D1591496

The Drinking Water Dictionary

James M. Symons
Editor and Project Supervisor
Cullen Distinguished Professor Emeritus of
Civil Engineering
University of Houston (retired)
Houston, Texas

Lee C. Bradley, Jr.
Associate Editor
Fort Worth Water Department
Fort Worth, Texas

Theodore C. Cleveland
Associate Editor
University of Houston
Houston, Texas

Technical Advisory Committee

J. Darrell Bakken
Indianapolis Water Company (retired)
Indianapolis, Indiana

Robert A. Bergman
CH2M Hill
Gainesville, Florida

A. Randolph Brown
United Water Resources Management
and Service Company
Jefferson City, Missouri

Richard J. Bull
Pacific Northwest National Laboratory
Richland, Washington

Zaid K. Chowdhury
Malcolm Pirnie, Inc.
Phoenix, Arizona

Gunther F. Craun
Gunther F. Craun & Associates
Staunton, Virginia

Charles D. Hertz
Philadelphia Suburban Water Company
Bryn Mawr, Pennsylvania

Debra E. Huffman
University of South Florida
St. Petersburg, Florida

Gregory J. Kirmeyer
Economics & Engineering Services
Bellevue, Washington

Stuart W. Krasner
Metropolitan Water District of
Southern California
La Verne, California

William H. Miller
Denver Water Department (retired)
Denver, Colorado

Peggy L. Murphy
Bureau of Water Works
Portland, Oregon

Douglas M. Owen
Malcolm Pirnie, Inc.
Carlsbad, California

Frederick W. Pontius
Pontius Consulting
Group
Lakewood, Colorado

Amy J. Purves
PlanGraphics, Inc.
Silver Spring, Maryland

Donald J. Reasoner
US Environmental Protection Agency
Cincinnati, Ohio

Stephen A. Tanner
Idaho Division of Environmental Quality
Coeur d'Alene, Idaho

Jack A. Weber
Montgomery Watson (retired)
Walnut Creek, California

Keh-Han Wang
University of Houston
Houston, Texas

Robert W. Weir
Denver Water Department (retired)
Denver, Colorado

The Drinking Water Dictionary

McGraw-Hill

New York Chicago San Francisco Lisbon London Madrid
Mexico City Milan New Delhi San Juan Seoul
Singapore Sydney Toronto

Ashland Community College Library
1400 College Drive
Ashland, KY 41101

72815

American Water Works Association
Dedicated to Safe Drinking Water

McGraw-Hill

A Division of The ***McGraw-Hill*** *Companies*

Copyright © 2001 by American Water Works Association. All rights reserved. Printed in the United States of America. Except as permitted under the United States Copyright Act of 1976, no part of this publication may be reproduced or distributed in any form or by any means, or stored in a data base or retrieval system, without the prior written permission of the publisher.

1 2 3 4 5 6 7 8 9 0 DOC/DOC 0 7 6 5 4 3 2 1

ISBN 0-07-137513-9

The sponsoring editor for this book was Larry S. Hager and the production supervisor was Sherri Souffrance.

Printed and bound by R. R. Donnelley & Sons Company.

This book was previously published in hardcover by American Water Works Association, copyright © 2000.

McGraw-Hill books are available at special quantity discounts for use as premiums and sales promotions, or for use in corporate training programs. For more information, please write to the Director of Special Sales, Professional Publishing, McGraw-Hill, Two Penn Plaza, New York, NY 10121-2298. Or contact your local bookstore.

This book is printed on recycled, acid-free paper containing a minimum of 50% recycled, de-inked fiber.

Contents

To my wife Joan, for her career-long support,
to my parents, for their life-long guidance and encouragement,
and to the waterworks profession that I have always tried to serve.

Foreword

The seed for development of this book was planted in early 1992 by Nancy M. Zeilig, Editor of the *Journal of the American Water Works Association,* when she found serious problems with outdated material in the 1981 (third) edition of the *Glossary, Water and Wastewater Control Engineering.* This reference, published jointly by the American Public Health Association, the American Society of Civil Engineers, the American Water Works Association, and the Water Pollution Control Federation (now Water Environment Federation), defined terms in common use in the field of drinking water, but the content needed updating.

AWWA management approved a proposal to rewrite, expand, and update the *Glossary* with a special focus on water supply terminology, and Nancy Zeilig recommended me to lead the project. AWWA recommended active members with various professional interests to join me in this work, and they, along with some participants I recommended, were recruited in 1993 to form the 22-person Technical Advisory Committee (TAC). Two TAC members, Lee Bradley and Theodore Cleveland, also agreed to be associate editors.

With the TAC formed, work started in earnest in early 1994. First, the previous edition of the *Glossary* was reviewed and terms not directly related to drinking water were deleted. Then, outdated definitions were updated, and arcane definitions were simplified.

Next, new entries were added from terms suggested by the TAC members (who provided most of the new definitions), from glossaries in references provided by AWWA, and from terminology encountered in technical articles and at professional conferences and meetings. Finally, in mid-1998, about 200 new entries were added to reflect the latest terminology in the drinking water field.

After obtaining definitions for all entries, the associate editors and I reviewed and edited the manuscript in preparation for submission to AWWA for copy and technical editing. At this time, pertinent formulas and illustrations were added, and units of measure—in both US customary and Système International (SI) units (where practical)—were gathered in a single list for ready reference.

I gratefully acknowledge the assistance of many helpers: Marty Allen, Appiah Amirtharajah, William Anderson, Richard Bannerot, Paul Barten, Mike Bauer, Thomas Bruns, O.K. Buros, Gary Carlson, Jack Cleasby, Dennis Clifford, Ovidiu Crisan, Denver Water, Peter Fu, Mirat Gurol, Charles Haas, Gerald Keeling, Doo Kim, Carolyn Kirkpatrick, Les Lampe, Brian Larson, Desmond Lawler, Minna Li, John Lienhard, Margaret Lingerfelt, Stuart Long, C. Scott MacDonald, James Mavis, Douglas McIlroy, Ross McKinney, Ginny Miller, Irving Moch Jr., Christine Moe, Brian Murphy, Michael O'Neill, Jennifer Orme Zavaleta, Junboum Park, Sal Pastor, Linda Symons Raab, William Rixey, Debbie Roberts-LaMountain, Joan Rose, Michael Schock, Mike Sclimenti, Ben Smith, Gerald Speitel Jr., Alan Stevens, Andrew Symons, George Symons, Joan Symons, James Taylor, Ross Walker, Shelia White, Mark Wiesner, Nancy Zeilig, and Karen Zicterman. These people assisted for varying amounts of time with tasks such as providing definitions, reviewing definitions, proofreading, typing copy, and photocopying and mailing parts of the manuscript.

In attempting to incorporate standard language for all definitions, the development team incorporated text from the Water Quality Association's *WQA Glossary of Terms.* I am grateful to Joseph F. Harrison and the WQA for permission to reprint this material.

Special recognition is given to Roger Eichhorn, University of Houston, who gave generously of his time to help with units, conversions between US customary and SI units, and definitions related to physics parameters.

Acknowledgment is gratefully given to those who worked on the development and production phases of this document: AWWA staffers Mary Kay Kozyra and Mindy Burke, senior technical editors; David Talley, technical editor; Scott Nakauchi-Hawn, production manager; and Michael Malgrande, production editor; Mart Kelle, proofreader; and Carol Magin, cover designer; as well as AWWA contractors Deborah Lynes, project development editor; Phil Murray, copy editor; and Diane Serafin, word processor.

My sincere thanks go to the TAC members, who took countless volunteer work hours from their busy schedules to make this book a reality and a success.

Every effort has been taken to ensure accuracy in this book. If, however, you would like to recommend a change in definition text, please mail or fax a copy of the suggestion form (printed at the back of the book) to AWWA, which will consider your input in improving future printings. This invitation constitutes permission to copy only the suggestion form.

In summary, *The Drinking Water Dictionary* is a unique reference book. It pulls together information from numerous sources, covering the many broad areas that make up the drinking water field.

James M. Symons
October 1998

The Drinking Water Dictionary

A

A *See* ampere *in the Units of Measure list.*
Å *See* angstrom *in the Units of Measure list.*
A–C pipe *See* asbestos–cement pipe.
A·h *See* ampere-hour *in the Units of Measure list.*
A/W ratio *See* air-to-water ratio.
AA *See* atomic absorption spectrophotometric method.
AADF *See* annual average daily flow.
AAEE *See* American Academy of Environmental Engineers.
AAES *See* American Association of Engineering Societies.
AAS Method *See* atomic absorption spectrophotometric method.
abandonment In legal terminology, the giving up, with the definite intent to do so, of (1) the right to use water for any purpose or (2) the method of using such water. A temporary cessation of such use or temporary giving up of such right does not constitute abandonment. Intent to abandon may be implied from acts of the one using the water, such as nonuse for a considerable period or diversion without beneficial use.
ABC *See* Association of Boards of Certification.
abiological Pertaining to (1) processes, actions, or chemical reactions that occur in the absence of biological activity or (2) events that are not biologically mediated.
abiotic Pertaining to the nonliving parts of a system.
abrasion number A number associated with granular activated carbon and used to define the resistance of the particles to degradation on handling. It is calculated by contacting an activated carbon sample with steel balls on a sieve column vibrator and determining the percentage ratio of the final mean particle diameter to the original mean particle diameter.
abrupt wave A translatory wave or increase in water depth in an open channel caused by a sudden change in flow conditions. *See also* hydraulic jump; sloping wave.
ABS *See* alkylbenzene sulfonate.
absolute Measured only by the fundamental units of mass, length, or time.
absolute filter rating Filter rating indicating that 99.9 percent (or essentially all) of the particles larger than a specified micrometre rating will be trapped on or within the filter.
absolute humidity A measure of the amount of water vapor held in a volume of air (in grams per cubic metre), frequently the atmosphere. *See also* relative humidity.
absolute ownership Complete ownership of water by one person. The term is used in the context of water rights.
absolute pressure The total pressure in a system, including both the pressure of the water and the pressure of the atmosphere. *Contrast with* gauge pressure.
absolute right In the context of water rights, a water-user right that cannot be lessened.
absolute temperature A temperature measurement relative to absolute zero (–273° Celsius, –459° Fahrenheit, 0 kelvin, or 0° Rankine). *See also* absolute zero.
absolute value The numerical value of a number without reference to the sign (i.e., plus or minus) associated with it.
absolute viscosity (μ) A numerical factor that represents a measure of a fluid's internal resistance to flow. The greater the resistance, the stickier the fluid and the greater the absolute viscosity. The absolute viscosity is also known as the coefficient of viscosity and the dynamic viscosity. It is calculated by dividing the shear stress (τ) by τ the velocity gradient normal to a fixed surface (dv/dy, the change in velocity per unit distance). In other words, $\mu = \tau/(dv/dy)$ in any consistent set of units. In US customary units, the absolute viscosity is expressed in terms of pounds-force seconds per square foot. In Système International units, it is expressed in terms of newton seconds per square metre or kilograms per metre second or pascal seconds.
absolute zero The temperature at which the molecules of an ideal gas are at rest, i.e., motionless. Common values for this temperature are –273° Celsius, –459° Fahrenheit, 0 kelvin, or 0° Rankine.
absorbed dose The amount of a chemical that enters the body of an exposed organism.
absorbing well A shaft or well driven through an impermeable stratum to allow water to drain through to a permeable one; also called a drain well, negative well, or dead well.
absorption (1) The penetration of one substance into the structure of another substance. (2) The penetration of a chemical

through the skin, wall of the gastrointestinal tract, or the lung into the systemic circulation (i.e., blood). *Contrast with* adsorption.

absorption analysis A technique in which measurement of an analyte is based on a decrease in signal from a radiation source. The type of absorption analysis is based on the wavelength of the electromagnetic radiation source (e.g., ultraviolet, visible, infrared).

absorption capacity The quantity of a soluble substance that can be absorbed by a given quantity of a solid substance.

absorption factor The fraction of a chemical making contact with an organism that is absorbed by the organism.

absorption spectroscopy A class of techniques in which the properties of molecules or atoms are studied by measuring the characteristics of electromagnetic spectra. Absorption is the selective removal of energy resulting from interaction with atoms. The absorbed energy promotes the transfer of an electron to a higher orbital.

Abyssinian well A tube with perforations above the pointed end, driven into strata of moderate hardness to obtain a supply of water. It is also called a well point.

AC *See* alternating current.

Acanthamoeba A genus of free-living amoebas that are found in soil and water habitats, including drinking water; that can infect a variety of mammals, including humans; and that can produce severe, even fatal disease. *Acanthamoeba* may cause keratitis (infection of the cornea) in contact lens wearers. The primary source in those cases is believed to be tap water used for cleaning the contact lenses and lens cases. *Acanthamoeba* cysts are large (15–28 micrometres) and should easily be removed by filtration; cysts are resistant to chlorine residuals found in adequately treated distribution water, but trophozoites are inactivated in such water.

accelerated depreciation An approach for depreciating an asset by amortizing its cost at a faster rate than would occur under the straight-line method. The three principal methods of accelerated depreciation are known as (1) sum of the year's digits, (2) double declining balance, and (3) units of production.

acceleration units In the US customary system, feet per second squared. In Système International, metres per second squared.

acceptable daily intake (ADI) An amount of a chemical that can be taken into the body without producing harm. It is generally calculated by dividing an experimentally determined no-adverse-effect level of a chemical in humans or experimental animals by a series of safety or uncertainty factors. ADI is also known as the reference dose.

acceptable risk The level of risk associated with minimal adverse effects. Acceptable health and environmental risks are usually determined by risk analysis. Cost may be an important consideration for establishing regulations and public policy. In situations where benefits may accrue, an acceptable risk may be the risk for which potential benefits outweigh adverse effects.

acceptance limits criteria Contaminant-specific analytical performance limits established by the US Environmental Protection Agency to judge the performance of laboratories analyzing contaminants under Safe Drinking Water Act requirements. Laboratories seeking certification to conduct analyses used to judge compliance under the Safe Drinking Water Act must demonstrate that they can quantify a contaminant within the contaminant's specified acceptance limits. These limits are usually expressed as a percentage range greater or less than the practical quantitation limit (e.g., ±40 percent at 0.005 milligrams per litre).

access time The cycle time required for a memory device to produce an output after receiving a request from the central processing unit.

accessible equipment Water treatment equipment that, when installed and operating, is exposable and available for proper and thorough cleaning and inspection using only simple tools such as a screwdriver, pliers, or an open-end wrench. *See also* readily accessible equipment.

accidental release prevention program A program intended to prevent and minimize the impact of accidental releases of hazardous substances to the environment. Organizations such as water and wastewater utilities, manufacturing facilities, and energy installations may have such programs. *See also* risk management plan.

accidental spill The unplanned release of substances, either directly or indirectly, in such quantities that substantial effects on receiving systems will be noted. Such a release is the result of accidents, acts of nature, or operational malfunctions.

acclimation Adjustment to a change in a system.

acclimation period The time that passes before a process is ready to perform at its best, sometimes called lag time.

accounts Records of the financial transactions of a certain property, person, or business.

accounts receivable Moneys not yet received by a business or person for goods delivered or services rendered.

accrual basis accounting A method of accounting in which revenues are recorded when they are earned and expenditures become liabilities for benefits received, even though payment for the expenditures or receipt of the revenue may take place in another accounting period.

accrued depreciation The monetary difference between the original cost of an article and its remaining value. *See also* depreciation.

accumulated deficiency The total amount less than the normal level of runoff in units of million gallons, acre-feet, or cubic metres from any stated time (such as the beginning of a drought, month, or calendar year) to date, as used in the design of impounding reservoirs. *See also* accumulated excess.

accumulated excess The total amount in excess of the normal level of runoff from any stated time (such as the beginning of the month or calendar year) to date. An accumulated excess is indicated by a plus sign. *See also* accumulated deficiency.

accumulation tank A vessel or tank that receives and stores product water for use on demand.

accuracy (1) The closeness of a result to the true value. (2) In the context of maps, the closeness of results of observations, computations, or estimates of graphic map features to their true values or positions. Relative accuracy is a measure of the accuracy of individual features on a map to other features; absolute accuracy is a measure of the accuracy of features compared to their true positions on the face of the earth. Mapping accuracy standards are generally stated in terms of an acceptable error that must be achieved and the proportion of measured features that must meet the criterion. In the case of some plotting and display devices, accuracy refers to tolerance in the display of graphic features relative to the original coordinate file. (3) In the context of a geographic information database, accuracy also takes into consideration the correctness of content (e.g., errors of commission and omission of features), the correctness in identification of features, the currency or temporal characteristics of the data, and the topologic integrity of graphic and nongraphic information. *See also* bias; precision.

acetaldehyde (CH_3CHO) An organic chemical formed during the disinfection of water. It is most commonly associated with the use of ozone (O_3) as a disinfectant.

$$CH_3\text{-}\overset{\overset{\displaystyle O}{\|}}{C}\text{-}H$$

Acetaldehyde

acetochlor An herbicide that was registered by the US Environmental Protection Agency's Office of Pesticide Programs in 1994. *See also* herbicide.

acid (1) A chemical substance that can donate a hydrogen ion, H^+, or a proton as it is commonly called. (2) Any chemical species that can accept an electron pair.

acid extractable concentration The concentration of a substance in solution after treatment of an unfiltered sample with hot, cold, or dilute mineral acid.

acid-fast stain A bacterial staining technique, also called the Ziehl–Neelsen stain (used primarily for bacteria of the genus *Mycobacterium*). In this technique, a solution of 95 percent ethanol with 3 percent hydrochloric acid (HCl) will remove the carbol fuchsin stain from all bacteria other than the mycobacteria and some closely related organisms.

acid precipitation Precipitation having a pH lower than the pH range commonly found in natural waters, caused by absorption from the atmosphere of sulfur dioxide gas (SO_2), which then forms sulfuric acid (H_2SO_4) in solution. It is sometimes called acid rain.

acid rain *See* acid precipitation.

acid sensitive Pertaining to a mixture that has a low capacity to neutralize hydrogen ions (buffering capacity) and can therefore be affected by acidic material, such as acid rain.

acid shock *See* pH shock.

acid-fast bacteria Bacteria belonging to the genus *Mycobacterium*. When stained by a procedure called the acid-fast stain, these organisms retain the dye carbol fuchsin when washed with 95 percent ethanol containing 3 percent hydrochloric acid.

acid-forming bacteria Microorganisms that can metabolize complex organic compounds under anaerobic conditions. This metabolic activity is the first step in the two-step anaerobic fermentation process leading to the production of methane. It may occur in the bottom muds of reservoirs.

acid-washed activated carbon Activated carbon that has been washed with an acid solution for the purpose of dissolving any ash in the activated carbon.

acidic *See* acidic solution.

acidic rain *See* acid precipitation.

acidic solution Water or soil that contains a sufficient amount of acid substances (hydrogen [H+] ions) to lower the pH to less than 7.0.

acidification *See* acidified.

acidified Pertaining to water in which acid has been added to lower the pH. In many analytic procedures, a sample will be acidified to a pH of less than 2 for preservation until actual analysis.

acidity A measure of the capacity of a water to neutralize strong base. In natural waters this capacity is usually attributable to acids such as $H_2CO_3^*$ and sometimes hydrogen ion (H^+). ($H_2CO_3^*$ is a hypothetical species that represents the molar concentration of carbonic acid (H_2CO_3) plus aqueous carbon dioxide; because very little H_2CO_3 is actually in solution, $H_2CO_3^*$ is roughly equivalent to aqueous carbon dioxide.) Acidity is usually expressed in terms of milligrams of calcium carbonate per litre (mg $CaCO_3$/L).

acidized Pertaining to a solution or material to which acid has been added.

acquired immune deficiency syndrome (AIDS) A disease caused by the human immunodeficiency virus (HIV) that leads to loss of the immune functions and makes individuals susceptible to a variety of diseases. HIV may be found in blood and other human excretions but is not found in drinking water.

acquired immunity Resistance to a pathogen or foreign substance acquired as a result of previous exposure.

acquisition The process of attaining ownership and operation of a small water system by a larger water utility or regulatory agency.

acre *See the Units of Measure list.*

acre-foot *See the Units of Measure list.*

acrylamide ($CH_2CHCONH_2$) An organic monomer used as a starting material for polymers that are used as coagulants or filter aids. Its concentration in finished drinking water is controlled by limiting the allowable dose of polymer that can be added to water.

$$H_2C{=}CHCONH_2$$

Acrylamide

acrylic resin A plastic material, tradename Lucite®.

ACS *See* American Chemical Society.

actinometer A tool (i.e., instrument or chemical) for measuring the power of radiant energy, especially in the visible or ultraviolet spectral regions. *See also* chemical actinometer.

actinometry The determination of irradiation intensity through the use of a chemical actinometer such as potassium ferric oxalate ($K_3Fe(C_2O_4)_3{\cdot}3H_2O$). *See also* chemical actinometer.

Actinomycetales An order of filamentous prokaryotic organisms (bacteria) that have moldlike characteristics and are considered transitional forms. This order includes eight families and a number of lesser known genera in addition to better known genera such as *Actinomyces*, *Nocardia*, *Streptomyces*, *Micromonospora*, and *Mycobacterium*.

Actinomycetes A group of bacteria that occupy an uncertain position between the bacteria and the true fungi. This group includes both anaerobic and aerobic forms, and morphologies range from filamentous forms that show true branching to

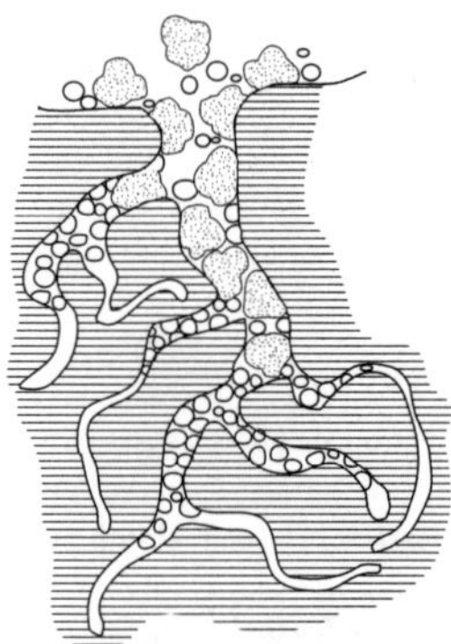

Idealized pore sturcture of activated carbon

club-shaped rods to others that demonstrate a large amount of pleomorphism during different life stages. This group includes many organisms that produce antibiotic substances. *Actinomycetes* can be present in soil, lake, and river muds. Certain *Actinomycetes* produce geosmin, 2-methylisoborneol, or both, compounds that produce earthy–musty odors at very low levels. *See also* geosmin; 2-methylisoborneol.

action level A specified concentration of a contaminant in water that determines treatment requirements or actions that a water system is required to complete to control the contaminant (such as further treatment or further monitoring, or both) to comply with a drinking water regulation.

activated alumina A charged form of aluminum, used in combination with a synthetic, porous media in an ion exchange process, for removing charged contaminants from water. It is typically used for fluoride removal. *See also* ion exchange.

activated carbon A form of particulate carbon (a crude form of graphite) with increased surface area to enhance adsorption of soluble contaminants. The primary source of activated carbon is bituminous coal, which is activated through a combustion process. Activated carbon is sometimes mistakenly called charcoal or just carbon.

activated carbon adsorption treatment A treatment process using either powdered activated carbon or a granulated form of activated carbon—granular activated carbon—to remove soluble contaminants from water by adsorption. Powdered activated carbon is added as a chemical slurry and is removed along with chemical sludges after sedimentation; granular activated carbon is contained in separate contactors or as filter media. *See also* activated carbon; granular activated carbon; powdered activated carbon.

activated carbon canister *See* chemical cartridge.

activated carbon reactivation *See* activated carbon regeneration.

activated carbon regeneration The process of restoring the adsorption capacity of granular activated carbon or (rarely) powdered activated carbon by thermal means. Used, or spent, activated carbon is removed from the process, dewatered, and combusted in furnaces in the absence of oxygen to remove adsorbed contaminants and restore the microporous structure (i.e., to increase surface area) for adsorption. *See also* granular activated carbon; powdered activated carbon.

activated carbon usage rate (CUR) A measure of the capacity of granular activated carbon or powdered activated carbon to remove a contaminant to a specified level. Usage rates are often expressed in terms of weight of activated carbon used per unit volume treated. *See also* activated carbon adsorption treatment; granular activated carbon; powdered activated carbon.

activated silica A negatively charged colloid used primarily as an aid to coagulation. Activated silica is formed from a reaction between sodium silicate ($Na_2Si_4O_9$) and acid to form a supersaturated gel, which is then diluted. *See also* coagulation.

activation (1) The process of producing a highly porous structure in carbon by exposing the carbon to high temperatures in the presence of steam. (2) The process of making a solid material more capable of a desirable selective action. *See also* activated carbon.

activation polarization A corrosion phenomenon in which the electric potential or net current, and therefore the rate of corrosion, is controlled by the activation energy of the elements participating in the reaction (e.g., iron). An activation polarization diagram is often used to determine the electron transfer step that controls the reaction. *See also* corrosion.

active humoral immunity Immunity attained either naturally by infection, with or without clinical symptoms, or artificially by inoculation of the agent, fractions of the agent, or products of the agent. Such immunity usually lasts for years.

active power (*P*) The time average over one period of the instantaneous power, measured in watts (W). For a single-phase circuit,

$$P = V \times I \times \text{PF}$$

Where:

V = root mean square (rms) voltage, in V

I = rms current, in A

PF = power factor

For a sinusoidal three-phase symmetrical and balanced circuit,

$$P = 3^{0.5} \times V \times I \times \text{PF}$$

Where:

V = line-to-line rms voltage, in V

I = rms current, in A

PF = power factor

The latter equation gives the total active power for all three phases.

active transport An energy-expending mechanism by which a cell moves a chemical across the cell membrane from a point of lower concentration to a point of higher concentration against the diffusion gradient.

active water *See* aggressive water.

activity coefficient A fractional number that, when multiplied by the molar concentration of a substance in solution, yields the chemical activity. This value gives an idea of how much interaction occurs between molecules at higher concentrations. The value of chemical (thermodynamic) activity replaces the actual molar concentration in mathematical expressions for the equilibrium constant, thereby eliminating the effect of concentration on the equilibrium constant. *See also* chemical activity.

activity network diagram A diagram used for planning the most appropriate schedule for completion of any complex task and all of its related subtasks.

actual evaporation The amount of water loss that can be attributed to evaporation based on measured values of precipitation, runoff, and estimated or calculated soil moisture. It represents the volume of water that evaporates over a given time period based on a water budget calculation using the following formula:

$$AE = P_i - R_o + S_m$$

Where:

AE = actual evaporation

P_i = the cumulative precipitation for the time period (typically expressed as a length)

R_o = the cumulative runoff for the same time period (expressed as a length)

S_m = the cumulative soil moisture (also expressed as a length)

Actual evaporation is different from evaporation measured using an evaporation pan.

actual evapotranspiration The evapotranspiration that actually occurs under particular climatic and soil moisture conditions.

actual groundwater velocity The effective or field velocity of groundwater percolating through water-bearing material. It is measured by the volume of groundwater passing through a unit cross-sectional area in unit time divided by the effective porosity. It is also called average linear velocity, true groundwater velocity, field groundwater velocity, seepage velocity.

acute Pertaining to an intense effect of a chemical or infectious agent. Acute effects are generally closely associated with high doses of the causative agent. They are not necessarily confined to brief exposure to the agent.

acute exposure A single exposure to a toxic substance that results in severe biological harm or death. An acute exposure is usually characterized as lasting no longer than a day.

acute gastroenteritis (1) Acute viral gastroenteritis presents as a sporadic or epidemic illness. Several enteropathogenic viruses (e.g., rotaviruses) affect primarily infants and young children, resulting in a diarrheal illness that may be severe enough to produce dehydration; other enteric viruses (e.g., Norwalk-like viruses) affect primarily older children and adults and cause self-limiting (recovery without treatment) sporadic illness or outbreaks. (2) Acute bacterial gastroenteritis is characterized by diarrhea and sometimes vomiting and may be mild or severe, resulting in dehydration and occasionally septicemia if high levels of the causative agent or its toxins occur in the circulating blood. *See also* acute gastroenteritis of undetermined etiology; diarrhea.

acute gastroenteritis of undetermined etiology (AGI) A designation used to describe cases of illness in outbreaks where no specific etiologic (causative) agent has been identified. In some instances no etiologic agent is identified, even after exhaustive laboratory analysis, but in most cases a lack of identification is a result of limited laboratory analysis or lack of collection of appropriate or timely clinical specimens during the epidemiologic investigation. Although in many waterborne outbreaks, the symptoms of acute gastroenteritis of undetermined etiology may suggest a viral etiology, the illness may also be of bacterial or protozoan etiology.

acute poisoning Poisoning that yields a rapid and intense response. It is generally associated with a single dose.

acute respiratory diseases Immediate responses of the respiratory system to infectious agents, allergens, or irritant chemicals. These responses generally restrict the passage of air through the upper airways of the lung or reduce the area for exchange of oxygen (O_2) and carbon dioxide (CO_2) between inhaled air and blood. Examples include bronchitis, chloramine-induced upper respiratory tract irritation, hay fever, and the common cold.

acute toxicity (1) The ability of a substance to cause poisonous effects resulting in severe biological harm or death soon after a single exposure or dose. (2) Any severe poisonous effect resulting from a single short-term exposure to a toxic substance.

acute violation Failure of a water system to meet a Safe Drinking Water Act regulation, resulting in an immediate public health risk. The US Environmental Protection Agency specifies which violations are acute when it sets drinking water regulations, and it requires extensive public notification when they occur. *See also* public notification.

ACWA *See* Association of California Water Agencies.

ad hoc query An operator-initiated request for information to be extracted from a database.

ad valorem tax A state or local tax based on the assessed value of real or personal property.

ADA *See* Americans With Disabilities Act.

adaptation (1) The condition of showing fitness for a particular environment, as applied to characteristics of a structure, function, or entire organism. (2) A modification of a species that makes it more fit for reproduction, existence, or both under the conditions of its environment.

additional bonds test A statement in a bond indenture that stipulates requirements to be met before additional bonds can be issued with a parity claim on revenues already pledged. It ensures that additional bonds are not issued unless historical and projected revenues indicate sufficient revenue exists to avoid dilution of coverage on outstanding bonds.

additive A chemical agent added to meet treatment goals (e.g., coagulants, chemicals to adjust pH, and corrosion inhibitors).

additive effect The toxicological effect of two chemicals given in combination, equivalent to the sum of their independent effects. Additivity can be defined in terms of either the intensity of the effects or the frequency in which a stochastic response is observed within a population of individuals at risk.

address matching The process of relating street addresses to point locations or areas such as census blocks, tracts, administrative units, buildings where permits are pending, or points of emergency response incidents.

adduct A covalent product of a reaction between a chemical or one or more of its metabolites and biological constituents. This term is most often applied to the product of reactions between reactive chemicals and cellular macromolecules such as nucleic acids or proteins.

adenoma A benign tumor that has originated from epithelial cells. It is produced in a variety of organs such as the pituitary gland, alveoli of the lung, liver, and kidney. Adenomas are recognizable because they maintain the normal structures of the cells from which they are derived.

adenosine triphosphate (ATP) ($C_{10}H_{16}N_5O_{13}P_3$) A compound found in all living cells in which energy is stored in high-energy phosphate bonds. Its components are the purine adenine, D-ribose, and three phosphoric acid groups.

adenovirus A large icosahedral deoxyribonucleic acid (DNA) virus, approximately 70 nanometres across. At least 41 different types are known, many of them associated with respiratory effects. Subtypes 40 and 41 are intestinal, associated with gastrointestinal disease, and waterborne. At present (1998) they are difficult to cultivate in cell culture.

ADF *See* average daily flow.

adhesion (1) Attachment or binding of unlike materials to one another. (2) Attachment of microorganisms to solid surfaces during biofilm formation. Adhesion may be reversible or irreversible in either case.

ADI *See* acceptable daily intake.

adiabatic expansion The expansion (increase) in volume of an air mass that occurs when the mass rises and expands without exchanging heat with its surroundings.

ADID *See* advance identification.

adipate A class of synthetic organic chemicals with various industrial uses, including use as a plasticizer. Di(2-ethylhexyl)adipate—(CH_2 $CH_2COOCH_2CH(C_2H_5)C_4H_9)_2$—is an adipate that is regulated by the US Environmental Protection Agency. *See also* di(2-ethylhexyl)adipate; plasticizer; synthetic organic chemical.

adjacent wetlands Wetlands separated from other waters of the United States by constructed dikes or barriers, natural river berms, beach dunes, and the like.

adjusted r^2 A statistical term describing the goodness of fit of an equation to a given data set while considering the number of parameters included in the equation.

adjusted water budget A quantity of water used to maintain a landscape based on evapotranspiration and area, adjusted to reflect an efficiency standard.

adjustment factor In the context of water-efficient landscapes, a decimal fraction used to modify reference evapotranspiration to reflect an efficiency standard.

administration The collective management of governmental, business, or institutional affairs.

Administrative Procedure Act (APA) A collection of requirements codified in Title 5, Sections 551-59, 701-06-1305, 3105, 3344, 5372, and 7521 of the US Code that federal agencies must follow when establishing regulations.

administrative order A legally binding compliance order that can be issued by the US Environmental Protection Agency under the Safe Drinking Water Act to the owner of a public water system that is not in compliance with a national primary drinking water regulation.

administrative water law A subcategory of administrative law that focuses on water rights.

adsorbable organic halogen (AOX) *See* total organic halogen.

adsorbate A solid, liquid, or gas substance that is adsorbed as molecules, atoms, or ions.

adsorbent Any material that can be used to adsorb substances on its surface (e.g., activated carbon).

adsorber A vessel designed to hold any adsorbent.

adsorption (1) The attraction and adhesion of molecules of a gas, liquid, or dissolved substance to a surface. Adsorption is generally a passive and reversible process. Granular or powdered activated carbon is often used as an adsorption medium. (2) The interaction of an analyte with the surface of a matrix. This interaction can form the basis for the extraction of analytes from water or the chromatographic separation of compounds. *See also* activated carbon adsorption treatment.

adsorption isotherm The plotted output of a test to evaluate the extent of adsorption determined at a constant temperature by varying the amount of activated carbon used or the concentration of the impurity in contact with the activated carbon. This test produces data that, when graphed, yields a line called the adsorption isotherm.

adsorption pores (1) The finest pores in an adsorbent's structure. (2) Pores that have adsorption capabilities.

adsorption water Water held on the surface of solid particles by molecular forces, with emission of heat (heat of wetting) taking place.

adsorption–destabilization A mechanism for the coagulation of particles in which counterions are adsorbed on the surface of nonsettling (stable) particles; thus, the particles can approach one another close enough to stick together and lose their stability (i.e., become settleable). *See also* bridging; double-layer compression; sweep-floc coagulation.

advance for construction An advance payment made by or on behalf of customers or others for the purpose of construction, to be refunded either wholly or in part. When applicants are refunded the entire amount to which they are entitled according to the agreement or rule under which the advance was made, the balance, if any, remaining in this account shall be credited to contribution in aid of construction.

advance identification (ADID) A planning process under which the US Environmental Protection Agency, in cooperation with the US Army Corps of Engineers and after consultation with the state, identifies wetlands and other waters of the United States that are either generally suitable or unsuitable for the discharge of dredged and fill material prior to the receipt of the Clean Water Act Section 404 permit application.

advance refunding bonds Bonds issued to replace an outstanding bond issue prior to the date on which the outstanding bond becomes due or callable. Proceeds of the advance refunding bonds are deposited in escrow with a fiduciary, invested in US treasury bonds or other authorized securities, and used to redeem the underlying bonds at maturity or the call date and to pay interest on the bonds being refunded or the advance refunding bonds.

Advanced Notice of Proposed Rulemaking (ANPRM) A notice issued by a federal agency and published in the *Federal Register* announcing the intent of the agency to develop a proposed rule.

advanced oxidation process (AOP) A process in which the oxidative capacity of a parent compound is modified to make oxidation–reduction reactions more rapid or complete. For example, hydrogen peroxide (H_2O_2) may be added to ozone (O_3), or hydrogen peroxide may be photolyzed by ultraviolet light, to generate hydroxyl radical, which are strong, nonspecific oxidizing agents. *See also* hydrogen peroxide; hydroxyl radical; ozone; ultraviolet light.

advanced treatment plant (ATP) A treatment facility using treatment processes that provide treatment to a higher level than that considered conventional. For example, a conventional surface water treatment process includes coagulation, flocculation, sedimentation, and filtration. An advanced treatment plant may include processes such as ozonation, granular activated carbon adsorption treatment, or both.

advection The transport of dissolved or suspended mass by motion of the host fluid. The motion is typically caused by pressure, density, or thermal gradients. The analogous term, convection, is often reserved to describe advection that results from density or thermal gradients.

advection dispersion equation A mathematical relationship to estimate the concentration of a given solute based on the flow (advection) and mixing (dispersion) in the aquifer. The estimate of dispersion includes effects of both molecular diffusion and mechanical mixing caused by shear forces at boundary layers. This equation is given as

$$\frac{\partial C}{\partial t} = D\frac{\partial^2 C}{\partial^2 x} - V\frac{\partial C}{\partial x}$$

Where:

C = concentration of species in water phase, in g/m^3

D = dispersion coefficient, in m^2/s

V = average linear velocity, in m/s

X = distance, in m

adverse effect An effect that is judged to compromise the normal physiology of an organism. In the context of regulatory decisions, the term is used to differentiate between effects that clearly affect health and responses unlinked to health effects or of unknown significance.

adverse reaction An effect of a chemical other than the intended effect, posing some threat to health or life. The term is usually applied in the context of drug use.

advisory group A group of people representing various interests, points of view, and fields of expertise. The group's purpose is to provide advice, credibility (sometimes), input, structured feedback, and support to an agency's projects or policies. The group may be standing or ad hoc, as well as handpicked or chosen from interested applicants.

advocacy The act of speaking or writing in support of an individual or thing.

ADWF *See* average dry weather flow.

AEESP *See* Association of Environmental Engineering and Science Professors.

aeolian deposit Soil deposited by wind, also called a wind deposit.

aeration A gas transfer unit process that allows for the absorption of gas (frequently oxygen) by water.

aeration zone The zone between the land surface and the water table. It includes the root zone, intermediate zone, and capillary fringe. The pore spaces contain water, as well as air and other gases. Saturated bodies, such as perched groundwater, may exist in the aeration zone, which is also called the vadose zone or the unsaturated zone.

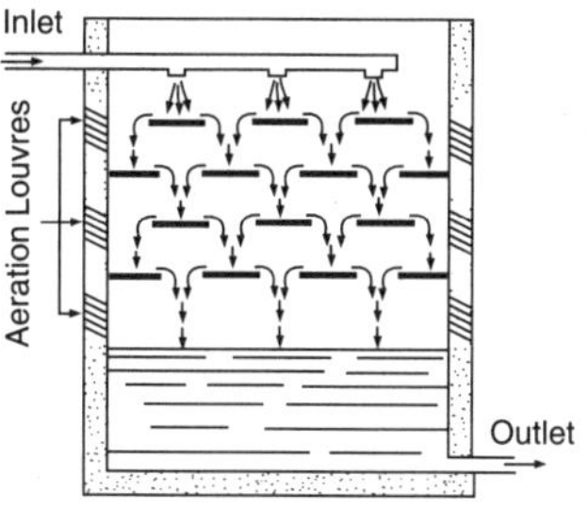

One type of aerator

aerator A treatment device that brings air into contact with water for the purposes of transferring gases (e.g., oxygen) from the air into the liquid phase. When this type of device is used for transferring contaminants from the liquid to the gaseous phase, it is called a stripper. *See also* air-stripping; diffused aeration; tray aerator.

aerial photography A method of obtaining photographic images taken from an airborne platform, commonly an airplane or satellite. Information derived from these photographs can be used to determine watershed characteristics, land use, topographic contours, geologic features, and artificial features in a given area.

Aerobacter aerogenes *See Enterobacter aerogenes.*

aerobe Any organism requiring free oxygen (O_2) or air to maintain its life processes.

aerobic (1) In reference to microorganisms, pertaining to the requirement for oxygen for growth. (2) In reference to environmental conditions, pertaining to the presence of oxygen.

aerobic bacteria *See* aerobe.

aerobic condition An environmental condition in which oxygen is available. In water, this means that dissolved oxygen is present.

aerohydrous Pertaining to both air and water. This term applies to minerals containing water in pores or cavities.

Aeromonas hydrophila One species of aerobic, straight-rod or coccobacillus shaped, gram-negative, motile (single polar flagellum) bacterium found in aquatic habitats. This bacterium grows well at 72° to 82° Fahrenheit (22° to 28° Celsius) and at 98.6° Fahrenheit (37° Celsius). It grows on a wide range of culture media and can be isolated on membrane filters of the type used for detection of coliform bacteria in water. *A. hydrophila* can cause wound infections and septicemia and has been associated with diarrheal illness worldwide. Other species of *Aeromonas* are pathogenic for fish and amphibians. *A. hydrophila* may regrow in treated distribution water that contains organic carbon and a low disinfectant residual and that has a long residence time in the system or in a storage reservoir or tank.

AES *See* atomic emission spectroscopy.

aesthetic Pertaining to a quality of water that is determined by the senses, e.g., color, taste, or odor.

aesthetic contaminants *See* aesthetic.

affinity constant A measure of the strength of association between two chemicals. In formal terms, it is the inverse of the dissociation constant. In toxicology it applies to chemicals that act through association with a biological molecule by noncovalent linkages; the biological molecule is referred to as the receptor.

affinity diagram A method of gathering large amounts of language data (ideas, opinions, issues) and visually organizing it into groupings based on natural relationships between each item.

affirmative action A program or plan designed to increase the number of minorities in the workplace.

affluent stream A stream or river flowing into a larger river or into a lake; a tributary.

aflatoxin A family of mycotoxins found in certain strains of *Aspergillis flavus*. Aflatoxin B1 is the most potent of these toxins. It is a hepatotoxin and among the most potent carcinogens known. Other aflatoxins are designated B2, G1, and G2, primarily on the basis of their fluorescent properties, either blue (B) or green (G). Metabolites of aflatoxin B1 are frequently designated M1, M2, and so on. The structural formulas for various aflatoxins are as follows:

aflatoxin	$C_xH_yO_z$, where *x* is typically 16 or 17; *y* is typically 10, 12, or 14; and *z* is typically 6, 7, or 8
aflatoxin B1	$C_{17}H_{12}O_6$
aflatoxin B2	$C_{17}H_{14}O_6$
aflatoxin B2a	$C_{17}H_{14}O_7$
aflatoxin G1	$C_{17}H_{12}O_7$
aflatoxin G2	$C_{17}H_{14}O_7$
aflatoxin G2a	$C_{17}H_{14}O_8$
aflatoxin M1	$C_{17}H_{12}O_7$
aflatoxin M2	$C_{17}H_{14}O_7$
aflatoxin P1	$C_{16}H_{10}O_6$
aflatoxin Q1	$C_{17}H_{12}O_7$

AFM *See* atomic force microscopy.

after-precipitation The continued precipitation of a chemical compound after leaving the sedimentation basin or solids contact basin. This process can cause scale formation on the filter media and in the distribution system.

after-the-fact (ATF) permit A permit issued under Section 404 of the Clean Water Act to allow dredged or fill material that accrued prior to permit application to be discharged into a wetlands or other waters of the United States.

afterburner A device that oxidizes compounds in the gaseous phase under conditions of high temperature. An afterburner can be used following an incinerator to convert undesirable gaseous by-products of the incineration process, such as carbon monoxide (CO) and hydrocarbons (C_xH_y), to carbon dioxide (CO_2) and water.

aftergrowth An increase in bacterial density in treated distribution water caused by growth of bacteria released from the pipe-wall biofilm and sediments, generally more apparent in areas where the disinfectant residual has been depleted. Aftergrowth is also called regrowth. *See also* regrowth.

AGA *See* American Gas Association.

agar A gelatinous material recovered from seaweed that is used to solidify media for microbiological assays.

age tank A tank used to store a chemical solution of known concentration for feed to a chemical feeder. It is also called a day tank.

agglomeration The collecting or coalescence of dispersed suspended matter into larger masses or flocs that can settle and be filtered from water.

agglutination The microscopic or macroscopic clumping of microorganisms. This process allows researchers to identify organisms by the formation of organism–antibody complexes that create a lattice or network.

aggradation The geological process of building up a surface by the accumulation of deposits.

aggrading river A river that is building up its valley bottom by deposition of material.

aggregate A collection of soil grains or particles gathered into a mass and behaving mechanically as a unit.

aggregate volume index (AVI) A ratio of total aggregate unit volume to the fraction of that volume occupied by dry sludge solids. This ratio is determined by performing sludge-thickening studies using batch cylindrical columns. The densities of sludge particles and aggregates, also measured during the test, are necessary to calculate the aggregate volume index. The AVI is an indication of the dewaterability of the sludge; higher aggregate volume indexes indicate poorer dewaterability.

aggressive Pertaining to a corrosive water that will deteriorate material such as distribution piping.

aggressive water Water having corrosive qualities. *See also* aggressive.

AGI *See* acute gastroenteritis of undetermined etiology.

agitator A mechanical apparatus for mixing, aerating, or both; a device for creating turbulence.

agricultural chemicals Chemicals used during various stages of agriculture to (1) control pests and weeds and (2) promote desired plant growth. *See also* fertilizer; fumigant; herbicide; pesticide.

agricultural drainage The runoff of water from farmed areas that travels through surface waterways and ditches or via subsurface tile drains and conduits.

agricultural reuse The use of treated wastewater for agricultural purposes, usually irrigation, to preserve drinking water supplies for human consumption.

agrochemicals Synthetic chemicals (pesticides and fertilizers) used in agricultural production.

Ah locus A location on deoxyribonucleic acid where the dioxin-activated receptor binds to activate synthesis of a variety of proteins. Prominent among these proteins are specific members of the cytochrome P-450 family of enzymes that are involved in the metabolism of chemicals that are foreign to the body.

Ah receptor A protein complex that binds to certain chemicals to induce a specific constellation of responses in the cell. It includes the induction of cytochromes in the 1A family and some changes in the rates of cell replication for a variety of tissues. Activation of the receptor seems clearly associated with the toxicological properties of chemicals in the dibenzodioxin, dibenzofuran, and, to a lesser degree, the polychlorinated biphenyl groups. Most effects of these chemicals seem to involve catalyzing the binding of the receptor to a variety of nuclear proteins, which in turn acts to stimulate or inhibit transcription of genes. Which of these responses is responsible for the carcinogenic and development effects of this class of compounds is not clear.

AI *See* artificial intelligence.

AIChE *See* American Institute of Chemical Engineers.

aid A chemical used to enhance and improve the performance of a physical–chemical process. For example, a coagulant aid is a chemical used in addition to the primary coagulant to improve performance; a filter aid is a chemical added to the filtration process to improve particle capture. *See also* coagulant.

AIDIS *See* Inter-American Association of Sanitary Engineering & Environmental Sciences.

AIDS *See* acquired immune deficiency syndrome.

AIEE *See* American Institute of Electrical Engineers.

AIHA *See* American Industrial Hygiene Association.

air backwash A process for cleaning filtration media in which air is introduced into a liquid backwash flow to assist in dislodging particles entrapped in the media. Air backwash is typically used for backwashing either pressure or gravity media filters. *See also* backwash.

air-and-vacuum valve An air valve that permits air to enter an empty pipe to counteract a vacuum and prevent accumulated air from escaping. This type of valve is also called a vacuum valve.

air binding (1) The clogging of a filter, pipe, or pump as a result of the presence of air released from water. Air can prevent the passage of water during the filtration process and can cause the loss of filter media during the backwash process. (2) An increase in the pressure of air trapped in soil interstices, which decreases the rate of water infiltration into the soil.

air chamber A closed pipe chamber installed on the discharge line of a reciprocating pump to take up irregularities in hydraulic conditions and relieve the pump of shocks caused by the pulsating flow.

air check A device that allows water, but not air, to pass through it.

air diffuser Devices of varied design that transfer (dissolve) air into a liquid, frequently water.

air diffusion The process of transferring (dissolving) air into a liquid, frequently water. *See also* air diffuser.

air dryer A device used to dry out (desiccate) air by removing the water vapor. An air dryer is used during the generation of ozone (O_3) from air to produce higher concentrations of ozone and lessen the production of corrosive nitrous oxides.

air gap A dedicated air space between a pressurized water supply and a source of contamination, used to ensure that an incompatible liquid or contamination source is physically disconnected from the piping system and therefore cannot be siphoned into the system. *See also* backflow prevention; siphon.

air lock A condition in which accumulated air in a high point of a piping system is sufficient to reduce or block the flow of water.

air padding The process of pumping dry air (dew point –40° Fahrenheit or –40° Celsius) into a container to force a liquefied gas such as chlorine out of a container or to assist with the withdrawal of a liquid.

air pocket A location within a pipeline or filtering medium in which air has collected. Air pockets interfere with the flow of a liquid and result in the loss of a siphon in pipelines. In filters, air pockets interfere with filter performance. They develop when pressure loss within a filter results in a localized pressure that is less than atmospheric pressure. Dissolved gases are released from the solution to form air pockets within the media. *See also* air binding.

air release valve *See* air relief valve.

air relief valve An air valve placed at the summit of a pipeline (1) to release air automatically and prevent air binding and pressure buildup or (2) to allow air to enter a line if the internal pressure becomes less than that of the atmosphere.

air scouring The practice of admitting air through the underdrain system to ensure complete cleaning of media during filter backwash. *See also* air backwash.

air scrubbing A process that uses a liquid spray to oxidize and subsequently remove undesirable contaminants (e.g., particulates, acid gases such as carbon dioxide (CO_2)) from the gaseous phase. Scrubbants can include either water or an agent such as sodium hydroxide (NaOH).

air-to-water (A/W) ratio A design criterion used in air-stripping to indicate the volume of air required per volume of water to effectively remove volatile contaminants.

air vent An opening in a penstock or other pipeline, covered tank, or well that allows inflow of air.

air wash *See* air backwash.

air-bound Pertaining to a pipe or pump obstructed because of air entrapped at a high point.

air-displacement pump A displacement pump in which compressed air, rather than pistons or plungers, is used to force liquid through a cylinder. Such a pump is also called an air-chamber pump.

air-fed ozonator A device for producing ozone (O_3) that uses dry air, rather than dry oxygen (O_2), as the feed material. An electrical discharge through the dry air produces the ozone.

air-gap device Any mechanism that uses the principle of separation or an air gap to prevent a cross-connection or physical connection where potable water could be contaminated. *See also* air gap.

air-gap fitting A physical device engineered to produce an air gap. *See also* air gap; air-gap device.

air-lift pump A pump based on the air-lift principle in which fine-pressurized air bubbles are released in water. These fine bubbles lower the density of the air–water mixture, allowing the weight of the more dense surrounding water to push the air–water mixture up the pipe.

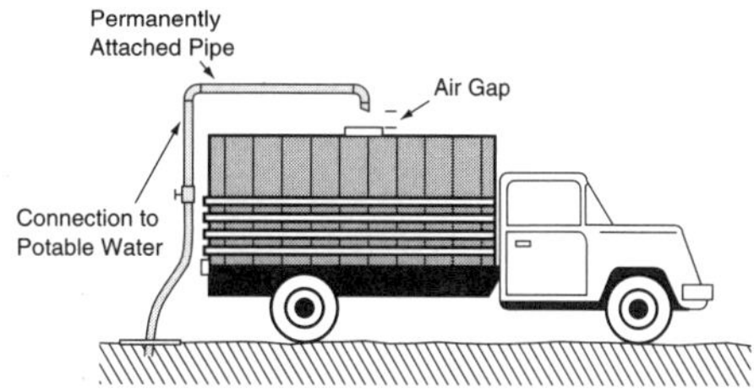

Water truck cross-connection prevented by air gap

air-line correction In wire sounding, a correction applied to that part of the line above the water surface when large vertical angles are induced by high velocities, great depth, insufficient sounding weight, or any combination thereof. *See also* wet-line correction.

air-purifying respirator A respirator that uses disposable, chemically active cartridges or high-efficiency particulate filters that remove harmful contaminants from the air.

air-stripping A process that removes volatile compounds from a liquid phase by passing air through the liquid. The process uses the principles of Henry's law to transfer volatile pollutants from a solution of high concentration into an airstream of lower concentration. Air can be diffused into a bulk liquid stream or can be introduced into packed towers by compressors, countercurrent to the water flow. *See also* packed tower aeration; stripping.

air-surging A well development technique that uses compressed air to remove fines (clay, sand, silt) that can clog the screen and filter pack of a well.

air–water wash A method of backwashing granular filter media in which both air and water are used. The air is entrained under pressure into the backwash water in the underlying media support structure and is released as the water flows upwardly through the granular media. The purpose of the air is to provide additional energy and some buoyancy that increases the scouring action and enhances the release of particles attached to the granular media. *See also* backwash.

ALA *See* aminolevulinic acid.

alachlor (2-chloro-2'-6'-diethyl-*N*-(methoxymethyl)-acetanilide) An herbicide used primarily on corn and soybeans. It is slightly soluble in water and is regulated by the US Environmental Protection Agency.

CH_3CH_2
N–CH_2OCH_3
C–CH_2Cl
O
CH_3CH_2

Alachlor

alachlor ethane sulfonic acid (ESA) A sulfonic acid (SO_3H) degradate of alachlor ($C_{14}H_{20}O_2NCl$). *See also* alachlor.

ALAD *See* aminolevulinic acid dehydrase.

alarm contact A switch that operates when some preset low, high, or abnormal condition exists.

albedo Reflective power, specifically, the fraction of incident light or electromagnetic radiation that is reflected by a surface or body (such as the moon or a cloud).

alcohol A class of organic compounds that have a hydroxyl (OH) functional group attached to one of the carbon atoms in the structure.

R-OH

Alcohol, where R = carbon-containing group

aldehyde A class of organic compounds that have a carbonyl (C=O) functional group on the first or last carbon atom in the chemical structure. Some aldehydes are created during the reactions of oxidants used as disinfectants, particularly ozone (O_3), with natural organic matter. *See also* disinfection by-product.

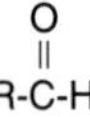

Aldehyde, where R = carbon-containing group

aldicarb ($CH_3SC(CH_3)_2HC{:}NOCONHCH_3$) The common name for 2-methyl-2-(methylthio) propionaldehyde *O*-(methylcarbamoyl) oxime, a synthetic organic chemical used as an insecticide or nematocide in soil. As of this writing, it is on the regulatory schedule of the US Environmental Protection Agency. *See also* insecticide; nematocide.

CH_3 O
CH_3-S-C-CH=N-O-C-NH-CH_3
CH_3

Aldicarb

aldicarb sulfone ($CH_3SO_2C(CH_3)_2HC{:}NOCONHCH_3$) Another name for aldoxycarb, which is the common name for 2-methyl-2-(methylsulfonyl)propanal *O*{(methylamino)carbonyl} oxime, a synthetic organic chemical used as a nematocide or insecticide. In addition, aldicarb sulfone is the final oxidation by-product of aldicarb. As of this writing, it is on the regulatory schedule of the US Environmental Protection Agency. *See also* aldicarb; insecticide; nematocide.

O CH_3 O
CH_3-S-C-CH=N-O-C-NH-CH_3
O CH_3

Aldicarb sulfone

aldicarb sulfoxide ($CH_3SOC(CH_3)_2HC{:}NOCONHCH_3$) The intermediate oxidation by-product of aldicarb. Aldicarb sulfoxide can be further oxidized to aldicarb sulfone. As of this writing, it is on the regulatory schedule of the US Environmental Protection Agency. *See also* aldicarb; aldicarb sulfone.

O CH_3 O
CH_3-S-C-CH=N-O-C-NH-CH_3
CH_3

Aldicarb sulfoxide

aldoacid A class of organic compounds that have both an aldehyde (HC=O) and carboxylic acid (COOH) functional group in the chemical structure. An aldoacid is also referred to as an aldehyde-acid. Some aldoacids are created during the reactions of oxidants used as disinfectants, particularly ozone Os, with natural organic matter. *See also* disinfection by-product.

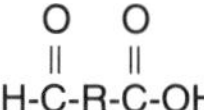

Aldoacid, where R = carbon-containing group

aldoketone A class of organic compounds that have both an aldehyde (HC=O) and carbonyl (C=O) functional group in the chemical structure. Some aldoketones are created during the reactions of oxidants used as disinfectants, particularly ozone (O_3), with natural organic matter. *See also* disinfection by-product.

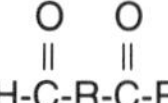

Aldoketone, where R = carbon-containing group

aldrin ($C_{12}H_8Cl_6$) The assigned common name for an insecticidal product containing 95 percent or more of 1,2,3,4,10,10-hexachloro-1,4,4a,5,8,8a-hexahydro-*exo*-1,4-*endo*-5,8-dimethanonaphthalene. *See also* insecticide.

alga *See* algae.

algae The simplest plants that contain chlorophyll and require sunlight; they vary from microscopic forms to giant seaweed. In drinking water sources, blooms of microscopic forms (phytoplankton) cause taste-and-odor problems.

algal assay An analytic procedure that employs specified nutrients and algal inoculums to identify which algal nutrient is limiting in water bodies.

algal bloom A sudden, dramatic growth of microscopic and macroscopic plant life, such as blue-green or green algae, that develops in lakes and reservoirs.

algal harvesting A process for recovering algal cell mass from suspension.

algal inhibition The process of slowing down or stopping algal growth.

algal mat A surface layer of dead algae.

algicide Any substance or chemical specially formulated to kill or control algae.

algogenic organic matter Natural organic matter derived from algae. Algogenic organic matter can contribute to both the bacterial regrowth potential and the disinfection by-product formation potential of a water. *See also* algae; bacterial regrowth; disinfection by-product; disinfection by-product formation potential; natural organic matter.

ALGOL (algorithmic language) A high-level programming language used to code mathematical problems.

algorithm An arrangement of logical steps to perform a certain operation within a computer program.

algorithmic language *See* ALGOL.

aliphatic *See* aliphatic compound.

aliphatic aldehyde An organic compound in which the aldehyde group (HC=O) is connected to a branched or unbranched open chain of carbon atoms rather than a ring. Some aldehydes are created during the reactions of oxidants used as disinfectants, particularly ozone (O_3), with natural organic matter. *See also* disinfection by-product.

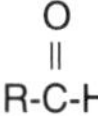

Aliphatic aldehyde, where R is an open chain of carbon atoms rather than a ring

aliphatic compound An organic compound with carbon atoms arranged in a branched or unbranched open chain rather than a ring. If it contains double bonds, it is said to be unsaturated; if not, it is saturated.

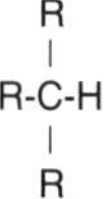

Aliphatic compound, where R is a hydrogen atom or an open chain of carbon atoms rather than a ring

aliphatic hydroxy acid (R-COOH) An organic acid with carbon atoms arranged in open chains, branched or unbranched, rather than rings.

aliphatic-type polyamide membrane A type of synthetic organic membrane with amide-linked open chains of carbon atoms, usually manufactured as a thin film membrane composite flat sheet. The thin film composite type often has a polyamide barrier layer supported by a microporous substrate (e.g., polysulfone). Variations of the membrane types are commonly used for spiral wound reverse osmosis. Typically these membranes exhibit good pH range tolerance but are intolerant to strong oxidants.

aliquot A representative portion of a sample, often an equally divided portion.

alizarin-visual test A laboratory procedure for determining the fluoride concentration in water.

alkali metals A class of soft metals in group IA of the periodic table with closely related outermost electron configurations that react rapidly with water, evolving hydrogen gas and forming an alkaline (basic) solution containing a hydroxide of the formula MOH, where M represents the metal lithium, sodium, potassium, rubidium, francium, or cesium.

alkaline *See* alkaline solution.

alkaline soil Soil having a pH greater than 7.0.

alkaline solution A solution that contains significant numbers of hydroxyl (OH^-) ions such that the pH is greater than 7.0. Such a solution is also called basic water.

alkaline water *See* alkaline solution.

alkaline-earth metals Metals in group IIA of the periodic table: beryllium, magnesium, calcium, strontium, and barium.

alkalinity A measure of the capacity of a water to neutralize strong acid; in natural waters this capacity is usually attributable to bases such as bicarbonate (HCO_3^-), carbonate (CO_3^{2-}), and hydroxide (OH^-) and to a lesser extent silicates, borates, ammonia (NH_3), phosphates, and organic bases. It is expressed in milligrams of equivalent calcium carbonate per litre (mg $CaCO_3$/L).

alkalinity test One of several analytic methods used to determine the acid-neutralizing capacity of a sample. Typically, a water sample is titrated with a standard solution of acid to a specific end point. End points can be detected by titrating potentiometrically or to a color change with an indicator dye.

alkane (C_iH_{2i+2}, where $i \geq 1$) An organic chemical in which the basic building group is ($-CH_2-$). It does not contain double bond (saturated) compounds. CH

$$CH_3\text{-}(CH_2)_i\text{-}CH_3$$

Alkane, where $i \geq -1$ ($i = -1$ corresponds to CH_4)

alkylbenzene sulfonate (ABS) ($R\text{–}C_6H_4SO_3^-$) A class of branched-chain sulfonate type of synthetic detergents. Usually it is a dodecylbenzene ($C_{12}H_{25}C_6H_4SO_3^-$) or tridecylbenzene sulfonate. In actuality, tridecylbenzenesulfonate is not a true compound but rather a mixture of C_{12} and C_{15} alkyl benzene sulfonates that approximates C_{13}.

R
C
H-C C-H
H-C C-H
C
SO_3^-

Alkylbenzene sulfonate, where R is a branched chain of 11 to 15 carbon atoms

allele One of two copies of a genetic locus in a diploid organism. One copy exists on one chromosome of a pair, the other on the other chromosome. These are not necessarily identical because one is inherited from each parent. This is a property common to all chromosomes except the sex chromosomes. One allele, if expressed, is all that is required to express a particular characteristic if it is a dominant trait. Therefore, if damage is induced in one allele of a genetic locus, it does not necessarily produce a harmful effect (although it does increase the probability of certain types of health effects, such as cancer). However, mutation at both loci can result in serious health effects if the cell survives (i.e., if the effects are not lethal).

allergy A response to a chemical that is mediated through the immune system. A prior exposure is required to sensitize the individual. Such effects are generally produced in a relatively small number of individuals within a population. They can range from mild skin reactions or hay fever to severe and unpredictable, even life-threatening, reactions. These reactions are difficult to predict from animal studies.

alligator teeth V-notched weirs in a short (3–6 feet [1–2 metres]) U-shaped effluent structure for a sedimentation basin. Frequently these structures cover the entire end of the basin.

allochthonous Pertaining to turbidity-causing material that originates from outside and is carried into a lake or other surface water. Such material includes humus, silt, organic detritus, colloidal matter, and plants and animals.

aluminum (Al) A metallic element. Aluminum is the most abundant metal in the earth's crust; it does not occur free in nature.

alluvia *See* alluvial deposit.

alluvial Pertaining to soil materials deposited by running water.

alluvial aquifer A water-bearing geologic unit composed of material deposited by flowing rivers. An alluvial aquifer is usually a good source of easily exploited groundwater when the aquifer is adjacent to a flowing stream.

alluvial deposit Mud, sand, or gravel deposited by water flowing over land that is not usually submerged. Alluvial deposits may occur after a heavy rainstorm.

alluvial fan A landform created in relatively dry climates when large quantities of sediments are deposited to the dry valley floors by seasonal rivers draining nearby mountains. These landforms are often the only water-bearing unconsolidated deposits in these climates.

alluvial plain A plain formed by alluvial material eroded from areas of higher elevation.

alluvial river A river that has formed its channel by the process of aggradation. The sediment that it carries is similar to that in its bed and banks.

alluvial-slope spring A spring occurring on the lower slope of an alluvial cone at the point where the water table slope and surface gradient are equal. Such a spring is also called a border spring. *See also* boundary spring.

alluvial terrace A terrace, usually adjacent to a river valley, that was originally deposited by stream action and, in later geologic periods, was cut through by the stream and thus left at some distance above the streambed.

alluvium Sediments deposited by streams or running water. *See also* alluvial deposit.

alpha (α) particle A positively charged particle emitted by certain radioactive materials. It consists of two neutrons and two protons and is identical to the nucleus of a helium atom. It is the least penetrating of the three common forms of radiation—alpha, beta (β), gamma (γ)—and is stopped by a sheet of paper. *See also* radiation; radionuclide.

alpha decay A radioactive process in which an alpha particle is emitted from the nucleus of an atom, decreasing the atomic number of that atom by two.

alpha value A commonly used variable name to indicate the level of statistical significance for a test of a hypothesis.

alphanumeric variable A variable, as found within a computer program, that includes alphabetical as well as numeric characters.

alternating current (AC) A periodic current for which the average value over a period is zero. AC is measured in amperes.

alternating system A multiple-unit system that functions with one or more units in service and one or more on standby.

alternative disinfectant A disinfectant used in place of chlorine, typically to minimize the formation of chlorination disinfection by-products. Common alternative disinfectants include

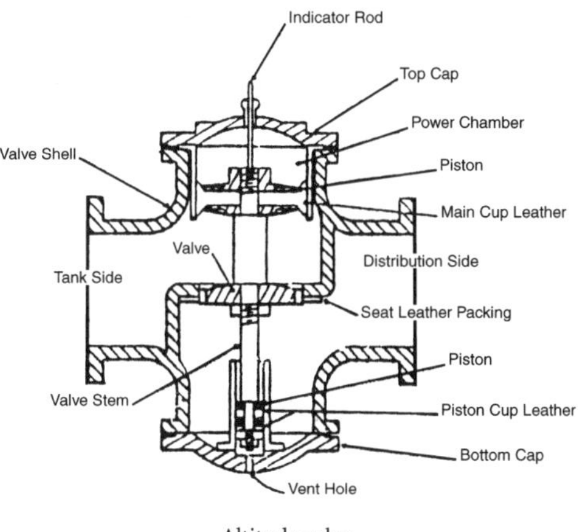

Altitude valve

chloramines (NH_xCl_y, where $x = 0–2$, $y = 1–3$), chlorine dioxide (ClO_2), and ozone (O_3). *See also* chloramines; chlorine dioxide; ozone.

alternative dispute resolution A class of techniques designed to result in "win-win" resolutions to disputes. Techniques include facilitated discussion, interest-based bargaining, interest-based negotiation, mediation, conflict resolution, and arbitration.

alternative funding mechanism A method for funding a state or federal drinking water program with funds other than general tax revenue, such as operating fees, plan and specification reviews, and inspection fees.

altitude valve A valve that automatically shuts off the flow into an elevated tank when the water level in the tank reaches a predetermined level. The valve automatically opens when the pressure in the distribution system drops to less than the pressure in the tank. An altitude valve is sometimes called an altitude-control valve.

ALU *See* arithmetic and logic unit.

alum ($Al_2(SO_4)_3$·14 H_2O) The common name for aluminum sulfate, a chemical used in the coagulation process to remove particles from water. *See also* aluminum sulfate; coagulation.

alum sludge Amorphous solids that result from the addition of alum as a coagulant during water treatment. Alum sludge is comprised both of hydrolyzed floc and particles that are entrapped in the hydrolyzed solids. Sludge is typically collected from the bottom of sedimentation basins and in settled backwash water from the filtration processes. *See also* alum; backwash.

alumino silicate *See* gel zeolite.

aluminum sulfate ($Al_2(SO_4)_3$) An inorganic compound commonly used as a coagulant in water treatment. It contains waters of hydration, $Al_2(SO_4)_3 \cdot XH_2O$ (where X is a variable number). Aluminum sulfate is often called alum.

aluminum silicate *See* pumicite.

Alzheimer's disease A brain disease of unknown origin that results in memory loss, among other symptoms.

AM *See* asset management; automated mapping system.

AM/FM system *See* automated mapping/facilities management system.

AM/FM/GIS *See* automated mapping/facilities management/geographic information system.

ambient Surrounding. For example, a test performed at ambient temperature would be performed at the temperature surrounding the apparatus, i.e., laboratory temperature if the equipment were in a laboratory.

ambient temperature Temperature of the surrounding air (or other medium), for example, the temperature of a room where chemical-feeding equipment is installed.

ambient water quality standards Provisions of state or federal law set under the authority of the Clean Water Act that consist of a designated use or uses for the waters within the state or of the United States and water quality criteria for these waters based on the designated use. Ambient water quality standards protect public health and welfare, enhance the quality of surface waters, and serve the purposes of the Clean Water Act.

amebiasis *See* waterborne disease.

American Academy of Environmental Engineers (AAEE) A professional association that encourages excellence in engineering and provides certification for qualified environmental engineers.

American Association of Engineering Societies (AAES) A professional organization founded in 1980 as the successor to the Engineers Joint Council to foster cooperation among engineering societies. It is composed of 31 member societies.

American Chemical Society (ACS) A professional society for chemists and chemical engineers.

American Gas Association (AGA) A professional organization of gas producers.

American Industrial Hygiene Association (AIHA) An association whose purpose is to promote the field of industrial hygiene, to provide education and training, and to represent the interests of industrial hygienists and those they serve.

American Institute of Chemical Engineers (AIChE) A professional organization of chemical engineers.

American Institute of Electrical Engineers (AIEE) A professional organization of electrical engineers.

American National Standards Institute (ANSI) An association of individuals and organizations involved in voluntary consensus standards development.

American Public Health Association (APHA) An association representing the variety of disciplines in the public health profession.

American Public Works Association (APWA) A professional organization whose members are interested in public works, drinking water, wastewater, electricity, gas, storm flow, and the like.

American Society for Testing and Materials (ASTM) An organization founded in 1898 that establishes standards for materials, products, systems, and services through technical committees. More than 9,000 standard test methods, specifications, classifications, definitions, and recommended practices are now in use.

American Society of Civil Engineers (ASCE) A professional organization of engineers interested in the practice of civil and environmental engineering.

American Society of Mechanical Engineers (ASME) A professional association representing mechanical engineers.

American Society of Safety Engineers (ASSE) A group of safety professionals dedicated to advancing the interests of safety professionals and fostering the well-being and professional development of its members.

American Standard Code for Information Interchange (ASCII) A code established by the American National Standards Institute that assigns 8-bit codes to all standard keyboard characters to provide compatibility for data communications.

American Standard fittings The standardized types and dimensions of the various malleable iron, cast-iron, steel, or other metal pipe fittings as set forth in standards published by the American National Standards Institute.

American Water Resources Association (AWRA) An organization dedicated to the advancement of interdisciplinary water resources research, planning, management, development, and education.

American Water Works Association (AWWA) A professional association representing the drinking water supply profession.

American Water Works Association Research Foundation (AWWARF) The only nongovernmental organization that sponsors research for the drinking water profession. It is independent of, but related to, the American Water Works Association.

American Rule A groundwater ownership doctrine that states that the landowner has the right to use only a reasonable amount of groundwater.

Ames test A test for mutagenic activity using a series of genetically engineered strains of *Salmonella*. It is frequently used in combination with drug-metabolizing enzymes isolated from rodent livers, usually in the form of a supernatant liquor from a $9{,}000 \times g$ centrifugation of liver homogenates (the so-called S9 fraction). This test is most frequently employed as a prescreen for potential carcinogenic activity. It is also utilized to determine whether a chemical that has been shown to induce cancer may be acting through a genotoxic mechanism. *See also* carcinogenesis bioassay.

amine ($R_{3-x}NH_x$, where $x = 0, 1,$ or 2) Any of a class of organic compounds of nitrogen that may be considered as derived from ammonia (NH_3) by replacing one or more of the hydrogen atoms with alkyl (C_nH_{2n+1}) groups. The amine is primary ($R\text{-}NH_2$), secondary ($R_1\text{-}NH\text{-}R_2$), or tertiary ($R_1\text{-}N(R_2)\text{-}R_3$) depending on whether one, two, or three of the hydrogen atoms are replaced, respectively. *See also* alkane; organic compound.

amino acid An important class of organic compounds that form building blocks of proteins. The 20 standard (alpha) amino acids have the general formula $R\text{-}NH_2$.

```
R-CH-COOH
     |
    NH2
```

Amino acid, where R is a carbon-containing group or a hydrogen atom

aminolevulinic acid (ALA) ($NH_2\text{-}CH_2\text{-}CO\text{-}CH_2\text{-}CH_2\text{-}COOH$) An amino acid precursor of heme proteins with the chemical name 5-amino-4-oxopentanoic acid. Heme is the part of certain proteins that interacts with oxygen (e.g., hemoglobin and cytochromes). It is used as a biomarker, most often for lead exposure.

aminolevulinic acid dehydrase (ALAD) The enzyme that catalyzes the conversion of aminolevulinic acid to porphobilinogen, an intermediate in the synthesis of heme. It is inhibited by lead. It is also called a dehydratase.

ammeter An instrument for measuring amperes (electrical current).

ammonia (NH_3) An inorganic gas commonly detected in wastewater. Ammonia may be present in drinking water that has been chloraminated, as well as in source waters.

ammonia nitrogen ($NH_3\text{-}N$) A common way to report a concentration of ammonia (expressed as nitrogen). The concentration of ammonia multiplied by 17/14 yields the concentration of ammonia nitrogen. Ammonia nitrogen is sometimes called ammonia as nitrogen.

ammoniator An apparatus used for applying ammonia (NH_3) or ammonium (NH_4^+) compounds to water.

ammonium (NH_4^+) One form of nitrogen usable by plants.

ammonium hydroxide (NH_4OH) A strongly basic inorganic solution formed when ammonia (NH_3) dissolves in water. It is often used as a source of ammonia in the formation of chloramines.

amoeba A single-celled protozoan that is widely found in fresh water and salt water. Some types of amoeba cause diseases such as amoebic dysentery. *See also* protozoa.

amoebal pathogen Any amoebal protozoan that can cause infection and illness in a host species or an accidental host. Such pathogens may be parasitic (require a host species to complete their life cycle) or free-living (do not require a host species to complete their life cycle).

amorphous Noncrystalline; having no ordered molecular structure.

amortization (1) The gradual reduction, redemption, or liquidation of the balance of an account according to a specified schedule of times and amounts. (2) A provision for the extinguishment of a debt by means of a sinking fund.

amperage The strength of an electric current, measured in amperes (coulombs per second); the amount of electric current flow, analogous to the flow of water in gallons per minute or litres per second.

ampere (A) *See the Units of Measure list.*

ampere-hour (A · h) *See the Units of Measure list.*

amperometric Relating to or being a chemical titration in which the measurement of the electric current flowing under an applied potential difference between two electrodes in a solution is used for detecting the end point.

amperometric methods Any of the analytical methods based on an electrochemical neutralization reaction of species in solution. Electric current is measured as it passes through a polarography cell. The most common example in water analysis is the determination of chlorine residual by amperometric titration. In this case, chlorine is an oxidizing species that is neutralized with a reducing agent, such as phenylarsene oxide (C_6H_5AsO).

amperometric titration Volumetric analysis in which the end point is based on the potential–current behavior of an electrode in solution. A common example is the determination of chlorine residual.

amperometric titrator A laboratory analytic device that is used for the determination of oxidants, chlorine, chloramine, and so on in water.

amphoteric Pertaining to a compound that has the ability to act as either an acid or a base. Amino acids are examples of such compounds.

AMR *See* automatic meter reading.

AMU *See* atomic mass unit.

AMW *See* apparent molecular weight.

AMWA *See* Association of Metropolitan Water Agencies.

Anabaena A genus of filamentous Cyanobacteria (formerly blue-green algae) characterized as resembling a string of beads. Various species in this genus are common to lakes and ponds. Dense blooms of these organisms may occur during the spring-to-fall period, contributing to taste-and-odor problems in lakes and reservoirs used for water supply.

anadromous Pertaining to migrating fish growing in the sea and returning to freshwater streams to spawn.

anaerobe *See* anaerobic.

anaerobic (1) In reference to microorganisms, pertaining to organisms that can grow in the absence of oxygen (anaerobes). (2) In reference to the environment, pertaining to the condition in which oxygen is absent or greatly reduced in concentration.

anaerobic condition An environmental condition in which oxygen is not available as a terminal electron acceptor. Under such a condition, compounds such as iron, manganese, and sulfur are reduced to dissolved constituents that must be removed in the treatment process.

anaerobic organism *See* anaerobic.

anal–oral route of exposure *See* fecal–oral transmission.

analog (1) Pertaining to a medium or mode in which data are represented by continuously variable quantities such as amplitude, frequency, shape, or position. Hard copy or screen displays of maps, drawings, records, and photographs are analog images. In analog phone communications, voices are transmitted as an electrical signal that continuously varies in frequency and amplitude. (2) Pertaining to the readout of an instrument by a pointer (or other indicating means) against a dial or scale. *Contrast with* digital.

analog controller A device used for proportion–integral–derivative deviation control. The device operates on analog values (voltages, currents, or air pressures).

analysis (1) An examination of a sample for its microbiological or chemical constituents. (2) An examination of data, fundamental principles, or operations in whole or in parts to determine their nature, proportion, functions, and interrelationships.

analysis of variance (ANOVA) A commonly used statistical procedure to compare the means of several normally distributed data sets to test the hypothesis that the data sets are statistically similar.

analyte The substance for which an analysis is performed. For example, if an analysis for the concentration of calcium in water was performed, calcium would be the analyte.

analytic balance A sensitive balance (device for weighing substances) used to make precise weight measurements.

analytic epidemiologic study An epidemiologic study that investigates the possible causal association between the disease or health status of individuals within a group. In contrast to ecologic studies, analytic studies obtain information about each individual's disease status, his or her exposures to possible risk factors and confounding variables, and other individual demographic characteristics. *See also* case-control epidemiologic study; cohort epidemiologic study; ecologic epidemiologic study.

analytic grade A characteristic of a chemical indicating a level of purity high enough to permit its use in chemical analysis. *See also* reagent grade water.

analytic method An analysis for which the description is sufficiently detailed to be set up in a laboratory. This type of method is less detailed than a standard operating procedure.

analytic standard A substance of known purity and concentration used to calibrate an analytic method.

analytic triangulation A process used in photogrammetric mapping that establishes the mathematical relationships among control points and adjusts them as a network of location values. Analytical triangulation involves a computer-aided extension of the control points in order to mathematically densify the control network and provide a positionally accurate structure for the photographs. *See also* control point.

analyzer A device that conducts periodic or continuous measurements of some factor such as chlorine or fluoride concentration or turbidity. Analyzers operate by one of several methods, including photocells, conductivity, or complex instrumentation.

anaphylaxis An acute allergic response; also referred to as immediate hypersensitivity. If severe enough, such a reaction can be life-threatening.

anchor ice Ice formed below the surface of a stream or other body of water, on the bed, or on a submerged body or structure.

anchorage A large and heavy block of material installed along a pipe for the purpose of resisting forces caused by the weight of the pipeline, the change in direction of the fluid flowing therein, or expansion or contraction in the pipeline caused by changes in temperature. Anchorage is also called an anchor or anchor block. When resisting forces are caused by change of direction, anchorage is usually called a thrust block.

ancillary charge A separate charge for ancillary services that is not included in costs for general water service. Often in providing water service, the utility must perform these ancillary services, which often benefit only the individual customer using the services and have no systemwide benefit.

anemia A condition in which the red blood cell concentrations in blood are reduced. Anemias can arise from a variety of causes including nutritional deficiencies (e.g., lack of iron) or the effects of toxic chemicals. The most serious type of chemically induced anemia is referred to as aplastic anemia, in which the bone marrow's ability to synthesize red blood cells is obliterated. Hemolytic anemia arises from increased destruction of existing red blood cells by a process referred to as hemolysis.

anemometer An instrument for measuring the force or velocity of wind; a wind gauge.

aneuploidy Loss of a chromosome following cell division.

angiosarcoma A malignant tumor formed from cells of blood vessels. It is frequently referred to by the organ in which it occurs (e.g., hepatic or liver angiosarcoma). However, an angiosarcoma is a tumor of the blood vessels, not of the tissue in which it is found.

angle gate valve *See* angle valve.

angle valve A 90° service fitting with a valve incorporated, used in setting water meters to allow a cutoff and direction change from vertical to horizontal.

angstrom (Å) *See the Units of Measure list.*

angular acceleration The rate of change in angular velocity, in degrees (or radians) per second squared.

angular frequency (ω) For a periodic function, the total angle per unit time.

$$\omega = 2 \times \pi \times f$$

Where:

f = the function's frequency, in hertz

Angular frequency is measured in units of degrees (or radians) per second. For example, if a periodic function had a frequency of 10 hertz, its angular frequency would be 3,600 degrees/second (62.83 radians/second).

angular velocity The rate of circular movement, in degrees (or radians per second).

animal bioassay *See* carcinogenesis bioassay.

animal infectivity The exposure of animals, generally mice, to a particular biological insult (bacterial, viral, protozoan) in an effort to produce a recognizable disease (i.e., histological changes, morbidity, mortality).

animal study An investigation using animals as surrogates for humans, on the expectation that results in animals are pertinent to humans.

anion (X^{-c}, where $c \geq 1$) A negatively charged atom or molecule that forms when an atom acquires one or more extra electrons. Anions may be present in solids and in solution in water or other solvents.

anion exchange A process in which anion contaminants are removed from a liquid phase by contacting a synthetic, porous medium or resin that is coated with other anions. The anions on the medium are exchanged for the anion contaminants. When the medium is depleted of the exchanging anions, it is backwashed with a concentrated solution to restore the bed with the exchange anions and to flush the contaminant anions for subsequent disposal. For exhaustion, the chemical formula is given as follows (where R represents the resin):

$$\overline{R_4N^+OH^-} + NaNO_3 \rightarrow \overline{R_4N^+NO_3} + NaOH$$

For regeneration, the chemical formula is as follows:

$$\overline{R_4N^+NO_3} + NaOH \text{ (high concentration)} \rightarrow \overline{R_4N^+OH^-} + NaNO_3$$

See also anion; ion exchange

anion membrane An electrodialysis membrane that allows the passage of cations (but not anions) and is practically impermeable to water under typical electrolysis system working conditions. An anion membrane is also called an anion transfer membrane.

anionic Having a negative ionic charge.

anionic polyelectrolyte *See* anionic polymer.

anionic polymer A negatively charged polymeric compound used to assist in removing particles from water. Anionic polymers are most typically used as flocculant aids; they bridge floc particles and thereby generate larger particles that can be removed by sedimentation, filtration, or both. *See also* flocculation; polymer.

anisotropic Pertaining to crystals for which the index of refraction varies with the direction of the incident light. This is true of most crystals, e.g., calcite (Iceland spar); it is not true of isometric (cubic) crystals, which are isotropic. *See also* isotropic.

anisotropic hydraulic conductivity Within a permeable medium, such as an aquifer, hydraulic conductivity that is dependent on the direction of measurement. For example, in many soils with high percentages of clay, the horizontal hydraulic conductivities are significantly greater than vertical hydraulic conductivities, and flow velocities are greater horizontally than vertically.

anisotropic membrane A membrane with nonuniform structure in cross section. Typically the support substructure has pores much larger than the barrier layer.

anisotropy A condition in which one or more of the hydraulic properties of a porous medium vary according to the direction of measurement. *See also* anisotropic; hydraulic conductivity.

ANN *See* neural network modeling.

anneal To reassociate or join together deoxyribonucleic acid or ribonucleic acid by base pairing two single-stranded sequences.

annexation The process of extending the service area of a water district or municipality to adjacent property owner(s) or a small water utility in order to provide service.

annotation The alphanumeric text or labels plotted graphically on a map, such as street names, place names, identification numbers, and dimensions.

annual average daily flow (AADF) The average of daily flows for a 12-month period. This value may also be determined by dividing the total volume of flow for the year by 365, the number of days in a year (AADF = annual flow/365).

annual flood The maximum 24-hour average rate of flow occurring in a stream or river during any period of 12 consecutive months. Commonly, the 12-month period is considered to run from October 1 of one year through September 30 of the following year.

annual load factor The load factor taken over an entire year. *See also* load factor.

annual variation The general pattern of a particular parameter throughout the year, obtained by plotting the normal values of the parameter for each month and connecting the points by a smooth curve.

annular reactor A small treatability reactor used to simulate the impact of biological treatment on biological regrowth in the distribution system. Treated water flows through cylindrical reactors with a volume of approximately 1 litre (0.25 gallons) that are stirred, creating internal liquid circulation through four draft tubes in the reactor. Hydraulic conditions in the draft tubes can be varied by changing the rotor speed, thereby simulating various conditions in this simulated distribution system.

annular space The space between the outside of a well casing and the drilled hole.

anode (1) In electrochemical or metal corrosion reactions, the site at which electrons are removed (flowing to the cathode) and dissolution of the metal occurs (releasing metal ions) as a result of the oxidation half-reaction of ions or molecules at the site. The anode is positively charged because of the movement of electrons away from it. (2) A positively charged electrode of an electrolysis cell, attracting negatively charged ions. Electrons flow away from the anode and negatively charged ions are oxidized. *See also* cathode.

anodic Pertaining to a site on a metal surface where the oxidation half-reaction occurs, in which metal ions tend to be released and dissolve and electrons tend to flow away to the cathode.

anodic stripping voltametry (ASV) An electrochemical technique used in the analysis of metals. Metals are plated onto an electrode and then oxidized back into solution at a characteristic voltage. The current generated is a function of the concentration. This technique can also be used to speciate metals.

anomaly A deviation from a norm for which an explanation is not apparent on the basis of available data.

anomorphic zone The zone of rock flowage, especially characterized by silicatization involving decarbonation, dehydration, and deoxidation. *See also* aeration zone; katamorphic zone; weathering zone.

ANOVA *See* analysis of variance.

anoxia A condition in which the blood oxygen level within the body is less than normal. It may be caused by lowered oxygen content of the air, obstruction of the respiratory tract, depressed breathing rate, chemically induced interference with oxygen (O_2) transport (e.g., carbon monoxide, CO), chemically induced interference with oxygen utilization (e.g., cyanide, CN^-), or interference with blood flow to a particular part of the body.

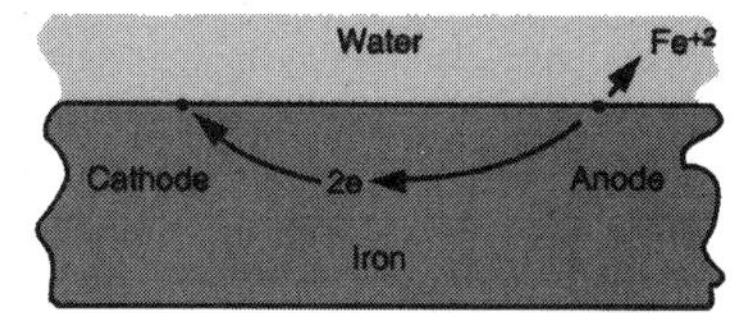

A. Minor Variations Cause Electric Current to Develop

$H_2O \rightleftharpoons H^+ + OH^-$

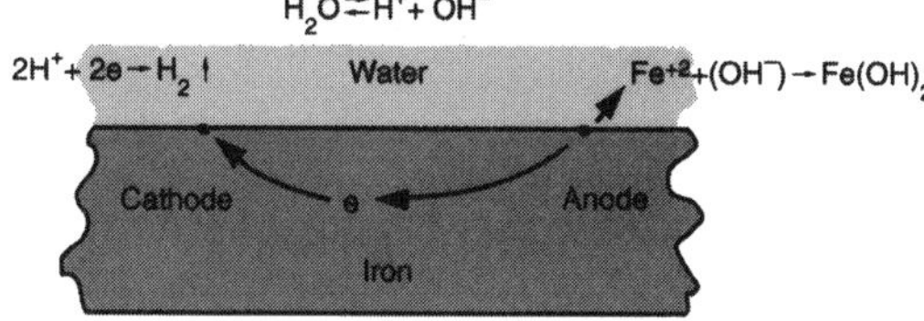

B. Chemical Reactions in Water Balance Those in Iron

Anode and cathode in corrosion

anoxic Lacking in oxygen. This term is synonymous with anaerobic.

anoxic condition A state in which free dissolved oxygen is absent but in which oxygen is available as a terminal electron acceptor through compounds such as nitrate (NO_3^-) or sulfate (SO_4^{2-}).

ANPRM *See* Advanced Notice of Proposed Rulemaking.

ANSI *See* American National Standards Institute.

antagonism In toxicology, a circumstance where exposure to one chemical blocks the effects of another chemical. This term is generally not useful unless the basis of the interaction is known. Antagonism is produced in several ways. A chemical can influence the distribution of a second chemical by interfering with transport processes. An effect can be elicited by one chemical that is the physiological opposite of the effect produced by a second chemical. A chemical can directly interfere with the binding of a second chemical to a receptor site on a protein that is responsible for producing the effect.

antecedent moisture The soil moisture present before a particular precipitation event. It is also called the antecedent-precipitation index.

anthracite *See* anthracite coal.

anthracite coal A particulate form of coal that is used in granular media filters to remove particles from water. Anthracite coal is typically used in dual-media filters in combination with sand. *See also* anthracite coal–sand filter; dual-media filter.

anthracite coal–sand filter A granular filter in which a layer of crushed anthracite coal of a specified size is placed over a layer of sand of a specified size. Such a filter is sometimes called a dual-media filter. These filters can often perform satisfactorily when operated at higher filtration rates than all-sand filters. *See also* dual-media filter.

anthropogenic Pertaining to or involving the impact of humans. *See also* xenobiotic.

antisiphon device *See* vacuum breaker.

antibiosis An antagonistic action of one or more microorganisms to other organisms.

antibiotic A chemotherapeutic chemical compound that is produced by the metabolism of a living organism. Fungi, *Actinomycetes*, and molds have been prolific sources of antibiotics.

antibiotic resistance A characteristic of a microorganism that allows it to neutralize the effect of an antibiotic compound. Antibiotic resistance in bacteria is mediated by (1) genes present in the bacterial genome or (2) extrachromosomal genetic elements (R plasmids) in the cytoplasm that are replicable and transferable (through a process called resistance gene transfer) to other bacteria. When an organism is resistant to two or more antibiotics, it is said to be multiply antibiotic resistant.

antibody A protein in mammals induced by the presence of a foreign substance called an antigen. The specificity of the antigen–antibody binding mechanism makes antibodies useful for identification of microorganisms.

antichlors Reagents, such as sulfur dioxide (SO_2), sodium bisulfite ($NaHSO_3$), and sodium thiosulfate ($NaS_2O_3{\cdot}5H_2O$), which can be used to remove excess chlorine residuals from water by conversion to an inert salt.

anticholinergic A drug or toxin that blocks the effects of acetylcholine, a chemical that is responsible for communication between nerve cells and between nerve cells and other cells within the body.

anticholinesterase A drug or chemical (usually either an organophosphorus or carbamate pesticide) that acts by inhibiting the enzyme that destroys acetylcholine. Acetylcholine serves as a chemical signal between nerves and muscles. This signal cannot be terminated rapidly enough without the activity of this enzyme. A high degree of inhibition is fatal.

anticipated yield The predicted yield of a production well based on an aquifer pumping test but neglecting inefficiencies that will arise after construction and operation.

anticlinal spring A contact spring occurring along the surface outcrop of an anticline (a folded geologic structure where the layer of interest is hill shaped), from a pervious stratum overlying a less pervious stratum.

anticorrosion treatment Treatment to reduce or eliminate the corrosion-producing or aggressive qualities of a water.

antigen A chemical, usually either a protein or a small molecule in combination with a protein, that stimulates the development of antibodies. Chemicals that covalently bind with proteins can often serve as antigens. This can set up a sensitization reaction on subsequent exposures to the chemicals that can result in toxicity that is mediated through the immune system.

antigen capture A system using support-bound antibodies to specifically bind and capture microorganisms, such as hepatitis A virus, from an environmental sample. The system is able to recover small numbers of microorganisms.

antigenic determinant A specific site on an antigen where an antibody binds. *See also* antibody; antigen.

antimetabolite A chemical or drug that closely resembles a normal metabolite structurally. These chemicals exert an effect by blocking reactions that the normal metabolite will participate in, or they may be metabolized to a product that cannot be effectively utilized by the organism.

antimony (Sb) A naturally occurring trivalent or pentavalent metalloid used as a constituent of metal in the manufacture of flame retardants, ceramics, glass, pesticides, and tin–antimony solder, as well as in medicine. It is regulated by the US Environmental Protection Agency.

antioxidant A chemical that is more easily oxidized than the material being protected. Classic examples in the biology of nutrition are vitamin E and ascorbic acid. These vitamins, by oxidizing easily, prevent oxidative damage within the body, although in certain combinations with metals ascorbic acid can act to generate oxygen radicals. Also, a variety of synthetic antioxidants are used as food additives, such as butylated hydroxytoluene (commonly known as BHT).

antiscalant A chemical that inhibits or delays precipitation and subsequent scale formation of sparingly soluble inorganic salts and silica.

antithyroid Pertaining to a chemical effect that decreases the normal function of the thyroid gland. A wide variety of mechanisms can result, ranging from interference with the normal utilization of iodine to the synthesis and release of thyroid hormones.

AOC *See* assimilable organic carbon.

AOP *See* advanced oxidation process.

AOX *See* total organic halogen.

APA *See* Administrative Procedure Act.

APD *See* apparent particle diameter.

aperiodic Not occurring regularly, but rather taking place at unequal intervals of time.

APHA *See* American Public Health Association.

aphotic Lacking light, as in the deeper part of a reservoir.

aplastic anemia *See* anemia.

apoptosis An active process of inducing cell death. It is also referred to as programmed cell death. Most organs (except the brain) have a continual process for replacing damaged cells. Death arises not from the aging of the cell per se, but through the recognition of specific types of damage and the generation of specific cellular signals that cause the cell to die. A similar process occurs during development of tissues when excess cells have been generated and need to be disposed of in the normal maturation process. Some chemicals that induce certain adverse effects (cancer and developmental effects, in particular) may act by suppressing apoptosis, allowing a damaged cell to remain alive and divide. Such cells could probably contain mutations that lead to malignancy. *See also* necrosis.

apparent cohesion Cohesion of moist soils caused by surface tension in capillary interstices. It disappears on immersion of the soil.

apparent color The color caused by substances in solution as well as that caused by suspended matter. Apparent color is determined on an original sample without filtration or centrifugation. In some waters, colloidal or suspended material may contribute to color. In such cases both true color and apparent color should be determined. *See also* true color.

apparent density (1) In soil mechanics, the mass per unit volume of oven-dried soil, pore space included. (2) The mass per unit volume of activated carbon.

apparent groundwater velocity The apparent distance covered by groundwater per unit time, computed by dividing the volume of flowing water per unit time by the cross-sectional area taken perpendicular to the streamlines. It is also called the discharge velocity.

apparent molecular weight (AMW) A measure of molecular weight based on the size of the molecule. Filters with pores of different sizes are calibrated with proteins of known molecular weight. Test molecules that pass through the same filter are assumed to have the same molecular weight as the calibration protein. The unit of this measure is the dalton.

apparent particle diameter (APD) The opening in a screen mesh used for sizing granular particles. The units of measure are millimetres.

apparent power (|S|) The magnitude of the complex power:

$$|S| = (P^2 + Q^2)^{0.5}$$

Where:

P = the active power

Q = the reactive power

For a sinusoidal single-phase circuit, the apparent power is given by

$$|S| = V \times I$$

Where:

V = root mean square (rms) voltage, in V

I = rms current, in A

For a sinusoidal three-phase circuit that is symmetrical and balanced, the apparent power is given by

$$|S| = (3)^{0.5} \times V \times I$$

Where:

V = the line-to-line rms voltage, in V

I = the rms current, in A

Apparent power is a scalar quantity and is customarily measured in units of volt-amperes.

apparent specific gravity The ratio of the mass of a unit volume of oven-dried soil to the mass of an equal volume of water under standard conditions. The term may be applied to either oven-dried undisturbed field samples or to samples manipulated in the laboratory. Referring to treatment by granular activated carbon (GAC) adsorption, the apparent specific gravity of the GAC is used to indicate the success of the regeneration process. *See also* apparent density.

apparent volume of distribution (*V*) The dose or concentration in a particular body fluid (usually blood) extrapolated to the time the drug was administered. In pharmacokinetics (or toxicokinetics), this quantity is used to describe the distribution of a drug within the body in abstract terms. A highly lipid-soluble chemical will have a very large volume of distribution (one that can actually be much greater than the total body volume) because the chemical will be concentrated in the fat, leaving a low concentration in the blood. This relationship is used in medicine primarily to calculate the dose needed to achieve a certain blood level in the therapeutic range. This concept is inherent in all the simple compartmental models used in pharmacokinetics.

application rate The rate of delivery by an irrigation circuit, in inches or centimetres for sprinkler irrigation and in gallons or litres for drip irrigation.

application software Software developed for a specific, practical application. Examples are word processing, database management, network solution, mapping, and illustration.

applications program *See* application software.

applied climatology The study of weather and its effects on human activities. *See also* climatology.

applied pressure The feedwater hydraulic pressure applied to a process. In pressure-driven membrane separation, applied pressure is equal to the hydraulic feedwater pressure minus the permeate pressure.

appraisal The value placed on a firm's assets by an estimator or assessor considering age, wear, condition, obsolescence, and marketability, usually to establish a value for the purpose of sale or taxation; the act of appraising the value of assets.

appraisal inventory A detailed list of the individual items constituting an assembled property, with or without unit costs or prices applied.

appreciation The increase in monetary value that is attributed to assets (especially land) from one period to the next, resulting from increased demand or marketability.

appropriate To take the legal actions necessary to establish the right to take water from a natural stream or aquifer for beneficial use.

appropriation (1) Federal funding limits, set under the jurisdiction of the House and Senate Committees on Appropriations, up to which, if approved by the US Congress, federal agencies can legally incur obligations and make payments out of the US Treasury for specified purposes. (2) The setting aside moneys for a specific use.

appropriation doctrine A legal concept developed in the western United States whereby a right to surface water is obtained simply by taking water and applying it to a beneficial use, with no limitations on the place of use or origin of the water. In times of shortage, the appropriation most junior in time must cease using water, if necessary, in order to satisfy more senior appropriations—hence the expression "first in time, first in right."

appropriation system A means of regulating water use by issuing permits so that overdrafts cannot occur. This approach is also called a permit system.

appropriative right Water rights to or ownership of a water supply that is acquired by diverting and putting the water to beneficial use following procedures established by state statutes or courts.

appropriator of water rights One who diverts and puts to beneficial use the water of a stream or other body of water under a water right obtained through appropriation.

appurtenances Machinery, appliances, structures, and other parts of a main structure that are necessary to allow it to operate as intended but are not considered part of the main structure.

apron A floor or lining of concrete, timber, or other resistant material placed at the toe of a dam, bottom of a spillway, chute, and so on to protect the surface from erosion from falling water or turbulent flow.

APWA *See* American Public Works Association.

aquaculture A process for removing contaminants from water through the use of aquatic plants (e.g., water hyacinths) in pond contaminants. The contaminants are either synthesized by, or bioaccumulated in, the aquatic plants, which ultimately are harvested for disposal. Aquaculture facilities are significantly more land-intensive than mechanical treatment methods but are simpler to operate.

aquagenic organic matter Natural organic matter originating from aquatic media. *See also* natural organic matter.

aquatic Associated with water.

aquatic animal A member of the animal kingdom that lives in water, e.g., fish, zebra mussels, nematodes.

aquatic ecosystem (1) A basic ecological unit composed of living and nonliving elements interacting in an aqueous mixture. (2) Waters, including wetlands, that serve as habitat for interrelated and interacting communities and populations of plants and animals.

aquatic environment *See* aquatic ecosystem.

aquatic fulvic acid A complex organic compound of unknown specific structure that leaches from decaying vegetation. Aquatic fulvic acids are the cause of most of the visible brownish color in some waters. Although nontoxic, they are major precursors of disinfection by-products. They make up most of the natural organic matter in water and are lower in molecular weight than aquatic humic acids. *See also* aquatic humic acid; disinfection by-product precursor; natural organic matter.

aquatic growth The aggregate of floating and attached organisms in a body of water; plankton.

aquatic habitat *See* aquatic ecosystem.

aquatic humic acid A complex organic compound of unknown specific structure that leaches from decaying vegetation. Aquatic humic acids contribute to the visible brownish color in some waters. Although nontoxic, they are major precursors of disinfection by-products. They make up about 10 percent of the natural organic matter in water and are higher in molecular weight than aquatic fulvic acids. *See also* aquatic fulvic acid; disinfection by-product precursor; natural organic matter.

aquatic humic substance A complex organic compound of unknown specific structure that can leach from decaying vegetation. Aquatic humic acids and aquatic fulvic acids fall in this category. Aquatic humic substances make up a significant fraction of the natural organic matter in water and are major precursors of disinfection by-products. *See also* aquatic fulvic acid; aquatic humic acid; disinfection by-product precursor; natural organic matter.

aquatic life All forms of animal and plant life that live in water.

aquatic plant A member of the plant kingdom that lives in water, e.g., algae, phytoplankton.

aqueduct A conduit, at or above ground level, usually of considerable size and open to the air, used to convey water by gravity flow.

aqueous Made up of, similar to, or containing water; watery.

aqueous vapor The gaseous form of water. *See also* water vapor.

aquiclude A low-permeability geologic unit that forms either the upper or lower boundary of a groundwater flow system.

aquifer A geologic formation, group of formations, or part of a formation that is saturated and sufficiently permeable to transmit economic quantities of water to wells and springs.

aquifer recharge area That land above an aquifer that contributes water to it.

aquifer safe yield The quantity of naturally occurring groundwater that can be economically and legally withdrawn from an aquifer on a sustained basis without impairing the native groundwater quality or creating an undesirable effect. It cannot exceed the increased recharge or leakage from adjacent formations plus the decreased natural discharge resulting from decline in head caused by pumping. *See also* groundwater safe yield.

aquifer storage and recovery (ASR) A management strategy in which excess surface water is treated and artificially recharged to an aquifer system for later withdrawal when surface water is in short supply.

aquifer storage recovery (ASR) well A well designed to inject treated water into an aquifer during periods of surplus water and to extract the treated water during periods of great water demand. The aquifer is used as a storage facility in this situation.

aquifer test A test made by pumping a well over a period of time and observing changes in hydraulic head in the aquifer. The test is used to determine the capacity of the well and the hydraulic characteristics of the aquifer.

aquifer transmissivity A measure of the ability of an aquifer to transmit water per unit width of the aquifer.

aquitard A low-permeability geologic unit that can store groundwater and also transmit it slowly from one aquifer to another.

arbitrage A buying of bills of exchange, stocks, bonds, and so on in one market and selling them at a profit in another. Note that, except for certain temporary periods defined in the federal tax code, governments normally are not permitted to reinvest the proceeds of tax-exempt debt in higher-yielding taxable securities. Even during these temporary periods, interest earnings in excess of interest expense must be rebated to the federal government.

arbitrage bond Any bond issue for which any portion of the proceeds is reasonably expected (at the time of issuance) to be used directly or indirectly to acquire higher-yielding investments or to replace funds that were used directly or indirectly to acquire higher-yielding investments (per Internal Revenue Service Code, Sec. 148).

arbitration The settlement of a dispute by one or more persons chosen to reach a decision after hearing both sides.

arbitration, binding *See* binding arbitration.

arbitration, nonbinding *See* nonbinding arbitration.

arch A curved structure supporting weight, for example, an arch dam or the curved top of a conduit.

arch dam A curved masonry dam that depends principally on arch action for stability. Such a dam is also called an arched dam or single-arch dam.

arch gravity dam A curved solid-masonry dam that depends on both arch action and gravity action for stability.

archie A method of searching files on anonymous file transfer protocol servers. *See also* file transfer protocol.

Archimedes principle The principle of buoyancy, stating that the resultant force on a wholly or partly submerged body is equal to the weight of the fluid displaced and acts vertically upward through the center of gravity of the displaced fluid.

Archimedes' screw A screw pump consisting of an inclined shaft carrying one or more helices (screw flights) that are rotated, with little clearance, in a circular or semicircular conduit, thus lifting water.

arching A condition that occurs when dry chemicals bridge the opening from the hopper to the dry feeder, preventing the discharge of chemicals from the feeder.

archive A place where public documents and records, private documents and records, or both, are kept to preserve and safeguard them.

area A bounded, continuous two-dimensional object (the boundary may or may not be included). Numerical units of area are expressed in terms of length squared (square feet or square metres). For example, the amount of land a parking lot occupies is an area.

area of diversion That portion of an adjacent area beyond the normal groundwater or watershed divide that contributes water to the groundwater basin or watershed under discussion.

area of influence The land area that has the same horizontal extent as the part of the water table or other piezometric surface that is perceptibly lowered by the withdrawal of the water through a well at a given rate. The area of influence for a given rate of discharge may vary with the period of withdrawal and the rate of recharge.

area takeoff A calculation of area based on measurements from plans drawn to scale.

area–capacity curve A graph showing the relationship between the surface area in a reservoir, lake, or impoundment and the corresponding volume.

areal cover A measure of dominance that defines the degree to which aboveground portions of plants cover the ground surface. The total areal cover for all strata combined in a community or a single stratum can exceed 100 percent because (1) most plant communities consist of two or more vegetative strata, (2) areal cover is estimated for each vegetative layer, and (3) foliage within a single layer may overlap.

areal standard unit count (asu) A unit of measurement used in the evaluation of the number of aquatic plankton, frequently algae, in water (this number is sometimes called the standing crop). A small volume of water is examined microscopically and the number of areal standard units counted. One areal standard unit is equal to four small squares in a Whipple grid at a magnification of 200. Areal standard units represent the number per unit volume. *See also* Whipple grid.

ARI *See* average rainfall intensity.

arid Pertaining to climatic conditions or a soil that lacks humidity.

arid climate A climate characterized by less than 10 inches (25 centimetres) of annual rainfall.

ARIMA *See* autoregressive integrated moving average forecasting method.

arithmetic and logic unit (ALU) A central part of the computer within the central processing unit that performs numerical operations fundamental to the execution of programs and other functions.

arithmetic average *See* arithmetic mean.

arithmetic mean An average value of a data set calculated by determining the sum of all the numbers in the data set and dividing by the number of data values within the set.

arithmetic operators Symbols representing various arithmetic operation, such as "+" for addition.

arithmetic scale A series of equally spaced intervals (marks or lines), usually marked along the side and bottom of a graph, that represent the range of values of the data being presented.

aromatic heterocycle A molecule having an aromatic ring composed of both carbon and heteroatoms, chiefly oxygen, sulfur, and nitrogen. *See also* aromatic hydrocarbon.

Pyridine, a nitrogen-containing aromatic heterocycle

aromatic hydrocarbon A six-carbon ring with alternating single and double bonds. *See also* polynuclear aromatic hydrocarbon.

Benzene, an example of an aromatic hydrocarbon, where one to several carbon-containing groups can replace hydrogen atoms to form other compounds

aromatic polyamide membrane A type of synthetic organic membrane with amide-linked closed rings of carbon atoms, each having one free valence bond site, usually manufactured as either an asymmetric hollow fiber or a thin film composite flat sheet. The hollow fiber type is typically asymmetric, with the supporting and "rejecting" layers made of aromatic polyamide. The thin film composite type often has a cross-linked polyamide barrier layer supported by a microporous substrate (e.g., polysulfone). Variations of the membrane types are commonly used for reverse osmosis and nanofiltration. Typically these membranes exhibit good pH range tolerance but are susceptible to attack by strong oxidants.

aromatic sulfonate (Ar-$(SO_3^-)_x$) [where Ar is an aromatic hydrocarbon—e.g., benzene, C_6H_6; naphthalene, $C_{10}H_8$; or anthraquinone, $C_6H_4(CO)_2C_6H_4$)—that may be substituted, e.g., with an amino (NH_2), nitro (NO_2), or hydroxy (OH) group, and where x typically equals 1 to 3)]

A type of synthetic organic chemical produced and applied in the chemical industry. Aromatic sulfonates are employed in the manufacture of azo dyestuffs (R-$(N{=}N)_y$, where $y = 1$–4), optical brighteners, ion exchange resins, plasticizers, pharmaceuticals, and fluorescent whitening agents for laundry products. They have been found as contaminants in river water.

aromaticity The condition in which a stable electron shell configuration in organic molecules, especially those related to benzene, renders aromatic rings very stable. Aromaticity is also known as resonance stabilization or delocalization. In nuclear magnetic resonance spectroscopy, aromatic rings produce a clear and dramatic shielding effect that assists the spectroscopist in assessing the aromaticity of a sample. *See also* aromatic hydrocarbon; benzene; nuclear magnetic resonance; resonance.

array In a pressure-driven membrane separation system, an arrangement of membrane elements or pressure vessels by stage. For example, a 4:2:1 pressure vessel array contains four pressure vessels in the first stage, two in the second stage, and one in the third stage. In a multiple-staged array, the concentrate from the first stage is the feed to the second stage, the concentrate from the second stage is the feed to the third stage, and so on.

arrhythmia An irregular beat of any of the heart chambers. Some arrhythmias are relatively benign (e.g., atrial flutter); others can be fatal (e.g., ventricular fibrillation). Arrhythmia is a concern for drinking water containing barium (Ba^{2+}).

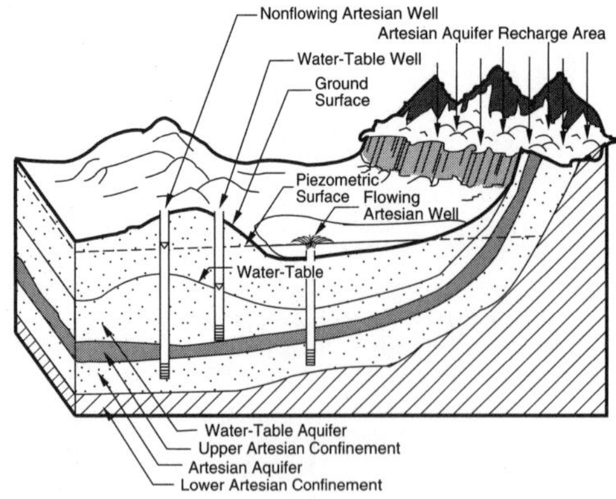

Artesian aquifer

arroyo A stream channel or gully, usually rather small, walled with steep banks and dry much of the time.

arsenic (As) An inorganic contaminant that is found in water supplies primarily as a result of natural geologic formations and has been associated with adverse health effects. It is regulated by the US Environmental Protection Agency.

artesian Pertaining to groundwater, a well, or an underground basin in which the water is under greater-than-atmospheric pressure and will rise higher than the level of its upper confining surface if given an opportunity to do so.

artesian aquifer An aquifer confined between less permeable materials from which water will rise above the bottom of the overlying confining bed if afforded an opportunity to do so. Such an aquifer is also called a confined aquifer.

artesian discharge The rate of discharge of water from a flowing well.

artesian head The distance, whether above or below the land surface, to which the water in an artesian aquifer or groundwater basin would rise if free to do so.

artesian pressure The pressure exerted by groundwater against an overlying impermeable or less permeable formation when the free water surface of the groundwater stands at a higher level than the bottom of the overlying formation.

artesian spring A spring from which water issues under pressure through some fissure or other opening in the confining formation above the aquifer.

artesian well A well that flows freely, without pumping, as a result of the piezometric surface of the aquifer area being at a higher elevation than the well discharge. Artesian wells are found in mountainous areas in which the aquifer supply is confined between two impervious layers, permitting a pressurized system to exist at a higher elevation than the well discharge.

artesian well water *See* bottled artesian water.

artifact A data point that does not fit into an experimental result properly. For example, if something went wrong during a portion of a test and was then corrected, the data point collected during the problem portion of the test would be called an artifact.

artificial intelligence (AI) The feature or property of a machine that allows it to process information based on reason and to learn new information (or update existing reasoning logic) based on additional new information processed by the machine. Expert systems are based on artificial intelligence in which available knowledge about a process is entered into the model and reasoning logics are assigned; on the basis of these logics, the machine can arrive at conclusions when new or different scenarios are presented to it. *See also* expert systems.

artificial neural network (ANN) modeling *See* neural network modeling.

artificial recharge The process of intentionally adding water to an aquifer by injection or infiltration. Dug basins, injection wells, or the simple spreading of water across the land surface are all means of artificial recharge.

artificial watercourse A surface watercourse constructed by human agencies.

artificial wetland A wetland created by the activities of humans, either accidentally or purposefully.

artificially created beta particle and photon emitter All radionuclides emitting beta particles, photons, or both that are listed in Maximum Permissible Body Burdens and Maximum Permissible Concentration of Radionuclides in Air or Water for Occupational Exposure, National Bureau of Standards Handbook 69 as amended August 1963 (Washington, D.C.: US Department of Commerce), except the daughter products of thorium-232, uranium-235, and uranium-238.

As *See* arsenic.

as $CaCO_3$ *See* as calcium carbonate.

as calcium carbonate (as $CaCO_3$) A phrase used in expressing particular concentrations for one or more chemicals in terms of an equivalent concentration of calcium carbonate ($CaCO_3$). The particular concentrations can include the following: (1) hardness; (2) acidity; or (3) carbon dioxide (CO_2), carbonate (CO_3^{2-}), bicarbonate (HCO_3^-) hydroxide (OH^-), or total alkalinity. The resulting concentration is expressed in milligrams per litre as calcium carbonate equivalent. Expressing these values in terms of calcium carbonate facilitates comparisons of values. For example, to convert a hardness concentration caused by calcium (Ca^{2+}) to an equivalent calcium carbonate concentration, one can multiply the calcium concentration by 2.50, based on the following equation:

$$\frac{\dfrac{50 \text{ milligrams } CaCO_3}{\text{per milliequivalent}} \times \dfrac{2 \text{ milliequivalents}}{\text{per millimole}}}{40 \text{ milligrams } Ca^{2+} \text{ per millimole}} = 2.50$$

Hence, for example, 50 milligrams per litre of calcium ions can be expressed as 125 milligrams per litre as calcium carbonate.

as-built plan A revised site plan reflecting the actual conditions of a landscape installation.

asbestos fiber A fibrous silicate mineral generally ranging from 0.03 to 0.10 micrometres in diameter and less than 10 micrometres in length. Asbestos fibers, from both natural and human-made sources, have been associated with adverse health effects when inhaled. Very little information is available regarding the adverse health effects of ingesting asbestos fibers, and studies are under way. Asbestos fibers are regulated by the US Environmental Protection Agency.

asbestos–cement (A–C) pipe A water main material made from a mixture of cement and asbestos fibers.

ASCE *See* American Society of Civil Engineers.

ASCII *See* American Standard Code for Information Interchange.

ASDWA *See* Association of State Drinking Water Administrators.

aseptic Free from living pathogens or organisms causing fermentation or putrefaction; sterile.

ash The mineral oxide constituents of activated carbon. Ash is normally measured on a weight percent basis after a given amount of sample is oxidized.

Asian-Pacific Group (ASPAC) The association representing the Asian and Pacific Rim members of the International Water Supply Association.

Asiatic clam A fresh water clam, *Corbicula fluminea*, that was introduced into US waters from southeast Asia in 1938. It is now present in almost all surface waters south of 40° latitude and causes problems by clogging intakes and mechanical systems.

ASME *See* American Society of Mechanical Engineers.

ASPAC *See* Asian-Pacific Group.

aspartic acid ($COOHCH_2CH(NH_2)COOH$) An organic compound (amino acid) containing two -COOH groups and one -NH_2 group. It is a precursor of a class of disinfection by-products called haloacetonitriles. Free chlorine reacts with aspartic acid to form some members of the haloacetonitrile group. *See also* amino acid; disinfection by-product; haloacetonitrile.

asphyxia A lack of oxygen, causing unconsciousness or death.

aspirate To remove a fluid from a container by suction.

aspirator A feeder using a hydraulic device that creates suction. The suction draws the material to be fed into the flow of the water.

aspirator feeder *See* aspirator.

ASR *See* aquifer storage and recovery.

assay A test for a particular chemical or effect.

ASSE (1) *See* American Society of Safety Engineers. (2) *See* Association of State Sanitary Engineers.

assembler *See* assembly language.

assembly An assemblance composed of one or more approved body components and including approved shutoff valves.

assembly language A low-level programming language that is unique to specific computers. It utilizes mnemonic instructions rather than binary numbers to represent machine language instructions.

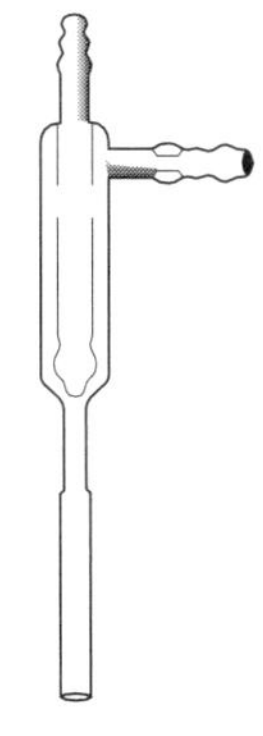

Aspirator

assessment district A specific area of land and property that, under the Municipal Improvement Act of 1913 and Improvement Bond Act of 1915, will be assessed a pro rata share of the cost of improved water service to that area. The debt incurred to make this improvement is secured by direct lien against the assessed property that receives the benefit. The assessment can be apportioned on any basis that reasonably defines the benefits: area served, units of water use, and so forth. In the context of development, an assessment district is sometimes referred to as a community facilities district.

assessment district bond A bond issued to finance required improvements to an assessment district under procedures specified by the Municipal Improvement Act of 1913 and Improvement Bond Act of 1915. The debt is secured by direct lien against a piece of property that will receive the special benefit and upon which an assessment is made. Assessments can be apportioned on any basis that reasonably defines the benefits: area, units of use, and so forth. Assessments are most often collected by means of an ad valorem tax, usually a percentage of true or market value. *See also* assessment district.

asset management (AM) The systematic care, disposal, maintenance, or replacement of the total resources of a business, including cash, notes, accounts receivable, inventories, securities, machinery, fixtures, real estate, or any source of wealth having economic value to the business.

assets The entire resources of a government entity or person, or private business, both tangible and intangible.

assimilable organic carbon (AOC) The fraction of organic carbon that can be used by specific microorganisms and converted to cell weight. AOC also represents a potential for biological regrowth in distribution systems. Ozone (O_3) can convert organic matter in water to assimilable organic carbon, whereas biological filtration can reduce the AOC level. *See also* biological filtration.

assimilation The transformation and incorporation of absorbed materials into components of the organism.

assimilative capacity of an aquifer The ability of an aquifer to receive an unacceptable-quality water at one point while being able to supply an acceptable quality of water further downgradient. This term is important when an aquifer system is used to treat water by exploiting both the filtration effect and dilution effect in porous media.

association The dependence between two or more events, characteristics, or variables (e.g., exposure and disease). Variables are associated if one is more (or less) common in the presence of the other. Events or variables are also associated when they occur more frequently together than expected by chance. The terms *association* and *relationship* are often used interchangeably for describing epidemiologic associations. *Association* does not necessarily imply a causal relationship between the events or variables. *See also* statistically significant.

Association of Boards of Certification (ABC) A group that promotes water and wastewater operator certification through model state legislation and national testing services.

Association of California Water Agencies (ACWA) An organization headquartered in Sacramento, Calif., that represents municipal and agricultural water agencies in that state on both a national and statewide basis.

Association of Environmental Engineering and Science Professors (AEESP) A group of college and university faculty who teach environmental engineering and environmental science.

Association of Metropolitan Water Agencies (AMWA) An association representing large municipal public water agencies.

Association of State Drinking Water Administrators (ASDWA) An association representing state drinking water regulatory agencies.

Association of State Sanitary Engineers (ASSE) A defunct organization that represented sanitary engineers employed by state regulatory agencies.

ASTM *See* American Society for Testing and Materials.

ASTM grade water *See* reagent grade water.

astrovirus An enteric virus (30 nanometres across) associated with waterborne diarrhea. It is identified by its characteristic star-like appearance in electron photomicrographs.

asu *See* areal standard unit count.

ASV *See* anodic stripping voltametry.

asymmetric Not similar in form, shape, or size arrangement, of parts on opposite sides of a line, point, or plane.

asymmetric membrane A type of membrane with a structure in which separation takes place in a thin microporous or dense permselective barrier layer supported by a less dense, more porous substrate of the same chemical composition. *See also* anisotropic membrane.

asymptomatic Pertaining to the presence of infection in a host without recognizable clinical signs or symptoms. An asymptomatic infection is also referred to as an inapparent infection or a subclinical illness or disease.

asymptotic Pertaining to the outcome of a process that gradually approaches but never reaches a goal or some absolute value.

asynchronous Pertaining to a transmission of data at irregular intervals to a computer. The data are preceded by a start bit and followed by a stop bit with no regard to timing.

asynchronous communication The transmission of data via serial lines without a specific timing pattern.

at wt *See* atomic weight.

ATF permit *See* after-the-fact permit.

atherosclerosis A chronic disease that is typified by the formation of lesions in the blood vessels. It usually involves damage to the endothelial cells that make up the blood vessel. As the lesion develops, plaques of fat deposits develop that can occlude the vessel.

atm *See* atmosphere *in the Units of Measure list.*

atmometer An instrument for measuring evaporation. An atmometer is also called an atmidometer or evaporimeter.

atmosphere The gaseous envelope (air) surrounding the earth. It consists of oxygen (O_2), nitrogen (N_2), and other gases, extends to a height of about 22,000 miles (35,400 kilometres), and rotates with the earth. *See also* atmosphere *in the Units of Measure list.*

atmospheric hazard A risk posed by constituents in the air at a particular location. For example, low oxygen (O_2) concentrations or the presence of toxic or flammable gases, mist, and dust can pose risks ranging from disorientation to death for individuals exposed.

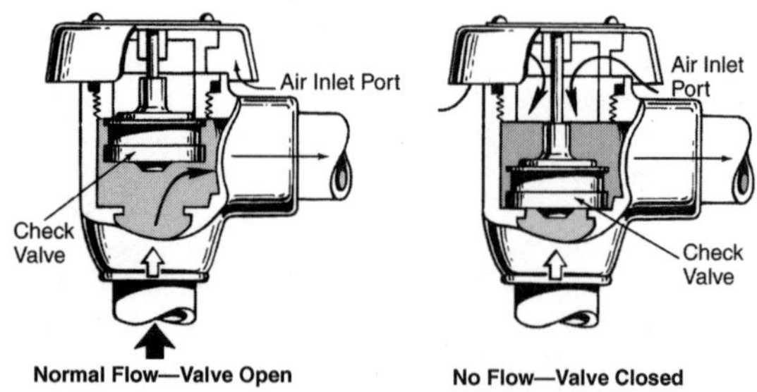

Atmospheric vacuum breaker

atmospheric moisture Water as it occurs in various forms in the atmosphere.

atmospheric pressure The pressure exerted by the atmosphere at any point. It decreases as the elevation above sea level increases.

atmospheric vacuum breaker (AVB) A device consisting of a float check, a check seat, an air inlet port, and possibly a shutoff valve immediately upstream, designed to allow air to enter the downstream water line to prevent backsiphonage. An AVB must never be subjected to a backpressure condition or have a downstream shutoff valve. Backpressure from, say, an irrigation system into the plumbing line will defeat the purpose of the AVB. A downstream valve, if closed, can create a vacuum, allowing water between the AVB and the closed valve to be pulled back into the supply system when the system is shut off.

atmospheric water Water in the atmosphere in gaseous, liquid, or solid state.

atom The smallest possible component of an element. In the classical model, an atom is composed of a nucleus (made up of one or more protons and two or more neutrons—except for hydrogen, which may have no neutrons) and one or more electrons that revolve around the nucleus. Atoms come together to form molecules. An atom is the smallest particle of an element that still retains the character of that element.

atomic absorption spectrometric method *See* atomic absorption spectrophotometric method.

atomic absorption spectrophotometer A spectrophotometer used to determine the concentrations of metals in water and other types of samples.

atomic absorption spectrophotometric method (AAS Method or AA) An analytic technique used to identify the constituents of a sample by detecting which frequencies of light the sample absorbs. This is the technique usually used to measure the concentrations of metals in water.

atomic absorption spectrophotometry *See* atomic absorption spectrophotometric method.

atomic absorption spectrophotometry, flame *See* flame atomic absorption spectrophotometry.

atomic absorption spectrophotometry, graphite furnace *See* graphite furnace atomic absorption spectrophotometry.

atomic absorption spectroscopy A common name for the atomic absorption spectrophotometric method. *See also* atomic absorption spectrophotometric method.

atomic emission The emission of photons when compounds are energized into excited atoms in a hot gas. The photons are emitted as the electrons return to the ground state and can be detected by a spectrophotometer, thus identifying the element.

atomic emission spectroscopy (AES) An instrumental technique for measuring metals based on the intensity of radiation formed from atoms in an excited state. An example is inductively coupled plasma spectroscopy, a technique that requires little sample preparation and can be very sensitive for certain elements.

atomic force microscopy (AFM) A type of nonoptical microscopy, also known as atomic resolution microscopy. Atomic force microscopy is a means of measuring differences in electrical potentials at the atomic forces level using a specially coated probe tip that rides across the surface of a specimen. The electrical potentials are converted to digitized signals that are used to generate a computer picture of the specimen surface.

atomic mass unit (AMU) A mass quantity equivalent to one-twelfth the atomic weight of carbon.

atomic number The number of protons in the nucleus of an atom.

atomic resolution microscopy *See* atomic force microscopy.

atomic weight (at wt) Approximately the number of protons and neutrons in the nucleus of an element.

ATP *See* adenosine triphosphate; advanced treatment plant.

atrazine (2-chloro-4-ethylamino-6-isopropyl amino-1,3,5-triazine) An herbicide and plant growth regulator used primarily on corn and soybeans. It is slightly soluble in water and is regulated by the US Environmental Protection Agency.

Cl
N N CH3
C2H5 NH— N —NH—CH3
CH3

Atrazine

atrazine-desethyl A degradation product of atrazine ($C_8H_{14}N_5Cl$). *See also* atrazine; degradation.

attack rate A rate, often expressed as a percentage (cases per 100), that describes the cumulative incidence of disease or illness in a particular group observed for a limited time and under special circumstances (i.e., during an epidemic or outbreak). The period of time used for observation of cases varies, but it begins at the time of exposure and continues over an interval that allows for the occurrence of all possible cases of illness attributable to the exposure. A secondary attack rate in communicable disease outbreaks refers to cases among familial, institutional, or other contacts following exposure to a primary case; secondary cases may be restricted to susceptible contacts.

atto Prefix meaning 10^{-18}, used in Système International.

attractive forces Forces that are created by dipoles on the surfaces of particles, called van der Waal forces. Their magnitude decreases markedly as the separation distance of the particles increases. They are important in the coagulation of particles in that they hold the particles together once they are close enough to each other.

attributable risk The rate of a disease or other outcome in exposed individuals that may be attributed to the exposure in question. This term is often used to denote several other risk measures, including rate difference, population excess rate, attributable fraction in the population, and attributable fraction among the exposed. To avoid confusion with similar and other measures of risk when evaluating epidemiologic studies, readers should carefully search for the investigator's definition of this term when it is used.

attribute A type of nongraphic data that describes the entities represented by graphic elements. The term is frequently used to cover all types of nongraphic, usually alphanumeric, data that are linked to a map element. For example, a map depicting parcels, each identified by a parcel number, may be linked to an attribute data file containing information about ownership, land use, and appraised value. In the context of engineered facilities, attribute data describe the facilities themselves, recording the characteristics of individual devices such as the type, size, material, and manufacturer.

attribute data A set of data that describes the characteristics of real-world objects and is usually represented alphanumerically.

attrition The gradual lessening of the capacity or effectiveness of media. This phenomenon may occur because of sacrificial properties of the media, friction, or chemical attack of the media.

audit A formal process in which an outside party examines the financial records and accounts of a business or public agency to verify their correctness.

auger hole test An aquifer test made in shallow auger holes to determine hydraulic properties of the aquifer or overlying soils.

augmentation plan A court-approved plan that allows a lower-priority water user to divert water out of priority if certain conditions are met: (1) adequate replacement is made to the stream system and to the water right that is affected and, in some cases, (2) injury to the water rights of other users is avoided.

augmentation source The supply of water that is used to replace any depletions in an augmentation plan. *See also* augmentation plan.

Australian Water and Wastewater Association (AW&WA) The association representing the Australian drinking water and wastewater professions.

autecology The study of the relationships between a species or an individual and its environment.

authority The power to give a command, enforce obedience, make final decisions, and take action.

authority bond A bond payable from the revenues of a specific authority. Because authorities usually have no revenues other than charges for services, their bonds are ordinarily revenue bonds.

authorization *See* authorizing legislation.

authorizing legislation Federal legislation that establishes or continues the operation of a federal program or agency, either indefinitely or for a specified period of time, or that sanctions a

Ashland Community College Library
1400 College Drive
Ashland, KY 41101
72815

particular type of federal monetary obligation or expenditure within a program.

autochthonous Pertaining to turbidity-causing material produced within a lake or other surface waters.

autoclave A device that sterilizes laboratory and microbial media by using pressurized steam.

autoclaved Sterilized with steam at elevated temperatures.

autocorrelation The autocorrelation function is used in regression analysis of time-series data (e.g., monthly water consumption) to measure the strength of the relationship of differences in the variable's terms separated by a specific number (n) of time periods. The strength of the measurement is given by the autocorrelation coefficient, r, which can take on positive or negative values from –1.0 to 1.0. The higher the positive or negative values, the stronger the correlation. If $n = 1$, the paired terms (C_t, C_{t-1}) are one period apart and the autocorrelation is a first-order autocorrelation. Correlating pairs two periods apart, $n = 2$, yields a second-order autocorrelation, and so forth.When pairs 12 monthly periods apart ($n = 12$)—one year apart—are being correlated, if the correlation of paired differences is high ($r \geq 0.75$), strong evidence exists of a seasonal pattern causing the differences because the pattern of data is recurring at 12-month intervals. Once the time-dependent pattern has been identified, it can be filtered out of the data series to achieve a stationary state in which no further significant autocorrelation exists. The filtering out process is done by taking second differences (for seasonality, taking the differences of the first differences, 12 periods apart) or by using autoregressive analysis to quantify the relationships. *See also* autoregressive term; Durbin–Watson test; seasonal analysis; serial correlation.

autocrine A mechanism of cellular control that is internal to a particular cell (i.e., the cell regulates itself).

autoimmunity An allergic response directed toward the self. Autoimmunity is the apparent basis for many chronic diseases. These activities can be triggered by chronic exposure to toxic chemicals, as well as other causes.

automated laboratory instrument An instrument that automatically runs tests on input samples. These instruments may perform scaling, linearization, and other necessary calculations, as well as providing data storage and limited reporting.

automated mapping (AM) system A computer system used to draw and produce maps that can be viewed on a terminal screen or on paper. Automated mapping systems (also known as computer mapping) are designed for efficient graphic data processing and display; they have little or no geographic analysis capability and a limited ability to store and manipulate any data other than graphic images.

automated mapping/facilities management (AM/FM) system A computer system that integrates automated mapping tools with facilities management capabilities, principally distinguished from automated mapping by the addition of database management capabilities and a linkage of attribute data to graphic entities. Automated mapping/facilities management systems can also be categorized as geographic information systems that emphasize applications related to managing utility networks or infrastructure. Typical automated mapping/facilities management applications include managing geographically distributed

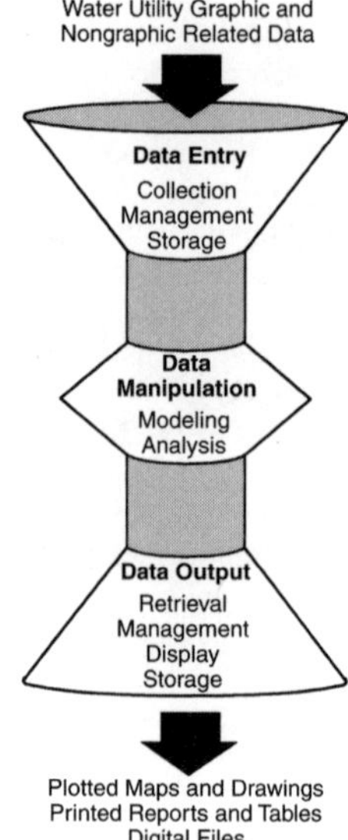

Elements of an AM/FM/GIS

facilities, maintaining facilities inventories, managing operations, overlaying combinations of features and recording resulting conditions, analyzing flows or other characteristics of networks, and defining districts to satisfy specified criteria.

automated mapping/facilities management/geographic information system (AM/FM/GIS) The combination of the terms automated mapping/facilities management (AM/FM) and geographic information systems (GIS) is often used to describe a multipurpose geographic information system that serves several agencies or departments and satisfies application requirements in several areas, such as infrastructure management, property, appraisal, planning functions, and development tracking. *See also* automated mapping/facilities management (AM/FM) system; geographic information system.

automatic controller A device that automatically executes the decision making necessary to determine the proper adjustment of a final control element.

automatic gate A gate that operates without human assistance when prescribed conditions are met.

automatic meter reading (AMR) The electronic reading of customers' meters from a remote central location. Data from meters are transmitted via telephone or cable television lines or radio frequency to a central computer for billing purposes.

automatic recording gauge An automatic instrument for continuously measuring and recording graphically; also called a register.

automatic sampling The process of collecting samples of prescribed volume over a defined time period by an apparatus designed to operate remotely without direct manual control. *See also* composite sampling.

automatic valve A valve that opens or closes without human assistance when prescribed conditions are met.

automatic water softener A water softener that is equipped with a clock timer that automatically initiates the backwash process, regeneration process, or both at certain preset intervals of time. All operations, including bypass of treated or untreated water (depending on design), backwashing, brining, rinsing, and returning the unit to service are performed automatically.

automation The use of electronic devices to control or perform a process or task. With respect to management of water utilities, areas of automation typically include process control, design and engineering, facilities mapping and analyses, hydraulic flow and hydraulic modeling, customer complaints handling and work order management, meter management, and customer billing. *See also* automated mapping/facilities management/geographic information system; computer aided design and drafting; supervisory control and data acquisition.

autoxidation A spontaneous process in which a compound loses electrons (i.e., is oxidized)—usually in the presence of oxygen, but other compounds can act as the electron acceptors. This term is usually associated with compounds that are chemically unstable in the presence of air or oxygen.

autoregressive integrated moving average (ARIMA) forecasting method An extension of the autoregressive method that includes three tools for capturing the statistical relationship among differences. The first tool addresses the autocorrelation of first differences in the original time-series data. The second tool specifies the order of the autocorrelation. The third tool applies a moving average to the forecast errors to improve the current forecast. The moving average can be first-order, using the most recent forecast error, or second-order, using the forecast error from the two most recent periods, and so on. The ARIMA model is referred to as a univariate model because it requires no data other than the single variable that is being forecast; however, various parameters for differencing and order are required. The model extracts from the dependent variable any patterns of seasonality, trend, and recurring cycle that can be identified and uses these patterns to make forecasts. *See also* autocorrelation; autoregressive term; serial correlation; weighted average.

autoregressive term A coefficient derived in regression analysis of time-series data (e.g., monthly water consumption) that quantifies the time dependency of random errors. The assumption in this type of analysis is that the effects of changes in the variables are not instantaneous, that they carry forward to future periods. By using autoregressive analysis, the errors of prior observations are included as variables in the regression analysis to identify the relationship between the errors or residuals in one period and the variable values in future periods. Because each coefficient that is identified will generate different residuals throughout the time series, an iterative process is used to find the coefficient with the most stable or stationary results. The most popular form of this analysis is the first-order autoregressive process, which measures the effect of residuals one period apart, but higher-order autoregressive models can also be used. The key value of this type of analysis is that, by identification of the autocorrelation, the coefficients of other variables take on more accurate values. *See also* autocorrelation; Durbin–Watson test; serial correlation.

autosampler A device that allows automated introduction of a sample or sample extract into an instrument.

autotroph An organism that obtains energy by the oxidation of inorganic compounds.

autotrophic Pertaining to organisms capable of synthesizing organic nutrients directly from simple inorganic substances such as carbon dioxide (CO_2), nitrogen (N), and nitrate ion (NO_3^-). *See also* autotroph.

autotrophic bacteria Bacteria that derive energy for growth and metabolism from oxidation–reduction reactions in which inorganic substances are used as hydrogen donors and carbon dioxide (CO_2) is the carbon source for biosynthesis. Such bacteria are also called chemoautotrophs.

auxiliary scour *See* filter agitation.

auxiliary source A source of water supply that (1) is not normally used but has been approved for use by agencies having the jurisdiction or primacy and (2) has been or may be developed for use when the normal sources are inadequate.

auxiliary storage device A device used to maintain large volumes of digital data and software that are directly accessible through high-speed connections to a processing unit. Examples include disk and tape drives. Such devices are sometimes called mass storage devices.

auxiliary tank valve In a chlorination system, a union or yoke-type valve connected to the chlorine container or cylinder. It acts as a shutoff valve in case the container valve is defective.

auxiliary water supply Any water supply on or available to the premises other than the purveyor's approved public water supply. These auxiliary waters may include (1) water from another purveyor's public potable water supply or any natural source(s), such as a well, lake, spring, river, stream, harbor, and so forth or (2) used waters or industrial fluids. These waters may be contaminated or polluted, or they may be objectionable and constitute an unacceptable water source over which the water purveyor does not have sanitary control.

availability charge A limited-use charge made by a water utility to a property owner between the time when water service is made available to the property and the time when the property connects to the utility's facilities and starts using the service.

available chlorine A measure of the amount of chlorine in chlorinated lime, hypochlorite compounds, chloramines, and other materials that are used for disinfection as compared to the amount in elemental (liquid or gaseous) chlorine. For example, the molecular weight of monochloramine (NH_2Cl) is 51.5 and each mole is equivalent to 71 g of available chlorine (Cl^{1+} undergoes two-electron addition (reduction) to Cl^{1-}–$2 \times 35.5 = 71$). Thus, monochloramine is equivalent to 1.38 (71/51.5) grams of available chlorine per gram.

available expansion The vertical distance from the media surface to the underside of a trough in a granular filter. This distance is also called freeboard.

available fresh water The quantity of renewable water available for human use. It is sometimes called the renewable water supply.

available water supply The quantity of water in a source such as a stream or groundwater basin in excess of the quantity necessary to supply valid prior rights and demands.

AVB *See* atmospheric vacuum breaker.

avdp *See* avoirdupois.

average *See* arithmetic mean.

average annual flood A flood discharge equal to the mean of the discharges of all of the maximum annual floods during the period of record.

average cost The total cost divided by the total number of units associated with the total cost.

average daily demand The average amount of water necessary in a 24-hour time frame to meet all needs of all customers. It is determined by dividing annual usage by the total number of days in the year.

average daily flow (ADF) A measurement of the total amount of flow past a point over a day or over a period of several or more days divided by the number of days in that period. Mathematically, it is the sum of all daily flows divided by the total number of daily flows used.

average dry weather flow (ADWF) In rivers and streams, the normal flow of water made up primarily of water that seeps into the ground.

average efficiency The efficiency values of a machine or mechanical device averaged out over the range of loads through which the machine operates.

average feed concentration The arithmetic average feedwater composition of a substance. In pressure-driven membrane separation,

$$\text{average feed concentration} = \frac{(C_f + C_c)}{2}$$

Where:

C_f = feed concentration, in milligrams per litre

C_c = concentrate concentration, in milligrams per litre

For reverse osmosis, a log-mean feed concentration is sometimes considered instead of the arithmetic average feed concentration to better simulate conditions in a concentrate-staged system. *See also* log-mean feed concentration.

average flow The arithmetic average of flows measured at a given point. *See also* mean flow; normal flow.

average flow rate The average of the instantaneous flow rates over a given period of time, such as a day.

average free available chlorine concentration The sum of the free chlorine dose and the free chlorine residual divided by two. This value is used in disinfection by-product formation tests to indicate the average free chlorine concentration during the time of the test.

average groundwater velocity The average distance traveled per unit of time by a body of groundwater. It is equal to the total volume of groundwater passing through a unit cross-sectional area in unit time divided by the porosity of the water-transmitting material.

average linear velocity *See* actual groundwater velocity.

average rainfall intensity (ARI) A measurement of rainfall, expressed in inches or centimetres of rain per unit of time.

average streamflow The average rate of discharge during a specified period. *See also* mean flow.

average velocity The average velocity of a stream flowing in a channel or conduit at a given cross section or in a given reach. It is equal to the discharge divided by the cross-sectional area of the section or the average cross-sectional area of the reach. Average velocity is also called mean velocity.

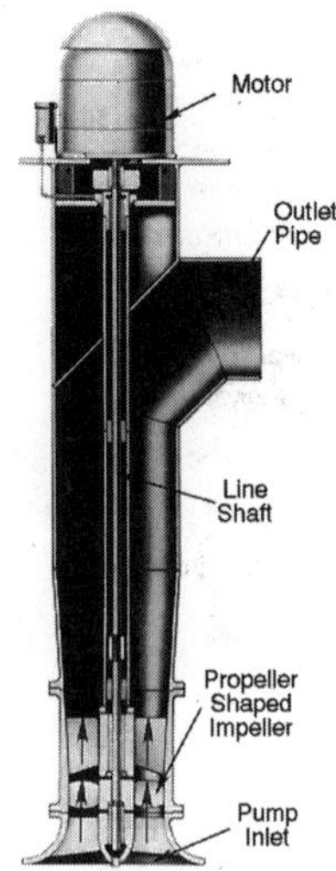

Axial flow pump

average year A year for which the observed quantities of hydrologic phenomena, such as precipitation, evaporation, temperature, and streamflow, approximate the mathematical mean of such observed quantities for a considerably longer period. *Normal year* is a better term.

average-day demand Annual water consumption divided by 365 days in a year; the average water demand a given water distribution system experiences over a one-day period. The average-day demand applies to a specific time period over which the average is calculated.

AVI *See* aggregate volume index.

Avogadro number The number of molecules in a mole of any molecular substance, or the number of atoms in a mole of any substance in atomic form, its value being 6.023×10^{23}. For example, 1 mole (16 grams) of oxygen atoms contains 6.023×10^{23} oxygen atoms.

avoided cost The incremental savings associated with not having to produce additional units of water or water service. Avoided cost can be used to compare demand management and supply management options and to encourage utilities to seek out least-cost alternatives for meeting future water needs.

avoirdupois (avdp) The weight of an object in the English system in which a pound equals 16 ounces. This weight is not given in Système International units. To convert to SI units, multiply pounds by 454 to obtain grams and ounces by 28.4 to obtain grams.

AW&WA *See* Australian Water and Wastewater Association.

AWRA *See* American Water Resources Association.

AWWA *See* American Water Works Association.

AWWA standard The minimum requirements for water supply products or procedures established by the American Water

Works Association through a consensus process that involves representatives from different interest groups, including manufacturers, water utilities, consultants, governmental agencies, and others. AWWA standards are not legally binding and are not specifications, but they may be referenced in specifications of water utilities or others who are acquiring a particular type of product or procedure.

AWWARF *See* American Water Works Association Research Foundation.

axenic culture A culture free of all living organisms except those of a single species. This concept gave rise to the development of pure culture techniques in bacteriology.

axial flow *See* longitudinal flow.

axial flow pump A type of centrifugal pump that develops most of its head by the propelling or lifting action of the vanes on the liquid. Such a pump is also called a propeller pump.

axial to impeller The direction of the discharge of the flow—parallel to the impeller shaft—with propeller pumps, as opposed to the discharge being perpendicular to the impeller shaft with centrifugal pumps.

axis of impeller An imaginary line running along the center of an impeller shaft.

B

B cell A lymphocyte (white blood cell) responsible for synthesizing circulating antibodies.

B. coli A member of the coliform group, now classified as *E. coli*. *See also* coliform group bacteria.

B6C3F1 mouse The standard mouse used by the National Toxicology Program in the United States. It is a hybrid derived from crossing male C3H mice with female C57BL mice. These mice inherit the very high background incidence of liver tumors (hepatocellular adenomas and carcinomas) from the tumor-prone C3H parent. The tendency for liver tumors is particularly evident in the males.

BAA *See* bromoacetic acid.

BAC *See* biologically enhanced activated carbon.

backblowing A reversal of the flow of water under pressure in a conduit or well to remove clogging material. *See also* backwash.

backfill The material placed in an excavation after construction or repair of water facilities.

backflow A hydraulic condition, caused by a difference in pressures, that causes nonpotable water or other fluid to flow into a potable water system.

backflow connection In plumbing, any arrangement whereby backflow can occur. It is also called an interconnection or cross-connection. *See also* backflow; cross-connection.

backflow preventer A device for a water supply pipe to prevent the backflow of water into the water supply system from the connections on its outlet end. *See also* check valve; vacuum breaker.

backflow prevention The use of a system or a device such as a double gate, double check assembly, or air gap to separate a potable water system from a system of water of unknown quality or source. The device or system prevents water from flowing back into a protected system when the pressure in the protected system drops lower than in the system of unknown quality. *See also* backflow-prevention device.

backflow-prevention device Any effective device, method, or construction used to prevent backflow into a potable water system. *See also* air gap; backflow prevention; check valve.

backflushing The reversing of flow direction through a filter, ion exchange column, or membrane to remove particles for cleaning purposes. The backflush source can be feedwater, treated water, air, or some other fluid. *See also* backwash.

background concentration The quantity of gaseous liquid or solid material in relation to a weight or a volume of other substances in which the material is mixed, suspended, or dissolved as a result of natural processes.

background correction, deuterium *See* deuterium background correction.

background correction, Smith–Hieftje *See* Smith–Hieftje background correction.

background correction, Zeeman *See* Zeeman background correction.

background level In toxic substance monitoring, the average presence of a substance in the environment, formerly referred to as naturally occurring phenomena.

background organic matter (BOM) Natural organic matter in a mixture with one or more specific organic chemicals but at a much higher concentration than those chemicals. Background organic matter frequently interferes with the adsorption of the specific organic compound on solid adsorbents. *See also* fouling; natural organic matter; preloading.

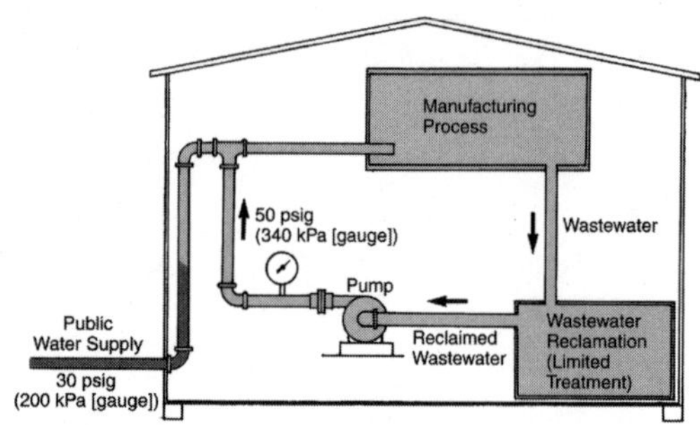

Backflow

background radiation Radiation extraneous to an experiment.

backhoe A track-mounted or wheeled machine hydraulically controlled and operated to excavate for water facilities construction or repair.

backpressure A pressure that can cause water to backflow into the water supply when a user's water system is at a higher pressure than the public water system.

backpressure valve A valve provided with a disk hinged on the edge so that it opens in the direction of normal flow and closes with reversal of flow. Such a valve is frequently called a check valve.

back-pullout An end-suction pump design that allows a portion of the pump casing to be removed without disturbing the suction or discharge pipes.

backsiphonage A form of backflow caused by a negative or subatmospheric pressure within a water system. *See also* backflow; vacuum breaker.

backup system Any part of a water system that is used as an auxiliary or alternate, e.g., a diesel engine backing up an electric motor.

backward elimination A method used for statistical regression analyses in which all variables are initially included and the least important ones are eliminated in sequence.

backwash (1) The process of cleansing filter media of particles that have been removed during the filtration or adsorption process. Backwashing involves reversing the flow through a filter to dislodge the particles. Backwash water is either treated and returned to the plant influent or is disposed. (2) The passing of a fluid (feedwater, treated water, air, or other fluid) through an ion exchange column or some types of microporous membranes for cleaning purposes to remove particles.

backwash rate The velocity of water used in the backwash process. The unit of measure is a flow rate per unit area. *See also* backwash.

backwash stage A specific action in the sequence of actions that form the backwash process. Examples of backwash stages include filter drawdown, initial media fluidization, backwashing at one or more rates, and filter-to-waste. *See also* backwash; filter drawdown; filter-to-waste; fluidization.

backwashing *See* backwash.

backwater (1) The increased depth of water upstream from a dam or obstruction in a stream channel caused by the existence of such obstruction. (2) The body of relatively still water in coves or covering low-lying areas and having access to the main body of water. (3) A water reserve obtained at high tide and discharged at low tide. *See also* backwater curve.

backwater curve The longitudinal shape of the water surface in a stream or open conduit where such water surface is raised or lowered from its normal level by a natural or artificial constriction. In uniform channels, the curve is concave upward, the velocities decrease in a downstream direction, and the flow is nonuniform.

bacteria (bacterium) Microscopic unicellular organisms having a rigid cell wall. They lack a nuclear membrane, a mitotic system, and mitochondria; possess only a single piece of chromosomal deoxyribonucleic acid; and divide by binary fission. Most are nonphotosynthetic; photosynthetic forms do not contain plant-type chlorophyll.

bacterial aftergrowth The growth of bacteria in treated water after that water reaches the distribution system. *See also* bacterial regrowth.

bacterial examination An analysis of water for bacteria, conducted qualitatively, quantitatively, or both. Drinking water is examined to determine the presence, number, and identity of bacteria. This process is also called a bacterial analysis. *See also* bacteriological count.

bacterial gastroenteritis *See* waterborne disease.

bacterial mutagenesis Mutagenic activity in bacteria (e.g., *Salmonella* in the Ames test) that are being used to prescreen chemicals for such activity. Positive results in this type of test heighten concern that a carcinogenic chemical may have the capability of acting as a genotoxic agent. A positive response in these assays suggests that an untested chemical may be carcinogenic. These data cannot substitute for carcinogenesis tests in either a qualitative or a quantitative way.

bacterial nutrient An organic or inorganic compound used by bacteria to meet their nutritional requirements for survival, growth, and multiplication. *See also* assimilable organic carbon; biodegradable organic carbon.

bacterial regrowth The presence of a persistent population of bacteria in a water distribution system. Such regrowth may lead to problems in the compliance monitoring of total coliform bacteria.

bacterial regrowth potential An assessment of the potential for bacteria present in treated drinking water to increase in density in the absence of a disinfectant residual.

bactericide Any substance or agent that kills bacteria.

bacteriological analysis The examination of an environmental, water, or food sample for the presence of bacteria. The analysis may be to detect or enumerate either specific types of bacteria, the general heterotroph population density, or both. It may involve a variety of methods, including microscopic examination, culture techniques, and genetic, immunologic, or enzyme probes.

bacteriological corrosion Corrosion that results from the byproducts of sulfate-reducing bacteria in media of very low or no oxygen content.

bacteriological count A quantification of the number of bacteria per unit volume in a water sample. *See also* heterotrophic plate count; indicated number; most probable number; plate count.

bacteriology A subset of microbiology focusing on the study of small, single-celled prokaryotic organisms.

bacteriophage A group of viruses that infect and grow in bacteria, such as coliphages that grow in *E. coli.* Following replication in the host bacterium, new bacteriophages are released by lysis of the host cell. Bacteriophages can be used as surrogates or models in place of human enteric viruses during water treatment testing and can be potential indicators of pathogens. Examples of common bacteriophages are MS2 coliphages and f-specific phages. A bacteriophage is also called a phage.

bacteriostatic Having the ability to inhibit the growth of bacteria without destroying the bacteria.

***Bacteroides* species phage** A phage specific to the bacterium *Bacteroides* that is common in the feces of humans and grows under anaerobic conditions. Detection of this phage is an indicator of polluted water.

BAF *See* biologically active filtration.

baffle A metal, wooden, or plastic plate installed in a flow of water to slow the water velocity and provide a uniform distribution of flow.

baffle wall A physical barrier designed to modify the flow characteristics into a treatment basin, alter the distribution of the flow stream, and minimize short-circuiting. For example, a sequence of solid baffle walls may be used to approximate plug flow conditions, or a perforated baffle wall may be used to improve the flow distribution into a basin. *See also* dispersion; perforated baffle wall; plug flow; sedimentation basin.

baffled-channel system A system that includes baffles to minimize short-circuiting or to provide flocculation as a result of head loss through the baffled system. *See also* baffling; flocculation.

baffling The process of adding physical barriers to modify the distribution of flow and alter the distribution of the flow stream. *See also* baffle wall.

bail-down test A type of slug (aquifer) test performed by using a bailer to remove a measured volume of water from a small-diameter well.

bailer A device used to withdraw a water sample from a small-diameter well. Typically a piece of pipe attached to a wire with a check valve in the bottom is used as a bailer.

balance An instrument used to measure weight.

balance sheet A financial statement summarizing the assets, liabilities, and net worth of a business or an individual at a given time. A balance sheet is so called because the sum of the assets equals the total of the liabilities plus the net worth.

balanced Pertaining to a chemical equation such that, for each element in the equation, the same number of atoms shown on the right side of the equation are shown on the left side.

balanced flow A flow pattern that is controlled to achieve the flow specified for that water treatment system.

balanced gate A gate in a reservoir outlet, conduit, or similar structure that operates automatically to release water and that usually is controlled by the head of water behind the gate.

balanced mechanical seal A spring-loaded or rubberized seal designed to reduce the leakage between a pump casing and the rotating shaft, used rather than a packing gland and pump packing material.

balancing reservoir A holding basin in which variations in flow and composition of a liquid are averaged. Such a basin is used to provide a flow of reasonably uniform volume and composition to a treatment plant. It is also called an equalizing basin.

ball joint A flexible pipe joint made in the shape of a ball or sphere.

ball valve A valve with the closing and opening mechanism formed in the shape of a ball with a hole. The valve is opened by rotating the ball so the hole is parallel to the flow, allowing it to pass. The valve is closed when the hole is perpendicular to the flow.

BAN *See* bromoacetonitrile.

band A thin circular metal strip with a gasket inside used to repair a leaking pipe by applying compression around the pipe by bolts attached to each end of the metal strip. It is also called a repair clamp, full circle repair clamp, full circle repair band, or a stainless-steel band.

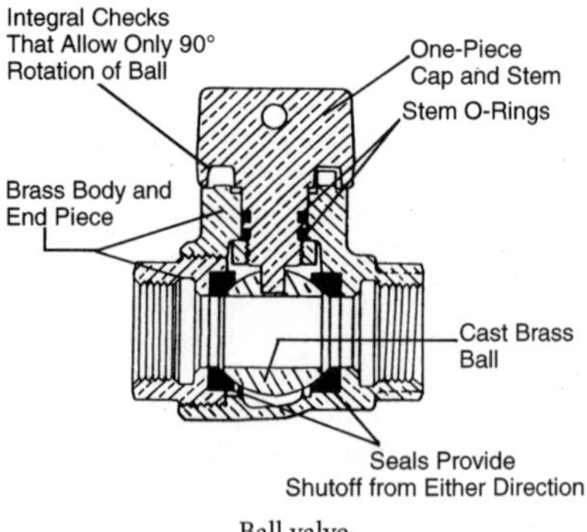

Ball valve

banded steel pipe A steel pipe for which the strength has been increased by the use of bands shrunk around the shell. Such pipe is used under extremely high heads for which the shell thickness ordinarily required would be greater than about 1¼ inches (3 centimetres).

bandwidth A measure of data transmission speed. For example, a bandwidth of 10 megabytes means the line can transfer data at the rate of 10 million binary digits per second.

bank (1) The continuous margin along a river or stream where all upland vegetation ceases. (2) The elevation of land that confines waters of a stream to their natural channel in their normal course of flow. (3) The rising land bordering a river, lake, or sea. (4) In a pressure-driven membrane treatment system, a grouping of membrane modules or pressure vessels in a common control unit or stage.

bank instrument A legal document that sets up the specifics of a bank. For example, a memorandum of agreement is a bank instrument.

bank protection Riprap, paving, brush, concrete, or other material placed to prevent erosion on a stream, reservoir, or lake shore. It usually extends to and beyond the thalweg of the channel.

bank sponsor A person or entity that financially supports or funds the mitigation bank.

bank storage Groundwater temporarily stored in sediments adjacent to a stream channel caused by a rise in stream elevation during flooding. It is a significant storage component in controlled streams and in arid climates.

bar (1) An alluvial deposit or bank of sand, gravel, or other material at the mouth of a stream or at any point in the stream itself that obstructs flow or navigation; in the case of a bay or inlet, a spit that extends completely across the waterway. (2) *See the Units of Measure list.*

bar code An identification code made up of lines of varying thickness that are optically read and most commonly used for inventory and work time tracking.

bar screen A series of straight steel bars welded at their ends to horizontal steel beams, forming a grid. Bar screens are placed on intakes or in waterways to remove large debris.

barium (Ba) An alkaline-earth metal. In nature, barium is found in the ores of barite and witherite. Its atomic number is 56, and it is bivalent. Barium has various industrial uses and is regulated by the US Environmental Protection Agency. *See also* alkaline-earth metals.

barometric efficiency (of an aquifer) The ratio of groundwater elevation change in a confined aquifer to change in atmospheric pressure. It is significant in estimating water reserves and hydraulic performance of wells in confined aquifers.

barometric pressure The force per unit area exerted by air. The measuring device is a barometer. The numerical value of the barometric pressure is a function of the air temperature, relative humidity, and the elevation above sea level of the instrument. Typical units are inches of mercury, pounds per square inch, millibars, atmospheres, and kilopascals.

barothermograph An instrument that records simultaneous barometric pressure and temperature on the same chart.

barrel (bbl) *See the Units of Measure list.*

barrier boundary (1) A groundwater flow boundary caused by an abrupt change in hydraulic conductivity of the aquifer. (2) An engineered low-hydraulic-conductivity unit used to isolate a contaminated part of an aquifer. Examples include slurry walls, sheet pile walls, and grout curtains.

barrier spring (1) A spring formed above a fault between a raised bedrock block and a depressed block covered by a thick deposit of alluvium. (2) A spring that occurs when a raising of the confining bed forces the groundwater to rise to the ground surface.

base In practical terms, a chemical that, when dissolved in water, yields a pH greater than 7. The Bronsted–Lowry concept defines a base by its function as a proton acceptor. The Lewis theory describes a base as a species that can donate its lone pair of electrons.

base exchange *See* cation exchange.

base exchange process A process permitting an exchange of positive ions on a prepared medium. *See also* cation exchange.

base flow (BF) That part of the stream discharge that is not attributable to direct runoff from precipitation or melting snow. It is usually sustained by water draining from natural storage in groundwater bodies, lakes, or swamps.

base flow recession The declining rate of discharge observed for an extended period in a stream fed only by base flow. Typically the recession can be described by an exponential function.

base flow-recession hydrograph A hydrograph depicting the base flow-recession curve. It is interpreted using an exponential model to determine the annual recharge to a groundwater basin.

base load The minimum use of something (water, electricity, etc.) over a given period of time.

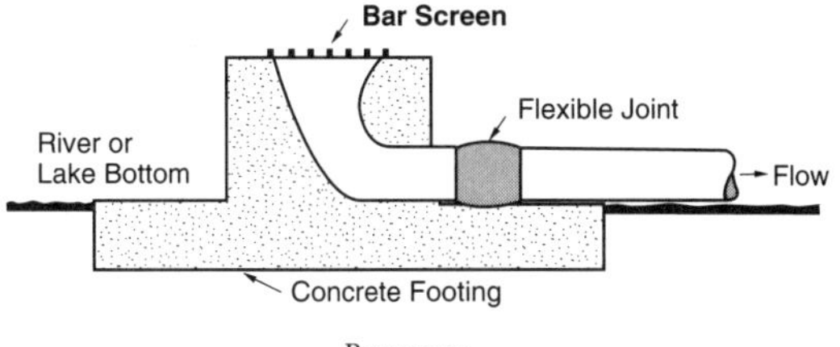

Bar screen

base map A map of locations of basic mapped data that remain constant.

base metal A metal (such as iron) that reacts with dilute hydrochloric acid (HCl) to form hydrogen. *See also* noble metal.

base peak The most abundant mass/charge ratio peak in a mass spectrum being used for organic analysis. In most cases, the peaks in a mass spectrum are normalized to the size of the base peak; therefore, the base peak will be the largest peak in the mass spectrum.

base plate A flat steel section to which an electric motor and pump are bolted, used to provide a rigid and level platform for installing the pump–motor unit on a foundation.

base runoff Sustained or dry weather flow. In most streams base runoff is composed largely of groundwater runoff.

base-catalyzed hydrolysis A chemical reaction that is catalyzed by hydroxyl ions (OH^-) in which water reacts with another substance to form two or more new substances. A number of disinfection by-products can undergo base-catalyzed hydrolysis (e.g., 1,1,1-trichloropropanone can be converted to chloroform by this process). *See also* catalysis.

base–extra-capacity *See* base–extra-capacity approach to rate structure design.

base–extra-capacity approach to rate structure design A rate structure design method in which the costs of service are assigned to four cost components: (1) base costs that include operations and maintenance (O&M) and capital costs that tend to vary with the total quantity of water used and costs that relate to average load conditions; (2) extra capacity costs that include O&M and capital costs associated with meeting rate-of-use requirements in excess of average use, usually expressed in terms of meeting maximum-day and maximum-hour demand; (3) customer costs that relate to serving customers irrespective of the amount or rate of water use; and (4) direct costs incurred solely for the purpose of fire protection.

baseline The condition of a natural community or environment at the starting point (e.g., prior to restoration activities) that will be used to track future changes.

baseline data The historical water usage of the water provider's customers, either in total or for the selected customer sample.

baseline survey A detailed assessment of the environmentally significant ambient chemical, physical, and biological conditions existing in a specific geographical area before superimposing additional human activity.

baseline test The process of taking a measurement that represents a starting or initial level of ability. Such a measurement is typically used for health tests.

base-pair substitution An alteration in a deoxyribonucleic acid (DNA) sequence that results from the substitution of one normal base (i.e., adenine, guanine, thymine, or cytosine) for another in DNA. When this DNA is replicated during cell division, a new complementary base will be inserted into the new strand; thus, the substitution comes as a pair. Such an alteration will be passed on as a mutation to subsequent generations of the affected cell. If this mutation results in changing the amino acid sequence in the protein coded for in this stretch of DNA, it may alter the protein's function or the regulation of its activity by the cell. Such mutations may be lethal or may result in abnormal development or function of a cell. A series of such mutations can give rise to a cancer cell.

BASIC (Beginner's All-Purpose Symbolic Instruction Code) A computer programming language in which statements are translated and executed rather than being compiled prior to execution.

basic In practical terms, pertaining to a solution with a pH greater than 7.

basic data Records of observations and measurements of physical facts, occurrences, and conditions, as they have occurred, excluding any material or information developed by means of computation or estimate. In the strictest sense, basic data include only the recorded notes of observations and measurements, although in general use it is taken to include computations or estimates necessary to present a clear statement of facts, occurrences, and conditions.

basic hydrologic data Records of (1) measurements and observations of the quantity per unit of time of precipitation, including snowfall; streamflow; evaporation from water; the elevation of natural water planes, both surface and underground, and the change thereof from time to time; and (2) any related phenomena and natural conditions necessary to allow estimates to be made of the past, present, and probable future occurrence of water. The unit of time involved in recording the quantities may vary from seconds to years, depending on custom, convenience, and method of compilation and publication of the data.

basic polyaluminum chloride (BPACl) A preformed inorganic polymeric coagulant, used in place of or in conjunction with a metal coagulant like alum. *See also* polyaluminum chloride.

basic solution A solution that contains significant numbers of OH^- ions and has a pH greater than 7.

basic stage flood An arbitrarily selected rate of flow of a stream used as the lower limit in selecting floods to be analyzed, usually taken as the minimum annual flood.

basic water *See* basic solution.

basin (1) A natural or artificially created space or structure, on the surface or underground, that has a shape and character of confining material that enable it to hold water. This term is sometimes used to describe a receptacle midway in size between a reservoir and a tank. (2) The surface area within a given drainage system. (3) An area upstream from a subsurface or surface obstruction to the flow of water. (4) A shallow tank or depression through which liquids may be passed or in which they are detained for treatment or storage. *See also* tank.

BAT *See* best available technology.

batch mode A type of process operation in which the feed input and process output flows are not continuous. *Contrast with* continuous mode.

batch process A process for treatment in which a reactor or tank is filled, the water is treated or a chemical solution is prepared, and the tank is emptied. The tank may then be refilled, and the process may then be repeated. *See also* batch treatment.

batch processing A mode of processing in which the user is not in direct communication with the processing unit while a program is being executed. In batch, or off-line, processing, the user submits a job request to the processing unit for execution when time is available. The job is placed in a batch queue, a temporary storage area. The job contains information necessary to execute the program, the destination of products to be generated, and identification of the user.

batch treatment A method in which a fixed quantity of water is processed through a single treatment device in a single vessel. *See also* batch process.

bathing As interpreted by the US Environmental Protection Agency, the use of water for personal hygiene purposes in a home, business setting, school, and so forth. The term *bathing* does not refer to swimming in an open canal or to incidental, casual contact with water from an open canal in connection with outdoor activities (such as agricultural work, canal maintenance, or lawn and garden care). *See also* human consumption.

battery A device for producing direct current electric current from a chemical reaction. In a storage battery, the process may be reversed, with current flowing into the battery, thus reversing the chemical reaction and recharging the battery.

battery backup Power supplied by battery to components requiring power to retain stored information. Clocks and calendars used in system setup are supplied with battery backup.

battery of wells A group of wells from which water is drawn by a single pump or other lifting device. It is also called a gang of wells.

baud *See the Units of Measure list.*

Baumé (Bé) degree A unit of measure used in two scales of density that are related to the specific gravity of a solution. In the heavy Baumé scale, 0°Bé corresponds to a specific gravity of 1.000 (water at 4° Celsius). In the light Baumé scale, 0°Bé is equivalent to a solution of 10 percent sodium chloride (NaCl). Aqueous solutions of ammonium hydroxide (NH_4OH) used in water treatment (aqua ammonia) are often specified in terms of degrees Baumé.

bay region epoxide A reactive intermediate that forms in a chemical structure made up of (1) ring substitutions that are ortho (adjacent) to one another on a third ring or (2) analogous substitutions on two aliphatic carbons. The diol-epoxides of certain polyaromatic hydrocarbons (e.g., benzo(a)pyrene) are the metabolites that are responsible for the carcinogenic effects of those hydrocarbons, as opposed to other epoxide intermediates (e.g., K-region epoxides). It is the active form of the carcinogenic polyaromatic hydrocarbons.

bay salt A relatively coarse salt made from seawater.

Bayesian principle A statistical principle used to select the strategy with the greatest payoff from among several available ones. The procedure involves calculating the probability of occurrences for each strategy.

Bazin roughness coefficient The roughness coefficient, *m*, in Bazin's discharge coefficient formula:

$$C = \frac{157.6}{1 + m/R^{0.5}}$$

Where:

C = open-channel discharge coefficient

m = a coefficient of channel roughness, the value of which varies from 0.100 to 3.20

R = hydraulic radius of the channel, in feet

The value of C is then used as the discharge coefficient in the Chezy open-channel formula:

$$V = C(RS^{0.5})$$

Where:

V = velocity of flow, in ft/s

R = hydraulic radius of the channel, in ft

S = friction slope (slope of the energy gradeline), dimensionless

BBDR modeling *See* biologically based dose-response modeling.

bbl *See* barrel *in the Units of Measure list.*

BCAA *See* bromochloroacetic acid.

BCAN *See* bromochloroacetonitrile.

BCD *See* binary coded decimal.

BCF *See* brine consumption factor.

BCSP *See* Board of Certified Safety Professionals.

BDCAA *See* bromodichloroacetic acid.

BDCM *See* bromodichloromethane.

BDOC *See* biodegradable organic carbon.

BDOM *See* biodegradable organic matter.

Bé *See* Baumé (Bé) degree.

bead count A method of evaluating the physical condition (quality) of the resin in a bed by determining the percentages of whole, cracked, or broken beads in a wet sample of the resin.

beaker A container with an open top, vertical sides, and a pouring lip used for mixing chemicals.

bearer bond A security that has no identification as to owner. It is presumed to be owned by the bearer of the bond. Such bonds are no longer issued since Congress required full registration of all new issues, effective July 1, 1983. Notes are still bearer designated.

bearing Usually stainless-steel balls set in grooved rings over one or both ends of a pump shaft to reduce friction. Pump bearings are usually regreasable; smaller motor bearings may be sealed bearings.

beaver fever *See Giardia lamblia.*

becquerel (Bq) *See the Units of Measure list.*

bed (1) The bottom of a watercourse or any body of water. (2) The weight of ion exchange resin or other media through which the water passes in the process of water treatment.

bed depth The height of the medium (excluding support material) in a bed, usually expressed in inches or centimetres.

bed expansion The effect produced during backwashing when the filter medium becomes separated and rises in the tank or column. It is usually expressed as a percentage increase of bed depth, such as 25 percent, 50 percent, or 75 percent.

bed life The time it takes for a bed of adsorbent to lose its adsorptive capacity. When this loss occurs, the bed must be replaced with fresh adsorbent or regenerated.

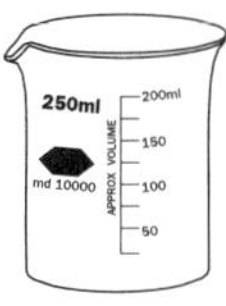

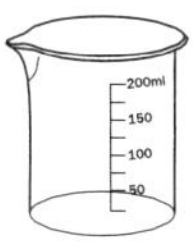

Beakers

bed load Sediment that moves by sliding, rolling, or skipping on or very near the stream bed; sediment that is moved by tractive forces, gravitational forces, or both, but at velocities less than that of adjacent flow.

bed material In a stream system, the geologic formations and the alluvial deposits through which the stream channel is cut.

bed material sampler A device for sampling bed material.

bed volume (BV) The volume occupied by a bed of porous media or resin used for filtration, adsorption, or ion exchange. The number of bed volumes is a unitless measure that indicates the capacity of a system to remove contaminants. The product of the number of bed volumes and the empty bed contact time is the operation time. *See also* breakthrough; empty bed contact time; ion exchange.

bedding The earth or other materials on which a pipe or conduit is supported.

bedding compaction The consolidation of bedding material to allow the bedding to withstand loads. Pipe bedding may be compacted to prevent pipe movement.

bed-load discharge The quantity of bed load passing a given point in a unit of time.

bed-load sampler A device for sampling the bed load.

bedrock Solid rock underlying unconsolidated surficial cover such as soil.

beef extract A by-product of beef processing available as a powder, made up of proteins, used in 1 to 10 percent solutions to recover and desorb viruses from filters, sludges, or soils.

Beginner's All-Purpose Symbolic Instruction Code *See* BASIC.

behead To cut off and capture by erosion, as sometimes occurs when the upper portion of one stream is captured by another. *See also* beheaded stream; river piracy.

beheaded stream A stream that has been separated from its original headwater drainage area by a second stream, which had greater erosive power, allowing it to drain the headwater area.

Belanger's critical velocity The flow velocity in a channel when critical flow occurs; usually called critical velocity. In equation form, for rectangular channels,

$$V_c = (gY_c)^{0.5}$$

Where (in any consistent set of units):

V_c = Belanger's critical velocity

g = the gravity constant

Y_c = the critical water depth

See also critical depth; critical flow condition; critical velocity.

bell (1) In a pipe fitting, the recessed, overenlarged end of a pipe into which the end fits; also called a hub. (2) In plumbing, the expanded spigot end of a wiped joint.

bell joint clamp A repair clamp that can be installed over a water line joint that compresses a gasket against the bell and spigot to seal a leaking joint or gasket.

bell-and-spigot joint A form of joint used on pipes that has an enlarged diameter or bell at one end and a spigot at the other that fits into and is laid in the bell. The joint is then made tight by lead, cement, a rubber O-ring, or other jointing compounds or materials.

bellmouth An expanding rounded entrance to a pipe or orifice.

bellmouth inlet A converging, funnel-shaped, or bellmouthed entrance to a pipe or conduit, used to facilitate entry of water. It is also called flaring.

bellmouthed orifice An orifice with a short tube on the outside that flares or increases in diameter, away from the opening in the orifice plate.

bellows gauge A device for measuring pressure in which the pressure on a bellows, with the end plate attached to a spring, causes a measurable movement of the plate.

belt filter *See* belt filter press.

belt filter press A piece of mechanical equipment designed to dewater sludge by applying pressure between two continuously rotating horizontal belts. The belts are supported by rollers that rotate at a fixed speed. Sludge is fed onto a lower belt, and water is squeezed from the sludge as it is pressed between the lower and upper belt. The filtrate is often either discharged to a sanitary collection system or treated and recycled to the plant influent. The dewatered sludge is typically applied to land or sent to a landfill. *See also* dewatering of sludge; filtrate; land application; landfill.

belt-line layout *See* gridiron layout.

benchmark (1) A reference point from which measurements can be made. (2) A standard or point of reference in judging or measuring quality or value.

benchmark dose The dose that has been derived based on actual dose-response data from a mathematical model and that serves as a common point of departure for calculating a "safe" dose. The most common method of arriving at a benchmark dose is to apply the multistage model to a stochastic data set to derive the upper 95 percent confidence limit on a dose that would be predicted to give a 1 percent or 10 percent response rate (i.e., a response relatively close to the experimental data rather than one extrapolated over orders of magnitude). Then appropriate uncertainty factors are applied in much the same way that they are applied to no-observed-adverse-effect levels (NOAELs) or lowest-observed-adverse-effect levels (LOAELs). The advantage of this methodology is that it uses all the data in a data set and escapes the dilemma posed by the NOAEL and LOAEL, which are determined, in part, by the spacing between the doses used in the study and the study's statistical power (i.e., the numbers of animals used). *See also* multistage model; uncertainty factor.

bench-scale study An experimental study to evaluate the performance of unit processes performed in laboratory surroundings. For example, a jar test to evaluate the coagulation, flocculation, and sedimentation processes is performed on a batch of water, and the apparatus is of a size that will fit on a laboratory "bench." *See also* jar test.

bend A directional-change length of pipe, normally at a designated angle of 90°, 45°, 22½°, or 11¼°—known respectively as a ¼ bend, ⅛ bend, 1/16 bend, or 1/32 bend—but able to be manufactured in a specific angle in some types of materials.

beneficial rainfall *See* effective precipitation.

beneficial use A useful purpose for a given amount of water. A water user may be legally required to meet the criteria for beneficial use or risk losing the right to that water.

benefit–cost analysis A type of analysis often used in making decisions among alternatives, such as different sources of water, different types of equipment, different water treatment processes, and so forth. Benefit–cost analysis is most frequently applied to capital-spending decisions, although the process can be applied to cash flows of any type. The process is to calculate the net present values of alternative cash flows for the benefits (cost reduction, avoided capital, and so on) and compare them with the costs (new capital, processes, and so on). When the benefit-to-cost ratio is 1.0 or more, the benefits have a greater value than costs, and the project has at least passed this hurdle on the way to being undertaken. *See also* net present value; return on investment.

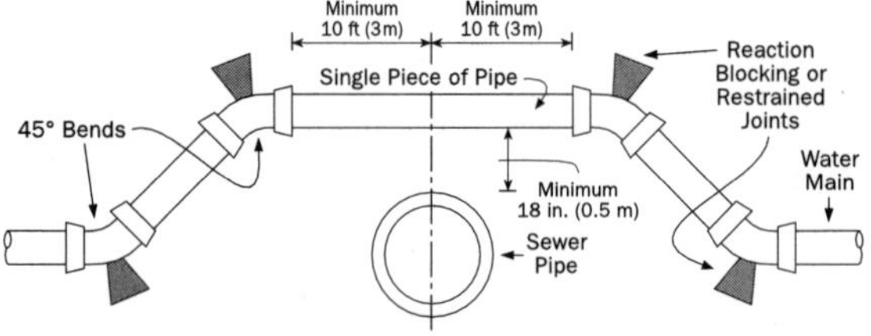

Bends in a water main

benefit–cost ratio *See* benefit–cost analysis.

benefits The advantages, tangible or intangible, gained by installing or constructing a system or works for one or more given purposes. Benefits may also be used as a measure for justifying projects proposed for construction at public expense.

benthic Relating to the bottom or bottom environment of a body of water.

benthos Organisms living on the bottom of bodies of water.

bentonite A form of clay that has been studied extensively with respect to the coagulation process. Bentonite can be added in waters of low turbidity to improve coagulation and flocculation and thereby improve particle removal in the source water. *See also* coagulation.

benzene (C_6H_6) An aromatic hydrocarbon used as a solvent. It is a volatile organic chemical and is regulated by the US Environmental Protection Agency. *See also* aromatic hydrocarbon; solvent; volatile organic compound.

Benzene

benzene–toluene–xylene (BTX) (C_6H_6, benzene; $C_6H_5CH_3$, toluene; $C_6H_4(CH_3)_2$, xylene) The volatile organic compounds measured in gasoline.

benzo(a)pyrene ($C_{20}H_{12}$) A polynuclear (five-ring) aromatic hydrocarbon found in coal tar, in cigarette smoke, and in the atmosphere as a product of incomplete combustion. It is regulated by the US Environmental Protection Agency. *See also* coal tar; polynuclear aromatic hydrocarbon.

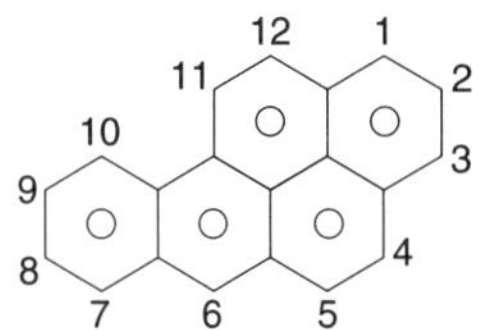

Benzo(a)pyrene

berm ditch A ditch constructed along a berm of ground to convey runoff of surface water.

Bernoulli law A physical law of hydraulics that states that under conditions of uniform steady flow of water in a conduit or stream channel, the sum of the velocity head, pressure head, and head resulting from elevation at any given point along the conduit or channel is equal to the sum of these three heads at any other point along the conduit or channel, taking into account the losses in head between the two points due to friction. In equation form, Bernoulli law may be expressed as follows for upstream point 1 and downstream point 2:

$$\frac{P_1}{\gamma} + \frac{V_1^2}{2g} + Z_1 = \frac{P_2}{\gamma} + \frac{V_2^2}{2g} + Z_2 + h_{f1-2} = \text{constant}$$

Where (in any consistent set of units):

P_1 = pressure at point 1
P_2 = pressure at point 2
V_1 = velocity at point 1
V_2 = velocity at point 2
Z_1 = elevation at point 1
Z_2 = elevation at point 2
γ = specific weight of the fluid
h_{f1-2} = amount of head loss caused by friction between points 1 and 2

beryllium (Be) An alkaline-earth metal. In nature, beryllium is found in the ores of beryl. Beryllium has various industrial uses. Its atomic number is 4. It is regulated by the US Environmental Protection Agency. *See also* alkaline-earth metals

best available technology (BAT) (1) Under Section 1412(b)(4)(D) of the Safe Drinking Water Act, the best technology, treatment techniques, or other means that the US Environmental Protection Agency administrator finds available to water systems—after examination for efficacy under field conditions, not solely under laboratory conditions, and taking cost into consideration—for meeting National Primary Drinking Water Standards. For the purposes of setting maximum contaminant levels for synthetic organic chemicals, any best available technology must be at least as effective as granular activated carbon. Water systems are not required to use the best available technology set under Section 1412 to meet a regulation; they may use any technology acceptable to the state primacy agency. (2) Under Section 1415(9)(1)(A) of the Safe Drinking Water Act, the best technology, treatment techniques, or other means that the US Environmental Protection Agency administrator finds available (taking costs into consideration). To be eligible for a variance, a water system must agree to apply the best available technology designated under Section 1415 prior to applying for a variance.

Best available technology designations by the US Environmental Protection Agency under Sections 1412 and 1415 are separate and may or may not be the same. *See also* variance.

best management practices (BMPs) Conservation practices or systems of practices and management measures that (1) reduce water quality degradation caused by nutrients, animal waste, toxics, and sediment, as well as control soil loss; and (2) minimize adverse impacts on surface water, groundwater flow, and circulation patterns and on the biological, chemical, and physical characteristics of wetlands.

BET method *See* Brunauer-Emmet-Teller method.

β-D-galactosidase *See under* galactosidase.

β-D-galactoside *See under* galactoside.

β-glucuronidase *See under* glucuronidase.

β-D-glucuronidase *See under* glucuronidase.

β-D-glucuronide *See under* glucuronide.

beta (β) particle A directly ionizing product of radioactive decay that can be positively or negatively charged. Commonly used radioisotopes, such as ^{14}C, ^{3}H, and ^{32}P emit only a single type of ionizing particle, the beta particle. Negatively charged beta particles are equivalent to high-speed electrons.

beta ratio The ratio of the diameter of the constriction (or orifice) to the pipe diameter in a differential pressure-type flowmeter.

beta site The location for the testing of a new program or hardware.

beta value A commonly used statistical term to indicate the significance of a coefficient of regression.

beta version The first version of a product, usually in a test mode prior to commercial use.

BeV *See* billion electron volts *in the Units of Measure list.*

BF *See* base flow.

bgd *See* billion gallons per day *in the Units of Measure list.*

BGM cells *See* Buffalo Green monkey cells.

bias (1) A systematic error in an analytical method. Bias can be positive or negative. (2) A fixed amount that is always added or subtracted to a measurement. For example, if a tank level is measured as water depth by a level sensor but water surface elevation is needed for operations, then a bias equal to the elevation of the bottom of the tank can be added to the measurement of depth. This would result in the readout indicating the elevation of the water surface. (3) In epidemiology, an error or effect that may produce associations or measures of risk that depart systematically from the true value; to be distinguished from random error. Bias must be avoided during the study or controlled in the data analysis to ensure the internal validity of an epidemiologic study. Most common types of bias in epidemiologic studies are selection, observation, confounding, and misclassification bias.

bib valve A valve closed by screwing down a leather- or fiber-washered disk onto a seat in the valve body.

bicarbonate (HCO_3^-) An inorganic monovalent anion usually found in natural water. *See also* bicarbonate alkalinity.

bicarbonate alkalinity The acid-neutralizing capability caused by bicarbonate ions (HCO_3^-). At a normal pH of drinking water (pH < 9), the anions of calcium carbonate ($CaCO_3$) and magnesium carbonate ($MgCO_3$) are present primarily as bicarbonate ions, therefore, the concentration of alkalinity and bicarbonate ion are equal. *See also* alkalinity.

bicarbonate hardness The hardness of a water caused by the presence of calcium bicarbonate ($Ca(HCO_3)_2$) and magnesium bicarbonate ($Mg(HCO_3)_2$). It is usually the major component of carbonate hardness or total hardness. Bicarbonate hardness is often referred to simply as carbonate hardness. *See also* carbonate hardness; total hardness.

bifurcate To divide into two branches.

bifurcation gate A gate located at the point where a conduit is divided into two branches to divert the flow into either branch or allow flow into both branches.

bile A fluid that aids in the absorption of dietary fat. It is formed in canniculi passing between hepatocytes (liver parenchymal cells) by active transport of various solutes in blood, including bilirubin, bile acids, and their conjugates. Bile is stored in the gallbladder and secreted into the intestine in response to dietary fat. The bile is a common excretory mechanism for conjugated chemicals, particularly amphipathic chemicals (i.e., chemicals that have both polar and nonpolar character). Processing of conjugates in the intestine by endogenous bacteria can, however, free the parent chemical or one of its metabolites for reabsorption into the systemic circulation. A metabolite of trichloroethylene, trichloroethanol glucuronide, is secreted into the bile.

bile acid Acidic metabolites of cholesterol, generally found in bile as conjugates. *See also* bile.

bile salt The salts of bile acids. *See also* bile acid.

bilirubin A bile pigment that is a breakdown product of heme. It is usually excreted into the bile as a glucuronide conjugate, but it is extensively processed by intestinal flora, reabsorbed, reconjugated, and secreted. *See also* bile; heme.

biliverdin A green bile pigment that is the initial breakdown product of hemoglobin. It is converted to bilirubin by reduction of a methine bridge. *See also* bile; bilirubin; hemoglobin.

bill frequency analysis An analysis of water bills, usually by customer class, to develop a distribution of how many customers use increasing amounts of water. These amounts are specified in terms of consumption blocks. Initially consumption blocks represent one unit (usually 1,000 gallons or 1,000 litres), but they are frequently summarized in 5- or 10-unit increments or according to water rate blocks. The purpose of the bill frequency analysis is usually to select break points for increasing block rates or to analyze historical consumption within existing rate blocks. *See also* inclining block rate.

bill of lading A contract issued to a shipper by a transportation agency. It lists the goods shipped, acknowledges their receipt, and promises delivery to the person named.

billing The act of issuing a statement, of charges for goods and services delivered. The statement is usually itemized.

billing system The system that takes data from customers' meters to produce bills and track payments.

billion electron volts (BeV) *See the Units of Measure list.*

billion gallons per day (bgd) *See the Units of Measure list.*

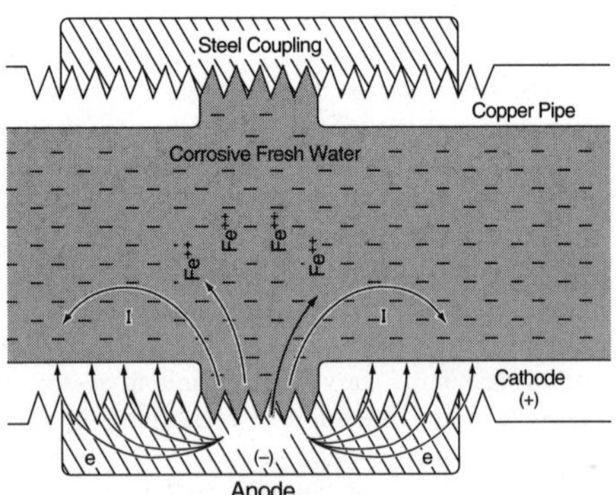

Bimetal couple

bimetal couple A type of corrosion in which two different metals or alloys are in contact with each other in a common medium.

bimonthly user charges The charges made to the users of water service through the general water rate structures of the utility for the utility's share of the cost of servicing the water service requirements.

binary Having two possible states or values (e.g., on and off). Computer instructions and data are represented as sequences of binary digits.

binary coded decimal (BCD) A positional value code for which each decimal digit is binary coded into 4-bit segments.

binary digit (bit) The basic element of a binary code. A bit is a unit of computer information represented by an electrical pulse (binary value 1) or no pulse (binary value 0) or a similar code equivalent to the result of a choice between two alternatives (e.g., yes or no).

binary digits per second *See the Units of Measure list.*

binary synchronous control (BSC) A protocol used for synchronous communication, often referred to as bisynch or bisynchronous protocol. This protocol, originally developed by IBM, was adopted by many mainframe and minicomputer vendors in the 1970s and early 1980s. Binary synchronous control is classified as a character-oriented protocol because a data transmission consists of frames of characters (8-bit strings), each of which represents a control signal or a portion of the message. Significant disadvantages of this protocol for interactive processing in complex networks led to the development of bit-oriented protocols in the late 1970s. *See also* synchronous communication.

binding arbitration A quasi-judicial process in which the parties agree to submit an unresolved dispute for binding agreement to a neutral third party. The arbitrator makes a decision after both parties submit their positions.

binding region The site on which a natural ligand, drug, or other chemical binds to a macromolecule. These binding sites can be specific or nonspecific. Specific sites are those in which a biological response can be expected as a result of binding. This term can also refer to sites on deoxyribonucleic acid where protein effectors (e.g., transcription factors) control the expression of particular genes.

binomial distribution A type of nonparametric data distribution.

bioaccumulation (1) A process by which a chemical becomes concentrated within organisms within a food chain. With organic chemicals, the extent of bioaccumulation is related to the chemical's lipid solubility (i.e., its oil–water partition coefficient) and its resistance to physical, chemical, and metabolic degradation. Inorganic chemicals can also be bioaccumulated, particularly some metals that are incorporated into bone (e.g., lead) or are complexed by proteins in particular organs (e.g., cadmium). Ambient water quality standards explicitly consider bioaccumulation in the human food chain; this is one reason why these standards vary from drinking water standards. (2) The uptake and retention of substances by an organism from its environment. *See also* ambient water quality standards.

bioassay A technique that uses a change in biological activity (stimulatory, inhibitory, or toxic) of a test organism to qualitatively or quantitatively assess the effects of exposure to a defined chemical compound; a defined mixture of compounds; or a complex, undefined mixture.

bioavailability A measure of the extent to which a chemical is available for systemic absorption by an organism. Bioavailability varies by chemical form (e.g., depending on the oxidation states of metals or metalloids), the matrix in which the chemical occurs, and the route of exposure (i.e., skin, gastrointestinal tract, or lung).

biocenosis A group of plants and animals living under the same conditions in the same space. It is also called a biotic community.

biochemical Pertaining to chemical change resulting from biological action.

biochemical action A chemical change resulting from the metabolism of living organisms.

biochemical fingerprinting The process of identifying microorganisms based on a matrix system of biochemical reactions by bacteria, e.g., glucose fermentation.

biochemical oxygen demand (BOD) A measure of the quantity of oxygen used in the biochemical oxidation of organic matter in a specified time and at a specific temperature. It is not related to the oxygen requirements in chemical combustion; it is determined entirely by the availability of the organic matter as a biological food and by the amount of oxygen the microorganisms use during oxidation. BOD serves as a surrogate for the biodegradable organic content (BDOC) in water because it is proportional to the BDOC.

biochemistry The branch of chemistry that deals with the chemical reactions involved in the life processes of plants and animals.

biocide A chemical that can kill or inhibit the growth of living organisms, such as bacteria, fungi, molds, and slimes.

bioconcentration The physical, chemical, and biological processes that act to increase the concentrations of a chemical in an organism relative to its environment. Bioconcentration can occur by passive or active processes. Passive processes are governed primarily by the oil–water partition coefficient or specific complexation within a compartment such as the bone matrix. Active processes require the expenditure of metabolic energy.

biodegradability The susceptibility of a substance to decomposition by microorganisms.

biodegradable Subject to degradation (breakdown) into simpler substances by biological action.

biodegradable organic carbon (BDOC or BOC) That portion of the organic carbon in water that can be mineralized by heterotrophic microorganisms. Ozone (O_3) can convert organic matter in water to biodegradable organic carbon, whereas biological filtration can reduce the biodegradable organic carbon level. *See also* biological filtration.

biodegradable organic matter (BDOM or BOM) That portion of the organic matter in water that can be degraded by microorganisms. Biodegradable organic carbon is a subset of biodegradable organic matter. *See also* biodegradable organic carbon.

biodegradation The breakdown of organic matter by microorganisms.

biofilm A layer of microorganisms held together by a matrix of organic polymers and found at the interface between water and a solid substrate in an aquatic environment. Biofilms in natural environments are heterogeneous and contain a range of microenvironments conducive to development of a diverse bacterial population. A biofilm can be detrimental to a process (say, by fouling a water treatment membrane) or beneficial (as in a biological treatment system). *Contrast with* suspended growth.

biofilter *See* biologically active filter.

biofiltered Removed by means of a biologically active filter. *See also* biological filtration; biologically active filter.

biofouling A phenomenon in which the performance of a unit process is compromised by biological growth. For example, biofouling can occur in a membrane process when microorganisms grow on the membrane surface and result in a premature and excessive loss of flow rate, or flux, through the membrane.

biogeochemical cycling The continuous and regularly occurring circulation of chemicals among atmosphere, biosphere, hydrosphere, and lithosphere.

biohazard An infectious agent presenting a risk or potential risk to human health, either directly through infection or indirectly through disruption of the environment. *See also* bloodborne pathogens; hazardous material.

bioindicator An organism that produces an observable response on exposure to a given substance.

biologic variation The totality of differences in response to a toxic or infectious insult that can be observed within representatives of a single species. Some of these differences are caused by differences in the genetics of an individual. Other differences can result from differences in the environment. For example, exposure to other agents or foods can alter the metabolism of a chemical; for another example, preexisting diseases can alter the susceptibility of an individual to chemical, physical, or microbial agents.

biological activated carbon *See* biologically enhanced activated carbon.

Biohazard warning symbol

biological analysis The examination of an environmental sample for the presence of biological agents, including macroscopic forms such as nematodes, worms, insects, and plants, as well as microscopic forms such as bacteria, fungi, algae, protozoa, and viruses.

biological denitrification The transformation of nitrate nitrogen (NO_3^-–N) to nitrogen gas (N_2) by microorganisms in an oxygen-free (anoxic) environment and in the presence of an electron donor (usually organic carbon) to drive the reaction:

$$2\ NO_3^- \rightarrow N_2 + 3\ O_2 + 2\ e^-$$

biological filtration The process of filtering water through a filter medium that has been allowed to develop a microbial biofilm that assists in the removal of fine particulate and dissolved organic materials.

biological fouling The clogging of a filtration medium as a result of microbiological growth. It is typically associated with (1) a loss of flux and potential filtrate quality degradation in membrane filters or (2) growth on a groundwater well casing. It is also called biofouling. *See also* flux; membrane filter.

biological growth The activity and growth of any and all living organisms.

biological indicator *See* bioindicator; indicator organism.

biological iron Iron involved in biological processes. Iron is involved in numerous oxidation–reduction reactions that are important in biological processes. For example, consider iron in the ferredoxins involved in the reduction of hydrogen and its incorporation into carbohydrates during photosynthesis; the role of iron in functional groups of most enzymes of the Krebs cycle and as an electron carrier in the cytochromes; and the central role of iron in the heme molecule that permits oxygen and electron transport essential to respiration in all vertebrates.

biological oxidation The process by which living organisms in the presence of oxygen convert organic matter into a more stable or a mineral form.

biological pest control The process of controlling nuisance insects and plants through the use of parasitic or predatory organisms.

biological reactor A vessel or chamber with a controlled volume in which biological reactions proceed. *See also* biological treatment processes.

biological regrowth The phenomenon of biological growth in treated water, typically in the distribution system. *See also* bacterial regrowth.

biological slime The gelatinous film formed by microbial growths covering the medium or spanning the interstices of a biological bed. Biological slime is also called microbial film.

biological stability A condition in which the nutrient status of treated drinking water is such that the water will not support (or will support only minimally) the growth of bacteria.

biological treatment processes Treatment processes that achieve contaminant removal through biological means. Conventional physicochemical processes, such as filtration, can remove biodegradable compounds if permitted to become biologically active by allowing microbial growth on the filtering medium. Biological treatment processes can also be designed to remove specific contaminants, such as nitrate (NO_3^-), in either fixed-film or suspended-growth biological reactors. *See also* fixed film; suspended growth.

biologically active carbon *See* biologically enhanced activated carbon.

biologically active filter Granular media (e.g., activated carbon, anthracite coal, or sand) that has developed a biofilm capable of degrading organic matter, ammonia (NH_3), or both. *See also* biologically enhanced activated carbon.

biologically active filtration (BAF) The removal of biodegradable organic matter by passing water over a fixed medium covered with a biofilm.

biologically available The physical or chemical form of a substance that can be directly used by an organism.

biologically based dose-response (BBDR) modeling Modeling of toxicological risks from a chemical by individually considering the contributions of biological variables to the response of interest. For example, in modeling the effects of a putative chemical carcinogen in humans based on animal data, the relative rates of activation and deactivation of the chemical by metabolism, as well as the kinetics of the chemical and any active metabolites, would be formally incorporated into a comparative pharmacokinetic model. This would normalize dosimetry between species. Then those biological responses, such as the chemical's genotoxic activity versus cytotoxic effects, which are induced by either the parent compound or its metabolites, would be individually factored into the model. In addition, any inherent differences in the responses of different species to these variables might be included, if known. Such models will inherently include more information about biological variability both within and across species and when the appropriate information is available, place the risks from exposure to a chemical in a much better context. The data demands from such an approach are much greater than for current risk assessment methods. Therefore, the use of such models will probably be reserved for those cases where competing risks (i.e., health risks from competing processes or high economic impacts) compel a more accurate understanding of the risks than provided by standard methodologies. *See also* linearized multistage model; log-probit model; logit model; multihit model; multistage model; one-hit model; probit model; Weibull model.

biologically enhanced activated carbon (BAC) Granular activated carbon (GAC) used as a treatment medium in which a microbial population is permitted to grow on the GAC surface. The GAC removes contaminants through adsorption, whereas the microbial population removes biodegradable components. The process is often coupled with ozone (O_3) pretreatment, which creates more biodegradable compounds and increases the biological activity on the GAC, but frequently decreases organic adsorbability. *See also* adsorption; granular activated carbon; ozone.

biologically refractory Difficult to biodegrade. *See also* biodegradability.

biologically significant Having the ability to measurably affect an organism.

biologically stable water Treated water that does not promote (or support to a significant extent) the growth of microorganisms during its distribution.

biology The science of living things. It is subdivided into specialties such as zoology, botany, and microbiology.

biomagnification A process by which a compound increases in concentration in an organism's tissue as the food level increases. Organisms at the top of the food chain have an unusually high concentration of the compound. Compounds that tend to biomagnify have a high affinity for the lipids (fats) of organisms. *See also* bioaccumulation; bioconcentration.

biomass (1) The total weight of biological matter, including any attached extracellular polymeric materials. (2) A material that is or was a living organism or was excreted from a microorganism.

biome A region or group of regions with specific climatic and other environmental conditions supporting similar flora and fauna.

biometry The study of numerical data concerning biological observations and phenomena via the application of statistical methods.

bioproductivity The quantity of organic material produced by the organisms in an ecosystem during a given period of time.

bioregeneration A process in which sites previously covered in an adsorbing medium are regenerated, or made available to perform future adsorption, through biological means. In bioregeneration, microorganisms residing on the external surface of the medium biodegrade contaminants that have been previously adsorbed, thereby regenerating the site for further adsorption. Bioregeneration is often associated with granular activated carbon adsorption treatment. It is also called bioreactivation.

bioremediation A process of adding nutrients to speed up processes in which microorganisms break down organic contaminants into harmless compounds.

biostat A substance that inhibits biological growth without destruction of the biomass.

biostatistics The application of statistics to biomedical issues. It is also called medical statistics.

biosurfactant A surface active chemical agent produced by a microorganism.

biota Living organisms, including bacteria, plants, and animals, in a given ecosystem.

biotic community *See* biocoenosis.

biotic influence Biological influence on life, in contrast to climatic influence.

biotope A limited space characterized by certain ecological conditions and inhabited by one or more plant or animal species.

biotransformation The metabolism of a chemical. Metabolism can either detoxify or actually mediate the toxicity of a chemical. Metabolic reactions in mammalian systems are frequently placed into two categories, referred to as phase I and II reactions. Phase I reactions generally result in the production of reactive intermediates that are often the mediators of toxic responses. Phase II reactions generally involve conjugation reactions with moieties (chemical groups) that increase the water solubility of the chemical and thus make it easier to eliminate from the body. These latter reactions most frequently lead to reductions in toxicity, but definite exceptions to this rule exist.

biotreatment A treatment process using biological methods to remove contaminants. *See also* biological treatment processes; biologically active filtration; biologically enhanced activated carbon; bioremediation.

Birge-Eckman dredge A sampler designed for collecting grab samples from the top inch or so (2–3 centimetres) of the bottom sediment of a lake or river.

birth defect A structural or functional abnormality that is induced in the womb and is apparent at birth or expressed later in life as a clear pattern of symptomology.

bit *See* binary digit.

bits per second (bps) *See* binary digits per second *in the Units of Measure list.*

bittern *See* mother liquor.

bituminous-based activated carbon A form of granular activated carbon or powdered activated carbon that is produced using bituminous coal as the raw material. *See also* granular activated carbon; powdered activated carbon.

bivalent ion An ion that has a valence charge of two. The charge can be positive or negative.

BK *See* bromoketone.

black alum Alum used as a coagulant that contains a small percentage of activated carbon.

black water An off-color condition that develops in drinking water from the presence of excess manganese that has been oxidized. Normally, less than 0.005 milligrams per litre of manganese will not cause black water; however, as levels approach 0.10 milligrams per litre, black water is more likely to occur.

blackfoot disease A circulatory disease that has been associated with arsenic exposures in Taiwan.

bladder pump A device for lifting water in a well by alternately filling a submerged bladder and then squeezing the bladder to force the water toward the surface. This type of pump is frequently used for contaminant characterization because of its ability to precisely pump small volumes of sample water.

blank A sample or result representing a matrix with all constituents except the analyte. A blank is analyzed in the same manner as standards and samples.

Blastocystis An intestinal protozoan parasite transmitted through the fecal–oral route by contaminated food or water.

bleach A strong oxidizing agent and disinfectant formulated to break down organic matter and destroy biological organisms. The term *bleach* commonly refers to a 5.25 percent nominal solution of sodium hypochlorite ($NaOCl$, household bleach) that is equivalent to 3 percent to 5 percent available free chlorine (strength varies by shelf life). Sodium hypochlorite is also available commercially in concentrations of between 5 percent and 15 percent available chlorine. Dry bleach is a dry calcium hypochlorite ($Ca(OCl)_2$) with 99.2 weight percent available chlorine.

bleaching powder A combination of slaked lime and chlorine.

bleed (1) To drain a liquid or gas, as in bleeding accumulated air from a water line or draining a trap or a container of accumulated water. (2) The exuding, percolation, or seeping of a liquid through a surface.

blending The process of mixing flow from more than one source into a single flow stream. Blending can be used to lessen the concentrations of compounds to desired levels by mixing a treated flow stream with one that is untreated or partially treated.

blind flange A pipe flange used to close the end of a pipe. It produces a blind end that is also called a dead end.

blind spots Places in the filter medium or membrane where no filtration takes place.

blind study In epidemiology, an experimental study in which either the investigators, study participants, or both (a double blind study) do not know to which group—control or study—a participant has been assigned. This information can also be withheld from laboratory analysts (a triple blind study). The intent is to eliminate biases or prejudices of the investigators and study participants.

blinding The reduction or shutting off of flow due to filter medium or membrane fouling. *See also* fouling.

block In a pressure-driven membrane treatment system, a grouping of membranes and pressure vessels in a single unit having common control.

block rate (1) Inverted: A rate structure of increasing price per unit of water delivered after the first unit. (2) Declining: A rate structure of reducing price per unit of water delivered after the first unit.

blocking A volume of concrete poured behind and around a bend, tee, or directional change that resists the forces created by the liquid moving inside the pipe. A vertical bend may require blocking under and around the pipe, with straps or reinforcement around the bend to hold down the bend and resist the upward forces when the depth above the bend is small. An instance of blocking is also called a thrust block.

blood–brain barrier Tight junctions that exist between the endothelial cells of vessels that supply the brain. Similar tight junctions are found between the ependymal cells lining the choroid plexus, a structure involved in the formation of cerebrospinal fluid. These two barriers make it necessary for chemicals to pass through cells before they can gain access to the brain. Therefore, the brain is more protected than other organs against polar chemicals that are present in the blood. Lipid-soluble chemicals pass through the blood–brain barrier very easily.

blood–testis barrier A poorly defined diffusional barrier that limits access of polar chemicals to the germ cells. The permeability of this barrier differs significantly between species. As a consequence, the prediction of a chemical's access to germ cells based on simple consideration of physical and chemical properties is problematic at best.

blood–urea nitrogen (BUN) A convenient measure of chemically induced renal damage in toxicological studies. Urea is a by-product of protein catabolism and provides a means of eliminating metabolic ammonia (NH_3) from the body. Urea is eliminated by the kidney, so compromised renal function leads to increased concentrations of urea in blood.

bloodborne pathogens Potentially infectious bloodborne materials, including non-A hepatitis, non-B hepatitis, hepatitis B, and delta hepatitis viruses, as well as human immunodeficiency virus. The two most significant bloodborne pathogens are hepatitis B and human immunodeficiency virus. The skin (particularly damaged skin), eye, or mucous membranes may become entry points.

bloom A large, undesirable growth of algae or diatoms in a water supply.

blowby (1) The technique sometimes used for recycling concentrate (e.g., membrane reject) back to the feed. (2) Contaminant leakage through or by the water treatment device.

blowdown (1) The continuous or intermittent removal of a portion of any process flow to maintain the constituents of the flow within desired levels. (2) The water discharged from a boiler, cooling tower, or membrane water treatment system to dispose accumulated dissolved solids.

blowoff (1) A connection at the low point on a pipeline used to drain or flush the pipe. (2) A device on an enclosed tank such as a water heater that opens at a specific pressure to prevent excessive pressure from occurring inside the tank.

blowoff valve A valve installed in a low point or depression on a pipeline to allow drainage of the line. It is also called a washout valve.

blowout A condition of bursting under hydrostatic pressure, e.g., a conduit, dam, or canal blowout.

blue copperas ($CuSO_4$) Copper sulfate.

blue stone ($CuSO_4$) A common name for copper sulfate.

blue vitriol ($CuSO_4$) A water supply term for copper sulfate.

blue-baby syndrome A condition in which infants have markedly elevated concentrations of methemoglobin. Methemoglobin is a form of hemoglobin in which the iron has been oxidized to the ferric state. In this form, it is incapable of acting as a carrier of oxygen in the blood. If not corrected, blue-baby syndrome can result in asphyxiation. Having small amounts of methemoglobin in the blood is normal. The level can be markedly elevated, however, when an individual is exposed to chemicals that directly oxidize hemoglobin or participate in reactions that produce such an oxidant; nitrite (NO_2^-) and chlorite (ClO_2^-) are examples of chemicals that can produce this effect.

blue-green algae A common name for Cyanobacteria. These microorganisms have been referred to as algae and bacteria in the literature. They belong to a class of living organisms called prokaryotes that are primitive in that they have no organized nucleus. Certain blue-green algae release odorous chemicals, such as geosmin and 2-methylisoborneal, with an earthy–musty smell. *See also* Cyanobacteria; geosmin; 2-methylisoborneal; prokaryote.

BMPs *See* best management practices.

board feet *See* feet board measure *in the Units of Measure list.*

Board of Certified Safety Professionals (BCSP) A peer certification organization founded in 1969. The Board's energies are directed toward evaluating academic and professional experience qualifications of safety professionals, administering examinations, and issuing certificates of qualification to those meeting its criteria.

BOC *See* biodegradable organic carbon.

BOD *See* biochemical oxygen demand.

body burden The total amount of a chemical that is in the body. This concept is most useful when one is referring to metals or very poorly metabolized nonpolar chemicals that accumulate in the body with repeated exposure. In these cases, chronic toxicity parallels the body burden because the compounds are not metabolized. Frequently, a reservoir may be established in a particular body compartment that acts to buffer and maintain concentrations of the chemical in blood (examples are lead in bone and dioxin in fat). The term is not particularly useful when one is referring to chemicals that produce cumulative damage without accumulating in the body. The best examples of this latter group of chemicals are organic chemicals that interact with deoxyribonucleic acid to produce point mutations.

body feed In diatomaceous earth filters, the continuous addition of diatomaceous earth during the filtering cycle to provide a fresh filtering surface as the suspended material clogs the precoat.

body water The portion of the body that is water, generally about 80 percent of the total body weight (the percentage decreases with increasing body fat). Water is, however, compartmentalized in the body into extracellular and intracellular compartments. The extracellular water is divided into vascular water (i.e., within blood vessels) and interstitial water (between cells). Interstitial water in some organs (e.g., the brain) is very isolated from vascular water, whereas in other areas (e.g., the peritoneal cavity) relatively rapid mixing of vascular and interstitial water occurs. Chemicals in the body have varying access to these different water compartments. Nonpolar, or uncharged, molecules move between these compartments with ease, whereas polar, or charged, molecules move with difficulty unless they have very small molecular weights (lower than 100) or are transported by a carrier mechanism.

body weight (Bw) The weight of an animal or a person. It is used in the calculation of safe doses of chemicals, expressed in terms of milligrams of the chemical per kilogram of body weight.

bog Permanently wet terrain characterized by an acidic soil that produces peat and dominated by sphagnum and shrubs of the *Ericaceae* family.

boil A rise in the water surface caused by the turbulent upward movement of water.

boil water advisory A public notice issued by a federal, state, or local health department via broadcast media that informs users of a public water system that their drinking water is, or potentially is, unsafe microbiologically and that the drinking water should be boiled before use. To kill waterborne pathogens, water at sea level should be brought to a rolling boil, held in that condition for 1 minute, and then allowed to cool (protected from further contamination) before consumption.

boil water order *See* boil water advisory.

Boltzmann's constant A mathematical constant used in Ludwig Boltzmann's kinetic theory of gases and the rules governing their viscosity and diffusion. The Boltzmann constant equals the gas constant (8.314×10^7 ergs per mole degree) divided by Avogadro number (6.02×10^{23} molecules per mole), which equals 1.3804×10^{-16} ergs per degree molecule.

bolus dose A dose of chemical given to an animal in a single volume, instead over an extended period as would be the case if the contaminant were in drinking water.

BOM *See* background organic matter; biodegradable organic matter.

bond (1) A long-term debt instrument (covering usually 20 to 30 years) that includes a promise to pay a specified sum of money (the principal) at a specified time in the future. Periodic interest payments (usually tax-free for municipal or governmental entities) on the face value of the bonds are made at a specified rate and at specified dates (usually semiannually). A number of basic bond types exist: general obligation bonds are secured by the full faith and credit of an issuing entity and are backed by the taxing authority of that entity; revenue bonds are secured by the revenue-producing ability of the entity; special assessment bonds (under the Improvement Acts of 1911, 1913, and 1915) are secured by a direct lien against a piece of property in the assessment district for specific public improvements. Many other variations of long-term bond debt instruments exist: certificates of participation (lease revenue bonds), zero coupon bonds, limited obligation bonds, variable rate bonds, adjustable rate bonds, and moral obligation bonds. (2) *See* chemical bond.

bond anticipation note A short-term interest-bearing note issued by a governmental agency in anticipation of bonds to be issued at a later date. The note is retired from proceeds of the bond issue to which it is related.

bond council A council typically made up of one or more attorneys who ensure that all of the legal aspects of a bond issued as a means of borrowing money for capital programs are properly conducted and stated in the debt agreement.

bond covenant A binding agreement to perform actions or meet conditions. Bond issues include various covenants in addition to the promise to make periodic interest payments and retire the debt according to a prescribed schedule. Maintaining a minimum ratio of net revenues to debt service payments (a debt service coverage ratio) is a common covenant.

bond discount The excess of the face value of a bond over the price for which it is acquired or sold.

bond fund The proceeds realized or received from a bond issue after payment of the costs of issuing and underwriting. Bond funds are typically designated to be used for one or more specific projects over a prescribed period of time.

bond indenture The legal instrument, as adopted by the issuer, that concerns a bond issue. Typically, such instruments contain pledges to bondholders regarding the circumstances of payment of principal and interest, operation, flow of funds, and further debt issuance.

bond ordinance *See* bond indenture.

bond premium The excess of the price at which a bond is acquired or sold over its face value.

bond resolution *See* bond indenture.

bonded debt That portion of the indebtedness of an enterprise represented by outstanding bonds.

bonds payable The face value of bonds issued and unpaid.

bone char A black pigment substance, with a carbon content of about 10 percent, made by carbonizing animal bones. Bone char is used for decolorizing sugar and in water treatment. In water treatment, it has been used as a selective anion exchanger for fluoride and arsenic control.

bone marrow The soft organic material that fills the cavities of bones. In adults this is the site in which blood cells are formed. It is the target of certain toxic chemicals, most notably benzene. Benzene can completely suppress blood cell production, leading to aplastic anemia, a fatal disease. It can also cause loss of control over the production of white blood cells, leading to leukemia.

bonnet The cover on a gate valve.

BOO *See* build-own-operate.

boom A floating structure, usually made of timber or logs, used (1) to protect the face of a dam or other structure built in or on water from damage by wave action or by floating material being dashed against it by the waves or (2) to deflect floating material away from the dam or other structure. A boom is also used to contain floating materials, as in the case of oil spills.

booster chlorination The use of chlorination facilities in a distribution system to increase (boost) the chlorine (HOCl) residual after an initial decay downstream of the treatment plant. Booster chlorination is used to lower initial dosages at the treatment plant while maintaining residuals at the far ends of the distribution system.

booster pump A pump used (1) to increase the pressure of the fluid on the discharge side of a pump that provides pressure in a closed distribution system and (2) to lift or boost water from a lower pressure plane to a higher pressure plane.

booster station (1) A pumping station in a water distribution system that is used to increase the pressure in the mains on the discharge side of the pumps. (2) A station that pumps water from low-level or ground storage to a distribution system.

borate A salt or ester of a boric acid. The salts of boric acid are quite complicated and, at least in aqueous solution, never contain the simple anion BO_3^{3-}. For example, the anion that occurs in borax (a sodium borate) is $(B_4O_5(OH)_4)^{2-}$.

borate buffer A solution of boric acid (H_3BO_3) and sodium hydroxide (NaOH) used to maintain a constant pH of 8 during simulated distribution system (or other) testing. *See also* simulated distribution system test.

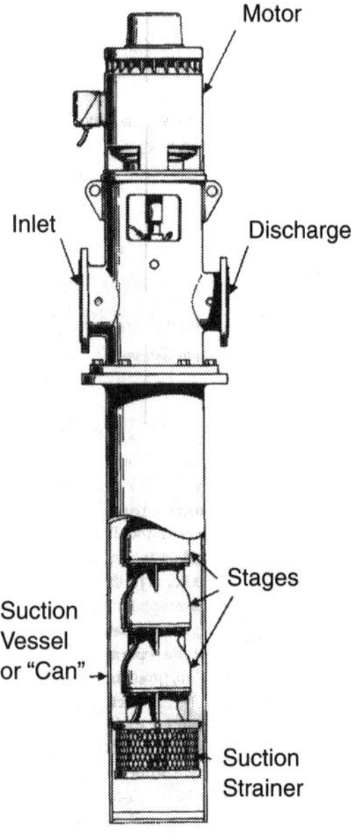

Booster pump

bore (1) A wave of water, such as a tidal wave, having a nearly vertical front and advancing upstream as a result of high tides in certain estuaries; a similar wave advancing downstream as the result of a cloudburst or the sudden release of a large volume of water from a reservoir. A bore is analogous to a hydraulic jump in that it represents the limiting condition of the surface curve in which the curve tends to become perpendicular to the bed of the stream. (2) A standing wave that advances upstream in an open conduit from a point where the flow has suddenly been stopped. (3) A tunnel, especially during its construction. *See also* hydraulic bore; suction wave.

bored well A well that is excavated by means of an auger (hand or power), as distinguished from one that is dug, drilled, or driven.

borehole geochemical probe A water quality monitoring device that is lowered into a well on a cable and can make direct measurements of physical parameters such as pH, E_H, temperature, and specific conductivity. Ion-sensitive electrodes can be added to allow for direct measurement of certain chemical parameters for which oxidation–reduction (redox) couples exist, e.g., ferrous iron.

borehole geophysics A subdiscipline of geophysics developed around the lowering of various probes into wells. The probes are used to infer hydraulic, chemical, and physical properties from various physical, electrical, radioactive, and chemical measurements made in the borehole.

boring (1) Subsurface investigations performed by drilling down to the desired depth, removing samples of the material penetrated so that it can be examined at the surface, recording the elevation at which changes in material are found, obtaining samples in a disturbed or undisturbed condition from the various strata, and preparing a log or chart of the boring data. Borings may be classified as follows: (a) soil auger boring, (b) churn drilling, (c) rotary drilling, and (d) core boring. (2) Holes made by boring. (3) Material removed by boring.

boring sample Material recovered from a boring and used to indicate the character of the formation penetrated by the boring.

boron (B) A trivalent metallic element (atomic number 5) found in nature only in combination with other elements.

borosilicate glass A type of heat-resistant glass used for labware.

borrow pit A bank or pit from which earth is taken for use in filling or embanking.

bottled artesian water Bottled water that comes from a well drilled into a confined aquifer in which the water level stands at some height above the top of the aquifer, according to the May 13, 1996, Food and Drug Administration rules for the labeling of bottled water. Bottled artesian water is also referred to as artesian well water.

bottled groundwater Bottled water from a subsurface saturated zone that is under a pressure greater than or equal to atmospheric pressure, according to the May 13, 1996, Food and Drug Administration rules for the labeling of bottled water. The groundwater must not be under the direct influence of surface water.

bottled mineral water According to the May 13, 1996, Food and Drug Administration rules for the labeling of bottled water, water that meets the following criteria: contains no less than 250 milligrams per litre of total dissolved solids (determined by evaporating to dryness and weighing the residue); comes from a source tapped at one or more boreholes or springs; and originates from a geologically and physically protected underground water source. *See also* total dissolved solids.

bottled purified water Bottled water that has been produced by distillation, deionization, reverse osmosis, or other suitable process that meets the definition of purified water in the United States *Pharmacopeia,* 23rd revision, Jan. 1, 1995, according to the May 13, 1996, Food and Drug Administration rules for the labeling of bottled water. Bottled purified water is also called demineralized water. Alternatively, bottled water may be called deionized water if it has been processed by deionization; distilled water if it has been processed by distillation; reverse osmosis water if it has been processed by reverse osmosis; and so forth.

bottled sparkling water Bottled water that, after treatment and possible replacement of carbon dioxide (CO_2), contains the same amount of carbon dioxide that it had when it was taken from the source, according to the May 13, 1996, Food and Drug Administration rules for the labeling of bottled water.

bottled spring water Bottled water that is collected from an underground formation from which water flows naturally to the surface of the earth, according to the May 13, 1996, Food and Drug Administration rules for the labeling of bottled water. The water must be collected at the spring or through a borehole tapping the underground formation feeding the spring. A natural force must cause the water to flow to the surface, and the location of the spring must be identified.

bottled sterile water Bottled water that meets the requirements of the sterility tests in the United States *Pharmacopeia,* 23rd revision, Jan. 1, 1995, according to the May 13, 1996, Food and Drug Administration rules for the labeling of bottled water. Such water is also called sterilized water.

bottled water Water intended for human consumption that is sealed in bottles or other containers with no added ingredients except a disinfectant, if necessary, to ensure microbial quality. Bottled water is regulated in the United States as a food by the Food and Drug Administration.

bottled well water Bottled water from a hole—bored, drilled, or otherwise constructed in the ground—that taps the water of an aquifer, according to the May 13, 1996, Food and Drug Administration rules for the labeling of bottled water.

bottom contraction The reduction in the area of overflowing water caused by the crest of a weir contracting the nappe.

bottom sediment Those sediments making up the bed of a body of still or running water.

bottom water level (BWL) (1) The lower level in a tank or reservoir used to set a level at which to turn on a pump or open an altitude valve. (2) The level beneath an oil or gas production formation where water is found.

bound water (1) Water held strongly on the surface or in the interior of colloidal particles. (2) Water associated with the hydration of crystalline compounds. *See also* deliquescence; hydrophilic; hydroscopic.

boundary layer (1) The region near a solid surface where fluid motion and transport are negatively affected by that surface. (2) For membrane water treatment, a thin layer at the membrane surface where water velocities are significantly lower than in the bulk stream flow.

boundary spring A spring occurring on the lower slope of an alluvial cone at the point where the water table slope and surface gradient are equal. Such springs are usually located on the boundary between two formations, the upstream one being the more pervious and having a flatter slope to its water table. A boundary spring is also called an alluvial-slope spring.

bovine serum albumin (BSA) A blood protein preparation obtained from the serum of cattle. A dilute (usually 5 percent) filter-sterilized solution of bovine serum albumin is used as an enrichment in media for culturing a variety of microorganisms and tissue cells.

bowl, pump *See* pump bowl.

box plot A graphical representation of statistical parameters of a data set. Various indicators in a box plot correspond to various quartiles and the mean of the data distribution. *See also* box-and-whisker plot.

box-and-whisker plot A graphic presentation of the statistical analysis of a data set in which the bottom and top of the box equal the twenty-fifth and seventy-fifth percentile values, respectively; an additional line within the box corresponds to the median level; lines perpendicular to and above or below the box indicate the maximum or minimum values excluding statistical outliers; and outliers are shown as separate points. Other variations are possible on this representation in which (1) a dotted line within the box corresponds to the mean level or (2) the sides of the box can be notched to indicate the 95 percent confidence interval for the median.

Box–Jenkins method *See* autoregressive integrated moving average forecasting method.

BPACl *See* basic polyaluminum chloride.

bps *See* binary digits per second *in the Units of Measure list.*

Bq *See* becquerel *in the Units of Measure list.*

brackish water Water having a mineral content in the range between fresh water and seawater. In water-desalting practice, brackish water is generally considered to be water containing 1,000 to 10,000 milligrams per litre of total dissolved solids.

brake horsepower (bhp) The power supplied to a pump by a motor, expressed in units of horsepower (watts). *See also* motor horsepower; water horsepower.

branch A special form of cast-iron pipe used for making connections to water mains. The various types are called T, Y, T–Y, double Y, and V branches, according to their respective shapes.

brass A metal alloy of copper, zinc, and usually some lead. Brass is harder and stronger than copper because of its zinc content; lead contributes malleability and ductility.

brazing The process of joining metal parts with an alloy that melts at a temperature higher than 800° Fahrenheit (427° Celsius) but lower than the melting temperature of the metal parts to be joined.

break tank A storage device that is used for hydraulic isolation and surge protection from a pressurized system for which the water surface is at atmospheric pressure. For example, a tank receiving pumped well water for which the contents are either repumped or flow by gravity to a subsequent process is a break tank.

breakaway point The rate of flow at which a water meter begins to register, or the lowest rate of flow at which a definite movement of the register occurs.

breakeven analysis A method of analyzing costs and revenues over a range of treated water volumes to determine the volume point at which revenues and total costs are equal. The method assumes that a portion of costs is fixed (does not vary with volume) and that revenue and a portion of costs vary directly with volume. The basic equation is:

$$\text{breakeven volume} = \frac{\text{fixed costs}}{\text{unit price} - \text{variable cost per unit}}$$

breakpoint The point at which the chlorine dosage has satisfied the chlorine demand exerted by ammonia (NH_3).

breakpoint chlorination A process by which ammonia (NH_3) is removed from water through the continuous addition of chlorine (Cl_2 or HOCl). As chlorine is added, monochloramine (NH_2Cl) is initially formed, followed by dichloramine ($NHCl_2$) and trichloramine (nitrogen trichloride—NCl_3). At the breakpoint, the nitrogen has been completely released and any additional chlorine added is measured as a free, rather than combined, chlorine residual. *See also* chloramines; combined chlorine; dichloramine; trichloramine.

breakpoint curve A graphical representation of chemical relationships that exist as varying amounts of chlorine are added to water containing small amounts of ammonia-nitrogen (NH_3-N). Prior to the breakpoint, chlorine is present in the form of chloramines; subsequently, free chlorine is available. *See also* breakpoint; breakpoint chlorination.

breakthrough (1) The point in a filtering cycle at which turbidity-causing material starts to pass through the filter. (2) The time in the cycle of a treatment bed when an increase, sometimes defined as an unacceptable increase, in the effluent concentration occurs for the contaminant being controlled.

breakthrough capacity The capacity of an ion exchange column at a fixed regeneration level. It is usually expressed in terms of kilograms per cubic foot or kilograms per cubic metre of ion exchange resin.

bridge A device for interconnecting two local area networks. The bridge can be used for extending the range of a local area network and to filter the messages that are passed from one portion of a local area network to another.

bridging (1) In water softening, the caking in a dry-salt brine tank that causes the liquid or brine beneath the dry salt to fail to become saturated. The result of bridging is insufficient salt in the regenerant solution to properly regenerate the cation resin. (2) The ability of particles to form a crustlike film over void spaces within a filter medium or membrane. (3) A condition in which dry chemicals adhere in an arch-like shape and do not fall out of a storage tank as intended. (4) A coagulation mechanism in which long chains of organic polymers added as coagulants bridge the distance between small particles, holding them together to create larger particles that will settle readily. *See also* adsorption–destabilization; arching; double-layer compression; sweep-floc coagulation.

brightfield microscope Another term for a light microscope; because the object being viewed is illuminated by the light source, the field of view is bright, hence the term *brightfield*.

brightfield microscopy The most common form of light microscopy, in which the light source is emitted upward from beneath the specimen to the lens and the oculars. With brightfield illumination, specimen images appear dark against a bright background.

brine (1) A concentrated salt solution, generally containing sodium, chloride, and other ions typically having a concentration of 3 weight percent or more. (2) A concentrated salt solution remaining after desalting brackish or seawaters. For a brackish water membrane desalting system, the use of the word *concentrate* or *reject* is commonly preferred over *brine*. (3) A concentrated potassium chloride (KCl) or sodium chloride (NaCl) solution used in the regeneration stage of either cation or anion exchange water treatment devices. Sodium chloride brine saturation in an ion exchange softening brine tank is about 26 percent NaCl by weight at 60° Fahrenheit (15.5° Celsius). *See also* salt water.

brine collector A device used to gather and retrieve brine from a brine tank or ion exchange bed. *See also* collector.

brine consumption factor (BCF) In ion exchange, the equivalents of regenerant required divided by the equivalents of target ion removed. In nitrate (NO_3^-) removal, the BCF is sometimes called the brine use factor. *See also* ion exchange.

brine disposal The process of ultimately discharging a reject stream, most often associated with desalination or membrane processes, to the environment. Brine disposal often involves ocean discharge, deep well injection, or discharge to another receiving medium for which the mineral concentration is at least as high as the brine's concentration. *See also* deep-well injection.

brine draw The process of drawing a brine solution into a cation or anion exchange water treatment device during regeneration.

brine ejector (eductor) A device used to draw (or educt) brine from a brine tank and force (or eject) it into a either a cation or anion water treatment device. It is usually a component of the brine tank's control valve.

brine use factor (BUF) *See* brine consumption factor.

brinelling Tiny indentations (dents) high on the shoulder of the bearing race or bearing. Brinelling is a type of bearing failure.

brining In ion exchange softening, the passing of a concentrated salt (sodium chloride, NaCl) solution through the resin bed for regeneration.

British thermal unit (Btu) *See the Units of Measure list.*

Broad Street pump A pump associated with John Snow's discovery in 1854 that cholera is conveyed by water. The well and pump were located near the corner of what is now Broadwick and Lexington Streets in Soho, London. In late August and early September 1854, Snow noted that within 250 yards (230 metres) "there were upwards of five hundred fatal attacks of cholera in ten days" and "there were no other outbreaks of or increase in cholera in that part of London except amongst those who habitually used the pump." He drew a map of the district showing the distribution of the deaths and tabulated the rise and fall of the epidemic. Snow presented his evidence the evening of Sept. 7, 1854, to the Board of Guardians, and the pump handle was removed the following day.

broadband network A communications facility that operates with uniform efficiency over a wide band of frequencies and has the capability to transmit multiple channels simultaneously over the same physical connector.

broad-base terrace A long ridge of earth, usually 1 to 3 feet (0.3 to 1 metre) high and 15 to 30 feet (4.5 to 9 metres) wide, with gently sloping sides, a rounded crown, and a broad, shallow channel along the upper side, constructed to control erosion by diverting runoff along the contour at low velocity instead of permitting it to rush down the slope.

broad-crested weir A weir having a substantial width of crest in the direction parallel to the direction of flow of water over it. This type of weir supports the nappe for an appreciable length and produces no bottom contraction of the nappe. It is also called a wide-crested weir.

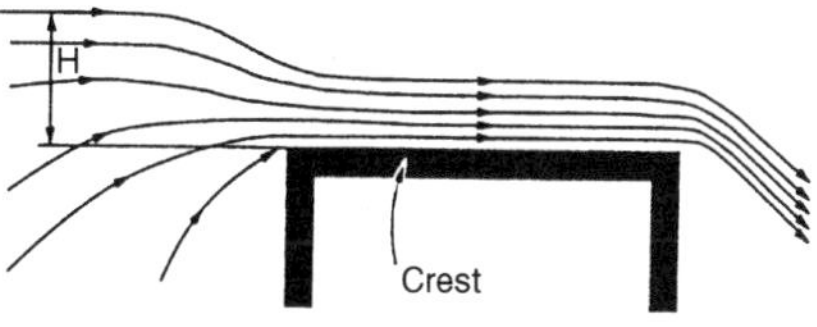

Broad-crested weir

broad-screen analysis A chemical analysis in which unknown compounds are being sought.

bromamines Chemicals formed during the mixing of chlorine (Cl_2 or HOCl), ammonia (NH_3), and the bromide ion. They can be produced during the chloramination of bromide-containing waters. *See also* chloramines.

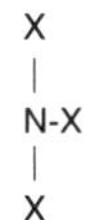

An example of bromamines where one to three of the Xs are bromine atoms and the rest are hydrogen or chlorine atoms

bromate *See* bromate ion.

bromate ion (BrO_3^-) The highest oxidation state of the bromide ion (Br^-). The bromate ion can be formed during the ozonation of bromide-containing waters.

bromide (Br^-) An inorganic ion found in surface water and groundwater and caused by (1) seawater intrusion, (2) the impact of connate water, or (3) industrial and oil-field brine discharges. When oxidized by chlorine (Cl_2 or HOCl) or ozone (O_3), it can result in the formation of bromide-substituted disinfection by-products. *See also* connate water; disinfection by-product; seawater intrusion.

bromide-to-FAC ratio (Br^-/Cl^+) The molar ratio of the water bromide level to the average free available chlorine (FAC)—$([FAC_0]+[FAC_t])/2$, where FAC_0 and FAC_t are the FAC values at the beginning and end of the contact period, respectively—concentration during the chlorine contact period. This ratio gives an indication of the relative amounts of hypobromous acid (HOBr) and hypochlorous acid (HOCl) available for trihalomethane formation, which influences the degree of bromine substitution. *See also* trihalomethane.

bromide-to-DOC (dissolved organic carbon) ratio *See* bromide-to-TOC ratio.

bromide-to-TOC ratio (Br^-/TOC) The weight ratio of the water bromide level to the total organic carbon (TOC) concentration at the point of chlorination. Because coagulation and granular activated carbon filtration can remove organic carbon but not bromide, this ratio increases during treatment and can result in a shift to more bromine-substituted disinfection by-products upon chlorination. *See also* disinfection by-product.

bromide utilization A measure of the degree of bromine substitution in trihalomethanes, where the molar sum of bromine in the individual trihalomethanes is divided by the molar amount of the water bromide. This value provides an indication of the percentage of bromide that is substituted into trihalomethanes. *See also* trihalomethane.

brominated alcohol *See* bromohydrin.

brominated organic *See* bromine-substituted organic.

bromine (Br_2) The oxidized form of the bromide ion (Br^-). In water, bromine is present as hypobromous acid (HOBr) and the hypobromite ion (OBr^-). *See also* bromide; hypobromite ion; hypobromous acid.

bromine atom (Br) A halogen atom that can be substituted into organic molecules during chlorination or ozonation of bromide-containing waters, thus forming disinfection by-products. Bromine atoms are also present in a number of synthetic organic chemicals. *See also* bromide; disinfection by-product; halogen; synthetic organic chemical.

bromine incorporation factor *See* bromine incorporation factor for haloacetic acids; bromine incorporation factor for trihalomethanes.

bromine incorporation factor for haloacetic acids [$n'(x/y)$] A measure of the degree of bromine substitution in haloacetic acids (HAAs), where the molar sum of bromine in all the individual HAAs is divided by the molar amount of the total HAAs. Because typically, not all nine HAAs are measured, the bromine incorporation factor $n'(x/y)$ is calculated for the molar sum of bromine in the measured bromine-containing HAAs (x measured species) divided by the molar amount of the total measured HAAs (y total measured species). For example, when the five HAA species (two of which contain bromine) that make up HAA5 have been measured, $x = 2$ and $y = 5$. *See also* bromine incorporation factor for trihalomethanes; haloacetic acid.

bromine incorporation factor for trihalomethanes (n) A measure of the degree of bromine substitution in trihalomethanes, where the molar sum of bromine in all the individual trihalomethanes is divided by the molar amount of the total trihalomethanes. This value can vary from 0 (corresponding to all chloroform) to 3 (corresponding to all bromoform). *See also* trihalomethane.

bromine-substituted Pertaining to a molecule into which a bromine atom has been introduced.

bromine-substituted by-product A disinfection by-product containing bromine atom(s). *See also* disinfection by-product.

bromine-substituted organic An organic molecule into which a bromine atom has been introduced.

bromine-substituted trihalomethanes, total (TTHM-Br) *See* total trihalomethanes bromine.

bromism A syndrome produced by chronic intakes of bromide. This poisoning is marked by signs of central nervous system depression. It was observed when bromides were commonly used as sedatives and hypnotics. Suspension of use corrects the problem. Bromism is unlikely to be associated with the milligram quantities of bromide that are frequently found in drinking water. The syndrome is also called brominism.

bromoacetic acid (BAA) (CX_3COOH, where one to three of the X atoms are bromine atoms and the remaining are chlorine and/or hydrogen) A haloacetic acid containing bromine atom(s). *See also* haloacetic acid.

```
 X O
 | ||
X-C-C-OH
 |
 X
```

Bromoacetic acid, where one to three of the Xs are bromine atoms and the rest are hydrogen atoms

bromoacetonitrile (BAN) ($CX_3C{\equiv}N$, where one to three of the X atoms are bromine atoms and the remaining are chlorine and/or hydrogen) A haloacetonitrile containing bromine atom(s). *See also* haloacetonitrile.

```
  X
  |
X-C-C≡N
  |
  X
```

Bromoacetonitrile, where one to three of the Xs are bromine atoms and the rest are hydrogen atoms

bromobenzene (C_6H_5Br) A synthetic organic chemical used as a solvent. *See also* solvent; synthetic organic chemical.

bromochloroacetic acid (BCAA) (CHBrClCOOH) A haloacetic acid containing one bromine atom and one chlorine atom. *See also* haloacetic acid.

```
  BrO
  | ||
Cl-C-C-OH
  |
  H
```

Bromochloroacetic acid

bromochloroacetonitrile (BCAN) (CHBrClC≡N) A haloacetonitrile containing one bromine atom and one chlorine atom. *See also* haloacetonitrile.

bromodeoxyuridine (BrdU) A pyrimidine analog that is used in toxicology and related sciences to detect cells that divide. It is used in place of tritiated thymidine. It is injected into animals or administered over an extended period via miniature pumps. It is also frequently employed in cell cultures. BrdU is incorporated in the deoxyribonucleic acid when cells replicate. BrdU so incorporated can be detected in histological sections by using an antibody and immunohistochemical techniques. This detection provides a very important parameter for use in nonlinear low-dose extrapolation models for carcinogenesis.

bromodichloroacetic acid (BDCAA) ($CBrCl_2COOH$) A haloacetic acid containing one bromine atom and two chlorine atoms. *See also* haloacetic acid.

```
  BrO
  | ||
Cl-C-C-OH
  |
  Cl
```

Bromodichloroacetic acid

bromodichloromethane (BCDM) ($CHCl_2Br$) A trihalomethane containing one bromine atom and two chlorine atoms. It is a commonly formed trihalomethane, even in waters with low bromide levels. *See also* trihalomethane.

bromoform ($CHBr_3$) A common name for tribromomethane, a trihalomethane containing only bromine atoms. It is typically formed to a significant extent only in waters containing moderate amounts of bromide. *See also* trihalomethane.

bromohydrin A class of organic compounds that are bromine-containing alcohols. Specifically, 3-bromo-2-methyl-2-butanol has been identified as a possible by-product from the ozonation of bromide-containing waters.

```
     HO Br
     |  |
CH3-C-C-CH3
     |  |
    CH3 H
```

3-bromo-2-methyl-2-butanol, an example of bromohydrin

bromoketone (BK) (R_1-CO-R_2, where one or both R groups contain one or more bromine atoms) A haloketone containing bromine atoms. *See also* haloketone.

```
  X O H
  | || |
X-C-C-C-H
  |   |
  X   H
```

Bromoketone, here one to three of the Xs are bromine atoms and the rest are hydrogen atoms

bromopicrin (CBr_3NO_2) A common name for tribromonitromethane, a disinfection by-product formed during the chlorination or ozonation of bromide-containing waters. *See also* disinfection by-product.

```
   Br
   |
Br-C-NO2
   |
   Br
```

Bromopicrin

bronze Any of various copper-base metal alloys that may or may not include tin, lead, or other metals in small amounts. Bronze is a commonly used material in water utility products such as water meters and valves.

Brownian motion The random movement of particles suspended in a fluid medium resulting from molecular bombardment. Brownian motion accounts for the diffusion mechanism in particle removal by filtration and in perikinetic flocculation. *See also* diffusion; perikinetic flocculation.

browser (1) A World Wide Web client. (2) An information retrieval tool. *See also* World Wide Web.

Brunauer-Emmet-Teller (BET) method A laboratory procedure used to determine the surface area of granular activated carbon (GAC). The method is based on the adsorption of nitrogen to the surface of the GAC. Based on the vapor pressure and equilibrium pressure of nitrogen, the amount of nitrogen adsorbed, and a linearized form of the Brunauer-Emmet-Teller equation, the GAC surface area available to form a monolayer of nitrogen can be calculated based on an assumption of the area of an adsorbed nitrogen molecule. The BET equation is:

$$\frac{q}{Q} = \frac{BC}{\left[(C_s - C)(1 + B - 1)\left(\frac{C}{C_s}\right)\right]}$$

Where (in any consistent set of units):

q = number of moles of adsorbate per mass of adsorbent at equilibrium

Q = maximum number of moles adsorbed per mass of adsorbent when the surface sites are saturated with adsorbate

B = dimensionless constant related to the difference in free energy between adsorbate on the first and successive layers

C = concentration of adsorbate in solution in milligrams per litre

C_s = saturation concentration of the adsorbate in solution in milligrams per litre

BSA *See* bovine serum albumin.

BSC *See* binary synchronous control.

Btu *See* British thermal unit *in the Units of Measure list.*

BTX *See* benzene–toluene–xylene.

bu *See* bushel *in the Units of Measure list.*

bubble memory Magnetic regions in thin film crystal that maintain data when electric power is shut off.

bubble point The required gas pressure point to fully displace liquid from the pores of a wetted filtration material, such as from a microfiltration membrane. The bubble point is also called the bubbling pressure or gas entry pressure.

bubble size The average diameter of the gas (e.g., ozone [O_3] or air) bubbles discharged from a diffuser at the bottom of the contactor in an ozone or aeration system. Generally, the finer the bubble size and the longer the bubble resides (contacts) within the water, the greater the transfer of air or ozone to the water.

bubbler A type of sprinkler head that delivers a relatively large volume of water to a level area where standing water gradually infiltrates into the soil. The flow rate is large relative to the area to which the water is delivered. Bubblers are used to irrigate trees and shrubs.

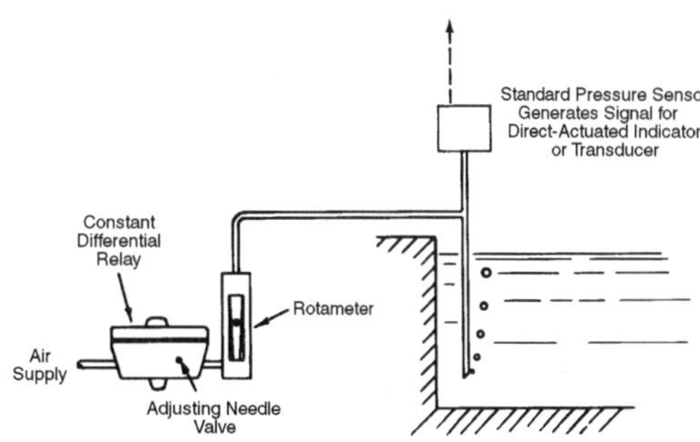

Bubbler-tube level indicator

bubbler-tube level indicator A gauge to monitor the water level by using a continuous air supply bubbled through the water at a set elevation to supply a pressure that can be recorded.

bubbler-tube level-sensing transmitter A electronic device that sends a signal from a water-level monitoring gauge to a remote receiver through a communications network for water-level recording.

Buchner funnel A laboratory filter consisting of a funnel with a perforated bottom. A disposable filter is placed atop the perforated bottom before liquid is filtered. A Buchner funnel is used to evaluate sludge dewaterability.

budget A plan adjusting expenses for a given time period to the estimated income for that same period.

budgeting The process of (1) identifying detailed operating and capital expenditures of all types required for efficient operations at a defined level of volume and service, and (2) matching revenues to the level of expenditures through adjustments of rates and charges, allowing for necessary levels of dedicated and available cash reserves.

BUF *See* brine use factor.

Buffalo Green monkey (BGM) cells A type of cell used for cell cultures that can be cultured in the laboratory continuously and is excellent for growing some of the human enteric viruses.

buffer A chemical substance that stabilizes the pH value of solutions. *See also* buffer solution.

buffer action The action of certain ions in solution in opposing a change in hydrogen ion concentration.

buffer capacity The concentration of a strong base (in moles per litre) that causes a unit change in pH when added to a solution. Buffer capacity is also known as buffer intensity.

buffer intensity *See* buffer capacity.

buffer solution A solution containing one or more substances that, in combination, resist any marked change in pH following addition of moderate amounts of either strong acid or strong base.

Buchner funnel

buffer strip A strip of grass or other close-growing vegetation that separates a waterway from an intensive land use area. Buffer strips are also referred to as filter strips, vegetated filter strips, and grassed buffers.

buffered Able to resist changes in pH.

bug An error in either a software program or the hardware. *See also* debug.

build-own-operate (BOO) A process enabling privately owned companies to finance, design, construct, and operate water or wastewater facilities over a long term for a municipality or other governmental agency. Build-own-operate negates the need for government agencies to obtain capital funds for construction and privatizes the utility operation.

building footprint The dimensions and area of a building foundation as represented in a two-dimensional plan.

building service The water pipe from the public main to a building. Building services are owned and maintained by the building owner or the distribution or collection agency, depending on local regulations.

buildup An increase of groundwater elevation around a recharge well or spreading basin. Buildup is the opposite of drawdown.

bulk density The mass per standard volume (usually kilograms per cubic metre or pounds per cubic foot) of material as it would be shipped from the supplier to the treatment plant.

bulk modulus of elasticity of water (E_v) The pressure required to produce a small change in water volume. In equation form,

$$E_v = -(dP)/(dV/V)$$

Where (in any consistent set of units):

E_v = the bulk modulus of elasticity of water
dP = the change in pressure
dV = the change in volume
V = the initial volume

bulkhead (1) A structure of wood, stone, or concrete erected along the shore of a water body to arrest wave action or along a steep embankment to control erosion. (2) A permanent or movable wall closely fitting into and across a waterway and intended to hold back earth or water. *See also* groin.

bulletin board system A means of disseminating information to individuals or groups through electronic mail.

bump joint A type of joint used to connect riveted or welded steel pipes of large diameter, usually 24 inches (0.6 metres) or more. One end of the length of pipe is flared out, and the end of the adjoining pipe is shaped so that it fits snugly into the flared end. Bends up to 5° per length of pipe can be made readily with such a joint.

BUN *See* blood–urea nitrogen.

buoyancy The upward force exerted by the fluid in which a body is immersed.

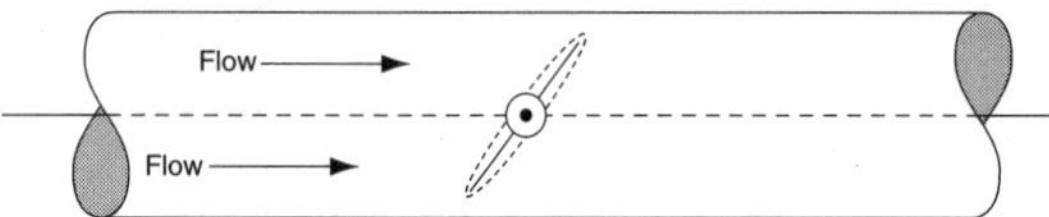

Cross section of a butterfly valve

burden The amount of indirect costs that are incurred and applied (burdened) to direct cost. Direct cost is usually direct labor (labor hours applied directly to the production of a product), but sometimes machine hours or other measures of direct cost are used if they are more closely related to the product produced. Typical burden costs are, for example, clerical support, material handling, maintenance, and rent. Burden rates (indirect cost divided by direct cost) are often developed by department, cost center, plant, function, or entity and applied to direct cost to estimate the total cost of producing a product through that cost center. Burden is sometimes called overhead.

burette A graduated glass tube fitted with a stopcock. It is used to dispense solutions during titration.

buried channel A former stream channel that has been filled with alluvial or glacial deposits and later covered by other material so that little surface indication of its existence or location exists.

buried gate valve A valve installed directly in the ground with a valve box extending from the valve to the ground or street surface to provide operating access.

burn, first to third degree Damage to the skin caused by various sources of heat, resulting in skin damage of increasing severity measured in degrees. A first-degree burn shows redness in unbroken skin. Indications of a second-degree burn are blistering and some breaks in the skin. Third-degree burns range from blistered, missing skin to damage to underlying tissues.

burner A high-temperature heating device that uses natural or bottled gas. It is also called a Bunsen burner.

burnt lime *See* lime.

bury length The depth from the surface of the ground to the bottom of the pipe to which a hydrant is connected.

bus A parallel data path that carries data at high speed between certain components of a computer system. The information-carrying capacity of a bus depends on its width, or the number of bits that can be transferred simultaneously. Mainframe buses usually are 24, 32, 48, or 64 bits wide, whereas minicomputer and microcomputer buses are typically 8, 16, or 32 bits wide.

bushel (bu) *See the Units of Measure list.*

bushing A short tube threaded inside and outside that screws into a pipe fitting to reduce its size.

butane number The volume of butane ($CH_3CH_2CH_2CH_3$) adsorbed per unit weight of activated carbon after air saturated with butane is passed through an activated carbon bed at a given temperature and pressure.

butterfly gate A gate that opens like a damper turning on a shaft inside the pipe. *See also* butterfly valve.

butterfly valve A valve that operates by the disk rotating on a shaft or stem through the center section of the valve body, closed when the disk is perpendicular to the flow and pressed against the seat around the outer edge, and fully open when unseated to a position such that the disk is parallel to the flow.

BV *See* bed volume.

Bw *See* body weight.

BWL *See* bottom water level.

bypass channel A channel, such as a diversion cut or spillway, formed around the side of a reservoir past the end of the dam to convey flood discharge from the stream above the reservoir into the stream below the dam.

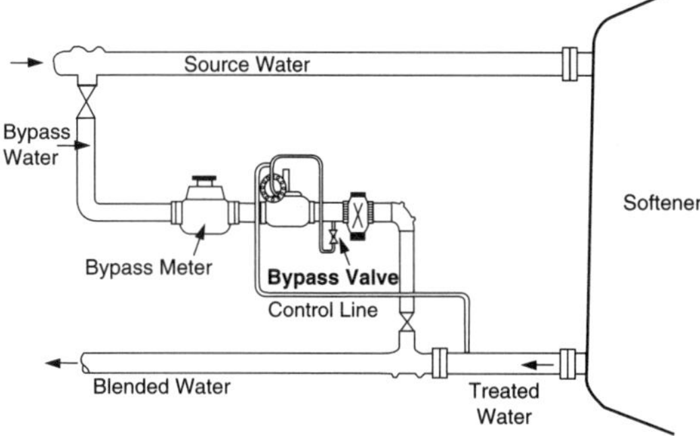

Bypass valve

bypass flow Water that is allowed to move past a storage facility or diversion structure.

bypass valve A small valve that is connected on either side of a large valve, used to equalize the pressure on both sides of the larger valve to allow easier operation of the larger valve, to control the filling of the water main on the downstream side of the large valve, or to blend source water and treated water.

bypassing The diversion of a flow of water around processes or structures through which the flow normally passes.

by-product A compound that is not the compound of interest but is also formed during a chemical reaction. Many by-products are formed during water treatment, but not all have yet been identified.

by-product, disinfection *See* disinfection by-product.

by-product, oxidation *See* oxidation by-product.

by-product, ozonation *See* ozonation by-product.

byte A group of adjacent binary digits that a computer stores and retrieves as a unit. A byte is commonly eight bits of data.

C

C *See* coulomb *in the Units of Measure list.*

C *See* C factor; Chezy resistance factor; residual disinfectant concentration.

c The speed of light, 9.81×10^{10} feet per second (2.99×10^{8} metres per second) in a vacuum.

C. cayatensis *See Cyclospora cayatensis.*

C factor (1) A coefficient representing the relative roughness of a pipe in the Hazen–Williams formula for water flow in a pipe. Pipes with low Hazen–Williams *C* factors have greater resistance to flow than pipes with large Hazen–Williams *C* factors. (2) The coefficient in the Chezy open-channel flow equation that represents the frictional effects of the channel on the flow. (3) A coefficient of proportionality in weir flow equations and orifice flow equations. (4) The coefficient of proportionality in the rational formula for peak flow prediction from the rainfall–runoff process. Usually the term is modified by the particular formula, as in "the Hazen–Williams *C* factor," "Chezy's *C*," or "the rational *C* coefficient." *See also* Chezy open-channel formula; Chezy resistance factor; Hazen–Williams formula; Manning formula.

C. fruendii *See Citrobacter fruendii.*

C. jejuni *See Campylobacter.*

C language A third-generation source code programming language often used to develop application software for computers.

C. perfringens *See Clostridium perfringens.*

$\boldsymbol{C \times T}$ The product of disinfectant concentration (in milligrams per litre) determined before or at the first customer and the corresponding disinfectant contact time (in minutes). It is also called the *CT* value. Units are milligram minutes per litre. *See also* $C \times T_{99.9}$.

$\boldsymbol{C \times T_{99.9}}$ The $C \times T$ value required for 99.9 percent (3-log) inactivation of *Giardia lamblia* cysts. These values appear for a variety of disinfectants and conditions in Tables 1.1–1.6, 2.1, and 3.1 of Section 141.74(b)(3) in the Code of Federal Regulations. Units are milligram minutes per litre.

$\mathbf{C^{13}}$ *See* carbon-13 (^{13}C) nuclear magnetic resonance.

CA *See* conventional aeration.

CA membrane *See* cellulose acetate membrane.

$\boldsymbol{C_a}$ *See* Cauchy number.

CAA *See* Clean Air Act.

CAB (cellulose acetate blend) membrane *See* cellulose acetate membrane.

cable-tool drilling A method of drilling wells by the use of cable tools. The hole is drilled by a heavy bit, which is alternately raised by a cable and allowed to drop, breaking and crushing the material that it strikes. Such material is removed from the hole by bailing or sand pumping.

cache memory Temporary storage for frequently accessed computer data and programs. When a program is executed by the central processing unit, the cache is first examined for the required data or instructions; auxiliary storage devices are accessed only if the information is not found in the cache.

CAD *See* computer-aided dispatch; computer-aided drafting.

CAD/CAM *See* computer-aided design/computer-aided manufacturing.

cadastral features Map elements created from legal descriptions that depict boundaries of property ownership and other rights to land.

cadastral map A map composed of cadastral features.

cadastre A public register or survey of a geodetic framework that defines or reestablishes boundaries of public land, private land, or both for purposes of ownership, taxation, or both.

CADD *See* computer-aided design and drafting.

cadmium (Cd) A malleable, ductile, bivalent metallic element (atomic number 48). In nature, cadmium is found in the ores of greenockite. It has various industrial uses, especially in protective platings and bearing metals. It is regulated by the US Environmental Protection Agency.

CAG *See* Carcinogen Assessment Group.

cage screen A cage-shaped screen built of bars, rods, or mesh to catch debris. A cage screen is arranged for lowering into water and for raising when the screen is to be cleaned.

caisson A large-diameter, watertight chamber in which work can be done below the surface of the earth or water.

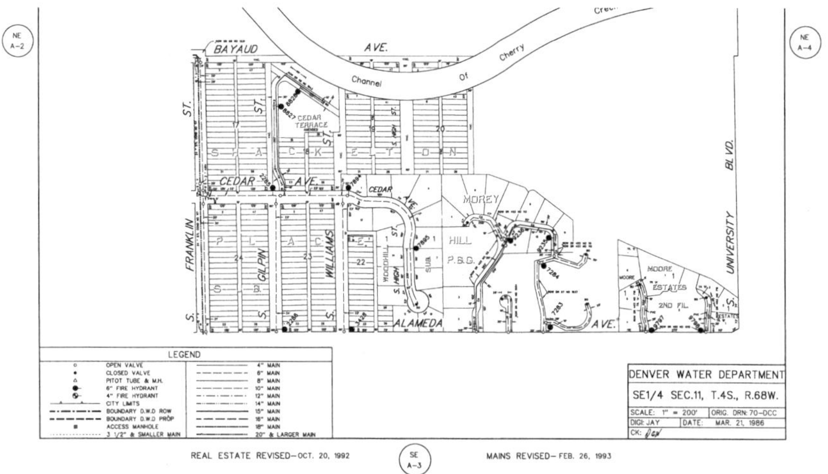

Cadastral map with specifications for distribution system

cake *See* sludge cake.

cake space The volumetric space available in a filter to support the formation of a sludge cake.

calcareous Pertaining to material containing calcium carbonate ($CaCO_3$) in sufficient concentrations to effervesce visibly with cold dilute (0.1 molar) hydrochloric acid (HCl).

calcareous spring A spring in which the water contains a considerable quantity of calcium carbonate ($CaCO_3$) in solution.

calcite ($CaCO_3$) The most common form of natural calcium carbonate ($CaCO_3$). Calcite is an essential ingredient in limestone, marble, and chalk. Calcium carbonate (limestone, calcite, whiting, chalk) is sometimes termed unburned lime. It is widely used as an agricultural lime and is also used in neutralization, stabilization, or prevention of corrosion in water treatment. *See also* calcium carbonate; lime; limestone.

calcite contactor A layer or bed of crushed limestone through which water passes. The pH, alkalinity, and calcium content of the water are normally increased by the dissolving calcium carbonate ($CaCO_3$).

calcite saturation index (CSI) The saturation index for calcium carbonate ($CaCO_3$). For any solid, a general saturation index can be defined as

$$SI = \log_{10} (Q/K_{sp})$$

Where:

SI = the saturation index

Q = the product of the ion activities

K_{sp} = the solubility constant for the given temperature

Interpreting this SI value is analogous to interpreting the Langelier saturation index: $SI = 0$ indicates saturation equilibrium; $SI < 0$ indicates undersaturation; and $SI > 0$ indicates supersaturation. In the preceding equation, ion concentrations can be substituted for ion activities in expressing Q, and the solubility constant can be corrected for the ionic strength (yielding K'_{sp}). The activities (and concentrations) represent those of the free ions, not the total analyzed concentrations. Thus, for calcium carbonate, the preceding equation translates into

$$CSI = \log_{10}([Ca^{2+}][CO_3^{2-}]/K'_{sp})$$

Where:

$[Ca^{2+}]$ = concentration of calcium ions, in moles per litre

$[CO_3^{2-}]$ = concentration of carbonate ions, in moles per litre

calcium (Ca) One of the principal elements making up the earth's crust. The presence of calcium in water is a factor contributing to the formation of scale and insoluble soap curds that are a means of clearly identifying hard water. *See also* hard water.

calcium carbonate ($CaCO_3$) A colorless or white crystalline compound that occurs naturally as chalk, limestone, marble, and other forms. Pure calcium carbonate exists in two distinct crystalline forms: the trigonal solid, calcite; and the orthorhombic solid, aragonite. $CaCO_3$ is a sparingly soluble salt, the solubility of which decreases with increasing temperature. It has the potential to cause scaling if it is concentrated to supersaturation.

calcium carbonate equivalent An expression of the concentration of specified constituents in water in terms of the equivalent value to calcium carbonate ($CaCO_3$). For example, the hardness in water that is caused by calcium, magnesium, and other ions is usually described in terms of the calcium carbonate equivalent. For example, the concentration of calcium ions (i.e., $[Ca^{2+}]$) can be multiplied by 100/40 (the ratio of $CaCO_3$'s molecular weight to Ca^{2+}'s atomic weight) to give $[Ca^{2+}]$ as equivalent $CaCO_3$.

calcium carbonate precipitation potential (CCPP) The amount of solid calcium carbonate ($CaCO_3$) that will precipitate or dissolve as water equilibrates. *See also* calcium carbonate precipitation potential index.

calcium carbonate precipitation potential (CCPP) index An index that predicts the milligrams per litre of calcium carbonate ($CaCO_3$) that should dissolve or precipitate with a particular water. The equation used to determine this index is:

$$CCPP = 50{,}000([Alk]_i - [Alk]_{eq})$$

Where:

CCPP = calcium carbonate precipitation potential, in milligrams per litre as $CaCO_3$

50,000 = a unit conversion factor

$[Alk]_i$ = measured total alkalinity of a given water, in milligrams per litre as $CaCO_3$

$[Alk]_{eq}$ = total alkalinity that the water would have in equilibrium, measured as milligrams per litre as $CaCO_3$

calcium carbonate saturation index *See* Langelier saturation index.

calcium chloride ($CaCl_2$) A soluble salt, some uses of which are similar to those of sodium chloride (NaCl). Because its most striking property is its ability to draw moisture from the air and thereby dissolve itself, it is often used as an air dryer and a deicing salt.

calcium hardness The portion of total hardness caused by calcium compounds such as calcium carbonate ($CaCO_3$) and calcium sulfate ($CaSO_4$).

calcium hydrate *See* hydrated lime.

calcium hydroxide ($CaOH_2$) *See* hydrated lime.

calcium hypochlorite ($Ca(OCl)_2$) A chemical compound used as a bleach or disinfecting agent and a source of chlorine in water treatment. Commercial grades contain 70 percent available chlorine (99.2 percent available chlorine for the pure chemical). Calcium hypochlorite is specifically useful because it is stable as a dry powder and can be formed into pellets. *See also* disinfection.

calcium oxide (CaO) *See* lime.

calcium sulfate ($CaSO_4$) A sparingly soluble salt, the solubility of which decreases with increasing temperature. It is called gypsum in its hydrated form. It is a potential source of scaling in desalting systems if it is concentrated to supersaturation.

Caldwell–Lawrence diagram A diagram illustrating a series of relationships associated with the chemical equilibrium of calcium carbonate ($CaCO_3$) as a function of pH, alkalinity, and calcium hardness for a solution of given temperature and ionic strength. *See also* alkalinity; calcium carbonate; calcium hardness.

caliber (1) The diameter of a round body, especially the internal diameter of a hollow cylinder. (2) The diameter of a bullet or other projectile or the diameter of a bore of a gun. In US customary units, caliber is usually expressed in hundredths or thousandths of an inch and typically written as a decimal fraction (e.g., 0.32). In Système International units, calibers are expressed in units of millimetres.

calibrate To adjust a measuring instrument so that it gives the correct result with a known concentration or sample.

calibration A procedure that adjusts or checks the accuracy of an instrument by comparison with a standard or reference.

caliche (1) A hard deposit, consisting mostly of calcium carbonate ($CaCO_3$) or of gravel and sand cemented by calcium carbonate, found in the subsoil in arid regions. The deposit may range from several inches to several feet in thickness and is presumed to have been created by the evaporation of mineral-laden capillary water, leaving a residue that served as a cementing material. (2) Crude sodium nitrate ($NaNO_3$) in Chilean deposits.

Calicivirus A single-stranded ribonucleic acid virus. It is transmitted through the fecal–oral route and causes gastroenteritis. *See also* gastroenteritis.

calking (sometimes spelled "caulking") (1) The process of driving, pouring, or forcing lead, oakum, plastic, or other material into a joint to make it leakproof. (2) The materials used in the calking process.

calking tool A blunt offset chisel-like tool that is used to drive or press the lead ring poured in a bell-and-spigot-joint water pipe. The tool is placed against the lead and hit with a hammer. The tool has been adapted to be used with a pneumatic calking gun or hammer and is made in various widths to correspond to the width of the lead ring in the joint. *See also* bell-and-spigot joint.

call (1) The process of redeeming a bond or preferred stock before its normal maturity. (2) An option to buy (or call) a share of stock at a specified price within a specified period. (3) A demand that upstream water right holders with junior (more recent) priority dates than the calling right holder cease diverting water. The exercise of a senior (older) water right holder in calling for her or his water rights requires the junior water right holders to allow water to flow to the senior right holder.

call premium The amount of excess or par value that a company must pay when it calls a security.

callable bond A type of bond that permits the payment of the obligation before the stated maturity date by giving notice of redemption in a manner specified in the bond contract.

calorie (gram calorie) *See the Units of Measure list.*

calyx *See* lime.

CAM *See* computer-assisted manufacturing; computer-assisted mapping.

CAMA *See* computer-assisted mass appraisal.

Campylobacter A genus of bacteria characterized as gram-negative, microaerophilic curved rods. Such bacteria are motile by means of a single polar flagellum, although some strains may have a flagellum at each pole. *Campylobacter jejuni* (*C. jejuni*) is a common bacterial cause of gastroenteritis in humans and has been implicated in waterborne outbreaks associated with municipal water systems.

CAN/CSA *See* Canadian Standards Association.

Canadian Standards Association (CAN/CSA) A standards organization that, among other things, provides water usage standards for water closets.

Canadian Water and Wastewater Association (CW&WA) An association representing Canadian water and wastewater utilities. The CW&WA is not affiliated with the American Water Works Association.

canal An artificial open channel or waterway constructed for one or more of the following purposes: transporting water, connecting two or more bodies of water, and serving as a waterway for watercraft.

canal lock A crest gate, the face of which is a section of a cylinder, that rotates about a horizontal axis downstream from the gate. The water pressure against the gate is concentrated in the axis, reducing friction in raising and lowering the gate. Such a gate is now called a tainter gate.

canal rapids An inclined conduit or structure used for conveying water at high velocity to lower levels. *See also* chute.

canal section (1) The shape of the cross section of a canal at right angles to its axis. (2) The data, either in graphic or tabular form, that describe the shape, dimensions, side and bank slopes, normal depth of water, slope of bottom, thickness and kind of lining material (if any), and other such characteristics of a canal at a given point.

canal seepage loss The loss of water from a canal by capillary action and percolation after the canal has reached a stable condition.

cancer A condition of having tumors that will prove fatal if allowed to run its natural course. Cancer cells have the properties of invasion and metastasis, but benign tumors frequently develop these properties, too, so the term is frequently more broadly used to refer to any tumor. This popular definition of cancer refers to almost any circumstance in which cells fail to respond to normal growth controls. However, some types of benign tumors practically never become malignant.

candela (cd) *See the Units of Measure list.*

candle *See the Units of Measure list.*

candlepower *See the Units of Measure list.*

canvass The process of blanketing an entire geographical area with news concerning a program. In the context of a water utility, canvassing involves the distribution of retrofit materials and the necessary door-to-door follow-up to ensure that items are installed, to assist customers with installation, or to retrieve unwanted materials.

CAP grade water *See* College of American Pathologists grade water.

cap (1) A fitting for the spigot or screw end of a pipe intended to protect the end or make it watertight. (2) In plumbing, a fitting for the spigot or screw end of a metal pipe.

capacitance A part of the total impedance of an electrical circuit tending to resist the flow of current. Capacitance can be added to cancel the effect of inductance. It is expressed in units of farads.

capacitive deionization (CDI) An electrically regenerated electrosorption process capable of desalting saline water. Feedwater passes between multiple parallel plates made of highly porous carbon material (carbon aerogel). The plates are placed under a low voltage (typically 1 to 5 volts), which causes ions in the water to move toward and collect at the plates (electrodes) of opposite charge. Periodic regeneration is accomplished by discontinuing the electric field or by reversing the polarity of the electrodes. Regeneration releases the ions from the electrodes to a concentrate stream for discharge from the system.

capacitor Two conductive plates separated by a nonconductor, used to create capacitance in a circuit. A capacitor is also called a condenser.

capacity (1) The flow rate that a pump is capable of producing. (2) A water utility's ability to have resources available to meet the water service needs of its customers. In this context, capacity is the combination of plant- and service-related activities necessary to meet the quantity, quality, peak loads, and other service needs of the various customers or classes of customers served by the utility.

capacity charge A one-time charge assessed against new customers (applicants for service) or developers as a means of recovering all or part of the costs of existing or additional system capacity constructed for their use. It is a means, devised in the 1970s and 1980s, to assess the cost of growth to those who cause the costs to be incurred. The total capacity charge often has a buy-in component for existing unused capacity and a new facilities component for specific facilities to be constructed. The fees collected are generally maintained in a separate fund or restricted asset account to be used only for expenditures directly related to the purpose for which the funds are assessed. Capacity charges are known by many different names: capital recovery charges, connection charges, system development charges, facility expansion charges, and impact fees.

capacity curve In the case of a reservoir, the relationship between the elevation of the water surface in the reservoir and the volume of water below that elevation. It is also called the storage-capacity curve.

capillary action The action of a material (e.g., filter paper) or an object, containing minute openings or passages and partly immersed in water, in drawing water up into the openings or passages to a level above the free water surface.

capillary capacity *See* field capacity.

capillary chromatography A separation technique in which a thin film of material called the stationary phase is coated inside a long, narrow tube. This technique is characterized by an ability to separate complex multicomponent solutions. Capillary chromatography columns used in gas chromatography range from 50 feet (15 metres) to over 325 feet (100 metres) in length.

capillary electrophoresis *See* capillary zone electrophoresis.

capillary flow *See* film flow.

capillary force The force, caused by molecular attraction between soil particles and water, that acts on water in the unsaturated zone of soil.

capillary fringe The zone immediately above the water table where water is drawn upward by capillary forces.

capillary head The difference in elevation between the ultimate position of the meniscus and the position of the meniscus at a given time. The head is zero when the water has risen to its ultimate position.

capillary interstice An interstice that is (1) small enough for water to be held in it at appreciable height in a capillary fringe above a water table or hydrostatic pressure level and (2) large enough to preclude molecular attraction of its own walls from spanning the interstice. Movement of water through saturated capillary interstices conforms to the laws of streamline flow acting under (1) hydrostatic pressure (percolation) or (2) the resultant of gravity and unbalanced film pressure.

capillary ion electrophoresis A technique that is applicable to the analysis of ions in solution. The technique has been used for milligrams-per-litre and lower concentrations of ions such as nitrate, bromide, and chloride in water. Analytic methods using this technique are characterized by little sample preparation and rapid analysis. Organic anions have also been analyzed with this technique. *See also* capillary zone electrophoresis.

capillary lift The height to which water or other liquid rises in a capillary tube.

capillary migration The movement of water through a rock or soil, as produced by molecular attraction of the rock or soil material for the water.

capillary moisture Available soil moisture easily abstracted by roots of plants.

capillary movement *See* film flow.

capillary opening An opening small enough in cross-sectional area to create a condition in which capillary flow can occur.

capillary percolation Percolation through capillary interstices. *See also* percolation.

capillary suction test A test in which a water treatment sludge is placed on a special blotting paper. The time for the wetted area to move a set distance, often 1 centimetre, is an indication of the dewatering characteristics of the sludge. *See also* capillary suction time.

capillary suction time (CST) The time required, in seconds, for the wetted area in a special blotting paper to travel a set distance (normally 1 centimetre) as a result of capillary action after a water treatment sludge is placed in the sample holder. The time measurement is accomplished by an automatic timer that starts recording when the wetted surface reaches the first probe and stops recording when the wetted surface reaches the second probe. Capillary suction time is an indication of the dewatering characteristics of water treatment sludge. Larger capillary suction time values are an indication of poorer dewaterability. Capillary suction time values decrease with chemical conditioning of the sludge.

capillary tube A tube with an interior area of such small size that water within it is raised or moved by capillary force.

capillary water Water retained in the soil by capillary forces.

capillary zone The zone in which soil water is held by capillary forces. *See also* capillary migration.

capillary zone electrophoresis A technique in which charged particles are separated because of differential migration in an electric field. Although electrophoresis has traditionally been used in the separation of large biomolecules, capillary zone electrophoresis has emerged in recent years as a powerful tool in a wider range of applications, e.g., the separation of proteins, peptides, amino acids, and carboxylic acids.

capita Person.

capital The net worth of a business; the amount by which the liabilities are exceeded by the assets.

capital asset An asset of a relatively permanent nature. *See also* fixed asset.

capital cost A cost (usually associated with long-term debt) of financing construction and equipment. Capital cost is usually a fixed, one-time expense that is independent of the amount of water produced. It is also called capital expenditure or capital outlay.

capital expenditure *See* capital cost.

capital grant A grant that is restricted by the grantor for the acquisition and/or construction of fixed assets.

capital improvement plan (CIP) A plan, updated or compiled annually by most utilities, that identifies facility requirements over an extended period, often 20 years or more. The capital improvement plan is often a part of or stems from a water system master plan that combines water demand projections with supply alternatives and facility requirements. It is different from an asset replacement program concentrating on smaller assets that are generally funded from current revenues or from funds established for replacement. The capital improvement plan, by contrast, is focused on larger expenditures that are often debt funded but are sometimes funded from capacity charges or current revenues.

capital program A set of planned major improvements to be accomplished by a business or government within a set period of time.

capital recovery charge *See* capacity charge.

capitalized interest cost The interest cost incurred when a utility, during the construction of major debt funded facilities, opts to capitalize the interest on debt for 1 or 2 years. By so doing, the utility can use bond proceeds rather than current revenues to pay the interest costs during the period before the facilities begin to produce product and possibly generate revenue.

capsid A regular, shell-like structure, composed of aggregated protein subunits, that encloses the nucleic acid component of viruses.

capture zone The up-gradient and down-gradient areas of an aquifer that drain into a particular well. The delineation of capture zones is used extensively in wellhead protection planning and in contaminant recovery.

carbofuran The common name for 2,3-dihydro-2,2-dimethyl-7-benzofuranylmethylcarbamate, a pesticide designed to combat corn rootworm and rice water weevil. It is regulated by the US Environmental Protection Agency. *See also* pesticide.

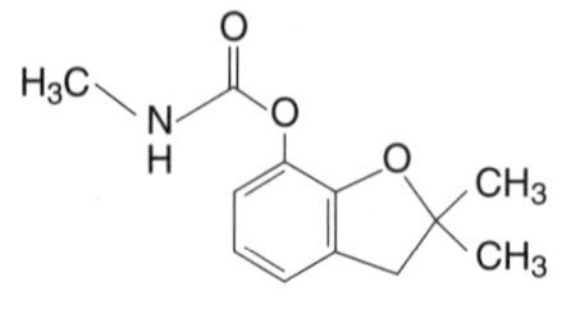

Carbofuran

carbon (C) (1) A chemical element essential for growth. (2) A solid material that, when activated, may be used to adsorb contaminants.

carbon chloroform extract (CCE) The residue from a carbon chloroform extraction test. *See also* carbon chloroform extraction.

carbon chloroform extraction An outdated method for assessing organic pollutants in water. It consists of adsorbing organic material onto activated carbon, then extracting the activated carbon with chloroform ($CHCl_3$). The chloroform is then evaporated and the residue weighed. The residue is called the carbon chloroform extract (CCE).

carbon dioxide (CO_2) A colorless, odorless, incombustible gas that is a normal component of natural waters. It may enter surface water and groundwater by absorption from the atmosphere or biological oxidation of organic matter.

carbon reactivation *See* activated carbon regeneration.

carbon regeneration *See* activated carbon regeneration.

carbon tetrachloride (CCl_4) A common name for tetrachloromethane, a volatile organic compound with various industrial uses (e.g., as a solvent). It is not a disinfection by-product. *See also* volatile organic compound.

carbon tetrachloride activity The maximum percentage increase in weight of an activated carbon bed after air saturated with carbon tetrachloride (CCl_4) is passed through it at a given temperature.

carbon usage rate (CUR) *See* activated carbon usage rate.

carbon-13 (^{13}C) nuclear magnetic resonance A spectrometric technique that can provide structural information about organic compounds. The adsorption of electromagnetic radiation depends on the nuclei present in a compound. A nuclear magnetic resonance spectrum plots the frequencies of absorption versus the peak intensity. The technique is similar to proton nuclear magnetic resonance (i.e., ^{1}H nuclear magnetic resonance), yet it became popular only after Fourier transform instruments became widely available. When used in conjunction with other electromagnetic techniques, carbon-13 nuclear magnetic resonance can be a powerful tool in the elucidation of molecular structure. *See also* proton nuclear magnetic resonance.

carbonaceous Containing carbon and derived from organic substances such as coal, coconut shells, and wood.

carbonate (CO_3^{2-}) A divalent negatively charged anion.

carbonate alkalinity Alkalinity caused by carbonate ions (CO_3^{2-}) and expressed in terms of milligrams of equivalent calcium carbonate ($CaCO_3$) per litre. *See also* calcium carbonate equivalent.

carbonate hardness Hardness caused by the presence of carbonates and bicarbonates of calcium and magnesium in water. Such hardness may be removed to the limit of solubility by boiling the water. When the hardness is numerically greater than the sum of the carbonate alkalinity and the bicarbonate alkalinity, that amount of hardness that is equivalent to the total alkalinity is called carbonate hardness and is expressed in milligrams of equivalent calcium carbonate ($CaCO_3$) per litre. Carbonate hardness was previously called temporary hardness. *See also* calcium carbonate equivalent; hard water; hardness.

carbonate system A system of chemical relationships in natural waters that performs important functions related to acid–base chemistry, buffering capacity, metal complexation, solids formation, and biological metabolism. Species comprising the carbonate system include carbon dioxide (CO_2), carbonic acid (H_2CO_3), bicarbonate (HCO_3^-), carbonate (CO_3^{2-}), hydroxide (OH^-), hydrogen ion (H^+), and calcium carbonate ($CaCO_3$). *See also* alkalinity; carbonate alkalinity; carbonate hardness.

carbonated spring A spring of water containing carbon dioxide (CO_2) gas. Such springs are very common, especially in volcanic regions.

carbonation The diffusion of carbon dioxide (CO_2) gas through a liquid to render the liquid stable with respect to precipitation or dissolution of alkaline constituents. *See also* recarbonation.

carbonator A device for the carbonation or recarbonation of water.

carbonyl (C=O) An organic functional group composed of a carbon and oxygen atom double bonded together. It is present in aldehydes, ketones, organic acids, and sugars. *See also* aldehyde; ketone; organic acid.

carboxyhemoglobin The complex between carbon monoxide (CO) and hemoglobin. Carbon monoxide competes effectively with oxygen at its binding site on hemoglobin. Consequently, carbon monoxide poisons individuals by sharply reducing the oxygen-carrying capacity of the blood.

carboxylic acid (R-COOH) Any of a broad array of organic acids composed chiefly of hydrocarbon groups, usually in a straight chain, terminating in a carboxyl group (COOH). They contribute cation exchange ability to some resins. Carboxylic acids are sometimes called weak acid cation exchangers. *See also* organic acid.

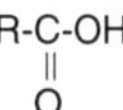

Carboxylic acid, where R is a carbon-containing group

carcinogen A chemical, physical, or microbial agent that is capable of inducing cancer. The mechanisms by which these agents act can vary widely. These differences seriously affect the extrapolation of risks to low exposure levels. Some carcinogens can alter a deoxyribonucleic acid (DNA) sequence by causing errors in its replication or by incorporating new material into the genome that is inappropriately replicated in the infected cell. These types of events are essentially irreversible and are assumed to accumulate during a lifetime. Therefore, a linear relationship between dose and response is assumed for estimating risks at low doses. Such agents are called genotoxic carcinogens. Agents can also produce cancer by altering the control of gene function without being directly involved in producing a permanent error in the DNA sequence. The effects of these agents would be considered largely, if not entirely, reversible once the exposure is removed. These are referred to as epigenetic or nongenotoxic carcinogens because their effects are reversible. Usually these agents affect the relative rates at which normal and transformed cells divide within a particular tissue. Chemicals that induce death of normal cells can also produce cancer by increasing the rate at which cells within the organ divide. A higher degree of cell turnover increases the probability of mistakes being made in DNA replication. Moreover, the killing of normal cells provides a selective growth advantage to abnormal cells that may not be sensitive to the cytotoxic effects of the chemical. Generally, chemicals that produce cancer in this way should be treated as toxins rather than as carcinogens. *See also* tumor initiator; tumor promoter.

Carcinogen Assessment Group (CAG) A group administratively housed in the Office of Health and Environmental Assessment within the Office of Research and Development of the US Environmental Protection Agency that has developed and promulgated the accepted methods of identifying hazards and assessing risks associated with putative carcinogens found in various environmental media. The principles laid out in their documents form the basis of the US Environmental Protection Agency's policy for developing standards, guidelines, regulations, and maximum contaminant levels for these chemicals.

carcinogenesis bioassay A test of the cancer-causing properties of chemical, physical, or microbial agents. The term is applied loosely to a wide variety of in vivo and in vitro assay systems. Technically, however, it should be applied only to assays that actually measure cancer as an end point. The studies conducted by the National Toxicology Program (NTP) represent the cornerstone of cancer bioassays conducted today. The experimental design for these studies is standardized and involves administration of the test substance for a major portion of the test animals' lifetimes. The standard animals used in these tests are currently the F344 rat and the B6C3F1 mouse, although exceptions are based on the needs of particular experiments. The NTP bioassay is the most frequent source of data used for establishing regulatory limits on chemicals. The main reason for this is the general high quality of these studies relative to more limited tests that are reported in the general scientific literature. However, the application of data to estimating cancer risk across species and to low doses is controversial. The lay public does not always recognize that such extrapolations are not really justifiable on strictly scientific grounds because it is very rare that these extrapolations are confirmed, particularly the low-dose extrapolations. This extrapolation usually means taking data from an experiment conducted in a limited number of animals (typically 50 animals from each sex of each species) to estimate doses that would produce cancer at a rate as low as one extra cancer death in a population of one million people per lifetime. However, the sensitivity of the actual NTP bioassay is usually about one extra cancer death per population of 10 people per lifetime. The high doses and small numbers of animals are practical limitations of the bioassay both from a cost standpoint and because of the almost impossible logistical difficulties that arise from attempting studies with the appropriate sensitivity. Because the technical community generally regards the cancer bioassay in animals as a qualitative test, the standard practice has been to use doses that approximate the maximally tolerated dose for the animals. At such doses, almost by definition, a variety of effects can occur that are unique to high doses, almost certainly affecting the outcome of the bioassay. Therefore, as already noted, the extension of these data to low doses is highly questionable. The extrapolation does provide a standard means of ranking carcinogens, but this must be recognized as largely a governmental policy approach based on some hypothetical and general conception of chemical carcinogenesis. Once implemented, the policy does have the logical benefit of establishing a way of essentially dismissing those chemicals that occur at concentrations much less than some hypothetical risk considered to be of regulatory importance. Implicit in this philosophy is the concept that sufficient economic interest exists to drive the development of data that are more appropriate for assessing the risks of chemicals in instances where human exposures approach some predetermined level of concern. Unfortunately, such data are not commonly developed, partly because of the expense of these more elaborate studies, particularly if the agent involved is generally ubiquitous in the community as opposed to being the responsibility of a small commercial sector. A second major problem is that such research almost invariably requires a chemical-by-chemical approach rather than a standardized methodology. As of 1996, a consensus seems to be emerging that the linear models are probably not appropriate for chemicals that are not mutagenic. The type of experimental data that would be needed to demonstrate that an alternative mechanism is actually operative has yet to be defined, however. *See also* B6C3F1 mouse; F344 rat; linearized multistage model; multihit model; multistage model; one-hit model; probit model; Weibull model.

carcinogenic Able to produce carcinoma or a cancer arising from epithelial cells. The term is also commonly used to describe any process or agent that produces cancer, regardless of the cell of origin.

carcinogenicity The power or ability of an agent to produce cancer. The agent can be a virus, a chemical, or a physical agent. The hepatitis B virus is an example of a viral carcinogen, benzo(a)pyrene is a chemical carcinogen, and radiation is a physical agent.

carcinoma A malignant growth arising from epithelial cells.

cardiomyopathy A general diagnostic term applied to primary disease of the muscles of the heart. Frequently, the term is applied to heart muscle disease for which the cause is obscure or unknown.

cardiopulmonary resuscitation (CPR) A first aid procedure used to maintain breathing and blood circulation for an injured person.

cardiovascular disease A group of diseases that affect either the heart or vasculature. The diseases range from the highly specific (e.g., myocardial infarct) to descriptions of symptoms (e.g., angina, hypertension). The diseases can arise from physical damage to critical organs such as the heart, lungs, or kidneys, or they may arise from changed regulation of the function of the heart or vasculature. Cardiovascular disease includes blocked (infarcted) blood flow to a variety of organs, most notably the brain (e.g., cerebrovascular accidents or strokes, thrombophlebitis). As can be imagined, these diseases have a wide variety of causes. The one cardiovascular disease that has been most frequently associated with drinking water quality is sudden death syndrome, which primarily reflects strokes.

cardiovascular system The heart and blood vessels. *See also* vascular system.

carrier (1) A person or animal that harbors a specific infectious agent and serves as a potential source of infection. Clinical disease is absent in the carrier, and the carrier state may be of a long or short duration. (2) An organization supplying insurance coverage, such as an insurance company, self-insurance group, or state fund.

carrying capacity The maximum rate of flow that a conduit, channel, or other hydraulic structure is capable of passing.

carryover (1) The entrainment of liquid or solid particles in the vapor evolved by a boiling liquid. (2) The entrained particles in a boiling liquid's vapor.

Cartesian coordinate system A coordinate system in which the location of a point on a plane is expressed by (1) two coordinates (x, y) that measure the point's distance from two intersecting, often perpendicular, straight-line axes along a line parallel to the other axis or (2) three coordinates (x, y, z) that locate a point in space by its distance from three fixed planes that intersect one another at right angles.

cartography The art and science of making maps.

cartridge A removable cylindrical water treatment separations device installed, alone or with others, in a pressure vessel (housing); such as (1) A micrometre-pore-size filter element installed in a cartridge filter pressure vessel or (2) a spiral-wound

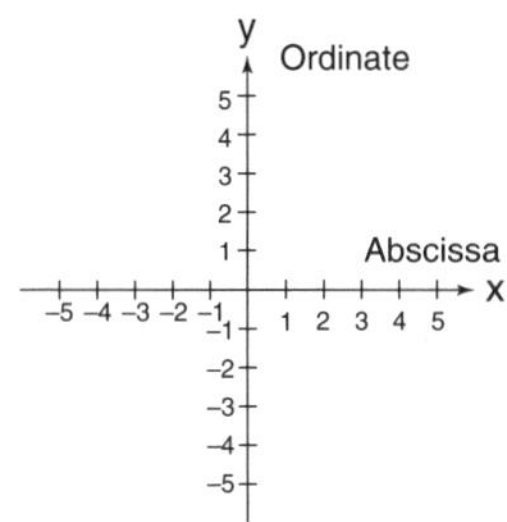

Cartesian coordinate system

membrane element housed in a pressure vessel of a membrane treatment system.

cartridge filter A filtration device that has a pressure vessel containing one or more cartridges of a specified nominal (or sometimes absolute) pore size rating used to remove particles from a process stream. *See also* cartridge.

CAS registration number *See* Chemical Abstracts Service number.

CAS# *See* Chemical Abstracts Service number.

cascade (1) A stretch of a stream, intermediate between a rapids and a waterfall, where the drop in elevation of the stream bed is considerable but not sufficient to cause the water to fall vertically. (2) A sudden drop installed in a waterway to produce agitation and aeration of the liquid flowing over it.

cascade aerator An aerating device built in the form of either steps or an inclined plane, on which are placed staggered projections arranged to break up the water and bring it into contact with air.

CASE *See* computer-aided software engineering.

case (1) In epidemiology, a person identified as having a particular disease or health condition under investigation. A variety of criteria may be used to identify a case, and the case definition for an epidemiologic investigation may not necessarily be the same as a clinical definition. (2) In a database, a complete set of information in the database, i.e., a row of data.

case fatality rate The percentage of individuals diagnosed with a specific disease who die as the result of that disease. This rate is different from mortality rate, which considers the entire population with and without disease.

case-control epidemiologic study A study in which individuals enter on the basis of their disease status without knowledge of their exposure status. A single disease or health outcome (e.g., giardiasis, lung cancer, bladder cancer, blood lipid levels) is selected for study, and two groups are assembled: those with the disease or condition and those without. Information about any number of individual exposures or other characteristics (e.g., smoking, use of chlorinated or unchlorinated water, radon exposures) can be obtained and studied to determine possible associations (an odds ratio) with the disease. Such a study is also called a case-comparison epidemiology study or a case-referent epidemiology study. *See also* odds ratio.

case-referent epidemiology study *See* case-control epidemiologic study.

cash basis The basis of accounting under which revenues are recorded when cash is received and expenditures are recorded when cash is disbursed.

cash requirements method A method of accounting used for ratemaking analysis by an estimated 90 percent of all not-for-profit water utilities. The major characteristic of this method is that the net revenue requirement that must be funded by water rates is based on actual (or estimated) cash expenditures and actual cash receipts only, which by definition exclude depreciation, accrued expenses, and accrued revenues. This method, which can be contrasted with the utility method, also excludes return on investment as a cost to be recovered by rates. *See also* net revenue requirement; rate of return; return on investment; utility method.

cash reserve Cash set aside in a bank account by businesses to pay operating expenses in winter months when expenses often exceed revenues, to pay claims not reimbursed from insurance, or to cover any other emergency or unplanned need for cash. Reported cash reserves should exclude cash that is restricted for specific purposes, such as cash that has accumulated in a sinking fund for asset replacement or that has been collected from capacity charges to pay for new facilities. Cash reserves are generally targeted to cover 3 to 6 months of operating expenses.

Casil index (CI) A measure of the corrosiveness of water based on a cation–anion balance. As the number of cations decreases, the index decreases, indicating more corrosive conditions. As the number of anions increases, the index decreases, again indicating more corrosive conditions. A decrease in pH through the addition of acid will increase the number of anions, which will lower the index, thus indicating increasing corrosiveness. The formula for determining the Casil index is

$$CI = \frac{\text{calcium ions} + \text{magnesium ions} + \text{silica ions} - \text{anions}}{2}$$

All ions are expressed in terms of milliequivalents per litre.

casing A solid piece of pipe used to keep a well open in unstable materials.

casing head In well boring, a heavy weight of iron screwed into the top of a string of casing to take the blows produced by driving the pipe.

casing shoe A rigid annular fitting placed at the lower end of a metal well casing, commonly with a cutting edge on the bottom.

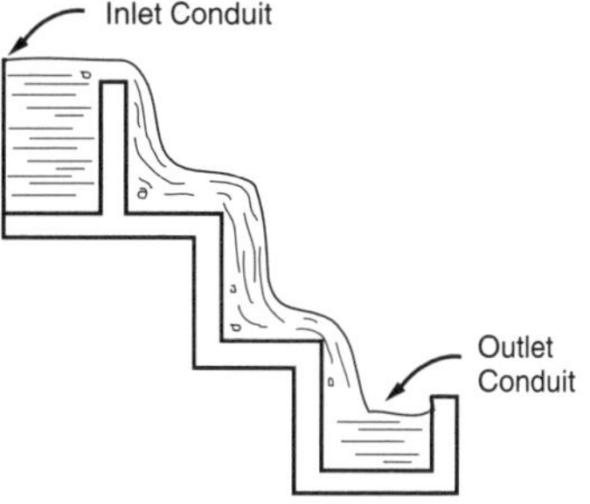

Stairway-type cascade aerator

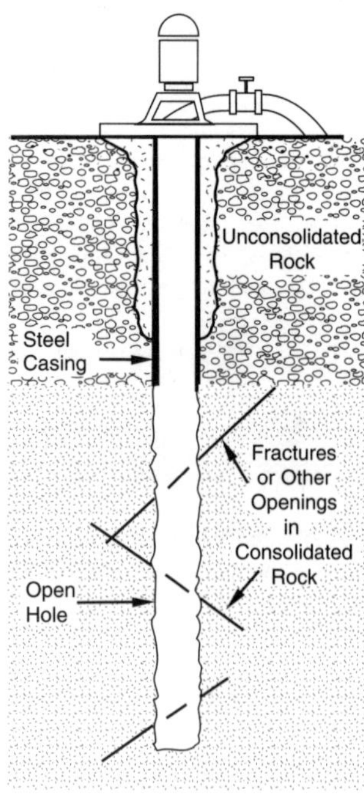

Casing

cast iron (CI) A heavy brittle alloy of iron, carbon, and silicon that is normally cast in specific shapes and has the ability to withstand shocks. It is used for such construction materials as piping, valves, and fittings.

Cast Iron Pipe Research Association (CIPRA) *See* Ductile Iron Pipe Research Association.

cast-iron pipe A pipe made from pig iron cast in a revolving, water-cooled mold, or in a stationary, cylindrical sand mold with a round, central core.

casting The shaping of an object such as a tee or valve body by pouring melted cast iron or ductile iron into a mold, usually a sand mold, and letting it harden.

catalase An enzyme that catalyzes the degradation of hydrogen peroxide (H_2O_2).

catalysis The speeding up (in most cases) of a chemical reaction by the addition of a specific substance, the catalyst. Although the catalyst causes the reaction to speed up, the catalyst is not changed chemically in any way.

catalyst Any substance of which a fractional percentage notably affects the rate of a chemical reaction, without the substance itself being consumed or undergoing a chemical change. Most catalysts accelerate reactions, but a few retard them (the latter substances are called negative catalysts or inhibitors).

catalyst filter A class of media bed filters that contain manganese-treated greensand, zeolites, or pumicites. Such filters are sometimes called oxidizing filters.

catalyst media Those filter media that can cause certain reactions to occur in water treatment, e.g., activated carbon, calcite, manganese greensand, magnesium oxides, and dissimilar metal alloys.

catalytic filtration A method of filtration in which a contaminant is removed through the catalytic effect of a coating on the filtering media. For example, iron and manganese can be removed through greensand filters, in which the natural zeolite medium catalyzes the oxidation and removal of iron and manganese.

catalytic ozone destructor A type of ozone (O_3) destruction unit that uses a catalyst to enhance the performance of the off-gas treatment system.

catalytic photochemical oxidation A method of accelerating and increasing the oxidation of compounds by using a catalyst together with a strong oxidant. For example, ultraviolet absorbance can be used together with ozone to rapidly degrade specific compounds.

catalytically destroy To break down a substance through catalysis. *See also* catalysis.

catalyze To act as a catalyst; to speed up a chemical reaction.

catalyzed oxidation The enhancement of chemical or biological oxidation by the addition of agents called catalysts, which promote higher rates of reaction.

cataract An opacity of the crystalline lens of the eye.

cataract action The digging back action of rapid current on a river bed.

catch basin A chamber or well, usually built at the curbline of a street, that admits surface water for discharge into a stormwater drain. *See also* curb inlet.

catch-can test A measurement of precipitation from a sprinkler system, taken by placing graduated containers at evenly spaced intervals throughout an irrigated area.

catchment The area tributary to a lake or stream. *See also* catchment area; drainage area; river basin; watershed.

catchment area (1) The intake area of an aquifer, as well as all areas that contribute surface water to the intake area. (2) In tropical island zones, a hard-surfaced area on which rain is collected and then conducted to a reservoir. (3) The area tributary to a lake or stream. A catchment area is also called a catchment basin. *See also* drainage area; intake area; watershed.

catchment basin *See* catchment area.

caterpillar gate A roller-bearing crest gate equipped on each side with a continuous chain of caterpillar rollers that run in a vertical groove and travel around the gate as it is raised or lowered.

cathode (1) The pole of an electrolytic cell that attracts positively charged particles or ions (cations). It is the electrode where reduction (and practically no corrosion) occurs. It is the opposite of the anode. (2) The negatively charged electrode of an electrodialysis cell.

cathode ray tube (CRT) A phosphorescent screen commonly used for displaying computer data, usually 20 to 24 lines of data with 80 to 132 characters per line.

cathodic corrosion An unusual condition (especially in the case of aluminum, zinc, and lead) in which corrosion is accelerated at the cathode because the cathodic reaction—in which cations are attracted to the negatively charged electrode—creates an alkaline condition that is corrosive to certain metals.

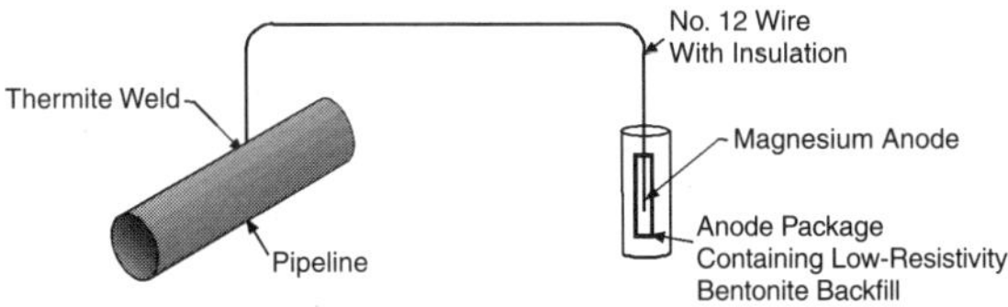

External cathodic protection using sacrificial anodes

cathodic protection Reduction or elimination of corrosion by making the metal a cathode by means of an impressed direct current or attachment to a sacrificial anode (usually magnesium, aluminum, or zinc).

cation A positively charged ion (e.g., H^+ or Zn^{2+}) or radical (as NH_4^+) that migrates toward the cathode.

cation exchange A process in which cation contaminants are removed from a liquid phase by contacting a synthetic, porous medium or resin that is coated with different cation species. The cations on the medium are exchanged for the cation contaminants. When the medium is depleted of the exchanging cations, it is regenerated with a concentrated solution to restore the bed with the exchanging cations and flush the contaminant cations for subsequent disposal. *See also* cation; ion exchange.

cation exchange capacity A measure of the ability of soil to absorb cations.

cation exchange material A material that releases nontroublesome ions into water in exchange for hardness-causing or other ions.

cation exchange resin A synthetic material possessing reversible ion exchange ability for cations. Sulfonated polystyrene copolymer divinylbenzene exchange resin is used almost exclusively today in ion exchange water softeners. A cation exchange resin may be called a cation exchanger or a base exchanger. *See also* ion exchanger.

cation exchange water softener An equipment unit capable of reducing water hardness by the cation exchange process.

cation exchanger *See* cation exchange resin.

cation membrane An electrodialysis membrane that allows the passage of anions (but not cations) and is practically impermeable to water under typical electrodialysis system working pressure.

cation transfer membrane *See* cation membrane.

cationic Having a positive ionic charge.

cationic flocculant A polyelectrolyte with a net positive electrical change.

cationic polyelectrolyte A polyelectrolyte that forms positively charged ions when dissolved in water.

cationic polymer A polymeric substance with a net positive charge, used in coagulation, flocculation, or filtration processes to improve the removal of negatively charged particles from natural waters. Cationic polymers can be used to destabilize negatively charged particles or to form a bridge between destabilized floc. *See also* charge neutralization; floc.

Cauchy number (C_a) A numerical quantity used as an index to characterize the type of flow in a hydraulic structure in which compression forces, in conjunction with the resisting force of inertia, influence the motion of the liquid. Cauchy number is the ratio of inertia forces to compression forces (in any consistent set of units):

$$C_a = rV^2/E$$

Where:

V = the characteristic velocity of the fluid

E = the bulk modulus of elasticity of the liquid

r = the density

The number is useful in the study of water hammer.

caulking *See* calking.

caulking tool *See* calking tool.

causal association An association, as observed in epidemiological studies, indicating that sufficient evidence is available to make a causal inference (i.e., that the exposure is a cause of the disease rather than the disease merely being associated with exposure). Guidelines used to evaluate causality include consideration of the association's biological plausibility, coherence, and specificity of effect; the consistency of results of studies in different geographic areas; whether a dose-response relationship was observed; and magnitude and temporal nature strength of the association. *See also* association; causality.

causality In epidemiology, the relation of causes to their effects; causes may be necessary, sufficient, or both. Criteria are available to help determine whether epidemiologic associations may be causal. Establishing whether an epidemiologic association is causal usually requires evidence outside the field of epidemiology; such a process is not performed without considerable debate among epidemiologists.

cause-and-effect diagram A graphic showing the relationship between cause and effect. It is an investigative tool that organizes randomly connected causes.

causeway A raised road across wet or marshy ground or across water.

caustic (1) Caustic soda (NaOH) or any compound chemically similar to caustic soda. (2) Any substance capable of burning or destroying animal flesh or tissue. The term is usually applied to strong bases. *See also* caustic soda.

caustic alkalinity The alkalinity caused by hydroxyl ions (OH^-). *See also* hydroxide alkalinity.

caustic corrosion Deterioration of a material caused by exposure to caustic soda (sodium hydroxide, NaOH). Caustic soda will rapidly attack zinc, tin, aluminum, bronze, and brass, so exposure of these materials to caustic soda is not advised.

caustic lime *See* hydrated lime.

caustic soda (NaOH) Sodium hydroxide, a strongly alkaline chemical used for pH adjustment, water softening, anion exchange demineralizer regeneration, and other purposes. It is sometimes called caustic.

caustic soda softening A process to remove hardness from waters by using caustic soda (NaOH). Caustic soda softening is effective when the source water contains insufficient carbonate hardness to react with lime. *See also* carbonate hardness.

cave-in An occurrence in an excavation where the sidewalls slough off into a trench or hole.

cavern flow The movement of water through large openings in rocks, such as caves, or through coarse sorted granular materials.

cavitation The formation and sudden collapse of vapor bubbles in a liquid, usually resulting from local low pressures, as on the trailing edge of a propeller. This phenomenon develops a momentary high local pressure that can mechanically destroy a portion of a surface on which the bubbles collapse. Cavitation can occur in pumps when the suction side has insufficient head for the current operating conditions.

CBFP *See* continuous belt filter press.

CBO *See* Congressional Budget Office.

CCE *See* carbon chloroform extract.

ccf *See* 100 cubic feet *in the Units of Measure list.*

CCL *See* Drinking Water Contaminant Candidate List.

CCML method *See* Cohen censored maximum likelihood method.

CCP *See* Composite Correction Program.

CCPP *See* calcium carbonate precipitation potential.

CCPP index *See* calcium carbonate precipitation potential index.

CCR *See* consumer confidence report.

C_D *See* drag coefficient.

CD *See* compact disk.

CD-ROM *See* compact disk, read-only memory.

cd *See* candela *in the Units of Measure list.*

CDA membrane *See* cellulose diacetate membrane.

CDC *See* Centers for Disease Control and Prevention.

CDI *See* capacitive deionization.

c-deoxyribonucleic acid (c-DNA) probe hybridization The use of complementary deoxyribonucleic acid clones made from ribonucleic acid to detect target gene sequences.

c-DNA probe hybridization *See* c-deoxyribonucleic acid probe hybridization.

celerity The velocity of propagation of a wave through a fluid medium relative to the undisturbed velocity of the fluid through which the disturbance is moving. *See also* wave velocity.

cell culture In vitro growth and replication of cells in suspension or on surfaces.

cell line Repeated subcultured cells derived from a primary cell culture. A cell line has unlimited life (immortalization). Cell replication ceases when a single layer (monolayer) of cells occupies all of the surface area of the vessel.

cell membrane A lipid bilayer with imbedded structural and functional proteins that encloses the contents of a cell. The cell membrane carries on many specialized cellular functions. From a toxicological point of view, it is perceived as a lipid barrier to the free diffusion of polar chemicals, especially those with molecular weights greater than 100. In tissues where tight junctions between cells exist, such as the skin and intestinal epithelium, the resulting membrane forms an effective barrier for systemic absorption of many chemicals that contact these surfaces.

cell organelle A class of organized structures found inside cells and having specialized functions. Examples include mitochondria, which are involved in energy metabolism; and endoplasmic reticulum, which is involved in protein synthesis (rough endoplasmic reticulum) or in drug or chemical metabolism (smooth endoplasmic reticulum).

cell pair In an electrodialysis process, a combination of a cation membrane, anion membrane, demineralized water flow spacer, and concentrate flow spacer. Multiple cell pairs are commonly placed in membrane stack(s).

cell pair resistance The electrical resistance of one electrodialysis cell pair.

cell transformation A process in which a cell takes on one or more properties that are associated with cancer. An example would be a loss of sensitivity to growth controls exerted when cells contact one another. Cell transformation assays are used as in vitro methods for detecting putative carcinogenic effects of a chemical, physical, or viral agent.

cell-associated hepatitis A virus (HAV) A hepatitis A virus not causing cytopathic effects in cell cultures.

cell-mediated immunity Immunity in which small lymphocytes derived from the thymus gland (T-lymphocytes) play a major role. This process is responsible for resistance to infectious agents; it can play a role in the resistance to cancer; it participates in delayed hypersensitivity reactions; it may be involved in certain autoimmune diseases and rejection of transplanted tissues; and it plays a role in certain allergic reactions. *See also* humoral immunity; immunity.

cellular macromolecule A high-molecular-weight biomolecule that includes proteins and nucleic acids. The toxicity and carcinogenicity of many chemicals are mediated by interaction with macromolecules, frequently by the reactions of an electrophilic metabolite with nucleophilic centers on these molecules. In the case of proteins, this interaction frequently inactivates the protein, inhibiting its normal function. In the case of deoxyribonucleic acid, damage induced by such reactions may lead to mutation and toxicological effects that result from the mutation (e.g., cancer or developmental toxicities). Noncovalent interactions of chemicals with proteins (e.g., enzymes, receptors, or both) that can play a role in *acute* toxic reactions induce adaptations that give rise to *chronic* disease. Chronic disease can also result from prolonged exposure to these chemicals.

cellulose acetate (CA) membrane (1) A reverse osmosis or other type of membrane consisting of cellulose with attached acetyl groups and having an asymmetric structure. (2) One of various types of membranes with cellulosic composition; commonly, a cellulose acetate blend membrane having approximately 2.5 acetyl groups per repeating carbon ring, produced using a blend of cellulose diacetate and cellulose triacetate.

cellulose acetate blend (CAB) membrane *See* cellulose acetate membrane.

cellulose diacetate (CDA) membrane A type of cellulose acetate membrane having approximately two acetyl groups per repeating carbon ring. *See also* cellulose acetate membrane.

cellulose ion exchanger A cellulose-based organic material that has been cross-linked and then modified with either anion or cation groups capable of selective ion exchange. Cellulose materials have some natural weak acid functionality.

cellulose triacetate (CTA) membrane A type of cellulose acetate membrane having approximately three acetyl groups per repeating carbon ring. *See also* cellulose acetate membrane.

cellulosic membrane A membrane with cellulose-based composition. *See also* cellulose acetate membrane.

Celsius degree *See* degree Celsius *in the Units of Measure list.*

Celsius scale The international name for the centigrade scale of temperature, on which the freezing point and boiling point are

0° and 100°, respectively, at a barometric pressure of 29.92 inches of mercury (1.013×10^5 Pascals). The scale was invented in 1742 by Anders Celsius, a Swedish astronomer.

cement grout A mixture of cement and water that is used to seal openings in well construction.

cement-lined pipe A pipe for which the interior has been lined with a smooth, dense, thin layer of cement mortar to reduce corrosion and increase smoothness.

cement-mortar lining A mixture of portland cement, sand, and water used to line pipe to prevent internal corrosion and maintain flow coefficients over the life of the system. The lining may be applied in the factory or in the field after cleaning in place. A seal coat of asphaltic material may be applied to the uncured lining at the purchaser's option.

cementation The process by which voids in a sediment become filled with precipitated minerals. Cementation of aquifer materials near a well caused by chemical imbalances can drastically effect the hydraulic performance of the well.

cemented soil A soil held together by calcium carbonate ($CaCO_3$, cement), such that a hand-sized sample cannot be crushed into particles or powder by finger pressure.

censored data A set of data from which selected values have been removed based on specified criteria. A national occurrence data set for a contaminant and from which all data less than the maximum detection limit have been removed is an example of censored data.

census tract A small, relatively permanent statistical subdivision of a county. Census tracts usually have between 2,500 and 8,000 persons and, when first identified, are designed to be homogenous with respect to population characteristics, economic status, and living conditions. Census tracts are identified for all metropolitan areas and other densely populated counties following US Census Bureau guidelines; more than 3,000 census tracts have been established in 221 counties outside metropolitan areas.

census-tract analysis An analysis in which health statistics, including morbidity and mortality, are compiled by census tract and compared because other population characteristics and statistics are available for the census tract. *See also* morbidity rate; mortality rate.

Center for Environmental Research Information (CERI) The office within the US Environmental Protection Agency's Office of Research and Development responsible for dissemination of information and reports on US Environmental Protection Agency research.

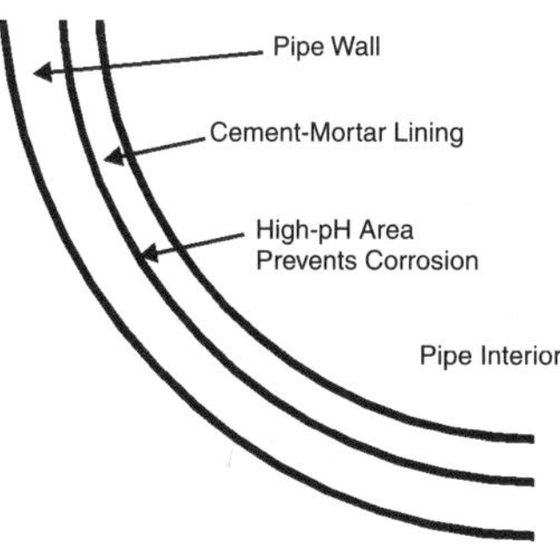

Pipe with cement-mortar lining

center of buoyancy The center of gravity of the space occupied in a body of water by a floating object.

center of flotation The centroid of the water plane area of a floating body.

center of gravity (1) The point in a body through which the resultant of the parallel forces of gravity, acting on all particles of the body, passes, no matter what the position in which the body is held. (2) The point in a plane surface through which the resultant of a force uniformly applied over such surface might be considered to act.

center of pressure The point of application of the resultant of all normal pressures acting on a surface.

Centers for Disease Control and Prevention (CDC) An operating health agency within the US Public Health Service responsible for protecting public health by providing leadership and direction in the prevention and control of diseases and other preventable conditions and by responding to public health emergencies.

centi (c) Prefix meaning 10^{-2}, used in Système International. Its use should be avoided except in the measurement of area and volume.

centigrade *See* Celsius scale.

centimetre *See the Units of Measure list.*

centimetre, gram, second (CGS) system A metric system of physical measurements in which the fundamental units of length, mass, and time are the centimetre, the gram, and the mean solar second. This system is not the Système International system.

central irrigation control A computerized system for programming irrigation controllers from a central location. It uses personal computer and radio waves or hard wiring to send program information to controllers in the field.

central nervous system (CNS) The portion of the nervous system that consists of the brain and spinal cord.

central office access unit (COAU) A piece of equipment installed in the telephone central office to facilitate meter reading via polling of meter interface units at customers' premises.

central processing unit (CPU) The hardware that directs and supervises all of a computer's functions.

central tendency A representation of the average behavior of a data set. The central tendency is often described by mean, median, and mode of the data distribution.

centralized processing Centralization of a computer system's "intelligence" and central execution of all instructions and functions of a system.

centralized system A system that is controlled by a central operation.

centrate The water that is separated from sludge and discharged from a centrifuge.

centrifugal force The force that makes rotating bodies move away from the center of the rotation.

centrifugal pump A pump that has a center suction and an outward discharge that uses rotation of an impeller on a shaft inside a casing to provide a velocity or outward force to the water, thus providing a lift or pressure.

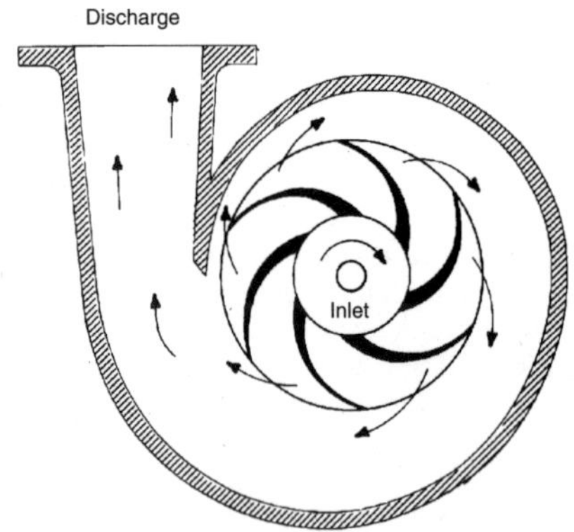

Centrifugal pump

centrifugal screw pump A centrifugal pump having a screw-type impeller. Such a pump may be of an axial flow type or combined axial and radial flow type.

centrifugation In water treatment, a method of dewatering sludge via a mechanical device (centrifuge), which spins the sludge at a high speed.

centrifuge A mechanical device used for dewatering liquid sludges by means of imparting centrifugal force through rotation. The centrifuge bowl rotates at a high velocity, forcing water to the periphery as dewatered sludge is transported in the opposite direction with a screw conveyer.

centrilobular Pertaining to the hepatocytes that surround the central vein of a liver lobule. Certain liver toxins induce damage (necrosis) in this area, e.g., carbon tetrachloride (CCl_4). This is frequently due to the low oxygen tension in that area relative to the oxygen tension observed surrounding liver parenchymal cells near the periportal region of the lobule (i.e., the region of lobules where the hepatic artery and portal vein deliver oxygenated blood to the liver cells).

centrilobular lesion Damage to hepatocytes (parechmymal cells) that surround the terminal hepatic vein (or central vein) of the liver lobule. In newer terminology, this is usually referred to as "Zone 3 damage." The lesion is usually necrosis (i.e., dead cells). It is a common site of damage for chemicals such as carbon tetrachloride (CCl_4) and chloroform ($CHCl_3$).

centrilobular necrosis The death of cells in the area surrounding a branch of the central vein of a liver lobule. *See also* centrilobular.

centripetal drainage Drainage more or less radially inward toward a central point.

centroid of flow The midpoint in a channel of uniform flow.

CEQ *See* Council on Environmental Quality.

ceramic membrane An inorganic membrane composed of ceramic material commonly containing a rejecting-layer surface coating, also of inorganic composition, e.g., a tubular ceramic ultrafiltration membrane with an alumina–zirconia barrier layer.

Ceratium Phytoplankton of the class *Dinoflagellata*, phylum *Mastigophora*, that are unicellular and flagellated and have spinous projections. Blooms of these and other organisms in reservoirs or lakes contribute to taste-and-odor problems; *Ceratium* in abundance has a particularly obnoxious odor.

CERCLA *See* Comprehensive Environmental Response, Compensation, and Liability Act.

cerebrospinal fluid The fluid, formed by a specialized structure called the choroid plexus, that flows slowly through the ventricles of the brain and down through the spinal column. This structure forms a compartment that is separated from the blood by a selectively permeable membrane that generally does not allow free passage of polar and high-molecular-weight chemicals unless some type of specific carrier mechanism exists for the chemical. *See also* blood–brain barrier.

CERF *See* Civil Engineering Research Foundation.

CERI *See* Center for Environmental Research Information.

certificate of participation (COP) A popular means of financing long-term capital requirements in most states. Generally, a nonprofit entity or authority is established that borrows the funds needed to construct a facility. The authority issues certificates of participation, which are equivalent to tax-exempt bonds, to the public, and it controls the funds that are collected to pay for construction and startup. The authority then leases the facilities to the utility or utilities that initiated the process for a period coincident with the maturity of the certificates of participation. The lease payments cover the debt service on the certificates of participation and the operating costs of the lessor authority. Ownership transfers to the lessee at the termination of the lease. The popularity of certificates of participation stems primarily from the fewer restrictions—in terms of debt limitations, voter approvals, and reserve funds—that are tied to general obligation or revenue bonds. *See also* Joint Powers Authority.

certification A program to substantiate the capabilities of personnel by documentation of their experience and learning in a defined area of endeavor.

certified backflow-prevention assembly tester A person who is certified by the approving authority to test, repair, and maintain backflow-prevention assemblies.

certified laboratory A laboratory that meets specified requirements set by the state or US Environmental Protection Agency and has been approved by such agencies to conduct laboratory analyses for compliance reporting under the Safe Drinking Water Act.

certified reference material A stable, homogenous, and well-characterized substance having a chemical and physical composition essentially identical or very similar to the test samples in question. It is used to eliminate or reduce the matrix effect that might be present in the actual samples to be analyzed. *See also* matrix effect.

certified utility safety administrator (CUSA) As certified by the National Safety Council Utilities Division, an individual who meets experience and educational criteria and successfully passes a written examination. CUSAs can achieve specialized certification by passing additional examinations that concentrate on the specifics of a certain type of utility safety, such as water, electric, gas, or communications.

CFM *See* cubic feet per minute *in the Units of Measure list.*

CFR *See* Code of Federal Regulations.

CFS *See* cubic feet per second *in the Units of Measure list.*
CFSTR *See* continuous-flow stirred-tank reactor.
cfu *See* colony-forming unit *in the Units of Measure list.*
CGR *See* coliform growth response.
CGS system *See* centimetre, gram, second system.

chain-and-flight sludge collector A device for removing sludge from a rectangular basin by a scraping mechanism. Collected sludge is scraped into a hopper, from which it typically is discharged to solids processing by means of a screw conveyer. The scrapers, or flights, are attached to chains that are driven along the length of the basin by rotating cogs located at the ends of the basin. The chains move longitudinally along the basin bottom and across the top of the basin in a circular fashion. *See also* screw conveyer.

chain bucket A continuous chain equipped with buckets and mounted on a scow. It is also called a ladder dredge.

chain gauge A gauge consisting of a tagged or indexed chain, tape, or other line attached to a weight that is lowered from a boat or platform to touch the water surface, whereupon the gauge height is determined based on the tag or index indicator on the boat deck or platform. *See also* tape gauge.

chain of custody (1) The sequence of persons handling a sample, as specified in a document accompanying a sample when certifying sampling information and the person with control of the sample at any given time is important. A chain-of-custody document is an important part of a quality assurance program when rigorous data validation is required. (2) The document that specifies this sequence of persons having custody.

challenge water Water specifically prepared for testing the performance of water treatment equipment products. Challenge water for each type of equipment is specifically defined in individual equipment testing standards, such as those established by the Water Quality Association and NSF International.

chalybeate spring A spring in which the water contains a considerable amount of iron compounds, especially the sulfate, $Fe_2(SO_4)_3$, in solution.

change of state The process by which a substance passes from one phase to another among the solid, liquid, and gaseous states and in which marked changes in the substance's physical properties and molecular structure occur.

channel (1) A perceptible natural or artificial waterway that periodically or continuously contains moving water or that forms a connecting link between two bodies of water. It has a definite bed and banks that confine the water. (2) The deep portion of a river or waterway where the main current flows. (3) The part of a body of water deep enough to be used for navigation through an area otherwise too shallow for navigation. (4) Informally, a more or less linear conduit of substantial size in cavernous limestones or lava racks. *See also* buried channel; bypass channel; open channel.

channel of approach That reach of a channel immediately upstream from a control structure such as a dam, weir, sill, or bridge.

channel inflow Water that, at any instant, is flowing into a channel system from surface flow, subsurface flow, base flow, or rainfall directly on the channel.

channel line The route of strongest flow of a river. It usually coincides, or nearly coincides, with the thalweg.

channel loss The loss of water from a channel by capillary action and percolation.

channel roughness Variation from smooth texture in a channel, including the extra roughness caused by local expansion or contraction and obstacles, as well as the roughness of the streambed proper. This roughness is expressed in terms of a roughness coefficient in the velocity formulas.

channel spring A spring occurring on the banks of a stream that has cut a channel below the water table.

channel storage The in-channel storage volume, which depends on the stage of the water surface in the channel.

channeling The flow of water or regenerant through a limited number of passages in a filter or ion exchanger bed, as opposed to the usual distributed flow through all passages in the bed. Channeling may be caused by fouling of the bed and the resulting plugging of many passages; poor distributor design; low flow rates; faulty operations procedures; insufficient backwash; or other causes.

channel-phase runoff That phase of runoff that takes place in the channel and is governed by the laws of channel hydraulics.

chapelet (1) A machine for raising water, or for dredging, that uses buckets on a continuous chain passing between two rotating sprocket wheels. (2) A chain pump having buttons or disks at intervals along its chain.

characteristic speed A speed or velocity of revolution, expressed in revolutions per minute, at which the runner of a given type of turbine would operate if it were so reduced in size and proportion that it would develop 1 horsepower under a 1-foot head (1.341 kilowatts under a 0.305-metre head). *See also* specific speed.

characteristic type The speed, in revolutions per minute, at which a waterwheel would run if it were reduced proportionally in all dimensions to develop 1 horsepower under a 1-foot head (1.341 kilowatts under a 0.305-metre head). This speed is also called characteristic speed or specific speed.

charcoal An adsorbent carbon product that has about one-third the surface area of activated carbon. It is often confused with activated carbon. *See also* granular activated carbon.

charge (1) An excess or deficiency of electrons in a body. (2) The cost for goods or services rendered.

charge neutralization A process under which a positively charged material (coagulant) is added to a negatively charged particle suspension to neutralize, or destabilize, particles. After charge neutralization, particles can agglomerate and settle. *See also* adsorption–destabilization; bridging; double-layer compression; sweep-floc coagulation.

check dam A low, uncontrolled dam used for a variety of purposes including (1) the spreading of water in shallow basins for artificial recharge to underlying aquifers, (2) dissipation of energy in flood channels, and (3) as a weir for maintaining a water depth suitable for navigation.

check sample A sample taken to verify the results of a previous sample. Under the Total Coliform Rule, water utilities are required to take check samples after a routine positive coliform sample is detected.

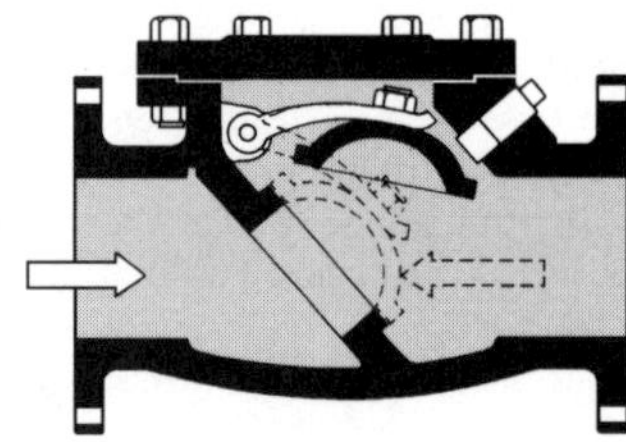

Check valve

check valve A valve that allows flow in one direction and that closes when the flow tries to reverse, used on pump discharge lines for pump protection and on building services to prevent draining of plumbing lines back into the distribution system. A single check valve is not a backflow prevention device; that requires a double check assembly that can be tested for positive closing.

checklist approach A method of establishing water efficiency standards by prescribing specific criteria for landscape design, installation, and management.

chelating agent A chemical or complex that interacts with an ion, usually a metal, causing the ion to join that chemical or complex by both ordinary and coordinate valence forces. Such linkages result in the formation of one or more heterocyclic rings in which the metal atom is part of the ring. Commercially available chelating agents may be used to remove traces of metal ions. *See also* sequestering agent.

chelating ion exchanger A special selective resin that will adsorb one metal ion to the exclusion of any other ion present in a stream of water. *See also* chelation.

chelation A process in which metal ions are reacted with a ligand to form a cyclic compound. Such complexes are used to bind certain ions for specific purposes in analytical methods. An example is the use of ethylenediaminetetraacetic acid as a chelating agent in determining the hardness in water.

chemical (1) A specific association of atoms into a single entity or molecule that has characteristic properties of form and structure under a given set of environmental conditions. (2) A substance used or produced in a reaction.

Chemical Abstract Service number (CAS#) A number assigned by the Chemical Abstracts Service to a chemical substance that is manufactured or processed in the United States. The US Environmental Protection Agency compiles and keeps these numbers current under the Toxic Substances Control Act.

chemical actinometer An actinometer utilizing a chemical reaction that has been accurately investigated in terms of its photochemical behavior. For example, the decomposition of oxalic acid ($H_2C_2O_4$) in the presence of uranyl (UO_2^{2+}) sulfate is a common actinometric reaction. Potassium ferric oxalate ($K_3Fe(C_2O_4)_3 \cdot 3H_2O$) is another common chemical actinometer. *See also* actinometer.

chemical activity The molar concentration of a chemical species multiplied by its activity coefficient. For electrolytes in very dilute aqueous solution, activity is very nearly equal to molal or molar concentration. *See also* molality; molarity.

chemical analysis An examination of a sample for its constituents. The examination may be quantitative, qualitative, or both. In a chemical analysis of water, chemical constituents other than water are typically present in concentrations ranging from milligrams per litre to micrograms per litre.

chemical antagonism Neutralization of the effects of one compound as a result of a chemical reaction involving that compound. For example, large cations and large anions tend to react with one another at neutral pH to form an insoluble precipitate. Such reactions can interfere with the absorption of the chemicals from the gastrointestinal tract. Relatively few examples of such reactions exist within tissues. Mercury and selenium form insoluble complexes in tissues, and this limits the toxicity of both elements. *See also* antagonism.

chemical bond The force that holds atoms together within molecules. A chemical bond is formed when a chemical reaction takes place. Two types of chemical bonds are ionic bonds and covalent bonds.

chemical carcinogen A chemical that is capable of inducing cancer in humans or animals. *See also* carcinogen.

chemical cartridge The filtering element of a respirator. It contains a compound-specific medium that absorbs or filters out hazardous fumes or dust.

chemical coagulation The destabilization and initial aggregation of colloidal and finely divided suspended matter by the addition of an inorganic coagulant. *See also* flocculation.

chemical dissolving box A box or small tank for dissolving chemicals before their introduction into the liquid to be treated. *See also* chemical solution tank.

chemical dose A specific quantity of chemical applied to a specific quantity of fluid for a specific purpose. *See also* dose.

chemical equation A shorthand way, using chemical formulas, of writing the reaction that takes place when chemicals are brought together. The left side of the equation indicates the chemicals brought together (the reactants); the arrow indicates which direction the reaction occurs (a bidirectional arrow indicating a reversible reaction); and the right side of the equation indicates the results (the products) of the chemical reaction.

chemical equilibrium The condition that exists when no net transfer of weight, energy, or both exists between the components of a system. This is the condition in a reversible chemical reaction when the rate of the forward reaction equals the rate of the reverse reaction.

chemical equivalent The weight, in grams, of a substance that combines with or displaces 1 gram of hydrogen. It is found by dividing the substance's formula weight by its valence. For example, calcium has a molecular weight of 40 and a valence of +2; thus, its chemical equivalent is 20 grams per mole.

chemical feed pump A pump used to apply chemicals to water. Pumps are used to apply chemicals into conduits under pressure and to control the chemical feed rate relatively precisely.

chemical feeder A device for dispensing a chemical at a predetermined rate for the treatment of water. A change in rate of feed may be effected manually or automatically by flow-rate changes. Feeders are designed for solids, liquids, or gases.

chemical formula *See* formula.

chemical gas feeder A feeder for dispensing a chemical in the gaseous state. The rate is usually graduated in gravimetric terms. Such devices may have proprietary names.

chemical gauging A method of measuring the flow or discharge of water in which a solution containing certain chemicals in known concentration is added at a given rate to a stream of water. Opportunity is afforded for complete mixing, and samples of the water are then taken and analyzed to determine the quantity of chemical in the resulting mixed flow. By computation, using the figures for the quantity and rate at which the chemical is added and the concentration of the chemical in the resulting mixture, one can determine the dilution and thus the discharge. This method is used most frequently when direct measurement of the velocity of flow is difficult or impossible. *See also* salt-velocity method.

chemical grout A material used to seal cracks or joints that uses a catalyst to cause the material to develop specific desired properties.

chemical handling and feeding A system designed to load, store, and apply chemicals for drinking water treatment. Components include storage tanks, day tanks, dilution water, feed pumps, flowmeters, and valves. *See also* day tank.

chemical ionization (CI) A mode of operation in mass spectrometry in which analyte molecules are bombarded with a reagent gas such as methane (CH_4). Molecules can be selectively ionized and often produce little fragmentation relative to mass spectra produced in the electron impact mode. The determination of molecular weight is often possible when chemical ionization mass spectra are being generated. *See also* electron impact ionization.

chemical lime (CaO) *See* calcium oxide.

Chemical Monitoring Reform An effort undertaken by the Office of Ground Water and Drinking Water of the US Environmental Protection Agency (USEPA) from 1992 to 1996 to revise the standardized monitoring framework to reduce unnecessary monitoring requirements and use chemical-monitoring resources more efficiently. While USEPA was working with state primacy agencies on Chemical Monitoring Reform, Congress enacted the 1996 amendments to the Safe Drinking Water Act (SDWA), which reflected a number of the issues being addressed. (See 63 *Federal Register* 40709 for additional background.) USEPA incorporated its Chemical Monitoring Reform effort into a broader effort under the 1996 SDWA amendments; this broader effort falls under the heading of Chemical Monitoring Revisions. *See also* Chemical Monitoring Revisions; standardized monitoring framework.

Chemical Monitoring Revisions An ongoing effort by the US Environmental Protection Agency's Office of Ground Water and Drinking Water to revise chemical-monitoring requirements to reduce unnecessary monitoring requirements, use chemical-monitoring resources more efficiently, and satisfy provisions of the 1996 Safe Drinking Water Act amendments to revise and update chemical-monitoring requirements.

chemical oxidation The process of using an oxidizing chemical to remove or change some contaminant in water by removing electrons.

chemical oxygen demand (COD) An operationally defined surrogate parameter in water analysis that is a measure of the constituents susceptible to oxidation by strong chemical oxidants.

chemical and physical analysis A series of examinations incorporating chemical and physical parameters. Physical analyses include parameters such as color, turbidity, and conductivity, which result from a combination of chemical constituents. In a chemical analysis of water, chemical constituents other than water are typically present in concentrations ranging from milligrams per litre to micrograms per litre.

chemical precipitation The process of generating solids from the soluble phase by changing the equilibrium conditions of a solution. By adding chemicals that react with the soluble material to be removed, or by changing pH, one can alter the solubility of constituents, resulting in the formation of solids that subsequently can be removed by physical processes (e.g., sedimentation, filtration).

chemical precipitation softening *See* lime softening; lime–soda ash softening.

chemical proportioner A device for dispensing a chemical into water in proportion to the flow. *See also* chemical feeder.

chemical reaction A process that occurs when atoms of certain elements are brought together and combine to form molecules, or when molecules are broken down into individual atoms.

chemical reagent A chemical added to a system to induce a chemical reaction.

chemical solution tank A tank in which chemicals are added in solution before they are used in a water treatment process.

chemical stability Resistance to attack by chemical action. This term is often applied to the resistance of ion exchange resins to breakdown caused by contact with aggressive solutions.

chemical tank A tank in which chemicals are stored before use in a water treatment process.

Chemical Transportation Emergency Center (CHEMTREC) A 24-hour-a-day emergency telephone service that provides information on hazardous chemical spills in transport. CHEMTREC is an emergency service operation, free of charge to fire, police, and other emergency agencies. The phone number is (800) 424-9300.

chemical treatment Any water treatment process involving the addition of chemicals to obtain a desired result, such as precipitation, coagulation, flocculation, sludge conditioning, or disinfection.

chemihydrometry A form of chemical gauging to determine the flow or discharge of water. *See also* chemical gauging; salt-velocity method.

chemisorption A process related to adsorption in which atoms or molecules of reacting substances are held to the surface atoms of a catalyst by electrostatic forces having about the same strength as chemical bonds. Chemisorption differs from physical adsorption chiefly in the strength of bonding, which is much greater in chemisorption. *See also* adsorption.

chemistry A science concerned with the composition, properties, and reactions of matter.

chemocline A zone of a lake, reservoir, or impoundment in which the concentration of dissolved substances changes rapidly with depth.

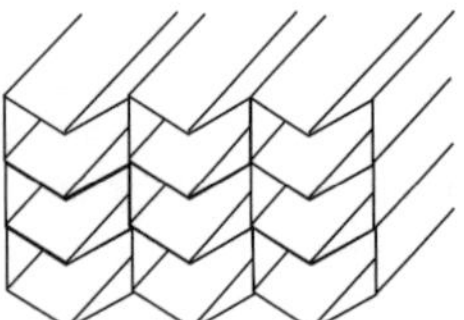

Chevron tube settler module

chemolithotroph A type of bacterial organism that obtains its energy by chemical oxidation of inorganic substances and that has the ability to fix carbon dioxide (CO_2) as its sole source of carbon. *See also* autotroph.

chemoorganotroph A type of bacterial organism that obtains energy and carbon through the oxidation of organic compounds.

chemosynthetic bacteria Bacteria that synthesize organic compounds by using energy derived from the oxidation of organic and inorganic materials without the aid of light.

CHEMTREC *See* Chemical Transportation Emergency Center.

chevron tube settler A type of tube settler with a design in which the cross-sectional area approximates a V shape. *See also* tube settler.

Chezy open-channel formula A basic hydraulic formula developed by Antione Chezy in 1775 for determining the flow of water in open channels:

$$V = C\sqrt{RS}$$

Where:

V = flow velocity, in feet per second (metres per second)

C = Chezy resistance factor, in feet$^{1/2}$ per second (metres$^{1/2}$ per second)

R = hydraulic radius of the channel, in feet (metres)

S = channel slope (for uniform flow) or the energy slope (for nonuniform flow), dimensionless

See also Chezy resistance factor; Kutter roughness coefficient; Manning formula; Manning roughness coefficient.

Chezy resistance factor (*C*) A coefficient used in the Chezy open-channel formula. Two formulas can be used to determine its value. The first is the Ganguillet-Kutter formula:

$$C = \frac{41.65 + \frac{0.00281}{S} + \frac{1.811}{m}}{1 + \left(41.65 + \frac{0.00281}{S}\right)\frac{m}{\sqrt{R}}} \qquad C = \frac{23 + \frac{0.00155}{S} + \frac{1}{m}}{1 + \left(23 + \frac{0.00155}{S}\right)\frac{m}{\sqrt{R}}}$$

US customary units — Système International units

Where:

C = Chezy resistance factor, in feet$^{1/2}$ per second (metres$^{1/2}$ per second)

S = channel slope (for uniform flow) or energy slope (for nonuniform flow), dimensionless

m = Kutter roughness coefficient, dimensionless

R = hydraulic radius, in feet (metres)

The second formula is

$$C = \frac{1.486}{n}R^{1/6} \qquad C = \frac{1}{n}R^{1/6}$$

US customary units — Système International units

Where:

n = Manning roughness coefficient (generally very close numerically to the value of the Kutter roughness coefficient), dimensionless

R = hydraulic radius, in feet (metres)

See also Chezy open-channel formula; Kutter formula; Kutter roughness coefficient.

chi-square distribution The distribution of sample variance in a normally distributed data set. Sample variances are calculated from the following equation:

$$S^2 = \sum\frac{\chi_i - \bar{\chi}}{n-1}$$

Where:

S = sample variance

χ_i = ith observation from a normally distributed data set containing n observations

$\bar{\chi}$ = average of the observed data set

See also normal distribution.

chi-square test A statistical procedure to determine the level of significance of the difference between two non-normally distributed data sets.

Chick–Watson model A model of chemical inactivation of microorganisms with the reaction rate of inactivation dependent on the relative concentrations of microorganisms and changes in disinfectant concentration during a specific time interval. It is expressed as follows:

$$\log(N/N_0) = -kC^n t$$

Where:

N = the number of organisms per unit volume recovered at time t

N_0 = the number of organisms per unit volume at time 0

k = the pseudo first-order reaction rate constant

C = the concentration of disinfectant, in milligrams per litre

n = the coefficient of dilution, an empirical factor, usually assumed to be 1.0

See also Hom model; modified Hom model.

chiller A component (generally a heat exchanger) designed to remove heat from a gas or liquid stream.

chloracne A form of skin eruption (i.e., acne) that is induced by some, but not all, chlorinated compounds. The effect is most commonly associated with 2,3,7,8-tetrachlorodibenzodioxin (commonly called dioxin) and compounds with similar toxicological effects. *See also* dioxin.

chloral hydrate ($CCl_3CH(OH)_2$) The common name for hydrated trichloroacetaldehyde, a disinfection by-product produced at low levels during chlorination. It is medically used as a sedative. It is also known as chloral. *See also* disinfection by-product.

chloramination The process of disinfecting water with chloramines. *See also* chloramines.

chloramines Disinfectants produced from the mixing of chlorine (Cl_2) and ammonia (NH_3). The general formula is NH_xCl_y, where x can be 0, 1, or 2; y can be 1, 2, or 3. Typically, monochloramine (NH_2Cl) and a small percentage of dichloramine ($NHCl_2$) are formed, depending on the pH and the chlorine-to-ammonia ratio that reacts. Under certain circumstances, nitrogen trichloride (trichloramine, $NHCl_3$) can be formed. In the presence of the bromide ion, bromamines can be produced. In the presence of organic nitrogen, organic chloramines can also form; however, organic chloramines are not considered to be disinfectants. *See also* bromamines; dichloramine; monochloramine; organic chloramines.

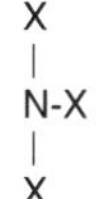

Chloramines, where one to three of the Xs are chlorine atoms and the rest are hydrogen atoms

chlorate ion (ClO_3^-) A disinfection by-product formed during the treatment of water with chlorine dioxide (ClO_2). The chlorate ion is also produced during the decomposition of bleach (NaOCl) in stock solutions. Furthermore, ozonation of water with a chlorine residual can yield chlorate ions. *See also* bleach; chlorine dioxide; chlorite ion; hypochlorite ion; liquid chlorine.

chlordane ($C_{10}H_6Cl_8$) The common name for 1,2,4,5,6,7,8,8-octachloro-4,7-methano 3a,4,7,7a-tetrahydroindane or 1,2,4,5,6,7,8,8-octachloro-2,3,3a,4, 7,7a-hexahydro 4,7-methanoindane. Chlordane is an insecticide and fumigant regulated by the US Environmental Protection Agency. *See also* fumigant; insecticide.

Chlordane

chloride (Cl^-) One of the major anions commonly found in water and wastewater. Its presence is often determined by ion chromatographic or volumetric analysis. Consumers who drink water with concentrations of chloride exceeding a secondary maximum contaminant level of 250 milligrams per litre may notice a salty taste.

chlorinated (1) Pertaining to a water to which chlorine was added as a disinfectant or oxidant. (2) Pertaining to a chlorine-containing organic compound that may be present in water. For example, chlorinated organic solvents are among the most commonly detected groundwater contaminants. The preferred term is *chlorine-substituted. See also* chlorine-substituted.

chlorinated copperas ($FeSO_4 + FeCl_3$) A solution of ferrous sulfate ($FeSO_4$) and ferric chloride ($FeCl_3$) produced by chlorinating a solution of ferrous sulfate.

chlorinated lime (approximately $CaCl(ClO){\cdot}4H_2O$) A water treatment chemical that is used primarily for disinfection, mostly in swimming pools, and also has buffering capacity. Chlorinated lime is prepared by chlorinating slaked lime ($Ca(OH)_2$); it decomposes in water and releases 39 percent available chlorine for disinfecting action. *See also* available chlorine; buffer; hydrated lime.

chlorinated organic *See* chlorine-substituted organic.

chlorinated polyvinyl chloride (CPVC) A modified form of polyvinyl chloride, a synthetic resin that is composed of polymerized vinyl chloride and that, when plasticized or softened with other chemicals, has some rubber-like properties. Vinyl chloride ($CH_2{:}CHCl$) is derived from acetylene ($CH{\equiv}CH$) and anhydrous hydrochloric acid (HCl). A chlorinated polyvinyl chloride molecule has more chlorine atoms per repeating monomer unit than does the polyvinyl chloride molecule. This extra chlorine is responsible for CPVC's strength at high temperatures, as well as other properties useful in industrial piping applications.

chlorination An oxidation process that is initiated through the addition of chlorine. In chlorination, chlorine oxidizes microbiological material, organic compounds, and inorganic compounds. Chlorination is the principal form of disinfection in US water supplies. *See also* chlorine; disinfection.

chlorination chamber A detention basin provided primarily to secure the diffusion of chlorine through the liquid. It is also called a chlorine contact chamber.

chlorination of wells The addition of chlorine and mixing energy to routinely disinfect production wells.

chlorinator Any device that is used to add chlorine to water.

chlorine A chemical used as a disinfectant and oxidizing agent. Gaseous molecular chlorine (Cl_2), when introduced into water, is converted into hypochlorous acid (HOCl) and the hypochlorite ion (OCl^-); the ratio of the two substances is dependent on the pH of the solution ($HOCl \Leftrightarrow OCl^- + H^+$). Chlorine is also commercially available in liquid form as a hypochlorite ion solution. Individual chlorine atoms are halogen atoms that can be substituted into organic molecules during chlorination, thus forming disinfection by-products. Chlorine atoms are also present in a number of pesticides and other synthetic organic chemicals. *See also* disinfection by-product; halogen; hypochlorite ion; hypochlorous acid; liquid chlorine; pesticide; synthetic organic chemical.

chlorine, combined *See* combined chlorine.

chlorine, free *See* free available chlorine.

chlorine contact chamber *See* chlorination chamber.

chlorine cylinder A container for storing chlorine gas (Cl_2) under pressure. Chlorine cylinders are typically made of steel and are commercially available in 150-lb (68-kilogram), 1-ton (907-kilogram), and 15- to 17-ton (13,608- to 15,422-kilogram) tank trucks, as well as 16- to 90-ton (14,515- to 81,647-kilogram) railroad cars.

chlorine demand The quantity of chlorine consumed in a specified time period by reaction with substances present in water that exert an oxidant demand (e.g., natural organic matter, ammonia [NH_3], hydrogen sulfide [H_2S]). The chlorine

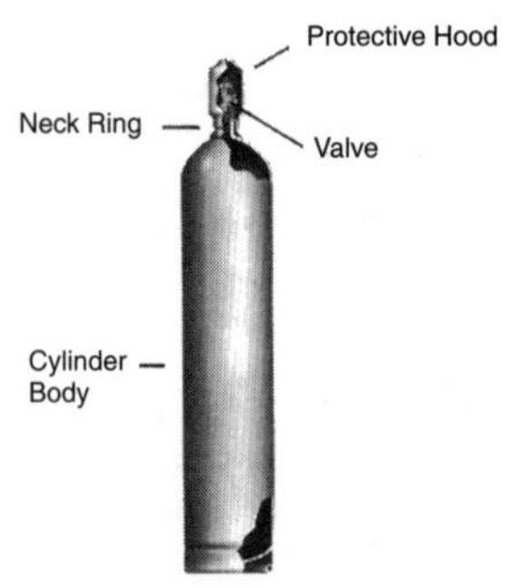

Chlorine cylinder

demand for a given water varies with both contact time and temperature. *See also* oxidant demand.

chlorine dioxide (ClO_2) A red-yellow gas that is very reactive and unstable. It is a strong oxidizing agent and is also used as a disinfectant. Chlorine dioxide decomposes in water to yield the chlorite ion (ClO_2^-) and, to a lesser extent, the chlorate ion (ClO_3^-). *See also* chlorate ion; chlorite ion.

chlorine dose The amount of chlorine (Cl_2) applied to a liquid, usually expressed in milligrams per litre or pounds per million gallons. *See also* chlorine demand; combined available chlorine; combined available residual chlorine; free available chlorine; free available residual chlorine.

chlorine feed rate The rate at which chlorine (Cl_2) is fed into a water to achieve the desired dosage. Chlorine can be fed as a gas, through an evaporator, or as a liquid when available as liquid sodium hypochlorite (NaOCl). *See also* evaporator.

chlorine feed room A room in a water treatment facility specially designed for chlorine storage and feed. Because of the hazardous nature of chlorine gas, chlorine must be stored separately from other chemicals in rooms that are designed to provide quick exit, gas containment, alarming, and possibly medical treatment in times of accidental releases.

chlorine hydrate ($Cl_2 \cdot 8H_2O$) A yellowish ice formed in a chlorinator when chlorine gas comes in contact with water at 49° Fahrenheit (9.4° Celsius) or lower. It is frequently detrimental to the performance of a chlorinator if formed in quantities sufficient (1) to interfere with the safe operation of float controls or (2) to plug openings essential to flow indication, control, or rate of application. Chlorine hydrate is also called chlorine ice.

chlorine ice *See* chlorine hydrate.

chlorine removal The process of removing chlorine from solution. Chlorine removal may be required (1) to reduce the concentration in the distribution system when high dosages are applied to destroy tastes and odors, (2) to reduce the potential toxic effect of a chlorinated discharge to aquatic life, (3) to protect downstream processes, such as zeolite softeners, from the oxidative effects of chlorine, or (4) to reduce chlorinous tastes in drinking water (particularly bottled water). Chlorine removal is typically achieved through chemical addition, and the most commonly used chemical is sulfur dioxide (SO_2).

chlorine requirement The amount of chlorine that is needed for a particular purpose. Some reasons for adding chlorine are to reduce the number of coliform bacteria, to obtain a particular chlorine residual, or to oxidize some substance in the water. In each case a definite dosage of chlorine (i.e., the chlorine requirement) will be necessary.

chlorine residual A concentration of chlorine species present in water after the oxidant demand has been satisfied. Chlorine residual can be determined and expressed in a number of ways. Commonly used analytical techniques include a colorimetric method using the reagent N,N-diethyl-p-phenylenediamine and a volumetric method using an amperometric titration. The concentration is often expressed in terms of free chlorine, total chlorine, or combined chlorine. *See also* amperometric titration; N,N-diethyl-p-phenylenediamine methods.

chlorine-to-ammonia–nitrogen ratio (Cl_2/NH_3–N) The ratio (typically on a weight basis) of the chlorine and ammonia doses during chloramination. During breakpoint chlorination of ammonia-containing waters, a theoretical Cl_2/NH_3–N ratio of 7.6:1 milligrams/milligram is required. *See also* breakpoint curve.

chlorine–ammonia process The application of chlorine (Cl_2) and ammonia (NH_3) to water, or of chlorine to water containing ammonia, in such ratios as to produce chloramines.

chlorine–ammonia treatment *See* chlorine–ammonia process.

chlorine-demand–free water A type of water with a very low oxidant demand. When chlorine is added to this type of water, none of the chlorine reacts and the chlorine residual is equal to that expected from simply diluting the chlorine.

chlorine-gas scrubbing system A mechanical system designed to collect and remove chlorine gas during a release. Sodium hydroxide (caustic solution, NaOH) is most commonly used to convert chlorine gas to salt (NaCl) and water. In the scrubber, the gas is directed upwardly as the scrubbant (caustic) solution is sprayed downward in a countercurrent fashion to improve contact and reaction. The components of such a system include compressors (fans), scrubber(s), feed chemicals, and recirculation pumps. The scrubbed gases may be released to the atmosphere or recirculated into the contaminated room until the chlorine gas concentration is reduced to allowable levels.

chlorine-substituted Pertaining to a molecule in which a chlorine atom has been introduced.

chlorine-substituted by-product A disinfection by-product containing chlorine atoms. *See also* disinfection by-product.

chlorine-substituted organic An organic molecule into which a chlorine atom has been introduced.

chlorine-substituted trihalomethanes, total (TTHM-Cl) *See* total trihalomethane chlorine.

chlorinity (1) The relative concentration of halides in salt water (typically, seawater). (2) The weight of halides in seawater, usually reported in grams per kilogram or parts per thousand of chlorine in seawater. *See also* chlorosity; salinity.

chlorite ion (ClO_2^-) A disinfection by-product formed during treatment of water with chlorine dioxide (ClO_2). *See also* chlorate ion; chlorine dioxide.

3-chloro-4-(dichloromethyl)-5-hydroxy-2(5H)-furanone (MX) A potent bacterial mutagen that was first identified in the chlorinated effluents of pulp mills. Initially, the chemical structure was unknown, so the compound was referred to as mutagen X (i.e., MX). Subsequently, this compound has been identified in a number of chlorinated drinking water samples. Although 3-chloro-4-(dichloromethyl)-5-hydroxy-2(5H)-furanone has been shown to contribute significantly to the mutagenic activity of a number of chlorinated drinking water samples, it was present in typical chlorinated drinking waters at concentrations of less than 100 nanograms per litre. This compound has not been evaluated for carcinogenic potential. *See also* mutagen.

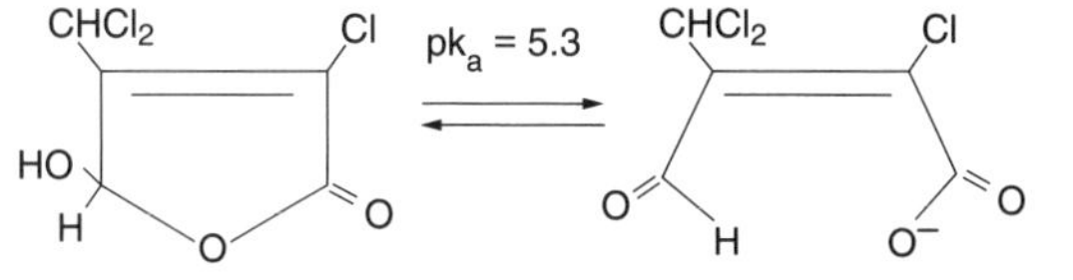

3-chloro-4-(dichloromethyl)-5-hydroxy-2(5H)-furanone (MX) showing the transition between the ring and tantomeric forms

chlorobenzene ($C_6H_{6-x}Cl_x$) A volatile organic compound with various industrial uses (e.g., as a solvent). It is not a disinfection by-product. *See also* volatile organic compound.

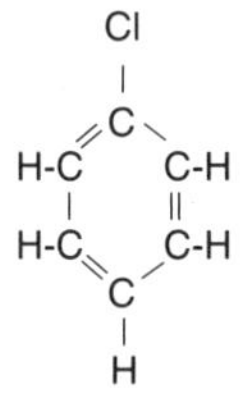

Chlorobenzene

chloroform ($CHCl_3$) The common name for trichloromethane, a trihalomethane containing only chlorine atoms. Typically, this is the principal trihalomethane formed during chlorination, except for in waters containing moderate amounts of bromide. *See also* trihalomethane.

chlorophenol ($C_6Cl_{5-x}H_xOH$, where x is 0 to 4) (1) The common name for chlorohydroxybenzene. The chlorine atom can be in one of three isomeric positions with respect to the hydroxyl group. Chlorophenol is produced during the chlorination of phenol. It may also be formed during the chlorination of natural organic matter. *See also* chlorophenols.

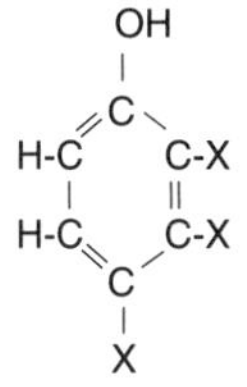

Monochlorophenol, a member of the chlorophenol class of compounds, where one of the Xs is a chlorine atom and the rest are hydrogen atoms

chlorophenolic Pertaining to chlorine-substituted phenolic compounds. *See also* chlorophenols.

chlorophenols A class of disinfection by-products produced during the chlorination of phenol. They may also be formed during the chlorination of natural organic matter. The disinfection by-products include mono-, di-, and trichlorinated species (C_6H_4ClOH, $C_6H_3Cl_2OH$, and $C_6H_2Cl_3OH$, respectively). Pentachlorophenol (C_6Cl_5OH), though, is a synthetic organic compound, not a disinfection by-product. Chlorophenols are more odorous than the unreacted phenol. *See also* disinfection by-product; pentachlorophenol.

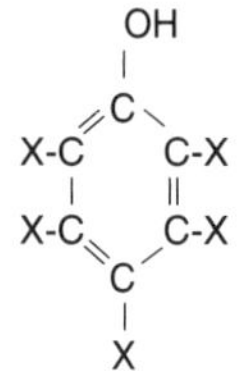

Chlorophenols, where one to five of the Xs are chlorine atoms and the rest are hydrogen atoms

chlorophenoxy herbicides ($(CH_3)_yCl_xH_{5-x-y}ORCOOH$, where y is 0 to 1 and x is 1 to 3, except for erbon and falone) A class of herbicides that may be found in domestic water supplies and that cause adverse health effects. Two widely used chlorophenoxy herbicides are 2,4-dichlorophenoxy acetic acid ($Cl_2C_6H_3OCH_2COOH$) and 2,4,5-trichlorophenoxy propionic acid ($Cl_3C_6H_2OCH(CH_3)COOH$). *See also* 2,4-dichlorophenol; silvex; 2,4,5-trichlorophenoxy propionic acid.

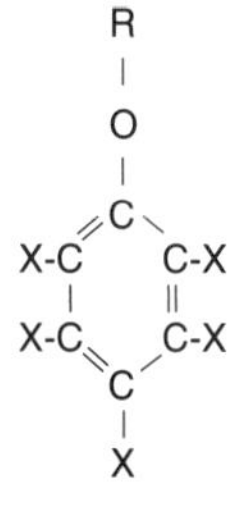

Chlorophenoxy herbicide, where Xs are chlorine and hydrogen atoms and R is an organic acid

chlorophyll a One of a group of special green pigments found in plants, algae, and some bacteria that assist in the process of photosynthesis. In photosynthesis, light is converted into chemical energy that is subsequently used to reduce carbon dioxide (CO_2) to organic compounds in the absence of light. Chlorophyll a is the principal chlorophyll in higher plants. *See also* chlorophyll b.

chlorophyll b One of a group of special green pigments found in plants, algae, and some bacteria that assist in the process of photosynthesis. Chlorophyll a and b are the most common chlorophylls in plants. *See also* chlorophyll a.

chloropicrin (CCl_3NO_2) The common name for trichloronitromethane, a disinfection by-product formed at trace levels during chlorination. Industrially, chloropicrin is used as a pesticide, as a fungicide, and in tear gas. *See also* disinfection by-product; pesticide.

chloroplatinate (Co-Pt) unit *See* color unit *in the Units of Measure list.*

chlororganic Pertaining to chlorine-substituted disinfection by-products and a number of synthetic organic compounds. *See also* disinfection by-product; synthetic organic compound.

chlorosity Chlorinity multiplied by the density of the solution water at 68° Fahrenheit (20° Celsius). *See also* chlorinity; salinity.

cholera *See* waterborne disease.

cholestasis Stoppage of the flow of bile. It can be caused by inhibition of the transport of bile salts by chemicals or some disease processes. It may also be caused by obstruction of the bile ducts (e.g., by formation of stones). In either of these cases it leads to jaundice (yellowing of the skin).

cholesterol ($C_{27}H_{45}OH$) A fat-like steroid alcohol that serves as an integral part of biological membranes and is a precursor for a variety of steroid hormones. Excessive amounts of cholesterol in blood have been associated with a variety of diseases, including atherosclerosis.

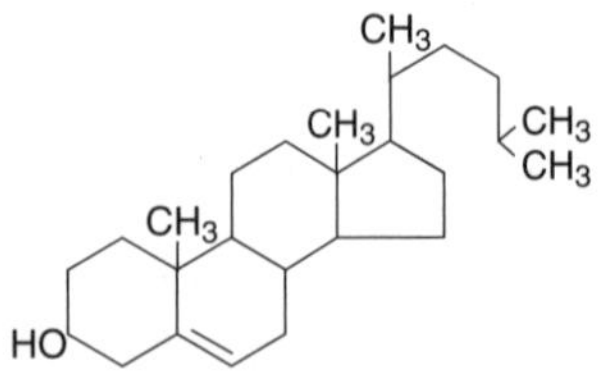

Cholesterol

cholic acid ($C_{23}H_{49}O_3COOH$) An oxidized form of cholesterol conjugated to glycine (NH_2CH_2COOH) or taurine ($NH_2CH_2CH_2SO_3H$) to form one of the more important bile salts.

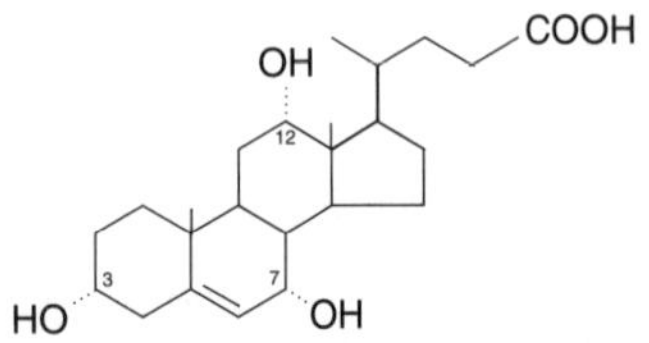

Cholic acid

cholinesterase A hydrolytic enzyme that hydrolyzes esters of choline. The term applies to a wide variety of such enzymes. Toxicologically, the most important of these enzymes is acetylcholinesterase, the enzyme that inactivates acetylcholine in the synaptic clefts formed by nerves that secrete this neurotransmitter. Inhibition of this enzyme is responsible for the acute neurotoxic symptoms associated with poisoning by organophosphorus pesticides. *See also* organophosphorus pesticide.

cholinesterase inhibitors Toxic compounds that act by inhibiting cholinesterases. The most important inhibitors found in the environment are the organophosphorus pesticides. They produce acute neurotoxic effects by virtue of their inhibition of a particular cholinesterase, acetylcholinesterase. *See also* cholinesterase; organophosphorus pesticide.

chromatin A combination of proteins and deoxyribonucleic acid. The proteins are a complex group of basic proteins referred to as histones and are responsible for the staining properties of the chromosomes. Chromatin exists in two basic states, a tightly coiled form and an open form that is actively involved in the transcription of ribonucleic acid used in the synthesis of proteins.

chromatographic behavior Movement of chemical components through a sorption bed in a way that is similar to movement through a chromatograph during an analytic method using chromatographic principles. In granular activated carbon adsorption or in ion exchange, some components of a mixture will have more affinity for the sorbent than others. The more strongly sorbed component will push a less strongly sorbed component out of the sorption bed first, causing a "peak" of concentration of the less strongly sorbed component in the effluent at the time it exits the column. *See also* chromatography.

chromatography A group of separation techniques based on the distribution of analytes between two phases. Separation is based on the difference in an analyte's affinity for the stationary phase relative to that for the mobile phase. Many analyses of organic compounds in water involve such chromatographic techniques as gas chromatography or liquid chromatography.

chromium (Cr) Chromium is a blue-white metallic element (atomic number 24) with various industrial uses, e.g., in alloys and electroplating. It is regulated by the US Environmental Protection Agency.

chromogenic (or fluorogenic) substrate technology Bacteriological testing technology that uses hydrolyzable substrates linked to a chromogen (or fluorogen) molecule. The test is based on the target bacteria having the specific enzyme(s) to hydrolyze the substrate-chromogen or substrate-fluorogen molecules, resulting in the release of a colored compound or a fluorogenic compound. When exposed to ultraviolet light of appropriate wavelength, colored compounds (chromogens) impart a characteristic color to the culture suspension, and fluorogenic compounds (fluorogens) impart a bright, colored fluorescence. Examples of these substrates are: *ortho*-nitrophenyl-β-D-galactopyranoside (ONPG), which, when hydrolyzed by the enzyme β-galactosidase, releases the yellow chromogen *ortho*-nitrophenol; and 4-methylumbelliferyl-β-D-glucuronide (MUG), which, when hydrolyzed by the enzyme β-glucuronidase, releases 4-methylumbelliferyl, which gives a bright bluish fluorescence when exposed to long-wavelength (366 nanometres) ultraviolet light. *See also* *ortho*-nitrophenyl-β-D-galactopyranoside; 4-methylumbelliferyl-β-D-glucuronide.

chromosomal abnormality Damage to chromosomes as indicated by breaks in the chromosome, altered banding patterns (indicating translocations or deletions of a portion of the chromosome), and increased or decreased numbers of chromosomes in a cell.

chromosomal deoxyribonucleic acid Deoxyribonucleic acid that contains the necessary sequences for cell growth, maintenance, and reproduction. Chromosomal deoxyribonucleic acid must be present in a cell for it to survive.

chronic Pertaining to symptoms that continue for a long time as a result of a single exposure or repeated exposures over a long period of time.

chronic exposure Exposure to an agent for an extended period of time. Although frequently referred to in a nonspecific way, this term ordinarily refers to periods of time in excess of 1 year in most animal species.

chronic toxicity A condition in which an adverse effect persists for an extended period of time. The term is more loosely used to describe the effects of a chemical that is administered over a long period of time. However, chronic toxicity can arise from short-term exposures to chemicals if the effect is irreversible.

chronicity index An expression of the potential of a chemical to produce cumulative effects with repeated exposure. The index is calculated as the ratio of the daily dose necessary to produce a chronic toxicity relative to the dose that is required to produce acute toxicity.

chronotropic action Alteration of the rate at which the heart beats as a result of chemical effects.

churn drill A type of drill used in cable-tool drilling.

chute (1) An inclined conduit or structure used for conveying water, other liquids, or granular material at high velocity to lower levels. (2) An inclined drop or fall. (3) The narrow, usually shorter channel around an island in a river. (4) A short, straight channel that bypasses a long bend in a river, formed by the river breaking through a narrow land area between two adjacent bends. *See also* drop.

CI *See* Casil index; cast iron; chemical ionization.

Ci *See* curie *in the Units of Measure section.*

CID *See Cryptosporidium* inactivation device.

ciénega An area where the water table is at or near the surface of the ground. Standing water occurs in depressions in the area, which is covered with grass or sometimes with heavy vegetation. The term is usually applied to areas ranging in size from several hundred square feet to several thousand square feet (several hundred square metres to several thousand square metres). Sometimes springs or small streams originate in the cienega and flow from it for short distances.

CIP *See* clean-in-place system; capital improvement plan.

Cipolletti weir A contracted weir of trapezoidal shape, in which the sides of the notch are given a slope of one part horizontal to four parts vertical to compensate as much as possible for the effect of end contractions.

CIPRA *See* Ductile Iron Pipe Research Association.

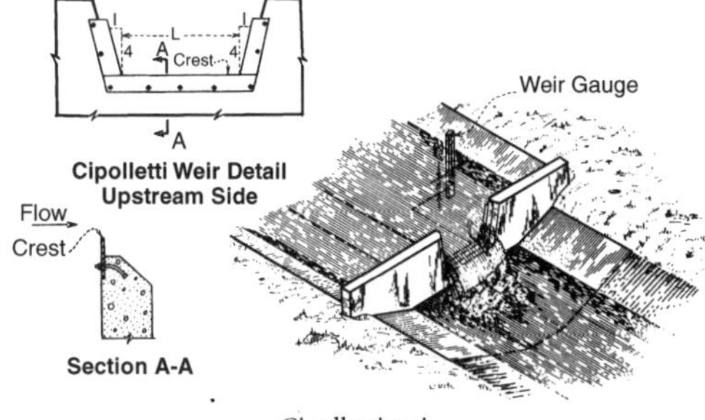

Cipolletti weir

circle of influence The circular outer edge of a depression produced in the water table by the pumping of water from a well. *See also* cone of depression; drawdown; well cone of influence; zone of influence.

circuit (1) The complete path of an electric current, including the generating apparatus or other source. (2) A specific segment or section of that complete path.

circuit breaker A device that functions both as a current overload protective device and as a switch.

circuit rider A person providing on-site technical assistance in the operation of a number of small water systems within a prescribed geographical area.

circular clarifier A unit process device used for removing particles from solution by settling. In a circular clarifier, the influent is most commonly fed to the unit from the center and the flow is directed radially to effluent weirs at the periphery of the clarifier. Collected particles, or sludge, are continuously removed from the bottom of the clarifier with a scraper mechanism.

circulation loop The portion of some types of membrane process systems where some or all of the concentrate stream is continuously being recycled and blended with incoming water to produce the membrane feedwater. This loop is used to increase the crossflow velocity across the membrane surface.

circumference The distance measured around the outside edge of a circle.

circumferential flow The flow of water or other liquid parallel to the circumference or periphery of a circular tank or other circular structure. It is also called peripheral flow.

cirrhosis A disease state in which the normal microscopic lobular structure of the liver is replaced by fibrosis and nodular regeneration within the liver.

CIS *See* customer information system.

cis Indicating a geometric isomer in which two groups attached to the two carbons in a double bond are at adjacent corners of the molecule. *See also* geometric isomers; *trans.*

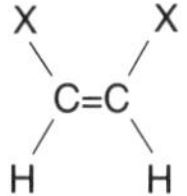

Example of *cis* isomer

***cis*-1,2-dichloroethylene** *See under* dichloroethylene.

CISC *See* complex-instruction-set computer.

cistern A small, covered tank, usually placed underground, in which rainwater is stored for household purposes.

citric acid ($C_3H_4(OH)(COOH)_3 \cdot H_2O$) An acidic chemical, commercially available as colorless crystals or powder, that is highly soluble and used in water treatment for pH adjustment, as a sequestering agent, as a cleaning agent, and for other purposes.

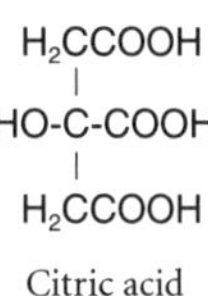

Citric acid

Citrobacter A genus of gram-negative bacteria in the family Enterobacteriaceae. Members of this genus are closely related to the coliform bacteria but either are unable to ferment lactose or do so only very slowly.

***Citrobacter fruendii* (*C. fruendii*)** A species of *Citrobacter* occasionally found in coliform biofilm problems in drinking water distribution systems. *See also Citrobacter.*

city council A municipality's elected legislative body.

civil action Legal action taken against a party that has violated an environmental law but has not committed a criminal offense.

Civil Engineering Research Foundation (CERF) A subunit of the American Society of Civil Engineers that facilitates, coordinates, and integrates research in the civil engineering field.

clamp A thin, circular metal strip with a gasket inside used to repair a leaking pipe by applying compression around the pipe by bolts attached to the metal strip.

Clara cell A nonciliated cell in the small bronchioles of the lung. The function of these cells is not well understood, but they contain most of the lung's capacity for metabolizing foreign chemicals. As a result, they are frequently injured by chemicals that are metabolized to reactive intermediates.

clarification Any process or combination of processes that reduces the amount of suspended matter in water.

clarifier *See* sedimentation basin.

clarity Clearness of liquid, as measured by a variety of methods.

class, pipe *See* pipe class.

classification *See* hydraulic classification; pipe class.

clastogenesis Chromosomal aberrations consisting of breaks in a chromosome or translocation of portions of one chromosome to another. These events are frequently associated with the cause of congenital abnormalities, birth defects, or cancer.

clay (1) Soil consisting of inorganic material for which the grains have diameters smaller than 0.002 millimetres. (2) A mixture of earthy matter formed by the decay of certain minerals.

clay pipe A pipe made of clay and baked in a kiln. Such a pipe is also called a tile. *See also* vitrified clay pipe.

clay soil A soil containing more than 40 percent clay but less than 45 percent sand and less than 40 percent silt.

claypan A stratum or horizon of accumulated stiff, compact, and relatively impervious clay. It is not cemented, but it may interfere with water movement.

Clean Air Act (CAA) The principal US law intended to protect human health and the environment by controlling ambient, or outdoor, air pollution through controlling pollutants at their source. It is codified generally as 42 U.S.C. 7401–7626.

Clean Water Act (CWA) The US law, codified generally as 33 U.S.C. 1251–1387, that establishes a regulatory and enforcement program administered by the US Environmental Protection Agency to control pollutant discharges into US waters.

clean river A river that has no obvious evidence of pollution and from which wholesome drinking water can be obtained by practicable methods of water purification.

clean room A room for a precise operation that requires a contamination-free environment. The room is maintained at a high level of cleanliness by special means (e.g., activated carbon filtered air).

clean water reservoir A holding basin where treated water is stored for delivery to customers.

clean-in-place (CIP) system Facilities used for in situ cleaning. *See also* cleaning-in-place; membrane cleaning.

cleaning The flushing of an item such as a pipe or tank by spraying water or running water through the item with enough velocity to remove debris or contaminants. The initial flush may contain a chlorine solution for disinfection, and a chlorine solution may be held in the facility for a specific time prior to flushing.

cleaning system Tank(s), pump(s), and other equipment used to prepare and recirculate membrane cleaning and, commonly, storage solutions through the membrane modules of a membrane water treatment system.

cleaning wye A wye fitting, usually with an oversized branch, installed on a water main to allow the insertion of a pig for cleaning water mains.

cleanout Any structure or device that is designed to provide access for the purpose of removing deposited or accumulated materials.

cleaning-in-place A chemical cleaning process in which the membranes in a membrane water treatment system (1) are not removed from their housings (pressure vessels) or the system and (2) are cleaned by being exposed to cleaning solution(s), which are commonly recirculated through the cleaning system and membranes. *See also* cleaning system; membrane cleaning.

clear acrylic The generic name for the tradename Plexiglas®.

clear water iron *See* ferrous iron.

clearance The rate at which a chemical is cleared from a body compartment, generally blood. This is a formal pharmacokinetic term. It is expressed in units of millilitres of plasma cleared completely of chemical (or drug) per unit time.

clear-water basin *See* clearwell.

clear-water reservoir *See* clearwell.

clearwell A tank or vessel used for storing treated water. Typical examples of storage needs include (1) finished water storage to prevent the need to vary the rate of filtration with variations in distribution system demand, and (2) backwash water for filters. Clearwells are located on-site at a water treatment plant. A clearwell is also called a filtered-water reservoir. *See also* backwash.

cleat A device to join together the crossbraces and wales in a timber-shoring setup.

client–server computing The performance of tasks by one or more available hardware entities, software entities, or both in a distributed system at the request of another entity.

cliff spring A spring occurring at the base of a cliff, where the water table is intersected by the face of the cliff. *See also* contact spring.

climate (1) The total of meteorological phenomena that combine to characterize the average and extreme conditions of the atmosphere at any specified place on the earth's surface. *See also* weather. (2) The collective state of the atmosphere at a given place or over a given area within a specified period of time; weather averaged on the daily, monthly, seasonal, and yearly basis.

climatic Pertaining to climate.

climatic cycle Actual or supposed recurrences of such weather phenomena as wet and dry years or hot and cold years at more or less regular intervals, in response to long-range terrestrial and solar influences.

climatic province An area characterized by a general similarity of climate throughout.

climatic variation The gradual change in the climate of a given locality occurring over an extended period of time, usually measured in centuries.

climatic year A continuous 12-month period during which a complete annual cycle occurs, arbitrarily selected from the presentation of data relative to hydrologic or meteorologic phenomena. The US Geological Survey uses the period October 1 to September 30 in the publication of its records of stream flow. A climatic year is also called a water year.

climatology The science dealing with the collective state of the atmosphere over a given area within a specified period of time. It is a subdivision of meteorology that deals with the average or normal state of the atmosphere. The distinction between meteorology and climatology is indefinite, and the two terms are often used interchangeably.

cline A graded series of morphological or physiological characteristics exhibited by a natural group, such as a species, along a line of environmental transition (e.g., in terms of climate, soil) or geographic transition.

clinical trial Testing a drug's effectiveness in the treatment of a particular disease state. During this period the drug's side effects are monitored as its effectiveness as a therapeutic agent is monitored. This term has also been applied to studies of the potentially adverse effects of nondrug chemicals in humans. Generally, these types of studies are limited to circumstances that would not be anticipated to produce irreversible damage to the individuals participating in the study. For example, the ability of humans to metabolize a particular chemical might be examined at doses much lower than those that would produce toxicity. *See also* experimental epidemiologic study.

clinoptilolite A naturally occurring zeolite that exchanges ammonium ions (NH_4^+) in preference to sodium, magnesium, and calcium ions. It is sometimes used as an ion exchange resin in water treatment.

CLLE *See* continuous liquid–liquid extraction.

clogging An impeding of flow, typically as a result of particle blockage. Clogging can occur in (1) filters as a result of excessive particle removal prior to backwash or mudball formation or (2) piping as a result of particles or highly viscous liquids. *See also* backwash; mudball.

clonal expansion An increase in the population of cells carrying the same genetic characteristics (i.e., genotype). The term is frequently applied to the division of a cell initiated for cancer into a population of cells containing the same mutation. The expansion of this population dramatically increases the probability that a group of these cells will undergo additional mutations to yield a malignant tumor. Clonal expansion of a group of initiated cells can be induced by a group of carcinogenic chemicals referred to as tumor promoters. *See also* tumor initiator; tumor promoter.

Closed impeller

clone (1) A copy or lookalike computer. Computers have evolved toward universal compatibility, such that most programs are not computer brand specific. (2) A virus or cell descended from a single common ancestor.

cloning The entire process of isolating, identifying, and manipulating genes or cells in recombinant deoxyribonucleic acid research.

close nipple A nipple having no unthreaded length between the two threads. *See also* nipple.

close-coupled pump A pump directly connected to its power unit without any reduction gearing or shafting. *See also* direct-connected pump.

closed basin (1) A basin draining to some depression or pond within its area, from which water is usually lost only by evaporation or percolation. (2) A basin without a surface outlet for precipitation. *See also* sink.

closed centrifugal pump A centrifugal pump having its impeller built with the vanes enclosed within circular disks.

closed conduit Any closed duct for conveying fluids.

closed fire line An unmetered connection to the distribution system to provide water for a fire protection or fire sprinkler system without hose connections or fire hydrants.

closed impeller An impeller having the sidewalls extended from the outer circumference of the suction opening to the vane tips.

closed system A distribution system that has no elevated storage and uses continuously running pumps to provide pressure and meet water usage demands.

closed-loop stripping analysis (CLSA) An extraction technique useful for the isolation of volatile and semivolatile organic compounds, typically in the range of nanograms per litre to micrograms per litre. An inert gas is bubbled into a water sample, and the headspace is continuously recycled through an adsorbent trap. Analytes are desorbed chemically or thermally and injected into a gas chromatographic system. The closed-loop stripping analysis technique is a very sensitive way of determining volatile compounds such as 2-methylisoborneol and geosmin, which are known to cause taste-and-odor problems in drinking water.

closing dike A structure built across the branch channel of a river to stop or reduce the flow entering that channel.

Clostridium A genus of anaerobic, gram-positive, spore-forming bacteria. Many organisms in this genus produce powerful toxins involved in human diseases.

***Clostridium perfringens* (*C. perfringens*)** An anaerobic, spore-forming bacillus that is a normal inhabitant of the human gastrointestinal tract. It has limited use as an indicator of fecal pollution of water. This organism is also important as a cause of gas gangrene in wound infections and food poisoning.

cloth A type of woven filter septum made from natural or synthetic yarns.

cloudburst (1) A rainstorm of extraordinary intensity and relatively short duration that usually occurs over a rather small area. Such storms may result in severe floods. (2) Quantitatively, a rainstorm in which the rate of rainfall is at least 3.94 inches (100 millimetres) per hour (more than 10 times the rate used by the National Weather Service in defining a heavy rainfall). *See also* thermal-convection storm.

cloudiness The visible effect in water caused by material in suspension (e.g., turbidity). *See also* turbidity.

CLSA *See* closed-loop stripping analysis.

clumping The formation of media agglomerations or resin clumps within an operating filter or ion exchange bed caused by organic fouling or electrostatic charges.

cluster Typically a group of closely interconnected computers that are configured so that they share the same operating system and peripheral devices.

CMC *See* critical micelle concentration.

CMF *See* completely mixed flow.

CMOS *See* complementary metal oxide semiconductor.

CNS *See* central nervous system.

cocarcinogenesis The interaction of a chemical, physical, or biological process with other factors that cause cancer to increase the magnitude of a carcinogenic response induced by a second agent. A wide variety of mechanisms can fall into the category of cocarcinogens, ranging from chemicals that are true tumor promoters to chemicals that modify the metabolism of a carcinogen to increase its activity as a carcinogen.

coagulant A chemical added to water that has suspended and colloidal solids to destabilize particles, allowing subsequent floc formation and removal by sedimentation, filtration, or both (e.g., the use of alum or iron salts for removing turbidity in a water treatment process).

coagulant aid A chemical added during coagulation to improve the process by stimulating floc formation or by strengthening the floc so it holds together better. Such a chemical is also called a flocculant aid. *See also* coagulant; flocculant.

coagulate To combine or aggregate small particles into larger masses or clumps by chemical or physical means.

coagulated–settled water Treated water, prior to filtration, to which one or more coagulant chemicals was added to form particles that subsequently settled. *See also* coagulation.

coagulation The process of destabilizing charges on particles in water by adding chemicals (coagulants). Natural particles in water have negative charges that repel other material and thereby keep it in suspension. In coagulation, positively charged chemicals are added to neutralize or destablize these charges and allow the particles to accumulate and be removed by physical processes such as sedimentation or filtration. Commonly used coagulants include aluminum and iron salts and cationic polymers. *See also* aluminum sulfate; cationic polymer; ferric chloride; flocculation.

coagulation basin A basin used for the coagulation of suspended or colloidal matter. In this basin, the liquid is mixed gently to induce agglomeration, with a consequent increase in the settling velocity of particulates.

coal A solid, carbon-based material formed naturally through heat and pressure. Coal is the raw material for anthracite filter media used in filtration and for some activated carbons. *See also* activated carbon; anthracite coal; filtration.

coal tar A black, viscous liquid (or semisolid) obtained by destructive distillation of bituminous coal. Among its uses, coal tar can be applied for waterproofing and pipecoating.

coalescence The union or growing together of colloidal particles into a group or larger unit as a result of molecular attraction on the surfaces of the particles.

coarse rack A rack with relatively wide spaces between bars, usually 1 inch (2.5 centimetres) or more.

coarse sand Sediment particles having diameters between 0.500 and 1.000 millimetre.

coarse screen A mesh or bar screen in which the openings are greater than 1 inch (2.5 centimetres) in least dimension. *See also* coarse rack.

Coastal Zone Management Act (CZMA) The US law, codified generally as 16 U.S.C. 1451–1464, that requires all federal agencies and permittees who conduct activities that affect a state's coastal zone to comply with an approved state coastal zone management program.

coastline The line that separates the land surface and the water surface of the sea or ocean. Strictly speaking, it is a strip of land surface rather than a line and is considered somewhat wider than the shoreline. *See also* shoreline.

coating A material applied to the inside or outside of a pipe, valve, or other fixture to protect it, primarily against corrosion. Coatings may be formed of various materials.

COAU *See* central office access unit.

COBOL (common business-oriented language) A source-code programming language.

cocci Sphere-shaped bacteria.

Coccidia A general name for fungi of the genus *Coccidioides*. *Coccidioides immitis* causes an infection called coccidioidomycosis, also known as valley fever, desert rheumatism, and San Joaquin fever.

coccidian parasite A sporozoan that alternates between asexual and sexual generation and requires only one host. It is a parasite of the intestinal epithelium. *See also* Sporozoa.

coccidioidomycosis An acute, self-limiting respiratory disease, or a progressive chronic infection that disseminates to produce lesions in any organ of the body, caused by the fungus *Coccidioides immitis*.

cock A device for regulating or stopping the flow in a pipe, consisting of a plug that may be rotated in a body that has ports corresponding to those in the plug.

cocktail An organic solution used in the analysis of radioisotopes. For example, in the determination of radon in water, a scintillation cocktail is used to extract radon from the water sample. *See also* scintillation cocktail.

coconut-based activated carbon Activated carbon, used for adsorption, that is produced from coconut shells. Although coconuts are a relatively inexpensive raw material, the use of coconuts for activated carbon may be sensitive to available supply.

cocurrent In the same direction as the water flow.

cocurrent flow *See* downflow.

cocurrent operation The operation of two components of a unit process in a parallel fashion, e.g., when a feed gas is injected into a vessel in the same direction as the fluid flow.

COD *See* chemical oxygen demand.

Code of Federal Regulations (CFR) The United States government publication that codifies the general and permanent rules published in the *Federal Register* by the executive departments and agencies of the federal government.

codisposal of sludge The simultaneous disposal of sludges generated from two unique processes, resulting in a more efficient operation than disposing the sludges individually. For example, water treatment plant sludges can be codisposed with wastewater sludge by discharging the water treatment plant sludge to the sanitary sewer for processing at the wastewater plant. Wastewater sludge, when combined with organically derived wastes such as lawn clippings and leaves, can be codisposed in a composting operation.

codescent debris cone A debris cone composed of two or more debris cones formed by debris contributed by large streams. The lower parts of the cones merge. The space between the apexes of the smaller cones is filled by debris from small intervening streams. The small stream may contribute enough debris to form the apex of a major debris cone.

codon A single frame of three sequential bases in one polynucleotide chain of deoxyribonucleic acid or ribonucleic acid that specifies the insertion of a particular amino acid at a specific location during protein synthesis. The frame for the three bases in each codon is set by an initiation site for transcription and translation.

codon rearrangement Multiple transpositions of codons between two strands of deoxyribonucleic acid. This can result in a different sequence of amino acids being synthesized. If these changes occur in a catalytic or regulatory site in the protein, changes in the functional properties of the protein will occur.

COE *See* US Army Corps of Engineers.

coefficient A numerical quantity determined by experimental or analytical methods and used in a formula that expresses the relationship between two or more variables.

coefficient of determination (r^2) A commonly used statistical measure of the goodness of fit of a linear model. An r^2 value of 1 represents a perfect linear relationship between an independent variable, *X*, and a dependent variable, *Y*. An r^2 value of 0.9 means that the equation explains 90 percent of the variability in the dependent variable. An r^2 value of 0 indicates no linear relationship between *X* and *Y*. The equation for the coefficient of determination is

$$r^2 = \frac{\text{regression sum of squares}}{\text{total sum of squares}}$$

$$= \frac{\sum_{i=1}^{N}\left(\hat{Y}_i - \bar{Y}\right)^2}{\sum_{i=1}^{N}\left(Y_i - \bar{Y}\right)^2}$$

Where:

Y_i = measured value

$\hat{Y}_i$ = predicted value

$\bar{Y}$ = average of measured values

N = number of measured values

Suppose, for example, in water demand forecasting, that the value of r^2 is 0.90 for the regression of residential daily water volume use (the dependent variable) on the independent variables of lot size, normal precipitation, and deviation of temperature from normal temperature. In this case, these three variables explain 90 percent of the variation in residential daily water volume use; the remaining variation (10 percent) is said to be residual or unexplained variation. The objectives are (1) to explain most of the variation with as few logically sound independent variables as possible and (2) to have the unexplained variation be randomly distributed (i.e., without a pattern). The coefficient of determination should not be the only test of adequacy in regression analysis. At minimum, the standard error of the estimate, the mean absolute percentage error, a test for heteroskedasticity, and the standard errors of the coefficients should also be analyzed. *See also* heteroskedasticity; mean absolute percentage error; standard error of the coefficient; standard error of the estimate.

coefficient of fineness The ratio of suspended solids to turbidity. It is a measure of the size of particles causing turbidity; a higher particle size means a higher coefficient of fineness.

coefficient of regime The ratio of the maximum daily flow in 1 year to the minimum daily flow in the same year.

coefficient of viscosity *See* absolute viscosity.

cofferdam A temporary structure built around a construction site to allow the removal of water and to permit free access to the area within. It may take various forms, such as an earth embankment, a single row of steel or timber sheet piling, or a double row of sheet piling with the space between filled with impermeable material.

cogeneration The use of an energy source for multiple purposes. For example, when natural gas is used to produce steam to drive turbines and produce electricity, the waste heat from the process can be used in other energy applications.

COGO *See* coordinate geometry.

Cohen censored maximum likelihood (CCML) method A statistical procedure to determine the mean and variance of a normal distribution when the data set is either left or right censored, i.e., biased toward left or right.

cohesion The force of molecular attraction that exists between the particles of any substance and tends to hold them together.

cohesive soil Clay or soil with a high clay content, which provides cohesive strength. Cohesive soil does not crumble; it is plastic when moist, hard to break when dry.

cohort A designated group of persons who are assembled based on a common characteristic (e.g., age, period of birth, exposure) and followed or traced over a period of time. In epidemiology, the cohort is traced to determine a health-related outcome, morbidity, or death.

cohort epidemiologic study A type of epidemiologic study in which individual subjects are selected on the basis of their exposure status (e.g., chlorinated or unchlorinated water;

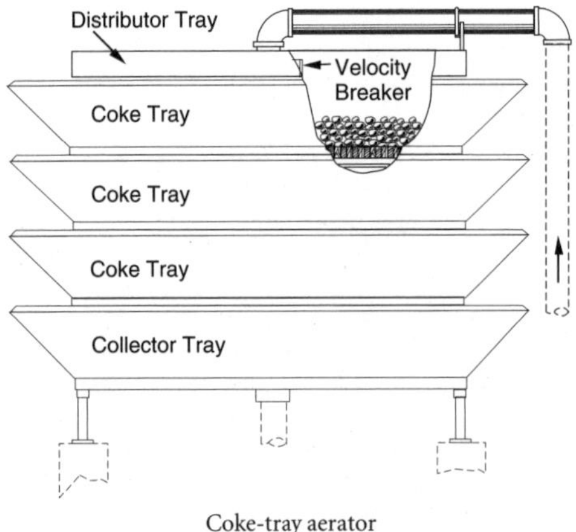

Coke-tray aerator

high, moderate, low, or no levels of radon in the home) without knowledge of their disease status. These studies are also called follow-up studies. Any number of health-related outcomes or diseases can be ascertained in a cohort study. Morbidity or mortality rates are determined for each group in the cohort and compared (in terms of rate ratio, rate difference) among the various exposure categories. There are two types of cohort studies: prospective and retrospective. In a prospective cohort study, the disease has not occurred at the time the exposed and unexposed groups are defined by the investigator. In a retrospective cohort study, the disease has occurred at the time these groups are defined. *See also* morbidity rate; mortality rate.

co-ion An ion of similar charge to another.

coke-tray aerator An aerator in which the water is passed or sprayed through coke-filled trays.

cold spring (1) A nonthermal spring in which the water has a temperature appreciably lower than the mean annual temperature of the atmosphere in the vicinity of the spring. (2) In areas of thermal springs, any nonthermal spring.

cold sterilization The use of submicrometre-pore-diameter filtration to screen out bacteria from a water or fluid.

cold weather operation Modification of operating procedures to account for cold weather conditions. For example, chemical dosages and settling rates in conventional water treatment will be affected by colder weather. Higher coagulant dosages may be required because of slower reaction kinetics, and settling performance may decrease as a result of lower differential density between particles and water (because the density of water decreases with temperature above 39° Fahrenheit (4° Celsius) as does the viscosity). *See also* conventional water treatment.

coli-aerogenes bacteria A group of bacteria predominantly inhabiting the intestines of humans or warm-blooded animals but also occasionally found elsewhere. *See also* coliform group bacteria.

coliform bacteria Bacteria of the family Enterobacteriaceae, commonly found in the intestinal tracts of warm-blooded animals. In sanitary bacteriology, these organisms are defined as all aerobic and facultative anaerobic, gram-negative, nonspore-forming, rod-shaped bacteria that ferment lactose with gas and acid formation within 48 hours at 95° Fahrenheit (35° Celsius). *See also* coliform group bacteria; *Escherichia coli*; fecal coliform.

coliform group bacteri A group of related bacteria comprised of the genera *Escherichia*, *Enterobacter*, and *Klebsiella*, commonly associated with the intestinal tracts of warm-blooded animals and shed in their fecal material.

coliform growth response (CGR) A bioassay method developed to assess the potential for a potable water to support the growth of coliform bacteria. The bioassay organism used is generally a strain of *Enterobacter cloacae*.

coliform regrowth potential *See* coliform growth response.

Coliform Rule *See* Total Coliform Rule.

coliphage A bacterial virus, also known as a bacteriophage, that uses *Escherichia coli* as its host cell.

collagen A proteinaceous material that makes up the basic structure of most tissues and organs. Basement membranes and various supportive tissue (e.g., tendons) are composed largely of collagen, which is characterized by a high proportion of its structure being made up of proline residues.

collapse pulsing backwash A method of cleaning a filter by introducing air and subcritical fluidization backwash water, causing the formation and collapse of air pockets within the bed, a condition called collapse pulsing. This condition enhances the abrasion between the media grains for removing retained particles. *See also* subcritical fluidization backwash.

collar (1) A projection secured to the outer surface of a hydraulic structure to increase the resistance to seepage or flow of water along the outside of the structure. It also acts as a rigid anchor when it is embedded in concrete. (2) A weight of concrete around the joint between vitrified-tile pipes and cast-iron pipes, where the socket cannot be forced into the bell, to provide a watertight joint. (3) A short sleeve or cylinder of any material, with diameter sufficiently large to permit two pipes of lesser diameter to be connected by sliding their ends into opposite ends of the collar and sealing with a watertight substance such as a rubber ring, cement mortar, or bituminous compounds.

collateral material (1) Multiple public relations tools developed for simultaneous or serial use to meet one or more particular objective(s). (2) Materials that deal with related issues.

collection lysimeter A device that collects water samples from the unsaturated zone using mild negative pressures. It is also known as a suction lysimeter. It should not be confused with a lysimeter cell that contains soil and vegetation and is used for measuring evapotranspiration.

collective bargaining Negotiation between organized (unionized) workers and their employers for coming to an agreement (contract) on benefits, wages, and working conditions.

collector A device or system designed to collect backwash water from a filter or ion exchange bed. A collector may also be used as an upper distributor to spread the flow of water in downflow column operation. *See also* distributor.

collector well A well, usually located near a river, with a vertical vault that connects a series of horizontal shafts (the collectors) arranged in a radial fashion. The groundwater enters the shafts and flows to the vault, where it is collected. Collector wells can produce enormous quantities of water under ideal conditions when the adjacent river is running at high stages.

College of American Pathologists (CAP) grade water Water that meets standards established by the College of American Pathologists. The standards cover three types of laboratory grade water: clinical, cell or tissue, and cultural.

collimated light Electromagnetic radiation for which the waves have been rendered parallel. In a spectrometer, a mirror may serve to collimate radiation prior to dispersion in a prism. *See also* spectrometer.

collision efficiency factor A measure of the efficiency of the collision of destabilized particles in forming larger particles in the flocculation process. After particles become destabilized in the coagulation process, they can be brought together to agglomerate into larger particles via flocculation. *See also* flocculation.

colloid A small, discrete solid particle in water that is suspended (not dissolved) and will not settle by gravity because of molecular bombardment.

colloidal matter Finely divided solids that will not settle but may be removed by coagulation or membrane filtration. *See also* colloid.

colloidal solid A finely divided solid that will not settle out of water for very long periods of time unless the coagulation–flocculation process is used. *See also* colloid.

colluvial Consisting in part of alluvium and in part of angular fragments of the original rocks. Colluvial soil is formed in regions of precipitous topography and is made up of fragments of rocks detached from the slopes above. Talus slopes, cliff debris, and other heterogeneous rock detritus are examples of colluvial soil.

colluvial deposit A heterogeneous deposit of rock waste, such as talus, cliff, and avalanche accumulations, that results from the transporting action of gravity. *See also* colluvial.

colluvial soil Soil formed when running water or wind transports sedentary soil from one location and deposits it in another, where it intermingles with other sedentary soil. Such soil forms a large portion of the soil found on rolling and hilly upland areas.

colluviarium An access opening in an aqueduct, used for maintenance and ventilation.

colmatage In a pressure-driven membrane treatment system, the portion of water flux decline that is reversible by chemical cleaning or hydrodynamic means.

colon bacilli Bacteria residing normally in the intestinal tract. These bacteria are not necessarily dangerous if present in drinking water, but they are indicators of other possible pathogens. *See also* coliform bacteria.

colony (1) A localized population of individuals, animals, or plants. (2) A cluster of microorganisms in or on a nutrient medium that was derived from a single microorganism.

colony counter An instrument used to count bacterial colonies for the standard plate count test.

colony-forming unit (cfu) *See the Units of Measure list.*

color A physical characteristic describing the appearance of water (different from turbidity, which is the cloudiness of water). Color is frequently caused by fulvic and humic acids.

color body Those complex molecules that impart color (usually undesirable) to a solution.

color comparator A device used for tests such as chlorine residual or pH tests. Personnel can determine concentrations of constituents by visually comparing a permanent standard (usually sealed in glass or plastic) and a water sample.

color removal The process of removing or reducing color in water. Color is undesirable aesthetically and is related to the abundance of natural organic matter in a water supply. Color removal can be achieved through physical–chemical processes such as coagulation, adsorption processes such as granular activated carbon or powdered activated carbon adsorption, or oxidation processes such as chlorination or ozonation. While color is removed, the natural organic matter causing the color is only transformed, not typically removed, by oxidation processes.

color throw The discharge of color into the effluent of an ion exchange or filter media system by any component. This usually occurs after an extended standing period that allows slowly soluble colored matter to accumulate in the water. A color throw may result from the leaching of color bodies from an ion exchange resin into the water.

color unit (cu) *See the Units of Measure list.*

colorimeter A device used for absorption analysis. In some cases the human eye can serve as the detector in visual methods, such as those in which Nessler tubes are used. Another technique involves a filter photometer, an instrument that uses filters to produce a wide band of wavelengths suitable to the particular determination. *See also* absorption analysis.

colorimetric analysis The process of forming a group of absorption analyses in which a fairly broad band of radiation, usually in the visible wavelengths, is used. These techniques find wide application in water analysis. For example, the determination of chlorine residual by the N,N-diethyl-p-phenylenediamine method uses a visual comparison of samples and standards. Nessler tubes and photometers are also commonly used in colorimetric analyses. Colorimetric analysis is also called colorimetric measurement or the colorimetric method. *See also* absorption analysis.

colorimetry The measurement of color naturally present in samples or developed in them by the addition of reagents. The color is measured in terms of the absorption of a particular wavelength of light.

column A vessel, usually a cylindrical and vertical tank, with an inlet at one end and an outlet at the other end, with some means of holding the medium in place so that a stream of water passing through it is processed.

column operation A process in which a solution to be treated is passed through a column in either an upflow or downflow pattern.

COM port *See* communications port.

combination bond A government-issued bond that is payable from the revenues of a governmental enterprise but is also backed by the full faith and credit of the government.

combination starter An electric motor controller with a manual disconnect, fuses, a circuit breaker, or an electromagnetic contactor. It may also include an overload relay, phase failure, and a lightning arrestor.

combination well A well system consisting of an open well and one or more wells or infiltration galleries that are connected to it.

combined available chlorine The total chlorine, present as chloramine or other derivatives, that is present in a water and still available for disinfection and for oxidation of organic matter. The combined chlorine compounds are more stable than free chlorine forms, but they are somewhat slower in action. *See also* chloramines.

combined available residual chlorine The concentration of residual chlorine that is combined with ammonia (NH_3), organic nitrogen, or both in water as a chloramine (or other chloro derivative) yet is still available to oxidize organic matter and help kill bacteria. *See also* chloramines.

combined chlorine The sum of the species composed of free chlorine and ammonia (NH_3), including monochloramine (NH_2Cl), dichloramine ($NHCl_2$), and trichloramine (nitrogen trichloride, $NHCl_3$). Dichloramine is the strongest disinfectant of these species, but it has less oxidative capacity than free chlorine. Di- and trichloramine can result in tastes and odors. *See also* chloramines.

combined moisture Moisture in combination with organic or inorganic matter.

combined residual A compound of an additive (such as chlorine) that has combined with something else and that remains in the water. Chloramines, where the chlorine has combined with ammonia, are combined residuals. *See also* chloramines.

combined residual chlorination The application of chlorine to water to produce combined available residual chlorine. This residual can be made up of monochloramine (NH_2Cl), dichloramine ($NHCl_2$), and nitrogen trichloride ($NHCl_3$). *See also* chloramines.

combined water Soil water that is held in chemical combination and remains after hygroscopic water evaporates. It will not evaporate and is driven off only by heating.

combining weight The formula weight of a dissolved species divided by its electrical charge.

combustible Pertaining to materials that have the ability to catch fire and burn.

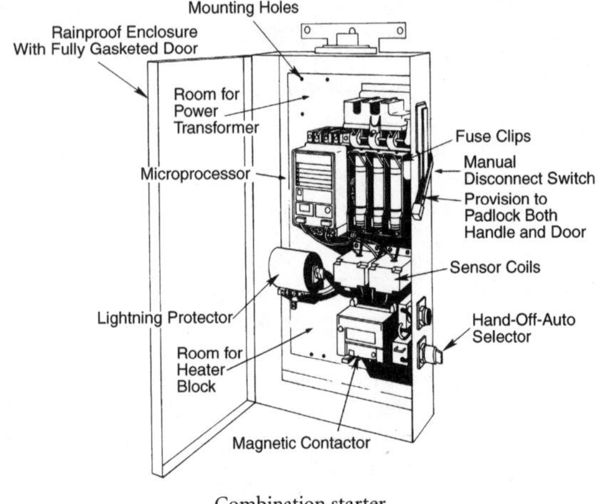

Combination starter

commercial water conservation Wise water use by business customers.

commercial water use The use of potable water by business customers (as opposed to residential, industrial, or public customers) such as stores, motels, shopping centers, gas stations, laundries, offices, and services.

commodity cost Operations and maintenance costs that vary with the quantity of water produced, such as purchased water costs, chemicals, direct power costs, and storage or other costs that vary with average daily demand.

commodity demand approach to rate structure design A rate structure design method in which the costs of service are assigned to four cost components: commodity costs, demand costs, customer costs, and direct fire protection costs. Commodity costs include the operations and maintenance and capital costs that vary with the quantity of water produced, such as purchased water costs, chemicals, direct power costs, and storage or other costs that vary with average daily demand. Demand costs are associated with the facilities that meet peak rates of use, or demands, placed on the water system (e.g., maximum-day or maximum-hour demands). Customer costs are those that relate to serving customers irrespective of the amount or rate of water use. Direct fire protection costs are those incurred solely for the purpose of fire protection. This rate is also called the commodity demand rate. *See also* base–extra-capacity approach to rate structure design.

common business-oriented language *See* COBOL.

common law That body of law developed in England prior to the establishment of the United States. It refers principally to rights and privileges and, though generally followed in the United States, has in some of its applications been abrogated or modified, as in the case of riparian rights to water in some jurisdictions in the United States.

common law doctrine A legal doctrine that arose out of the English system of law.

common logarithm The exponent in the representation of a number as a power of 10. For example, the common logarithm of 100 is 2, because 10 raised to the power of 2 equals 100.

common salt (NaCl) Sodium chloride, a white or colorless crystalline compound that occurs abundantly in nature (present as 2.6 percent of seawater) and in animal fluids. Sodium chloride is used in water treatment to regenerate cation exchange water softeners and some dealkalizer systems. It is also called table salt or common table salt. *See also* salt.

common source epidemic An outbreak of illness in a group of persons that is caused by a common exposure. When the exposure is brief, cases develop within one incubation period of the disease.

common-ion effect The decrease in the solubility of a salt dissolved in water caused by the presence of ions of the salt already contained in the water.

communicable disease An illness caused by a specific infectious agent or its toxic products and transmitted from an infected person, animal, or the environment (e.g., through water, food, fomites) to a susceptible host. Transmission can be direct or indirect.

communications port (COM port) A computer communications channel that uses the serial transmission of data from one device or computer to another. Most computers use as many as four communications ports.

Community Right to Know A provision of Title III of Superfund Amendments and Reauthorization, requiring companies that keep hazardous materials on-site in sufficient quantities to report to local officials, such as fire departments or local emergency planning agencies.

Community Water Supply Study (CWSS) A comprehensive survey of 969 US water supplies conducted by the US Public Health Service (USPHS) to determine whether the US consumer's drinking water met the 1962 USPHS standards. The study was initiated in 1969 and results were released in 1970 in two reports: (1) *Community Water Supply Study: Analysis of National Survey Findings.* 1970. NTIS PB214982. Springfield, Va.: USPHS; and (2) *Community Water Supply Study: Significance of National Findings.* 1970. NTIS PB215198/BE. Springfield, Va.: USPHS.

community Any naturally occurring group of organisms occupying a common environment.

community system *See* community water system.

community water system (CWS) A public water system serving at least 15 service connections used by year-round residents or regularly serving at least 25 year-round residents. *See also* public water system.

compact A formal agreement between states relating to the use of water in a stream or river that flows across state boundaries.

compact call The requirement that an upstream state stop or reduce diversions of water from a river system that is the subject of the compact, in order to satisfy the downstream state's compact entitlements.

compact disk (CD) A computer peripheral device used to optically store large amounts of digitized text, data, images, and sound. One disk typically holds in excess of 600 megabytes of information.

compact disk, read-only memory (CD-ROM) A specialized computer peripheral device used to store large amounts of data in a read-only medium. CD-ROM disks store digitized text, data, images, and sound.

compaction For a pressure-driven membrane process, a physical reduction in membrane thickness, deformation of the membrane, or both. Compaction is dependent on pressure, temperature, and operating time and is caused by transmembrane pressure resulting in decreased membrane permeability. It is typically associated with some irreversible water flux decline.

compartment A chamber or other enclosed or partially enclosed division of a structure, generally underground, that may or may not be windowless. The term is usually applied to a space that is a subdivision of another space and is used for a specialized purpose in connection with some other space or chamber. The term is usually applied to smaller spaces than is the term *chamber* and to larger spaces than *access hole* and *vault.*

compartmental model A type of pharmacokinetic model that treats the body as abstract compartments that are expressed as volumes. The sizes of compartments are estimated based on concentration measures in some body fluid (e.g., blood or urine). The sizes of various compartments are also estimated on the basis of the behavior of the chemical in the sampled fluid over time. The compartments do not actually correspond to specific body compartments.

compensated hardness A calculated value based on the total hardness, the magnesium-to-calcium ratio, and the sodium, iron, and manganese concentrations in a water. This value is used to correct for the reduction in hardness removal capacity of a cation exchange water softener as caused by the presence of these substances. No single method of calculation has been uniformly accurate.

compensation ratio A value that establishes the types and levels of allowable trades of bank currency for permitted wetland impacts.

competent person An employee who possesses the following: (1) the ability to identify existing and predictable hazards in the surroundings, (2) the ability to identify working conditions that are unsanitary, hazardous, or dangerous to employees, and (3) authorization to take prompt corrective action to eliminate these problems.

competition (1) Rivalry in business for markets and customers in an effort to increase profits. Economists generally regard price competition as the only true form of competition. (2) The struggle among individual organisms for water, food, space, and so on when the available supply is limited.

competitive adsorption A phenomenon in which the adsorption of a target compound is hindered by the presence of another compound. These latter, or competing, compounds either are preferentially adsorbed on the medium or, in the case of granular activated carbon, may also occupy the pore space through which the target compound must pass to be adsorbed. These conditions prevent the target compound from being effectively removed from the liquid phase. *See also* fouling; granular activated carbon.

competitive index A listing of costs incurred by other organizations performing like functions to determine if those functions are being accomplished in-house at a comparable price.

compiler A computer program that translates a source code language into machine code, which is the binary representation of the program, for execution by a computer.

complaint investigation A professionally conducted investigation of a customer's water quality complaint.

complement A complex set of proteins that interacts with the antigen–antibody complex and attaches to cells and causes lysis.

complementary metal oxide semiconductor (CMOS) A semiconductor device that, by design, has a high noise and voltage tolerance and uses very little power. This advanced design is currently used in most microprocessors.

complete diversion The taking or removing of water from one location in a natural drainage area and the discharge of it into another drainage.

complete treatment A method of treating water that consists of the addition of coagulant chemicals, flash mixing, coagulation–flocculation, sedimentation, filtration, and disinfection. This approach is also called conventional filtration or conventional water treatment.

complete uncertainty analysis The process of using ranges and distributions for the diverse inputs in a risk assessment analysis and incorporating these variations into the final result, either analytically or by using various computer methods, such as Monte Carlo simulation. In a risk assessment, a number of inputs exist, such as water consumption, contaminant concentration, potency of the contaminant, and risk mediators (e.g., smoking, alcohol, diet, age). As with a sensitivity analysis for a model, obtaining a measure of the degree to which different values of the inputs will impact the computed population or individual risk of an adverse event (illness, death, and so forth) is useful in risk assessment. This overall process is the essence of performing an uncertainty analysis. In recent applications, the various input variables have been placed into two categories: those possessing variability and those possessing uncertainty. Variability characterizes those input variables that are intrinsic properties of the population and for which no amount of effort will lessen the variability (e.g., water consumption, body weight, distribution of contaminant concentration). Uncertainty characterizes those input variables that, theoretically at least, can be obtained more precisely with greater effort (such as potency of the contaminant). Unfortunately, components of uncertainty exist within variability input variables (and probably vice versa). For example, in a given situation, because only a finite sample of persons are consuming water, only an estimate of the true water consumption distribution is available. *See also* health effects risk analysis; Monte Carlo analysis; sensitivity analysis.

completed test The third major step of the multiple-tube fermentation test. It confirms that positive results from the presumptive test are caused by coliform bacteria. *See also* confirmed test; presumptive test.

completely mixed flow (CMF) A liquid stream in which a heterogeneous mixture of constituents exists throughout the stream.

complex power (*S*) A complex quantity, customarily measured in volt-amperes, that has the active power as the real component and the reactive power as the imaginary component:

$$S = P + \sqrt{-1}$$

Where:

P = the active power, in volt-amperes

Q = the reactive power, in volt-amperes

complexation The inactivation of an ion by addition of a reagent that combines with it and, in effect, prevents it from participating in other reactions. Complexation is also called sequestration.

complexes Compounds formed by the union of two or more simple salts.

complex-instruction-set computer (CISC) A computer containing a processor designed to sequentially run variable-length instructions, many of which require several clock cycles. *See also* reduced-instruction-set computer.

compliance The act of meeting the requirements of a national primary drinking water regulation or other regulation.

compliance cycle The 9-year calendar year cycle during which public water systems must monitor. Each compliance cycle consists of three 3-year compliance periods. The first compliance cycle began Jan. 1, 1993, and ends Dec. 31, 2001; the second begins Jan. 1, 2002, and ends Dec. 31, 2010; the third begins Jan. 1, 2011, and ends Dec. 31, 2019, and so on. *See also* compliance period.

compliance monitoring Monitoring required by the US Environmental Protection Agency for the purpose of demonstrating compliance with a regulation.

compliance period A 3-year calendar period within a compliance cycle. Each compliance cycle has three 3-year compliance periods. Within the first compliance cycle, the first compliance period began Jan. 1, 1993, and ended Dec. 31, 1995; the second began Jan. 1, 1996, and ended Dec. 31, 1998; the third began Jan. 1, 1999, and ends Dec. 31, 2001. *See also* compliance cycle.

compliance safety health officer (CSHO) A field representative of the Occupational Safety and Heath Administration who reviews the types of conditions in the workplace likely to be encountered by employees, including work processes, equipment, and machinery involved and the hazards associated with them.

compliance support program An American Water Works Association program designed to provide guidance to its local sections on providing assistance to small systems in order to improve compliance with the drinking water regulations.

composite (1) The act of adding one sample to one or more samples for the purpose of obtaining a final sample that is representative of a certain period of time. (2) A type of sample to which more than one subsample has been added. The purpose of a composite is to obtain a sample that is representative of a period of time or series of subsamples.

Composite Correction Program (CCP) A two-stage program for evaluating the performance of a water treatment plant. The first stage is a comprehensive performance evaluation, which can be followed by comprehensive technical assistance. The Composite Correction Program was developed for the US Environmental Protection Agency to improve performance at wastewater treatment plants. It has been applied to drinking water treatment plants to determine whether treatment performance can be optimized without major capital improvement.

composite membrane A semipermeable membrane used in water treatment, consisting of a rejecting barrier layer of one chemical composition (usually a type of polymer) supported by one or more layers of porous materials with different composition(s).

composite rate of depreciation A single depreciation rate to be applied against the total depreciable assets to determine the depreciation for an accounting period. The use of a composite rate is based on the assumption that the depreciation expense is the same as would be obtained by applying proper individual rates of depreciation against the cost of each unit and totaling the individual amounts.

composite sample A type of sample to which more than one subsample has been added in order to obtain a sample that is representative of a period of time or series of subsamples.

composite sampling The taking of discrete individual samples, which are then combined to represent the integrated composition of the sample source. *See also* automatic sampling; depth-integrated sample.

composite unit hydrograph A unit hydrograph for a large area, constructed from unit hydrographs for the subareas that compose the large area. The composite unit hydrograph is the sum of the individual unit hydrograph discharge values at time values that have been shifted to reflect the time of travel from the outlets of the subareas to the outlet of the large area.

compositing The process of combining discrete individual water samples prior to performing a water quality analysis. *See also* composite sampling.

compost The product of thermophilic biological oxidation of organic materials.

composting The combining of biologically active sludge, such as wastewater sludge, with "green" wastes, such as lawn clippings or leaves, to lessen sludge volume by biological activity and generate a stable, reusable organic solid additive. *See also* windrow sludge composting.

compound A substance composed of at least two elements in specific proportions.

compound hydrograph The hydrograph of an intermittent storm when the flow resulting from one substorm continues during the next substorm.

compound loop control The use of both feedforward and feedback control loops for the same final control element.

compound meter A water-measuring device that has both a small meter and a large meter and alternative flow paths. Such a meter is used where flow demands vary or fluctuate beyond the range of one type of meter, and the flow is diverted to the alternate meter when it varies outside the range of either meter.

compound microscope A microscope with two or more lenses.

compound pipe (1) A pipeline made up of two or more pipes of different diameters. (2) Two or more pipes of the same or different diameters connected in parallel pipelines to form a loop.

compound tube A tube made up of several shorter tubes of various diameters, of various longitudinal cross sections, or both.

comprehensive coliform monitoring plan A plan required by the Total Coliform Rule, developed by a public water system, and approved by the state primacy agency that specifies the sampling locations and monitoring protocol to be used to comply with the monitoring requirements of the Total Coliform Rule.

Comprehensive Environmental Response, Compensation, and Liability Act (CERCLA) The US law, codified generally as 42 U.S.C. 9601–9675, that establishes a regulatory and enforcement program administered by the US Environmental Protection Agency to clean up abandoned hazardous waste sites. It is also known as "Superfund."

comprehensive performance evaluation (CPE) A systematic review and analysis of a water treatment plant's performance without major capital improvements. It is the first part of a two-stage composite correction program, which was developed for the US Environmental Protection Agency to improve performance at wastewater treatment plants. It has been applied to drinking water treatment plants in conjunction with the Surface Water Treatment Rule.

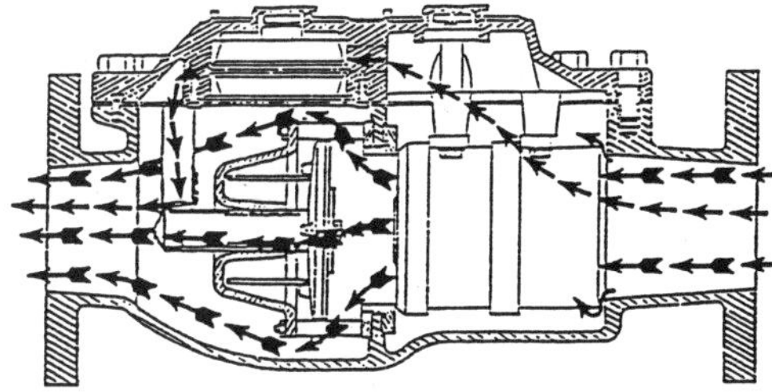

Compound meter at full flow. The bulk of the water passes through a turbine element, with a minimal flow through a disk element.

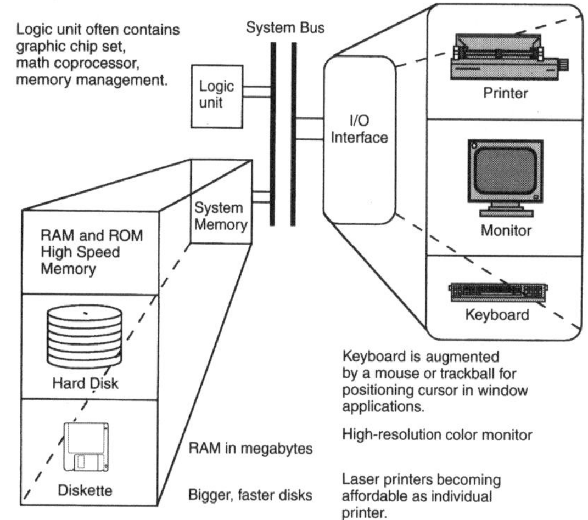

Typical personal computer

comprehensive plan A method for formulating beforehand all relevant details—including schedule, methods of proceeding, finances, and so on—for creating or improving a system or physical plant to provide a specific service.

comprehensive planning A process that results in widespread community planning.

comprehensive technical assistance (CTA) The second step in what is known as the Composite Correction Program. CTA involves correcting the problems identified during the comprehensive performance evaluation that limit treatment plant performance. The major purpose of the CTA is to optimize the performance of existing facilities and to train plant staff and administrators to maintain long-term, optimal plant performance. *See also* comprehensive performance evaluation.

compressibility The degree of physical change in suspended solids or filter cake when subjected to pressure.

compression settling The settling of particles at very high concentrations, as would occur near the bottom of a settling basin. The particles actually touch each other, and settling can occur only by the compression of the compacting mass. *See also* type IV settling.

compression-type hydrant A hydrant that opens against the flow of water by the movement of the operating stem and in which the water pressure tends to keep the main valve closed.

comptroller A person in charge of finances or expenditures in a business or institution.

computer A machine that is capable of executing a set of instructions of data stored in digital form.

computer mapping *See* automated mapping system.

computer model A means of defining a system using computer programs.

computer program An arrangement of instructions for computational operation to achieve a set of desired results.

computer-aided design/computer-aided manufacturing (CAD/CAM) The use of computers in the design and control of manufacturing processes.

computer-aided design and drafting (CADD) Software that supports civil, mechanical, and other engineering design activities, including interactive graphics display, engineering calculation and analysis, and limited attribute processing. In addition, computer-aided design and drafting systems enable some modeling of graphic relationships and analysis of logical relationships.

computer-aided dispatch (CAD) A group of automated tools used by organizations to identify the physical location of a query for service or emergency assistance and to assist in identifying the most appropriate vehicle and route of response.

computer-aided drafting (CAD) The use of computers to automate drafting functions, storing drawing elements as graphic features with *x*, *y*, *z* coordinates in digital form. Computer-aided drafting allows data to be entered and modified through digitizing or manipulation of a screen cursor via a puck or mouse. Very limited attribute data or analytic capabilities are typically included.

computer-aided modeling system A complex system of simulation modeling that is enhanced by the application of pre- and postprocessors to aid the experienced or inexperienced user in setting up and running the model. The user must have great expertise in the topic for which the model is designed.

computer-aided software engineering (CASE) A set of programs and techniques designed to improve programmer productivity.

computer-assisted manufacturing (CAM) The use of computers in the control of manufacturing processes.

computer-assisted mapping (CAM) The use of computers in the creation and production of maps. Map features are stored in digital form as elements with coordinate values.

computer-assisted mass appraisal (CAMA) A computer-aided analysis of data describing property characteristics, used in establishing property values for tax assessment.

computer-based modeling The simulation of an activity or condition through the use of equations developed from a collection of data.

concave stream bank A stream bank that has its center of curvature toward the channel.

concentrate The concentrated solution containing constituents removed or separated from the feedwater by a membrane water treatment system. It is commonly in the form of a continuous flow stream. Concentrate is also called reject, brine, retentate, or blowdown, depending on the specific membrane process.

concentrate disposal The process, method, or facilities used to dispose of the concentrate generated by a membrane treatment system.

concentrate recycle A technique for increasing the amount of product water from reverse osmosis and electrodialysis systems by recycling a fraction of the concentrate stream back through the membrane or membrane stack.

concentrate staging For a pressure-driven membrane process, the membrane module arrangement such that the concentrate from one stage is further processed in a following stage, nearly always without an additional pumping step prior to the following stage. *See also* array; hydraulic staging.

concentrate stream (1) For pressure-driven membrane treatment systems, the stream into which rejected materials are concentrated, including ions in the case of reverse osmosis and nanofiltration. (2) For electrical-driven membrane treatment systems (e.g., electrodialysis), the stream in a membrane stack into which ions are transferred and concentrated.

concentrated fall The fall that is concentrated at one point on the stream in a hydroelectric power development, e.g., the fall a dam uses to generate power.

concentrated solution A solution that contains a relatively high quantity of the solute.

concentration A means of expressing the strength of a solution. A common way to express results of water analyses is by weight of a solute per unit volume of the solution, e.g., milligrams per litre or micrograms per litre. Molarity (moles per litre) or normality (equivalent weights per litre) are other ways of expressing solute concentration.

concentration cell A corrosion cell involving two identical electrode materials, with corrosion resulting from mechanical, physical, or chemical differences of the environments adjacent to the two electrodes.

concentration cell corrosion A form of localized corrosion that can form deep pits and tubercles.

concentration factor (1) An expression of the enhancement in concentration that is possible (or has occurred) during sample preparation. It can refer to the enhancement of either the solute or solvent. In the case of solutes, the concentration factor is the ratio of concentration in the final extract to that in the original sample. (2) In a membrane treatment process, the concentrate solute concentration divided by the feed solute concentration.

concentration polarization (1) A corrosion phenomenon in which a transport step for movement of the dissolved species to or from the surface sites is limiting, thus controlling the rate of corrosion. The polarization, or difference in potential, in a corroding metal system is related to concentration differences between the surface and the bulk solution. (2) In a membrane treatment process, the phenomenon in which retained solutes accumulate at the membrane surface in concentrations greater than the bulk stream. *See also* boundary layer.

concentration tank A settling tank of relatively short detention period in which sludge is concentrated by sedimentation or flotation before treatment, dewatering, or disposal. *See also* sedimentation basin.

concentration time (1) The period of time required for storm runoff to flow from the most remote point of a catchment or drainage area to the outlet or point under consideration. It is not a constant, but rather varies with the depth of flow and the condition of the channel. (2) The time at which the rate of runoff equals the rate of rainfall of a storm of uniform intensity.

concentration units The amount of a substance per unit volume: in the US customary system, milligrams per litre; in Système International, moles per cubic metre.

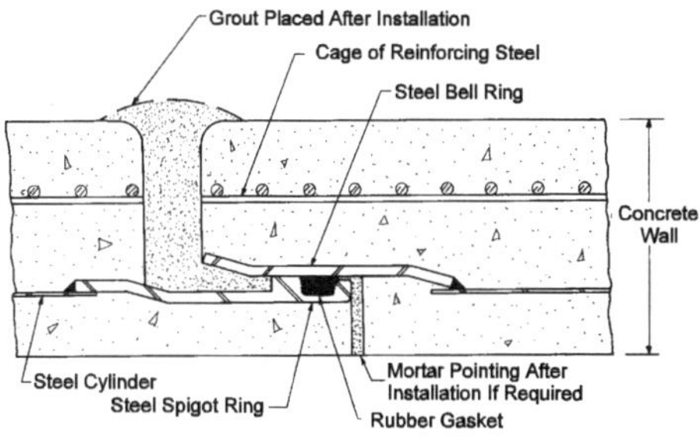

Reinforced Concrete Cylinder Pipe

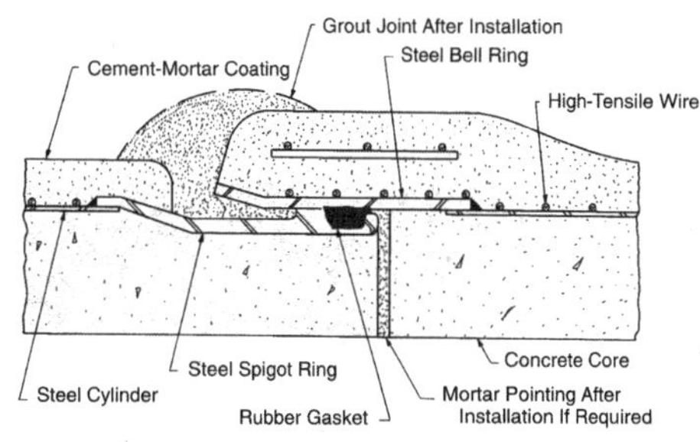

Prestressed Concrete Lined-Cylinder Pipe

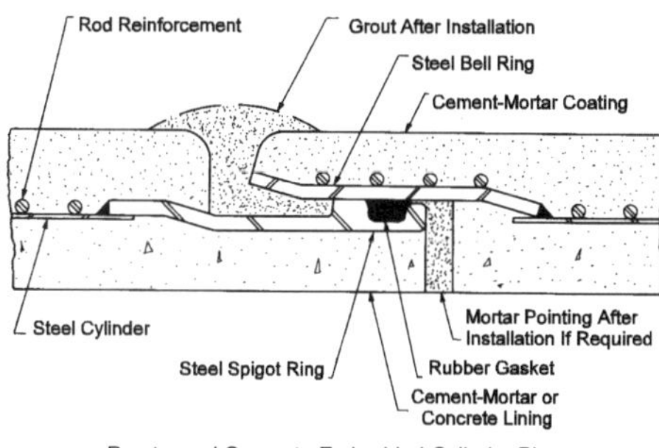

Prestressed Concrete Embedded-Cylinder Pipe

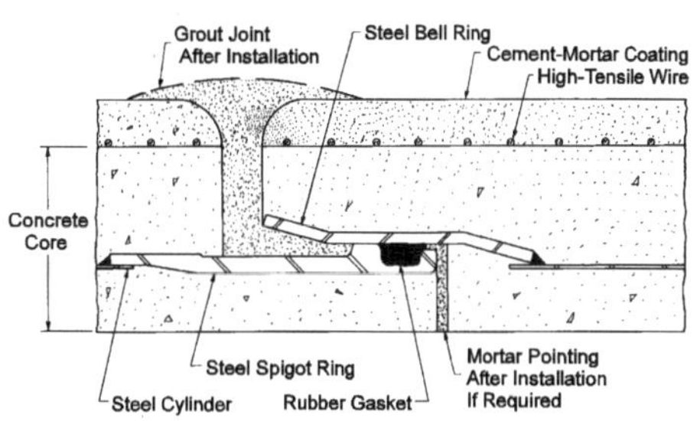

Bar-Wrapped Concrete Cylinder Pipe

Concrete pressure pipe

concentrator A solids-contact unit used to decrease water content of a sludge or slurry.

concrete pipe A conduit manufactured from portland cement and aggregates combined with different forms of reinforcement, used to convey water.

concrete pressure pipe A conduit or pipe manufactured from portland cement and aggregate, reinforced with steel wire or bar. Four of five configurations incorporate a steel cylinder; the fifth type does not. Reinforced concrete cylinder pipe consists of a concrete core, an embedded steel cylinder, and mild steel reinforcement cast into the pipe wall. Prestressed concrete cylinder pipe includes two types, consisting of a steel cylinder either lined with or embedded in structural concrete, wrapped by prestressing wire, and coated in portland cement mortar or concrete. Bar-wrapped concrete cylinder pipe consists of a steel cylinder lined with portland cement mortar or concrete and wrapped with a mild steel bar, all coated with portland cement mortar. Reinforced noncyclinder pipe consists of a concrete core reinforced with one or more steel bar cages; its use is limited to low-pressure applications. Concrete cylinder pipes are joined by steel joint rings, while noncylinder pipes may have steel or cast concrete joint rings.

condensate (1) Condensed vapors from a heat exchanger, also called distillate. (2) The product from a distillation water treatment process.

condensation (1) Conversion from the vapor state to the liquid state, such as the production of condensate or distillate from a distillation plant. This is caused by the formation of water droplets around condensation nuclei or on surfaces when a saturated air mass contacts these surfaces. A common example is the formation of water droplets on the outside of windows in air-conditioned buildings in humid climates. (2) A chemical reaction with two or more molecules joining to form a more complex compound, such as in the case of polymerization.

condensation nucleus A particle on which condensation of water vapor begins in the free atmosphere, where it invariably takes place on hygroscopic dust or hygroscopic gases. The common sources of such dust and gases are sea salt, products of combustion, and dust blown from the earth's surface.

condensed time A foreshortened time base used in developing an infiltration capacity curve in hydrological studies. Time condensation is applied for a period during a storm when the rain intensity is less than infiltration capacity. This is necessary to meet the requirement that the volume of infiltration under the curve of infiltration capacity must equal the observed volume of infiltration.

condensor *See* capacitor.

conditional water right The legal preservation of a priority date that provides a water user time to develop a water right while preserving a more senior date. It becomes an absolute right when the water is actually put to beneficial use. *See also* beneficial use.

conditioned water Any water that has been treated by one or more processes (e.g., adsorption, deionization, reverse osmosis) to improve the water's usefulness, aesthetic quality, or both by reducing undesirable substances (e.g., iron, hardness) or undesirable conditions (e.g., color, taste, odor).

conductance A measure of the conducting power of a solution, equal to the reciprocal of the resistance.

conductance units In the US customary system, mhos. In Système International, siemens; seconds cubed amperes squared per square metre per kilogram.

conductivity (specific conductance) A measure of the ability of a solution to conduct electrical current. Its value is inversely proportional to the solution's electrical resistance. The conductivity value is commonly used in water-desalting processes as a means to evaluate desalting efficiency and to estimate the total dissolved solids concentration; the conductivity value of a water sample is multiplied by an empirical factor representative of the typical total dissolved solids/conductivity ratio for the specific type of water. The units of conductivity are often reported as micromhos per centimetre at 25° Celsius, but this is not a Système International unit; multiplying such a value by 10^{-4} converts the value to units of siemens per metre.

conductivity bridge A means of measuring conductivity whereby a conductivity cell forms one arm of a Wheatstone bridge, a standard fixed resistance forms another arm, and a calibrated slide wire resistance with end coils provides the remaining two arms. A high-frequency alternating current is supplied to the bridge.

conductivity detector A type of detector frequently used in conjunction with an ion chromatograph. Ions present in the elution solvent conduct an electrical current that is converted into a detector response. The conductivity measured is proportional to analyte concentration.

conductivity units In the US customary system, micromhos per centimetre. In Système International, siemens per metre; seconds cubed amperes squared per metre per kilogram.

conductor A substance that permits the flow of electricity, especially one that conducts electricity with ease.

conductor casing The outer casing of a well. The purpose of this casing is to prevent contaminants from surface waters or shallow groundwaters from entering a well.

conduit Any artificial or natural duct, either open or closed, for conveying fluids.

cone of depression The depressed groundwater elevations that form around a well during pumping, caused by the loss of water from storage in the nearby aquifer materials. Also called cone of influence. *See also* circle of influence.; drawdown; well cone of influence; zone of influence.

cone valve A valve with a housing and tapered plug containing a hole that is operated by turning the tapered section. The valve is open when the hole is parallel with the flow and closed when the hole is perpendicular to the flow. A cone valve is designed so that the tapered section can be lifted slightly for easier rotation and forced back against the housing for closing.

confidence interval An interval constructed around the mean of sample data that includes the true population mean for a specified percentage of repeated samples taken. The data are assumed to be random and normally distributed, so the percentage of

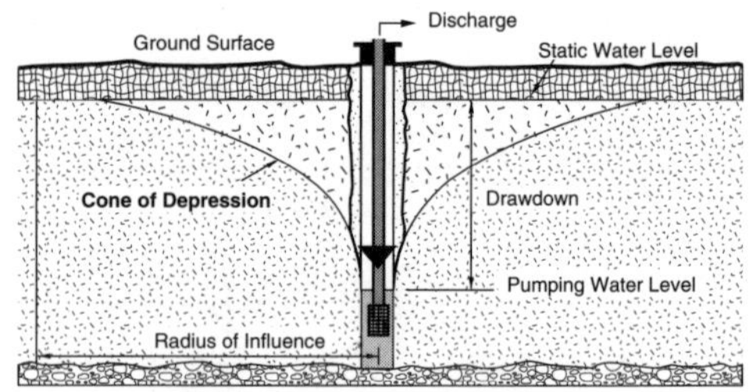

Cone of depression

confidence is based on the standard deviation of the sample data. A confidence interval of two standard deviations on either side of the sample mean will include the population mean in about 95 percent of repeated samples. This same statistical method can be applied to the predicted values in multiple regression analysis, such that the standard error of the estimate is substituted for the standard deviation. In epidemiology, the confidence interval is preferred over the *p*-value because more information is conveyed about the magnitude of an association and its range of values (confidence limits). *See also* confidence limit; odds ratio; probability value; standard deviation; standard error of the coefficient; standard error of the estimate; t-test.

confidence level *See* confidence interval.

confidence limit An upper or lower bound specified by the confidence interval surrounding a statistical mean. *See also* confidence interval.

confined aquifer An aquifer overlaid by a confining bed that has significantly lower hydraulic conductivity than the aquifer.

confined groundwater Water contained in a confined aquifer. Pore water pressure exceeds atmospheric pressure at the top of the confined aquifer. *See also* confined aquifer.

confined space (1) A space that is configured so that an employee can bodily enter and perform assigned work but that has limited or restricted means for entry and exit and is not designed for continuous employee occupancy. (2) A space defined by the concurrent existence of the following conditions: (a) existing ventilation is insufficient to remove any dangerous air contamination, oxygen deficiency, or both that may exist or develop, and (b) ready access or egress (getting out) for the removal of a suddenly disabled employee (operator) is difficult because of the location of the opening(s), size of the opening(s), or both. *See also* dangerous air contamination; oxygen deficiency.

confined space entry The passing of an employee through an opening into a permit-required confined space. The purpose of such entry includes work activities in that space. Entry is considered to have occurred as soon as any part of the entrant's body breaks the plane of an opening into the space.

confined water *See* confined groundwater.

confining bed A geologic unit of low hydraulic conductivity that is adjacent to one or more aquifers. It may lie above or below the adjacent aquifer.

confining stratum An impervious stratum or confining layer directly above or below a stratum or layer bearing water. *See also* confining bed.

confirmed test The second major step of the multiple-tube fermentation test. It confirms that positive results from the presumptive test are caused by coliform bacteria. *See also* completed test; presumptive test.

confluence A junction or flowing together of streams; the place where streams meet.

confluent Flowing together to form one stream.

confluent growth The growth of bacterial colonies that are not discretely separated on a growth medium.

confluent stream A stream that unites with another; a fork or branch of a river, especially when the streams are nearly equal in size.

confounding bias A potential source of error in evaluating an epidemiologic association between exposure and disease. Confounding bias must be avoided because it may incorrectly convey the appearance of such an association. For example, the confounding characteristic rather than the putative exposure may be responsible for all or much of the observed risk. This bias results when one or more variables that cause the disease exist and are also associated with exposure. Confounding bias is potentially present in all studies and does not result from an error of the investigator. This bias can be minimized if certain procedures are followed during the study design (matching, randomization) and during data analysis (multivariate analysis, stratification).

confounding factor A variable other than the specific exposure being evaluated that is associated with both the exposure and outcome of interest and may bias an analysis of the effect of that exposure. *See also* confounding bias.

congenital malformation An abnormal structure in certain limbs or organs that was present at birth. Increased risks of some of these abnormalities appear to be caused by microbial, chemical, and physical agents in the environment.

Congressional Budget Office (CBO) A government office established by the Congressional Budget Act of 1974 (2 U.S.C. 601) to provide the US Congress with basic budget data and analyses of alternative fiscal, budgetary, and programmatic policy issues. The Congressional Budget Office provides Congress with biannual forecasts of the economy and analyses of economic trends and alternative fiscal policies; prepares a 5-year cost estimate for carrying out any public bill or resolution reported by congressional committees; is responsible for furnishing the House and Senate budget committees by February 15 of each year with a report that discusses alternative spending and revenue levels and alternative allocations among major programs and functional categories; and undertakes studies requested by the Congress on budget-related issues.

Congressional Record The US government publication published each day Congress is in session to document the deliberations, actions, and proceedings of the US Congress.

conical plug valve A valve in which the moving plug is conical. This type of valve is opened by unscrewing the plug from the seat and turning it through an angle of 90°. It is also called a cone valve.

conjugate *See* conjugation reaction.

conjugate base A chemical species that can accept a proton and is related to an acid by the difference of that proton (e.g., chloride ion [Cl^-] is the conjugate base of hydrochloric acid [HCl]).

conjugate depths The depths before and after a hydraulic jump. These depths are also called sequent depths.

conjugation reaction Generally a detoxification reaction that occurs in an organism exposed to a chemical or its active metabolite. Such a reaction usually involves the biochemical addition of a conjugate (metabolite of normal biological constituents) to increase the water solubility of the chemical, thereby increasing the chemical's rate of elimination via the urine. Common conjugates include glucuronides, glycine, sulfate, glutathione, and taurine.

conjunctive use A water supply management concept in which groundwater and surface water supplies are managed jointly to produce a larger and more stable economic yield of water than operating the two supplies independently. One such scheme may be to use surface water supplies as a base supply and then supplement with groundwater in times of drought.

conjunctive water management The management of both surface water and groundwater resources in a region where the surface water and groundwater systems are hydraulically interactive, in order to satisfy water demands and place limited stress on either resource.

connate seawater Ancient seawater entrapped in sediments at the time of their deposition. Connate seawater can be a source of bromide ions. *See also* bromide; connate water.

connate water Subsurface water that was not buried with a geologic unit but has been out of contact with the atmosphere for a large portion of a geologic period.

connection band A collar or coupling that fits over adjacent ends of pipe to be joined and that, when drawn tight, holds the pipe together either by friction or by mechanical bond.

connection charge A charge assessed by the utility to recover the cost of connecting the customer's service line to the utility's facilities, typically covering only the direct and indirect costs of physically tying the service line into the main. In this case, the payment is considered a service charge. When this charge includes system backbone capacity costs in addition to direct hook-up costs, it is considered a contribution of capital by the customer or other agency applying for service. Such an all-inclusive connection charge is also called a system development charge.

connection fee *See* connection charge.

connectivity The establishment of logical relationships or topology between graphic features, for instance, linking a water main junction to its connecting pipes.

consecutive systems Two water systems that are interconnected.

consent order A legally binding agreement between water utility owner(s) and a regulatory agency prescribing a set time period for the water utility to complete prescribed tasks.

consequent stream A stream for which the course has been controlled in terms of direction and location by the general slope of the surface topography.

conservation The practice of protecting against the loss or waste of natural resources.

conservation district A governmental entity that acts to protect soil and water.

conservation master plan A plan, usually an integral component of a water system master plan, that defines the demand management measures available to a utility, the cost of implementing them, the benefit–cost relationship (ratio) of each measure, and a recommended set of the most cost-effective conservation measures in order to achieve a prescribed amount of conservation over an extended period, usually 20 years.

conservation of energy The principle that the total amount of energy in an isolated system remains unchanged while internal changes of any kind occur. In hydraulics, this principle dictates that the energy of water at any point in a waterway is equivalent to the energy at any other point plus or minus friction losses.

conservation rate An increased rate of charge for water use over an established amount per billing cycle to discourage overuse and waste.

conservation storage Storage of water for future useful purposes such as municipal supply, power, or irrigation.

conservative constituent A chemical that essentially maintains its concentration in passing through a unit treatment process. For example, coagulation can remove turbidity and organic carbon, but it does not reduce the bromide ion concentration; thus, bromide ion is called a conservative constituent in this circumstance.

consolidated formation A geologic material for which the particles are stratified (layered), cemented, or firmly packed together (hard rock). Such a formation usually occurs at a depth below the ground surface. *See also* unconsolidated formation.

consolidation (1) In geology, any or all processes by which loose, soft, or liquid earth materials become firm and coherent. (2) A merger or acquisition of one or more small systems, usually by a larger system.

consolidation sedimentation (1) Any or all processes by which loose, soft, or liquid earth materials become firm and coherent. (2) In soil mechanics, the adjustment of saturated soil in response to increased load, involving removal of water from pores by an increase in pressure and a decrease in void ratio.

consortium In microbiology, a heterogeneous group of microorganisms found to be associated with a biochemical process. For example, in the biodegradation of a complex hydrocarbon mixture such as crude oil, a variety of microorganisms, including different types of bacteria and fungi, may be involved to achieve complete mineralization of the crude oil.

constant (1) A universally accepted value for a certain parameter. (2) An unchanging number with or without fractional values.

constant-rate filtration The most common method of filter operation, in which the filters are operated at a constant rate. The rate is controlled either by (1) an effluent valve that is gradually opened as head loss increases throughout the filter run or (2) a variable water surface above the filter media that gradually rises to increase the applied head as head loss accumulates.

constant-speed pump A centrifugal pump designed to operate at a constant speed. For a given system head loss, the pump will deliver a constant flow rate. As head loss increases, flow rate will decrease, and vice versa. The extent to which the flow rate will increase or decrease as system head loss fluctuates is dependent on the shape of the pump curve. *See also* pump characteristic curve.

constant spring A spring in which variation in discharge from maximum to minimum does not exceed one-third of the average discharge.

constant-flow stirred-tank reactor *See* continuous-flow stirred-tank reactor.

constriction In a waterway, an obstruction that confines the flow to a narrower section or to a smaller area, thus throttling the flow. Natural gorges, bridge piers, weirs, and orifices are examples of constrictions. This term is sometimes used synonymously with the term *contraction*.

constructed conveyance In broad terms, for the purposes of defining a public water system under the Safe Drinking Water Act (SDWA), any artificial conduit, such as a ditch, culvert, waterway, flume, mine drain, or canal. A constructed conveyance does not include water that is delivered by bottle, other package unit, vending machine, or cooler; nor does it include water that is trucked or delivered by a similar vehicle. Water bodies or waterways that occur naturally but that are altered by humans may, in some cases, be constructed conveyances. The US Environmental Protection Agency issued guidelines, published on Aug. 5, 1998 (63 *Federal Register* 41939–41946), regarding interpretation of the definition of *public water system*, including *constructed conveyance*, which is defined in Section 1401(4) of the SDWA. *See also* public water system.

construction The overall activities associated with assembling and building components of a facility.

construction joint A temporary joint along a plane or surface in a structure composed of homogenous material such as earth or concrete. This joint allows for ceasing to place additional material for a certain amount of time, such as overnight or several days, and allows for forming complex concrete pours.

construction work in progress (CWIP) A utility's investment in facilities under construction but not yet dedicated to service. The inclusion of CWIP in the rate base varies from one regulatory agency to another.

consumer confidence report (CCR) An annual report that community water systems must deliver to their customers summarizing information regarding source water, detected drinking water contaminants, and other information. A CCR is intended to educate and inform the public regarding the quality of their drinking water. The US Environmental Protection Agency established requirements for issuance of consumer confidence reports on Aug. 19, 1998 (63 *Federal Register* 44512–44536). Community water systems must issue their first CCR under this rule by Oct. 19, 1999, covering calendar year 1998. The second CCR must be issued by July 1, 2000, covering calendar year 1999. Subsequent CCRs must be distributed annually by July 1, covering the previous full calendar year.

Consumer Products Safety Commission (CPSC) An independent US government agency established by the Consumer Products Safety Act (15 U.S.C 2051 et seq.). The commission consists of five commissioners, appointed by the president of the United States with advice and consent of the US Senate, one of whom is appointed chair. The commission protects the public against unreasonable risks of injury from consumer products; assists consumers in evaluating the comparative safety of consumer products; develops uniform safety standards for consumer products and minimizes conflicting state and local regulations; and promotes research and investigation into the causes and prevention of product-related deaths, illnesses, and injuries.

consumptive use The loss of irrigation water to the atmosphere via transpiration or evaporation. *See also* evapotranspiration.

consumptive waste The water that returns to the atmosphere without benefiting people.

contact chamber Any large tank in which water can be mixed with and allowed to react with a disinfectant or other chemical agent.

contact dermatitis An inflammatory reaction of the skin that has as its basis an allergic reaction. The response can be a result of a chemical or biochemical that comes in contact with the skin.

contact filter A filter used in a water treatment plant for the partial removal of turbidity before final filtration. *See also* preliminary filter.

contact flocculation A process in which coagulated water is passed through a coarse-media or gravel bed that acts as a flocculation system, thereby reducing the time to produce a settleable or filterable floc. Reducing the time may be important for plants treating very cold water, portable plants, pressure plants, or plants treating water with low total dissolved solids that do not respond readily to metal salt coagulants. *See also* flocculation.

contact roughing filter A filter used in a water treatment plant for the partial removal of turbidity before final filtration. *See also* preliminary filter.

contact spring A spring forming at a lithologic contact when a more permeable unit overlies a less permeable unit. *See also* cliff spring.

contact tank A tank used in water treatment to promote contact between treatment chemicals or other materials and the body of liquid treated.

contact time (CT) The time in which a chemical or constituent is in contact with another reacting chemical or constituent. The contact time in a basin or storage vessel can be expressed on a theoretical basis (i.e., the volume divided by the flow rate) or on a measured basis using tracer studies that account for contactor hydrodynamics. Contact time is not to be confused with $C \times T$. *See also* contactor hydrodynamics; $C \times T$; empty bed contact time.

contactor (1) A vertical pressure vessel used to hold an activated carbon bed. (2) An electric switch, usually magnetically operated. *See also* reactor.

contactor hydrodynamics The hydraulic behavior and pathways of a fluid in a reactor or contactor. The hydrodynamics dictate the extent of short-circuiting or the extent to which the reactor behaves as a plug flow system. The hydrodynamics change as flow rate increases or decreases in a fixed-volume contactor. *See also* plug-flow reactor; short-circuiting.

container valve The valve mounted on a chlorine container or cylinder.

contaminant Any physical, chemical, biological, or radiological substance or matter in water.

Contaminant Candidate List *See* Drinking Water Contaminant Candidate List.

Contaminant Selection List *See* Drinking Water Contaminant Candidate List.

contaminant source inventory A record of the activities on a watershed or aquifer recharge area that have a potential to contaminate water.

contamination Any introduction into water of microorganisms, chemicals, wastes, or wastewater in concentrations that make the water unfit for its intended use.

contingency management An approach to handling adverse conditions that (1) emphasizes that no one best way exists at all times and in all situations and (2) recognizes differing abilities of people at various times and in actual circumstances. This approach is also called situational management.

contingency plan A document that details the intended actions of a water utility under specified adverse conditions.

contingency planning A process wherein specific alternatives are specified to provide direction in the event of unforeseen occurrences.

continuing education A planned or mandated program in which those in a profession or occupation attend formal classes, seminars, training sessions, and so on to keep informed of new developments in their specific discipline.

continuity equation An axiom stating that the rate of flow past one section of a conduit is equal to the rate of flow past another section of the same conduit plus or minus any additions or subtractions of water flow between the two sections. *See also* storage equation.

continuous air supply A grade D air that is supplied to a worker who is wearing a hood or helmet while working in a hazardous environment. *See also* grade D air.

continuous belt filter press (CBFP) A belt filter press to which sludge is fed continuously. *See also* belt filter press.

continuous control *See* proportional integral derivative control.

continuous flow operation The process wherein a continuous and steady flow of water is processed for treatment through the media. *Contrast to* intermittent flow operation.

continuous immunomagnetic collection An antibody capture technique used for the selective separation of viruses and protozoa. The process involves the capture of target organisms by a specific antibody attached to a magnetic bead.

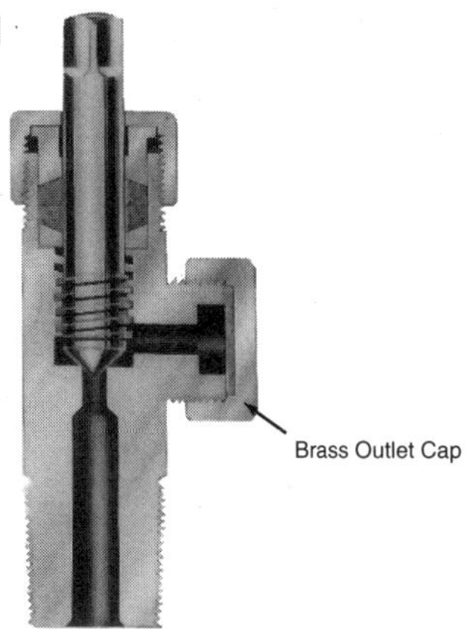

Container valve

continuous interstice An interstice that occurs in (1) granular material, (2) a fracture in a rock formation, or (3) a tubular opening in soluble rocks or lava and that is connected with other interstices.

continuous liquid–liquid extraction (CLLE) An isolation technique based on the distribution of analytes between two partially soluble solvents. In continuous extraction, an organic solvent phase is recycled so that analytes have multiple opportunities to partition from the aqueous phase. Continuous extractors can be used to extract contaminants from very large volumes of water and for samples that may cause emulsions or other problems if the sample were shaken.

continuous mode A mode of water treatment operation in which the feedwater input and process output flow(s) are continuous. *Contrast with* batch mode.

continuous sample A sample taken from water that flows from a particular place in a plant to the location where samples are collected for testing. Personnel may obtain grab or composite samples from this continuous stream. Frequently, several taps (faucets) will flow continuously in the laboratory to provide test samples from various places in a water treatment plant. Possible changes in water quality may occur during travel to the sampling tap. *See also* composite sample; grab sample.

continuous sludge-removal tank A sedimentation tank equipped to permit the continuous removal of sludge.

continuous stream A stream that does not have interruptions in course. It may be perennial, intermittent, or ephemeral, but it does not habitually have wet and dry stretches.

continuous-backwash filter A modification of a rapid granular filter in which the filter bed is divided into multiple compartments with the filtrate flowing to a common effluent channel. A traveling backwash system equipped with a backwash collection system washes each cell in succession as it continuously or intermittently traverses the length of the filter.

continuous-flow pump A displacement pump within which the direction of the flow of water is not changed or reversed.

continuous-flow tank A tank through which liquid flows continuously at its normal rate of flow, as distinguished from a fill-and-draw or batch system.

continuous-flow stirred-tank reactor (CFSTR) A reactor in which the contents are completely mixed on a continuous basis. In theory, flow introduced to the continuous-flow stirred-tank reactor is immediately mixed to a uniform concentration upon entering the reactor, and therefore fractions of the flow will be exiting as soon as they enter. Steady-state conditions prevail, and no accumulation of constituents occurs in the reactor with time. Continuous-flow stirred-tank reactors are a fundamental component of reactor theory, although few true applications exist in water treatment. A rapid mix basin for adding coagulant chemicals, if mechanically mixed at high intensity, is one example of a treatment process that approximates this behavior.

continuous-stave pipe A pipeline made of wooden staves, constructed so that the staves—which form the barrel of the pipe—are placed in a continuous line from one end of the pipeline to the other. Such a pipe is assembled in place in the field with staggered joints and held to position and tightness by encircling steel bands.

continuously stirred tank reactor (CSTR) *See* continuous-flow stirred-tank reactor.

contour A line of equal elevation above a specified datum, usually mean sea level.

contour basin A basin made by levees or borders following contours, with occasional cross levees.

contour interval The difference in elevation between adjacent contours on a map.

contour line A line joining points having or representing equal elevations.

contour map A map showing the configuration of the surface by means of contour lines drawn at regular intervals of elevation, such as one for every 10 metres. A crowding of the contour lines indicates steepness.

contour plot *See* contour map.

contract demand The water quantity authorized by contract between a water utility and a large-use customer who requires a significant amount of the total capacity of the utility. The agreement fixes the terms and conditions under which the water utility will provide water to the customer. Such an agreement has also been called contract capacity.

contract operation and management The provision of water utility services by a private company under contract (rather than another water system).

contracted weir A rectangular notched weir that has (1) a crest width narrower than the channel across which it is installed and (2) vertical sides extending above the upstream water level. These features produce a contraction in the stream of water as it leaves the notch.

contracted-opening discharge measurement A determination (by indirect measurement) of peak discharge following a flood based on a field survey of high-water marks and channel and bridge geometry at a bridge constriction. Discharge is computed on the basis of an evaluation of energy changes between the approach section and the downstream side of the constriction.

contracting reach A reach of channel wherein flow is accelerating. The velocity head at a downstream cross section of the reach exceeds the velocity head at an upstream cross section.

contraction (1) The extent to which the cross-sectional area of a jet, nappe, or stream is decreased after passing an orifice, weir, or notch. (2) The reduction in cross-sectional area of a conduit along its longitudinal axis. *See also* constriction.

contraction coefficient A coefficient in a formula for calculating the discharge of a weir, orifice, or other constriction in a waterway, introduced to correct the result because the water stream's cross-sectional area will contract to a size smaller than the nominal area of the constriction after the fluid passes the constriction. The contraction coefficient is equal to the ratio of the smallest cross-sectional area of the fluid after passing the constriction to the nominal area of the constriction.

contribution in aid of construction An amount of money, services, or property that a water utility receives at no cost from any person, government agency, or other entity. It represents an addition or transfer to the capital of the utility and is used to offset the acquisition, improvement, or construction costs of any utility property, facilities, or equipment used to provide utility services to the public. It includes amounts transferred from advances for construction representing (1) any unrefunded balances of expired refund contracts or (2) discounts resulting from the termination of refund contracts. Contributions received from government agencies and others for relocating water mains or other plant facilities are also included.

control (1) A condition in which specific quality criteria have been achieved in a laboratory analysis. For example, an analysis is said to be in control when a series of standards or samples have met a set of conditions such as precision or accuracy. (2) A type of sample used to assess the quality of an analytical process. (3) A section or reach of an open conduit or stream channel where artificial or natural conditions, such as the presence of a dam or a stretch of rapids, make the water level upstream a stable indicator of the discharge. Controls may be complete or partial. Complete control exists if the elevation of the water surface upstream of the control is entirely independent of fluctuations of the water level downstream from it. Partial control exists if downstream fluctuations have some effect on the upstream water level. (4) A waterway cross section that serves as a bottleneck for a given flow and that determines the energy head required to produce the flow. In the case of open channels, the control is the point where the flow is at critical depth; hydraulic conditions upstream of this point are wholly dependent on the characteristics of the control section and are entirely independent of hydraulic conditions downstream of the point. In the case of closed conduits, the control is the point where the hydrostatic pressure in the conduit and the cross-sectional area of flow are definitely fixed, except in cases where the flow is limited at some other point by a hydrostatic pressure equal to the greatest vacuum that can be maintained unbroken at that other point.

control action The particular mathematical relationship between an automatic controller's input and output. This term is synonymous with the term *control mode.*

control character A specific character imbedded in a data stream. Examples include uppercase (capital) or line-feed control characters. Such characters specify functions or operations such as capital line feed, underline, or highlight and are used to direct a video display or printer.

control chart A plot of analytical results that illustrate the state of statistical control exhibited over a series of measurements. A control chart is typically one of two types: a property chart or a precision chart. Sometimes called Shewhart charts, these plots are important tools in quality assurance.

control float A float installed in a tank or body of liquid to control the pumps.

control flume A flume arranged for measuring the flow of water, generally including a constricted section wherein a minimum energy head exists at all stages.

control group A set of substances or organisms used to compare with a similar set that has undergone one or more modifications.

control limits Specifications on a control chart beyond which a given point in a system would be unlikely to be in statistical control. The value, calculation, or both of a control limit depends on the type of control chart.

control lines *See* control limits.

control loop (1) The combination of all instruments and devices that directly sense a condition or control a field element. (2) The path through the control system between the sensor that measures a process variable and the controller that controls or adjusts the process variable.

control mode *See* control action.

control point A reference point used to perform absolute orientation, or georeferencing, of graphic information system data. Control points can be used for horizontal orientation, vertical orientation, or both; they can be derived from field surveys, aerotriangulation, or documents published by recognized institutions such as the National Geodetic Survey, US Geological Survey, Department of Transportation, and other federal, state, or local government agencies.

control reach A section or reach of an open conduit or stream channel where artificial or natural conditions, such as the presence of a dam or a stretch of rapids, make the water level upstream a stable indicator of the discharge. *See also* control.

control relay An electromechanical device used to switch one or more electrical control signals in response to a trigger signal. Control relays contain a switch of one or more poles that is operated by an electromagnet. They are usually small because they are designed to carry control signals rather than power.

control section A waterway cross section that serves as the bottleneck for a given flow and determines the energy head required to produce the flow. *See also* control.

control system A system that senses and controls its own operation on a close, continuous basis. Many different types of control algorithms exist, including proportional, integral, and derivative control. *See also* derivative action; integral action; proportional action.

control works (1) All the structures, devices, and so on located at the head or diversion point of a conduit or canal. A control works is practically the same thing as a diversion works (i.e., an intake heading). (2) Structures and reservoirs constructed to reduce the flood peaks on streams subject to damaging floods.

controlled discharge Regulation of effluent flow rates to correspond with flow variations in receiving waters in order to maintain established water quality.

controlled globe valve An electrically or hydraulically remotely operated globe valve. *See also* globe valve.

controlled variable A quantity or condition that is measured and controlled.

controller A device that controls the starting, stopping, or operation of a device or piece of equipment.

convection In physics, mass motions within a fluid, resulting in transport and mixing of the properties of that fluid. Convection is caused by the force of gravity and by differences in density caused by nonuniform temperature.

convection current An ascending or descending (vertical) movement of water in a settling basin or lake, caused by differences of temperature or density (when the density of an upper stratum becomes greater than that of the underlying stratum).

convective precipitation Precipitation resulting from vertical movement of moisture-laden air that, on rising, cools and precipitates its moisture.

conventional aeration **(CA)** The process of adding air to a treatment basin by mechanical means (surface aerators) or by submerged air diffusers that are supplied by compressors.

conventional filtration *See* conventional water treatment; direct filtration; in-line filtration.

conventional filtration treatment *See* conventional water treatment.

conventional treatment *See* conventional water treatment.

conventional water treatment The use of coagulation, flocculation, sedimentation, filtration, and disinfection, together as sequential unit processes, in water treatment. This process is also called complete treatment.

convergence The coming together of two or more mathematical functions to a single point. The point of convergence is commonly referred to as the solution.

converging tube A tube for which diameter decreases, usually at a uniform rate, along its longitudinal axis from the end at which the liquid enters.

conversion (1) The process of converting nondigital data from hard copy maps or other sources to digital format. This term usually refers to the capture of graphic features and their incorporation into a geographic information system database; it also applies to the conversion of digital data from one format to another. (2) A short transition section in a conduit uniting two other sections that have different hydraulic elements; a transition. (3) In a membrane water treatment system, the amount of permeate, filtrate, or product flow divided by the amount of feedwater, commonly expressed as a percentage. In this instance, conversion is also called recovery.

conversion factor A numerical constant by which a quantity with its value expressed in units of one kind is multiplied to express the value in units of another kind.

converter Generally, a direct current generator driven by an alternating current motor.

convex stream bank A stream bank that has its center of curvature away from the channel.

conveyance loss Water lost during conveyance (via a pipe, channel, conduit, or ditch) as a result of leakage or evaporation.

coolant A fluid used to cool mechanical equipment and prevent overheating. Coolants include water or other solutions that have higher coefficients of specific heat (i.e., can absorb more heat energy for a fixed increase in temperature).

cooling coil A coil of pipe or tubing containing a stream of hot fluid that is cooled by the transfer of heat to a cold fluid outside the coil. Conversely, the coil may contain a cold fluid to cool a hot fluid in which the coil is immersed.

cooling tower A hollow, vertical structure with internal baffles to break up falling water so that it is cooled by upward-flowing air and by evaporation of water.

cooling water Water used to reduce temperature.

cool-season turfgrass Turfgrass that does not ordinarily lose its color unless the average air temperature drops below 32° Fahrenheit (0° Celsius) for an extended period. It is not usually damaged by subfreezing temperatures. Cool-season grasses grow actively in the cool weather of spring and fall and slowly in summer heat.

cooperative A group of small water systems joining together for the purpose of buying drinking water supplies—such as water treatment chemicals, equipment, and services—at a reduced cost.

coordinate geometry (COGO) The common shorthand term for the technique of using mathematical algorithms to compute coordinates from geometric descriptions such as bearings and distances. The coordinates are stored and used to generate graphic map displays.

coordinate system A framework used to define the position of a point, line, curve, or plane, as well as derivative map features, within a two- or three-dimensional space.

coordinated phosphate treatment A boiler treatment process using phosphate buffers to avoid the presence of hydroxyl alkalinity.

COP *See* certificate of participation.

copper (Cu) A common reddish metallic element (atomic number 29). In nature, copper is found in the ores of azurite, azurmalachte, chalcocite, chalcopyrite, covellite, cuprite, and malachite. It is ductile and malleable and is used to manufacture pipe and other products used in water and treatment. Copper has various industrial uses and, as copper sulfate, is used to control algal growths. It is regulated by the US Environmental Protection Agency. *See also* copper sulfate.

copper pipe *See* copper tubing.

copper service A small water line constructed with copper material extending from the water main to the building plumbing.

copper sulfate ($CuSO_4 \cdot 5H_2O$) A chemical prepared from copper and sulfuric acid (H_2SO_4). It is usually used to control algal growths. Copper sulfate is also called blue vitriol, blue copperas, bluestone, and cupric sulfate.

copper tubing A conduit made from copper that has been heated and forced through dies to form a hollow cylindrical body of a required size for water services or plumbing pipes.

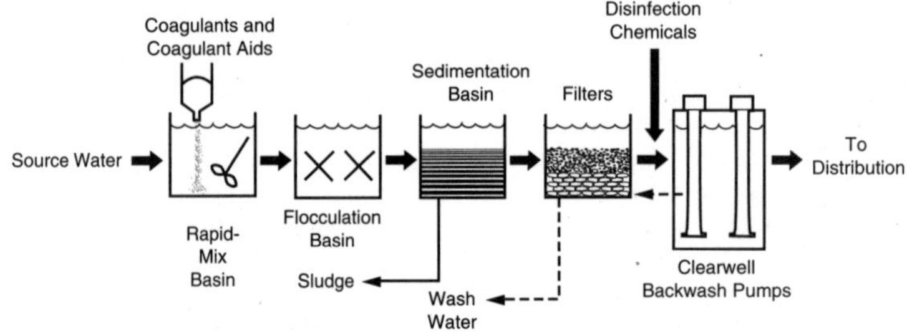

Conventional water treatment plant

copperas ($FeSO_4 \cdot 7H_2O$) A common name for ferrous sulfate heptahydrate.

copper–copper sulfate electrode (CSE) A copper rod partially immersed in a copper sulfate solution inside a cylindrical tube with a porous bottom, creating the cathodic half of a galvanic cell. The copper sulfate provides excellent soil contact, much better than would a bare copper rod. A reference electrode made with this cell is used in making field measurements of pipe-to-soil potentials.

COPPERSOL (copper solubility) A straightforward program that develops solubility diagrams for copper, considering carbonate, hydroxide, and phosphate complexes at a given temperature and ionic strength. The expected solids formed at equilibrium are also determined.

Co-Pt unit *See* chloroplatinate unit *in the Units of Measure list.*

core (1) A small cylindrical sample of rock removed by a core drill. (2) A section constructed in the center of an earth-fill or rock-fill dam along its axis to prevent the passage of water through the dam. When the section is made of concrete, it is usually called a core wall. It may also consist of puddled clay, timber, rolled or tamped earth, or fine material deposited underwater.

core boring A boring from which a sample of rock, and occasionally soil, is obtained by the use of a drill bit that cuts out a core of the material. The size of the core usually varies from 1⅛ to 2¼ inches (28.6 to 57.2 millimetres); however, calyx drills are sometimes used to take 12-inch to 48-inch (305-millimetre to 1,220-millimetre) cores. The term *core boring* is also applied to similar cores taken from concrete or other materials in place for testing purposes.

core drill An instrument for boring holes in rock or other material to obtain a cylindrical sample of the material passed through.

core drilling A method of drilling with a hollow bit and core barrel to obtain a rock core.

core sample A sample of the medium obtained to represent the entire bed depth when the bed is being analyzed for capacity or usefulness. A hollow tube is sent down through the bed to extract the sample.

core wall A wall of masonry, sheet piling, or puddled clay built inside a dam or embankment and usually along the axis to provide resistance to seepage.

coring The mechanical cultivation of turfgrass using hollow tines to remove cores of turf. This approach improves soil texture and increases air and water movement.

corona An electrical discharge effect that causes ionization of oxygen and the formation of ozone (O_3).

corona discharge A discharge of electricity causing a faint glow adjacent to the surface of an electrical conductor and, similarly, adjacent to the dielectrics in an ozone (O_3) generator during ozone production. Corona discharge results from electrical discharge and indicates the ionization of oxygen and formation of ozone in the surrounding air.

coronavirus A human respiratory virus; a single-stranded ribonucleic acid virus, 70–120 nanometres in size, that replicates in the cytoplasm.

corporation cock A valve for joining a service pipe to a street water main. It is usually owned and operated by the water utility or department. It cannot be operated from the surface. A corporation cock is also called a corporation stop or ferrule.

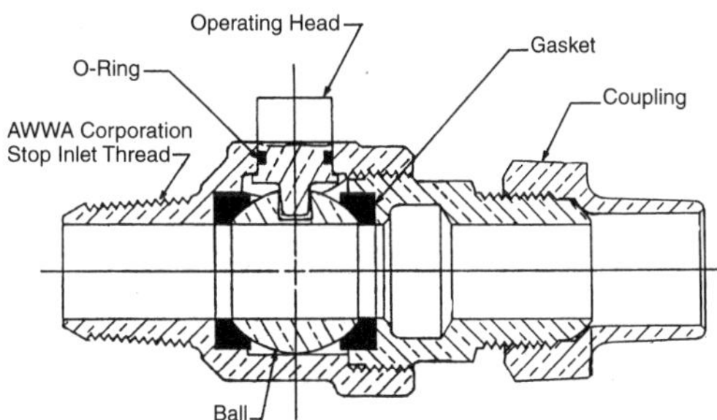

Ball-style corporation cock

corporation stop *See* corporation cock.

Corps *See* US Army Corps of Engineers.

correlation A statistical term indicating the relationship between two variables.

correlation coefficient A measure of closeness of a regression line to the data set, often symbolized by *r* and equal to the square root of the coefficient of determination.

correlation matrix A matrix containing correlation coefficients between pairs of variables listed in the first row and the first column.

correlative right A doctrine of water ownership stating that water use during a drought will be shared among users proportional to the areal extent of the land owned by the competing users.

corrode To destroy gradually by electrochemical action.

corrosion The gradual deterioration or destruction of a substance (usually a metal) or its properties as a result of a reaction with thc substance's surroundings.

corrosion cell The combination of an anode, a cathode, an external circuit (or connection between the anode and cathode), and an internal circuit (or conducting solution between the anode and cathode) that can cause corrosion. The electrons generated at the anode migrate to the cathode, and the positive ions generated at the anode migrate to the cathode. Different forms of corrosion can occur as a result of the distribution of anodic and cathodic areas over the corroding material.

corrosion control (1) Any water treatment method that keeps the metallic ions of a material, typically a conduit, from going into solution, e.g., increasing the pH of the water, removing free oxygen from the water, or controlling the carbonate (CO_3^-) balance of the water. (2) The sequestration of metallic ions and the formation of protective films on metal surfaces by chemical treatment.

corrosion index A prediction of the corrosivity of a water. In practice, no index has been developed that adequately predicts the corrosiveness of all waters. Most indexes do not predict corrosion per se, they predict the tendency for a water to dissolve or precipitate a solid, such as calcium carbonate ($CaCO_3$).

corrosion inhibitor A substance that slows corrosion by forming a protective film on the interior surface of pipes and tanks.

corrosion prevention The act of reducing or eliminating deterioration caused by corrosion. Three methods of corrosion prevention exist: (1) selecting and using an inert material or one that is highly resistant to deterioration, (2) providing a coating on the material of concern such that the metal is separated from the aqueous environment, and (3) altering the water composition such that deterioration is lessened.

corrosion rate A measure of the loss of metal from a surface caused by corrosion over time. It is the speed, usually an average, with which corrosion progresses. Corrosion rates may be curvilinear or linear with time, although they are often expressed as though they were linear. They are often reported in units of milligrams per square decimetre per day for weight changes, or mils per year for thickness changes.

corrosion scale The material formed on the surface of the pipe as a result of corrosion or corrosion control treatment. Scale may result from precipitation of a constituent present in the water, such as calcium carbonate ($CaCO_3$), or from the interaction of the metal and the water during passivation. Scales may be very thin and uniform over the surface of the pipe, or they may be relatively thick, thereby decreasing pipe diameter and increasing roughness.

corrosion treatment Treatment to minimize the loss of metal from the pipe or appurtenance, the uptake of the metal by the water during delivery to consumers, or both. Two general corrosion treatment approaches exist: precipitation and passivation. Precipitation involves forming a precipitate in the potable supply that deposits onto the pipe wall to create a protective coating. Passivation involves causing the pipe material and the potable supply to interact in such a way that metal compounds are formed on the pipe surface, creating a film of less soluble material.

corrosion-resistant material A material that resists corrosion after prolonged placement in the environment in which the material was intended to be used. Corrosion-resistant materials do not contribute unacceptable amounts of corroded material into the processed water.

corrosive Tending to deteriorate material, such as pipe, through electrochemical processes.

corrosivity An indication of the corrosiveness of a water. The corrosiveness of a water is described by the water's pH, alkalinity, hardness, temperature, total dissolved solids, dissolved oxygen concentration, and Langelier saturation index. *See also* corrosion index; Langelier saturation index.

cosmic water Water that comes in from space with meteorites.

cosmic-ray-produced nuclides Isotopes, usually radioactive, such as tritium, beryllium-7, and carbon-14, formed by high-energy space particles (e.g., electrons and nuclei) or cosmic rays, interacting with certain atmospheric and terrestrial elements.

cost–benefit analysis *See* benefit–cost analysis.

cost–benefit ratio *See* benefit–cost analysis.

cost of capital The long-term cost of borrowing to a water utility (expressed as a percent). For a not-for-profit utility, the cost of capital is usually the average interest rate (weighted by size) for all outstanding bond issues. For a for-profit utility, the cost of capital is usually calculated as the weighted average (in proportion to balance sheet accounts) of both long-term debt and equity (stock) used to fund the capital requirements. The cost of capital is generally in the 5 to 10 percent range, depending on the amount of inflation in the interest rate; the required return on equity is generally in the 10 to 15 percent range, which is the expected stockholder's return on invested capital. *See also* rate of return; return on investment.

cost of service The total cost of providing any service, such as drinking water, to customers.

cost-of-service analysis A process of determining how to assign costs to various groups of customers for services provided. The purpose of a cost of service analysis is to determine an equitable basis for allocating the costs of providing water to those who receive the service. Fairness requires allocation of costs among customer groups commensurate with their service requirements for total volume of water, peak rates of use, location, and other factors. The base–extra-capacity approach to rate structure design is one method of allocating the cost of service. *See also* base–extra-capacity approach to rate structure design.

cost sharing A publicly financed program through which society, as the beneficiary of environmental protection, shares part of the cost of pollution control with those who must actually install the controls.

coulomb (C) *See the Units of Measure list.*

coulometric cell An apparatus used in a coulometric titration. It consists of a sealed vessel containing a sample and two electrodes. One of the electrodes generates a reagent that reacts with the analyte; the second electrode completes the circuit.

coulometric titration A titration method measuring the quantity of electricity passed during an electron exchange involving the substance being determined.

Council on Environmental Quality (CEQ) A three-member council appointed by the president of the United States that formulates and recommends national policies to the president to promote improving the quality of the environment.

countercurrent efficiency The advantage of a granular activated carbon (GAC) adsorber in permitting partially spent GAC to adsorb impurities before the semiprocessed stream comes in contact with fresh GAC. This allows the maximum capacity of the GAC to be used.

countercurrent flow *See* countercurrent operation; upflow.

countercurrent operation The operation of two components of a unit process in a manner such that they flow in opposite directions. For example, injecting a feed gas into a vessel in the opposite direction as the fluid flow or sludge flowing in the opposite direction to water are countercurrent operations.

counterion An ion of opposite charge with respect to another ion.

counts per minute (cpm) In an examination of water for radioactivity, the number of disintegrations measured in 1 minute after correction for background; a net count rate. The net count rate is used in the calculation of a radioisotope's concentration.

coupling (1) A device to connect the ends of two pipes. (2) A compression fitting to attach the ends of service pipe or tubing. (3) A steel or cast-iron sleeve with a gasket and retainer glands on each end bolted together across the sleeve to apply compression against the ends of two connecting water mains.

coupon A piece of metal or other material used to evaluate the rate of corrosion or deterioration due to exposure to the water of interest. *Coupon* is a broad term and may include a length of

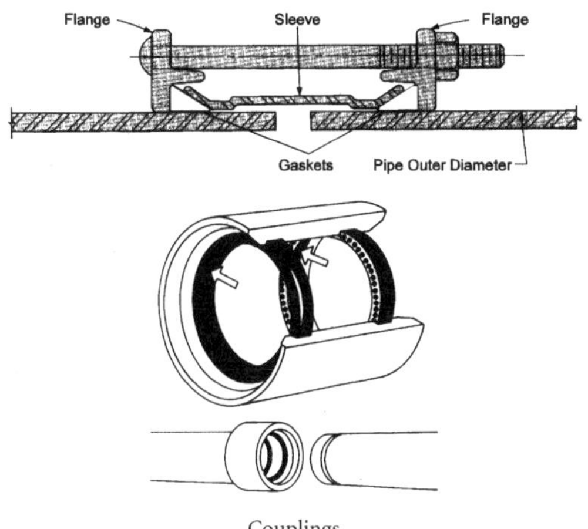

Couplings

pipe from the water distribution system or a specially prepared, preweighed pipe segment used in pipe loop testing. Coupons may be visually inspected, weighed to determine metal loss, chemically tested to determine scale composition, physically measured to determine depth of pits, or subjected to a variety of other tests including x-ray diffraction and scanning electron microscopy.

coupon bond A security that has no identification as to owner. It is presumed to be owned by the bearer of the bond. Such bonds are no longer issued since Congress required full registration of all new issues, effective July 1, 1983. Notes are still bearer designated.

coupon rate The interest rate specified on interest coupons attached to a bond. The coupon rate is also called the nominal interest rate.

coupon test A method of determining the rate of corrosion or scale formation by placing metal strips (coupons) of a known weight in the pipe.

covalent bond A type of chemical bond in which electrons are shared. *See also* ionic bond.

cover (1) The depth of earth material from the top of a water main to the ground surface to prevent freezing of the water in the pipe. (2) The roof, fixed or flexible, over a reservoir in the distribution system.

covered storage A water reservoir or tank that is enclosed, including bottom, sides, and top, to prevent contamination.

cows A type of construction of Japanese origin, used for toe protection and for groins, in which side posts tied together project above like a pair of horns. Several types exist, such as diamond cows and sheet cows.

coxsackie virus An *Enterovirus* classified as a picornavirus, about 20 to 30 nanometres in diameter, containing single-stranded ribonucleic acid. Humans appear to be the only natural host. Coxsackie viruses have been implicated in a variety of diseases, from common-cold symptoms to aseptic meningitis and pericarditis to fatal myocarditis of the newborn. Twenty-four group A and six group B serotypes can be distinguished by type-specific antigens on the virion surface.

CPE *See* comprehensive performance evaluation; cytopathic effect.

cpm *See* counts per minute.

CPR *See* cardiopulmonary resuscitation.

cps *See* cycles per second *in the Units of Measure list.*

CPSC *See* Consumer Products Safety Commission.

cpu/CPU *See* central processing unit; chloroplatinate unit *in the Units of Measure list*; cytopathic unit.

CPVC *See* chlorinated polyvinyl chloride.

cradle (1) One of a series of piers supporting a pipe laid above or below ground, spaced so that the pipe itself furnishes its own structural bridge. (2) A continuous concrete bed placed in the bottom of a trench to support and partially envelope a pipe, particularly where the trench is in soft ground or where bearing support is inadequate to carry the load of pipe and backfill material.

crank-and-flywheel pump A steam-driven reciprocating pump with a crankshaft on which a flywheel is mounted for storing energy during the early part of the stroke and imparting it to the liquid piston or plunger during the latter part of the stroke after the steam is cut off in the steam cylinder.

created wetland A persistent upland or shallow water area converted into a wetland through human activity.

creation Actions performed that establish nontidal wetlands on upland sites.

creek (1) A small stream of water that serves as the natural drainage course for a drainage basin of small size. The term is a relative one as to size; some creeks in a humid region would be called rivers if they occurred in an arid region. (2) A small tidal channel through a coastal marsh. (3) The short arm of a stream.

creep The movement of water under or around a structure built on permeable foundations. *See also* piping.

Crenothrix polyspora Filamentous, sheathed, iron bacteria that are considered nuisance bacteria because the slimes they produce create problems in water treatment and distribution systems, particularly in waters for industrial use, such as cooling and boiler waters. They metabolize reduced iron in their environment and deposit it as hydrated ferric oxide on or in their mucilaginous secretions.

cresol red An acid–base indicator dye. The color of the solution changes as the pH changes: pH 2–3, orange or amber; pH 7.2, yellow; pH 8.8, red.

crest (1) The top of a dam, dike, spillway, or weir, to which water must rise before passing over the structure. (2) The highest elevation reached by floodwaters flowing in a channel.

crest control A control method in which a device on the crest of the spillway of a dam is used to raise or lower the crest when variations in flow of the stream occur. The device may consist of flashboards, a radial drum, a tilting gate, or a bear-trap crest.

crest gate A gate installed on the crest of the spillway of a dam, operated to vary the discharge over the spillway.

crevasse A break in a levee through which flood water discharges.

crib weir A low diversion weir built of log cribs filled with rock.

crisis response chart An emergency planning tool to show alternate job duties in an organization in times of an emergency or disaster.

critical bed depth The depth of granular activated carbon (GAC) that is partially spent in a GAC adsorption bed. This depth lies between the fresh GAC and the spent GAC, defining the zone where adsorption takes place to the extent required to reach a selected effluent concentration target. The zone defined by this depth is equal to or less than the mass transfer zone, depending on the target removal. The critical bed depth is sometimes simply called the critical depth. *See also* mass transfer zone.

critical depth (1) The depth of water flowing in an open channel or partially filled conduit corresponding to one of the recognized critical velocities. (2) *See* critical bed depth. *See also* critical velocity.

critical drawdown period A parameter associated with the water supply characteristics of a lake or reservoir, equal to the length of time between the beginning of reservoir drawdown and the time when the lowest useful water surface elevation has been reached.

critical drought The design drought, or drought of record, for a surface water impounding reservoir or lake.

critical flow condition The state of flow at which the specific energy is a minimum for a given discharge. In equation form,

$$V_c = \sqrt{gD_c}$$

Where (in any consistent set of units):

V_c = critical velocity
g = gravitational constant
D_c = hydraulic depth = A_c/B_c
A_c = cross-sectional area of the flow
B_c = surface width of the open channel

For a rectangular channel,

$$V_c = \sqrt{gY_c}$$

Where (in any consistent set of units):

Y_c = critical water depth

See also Belanger's critical velocity; critical velocity; specific energy.

critical gradient (1) The channel slope for a given discharge where flow and critical flow occur at the same depth of flow. (2) The minimum grade of a conduit that will produce critical flow.

critical level A reference line representing the level of the check valve seat within a backsiphonage control unit. It is used to establish the height of the unit above the highest outlet or flood level rim.

critical micelle concentration (CMC) The surfactant concentration at which monomers begin to assemble in ordered, colloidal aggregates (micelles). At surfactant concentrations less than a compound-specific threshold value, surfactant molecules exist predominantly in monomeric form, with some fraction being adsorbed at system interfaces.

critical moisture content For a saturated soil, the moisture content at which—when a sample is subjected to shear (either confined or unconfined specimens)—the tendency of the sample is neither to expel nor to suck in water.

critical point drying A procedure used in preparing specimens for scanning electron microscopy. It eliminates forms of distortion such as shrinkage and wrinkles that ordinarily might occur if a specimen is air dried. *See also* scanning electron microscopy.

critical pressure The minimum pressure necessary to liquefy a gas that is at critical temperature.

critical slope *See* critical gradient.

critical temperature The temperature above which a gas cannot be liquefied solely by an increase in pressure.

critical velocity The velocity in an open channel for which the flow reaches the critical flow condition. This critical velocity is equal to the velocity at which a long wave of low amplitude propagates itself in water with a depth equal to the critical depth. *See also* Belanger's critical velocity; critical depth; critical flow condition.

critical-depth discharge measurement A determination (by indirect measurement) of peak discharge following a flood based on a field survey of high-water marks and channel and control-section geometry. The discharge is computed on the basis of critical flow (flow at minimum energy) theory.

crop coefficient (K_c) A factor used to adjust reference evapotranspiration and calculate water requirements for a given plant species.

cross A pipe fitting with four branches arranged in pairs, with each pair forming one axis and the axes at right angles.

cross contamination (1) Contamination that occurs in a mixed bed deionizer unit when anion and cation resins are mixed together after regeneration as a result of a system malfunction. (2) The intermixing of two water streams that results in unacceptable water quality for a given purpose.

cross leakage (1) Water leakage between the demineralized and concentrate streams in a membrane stack used in the electrodialysis process. (2) In a electrodialysis process, the permeation of water through the membrane from the concentrate stream to the demineralized stream in the membrane stack.

cross multiplication A method used to determine if two ratios are in proportion. In this method, the numerator of the first ratio is multiplied by the denominator of the second ratio. Similarly, the denominator of the first ratio is multiplied by the numerator of the second ratio. If the products of both multiplications are the same, the two ratios are in proportion to each other.

cross section A plane that is perpendicular to an axis, such as the axis of flow.

cross-connection The physical connection of a safe or potable water supply with another water supply of unknown or contaminated quality or such that the potable water could be contaminated or polluted.

cross-connection control The enforcement of an ordinance or other legal statement regulating cross-connections.

cross-linkage The bonding of linear polymers into a resinous product. Cross-linking with a material such as divinylbenzene produces a tridimensional exchanger product. The degree of cross-linkage is a factor of the resin's ability to withstand chemical oxidation. Standard softening resin is usually 8 percent cross-linked with divinylbenzene. Anion resins can be from 2 percent to 8 percent cross-linked. Acrylic compounds can also be used instead of divinylbenzene for cross-linking. *See also* cross-linking; polymer.

cross-linked Pertaining to chains of polymer molecules attached by cross-linking.

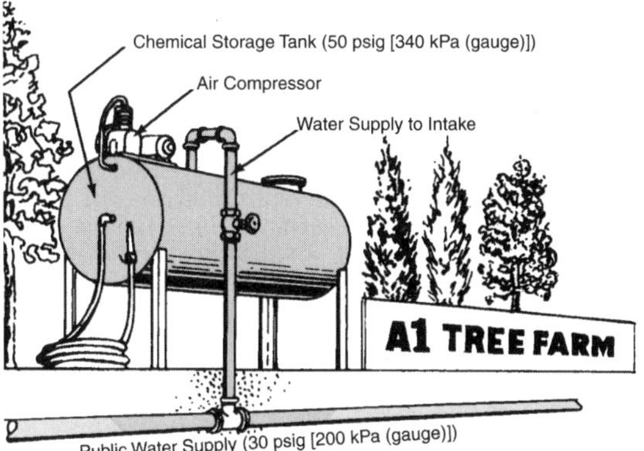

Cross-connection

cross-linking The attachment of two chains of polymer molecules by bridges composed of either an element, a group, or a compound that join certain carbon atoms of the chains by primary chemical bonds. Cross-linking increases the chains' strength, chemical resistance, or both. Cross-linking is used in the manufacture of some types of ion exchange resins and polyamide composite membranes. Cross-linking also occurs in nature, for example, in substances made up of polypeptide chains that are joined by the disulfide bonds of the cystine residues, as in certain proteins. Cross-linking can be effected artificially, either by the addition of a chemical substance (cross-linking agent) and exposing the mixture to heat or by subjecting the polymer to high-energy radiation (e.g., vulcanization of rubber with sulfur or organic peroxides). *See also* polymer; protein.

cross-sectional area The area of a plane in a tank or vessel that is at a right angle to the direction of the flow. The cross-sectional area is expressed, for example, in square feet or square metres.

cross-sectional bed area The area of activated carbon in a filter bed through which the stream flow is perpendicular.

cross-sectional epidemiologic study An analytical epidemiologic study in which information about disease status and exposures relate to the same period of time. The time sequence between exposure and disease status cannot necessarily be inferred.

crossbrace A horizontal member of a shoring system installed across to the sides of the excavation, the ends of which bear against either uprights or wales.

crossflow (1) An operating condition, in some types of pressure-driven membrane water treatment processes, where the feedwater and resulting concentrate or retentate pass tangentially across the surface of the membrane and the permeate passes in a perpendicular direction through the membrane barrier. With hollow fiber membranes, crossflow commonly means that pressurized feed passes tangentially through the inner lumen of the membrane fibers and that the permeate flow direction is inside out, or lumen to shell side (opposite the permeate flow direction, as in a transverse-flow hollow fiber membrane). (2) Flow that is at right angles to an inclined plate settler. *See also* hollow fiber membrane; inclined plate separator; transverse flow.

crossflow filtration A type of filtration that uses the shear force of tangential flow across the membrane surface (during suspension recirculation) to keep the particle buildup to a minimum. However, particle boundary layers cannot be completely eliminated by a crossflow of low fluid velocity that exists at the membrane surface. Ultrafiltration and reverse osmosis are examples of crossflow filtration. Crossflow filtration is also called tangential flow filtration. *See also* normal flow filtration.

crossing (1) The portion of a river between two bends, in one of which the flow is clockwise and in the other the flow is counterclockwise. (2) The relatively short and shallow length of a river between bends. (3) In rivers, flow over a shoal or bar between successive bendways.

crown The top of a levee, embankment, roadway, or pipe.

crown gate A canal-lock head gate.

CRT *See* cathode ray tube.

crude mortality rate *See* death rate.

crude rate A measure of morbidity and mortality computed for the whole population. A crude rate does not take into account population characteristics (confounding variables)—such as age, race, or gender—that may also affect morbidity or mortality; this feature is a disadvantage. Crude rates should not be used to compare populations in different geographic areas or time periods. Rates should be standardized, or specific rates should be computed; e.g., age-specific rates can be compared for the various age groups. *See also* standardized mortality rate; standardized rate.

Crustacea Aquatic arthropods having a body covered with a hardshell or crust, such as lobsters, shrimps, crabs, and barnacles.

cryptosporidiosis A protozoan infection (via *Cryptosporidium spp.*) of the epithelial cells of the gastrointestinal, biliary, and respiratory tracts of humans and other vertebrates, including birds, reptiles, rodents, cats, dogs, sheep, and cattle. In humans, the average incubation period is 7 days and the infectious dose is low. The primary symptom is profuse, watery diarrhea with cramping abdominal pain, but infection is often asymptomatic. In immunologically healthy individuals, symptoms usually end in 30 days; however, immunocompromised persons, especially AIDS patients, may be unable to clear the infection and illness may be prolonged, contributing to death. Cryptosporidiosis occurs worldwide, with children, animal handlers, people who have close personal contact with infected persons, and travelers most likely to be infected. Person-to-person transmission occurs by hand-to-mouth transfer of oocysts from feces of an infected person, especially in institutions and day-care centers. Animal-to-person contact and waterborne transmission are also important. In 1993, contamination of the Milwaukee, Wis., water system caused more than 400,000 cases of this disease; over 1,000 people were hospitalized. *See also Cryptosporidium*; oocyst; waterborne disease.

Cryptosporidium A widespread intestinal coccidian protozoan parasite about 3.5 micrometres in diameter, causing diarrhea and capable of infecting humans, birds, fish, and snakes. It is responsible for waterborne disease outbreaks.

Culture tube

***Cryptosporidium* inactivation device (CID)** An ultraviolet unit specifically designed to inactivate *Cryptosporidium* oocysts. It consists of two chambers, each with a nominal 2-micrometre pore diameter filter and three 85-watt low-pressure mercury lamps. The oocysts are captured on the first filter, and the ultraviolet dosage is established. The oocysts are then backflushed into the second chamber, where they are trapped until properly irradiated.

crystallizer An evaporator used to concentrate salts, producing crystalline solids. Commonly, feed salt is combined with recirculating brine and passes to a shell-and-tube heat exchanger for heating and subsequent discharge to a vortex chamber, where evaporation takes place and crystals form. The remaining brine is recirculated. Crystallizers are sometimes used in zero-discharge wastewater treatment facilities and for concentrate disposal in some small-scale desalting systems. *See also* zero discharge.

CSE *See* copper–copper sulfate electrode.

CSHO *See* compliance safety health officer.

CSI *See* calcite saturation index.

CSP *See* compliance support program.

CST *See* capillary suction time.

CSTR *See* continuously stirred tank reactor.

CT *See* contact time.

CTA *See* cellulose acetate membrane; cellulose triacetate membrane; comprehensive technical assistance.

cu *See* color unit.

cubature The cubic content or volume of tidal water as determined by computing the discharges and average currents of a cross section of a waterway from the tidal rises and falls upstream of the section. Cubature is essentially the arithmetic integration of the equation of continuity.

cube A cubic foot (0.028 cubic metres) of cation resin or filter medium.

cubic feet (ft^3) *See the Units of Measure list.*

cubic feet per hour (ft^3/h) *See the Units of Measure list.*

cubic feet per minute (ft^3/min) *See the Units of Measure list.*

cubic feet per second (ft^3/s) *See the Units of Measure list.*

cubic inch ($in.^3$) *See the Units of Measure list.*

cubic metre (m^3) *See the Units of Measure list.*

cubic spline A mathematical procedure to draw a smooth line through a series of observational data. A cubic spline can be roughly defined as a running interpolation of cubic polynomials.

cubic yard (yd^3) *See the Units of Measure list.*

culture The growth of living microorganisms in a medium.

culture medium A nutrient substance used to grow (culture) microorganisms. It may be a liquid (broth) or a solid medium made by adding agar as a solidifying agent.

culture tube A hollow, slender glass tube with an open top and a rounded bottom used in microbiological testing procedures such as the multiple-tube fermentation test.

culvert (1) A closed conduit for the free passage of surface drainage water under a highway, railroad, canal, or other embankment. (2) In highway usage, a bridge waterway structure having a span of less than 20 feet (6.1 metres).

culvert flow discharge measurement A determination (by indirect measurement) of peak discharge following a flood based on a field survey of high water marks and channel and culvert geometry. Discharge is computed on the basis of an evaluation of energy changes between the approach section and the control section.

cumulative data function A data function that determines the cumulative or integrated value of dependent variable y for a given value of the independent variable x.

cumulative effects (1) Effects that are irreversible or slowly reversible such that repeated exposure to a chemical or infectious agent leads to increasing severity of an adverse health effect. The chemical can be in the system for a very short time. (2) Health effects that can arise from the accumulation of a toxic form of a chemical in a tissue with repeated exposure. By definition, the chemical is poorly eliminated from the tissue.

cumulative exposure The summation of exposures of an organism to a chemical over a period of time.

cumulative impact analysis A review of the cumulative economic, environmental, and social impacts of proposed water activities and projects associated with development within a drainage area or ecosystem.

cumulative runoff The total volume of runoff over a specified period of time. Successive summations are frequently plotted against time to produce a mass curve. *See also* mass diagram.

cumulose deposit A deposit formed from the accumulation of organic matter, with small amounts of rock waste included.

cupric sulfate *See* copper sulfate.

cuprichloramine A mixture of copper sulfate ($CuSO_4{\cdot}5H_2O$), ammonia (NH_3), and chlorine (Cl_2), used as an algicide.

CUR *See* activated carbon usage rate.

curb A precast concrete liner for a dug well.

curb cock *See* curb stop and box.

curb inlet An intake structure to allow stormwater to enter a stormsewer from a roadway gutter. It is configured to the shape of the curb and gutter to provide for easy installation and efficient operation. *See also* catch basin.

curb stop *See* curb stop and box.

curb stop and box A shutoff valve in a water service line buried near the curb of a customer's premises, with a box or housing over the valve extending to the surface of the ground for access to the valve.

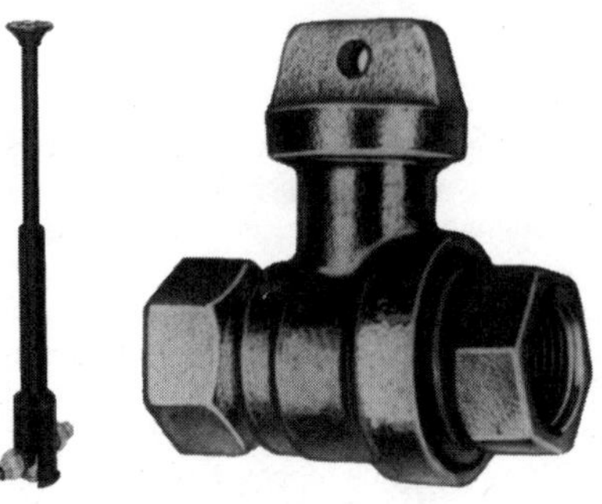

Curb stop and box

cured-in-place lining A cement–mortar mixture applied to the interior of steel or iron pipe water mains after they have been installed. The lining is spun onto the interior by centrifugal force and then worked with a cone-shaped trowel or spiral rotating trowel to smooth the surface of the lining. The lining can be applied to either new or old unlined pipe after that pipe has been scraped and cleaned.

curie (Ci) *See the Units of Measure list.*

curing time (for residuals) The period of time required, following treatment, for residuals to be suitable for ultimate disposal.

current The speed of electrical charge movement, or the "flow rate," of electricity, measured in amperes. *See also* Ohm's law.

current density As used in desalting, amperage per unit area flowing through a stack; current flow through a stack per unit cross-sectional area.

current density units In the US customary system, amperes per square foot. In Système International, amperes per square metre.

current diagram A graphic representation showing the velocities of flood and ebb currents and times of slack and strength over a considerable stretch of the channel of a tidal waterway. The times relate to the tide or current phases at some reference station.

current difference The difference between the time of slack water or strength of current in a locality and the time of the corresponding phase of the current at a reference station for which predictions are given in a current table.

current efficiency In desalting processes, the theoretical power required for a particular job outcome compared to (i.e., divided by) actual power consumed.

current meter A water-measuring device that uses the velocity of water flowing through the device to turn a turbine wheel, which in turn reflects the volume of water passing through the device.

current pole A pole used in observing the direction and velocity of water current. The standard pole used by the US Coast and Geodetic Survey is about 3 inches (76 millimetres) in diameter and 15 feet (4.6 metre) long and is weighted at one end to float upright with roughly the top 1 foot (0.3 metre) out of the water. As the pole is carried by the current out from an observing vessel, its direction and the amount of line passing over a fixed reference point during a specified interval indicate the direction and velocity of the current.

current regulator A device that automatically holds electric current within certain limits.

current span The amount of electric current moving through a pipe, calculated based on a measurement of the difference in voltage between two locations along the pipeline.

current table A table that gives daily predictions of the times of slack water and of the times and velocities of the flood and ebb maximums, usually supplemented by current differences and constants by which additional predictions can be obtained for other places.

current-meter rating The operation of developing for a given current meter the relationship between the revolutions of its wheel or screw per unit of time (usually 1 minute) and the velocity at which the meter travels through still water. With such a relationship established in graphical, algebraic, or tabular form, the velocity of water passing the meter when it is held in a stationary position can be determined.

curtain A wall or diaphragm of concrete or steel, or a trench filled with clay slurry or impervious earth, extending into the foundation of a dam or levee and either making a watertight connection with the dam or its impervious facing or extending into the body of the dam a considerable distance. Its purpose is to prevent or reduce the passage of water under the dam and the foundation material or through the upper layers of the foundation material. A curtain is sometimes called a cutoff.

curtain drain A drain constructed at the upper end of an area to be drained to intercept surface water or groundwater flowing toward the protected area from higher ground and carry it away. A curtain drain is also called an intercepting drain.

curtain wall A deep cutoff wall used under an overflow masonry dam on pervious foundations to stop underflow by lengthening the path of travel of underflowing water and thereby providing resistance to seepage greater than the hydrostatic head of the water impounded by the dam.

curvature factor A numerical quantity used to express the energy loss caused by one or more sharp curves in a pipeline in terms of the velocity head of water flowing in the line.

curve (1) A graph representing the changes in value of physical or statistical quantities. (2) A continuous bending.

curve fitting A mathematical procedure to develop an equation defining a series of observations. Curve fitting is often performed by the method of least squares to define the parameters of the assumed equation to best describe the observations.

CUSA *See* certified utility safety administrator.

customer class A grouping of individual customers to a specific category such as residential, commercial, or industrial users for rate-making purposes.

customer costs Costs that relate to serving customers irrespective of the amount or rate of water use.

customer information system (CIS) A system that manages, processes, and controls information for each utility customer, the customer's premises, the individual facilities that serve that customer, water usage, and accounting. The system supports and controls interactions with the customers, most notably meter reading and billing.

customer service All elements of providing water that meets all health standards to a consumer: delivery, sufficient pressure, measuring the amount delivered, billing at regular intervals, and communicating with the customer to understand service needs.

customer service system A system that develops and generates a response to customers' problems or inquiries.

cut-in repair coupling A coupling designed with a middle ring, two sleeves, and end rings bolted together with gaskets, such that parts can be inserted in a small gap in a water main and then reconnected to cover the gap. This type of coupling is normally used where a bell on the pipe had to be removed to repair a leak near the joint.

cutaneous irritation An inflammatory response of the superficial cells of the skin.

cutoff (1) A more direct channel, formed either naturally or artificially, connecting two points on a stream channel, shortening the original length of the channel and increasing its slope. (2) *See* curtain. (3) Water stop.

cutoff ratio The ratio of the length of a cutoff to the original length of the river.

cutoff trench A trench excavated below the normal base of a dam or other structure and filled with relatively impervious material to reduce percolation under the structure.

cutoff wall A thin wall or footing constructed downward from, under, or around the head wall and lip wall of a dam to provide resistance to seepage.

cutwater The angular edge of a bridge pier, shaped to lessen to the flow of water, ice, debris, and so on.

CWA *See* Clean Water Act.

CW&WA *See* Canadian Water and Wastewater Association.

CWIP *See* construction work in progress.

CWS *See* community water system.

CWSS *See* Community Water Supply Study.

cwt *See* hundredweight *in the Units of Measure list.*

cyanazine (C_2H_5NH-$C_3N_3(Cl)$-$NHC(CH_3)_2CN$) The common name for the herbicide 2-((4-chloro-6-(ethylamino)-*s*-triazin-2-yl)amino)-2-methylproprionitrile. *See also* herbicide.

cyanide (CN^-) An inorganic ion. Hydrogen cyanide (HCN) has various industrial uses, including use as a rodenticide or pesticide. Sodium cyanide (NaCN) is used in, for example, the extraction of gold and silver from ores, electroplating, and insecticides or fumigants. Cyanide is regulated by the US Environmental Protection Agency. *See also* fumigant; insecticide; pesticide; rodenticide.

Cyanobacteria The name given to organisms formerly referred to as blue-green algae. Many of the Cyanobacteria cause taste-and-odor problems in lakes and impoundments used as sources for drinking water.

cyanogen bromide (CNBr) A disinfection by-product formed during the ozonation or chloramination of bromide-containing waters. *See also* disinfection by-product.

cyanogen chloride (CNCl) A disinfection by-product formed in low levels during chloramination. It can also be produced during the chlorination of cyanide-containing waters, though alkaline chlorination will destroy the cyanogen chloride. Cyanogen chloride has an industrial use in tear gas. *See also* disinfection by-product.

cyanogen halide (CNX) A class of disinfection by-products produced primarily during chloramination. *See also* cyanogen bromide; cyanogen chloride.

cyanosis A dusty bluish or purplish discoloration of the skin or mucous membranes caused by insufficient oxygen delivery to tissues. One potential cause associated with drinking water is production of methemoglobinemia by excessive levels of nitrates (after conversion to nitrite) or chlorite.

cycle In alternating current electrical systems, one complete change in direction of current flow from zero to maximum to zero in one direction, then to maximum and back to zero in the opposite direction.

cycles of concentration The number of times an initial solids concentration in a liquid is increased by removal of liquid volume (and weight), thus increasing the solids. For example, if water has a solids concentration of 500 milligrams per litre and is evaporated to half the original liquid volume, the solids concentration becomes 1,000 milligrams per litre, or twice the original concentration. If the water is evaporated to one-fourth of its original liquid volume, the solids concentration becomes 2,000 milligrams per litre, or four times the original concentration. For boiler water applications, the number of cycles of concentration is the ratio of circulating water solids concentration to makeup water solids concentration. The number of cycles of concentration for boiler water is controlled by the blowdown rate of the circulating water from the system and makeup water quantity and quality; it is limited by the solubility of solids in the system and the use of chemical additives (e.g., acid, scale inhibitors). To reduce the cycles of concentration, the blowdown rate is increased; to increase the cycles of concentration, the blowdown rate is reduced. *See also* blowdown.

cycle of fluctuation The total time occupied by a period of rise and a succeeding period of decline of a water table. The most common kinds of cycles are daily, annual, and secular (or cycles extending over longer periods). A cycle of fluctuation is also called a phreatic cycle.

cycles per second (cps) *See the Units of Measure list.*

cyclic depletion Withdrawal of water at a rate in excess of the average rate of supply over a secular cycle. *See also* seasonal depletion.

cyclic recovery The gradual rise in elevation of the water table caused by additions of water, usually extending over several years and following a period of cyclic depletion.

cyclic storage The accumulation in storage of water in a reservoir during one or several years of greater-than-average supply, the holding of such water for 1 or more years, and the releasing of it for use during a period of years when the supply would otherwise be insufficient. Cyclic storage is also called overyear storage.

cyclone degritter A centrifugal sand-and-grit removal device.

cyclone separator A conical unit used for separating particles by centrifugal force. It is also called a cyclone precipitator. *See also* hydroclone.

cyclonite ($C_3H_6O_6N_6$) The common name for the explosive symtrimethylene trinitramine; also known as RDX.

Cyclospora A family of protozoan organisms in the subclass *Coccidia* that are distributed worldwide. They infect both immunocompetent and immunosuppressed children and

Cyclone separator

adults, causing explosive, watery diarrhea (usually self-limiting, even in immunosuppressed subjects), anorexia, fatigue, and weight loss. The organisms are thought to be waterborne with some seasonal incidence. *See also Cyclospora cayatensis.*

Cyclospora cayatensis A coccidian parasite distributed worldwide, related to *Cryptosporidium parvum*. Oocysts of *Cyclospora cayatensis* are about 9 to 10 micrometres in size, about twice the size of *Cryptosporidium* oocysts. The organism infects both immunocompetent and immunosuppressed children and adults. *See also Cyclospora.*

cycrophilic *See* psychrophilic range.

cylinder A metal container to hold pressurized gas, such as chlorine for disinfecting in water.

cylinder gate A gate installed in a dam to control the flow from a reservoir. It consists of a steel cylinder that is open at the top and bottom and has a balanced water pressure on the inside and outside surfaces. The cylinder forms a seat over the opening and closes it when seated.

cylinder valve *See* container valve.

cyst The infectious stage for *Giardia*, 7 to 10 micrometres long and refractile to light when viewed with a brightfield microscope. *See also* oocyst.

cysteine conjugate-lyase An enzyme that catalyzes the cleavage of chlorovinyl-cysteine conjugates on the cysteine ($HSCH_2CH(NH_2)COOH$) side of the sulfur atom, yielding several reactive metabolites (e.g., thioketenes) that can react with nucleophilic centers on proteins and nucleic acids. Researchers have proposed that this pathway is responsible for toxic and carcinogenic effects of trichloroethylene ($CHCl{:}CCl_2$) and tetrachloroethylene ($Cl_2C{:}CCl_2$) in the kidney.

cytochrome b_5 reductase A flavoprotein responsible for transferring electrons from nicotinamide-adenine dinucleodide phosphate to cytochrome b_5 to initiate the microsomal metabolism of foreign and endogenous chemicals.

cytochrome oxidase The terminal oxidase of the mitochondrial electron transport system. It is most generally studied as a complex of cytochromes a + a_3. The mitochondrial electron chain serves to transfer electrons from substrates to oxygen in a way that captures the energy in the form of adenosine triphosphate that can be used for cellular functions.

cytochrome P-450 A family of hemeproteins that are localized in the smooth endoplasmic reticulum of cells (particularly the liver) and are thought to be responsible for the metabolism of chemicals foreign to the body. Many of these enzymes also have endogenous substrates that participate in a variety of cellular functions. The different isoforms of P-450 have a wide variety of substrate specificities. Moreover, specific isoforms of P-450 are frequently increased in concentration in response to exposure to a chemical. This may result in an increased or decreased toxic response to subsequent exposures to the chemical, depending on whether the P-450 metabolizes the chemical to a more or less toxic form.

cytopathic Pertaining to cell death or morphological changes within a cell.

cytopathic effect (CPE) An effect consisting of morphologic alterations of host cells, usually resulting in cell death.

cytopathic unit (cpu) A unit used to measure changes in cells (i.e., tissue culture) caused by microorganism infection.

cytosol The fluid matrix in the interior of cells. The internal constituents of a cell are said to be in the cytosol if they are not associated with the clearly particulate bodies within the cell (e.g., nucleus, mitochondria, peroxisomes, endoplasmic reticulum). Sometimes this definition is an operational one, depending on how the constituent distributes when cells are lysed and the contents analyzed by ultracentrifugation. The operational definition should not be confused with what occurs in vivo.

cytotoxic *See* cytotoxicity.

cytotoxic effects *See* cytotoxicity.

cytotoxicity Lethal injury to cells. In addition to compromising the function of the tissue or organ in which these cells occur, the repeated induction of cytotoxic effects in vivo can, through repeated activation of repair processes, contribute to the development of chronic and less reversible effects in a tissue (e.g., fibrosis, cirrhosis of the liver, cancer).

CZMA *See* Coastal Zone Management Act.

D

D *See* dalton *in the Units of Measure list.*

2,4-D *See* 2,4-dichlorophenoxyacetic acid.

d *See* day *in the Units of Measure list.*

D/DBP *See* disinfectant/disinfection by-products.

D/DBP Rule *See* Disinfectant/Disinfection By-Products Rule.

da *See* deka; darcy *in the Units of Measure list.*

DAF *See* dissolved air flotation.

daily flood peak The maximum mean daily discharge occurring in a stream during a given flood event.

daily flow The volume of water that passes through a plant in 1 day (24 hours). It is more precisely called daily flow volume.

dalapon (CH_3CCl_2COOH) The common name for 2,2-dichloropropionic acid. Dalapon is an herbicide regulated by the US Environmental Protection Agency.

dalton (D) *See the Units of Measure list.*

dam A barrier constructed across a watercourse for the purpose of creating a reservoir, diverting water into a conduit or channel, creating a head that can be used to generate power, improving river navigability, or a combination of these goals. Dams are classed as fixed or movable and may be of such types as gravity, arch, earth, rock fill, or combinations.

dam site A location where the topographical and other physical conditions are favorable for the construction of a dam, or any site where a dam has been built, is being built, or is contemplated.

dam toe The downstream edge at the base of a dam.

dam top Usually a top of a roadway, a top or earth top of a parapet, top of a gate, a top of nonoverflow masonry, or a top of a spillway. (Dams have different designs and thus different tops.)

dangerous air contamination An atmosphere presenting a threat of death, injury, acute illness, or disablement due to the presence of flammable or explosive toxic or otherwise injurious or incapacitating substances: (1) Dangerous air contamination caused by the flammability of a gas or vapor is defined as an atmospheric concentration of the gas or vapor greater than 10 percent of its lower explosive (lower flammable) limit. (2) Dangerous air contamination caused by a combustible particulate is defined as an atmospheric concentration greater than 10 percent of the minimum explosive concentration of the particulate. (3) Dangerous air contamination caused by the toxicity of a substance is defined as an atmospheric concentration immediately hazardous to life or health.

Daphnia A microscopic crustacean found in water.

DAPI/PI *See* 4'6-diamidino-2-phenylindole/propidium iodide.

Darcian velocity *See* specific discharge.

Darcy velocity *See* specific discharge.

Darcy law An equation used to compute the discharge of water flowing through an aquifer. The law states that the discharge is equal to the product of hydraulic conductivity, cross-sectional area perpendicular to flow, and hydraulic gradient. It is the fundamental equation of motion used in nearly all groundwater computations.

$$Q = KA(\Delta h/\Delta l)$$

Where (in any consistent set of units):

Q = the flow
K = the hydraulic conductivity
A = the cross-sectional area perpendicular to the flow
Δh = the difference in hydraulic head
Δl = the length traveled

The term $\Delta h/\Delta l$ is called the hydraulic gradient.

Darcy–Weisbach equation *See* Darcy–Weisbach formula.

Darcy–Weisbach formula The formula for calculating friction-induced head loss in pipes:

$$h_L = f(L/D)(V^2/2g)$$

Where (in any consistent set of units):

h_L = the head loss
f = the friction factor, dimensionless
L = the length of pipe
D = the diameter of the pipe
V = an average velocity
g = the gravity constant

Darcy–Weisbach roughness coefficient The roughness coefficient (f) in the Darcy–Weisbach formula. It is dimensionless. *See also* Darcy–Weisbach formula.

data annotation Textual annotation on a map or graphic display that is derived from a feature attribute or database element.

data collection unit (DCU) A device used in some automatic meter-reading systems to collect and collate data for input to the billing system.

data conversion The transformation of existing manual maps and data or other data such as charts to computer-readable format.

data dictionary A database that serves as a catalog containing information about map features or attributes. The catalog may define data file and element names, sources, accuracy, date of entry or update, and other characteristics of the data and their sources.

data dispersion (1) Dissimilarity between two sets of data. (2) The archaic term used by statisticians for variance. *See also* variance.

data flow diagram A tool used in conducting structured designs that illustrate the flow of data between processes in a system.

data format A specific order in which data are organized.

data layer A specific homogeneous facility, such as a water distribution network with all its appurtenances (valves, hydrants, and so forth), or utility mapped over base map data.

data questionnaire A questionnaire organized to collect data from various processes, including water treatment plants and distribution systems.

database (1) A large collection of data, organized so that it can be expanded, updated, and retrieved rapidly for various uses; a specific grouping of data within the structure of a database management system that has been defined by a particular use, user, system, or program. The data are organized so that they can be expanded. A database consolidates many records that were previously stored in separate files so that a common pool of data records serves as a single central file for many processing applications. (2) A set of interrelated records that are stored on a device, frequently a computer, for direct access by multiple users for subsequent analysis. It is highly structured to minimize redundant data.

database management system (DBMS) A software package providing the ability to store and retrieve information.

dateometer A small calendar disk attached to motors and equipment to indicate the year in which the last maintenance service was performed.

datum (1) An agreed upon standard point or plane of stated elevation, noted by permanent bench marks on some solid, immovable structure, from which elevations are measured or to which they are referred. (2) Any position or element in relation to which others are determined, e.g., the horizontal control system used in map making. (3) Any numerical or geometrical quantity or set of such quantities that may serve as a reference or base for other quantities.

datum line A line from which heights and depths are calculated or measured. It is also called a datum plane or datum level.

day (d) *See the Units of Measure list.*

day tank A treatment chemical storage vessel that contains a diluted concentration in a feed volume suitable for a short period, typically from 1 to 3 days. For example, dry or viscous polymers are often diluted, or "aged," in day tanks prior to application. A day tank is also called an age tank. *See also* polymer aging.

dB *See* decibel *in the Units of Measure list.*

DBAA *See* dibromoacetic acid.

DBAN *See* dibromoacetonitrile.

DBCAA *See* dibromochloroacetic acid.

DBCM *See* dibromochloromethane.

DBCP *See* 1,2-dibromo-3-chloropropane.

DBMS *See* database management system.

DBP *See* disinfection by-product.

DBP_0 *See* instantaneous disinfection by-product concentration.

DBPFP *See* disinfection by-product formation potential.

DBPP *See* disinfection by-product precursor.

DBPRAM *See* disinfection by-product regulatory assessment model.

DBP_t *See* terminal disinfection by-product concentration.

DC *See* direct current.

DCAA *See* dichloroacetic acid.

DCAN *See* dichloroacetonitrile.

DCPA *See* dimethyl-2,3,5,6-tetrachloroterephthalate.

DCS *See* distributed control system.

DCU *See* data collection unit.

DCVA *See* double check valve assembly.

DDC *See* direct digital control.

DDD *See* dichlorodiphenyl dichloroethane.

DDE *See* dichlorodiphenyl dichloroethylene.

DDT *See* dichlorodiphenyl trichloroethane.

DE *See* diatomaceous earth; district engineer.

DE filter *See* diatomaceous earth filter.

deactivation The loss of catalytic activity by an absorbent or a catalyst.

dead end A section of a water distribution system that is not connected to another section of pipe by means of a connecting loop. Such portions of a distribution system can experience lower flows than surrounding portions, which can lead to water quality problems caused by somewhat stagnant water. Examples of problems include tastes or odors, bacteriological growth, loss of chlorine residual, or any combination of these.

dead storage Storage below the lowest outlet levels of a reservoir, not available for use.

dead time Any definite time period required for any process material to propagate or travel between two different locations. Dead time can occur in the measuring system, the controller, or any of the process streams.

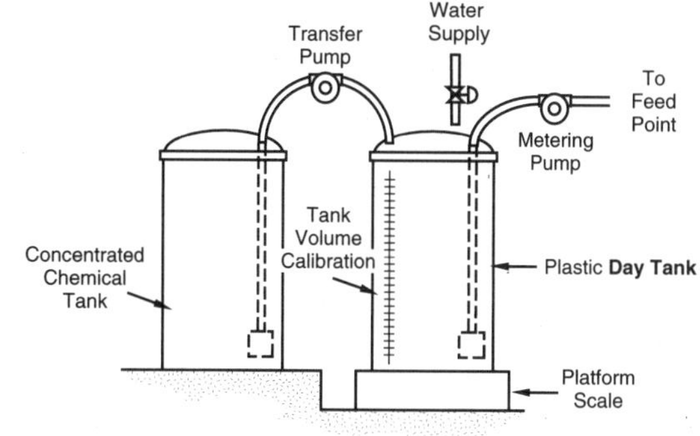

Day tank

dead water (1) Standing or still water. (2) Water in a boiler, settling tank, or similar equipment that fails to circulate to the extent required for proper functioning of the equipment.

dead well A shaft or well driven through an impermeable stratum to allow water to drain through to a permeable one. A dead well is also called an absorbing well, drain well, or negative well.

dead-end filtration A flow pattern in which all water flows through the medium or membrane (as opposed to crossflow filtration), thus allowing a buildup of a particulate layer on or near the surface of the medium and requiring periodic backwashing, repeated cleaning, or cartridge replacement.

dead-end flow A flow pattern in which all the feedwater passes through a filtration medium during service operating mode.

deadline The date and time when a program, project, or product must be completed.

deaerate To reduce or eliminate oxygen or air from a liquid or from granular activated carbon.

deaeration *See* deaerate.

deaerator A device that removes oxygen or air from a liquid. *See also* degasifier.

dealkalization Any process for the reduction of alkalinity in a water supply. Dealkalization is generally accomplished by a chemical feed process or combined cation and anion ion exchange systems.

death rate A rate expressing the proportion of a population who die of a certain disease or of all causes in a given period, usually a calendar year. The denominator is the average total population in which the deaths occurred. The rate is usually multiplied by 1,000 to produce a rate per 1,000 (another convenient multiplier of 10 can also be used). Morbidity can also be expressed in this manner, i.e., crude morbidity rate. The death rate is sometimes called the crude mortality rate. *See also* crude rate; standardized mortality rate; standardized rate.

debenture A long-term debt instrument that is not secured by a mortgage on specific property.

debris Generally solid wastes from natural and artificial sources deposited indiscriminately on land and water. *See also* detritus; flotsam; jetsam; litter; trash.

debris basin A basin formed behind a low dam or excavated in a stream channel to trap debris or bed load carried by mountain torrents. Such a basin requires periodic cleaning of debris by excavation to restore capacity.

debris cone A fan-shaped deposit of soil, sand, gravel, and boulders deposited at the foot of a range of hills or mountains by a stream that flows from the higher area or where the velocity of the stream is reduced sufficiently to cause such deposits.

debris dam A fixed dam built across a stream channel to catch and retain debris such as sand, gravel, silt, or driftwood.

debris flow An unsorted mixture of water, rock, soil, and earth material that moves downslope as a mass. Such flows are common in areas where vegetation is lacking and surface water runoff reaches drainageways rapidly.

debris rack *See* bar screen.

decant To draw off the liquid from a basin or tank without stirring up the sediment in the bottom.

decantation The process of drawing off a supernatant liquid without disturbing the underlying lower liquid layers and the precipitate.

decarbonate To reduce or eliminate carbon dioxide (CO_2) from a liquid.

decarbonation *See* decarbonate.

decarbonator A device that removes carbon dioxide (CO_2) from a liquid. *See also* degasifier.

decationize The exchange of cations for hydrogen ions by a strong acid cation exchanger operated in the hydrogen cycle.

decentralized processing A distribution of computer processing tasks among separate processors.

dechlorination The process of removing chlorine ($HOCl$, OCl^-) from solution. Dechlorination may be required (1) when high dosages are provided to destroy tastes and odors or to reduce their concentration in the distribution system, (2) to reduce the potential toxic effect of a chlorinated discharge to aquatic life, (3) to protect downstream processes, such as zeolite softeners, from the oxidative effects of chlorine, or (4) to reduce chlorinous tastes in drinking water (particularly bottled water). Dechlorination is typically achieved through chemical addition of a reducing agent, the most common of which is sulfur dioxide (SO_2).

dechlorination agent A chemical used to reduce the chlorine residual for the purpose of sample preservation. Commonly used dechlorination agents include sodium thiosulfate ($Na_2S_2O_3 \cdot 5H_2O$), sodium sulfite (Na_2SO_3), ammonium chloride (NH_4Cl), and ascorbic acid ($C_6H_8O_6$). *See also* dechlorination.

deci Prefix meaning 10^{-1} used in Système International. Its use should be avoided except for the measure of area and volume.

decibel (dB) *See the Units of Measure list.*

decilitre *See the Units of Measure list.*

decision tree A decision matrix developed by the US Environmental Protection Agency that estimates the number or percentage of public water systems that are affected by a new regulation and would be expected to choose any one of a predetermined set of alternatives for complying with the regulation. The decision matrix is used to develop national cost estimates for meeting a new regulation by multiplying the number of systems expected to choose a particular compliance option by the unit cost of that option and then summing across all options considered.

declining block pricing A rate design wherein the unit price of water to all classes of users declines with each succeeding block of use, resulting in both incremental and average cost of water decreasing with increased customer use. This was the traditional rate structure in the United States for many years. It recognizes the lower unit cost of water associated with the relatively level demands of large industrial customers compared to the peaking demands of, for example, residential customers. Because of scarcity of water in many areas, a desire to encourage customers to conserve the resource, and escalating capital costs, many utilities have converted the declining block rate structure into single unit rates, seasonal rates, and inclining block rates.

declining block rate *See* declining block pricing.

declining rate filtration A filtration process in which the filter rate gradually decreases throughout the course of the filter run. A flow-restricting orifice is used in the effluent piping to control the maximum rate when the filter is clean. As the filter clogs with solids, resistance through the filter bed increases, which causes the flow rate to decrease as flow is shifted to other, cleaner filters in the treatment plant.

decolorization The process of removing color bodies such as tannins or humic acid from water by means of oxidation, coagulation–filtration, adsorption, or ion exchange.

decomposition The conversion of chemically unstable materials to more stable forms by chemical or biological action. If organic matter decays when no oxygen is present (i.e., under anaerobic conditions or putrefaction), undesirable tastes and odors are produced. Decay of organic matter when oxygen is present (i.e., under aerobic conditions) tends to produce much less objectionable tastes and odors.

decreasing block rate *See* declining block pricing.

decross-linking The alteration of an ion exchange resin structure by destruction of the cross-linked polymer (such as divinylbenzene) as a result of very aggressive chemical attack—by chlorine (HOCl, OCl^-), ozone (O_3), or hydrogen peroxide (H_2O_2), for example—or heat. Decross-linking causes increased moisture content in an ion exchange resin and physical swelling of the beads.

dedicated capacity The portion of the water utility's total capacity that is set aside, or dedicated, for use by an individual large-use customer or group (class) of customers whose total use is a significant part of the utility's total capacity requirement.

dedicated metering Metering of water service based on a single type of use, as in separate metering for landscape irrigation only.

DEE *See* diethyl ether.

deep-bed filtration A type of filter in which the medium is deeper than a standard design of 2 to 3 feet (0.6 to 0.9 metre). A deeper bed is often combined with larger filter media to allow deeper penetration of particles, thereby more effectively using the full volume of the bed in collecting particles. Deep-bed filters are also used with ozonation to provide a larger surface area for removing biodegradable components formed by ozone.

deep injection well A well discharging under pressure to a deep subsurface stratum. Such a well is often used for disposal of liquid waste streams to a suitable confined poor-water-quality aquifer that is generally considered unusable for other purposes.

deep percolation The movement of groundwater through deeply buried permeable rocks.

deep seepage That portion of the runoff that escapes from a reservoir through the underlying earth or rock strata, below any possible intercepting cutoff constructed at the dam.

deep well A drilled well for the extraction of water, salt, gas, oil, or other minerals.

deep-well injection The disposal of waste by pumping into a deep well discharging to an aquifer that is not a water supply.

deep-discount bond A bond that typically does not pay any annual interest; instead, the bondholder is rewarded in appreciation of the principal payable. *See also* zero-coupon bond.

deep-well pump A pump used for lifting water from deep wells. The pumping mechanism is usually installed within the well at a considerable distance below the surface. The pump may be of a reciprocating or centrifugal type.

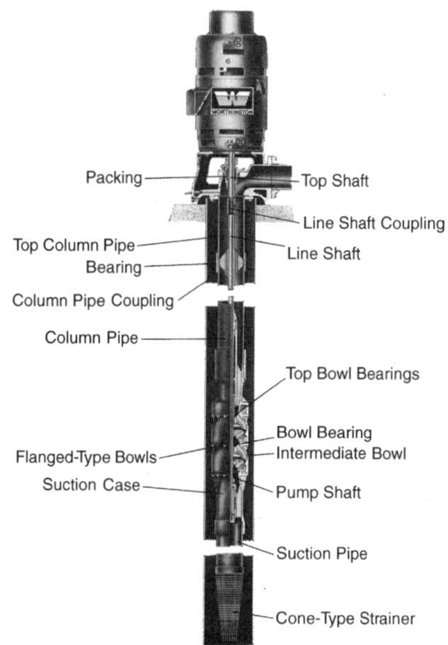

Deep-well pump

deep-well turbine pump A centrifugal pump adapted for deep-well use and consisting of a series of stages. Each stage comprises a set of vanes in a case or bowl, and the number of stages increases with the operating head.

defecate To excrete waste matter from the intestinal tract.

deferred serial bond A serial bond in which the first installment does not fall due for 2 or more years from the date of issue.

deferrization In water treatment, the removal of soluble compounds of iron from water.

deficiency (1) The amount or amounts by which quantities fall short of a given demand. (2) The amount by which the natural flow of a stream or other source of water supply fails to meet the demand for irrigation, hydroelectric power, domestic consumption, or other purposes. (3) The amount by which the precipitation during a given period falls short of the normal precipitation for that period.

defined-substrate technology (DST) A microbiological testing technology based on the ability of coliform bacteria to produce an enzyme that reacts with a substrate, ortho-nitrophenyl-β-D-galactopyranoside, to release a compound that produces a yellow color. When used in conjunction with another substrate, 4-methylumbelliferyl-β-D-glucuronmide, the technique can be used to detect both total coliform bacteria and *E. coli* within 24 hours. Commercial kits using this technology are available.

defluoridation The removal of fluoride through treatment by lime softening of high-magnesium water, ion exchange, or activated alumina. *See also* activated alumina; ion exchange; lime softening.

defoliant An herbicide (e.g., 2,4-dichlorophenoxyacetic acid) that removes leaves from trees and growing plants. *See also* 2,4-dichlorophenoxyacetic acid; herbicide.

deformable Pertaining to suspended solids that extrude into the interstices of a filter cake and cause rapid filter plugging.

degasification The removal of dissolved gases from water to reduce their impact on water quality, filter operation (via air binding), pump cavitation, corrosion, or other parameters. Degasification is accomplished by mechanical methods (e.g., a degasifier or venturi), chemical methods, or a combination of both. *See also* air binding; cavitation; corrosion; degasifier; stripping.

degasifier A device that removes dissolved gases, such as carbon dioxide (CO_2) or hydrogen sulfide (H_2S), from water. The removal rate is dependent on the concentration of the gas in the liquid phase compared to the equilibrium concentration of the gas in the liquid phase for the partial pressure of the gas in the gas phase, as determined by Henry's law. Commonly, the water falls downward by gravity through packing media, and air passes in a countercurrent direction driven by forced or induced draft fans or blowers. The air removes (strips) the gas or gases from the water. *See also* deaerator; decarbonator; Henry's law; stripping.

degasify To reduce the concentration of one or more gases and other volatile substances from a liquid.

degassing *See* degasification.

degradation (1) The loss of capacity, reduction of resin particle size, excessive swelling of resin particles, or any combination of these factors resulting in a lessening of the ion exchange capabilities of a resin. This may occur as a result of the type of service for which the resin was used, the solution concentrations used, heat, or aggressive operating conditions. (2) A decrease in the quality of a water. (3) A reduction in performance of a unit process used to treat water. (4) The changing of an organic material or substrate into smaller, less complex compounds, principally by biological means.

degrade To convert complex materials to simpler compounds by biological, chemical, or physical treatment. *See also* biodegradable.

degraded wetland A wetland altered by human action through impairment of some physical or chemical property in a way that results in reduced habitat value or other reduced functions (e.g., by flood storage).

degrading river A section of a stream or river where the bottom elevation is declining because of erosion. *See also* aggrading river.

degree (1) On the Celsius temperature scale, $1/100$ of the interval from the freezing point to the boiling point of water under standard conditions; on the Fahrenheit scale, $1/800$ of this interval. (2) A unit of angular measure; the central angle subtended by $1/360$ of the circumference of a circle.

degree Celsius (°C) *See the Units of Measure list.*

degree Fahrenheit (°F) *See the Units of Measure list.*

degree, Baumé (Bé) *See* Baumé degree.

degree, temperature *See* temperature degree.

dehydration (1) The chemical or physical process whereby water in chemical or physical combination with other matter is removed. (2) The geologic process in the weathering of rocks or soils whereby water is given up and a new mineral compound formed.

deionization The removal of all ionized minerals and salts (both organic and inorganic) from a solution by a two-phase ion exchange procedure. First, positively charged ions are removed by a cation exchange resin in exchange for a chemically equivalent amount of hydrogen (H^+) ions. Second, negatively charged ions are removed by an anion exchange resin for a chemically equivalent amount of hydroxide (OH^-) ions. The hydrogen and hydroxide ions introduced in this process unite to form water molecules. This process is also called demineralization by ion exchange. *See also* demineralization; demineralize.

deionize *See* deionization.

deionized (DI) water A type of reagent water used in the laboratory to prepare blanks, standards, and so on. It is typically prepared in the laboratory by passing tap water through a series of ion exchange cartridges. Many of the major constituents in tap water, such as minerals, are essentially removed during deionization.

deionizer A device used to remove all dissolved inorganic ions from water.

DEIS *See* Draft Environmental Impact Statement.

deka (da) Prefix meaning 10^1 used in Système International. Its use should be avoided except for the measure of area and volume.

delayed hypersensitivity Sensitivity to a chemical or biological agent that develops only after prior exposure. It essentially pertains to an allergic reaction to the agent, mediated via the immune response.

delayed neurotoxicity (1) A specific neuropathy that is produced by certain organophosphorus compounds, including some (but not all) pesticides that have been developed in this class. This neuropathy takes some time to develop and is to be distinguished from the effects of these compounds on the inhibition of acetylcholinesterase; such inhibition is the primary effect responsible for the compounds' neurotoxic effects in insects and is largely responsible for the acute effects of organophosphorus pesticides. The delayed neurotoxic effect is produced by those organophosphorus pesticides (a) in which the alkyl side chains are hydrolyzed away before the esteratic bond between the pesticide and enzymes can hydrolyze and (b) that inhibit the so-called neurotoxic esterase. (2) A toxicity to the nervous system that takes some time to develop, e.g., the peripheral neuropathies produced by acrylamide ($CH_2CHCONH_2$) or n-hexane ($CH_3(CH_2)_4CH_3$).

delayed toxicity An adverse chemically induced effect that has some minimum latent period before it becomes manifest in terms of overt symptomatology. In some ways this delay is more apparent than real. For example, the action of a carcinogen to induce mutation in a cell occurs rapidly, but the development of a malignant tumor as a result of that effect on a single cell will take years or even decades before it can be detected.

delimiter A special character designating the beginning or end of a field or string of characters. Such a character is commonly used in data strings.

deliquescence The ability of a dry solid to become a saturated solution by absorbing water from the air.

delivered water A public or private utility water plant's finished product water that is carried through a water main network of pipes and arrives at the points of use (homes, institutions, and business facilities).

delta (1) An alluvial deposit, often in the shape of the Greek letter delta (Δ), from which it derives its name, that is formed where a stream drops its debris on entering a body of quieter water. (2) The terminal deposit of a river.

delta agent *See* delta virus.

delta E^o (ΔE^o) *See* standard cell potential.

delta P (ΔP) The pressure drop or loss by flowing fluid in a pressurized system due to the velocity and turbulence of the flowing fluid, restrictions the fluid flows through, and the roughness of wetted surfaces. *See also* pressure drop.

delta virus A virus found most frequently in patients who receive massive blood transfusions, in intravenous drug users, and in close personal contacts of infected subjects. Infection with the delta virus is dependent on co-infection with hepatitis B and the production of hepatitis B surface antigen. The delta virus appears to be a defective virus that can interfere with replication of the hepatitis B virus. It is also known as the delta agent.

deluge shower A safety device used to wash chemicals off the body quickly.

DEM *See* digital elevation model.

demand (1) The amount of electric power that may be required during a certain time interval. (2) The amount of water used by customers during a certain time interval from a water system.

demand costs Costs that are associated with the facilities that meet peak rates of water use, such as maximum-day, maximum-hour, or other rates.

demand curve A curve that defines the amount of water that customers are willing to buy at different rates, all other things being equal. The demand for water is usually defined by means of an econometric model (regression analysis) that analyzes historical demand and allows for other influences on demand, such as number of occupants (by age), household income, lot size, topography, and weather.

demand factor The ratio of the peak or minimum demand to the average demand.

demand forecasting The process of anticipating amounts of water required to meet customers' maximum hourly, daily, and annual requirements.

demand initiated regeneration (DIR) A method of automatically initiating regeneration or recycling in filters, deionizers, or softeners after a predetermined metered volume of water has been processed. In a softener or deionizer, regeneration may be triggered automatically based on a sensor reading. All operations, including bypass (of hard or soft water depending on design), backwashing, brining, rinsing, and returning the unit to service, are initiated and performed automatically in response to the demand for treated water.

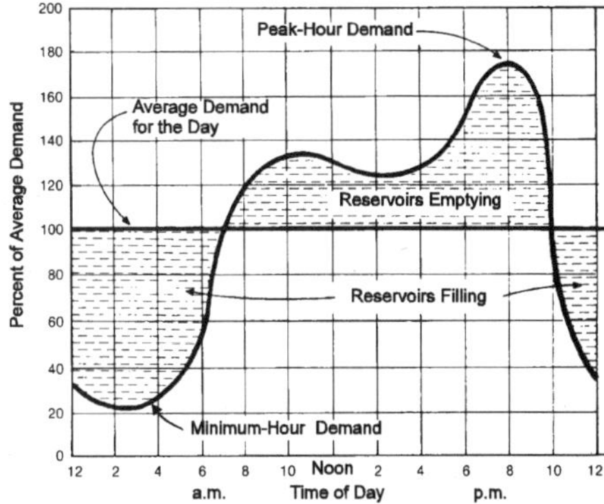

Daily system demand

demand management The use of measures that improve efficiency by lessening water use or altering patterns of water use. Some demand management measures can be implemented by consumers on their own, whereas others can be implemented through utility-sponsored programs. Examples of demand management include conservation-oriented pricing, water fixture plumbing standards and retrofitting, water-efficient landscaping, changes in water use practices, and public education.

demand meter (1) An instrument that measures the average power of a load during some specific interval. (2) A meter that will record the maximum rate of water usage over a specific interval.

demand uncertainty A margin of safety included in demand forecasting to cover the variability of the combination of those factors that are most likely to affect the short- or long-term forecast being made. Water demand forecasts are affected in the short run by such factors as weather departures from normal; water rate changes; conservation measures that are implemented; water shortage (drought) restrictions; and various economic variables that include changes in unemployment, household income, housing starts, and tourist visits. In the long run, the reliability of demand forecasts is affected by additional forces, such as changes in population growth, mix of household formation, family size, naturally occurring conservation effects (mostly from more efficient water fixtures), and supply availability.

demanganization In water treatment, the removal of compounds of manganese from water.

demineralization The removal of dissolved minerals and inorganic constituents from water to meet water quality standards, improve aesthetic quality, or reduce corrosivity. Commonly used treatment techniques for demineralization include chemical coagulation or lime softening for heavy metals and activated alumina. *See also* activated alumina; coagulation; deionization; desalting; ion exchange; lime softening.

demineralize To reduce the concentrations of minerals and inorganic constituents from water by ion exchange deionization or desalting processes such as distillation, electrodialysis, and reverse osmosis. *See also* deionization; desalting.

demineralized stream (1) A water stream from which minerals have been removed. (2) For an electrical-driven membrane treatment system (e.g., electrodialysis), the stream in a membrane stack in which the concentrations of ions have been reduced as a result of ion transfer through the membranes to the concentration stream.

demineralized water *See* bottled purified water.

demineralizer A device that demineralizes water.

***deminimus* risk** A risk that is so low as to be considered negligible.

denitrification The conversion of nitrite (NO_2^-) and nitrate (NO_3^-) to molecular nitrogen (N_2), nitrogen dioxide (NO_2), or a mixture of these two gases. Denitrification is accomplished by biological means under reducing conditions in the absence of free dissolved oxygen (i.e., via anaerobic respiration). *See also* biological denitrification; nitrate removal.

denominator The part of a numerical fraction below the line. A fraction indicates division of the numerator by the denominator. *See also* numerator.

dense, nonaqueous phase liquid (DNAPL) A liquid compound that is immiscible with water and for which the density is greater than water. Chlorinated solvents (e.g., tetrachloroethylene) and related compounds are examples of dense, nonaqueous phase liquids.

density (1) The mass of a substance per unit volume. (2) The number of bits or characters that can be stored on a specific length of recording surface.

density current A flow pattern that is moderated by the different densities of at least two liquids in a mixture. For example, salt water in tidal-influenced rivers establishes a countercurrent flow beneath the freshwater flow. Temperature can also influence density; as water temperature increases in settling basins, density currents due to the lower-temperature influent water can disturb settling patterns. A density current is also called a convection current.

density flow That movement of water near or on the surface of a reservoir, as an underflow or interflow, caused by the introduction of water that is different in terms of temperature, salinity, suspended solids content, or some other quality and so has a different density.

density of coliforms The concentration of organisms per unit volume of water examined. For drinking water analysis, the standard sample volume is 100 millilitres. In food analysis, the density is expressed in terms of the number of organisms per gram.

density of snow The ratio of (1) the volume that a given quantity of snow would occupy if it were reduced to water to (2) the volume of snow, expressed as a percentage. When a snow sampler is used, the density of snow is the ratio of the scale reading of water on the sampler to the length of the snow core or sample, expressed as a percentage. This value is numerically equal to the specific gravity of the snow.

density stratification The formation of identifiable layers of different densities in bodies of water. *See also* thermal stratification.

density units (1) In the US customary system, pounds mass per cubic foot or slugs per cubic foot. In Système International, kilograms per cubic metre or grams per cubic centimetre. (2) In the US customary system, bits per inch or track.

dental (denticle) *See* dentated sill.

dental fluorosis A condition resulting from excessive intake of fluoride. At low exposures it is manifested as mottling of the teeth that becomes quite marked as exposure increases. As the severity of this condition increases, the teeth become more brittle and subject to breakage.

dentated Having toothlike projections.

dentated sill A notched sill installed at the end of an apron to reduce the velocity of flowing water and the resulting scour below the apron or to promote aeration.

deoxygenation The depletion of the dissolved oxygen in a liquid either under natural conditions associated with the biochemical oxidation of organic matter present or by addition of chemical reducing agents.

deoxyribonucleic acid (DNA) A polymeric chemical that encodes genetic information that is transmitted between generations of cells. The code is rendered in the form of four purine and pyrmidine bases—adenine ($C_5H_5N_5$), thymine ($C_5H_6N_2O_2$), guanine ($C_5H_5N_5O$), and cytosine ($C_4H_5N_3O$)—placed in specific sequences referred to as codons (a three-base unit) that code for specific amino acids. The sequence of codons determines the sequence of amino acids in proteins that are synthesized from this template. *See also* codon; c-deoxyribonucleic acid probe hybridization.

deoxyribonucleic acid adduct A chemical metabolite that is covalently bound to deoxyribonucleic acid. Adducted deoxyribonucleic acid can either be repaired or cause insertion of an inappropriate base in the opposing or neighboring bases; the latter case leads to a mutation. If the mutation occurs in a critical part of the deoxyribonucleic acid, it will produce a protein product that has altered functional capabilities. This results in a cell that inappropriately expresses proteins involved in the regulation of the cell cycle; it can be responsible for a variety of toxic effects, including developmental toxicities and cancer.

deoxyribonucleic acid repair The process for correcting errors that have been introduced into deoxyribonucleic acid. This process is catalyzed by a wide variety of enzyme systems that recognize certain types of alterations in deoxyribonucleic acid.

Department of Agriculture (USDA) An agency of the US government's executive branch responsible for improving and maintaining farm income and expanding markets abroad for agriculture products. It works to enhance the environment and maintain production capacity by helping landowners protect the soil, water forests, and other natural resources. It administers national programs for rural development, credit, conservation, and research, and it safeguards and ensures standards of quality in the daily food supply through inspection and grading services.

Department of Commerce (DOC) An agency of the US government's executive branch responsible for encouraging, serving, and promoting US international trade, economic growth, and technological advancement. It administers programs to prevent unfair foreign trade competition; provides social and economic statistics and analyses for business and government planners; provides research and support for the increased use of scientific, engineering, and technological development; works to improve understanding and beneficial use of the earth's physical environment and oceanic resources; grants patents and registers trademarks; develops policies and conducts research on telecommunications; provides assistance to promote development; promotes travel to the United States by residents of foreign countries; and assists in the growth of minority businesses.

Department of Interior (DOI) An agency of the US government's executive branch responsible for most nationally owned public lands and natural resources. It fosters the sound use of land and water resources; protects fish, wildlife, and biological diversity; preserves the environmental and cultural values of national parks and historic places; and assesses mineral resources and works to ensure that their development is in the best interest of all citizens by encouraging stewardship and citizen participation in their care. The department also has a major responsibility for American Indian reservation communities and for island territories under US administration.

Department of Labor (DOL) An agency of the US government's executive branch that controls the following departments: Occupational Health and Safety Administration, National Institute for Occupational Safety and Health, Bureau of Labor Statistics, Mine Safety and Health Administration, Pension and Welfare Benefits Administration, Veteran's Employment and Training Service, and Employment Standards Administration.

Department of Natural Resources (DNR) The agency within a state government responsible for protection of natural resources.

Department of Transportation (DOT) The department of the US government responsible for railroads, motor carriers, hazardous materials, and vehicle and highway safety.

Department of Transportation hazard class (DOT hazard class) One of the classes (numbered 1–9) indicating the type of hazardous material being transported. The classes are as follows: class 1, explosive; class 2, gases; class 3, flammable liquid; class 4, flammable solid; class 5, oxidizer; class 6, poison (other than gases); class 7, radioactive material; class 8, corrosive material; class 9, miscellaneous hazard.

departure The difference between any single observation and the normal.

dependable capacity The capacity of a utility that can be relied on for service during all but exceptional circumstances.

dependable yield, groundwater *See* groundwater safe yield.

dependable yield, river *See* river safe yield.

dependable yield, surface water *See* impoundment safe yield.

dependent variable An outcome or variable for which the magnitude is dependent on other variables. In graphical operations the ordinate (y-axis) variable is generally referred to as the dependent variable because its value depends on the value of abscissa (x-axis) variable. *See also* regression analysis.

depletion (1) The continued withdrawal of water from a stream or from a surface water or groundwater reservoir or basin at a rate greater than the rate of replenishment. (2) The exhaustion of natural resources, usually in connection with commercial exploitation and usually recorded in monetary terms. *See also* streamflow depletion.

depletion curve (1) That part of a hydrograph extending from the point of termination of the recession curve to the point of a subsequent rise or alteration of inflow when additional water becomes available for stream flow. (2) That part of the hydrograph representing the rate of water flow or seepage from groundwater storage into the stream channels.

depletion hydrograph A hydrograph of discharge from water bodies depicting the drainage from groundwater, swamps, and lakes during long rainless periods of no recharge. The rate of this drainage is an indication of the rate at which storage in those water bodies is being depleted; the rates of drainage (as well as storage) decrease with time.

deposit Material left in a new position by a transporting agent such as gravity, human activity, ice, water, or wind.

deposition (1) The material that collects on the inside surface of a distribution system pipe as a result of suspended material in the water. (2) The material that forms in a water heater or utensil when water is heated.

depot A facility, such as a utility customer service office, mall, fire station, or park, that serves as a distribution center for water conservation retrofit materials.

depreciation An accounting method that allows the original cost of an asset (e.g., building, equipment, pipeline) to be deducted over its useful life as an expense on the annual income statement of the utility. (Land is not depreciated.) In for-profit entities, depreciation is an allowable tax deduction in deriving taxable income. Because the asset is paid for with cash or borrowed funds at the time of purchase, depreciation is a noncash expense item each year during the asset life that many utilities use as a convenient means of allocating funds for equipment replacement. Physical damage to an asset or obsolescence can cause the asset to be written off long before its expected useful life; on the other hand, if an asset lasts longer than the expected life, accounting depreciation simply stops when the total cost is written off.

depression head The magnitude of the lowering of the water surface in a well, and of the water table or piezometric surface adjacent to the well, resulting from the withdrawal of water from the well by pumping. *See also* drawdown.

depression spring A spring formed when the water table reaches a land surface formed by a depression in the local topography.

depression storage Precipitation that collects in puddles on the land surface.

depth filtration A filtration process in which water flows through progressively smaller pore spaces in a filter media bed. Depth filters are designed to entrap particles throughout the mass of filter media, as opposed to a surface filter, where only the surface layer does the actual filtering. Depth filtration is also called multimedia filtration.

depth of flotation The vertical distance from (1) the surface of the water in which a body is floating to (2) the point on the body farthest below the water surface.

depth profile A graphic representation of (1) the change with depth in a physicochemical characteristic of a body of water or (2) the varying depth of a body of water.

depth of runoff The total volume of runoff from a drainage basin divided by the basin's area. For convenience in comparing runoff with precipitation, the term is usually expressed in inches of depth over the drainage area during a given period of time.

depth–area–duration curve A set of curves that show the variation in depth of rainfall for a given design event (e.g., a 100-year event) as a function of the surface area covered by the storm and the duration of precipitation.

depth-integrated sample A sample that represents the water-suspended sediment mixture throughout the water column so that the contribution to the sample from each point is proportional to the stream velocity at that point. A depth-integrated sample is also called an integrated sample.

derivative action The control mode that varies the controller's output in proportion to the derivative (rate of change) of the controlled variable. Derivative action is also called rate action or derivative control. *See also* proportional integral derivative control.

derivative control *See* derivative action.

derivatization The process by which a chemical substance is derived from another substance. For example, esterification is a type of derivatization reaction in which an organic acid is converted into an ester. *See also* esterification.

dermal toxicity test A test of a chemical's potential to produce corrosive damage to the skin or to produce hypersensitivity reactions in the skin.

dermatitis Typically a redness and swelling of the skin.

desalinate *See* desalting.

desalination *See* desalting.

desalination plant A water treatment plant that conducts a desalting process.

desalination ratio In the electrodialysis process, the salt concentration in the demineralized stream divided by its concentration in the feed stream.

desalinization *See* desalting.

desalt *See* desalting.

desalting A water treatment process, such as distillation, reverse osmosis, or electrodialysis, that removes dissolved mineral salts and other dissolved solids from water.

desander A device for removing sand from drilling fluids during water well construction.

descaling The removal of encrustations within a pipe either by mechanical means such as flushing or pigging or by a chemical interaction with the scale that results in dissolution.

descending block rate *See* declining block rate.

descriptive epidemiologic study A study of population statistics pertaining to health. Descriptive studies primarily summarize routinely collected health-related data (e.g., cancer incidence rates) to reveal patterns of disease occurrence and observations about the association of disease with certain characteristics, including age, gender, race, occupation, geographic location, and time of occurrence. Information is not available from the individuals within the population studied about their exposures, risk factors, and disease status. *See also* ecologic epidemiology study.

descriptive statistics A set of statistical parameters describing various features of a data set.

desiccant A drying agent that is capable of removing or absorbing moisture from the atmosphere in a small enclosure.

desiccation A process used to remove virtually all moisture from air.

desiccator A tightly sealed container used to cool heated items before they are weighed. This prevents the items from picking up moisture from the air and increasing in weight.

design (1) The process of developing plans and specifications for the construction and implementation of an engineering system. (2) The result of that process.

design analysis In engineering reports, the tabulation and consideration of the physical data, present requirements, and probable future requirements pertaining to an engineering project. Such analysis should include the main features and principles of the design.

design criteria (1) Engineering guidelines specifying construction details and materials. (2) Objectives, results, or limits that must be met by a facility, structure, or process in performance of its intended functions. *See also* design loading.

design flood (1) The largest flow that a reservoir, channel, or other works can accommodate without damage or with limited damage. (2) The flood adopted for use in determining the hydraulic proportions of a structure, such as the outlet works of a dam, the height of a dam or levee, or the maximum water level in a reservoir. A design flood is also called a plan flood.

design life The projected or expected years of service for a major water utility component, or portion thereof, as developed by the designer and verified by the utility or owner.

design loading The flow rates and constituent concentrations that determine the design of a process unit or facility necessary for proper operation.

design point The mark on the head–capacity curve of a pump characteristics curve that indicates the head and capacity at which the pump is intended to operate for best efficiency in a particular installation.

design standard A commonly accepted standard professional engineering practice used to ensure that project construction accurately reflects design goals.

design storm (1) The storm for which a hydraulic structure such as a bridge, culvert, or dam is designed. (2) The rainfall estimate corresponding to an enveloping depth–area–duration curve for the selected frequency of the storm.

design–build An alternative method of designing and constructing a facility in which one entity both designs and constructs the plant. Under some circumstances, this approach can accelerate the schedule and may lower costs compared to a traditional approach, in which the facility is designed by one entity and then bid so that a contractor constructs the project. However, the owner often accepts more risk in a design–build approach.

desilting basin A settling basin for removal of silt from river or stormwater flows. *See also* sedimentation basin.

desilting works Basins installed just below the diversion structures of canals to remove the bed loads, suspended sand loads, and some of the heavier silt loads from the water.

desorption The movement of a previously adsorbed constituent into a liquid phase. Desorption can occur (1) when the liquid phase concentration decreases significantly, resulting in a driving force from the high-concentration adsorption surface to the low-concentration solution, or (2) when a more strongly adsorbed chemical is introduced into the liquid stream. *See also* adsorption.

destabilization of pipe film The chemical dissolution or physical separation of a film from the pipe surface. The film may have been placed on the pipe during the manufacturing process, it may develop as a result of the corrosion process, it may result from precipitation of a chemical constituent in the water onto the pipe, or it may result from addition of a corrosion inhibitor. Water quality conditions, flow velocity changes, changes of flow direction, physical abrasion, vibration, or pipe movement may cause the film to dissolve, loosen, or detach.

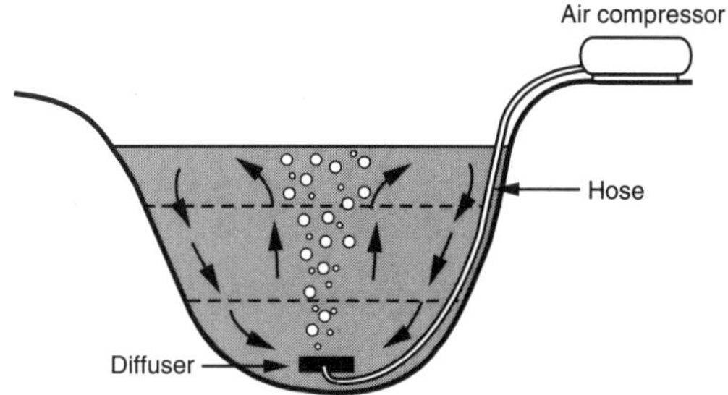

Destratification using compressed air

destratification The mixing of established, layered zones within a medium. Thermal layers that produce fixed aerobic and anaerobic zones in reservoirs can be destratified by mechanical means or temperature changes. Filter media that are layered as a result of particle size or density differences can be destratified by intense agitation during backwashing. *See also* aerobic condition; anaerobic condition; backwash; stratification; thermal stratification.

Desulfovibrio A genus of anaerobic bacteria found in water, sediments, and soils, characterized by an ability to use sulfate (SO_4^{2-}) as an electron acceptor, producing hydrogen sulfide (H_2S) as the final reduction product. *See also* sulfate-reducing bacteria.

detection lag The time period between the moment a change is made and the moment when such a change is finally sensed by the associated measuring instrument.

detection limit *See* limit of detection.

detector check A check valve that is used on a fire system and has a small bypass meter or indicator system that shows usage or that water has passed through the check valve.

detention dam A dam, usually small, constructed to impound or retard surface runoff temporarily. It is also used to bring about the deposition of soil being carried away by runoff of surface water. *See also* soil-saving dam.

detention reservoir A reservoir in which water is stored for a relatively brief period of time, with part of the water being retained until the stream can safely carry the ordinary flow plus the released water. Such reservoirs usually have outlets without control gates and are used for flood regulation. A detention reservoir is also called a retarding reservoir.

detention tank Any vessel used to hold flow to even out surges, allow sedimentation, or provide time for physical, chemical, or biological reactions. *See also* contact tank; equalization; sedimentation basin.

detention time The average length of time a drop of water or a suspended particle remains in a tank or chamber. Mathematically, it is the volume of water in the tank divided by the flow rate through the tank.

detergent Any material with cleansing powers: soaps, synthetic detergents, artificial alkaline materials, solvents, and abrasives. In common domestic usage, the term is often used to refer to synthetic detergents.

deterioration A change in the quality of a sample over a period of time as a result of improper preservation techniques.

determinant of water quality *See* water quality parameter.

determination The quantitative analysis of a specific analyte, e.g., a method for the determination of chloroform ($CHCl_3$) in water.

deterministic process A phenomenon determined by antecedent causes, with the same causes consistently producing the same effects.

detoxication Generally, the inactivation of a chemical as a toxin. More specifically, this term usually refers to the metabolism of the chemical to a form that is less toxic to the organism. However, metabolism is also frequently responsible for producing the toxic form of the chemical. Therefore, the terms *detoxication* and *metabolism* should not be used interchangeably.

detritus Finely divided, nonliving, settleable material that is suspended in water.

deuterated internal standard A compound that has a known proportion of an isotope and can be used to calibrate a mass spectrometer. For the purposes of calibration, a known number of hydrogen atoms in a compound are replaced with deuterium. The deuterated compounds behave in a chromatograph very much like the compound with a natural proportion of the isotope. However, the mass/charge ratio can be distinguished in a mass spectrometer as a result of the increase in the mass/charge ratio of the deuterated compound. *See also* mass spectrometer.

deuterium A hydrogen isotope having an atomic weight of 2.0147.

deuterium background correction An instrumental technique of compensating for interferences in atomic absorption spectrophotometry. In this approach, a deuterium arc lamp is commonly used as a continuum source for the lower ultraviolet wavelengths. Another continuum source is required for visible (higher) wavelengths.

deuterium oxide (D_2O) The chemical name for heavy water, which is similar in structure and composition to ordinary water except that the hydrogen present has an atomic weight of 2.0147 instead of 1.008.

developed length The length of a line of pipe along the center line of the pipe and fittings.

developed water A legal term that applies to (1) groundwater that is artificially brought to the surface or to the land and that otherwise would have run to waste or not have appeared in any known source, or (2) an artificially induced flow of water in a stream.

development impact fee (DIF) A fee that represents a pro rata share of an agency's cost to provide additions to the infrastructure to serve new development. The DIF is paid by the customers who benefit from the improvements. For existing customers, it is paid at the time the project is approved. For new customers, the fee is usually paid by developers, who pass the cost along to customers in the purchase price of the homes. To qualify as a fee (rather than as a tax), the levy must represent costs reasonably borne for the benefits provided, and the money collected must be used for the purposes for which the fee was established. *See also* capacity charge.

developmental toxicities Functional as well as structural abnormalities that can be induced by physical and chemical agents during normal differentiation and maturation of an organism. They include teratogenic effects. *See also* teratogenic effect.

deviation The difference between the controlled variable and the controller set point. Deviation is also known as process error.

device driver A program that provides an interface between the operating system and application software to support output on a specific peripheral device. The device driver translates commands issued through application software into instructions that a device, such as a plotter, can interpret to perform a certain function.

device server A hardware component used to connect multiple nonintelligent devices—such as nonintelligent terminals, query stations, plotters, and printers—to a local area network. A device server contains computer processors dedicated to supervising communications between the devices and the network.

dew point The temperature of an air mass when water condensation will begin.

dew point depression The difference, in degrees, between the prevailing temperature and the current dew point.

dew point hygrometer A hygrometer for indicating the dew point, from which the relative humidity can be calculated when the air temperature is known.

dewater *See* dewatering of sludge.

dewaterability The characteristic of a sludge relating to the ability to have water extracted and thus produce a higher solids concentration. As the dewaterability increases, the ability to generate a lower-volume, higher-solids-concentration sludge also increases. Sludges can be dewatered by gravity (e.g., a sludge-drying bed) or mechanical means (e.g., a belt filter press). *See also* belt filter press; sludge; sludge-drying bed.

dewatered sludge The solid residue remaining after removal of water from a wet sludge by draining or filtering. Dewatering is distinguished from thickening in that dewatered sludge may be transported by solids-handling procedures.

dewatering The process of partially removing water. It may refer to the removal of water from a basin, tank, reservoir, or other storage unit, or to the separation of water from solid material.

dewatering of reservoirs A physical method for controlling aquatic plants in which a water body is completely or partially drained and the plants allowed to die.

dewatering of sludge The removal of water in sludge to reduce sludge volume and facilitate disposal. Sludge can be dewatered by mechanical means (e.g., a belt filter press or centrifuge) or by gravity (e.g., a sludge-drying bed). *See also* belt filter press; centrifuge; sludge-drying bed.

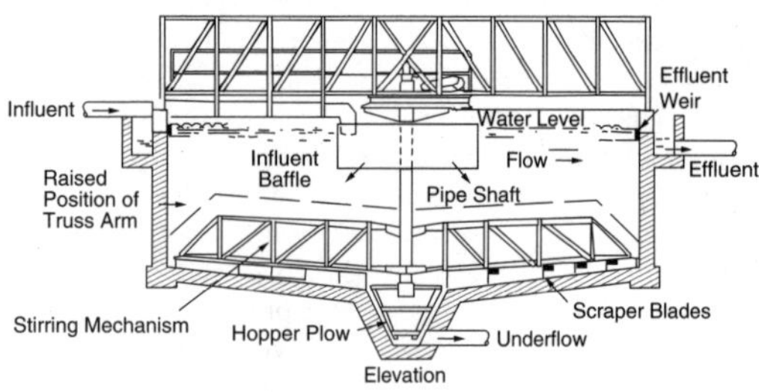

Thickener to aid dewatering of sludge

dezincification The parting of zinc from an alloy. In some cases, zinc is lost, leaving a weak, brittle, porous, copper-rich residue behind.

DF *See* domestic flow.

DHAN *See* dihaloacetonitrile.

DI water *See* deionized water.

di(2-ethylhexyl) adipate ($(CH_2CH_2COOCH_2CH(C_2H_5)C_4H_9)_2$) A synthetic organic chemical used as a plasticizer or solvent. It is regulated by the US Environmental Protection Agency. *See also* plasticizer; solvent; synthetic organic chemical.

di(2-ethylhexyl) phthalate ($C_6H_4(COOCH_2CH(C_2H_5)C_4H_9)_2$) A synthetic organic chemical used as a plasticizer for many resins and elastomers. It is regulated by the US Environmental Protection Agency. *See also* plasticizer; synthetic organic chemical.

dialdehyde A class of organic compounds that have two aldehyde (HC=O) functional groups in the chemical structure (e.g., glyoxal). Some dialdehydes are created during the reactions of oxidants used as disinfectants, particularly ozone (O_3), with natural organic matter. *See also* aldehyde; glyoxal; ozonation by-product.

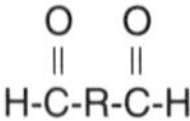

Dialdehyde (in the simplest dialdehyde, there is no R group)

dialysate The stream being depleted of salt in electrodialysis.

dialysis (1) The separation of small molecules from larger molecules in a solution by means of a semipermeable membrane. It occurs as a result of differing diffusion rates and is driven by concentration differences across the membrane barrier. (2) A process that takes advantage of the selective permeability of certain membranes to separate a solute from solution. In medicine, dialysis is used to remove waste products of metabolism from the blood of patients who are in renal failure. It is also frequently used to remove chemicals from the blood of poisoned individuals. (3) A membrane technique that can be used to transport analytes from one liquid to another. The major driving force for transport of solutes is the concentration gradient. Dialysis membranes have typically been used to remove salts and low-molecular-weight solutes from solution. In addition, dialysis membranes have been used to concentrate higher-molecular-weight organic matter and biological matter, such as viruses.

diameter The length of a straight line measured through the center of a circle from one side to the other.

4'6-diamidino-2-phenylindole ($C_{16}H_{15}N_5{\cdot}2HCl$)/propidium iodide ($C_{27}H_{34}N_4I_2$) (DAPI)/(PI) Two fluorogenic vital dyes used to assess the viability of *Cryptosporidium* oocysts in vitro. If an oocyst is 4'6-diamidino-2-phenylindole positive, it is assumed to be viable; if it is propidium iodide positive, it is assumed to be dead.

diaphragm A thin, flexible partition (disk) that is supported at the edges and is used to transmit pressure from one substance to another while keeping them from direct contact.

diaphragm float A float mounted on a truck that runs on rails along a straight and uniform section of a channel in which the velocity is to be gauged. It can be lowered quickly into the

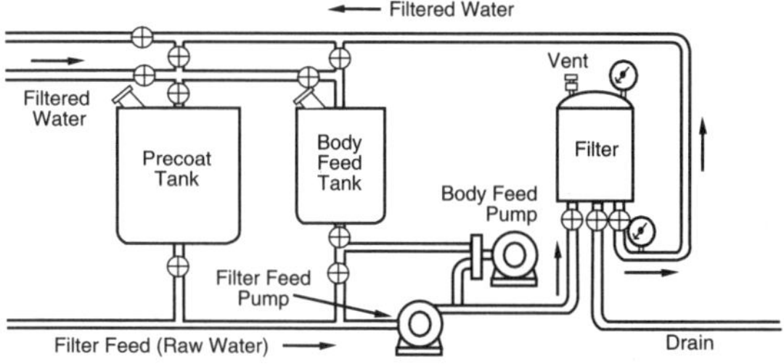

Diatomaceous earth filtration system

water, made to occupy nearly the whole area of the channel during the run, and lifted out at the end. Measuring the float's velocity yields the water velocity.

diaphragm-type metering pump A pump in which a flexible rubber, plastic, or metal diaphragm is fastened at the edges in a vertical cylinder. As the diaphragm is pulled back, suction is exerted and the liquid is drawn into the pump. When it is pushed forward, the liquid is discharged.

diapositive A positive photographic transparency (the mirror image of a photograph negative).

diarrhea An abnormally frequent discharge of fluid fecal matter from the bowel, sometimes accompanied by vomiting and fever (from the Greek *dia*, meaning "through," and *rhoia*, meaning "a flow"). It is a symptom of infection by bacterial, viral, and parasitic agents that cause enteric diseases (e.g., cholera, shigellosis, salmonellosis, yersiniosis, giardiasis, campylobacteriosis, cryptosporidiosis, and viral gastroenteritis). Diarrheal illness can be divided into several clinical syndromes: (1) simple diarrhea, managed by oral rehydration without specific identification of etiology; (2) dysentery, with scanty stools containing blood, mucous, or both; (3) persistent diarrhea of 14 or more days; (4) severe diarrhea, as seen in cholera; (5) minimal diarrhea associated with persistent vomiting, typical of viral gastroenteritis; and (6) hemorrhagic colitis, with watery diarrhea containing gross blood but no fever. *See also* gastroenteritis.

diatom A type of algae that is commonly found in freshwater and marine environments. Thousands of species, which are characterized by a cell wall composed of polymerized silica (exoskeleton), have been identified. Large deposits of diatoms are mined as diatomaceous earth, which is used in specific situations as filter media in water treatment. In addition, certain types of diatoms can contribute to taste-and-odor problems in drinking water supplies. *See also* diatomaceous earth.

diatomaceous earth (DE) The microscopic remnants of the discarded outer surface of diatoms. DE is also called fuller's earth. It is the medium most commonly used in precoat filtration. *See also* diatomaceous earth filter; diatomaceous earth filtration.

diatomaceous earth (DE) filter A pressure filter using a medium made from diatoms. The water is forced through the diatomaceous earth by pumping.

diatomaceous earth (DE) filtration A filtration method in which diatomaceous earth is used as the filtering medium. Initially, a 1/8- to 3/16-inch (2- to 5-millimetre) thick layer, or precoat, is applied to a septum or filter element. During operation, diatomaceous earth is fed continuously until a terminal head loss is reached, after which the filter influent is shut off and the diatomaceous earth layer falls off and is discharged. This type of filtration is used most frequently for swimming pools and is best applied to source waters that have consistently low turbidity. *See also* surface filtration.

diatomic molecule A molecule containing only two atoms, such as hydrogen as H_2 or oxygen as O_2.

diatomite A processed natural material, chiefly the skeletons of diatoms, used as a filter medium.

diatomite filtration *See* diatomaceous earth filtration.

diazinon (**$((CH_3)_2CHC_4N_2H(CH_3)O)PS(OC_2H_5)_2$**) The generic name for the insecticide *O,O*-diethyl *O*-(2-isopropyl-4-methyl-6-pyrimidinyl)phosphorothioate. *See also* insecticide.

diazomethane (**$H_2C{=}N^+{=}N^-$**) A methylating agent used in organic synthesis. Diazomethane is used to derivatize haloacetic acids, converting them to their corresponding methyl esters, so that they can be analyzed by gas chromatography. *See also* haloacetic acid.

dibromamine (**$NHBr_2$**) The brominated analog of dichloramine ($NHCl_2$), which can be formed during the chloramination of bromide-containing water. *See also* bromamines.

dibromoacetic acid (DBAA) (**$CHBr_2COOH$**) A haloacetic acid containing two bromine atoms. It is formed to a significant extent during the chlorination of waters containing moderate amounts of bromide (Br^-). It may also be formed during the ozonation of bromide-containing waters. *See also* haloacetic acid.

```
   Br O
   |  ||
Br-C-C-OH
   |
   Cl
```

Dibromoacetic acid

dibromoacetonitrile (DBAN) (**$CHBr_2C{\equiv}N$**) A haloacetonitrile containing two bromine atoms. It is formed to a significant extent during the chlorination of waters containing moderate amounts of bromide (Br^-). It may also be formed during the ozonation of bromide-containing waters. *See also* haloacetonitrile.

dibromochloroacetic acid (DBCAA) (**$CBr_2ClCOOH$**) A haloacetic acid containing two bromine atoms and one chlorine atom. *See also* haloacetic acid.

dibromochloromethane (DBCM) (**$CHClBr_2$**) A trihalomethane containing two bromine atoms and one chlorine atom. *See also* trihalomethane.

1,2-dibromo-3-chloropropane (DBCP) (**$CH_2BrCHBrCH_2Cl$**) A volatile organic chemical used as a soil fumigant for nematode control on crops. It has been detected in groundwater. The US Environmental Protection Agency has canceled all uses for this pesticide, which is also called dibromochloropropane.

DIC *See* dissolved inorganic carbon.

dichloramine (**$NHCl_2$**) A disinfectant produced from the mixing of chlorine and ammonia (NH_3). Typically, this chloramine species is produced to a lesser extent than monochloramine (NH_2Cl). However, chloramination at a low pH (e.g., 7), high chlorine-to-ammonia–nitrogen ratios (e.g., 7:1 by weight), or both results in higher levels of dichloramine. *See also* chloramines; chlorine-to-ammonia–nitrogen ratio; monochloramine.

dichloroacetic acid (DCAA) ($CHCl_2COOH$) A haloacetic acid containing two chlorine atoms. Typically, this substance and trichloroacetic acid are the principal haloacetic acids formed during chlorination, except for in waters containing moderate amounts of bromide (Br^-). *See also* haloacetic acid; trichloroacetic acid.

dichloroacetonitrile (DCAN) ($CHCl_2C{\equiv}N$) A haloacetonitrile containing two chlorine atoms. Typically, this is the principal haloacetonitrile formed during chlorination, except for in waters containing moderate amounts of bromide (Br^-). *See also* haloacetonitrile.

dichlorobenzene ($C_6H_4Cl_2$) An aromatic hydrocarbon with two chlorine atoms attached to a benzene ring. Two dichlorobenzene isomers (*ortho-* and *para-*) are regulated by the US Environmental Protection Agency. *See also* aromatic hydrocarbon; *ortho*-dichlorobenzene; *para*-dichlorobenzene.

***ortho*-dichlorobenzene ($C_6H_4Cl_2$)** The common name for 1,2-dichlorobenzene. This volatile organic chemical has various industrial uses, including use as a solvent, fumigant, or insecticide. It is regulated by the US Environmental Protection Agency. *See also* fumigant; insecticide; solvent; volatile organic compound.

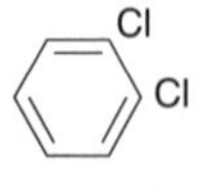

Ortho-dichlorobenzene

***para*-dichlorobenzene ($C_6H_4Cl_2$)** The common name for 1,4-dichlorobenzene. This volatile organic chemical has various industrial uses, including use as an insecticide, germicide, or fumigant. It is regulated by the US Environmental Protection Agency. *See also* fumigant; germicide; insecticide; volatile organic compound.

Para-dichlorobenzene

dichlorodiphenyl dichloroethane (DDD) ($(ClC_6H_4)_2CHCHCl_2$) The generic name for 1,1-dichloro-2,2 bis-(*para*-chlorophenyl) ethane, an insecticide (now discontinued) that was formerly applied to many fruits and vegetables. It is also referred to as tetrachlorodiphenylethane. *See also* insecticide.

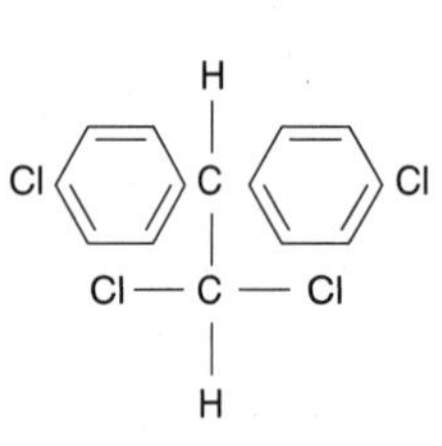

Dichlorodiphenyl dichloroethane

dichlorodiphenyl dichloroethylene (DDE) ($(ClC_6H_4)_2C{:}CCl_2$) A degradation product of dichlorodiphenyl trichloroethane that has been found as an impurity in that substance's residues. *See also* dichlorodiphenyl trichloroethane.

dichlorodiphenyl trichloroethane (DDT) ($(ClC_6H_4)_2CH(CCl_3)$) A generic name for 1,1,1-trichloro-2,2-bis (*para*-chlorophenyl) ethane, an insecticide–pesticide that is not biodegradable and is ecologically damaging. Its agricultural use in the United States was prohibited in 1973. *See also* dichlorodiphenyl dichloroethylene; insecticide; pesticide.

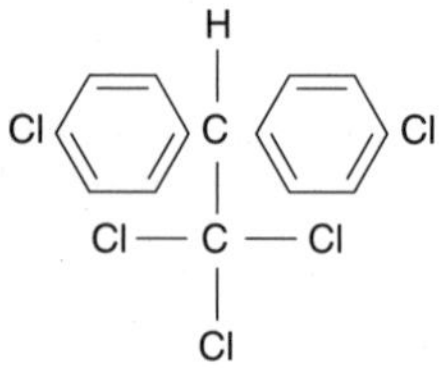

Dichlorodiphenyl trichloroethane

1,1-dichloroethane (CH_3CHCl_2) A volatile organic chemical used as a solvent. *See also* solvent; volatile organic compound.

1,2-dichloroethane ($ClCH_2CH_2Cl$) A volatile organic chemical that has various industrial uses, including use as a solvent. It is regulated by the US Environmental Protection Agency. It is also called ethylene dichloride. *See also* solvent; volatile organic compound.

1,1-dichloroethylene ($Cl_2C{:}CH_2$) A volatile organic compound that is regulated by the US Environmental Protection Agency. *See also* volatile organic compound.

1,2-dichloroethylene ($ClHC{:}CHCl$) A volatile organic chemical that has various industrial uses, including use as a solvent. It exists as *cis* and *trans* isomers. Both isomers are regulated by the US Environmental Protection Agency. *See also* isomer; solvent; volatile organic compound.

***cis*-1,2-dichloroethylene ($ClHC{:}CHCl$)** An isomer of 1,2-dichloroethylene in which the chlorine atoms are at adjacent corners of the molecule. *See also cis*; 1,2-dichloroethylene; isomer.

***trans*-1,2-dichloroethylene ($ClHC{:}CHCl$)** An isomer of 1,2-dichloroethylene in which the chlorine atoms are at opposite corners of the molecule. *See also* 1,2-dichloroethylene; isomer; *trans*.

dichloromethane (CH_2Cl_2) The legal label name for methylene chloride, a volatile organic chemical primarily used as a solvent. It is regulated by the US Environmental Protection Agency. *See also* solvent; volatile organic compound.

2,4-dichlorophenol ($Cl_2C_6H_3OH$) A synthetic organic chemical used in organic synthesis. In addition, 2,4-dichlorophenol can be formed during the chlorination of water. *See also* chlorophenol; synthetic organic compound.

2,4-dichlorophenoxyacetic acid (2,4-D) ($Cl_2C_6H_3OCH_2COOH$) A selective weed killer and defoliant. Is is regulated by the US Environmental Protection Agency. *See also* defoliant.

1,2-dichloropropane ($CH_3CHClCH_2Cl$) A volatile organic chemical with various industrial uses, including use as a solvent or nematocide. It is regulated by the US Environmental Protection

Agency. It is also known as propylene dichloride. *See also* nematocide; solvent; volatile organic compound.

1,3-dichloropropane ($CH_2ClCH_2CH_2Cl$) A volatile organic chemical. *See also* volatile organic compound.

2,2-dichloropropane ($CH_3CCl_2CH_3$) A volatile organic chemical. *See also* volatile organic compound.

1,1-dichloropropene (CCl_2:$CHCH_3$) A volatile organic chemical. *See also* volatile organic compound.

1,3-dichloropropene ($CHCl$:$CHCH_2Cl$) A soil fumigant. *See also* fumigant.

dieldrin ($C_{12}H_{10}OCl_6$) The generic name for an insecticidal product containing not less than 85 percent 1,2,3,4,10,10-hexachloro-6,7-epoxy-1,4,4a,5,6,7,8,8a-octahydro-1,4-endo,exo-5,8-dimethanonapthalene and not more than 15 percent active related compounds. Dieldrin is a stereoisomer of endrin. *See also* endrin.

dielectric constant An electrical property related to the polarizability of a medium such as water. Measurements are made in groundwater, and changes in the dielectric constant compared to a baseline indicate the presence of contaminants.

diene An aliphatic hydrocarbon that contains two double bonds (e.g., 1,3-butadiene: $H_2C{=}CHHC{=}CH_2$). A diene is also referred to as a diolefin. *See also* aliphatic compound; hydrocarbon.

diethyl ether (DEE) ($(C_2H_5)_2O$) An ether used as a solvent in analytical chemistry to extract polar organic compounds from water. It is also known as an ethyl ether. *See also* ether; solvent.

```
  H H   H H
  | |   | |
H-C-C-O-C-C-H
  | |   | |
  H H   H H
```

Diethyl ether

N,N-diethyl-p-phenylenediamine (DPD) ($(C_2H_5)_2NC_6H_4NH_2$) A reagent used in the determination of a chlorine residual. A DPD method is commonly used in the field as part of a colorimetric test kit. Variations of the DPD method can be performed in the laboratory in the form of a titration with ferrous ammonium sulfate or a spectrophotometric method. *See also* N,N-diethyl-p-phenylenediamine–ferrous ammonium sulfate test.

N,N-diethyl-p-phenylenediamine (DPD) methods Methods for the determination of chlorine residual in water using N,N-diethyl-p-phenylenediamine as a reagent. These methods are available for both volumetric (titration) and colorimetric determinations. They are generally simpler than amperometric methods. *See also* amperometric methods; N,N-diethyl-p-phenylenediamine.

N,N-diethyl-p-phenylenediamine–ferrous ammonium sulfate (DPD-FAS) ($Fe(SO_4)\cdot(NH_4)_2SO_4\cdot6H_2O$) test A test for the determination of chlorine residual. In this test, N,N-diethyl-p-phenylenediamine is used in the laboratory with a titration of ferrous ammonium sulfate. *See also* N,N-diethyl-p-phenylenediamine.

DIF *See* development impact fee.

differential gauge A pressure gauge used to measure the difference in pressure between two points in a pipe or receptacle containing a liquid.

differential intake An intake formed by a hollow, low-diversion dam with a longitudinal slot for intercepting a part of the stream flow and conducting it to a pipeline at one or both ends of the dam.

differential plunger pump A reciprocating pump with a plunger designed so that it draws the liquid into the cylinder on the upward stroke but has double action on the discharge stroke. *See also* double-action pump.

differential pressure The relative difference in pressure between two pressure sources or across some restriction, e.g., across a differential measuring meter that includes an orifice plate or a venturi tube. A differential pressure representing head loss due to pipe friction can be measured with a rigid U-tube connected at two points along a water line by long, flexible, small-diameter connecting tubes.

differential pressure (DP) cell An instrument used for determining the difference in pressure between two hydraulically connected locations. For example, a differential pressure cell could be used to determine head loss through a filter. *See also* differential pressure gauge; head loss.

differential pressure gauge An apparatus used to measure the pressure difference between two points in a system. The primary application is for the calculation of flow when the device senses a pressure differential in a venturi-type flow element. The difference in pressure between the upstream entrance to the inlet cone, on one hand, and the throat of the cone, on the other, is proportional to the flow through the element; this difference is also commonly used to indicate head loss across a filter element. Most elements utilize electronic signals generated by strain-gauge–type elements mounted on a metallic diaphragm.

differential pulse polarography (DPP) A technique used in the electrochemical analysis of a variety of analytes, such as metal cations and anions. It is based on the difference in current generated between two polarographic cells.

differential settling A method of flocculation in which particle contact is promoted as a result of the different settling velocities of various particles. *See also* orthokinetic; flocculation; perikinetic flocculation.

differential surge tank A combination of an orifice surge tank and a normal surge tank working together to dampen surges or control sudden pressure changes in a water distribution system or transmission main.

differential thermal analysis (DTA) One of a group of techniques used in analyzing the manner in which materials change when heated. It can be used to study phase transitions in materials. In differential thermal analysis, the behavior of a sample when heated is compared to the analogous behavior of a reference material. The technique is most often applied to the characteristics of materials, especially polymers, but could also be used to determine the specific heat of water.

diffraction A modification of the behavior of light or other radiation as it passes by the edge of an aperture. It is an interference phenomenon caused by the wave nature of radiation; the rays of light appear to be deflected, and they produce light and dark bands on a surface. Some laboratory instruments, such as spectrophotometers, use diffraction gratings as part of

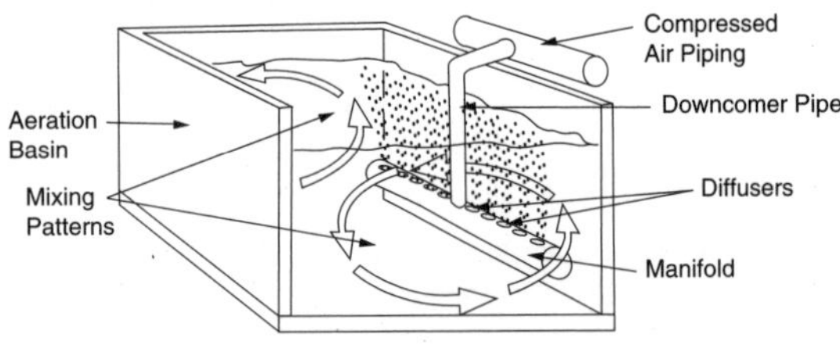

Diffusion aerator

their optical system to disperse the radiation. Selected wavelengths of radiation can then be used in spectroscopic analysis.

diffused aeration The dissolution of air or oxygen by the use of coarse-bubble spargers or fine-bubble porous disks or tubes. Examples of uses for diffused aeration include removing volatile compounds from water and increasing the oxygen concentration in water.

diffused air Small air bubbles formed below the surface of a liquid to transfer oxygen to the liquid.

diffused air aeration Aeration produced in a liquid by air passed through a diffuser.

diffused surface water (1) Floodwater that has escaped from a stream channel. (2) Water derived from rainfall, melting snow, seepage, or springs that is on its way to a stream but has not reached a definite channel.

diffuser (1) A section of a perforated pipe or porous plates used to inject a gas, such as carbon dioxide (CO_2) or air, under pressure into water. (2) A type of pump.

diffuser plate A porous plate used in aeration tanks to diffuse air or other gases in various water treatment processes. *See also* diffuser.

diffuser tube An air tube used in aeration tanks to diffuse air or other gases in various water and treatment processes. *See also* diffuser.

diffuser vane A vane installed within a pump casing (on diffuser centrifugal pumps) to change velocity head to pressure head.

diffusing pit A form of diffusing well.

diffusing well A well into which water is injected to restore water to an aquifer. *See also* inverted well.

diffusion A process whereby molecules or particles move and intermix because of a concentration gradient driving force; the movement of a compound within a medium or from one medium to another. For example, longitudinal diffusion refers to the movement of a compound in a conduit at a speed either faster or slower than the mean velocity of the solution, whereas boundary layer diffusion refers to the movement of a compound to or from a solution through a boundary layer surrounding a particulate medium. Molecular diffusion is quantified by Fick's law. *See also* Fick's law of diffusion.

diffusion aerator An aerator that blows air under low pressure through submerged porous plates, perforated pipes, or other devices so that small air bubbles rise through the water continuously.

diffusion feeder A chemical feed system in which chemicals are added to a water stream in controlled quantities. These chemicals provide pipeline or metal surface protection, disinfection, or both. A diffusion feeder is designed in such a way that a small stream of water is diverted through a tank so that the water flows over the chemical material; a small amount of the chemical is then diffused (dissolved) into the water and carried back to the main water line. A diffusion feeder is also called a bypass feeder.

diffusion van A fixed or removable casting in a pump between the impeller and the casing with liquid passages designed to convert velocity head to pressure head.

diffusion well A well constructed to conduct surface water or other surplus water into an aquifer to increase the groundwater supply. Such a well is also called a recharge well.

digital Pertaining to the discrete numerical representation of information. *Contrast with* analog.

digital capture device A device that converts an analog image, such as a hard copy map or aerial photograph, to digital form by optically reading the tone or color variation of the analog image and converting it to a raster form suitable for display and analysis.

digital computer model (1) A model of groundwater or surface water flow, contaminant transport, or both such that the aquifer or stream is described by numerical equations that have specified values for boundary conditions and are solved using a digital computer. (2) A model of pipeline flow, contaminant transport, or both for which the distribution system is described by hydraulic equations (with specified values for forcing functions) and storage conditions. These equations are solved using a digital computer. (3) A model of any water-related system that is described by numerical equations and solved on a digital computer. For instance, a spreadsheet that does $C \times T$ calculations for disinfection is a digital computer model.

digital elevation model (DEM) A digital cartographic representation of the surface of the earth or a subsurface feature through a series of three-dimensional coordinate values. Such a model is also called a digital terrain model.

digital line graph (DLG) A standard file structure for cartographic digital data established and used by the US Geological Survey that contains point coordinates describing planimetric and contour data such as boundaries, drainage lines, transportation routes, and features.

digital readout A readout of an instrument in the form of a direct, numerical display of the measured value.

digital service unit A peripheral device that enables data transmission between computer devices via an outside communication carrier. These units connect devices across digital lines that support both voice and data transmission.

digital terrain model (DTM) *See* digital elevation model.

digitization The process of converting analog data, such as an analog map, into a digital format for use by a computer program.

digitizer *See* digitizing tablet.

digitizing A technique used to convert graphic information from analog to digital form, in which maps are retraced by a person using a digitizing tablet and cursor. The map to be digitized is mounted on a digitizing tablet. The scale, angle of rotation, and area coverage are registered with the computer system through a standard procedure. The user then digitizes each map feature by pointing the cursor and pushing appropriate buttons.

digitizing table A large, free-standing digitizing tablet.

digitizing tablet A peripheral device used to convert graphic information from analog to digital form for use in a computer. The user moves a cursor to select positional points on a hard copy map mounted on the surface of the digitizing tablet; these points are transmitted by electrical signals from the cursor to the tablet. Conductors in the tablet receive the signals and convert them to relative tablet positions. A tablet may be placed on a horizontal surface such as a table.

dihaloacetonitrile (DHAN) ($CHX_2C{\equiv}N$, where X is Br, Cl, or both) A haloacetonitrile containing two halogen atoms (normally chlorine (HOCl, OCl^-), bromine (Br), or both). *See also* haloacetonitrile.

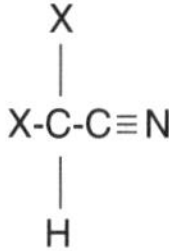

Dihaloacetonitrile, where the Xs are chlorine atoms, bromine atoms, or both

dihydrogen oxide (H_2O) Water.

dike (1) An embankment constructed to prevent overflow of water from a stream or other body of water. (2) An embankment constructed to retain water in a reservoir. The term *dam* is usually used for a structure constructed across a watercourse or stream channel, and *dike* is used for one constructed solely on dry ground. (3) A vertical or steeply inclined wall of igneous rock that has been forced into a fissure in a molten condition and has consolidated there. Dikes may obstruct the passage of groundwater.

diketone A class of organic compounds that have two ketone (R-(C=O)–R) functional groups in the chemical structure (e.g., dimethyl glyoxal). Some diketones are created during the reactions of oxidants used as disinfectants, particularly ozone (O_3), with natural organic matter. *See also* dimethyl glyoxal; ketone; ozonation by-product.

Diketone (in some diketones, there is no R group in between the two carbonyl groups)

dilute stream *See* demineralized stream.

dilution The act of adding more solvent or water to a given solution to make the solution less concentrated. Sometimes this is done to attain the proper concentration, sometimes to make the solution easier to handle.

dilution bottle A heat-resistant glass bottle used for diluting bacteriological samples before analysis. It is also called a milk dilution bottle or French square.

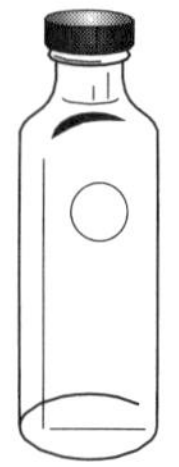

Dilution bottle

dilution gauging A method of measuring the flow of water by introducing a constant flow of a solution of known concentration for a sufficient length of time at one section of a water conduit and then determining the resulting dilution of this solution at another downstream section. Chemical gauging, electrochemical gauging, and radioactive solution gauging are different forms of dilution gauging.

dilution method A technique for determining the discharge in a flowing stream by dilution of a tracer injected at a known concentration and flow rate. The method assumes that no tracer compound is in the water upstream of the injection point and that complete mixing occurs between the injection point and the measuring point. The formula to compute discharge is:

$$Q_S = Q_T \frac{C_T - C_S}{C_S}$$

Where (in any consistent set of units):

Q_S = the volumetric flow rate in the stream
C_T = the concentration of the tracer in the injection fluid
Q_T = the injection fluid volumetric flow rate
C_S = the measured concentration in the stream downstream of the injection point

DIME file *See* dual independent map encoding file.

dimensionless unit hydrograph A unit hydrograph for which the discharge values are expressed as the ratio of actual discharge to the peak value and the time values are expressed as the ratio of actual time to the time where the peak discharge value occurs. The hydrograph is called dimensionless because the discharge ratio and time ratio have no units. *See also* unit hydrograph.

dimer An oligomer (a polymer molecule consisting of only a few monomer units) composed of two molecules of the same chemical composition. *See also* polymer.

dimethyl glyoxal ($CH_3COCOCH_3$) A diketone created during the reaction of ozone (O_3) with natural organic matter. *See also* diketone; ozonation by-product.

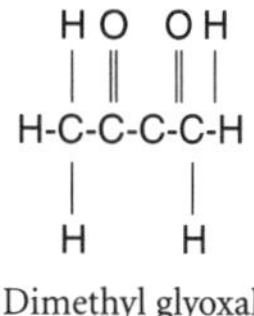

Dimethyl glyoxal

dimethyl-2,3,5,6-tetrachloroterephthalate (DCPA) ($C_6Cl_4(COOCH_3)_2$) An herbicide.

dimethyl-2,3,5,6-tetrachloroterephthalate (DCPA) di-acid degradate A degradation product of DCPA ($C_6Cl_4(COOCH_3)_2$), which is an herbicide. *See also* herbicide.

dimethyl-2,3,5,6-tetrachloroterephthalate (DCPA) mono-acid degradate A degradation product of DCPA ($C_6Cl_4(COOCH_3)_2$), which is an herbicide. *See also* herbicide.

dimictic Pertaining to lakes and reservoirs that freeze over and normally go through two stratification and two mixing cycles within a year.

dimple spring *See* depression spring.

2,4-dinitrophenol ($C_6H_3OH(NO_2)_2$) A synthetic organic chemical with various industrial uses, including use in dyes, preservation of lumber, manufacture of photographic developer, and explosives manufacture. *See also* synthetic organic chemical.

2,4-dinitrotoluene ($C_6H_3CH_3(NO_2)_2$) A synthetic organic chemical with various industrial uses, e.g., in organic synthesis, dyes, and explosives. *See also* synthetic organic chemical.

2,6-dinitrotoluene ($C_6H_3CH_3(NO_2)_2$) A synthetic organic chemical. *See also* synthetic organic chemical.

dinoseb ($CH_3(C_2H_5)CHC_6H_2(NO_2)_2OH$) The legal label name for 2,4-dinitro-6-sec-butylphenol, a synthetic organic chemical used as an insecticide, ovicide, and herbicide. It is regulated by the US Environmental Protection Agency. *See also* herbicide; insecticide; ovicide; synthetic organic chemical.

Dinoseb

dioxin Typically, a specific dioxin, i.e., 2,3,7,8-tetrachlorodibenzo-p-dioxin, which is a chlorinated hydrocarbon that occurs as an impurity in the herbicide 2-(2,4,5-trichlorophenoxy)propionic acid. In addition, other environmental sources exist for this highly toxic and persistent contaminant, which is regulated by the US Environmental Protection Agency.

Dioxin, where the Xs are chlorine atoms, hydrogen atoms, or both

DIP *See* ductile iron pipe.

1,2-diphenylhydrazine ($C_6H_5NHNHC_6H_5$) A synthetic organic chemical. *See also* synthetic organic chemical.

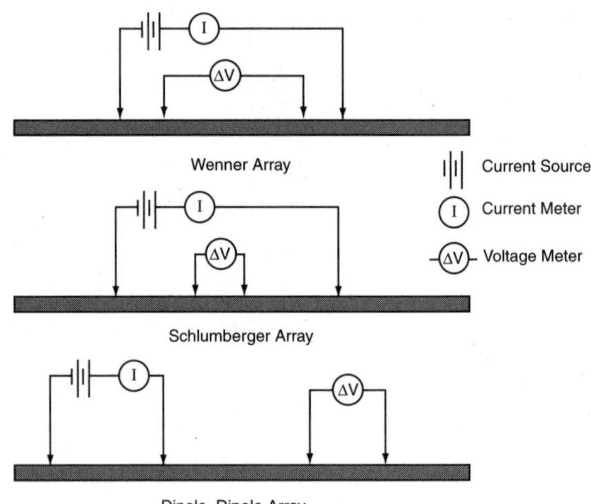

Dipole arrays

dipole array A particular arrangement of electrodes used in electrical resistivity surveys. Three configurations are in common usage: Wenner array, Schlumberger array, dipole–dipole array. The arrays differ principally in electrode spacing and relative orientation.

DIPRA *See* Ductile Iron Pipe Research Association.

diquat ($(C_5H_4NCH_2)_2Br_2$) The commercial name for 6,7-dihydrodipyrido(1,2-a:2-,1-c)pyrazidinium salt (1,1-ethylene 2,2-di-pyridium dibromide), an herbicide and plant growth regulator. It is regulated by the US Environmental Protection Agency. *See also* herbicide.

Diquat

DIR *See* demand initiated regeneration.

direct current (DC) A unidirectional electrical current that does not change in value over time (or changes so little that the change can be neglected). It is measured in amperes.

direct debt The debt an entity has incurred in its own name or assumed through the annexation of territory or consolidation with another entity.

direct digital control (DDC) A type of control in which a digital controller (usually a minicomputer) is directly connected to the controlled devices. *See also* distributed control system.

direct filtration A method of filtration in which coagulated and flocculated particles are applied directly to a filtering medium from the flocculation basin without settling. It is not to be confused with direct in-line filtration. *See also* coagulation; direct in-line filtration; filtration; flocculation.

direct fire costs Those costs incurred solely for the purpose of fire protection.

direct fire pressure The pressure required in the pipes of a water distribution system to provide adequate fire streams without the use of fire department pumpers when the fire hose is connected directly to the hydrant.

direct flow (1) *See* dead-end flow. (2) Water diverted from a stream or river without interruption between diversion and use except for incidental purposes, such as filtration or settling.

direct in-line filtration A method of filtration in which coagulated particles are flocculated "in-line" prior to direct application to a filtering medium. Neither a dedicated basin for flocculation nor settling is provided. Direct in-line filtration is not to be confused with direct filtration (which uses a dedicated flocculation basin). It is also called in-line filtration. *See also* direct filtration; filtration; flocculation.

direct labor Labor directly expended or applied in productive operations, as distinguished from that not directly connected with a productive process.

direct material A material that can be identified with a particular product or service and the cost of which can be charged directly to that product or service.

direct oxidation The direct combination of substances with oxidants, e.g., oxidation of substances in drinking water by the direct application of oxidizing agents such as chlorine.

direct precipitation Water that falls directly onto a lake, reservoir, or stream without passing through any land phase of the runoff cycle.

direct reuse The use of reclaimed water for nonpotable or potable purposes without first discharging to a water supply source. Direct reuse for a public drinking water supply is not accepted by the American Water Works Association or regulatory agencies.

direct right *See* direct flow; *see also* dead-end flow.

direct runoff Water that flows over the ground surface or through the ground directly into streams, rivers, or lakes.

direct-acting reciprocating pump A steam-driven reciprocating pump in which the steam piston is directly connected to the liquid piston or plunger through the piston rod.

direct-connected pump A power pump directly connected to the power unit that operates it by some form of coupling or clutch. Such a pump is also called a close-coupled pump.

Dirichlet condition A boundary condition common in groundwater modeling such that the head is known at the boundary of the flow field.

discharge The volume of water flowing past a specific point in a water system in a given period of time. Typical units are cubic metres per second, gallons per minute, million gallons per day, and cubic feet per second.

discharge area The cross-sectional area of a waterway, used to compute the discharge of a stream, pipe, conduit, or other carrying system.

discharge capacity The maximum rate of flow that a conduit, channel, or other hydraulic structure is capable of passing.

discharge coefficient A coefficient by which the theoretical discharge of a fluid through an orifice, weir, nozzle, or other passage must be multiplied to obtain the actual discharge.

discharge curve A curve that expresses the relationship between the discharge of a stream or open conduit at a given location and the stage or elevation of the liquid surface at or near that location. It is also called a rating curve or discharge rating curve.

discharge head The pressure measured at the center line of a pump discharge and very close to the discharge flange, converted into metres or feet of head. The pressure is measured from the center line of the pump to the hydraulic grade line of the water in the discharge pipe.

discharge hydrograph A hydrograph of the discharge or flow of a stream or conduit.

discharge measurement (1) A determination of the quantity of water flowing per unit of time in a stream channel, conduit, or orifice at a given point by means of a current meter, rod float, weir, pitot tube, or other measuring device or method. The operation includes not only the measurement of the velocity of water and the cross-sectional area of the stream of water, but also the necessary subsequent computations. (2) The numerical results of a measurement of discharge, expressed in appropriate units. *See also* stream gauging.

discharge piping The piping system or pipe header through which water flows from the outlet side of a pump, pump station, or reservoir.

discharge rate The volume of flow of water over a given time discharged by a pump, fire hydrant, hose bib, spillway, or other facility past a given point, usually the outlet. The flow is

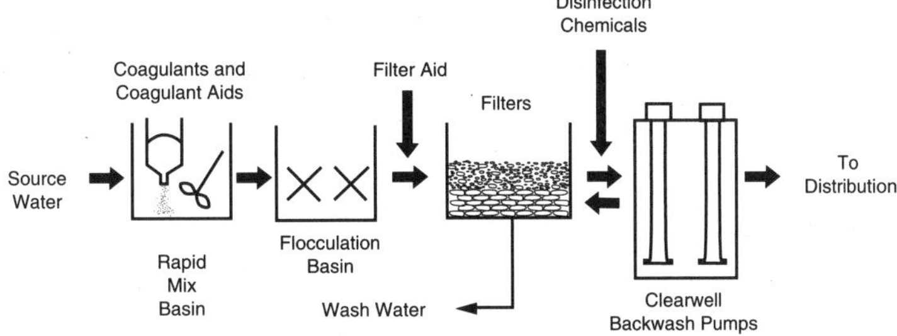

Direct filtration plant

expressed in units of volume per time, such as cubic metres per day, litres per minute, cubic feet per second, or gallons per minute.

discharge rating curve A curve showing the relationship between the discharge of a gate, meter, or other hydraulic structure or instrument and the pertinent hydraulic conditions that affect the discharge.

discharge table A table showing the relationship between the gauge height and the discharge of a stream or conduit at a given gauging station.

discharge valve A valve located on the outlet side of a pump or on the piping system where water flows from a reservoir.

discharge velocity *See* specific discharge.

disconnect switch A hand-operated switch for isolating electrical power from a device to permit safe maintenance on the device. The switch must open all power conductors running to the load and is always lockable in the off (or open) position. It is not necessarily the same as a loss-of-signal switch, which is used to prevent operation of electrical equipment. *See also* loss-of-signal switch.

discontinuous interstice A small open space or interstice that is not connected with another interstice. It usually occurs in lava or other effusive igneous rock of a vesicular texture. A rock that has only discontinuous interstices is not permeable.

discount As applied to the securities issued or assumed by a utility, the excess of the par or face value of the securities plus interest or dividends accrued at the date of the sale over the cash value of the consideration received from their sale.

discount rate A rate that it used to restate, or discount, future cash flows to their net present value. It usually represents the long-term cost of funds for a project or program. The discount rate can also be thought of as the interest-earning rate on an investment instrument, such as 6.0 percent on a long-term bond or 5.0 percent on short-term deposits in a financial institution. The discount rate includes both an inflation component and a real return-on-investment component. The real component is the long-term real return on the investment, usually calculated at about 3.0 percent, and the balance of the return rate is inflation. *See also* inflation; net present value; return on investment.

discounted cash-flow (DCF) model A model, often used in rate making, for estimating the investor required rate of return on common equity. By definition, the DCF model contends that the market price of a common stock is equal to the cumulative present value of all future cash flows to investors produced by said common stock.

discrete dynode detector A type of electron multiplier consisting of an array of individual electrodes, or dynodes, that are used in mass spectrometers to carry out the electron multiplication process and detect the ions passed by the mass spectrometer. Use of the discrete dynode detector is gradually replacing use of the channel electron multiplier detector in inductively coupled plasma–mass spectrometry. The discrete dynode detector's higher sensitivity, longer lifetime, and wider dynamic range are all desirable features. *See also* inductively coupled plasma–mass spectrometry.

discrete sedimentation Sedimentation in which the removal of suspended solids is a function only of terminal settling velocity. It is also called free settling or type I settling. *See also* flocculant; type III settling.

disease Any change from a state of health; an interruption in the normal functioning of a body structure.

disease-causing bacteria Pathogenic bacteria capable of infecting a host and causing disease.

disinfectant An agent that destroys or inactivates harmful microorganisms. *See also* disinfection.

disinfectant contact time (*T*) The time in minutes that it takes for water to move from the point of disinfectant application or the previous point of disinfectant residual measurement to a point before or at the point where residual disinfectant concentration (*C*) is measured. Disinfectant contact time in pipelines must be calculated based on plug flow by dividing the internal volume of the pipe by the maximum hourly flow rate through that pipe. Disinfectant contact time within mixing basins and storage reservoirs must be determined by tracer studies or an equivalent demonstration. *See also* $C \times T$.

disinfectant/disinfection by-products (D/DBP) A group of disinfectants and the chemical by-products resulting from the application of those disinfectants that have been selected by the US Environmental Protection Agency for regulation.

Disinfectant/Disinfection By-Products (D/DBP) Rule A national primary drinking water regulation promulgated by the US Environmental Protection Agency to regulate drinking water disinfectants and by-products of disinfection.

disinfectant-resistant pathogen *See* disinfection-resistant pathogen.

disinfection (1) The process of destroying or inactivating pathogenic organisms (bacteria, viruses, fungi, and protozoa) by either chemical or physical means. (2) In water treatment, the process in which water is exposed to a chemical disinfectant—chlorine ($HOCl$, OCl^-), chloramines ($NHCl_2$ or NH_2Cl), chlorine dioxide (ClO_2), iodine, or ozone (O_3)—for a specified time period to kill pathogenic organisms.

disinfection by-product (DBP) A chemical by-product of the disinfection process. Disinfection by-products are formed by the reaction of the disinfectant, natural organic matter, and the bromide ion (Br^-). Some disinfection by-products are formed through halogen (e.g., chlorine or bromine) substitution reactions; i.e., halogen-substituted by-products are produced. Other disinfection by-products are oxidation by-products of natural organic matter (e.g., aldehydes—RCHO). Concentrations are typically in the microgram-per-litre or nanogram-per-litre range. *See also* aldehyde; bromide; cyanogen halide; disinfection; dissolved organic halogen; haloacetic acid; halogen; halogen substitution; natural organic matter; oxidation by-product; ozonation by-product; trihalomethane.

disinfection by-product formation potential (DBPFP) The amount of disinfection by-products (DBPs) formed during a test in which a source or treated water is dosed with a relatively high amount of disinfectant (normally chlorine—$HOCl$, OCl^-) and is incubated (stored) under conditions that maximize DBP production (e.g., neutral to alkaline pH, warm water temperature, contact time of 4 to 7 days). This value is not a measure of the amount of DBPs that would form under normal drinking water treatment conditions, but rather an indirect measure of the amount of DBP precursors in a sample. If a water has a measurable level of DBPs prior to the formation potential test (e.g., in a prechlorinated sample), then the formation potential equals the terminal value measured at the end of the test minus the initial value. *See also* disinfection by-product; disinfection by-product precursor.

disinfection by-product precursor (DBPP) A substance that can be converted into a disinfection by-product during disinfection. Typically, most of these precursors are constituents of natural organic matter. In addition, the bromide ion (Br^-) is a precursor material. *See also* bromide; disinfection by-product; natural organic matter.

disinfection by-product regulatory assessment model (DBPRAM) A computer model used by the US Environmental Protection Agency in developing the Disinfectant/Disinfection By-Products Rule. The empirically based model assesses disinfection by-product formation based on source water quality and treatment plant operations, while simultaneously taking into account disinfection criteria and other water quality standards and objectives. *See also* Disinfectant/Disinfection By-Products Rule.

disinfection-resistant pathogen A microorganism, generally a bacterium, that shows less susceptibility to inactivation by a chemical disinfectant than related organisms or other strains of the same organism. Usually, the resistance shown is not great and the organism can be inactivated by increasing the exposure time at the same disinfectant concentration or by increasing the disinfectant concentration and maintaining the same exposure time. Complete resistance to a disinfectant is rare or nonexistent. Spores of spore-forming bacteria and *Mycobacterium* species are more disinfection resistant than other bacteria. Cysts and oocysts of pathogenic protozoa (*Giardia* and *Cryptosporidium*), as well as spores of fungi, are considered disinfection resistant because they are not inactivated by the disinfectant concentrations and contact times normally used in drinking water treatment processes. *See also Cryptosporidium*; cyst; *Giardia*; *Mycobacterium*; oocyst.

disintegration The breaking down of a substance into its component parts.

disk drive A peripheral mass storage device using magnetic or optical technology that enables direct and quick storage and retrieval of computer data.

disk friction Friction occurring in a turbine between (1) the water above and below the runner and (2) the stationary parts and upper and lower covers of the turbine.

disk operating system (DOS) A microcomputer operating system developed by IBM® and Microsoft® Corporation that has become a de facto standard in the microcomputer industry for IBM-compatible computers.

disk-type meter A water-measuring device containing a measuring chamber in which a disk is actuated by the passage of water. The nutating motion marks the discharge of a quantity of water through the measuring chamber and is translated into units of volume on a register dial.

dispersant A surface-active substance added to a suspending medium to promote and stabilize a dispersion. *See also* dispersion.

dispersed hepatitis A virus (HAV) A virus preparation such that individual virus particles are homogeneously dispersed in the preparation, as compared to clumping or cell-associated virus particles.

dispersion (1) The phenomenon in which a solute flowing in groundwater is mixed with uncontaminated water and becomes reduced in concentration. Dispersion is caused by differences in pore velocity and differences in flow paths at a small scale in the aquifer. A similar phenomenon is caused by turbulence in surface water systems. (2) A uniform and maximum separation of extremely fine particles, often of colloidal size.

dispersion index A measure of the short-circuiting of liquid through a continuous-flow tank. For example, the Morrill index, a common dispersion index, is the ratio of the time in which 90 percent of a unit volume of liquid passes through the tank to the time in which 10 percent of that unit volume passes through, t_{90}/t_{10}.

displacement device A toilet retrofit device, such as a dam, bag, bottle, and so forth, designed to displace toilet tank water.

displacement meter A water meter that measures the quantity of flow by recording the number of times a container of known volume is filled and emptied. Displacement meters are used primarily for relatively low flows.

displacement pump A device to move or lift water by using a piston to apply force directly to the water leaving a cylinder. Water is drawn into the cylinder through the inlet valve as the piston moves back, creating a partial vacuum with the outlet valve closed. The force is then applied to the water, which is forced out of the outlet valve at a higher pressure or lift.

displacement time The average period of detaining or holding a moving liquid in a tank or channel (tank volume less dead space, divided by rate of discharge). It is sometimes called the displacement period.

displacement velocity In settling tanks, the rate of displacement of the contents by inflowing liquid.

display (1) A visual exhibit designed to catch the eye. Objectives may include creating or maintaining a specific perception of the organization, maintaining contact with important audiences, or generating requests for more information. (2) An electronic device that temporarily presents visual information, as in a display for a computer.

disposable component Any component of a piece of water treatment equipment or water treatment system that is manufactured to be disposed of instead of repaired or reused, e.g., a cartridge filter element.

disposal site That portion of the waters of the United States where specific disposal activities are permitted, consisting of a bottom surface area and any overlying volume of water. In the case of wetlands on which surface water is not present, the disposal site consists of the wetlands surface area.

disposal well A shallow well used to place surface runoff and treated water into aquifers. Disposal wells differ from aquifer storage and recovery wells in that a disposal well employs the treatment capacity of an aquifer to purify the water before any subsequent withdrawal. The aquifer is used as a storage and filtration facility in this case.

disposition of complaint The official completion of a complaint investigation, including an assessment of the customer's satisfaction.

dispositional antagonism *See* antagonism.

disproportionation The simultaneous oxidation and reduction of a substance, such that two molecules of the same compound can react to form two other molecules. For example,

$$Cl_2^0\ (aq) + H_2O \Leftrightarrow H^+ + HO(Cl^{+1}) + Cl^{-1}$$

Here, the valence of chlorine both increases (oxidation) and decreases (reduction).

dissociation The process by which a chemical combination breaks up into simpler constituents—such as atoms, groups of atoms, ions, or multiple different molecules—without any change in valence. Often this breakdown is reversible, as in the case of ionization. *See also* ionization.

dissociation constant (1) Most generally, the rate at which two reversibly bonded chemical species separate from one another. This constant depends very heavily on the matrix in which the chemicals are placed or found. Specifically, in biochemistry, pharmacology, and toxicology, the concentration of a drug or toxin in solution at which a macromolecular binding site is half-saturated by the chemical. An apparent dissociation constant (known as the Michaelis constant, K_m) can be determined by examining the effects of various concentrations of the substrate on the rate of its conversion to product or by the extent to which varying the concentrations of a ligand activates a response that is mediated by a receptor. The inverse of K_m is defined as the affinity (or association constant) of a drug or chemical for the macromolecular binding site. (2) A value describing the tendency of a molecule to ionize in a solution at a given temperature. For example, the dissociation constant of acetic acid (CH_3COOH) in water at 77° Fahrenheit (25° Celsius) is 1.76×10^{-5}, which corresponds to pK_a 4.8. Thus, at a pH of 4.8, 50 percent of the acetic acid is ionized and 50 percent is unionized. This type of dissociation constant is also called the ionization constant.

dissociation in water The splitting of molecules dissolved in water into positive and negative ions. It is more properly termed *ionization.*

dissolved air flotation (DAF) A process in which air is dissolved into water under high pressure and is subsequently released into the bottom of a treatment unit to float solids. Upon release, the lower pressure in the unit results in the formation of bubbles that collect particles as they rise to the surface. The floated particles are then skimmed for subsequent processing. This process is effective in removing low-density solids and algae.

dissolved carbon dioxide The carbon dioxide (CO_2) that is dissolved in a liquid medium, typically expressed in milligrams per litre. The saturation concentration is dependent on several factors, including partial pressure, temperature, and pH. *See also* partial pressure.

dissolved concentration The amount per unit volume of a constituent of a water sample filtered through a 0.45-micrometre pore-diameter membrane filter before analysis.

dissolved constituent *See* dissolved solids.

dissolved gases The sum of gaseous components that are dissolved in a liquid medium. Typical dissolved gases found in water include oxygen (O_2), nitrogen (N_2), carbon dioxide (CO_2), methane (CH_4), and hydrogen sulfide (H_2S), among others. High concentrations of dissolved gases can result in filter air binding and pump cavitation. *See also* air binding; cavitation.

dissolved inorganic carbon (DIC) The fraction of inorganic carbon (the carbonate (CO_3^{2-}), bicarbonate (HCO_3^-), and dissolved carbon dioxide (CO_2)) in water that passes through a 0.45-micrometre pore-diameter filter.

dissolved load The portion of the stream load that is carried in solution.

dissolved matter That portion of matter or solids, exclusive of gases, that is dispersed in water to produce a homogenous liquid. According to the definition used in the water treatment industry, dissolved matter is that portion of the total matter that will pass through a 0.45-micrometre pore-diameter membrane filter.

dissolved minerals Inorganic salts held in solution in potable water. Dissolved minerals make up a portion of the dissolved solids in water. The sizes of the particles are used to differentiate between dissolved and particulate matter, with various size ranges being used. The dissolved solids in a water sample are determined by filtering the water, evaporating it, and weighing the residue.

dissolved organic bromine (DOBr) A surrogate measurement of the total quantity of dissolved bromine-substituted organic material in a water sample. The presence of bromine-substituted organic molecules in source water is typically caused by synthetic organic chemicals, whereas in finished water it is typically caused by disinfection by-products and high-molecular-weight, partially halogenated aquatic humic substances. Dissolved organic bromine is also called total organic bromine, adsorbable organic bromine, or carbon adsorbable organic bromine. *See also* aquatic humic substance; disinfection by-product; halogen-substituted organic material; surrogate measurement; synthetic organic chemical.

dissolved organic carbon (DOC) That portion of the organic carbon in water that passes through a 0.45-micrometre pore-diameter filter. For most drinking water sources, the dissolved organic carbon fraction represents a very high percentage of the total organic carbon pool. It is composed of individual compounds as well as nonspecific humic material, although humic substances account for a large portion of dissolved organic matter in natural waters. Typically, the dissolved organic carbon level provides some indication of the amount of disinfection by-product precursors in a water source. After filtration, dissolved organic carbon is determined in the same manner as total organic carbon. Organic carbon concentrations should be reported as dissolved organic carbon only if the sample has been filtered through a 0.45-micrometre pore-diameter filter before analysis. *See also* disinfection by-product precursor; nonpurgeable organic carbon; particulate organic carbon; purgeable organic carbon; total organic carbon.

dissolved organic chlorine (DOCl) A surrogate measurement of the total quantity of dissolved chlorine-substituted organic material in a water sample. The presence of chlorine-substituted organic molecules in source water is typically caused by synthetic organic chemicals, whereas in finished water it is typically caused by disinfection by-products and high-molecular-weight, partially halogenated, aquatic humic substances. Dissolved organic chlorine is also called total organic chlorine, adsorbable organic chlorine, or carbon adsorbable organic chlorine. *See also* aquatic humic substance; disinfection by-product; halogen-substituted organic material; surrogate measurement; synthetic organic chemical.

dissolved organic halogen (DOX) *See* total organic halogen.

dissolved organic halogen formation potential (DOXFP) *See* total organic halogen formation potential.

dissolved organic matter (DOM) That portion of the organic matter in water that passes through a 0.45-micrometre pore-diameter filter. Dissolved organic carbon represents the carbon portion of this matter. For aquatic humic substances, the carbon level typically represents approximately 50 percent of the organic matter (the rest being hydrogen, oxygen, nitrogen, and sulfur). *See also* aquatic humic substances; dissolved organic carbon.

dissolved oxygen (DO) The concentration of oxygen in aqueous solution, which is often expressed in units of milligrams per litre. It is usually determined by one of two methods: a dissolved oxygen probe or Winkler titration.

dissolved solids In operational terms, the constituents in water that can pass through a 0.45-micrometre pore-diameter filter. *See also* total dissolved solids.

dissolved zinc (Zn) Generally, that portion of the total zinc in water that passes through a 0.45-micrometre pore-diameter filter. Zinc most commonly enters the domestic water supply from deterioration of galvanized iron or dezincification of brass.

distillate A liquid produced by condensation of vapors of that liquid, forming the product of a distillation process.

distillation (1) A purification process in which a liquid is evaporated and its vapor is condensed and collected. For water treatment, distillation is used as a desalting technique in such processes as multistage flash distillation, multiple-effect distillation, and vapor compression. (2) A group of techniques used to separate components of liquid mixtures for purification, isolation of analytes, or the minimization of interferences. All types of distillation are based on the equilibrium between liquid and vapor. The two most commonly used types are simple distillation and fractional distillation.

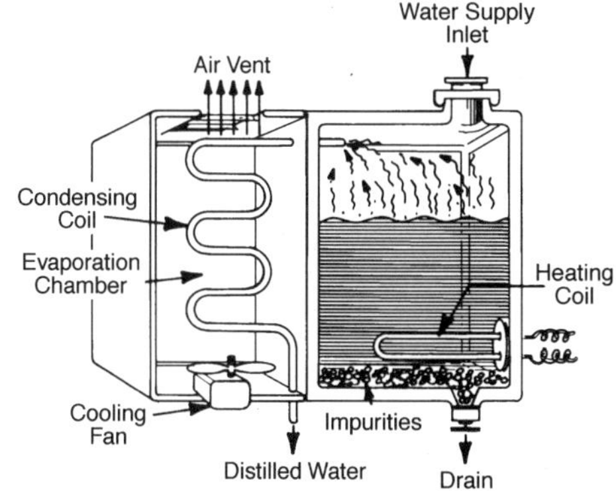

Point-of-use distillation system

distilled water A type of reagent water prepared by boiling a source water (usually tap water) and condensing the vapor for the purpose of purification. Many of the major constituents in tap water, such as minerals, are essentially removed during distillation. Distilled water is often prepared in the laboratory and can then be used to prepare standard solutions and blank samples.

distributed control system (DCS) The successor to direct digital control systems. In a DCS, many digital processors (usually microcomputers) are directly connected to the controlled devices.

distributed system A system in which processing tasks, data, or both are distributed among separate computers on a network.

distribution The system or process by which a commodity (e.g., water) is delivered to customers.

distribution coefficient The ratio of the concentration of a solute in phase A to the concentration of that same solute in phase B at equilibrium, in cases when the ratio is dependent on the initial concentration of the solute before partitioning occurs. The distribution coefficient is a way to express the ability of a particular method to extract a component from a matrix. It is similar to, but distinct from, the partition coefficient and distribution constant. *See also* distribution constant; partition coefficient.

distribution constant (*K*) The ratio of the concentration of a solute in phase A to the concentration of that same solute in phase B at equilibrium, in cases when the ratio is independent of the initial concentration of the solute before partitioning occurs. The distribution constant is a way to express the ability of a particular method to extract a component from a matrix. It is similar to the distribution coefficient but is the preferred term when the partitioning of the solute between phases is independent of the initial concentration of the solute before partitioning. *See also* distribution coefficient; partition coefficient.

distribution graph In hydrology, a unit hydrograph in which the ordinates of flow are expressed as percentages of the volume of the hydrograph. *See also* unit hydrograph.

distribution point A sampling point at which water quality is representative of that within the distribution system.

distribution reservoir A reservoir connected with the distribution system of a water supply project, used primarily to accommodate fluctuations in demand that occur over short periods (several hours to several days) and to provide local storage in case of emergency, such as a break in a main supply line or failure of a pumping plant.

distribution sample A sample of water taken from the distribution piping or a site served by the distribution system.

distribution system A system of conduits (laterals, distributaries, pipes, and their appurtenances) by which a primary water supply is distributed to consumers. The term applies particularly to the network of pipelines in the streets in a domestic water system.

distribution system characteristic curve A plot of the total dynamic head versus total discharge in a water distribution system. The plot shows the required head to supply water to the system at a particular total flow so that users will have sufficient water pressure. The plot considers all head losses in the system. The curve is used in conjunction with pump characteristic curves to determine optimal (minimum energy) pump operation strategies for different water demands and to evaluate elevated storage. *See also* pump characteristic curve.

distribution of toxicant The sites in the body where a chemical will locate given an oral, inhalation, or dermal exposure. Caution must be exercised in interpreting data purporting to describe a chemical's distribution. Frequently, data are provided on the distribution of a label attached to a compound. However, where the chemical and label can be separated by metabolism, this type of data provides very little useful information. Moreover, the site at which a chemical is deposited in the highest concentration does not necessarily reflect the organ in which the greatest effect will be observed. Data on the distribution of a chemical can provide insight on whether the chemical is likely to accumulate in the body, but this should not be considered a substitute for quantitative pharmacokinetic data.

distribution uniformity A measure of the efficiency of irrigation from above. It is calculated by analyzing the results of catch-can tests or by applying a formula to the dimensions and specifications of an irrigation plan.

distributor A fitting, usually installed at the top and bottom of the tank in a loose media system, that is designed to produce even flow through all sections of an ion exchanger or filter media bed and to function as a retainer of the media in the tank. It may also be called a diffuser. *See also* collector.

district engineer (DE) The principal engineer of a district of the US Army Corps of Engineers who is responsible for reviewing and issuing Clean Water Act Section 404 permits, unless the decision to issue a permit is elevated to a higher administrative level.

district metering The process of monitoring water flows into hydraulically discrete sections of a water distribution system to determine system integrity and the presence of leaks.

disturbed area An area where vegetation, soil, hydrology, or a combination of all three has been significantly altered, thereby making a wetlands determination difficult.

disturbed wetland A wetland directly or indirectly altered from a natural condition (including changes via natural perturbations) yet retaining some natural characteristics.

disulfoton ($(C_2H_5O)_2P(S)SCH_2CH_2SCH_2CH_3$) The generic name for the insecticide *O,O*-diethyl *S*-[2-(ethylthio)ethyl] phosphorodithioate. *See also* insecticide.

diuresis An increased excretion of urine.

diurnal (1) Occurring during a 24-hour period, as in diurnal variation. (2) Occurring during the daytime (as opposed to nighttime).

diurnal fluctuation (1) The cyclic rise and fall of the water table or stream flow during a 24-hour period in response to changes in evapotranspiration draft from groundwater. (2) Any daily variation in a groundwater characteristic, such as flow or dissolved solids.

diurnal demand curve A plot of water demand versus time for a 24-hour period. Time is plotted on the *x* axis and demand is plotted on the *y* axis. The curve depicts a typical period of time (average day, maximum day, minimum day, and so on) and is used to simulate the daily operation of the network, especially the cycling of system storage.

diuron ($C_6H_3Cl_2NHCON(CH_3)_2$) The generic name for the herbicide 3-(3,4-dichlorophenyl)-1,1-dimethylurea. *See also* herbicide.

divalent *See* divalent ion.

divalent ion An ion with two positive or negative electrical charges, such as ferrous ion (Fe^{2+}) or sulfate ion (SO_4^{2-}).

diverging tube A tube for which the diameter increases, usually at a uniform rate, along its longitudinal axis from the end at which the liquid enters.

diversion The use of part of a stream flow as a water supply.

diversion area That portion of an adjacent area beyond the normal groundwater or watershed divide that contributes water to the groundwater basin or watershed under discussion.

diversion box A type of diversion chamber.

diversion chamber A chamber in a water treatment plant that contains a device for drawing off all or part of a flow or for discharging portions of the total flow to various outlets.

diversion channel (1) An artificial channel constructed around a town or other point of high potential flood damages to divert flood water from the main channel for the critical reach. (2) A channel carrying water from a diversion dam. *See also* bypass channel.

diversion cut A channel cut around the side of a reservoir past the end of a dam to convey flood discharge from the stream above the reservoir into the stream below the dam; a bypass channel.

diversion dam A fixed dam built to divert part or all of the water from a stream out of and away from its course. *See also* overflow weir.

diversion gate A gate, having one of many different forms, that may be closed to divert flow from one channel to another.

diversion works Dams, pumping plants, and all appurtenant structures by means of which water is diverted from a stream or other body of water. *See also* headworks.

diversity index A nondimensional value relating the numbers of individuals of all species present to the number of species present at a site. The term is most commonly used in biological studies.

diverting weir *See* overflow weir.

divide The line that follows the ridges or summits forming the boundary of a drainage basin (watershed) separating one drainage basin from another. A divide is also called a watershed divide.

divining rod A stick, usually a forked branch of a certain kind of tree, that some people claim can be used to locate underground sources of water. This belief still persists in many sections of the United States, although no scientific basis has ever been found for it. A divining rod is also called a dowsing rod.

divinylbenzene (DVB) ($C_6H_4(CH{:}CH_2)_2$) A polymerization monomer used as a cross-linking agent by polymerization with styrene monomer in the manufacture of many synthetic ion exchange resin products. The degree of divinylbenzene cross-linkage is a factor in exchanger resistance to chemical oxidation. Standard cation resin usually contains about 8 percent divinylbenzene. Macroporous resins contain over 12 percent divinylbenzene cross-linking. *See also* cross-linking.

division box A device for splitting and directing discharge from a head box to two separate points of application. *See also* splitter box.

divisor A device used in soil conservation studies that controls the continuous separation and diversion of a representative fraction of the runoff from experimental areas. The diverted fraction is retained and analyzed to furnish a basis for calculating the total losses of soil and water from the area.

dL *See* decilitre *in the Units of Measure list.*

DLG *See* digital line graph.

DNA *See* deoxyribonucleic acid.

DNA adduct *See* deoxyribonucleic acid adduct.

DNA repair *See* deoxyribonucleic acid repair.

DNAPL *See* dense, nonaqueous phase liquid.

DNR *See* Department of Natural Resources.

DO *See* dissolved oxygen.

DOA *See* Department of Agriculture.

doble nozzle A nozzle, used in an impulse turbine, that is opened or closed by a pin or needle moving longitudinally along the axis of the jet and that may be adjusted to give a symmetrical jet of any effective area ranging downward from that obtained at full nozzle opening to that obtained at a fully closed position.

DOBr *See* dissolved organic bromine.

DOC *See* dissolved organic carbon; Department of Commerce.

DOCl *See* dissolved organic chlorine.

DOI *See* Department of Interior.

DOL *See* Department of Labor.

dolomite ($CaCO_3 \cdot MgCO_3$) An equimolar combination of calcium carbonate ($CaCO_3$) and magnesium carbonate ($MgCO_3$) that occurs in nature as a hard rock. Upon calcination it becomes dolomitic lime, containing 30 to 50 percent magnesium oxide (magnesia, MgO) and 50 to 70 percent calcium oxide (CaO).

dolomitic lime Lime containing 30 to 50 percent magnesium oxide (magnesia, MgO) and 50 to 70 percent calcium oxide (CaO).

DOM *See* dissolved organic matter.

domestic Pertaining to municipal (household) water services as opposed to commercial and industrial water. The term is sometimes used to include the commercial component.

domestic consumption The quantity, or quantity per capita, of water supplied in a municipality or district for domestic uses or purposes during a given period, usually 1 day. It is usually taken to include all municipal use of water, as well as any quantity wasted, lost, or otherwise unaccounted for.

domestic filter A small filter used in the home to purify a water supply. Such a filter is also called a household filter, point-of-use filter, or home treatment filter.

domestic flow (DF) The flow in a water distribution system required to meet the normal demand for residential and commercial usage. This flow excludes fire flow, industrial flow, and separate irrigation flow, which would be added to domestic flow to determine the total system flow.

domestic hot water Hot water that is generated from a public water supply at a residence.

domestic meter A water meter installed on a consumer's service line; a service meter.

domestic service The pipes to provide water service for a house. It is the same as a house service.

domestic use Water use by the public (home use).

dominance (1) In a wetland, the spatial extent of a plant species. (2) The condition of being (a) the most abundant single plant species in a vegetation stratum such that, when all species are ranked in descending order of abundance and cumulatively totaled, it immediately exceeds 50 percent of the total dominance measure (e.g., area cover or basal area) for the stratum; or (b) any additional species that accounts for 20 percent or more of the total dominance measure for the stratum.

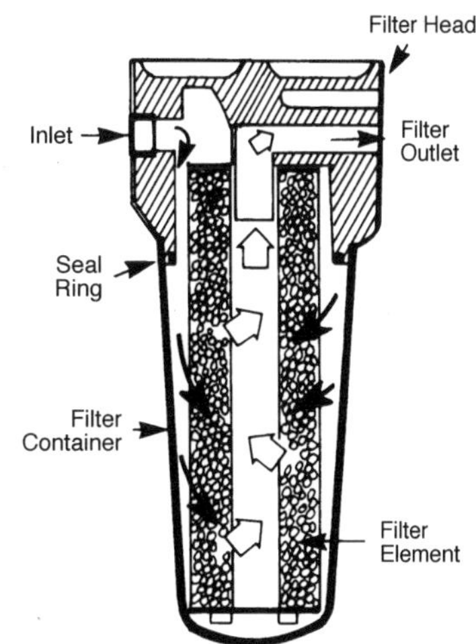

Domestic filter unit

dominance measure The means or method by which dominance is established, including area coverage and basal area. The total dominance measure is the sum total of the dominance measure values for all species that form a given stratum.

dominant lethal mutation An alteration in the base sequence of a developing organism's deoxyribonucleic acid that is incompatible with life. The site of the mutation occurs in the code of one or more proteins that have a function critical to maintaining life.

dominant resistance The condition of having a gene that codes for a factor that increases a cell's ability to survive certain insults. To be dominant, this gene need not be present on both alleles of deoxyribonucleic acid; one allele is sufficient. Plasmids with dominant resistance are frequently incorporated into bacteria in a way that can allow the bacteria to be selectively recovered from a medium containing other bacteria. For example, a mutated gene for ampicillin resistance can be placed in a bacterium to induce mutations; after enough time has passed, colonies of ampicillin-resistant bacteria can be detected based on the mutated gene.

dominant species In a wetland, a plant species in a given stratum that, when all plant species in that stratum are ranked in descending rank order and cumulatively totaled, either (1) immediately exceeds 50 percent of the total dominance measure or (2) accounts for 20 percent or more of the total dominance measure for the stratum.

donor strain The bacterial strain that contains the deoxyribonucleic acid that will be transferred to the recipient strain during mating.

72815
Ashland Community College Library
1400 College Drive
Ashland, KY 41101

Doppler-effect flowmeter An instrument for determining the velocity of fluid flow from the change in the observed frequency of an acoustic or electromagnetic wave. This change in frequency is due to relative motion between the observer and particles or discontinuities in the flowing fluid. Such an instrument that uses a wave of very high frequency is called a Doppler ultrasonic flowmeter.

DOS *See* disk operating system.

dosage A specified quantity of a material applied to a specified quantity of a second material, e.g., a certain number of milligrams per litre alum applied as a coagulant in water purification.

dose (1) The concentration of a chemical to which a person is exposed. This chemical can be measured as a gas, liquid, or solid. (2) The amount of chemical presented to an organism. A dose can be expressed in a variety of ways with varying degrees of accuracy. For example, a dose could be expressed as a concentration in drinking water, but this will yield varying doses depending on the amount of water that is consumed. Moreover, the dose that is consumed (the external dose) is only an indirect measure of the dose that reaches an active site other than the skin or mucous membranes in the body (the internal or systemic dose). Measures of external dose are inherently less accurate than measures of internal or systemic dose, except when the target tissue is the skin or mucous membranes. Therefore, in the following sequence, progressively greater accuracy of dose is achieved if the dose is expressed either as: the concentration in drinking water (least accurate); the dose per unit body weight or per unit surface area; the concentration of the parent compound in blood; the concentration of the parent compound in plasma water; the concentration of the active metabolite in blood plasma; the concentration of the active metabolite in the aqueous phase contacting a cell in a particular tissue (most accurate). The most accurate method of expressing dose is when the actual measurement of the amount of the active metabolite interacting with the receptor responsible for producing the effect (usually a macromolecule) is used. Knowledge of the dose at the target site is extremely important for interspecies extrapolations. The most common ways of comparing doses between species is on the basis of amount per unit body weight or amount per unit of body surface area. The latter convention is most rationally used when the action of the chemical is lessened by metabolism, because the extent of metabolism generally corresponds better to surface area rather than body weight. Thus, larger animals are considered less sensitive to a particular dose than smaller animals. This is the assumption that is made in assessing risks from chemical carcinogens. In addition, many carcinogens are activated by metabolism, further suggesting that small animals should be more sensitive than large animals. These general assumptions are, however, much better replaced by specific data that establish, first, which form of the chemical is responsible for the effect, and second, the relative rates of the formation and elimination of the active form. This allows accurate description of the "effective" doses in both the experimental animal and the human and avoids the use of assumptions. *See also* benchmark dose; lowest-observed-adverse-effect level; no-observed-adverse-effect level; reference dose.

dose equivalent The adsorbed dose from ionizing radiation, modified by factors that account for differences in biological effectiveness due to the type of radiation and its distribution in the body.

dose of radioactivity A quantity of ionizing radiation. This term is often used in the sense of an exposure dose, expressed in roentgens, which is a measure of the total amount of ionization that the quantity of radiation could produce in air.

dose rate The amount of chemical administered to an organism per unit time. In vivo, this rate is commonly expressed as milligrams per kilogram body weight per day. It could also be expressed as the area under the concentration versus-time curve for a body fluid such as blood or for an in vitro system.

dose rate of radioactivity The amount of ionizing (or nuclear) radiation to which an individual would be exposed per unit of time. It is usually expressed in terms of roentgens per hour or in submultiples of this unit, such as milliroentgens per hour. The dose rate of radioactivity is commonly used to indicate the level of radioactivity in a contaminated area.

dose response The effect of the dose of an agent administered or received on the incidence of an adverse health effect.

dose-effect curve A curve describing the relationship between the amount of chemical to which an organism is exposed and the frequency or intensity of the response that is produced. The relationship between dose and the frequency of response is the one most often dealt with in risk assessment.

dose-response evaluation A component of risk assessment that describes the quantitative relationship between the amount of exposure to a substance and the extent of toxic injury or disease.

dose-response experiment Any experiment, either biological or chemical, designed to obtain information on a dose response. The focus of such an experiment could be microbial inactivation data, optimal coagulant concentrations, or health effects.

dose-response model A formal mathematical formula that is applied to a set of data and describes the relationship between the dose of an agent that has been administered and the recipient's response. These models can take diverse forms, but the two most fundamental forms are graded dose-response models and the so-called all-or-none stochastic models. Graded models deal with the increasing intensity of response that is observed within an individual as the dose increases. All-or-none models are actually based on a distribution of sensitivities observed within a population; thus, at any one dose a certain proportion of the population would respond while the remaining individuals would not. In this case, as the dose increases, a progressively larger proportion of the population is assumed to respond. The all-or-none model is most frequently used in risk assessments and in extrapolating response information to doses considerably below the tested doses. The results of these extrapolations are expressed in terms of the number of deaths from a particular disease (e.g., cancer) within a population exposed for a given time. In either model type, uncertainty or safety factors are applied to account for deviations of the data from real-world responses.

dose-response relationship The quantitative relationship between the amount of exposure to a substance and the extent of toxic injury produced.

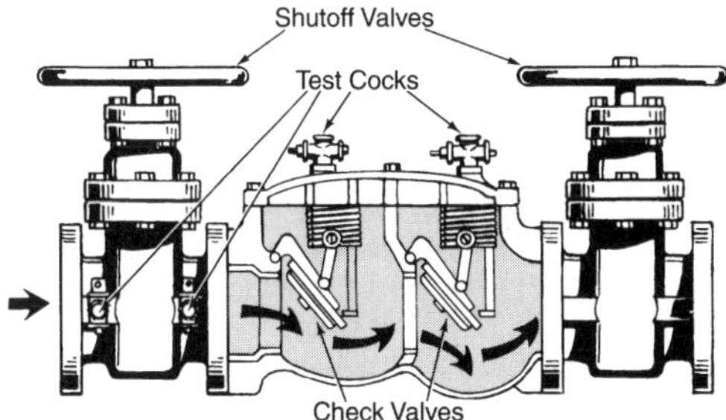

Double check valve assembly

dosimeter An instrument for measuring and registering total accumulated exposure to ionizing radiation.

dosing tank Any tank used in applying a dosage. Specifically, a dosing tank is used for intermittent application of a liquid to subsequent processes.

DOT *See* Department of Transportation.

DOT hazard class *See* Department of Transportation hazard class.

dot-blot A method of detecting viruses that involves immobilizing the nucleic acid of the virus onto nitrocellulose and detecting the nucleic acid by hybridizing it to a probe of complementary homologous nucleic acid.

double bond The binding together of two atoms by two pairs of electrons (rather than one pair as found in single bonds). Typical double bonds found in organic chemicals in water include carbon–carbon (C=C) and carbon–oxygen (C=O) double bonds.

double check valve assembly (DCVA) An assembly composed of two single, independently acting check valves. An approved assembly includes tightly closing shutoff valves located at each end of the assembly and suitable connections for testing the watertightness of each check valve.

double coagulation The application of coagulant to liquid at two separate points in the water treatment process.

double filtration The filtration of water through two or more slow sand filters in series, through a rapid granular filter and then through a slow sand filter, or through two rapid granular filters in series.

double gate, double check valve An assembly device consisting of two check valves with a gate valve on each end and with connections that provide points for testing to ensure positive shutoff of the check valves used for backflow prevention.

double mass curve A plot on arithmetic cross section paper of the cumulated values of one variable against the cumulated values of another or against the computed values of the same variable for a concurrent period of time.

double-action reciprocating pump A reciprocating pump for which the suction inlet admits water to both sides of the plunger or piston. This arrangement affords a more or less constant discharge.

double-layer compression A mechanism for destabilizing particles in which high–ionic strength solutions or solutions with high concentrations of total dissolved solids produce correspondingly high concentrations of counterions in the diffuse layer of particles. This mechanism reduces the range of repulsive interaction between similar colloidal particles, thus allowing attractive forces to dominate and coagulation to occur. For example, when a low–ionic strength river carrying colloidal particles flows into high–ionic strength seawater, the particles are destabilized by double-layer compression and settle out to form deltas. *See also* adsorption–destabilization; bridging; sweep-floc coagulation.

double-main system (1) A distribution system with mains on both sides of the street. (2) A system with separate water mains for domestic and fire-fighting use.

double-pan balance A balance in which a material is weighed by being placed on one pan while counterbalancing brass weights are placed on the other pan.

double-suction impeller An impeller with two suction inlets, one on each side of the impeller.

double-suction pump A centrifugal pump with suction pipes connected to the casing from both sides.

downflow The downward flow of water or a regenerant through an ion exchange or filter media bed during any phase of the operating cycle. This is the flow pattern found in conventional column operation: in at the top, out at the bottom of the column. The pattern may also be called co-current flow in ion exchange systems.

downflow softening The softening process in which water enters at the top of the softener bed column and passes downward through the cation resin and out the bottom. In this process, the brining also takes place in the same cocurrent direction.

downgradient The direction groundwater flows, similar in concept to *downstream* for surface water, such as a river.

downstream The direction toward which a stream flows.

downtime The time during which a machine, factory, conduit, or other item is shut down for repairs, replacement, or other work.

dowsing rod *See* divining rod.

DOX *See* total organic halogen.

DOXFP *See* total organic halogen formation potential.

DP cell *See* differential pressure cell.

DPD–FAS *See* N,N-diethyl-p-phenylenediamine–ferrous ammonium sulfate.

DPD *See* N,N-diethyl-p-phenylenediamine.

DPD methods *See* N,N-diethyl-p-phenylenediamine (DPD) methods.

DPP *See* differential pulse polarography.

dr *See* dram *in the Units of Measure list.*

drab A conduit or channel constructed to carry off, by gravity, liquids other than wastewater, including surplus underground water, stormwater, or surface water. It may be an open, lined or unlined ditch or a buried pipe.

draft (1) The act of drawing or removing water from a tank or reservoir. (2) The water that is drawn or removed from a tank or reservoir. (3) A written document prepared for review.

Draft Environmental Impact Statement (DEIS) A document that describes in great detail the effect a construction project will have on the local environment, prepared for review and consideration following the procedures established by regulatory agencies under the National Environmental Policy Act.

draft tube (1) An extension of the wheel passages in a hydraulic turbine from the point where the water leaves such passages down to the tailrace level. It may consist of pipes or passages constructed in the powerhouse structure. It allows the waterwheel and connected machinery to be constructed at higher elevations—usually above a possible high-water or flood level—without sacrificing any of the total height of power drop. It may also reduce velocity head losses in the water leaving the wheel.

draft tube loss The loss of energy, due to eddies and friction, that occurs when water passes through a draft tube.

drag The resistance offered by a liquid to the settlement or deposition of a suspended particle.

drag coefficient A measure of the resistance to sedimentation or flotation of a suspended particle, as influenced by the particle's size, shape, density, and terminal velocity. It is the ratio of the force per unit area to the stagnation pressure, and it is dimensionless. *See also* friction factor.

drain A pipe, conduit, or receptacle in a building that carries liquids by gravity to waste. The term is sometimes limited to referring to the disposal of liquids other than wastewater.

drain line A pipeline that is used to carry backwash water, regeneration wastes, or rinse water from a water treatment system to a drainage receptacle or waste system.

drain well *See* disposal well.

drainable sludge Sludge that can be dewatered readily by gravity.

drainage Any of several means of changing the hydrologic conditions of wetlands, including lowering groundwater or surface water levels through pumping or ditching or otherwise altering water flow patterns.

drainage area The area of a drainage basin or watershed, expressed in square metres, acres, square miles, or other units of area. A drainage area is also called a catchment area, drainage basin, watershed, or river basin.

drainage basin *See* drainage area.

drainage by well The removal of surplus or excess surface water or groundwater by (1) sinking wells to a porous formation in which the hydrostatic head is lower than that of the water on the surface to be drained and then carrying off the surface water through these wells, or (2) pumping the excess water into irrigation or drainage canals. When filled with stones, the well is called a dry well.

drainage canal A canal built and used primarily to convey water from an area where surface and soil conditions provide no natural outlet for precipitation.

drainage coefficient The discharge of an underdrainage system, expressed in inches (or centimetres) of depth of water that must be removed from the drainage area in 24 hours.

drainage divide The boundary between adjacent drainage basins. It is typically represented by a topographically high area.

drainage equilibrium A condition existing in an area where the quantity of water reaching the water table from all sources over a reasonable period of time just equals the quantity of water drained from the water table through all means during that time. The resulting position of the water table remains essentially unchanged over such a period of time.

drainage gallery A gallery in a masonry dam, parallel to the crest, that intercepts leakage from the water face and conducts it away from the downstream face.

drainage right The legal right of a landowner to dispose of excess or unwanted water that accumulates on his or her land over the land of others. Applicable common and statutory law and interpretations thereof vary widely among the various states.

drainage system (1) A system of conduits and structures for effecting drainage. (2) A surface stream or a body of impounded surface water, together with all surface streams and bodies of impounded surface water tributary to it.

drainage water (1) Water that has been collected by a drainage system and discharged into a natural watercourse. (2) Water flowing in a drain, derived from groundwater, surface water, or stormwater.

drainage well (1) A vertical shaft constructed in masonry dams to intercept seepage before it appears at the downstream face. (2) A well installed to drain surface water or stormwater into underground strata. It is also called a dry well. *See also* absorbing well.

drained Pertaining to a condition in which the level or volume of groundwater or surface water has been reduced or eliminated from an area by artificial means.

DRAM *See* dynamic random access memory.

dram (dr) *See the Units of Measure list.*

draw A natural depression or swale. This term also applies to a small watercourse in some parts of the United States.

drawback tank A tank used in some reverse osmosis systems (e.g., hollow-fiber seawater reverse osmosis systems) that provides permeate (desalted membrane product water) to the membrane modules during system depressurization. The natural phenomenon of osmosis causes permeate to flow backward through the membranes when there is a loss of feed pressure during shutdown; these vacuum conditions may cause damage if an adequate backflow water supply is not available.

drawdown (1) The drop in water elevation of a body of surface water as a result of the withdrawal of water. (2) The amount the water level in a well drops once pumping begins. In this case, drawdown equals the static water level minus the pumping water level. *See also* circle of influence; cone of depression; well cone of influence; zone of influence.

drawdown curve A graph of drawdown versus time or drawdown versus distance, used to determine hydraulic characteristics of water supply aquifers.

drawdown method A testing procedure in which water is pumped from a well and in which drawdowns in the same wells, nearby wells, or both, are recorded. Analysis of the time, drawdown, and distance relationships provides information on the hydraulic characteristics of the surrounding aquifer.

dredged material Material that is excavated or dredged from waters.

dredging (1) A physical method for controlling aquatic plants in which a dragline or similar mechanical equipment is used to remove plants and the bottom mud in which they are rooted. (2) The removal of sediment.

drenching shower A safety device used to wash chemicals off the body quickly.

drift (1) The difference between the actual value and the desired value (or set point) characteristics of proportional controllers that do not incorporate reset action. (2) The wandering of the controlled variable around its set point. Drift is also called offset.

drift barrier An artificial barrier built across a stream to catch driftwood. It may be of any form, from a simple wire fence to a barrier of massive piers with heavy cables strung between them.

drift tube A portion of a time-of-flight mass spectrometer in which ions are separated. *See also* time-of-flight mass spectrometer.

drill log A chronological record of the soil and rock formations that were encountered in the operation of sinking a well, with either their thickness or the elevations of the tops and bottoms of each formation given. *See also* well log.

drilled well A well that is constructed either by cable tool or rotary methods, usually to depths of 50 feet (165 metres) or more, and has the capacity to provide water for household, municipal, industrial, or irrigation requirements.

drilling Subsurface investigation performed by cutting holes down to the desired depth, with or without the removal of borings. *See also* boring.

drilling foam A bubbly drilling fluid made by injecting a small volume of water and surfactant (detergent) into an airstream. The foam is used to carry drilling cuttings to the surface, stabilize the borehole, and lift large volumes of water during drilling. Foam drilling is used only in unconfined aquifer formations.

Drinking Water Contaminant Candidate List (DWCCL) A list of contaminants issued by the US Environmental Protection Agency that are known or anticipated to occur in public water systems and that may require regulation under the Safe Drinking Water Act (Section 1412(b)(1)). The first Drinking Water Contaminant Candidate List was published March 2, 1998 (63 *Federal Register* 10274–10287), and the list will be updated every 5 years.

Drinking Water Research Division (DWRD) *See* Water Supply and Water Resources Division.

Drinking Water Standards Division (DWSD) The branch of the US Environmental Protection Agency's Office of Ground Water and Drinking Water that is responsible for development of national primary drinking water regulations.

drinking water Water that meets or exceeds all applicable federal, state, and local requirements concerning safety. Drinking water is also called potable water. *See also terms under* water.

drinking water equivalent level (DWEL) An estimated exposure (in milligrams per litre) that is interpreted to be protective against adverse health effects other than cancer over a lifetime of exposure, assuming all exposure is from drinking water ingestion.

drinking water priority list (DWPL) A list of contaminants that are known or anticipated to occur in public water systems and that may require regulation, published by the US Environmental Protection Agency as required by Section 1412(b)(3) of the Safe Drinking Water Act.

drinking water regulation A regulation set by a state or by the US Environmental Protection Agency that applies to public water systems. *See also* Primary Drinking Water Regulation; Secondary Drinking Water Regulations.

drinking water standard (DWS) A numerical limit established by a state or federal agency for contaminants that may be found in drinking water.

drinking water standard, primary *See* Primary Drinking Water Regulation.

drinking water standard, secondary *See* Secondary Drinking Water Regulations.

drinking water treatment facility The structures, equipment, and processes required to treat, convey, and distribute potable water.

drinking water treatment plant The central facility for the production of potable water, containing all treatment processes and appurtenances exclusive of the distribution system.

drip irrigation An irrigation method in which water is slowly applied through porous piping that has small openings. Drip irrigation reduces runoff and overall water consumption by avoiding the oversaturation that can result from irrigation methods using higher flow rates.

drip leg A small piece of pipe installed on a chlorine cylinder or container that prevents collected moisture from draining back into the container.

drive shoe A protecting end attached to the bottom of a drive pipe and casing.

driven well A shallow, usually small well (having a diameter of 1.5 to 3.0 inches [4 to 10 centimetres]) constructed without the aid of any drilling, boring, or jetting device by driving a series of connected pipe lengths into unconsolidated material to a water-bearing stratum.

driving The process of extending excavations horizontally.

drop (1) A structure in an open conduit or canal installed for the purpose of dropping the water to a lower level and dissipating its energy. It may be vertical or inclined; in the latter case it is usually called a chute. A drop is sometimes called a drop spillway. (2) The difference in water surface elevations upstream and downstream of a bridge or other constriction in a stream or conduit.

drop pipe The suction pipeline below a deep-well pump.

drop test A method to measure water leakage whereby in-system storage facilities or supplies are shut down and then the drop in the water level in the storage facilities or the drop in pressure in the system downstream from the facility or supply is measured.

drop-down curve The longitudinal shape of the water surface in a stream or open conduit upstream from a point where a sudden fall occurs, as when flowing water passes over a dam, weir, or other obstruction. In uniform channels the curve is convex upward. *See also* backwater curve; surface curve.

dropping head The decreasing head on an orifice or outlet pipe that may occur when inflow into a tank or reservoir is less than the outflow. The dropping head causes a continuing reduction in the rate of discharge.

drought A prolonged period of dry weather that causes a serious or prolonged deficiency or shortage of water.

drought price A set of special inclining block prices that are part of a comprehensive demand management program implemented by many utilities during periods of acute supply shortage (crisis situations). The total program might include public awareness, restrictions on use, monitoring of use, various indoor and outdoor conservation measures, and drought prices—all aimed at achieving a necessary reduction in demand. When the crisis is over, the drought prices are discontinued and replaced with a rate structure that reflects the new budget and supply conditions.

drowned weir A weir that, when in use, has the water level on the downstream side at an elevation equal to or higher than the weir crest. The rate of discharge is changed by the presence of the tailwater. A drowned weir is also called a submerged weir.

drug metabolism The conversion of a chemical to a series of products by enzymatic activity in an organism. In mammalian systems, the most prominent site of drug metabolism is in the smooth endoplasmic reticulum of the liver. However, metabolism of some chemicals occurs in other subcellular fractions and in other organs. This local metabolism can play an important role in the toxicology of a chemical. In some cases metabolism detoxifies the chemical, but in others it can convert the chemical to a reactive form that induces toxicity. For this reason, steps in the metabolism of drugs and other chemicals are divided into two phases. Phase 1 metabolism involves the production of a reactive metabolite, whereas phase 2 reactions tend to detoxify these reactive metabolites. The former reactions are conducted by cytochrome P-450 isoforms, whereas the latter phase involves various conjugation reactions with other biochemicals to form glucuronides, sulfates, glutathione conjugates, and so on.

drum gate A movable crest gate used at a dam to control spills or overflows. It is in the form of a sector of a circle hinged at the apex. The arc face effects a water seal with the edge of a recess into which the gate may be lowered. The gate is raised and held up by the pressure of water admitted to the recess from the headwater. It is lowered by closing the inlet port to the recess and draining the water from it.

dry barrel fire hydrant A device installed and spaced on a distribution system to supply water for fire protection, designed with a special fitting to connect to fire hoses or fire pumps and with a valve and drain system to allow the barrel of the hydrant to drain when the valve is turned off, protecting the hydrant from freezing.

dry chemical feeder A mechanical device that feeds chemicals in a dry form. Dry chemicals are typically loaded into a hopper, from which they are conveyed into a conduit for dilution or direct feed by a screw conveyer. The rate may be controlled volumetrically or gravimetrically (by weight). *See also* screw conveyer.

dry connection A connection to a water line installed when the line is not in service or during construction.

dry deposition Material deposited by the atmosphere on the earth's surface outside periods of precipitation.

dry feeder *See* dry chemical feeder.

dry pit An enclosure in which a pump or piece of mechanical equipment is not in contact with the solution being pumped. A dry pit is located adjacent to a wet well or basin that contains the liquid. A dry pit provides easy access for pump maintenance, although connections between the dry pit and wet well must be sealed properly to prevent leakage. *See also* wet well.

dry solids (DS) A measure of the mass of solids on a dry weight basis. In chemical analyses, a measurement of the dry solids in a solution yields a total solids concentration. The dry solids in a sludge provides an understanding of the amount of solids disposed of in a completely dewatered sludge. Certain land disposal requirements for sludge are based on the amount of dry solids per unit area per time.

dry spell (1) A period, usually of not less than 2 weeks duration, during which no measurable rainfall occurs in a certain place or region. When it extends to a month or more, a dry spell is considered to be a drought. (2) A period of not less than 4 days during which the daily maximum temperature remains at least 5°–7° Fahrenheit (3°–4° Celsius) greater than the normal maximum and the relative humidity continues at less than 50 percent.

dry strength The extent to which a soil is dry and crumbles into individual grains or fine powder on its own or under modest pressure.

dry suspended solids The weight of the suspended matter in a sample after drying for a specified time at a specified temperature.

dry weather flow The flow of water in a stream during dry weather, usually contributed entirely by groundwater. *See also* base flow; minimum flow.

dry weight capacity The capacity of a storage container expressed in terms of the dry weight of the chemical being stored.

dry well (1) A well that yields no water. (2) A dry compartment in a pumping station, near or below pumping level, where the pumps are located. (3) A common name for a drainage well. *See also* drainage well; pump pit.

dry year A year of drought during which the precipitation or stream flow is less than normal. Drought severity is usually expressed in terms of a probability or a return period.

dry-bulb temperature The temperature of air measured by a conventional thermometer.

dry-salt saturator tank A brine tank, usually full of undissolved salt and with saturated brine below the undissolved salt. This is the type of brine tank used with most automatically regenerated home softeners because it does not require refilling with salt as often as other tank types. *See also* wet-salt saturator tank.

drying bed *See* sludge-drying bed.

DS *See* dry solids.

DST *See* defined-substrate technology.

DTA *See* differential thermal analysis.

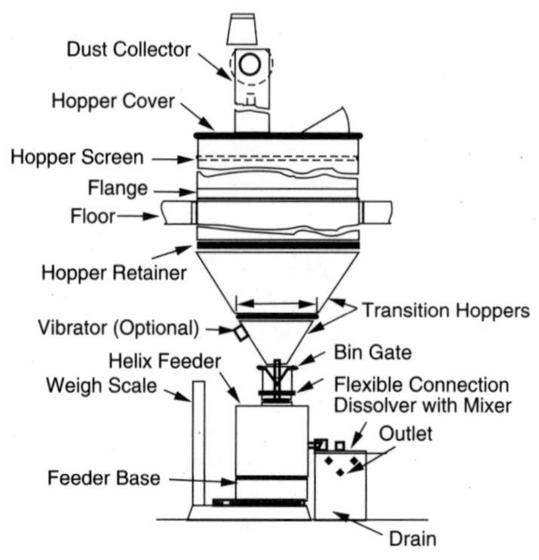

Dry chemical feed hoppers and dust collectors

DTM *See* digital terrain model.

dual coagulation A coagulation process that is used for more than one purpose. For example, coagulation can be used to remove both particles and naturally occurring organic material.

dual distribution system Two separate water piping systems distributing water to customers, one carrying potable water and the other conveying lesser-quality water (e.g., nonpotable reclaimed water) for reuse purposes.

dual programming The use of an irrigation controller to schedule the frequency and duration of irrigation cycles to meet varying water requirements of plants served by a system. Grouping plants and laying out irrigation circuits on the basis of similar water requirements will facilitate dual programming. If more than two groups are involved, the process is known as multiple programming.

dual independent map encoding (DIME) file A geographic file based on line segments produced by the US Bureau of the Census for each standard metropolitan statistical area in the United States. It is also called a geographic base file–dual independent map encoding (GBF–DIME) file when street segments are assigned geographic coordinates.

dual system *See* dual distribution system.

dual-flush device A toilet retrofit device designed to use only a portion of the toilet tank water for a liquid flush and the complete tank of water (full flush) to remove solids.

dual-media filter A filter containing two types of granular filtering media with different sizes and specific gravities to maintain media stratification after backwashing. Anthracite coal and sand are the most commonly used media in dual-media filters. *See also* anthracite coal; filter.

dual-media filtration A filtration method designed to operate at a filtration rate higher than conventional rapid granular filters by using two different types of filter media, usually sand and finely granulated anthracite.

dual-membrane system A water treatment membrane system composed of two different types of membranes in series, e.g., a surface water system that uses a microfiltration or ultrafiltration membrane system for particle removal followed by a nanofiltration or reverse osmosis system for dissolved material removal. A dual-membrane system is also called an integrated membrane system.

dual-purpose plant A plant with two production outputs. This term is commonly used to refer to a facility that is both producing electric power and desalting seawater. Waste heat from the electrical generating equipment is used (recovered) to lessen energy requirements for the water-desalting process facilities located at the same site.

duct A tube or channel for the conveyance of a fluid; a conduit.

ductile iron pipe (DIP) A water pressure pipe made from a material similar to cast iron, with the primary graphite in a nodular or spheroidal form as a result of the addition of an inoculant such as magnesium to molten iron of appropriate composition. These structural changes increase strength and flexibility.

Ductile Iron Pipe Research Association (DIPRA) A nonprofit corporation whose members are manufacturers of ductile iron pipe and fittings used by water, wastewater, and industrial facilities. The association's purpose is to provide research and engineering information to its members, utilities, industries, and the public relating to ductile iron pipe. Until 1980, DIPRA was known as the Cast Iron Pipe Research Association (CIPRA).

due diligence The efforts necessary to bring an intent to appropriate water to fruition; actions that demonstrate a good-faith intention to complete an undertaking within a reasonable time period.

duff The more or less firm organic layer in forests. It consists of fallen vegetative matter in the process of decomposition, including everything from the litter on the surface to the pure humus below.

dug well A shallow, large-diameter well constructed by excavating with power machinery or hand tools instead of drilling or driving. Typically a dug well is constructed for an individual residential water supply and yields considerably less than 100 gallons per minute (380 litres per minute).

dumb (1) Pertaining to the lack of intelligence of a database element (or a user workstation annotation incorporated into a map or engineering drawing) that exists only as a graphical element and is not linked to another graphical element or a corresponding attribute database. (2) Pertaining to workstations that have little or no local processing power and that rely on the processing power of another device, e.g., a mainframe or minicomputer.

dump power Power available in excess of primary and secondary power, usually for short periods of time during exceptionally favorable generating conditions. Dump, secondary, and off-peak power are not completely separable.

dump truck A vehicle used to haul earth material in the construction or maintenance of systems. It has a bed that will tilt to allow the material to be easily unloaded or dumped.

duplex pump A reciprocating pump consisting of two cylinders placed side by side and connected to the same suction and discharge pipe. The pistons move so that one exerts suction while the other exerts pressure, with the result that the discharge from the pump is continuous.

duplicate samples Samples obtained by dividing one sample into two or more identical subsamples.

Dupuit assumptions A set of assumptions used in the analysis of flow in unconfined aquifers. Solutions based on these assumptions are quite useful in practical problems. The assumptions are that (1) the hydraulic gradient is equal to the slope of the water table, and (2) the streamlines are horizontal and equipotential lines are vertical in the aquifer.

duration area curve A curve that shows the area beneath a duration curve at any value of the flow and is therefore the integral of duration with respect to stream flow. When the duration curve is plotted as a percentage of time, the resulting duration area curve shows the average flow available at less than a given discharge.

duration curve A graph showing the percentage of time that the given flows in a stream will be equaled or exceeded. The curve is generated based on a statistical study of historical stream flow records. *See also* intensity–duration frequency curve.

duration of inundation The length of time that water stands above the soil surface or that water fills most soil pores near the soil surface. Typically this term refers to a period during the growing season.

Durbin–Watson (DW) test The standard test used to determine whether serial correlation exists in a time series. The computed value of this test can range from 0, indicating extreme positive serial correlation, to 4.0, indicating extreme negative serial correlation. A Durbin–Watson value of 2.0 indicates no serial correlation.

duty of water The amount of water that, through careful use without waste and management, is reasonably required to be applied to a tract of land for a length of time adequate to produce the maximum quantity of crops that are ordinarily grown there.

DVB *See* divinylbenzene.

DW test *See* Durbin–Watson test.

DWCCL *See* Drinking Water Contaminant Candidate List.

DWEL *See* drinking water equivalent level.

DWPL *See* drinking water priority list.

DWRD *See* Water Supply and Water Resources Division.

DWS *See* drinking water standard.

DWSD *See* Drinking Water Standards Division.

dye A conservative, nonreactive substance, the concentration of which can be measured analytically, used as a tracer in water treatment applications. *See also* tracer.

dye dilution method Part of the dilution method that introduces a dye (sometimes fluorescent) as the tracer compound. *See also* dilution method.

dyke *See* dike.

dynamic discharge head The difference in height measured from the pump center line at the discharge of the pump to the point on the hydraulic grade line directly above it.

dynamic equilibrium (1) A condition wherein the recharge to an aquifer is equal to the amount of natural discharge. (2) A condition in which the inflows to a system exactly equal the outflows from the system.

dynamic head *See* total dynamic head.

dynamic membrane A membrane having a surface-layer coating of substances removed from the water such that this deposited layer also acts as a filtration barrier.

dynamic model A physical model of an object such that the dimensions, weights, and moments of inertia of the model are proportional to the full-sized object. Such a model is also said to have dynamic similarity.

dynamic pressure head When a pump is operating, the vertical distance from a reference point (such as a pump center line) to the hydraulic grade line.

dynamic random access memory (DRAM) The location where data are stored capacitively on a memory chip, requiring frequent recharge hundreds of times per second to avoid loss of the data. DRAM is a high-density, low-cost technology with operational speeds in the nanosecond range.

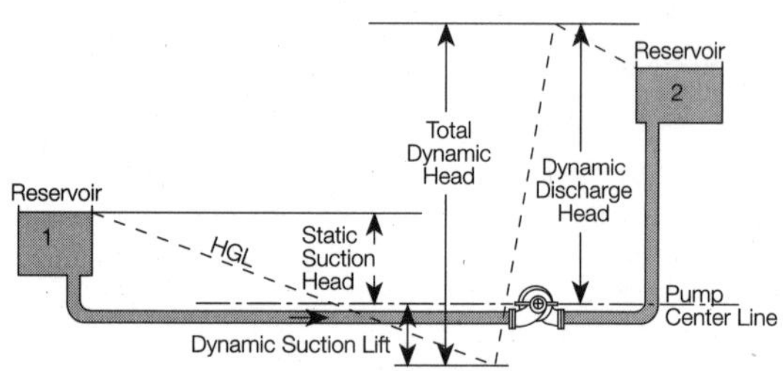

Dynamic suction lift for reservoir above pump

dynamic reaction An ion exchange reaction that takes place as the water moves past the exchange resin or resins.

dynamic-state extended-period simulation (EPS) A pipe network simulation comprising a series of steady-state runs that allow the user to view system operations over a period of time with defined intervals. EPS can model time-varying features, such as tank levels, head- or pressure-controlled valves, and the status of pumps and pipes.

dynamic suction head The distance from the pump center line at the suction of the pump to the point of the hydraulic grade line directly above it. Dynamic suction head exists only when the pump is below the piezometric surface of the water at the pump suction. When the pump is above the piezometric surface, the equivalent measurement is called dynamic suction lift.

dynamic suction lift The distance from the pump center line at the suction of the pump to the point on the hydraulic grade line directly below it. Dynamic suction lift exists only when the pump is above the piezometric surface of the water at the pump suction. When the pump is below the piezometric surface, the equivalent measurement is called dynamic suction head.

dynamic system A process or system in which motion occurs, as compared to static conditions with no motion. For example, an ion exchange system is considered a dynamic system because the continuous flow of the water to be treated creates continuous motion, as opposed to a static batch system in which the water does not move during the reaction process.

dynamic viscosity *See* absolute viscosity.

dynamic water pressure The water pressure at the inlet to a dynamic water-processing system.

dynamic water system *See* dynamic system.

dyne *See the Units of Measure list.*

dystrophic Pertaining to waters rich in humic materials, most present in colloidal form, and poor in nutrients and plankton.

E

e⁻ *See* electron.
E *See* exa.
E. cloacae *See Enterobacter cloacae.*
E. coli *See Escherichia coli.*
EA *See* environmental assessment.
EAP *See* emergency action plan.
early closure device A toilet retrofit device that takes advantage of the maximum water pressure in the tank provided by a full tank of water but closes the flapper early, saving a portion of water in the toilet tank.
earned surplus That part of a surplus that has arisen out of accumulated net income. It is also called retained earnings.
earth Unconsolidated material derived by weathering and erosion. In engineering practice, it is commonly called soil to distinguish it from rocks. It consists of gravel (boulders, cobbles, pebbles, and granules), sand, silt, and clay, including material of both inorganic and organic origin.
earth dam A dam for which the main section is composed principally of earth, gravel, sand, silt, and clay. An earth dam is sometimes called an earth-fill dam.
earth flow The moderately rapid flow of large masses of slippery earth material. It occurs along the edges of clay terraces in certain glaciated valleys and in some mountainous regions where shale bedrock occurs. It is also called landslide flow.
earth-dam paving Concrete slabs or stone blocks laid by hand, with or without mortar, on the water face of earth dams or levees to protect them against erosion by waves.
earthen reservoir A reservoir constructed in earth by using excavated material to form embankments. Such reservoirs may be unlined or lined with impervious materials.
earth-ground The primary electrical ground in a power system. All electrical power and lightning systems have conductors known as grounding conductors to carry errant currents safely into the earth, where the charge can be safely dissipated.
earthquake A slipping of the Earth's plates pushing against each other. When the stored energy overcomes the friction between the two plates, the slipping movement results in shock waves being sent out.
earth's gravitational constant (*g*) As adopted by the International Committee on Weights and Measures, 32.174 feet per second squared (9.80665 metres per second squared). It is also called the acceleration of gravity.
earthy–musty odor A smell associated with soil–dirt or mold–mildew. In drinking water, geosmin is a common source of earthy odors and 2-methylisoborneol is typically associated with musty smells. Both of these chemicals are produced by certain blue-green algae and *Actinomycetes*. *See also Actinomycetes*; blue-green algae; geosmin; 2-methylisoborneol.
easement An acquired legal right to the use of land owned by others. An easement is ordinarily more restricted in scope than a leasehold.
ebbing-and-flowing spring A spring that has periods of relatively large continuous discharge at more or less regular and frequent intervals. Such springs may be classified as perennial or intermittent. Although the discharge of such a spring somewhat resembles that of a geyser in its rhythmic action, such discharge is due to an entirely different cause, such as siphon action.
EBCDIC *See* Extended Binary Coded Decimal Interchange.
EBCT *See* empty-bed contact time.
ebullition The boiling of a liquid. Bubbles of saturated vapor form rapidly in the interior of the liquid and then rise to the surface and escape.
ECD *See* electron capture detector.
echo sounding A process of indirectly measuring depth by noting the time interval required for sound waves to go from a source of sound near the surface to the bottom and back again.
echovirus An acronym for the enteric cytopathogenic human orphan virus, a type of enterovirus that may cause diarrhea and aseptic meningitis.
ecokinetics The study of the rates of uptake and metabolism of specific compounds by living organisms in an ecosystem. The

term can also be applied to the apportioning of a compound among phases in an ecosystem.

ecologic epidemiologic study A study using descriptive epidemiologic data, vital statistics, and other information (e.g., general air pollution measurements for a region) to determine associations. In contrast to analytical epidemiologic studies, populations rather than individuals within these populations are studied, and any association between exposure and disease is indirect. These studies are used primarily to create hypotheses for further research by analytical epidemiologic studies.

ecological amplitude The range of one or more environmental conditions in which an organism or a process can function.

ecological balance The state of an ecosystem in which the relative populations of the system's components remain approximately constant.

ecological community An aggregation of naturally sustaining and interdependent organisms interacting in the same area.

ecological equivalence The condition in which two or more species can, because of their similarity, occupy the same ecological niche and thus be able to replace each other.

ecological equivalent One of two or more analogous species in similar environmental contexts (i.e., distantly related species) displaying closely similar adaptive mechanisms.

ecological fallacy An error in inferring that a population association observed in an ecologic or descriptive epidemiologic study applies to individual members of that population. An ecologic study does not link an individual's disease to exposures or confounding characteristics of that individual, nor does it link individual exposure and confounding characteristics to each other. These linkage failures are the source of special biases of ecological studies that are often referred to as the ecological fallacy.

ecological indicator An ecological community or organism so strictly associated with given environmental conditions that its presence is indicative of the existence of these conditions.

ecological niche The status of an organism within its community that determines its feeding activities and interactions with other organisms, including enemies.

ecologically sustainable development (ESD) A rate or level of development believed to enable an ecosystem to sustain itself.

ecology The study of the relationships between the environment and the living organisms present.

econometric forecasting The use of regression analysis in water demand forecasting, including (as independent variables) economic variables such as population, employment, household income, home value, and water prices. The resulting methods or models are also used to define the demand curve(s) for water and the price elasticity for projecting future demand.

econometrics The use of statistical and mathematical methods in the field of economics to develop and verify economic theories.

economic depreciation The loss in value of goods or services resulting from external economic conditions affecting the character or degree of utilization.

economic development A stage or step in growth or advancement that benefits a business, community, or organization.

economic development rate A water rate established to entice commercial and industrial customers either to stay in or to relocate into the utility's service area. Such a rate provides a temporary or permanent subsidy to the specific customers who qualify, justified by (1) the benefits accruing to the utility through additional water sales to new residents resulting from additional employment, or (2) the overall benefit accruing to the community as a whole from increased employment and growth in economic development. *See also* negotiated rate.

economic forecast A prediction of future production, distribution, and use of income, wealth, and commodities.

economic groundwater yield The maximum rate at which water can be withdrawn artificially from an aquifer throughout the foreseeable future without depleting the supply or altering the chemical character of the water to such an extent that withdrawal at this rate is no longer economically possible. The economic yield varies with economic conditions and other factors, such as recharge, natural discharge, and pumping head. The term may be applied with respect to the economic feasibility of withdrawal from the standpoint of the actual organization that withdraws water artificially or from the standpoint of the entire economy of a river valley or other larger area to which the aquifer contributes water.

economic impact analysis An evaluation of the economic impacts of a proposed or final rule prepared by the US Environmental Protection Agency. An economic impact analysis is prepared for minor rules (those with an economic impact of under $100 million), whereas a regulatory impact analysis is prepared for rules having a major impact (those with an economic impact of over $100 million). *See also* regulatory impact analysis.

economics The science that deals with the production, distribution, and consumption of wealth, as well as with the various related problems of labor, finance, taxation, and so on.

economy In the desalting processes, the ratio of the mass of distilled water produced per 1,000 British thermal units (1,055 kilojoules) of heat input.

economy of scale A reduction in unit cost resulting from a distribution of fixed costs over a larger number of units. Fixed costs are those necessary to and resulting from the implementation of an activity or project and not directly related to the actual production of a single unit.

ecosystem The composite balance of all living organisms and plants and their ambient environment (air, water, and solid phases) in a defined area.

ecosystem dynamics The interaction of characteristic and measurable processes within an ecosystem, such as succession, energy flow, nutrient cycling, and community metabolism.

ecosystem function Within an ecosystem, the manner in which energy flows and material production varies in cycles.

ecosystem structure The functionally important components of an ecosystem and their pattern of interrelations and spatial arrangements. This term is generally applied to organisms.

ecotone A transition zone where two adjacent communities meet and blend together. The conditions are intermediary between those of the two neighboring communities.

ecotoxicology A science that studies the response of the environment to insult by toxic substances.

Eductor

ecotype (1) A local variety or subspecies that has appeared as a result of and is adapted to a specific set of ecological conditions. (2) A group of organisms within a species that have morphological or physiological characteristics adapting them to their environment.

ED *See* effective diameter; electrodialysis.

ED_{10} The diameter of the d_{10} size particle in a particle size distribution. The ED_{10} represents the particle diameter for which 10 percent of the particles in a sample, by weight, have a smaller or equivalent diameter. The ED_{10}, or d_{10}, is typically used in defining the size characteristics of filter media, such as anthracite or sand. Common units are millimetres. *See also* effective diameter; particle size distribution.

EDB *See* ethylene dibromide.

eddy A circular movement occurring in flowing water, caused by currents set up in the water by obstructions or changes and irregularities in the banks or bottom of the channel.

EDF *See* Environmental Defense Fund.

edgematching The process of eliminating locational and content discrepancies in the representation of features at the edges of adjacent map sheets to create a continuous map.

EDI *See* electrodeionization.

EDM device *See* electronic distance-measuring device.

EDR *See* electrodialysis reversal.

EDTA *See* ethylenediaminetetraacetic acid.

eduction The process of bringing, sucking, or separating something out from something else, as in removing used filter sand from a filter box with a water slurry. *See also* eductor.

eductor A device used to mix a chemical with water. The water is forced through a constricted section of pipe (venturi) to create low pressure, which allows the chemical to be drawn into the stream of water or granular media to be removed from a vessel.

effect In water distillation, an evaporation–condensation step in which steam is condensed on one side of a heat transfer surface and salt water is evaporated on the other side using the available heat of condensation. *See also* multiple-effect distillation.

effect modifier A variable or characteristic that modifies the effect of an epidemiologic association (e.g., age is an effect modifier for many diseases).

effective corrosion inhibitor residual The corrosion inhibitor residual concentration that is capable of controlling corrosion or metal uptake to a target level. Corrosion inhibitors employ passivation of the metal surface as a means of corrosion control.

effective date The date a national primary drinking water regulation becomes enforceable.

effective diameter (ED) A criterion used in evaluating the size of a soil or particulate mixture. Often used for determining the size of filtration media in drinking water treatment, the effective diameter is the particle diameter in millimetres for which 10 percent of the medium (by weight) has a diameter less than or equal to that value. The effective diameter is determined by processing a sample through a series of sieves with progressively smaller openings and then measuring the mass of material retained by each sieve. The sieve opening diameter is plotted as a function of the percent mass less than or equal to that given sieve diameter. The tenth percentile value is the effective diameter, d_{10} or ED_{10}.

effective grain size *See* effective size.

effective groundwater velocity The actual or field velocity of groundwater percolating through water-bearing material. It is calculated as the volume of groundwater passing through a unit cross-sectional area in unit time divided by the effective porosity. It is also called actual groundwater velocity, field groundwater velocity, or true groundwater velocity.

effective head (1) The head available for producing energy in a hydroelectric plant after all losses from friction, eddies, the entrance, unrecovered velocity head at the mouth of a draft tube, and so on—except those chargeable to the turbine—are deducted. It is also called net head. (2) For reaction turbines, the difference between the total head at the entrance to the runner case and that in the tailrace at the draft tube. (3) For impulse turbines, the difference between the total head at the entrance to the nozzle and that at the lowest point of the pitch circle of the runner bucket.

effective height The total head against which a pump must work, measured in length units.

effective interest rate The rate of earnings on a bond investment based on the actual price paid for the bond, the coupon rate, the maturity date, and the length of time between interest dates.

effective opening The minimum cross-sectional area at the point of water supply discharge, measured or expressed in terms of the diameter of a circle or, if the opening is not circular, the diameter of a circle of equivalent cross-sectional area.

effective pore fraction The ratio of the porosity available for fluid flow to the bulk porosity of a sample.

effective porosity The volume of void spaces through which water or other fluids can travel in a porous material divided by the total volume of the material.

effective precipitable water In the context of rainfall, the greatest amount of precipitable water that can be removed from an atmospheric column by convective action.

effective precipitation Rainfall that offsets evapotranspiration losses during a given time period; rainfall that enters the soil and is available to plants when needed. In a given day, effective precipitation is less than or equal to the daily evapotranspiration.

effective rainfall Rain that produces surface runoff.

effective range That portion of the design range (usually the upper 90 percent) in which an instrument has acceptable accuracy. *See also* range; span.

effective size (ES) The granular medium particle diameter, d_{10}, for which 90 percent of a sample, by mass, has an equivalent or larger diameter. It is also called the effective grain size. Common units are millimetres. *See also* ED_{10}; effective diameter; uniformity coefficient.

effective storage The volume of water available for a designated purpose.

effective velocity *See* effective groundwater velocity.

effervescence The vigorous escape of small gas bubbles from a liquid, especially as a result of chemical action.

efficiency (1) A measure of the effectiveness of the operational performance of an ion exchanger, usually based on the ratio of output per unit of input. This ratio is often expressed as the amount of regenerant required to produce a unit of contaminant reduction capacity. Common units are pounds of salt per kilograin of hardness removed or kilograms of acid per kilogram of salt removed. (2) The percent reduction of contaminants that occurs with a specified medium volume and specified water contact time. (3) The value obtained (expressed as a percentage) by dividing the volume of product water produced by the total volume of feedwater fed to the particular unit or system. (4) A measure of the effectiveness of any operational performance. (5) The ratio of the resulting useful work to the energy expended to achieve the results. (6) The competence with which one completes one's duties. (7) The ability to produce a desired effect on a product, with a minimum of effort, expense, or waste. (8) The ratio of output power from a hydraulic machine to the input power. In pumps, mechanical efficiencies greater than 80 percent are typical. *See also* salt efficiency.

efficiency standard A value or criterion that establishes levels or conditions of water use in the ornamental landscape.

effluent Water flowing from a structure such as a treatment plant. *Contrast with* influent.

effluent exchange The practice of using wastewater effluent as a source of replacement of upstream water diversions.

effluent launder A trough that collects the water flowing from a basin and transports it to the effluent piping system.

effluent quality The physical, biological, and chemical characteristics of a liquid flowing from a basin, reservoir, pipe, or treatment plant.

effluent seepage The diffuse discharge of groundwater to ground surfaces or into a surface water body. *See also* leakage.

effluent stream A stream or stretch of stream that receives water from groundwater in the zone of saturation. The water surface of such a stream stands at a lower level than the water table or piezometric surface of the groundwater body from which it receives water.

effluent weir A weir at the outflow end of a sedimentation basin or other hydraulic structure.

efflux tube A tube inserted into an orifice to provide for outward flow of water.

EGL *See* Bernoulli law; energy grade line.

E_h An electric potential in an oxidation–reduction reaction. This potential (in volts, hydrogen scale), called a "redox" potential, represents the relative tendency of ions to acquire electrons. The more positive the value, the more oxidizing (electron-rejecting) the solution; more negative values represent more reducing (electron-acquiring) conditions. Because electron activity can vary over wide ranges, expressing the E_h value as the dimensionless pE is often more convenient. Because electron activity varies with the pH of water, it is often useful to refer to graphs of E_h (or pE) as a function of pH. In this manner, the range of thermodynamically stable species can be illustrated. *See also* oxidation–reduction potential; oxidation–reduction reaction.

EHEC *See* enterohemorrhagic *Escherichia coli.*

EI ionization *See* electron impact ionization.

EIA *See* enzyme-immuno-assay.

EIEC *See* enteroinvasive *Escherichia coli.*

18-megohm water A high-purity water that conducts electrical current poorly because of the lack of ionized impurities (electrolytes). It has an electrical resistivity of approximately 18 megohm-centimetres (180,000 ohm-metres) and a conductivity of 0.0556 micromhos per centimetre (0.00000556 siemens per metre) at a specified temperature, typically 77° Fahrenheit (25° Celsius). This type of water is also called ultrapure water.

einstein *See the Units of Measure list.*

EIS *See* Environmental Impact Statement.

EJC *See* Engineers Joint Council.

ejection The process of forcing something out or expelling it.

ejector The portion of a chlorination system that feeds the chlorine solution into a pipe under pressure.

ejector pump A shallow or deep-well pump operating on the venturi principle. An ejector pump is commonly referred to as a jet pump. *See also* jet pump; venturi principle.

El Tor (eltor) vibrato *See Vibrio cholerae.*

elastic demand Demand for a product (water) such that when a price increase (or decrease) is imposed, the resulting percentage change in quantity demanded is greater than the percentage change in price. Elasticity is defined as the percentage change in quantity demanded divided by the percentage change in price, which is always a negative term. With elastic demand, price elasticity is less than –1 (e.g., –1.1). The effect of elastic demand is that a utility's revenue will decrease with rate increases. Elastic demand rarely, if ever, applies to the water. *See also* inelastic demand; price elasticity.

elbow A pipe fitting that connects two pipes at an angle. The angle is always 90° unless another angle is stated. An elbow is also called an ell.

elbow draft tube A draft tube that makes a turn of approximately 90° after leaving the wheel, usually in order to save space or reduce the depth and quantity of excavation required.

ElCD *See* electrolytic conductivity detector.

electric charge density units In Système International, coulombs per cubic metre; ampere-seconds per cubic metre.

electric charge units In Système International, coulombs; ampere-seconds.

electric current units In Système International, amperes.

electric field strength units In Système International, volts per metre; metre-kilograms per second cubed per ampere.

electric flux density units In Système International, coulombs per square metre; ampere-seconds per square metre.

electric power Force or energy that is derived from electricity and is at work or can be put to work.

electric resistance units In Système International, ohms; metres squared kilograms per second cubed per ampere squared.

electric timer An electric control device that initiates regeneration of an automatic water softener or the recycle phase of a filter unit.

electric well log A record obtained by a traveling electrode used during a well investigation in rock. It is in the form of curves that represent the apparent values of the electric potential and electric resistivity or impedance of the rocks and their contained fluids throughout the uncased portions of a well.

electrical conductivity The reciprocal of the electrical resistance in ohms, measured between opposite faces of a 1-centimetre cube of an aqueous solution at a specified temperature. Units, in Système International, are siemens per metre, or square amperes cubic seconds per kilogram per cubic metre.

electrical double layer The electrically charged surface of a colloidal particle together with a diffuse surrounding layer of counterions. *See also* electrokinetic potential.

electrical ground (1) Any common point in an electrical circuit against which electrical properties are measured. (2) The portion of an electrical power circuit that is connected to earth.

electrical grounding The connection of an appliance, such as a pump motor or electrical drill, to the ground to prevent short circuits and to eliminate the impossibility that people will complete circuits and be fatally shocked by touching the appliance. *See also* earth-ground; ground fault interrupter.

electrical load The electrical demand for mechanical equipment, expressed in amperes, kilowatts, kilojoules, or horsepower.

electrical lockout device A piece of equipment that, when attached to a breaker, motor starter, or electrical switch, prevents the accidental startup of out-of-service equipment.

electrical logging A procedure used to determine the porosity (degree of spaces or voids) of formations during a search for water-bearing formations (aquifers). Electrical probes are lowered into wells, and an electrical current is induced at various depths. The measured resistance of various formations indicates the porosity of the material.

electrical potential units In Système International, volts; metres squared kilograms per seconds cubed per ampere.

electrical resistance model An analog groundwater flow model based on the flow of electricity through a circuit containing resistors and capacitors.

electrical sounding An earth resistivity survey made at one particular location by putting the electrode pairs of a dipole–dipole array, or the outer electrodes of a Schumberger array, progressively further apart. It shows the change of apparent resistivity with depth for local ground conditions. *See also* dipole array.

electrical stage In an electrodialysis system, a grouping of membrane cell pairs between one anode and one cathode.

electrical staging The addition of anode and cathode electrode pairs to an electrodialysis membrane stack to optimize the direct current electrical current efficiency. A single membrane stack may have one or more electrical stages.

electricity The movement of electrical charge from one location to another.

electrochemical gauging A method for measuring the flow of water based on the nearly linear relationship between the concentration and electrical conductivity of a salt solution added to the water as a tracer.

electrochemical generation The development of an electric current arising from the electric potential between anions and cations. *See also* anion; cation; current; electrode potential.

electrochemical instrumentation power positioner An energy-regulating device for battery-powered detection, sensing, or monitoring of equipment.

electrochemical potential The driving force for initiating a chemical reaction, giving rise to a flow of electrons. Corrosion can occur whenever electrochemical potential differs between two points on a metallic surface. Differences in potential can be caused by minor metallurgical or physical differences on a pipe surface, by a connection of old and new pipe, by a connection of dissimilar metals, by cracks in a pipe, and so on.

electrochemical reaction (1) A chemical reaction involving the transfer of electrons, involving both oxidation (the loss of electrons) and reduction (the gaining of electrons). (2) Chemical changes produced by electricity (electrolysis) or, alternatively, the production of electricity by chemical changes (galvanic action). In corrosion, a chemical reaction is accompanied by the flow of electrons through a metallic path. The electron flow may come from an external source and initiate the reaction (e.g., electrolysis caused by a direct current electric railway), or the electron flow may be caused by a chemical reaction (as in the galvanic action of a flashlight dry cell).

electrochemical series A list of metals with the standard electrode potentials given in volts. The size and sign of the electrode potential indicate how easily an element will take on or give up electrons, or corrode. Hydrogen is conventionally assigned a value of zero.

electrochemistry The study of chemical reactions involving electricity, such as when a metal goes into solution as an ion or reacts in water with another element to form a compound and results in a flow of electrons.

electrode A device through which electric current flows into or out of a cell, apparatus, or liquid electrolyte. *See also* anode; cathode.

electrode method Any analytical procedure that uses an electrode connected to a millivoltmeter to measure the concentration of a constituent in water.

electrode potential A measure of the difference in charge between metals in contact with a fluid or electrolyte. This potential can generate a current and is an important aspect in determining the rate of corrosion in piping.

electrodeionization (EDI) A demineralization process employing a specialized electrodialysis membrane process containing mixed bed ion exchange resin that is capable of producing high-purity water. Feedwater passes through ion exchange resin located in dilute-flow stream channels between parallel anion exchange membranes and cation exchange membranes in a direct current (DC) field. Feedwater ions move toward the electrodes of opposite charge and are transported through the ion exchange resin and ion selective membranes to concentrate-flow channels. The DC electrical field is operated to split water into hydrogen and hydroxyl ions near the membrane and resin bead surfaces that continuously regenerate the resin. The ions removed from the feedwater are discharged from the system in the concentrate flow stream.

electrodialysis (ED) A desalting process driven by an electrical potential difference between oppositely charged electrodes. Ions are transferred by direct electric current flow through cation and anion membranes, depending on ion charge, from a less concentrated solution to a more concentrated one, leaving a demineralized stream.

electrodialysis reversal (EDR) An electrodialysis process in which the electrical polarity of the electrodes is reversed on a set time cycle, thereby reversing the direction of ions in the system for membrane scaling and fouling control. *See also* electrodialysis.

electrokinetic potential The electrical difference between the firmly bound water layer surrounding a charged particle and the bulk water solution. This is often called the zeta potential. It can be computed operationally from the electrophoretic mobility. Zeta potential is used in water treatment as a way of optimizing coagulation.

electrolysis Chemical changes caused in an electrolyte by an electrical current. The US Environmental Protection Agency discourages the use of this word to mean "corrosion by stray currents."

electrolyte An ionic conductor (usually in aqueous solution).

electrolytic analysis The quantitative measurement of a chemical by (1) measuring the potential at an electrode or sensor or (2) passing a current through a liquid and measuring the quantity of material deposited on an electrode or the volume of gas released at an electrode.

electrolytic cell (1) A device in which the chemical decomposition of material causes an electric current to flow. (2) A device in which a chemical reaction occurs as a result of the flow of electric current. Chlorine and caustic (NaOH) are made from salt (NaCl) in this type of electrolytic cell.

electrolytic chlorine Chlorine produced by the electrolytic dissociation of hydrochloric acid (HCl) or one of its salts.

electrolytic conductivity detector (ElCD) A device used in the analysis of organic compounds by gas chromatography for the detection of individual organic compounds. This type of detector can be sensitive to a variety of compounds but is usually used in water analysis for the analysis of halogen-substituted compounds.

electrolytic corrosion cell An electrochemical cell in which an external direct current generates the corrosion. For this to occur, all elements of the electrochemical cell must be present: anode, cathode, a connection between the anode and cathode (internal circuit), and an electrochemical solution. *See also* anode; cathode.

electromagnetic conductivity A measurement of an induced electrical field in earth to determine the ability of the earth to conduct electricity. Electromagnetic conductivity is the inverse of electrical resistivity. It is important in water prospecting and contaminant surveys. It is also known as electric conductivity and terrain conductivity.

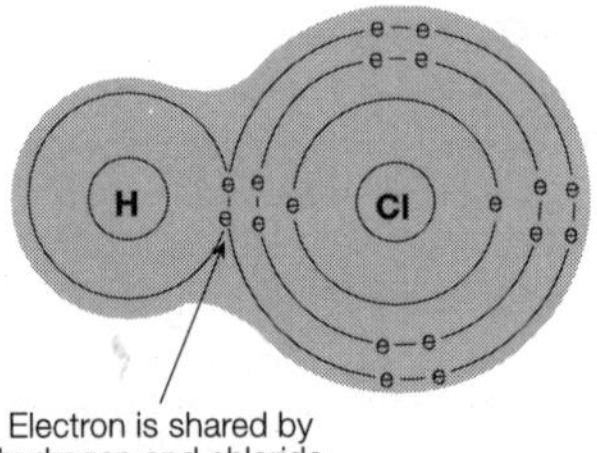
H
Cl
Electron is shared by
hydrogen and chloride

Electron shared in covalent bond for hydrogen chloride (HCl)

electromagnetics The study of the combined effects of electricity and magnetism.

electromechanical instrumentation power positioners Any number of combination electronic and mechanical devices designed to control or monitor electrical equipment such as variable-speed motors. Because they have electronic components, they should be maintained as electrical equipment.

electrometric titration A titration in which the end point is determined by observing the change of potential of an electrode immersed in the solution titrated.

electromotive force (emf) Basic electrical force, measured in volts.

electromotive force units In Système International, volts; metres squared kilograms per second cubed per ampere.

electromotive series A list of metals and alloys presented in the order of their tendency to corrode (or go into solution). It is also called the galvanic series. It is a practical application of the theoretical electrochemical series.

electron (e^-) One of the three elementary particles of an atom (along with protons and neutrons) in the simplified atom model. An electron is a tiny, negatively charged particle that orbits around the nucleus of an atom. The number of electrons in the outermost shell is one of the most important characteristics of an atom in determining how chemically active an element will be and with what other elements or compounds it will react. *See also* neutron; proton.

electron acceptor Any substance (a molecule or ion) that can accept electrons and thereby be reduced. For example, a Lewis acid is an electron acceptor (called an electrophile) that can combine with another molecule or ion by forming a covalent bond with two electrons from the second molecule or ion. *See also* covalent bond; electron donor; Lewis acid; oxidant.

electron beam process A process for disinfection that uses gamma radiation, sometimes from a cobalt-60 source.

electron capture detector (ECD) A device used in the analysis of organic compounds by gas chromatography. Most of the electron capture detectors used today make use of a radioactive foil coated with $nickel^{63}$. Electron capture detectors are especially sensitive to halogen-substituted compounds.

electron donor Any substance (a molecule or ion) that can donate electrons. For example, a Lewis base is an electron donator (called a nucleophile) that forms a covalent bond by donating a pair of electrons. *See also* covalent bond; electron acceptor; Lewis base.

electron microscope An instrument that uses electric or magnetic fields to focus a beam of electrons in order to form an enlarged image of an object on a fluorescent screen or photographic plate. An electron microscope can magnify images up to 50,000 times. The image of a virus is the result of the scattering of the electron beam by the electrons in the specimen. An electron microscope is also called a transmission electron microscope. *See also* scanning electron microscope.

electron microscopy The use of an electron microscope for counting typical viruses.

electron spin resonance (ESR) A spectroscopic technique based on the absorption of microwave radiation by electrons in the presence of a magnetic field. It is similar to nuclear magnetic resonance. Although most compounds do not exhibit an electron spin resonance spectrum because they have an even number of electrons, the technique has been used to study reactions of free radicals. Compounds that contain unpaired electrons can be studied by this technique.

electron volt (eV) *See the Units of Measure list.*

electron impact (EI) ionization A mode of operation in mass spectrometry in which analyte molecules are bombarded with an electron beam, typically at an energy of about 70 electron volts. This bombardment produces a fragmentation pattern that is characteristic of the analyte and can form a basis for the identification of unknown compounds. This is the most commonly used ionization mode in mass spectrometry. Large electronic databases are available that contain thousands of mass spectra determined in the electron impact mode. *See also* chemical ionization; mass spectrometry.

electronic distance-measuring (EDM) device A surveying instrument using laser and computer technology to digitally record distances and angles. It has replaced traditional optical surveying transits.

electronic meter reading (EMR) The use of hand-held data entry terminals, computers, or other electronic technology to increase the productivity of the meter reader.

electronic path The movement of electrons through a complete electrical circuit.

electroosmose treatment A process for the removal of objectionable salts from water by passing the water through diaphragms from cell to cell containing anode and cathode poles.

electrophilic intermediates Electron-deficient chemical metabolites that will readily react with nucleophilic centers in macromolecules. Such reactions lead to the destruction of the functional capabilities of proteins and can result in mutations in deoxyribonucleic acid. If a chemical is too electrophilic, however, reactions with water may prevent it from reaching the macromolecular targets.

electrophoresis The migration of electrically charged particles in solution or suspension in the presence of an applied electric field. *See also* electrophoretic mobility.

electrophoresis apparatus A device used to separate charged particles in an electric field. The charged particles, such as simple ions, proteins, or living cells, may be separated depending on the selected apparatus and technique. A given apparatus can also allow the use of several electrophoretic techniques, such as free-zone capillary electrophoresis, isoelectric focusing, and micellar electrokinetic capillary chromatography.

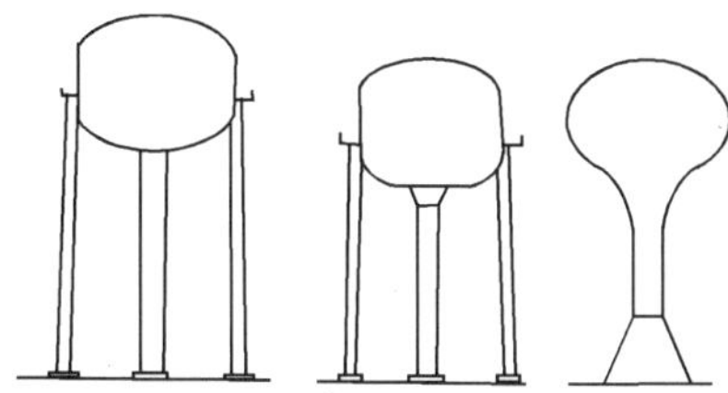

Cross-sectional views of some common elevated tank designs

electrophoretic mobility (EM) A measure of the ability of an ion to migrate in an electric field. The electrophoretic mobility is equal to the migration velocity per unit field strength. The units are often expressed as square centimetres per volt per second. This term is often used in colloid analysis and coagulation. Electrophoretic mobility is related to the electrokinetic potential on a particle. *See also* electrokinetic potential.

electrophotometer A photometer that uses different colored glass filters to produce the desired wavelengths for analyses. It is also called a filter photometer.

electropositive filter A filter composed of a cellulose–diatomaceous earth resin mix, used for concentrating viruses from large volumes of water. Such a filter is also called a zeta plus filter.

electrospray ionization (ESI) A technique used as an interface in liquid chromatography–mass spectrometry. In this approach, ions are formed from analyte molecules by evaporation from charged droplets in the mass spectrometer. The advantages of electrospray ionization over other types of interfaces include a combination of greater sensitivity and greater selectivity.

electrostatic plotter A peripheral device that produces hard copy graphic images in raster format.

electrostatic test A procedure for measuring or indicating the force that exists between charged bodies or particles. This force is expressed mathematically by Coulomb's law and can be measured or indicated with an electroscope.

electrowin To extract a metal from solution by electrolysis.

element (1) Any of more than 100 fundamental substances that consist of atoms of only one kind and that constitute all matter. (2) Any single discrete part of a control system. (3) *See* cartridge.

elemental analysis A method for determining the carbon, hydrogen, nitrogen, oxygen, sulfur, and silicon content of a sample. The method involves the combustion or pyrolysis of dry material. Elemental analyses are often performed during the characterization of the natural organic matter in water. *See also* natural organic matter.

elevated storage Water stored or the capacity to store water at an elevation. This elevated water provides pressure in a water pressure plane, thereby allowing for controlled pumping rates, water for peak usage and fire protection, and minimized service interruptions during equipment maintenance requirements.

elevated tank A water storage facility located on and supported by a tower constructed at an elevation to provide useful storage and pressure for a water pressure plane.

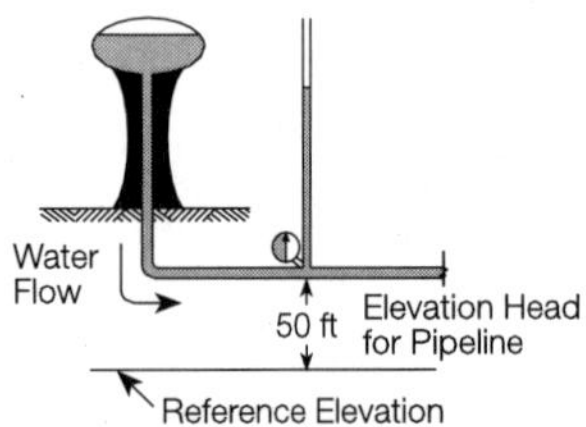

Elevation head

elevation charge A surcharge added to the base water rate for customers at different e levations to recover the incremental costs incurred for pumping and storage to serve each higher elevation band. Elevation bands may be set, for example, in increments of 100 feet (30.5 metres) or 500 feet (152.5 metres).

elevation head The energy per unit weight of a fluid due to the fluid's elevation above some reference point (called the reference datum). Elevation head is also called position head or potential head.

elevation view The view of an object showing the vertical dimensions and how the object would look rising from the ground. *See also* isometric view; plan view.

elimination half-life The time required to remove one-half of a chemical from the body. Frequently, this is approximated by the terminal half-life for removal of the chemical from the blood compartment after the distributive phase of chemical absorption has been completed. It is the inverse of the elimination rate constant for a chemical from the body. The longer the half-life, the slower the rate constant. Elimination can occur by metabolism, excretion of the chemical, or both.

elimination rate constant The inherent rate at which a chemical is removed from the body. This includes removal from various repositories within the body. Frequently, this constant is approximated by the terminal rate constant for chemical removal from the blood compartment after the distributive phase of drug or chemical absorption has been completed. Mathematically, the elimination rate constant is the reciprocal of the elimination half-life.

ELISA *See* enzyme-linked immunosorbent assay.

ell *See* elbow.

eluent A solution used to remove particles adsorbed to a solid. Solutions of proteins–peptides (e.g., beef extract, skim milk) or detergents (distilled water solution) at pH 7 to 10 are often used if viruses are to be removed.

elute To remove particles adsorbed to a solid. This is usually accomplished by treating the solid with a chemical that reverses adsorption.

elution (1) The process of separating or washing out adsorbed material, especially by use of a solvent. (2) In ion exchange, the stripping of ions from the medium by passing a more highly concentrated ionized solution through the ion exchanger bed.

elution curve A relationship illustrating the concentration of ions, in equivalents per litre, as a function of bed volume for the regeneration of ion exchange resin. The elution curve, similar to a breakthrough curve, provides an understanding of the regeneration level required to achieve a desired resin conversion and regenerant efficiency. *See also* bed volume; breakthrough; ion exchange; regeneration.

EM *See* electrophoretic mobility.

embedded knowledge-based system A complex simulation model that is enhanced by the application of rule-based pre- and post-processors. *See also* computer-aided modeling system.

embolism A blockage of a blood vessel. This is frequently the result of a blood clot that has broken free from one area of the circulatory system to be lodged in the fork of an artery. Such clots cause the acute damage associated with strokes, myocardial infarctions (i.e., heart attacks), or both.

embryolethality The process of producing death of an embryo.

embryotoxic A chemical or other agent that exerts greater toxicity for the embryo than for the pregnant woman or animal.

EMC *See* event mean concentration.

emergency action plan (EAP) A plan developed by a water utility that defines actions that will be taken in case of an emergency.

emergency planning The process of preparing and documenting an operations plan to be used when a natural or human-made disaster interrupts the delivery of goods or services for a period of time.

emergency preparedness plan A thought-out and tested procedure to deal with known hazards in a particular area, either artificial or natural.

emergency source A source of water supply that is not the regular source or auxiliary source but is developed during an emergency for temporary use.

emergent plant A rooted plant that has parts extending above a water surface, at least during portions of the year, but does not tolerate prolonged inundation.

emergent weed An aquatic plant, such as a cattail, that is rooted in the bottom mud of a water body but projects above the water surface.

emergent wetland That portion of a wetland dominated by erect, rooted, herbaceous vegetation as the uppermost vegetative stratum.

emf *See* electromotive force.

eminent domain The power of a government entity to seize property or take private property for the benefit of the public.

emission A venting of gas or radiation from a building or piece of equipment.

emission spectroscopy A chemical analytical technique used to determine metal elements in water by measuring the well-defined characteristic radiation given off by each respective element as the thermally excited element returns from an atomic vapor state to its fundamental state. *See also* atomic adsorption spectroscopy; inductively coupled plasma spectroscopy; spectroscopy.

emission standard A maximum level of contaminants allowed in airborne discharges based on applicable regulatory requirements that may be imposed by national, regional (state or province), or local agencies.

emitter A drip irrigation component that dispenses water to plants at a predictable rate, measured in gallons or litres per hour.

emphysema A disease of the lung that functionally results in a loss of the lung's normally high surface area for exchanging gases. This effect occurs in a variety of ways, but the process ultimately involves the coalescence of alveolar spaces. It is sometimes accompanied by fibrotic changes that further reduce the compliance of the lung (i.e., its ability to expand and contract and move air in and out of the alveoli). Individuals with emphysema are extremely sensitive to conditions that modify the availability of air to the lung or the exchange of oxygen across the membranes in the lung. The nose and lung are sensitive organs to orally administered compounds.

empirical equation A mathematical expression developed based on actual observations, rather than on theory alone, to predict the value of one variable based on several other variables.

empirical model *See* empirical equation.

empty bed contact time (EBCT) A standard convention or measure of the time during which a water to be treated is in contact with the treatment medium. The empty-bed contact time is calculated by dividing the empty volume in a contactor that will be occupied by the treatment medium by the flow rate

$$\text{EBCT} = V/Q$$

Where (in any consistent set of units):
V = the volume of the vessel
Q = the flow rate

Because the treatment medium, such as granular activated carbon, will occupy some volume, the empty-bed contact time overestimates the actual time that the flow resides in the contactor. *See also* granular activated carbon.

EMR *See* electronic meter reading.

EMSL *See* Environmental Monitoring Systems Laboratory.

emulsifying agent An agent capable of modifying the surface tension of emulsion droplets to prevent coalescence. Examples are soap and other surface-active agents, certain proteins and gums, water-soluble cellulose derivatives, and polyhydric alcohol esters and ethers.

emulsion A heterogeneous liquid mixture of two or more liquids not normally dissolved in one another but held in suspension (one in the other) by forceful agitation or by emulsifiers that modify the surface tension of the droplets to prevent coalescence. *See also* emulsifying agent.

emulsoid A colloid that is readily dispersed in a suitable medium and may be redispersed after coagulation.

encapsulate To enclose or seal off. For example, a tank being sandblasted to remove lead-base paint would be encapsulated.

encephalitis Inflammation of the brain. It may be caused by a specific viral infection (one of several arthropod-borne viruses), or it may be the result of a viral infection (e.g., influenza, herpes, chickenpox).

encrustation The built-up material on the inside of a potable water pipe or appurtenance caused by the precipitation of a mineral, such as calcium carbonate ($CaCO_3$); or by corrosion of an exposed metal, resulting in the formation of tubercles. Encrustations may result in a rough or smooth surface and normally restrict flow by reducing diameter or increasing roughness. *See also* tubercle.

end bell A device used to hold the rotor and stator of a motor in position.

end contraction (1) The extent of the reduction in width of the nappe due to a constriction caused by the ends of the weir notch. (2) The walls of a weir notch that does not extend across the entire width of the channel of approach.

end point The final state once a chemical reaction is completed. For example, suppose a chemical is added, drop by drop, to a sample until a certain color change (blue to clear, say) occurs. This new color condition is called the end point of the titration. In addition to a color change, an end point may be reached by the formation of a precipitate or the reaching of a specified pH. An end point may be detected by the use of an electronic device such as a pH meter.

Endangered Species Act (ESA) Public Law 93-205, enacted in 1973, creating a national program to prevent the extinction of imperiled animals and plants. The law was amended in 1984 (Public Law 98-327) and in 1988 (Public Law 100-478).

endangered species Plants and animals threatened by extinction, identified and protected under provisions of the national Endangered Species Act.

endangerment assessment A site-specific risk assessment of the actual or potential danger to human health or welfare and the environment from the release of hazardous substances or waste. The endangerment assessment document is prepared in support of the Comprehensive Environmental Response, Compensation, and Liability Act or the Resource Conservation and Recovery Act.

end-around baffles A series of physical obstacles within or between unit processes that reduce short-circuiting by requiring the flow to pass in sequence around the ends of the baffles, encouraging mixing. In addition, this arrangement may be used in multistage flocculation basins to ensure contact with the mixing devices in each stage. *See also* baffle; flocculation; over-and-under baffles.

endemic Pertaining to the constant presence of a disease, infection, or other health-related event in a community or geographic area.

endocrine An organ responsible for secreting hormones into the bloodstream to exert their effects on distant elements of the body. Examples include the thyroid gland, the pancreas, and adrenal glands.

endocrine disruptor An exogenous chemical substance or mixture that alters the function(s) of the endocrine system and thereby causes adverse effects to an organism, its progeny, or (sub)populations.

endogenous respiration Autooxidation by organisms in biological processes.

endoplasmic reticulum A membranous structure within the cell that has specialized functions in the synthesis and processing of proteins and in the metabolism of drugs and chemicals to which the organism is exposed. Increasing evidence exists that some of the systems found in the endoplasmic reticulum also have some role in the regulation of cellular activities. Drug and chemical metabolism occurs in the so-called smooth endoplasmic reticulum. Protein synthesis occurs in the rough endoplasmic reticulum. The particles adhering to rough endoplasmic reticulum are ribosomes that are responsible for specialized functions in protein synthesis.

endospore A dormant stage of the life cycle of certain groups of bacteria, notably the aerobic genus *Bacillus* and the anaerobic genus *Clostridium*, that are widely found in soils. Spores are

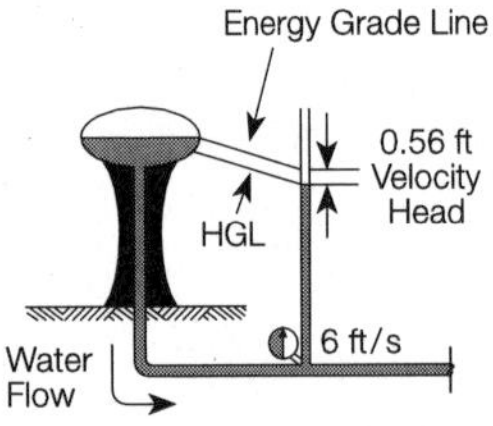

Energy grade line

highly resistant to desiccation (drying), heat (many can withstand boiling for 30 minutes), and chemicals. Endospores are small, generally about 0.5 × 1.0 × 1.5 micrometres, and they may be spherical, cylindrical, or ellipsoidal. They are commonly called spores. Because spore-forming bacteria generally occur in abundance in soils, they are carried into drinking source surface waters by surface runoff. Spores are removed as particulates by the coagulation, flocculation, and filtration processes in water treatment, as are *Cryptosporidium* oocysts. In addition, *Bacillus* species are not pathogenic to humans and pose no health threat. By evaluating the levels of indigenous spores in source water and after treatment steps, one can use spores to assess the effectiveness of water treatment processes. In the mid-1990s, the removal of spores or the aerobic bacilli by treatment processes was shown to parallel the removal of total particles (greater than 1 micrometre) and particles in the size range of 3.1 to 7 micrometres. The latter size range corresponds to the size range for *Cryptosporidium* oocysts; thus, removal of aerobic *Bacillus* spores may prove to be a useful surrogate for *Cryptosporidium* oocyst removal. If levels of indigenous spores are not high enough to follow through the water treatment processes, a suspension of *Bacillus subtilus* spores can be used in jar or bench test procedures. These bacteria will not grow in filter beds of a pilot plant or full-scale plant.

endospore detection method A simple bacteriological culture method for detecting endospores. It is much easier and quicker than the method used to monitor for the presence of *Giardia* cysts or *Crystosporidium* oocysts. The water sample to be examined is pasteurized by heating to 176° Fahrenheit (80° Celsius) and held at that temperature for 12 minutes to kill any vegetative *Bacillus* cells. The sample is then rapidly cooled in an ice bath. The spores are not killed by this procedure. Once the sample is cooled to 95° Fahrenheit (35° Celsius) or less, appropriate sample aliquots are filtered through 47-millimetre-diameter, 0.45-micrometre pore-diameter membrane filters, and each filter is transferred to the surface of a petri plate (60 × 15 millimetre diameter, loose lid) containing solidified nutrient agar plus trypan blue dye (15 milligrams per litre), and is incubated at 95° Fahrenheit (35° Celsius) for 20 to 22 hours. All colonies that arise on the membrane filter are considered to be spore formers and are counted and recorded.

endosymbiosis A type of symbiotic association in which one symbiont dwells within the other.

endothall ($C_8H_{10}O_5$) A generic name for 7-oxalobicyclo-[2.2.1]-heptane-2,3-dicarboxylic acid—a synthetic organic chemical used as a defoliant or herbicide. It is used as the sodium, potassium, or amine salt. It is regulated by the US Environmental Protection Agency. *See also* defoliant; herbicide; synthetic organic chemical.

Endothall

endotheliochorial placenta A placental type that has only the endothelial layer of the maternal tissue, so that the chorionic villi bathe in the maternal blood. This type of placenta is found in dogs and cats.

endothermic Pertaining to a chemical process in which heat is absorbed. For example, melting ice absorbs heat and therefore represents an endothermic process.

endotoxin A heat-resistant pyrogen (specifically a lipopolysaccharide) found in the cell walls of certain disease-producing bacteria.

endrin ($C_{12}H_8OCl_6$) A generic name for 1,2,3,4,10,10-hexachloro-6,7-epoxy-1,4,4a,5,6,7,8,8a-octahydro-1,4-endo-endo-5,8-dimethanonaphthalene. Endrin is a stereoisomer of dieldrin, both of which are insecticides. Endrin is regulated by the US Environmental Protection Agency. *See also* dieldrin; insecticide.

end-suction pump A centrifugal pump with the suction and discharge ports located at right angles to each other. Such a pump is also called a single-suction pump because the water enters the impeller from only one side.

energy In a mechanical sense, the capacity for doing work and overcoming resistance. Many other forms of energy exist: electrical, chemical, hydraulic, and so forth. *See also* power; work.

energy barrier The electrostatic force between two particles that must be overcome to allow particle flocculation and aggregation to occur. This force results from the interaction between the diffuse layers of the two particles, both of which are similarly charged. These charges are partially offset by attraction through van der Waals forces. The energy barrier is sometimes called the energy hill. *See also* van der Waal's forces.

energy budget A quantitative account of inputs, transformations, and outputs of energy in an ecosystem.

energy conservation A reduction in energy usage through the application of more efficient equipment, through varying equipment usage to take advantage of a stepped rate structure for electricity, through the use of variable-speed drives to reduce pump cycling (shutoff and restart), or through other operational measures.

energy density units In Système International, joules per cubic metre; kilograms per metre per second squared.

energy dispersive spectroscopy A group of techniques used to analyze the atomic structure of materials. For example, in the case of energy dispersive X-ray spectroscopy, a given element produces a characteristic peak or pattern of peaks. Dispersion involves the change in refractive index of a material as a function of wavelength. In laboratory instruments, dispersion of radiation often occurs by the use of a prism or diffraction grating. Normal dispersion occurs when the change in refractive index increases with increasing frequency (decreasing wavelength). When the reverse occurs, absorption takes place. The absorption of radiation by materials serves as the basis for a number of types of spectroscopic analyses.

energy dissipation The transformation of mechanical energy into heat energy. In fluids, this is accomplished by viscous shear. The rate of energy dissipation in flowing fluids varies with the scale and degree of the turbulence. Baffles, the hydraulic jump, and other damping methods are used to dissipate energy.

energy flow The one-way passage of energy (largely chemical) through any ecosystem, entering via photosynthesis, being exchanged through feeding interactions, and, at each stage, being reduced to heat.

energy grade line (EGL) A line joining the elevations of the energy heads; a line drawn above the hydraulic grade line by a distance equivalent to the velocity head of the flowing water at each section along a stream, channel, or conduit. It is sometimes called the energy gradient line or energy line.

energy gradient The slope of the energy line of a body of flowing water with reference to a datum plane.

energy gradient line The line representing the gradient that joins the elevations of the energy head. *See also* energy grade line.

energy head The height of the hydraulic grade line above the center line of a conduit plus the velocity head of the mean-velocity water in that section.

energy line *See* energy grade line.

energy management system A group of approximately 13 programs used by electric utilities for optimizing control of generation and transmission.

Energy Policy Act The 1992 federal law that states, among other things, that no toilet manufactured after Jan. 1, 1994, shall use more than 1.6 gallons (5.4 litres) per flush (except commercial toilets and blowout toilets) and that faucets and showerheads manufactured after the same date may not use more that 2.5 gallons (9.5 litres) per minute.

energy recovery A method whereby energy remaining from a process, which would otherwise be lost, is recovered to lessen the overall process energy requirement. In reverse osmosis desalting systems treating highly brackish or seawaters, energy recovery devices such as impulse turbines, turbopumps, turbochargers, and external work exchangers are commonly used to recover residual pressure from the membrane concentrate stream to help reduce system energy consumption.

energy source Thermal, chemical, mechanical, pneumatic, or hydraulic energy that is stored or used in the operation of equipment.

energy system The energy flow and the cycling of numerous elements and materials, including the energetic equivalent of the materials.

energy transfer process Any process that transfers energy from one component in an ecosystem to another.

energy units In the US customary system, British thermal units; foot-pounds force; horsepower hours; kilowatt-hours. In Système International, joules; newton metres; kilograms square metres per second squared.

Engineering drawing

enforcement Administrative or legal procedures and actions to require compliance with legislation or associated rules, regulations, or limitations.

engine A machine that converts energy derived from fuel into mechanical work. It operates independent of any other energy source. *Contrast with* motor.

engineer A person trained in science and concerned with putting scientific knowledge to practical uses and using that training to solve problems.

engineering The science concerned with putting scientific knowledge to practical uses. It is divided into different branches, such as civil, electrical, mechanical, chemical, industrial, and environmental engineering.

engineering drawing A scale drawing of a community's street layout, including the distribution system and sewer pipes. Typically, 1 inch (or 2.5 centimetres) on the map is equivalent to a certain number of feet (or metres). The drawing typically delineates pipe sizes and valves where pipes connect.

engineering structure A structure in which the various members or parts have been predetermined in size, shape, and composition, in order that they may withstand such physical forces or loads as may reasonably be expected to act on or against them.

Engineers Joint Council (EJC) An organization founded in 1945 to foster cooperation among engineering societies. It lost influence around 1967 when two major engineering societies dropped out. It was replaced in 1980 by the American

Association of Engineering Societies. *See also* American Association of Engineering Societies.

English rule (of capture) A groundwater ownership doctrine that holds that property owners have the right of absolute ownership of the groundwater beneath their land. This doctrine is also known as the right of capture.

Enhanced Surface Water Treatment Rule (ESWTR) A national primary drinking water regulation under development by the US Environmental Protection Agency to include *Cryptosporidium* within the scope of the Surface Water Treatment Rule.

enhanced coagulation The addition of excess coagulant for improved removal of disinfection by-product precursors by conventional coagulation–sedimentation–filtration treatment. In the Disinfectant/Disinfection By-Products Rule, the removal of total organic carbon is used as a performance indicator for the removal of disinfection by-product precursors. The Disinfectant/Disinfection By-Products Rule requires no conversion to optimized coagulation practices, but rather enhancement of an existing process to remove specified levels of total organic carbon based on influent water quality. Enhanced coagulation can also be used to remove arsenic during the coagulation process. *See also* coagulation; Disinfectant/Disinfection By-Products Rule; disinfection by-product precursor; optimized coagulation; total organic carbon.

enhanced softening The improved removal of disinfection by-product precursors by precipitative softening. In the Disinfectant/Disinfection By-Products Rule, the removal of total organic carbon is used as a performance indicator for the removal of disinfection by-product precursors. The removal of magnesium hardness can also be an indicator of enhanced softening for total organic carbon control. *See also* Disinfectant/Disinfection By-Products Rule; disinfection by-product precursor; enhanced coagulation; precipitative softening; total organic carbon.

enhanced wetland An existing wetland where human activity increases one or more values, often with an accompanying decline in other wetland values. Wetland values include, but are not limited to, creation of wetlands, aquatic productivity, pollutant assimilation, flood storage capacity, and wildlife habitat.

enhancement Actions performed to provide additional protection to, or create or improve the functions of, a nontidal wetlands.

enlargement loss The loss of head in a conduit resulting from eddy losses occasioned by a sudden change in velocity of water caused by enlargement in the cross-sectional area of the conduit.

enol A chemical compound containing both a double bond and a hydroxyl group (-OH), forming an intermediate and reversible product.

enolization A chemical process by which a carbonyl compound (CH–C=O) is converted to an enol (C=C=OH) under acidic conditions or to an enolate ion ($C{=}C{-}O^-$) under basic conditions. For example, in the haloform reaction, the enolization of acetone (CH_3-(C=O)-CH_3) yields the enolate $CH_2{=}(C{-}O^-){-}CH_3$ under alkaline conditions, the first step in the production of chloroform ($CHCl_3$).

enteric Of intestinal origin, especially with respect to wastes or bacteria.

enteric adenovirus *See* adenovirus.

enteric bacteria Gram-negative, non-spore-forming, facultative anaerobic, rod-shaped bacteria (bacilli) of the family Enterobacteriaceae found in the intestinal tracts of animals. This family of bacteria is broadly divided into three groups based on lactose utilization: the lactose fermenters, the coliforms (the genera *Escherichia, Enterobacter,* and *Klebsiella*); the lactose nonfermenters (the genera *Salmonella, Shigella,* and *Proteus*); and the slow lactose fermenters, the paracolon bacteria (organisms of the Bethesda–Ballerup and Arizona groups [genus *Citrobacter*], the Hafnia, and the Providencia).

enteric virus Any virus that inhabits the alimentary and gastrointestinal tracts. Many such viruses are stable in the feces and wastewater and can be transmitted through contaminated water supplies.

Enterobacter A genus of lactose-fermenting bacteria belonging to the family Enterobacteriaceae.

Enterobacter aerogenes One of the species of bacteria included in the coliform group. These bacteria are often of nonfecal origin.

Enterobacter cloacea A lactose-fermenting bacterium belonging to the family Enterobacteriaceae. This type of bacterium is one of the coliform group of bacteria used as sanitary indicators for drinking water quality.

***Enterobacter* species** A reference to bacteria identified as belonging to the genus *Enterobacter,* used when one is referring to similar organisms collectively or when incomplete characterization precludes species identification.

***Enterococcus* (*Enterococci*, pl.)** The name commonly used for bacteria of the genus *Streptococcus,* Lancefield Group D, also called the fecal streptococci. These organisms are generally found in human and animal feces.

enterohemorrhagic *Escherichia coli* (EHEC) *Escherichia coli* strains that have been associated with hemorrhagic colitis and that produce one or more verotoxins. *Escherichia coli* O157:H7 is an important pathogen that has been implicated in food-borne and waterborne disease and causes severe illnesses, such as hemorrhagic colitis, hemolytic uremic syndrome, and thrombocytopenic purpura. *See also Escherichia coli* O157:H7.

enterohepatic circulation A process in which natural metabolites (e.g., bilirubin, bile acids), drugs, chemicals, their metabolites, or combinations of these are absorbed from the intestine into the blood. Then they pass into the liver, where they may be concentrated in bile to be secreted into the intestine, where the chemical or its metabolite can be reabsorbed from the intestine. The process frequently involves the formation of various conjugates of the chemical or metabolite in the liver cell and hydrolysis or other degradation of the conjugate in the intestine by microflora.

enteroinvasive *Escherichia coli* (EIEC) *Escherichia coli* strains that produce an invasive dysentery type of diarrheal illness in humans.

enteropathogenic *Escherichia coli* (EPEC) *Escherichia coli* strains that cause diarrhea not related to (1) heat-labile or heat-stable enterotoxins, (2) shigella-like invasiveness, or (3) verotoxin production. These organisms are an important cause of infantile diarrhea in many developing countries.

enterotoxigenic *Escherichia coli* (ETEC) *Escherichia coli* strains that cause diarrhea by production of heat-labile enterotoxins, heat-stable enterotoxins, or both. ETEC are the most common cause of travelers' diarrhea, and infections are acquired primarily by ingestion of contaminated food or water.

enterovirus A taxonomical group of viruses that replicate initially in the cells of the enteric tract. They include polio viruses, coxsackie viruses, echoviruses, and the hepatitis A virus. These are small (22 nanometres across) ribonucleic acid viruses.

enterprise fund Any of various major activities for which separate sets of accounting books are often established to track costs and performance. In a city, for example, water, wastewater, and solid waste—as well as schools, streets, and law enforcement—could be separate enterprise funds, so that each has separate accounting to measure costs and revenues within the total city budget. Many enterprise funds are established and operated to be self-sufficient.

enthalpy A thermodynamic property defined as the sum of the internal energy plus the product of pressure and volume. In equation form,

$$H = U + pVf$$

Where:

H = the enthalpy, in British thermal units (kilojoules)

U = the sum of the internal energy, in British thermal units (kilojoules)

p = the pressure, in pounds force per square inch absolute (kilopascals)

V = the volume, in cubic feet (cubic metres)

f = a conversion factor = 144/778 for US customary units (1 for Système International)

The term pV is usually viewed as the flow work, i.e., the work associated with pushing a fluid along. Enthalpy is a useful property in open system analysis, i.e., analysis involving flow. It can be viewed as the sum of the thermal and mechanical energy associated with a given state.

entrain To trap bubbles in water either mechanically through turbulence or chemically through a reaction.

entrainment (1) The carryover of drops of liquid during a process such as distillation. (2) The trapping of bubbles produced in a liquid either mechanically through turbulence or chemically through a reaction.

entrainment separator A device to remove entrapped droplets from a vapor stream in a desalting process. It is also called a demister.

entrance head The head required to cause water to flow into a conduit or other structure.

entrance loss The head necessary to overcome resistance to the entrance of water into a conduit or other structure. This head is lost in eddies and friction at the inlet of the conduit.

entrance well A well or opening at the surface of the ground, constructed to receive surface water that is then conducted to a sewer. It is also called an inlet well.

entropy A thermodynamic property that is a measure of the disorder or uncertainty at the microscopic scale. On the macroscopic scale, it can be viewed as a measure of the degradation in the quality of energy. For example, the second law of thermodynamics puts a higher "value" on 1 joule of energy at 1,000° Celsius than on 1 joule of energy at 100° Celsius. On the macroscopic scale, the change in entropy for a given process can be defined as the integral of the reversible heat transferred divided by the absolute temperature.

entropy units In the US customary system, British thermal units per degree Rankine. In Système International, joules per kelvin; metres squared kilograms per second squared per kelvin.

entry permit A checklist that includes working conditions, hazard evaluations, equipment, and personnel requirements to be performed or assessed before entry into a confined space. The permit is to be reviewed and signed by a supervisor.

entry point A place where water enters the distribution system. Frequently, water is sampled at such points.

envelope The lipid-rich outer coat that covers the nucleocapsid (protein coat) of many but not all viruses.

envelope curve A smooth curve covering either all peak values or all trough values of certain quantities, such as rainfall or runoff, plotted against other factors, such as area or time. In general, none of the peak values goes above the curve in the former case, called the maximum envelope curve, and none of the minimum points falls below in the latter, called the minimum envelope curve.

Envirofacts Warehouse A series of databases operated by various program offices within the US Environmental Protection Agency. Access to the information is available via the Internet. One section of Envirofacts is the Safe Drinking Water Information System (SDWIS), which contains regulatory compliance information on public water supplies. Violations of regulations derived from the Safe Drinking Water Act are listed in SDWIS. *See also* Safe Drinking Water Information System.

environment A general description of the air, water, and land that support the life of an organism and receive its waste products. *See also* ecosystem.

environmental alteration The selection of an electrolytic backfill material to eliminate or reduce the rate of corrosion in buried pipes. It is also called trench improvement.

environmental assessment (EA) A brief document that provides sufficient information to a US Army Corps of Engineers district engineer on the potential environmental effects of a proposed action and, if appropriate, recommends whether to begin preparation of an environmental impact statement or issue a Finding of No Significant Impact.

environmental document A written material dealing with environmental subject matters, e.g., an environmental impact statement or an environmental assessment.

environmental engineering The application of the principles of engineering to maintain and improve the environment for the protection of human health, for the protection of nature's beneficial ecosystems, and for the enhancement of the quality of human life with respect to environment-related issues. The practice of environmental engineering as a profession includes several scientific disciplines, e.g., engineering, biology, chemistry, biochemistry, mathematics, microbiology, physics, ecology, epidemiology, and hydraulics. It entails design, construction, maintenance, operation, process control, and administration and management of facilities and structures utilized for the protection of public health and the environment, with particular reference to (1) the procurement, treatment, and distribution of public drinking water supplies, (2) the collection, treatment, and disposal of municipal and industrial wastewaters and stormwater runoff, (3) the collection, treatment, and disposal of municipal solid wastes (e.g., garbage, trash) and hazardous wastes, whether liquid or solid, (4) the treatment, abatement, and prevention of indoor and outdoor air pollution from whatever source, (5) the abatement of noise pollution, and (6) the control of the quality of surface waters, groundwaters, and watersheds. Environmental engineering was formerly called sanitary engineering.

environmental epidemiology The use of epidemiologic methods for assessing the human health effects that may be associated with biological, physical, and chemical contaminants in the environment. Indicators of disease, rather than the disease itself, are often studied. Frequently the exposures to environmental contaminants are low and small magnitudes of risk are observed.

environmental health The welfare of an individual as influenced by his or her environment, including the effects of biohazards, pollution, working and living conditions, and so on. *See also* industrial hygiene.

environmental impact A change in the environment resulting from specific actions or materials. *See also* product stewardship.

Environmental Impact Statement (EIS) An analysis required under Section 404 of the national Clean Water Act when a project would discharge dredged or fill material in waters of the United States. It is conducted to determine adverse effects on municipal water supplies, shellfish beds (including spawning and breeding beds), wildlife, aquatics, recreational areas, and so on.

Environmental Monitoring Systems Laboratory (EMSL) A laboratory of the US Environmental Protection Agency's Office of Research and Development responsible for developing analytical methods for environmental pollutants.

Environmental Protection Agency (EPA) A federal, state, or local unit of government established to consolidate programs dealing with air, water, solid waste, and other environmental concerns within one agency. The authority and scope of activity are determined by the legislation establishing the agency. *See also* regional office; US Environmental Protection Agency.

environmental quality The set of characteristics of the environment that generally indicate its desirability for human activity.

environmental stress A perturbation likely to cause an observable change in an ecosystem.

environmentally sensitive Having an awareness of human activities that may have an adverse effect on an ecosystem.

enzyme A protein (or occasionally a nucleic acid molecule) that catalyzes chemical reactions.

enzyme immunoassay (EIA) An assay that involves enzyme-labeled antibodies. Manufacturers produce enzyme immuno assay kits that contain everything necessary to perform a fixed number of assays. An enzyme immunoassay is commonly used for virus detection or for confirmation of the presence of viruses that had previously been detected via cell culture methods.

Ephemeral stream

enzyme immunoassay (EIA) kit *See* enzyme immunoassay.

enzyme induction A process by which a new enzyme is synthesized as a means of increasing enzyme activity within certain cells. The body uses this effect for a variety of regulatory purposes in responding to varying environmental conditions. A special case of enzyme induction in toxicology involves increased activity of enzymes responsible for the metabolism of chemicals that are foreign to the body. These responses are usually most prominent in the liver but are by no means confined to this organ.

enzyme-linked immunosorbent assay (ELISA) A test that involves the reaction of specific antibodies with target antigens (i.e., viruses). A positive reaction is measured by a colorimetric method whereby the enzyme–antibody–antigen complex, when present, will degrade a substrate for a color change.

EP *See* extraction procedure.

EPA *See* Environmental Protection Agency.

EPEC *See* enteropathogenic *Escherichia coli.*

ephemeral Pertaining to a transient or short-lived phenomenon.

ephemeral stream (1) A stream that flows only in direct response to precipitation. Such a stream receives no water from springs and no continued supply from melting snow or other surface sources. Its channel is above the water table at all times. (2) A stream or stretch of a stream that does not flow continuously during periods of as much as 1 month.

epichlorohydrin (chloropropylene oxide, C_3H_5OCl) A highly volatile, unstable liquid epoxide. It is a major raw material for epoxy and phenoxy resins and has other industrial uses. It is a treatment chemical that is regulated in drinking water under the Phase II Rule for synthetic organic contaminants and inorganic contaminants. *See also* epoxide; Phase II Rule; synthetic organic chemical.

```
  H   H H
  |   | |
H-C-O-C-C-Cl
  |___| |
        H
```

Epichlorohydrin

epidemic An occurrence of cases of disease in a community or geographic area clearly in excess of the number of cases normally found (or expected) in that population for a particular season or other specific time period. Most epidemiologists use the terms *epidemic* and *outbreak* interchangeably. When the excess disease is confined to a small or limited population or area, the term *outbreak* is preferred. An epidemic occurring over a very wide area and usually affecting a large proportion of the population is called a *pandemic.*

epidemic curve A graph showing the distribution of cases of disease by time of onset. From this curve, information can be obtained about the median incubation period and nature of the outbreak (e.g., point source or person-to-person spread, secondary spread).

epidemic, common source *See* common source epidemic.

epidemiologic study A study of human populations to identify causes of disease. Such studies often compare the health status of a group of persons who have been exposed to a suspect agent with that of a comparable nonexposed group. *See also* analytic epidemiologic study; case-controlled epidemiologic study; cohort epidemiologic study; cross-sectional epidemiologic study; descriptive epidemiologic study; ecologic epidemiologic study; experimental epidemiologic study; longitudinal epidemiologic study; observational epidemiologic study.

epidemiology The study of the determinants and distribution of injuries and diseases in human populations.

epidermis The superficial (outer) layer of skin cells.

epifluorescence microscope A microscope in which (1) the specimen to be observed is illuminated from above by short-wavelength (ultraviolet) light that excites dyes or naturally fluorescent material in the specimen, and (2) visible light is reflected upward from the specimen to the ocular. In transmission microscopy, light from a light source below passes through the specimen to the ocular.

epigenetic Pertaining to a process, response, or effect that is not mediated through modification or utilization of genetic information encoded in deoxyribonucleic acid.

epigenetic carcinogen A chemical or other agent that produces cancer without directly affecting deoxyribonucleic acid (DNA). Mechanisms by which a chemical could contribute to cancer without it or one of its metabolites binding to DNA include (1) stimulating rates of cell division by killing normal cells and providing a selective advantage to the growth of cells having altered genotypes that render them resistant to the chemical, or (2) directly stimulating division of these altered cells without similarly stimulating the division of normal cells. A third possibility identified in recent years is that certain chemicals prevent the death of cells with altered genotypes by an active process called apoptosis. Whereas genotoxic chemicals are assumed to increase the risk of cancer with a linear relationship to dose, epigenetic carcinogens are more likely to involve sublinear relationships with dose. This is most easily conceptualized with cytotoxic agents, for which no increase in cancer risk exists unless the dose has reached a level where some cell killing has occurred. *See also* apoptosis; genotoxic carcinogen.

epilimnion The upper water layer in a stratified lake. *See also* destratification; hypolimnion; metalimnion; stratification; thermocline.

epiphyte A plant that grows on other plants or nonliving structures without drawing its nourishment from them (as opposed to a parasite).

epiornithic Shared among members of a bird population; commonly refers to epidemic diseases.

epitheliochorial placenta A placental type that has all six layers interposed between maternal and fetal circulation. This placental type is found in pigs, horses, and donkeys.

epizootic An epidemic of disease in an animal population.

epoxidation A chemical process whereby an oxygen bridge is formed across a double bond.

epoxide An organic compound resulting from the union of an oxygen atom with two other atoms (usually carbon) that are joined in a ring. Epoxides contain a reactive group. They are three-membered cyclic ethers. Some epoxides may be created during the reactions of ozone (O_3) with natural organic matter. *See also* ether.

```
        O
  R    / \    R
   \  /   \  /
    C ----- C
   /         \
  R           R
```

Epoxide, where R is a carbon-containing group or hydrogen

EPROM *See* erasable programmable read-only memory.

EPS *See* dynamic-state extended-period simulation; extracellular polymeric substance.

EPTC *See S*-ethyl di-*N,N*-propylthiocarbamate.

eq *See* equivalent weight.

eq wt *See* equivalent weight.

eq/L *See* equivalents per litre *in the Units of Measure list.*

equal-energy depth One of two alternate depths in an open-channel cross section that can occur for a given flow rate and head, provided that the specific head exceeds the minimum needed for flow to occur.

equalization A means of providing a more uniform flow rate and composition for a water supply by using a reservoir that receives water from a pump or treatment system, evens out the incoming flow variation, and permits temporary water withdrawal in excess of the pump or treatment system capacity.

equalization storage The quantity of water from storage that enters the system during peak demand periods to augment other sources of supply. Equalization storage is replenished during off-peak demand periods.

equalizing basin *See* equalizing reservoir.

equalizing reservoir A reservoir interposed in a water supply system (or other hydraulic system) at any point between source and consumer to furnish elasticity of operation to the distribution system, so that different portions of the system may be more or less independent of each other. An equalizing reservoir is also called a balancing reservoir.

equilibrium (1) A condition in which reversible chemical reactions are simultaneously taking place in such a manner (i.e., at an equivalent rate) that no change in the net concentration of the substances involved in the chemical processes occurs. As a result of these simultaneous reactions, the substances involved in the reactions can be shown to be in a constant ratio to each other. (2) The state in which the action of multiple forces produces a steady balance or seeming lack of change. This may be caused by a true stop in action or by continuing actions that neutralize each other, resulting in no net change.

equilibrium constant The number prescribing the conditions of equilibrium for a given reversible chemical reaction.

equilibrium shift A change in the relative concentrations of reacting substances such that a different reaction or reaction rate is caused. For example, a change in the relative concentrations of sodium and calcium ions will dictate both the exchange rate and the selection of which ions will be adsorbed to and released from ion exchange resin beads.

equipluve A line on a rainfall map connecting places that have the same value of the pluviometric coefficient. *See also* pluviometric coefficient.

equipment Any mechanical device or supplies required to support the construction and operation of engineering facilities.

equipotential line A line drawn on a two-dimensional groundwater flow field map such that the total head is the same for all points along the line.

equipotential surface A geometric surface in a three-dimensional groundwater flow field such that the total hydraulic head is the same everywhere on the surface.

equivalence point The point at which the same number of equivalents of two reactants have been added (i.e., no excess of either reactant exists). During a titration, a suitable indicator is added to yield an end point that coincides with the equivalence point, although, as a rule, a difference between the end point and the equivalence point usually exists. In acid–base reactions, neutralization is achieved at the equivalence point. *See also* end point; equivalents per litre *in the Units of Measure list*; neutralization; titration.

equivalent (eq) *See* equivalent weight.

equivalent calcium carbonate *See* as calcium carbonate.

equivalent customer The designation for the number of fictitious customers whose water use through a ⅝-inch meter (single-family unit) is the same as the water use of a large-use customer. This calculation is based on a composite of all elements of cost differences between the single-family unit customer and the large-use customer. It is normally expressed as the ratio of the single-family unit customer to the large-use customer. For example, if the large-use customer is equivalent to 10 single-family use customers, the ratio is 10:1.

equivalent meter A means of expressing different sized water meters in terms of an equivalent number of "standard" size meters. The ⅝-inch or ¾-inch meter size is often used as the standard in terms of which all other meters are expressed. For example, a 1-inch meter is equivalent in capacity to two and one-half ⅝-inch meters. Equivalent meters are often used in ratemaking to assess customer monthly service charges based on meter size and to distribute capacity costs based on meter size in developing connection charges.

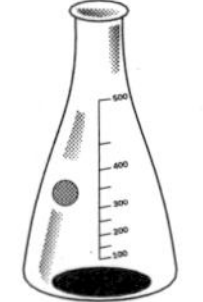

Erlenmeyer flask

equivalent pipes Two pipes or two systems of pipes in which the losses of head for equal rates of flow are the same. A single pipe is equivalent to a system of pipes when the same loss-of-head condition is satisfied.

equivalent residential unit A means of expressing different levels of residential customer demand for water in terms of an equivalent number of least-size residential demand accounts. If the base residential demand is 300 gallons (1,135 litres) per day per account, then other customer groups' accounts would be expressed in terms of the equivalent number of these least-size accounts. In this manner, the customer may be said to be paying for a proportionate share of the share of the capacity that customer is using. This concept is sometimes used to determine capacity costs for new accounts and to develop monthly service charges based on meter size. *See also* equivalent customer; equivalent meter.

equivalent weight A weight of a substance based on the formula weight and number of reactive protons associated with the substance. For an element, the equivalent weight is the atomic weight divided by the number of reactive protons. For example, calcium (Ca) has an atomic weight of 40 and has two reactive protons, so its equivalent weight is 40/2= 20. For a compound, the equivalent weight is the molecular weight divided by the number of reactive protons. For example, sodium chloride (NaCl) has a molecular weight of 58.5 and one reactive proton, so the equivalent weight is 58.5/1 = 58.5. Calcium carbonate ($CaCO_3$) has a molecular weight of 100 and two reactive protons (in the Ca^{2+} ion), so the equivalent weight is 100/2 = 50. Equivalent weight is also known as combining weight. *See also* oxidation number; valence.

equivalents per litre (eq/L) *See the Units of Measure list.*

erasable programmable read-only memory (EPROM) A read-only memory chip that can be erased, either electrically or by high-intensity ultraviolet light. After it is erased, the chip is reprogrammed in a read-only function mode.

Erlenmeyer flask A bell-shaped container used for heating and mixing chemicals and culture media.

erosion The wearing away of a material by physical means, often caused by the presence of abrasive particles in the stream.

erosion corrosion Deterioration of a surface by the abrasive action of moving fluids. This phenomenon is accelerated by the presence of solid particles or gas bubbles in suspension.

error A false or mistaken result in a study or experiment. In epidemiology, error refers to random error and bias (systematic error). *See also* bias; random error.

error control system A system for controlling equipment operation based on the difference between actual and desired values of the output. *See also* process error.

error estimates *See* standard error of estimate.

ES *See* effective size.

ESA *See* Endangered Species Act.

escape A wasteway installed in a conduit to discharge, when necessary, all or part of the conduit's flow.

escarpment A more or less continuous series of cliffs or sharp slopes oriented in the same direction, often at the edge of a plateau, resulting from erosion or faults.

***Escherichia coli* (*E. coli*)** A gram-negative, facultatively anaerobic, nonspore-forming bacillus commonly found in the intestinal tracts of humans and other warm-blooded animals. In sanitary bacteriology, *Escherichia coli* is considered the primary indicator of recent fecal pollution. *See also* coliform bacteria; fecal coliform.

***Escherichia coli* O157:H7** A strain of *E. coli*—serologically identified as O-antigen 157, H-antigen 7—that produces enterotoxin (Vero cytotoxin) and causes diarrheal illness. It has been identified as the disease agent in a variety of foodborne and waterborne disease outbreaks. This strain of *E. coli* can cause severe bloody diarrhea that may be followed by kidney problems due to hemolytic uremic syndrome.

ESD *See* ecologically sustainable development.

ESI *See* electrospray ionization.

ESR *See* electron spin resonance.

ester (R-CO-O-R') An organic compound formed by the reaction between a carboxylic acid and an alcohol with the elimination of a molecule of water.

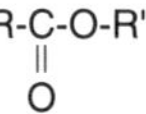

Ester, where R and R' are carbon-containing groups

esterification The process by which an ester is formed. The usual reaction is that of an acid (organic or inorganic) with an alcohol or other organic compound rich in hydroxyl (-OH) groups. Derivatization of haloacetic acids (to their methyl ester forms) with diazomethane is an esterification reaction. *See also* derivatization; diazomethane; ester.

estuary A passage in which the tide meets a river current, especially an arm of the sea at the lower end of a river; a firth.

ESWTR *See* Enhanced Surface Water Treatment Rule.

ET *See* evapotranspiration.

ETEC *See* enterotoxigenic *Escherichia coli*.

ether (R-O-R') A class of organic compounds in which an oxygen atom is interposed between two carbon atoms (organic groups) in the molecular structure.

R-O-R'

Ether, where R and R' are carbon-containing groups

ethology The study of the behavior of animals under natural conditions.

***S*-ethyl di-*N*,*N*-propylthiocarbamate (EPTC) ($C_2H_5SC(O)N(C_3H_7)_2$)** An herbicide. *See also* herbicide.

ethylbenzene ($C_6H_5C_2H_5$) A volatile organic compound used as a solvent. It is regulated by the US Environmental Protection Agency. *See also* solvent; volatile organic compound.

ethylene dibromide (EDB) ($BrCH_2CH_2Br$) A fumigant used for grains and tree crops. This chemical has other industrial applications as well. It is regulated in drinking water under the Phase II Rule for synthetic organic contaminants and inorganic contaminants. *See also* fumigant; Phase II Rule; synthetic organic chemical.

ethylene thiourea (ETU) ($NHCH_2CH_2NHCS$) The common name for 2-imidazolidinethione, a chemical with various industrial uses, including use as an insecticide or fungicide. *See also* fungicide; insecticide.

ethylenediaminetetraacetic acid (EDTA) ($(HOOCCH_2)_2NCH_2CH_2N(CH_2COOH)_2$) An organic chelating agent that forms very stable complexes with calcium, magnesium, and other divalent ions; is used as an analytical reagent (e.g., in hardness titration); and is in some detergents, cleaning agents, and scale preventatives.

etiologic agent The agent responsible for the postulated cause of a disease.

etiology The postulated causes of disease, specifically those that initiate the pathogenic mechanisms. *See also* pathogenesis.

ET_o *See* reference evapotranspiration.

ETU *See* ethylene thiourea.

eukaryote A cell in which the nucleoplasm is surrounded by a membranous envelope through at least part of each cell reproductive cycle; i.e., the cell has a true nucleus. This term is sometimes spelled eucaryote.

eukaryotic Pertaining to a eukaryote.

euphotic Characteristic of the upper part of a lake in which sufficient light penetrates for photosynthesis.

euploidy The condition of any cell that contains the proper or normal number of chromosomes.

eutrophic Pertaining to reservoirs and lakes that are rich in nutrients and very productive in terms of aquatic animal and plant life.

eutrophic lake *See* eutrophic reservoir.

eutrophic reservoir A reservoir rich in nutrients and characterized by large quantities of planktonic algae. In this type of reservoir, the input of oxygen-demanding materials has depleted dissolved oxygen, resulting in anoxic or anaerobic conditions, which potentially result in the formation of hydrogen sulfide (H_2S), release dissolved iron and manganese from sediments, lower the pH, and increase concentrations of carbon dioxide (CO_2). *See also* anaerobic condition; anoxic condition.

eutrophication An accelerated aging of a surface water body due to excess nutrients and sediments (over biological needs) entering the water body.

E_v *See* bulk modulus of elasticity of water.

eV *See* electron volt *in the Units of Measure list.*

evaporating dish A glass or porcelain dish in which samples are evaporated to dryness by high heat.

evaporation A process in which a liquid is changed by volatilizing (boiling) to a gaseous state at a set of temperature and pressure conditions. *See also* distillation; sublimation.

evaporation area The surface area of a body of water, and of any adjacent moist land to which water was supplied from the body of water, from which water is lost to the atmosphere by evaporation.

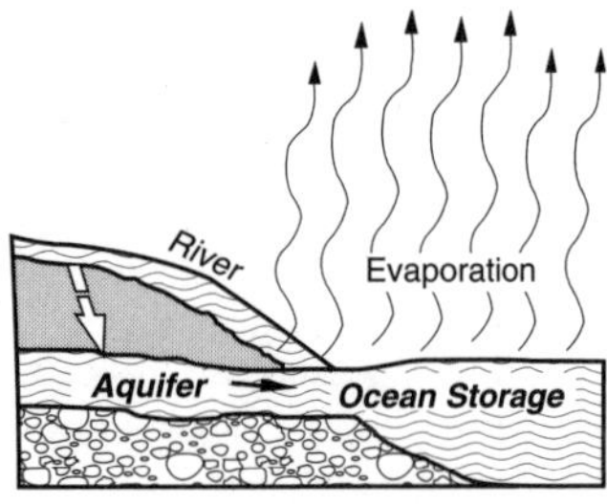

Evaporation

evaporation chamber That part of a distillation system in which water is changed into vapor.

evaporation discharge Discharge into the atmosphere of water, in the gaseous state, derived from the saturation zone. Evaporation discharge may be divided into vegetal discharge and soil discharge.

evaporation gauge A means of measuring evaporation. *See also* atmometer.

evaporation opportunity The ratio of the actual rate of evaporation to the potential rate. This ratio is also called the relative evaporation.

evaporation pan A pan used to hold water during observations for determining the quantity of evaporation at a given location. Such pans are of various sizes and shapes, the most common shape being circular or square. *See also* atmometer; floating pan; land pan; standard National Weather Service pan.

evaporation pond A constructed liquid containment with a free surface designed to promote evaporation, thereby reducing the volume of the liquid (e.g., a pond for disposing waste concentrate from a water-desalting process).

evaporation rate The quantity of water, expressed in terms of depth of liquid water, evaporated from a given water surface per unit of time. It is usually expressed in inches or millimetres of depth per day, month, or year.

evaporation, actual *See* actual evaporation.

evaporation, potential *See* potential evaporation.

evaporative concentration A technique used to increase the concentration of a solute by progressively vaporizing the solvent. For example, a Kuderna–Danish apparatus is used to concentrate analytes from an extraction solvent.

evaporative cooler A temperature-controlling unit that uses evaporation for cooling. An evaporative cooler works on the principle that energy is needed to evaporate water. Therefore, hot air is pumped from the affected area through water, and as the water evaporates the air is cooled. The cooled air is pumped back into the temperature-controlled area. These systems work most effectively in dry, arid regions, such as the southwestern United States, where evaporation rates are high.

evaporativity The potential rate of evaporation, as distinguished from the actual rate. *See also* potential evaporation.

evaporator A device in which a liquid is evaporated by the application of heat.

evaporimeter An instrument for measuring evaporation. *See also* atmometer.

evapotranspiration (ET) (1) The process of water vapor entering the atmosphere as a result of both evaporation and transpiration. (2) Water vapor given off by plants through their leaves.

evapotranspiration potential Water loss that would occur if a deficiency of water never existed in the soil for use by vegetation.

evapotranspiration tank A tank, filled with soil and provided with a water supply, in which representative plants are grown to determine the amount of water transpired and evaporated from the soil under observed climatic conditions. Such a tank is sometimes improperly referred to as a lysimeter.

evapotranspiration, actual *See* actual evapotranspiration.

evapotranspiration, potential *See* potential evapotranspiration.

event mean concentration (EMC) The average concentration of a contaminant in stormwater arising from a single storm event. The mean concentration is dependent on the sampling frequency and flow rate during each sampling event.

***ex parte* communication** Off-the-record communication between regulatory agency personnel responsible for a particular rulemaking and another party that could influence the content of the agency's rulemaking.

exa (E) A metric prefix meaning 10^{18}.

examination of water An investigation to determine the physical, chemical, and biological characteristics of water. *See also* water analysis.

excavation The removal of material to create a trench, hole, or cut in the earth's surface, resulting in a change in all or part of the elevation of a site.

exceedance probability The probability that a randomly selected event or sample exceeds a given value. For example, if the iron concentration is greater than 0.3 milligrams per litre in 100 out of 1,000 samples, the probability that a random sample has an iron content greater than 0.3 milligrams per litre is 10 percent.

excess rainfall That part of the rain of a given storm that falls at intensities exceeding the infiltration capacity and is thus available for direct runoff. *See also* net rainfall; storm flow.

excessive encrustation The buildup of precipitates on the inside of a pipeline such that pipe diameter is reduced, flow is blocked, and head loss increases. Calcium carbonate ($CaCO_3$), when present in concentrations greater than its solubility, is a common cause of encrustation. *See also* encrustation; tuberculation.

excess-lime treatment A modification of the lime–soda ash method of precipitative softening that uses additional lime to remove magnesium compounds.

exchange A process by which, under certain conditions, water may be diverted out of priority at one point by replacing a like amount of water at a location downstream.

exchange capacity *See* rated capacity.

exchange site A location on each bead of ion exchange resin that holds mobile ions available for exchange with other ions in the solution passing through the resin bed. In cation water softening, for example, mobile sodium ions located at the various exchange sites are exchanged for calcium or magnesium ions in the water being softened. Exchange sites are also called functional groups.

exchange tank *See* portable exchange tank.

exchange velocity The rate at which one ion is displaced in favor of another in an ion exchanger.

exchanger *See* ion exchanger.

excimer An excited dimer (for example, Xe_2). An excimer is formed when an electron from a corona discharge collides with a gas molecule and raises the molecule to an excited state; this excited molecule then collides with another molecule of gas to form an unstable excimer that spontaneously decays, generating ultraviolet light at a specific wavelength. The gases of primary interest for water and wastewater treatment (i.e., argon, krypton, xenon, fluorine, chlorine, bromine, and iodine) emit at wavelengths that either inactivate microorganisms, destroy organic pesticide molecules, or both. *See also* corona discharge; dimer; excited state; inactivation; pesticide; ultraviolet light.

excited state A higher-than-normal energy level (vibrational frequency) of the electrons of an atom, radical, or molecule resulting from absorption of photons (quanta) from a radiation source (e.g., an arc, flame, spark) in any wavelength of the electromagnetic spectrum. *See also* photon; radiation.

exclusion chromatography *See* high-performance liquid chromatography.

excystation The biological process of releasing sporozoites or trophozoites from a *Cryptosporidium* oocyst or a *Giardia* cyst in vivo or in vitro.

exemption A time extension allowed under the Safe Drinking Water Act for compliance with a national primary drinking water regulation. *See also* Variance.

exhaustion The state of an ion exchanger or other adsorbent that is no longer capable of useful ion exchange or adsorption as a result of the initial supply of available exchangeable ions being depleted or the adsorption sites being occupied. A unit that is exhausted requires regeneration to restore its capacity to treat water. *See also* regeneration.

exhibit *See* display.

exit loss The loss of head that occurs when a stream of water passes out of a hydraulic structure and assumes a lower velocity. Such loss is essentially caused by loss of velocity head.

exon The portion of a strand of deoxyribonucleic acid or m-ribonucleic acid that codes for a portion of a protein. *See also* intron.

exopolymer Any of several types of macromolecular polymeric materials released extracellularly by microbial cells. These polymeric materials may include polysaccharides, mucopolysaccharides, glycoproteins, proteins, and nucleic acids. Other terms such as *glycocalyx*, *slime*, *capsule*, and *sheath* have all been used in referring to microbial extracellular polymeric materials. *See also* excellular polymeric substance.

exopolysaccharide A complex molecule composed of monosaccharides linked together by glycosidic bonds that are released or extruded from the cells of microorganisms such as bacteria and algae. Exopolysaccharides are one of several types of extracellular polymeric substances excreted by microorganisms in biofilms. *See also* excellular polymeric substance.

exothermic Pertaining to a chemical process in which heat is released. For example, combustion is an exothermic process because heat is released.

exotic species A plant or animal that is not indigenous to a region, i.e., one that has been intentionally or accidentally introduced and is often persisting in the area.

expandable memory Computer memory that is used in addition to the disk operating system memory. It is composed of internal random access memory chips that act as a memory pool that can be mapped into one or more conventional memory areas. It requires a special driver and can be used by all processor types that have an expandable memory manager.

expanding reach A reach of a channel wherein the flow is decelerating. The velocity head at the downstream cross section is less than the velocity head at the upstream cross section.

expansion coupling A pipe coupling that permits relative movement of the two joined pipes.

expansion flow Fluid flow in a conduit where the cross-sectional area increases, resulting in a decrease in velocity and an increase in pressure.

expansion joint A specific location in a structure or pipeline designed to allow for fluctuations in length or size due to changes in temperature.

expenditure The act of using up or spending goods and services or money.

expense account An arrangement whereby an employer pays for certain expenses of employees in connection with work.

experimental epidemiologic study An epidemiologic study in which the investigator has control over the assignment of exposure for human subjects. Each study participant is assigned, usually at random, to an exposure classification. Because knowingly causing harm to individuals in such a study is clearly unethical, these studies are generally considered only in special circumstances—prophylactic trials to prevent disease or therapeutic trials to treat established disease processes. They may involve whole communities or groups of individuals. Experimental studies are sometimes referred to as intervention studies, clinical trials, or clinical epidemiology studies.

experimental matrix A design of an experiment with a set of independent and dependent variables called the matrix. It is used as part of a statistical approach to generating a maximum amount of information from a minimum number of experiments.

expert system A computer system using rule-based logic to simulate human intelligence.

explosimeter A device for measuring the concentration of potentially explosive fumes. It is also called a combustible-gas indicator.

explosive limit *See* lower explosion limit; upper explosion limit.

exponent An indication of the number of times a base number is to be multiplied by itself. For example, a base number of 3 with an exponent of 5, written 3^5, indicates that the base number is to be multiplied together five times: $3^5 = 3 \times 3 \times 3 \times 3 \times 3 = 243$.

exponential function A mathematical function containing exponential terms, e.g., e^x.

exponential increase An increase of some physical quantity according to an exponential function.

export file A file created by a software program to export data to another application.

exposure (1) The condition of being subject to legal liability. (2) Contact with a chemical, physical, or microbial agent. Being precise when exposure to various agents is discussed is important because different routes of exposure can have greatly different health impacts. Exposure can best be expressed in terms of the agent's free concentration at the site in the body where the agent exerts its toxic effect. This is preferred because many factors modify the delivery of an external exposure to a chemical or microbial agent to the active site. Practical considerations frequently require exposure to be expressed in terms of units of agent administered per unit of body weight or unit of body surface area. *See also* dose.

exposure assessment The determination or estimation (in qualitative or quantitative terms) of the magnitude, frequency, duration, route, and extent (number of people) of exposure to a chemical. *See also* dose.

exposure coefficient A quantitative value of the amount of contaminated medium contacted per day based on information on the frequency, mode, and magnitude of contact with the contaminated medium.

exposure control plan A strategy to limit legal liability or biological or chemical contact.

exposure determination The process of ascertaining the job classifications in which employees have occupational exposure to any hazard, as determined by listing all tasks (as well as procedures or groups that have closely related tasks) where exposure can occur.

exposure incident An instance of a person being exposed to an infectious material through open cuts, mucous membranes, eyes, nose, or indirect exposure by touching a contaminated object.

exposure level (chemical) The amount (concentration) of a chemical at the absorption surfaces of an organism.

exposure scenario A set of conditions or assumptions about sources, exposure pathways, concentrations of toxic chemicals, and populations (numbers, characteristics, and habits) that aid the investigator in evaluating and quantifying exposure in a given situation.

expression The dewatering of a filter cake or sludge, after gravity drainage, by physical compression of the filter cake or sludge.

Extended Binary Coded Decimal Interchange (EBCDIC) A standard that assigns 8-bit codes to standard keyboard characters. This is a data format common on IBM mainframe computers. *See also* American Standard Code for Information Interchange.

extended-period simulation (EPS), dynamic-state *See* dynamic-state extended-period simulation.

extension agent A government official who serves as a consultant to agriculture.

external corrosion Corrosion or deterioration of the outside of a pipe or appurtenance, i.e., the side of the pipe that does not carry or is not exposed to the potable water. External corrosion may be caused by soil or bedding conditions, surface characteristics, moisture conditions, grounding procedures, other utility structures, and so on. Cathodic protection systems are often used to mitigate external corrosion.

external exposure Exposure measured in terms external to the body. Examples include concentrations of a chemical in drinking water or a dose that is administered by a nonsystemic route (i.e., by stomach tube rather than intravenous injection).

external water treatment Preparation of the source water to be used for boiler feedwater or boiler makeup water outside the boiler. This preparation may include such steps as cation exchange softening, pH modification, or dealkalization. *See also* internal water treatment.

externality A cost or benefit that is associated with providing water service but is not typically reflected in the water utility's cost-of-service analysis for rates. Externalities are often associated with adverse environmental effects, such as reduced river flows or loss of animal habitat land. Including the cost (or benefit) of externalities in the cost-of-service for water would make the price of water reflect its true economic cost, but the practice raises policy and practical issues, especially with regard to equity, because the external costs are invariably difficult to accurately assess and are not paid out of pocket as cash disbursements.

extracellular polymeric substance (EPS) A substance released extracellularly by microbes, usually bacteria. These substances are primarily polysaccharides and exist as highly hydrated gels and fibers surrounding, encapsulating, or connecting a consortium of microbial species in an aquatic environment (i.e., a biofilm).

extracellular slime Any extracellular polymeric substance elaborated by micro-organisms that allows them to adhere to solid substrata. Extracellular slime may also be referred to as slime, glycocalyx, capsule, or sheath. Also called extracellular polymeric substance.

extracellular water The portion of the water found in an organism that exists outside of cellular membranes. In humans and other mammals, it includes water within the vasculature and in the interstitial spaces (i.e., spaces between cells). Specialized fluids, such as cerebrospinal fluid, are also included in the extracellular water.

extractable metal That part of a metal from a sediment, soil, or water sample that is found in the filtrate after treatment of the sample with hot or cold dilute mineral acid followed by filtration through a 0.45-micrometre pore-diameter filter.

extraction A separation technique used to increase the concentration of a solute, remove interferences from a matrix, or both. Many types of extraction exist and are frequently used in the chemical analysis of water. The extraction technique is used to isolate the analyte from the water matrix, increase the analyte concentration for better response from a detector, and minimize interferences that may hinder detection of the analyte.

extraction procedure (EP) A specific protocol for isolating an analyte for the purpose of increasing its concentration, minimizing interferences, or both.

extramedullary erythropoiesis The process of synthesizing the formed elements of the blood outside of the bone marrow. This occurs in mammals in utero and immediately after birth when a significant fraction of the blood cells are formed in the liver.

extrapolation (1) Projection of empirical or graphical trends to estimate the value of a dependent variable (y value) for a known value of the independent variable (x value) that is beyond the range of experimental observation. For engineering purposes, extrapolation of empirical or graphical trends is generally acceptable for a small range (less than or equal to 10 percent) of x values beyond the range of actual observations. (2) In risk assessment, the use of data from one species of animals to estimate risks in another without specific data justifying this use.

Eyewash station

The process of extending dose-response data to doses less than those that have actually been tested by some mathematical process is referred to as low-dose extrapolation. Frequently, the extent of this extrapolation spans several orders of magnitude (e.g., extrapolating from 1 in 1,000 to 1 in 10,000,000) between the doses tested and the environmental concentrations being evaluated.

extremely hazardous chemicals A list of chemicals designated by the US Environmental Protection Agency that require special reporting and handling to protect workers and communities.

eye irritation A process by which a chemical produces inflammation and corrosive damage to the eye.

eyewash A safety device used to wash chemicals from the eyes. The device, which resembles a drinking fountain, directs a gentle spray of water into each eye.

eyewash station A flushing fountain designed to wash debris or chemicals from the eyes at a flow rate of approximately 3 gallons (12 litres) per minute.

F

F *See* farad *in the Units of Measure list*; fertility plasmid.

F *See* Faraday constant.

f *See* femto; frequency.

FAB gas chromatography–mass spectrometry *See* fast atom bombardment gas chromatography–mass spectrometry.

FAC *See* facultative plant; free available chlorine.

FACA *See* Federal Advisory Committee Act.

facilitated diffusion A biological mechanism for accelerating the movement of substrates and other chemicals across cellular membranes. The process usually involves a protein that acts as a carrier molecule. This mechanism is particularly important for moving chemicals that are highly polar or have high molecular weights that preclude high rates of diffusion through the lipid environment of biological membranes.

facility expansion charge *See* capacity charge.

facility identifier A unique identifier, assigned to facilities such as personal access openings, valves, and fire hydrants, that is usually annotated on maps and entered into a geographic information system or relational database, to provide a unique key and link between database records pertaining to that facility.

facility maintenance system (FMS) A means of providing and recording preventive and scheduled equipment maintenance.

factored pooling A process of assigning water allocations by estimating the rates of water use and calculating allocations before consumption and billing.

factorial experimental design A test plan in which experimental parameters are selectively varied to optimize the amount of information that can be obtained on the effect of individual parameters without requiring that every unique test condition be studied.

FACU *See* facultative upland plant.

facultative Having the ability to live under different conditions, e.g., with or without free oxygen.

facultative bacteria Bacteria that can adapt themselves to grow and metabolize in both the presence and absence of oxygen.

facultative organism A microbe capable of adapting to either aerobic or anaerobic environments.

facultative plant (FAC) A plant species that is about equally likely to occur in a wetland or nonwetland environment; i.e., an estimated 34 to 66 percent of the members of that species are found in wetlands, and the rest are found in nonwetlands.

facultative upland plant (FACU) A plant species that usually occurs in nonwetlands (an estimated 67 to 99 percent of such plants are found there) but occasionally is found in wetlands (an estimated 1 to 33 percent).

facultative wetlands plant (FACW) A plant species that usually occurs in wetlands (an estimated 67 to 99 percent of such plants are found there) but occasionally is found in nonwetlands (an estimated 1 to 33 percent).

FACW *See* facultative wetlands plant.

Fahrenheit scale A temperature scale in which equal mixtures of water, ice, and salt produce a value of 0°. The point where water freezes was taken as 32° on this scale, which has 180 equal units between the freezing and boiling of pure water. This scale was invented by Daniel G. Fahrenheit (1686–1736), a German physicist. *See also* degree Fahrenheit *in the Units of Measure list.*

Fahrenheit degree *See the Units of Measure list.*

fall (1) A sudden difference in elevation in the bed of a stream of sufficient extent to cause the entire stream of water passing over it to drop vertically for some distance before resuming its course.

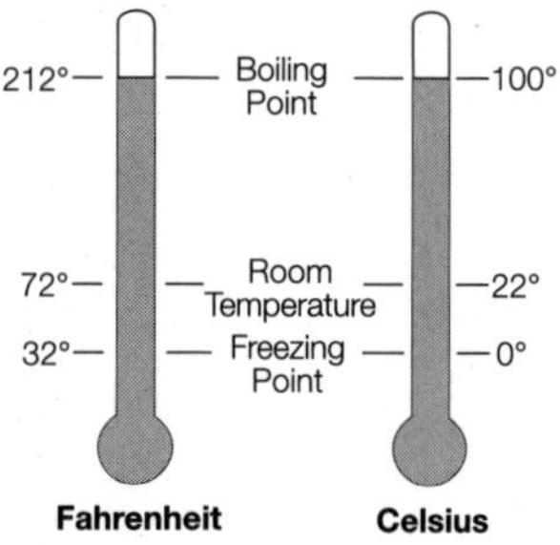

Fahrenheit and Celsius temperature scales

(2) The vertical distance between the water surface elevations of two points on a stream or conduit. *See also* drop; slope.

fall increaser A device connected with a draft tube to increase the discharge by reducing the tailwater pressure.

fall velocity The velocity applied to any rate of fall. *See also* standard fall velocity.

fallout Radioactive debris that resettles to the earth after a nuclear explosion. Fallout takes two forms. The first, local fallout, consists of the denser particles injected into the atmosphere by the explosion. These particles descend to the earth within 24 hours near the site of the detonation and in an area extending downwind for some distance (often hundreds of miles or kilometres), depending on meteorological conditions and the force of the detonation. The other form, worldwide fallout, consists of lighter particles that ascend into the upper troposphere and stratosphere and are distributed over a wide area of the earth by atmospheric circulation. These particles are brought to the earth, mainly by rain and snow, over periods ranging from months to years. When atmospheric testing of nuclear devices was permitted, fallout contaminated surface waters.

false positive A positive coliform result from a noncoliform bacterium. In testing a drinking water sample for the presence of coliform bacteria, some types of noncoliform bacteria may give a result that resembles a positive coliform result either in the presumptive step by the most probable number (MPN) method or in the membrane filter method. Further testing, called confirmation (for the MPN method) or verification (for the membrane filter method), is required to establish whether the positive result is real or a false positive. *See also* completed test; confirmed test; presumptive test.

FAME *See* fatty acid methyl ester.

FAP *See* fatty acid profile.

farad (F) *See the Units of Measure list.*

Faraday constant (*F*) The quantity of electricity that can deposit (or dissolve) 1 gram-equivalent weight of a substance during electrolysis. It is equivalent to 96,489 coulombs. Thus, 1 faraday of electricity will reduce 1 mole of silver ion (Ag^+) or 0.5 mole of copper ion (Cu^{2+}) to the metal form.

Faraday's law A law of electrochemistry stating that the amount of any substance dissolved or deposited in electrolysis is proportional to the total electric charge passed.

FAS *See* ferrous ammonium sulfate.

fascine A round bundle of brush from 1 to 3 feet (0.3 to 1 metre) in diameter and 10 to 20 feet (3 to 7 metres)—and occasionally up to 500 feet (17 metres)—long used to make a foundation mat; to make a revetment to protect a shore against erosion; to accumulate sand or silt on the bed of an estuary; or to make a mat to protect a streambed from scour around piers, bridge abutments, dike breeches, or similar points.

fast atom bombardment (FAB) gas chromatography–mass spectrometry An ionization technique used in the analysis of biological molecules and other materials of fairly high molecular weight, as well as in conjunction with liquid chromatography and capillary electrophoresis. The technique involves a beam of energetic atoms, such as argon, aimed at a sample on a direct insertion probe. Desorbed ions are then analyzed in the mass spectrometer. The technique can be used in either a static or continuous flow mode, with the latter mode being more sensitive.

fast rinse A process in which rinse water is applied to a softener bed at the end of brine regeneration at a faster rate of flow than that for which the brine was applied. Brine, because of its greater density, moves down through the bed more slowly than regular water. If rinsing were to continue at this slower rate until the chloride concentration had dropped to the acceptable level at which the softener unit could be returned to service, the time required would be excessive. After the higher concentrations of brine have passed from the unit, little is gained by prolonging the rinse time. Thus, the rinse rate during the last few minutes is increased approximately fivefold to complete the rinse cycle. This fast rinse quickly removes the last traces of chloride and significantly reduces the regeneration time.

fast-track design A method of project delivery that deviates from the usual approach in which a complete set of plans and specifications is prepared and bid to the contracting community. In a fast-track design, a design firm teams with a contractor; the contractor begins constructing the project according to partial plans, making design decisions "in the field," and communicating with the designer. A fast-track design is implemented to accelerate a project. The owner may not save money and typically sacrifices some control over the final quality of the project.

fathom (1) *See the Units of Measure list.* (2) To find the depth of something; to sound.

fathometer An instrument used in measuring the depth of water based on the time required for a sound wave to travel from the surface to the bottom and for its echo to be returned.

fatigue corrosion *See* stress corrosion.

fatty acid A carboxylic acid derived from or contained in an animal or vegetable fat or oil. A saturated fatty acid ($CH_3(CH_2)_iCOOH$, where $i = 0–20$) has the carbon atoms of the alkyl chain connected by single bonds, whereas an unsaturated fatty acid (like the monounsaturated fatty acid $CH_3(CH_2)_i CH{=}CH(CH_2)_jCOOH$, where $i + j \geq 14$) has one or more double bonds between the carbon atoms in the alkyl chain. *See also* carboxylic acid; saturated; unsaturated.

fatty acid methyl ester (FAME) (R–COOCH$_3$) A fatty acid with the active hydrogen replaced by the methyl group of methanol. The esterification of a fatty acid (R–COOH) by methanol (CH_3OH) yields the fatty acid methyl ester (R–$COOCH_3$). The methyl esters of fatty acids have higher vapor pressures than the corresponding acids and are more easily distilled. *See also* esterification; fatty acid.

fatty acid profile (FAP) The array of fatty acids that an organism (usually a bacterium) contains when grown in a pure culture under specified conditions. It can be used in a manner similar to a fingerprint for identification of the organism.

fatty acid profiling The process of extracting the fatty acids from a pure culture of a bacterial isolate, esterifying the fatty acids, and analyzing them using a gas chromatograph to generate a profile of the fatty acids that were extracted. *See also* esterification.

faucet aerator A device inserted into the faucet head (or a complete faucet head itself) that reduces faucet water flow while adding air to the water to create the impression of more water.

faucet restrictor A device inserted inside a faucet behind the particle trap screen that forces water through a small opening. Faucet restrictors are usually incorporated into low-flow faucet aerators.

fault spring A spring fed by juvenile water and deep groundwater, usually hot and mineralized, that rises through a deep-rooted fault or fissure. Groups are sometimes collectively known as thermal or hot springs. A fault spring is also called a fissure spring.

fault-dam spring A spring created by the occurrence of a fault of low permeability in alluvial material. The fault acts as a dam in backing up groundwater to the surface and causing that water to appear as a spring.

fauna The animals of a given region or period considered as a whole.

FB reactor *See* fluidized bed reactor.

fbm *See* feet board measure *in the Units of Measure list.*

FC *See* fecal coliform.

FCE *See* final control element.

***F*-curve** A graphical representation of a cumulative data function.

FDA *See* Food and Drug Administration.

feasibility determination A determination made by the US Environmental Protection Agency under Section 1412 of the Safe Drinking Water Act that a technology, treatment technique, or other means is feasible and is a best available technology under the terms of Section 1412. The Safe Drinking Water Act states that this determination must involve examining for efficacy under field conditions (not solely under laboratory conditions), and it must take cost into consideration.

feature A graphic element or entity in a digital database, e.g., a water valve symbol in a geographic information system database.

fecal coliform (FC) Members of the total coliform group of bacteria that are characterized by their ability to ferment lactose at 112.1° Fahrenheit (44.5° Celsius) and that are considered more specific indicators of fecal contamination than are coliforms that ferment lactose only at 95° Fahrenheit (35° Celsius). *Escherichia coli* and some *Klebsiella pneumoniae* strains are the principal fecal coliforms. *See also* coliform bacteria; *Escherichia coli.*

fecal contamination Contamination of soil or water by feces from warm-blooded animals.

fecal detection A determination of water contamination by fecal matter from warm-blooded animals through the identification of specific bacterial groups such as fecal coliform or fecal streptococci.

fecal indicator Fecal coliform (including *Escherichia coli*, fecal streptococci, or other bacterial groups), originating in humans or other warm-blooded animals, that indicate contamination by fecal matter.

fecal matter Matter (feces) containing or derived from animal or human bodily wastes that are discharged through the anus.

fecal streptococci *See* fecal streptococcus.

fecal streptococcus (FS) A gram-positive bacterium belonging to the genus *Streptococcus* (Lancefield Group D) that is generally found in human and animal feces. *See also Enterococcus.*

fecal–oral route of exposure *See* fecal–oral transmission.

fecal–oral transmission The conveyance of an infectious disease through contact with human or animal excreta that contain the infectious agent. The transmission of infection can be either direct or indirect. Contaminated drinking water can be one of several possible sources of infection, and an epidemiologic investigation is required to determine the source and mode of transmission. *See also* transmission of infection.

feces Excrement from the gastrointestinal tract, consisting of residue from food digestion and bacterial action.

fecundity index A measure of fertility in reproductive studies. It is formally defined as the percentage of copulations that result in pregnancy. *See also* fertility index.

Federal Advisory Committee Act (FACA) Public Law 92-463 (5 U.S.C. Appendix), enacted by the US Congress in 1972 to provide uniform standards for the operation of advisory committees within the US government's executive branch, to monitor the number of committees and their activities, and to ensure that the public will have access to the committees' deliberations.

Federal Emergency Management Agency (FEMA) A department of the US government responsible for emergency planning, preparedness, mitigation, response, and recovery, as well as for working with the states to plan, organize, and train for responding to disasters.

Federal Energy Regulatory Commission (FERC) An independent, five-member commission within the US Department of Energy that sets rates and charges for the transportation and sale of natural gas and for the transmission and sale of electricity; licenses hydroelectric power projects; and establishes rates or charges for the transportation of oil by pipeline, as well as the valuation of such pipelines.

Federal Geodetic Control Committee (FGCC) A committee established by the US government that has developed and published geodetic control standards and specifications for the United States describing the methods, equipment, and procedures used to achieve specified accuracy levels.

Federal Insecticide, Fungicide, and Rodenticide Act (FIFRA) The US law, codified generally as 7 U.S.C. 136–136y, that establishes a regulatory and enforcement program administered by the US Environmental Protection Agency to govern the use of pesticide products.

Federal Interagency Coordinating Committee on Digital Cartography (FICCDC) A group created by the federal Office of Management and Budget to promote the exchange of information and ideas on technology and methods for collecting and using digital cartographic data.

Federal Power Act (FPA) An act (16 U.S.C. 791a *et seq.*) amended several times since enactment in 1920, that established the Federal Power Commission, now the Federal Energy Regulatory Commission, to issue licenses for the construction and operation of hydroelectric facilities in any water over which the US Congress has jurisdiction.

Federal Register A US government document published every federal working day by the Office of the Federal Register. It provides official public notice of federal agencies' proposed regulations, final regulations, and other actions.

Federal Reporting Data System (FRDS) A computer database maintained by the US Environmental Protection Agency for

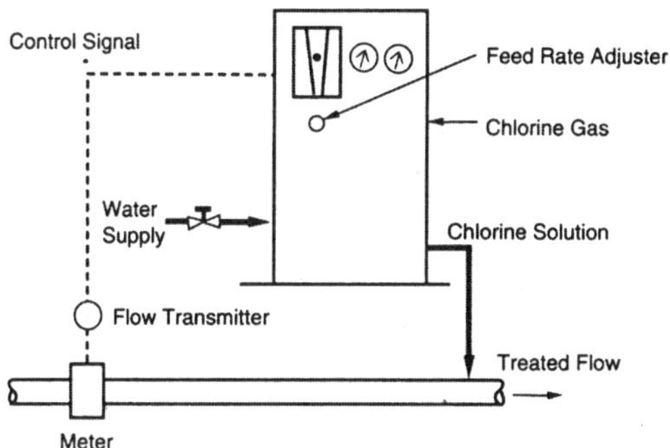

Feedforward control

tracking public water system compliance with national primary drinking water regulations.

Federal–State Toxicology and Regulatory Alliance Committee (FSTRAC) An ad hoc committee of federal and state toxicology and regulatory professionals that generally meets twice a year and is supported administratively by the Human Risk Assessment Branch of the US Environmental Protection Agency's Office of Science and Technology.

feed and bleed An ultrafiltration term borrowed from old-fashioned boiler operators. When applied to an ultrafilter design, it refers to the use of multiple stages of ultrafilter units such that the feedwater is controlled at a rate equal to the permeate plus concentrate flow rates and the reject water from the initial ultrafiltration stages is recirculated to subsequent stages.

feed pressure The pressure at which water is supplied to a water treatment device.

feed pump A pump used to add water treatment chemicals, either liquids or in solution, to the water being treated. These pumps are often very accurate in their delivery of what is being pumped.

feedwater The water to be treated that is fed into a given water treatment system.

feedback The circulating action between a sensor measuring a process variable and the controller that controls or adjusts the process variable. For example, chemical feeders are often controlled through links to rate-of-flow devices.

feedback control A type of control that measures the output of a process and adjusts one or more inputs to cause the output to assume a desired value or set point.

feedforward control A type of control that measures one input of a process and adjusts one or more other inputs to cause the output to assume a desired value.

feedwater treatment (1) Any chemical or other treatment of water to be used in boilers or manufacturing processes that requires removal of certain minerals, steps to make the water stable with respect to corrosion, or both. (2) Any treatment prior to a membrane process to remove substances that may plug, foul, or damage the membranes or may allow the membrane system to operate at a higher recovery or greater efficiency. *See also* pretreatment; water correction.

feet (ft) *See the Units of Measure list.*

feet board measure (fbm) *See the Units of Measure list.*

feet per hour (ft/h) *See the Units of Measure list.*

feet per minute (ft/min) *See the Units of Measure list.*

feet per second (ft/s) *See the Units of Measure list.*

feet per second squared (ft/s^2) *See the Units of Measure list.*

feet squared per second (ft^2/s) *See the Units of Measure list.*

Feigenbaum index A corrosion index related to hard-saline waters. It was developed in Israel for waters of the Negev Desert region. Tests led to the following corrosion index:

$$Y = AH + B[Cl^- + SO_4^{2-}]\exp(-1/AH) + C$$

Where:

$A = 0.00035$
$B = 0.34$
$C = 19.0$

$$H = \frac{[Ca^{2+}][HCO_3^-]^2}{[CO_2]}$$

(All concentrations are expressed as milligrams bicarbonate per litre.) Field tests indicate less corrosion with higher Y values; these test results may be generalized as follows: $Y > 500$, mild corrosion; $200 < Y < 500$, moderate corrosion; $Y < 200$, high corrosion. This method has not been widely tested.

FEIS *See* Final Environmental Impact Statement.

FEMA *See* Federal Emergency Management Agency.

female end Obsolete term; substitute *inside threaded connection* or *bell end. See also* male end.

femto (f) Prefix meaning 10^{-15} in Système International.

Fenton's reagent A ferrous salt (e.g., ferrous sulfate [$Fe(SO_4)$]) that acts as a catalyst for the production of hydroxyl radicals (•OH) from hydrogen peroxide (H_2O_2). During the reaction, the ferrous salt is oxidized to the ferric form and can participate in other chemical reactions, such as coagulation or metal complexation.

FEP *See* free erythrocyte porphyrin.

FERC *See* Federal Energy Regulatory Commission.

fermentation A process in which an organic substrate is broken down by fermentative metabolism of a microorganism culture (yeast or bacterium) to form stable intermediate metabolic products. Some examples of commercially important fermentation processes include wine making, beer brewing, production of ethyl alcohol, cheese making, and vinegar production.

fermentative respiration A form of anaerobic respiration in which metabolic products formed during catabolic breakdown of organic substrates serve as the final electron acceptors, forming stable fermentation products.

ferric chloride ($FeCl_3$) An iron salt used as a coagulant in water treatment. The iron has a valence of +3. *See also* coagulant; iron salt; natural organic matter.

ferric hydroxide ($Fe(OH)_3$) A coagulant precipitate used to remove turbidity, natural organic matter, and arsenic from water. Addition of iron salt coagulant to water results in ferric hydroxide formation. *See also* coagulant; iron salt; natural organic matter.

ferric iron (Fe^{+3}) Small solid iron particles containing trivalent iron, usually as gelatinous ferric hydroxide ($Fe(OH)_3$) or ferric oxide (Fe_2O_3), which are suspended in water and visible as "rusty water." Ferric iron can normally be removed by filtration. It is also called precipitated iron or iron +3.

ferric sulfate ($Fe_3(SO_4)_3$) An iron salt used as a coagulant in water treatment. The iron has a valence of +3. *See also* coagulant; iron salt; natural organic matter.

ferrous ammonium sulfate (FAS) ($Fe(SO_4)\cdot(NH_4)_2SO_4\cdot 6H_2O$) Part of the N,N-diethyl-p-phenylethylenediamine–ferrous ammonium sulfate method, which is used for determining the presence of chlorine. *See also* N,N-diethyl-p-phenylethylenediamine.

ferrous chloride ($FeCl_2$) A soluble iron salt. *See also* coagulant.

ferrous iron (Fe^{+2}) A reduced form of iron. When ferrous salts are added to water, the ferrous iron can act as a reducing agent. Ferrous iron is also referred to as iron(+2). *See also* ferrous salt; reducing agent.

ferrous oxide (FeO) A black powder that is insoluble in water.

ferrous salt A class of iron salts that includes ferrous chloride ($FeCl_2$) and ferrous sulfate ($Fe(SO_4)$). It is used in drinking water treatment as a coagulant, reducing agent, or both. *See also* coagulant; ferrous chloride; ferrous sulfate; reducing agent.

ferrous sulfate ($Fe(SO_4)$) An iron salt used as a coagulant in water treatment. The iron has a valence of +2. *See also* coagulant.

fertility (F) plasmid A circular piece of deoxyribonucleic acid (DNA) located outside the chromosome. It has a molecular weight of 6.3×10^7 and was originally found in *Escherichia coli*. It contains genes that code for its transfer by bacterial conjugation (bacterial mating), including genes for production of the fertility-pilli, which are hollow tubes of protein that extend from the bacterial cell by 2 to 20 micrometres and act as portals for DNA entry into the cell.

fertility index An index of the ability of male animals to impregnate female animals or the ability of female animals to conceive. The male fertility index is defined as the percentage of males impregnating females relative to the total number of males exposed to fertile, nonpregnant females. The female fertility index is the percentage of females who conceive after exposure to fertile males. *See also* fecundity index.

fertilization The result of genetic combination of a sperm with an ovum to form a zygote.

fertilizer A substance (manure or chemical mixture) used to make soil more fertile.

FGCC *See* Federal Geodetic Control Committee.

FIA *See* flow injection analysis.

fiber bundle For hollow fiber membrane elements or permeators, the membrane fiber assembly that can be removed from its pressure vessel.

fiber optics The use of fine glass fibers for transmitting signals by means of optical light.

fiber optics transmission The technique of transmitting data through an enclosed bundle of thin, transparent fibers of glass or plastic as pulses of light generated by lasers.

fiberglass A generic term for a plastic material containing glass fibers for reinforcement. The plastic is composed of a resin that is hardened by an exothermal chemical process. Fiberglass is often used as a material of construction when corrosion is a concern, such as for chemical storage tanks.

fibrosarcoma A sarcoma (i.e., tumor that has the appearance of embryonic connective tissue) that is derived from fibroblasts that produce collagen. These tumors are frequently highly malignant. Fibrosarcoma is a frequent finding in animal studies of chemicals.

fibrosis The formation of tissue that has a fibrous appearance. It is a frequent result of injury where fibrous tissue replaces the cells that are functionally important to a particular organ. The resulting tissue becomes more rigid and frequently restricts blood flow in and out of the tissue. In the lung it also reduces the ability of the lung to expand and contract, reducing the amount of air that can be exchanged with the environment. In the liver, fibrosis ultimately results in cirrhosis. Fibrosis could follow liver-damaging doses of organics (e.g., chloroform).

FICCDC *See* Federal Interagency Coordinating Committee on Digital Cartography.

Fick's law of diffusion A law of chemistry and physics stating that the rate of mass transfer of a substance by diffusion through a medium is proportional to the concentration gradient of the substance in the medium. Substances diffuse from a place of higher concentration to a place of lower concentration. In equation form,

$$\text{diffusive flux (mass per area per time)} = D\frac{\Delta C}{X}$$

Where (in any consistent set of units):

D = the coefficient of diffusion, in area per time

ΔC = the change in concentration of the substance from point A to point B in the medium, in mass per unit volume

X = the distance from point A to point B in the medium, in length

In engineering terms,

$$\text{diffusive flux (mass per area per time)} = k(C_A - C_B)$$

Where (in any consistent set of units):

k = the mass transfer coefficient, in length per time

C_A, C_B = the concentrations of the substance at points A and B in the medium, in mass per volume *See also* molecular diffusion.

fictive component A subset of a complex polymeric molecular structure that behaves similarly to the complex polymeric molecule in the context of a treatment process. For example, natural organic matter could be divided into several fictive components for modeling adsorption of natural organic matter on granular activated carbon.

FID *See* flame ionization detector.

field blank A water quality sample for which highly purified water is run through the field-sampling procedure and sent to the laboratory to detect whether any contamination of the samples is occurring during the sampling process.

field capacity (1) The maximum am ount of water that the unsaturated zone of a soil holds against the pull of gravity. It is the soil moisture content of a sample in which the force of gravity acting on the water is equal to the air–water interface tension in the sample. (2) The approximate quantity of water that can be permanently retained in the soil in opposition to the downward pull of gravity. It may be expressed in terms of percentage of dry weight or as a depth for a given depth of soil. The length of time required for a soil to reach field capacity varies considerably with various soils, being approximately 24–48 hours for sandy soils, 5–10 days for silt clay soils, and longer for clays. Field capacity is also called capillary capacity, field carrying capacity, maximum water-holding capacity, moisture-holding capacity, or normal moisture capacity. *See also* specific retention.

field carrying capacity *See* field capacity.

field groundwater velocity The actual or field velocity of groundwater percolating through water-bearing material. It is measured by the volume of groundwater passing through a unit cross-sectional area per unit time divided by the effective porosity. It is also called effective groundwater velocity, true groundwater velocity, actual groundwater velocity, or average linear velocity.

field moisture capacity *See* field capacity.

field permeability coefficient *See* hydraulic conductivity; permeability coefficient.

FIFRA *See* Federal Insecticide, Fungicide, and Rodenticide Act.

file transfer protocol (FTP) A method of transferring files to and from remote computers.

file transfer protocol (FTP) transfer File transfer protocol used in conjunction with transmission control protocol–Internet protocol.

fill material Any material used for the primary purpose of replacing an aquatic area with dry land or changing the bottom elevation of a water body. This term does not include any pollutant discharged into the water primarily to dispose of waste; that type of activity is regulated under Section 402 of the Clean Water Act.

filler gate A small gate, installed in a larger sliding gate, that admits water to fill the conduit or passage on the dry side of the larger gate and equalize the water pressure on both faces of the larger gate, thereby reducing the friction in opening the larger gate. A filler gate is also called a bypass gate or bypass valve.

film flow The movement of suspended water in any direction through a system of interconnecting films that adhere to the surfaces of solid particles or to the walls of fractures in the zone of aeration. In a capillary fringe, film flow may occur through interstices that are completely filled with water. Film flow is caused by an imbalance between gravitational forces, adhesion forces, and film pressure forces. The flow is laminar, and head loss varies directly with flow velocity. Film flow is also called capillary migration, capillary movement, or capillary flow.

film pressure The inward pull or pressure acting on the water–air surface of a film system of suspended water in soils composed principally of mobile or fringe water. The surface is curved, with the concavity toward the air space. The magnitude of the film pressure is controlled by surface tension and surface pressure. The magnitude of surface tension in suspended water is influenced by the chemical character of the soil solution and the temperature. The magnitude of the surface pressure in suspended water is controlled by the size and shape of the interstices between soil grains as established by soil texture and structure. In a capillary fringe, the additional controlling factor of solid–liquid contact angle exists.

filter (1) In the laboratory, a porous layer of paper, glass fiber, or cellulose acetate used to remove particulate matter from water samples and other chemical solutions. (2) The screening, removal, or both of harmful pollutants, as through respirators and dust masks. Many filters are compound specific. (3) A unit process containing a small-diameter medium, such as sand,

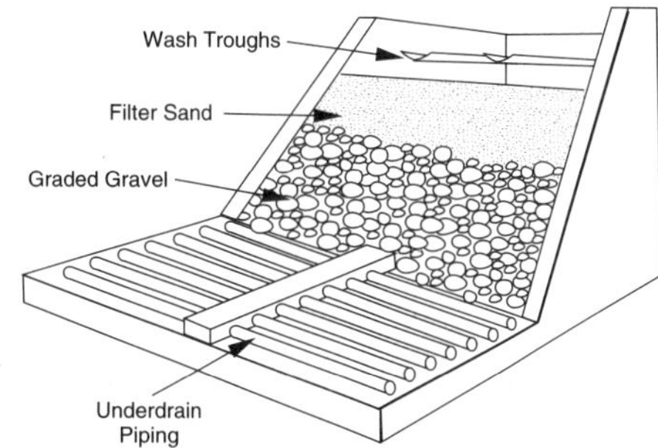

Filter (single medium)

that is designed to remove particulate matter from a liquid stream. Filters may operate by gravity or by externally applied pressure.

filter agitation A method used to achieve more effective cleaning of a filter bed. It usually involves using nozzles attached to a fixed or rotating pipe installed just above the filter media. Water or an air–water mixture is fed through the nozzles at high pressure to help agitate the media and break loose accumulated suspended matter. Filter agitation can also be called auxiliary scour or surface washing. *See also* surface wash.

filter aid An agent (such as diatomite) that improves filtering effectiveness in some way, such as by enhancing the retention of particles or increasing the permeability of the filter to water flow. A filter aid is either added to the suspensions to be filtered or placed on the filter as a layer through which the liquid must pass.

filter area The effective area, expressed in square feet or metres, through which water approaches the filter media. It is also called the surface area.

filter backwash A process in which retained particles on a filter are removed by reversing the flow direction. *See also* backwash.

filter backwash rate A measurement of the number of gallons (or litres) per unit time flowing upward (backward) through a square foot (or metre) of filter surface area. Mathematically, it is the backwash flow rate divided by the total filter area. It is typically expressed in gallons per square foot per minute or litres per square metre per second.

filter bed (1) A tank for water filtration that has a false filter bottom covered with granular media. (2) A pond with sand bedding, as in a sand filter or slow sand filter. (3) A type of bank revetment consisting of layers of filtering medium such that the particles gradually increase in size from the bottom upward. Such a filter allows the groundwater to flow freely, but it prevents even the smallest soil particles from being washed out.

filter bottom The underdrainage system for collecting the water that has passed through a rapid granular filter and for distributing the wash water that cleans the filtering medium.

filter box A rectangular filtering unit of a rapid granular filter plant.

filter cake The layer of fine material that deposits on the upgradient (upstream) side of any filter. Over time, the filter cake increases the head loss and decreases the performance of the filter. Sometimes the filter cake itself is the desirable end product of filtration. Filter cake formation can be problematic in cases where the filter cannot be cleaned easily, e.g., filter packs around well screens. The term *filter cake* is most often used to describe the dewatered product from a filter press. *See also* belt filter press; filter press; plate-and-frame filter press.

filter clogging The effect occurring when fine particles fill the voids of a rapid granular filter.

filter crib A water treatment works intake consisting of a wooden crib in an excavation in the bed of a stream, filled with gravel and covered with 3–4 feet (0.9–1.2 metres) of sand to the level of the streambed, into which a suction pipe is placed and connected to a pump on shore or elsewhere to draw filtered water for use as a potable supply.

filter drawdown The process of lowering the free water surface above filter media. Filter drawdown typically is performed prior to filter backwash.

filter efficiency The operating usefulness of a filter as measured by various criteria, such as percentage reduction in turbidity, bacteria, particles, *Giardia* cysts, or *Cryptosporidium* oocysts. It is sometimes expressed in terms of a log removal, with 90 percent removal being 1-log removal, 99 percent being 2-log removal, 99.9 percent being 3-log removal, and so on.

filter gallery A gallery provided in a water treatment plant for the installation of the conduits and valves and for a passageway to provide access to them. *See also* pipe gallery.

filter loading rate A measurement of the volume of water applied per unit time to each unit of filter surface area, i.e., flow rate into the filter divided by the total filter area. It is typically expressed in units of gallons per minute per square foot or metres per hour.

filter media The selected materials in a filter that form a barrier to the passage of filterable suspended solids. Filter designs include (1) loose media filters with particles lying in beds or loosely packed in column form in tank-type filters, or (2) cartridge-type filters that may contain membranes or fabric, fiber, bonded-ceramic, precoat, or cast solid-block filter media. The media used in some filters are chemically inert, such as sand, which performs only a mechanical filtration. Other filter media are multifunctional, chemically reactive media, such as calcite, granular activated carbon, magnesia, manganese dioxide (MnO_2), and manganese greensand.

filter operating table The table set in front of a filter on the operating floor of a rapid granular filtration plant. It supports all the equipment for the control and operation of that filter, usually including the loss-of-head gauges and rate-of-flow gauges.

filter pack A gravel or sand cylinder placed around the intake screen of a well to filter fines from the water before the water enters the well. It is used to protect the pump, as well as to prevent the well screen from clogging.

filter paper Paper with pore diameters usually between 5 and 10 micrometres, used to clarify chemical solutions, collect particulate matter, and separate solids from liquids.

filter plant The processes, devices, and structures used by a water treatment works for filtration of water.

filter press A piece of mechanical equipment designed to dewater sludge by applying pressure between two plates or belts. Dewatered sludge, or filter cake, is ultimately disposed of (e.g., via a landfill or acceptable alternative) and the filtrate is discharged to an acceptable disposal option or is treated for recycling. *See also* belt filter press; plate-and-frame filter press.

filter rate The rate of application of material to some process involving filtration, e.g., application of water to a rapid granular filter.

filter rate-of-flow controller A valve or orifice plate on the effluent line from a filter, used to control the flow rate through the filter. A valve can be modulated by a flow-measuring device to select the flow rate desired, whereas an orifice plate has a fixed opening and can control the maximum flow rate through the filter only when the filter is clean. *See also* Venturi meter.

filter rating *See* micron rating.

filter ripening A process by which granular media filter performance gradually improves at the beginning of a filter run as particles are deposited and act as collectors for subsequent particles applied to the filter. Filter ripening can last anywhere from a few minutes to more than an hour depending on the characteristics of the filter influent and the filter design. It is desirable for filters to ripen in less than 15 minutes because filtered water during the ripening process is often wasted.

filter run (1) The interval between the cleaning and washing operations of a rapid granular filter. (2) The interval between the changes of the filter medium on a sludge-dewatering filter.

filter sand Sand that is prepared according to detailed specifications for use in filters.

filter strainer A perforated device inserted in the underdrains of a rapid granular filter through which the filtered water is collected and through which the wash water is distributed when the filter is washed. It is sometimes called a strainer head.

filter tank The concrete or steel basin that contains the filter media, gravel support bed, underdrain, and wash water troughs.

filter underdrain A system in a filter designed to collect filtered water and evenly distribute backwash water. Typical underdrain designs include perforated pipe systems, precast or plastic blocks, strainers, and porous plates, among others. *See also* filter bottom.

filter wash The reversal of flow through a rapid granular filter to wash clogging material out of the filtering medium and relieve conditions causing loss of head. Filter wash is also called backwash.

filter washtrough A set of conduits, located above the filter media and open at the top, that are designed to collect filter

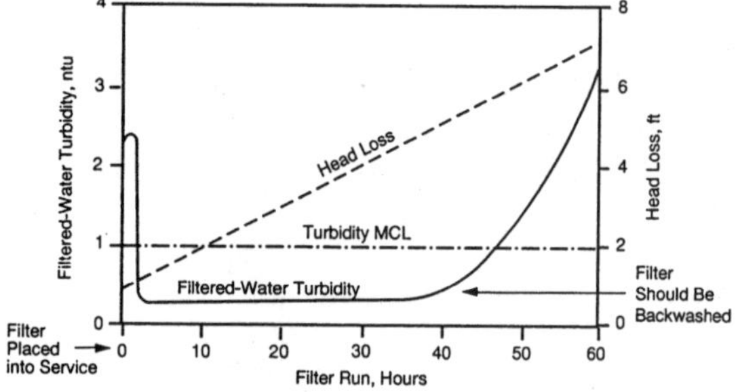

Performance over a filter run

backwash water. The bottoms of filter troughs should be at an elevation that permits the filter media to expand adequately during backwashing without coming in contact with the bottom of the trough.

filter, negatively charged *See* negatively charged filter.

filter, positively charged *See* positively charged filter.

filterable constituent *See* dissolved solids.

filterable residue *See* total dissolved solids.

filterable residue test *See* total dissolved solids test.

filtered-water reservoir A reservoir for the storage of filtered water in sufficient capacity to prevent the need for frequently varying the rate of filtration with variations in demands. It is also called a clearwell or clear-water reservoir.

filtering crucible A small porcelain container with holes in the bottom used in the total suspended solids test. It is also known as a Gooch crucible.

filtering medium Any material through which water is passed for the purpose of purification or treatment.

filter-pressed sludge Sludge that has been dewatered by being squeezed in filter presses.

filter-wash waste tank A reservoir or basin to collect filter wash water prior to further treatment and disposal.

filter-to-waste A practice of discharging filtered water directly to disposal immediately following backwashing. Filter-to-waste is typically performed for a period of 5 to 10 minutes following backwash, or until the filtered water is of acceptable quality. *See also* filter ripening.

filtrate The liquid that has passed through a filter.

filtration The removal of suspended materials in a fluid stream by passage of the fluid through a filter medium. *See also* water filtration plant.

filtration avoidance criteria for surface water Specific requirements established by the US Environmental Protection Agency that a public water system using surface water or groundwater under direct influence of surface water must meet to avoid installation of filtration as required by the Surface Water Treatment Rule.

filtration plant *See* filter plant.

filtration rate The hydraulic loading rate on a filter, typically expressed as a flow rate per unit area. It is typically expressed in units of gallons per minute per square foot or metres per hour.

filtration spring A spring that percolates from numerous small openings in permeable material. It is also called a seepage spring.

final control element (FCE) The actual device used to adjust a process. Motor speed controllers, chemical feeders, and control valves are common examples.

Final Environmental Impact Statement (FEIS) An environmental impact statement that has been reviewed and approved following the procedures established by regulatory agencies under the National Environmental Policy Act.

final rulemaking A rule promulgated by a US federal agency that has been prepared after consideration of public comments submitted in response to a proposed rulemaking.

Finding of No Significant Impact (FONSI) A formal determination by the US Army Corps of Engineers, based on an environmental assessment, that an action produces no significant impact on the quality of the human environment, thereby avoiding the requirement for an Environmental Impact Statement.

fine rack Generally, a rack that has clear spaces of 1 inch (2.5 centimetres) or less between its bars.

fine sand Sediment particles having diameters between 0.125 and 0.250 milllimetres.

fines The finer-grained particles of a mass of soil, sand, or gravel.

finger levee An L-shaped berm built alongside a stream channel to impound water but not entirely stop the flow of the stream. Such structures are commonly used in artificial groundwater recharge activities.

finished water Water that has passed through a water treatment plant, such that all the treatment processes are completed, or finished. This water is ready to be delivered to consumers. It is also called product water.

finite resource A natural resource, such as coal, oil, or water, that is definable and measurable.

finite-difference model A particular kind of digital computer model, typically based on a rectangular grid that sets the boundaries of the model and the nodes where the model will be solved. The mathematical expressions are usually based on truncated Taylor series expansions.

finite-element model A particular kind of digital computer model, typically based on a mesh formed of a number of polygonal cells or elements. The mathematical expressions are usually based on integrated descriptions of the modeled process.

fire cistern A cistern constructed for storing water to be used in fighting fires. The water may be stored to supplement the distribution system or to be used as an emergency supply in case the water mains serving the area should break.

fire demand The required fire flow and the duration for which it is needed, usually expressed in gallons or litres per minute for a certain number of hours. This term can also be used to denote the total quantity of water needed to deliver the required fire flow for the specified number of hours.

fire demand rate The rate of flow, usually expressed in gallons or litres per minute, that is needed at a specified residual pressure for fire fighting at a particular location or in a certain area. *See also* required fire flow.

fire department connection An inlet equipped with one or more couplings to which a fire hose can be attached and through which water can be delivered by fire department pumpers to sprinkler, standpipe, yard main, or other fire protection systems.

fire extinguisher (type A, B, or C) A device used to put out fires. Fire extinguishers are classified as to the types of fires they should be used on: A, trash, wood, and paper; B, liquids and grease; and C, electrical equipment.

fire flow The rate of flow, usually expressed in gallons or litres per minute, that can be delivered from a water distribution system at a specified residual pressure for fire fighting. When delivery is to fire department pumpers, the specified residual pressure is generally 20 pounds per square inch (138 kilopascals).

fire flow demand The quantity of water required specifically for fire fighting, generally determined by a fire department.

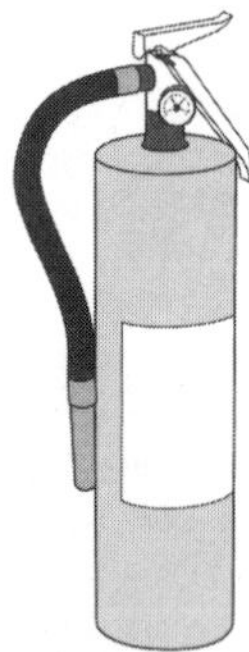
Fire extinguisher

fire hydrant A specifically designed fitting that serves as a connection point on a distribution system to provide water for fire protection directly through a fire hose or to a fire pumper.

fire line service charge A fire line connection charge that is separate from the domestic water charge based on the cost of service for providing water to the fire line. Fire lines are normally not metered; therefore, the cost is proportioned out according to the size of the fire line service per month or by other means of applying the cost, such as inch-foot (diameter and length) of the fire line.

fire pressure The pressure necessary in water mains at those times when water is used for fire fighting. This term applies to cases in which the pressure for fire fighting is increased to greater than the pressure normally maintained for general use.

fire protection The ability to provide water through a distribution system for fighting fires in addition to meeting the normal demands of water usage.

fire protection charges Charges made to recover the cost of providing both public and private fire protection service to the communities served by a utility. Such charges typically include the direct capital-related and maintenance costs for fire hydrants and private connections, as well as applicable indirect costs for supply, treatment, transmission, and distribution of water to the fire protection facilities.

fire pump A stationary pump installed solely to provide water for fire protection. Such a pump may be used to supply sprinkler, standpipe, yard main, and other types of fire protection systems.

fire service A connection to a water distribution system to provide water for a private fire sprinkler or fire protection system.

fire service connection A pipe extending from a main to supply a sprinkler, standpipe, yard main, or other fire protection system.

fire service detector check meter A special device for use on fire service connections consisting of a weighted check valve with a disk meter in a bypass. The disk meter measures small rates of flow only; the weighted check valve opens for large rates of flow so that the loss of head is relatively small, but these large rates of flow are not measured.

fire service meter A special type of meter causing a relatively small loss of head at high rates of flow, for use on fire service connections.

fire stream The stream of water issuing from a nozzle fastened to a fire hose that is used to extinguish fire. *See also* standard fire stream.

fire supply In municipal use, the quantity of water required for extinguishing fires, in addition to that required for domestic, industrial, and public use.

fire system A separate system of water pipes or mains and their appurtenances installed solely to furnish water for extinguishing fires.

fireplug An early form of fire hydrant, originally a wooden plug used to stop a hole bored through a wooden water pipe to allow water to be withdrawn for fire fighting. The term is still used to denote a fire hydrant.

firewall Computers set up on a network to prevent unauthorized access.

firm annual yield *See* firm yield.

firm pumping capacity The available capacity of a pumping system when the largest unit is out of service.

firm yield The ability of a reservoir, a water supply utility, or both to supply water.

first draw The water that immediately comes out when a tap is first opened. This water is likely to have the highest level of lead and copper contamination from plumbing.

first-draw residential lead sample A tap sample required by the Lead and Copper Rule consisting of the first 1 litre of water drawn from either a residential cold-water kitchen or bathroom faucet, where the water sampled has remained motionless in the pipe and faucet for at least 6 hours prior to sampling.

first-draw sample *See* first-draw residential lead sample.

first-draw tap sample *See* first-draw residential lead sample.

first-order kinetics *See* reaction order.

first-pass effect The fraction of a chemical that is cleared from the blood by a single pass through the liver. Liver metabolism can substantially reduce the amount of an ingested chemical that is introduced into the systemic circulation. This reduction occurs because the blood in the capillaries that contact the small intestine empty into the portal vein that takes the blood to the liver before it is returned to the heart.

fish bioassay Exposure of fish to various levels of a chemical under controlled conditions to determine safe and toxic levels of the test chemical.

fish elevator A type of fishway permitting fish to pass upstream around and over a dam. *See also* fish ladder.

fish ladder An inclined trough carrying water from above to below a dam at a velocity against which fish can easily swim. Various types exist, some with baffles to reduce the velocity of the water and some consisting of a series of boxes with water spilling down from one to the next. Their installation at dams is required in many states to allow fish to pass upstream to reach their spawning grounds.

fish screen A screen placed across the head of or inside an intake canal or pipeline to prevent fish from entering. In some parts of the United States, installations of such screens are required at head gates on all streams that fish inhabit.

fish toxicity (1) Impairment of fish development during one or more life stages as a result of prolonged exposure to a toxic material at sublethal levels. (2) The immediate death of fish exposed to a material at acutely toxic concentrations.

fishery A place where fish live and are able to reproduce.

fishway A device constructed in connection with a dam and usually consisting of a series of pools, one above the other, with low falls between. It allows fish to pass upstream and downstream over the dam. Fishways are necessary on many streams where fish periodically migrate upstream to the headwaters to spawn and the young fish come downstream again. One form is called a fish elevator, another a fish ladder. *See also* fish ladder.

fission The splitting of a heavy nucleus into two roughly equal parts (which are nuclei of lighter elements), accompanied by the release of a relatively large amount of energy and frequently one or more neutrons. Fission can occur spontaneously but is usually caused by the absorption of gamma rays, neutrons, or other particles. *Contrast with* fusion.

fissure spring A spring that issues from a large fissure. *See also* fault spring.

fissure water Water in open fractures, usually abundant only near the ground surface.

fissured soil Soil that has little cohesion and a preference to break along definite lines of fracture with little opposition (type C soil).

fitting A general classification of piping materials that join pipes together or connect to a pipe, including such items as tees, bends, wyes, corporations, plugs, caps, unions, and reducers. *See also* pipe fitting.

fittings, class *See* pipe class.

500 series methods A group of analytical methods published by the US Environmental Protection Agency for the analysis of drinking water.

fix To add chemicals in the field that prevent the water quality indicators of interest in a sample from changing before final measurements are performed later in the lab. *See also* sample preservation.

fixed asset A piece of permanent property, such as land, buildings, machinery, equipment, rights, and benefits (tangible and intangible), that is permanently used in the rendering of a service or in the production of a product.

fixed bed (1) The filter or ion exchange medium retained in a vessel. (2) A media bed that is "contained," that is, filled to the top or to the restraining barrier with filter media and is not capable of being expanded during backwashing.

fixed-bed column A treatment unit containing media that remain stationary during the course of treatment. Once the treatment capacity is exhausted, the media may be fluidized to remove contaminants or replaced. Examples of fixed-bed columns are gravity and pressure filters, ion exchange columns, and granular activated carbon filter columns. The water to be treated typically flows downward through the bed.

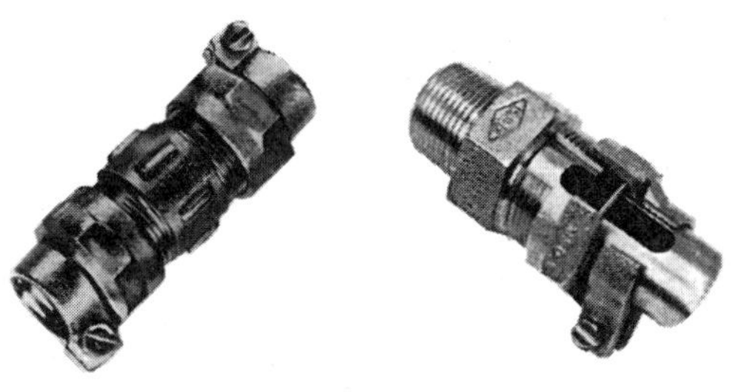

Compression fittings

fixed capital The investment represented by fixed assets.

fixed charge The carrying and operating cost of any business or project that continues to occur whether or not the business operates or produces anything. (2) A charge that cannot be escaped, shifted, or altered, such as interest, rent, taxes, and amortization.

fixed cost In economics, a cost that does not vary with the amount of product produced. Examples are rent, insurance, taxes, and most general and administrative expenses, such as executive salaries, personnel department costs, and public relations costs. *See also* breakeven analysis; semivariable costs; variable costs.

fixed dam A type of dam that has a fixed elevation for the outlet or overflow and has no provision such as gates for allowing releases. A fixed dam may have a movable crest arrangement, such as flashboards or tainter gates, to change elevation of the water surface. *See also* flashboard; tainter gate.

fixed distributor A distributor consisting of perforated pipes, notched troughs, sloping boards, or sprinkler nozzles, all of which remain stationary when the distributor is operating. *See also* distributor.

fixed film A type of microbiological population in which the microorganisms are attached to a medium through which a water to be treated flows. For example, a biologically active filter is a fixed film process that removes biodegradable contaminants. *See also* biological treatment processes.

fixed groundwater Water in saturated rocks so fine grained that the water is assumed to be permanently attached to the rock particles. Although the occurrence of water in such rock bodies is not fully understood, most of these formations in the zone of saturation will permit slow movement of water under a hydraulic gradient, although their well yields will be negligible.

fixed liability A liability that is to be paid 1 year or more after the date of a balance sheet.

fixed matter The residue (particulate material, dissolved material, or both) that remains behind (i.e., is immovable or fixed) despite action to remove it, e.g., the residue remaining after heating or burning a substance to drive off the volatile solids. *See also* fixed solids.

fixed moisture Moisture held in the soil below the hygroscopic limit.

fixed nozzle An immovable nozzle used in filter backwash to introduce backwash water or surface wash. Fixed nozzles can be connected to piping in the gravel support layer of a rapid granular filter to provide even distribution of backwash water over the entire area of the filter. Fixed nozzles also can be attached to a piping grid above the filter media to provide surface wash during the backwash cycle. *See also* backwash; rotary surface washer; surface wash.

fixed point arithmetic A method of storing and calculating numbers that makes use of a fixed location for the decimal point in the field of each number. Early computer programmers experienced difficulty in utilizing and keeping track of a

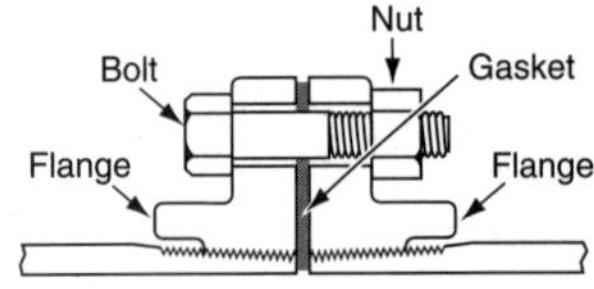

Flanged joint

fixed decimal; thus, almost all computer programmers now utilize floating point notation. *See also* floating point arithmetic.

fixed solids The residue of total suspended solids, dissolved solids, or both after ignition (burning) or heating of a water sample for a specified time at a specified temperature. This term is used in the laboratory analysis of the solids content of water. *See also* fixed matter; volatile solids.

fixed-grade node A node within a model network at which the pressure is specified. It is also known as a supply node.

fixture branch In plumbing, the water supply pipe between a fixture supply pipe and the water distribution pipe.

flagellate A microorganism that moves by the action of a tail-like projection.

flame arrester A device incorporating a fine-mesh wire screen or tube bundle inserted in a vent or pipe and designed to resist the flashback of flame.

flame atomic absorption spectrophotometry An instrumental technique used for the analysis of metals in which the sample is atomized in a flame. This approach is typically used for low milligrams per litre concentrations.

flame ionization detector (FID) A device used to produce a response from organic compounds as they elute from a gas chromatographic column. This electrical signal is converted into peaks on a chromatogram, which can be used in the identification and quantification of organic compounds. The detector is useful in water analysis, because it responds to most organic compounds and does not respond to water.

flame polished Melted by a flame to smooth out irregularities. Sharp or broken edges of glass (such as the end of a glass tube) can be rotated in a flame until the edge melts slightly and becomes smooth.

flaming The act of passing a flame over the end of a faucet in order to kill bacteria before taking a water sample for bacteriological sampling. The procedure is no longer recommended because it may damage the faucet.

flammable Pertaining to substances that burn easily, strongly, and at a rapid rate.

flammable liquid A liquid that has a flash point at a temperature less than 100° Fahrenheit (37.8° Celsius).

flange A projecting rim, edge, lip, or rib.

flanged joint A pipe joint made by flanges bolted together.

flanged pipe A pipe provided with flanges so that the ends can be joined together by means of bolts.

flap gate A gate that opens and closes by rotation around one or more hinges at the top side of the gate.

flap valve A valve that is hinged at one edge and that opens and shuts by rotating about the hinge. *See also* check valve.

flaring inlet A converging, funnel-shaped, or bellmouthed entrance to a pipe or conduit, used to facilitate entry of water. *See also* bellmouth.

flaring tool A shaping tool used to enlarge or flare the end of copper tubing to make a connection. To create and shape a flare, a flare nut is placed on the tubing and the flaring tool is hammered into the end of the tubing. The flare nut is then compressed against the flare as the flare is tightened onto a connecting fitting, making a watertight connection that will not pull apart under pressure.

flash distillation A distillation process in which hot incoming water at a high enough pressure to prevent boiling enters a chamber that is at reduced pressure, causing some of the water to "flash" (i.e., evaporate quickly into steam). For seawater desalting, multistage flash distillation is commonly used.

flash dryer A furnace used for drying sludge in which small particles of partly dewatered sludge are sprayed into a stream of air heated to the ignition temperature of the sludge's dry solids content.

flash evaporator A device used for flash distillation. *See also* flash distillation.

flash flood A flood of short duration with a relatively high peak rate of flow, usually resulting from a high-intensity rainfall over a small area.

flash mixer A device for quickly dispersing chemicals uniformly throughout a liquid.

flash mixing A method of mixing chemicals into solution by imparting a high velocity gradient to them, typically by mechanical means. In water treatment, flash mixing is typically associated with the addition of coagulant or softening chemicals at the beginning of the treatment process. *See also* velocity gradient.

flash point The minimum temperature at which a liquid gives off enough vapor to mix with air and allow a flame when the vapor is ignited.

flash range The difference between the maximum brine temperature and the temperature of the brine blown down from the last stage of a multistage flash evaporator.

flashboard A temporary barrier of relatively low height and usually constructed of horizontal wooden planks, placed along the crest of the spillway of a dam to allow the water surface in the reservoir to be raised above the spillway level to increase the storage capacity. It is constructed so that it can be readily removed or lowered or carried away by high flow or floods.

flashboard check gate A balanced crest gate that opens under a certain pressure on the flashboards that would otherwise cause the pins holding the flashboard in place to rupture.

flashy stream A stream in which flows collect rapidly from the steep slopes of the catchment, thereby causing flood peaks to occur soon after a rain. The flows in such a stream usually subside as rapidly as they increase. Rapidly varying turbidity, particle count, and particle size are also characteristics of this type of stream.

flask A container, often narrow at the top, used for holding liquids. Many types of flasks exist, each with its own specific name and use.

flat A level surface with little or no change in elevation; a level tract along the banks of a river.

flat fee A set fee charged to a water customer regardless of the amount of water used in a month. For example, a customer may be charged $10.00 a month for water, no matter how much is used. The fee may be based on a number of conditions, such as size of service pipe, number of water fixtures or outlets, land frontage, or other arbitrary measures, usually fixed by law.

flat rate *See* flat fee.

flat slope A conduit slope less than the critical slope for a particular discharge. Such a slope is also called a mild slope.

flat-crested weir A weir for which the crest is (1) horizontal in the direction of flow and (2) of appreciable length when compared with the depth of water passing over it.

flat-sheet membrane A membrane manufactured in flat-sheet form. Flat-sheet pressure-driven membranes used for water treatment are commonly assembled into spiral-wound or plate-and-frame configurations. *See also* membrane element, plate-and-frame membrane configuration; spiral wound.

flavor A combination of taste-and-odor sensations experienced when a sample is taken into the mouth. Flavor results from stimulation of the taste buds and the olfactory bulb. *See also* odor; taste.

flavor profile A systematic analysis of a water sample for the characteristics of taste or odor. The approach uses a panel of participants that have been trained and "calibrated" according to a protocol. Results from a flavor profile analysis can be helpful in the operation of a water treatment plant and in investigating customer inquiries.

flexible coupling (1) A joint or connection in a pipeline that will allow movement in the pipeline without failure in the joint. Such a coupling is sometimes called fault protection. (2) A section of pipe made from material that is formed, shaped, or manufactured to bend easily without failure. (3) A connection that transmits load or power between the shafts of a motor and pump and that allows for alignment variations caused by temperature changes.

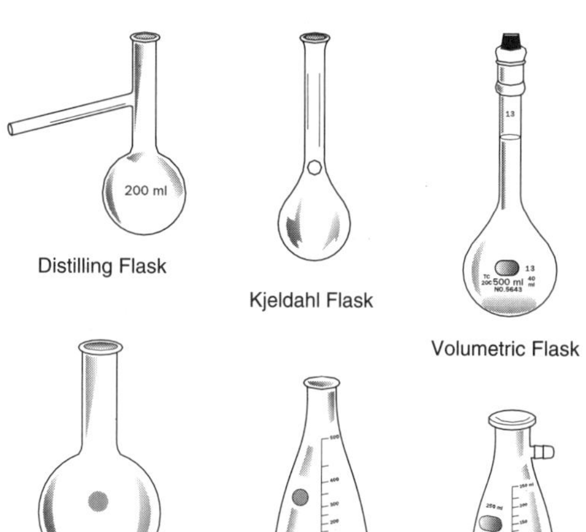

Flasks

flexible cover A cover on a process unit constructed in such a manner that the material assumes the shape of the volume to be contained. Flexible covers often are constructed of synthetic materials and are used on a finished water reservoir, "floating" on the surface of the water contained in the reservoir to protect the water from contamination.

flexible joint Any joint between two pipes that permits one of them to be deflected without disturbing the other pipe.

float The concentrated solids collected at the surface of a dissolved air floatation treatment unit. *See also* dissolved air floatation.

float control A float device, triggered by changing liquid levels, that activates, deactivates, or alternates equipment operation.

float gauge A device for measuring the elevation of the surface of a liquid. The actuating element of a float gauge is a buoyant float that rests on the surface of the liquid and rises or falls with it. The elevation of the surface is measured by a chain or tape attached to the float.

float gauging The act of measuring the velocity of water flowing in a stream or conduit by means of floats placed in the water.

float run The fixed distance over which any type of float is timed.

float switch An electrical switch operated by a float in a tank or reservoir and usually controlling the motor of a pump.

float tube (1) A tube for which the tube opening is held a certain level below the surface of the water by a float that goes up and down with the water level. (2) A vertical pipe connected to the wet well of a pumping station, in which a float is placed to actuate a float switch for controlling the pumps.

float valve A valve in which the closure to an opening, such as a plug or gate, is actuated by a float to control the flow into a tank.

floating pan An evaporation pan floating on a body of water.

floating point arithmetic A method of calculating that uses number quantities represented by a number called the mantissa in conjunction with a power or exponent of the number base. The actual number is obtained by multiplying the mantissa by the power of the number base. For example, 0.628475×10^2 is a floating point number; 62.8475 is the fixed point version of the same value. Computer programmers want to keep track of the decimal point easily; they want it in the same place every time. Hence, they prefer floating point arithmetic. Floating point arithmetic is also easier to use for large numbers. *See also* fixed point arithmetic.

floating on the system A method of operating a water storage facility such that daily flow into the facility approximately equals the average daily demand for water. When consumer demands for water are low, the storage facility will be filling. During periods of high demands, the facility will be emptying.

floating weed An aquatic plant, such as a water lily, that entirely or in part floats on the surface of the water.

floc Collections of smaller particles that have come together (agglomerated) into larger, more settleable particles as a result of the coagulation–flocculation process.

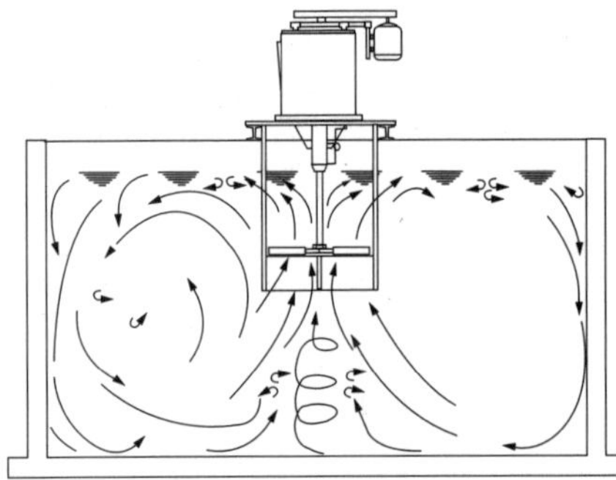

One type of flocculator, with mixing energy provided by a turbine mechanism

floc density A characteristic of floc that relates to settling velocity and volume of floc. The higher the floc density, the more rapidly the floc particles will settle and the lower the resulting sludge volume. The higher the ratio of particles (turbidity) removed to the coagulant dosage, the higher the floc density.

flocculant A water-soluble organic polyelectrolyte that is used alone or in conjunction with inorganic coagulants, such as aluminum or iron salts, to agglomerate solids present in water to form large, dense floc particles that settle rapidly. *See also* coagulant.

flocculant aid *See* coagulant aid.

flocculant settling The settling of flocculent particles in dilute suspension. The particles flocculate during settling; thus, they increase in size and settle at a faster velocity as time passes. *See also* type II settling.

flocculating agent *See* flocculant.

flocculating tank A tank used for the formation of floc by the gentle agitation of liquid suspensions, with or without the aid of chemicals.

flocculation The water treatment process following coagulation that uses gentle stirring to bring suspended particles together so they will form larger, more settleable clumps called floc.

flocculation agent A coagulating substance that, when added to water, forms a flocculant precipitate that will entrain suspended matter and expedite sedimentation. Examples are alum ($Al_2(SO_4)_3 \cdot 14H_2O$), ferrous sulfate ($FeSO_4$), and lime ($Ca(OH)_2$). *See also* coagulant.

flocculation limit The water content of a soil when it is in the condition of a deflocculated sediment (i.e., with individual soil particles configured generally parallel to each other microscopically).

flocculation ratio The void ratio of a soil when it is in the condition of a deflocculated sediment.

flocculator A device to enhance the formation of floc in a liquid. Mixing energy can be provided by mechanical means or head loss.

flood (1) A relatively high flow as measured by either gauge height or discharge quantity. (2) Any flow equal to or greater than a designated basic flow.

flood basin The part of a river valley that is outside the natural stream bank and is subject to flooding, particularly where the bank is higher than the valley floor and where a basin is formed between the bank and the sidehill.

flood benefit The value of a proposed flood protection works as estimated in terms of damage that would be avoided or other advantageous effects.

flood control The use of structural and nonstructural methods intended to reduce or eliminate the effects of flood events on people, buildings, or property.

flood control storage Storage of water during floods for later release as soon as channel capacities permit.

flood control works Structures and reservoirs constructed to reduce the flood peaks on streams subject to damaging floods.

flood event A series of flows of water in a stream constituting a distinct progressive rise and culminating in a peak, crest, or summit, together with the recession that follows the peak or crest.

flood flow The discharge of a stream during periods of flood.

flood forecasting The prediction of stream flow discharge events of specific flow rates. Forecasting takes into account drainage basin characteristics, rainfall duration and frequency information, and soil characteristics to predict the magnitude and frequency of flooding events.

flood frequency The frequency with which the maximum flood may be expected to occur at a site in any average interval of years. Frequency analysis defines the *n*-year flood as the flood that will, over a long period of time, be equaled or exceeded on the average once every *n* years. Thus, the 10-year flood would be expected to occur approximately 100 times in a period of 1,000 years; of these, 10 would be expected to reach the 100-year magnitude (i.e., they would be 100-year floods). Flood frequency is sometimes expressed in terms of a percentage of probability. For example, a probability of 1 percent would correspond to a 100-year flood; a probability of 10 percent would correspond to a 10-year flood. *See also* flood probability.

flood level The stage of a stream at a time of flood.

flood level rim That level from which liquid in plumbing fixtures, appliances, or vats could overflow to the floor when all drain and overflow openings built into the equipment are obstructed.

flood peak The maximum rate of flow, usually expressed in units of volume per unit time, that occurred during a flood.

flood probability The probability of a flood of a given size being equaled or exceeded in a given period. A probability of 1 percent corresponds to a 100-year flood; a probability of 10 percent corresponds to a 10-year flood. *See also* flood frequency.

flood protection works Structures built to protect lands and property from damage by floods.

flood relief channel A channel constructed to carry flood water in excess of the quantity that can be carried safely in the stream. Such a channel is also called a bypass channel or floodway.

flood routing The process of progressively determining the timing and shape of a flood wave at successive points along a river. *See also* flow routing; routing.

flood spreading The flooding of gravelly or otherwise relatively pervious lands to recharge a groundwater basin. *See also* spreading; water spreading.

flood source area That portion of a drainage basin where conditions of precipitation and cover, topography, or land use favor frequent flooding.

flood stage An arbitrarily fixed and generally accepted gauge height or elevation above which a rise in the water surface elevation is termed a flood. It is commonly fixed as the stage at which overflow of the normal banks or damage to property would begin.

flood wave A rise in stream flow to a crest in response to runoff generated by precipitation, as well as the stream flow's subsequent recession after the precipitation ends.

flooded Pertaining to a condition in which the soil surface is temporarily covered with flowing water from any source, such as streams overflowing their banks, runoff from adjacent or surrounding slopes, inflow from high tides, or any combination of sources.

floodplain (1) The area described by the perimeter of the probable limiting flood. (2) That portion of a river valley that has been covered with water when the river overflowed its banks at flood stage.

floodway *See* flood relief channel.

flora The plants of a particular region or period considered as a whole.

flotation A method for removing coagulated particles by introducing air to attach and float particles to the surface. Collected particles are skimmed and discharged to solids processing. *See also* dissolved air flotation.

flotsam Debris from natural sources or human activity that floats on oceans, lakes, and rivers. *See also* jetsam.

flow (1) The movement of a stream of water or other fluid from place to place; movement of silt, water, sand, or other material. (2) A fluid that is in motion. (3) The quantity or rate of movement of a fluid; the discharge; the total quantity carried by a stream. (4) To issue forth or discharge. (5) The liquid or amount of liquid per unit time passing a given point. *See also* discharge rate; Newtonian flow; Reynolds number.

flow augmentation The release of water stored in a reservoir or other impoundment to increase the natural flow of a stream.

flow cell A sensing element or combination of elements, such as electrodes, that is immersed in a flowing fluid and continuously measures some property of the fluid, such as pH, dissolved oxygen, or electrical conductivity.

flow control valve A cylindrical pressure-compensating valve installed to regulate the flow of water. Such a valve is rated in gallons or cubic metres per minute or day. *See also* filter rate-of-flow controller.

flow controller An in-line device or orifice fitting in a treatment device's piping that will regulate and control the flow of water or regenerant over a broad range of inlet water pressures. Some types are manually adjustable.

flow cytometry A method for measuring and then analyzing the signals that result as particles flow in a liquid stream through a beam of light. The instrument used for this method is known as a flow cytometer.

flow demand The flow required to satisfy demands on a system, such as fire demand.

flow diffuser A device to distribute flow and promote blending of two fluid streams.

flow duration curve A duration curve of stream flow. *See also* duration curve.

flow equalization The use of storage tanks to control a changing flow of water and make it nearly uniform with time.

flow index The slope of a flow curve of a standard liquid-limit test of soils, plotted with water content as the ordinate on an arithmetic scale and number of flows as the abscissa on a logarithmic scale. The flow index is also called the fluvial index.

flow injection analysis (FIA) A technique used in water analysis to automate some tests that are often conducted by traditional chemical methods. A water sample is injected into a continuously flowing stream of liquid, where the sample disperses in a reproducible manner. Reagents are mixed into the flowing stream and the solution flows through a detector, where the analyte is measured. In addition to automating traditional chemical methods, flow injection techniques tend to consume small volumes of reagents and sample.

flow level control The use of gates, dams, or other barriers in a flume, partially filled pipe, or stream to maintain the desired water depth in the channel.

flow line A path that fluid particles follow in a fluid system; e.g., the flow lines to a well are radial toward the well.

flow measurement A measurement of the quantity of water flowing through a given point in a given amount of time, such as gallons or cubic metres per day.

flow net A set of intersecting equipotential lines and flow lines representing two-dimensional steady flow through porous media.

flow rate A measure of the volume of water moving past a given point in a given period of time. *See also* average flow rate; instantaneous flow rate.

flow recording The process of documenting the quantity or rate of the flow of a fluid past a given point. The recording is normally accomplished automatically.

flow regulator A structure installed in a canal, conduit, or channel to control the flow of water at an intake or to control the water level in a canal, channel, or treatment unit. *See also* filter rate-of-flow controller; regulated flow.

flow routing A mathematical procedure for predicting the changing rate of flow as a function of time at various points along a watercourse. *See also* flood routing.

flow test (1) A measurement of water flowing in a distribution pipe, usually accomplished with a pitometer. (2) The measurement of water discharging from a fire hydrant at a certain pressure to determine the quantity of water available from the

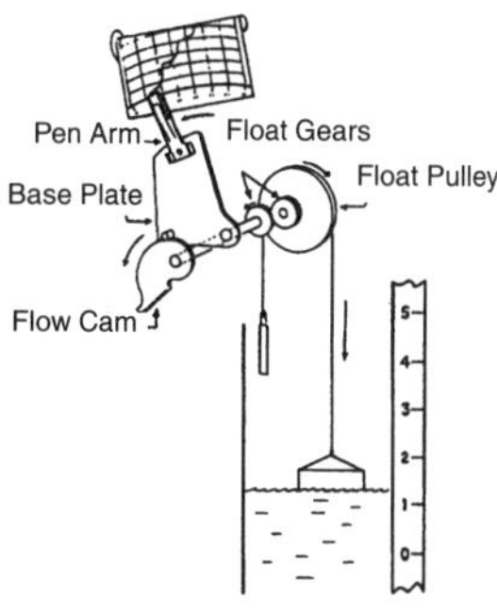

Flow recording

hydrant for fire fighting. (3) The measurement of water discharged from a pump at different heads to establish a flow curve for the pump.

flow tube One type of primary element used in a pressure differential meter. It measures flow velocity based on the amount of pressure drop through the tube. It is similar to a Venturi tube.

flow yield *See* recovery.

flow, steady *See* steady flow.

flow, unsteady *See* unsteady flow.

flowage The movement of liquid through a series of structures, such as a treatment plant. *See also* flow line.

flow-balancing tube For reverse osmosis water-desalting systems using multiple, parallel hollow fiber permeators, the specifically sized tubing or pipe section from each individual permeator that provides the needed pressure drop to equalize concentrate flow rates from each permeator to help balance permeate recovery from individual permeators.

flowchart A description of logical steps in a computer algorithm. *See also* flowsheet.

flowing sample A well-flushed sample, normally one that is collected from the water distribution system after the sample line has been flushed. Coliform sampling is normally done on a flowing sample. Flowing samples are normally used to gather general water quality characteristics about the system. *Contrast with* first-draw residential lead sample; standing sample.

flowing well A well that penetrates an aquifer in which the water has sufficient energy to make water flow from the well without pumping. *See also* artesian well.

flow-line aqueduct An aqueduct placed at such an elevation that the hydraulic gradient is lower than the crown of the aqueduct, so that the aqueduct never flows full.

flowmeter An instrument used to measure volume, pressure, and flow rate for water and chemicals in a liquid or gaseous state in a pipe or system.

flow-nozzle meter A water meter of the differential-medium type in which the flow through the primary element (nozzle) produces a pressure difference (or differential head), which the secondary element (float tube) then uses as an indication of the rate of flow.

flow-paced feed system A system for feeding chemicals in which the rate of application is proportional to the flow into which the chemicals are being fed. In this manner, the concentration of the chemical in the flow stream remains constant as flow changes.

flow-proportional composite *See* flow-proportional composite sample.

flow-proportional composite sample A sample obtained by (1) continuous pumping at a rate proportional to the flow, (2) mixing equal volumes of water collected at time intervals inversely proportional to the volume of flow, or (3) mixing volumes of water that are proportional to the flow and were collected during or at regular time intervals. This approach produces a sample that is representative of true conditions over the sampling time period.

flow-proportional control A method of controlling chemical feed rates by having the feed rate increase or decrease as the flow increases or decreases.

flowsheet A diagrammatic representation of the progression of steps in a process, showing their sequence and interdependence. A flowsheet is sometimes also called a flowchart.

flow-through time (1) The time required for a volume of liquid to pass through a basin, identified in terms of the time characteristic being measured, such as mean time, modal time, or minimum time. These various times may be estimated from tracer recovery curves. Dyes, electrolytes, and radioactive substances are used as tracers. The T_{10} time—the time at which 10 percent of the tracer has passed through the system—is important in calculating the $C \times T$ values for disinfection. (2) The average time required for a small volume of liquid to pass through a basin from inlet to outlet. *See also* $C \times T$; disinfectant contact time.

fluid (1) A substance having molecular particles that easily move and change their relative positions without separation of the mass and that easily yield to shear stress. (2) Capable of flowing; liquid or gaseous. *See also* gas; liquid.

fluid mechanics The science of motion of fluids, based on physical analysis and experimental verification.

fluid potential A measure of the energy per unit weight of water. It is often used to calculate the rate of water movement through some conduit. Fluid potential normally includes velocity potential, pressure potential, and gravitational (elevation) potential.

fluid velocity log A set of flow rate measurements in a borehole as a function of depth. This type of log can be interpreted to determine which strata in an aquifer can produce economically useful quantities of water.

fluidization The process of suspending a particulate medium such that the particles are mobile and are not continually in contact with each other. For example, filter media are fluidized during backwashing to remove entrapped particulates. *See also* backwash.

fluidized Pertaining to a mass of solid particles that is made to flow like a liquid by injection of water or gas. In water treatment, a bed of filter media is fluidized by backwashing water through the filter. *See also* backwash.

fluidized bed (FB) reactor A pressure vessel or tank that is designed for liquid–solid or gas–solid reactions. The liquid or gas moves upward through the solid particles at a velocity sufficient to suspend the individual particles in the fluid. Ion exchangers, granular activated carbon adsorbers, and some types of furnaces and kilns use fluidized bed reactors.

fluidized-bed granular activated carbon adsorption An adsorption process in which an upward flow maintains an approximate 10 percent expansion of the granular activated carbon media. This allows for the passage of suspended particles to minimize head loss and for extended operation without backwashing. A loss of fine granular activated carbon particles in the effluent is often a problem.

flume An open conduit of wood, masonry, or metal constructed on a grade and sometimes elevated. It is sometimes called an aqueduct.

fluorescein (resorcinolphthalein, $C_{20}H_{12}O_5$) An orange-red compound that exhibits intense fluorescence in alkaline solutions and is used to dye water in order to trace the water's course and movement. It is relatively harmless in low concentrations (though it will still fluoresce).

fluorescence The emission of light by excited molecules. This phenomenon is used as the basis for detecting certain compounds, such as symmetrically conjugated organic compounds like benzene. *See also* immunofluorescence.

fluorescence detector A device used in the analysis of trace organic compounds. It is composed of excitation and emission filters that can be specified or changed for the detection of specific types of compounds. These detectors are very sensitive to certain types of organics, such as aromatic compounds. Fluorescence detectors have been used in water analysis to determine the presence of certain compounds interfaced with a high-performance liquid chromatograph. *See also* fluorimeter.

fluorescence immunofluorescence *See* immunofluorescence.

fluorescence microscope A microscope used for observing and studying microscopic specimens that can absorb light (exciting light) of one wavelength and emit light (emitted light) of a longer wavelength. The longer-wavelength light emitted is called fluorescence.

fluorescent dye A substance used for measuring a feeble water current and determining the path of movement of currents and sediment. Such a dye must be stable, have a color strikingly different from that of the water, be harmless and cheap, and have the property of emitting radiation as a result of absorption of incident irradiation generated in a measuring device. *See also* fluorescein.

fluoridation The process of adding fluoride to water to help prevent tooth decay.

fluoride ion (F^-) A halide ion. Fluoride salts are added to drinking water for fluoridation. Fluoride is regulated by the US Environmental Protection Agency. *See also* fluoridation; halide.

Fluoride Rule The national primary drinking water regulation set for fluoride by the US Environmental Protection Agency.

fluorimeter An instrument that measures the visible light emitted when a fluorescent substance is irradiated with ultraviolet light. In water studies, fluorescent dyes are frequently used as tracers.

fluorochrome A chemical dye—e.g., fluorescein (resorcinolphthalein, $C_{20}H_{12}O_5$) and rhodamine B($C_{28}H_{31}ClN_2O_3$)—that emits colored light when excited by an appropriate wavelength of light. These dyes can be coupled readily to antibody molecules to form conjugated fluorescent dyes that can be used to specifically stain the antigenic reactive sites of the target microorganisms. These organisms can then be visualized using a fluorescence microscope with an excitation wavelength appropriate for the fluorescent dye attached to the antibodies. *See also* immunofluorescence.

fluoroelastomer Any elastomeric high polymer (an elastic substance resembling rubber) containing fluorine.

fluorogenic substrate technology *See* chromogenic substrate technology.

fluorosis The staining or pitting of the teeth caused by excessive amounts of fluoride in water.

fluosilicic acid (H_2SiF_6) *See* hydrofluosilicic acid.

fluorspar (CaF_2) Natural calcium fluoride. Fluorspar may be used as a direct source of fluoride for water fluoridation, in which case an alum ($Al_2(SO_4)_3 \cdot 14H_2O$) solution is used to dissolve the fluorspar. *See also* fluoridation.

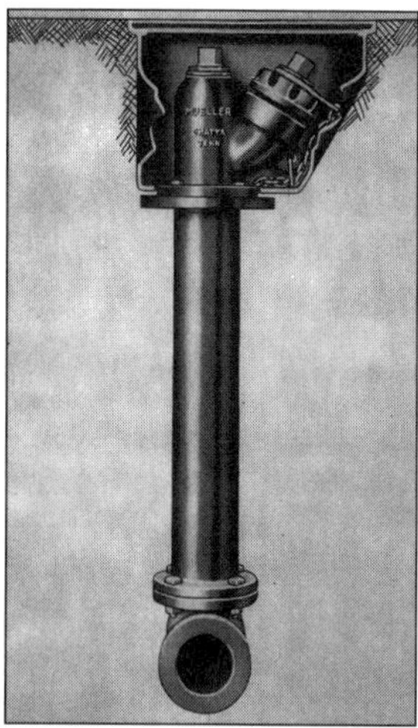

Flush hydrant

flush (1) To open a cold-water tap to clear out the water that may have been sitting for a long time in the pipes. In new homes, to flush a system means to send large volumes of water gushing through the unused pipes to remove loose particles of solder and flux. (2) To force large amounts of water through piping to clean out piping, tubing, or storage or process tanks.

flush hydrant A hydrant with outlets at or below grade.

flush valve A valve used to expel water and sediment from irrigation lines.

flush water The water that is discharged from a blowoff, fire hydrant, or open pipe to remove debris or discolored water from a distribution system. Such flushing normally takes place from a line being placed in service or from a line that has only one end connected.

flushing The act of running water through a distribution system or water main in order to remove debris, discolored water, or chemical solutions in order to clean the line or system.

flushing chamber A tank or reservoir used to store water used for flushing a line or sewer where quick release increases the volume of flow.

flushing program A systematic or preventative maintenance program designed to flush dead-end water mains or distribution systems to prevent discolored water or stale water from occurring. This program involves flowing or flushing designated fire hydrants on a routine or scheduled basis to replace water in the distribution system that has remained static over a period of time.

fluvial Of or pertaining to one or more rivers; produced by the action of a stream or river; growing, living, or existing in or near a stream or river.

fluvial deposit Sediment deposited by the action of streams. It is also called an alluvial deposit.

fluvial erosion Erosion caused by the action of streams.

fluvial index *See* flow index.

fluviation Collectively, all the numerous activities of streams.

fluvioglacial stream A stream that obtains its water and much of its load from melting glacial ice.

flux (1) A substance applied to surfaces that are to be joined to clean and free them from oxides and promote their union. Typically fluxes are used after cleaning copper pipe in preparation for soldering or brazing. Overuse or improper use of fluxes can cause corrosion. (2) The amount of a substance that passes a point in space per unit area per unit time. (3) For a membrane separation process, the flow rate of solvent (water) or salt (or specific solute) through the membrane area. For a pressure-driven membrane, when flux is not specified as a solvent or solute flux, it is commonly assumed to be a solvent flux. *See also* flux rate (solvent, water).

flux decline *See* flux rate decline.

flux rate (solvent, water) For a membrane separation process, the volume or mass of permeate passing through the membrane per unit area per unit time. Solvent (water) flux rate is commonly expressed in gallons per square foot per day, or cubic metres per square metre per second, or metres per second.

flux rate decline The decrease in flux rate over a time period. For a pressure-driven membrane process, flux rate decline is commonly used to quantify the hydraulic performance decline caused by membrane compaction, fouling, and other factors. It is calculated as the rate of decline of the temperature-corrected solvent (water) mass transfer coefficient over a specified time period. *See also* solvent (water) mass transfer coefficient; temperature correction factor.

fluxing lime *See* lime.

FMA (free mineral acidity) *See* mineral acidity.

FMS *See* facility maintenance system.

foam The frothy substance composed of an aggregation of bubbles on the surface of liquids and created by violent agitation or by the admission of air bubbles to liquid containing surface-active materials solid particles, or both. Foam is also called froth.

foci of infection Small areas of a cell culture that shows signs of virus replication. The initial sites (foci) of infection usually spread to infect the entire culture.

fog drip Water dripping to the ground from trees or other objects that have collected moisture from wind-blown fog. The dripping is in some instances as heavy as light rain.

follicular cell Generally, the epithelial cells that surround ovarian follicles. However, a follicle is simply a pouchlike depression or cavity, and can be found in a variety of other specialized structures within the body, e.g., hair follicles or thyroid follicles. The reproductive effects of chemicals involve follicular cells.

follow-up monitoring Monitoring required for lead, copper, and other water quality parameters after a public water system installs treatment.

fomites Articles contaminated by pathogens that convey infection. Toys, soiled diapers and clothes, drinking cups, and door handles are possible examples of fomites.

fonofos ($C_{10}H_{15}OPS_2$) The common name for the insecticide and soil fumigant ethylphosphonodithioic acid *O*-ethyl *S*-phenyl ester. *See also* fumigant; insecticide.

FONSI *See* Finding of No Significant Impact.

Food and Drug Administration (FDA) A US federal agency that, among other activities, regulates bottled water as a food.

Food Security Act (FSA) Public Law 99-198 (16 U.S.C. 3801–3845), amended several times since enactment in 1985, governing farming practices and programs. It is also called the Farm Bill.

food chain A set of relationships describing which organisms eat or are eaten by which other organisms (predator and prey). A food chain can be a means of concentrating certain types of toxic chemicals in the tissues of the predator chain. These chemicals are generally very nonpolar, such as the polychlorinated biphenyls, dioxins, and dibenzofurans. However, chemicals that may be sequestered by specific biological processes also bioaccumulate (e.g., lead and cadmium).

food web A series of interconnected food chains forming interlocking patterns.

foot *See* feet *in the Units of Measure list.*

foot valve A check valve placed on the lower section of a water system, such as on the suction side of a pump or at the bottom of a well with a submersible pump, to prevent water from draining out of the system or back into a well.

foot-pound, torque *See the Units of Measure list.*

foot-pound, work *See the Units of Measure list.*

forage fish Small fish eaten by predatory fishes as prey.

force The agent or cause that is capable of putting a mass in motion or alters the motion of a mass. In equation form (in any consistent set of units), force equals mass times acceleration.

force main A pressure pipe joining the pump discharge at a water-pumping station with a point of gravity flow.

force potential The sum of the kinetic energy, elevation energy, and pressure at a point in an aquifer. Because the kinetic energy (velocity head) is neglected in most groundwater problems, force potential is taken as equal to the hydraulic head times the acceleration of gravity times the density of the liquid (in any consistent set of units).

force units In the US customary system, pounds force. In Système International, newtons; kilogram-metres per second squared.

forced aeration The application of pressurized air below the surface of a liquid through diffusers or other devices to promote the formation of small bubbles.

forced draft degasifier A device for removing gases from water in which, commonly, the water is distributed at the top of a column and falls by gravity while air or other gas is blown in from the bottom, moves in an upward direction, and removes the volatile gases from the water. *See also* degasifier.

forebay A small reservoir or pond located at the head of a water treatment plant, used to store water in relatively small quantities to compensate for variations in flow that occur over short periods of time, usually not exceeding several hours.

forebay area (1) Any holding reservoir supplying water to a water treatment plant. (2) In groundwater hydrology, a free groundwater basin that serves as recharge area to an artesian basin.

forecasting The use of mathematical procedures for predicting behavior based on historical data. Forecasting can be applied to budgeting processes, statistical modeling, population growth, or any function requiring a predictive capability.

foreign water Water that is found in a particular stream or other body of water but that originated in another drainage basin. In some states, the legal rights to the use of such water are different from the rights to the use of stream water that originated within the stream's own drainage basin.

forest influence Effects of forest or brush on climate, water, and soil.

forested wetland That portion of a wetland dominated by woody vegetation greater than about 20 feet (6 metres) in height.

form loss of head A loss of head caused by the change in shape of the waterway, distinct from a loss of head caused by friction or roughness.

formaldehyde (HCHO) An aldehyde created during the reactions of oxidants used as disinfectants, particularly ozone (O_3), with natural organic matter. Typically, formaldehyde is the most prevalent aldehyde produced during the ozonation of drinking water. It is quite biodegradable and can readily be removed during biological filtration. *See also* aldehyde; biological filtration; ozonation by-product.

formalin-fixed cysts or oocysts Cysts or oocysts preserved in formalin storage media. Storage in formalin renders the organisms noninfectious.

format *See* data format.

formation A geologic group of rock or soil of similar consolidation that forms an identifiable lithologic unit. An aquifer composed of sand and gravel and bounded by clay layers is an example of a formation.

formation packer A set of inflatable bladders lowered into a well to vertically isolate different parts of its depth.

formation potential The future expectation of the creation of disinfection by-products (DBPs), measured individually or as a group, following the addition of a disinfectant to produce a target residual, in a given time and under a given set of environmental conditions (e.g., temperature, pH). When a formation potential is being reported, all conditions of the test should also be reported. Formation potential tests tend to produce near the maximum DBP concentration possible with whatever disinfectant is being tested. *See also* disinfection by-product formation potential; simulated distribution system test; total organic halogen formation potential; trihalomethane formation potential.

formazin turbidity unit (ftu) *See the Units of Measure list.*

formula (1) A mathematical function. (2) A shorthand way of writing the elements present in a molecule and how many atoms of each element are present in each of the molecules (e.g., carbon dioxide's formula is CO_2). Such a formula is also called a chemical formula.

formula weight *See* molecular weight.

fortified *See* spiked.

forward mutation A change in the naturally occurring genomic sequence. Such a mutation may lead to the synthesis of a protein with altered function, or it may have little effect on the organism. The main use of this term is in contrast to the genetically engineered organisms that are most frequently used in mutagenesis tests. Those organisms contain a mutation that makes them dependent on exogenous sources of nutrients (e.g., the histidine-dependent *Salmonella* strains used in the Ames test for bacterial mutagenicity). A reverse mutation may occur after exposure to a test chemical at the specific locus of the gene that makes the organism dependent on a nutrient. This event will render the organism capable of growing on media deficient in the nutrient. Reverse mutations at specific loci are much easier to detect than forward mutations because a selective medium simplifies detection of the mutant. *See also* mutagenesis; reverse mutation.

fossil water Interstitial water that was buried at the same time as was the original sediment.

fouling (1) The accumulation of undesirable foreign matter in a filter or ion exchange media bed, causing clogging of pores or coating of surfaces and inhibiting or limiting the proper operation of the bed and the treatment system. (2) The reduction in adsorption capacity of granular activated carbon as a result of other material occupying the adsorption surface. For example, granular activated carbon can be fouled by naturally occurring organic matter in drinking water, thereby limiting the absorption of specific microcontaminants. (3) In a membrane water treatment process, the deposition of material such as colloidal matter, microorganisms, and metal oxides on the membrane surface or in its pores, causing a change in membrane performance (e.g., flux decline). (4) A gelatinous, slimy accumulation resulting from the activity of organisms in water. This type of fouling may be found on concrete, masonry, and metal surfaces (tuberculation). *See also* preloading.

fouling index One of several types of empirically obtained values that is used to characterize membrane feedwaters and is indicative of the amount of particulate and colloidal matter present. Index values are calculated from empirical formulas using data obtained by filtering a water sample through a specified filter media disk under controlled conditions and measuring the change in flow rate over time. *See also* plugging factor; silt density index.

fouling potential In membrane process treatment, the degree to which the feed stream contains substances that may foul a membrane system and cause a decline in performance. Potential foulants include suspended and colloidal solids, metal oxides, microbial matter, organic compounds, and other solutes. The feedwater colloidal fouling potential for some membrane processes is sometimes estimated based on turbidity, the silt density index, or other empirical measurements. *See also* silt density index.

fountain A spring of water issuing from the earth; the source or head of a stream of water.

fountain aerator An aerator in the form of a fountain with tight horizontal basins of various diameters, the largest basin being found at the bottom. The ensemble resembles an ornamental fountain.

fountain flow The manner in which water flows from the open end of a vertical pipe, rising a certain distance and then spreading out in an umbrella-like shape.

fountain head The head in a saturated confined aquifer.

fountain jet The stream of water issuing from and over the horizontal circumference of a vertical pipe.

4-log removal The removal of 99.99 percent of a contaminant, usually microbes.

4-methylumbelliferyl-β-D-glucuronide *See under* methylumbelliferyl-β-D-glucuronide.

4'6-diamidino-2-phenylindole/propidium iodide *See under* diamidino-2-phenylindole/propidium iodide.

four-way valve A valve constructed with four waterways and with a movable element operated by a quarter turn to provide passage between either pair of adjacent waterways.

Fourier transform infrared (FTIR) analysis A spectroscopic technique that provides information on the molecular structure of a substance. Detectors employing this technique have been interfaced to a variety of separation devices, such as gas and liquid chromatographs. A computer is an important part of such an instrument setup; it allows for rapid recording of the spectra of organic compounds. However, these interface techniques often lack the sensitivity desired in the analysis of trace organic compounds in water.

fourneyron wheel An outward-flow waterwheel.

FPA *See* Federal Power Act.

fps *See* feet per second *in the Units of Measure list.*

F_r *See* Froude number.

fractal index A measure of the manner in which solids deposit on a surface. It can be used to predict sludge volume and density.

fractionate To separate into fractions or parts.

fracture spring A spring created via fracturing or jointing of a rock.

fracture trace The surface representation of a fracture zone, often with valleys and stream segments running along the zones. The trace may be a line of characteristic vegetation, a linear soil-moisture pattern, or a linear topographic feature such as a sag or cliff. *See also* fracture zone.

fracture zone A zone in geologic formations where fractures are plentiful. Wells located in these zones can be very productive water sources, whereas wells drilled between fracture zones may not yield any water.

fragment model A subdivision of a system that can be modeled independently of the larger system. A large, complicated system is sometimes broken down into fragment systems, such as pressure zones, to facilitate analysis.

framed dam A low, fixed dam that is generally built of timber framed to form a water face and supported by struts.

frame-shift mutation A change in the sequence of purine and pyrimidine bases within deoxyribonucleic acid (DNA) that results from the loss (or gain) of bases in combinations that are not divisible by 3. Essentially, this changes the reading frame of the sequence and in turn alters the arrangement of amino acids in the protein that is synthesized from this DNA template. *See also* codon; mutagenesis.

Francis turbine A reaction turbine of the radial inward-flow type.

Francis wheel An inward-flow waterwheel.

frank pathogen An infective agent for which multiplication in or on a host organism usually results in disease of the host. *See also* infectious agent; pathogenic organism; virulence.

***F*-ratio** A statistic used in regression analysis to test the significance of independent variables and the regression as a whole. The ratio is computed by dividing the amount of variance explained by the model by the amount of unexplained variance. The resulting ratio can then be compared to a standardized table of *F* values to determine the level of confidence that can be placed in the relationship. The standard *F*-test values are uniquely determined by the degrees of freedom in the numerator (the number of variables included) and in the denominator (the number of observations included in the analysis) and by the level of confidence that is desired for the test (90, 95, or 99 percent). A calculated *F*-ratio (or *F*-value) of about 3.0 is generally considered adequate at the 95 percent confidence level, with at least 30 observations to support the hypothesis that the regression relationship did not occur by chance.

frazil ice Small ice crystals that can block water intakes.

FRDS *See* Federal Reporting Data System.

free acid form The regenerated form of a weak acid cation exchanger. *See also* weak acid cation exchanger.

free ammonia Ammonia (NH_3) that is free in solution, not bound up as in chloramines. *See also* chloramine.

free available chlorine (FAC) The amount of chlorine available as dissolved gas (Cl_2), hypochlorous acid (HOCl), and hypochlorite ion (OCl^-), that is not combined with ammonia (NH_3) or other compounds in water. *See also* hypochlorite ion; hypochlorous acid.

free available chlorine residual The free chlorine residual formed once all the chlorine demand has been satisfied. The chlorine is not combined with other constituents in the water. *See also* chlorine demand.

free base form The regenerated form of a weak base anion exchanger. *See also* weak base anion exchanger.

free carbon dioxide Carbon dioxide (CO_2) present in water as gas or as carbonic acid (H_2CO_3) but not including carbon dioxide in combination, as in the carbonates or bicarbonates.

free chlorine *See* free available chlorine.

free chlorine residual *See* free available chlorine residual.

free convection Motion caused only by density differences within the fluid. It is also called gravitational convection.

free energy A measure of the thermodynamic driving energy for a chemical reaction. Free energy is also known as the Gibbs function. *See also* Gibbs free energy.

free erythrocyte porphyrin (FEP) Protoporphyrin IX found in the red blood cells. It is also known as free erythrocytic protoporphyrin. It is a precursor of heme, requiring only the insertion of iron to complete the synthesis. Iron is inserted by an enzyme called ferrochetalase that is inhibited by certain heavy metals, particularly lead. In lead-poisoned individuals, FEP accumulates in the immature red cells; this accumulation can easily be detected because protoporphyrin IX is fluorescent. Consequently, protoporphyrin IX is used as a sensitive and easily administered prescreen for excessive exposure to lead. *See also* heme.

free flow A condition of flow through or over a structure such that the flow is not affected by submergence of the structure or the existence of tailwater downstream of the structure.

free groundwater Groundwater in the interconnected interstices in the saturation zone. It extends down to the first impervious barrier and moves under the influence of gravity in the direction of the slope of the water table. *See also* phreatic water.

free ligand The fraction of a chemical that has dissociated from its binding site on a protein or chelating agent. The free ligand reaches equilibrium with the bound ligand. Interactions between a chemical and its binding site are considered to be specific if the interaction is responsible for inducing a response within the body. Nonspecific interactions also occur, however, and are generally thought of as a way of reducing the chemical's availability for distribution throughout the body. Ligands also form complexes with certain biological macromolecules in order to exert effects on the body. This latter type of complex is usually referred to as a drug-receptor complex. *See also* ligand; ligand binding site.

free mineral acidity (FMA) *See* mineral acidity.

free radical A chemical that possesses an unshared electron. Such chemicals are usually quite reactive. Free radicals that are generated in the metabolism of chemicals within the body are frequently held responsible for mediating toxic responses. One of the outcomes of the resulting free radical is lipid peroxidation. *See also* hydroxyl radical; lipid peroxidation.

free residual chlorination The application of chlorine to water to produce a free available chlorine residual equal to at least 80 percent of the total residual chlorine (the sum of free and combined available chlorine residual). *See also* free available chlorine.

free residual chlorine *See* free available chlorine residual.

free settling The settling of discrete, nonflocculent particles in dilute suspension such that no apparent flocculation or interaction occurs between particles. *See also* Stokes's law; type I settling.

free surface The boundary surface of a liquid that is in contact with the atmosphere. It is also called the free water surface.

free surface energy The free energy in a liquid surface produced by the unbalanced inward pull exerted by the underlying molecules on the layer of molecules at the surface. For purposes of calculation, a hypothetical tension equal to the free surface energy is assumed to be acting in all directions parallel to the surface. This mathematical equivalent of free surface energy is called surface tension.

free water Suspended water that constitutes films covering the surfaces of solid particles or the walls of fractures but is in excess of pellicular water. Free water can move in any direction under the pull of the resultant of the force of gravity and unbalanced film pressure. It is also called mobile water.

free weir A weir that is not submerged; a weir in which the tailwater is below the crest or where the flow is not in any way affected by tailwater.

freeboard (1) The vertical distance from the normal water surface to the top of a confining wall. (2) The vertical distance from the media surface to the underside of a trough in a granular filter. This distance is also called available expansion.

freeze-drying An effective method for long-term preservation of bacteria and other microorganisms. It involves the removal of water from frozen microorganism suspensions by sublimation under reduced pressure. It is also called lyophilization.

freeze–thaw cycles A process to separate solids from water in open beds by alternately freezing and thawing the mixture.

freeze–thaw sludge dewatering A process of improving the dewaterability of sludge by allowing the sludge to freeze. Freezing waste alum solids causes the water in the gelatinous material to crystallize and, upon thawing, leaves a granular solid of coffee-ground consistency. *See also* alum.

freezing A process in water desalting whereby salt water is cooled under controlled conditions to form ice crystals, remaining dissolved solids are washed out, and the separated ice is melted to produce fresh water.

freezing process In desalting operations, the separation of salt from water by the formation of salt-free ice crystals.

French drain An underground passageway for water through the interstices among stones placed loosely in a trench.

French square *See* dilution bottle.

frequency (f) (1) For a periodic quantity, the number of periods per unit time:

$$f = (1/T)$$

Where:

f = the frequency

T = the period (in units of time)

Frequency is measured in hertz. (2) The degree of regularity of (a) coverage of an area by surface water or (b) saturation of the soil. It is usually expressed as the number of years the soil is inundated or saturated during part of the growing season of the prevalent vegetation (e.g., 50 years per 100 years) or as an inundation frequency such as once every year, once every 2 years, or once every 5 years.

frequency analysis A method of evaluating vegetation in an area by establishing a transect and counting the occurrences of plant species at various sampling points along the transect.

frequency curve A graphical representation of the frequency of occurrence of specific events. The event that occurs most frequently is termed the mode. When the mode coincides with the mean value, the curve is symmetrical, but when it differs from the mean, the curve lacks symmetry and is termed a skew frequency curve. *See also* duration area curve; probability plot.

frequency distribution A description of the frequency of occurrence of certain values for a parameter within a data set. *See also* probability distribution.

frequency units In the US customary system, cycles per second. In Système International, hertz; number per second.

frequent flooding Flooding that is likely to occur often during usual weather conditions (i.e., has more than a 50 percent chance of occurring in any year or will occur more than 50 times in 100 years).

fresh water Water having a relatively low mineral content. Fresh water is generally considered to be water containing less than 500 or 1,000 milligrams per litre of total dissolved solids, depending on location.

freshwater lens An occurrence of fresh water found above salt water, commonly associated with coastal areas where aquifers contain both fresh and saline water.

fresh–salt water interface An idealized location in coastal aquifer systems where the inland, fresh groundwater meets the offshore, saline groundwater. Because fresh water is less dense than saline water, the interface has vertical as well as horizontal variation. Pumping in the freshwater zone can cause the interface to move inland.

freshet A great rise or overflowing of a stream caused by heavy rains or melted snow.

fretting corrosion *See* stress corrosion.

Freundlich adsorption constant (K_F) An indicator of the adsorptive capacity of an adsorbent for a solute. It appears in the Freundlich equation, which mathematically describes the observed dependence of the amount of solute adsorbed on an adsorbent such as activated carbon. This constant is empirically determined from a Freundlich isotherm, in which a plot of the logarithm of the loading (q, the amount of solute adsorbed per unit of adsorbent) versus the logarithm of the equilibrium concentration of the solute remaining in solution (C_e) in any consistent set of units yields a straight line with an intercept of log K_F. If the data are plotted on log–log paper, K_F equals the loading when $C_e = 1$. The Freundlich constant is useful for interpreting the relative strengths of the adsorption of compounds on the adsorbent when the experimental conditions of temperature, contact time, and pH value are the same. *See also* activated carbon; adsorption; adsorption isotherm; Freundlich isotherm.

Freundlich capacity coefficient *See* Freundlich adsorption constant.

Freundlich isotherm An adsorption relationship that describes the equilibrium state of adsorbate, adsorbent, and solution at a given temperature. The Freundlich isotherm takes the form of

$$q = K_F C_e^{1/n}$$

Where (in any consistent set of units):

q = the mass of adsorbate per mass of adsorbent at equilibrium

C_e = the concentration of the adsorbate in dilute solution when adsorption equilibrium is reached

K_F = the Freundlich adsorption constant

$1/n$ = the slope of the straight-line isotherm when q is plotted as a function of C_e on log–log paper *See also* Langmuir isotherm.

friability An expression of the ability of ion exchange beads to resist cracking under hydrostatic operation.

friction factor (C) A measure of the resistance to flow of fluid in a conduit as influenced by wall roughness. *See also* C factor.

friction head loss The head lost by water flowing in a stream or conduit as a result of the disturbances set up by the contact between the moving water and its containing conduit, as well as by intermolecular friction. In laminar flow, the head loss is approximately proportional to the first power of the velocity:

$$h_L = 64\nu\left(\frac{L}{D^2}\right)\left(\frac{V}{2g}\right)$$

Where (in any consistent set of units):

h_L = the friction head loss

L = the conduit length

D = the conduit diameter

V = the mean velocity

g = the gravitational constant

ν = the kinematic viscosity of the fluid

In turbulent flow, friction head loss is proportional to a higher power, approximately the square of the velocity:

$$h_L = f\left(\frac{L}{D}\right)\left(\frac{V^2}{2g}\right)$$

Where:

f = the friction coefficient

Although, strictly speaking, head losses such as those caused by bends, expansions, obstructions, and impacts are not included in this term, the usual practice is to include all such head losses under this term. Friction head loss is also called friction head.

friction loss *See* friction head loss; head loss.

friction slope The friction head or loss per unit length of conduit. For most conditions of flow, the friction slope coincides with the energy gradient, but where a distinction is made between energy losses caused by bends, expansions, impacts, and so on, a distinction must also be made between the friction slope and the energy gradient. In uniform channels, the friction slope is equal to the bed or surface slope only for uniform flow.

friction weir *See* free weir.

fringe water Free water occurring in the saturated capillary fringe above the water table and completely filling the smaller interstices. It may consist solely of capillary water, or it may be combined with water in transit to the water table. It moves with the capillary fringe as the latter rises and falls with the water table.

front of levee The side of the levee next to the river, facing away from the protected area. It is sometimes called the river side.

frontal precipitation Precipitation occurring at a boundary between pockets of cool and warm air.

front-end loader A wheeled or track-mounted tractor with a hydraulically operated bucket mounted on the front to move earth material and especially to load the material into trucks.

frost bottom A section of a meter case designed to break easily without damaging other parts of the meter when water freezes within the meter.

frost box A box surrounding a water meter or water pipe and containing some heat insulator such as mineral wool, excelsior, or sawdust, to prevent the water from freezing.

Froude number (F_r) A numerical quantity used as an index to characterize the type of flow in a hydraulic structure where the force of gravity (as the only force producing motion) acts in conjunction with the resisting force of inertia. It is equal to the square of a characteristic velocity of the system (the mean, surface, or maximum velocity) divided by the product of a characteristic linear dimension (such as diameter or depth) and the gravitational constant, all expressed in consistent units so that the combinations will be dimensionless. The Froude number is used in open-channel flow studies or in cases in which the free surface plays an essential role in influencing motion. It is given by the following equation:

$$F_r = \frac{V^2}{gL}$$

Where (in any consistent set of units):

F_r = Froude number

V = the characteristic velocity

g = the gravitational constant

L = the characteristic linear dimension

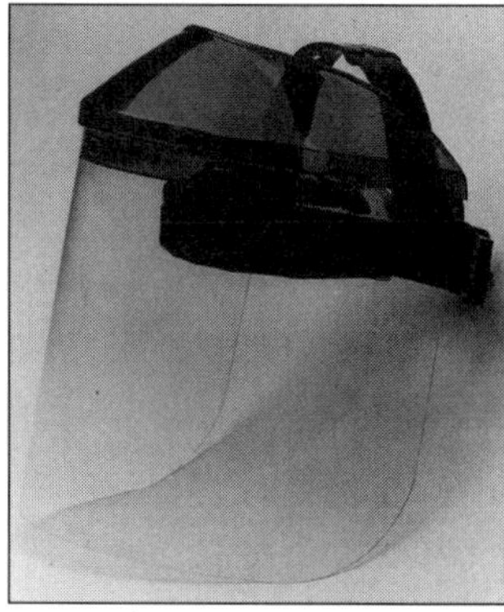

Full-face shield

FS *See* fecal streptococcus.

FSA *See* Food Security Act.

FSTRAC *See* Federal–State Toxicology and Regulatory Alliance Committee.

ft/h *See* feet per hour *in the Units of Measure list.*

F344 rat The Fischer 344 rat strain that is used by the National Toxicology Program in cancer bioassays. It remains the "standard" rat as of this writing. The F344 strain is an inbred albino rat (*Rattus norvegicus*) with a pedigree stretching back to the experiments of Maynie Rose Curtis of Columbia University. She began brother–sister matings with rat number 344 produced by the breeding pairs obtained from a commercial supplier of rats, Fischer, in 1920.

ft/min *See* feet per minute *in the Units of Measure list.*

ft/s *See* feet per second *in the Units of Measure list.*

ft/s^2 *See* feet per second squared *in the Units of Measure list.*

ft^2/s *See* feet squared per second *in the Units of Measure list.*

ft^3 *See* cubic feet *in the Units of Measure list.*

ft^3/h *See* cubic feet per hour *in the Units of Measure list.*

ft^3/min *See* cubic feet per minute *in the Units of Measure list.*

ft^3/s *See* cubic feet per second *in the Units of Measure list.*

FTIR analysis *See* Fourier transform infrared analysis.

FTP *See* file transfer protocol.

FTP transfer *See* file transfer protocol transfer.

ftu *See* formazin turbidity unit *in the Units of Measure list.*

fugitive water Leakage from impounding reservoirs or an irrigation system.

full duplex Pertaining to communications for which information can be simultaneously transmitted and received.

full fluidization backwash A method of backwashing in which the backwash water is added at a sufficient rate that the media grains separate from one another and become freely supported by the liquid. This allows dislodged particles to be fully released from the bed during backwash.

fuller's earth The microscopic remnants of the discarded outer surface of diatoms used in precoat filtration. Fuller's earth is also called diatomaceous earth.

full-face shield A shatterproof plastic shield worn to protect the face from flying particles and chemicals.

fulvic acid A complex organic compound of unknown specific structure that can be either soil derived or aquagenic in origin. When present in water, fulvic acids are major precursors of disinfection by-products. *See also* aquagenic organic matter; aquatic fulvic acid; disinfection by-product precursor.

fume hood A large enclosed cabinet equipped with a fan to vent fumes from a laboratory. Chemicals are mixed and heated under the hood to prevent fumes from spreading through the laboratory.

fumigant A toxic agent in vapor form that destroys rodents, insects, infectious organisms, and so on. It is a type of pesticide. *See also* pesticide.

function A role that a wetland serves: (1) reduction of pollutant loadings, including excess nutrients, sediment, and toxics; (2) attenuation of flood waters and stormwaters; (3) shoreline stabilization and erosion control; (4) breeding ground and habitat for many species of plants and wildlife, including fish, birds, and mammals, extending to threatened and endangered species and species in need of conservation; (5) food chain support; and (6) timber production.

function argument A parameter used to define a mathematical function. For example, if a function called CHEMCOST were defined to calculate the cost of a chemical feed system based on the feed rate of the chemical in pounds per day and the unit cost of the chemical in dollars per pound, the arguments of the CHEMCOST function would be chemical feed rate and chemical unit cost.

functional antagonism In pharmacology and toxicology, a general case where the effect of one chemical (or drug) is offset by the effect of a second chemical acting on a second physiological process.

functional assessment A valuation of physical, chemical, and biological processes or attributes in a wetland that are vital to the system's integrity.

functional equivalency The ability of a restored or created ecosystem to perform ecosystem services, such as flood water storage or pollutant detoxification, that are indistinguishable in effects from corresponding services performed by natural ecosystems.

functional group *See* exchange site.

functionally required area In a landscape design, a portion of an ornamental landscape intended to serve a specific function, e.g., for pedestrian traffic, sports, or recreational activities.

fund management The process whereby a group, organization, or individual administers sums of money set aside for a specific purpose. One example is a pension fund for retired employees.

funding bonds Bonds issued to retire outstanding floating debt and to eliminate deficits.

fungal *See* fungus.

fungi *See* fungus.

fungicide Any substance that kills or inhibits the growth of fungi. *See also* fungus.

fungus (pl. fungi) Eumycetes, eukaryotic organisms that are classified as plants, simple or complex, lacking chlorophyll and a vascular system. Simple forms are single celled (yeasts); higher forms have branched filaments and complicated life cycles (molds and mushrooms).

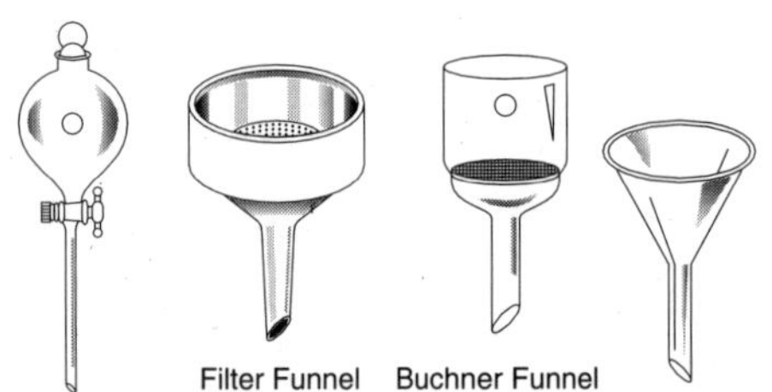

Funnels

funnel A utensil made of glass or plastic and used in the laboratory for pouring liquids into flasks and other containers.

fuse A device for limiting the amount of electrical current in a circuit. When an excessive current flows, the interior conductor of the fuse melts because of the heat generated by the current, thereby opening the circuit. Unlike a circuit breaker, a fuse cannot be reused; it must be replaced.

fused disconnect A switch device that allows electrical current to be interrupted automatically when a replaceable fuse in the switch fails. This device is used to isolate a circuit or piece of electrical equipment from a power source when the level of current is too high.

fusible plug A safety device on a chlorine cylinder below the outlet valve that melts at 158° Fahrenheit (70° Celsius) for protection from excessive pressure when the cylinder is heated.

fusion In physics, the process by which the nuclei of two light elements combine to form the nucleus of a heavier element and release substantial amounts of energy. *Contrast with* fission.

futile call In the context of water rights, a situation in which a more recent (junior) priority is allowed to continue to divert water in spite of a downstream senior call, when curtailing the more recent priority would not produce any more water for the senior call.

future capacity The capacity for service somewhat in excess of immediate requirements. It is built into a utility in anticipation of increased demands for service resulting from higher uses by existing customers or growth in the service area.

fuzzy logic An approach to approximate reasoning for which the rules of inference are approximate rather than exact. It can be used to manipulate incomplete, imprecise, or unreliable information and is used extensively in the field of artificial intelligence. *See also* artificial intelligence.

***F*-value** *See F*-ratio.

FWS *See* US Fish and Wildlife Service.

FY *See* fiscal year.

G

G *See* giga.

G *See* Gibbs free energy; universal gravitational constant; velocity gradient.

$\overline{G}$ *See* mean velocity gradient.

g *See* gram *in the Units of Measure list.*

g *See* earth's gravitational constant.

GA salt A very coarse grainy salt of ground alum size (hence the "GA") that was formerly used for regenerating zeolite water softeners and is sometimes used coincidentally with alum as a coagulant.

gabion A long basket lowered in place on a wire and filled with earth stones to serve the same purpose as a fascine. It is also called a pannier. *See also* fascine.

GAC *See* granular activated carbon.

gaining stream A stream for which the flow is partially or completely composed of water discharged from an underlying aquifer.

β-GAL *See* β-D-galactosidase.

gal/flush *See* gallons per flush *in the Units of Measure list.*

gal/ft² *See* gallons per square foot *in the Units of Measure list.*

β-D-galactosidase (β-GAL) An enzyme, sometimes called lactase, that catalyzes the breakdown of lactose into glucose and galactose. It is the basis for tests for detecting the coliform group bacteria using defined-substrate technology. *See also* defined-substrate technology.

β-D-galactoside A glycoside derived from galactose. It is a substrate for the enzyme β-D-galactosidase, part of the lactose operon (group of genes) of lactose-fermenting bacteria such as the coliform group.

Galileo number (N_{Ga}) The reciprocal of the sedimentation number. It is used in describing the sedimentation properties of particles and in calculating the minimum upflow velocity in a granular filter to produce fluidization.

$$N_{Ga} = \frac{\rho D^3 g(\rho_s - \rho)}{\mu^2}$$

Where (in any consistent set of units):

ρ = the fluid density

D = the particle diameter

g = the gravitational constant

ρ_s = the density of the particle

μ = the absolute viscosity

gallery (1) An underground structure designed and installed to collect subsurface water. (2) A passageway in a structure, such as a dam or a water treatment plant, used to obtain access to interior parts, to carry pipes, or to house machinery. (3) An underground conduit, reservoir, or passage.

Gallionella A bacterial genus of unicellular, colorless, curved, rod-shaped bacteria that secrete a ribbon-like stalk from the concave side of the cell. The stalk is characteristically twisted. These organisms are generally considered iron bacteria because of their occurrence in iron-bearing waters and because iron oxide (Fe_2O_3) is deposited in or on the stalks. They are considered nuisance bacteria because of their association with fouling in pipes that convey iron-containing groundwaters used for drinking water or industrial uses.

Gallionella ferruginea A bacterium designated as the type species of the genus *Gallionella*. It is one of several types of bacteria that use iron in their metabolism and are capable of depositing gelatinous ferric hydroxide ($Fe(OH_3)$). *See also Gallionella*; iron bacteria.

gallon *See the Units of Measure list.*

gallons per capita per day (gpcd) *See the Units of Measure list.*

gallons per day (gpd) *See the Units of Measure list.*

gallons per day per square foot (gpd/ft²) *See* gallons per square foot per day *in the Units of Measure list.*

gallons per flush (gal/flush) *See the Units of Measure list.*

gallons per hour (gph) *See the Units of Measure list.*

gallons per minute (gpm) *See the Units of Measure list.*

gallons per minute per square foot (gpm/ft²) *See the Units of Measure list.*

gallons per second (gps) *See the Units of Measure list.*

gallons per square foot (gal/ft²) *See the Units of Measure list.*

gallons per square foot per day (gfd) *See the Units of Measure list.*

gallons per year (gpy) *See the Units of Measure list.*

galvanic cell An electrolytic cell capable of producing electrical energy by electrochemical action. The decomposition of materials in the cell causes an electric current (electrons) to flow from cathode to anode.

galvanic corrosion Corrosion that is increased because of the current caused by a galvanic cell. It is sometimes called couple action.

galvanic coupling A cell consisting of two dissimilar metals in contact with each other (i.e., a coupling) and with a common electrolyte.

galvanic series A listing of metals and alloys according to their corrosion potential.

galvanization The process of coating a metal with zinc.

galvanize To coat a metal (especially iron or steel) with zinc.

galvannealing Extension of the iron content of a galvanized pipe all the way to the pipe surface. When properly performed, the galvanizing process produces a coating that gradually changes in composition from the pure iron of the pipe to the pure zinc surface that protects the pipe. Under certain manufacturing conditions, however, the iron content of the alloy layer extends all the way to the surface, compromising the protection normally afforded by the galvanizing processes. Among the manufacturing conditions that can cause galvannealing are (1) zinc dip that is too hot, (2) keeping the iron pipe in the zinc dip too long, or (3) certain impurities in the metals.

gametes *See* gametogenesis.

gametogenesis The formation of male and female sex cells (sperm or ova), which are also referred to as gametes.

gamma (γ) *See the Units of Measure list.*

gamma (γ) decay A nuclear reaction in which the nucleus of an atom emits a pulse of energy in the form of gamma radiation.

γ-glutamyltranspeptidase *See under* glutamyltranspeptidase.

gamma (γ) log A set of geophysical measurements of the background gamma radiation given off by a subsurface formation. The log can be interpreted to determine water-bearing and nonwater-bearing layers in an aquifer.

gamma (γ) ray High-energy, short-wavelength electromagnetic radiation emitted during radioactive decay. Energies of gamma rays are usually between 0.010 and 10 megaelectron volts. Gamma radiation usually accompanies alpha and beta emissions and always accompanies fission. Gamma rays are very penetrating and are best attenuated by dense materials such as lead and depleted uranium.

gamma–gamma (γ–γ) radiation log A set of geophysical measurements of the backscattered gamma radiation given off by a subsurface formation that is exposed to a gamma radiation source. The log can be interpreted to determine water-bearing and nonwater-bearing layers in an aquifer, as well as the porosity of the medium.

gang of wells A group of wells from which water is drawn by a single pump or other lifting device. A gang of wells is also called a battery of wells.

Gantt chart A means of graphically representing activities across a time scale, developed by Henry L. Gantt to assist with project management systems. Initially called bar charts, these graphic representations have since been renamed Gantt charts in honor of their inventor.

GAO *See* General Accounting Office.

gap junction A structure that passes through the plasma membrane of adjoining cells. These structures form channels that allow low-molecular-weight chemicals to pass between cells. They are thought to play an important role in communication between cells. The proteins that form the junctions disappear from the plasma membrane when cells undergo division. Gap junctional proteins are also lost when cells are treated with tumor promoters.

garnet A dense granular filter medium (specific gravity of 3.6–4.2) that is the bottom layer in multimedia, triple-media, or mixed media filters. Garnet is often used in rapid granular filters, with anthracite as the top layer, then sand, then garnet.

gas A state of matter characterized by very low density and viscosity (relative to liquids and solids); comparatively great expansion and contraction with changes in pressure and temperature; an ability to diffuse readily into other gases; and an ability to occupy almost completely uniformly the whole of any container. A perfect, or ideal, gas is one that closely conforms to the simple gas laws for expansion and contraction. (Use of the word *gas* for gasoline, natural gas, or the anesthetic nitrous oxide is acceptable only in informal communication.) *See also* gas constant.

gas chlorination The application of chlorine gas (Cl_2) to water.

gas chlorinator A device for adding gaseous chlorine (Cl_2) to water. Chlorine that is stored as a liquid in cylindrical containers evaporates into the gaseous phase, withdrawn from the cylinder by vacuum through a metering device (typically a rotometer), and combined with water for injection into the flowstream.

gas chromatograph An instrument commonly used to separate organic compounds at trace concentrations. The major components of such an instrument are a sample injector, an oven, and a detector. A chromatography column, which is in the oven, links the injector port and the detector.

gas chromatography (GC) A technique commonly used in the analysis of organic compounds in water. In most cases, an extract of a water sample is injected into a gas chromatograph.

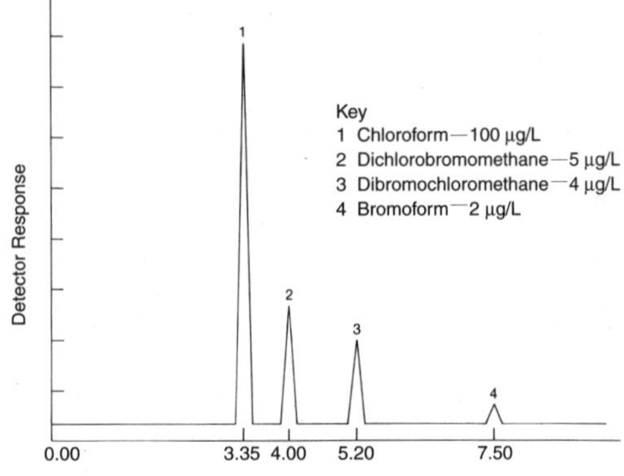

Sample readout from a gas chromatograph

Analytes are volatilized in an injector port and migrate as a gas through a chromatography column. The speed with which the analytes migrate depends on the relative affinity of the analyte for the stationary phase in the chromatography column. Compounds are identified based on their retention time in the column. *See also* stationary phase.

gas chromatography–electron capture detector (GC–ECD) A combination of analytic tools to separate and detect organic compounds in water. *See also* electron capture detector; gas chromatography.

gas chromatography–mass spectrometry (GC–MS) The combined use of a gas chromatograph and a mass spectrometer. This combination of instruments provides one of the most powerful techniques used in water analysis. Compounds that elute from a gas chromatographic column are detected in a mass spectrometer. The mass spectrometer serves as a powerful detection device that generates much more information about a compound than most other gas chromatographic detectors provide. *See also* mass spectrometry.

gas constant (*R*) The universal gas constant, equal to 0.08205 litre-atmospheres per mole-kelvin. In the ideal gas equation,

$$PV = nRT$$

Where:

P = pressure, in atmospheres
V = volume, in litres
n = concentration, in moles
T = temperature, kelvin

The gas constant is independent of pressure, temperature, or the number of moles in the sample; however, the numerical value of R depends on the units used to measure the parameters of the gas.

gas membrane *See* gas separation membrane.

gas production The creation of a gas by chemical or biological means.

gas separation membrane A membrane used in gas separation processes, such as one used for producing a gas higher in oxygen content than air.

gas solubility coefficient The mass of gas that is absorbed by a unit quantity (e.g., milligrams per litre) of water at a given temperature when the water is exposed to a pure atmosphere of the gas under a barometric pressure of 1 atmosphere (101.3 kilopascals). The gas solubility coefficient was formerly called the solubility coefficient.

gas trap A device, usually a bent section of drain line, in which enough water is retained to prevent the return or counterflow of obnoxious gases. *See also* gas vent.

gas vent A passage to permit the escape of gases.

gasket A ring of material used to make a joint or connection watertight.

gas-lift flow area An area under which the subsurface water is under pressure because of the presence of gas that would force the water to rise to the ground surface if afforded an opportunity to do so.

gas-phase chlorine dioxide (ClO_2) A strong oxidant that is produced by a chemical reaction between hydrochloric acid (HCl) and potassium chlorate ($KClO_3$). Chlorine dioxide is almost never used commercially as a gas because of its explosiveness. It is used only as a liquid in water treatment.

gastric lavage An artificial "washing" of the stomach. Fluid is pumped in and out of the stomach to remove its contents. The process is frequently used in cases of poisoning if the poison has been recently ingested. It is used as an alternative to emetics (drugs that induce vomiting) in poisoned patients who are unconscious.

gastroenteric illness An illness affecting the gastrointestinal system. Such an illness is often caused by microbial infection, but it may also result from nonmicrobial causes.

gastroenteritis An inflammation of the mucous membranes of both the stomach and intestine that may result in a relatively mild to severe illness with a wide range of symptoms, usually including vomiting, diarrhea, and fever. Symptoms are usually acute but may also be chronic. Most cases are caused by microbial agents, but chemicals can also be responsible. In reports of waterborne disease outbreaks, the designation "acute gastroenteritis of undetermined etiology" is used to describe cases of illness in outbreaks where no specific etiologic (causative) agent has been identified. In some instances no etiologic agent is identified even after exhaustive laboratory analysis, but this problem is usually due to limited laboratory analysis or a lack of appropriate or timely clinical specimens collected during the epidemiologic investigation. Although in many waterborne outbreaks the symptoms of acute gastroenteritis of undetermined etiology may suggest a viral etiology, the illness may also be of bacterial or protozoan etiology. Acute viral gastroenteritis presents as a sporadic or epidemic illness. Several enteropathogenic viruses (e.g., rotaviruses) affect primarily infants and young children, resulting in a diarrheal illness that may be severe enough to produce dehydration; other enteric viruses (e.g., Norwalk-like viruses) affect primarily older children and adults and cause self-limiting sporadic illness or outbreaks (i.e., illnesses for which recovery does not require treatment). Acute bacterial gastroenteritis is characterized by diarrhea and sometimes vomiting and may be mild to severe, resulting in dehydration and occasionally septicemia if high numbers of the causative agent or its toxins are in the circulating blood. *See also* acute gastroenteritis; diarrhea.

gastroenteritis, acute *See* acute gastroenteritis.

gastrointestinal absorption of toxicants The movement of chemicals from the lumenal to the serosal side of the gastrointestinal tract (the stomach and intestines). Such movement depends on the nature of the chemicals and the intrinsic structure of the membranes present in these tissues. Because they are epithelial tissues, cells in the gastrointestinal tract are joined by tight junctions. These junctions prevent chemicals from moving between cells, generally forcing them to pass through the cells that make up the barrier between the two sides of the gastrointestinal tract. Polar compounds of very low molecular weight (less than 100) pass freely through the cell membranes, as do nonpolar compounds of molecular weights as high as 1,000. This reflects the essentially lipid character of the cell membrane. Chemicals and toxicants can also move through the membrane by a variety of more specific mechanisms, however. For example, they may be substrates for various carrier molecules that are responsible for transporting nutrients across the membranes. Pinocytosis, an engulfment of a small bit of extracellular fluid followed by a pinching off of a segment of the membrane to the interior of the cell, is another mechanism that can account for absorption of higher-molecular-weight substances.

gastrointestinal excretion of toxicants The elimination of chemicals that are not absorbed by the body (e.g., many heavy metals)—as well as a number of chemicals secreted into the intestine by a variety of processes—through feces. A major means of such excretion for very polar compounds or compounds that have been bonded with polar substances (e.g., glutathione conjugates, glucuronides, sulfates, taurine, glycine) is by active transport into the bile. The bile is subsequently secreted into the intestine, and the chemical or its metabolite is eliminated with feces. Activities within the intestine can render these chemicals reabsorbable in the small intestine, however. *See also* enterohepatic circulation.

gate A movable, watertight barrier for the control of a liquid in a waterway.

gate chamber A structure erected to house a valve or regulating device and provide access to the valve for maintenance and repair. If the chamber has a superstructure, it is termed a gate house.

gate house A superstructure built at the headworks of a canal, conduit, dam, or powerhouse to house the gates. *See also* gate chamber.

gate lift A device for operating a gate, usually in a vertical direction.

gate stem The rod attached to a gate and by which the gate is opened or closed. The rod serves to lift the gate for opening or to push it down for closing.

gate valve A mechanical device used to turn on or shut off the flow of water in a distribution or piping system. It is operated by turning a stem that raises or lowers a disk. This disk covers the flow way, pressing against a seat when closed; it moves into a space above the flow way when open, providing an unrestricted flow.

gate vault *See* gate chamber.

gate-type hydrant A hydrant having one main valve consisting of a vertical disk that moves in a vertical plane across the valve seat and is held closed against pressure by a wedge nut.

gateway A means of allowing an interconnection between computer networks that have different communications protocols. It could be called a "simultaneous interpreter" between computers that use different languages.

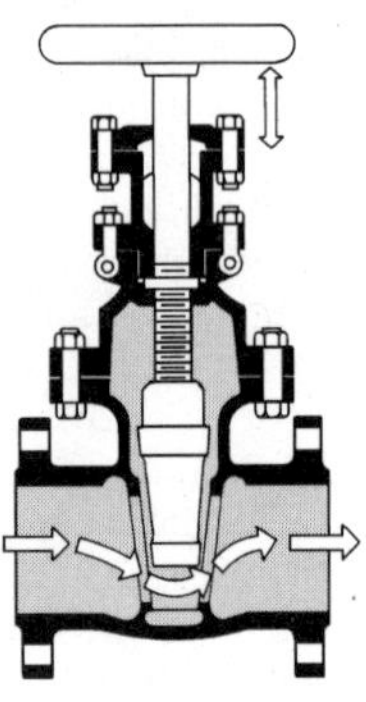

Gate valve

gauge (1) A device for indicating the magnitude or position of an element in specific units when such magnitude or position is subject to change. Examples of such elements are the elevation of a water surface, the velocity of flowing water, the pressure of water, the amount or intensity of precipitation, and the depth of snowfall. (2) The act or operation of registering or measuring the magnitude or position of something when these characteristics are subject to change. (3) The operation of determining the discharge in a waterway by using both discharge measurements and a record of stage.

gauge correlation A stage–discharge relation between the gauge height and discharge of a stream or conduit at a gauging station. This relation is shown on the rating curve or rating table for the station.

gauge datum The elevation corresponding to the zero value of a gauge above a certain datum.

gauge height The height of a water surface as measured on a gauge, the zero value of the gauge being referred to some datum. *See also* stage.

gauge pressure As measured by a gauge, the water pressure that is in excess of atmospheric pressure at the referenced elevation. Gauge pressure is not the total pressure. Total water pressure (absolute pressure) also includes the atmospheric pressure exerted on the water. Because atmospheric pressure is exerted everywhere, however (against the outside of the main as well as the inside, for example), it is generally not written into water system calculations. Gauge pressure is reported in pounds per square inch or kilopascals. *Contrast with* absolute pressure.

gauge, pipe *See* pipe gauge.

gauging The process of determining the quantity of water flowing per unit of time in a stream channel, conduit, or orifice at a given point by means of current meters, rod floats, weirs, pitot tubes, or other measuring devices or methods. *See also* discharge measurement; stream gauging.

gauging station A location on a stream or conduit where measurements of discharge are customarily made. The location includes a stretch of channel through which the flow is uniform and a control downstream from this stretch. The station usually has a recording or other gauge for measuring the elevation of the water surface in the channel or conduit.

Gaussian distribution The so-called normal distribution. It is bounded by a curve (the normal curve) that has as its maximum the mean value of a series of quantitative observations. The true normal curve will show an equal distribution of values greater than and smaller than the mean value (i.e., the mean value will approximate the median), and the probability that a particular value will be observed decreases the further that value diverges from the mean value. Values that are equally distant from the mean value in the positive and negative directions have equivalent probabilities of being observed. This distribution is a very important concept in statistics. One must be careful to apply appropriate statistical analyses to data sets if the distribution of data is not normal. *See also* chi-square distribution; Gumbel distribution; log-normal distribution.

Gear pump

gauze filtration A historical method for the recovery of viruses from water.

gauze number The number of openings per inch in a wire mesh. This number is used in specifying well screens and filter screens. For example, a mesh with a gauze number of 60 has openings that are 0.010 inches (0.25 millimetres) wide and 0.010 inches (0.25 millimetres) tall, or 60 openings per inch.

gavage Administration of a material by stomach tube.

GBF–DIME file *See* geographic base file–dual independent map encoding file.

GC *See* gas chromatography.

GC–ECD *See* gas chromatography–electron capture detector.

GC–MS *See* gas chromatography–mass spectrometry.

gear box A machine of different-diameter gears that connect a prime mover drive shaft to a driven piece of equipment and reduces the revolutions per minute. It is also called a gear reducer.

gear pump A rotary displacement pump that uses meshing gears that rotate in a housing, trapping water between the gear teeth in an inlet area, carrying it around to the outlet side, and imparting pressure. The meshing gears prevent water from returning to the inlet area.

gear reducer *See* gear box.

Geiger counter A device used to detect radiation by collecting and observing the pulse of ions created in an enclosed volume of gas by the passage of energetic ionizing particles or rays. It is used primarily for detecting and measuring beta and gamma radiation.

gel (1) A borehole lubricant used in drilling, similar to drilling mud. (2) A compound injected into aquifers to temporarily reduce the permeability in specific locations and thus direct flow in a desired direction. The technology is borrowed from the enhanced oil recovery industry and may have some value in controlling and recovering contaminated groundwater.

gel permeation chromatography (GPC) A separation technique based on the molecular weight of the analytes relative to the pore size of a stationary-phase gel. Because smaller molecules can enter the pores more easily than larger molecules, smaller molecules are retained longer in the column. This technique has been used in applications such as sample preparation and the determination of molecular weight distributions. Various materials have been used as the gel, such as dextran, agarose, and divinyl benzene/styrene.

gel resin A gel-based form of ion exchange resin. Its "pores" are on the molecular scale, so no "true" pores exist.

gel zeolite A synthetic sodium alumina silicate cation exchange product that was very widely used in residential water softeners prior to the development of divinyl benzene/styrene cation resin. It is also called siliceous gel zeolite. *See also* zeolite.

gelatinous Pertaining to suspended solids that are slimy, deformable, and capable of causing rapid filter plugging.

gene The functional unit of inherited information, expressed as a single trait. Genes are located on the chromosomes; each gene is coded for by a specific sequence of nucleotides in the deoxyribonucleic acid of the organism.

gene locus The location of a gene on a particular chromosome.

gene mapping A pictorial representation of the locations of various genes on a chromosome, as determined by laboratory techniques.

gene probe A piece of deoxyribonucleic acid that is specific to a target sequence and will bind to that target sequence. It can be labeled with enzymes for radioactive, colorimetric, or chemiluminescent detection.

gene splicing A process in the laboratory of cutting and inserting gene sequences along a chromosome.

gene-locus mutation Small alterations in a deoxyribonucleic acid (DNA) sequence. These mutations are created by the insertion of an inappropriate base in the newly synthesized or repaired strand of DNA (a base substitution mutation) or by small deletions or additions of bases. Gene-locus mutation is also referred to as point mutation. *See also* frame-shift mutation.

General Accounting Office (GAO) The investigative arm of the US Congress established by the Budget and Accounting Act of 1921 (31 U.S.C. 702). It is charged with examining all matters relating to the receipt and disbursement of public funds. GAO independently audits government agencies. It is under the control and direction of the comptroller general of the United States, who is appointed by the president of the United States with the advice and consent of the Senate for a term of 15 years.

general duty clause A contract provision that may be applied in order to cite a particular hazard when no standard applies. Its reference number in the Occupational Safety and Heath Administration regulations is CFR 29 1910.5(f).

general fund The primary fund (bank account) of a utility through which cash receipts and disbursements flow and in which the cash reserves of the utility reside. Although strict accounting rules are applied to the administration of funds, the general fund is generally unrestricted as to use, unlike special-purpose funds that can be used only for the purpose they were created (e.g., schools, streets, law enforcement).

general obligation (GO) bond A debt obligation issued by a government entity such as a state or municipality and backed by a pledge of the entity's full faith and credit. *See also* bond.

general permit An authorization that is issued on a nationwide or regional basis by the US Department of the Army or the US Environmental Protection Agency's National Pollution Discharge Elimination System for one or more categories of activities when (1) those activities are substantially similar in nature and cause only minimal individual and cumulative environmental impacts; or (2) the general permit would result in avoiding unnecessary duplication of regulatory control exercised by another federal, state, or local agency provided that the environmental consequences of the action have been determined to be individually and cumulatively minimal.

general-user bank A mitigation bank that can be used (i.e., allow withdrawal of credits) by a variety of unrelated entities.

generally regarded as safe (GRAS) A legal term defining a group of food additives with established use in the food industry under the Food, Drug, and Cosmetic Act. Some of these additives have been controversial because they have not been subjected to the formal safety testing required for new additives to food.

generator A piece of equipment used to transform rotary motion (e.g., the output of a diesel engine) into electric current.

genetic engineering The practice of modifying genomic deoxyribonucleic acid.

genetic mapping The process of determining the relative positions of genes on chromosomes.

genetic material Some quantity of deoxyribonucleic acid, the heritable information of living plants and animals.

genetic probe *See* gene probe.

genetic toxicology The study of the interaction of chemical or physical agents with the processes of heredity. Agents that produce damage to genes are responsible for a variety of diseases that are inherited, as well as for inducing cancer by inducing damage in somatic cells (i.e., nongerm cells).

genome The total deoxyribonucleic acid of an organism.

genotoxic Having the ability to induce damage to deoxyribonucleic acid (DNA) that results in a mutation upon replication of that DNA and division of that cell. In general, this term applies to a variety of mechanisms that can give rise to point mutations, as well as agents that produce overtly observable chromosomal damage (i.e., calstogens). Technically speaking, *genotoxic* is an inaccurate term because the gene will survive. Damage to DNA can also result in toxicity to the cell, but this type of activity is not specifically considered genotoxic. *See also* nongenotoxic.

genotoxic effect *See* genotoxic.

genotoxic activity The activity of a chemical, physical, or biological agent if that agent or one of its metabolites is capable of producing mutations that are successfully passed on to daughter cells. Such activity is known to be involved in the induction of cancer, birth defects, and other toxicities (though chemicals can produce these same effects by other mechanisms, too). Mutation is considered an irreversible event, so a linear relationship with dose at low response rates is assumed. *See also* carcinogen; nongenotoxic.

genotoxic carcinogen A chemical that induces cancer (1) by direct interaction with deoxyribonucleic acid (DNA), (2) through metabolites that induce damage to DNA, (3) or by indirectly giving rise to products that can damage DNA (e.g., by increasing oxygen radicals or reactive products of lipid preoxidation within the cell). Such damage increases the probability that errors will be made in the replication of DNA, resulting in a mutation that causes a protein with abnormal functional or control characteristics to be synthesized.

genotype (1) The entire genetic constitution of an individual. (2) An individual allele at a specific locus within an individual. *See also* phenotype.

genus A group of closely related species. Members of the same genus often have a number of common characteristics by which one can clearly see that the members are related.

geochemistry A science that deals with the chemical composition and chemical reactions of the substances that make up the crust and pore water of the earth and are contained in the pore water.

geocode A spatial index code (e.g., an address or parcel number) identifying unique points, lines, or areas that are stored in both graphic and nongraphic data. Geocodes are used extensively in geographic information system database management algorithms.

geodesy A branch of applied mathematics that determines by observation and measurement the shape and size of the earth or a large part of its surface, as well as the exact locations of points on its surface.

geodetic control A network of surveyed and monumented points on the earth's surface for which the locations are established in accordance with national accuracy standards.

geodetic head The static head or total head, without deduction for velocity head or losses.

geographic base file–dual independent map encoding file (GBF–DIME file) A geographic file based on line segments, produced by the US Bureau of the Census for each Standard Metropolitan Statistical Area in the United States. These files were the basis for developing the Census Bureau's Topological Integrated Geographic Encoding and Referencing file for urban areas.

geographic data Information that can be related to a physical location, such as a building, land parcel, or natural feature. Geographic data generally consist of both the graphic feature that indicates the spatial location and the attributes of that feature. The attributes may exist in the same data set as the graphic entities, depending on the data structure used, and may also exist in external tables that can be related to the feature by a unique identifier, such as parcel number, address, of facility identification.

geographic information system (GIS) A computer system that stores and links nongraphic attributes and geographically referenced data with graphic map features to allow a wide range of information processing and display operations, as well as map production, analysis, and modeling.

geographic reference system An *XYZ* coordinate system that locates specific places on the earth's surface for use in specifying points on a map.

geographically referenced data Data that contain some spatial reference that can be used to link graphic features and nongraphic attribute data in a geographic information system through geocodes.

geohydrology The study of the flow of water through the earth without considering the effects of the geology.

geological log A detailed description of all underground features discovered during the drilling of a well (e.g., depth, thickness, type of formation).

geology A branch of science devoted to the study of the earth. It primarily deals with the study of rocks, minerals, and soils, but certain aspects of water chemistry and atmospheric behavior are also in its domain.

geometric isomers Compounds that have the same groups attached to each carbon atom in a carbon–carbon double bond but for which the spatial arrangement is altered. The two spatial arrangements represent two different structural isomers of the same molecular formula. *See also cis*; isomer; *trans*.

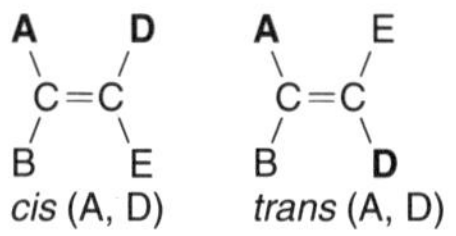

cis (A, D) *trans* (A, D)

Geometric isomers

geometric mean The *n*th root of the product of *n* given numbers. (It can also be calculated by taking the antilog of the mean of the logarithms of the individual numbers.) For example, the geometric mean of 6, 8, and 10 (three numbers, so $n = 3$) is the cube (third) root of $6 \times 8 \times 10$, or the cube root of $480 = 7.83$.

geomorphology The study of the evolution of land forms, not including the major forms of the earth's surface such as ocean basins and mountain chains.

geophysical log A record of the structure and composition of the earth encountered during drilling of a well or similar type of test hole or boring.

geophysical method Any of several indirect methods of determining the structure and composition of underground geological formations. Each method is based on using different types of scientific equipment and phenomena. The phenomena involved include the rate of travel of sound waves, electrical currents, and radio waves through the lithosphere, as well as variation in gravitational or magnetic force.

geophysiographic region A subdivision of the earth's relief according to geological landform.

geoprocessing and network analysis system A computer system incorporating geographic identifiers, or geocodes, used primarily for geographic analysis. The most common application is the assignment of planning, census, or statistical area identifiers to data records containing addresses via using a geographic reference file such as a dual independent map encoding file. *See also* geocode.

geosmin ($C_{12}H_{22}O$) The common name for *trans*-1,10-dimethyl-*trans*-9-decalol, an earthy-smelling chemical produced by certain blue-green algae and *Actinomycetes*. This odorous compound can be perceived at low nanogram-per-litre concentrations. Geosmin removal at a treatment plant requires either ozone (O_3) oxidation or activated carbon adsorption. *See also Actinomycetes*; blue-green algae.

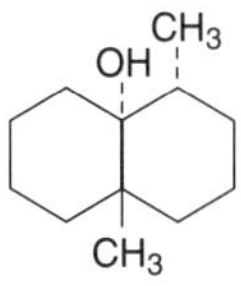

Geosmin

geothermal Having to do with the heat of the interior of the earth.

germ cell A cell that serves a sexual reproductive function. In humans these cells are sperm and eggs (ova). *See also* ova.

germicidal effect *See* germicidal treatment.

germicidal treatment Any treatment that kills microorganisms.

germicidal ultraviolet (UV) An ultraviolet light that peaks at a 253.7-nanometre wavelength and is of a wavelength between 200 and 300 nanometres. This range is known as the germicidal or short-wave ultraviolet band.

germicide A substance formulated to kill pathogens or microorganisms or to inactivate viruses. The germicidal properties of chlorine make it an effective disinfectant. *See also* disinfectant.

gestation index In reproductive toxicology, the percentage of pregnancies that result in the birth of live young.

GFAA *See* graphite furnace atomic absorption.

gfd *See* gallons per day per square foot *in the Units of Measure list.*

GFF *See* glass fiber filter.

GFI *See* ground fault interrupter.

Ghyben–Herzberg lens A sea-fed body of saline groundwater depressed by an overlying body of rain-fed groundwater.

Ghyben–Herzberg principle An equation relating the depth of a saltwater interface in a coastal aquifer to the height of the freshwater table above sea level. In equation form

$$z = \left(\frac{\rho_w}{\rho_s - \rho_w}\right)h$$

Where:

z = the depth of the saltwater interface below sea level, in feet (metres)

h = the elevation of the freshwater table above sea level, in feet (metres)

ρ_w = the freshwater density, in pounds mass per cubic foot (kilograms per cubic metre)

ρ_s = the saltwater density, in pounds mass per cubic foot (kilograms per cubic metre)

Giardia The genus name for a group of single-celled, flagellated, pathogenic protozoans found in a variety of vertebrates, including mammals, birds, and reptiles. These organisms exist either as trophozoites or as cysts, depending on the stage of the life cycle. *See also* cyst; trophozoite.

Giardia lamblia The species of *Giardia* that is a common cause of human diarrheal disease.

Giardia muris The species of *Giardia* found in rodents.

giardiasis A protozoan infection, principally of the upper small intestine, with *Giardia lamblia*. The median incubation period is 7 to 10 days, and the infectious dose is low—10 cysts have caused infection. Although infected persons are usually asymptomatic, infection may be associated with a variety of intestinal symptoms, such as abdominal cramps, chronic diarrhea, frequent loose and greasy stools, bloating, flatulence, fatigue, and weight loss. Malabsorption of fats or fat-soluble vitamins may occur. Occurrence is worldwide, with children found to be more frequently infected than adults. Person-to-person transmission occurs by hand-to-mouth transfer of cysts from the feces of an infected person, especially in institutions and day-care centers. Outbreaks often occur from ingestion of cysts in fecally contaminated water, especially unfiltered surface water, and less frequently from contaminated food. Reservoirs of infection include humans and wild and domestic animals. *See also* cyst, *Giardia lamblia*, waterborne disease.

Gibbs free energy (*G*) A function of state used to determine the spontaneity of a process; formerly called Gibbs free energy function. *See also* standard free-energy change; state function.

GIF *See* graphic interchange format.

giga (G) Prefix meaning 10^9 in Système International.

gigabyte *See the Units of Measure list.*

gigalitre (GL) *See the Units of Measure list.*

Girard turbine An impulse turbine with buckets shaped so that, upon leaving the wheel, the water has a relative velocity in the direction opposite to that of the wheel's motion.

GIS *See* geographic information system.

GL *See* gigalitre *in the Units of Measure list.*

glacial lake (1) A lake occurring in the bottom of a valley that was formed by glacial action. (2) A body of fresh water on the land surface impounded by a dam of glacial ice.

glacial stream A stream supplied by melting glacier ice.

glacier An extensive, usually slowly flowing body of ice formed on land in a cold region where more snow falls than is melted. The snow is transformed into ice by pressure and, together with frozen liquid water, constitutes the body of the glacier.

glacier burst A sudden release of a volume of water that has been impounded by a glacier.

glacier snow The compacted mountain snow that is in the intermediate stage between ordinary snow and glacier ice.

glaciofluvial stream A stream having melting glacial ice as the source of its water and much of its load.

gland (1) A mechanical device usually bolted to a pump casing to keep a mechanical seat or to adjust packing in a stuffing box between a pump shaft and pump casing, so as to reduce or control water leakage along the pump shaft. A small amount of leakage along a pump shaft through the packing is necessary for lubrication of the shaft. (2) A mechanical device to retain and compress packing around a valve stem to prevent flow between the valve stem and valve bonnet.

glass fiber filter (GFF) A filter made of uniform glass fibers with a pore diameter of 0.7 to 2.7 micrometres. Such filters are used to filter fine particles and algae while maintaining a high flow rate.

glass, pipe and fittings *See* pipe class.

glass-distilled water Water that has been freed from dissolved and suspended matter by distillation in an all-glass system.

Glauber salt ($Na_2SO_4 \cdot 10H_2O$) An anhydrous sodium sulfate salt compound.

glauconite ($K_{15}(Fe, Mg, Al)_{4-6}(Si, Al)_3O_{20}(OH)_4$ approx.) A natural dull-green iron potassium silicate mineral occurring abundantly in greensand. *See also* greensand.

gleyed Pertaining to a soil condition resulting from gleization (the weathering process that produces a gleisoil horizon) that is manifested by the presence of neutral gray, bluish, or greenish colors through the soil matrix or in mottles (spots or streaks) among other colors. Reduced iron is often involved.

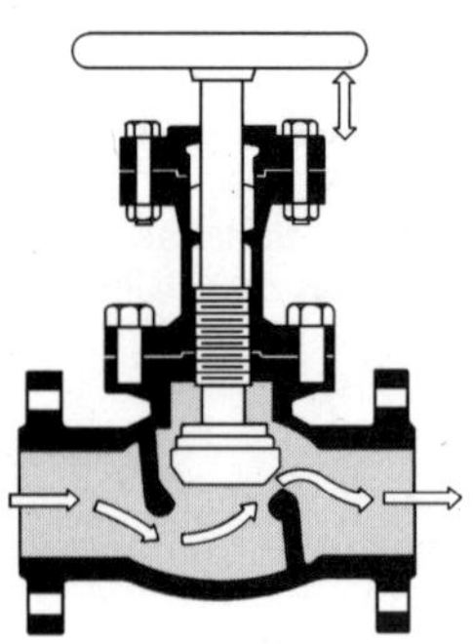

Globe valve

global adjustments Systemwide adjustments made to a particular parameter. An example is applying a systemwide percentage change in water consumption (in a computer simulation model) to simulate a uniform change in demand.

global positioning system (GPS) A constellation of satellites originally developed by the US Department of Defense as a navigational aid that in 1993 became available for geodetic control surveying. The satellites transmit signals that can be decoded by specially designed receivers to determine positions precisely (within centimetres). These measurements can be made at a fraction of the cost of traditional first-order surveying techniques. Eventually, the global positioning system will be replaced by an 18-satellite constellation that will allow 24-hour, all-weather operational capacity in both navigation and relative positioning.

globe valve A valve that has a round opening to let liquid pass and that closes when a stem is turned to press a disk against the round opening. Globe valves are used in plumbing where numerous openings and closings are anticipated.

globulin A plasma protein characterized by being "insoluble" in water but "soluble" in saline. This property results because salt, which is necessary to act as a counterion to charged groups on the protein, prevents ionic bonds from forming between molecules of the globulins; these bonds ordinarily cause the molecules to aggregate and precipitate. Note that the use of the term *soluble* is not precisely correct for proteins of this size; these proteins actually form a really fine suspension that is maintained by Brownian motion.

glomerulonephritis A variety of inflammatory processes that involve the capillary loops of the glomeruli in the kidney. The glomerulus is the structure within the kidney in which an ultrafiltrate of blood is formed that is eventually made into urine after the body reabsorbs a large fraction of the water and most nutrients. The inflammation is produced with chronic metal poisoning—mercury in particular but cadmium and lead as well. It tends to be an autoimmune response rather than a direct effect of these chemicals.

glory-hole spillway A vertical shaft having a funnel-shaped entrance and ending in an outlet tunnel, used to provide an overflow from a reservoir. It is also called a morning-glory spillway or shaft spillway.

β-glucuronidase An enzyme that is capable of hydrolyzing the glycosidic linkage between glucuronic acid and a chemical possessing a hydroxyl group. It occurs in a variety of tissues (e.g., spleen, liver, and endocrine glands) but is also secreted by a number of bacteria found in the colon. The bacterial enzyme can hydrolyze off the glucuronide bonded to a drug, making it available for reabsorption. *See also* enterohepatic circulation.

β-D-glucuronidase (GUD) An enzyme that catalyzes the hydrolysis of β-D-glucopyranosiduronic derivatives into their corresponding aglycons and D-glucuronic acid. The presence of this enzyme in most *Escherichia coli* isolates and some *Shigella* and *Salmonella* strains allows for a useful and specific test for these organisms.

glucuronic acid ($COOH(CHOH)_4CHO$) The acidic derivative of glucose (i.e., the aldehyde group has been oxidized to a carboxylic acid).

glucuronide A glycosidic complex of glucuronic acid with any chemical.

β-D-glucuronide A glycoside derived from glucuronic acid that is a substrate for the enzyme β-D-glucuronidase, an enzyme produced by 94–96 percent of *Escherichia coli* isolates tested.

glucuronyltransferase Any of a group of enzymes that are responsible for transferring glucuronic acid from uridine-5'-diphospho-D-glucuronic acid to form a glycosidic linkage with metabolites of foreign chemicals or biochemicals.

γ-glutamyltranspeptidase An enzyme that hydrolyzes the glutamic acid group from glutathione conjugates of various chemicals. It is a membrane-bound enzyme found in high concentrations in cells that exhibit absorptive or excretory functions (e.g., the bile ducts in the liver). This enzyme is also used as a marker for preneoplastic lesions in the livers of rats. Normal rat hepatocytes (parenchymal cells of the liver) do not express significant amounts of this enzyme. Hepatocytes that have been altered by treatment with genotoxic carcinogens will express much higher concentrations of this enzyme, allowing them to be detected with appropriate histochemical staining.

glutaraldehyde ($OHC(CH_2)_3CHO$) A chemical biocide used to sterilize high-purity water systems. For some types of reverse osmosis membrane systems, it may be used periodically at dilute concentrations to sterilize membranes and piping and as a membrane element storage (preservative) solution.

glutathione ($C_{10}H_{17}O_6N_3S$) A tripeptide of glycine, glutamic acid, and cysteine that has diverse roles in protecting cells from damage. Glutathione is frequently bonded with the metabolites of toxic chemicals. This most frequently results in inactivation of the chemical, but in special cases metabolites that are toxic, carcinogenic, or both (e.g., 1,2-dichloroethane) can be formed by conjugation with glutathione. The sulfur group of glutathione is used as an antioxidant for the cell because it easily forms disulfide bonds with itself or other thiol-containing biochemicals in the body. Processes exist to reduce these disulfide bonds by the reduced form of nicotinamide-adenine dinucleatide phosphate within the cell.

glutathione peroxidase An enzyme that uses glutathione to inactivate peroxides.

glutathione reductase An enzyme that regenerates glutathione that has been directly oxidized or oxidized by glutathione peroxidase.

glutathione-S-transferase Any of a group of enzymes that transfer glutathione to reactive metabolites of chemicals. This generally results in the detoxification of the chemical.

glycocalyx The extracellular slime that surrounds bacteria found in the biofilms that form on solid surfaces in contact with water. Such slimes consist of materials created by the bacteria and extruded from the cells. These materials are exopolymers that are primarily polysaccharide in nature. *See also* extracellular polymeric substance.

glycogen ($(C_6H_{10}O_5)_n$) A polymer of glucose that serves as a reserve for replenishing cellular and circulating levels of glucose. The levels and mobilization of glycogen in cells are intimately related and coordinated with other cellular functions through signaling pathways. Some of these pathways are responsive to insulin and a variety of autocrine (within the cell) and paracrine (affecting neighboring cells) factors. Human genetic diseases linked with the accumulation of glycogen are associated with an early induction of liver tumors.

glyoxal (OHCCHO) A dialdehyde created during the reactions of oxidants used as disinfectants, particularly ozone (O_3), with natural organic matter. Glyoxal is often the second most prevalent aldehyde produced (after formaldehyde) during the ozonation of drinking water. Glyoxal is not as biodegradable as formaldehyde, but it can readily be removed on a biologically active carbon filter. *See also* biologically enhanced activated carbon; dialdehyde; formaldehyde.

glyoxylic acid (OHCOOH) An aldoacid created during the reactions of oxidants used as disinfectants, particularly ozone (O_3), with natural organic matter. Glyoxylic acid can be removed during biological filtration. *See also* aldoacid; biological filtration.

glyphosate ($HO–CO–CH_2–NH–CH_2–PO–(OH)_2$) The common name for the isopropylamine salt of N-(phosphonomethyl)glycine. Glyphosate is a nonselective, postemergence herbicide. It is regulated by the US Environmental Protection Agency. *See also* herbicide.

```
  O              O
  ||             ||
HO-C-CH2-NH-CH2-P-OH
                 |
                 OH
```

Glyphosate

gnotobiotic Pertaining to an animal that has been raised in germ-free (bacteria-free) conditions. Because these animals have no resident microflora, they are usually more susceptible to infection than are normal animals.

GO bond *See* general obligation bond.

***G*⁰** *See* standard free energy.

go-devil A scraper with self-adjusting spring blades inserted in a pipeline and carried forward by the fluid pressure to clear away accumulations, tuberculations, and similar coverings. *See also* polypig.

gonadal function The capacity of the gonads to produce gametes (i.e., sperm and ova) and to maintain a variety of secondary sexual characteristics (e.g., beard in men, enlarged breasts in females). Gonadal function refers primarily to the production of gametes, but altered function will also have impacts on secondary sexual characteristics.

Gooch crucible *See* filtering crucible.

goodness of fit The degree to which observed or experimental data correspond to an assumed mathematical model to describe the data set. Goodness of fit is indicated by the magnitude of the coefficient of determination. *See also* coefficient of determination.

Goodpasture's disease An inflammation of the glomerular capillaries of the kidney that is associated with the production of antibodies to proteins in the basement membrane (i.e., the membrane to which cells attach in the formation of an organ). The syndrome has been associated with excessive exposure to solvents, but the causal pathway has not been worked out in detail.

gooseneck A sweeping bend in a service line resembling the shape of a goose's neck that will allow soil movement without damaging the service line.

gorge A narrow passage often formed by a river cutting through the wall of a canyon—especially one with steep, rocky walls—or through soft rock between mountains.

Government Printing Office (GPO) An agency of the US government established by the US Congress to execute orders for printing and binding placed by Congress and departments and establishments of the federal government. It prepares catalogs and distributes and sells government publications through mail orders and government bookstores.

government-owned water utility A water utility created by the state or other government agency's legislative action, with the mandate that the utility's purposes are public purposes and that its functions are essential governmental proprietary functions. Its primary purpose is to provide its designated area with potable water in an adequate supply and at a reasonable cost so that people of the area may improve their health, safety, and welfare. A government-owned water utility may be part of a municipal government operation, a county agency, or a regional authority, or it may take some other form appropriate for the service area.

GPC *See* gel permeation chromatography.

gpcd *See* gallons per capita per day *in the Units of Measure list.*

gpd *See* gallons per day *in the Units of Measure list.*

gpg *See* grains per gallon *in the Units of Measure list.*

gph *See* gallons per hour *in the Units of Measure list.*

gpm *See* gallons per minute *in the Units of Measure list.*

gpm/ft^2 *See* gallons per minute per square foot *in the Units of Measure list.*

GPO *See* Government Printing Office.

GPS *See* global positioning system.

gps *See* gallons per second *in the Units of Measure list.*

gpy *See* gallons per year *in the Units of Measure list.*

gr *See* grain *in the Units of Measure list.*

grab sample A discrete sample taken under specific circumstances at a given time and location. A grab sample is easier to take than a composite sample, but it allows only a "snapshot" of the water quality at the time the sample was taken. A grab sample is most representative when the water quality is fairly constant. Many of the methods for the compliance monitoring of drinking water require that grab samples be taken.

grad *See the Units of Measure list.*

gradation (1) The geologic process, including aggradation and degradation, of bringing the surface of the Earth, or any portion thereof, to a common level or uniform slope. (2) The process of bringing the bed of a stream to a gradient (sharper or gentler) at which the water flowing in it is just able to transport the material delivered to it.

grade (1) The elevation of the invert (or bottom) of a pipeline, canal, culvert, or similar conduit. (2) The inclination or slope of a pipeline, conduit, stream channel, or natural ground surface, usually expressed in terms of the ratio or percentage of number of units of vertical rise or fall per unit of horizontal distance. For example, a 0.5 percent grade would be a drop of 0.5 feet per 100 feet of pipe.

grade D air The type of air used in self-contained breathing apparatuses. It is also known as respirator air. The Compressed Gas Association sets standards for air grades A–J. Grade A is the ambient surrounding air, and Grade J is for very deep sea diving. Grade D air meets the following limits: hydrocarbons, 5 milligrams per cubic metre, carbon monoxide (CO), 20 parts per million; carbon dioxide (CO_2), 1,000 parts per million; oxygen (O_2), 19.5–23.5 percent; water vapor, none; no noticeable odors. *See also* self-contained breathing apparatus.

grade tunnel A waterway tunnel constructed with the elevation of the top above the hydraulic gradient.

graded response An individual response to chemical exposure that increases in intensity as the dose is increased. (This is in contrast to responses that are considered stochastic or all-or-none.) For example, an individual who was exposed to a renal toxin would experience increasing damage as the dose of the toxin increased. On the other hand, a lethal dose response curve describes the sensitivity of a population of individuals with varying sensitivity to the chemical.

graded stream A stream in which the channel has reached a stable form as a result of the stream's flow characteristics.

gradient The rate of change of any characteristic per unit of length, i.e., the slope. The term is usually applied to such characteristics as elevation, velocity, and pressure. *See also* slope; velocity gradient.

gradient ion chromatography A technique in which the eluant strength is increased during the separation of analytes on an ion exchange column. This is accomplished by mixing two or more elution solvents during the analytical run. A gradient approach typically reduces the analysis time compared to an isocratic approach, in which the elution solution is constant during analyte separation.

grading (1) The design of landscape contours to accommodate a site's uses and provide adequate storm drainage. (2) The process of constructing desired landscape gradients.

grading plan A plan drawn to scale that expresses the designed landscape gradient and elevation by showing contour lines or numeric notation of elevations.

gradual contraction A reduction in the cross-sectional area of a stream, channel, conduit, or other hydraulic structure that

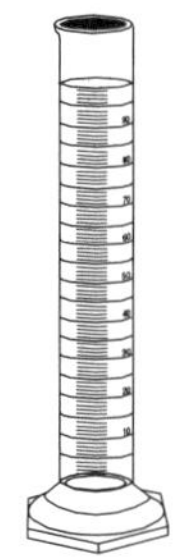
Graduated cylinder

occurs over a considerable distance relative to the dimensions of the channel or conduit, such as depth or diameter.

graduated cylinder A tall, cylindrical, glass or plastic container with a hexagonal base and a pouring lip. It is used for measuring liquids quickly without great accuracy. Graduations are marked on the side.

grain (gr) *See the Units of Measure list.*

grains per gallon (gpg) *See the Units of Measure list.*

gram (g) *See the Units of Measure list.*

gram-mole *See* gram molecular weight *in the Units of Measure list.*

gram-milliequivalent The equivalent weight (in grams) divided by 1,000.

Gram's stain A common staining procedure used to differentiate bacteria into gram-negative and gram-positive categories. *See also* coliform bacteria.

granular activated carbon (GAC) A form of particulate carbon manufactured with increased surface area per unit mass to enhance the adsorption of soluble contaminants. Granular activated carbon is used in fixed-bed contactors in water treatment and is removed and regenerated (reactivated) when the adsorption capacity is exhausted. In some applications, granular activated carbon can be used to support a biological population for stabilizing biodegradable organic material. *See also* biologically enhanced activated carbon; fixed-bed column; granular activated carbon capped filter; granular activated carbon–sand filter; powdered activated carbon.

granular activated carbon (GAC) capped filter A filter containing at least two different types of filtering media in which the top layer is granular activated carbon. The granular activated carbon provides some particle removal, but its primary purpose is to adsorb organic materials, including many taste-and-odor-causing compounds. These filters can also stabilize biodegradable organic materials if microorganisms are permitted to grow on the surface of the granular activated carbon. *See also* biologically enhanced activated carbon; granular activated carbon–sand filter; powdered activated carbon.

granular activated carbon–sand (GAC–sand) filter A dual-media filter in which the top layer is granular activated carbon and the bottom layer is sand. In retrofit applications, the granular activated carbon often replaces an anthracite layer to improve the adsorption of organic material, to stabilize biodegradable material when biologically active, or both. In most cases, the depth of the granular activated carbon is greater than in a granular activated carbon capped filter. *See also* granular activated carbon capped filter.

granular activated carbon (GAC) sandwich filter A modification of a slow sand filter for which a layer of granular activated carbon is sandwiched between two layers of sand in order to add an adsorptive removal capacity to the process of slow sand filtration and to further enhance the biological purification process. The operation and maintenance of a GAC sandwich filter should be identical with a conventional slow sand filter because the subsurface layer of GAC does not interfere with the normal sand cleaning operations. The upper sand layer acts as a normal slow sand filter, purifying the water by the combined processes of microstraining, settling, and biological degradation of dissolved and particulate organic matter. The GAC layer removes nonbiodegradable dissolved organics (such as pesticides, color, taste-and-odor-causing compounds, and precursors of disinfection by-products) and further enhances the biological decomposition of the natural organic matter. The bottom sand layer resumes the typical slow sand filter functions of ensuring the microbiological stability of the water and the high quality of the filtrate. The GAC sandwich technology comprises techniques of installing, removing, separating, and reprocessing the media; it was developed and patented by Thames Water Utilities.

granular media A material used for filtering water, consisting of grains of sand or other granular material.

granular soil A soil that looks and feels dry and breaks up into clumps with moderate pressure.

graphic Pertaining to digital descriptions of map features and logical geographic relationships among features, drawings, and images in a geographic information system. Graphic features or geometric forms are described by seven elements: points, lines, areas, grid cells, pixels, symbols, and annotation. A geographic information system uses graphic data to generate a map or cartographic "picture" on a display device (computer screen), paper, or other medium.

graphical user interface (GUI) A user interface to an application or database that provides a set of menus using combinations of icons, buttons, dialog areas, and scroll bars. The user needs little understanding of the operating system, software commands, or data structure in order to use the interface.

graphics Combinations of individual graphic features, generally referring to the contents of a file or database of graphic elements as opposed to tabular data. *See also* graphic.

graphite furnace *See* graphite furnace atomic absorption spectrophotometry.

graphite furnace atomic absorption (GFAA) *See* graphite furnace atomic absorption spectrophotometry.

graphite furnace atomic absorption (GFAA) spectrophotometry An electrothermal instrumental technique used for the analysis of metals. The sample is atomized in a graphite tube after the temperature is raised quickly. This approach is typically used for microgram-per-litre concentrations. Drinking water samples require a small volume of sample, typically 5–50 microlitres, and minimal sample preparation. The spectrophotometer used is composed of a power supply, an atomizer, and a temperature programmer. A small aliquot of the sample is injected into a graphite tube. The temperature of the graphite tube is raised rapidly to atomize the metal in the path of the light source.

grapple dredge A floating derrick with a clamshell, orange peel, or other type of grab bucket used for removing material below water in very deep water or in confined places.

GRAS *See* generally regarded as safe.

grassed waterway A vegetated natural waterway used to conduct the accumulated runoff from cultivated land or fields in a strip-crop system.

grating A screen consisting of two sets of parallel bars transverse to each other in the same plane.

gravel Rock fragments 0.08–3 inches (2–75 millimetres in diameter.

gravel bed *See* gravel support bed.

gravel pack Gravel surrounding a well intake screen, artificially placed ("packed") to aid the screen in filtering sand out of the aquifer. Gravel packs are usually needed in aquifers containing large proportions of fine-grained material.

gravel support bed A combination of various layers of different-sized gravel and coarse sand placed above the underdrain network to support filter or ion exchange media beds. The gravel support bed contributes greatly to the collection of product water and the even dispersal of backwash water flow. *See also* backwash.

gravel-wall well A type of well used in a water-bearing formation containing a large proportion of fine-grained material to permit the passage of water at low velocity. Gravel is introduced around the screen or intake section of the well to increase the specific capacity and to prevent extremely fine material from flowing into the well. Depending on the level of construction detail, such a well may be called a gravel well, gravel-envelope well, gravel-filled well, or gravel-packed well.

gravimetric Of or pertaining to measurement by weight.

gravimetric analysis The process of taking measurements based on the weight of a substance. For example, in the determination of total solids in water, a known volume of water is evaporated to dryness; the weight of the residue is then weighed and expressed in milligrams per litre.

gravimetric dry feeder *See* gravimetric feeder.

gravimetric feeder A chemical feeder that adds specific weights of dry chemical.

gravimetric measurement Measurement on the basis of weight.

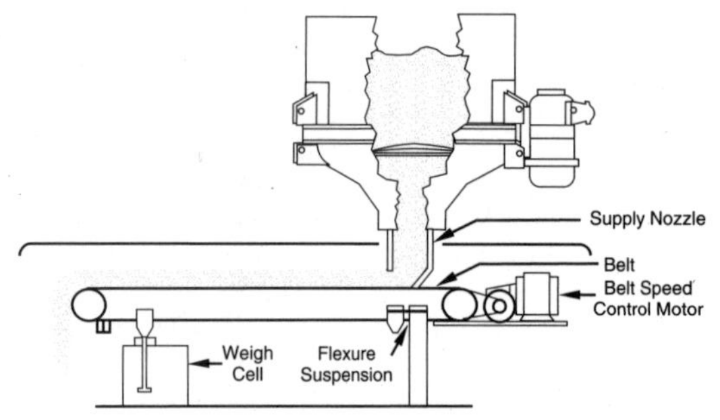

Gravimetric feeder

gravimetric procedure Any analytical procedure that uses the weight of a constituent to determine the constituent's concentration.

gravitational convection *See* free convection.

gravitational potential *See* gravity potential.

gravitational water Water that moves under the force of gravity.

gravity distribution A water supply that uses natural flow from a mountain reservoir or elevated tank to supply pressure.

gravity drainage The downward movement of water in the vadose zone caused by gravity.

gravity filter A rapid granular filter of the open type for which the operating level is placed near the hydraulic grade line of the influent and through which the water flows by gravity.

gravity potential A potential resulting from the position of groundwater or soil moisture above a datum.

gravity spring A spring in which water flows from permeable material or from openings in a rock formation entirely under the action of gravity.

gravity system (1) A system of conduits (open or closed) in which the liquid runs on descending gradients from source to outlet and for which no pumping is required. (2) A water distribution system in which no pumping is required.

gravity thickener A device used for removing solids from water by settling and removal. Gravity thickeners are typically circular sedimentation basins that may use rotating arms to assist in the thickening process, depending on the sludge concentration.

gravity thickening A method for reducing the water content of a sludge by settling.

gravity water A supply of water that is transported from its source to its place of use by means of gravity, as distinguished from a supply that is pumped. *See also* free water.

gravity water supply A potable water supply available to consumers without pumping.

gravity wave A displacement or wave that forms and propagates at the free surface of a body of fluid after that surface has been disturbed and the fluid particles have been displaced from their original positions. Such waves form because of the role played by the restoring force of gravity.

gray (Gy) *See Units of Measure list.*

graywater Untreated household used water—such as wash or rinse water from a sink, bathtub, or other household plumbing fixture, except a toilet—that does not contain human wastes.

grease An aggregate group of substances that can analytically be measured—along with oil—by extraction with trichlorotrifluoroethane. (The result of this extraction is called oil and grease.) Grease includes rendered animal fat, oily matter, and thick lubricants. *See also* oil.

green industry (1) The trades, professions, and disciplines related to landscape and irrigation research, design, installation, and management. (2) The renewables industry.

green rust Any of the solids in a class of mixed valence iron (II) and iron (III) [Fe(II)–Fe(III)] hydroxides that consist of alternating positively charged trioctahedral metal hydroxide layers and negatively charged interlayers of anions such as carbonate (CO_3^{2-}), sulfate (SO_4^{2-}), or chloride (Cl^-). Green rust is formed readily when iron corrodes in the presence of oxygen or when ferric oxyhydroxides (such as ferrihydride or lepidocrocite) are dissolved under reducing conditions in contact with dissolved ferrous iron species at near-neutral pH. Green rust solids have been reported as a corrosion product of cast-iron water pipes and may be an important intermediate phase in oxidation–reduction processes.

green vitriol ($FeSO_4$) A common name for ferrous sulfate (copperas).

greenhouse effect Warming of the Earth's surface resulting from the capacity of the atmosphere to transmit short-wave energy (visible and ultraviolet light) to the Earth's surface and to absorb and retain heat radiating from the surface.

greensand (1) A naturally occurring mineral that consists largely of dark greenish grains of glauconite and possesses ion exchange properties. Greensand was the original product used in commercial and home cation exchange water-softening units and was the base product for manufacturing manganese greensand zeolite products. (2) A filter that adsorbs soluble iron and manganese through the use of a sand medium coated with manganese dioxide (MnO_2). When chlorine (Cl_2) or potassium permanganate ($KMnO_4$) is added to the influent, the adsorbed iron and manganese are oxidized, thus "regenerating" the greensand filter.

greywater *See* graywater.

gridiron layout, gridiron system A system of distribution pipes in which, if one path is disturbed, alternative paths exist through which water can flow.

grit Sand-like particles mixed with debris and mud.

groin A wall, crib, row of piles, stone jetty, or other barrier projecting outward from the shore or bank into a stream or other body of water for such purposes as protecting the shore or bank from erosion, arresting sand movement along the shore, or concentrating the low flow of a stream into a smaller channel. A groin is also called a wing, dam, jetty, or spur. *See also* bulkhead.

gross alpha (α) particle activity The total radioactivity caused by alpha particle emission as inferred from measurements on a dry sample. It is regulated by the US Environmental Protection Agency.

gross alpha radiation *See* gross alpha particle activity.

gross available head (1) The total difference in elevation between the water surface at the point on a stream where water is diverted from the stream and the water surface at the point where the water is returned to the stream. (2) The amount of fall in a stream that is available for hydroelectric power development. Gross available head is also called gross head.

gross beta (β) particle activity The total radioactivity caused by beta particle emission as inferred from measurements on a dry sample. It is regulated by the US Environmental Protection Agency.

gross beta (β) radiation *See* gross beta particle activity.

gross duty of water The water requirement or duty measured at the point of diversion. It includes all water uses as well as all canal or conduit losses, seepage losses, evaporation and transpiration losses in the system, and all waste. It is also called the diversion duty of water or head-gate duty of water.

ground An electrical connection to the earth or to a large conductor that is known to be at the same potential as the earth.

ground air The gases in the interstices of the aeration zone that open directly or indirectly to the surface and therefore

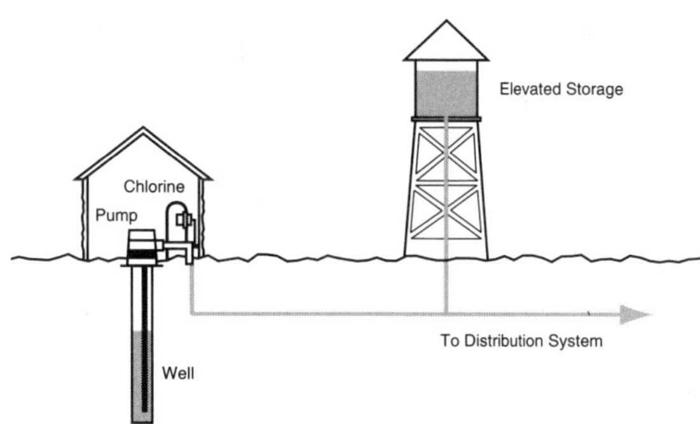

Simple groundwater system

"communicate" with the atmosphere. Ground air is also called soft air or subsurface air.

ground cover All plants except trees and grass (e.g., herbs, moss, and small shrubs) covering the soil, protecting it from temperature and humidity variations and from erosion.

ground, electrical *See* electrical ground.

ground fault circuit breaker *See* ground fault interrupter.

ground fault interrupter (GFI) A device that detects voltage leakage on an electrical line and discontinues voltage to prevent the user from receiving an electrical shock. Each outlet protected by a ground fault interrupter should be marked.

ground key valve A valve that can be closed or opened to full flow in a turn of the plug. Such a valve is generally used on water service pipes.

ground sills Underwater walls built at intervals across the bed of a channel in order to prevent excessive scour of the bed or to increase the width of the water surface.

ground storage A reservoir or tank for storage of water at or below ground level. It allows larger volumes of water to be stored relative to elevated storage, but generally ground storage water must be pumped to provide pressure unless the ground is higher than the pressure plane being served.

grounding An electrical wiring practice that protects people and equipment from electrical voltage and sends the charge to the ground.

ground-level storage In any distribution system, storage of water in a shallow tank for which the bottom is below or at the surface of the ground. Booster pumps are ordinarily required to raise water from ground-level storage.

ground-penetrating radar A surface geophysical measurement device that operates based on the transmission of repeated pulses of electromagnetic waves into the ground. Some of the radiated energy is reflected back to the surface, and the reflected signal is captured and processed.

ground–voltage gradient A record of the voltage differences between each two points in a series of locations along the ground surface.

groundwater The water contained in interconnected pores located (1) below the water table in an unconfined aquifer or (2) in a confined aquifer.

groundwater aquifer *See* aquifer.

groundwater artery A body of permeable material encased in a matrix of less permeable or impermeable material and saturated with water that is under pressure (usually artesian).

groundwater basin A groundwater reservoir that is separate from neighboring groundwater reservoirs. A groundwater basin could be separated from adjacent basins by geologic boundaries or by hydrologic boundaries.

groundwater cascade The descent path of groundwater on a steep hydraulic gradient to a lower and flatter water table slope. A cascade occurs below a groundwater barrier or dam and at the point of contact between less permeable material and more permeable material downslope.

groundwater classification A scheme to categorize groundwater according to its quality.

groundwater, confined *See* confined groundwater.

groundwater dam A geologic formation that is impermeable or has a low permeability and that occurs in such a position that it impedes the horizontal movement of groundwater and consequently causes a pronounced difference in the levels of the water table on opposite sides.

groundwater decrement All groundwater extracted from the groundwater reservoir beneath a given surface area by evaporation, transpiration, spring flow, effluent seepage, pumping from wells, and other underground loss or diversion.

groundwater depletion curve A graphical representation of groundwater recession after the flow created by direct runoff has passed. Such a curve is also known as a groundwater recession curve.

groundwater under the direct influence of surface water (GWUDI) Water defined by the US Environmental Protection Agency in the Surface Water Treatment Rule as any water beneath the surface of the ground that has (1) significant occurrence of insects or other macroorganisms, algae, organic debris, or large-diameter pathogens such as *Giardia lamblia*, or (2) significant and relatively rapid shifts in water characteristics—such as turbidity, temperature, conductivity, or pH—that closely correlate with climatological or surface water conditions. The Enhanced Surface Water Treatment Rule amends the first item of this definition to include *Cryptosporidium*.

groundwater discharge Discharge of water from the saturation zone directly onto the land surface, into a body of surface water, or into the atmosphere by means of springs, wells, infiltration galleries, or infiltration tunnels and other subterranean channels. It is also called phreatic water discharge.

groundwater discharge area (1) An area in which groundwater is discharged through springs, wells, or percolation or is evaporated from the soil or is transpired from vegetation. (2) The cross-sectional area of an aquifer through which groundwater is moving.

Groundwater Disinfection Rule (GWDR) A regulation of the US Environmental Protection Agency to establish disinfection requirements for public water systems using groundwater.

groundwater divide A boundary between two adjacent groundwater basins, represented by a high point in the water table or piezometric surface.

groundwater drain A drain that carries away groundwater; a drain laid below a pipe through wet ground to facilitate construction. A groundwater drain is also called an underdrain.

groundwater flow (1) The flow of water in an aquifer. (2) The movement of water through openings in sediment and rocks in the zone of saturation. (3) That portion of the discharge of a stream that is derived entirely from groundwater, through springs or seepage water. This flow is also called groundwater runoff.

groundwater hill A mound- or ridge-shaped feature of a water table or piezometric surface, usually produced by downward percolation of water to water-bearing deposits. Such a feature is also called a groundwater mound or groundwater ridge.

groundwater hydrology The branch of hydrology that deals with groundwater, its occurrence and movements, its replenishment and depletion, the properties of rocks that control groundwater movement and storage, and the methods of investigation and use of groundwater. Groundwater hydrology is sometimes called geohydrology or hydrogeology.

groundwater increment Water added to a groundwater reservoir from all sources: influent seepage from streams, rainfall, irrigation, and inflow of groundwater from outside the area under consideration.

groundwater infiltration (GWI) The seepage of groundwater into shafts and tunnels. It is desirable when the shafts are designed for collecting groundwater (e.g., skimmer wells, infiltration galleries) but undesirable when the shafts are designed for other purposes, such as mining or water transmission.

groundwater inventory A detailed estimate of the quantities of water added to the groundwater reservoir of a given area (groundwater increment) balanced against estimates of quantities abstracted or lost from the groundwater reservoir of the area (groundwater decrement). *See also* hydrologic inventory.

groundwater level The level below which the rock and subsoil, down to unknown depths, are saturated with water.

groundwater mining The long-term extraction of groundwater from an aquifer at a rate that far exceeds the natural and artificial replenishment rate of the aquifer. The intent in groundwater mining is to extract the water resource without any expectation of replacement.

groundwater model A tool to approximate the response of an aquifer to pumping, injection, contamination, and other natural and artificial hydraulic and hydrologic activities. The models can be very simple equations or complex computer programs, depending on the nature of the problem.

groundwater mound *See* groundwater hill.

groundwater, perched *See* perched groundwater.

groundwater piracy The tapping of water that originates in another surface drainage basin by a solution channel in limestone terrain.

groundwater pressure head At a given point, the hydrostatic pressure expressed as the height of a column of water that can be supported by the pressure. In a tightly cased well where no discharge occurs, this is equivalent to the distance that a column of water rises.

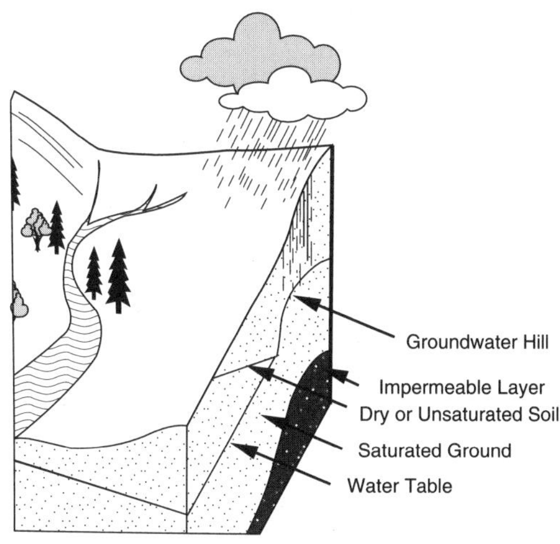

Groundwater hill

groundwater province An area characterized by a general similarity in the mode of occurrence of groundwater.

groundwater recession A lowering of the groundwater level of an area.

groundwater recharge The replenishment of a groundwater source through percolation, injection, or other means. In arid areas that depend on groundwater as a primary source, using highly treated wastewater or other nonpotable water sources for groundwater recharge is becoming more common.

groundwater reservoir A geologic formation that stores sufficient water to yield economically significant water to wells and springs. *See also* aquifer; groundwater basin.

groundwater ridge *See* groundwater hill.

groundwater runoff That portion of runoff that has infiltrated the groundwater system and has later been discharged into a stream channel as spring or seepage water. Groundwater runoff is the principal source of base or dry-weather flow for streams unregulated by surface storage, and such flow is sometimes called groundwater flow.

groundwater safe yield The amount of naturally occurring groundwater that can be economically and legally withdrawn from an aquifer on a sustained basis without creating an undesirable effect.

groundwater storage Water temporarily stored within the functional interstices of permeable rocks. *See also* groundwater reservoir.

groundwater storage curve A curve summing up the area under the groundwater depletion curve to show the volume of groundwater available for runoff at given rates of groundwater flow.

groundwater table (1) The upper surface of a body of unconfined groundwater. (2) The elevation of depth below the ground surface of such a water surface. In a confined aquifer, this elevation is defined by the static levels in wells that draw from the aquifer.

groundwater time of travel The elapsed time required by a fluid particle on a particular flow line to travel a prescribed distance. It is a common term in well-head protection planning. It is commonly expressed in years. *See also* flow line.

groundwater trench A trench-shaped depression of the water table caused by effluent seepage into a drainage ditch or a stream.

groundwater turbulent flow Turbulent flow that occurs under high velocities in large openings in the saturation zone, as in cavernous limestone or lava rock.

groundwater, unconfined *See* unconfined groundwater.

groundwater wave A wavelike movement of the water table in the direction of the latter's slope. It assumes the form of a groundwater mound or ridge. It is produced by a considerable addition of water to the water table in a relatively short time and over a relatively small area, and it flattens out as it progresses to lower levels. It is also called a phreatic wave.

groups The vertical columns of elements in the periodic table. *See also* periodic table.

grout curtain An underground wall created by injecting grout into the ground that subsequently hardens to become impermeable. It is designed to stop groundwater flow in a certain area.

grouting The placement of grout in the subsurface by drilling and then injection.

growing season The portion of the year when soil temperatures are above biologic zero (41° Fahrenheit [5° Celsius]). The following growing season months are assumed for each of the soil temperature regimes: thermic, February–October; mesic, March–October; frigid, May–September; cryic, June–August; pergelic, July–August; isohyperthermic, January–December; hyperthermic, February–December; isothermic, January–December; and isomesic, January–December.

growth retardation Slowed growth following exposure of a developing organism (either in utero or postnatally) to chemical or physical agents. In toxicology, growth retardation is often considered a nonspecific chemical effect that can be mediated by such effects as depressed maternal appetite. In the absence of more specific damage (i.e., malformations), these mediating effects do not play a role in the assessment of a chemical's safety if they are produced only at the same dose levels that result in overt maternal toxicity.

groyne *See* groin.

grubbing The operation of removing stumps and roots.

gsfd *See* gallons per day per square foot *in the Units of Measure list.*

guard gate An additional pair of gates installed in front of lock gates for emergency use in case the lock gates cannot be operated.

GUD *See* β-D-glucuronidase.

GUI *See* graphical user interface.

guidance Recommendations and suggestions issued by a federal agency in a report, document, statement, letter, or other communication that expresses the views of the agency. Such guidance is not a regulation subject to the public review and development process under the Administrative Procedure Act, and it is not federally enforceable.

guide vane (1) A fixed or removable casting in a pump between the impeller and the casing, with liquid passages designed to convert velocity head to pressure head. (2) A fixed or adjustable device intended to direct the flow of liquid in a conduit or channel.

gullet An open-channel collector, typically used in a filtration process, that collects backwash water for subsequent discharge. Gullets can be used with or without backwash troughs. If backwash troughs are used, the backwash water flows over weirs into the troughs that then discharge to the gullet. If backwash troughs are not used, the gullet has a weir over which backwash water flows directly for collection.

gulmite screw The screw that holds the swivel part of a yoke connector for a portable exchange tank in place.

Gumbel distribution A statistical distribution used in flood frequency analysis to determine the probability that a given flow will occur within a given time interval.

gut (1) A narrow passage or contracted strait connecting two bodies of water. (2) A channel in otherwise less deep water, generally formed by water in motion. (3) The intestinal tract of an organism.

GWDR *See* Groundwater Disinfection Rule.

GWI *See* groundwater infiltration.

GWUDI *See* groundwater under the direct influence of surface water.

Gy *See* gray *in the Units of Measure list.*

gypsum ($CaSO_4 \cdot 2H_2O$) A mineral composed of calcium sulfate ($CaSO_4$) with attached water molecules, or a rock primarily containing calcium sulfate. It is often used as a soil amendment to aid in building soil structure and permeability or in the preparation of gypsum cements. This term is sometimes used to refer to a type of scale consisting primarily of insoluble calcium sulfate that can form during a water-desalting process if the reject salt solution becomes concentrated beyond solubility. *See also* calcium sulfate.

H

H *See* henry *in the Units of Measure list*; hydrogen.
H *See* Henry's constant.
H^+ *See* hydrogen ion.
h *See* hour *in the Units of Measure list*; hecto.
h *See* height.
H_2O The chemical formula for water (dihydrogen oxide).
H_2O_2/Vis–UV light process *See* hydrogen peroxide/visible–ultraviolet light process.
H_3O^+ *See* hydrogen ion.
HA *See* Health Advisory.
ha *See* hectare *in the Units of Measure list.*
HAA *See* haloacetic acid.
HAA5 *See* sum of five haloacetic acids.
HAA6 *See* sum of six haloacetic acids.
HAAFP *See* haloacetic acid formation potential.
Haber's law A rule of thumb stating that the concentration of a toxin times the time of exposure provides an equivalent effective dose. Haber's law has been primarily applied to respiratory irritants. It provides a generally conservative means of predicting safe levels to protracted exposures from the effects seen with shorter exposures. The law, however, has very limited application, and many examples exist where it clearly does not apply. Haber's law should never be used to predict short-term impacts from long-term exposures to a toxicant.
habitat The place where an organism, such as an animal or plant, lives.
habitat protection Measures aimed at maintaining or enhancing the quality of a habitat for the well-being of the species that inhabit it.
habitat value The suitability of an area to support a given evaluation species.
HACCP *See* hazard analysis critical control point.
half-duplex Pertaining to the transmission of data in either direction but only in one direction at a time.
half-life (1) The time during which one-half of the atoms of a radionuclide undergo radioactive decay. It is the reciprocal of the rate constant for such decay. Half-lives of selected radionuclides are as follows: carbon-14, 5,730 years; hydrogen-3, 12.3 years; radon-222, 3.8 days. (2) A measure of the rate at which a chemical disappears from the body (i.e., biological half-life). The biological half-life can reflect the rates of several processes—at a minimum, the rate at which the chemical is biotransformed (i.e., metabolized) and the rate at which it is excreted. Metabolism and excretion through several pathways is not unusual.
halide (X^-) The ionic form of a halogen atom. *See also* halogen.
halite Rock salt, a mineral that is more than 95 percent sodium chloride (NaCl). It is also called native salt or fossil salt.
Hall detector A device used to detect specific types of compounds, such as halogen-substituted organics, as they elute from a gas chromatograph. R.C. Hall refined this type of detector in the early 1970s. This detector is also known as a Hall electrolytic conductivity detector or simply as an electrolytic conductivity detector. The electrolytic conductivity detector was originally developed by D.M. Coulson in the early 1960s.
hallucinogen Any of several chemical substances that are capable of altering sense perceptions, particularly sight, smell, and hearing. Hallucinogens are frequently referred to as mind-altering drugs. They can act by a variety of neurochemical mechanisms.
haloacetic acid (HAA) (CX_3COOH, where X = Cl, Br, H in various combinations) A class of disinfection by-products formed primarily during the chlorination of water containing natural organic matter. When bromide (Br^-) is present, a total of nine chlorine-, bromine-and-chlorine-, or bromine-substituted species may be formed. Trihalomethanes and haloacetic acids are the two most prevalent classes of by-products formed during chlorination; and subject to regulation under the Disinfectant/Disinfection By-Products Rule. *See also* chlorination; Disinfectant/Disinfection By-Products Rule; disinfection by-product; natural organic matter.
haloacetic acid formation potential (HAAYFP, where Y is the number of haloacetic acids measured) The amount of haloacetic acids formed during a test in which a source or treated water is dosed with a relatively high amount of disinfectant (normally

chlorine) and is incubated (stored) under conditions that maximize haloacetic acid production (e.g., pH less than or equal to 8, warm water temperature, contact time of 4 to 7 days). This value is not a measure of the amount of haloacetic acids that would form under normal drinking water treatment conditions, but rather an indirect measure of the amount of haloacetic acid precursors in a sample. If a water has a measurable level of haloacetic acids prior to the formation potential test (e.g., in a prechlorinated sample), then the formation potential equals the haloacetic acid concentrations measured at the end of the test at time t [$HAAY_t$] minus haloacetic acid concentrations measured at the start of the test at time 0 [$HAAY_0$]. *See also* disinfection by-product precursor; haloacetic acid.

haloacetic acids, sum of five *See* sum of five haloacetic acids.

haloacetic acids, sum of six *See* sum of six haloacetic acids.

haloacetonitrile (HAN) ($CX_3{\equiv}N$, where X = Cl, Br, H in various combinations) A class of disinfection by-products formed primarily during the chlorination of water containing natural organic matter. Some amino acids (e.g., from algae) have been found to be precursors to haloacetonitriles. When bromide (Br^-) is present, a total of four chlorine-chlorine-and-bromine-, or bromine-substituted species may be formed. Monitoring for haloacetonitriles was required in the Information Collection Rule. *See also* amino acid; chlorination; disinfection by-product; Information Collection Rule; natural organic matter; precursor.

haloform (CHX_3) Any of the trihalomethanes, one-carbon compounds containing three halogen atoms. *See also* halogen; trihalomethane.

haloform reaction Generally, the reaction of a methyl ketone (e.g., acetone [C_3H_6O]) with a halogen (e.g., chlorine) under base-catalyzed conditions. The initial product is an α,α,α-trihaloketone (e.g., 1,1,1-trichloroacetone). In basic solution, such trihaloketones undergo cleavage to yield as final products a haloform—e.g., chloroform ($CHCl_3$) and a carboxylic acid (R–COOH, where the number of carbon atoms ranges from 1 to 26). *See also* enolization; haloform; trihalomethane.

halogen Any of the chemical elements composing group VIIA of the periodic table: fluorine, chlorine, bromine, iodine, and astatine.

halogen substitution Incorporation of one of the halogen elements, usually chlorine or bromine, into a chemical compound. Typically the halogen atom is replacing a hydrogen atom. *See also* halogen.

halogen-demand-free phosphate buffer saline (HDFPBS) A specially prepared stock buffer solution used to conduct disinfection experiments with microorganisms. A halogen-demand-free buffer is defined as a buffer that loses no more than 10 percent of the initial halogen disinfection dose or 0.1 milligram per litre, whichever is less, during a holding time of 1 hour.

halogen-substituted by-product Typically, a disinfection by-product containing chlorine (from chlorination), bromine (from chlorination or ozonation of bromide-containing waters), or both. *See also* disinfection by-product.

halogen-substituted nitrile An organic compound containing the –CN grouping and one or more halogen atoms. Haloacetonitriles ($CX_3C{\equiv}N$, where X = Cl, Br, H in various combinations) are a type of halogen-substituted nitrile. *See also* haloacetonitrile; halogen.

halogen-substituted organic material An organic compound containing one or more halogen atoms. Some halogen-substituted organic compounds are disinfection by-products, whereas others are synthetic organic chemicals. *See also* disinfection by-product; halogen; organic compound; synthetic organic chemical.

halogenated by-product *See* halogen-substituted by-product.

halogenated nitrile *See* halogen-substituted nitrile.

halogenated organic material *See* halogen-substituted organic material.

halogenating species A chemical (e.g., hypochlorous acid [HOCl] or hypobromous acid [HOBr]) that can induce halogen substitution (i.e., the formation of halogen-substituted organic material). *See also* halogen substitution; halogen-substituted organic material; hypobromous acid; hypochlorous acid.

halogenation *See* halogen substitution.

haloketone (HK) (CX_3COCX_3, where X = Cl, Br, H in various combinations) A class of disinfection by-products formed primarily during the chlorination of water containing natural organic matter. When bromide (Br^-) is present, bromine-substituted species may be formed. The Information Collection Rule required monitoring of two haloketones (i.e., 1,1-dichloropropanone and 1,1,1-trichloropropanone). *See also* chlorination; disinfection by-product; Information Collection Rule; natural organic matter.

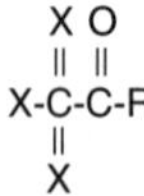

Haloketone, where one to three of the Xs are chlorine atoms, bromine atoms, or both, and the rest are hydrogen atoms; R is a methyl (CH_3) group in common disinfection by-products

halophilic Thriving in a salt environment.

halophilic bacteria Salt-tolerant bacteria often found in solar salt that has not been fully kiln dried or in salt that may have been exposed to unsanitary conditions over a long period of time.

halophyte A plant that can tolerate high concentrations of dissolved minerals, particularly salts of chlorine, iodine, and bromine. Halophytes are also called salt-loving plants.

HAN *See* haloacetonitrile.

hand-held data entry terminal (HDET) A portable device used to record meter readings in the field, which are then transferred electronically to the billing system.

haptenization The formation of a bond between a small molecular compound and a high-molecular-weight chemical (generally a protein) to form an antigenic substance (i.e., a compound to which the immune system will react by producing antibodies).

hard detergent A synthetic detergent that is resistant to biological attack. *See also* alkylbenzene sulfonate.

hard water Generally, water that requires considerable amounts of soap to produce a foam or lather and that also produces scale in hot water pipes, heaters, boilers, and other units in which the temperature of water is increased materially. With respect to hardness, waters have been classified as follows: 0–25 milligrams per litre as calcium carbonate (CaCO3), very soft; 25–75 milligrams per litre as $CaCO_3$, soft; 75–150 milligrams per litre as $CaCO_3$, moderately hard; 150–300 milligrams per litre as $CaCO_3$, hard; 300 milligrams per litre as $CaCO_3$ and up, very hard. *See also* hardness.

hard water scale *See* scale.

hardness A quality of water caused by divalent metallic cations and resulting in increased consumption of soap, deposition of scale in boilers, damage in some industrial processes, and sometimes objectionable taste. The principal hardness-causing cations are calcium, magnesium, strontium, ferrous iron, and manganese ions. Hardness may be determined by a standard laboratory titration procedure or computed from the amounts of calcium and magnesium expressed as equivalent calcium carbonate ($CaCO_3$). *See also* calcium carbonate equivalent; carbonate hardness; noncarbonate hardness.

hardness as calcium carbonate ($CaCO_3$) The value obtained when the hardness-forming salts are calculated in terms of equivalent quantities of calcium carbonate. This method of water analysis provides a common basis for comparison of different salts and compounds. *See also* calcium carbonate equivalent.

hardness leakage *See* leakage.

hardness number The resistance of a granular carbon to the degradation action of steel balls in a sieve column vibrator. It is calculated by using the weight of granular carbon retained on a particular sieve after the granular carbon has been in contact with the steel balls.

hardness sensor *See* sensor.

hardness, total *See* total hardness.

hardpan A layer of soil that has accumulated leached calcium-rich minerals from weathering and that is well cemented by the minerals and practically impenetrable by water. Caliche is another name used for hardpan.

hardpan spring A contact spring occurring above a layer of hardpan and flowing from a perched water table.

hardscape Landscaping constructed from nonliving materials, such as concrete, brick, and lumber.

hardware efficiency A value, expressed as a percentage or a decimal fraction, representing the portion of water applied by an irrigation system that benefits the intended plants.

Hardy Cross method A mathematical method of analyzing flow in a water distribution system, devised by Hardy Cross in 1936.

Hartmannella A genus of small, free-living amoebae characterized by moderately or well-developed ectoplasm, a vesicular nucleus, an ellipsoidal or cylindrical mitotic figure, and no polar caps. Cysts are rounded with smooth or, in one species, slightly wrinkled surfaces.

Hartmannella vermiformis A species of free-living amoebae, often found in potable water systems, that has been shown to be associated with the amplification of *Legionella pneumophila* (the Legionnaires' disease bacillus) in hot water tanks and hot water distribution systems.

harvesting A physical method for controlling aquatic plants in which the plants are pulled or cut and raked from the water body.

HAV *See* hepatitis A virus.

hazard abatement The correction of a known hazardous condition according to standards.

hazard analysis critical control point (HACCP) A point, identified during a systematic analysis of the potential hazards in a process, where hazards may be controlled and where such controls are deemed critical to consumer safety.

hazard communication The means or process of conveying needed information on materials and workplace dangers to employees. Such information includes standard operating procedures, personal protective equipment information, and material safety data sheets. This type of program is sometimes called a worker's right-to-know program.

hazard evaluation A component of risk assessment that involves gathering and evaluating data on the types of health injuries or diseases (e.g., cancer) that may be produced by a chemical or by the conditions of exposure under which injury or disease is produced.

hazard index The primary method for component-based noncancer risk assessment based on dose addition. It is defined as the weighted sum of the measures of exposure for the mixture of component chemicals. The weighting factor according to dose addition should be the relative toxic strength, also called potency. Calculating a hazard index is recommended for groups of toxicologically similar chemicals (i.e., chemicals usually viewed as having similarity in target organs such as liver, kidney, and so forth). The hazard index is calculated as:

$$\text{hazard index} = \sum_{1}^{n} \frac{\text{exposure level}}{\text{acceptable level characterized by a reference dose}}$$

where n is the number of component chemicals. More information can be obtained from *Federal Register*, 51:34014–34025 (Sept. 24, 1986). *See also* reference dose; toxicity equivalency factor.

hazard summary (1) A list of all hazards that an employee may be exposed to, such as chemicals and equipment. (2) A list of all hazards a community or business may be exposed to, such as earthquakes and tornadoes.

hazardous atmosphere An atmosphere that may contain one or more hazardous gases that could cause an employee to pass out, be injured, or possibly be killed.

hazardous material Any solid, liquid, or gas that is detrimental to the health and safety of a person.

hazardous material spill An uncontrolled release of any solid, liquid, or gas that is detrimental to the health and safety of a person.

Hazardous Materials Transportation Act (HMTA) A Public Law enacted in 1988 (49 U.S.C. Sec. 1801 *et seq.*) governing the transport of materials defined as hazardous under the act. Under this act, the US Department of Transportation has established an extensive and comprehensive regulatory program covering the classification, packaging, labeling, and transportation of hazardous materials.

hazardous substance Any material that is dangerous and poses a threat to public health or safety.

hazardous waste A by-product or leftover product from processes that are harmful to the environment and living creatures.

Hazardous Waste Operations and Emergency Response (HAZWOPER) Planning required by an employer under Occupational Safety and Health Agency standards 1910.120 and 1926.65 to determine how employees should react to hazardous material spills. The standards call for the emergency response team to receive training concerning the specific hazards of that location.

Hazen method An empirical method to estimate the hydraulic conductivity of a sediment on the basis of grain size.

Hazen unit (HU) of color A color standard for natural waters introduced by Allen Hazen in the 1890s. Currently, the standard method for the visual determination of color is based on the platinum–cobalt (Pt–Co) color unit. *See also* chloroplatinate unit *in the Units of Measure section.*

Hazen–Williams C A friction coefficient representing the degree of roughness of the inside surface of a water main. *See also* Hazen–Williams formula.

Hazen–Williams formula A formula used to calculate the flow in a water main by applying a roughness or friction coefficient as a factor. It was developed in 1902 by Gardner Williams and Allen Hazen. The value of the coefficient indicates the condition of the inside surface of the pipe. The Hazen–Williams formula for determining head loss is

$$h_f = k_1 \frac{LQ^{1.85}}{C^{1.85}D^{4.87}}$$

Where:

h_f = head loss, in feet (metres)

k_1 = 4.72, in units of $\text{seconds}^{1.85}$ per $\text{feet}^{0.68}$ in the US customary system (10.63, in units of $\text{seconds}^{1.85}$ per $\text{metre}^{0.68}$ in Système International)

L = pipe length, in feet (metres)

Q = flow rate, in cubic feet per second (cubic metres per second)

C = Hazen–Williams roughness coefficient

D = pipe diameter, in feet (metres)

The value of C ranges from 60 for corrugated steel to 150 for clean, new asbestos–cement pipe.

Hazen–Williams roughness coefficient *See* Hazen–Williams *C*.

hazmat *See* hazardous material.

HAZWOPER *See* Hazardous Waste Operations and Emergency Response.

HBV *See* hepatitis B virus.

HCV *See* hepatitis C virus.

HDB *See* hydraulic design basis.

HDET *See* hand-held data entry terminal.

HDFPBS *See* halogen-demand-free phosphate buffer saline.

HDPE *See* high-density polyethylene.

HDT *See* hydraulic detention time.

HDV *See* hepatitis D virus.

head The energy per unit weight of a liquid. In practical terms, head is the pressure at any given point in a water system. It may also be called pressure head or velocity head. Head is calculated as the pressure exerted by a hypothetical column of water standing at the height to which the free surface of water would rise above any point in a hydraulic system. Head is often measured in pounds per square inch or kilopascals. Head is sometimes expressed as the height of a column of water in feet or metres that would produce the corresponding pressure; this measurement may be called hydrostatic head.

head flume A flume, chute, trough, or lined channel constructed at the head of a gully or at a terrace outlet to prevent cutting or scouring by running water.

head gate A gate or valve at the entrance of a waterway such as a conduit, penstock, treatment process, or canal.

head gate duty of water *See* gross duty of water.

head increaser A device connected with a draft tube to increase the discharge by reducing the tailwater pressure.

head loss A reduction of water pressure (head) in a hydraulic or plumbing system. Head loss is a measure of (1) the resistance of a medium bed (or other water treatment system), a plumbing system, or both to the flow of the water through it, or (2) the amount of energy used by water in moving from one location to another. In water treatment technology, head loss is basically the same as pressure drop. *See also* head; pressure drop.

head race A channel that conducts water to a waterwheel; a forebay.

head radius The radius of the circular arc pattern of an overhead irrigation nozzle or sprayer.

headbox A chamber for mixing, collection, equalization, or a combination of these into which chemicals may be added so that the mixture may be distributed from the box to subsequent processing or treatment.

header A large pipe to which a series of smaller pipes are connected. Such a pipe is also called a manifold.

head-to-head spacing Spacing of irrigation nozzles so that the pattern of precipitation from one head completely overlaps the area between that head and an adjacent head.

heading The place at which a canal or pipeline diverts water from a stream or other body of water.

headpond A reservoir of water established at a given elevation to generate head for the production of energy. Headponds store water that can flow through turbines to generate electricity or can turn paddlewheels to operate equipment.

headspace The space between the top of a filled liquid and the cover of a container. Because some volatile organic chemicals and disinfection by-products are relatively volatile, sampling for such constituents usually requires the bottle to be filled headspace-free in order to prevent volatile constituents from leaving the liquid phase and entering the headspace air. *See also* disinfection by-product; volatile organic compound.

headspace analyzer A piece of instrumentation that can be used to analyze for volatile organic chemicals and selected disinfection

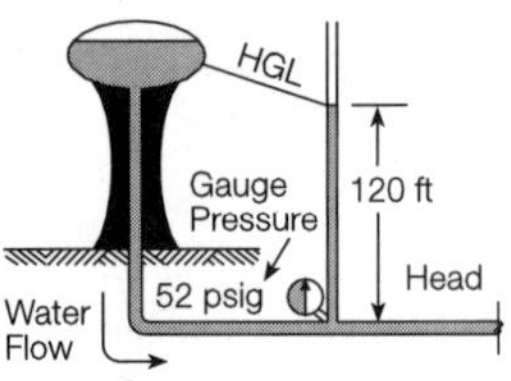

Head

by-products. Typically the sample is warmed in order to volatilize organic compounds from the liquid phase into the headspace air above the sample. The headspace air is then swept with a gas to a detector that can measure the presence of volatile organic chemicals that may be present in the sample. *See also* disinfection by-product; headspace; volatile organic compound.

headwater (1) The upper reaches of a stream near its source. (2) The region where groundwaters emerge to form a surface stream. (3) The water upstream from a structure.

headworks (1) All the structures and devices located at the head or diversion point of a conduit or canal. The term as used is practically synonymous with *diversion works* or *intake heading.* (2) The initial structures and devices of a water treatment plant.

Health Advisory (HA) A contaminant-specific document prepared and issued by the US Environmental Protection Agency for contaminants not regulated under the Safe Drinking Water Act. It provides technical guidance to public health officials on health effects, analytical methodologies, and treatment technologies in response to the public need for guidance during emergency situations involving drinking water contamination. *See also* guidance.

health contaminant Any substance or condition that may have any adverse effect on human health. Health contaminants in water are regulated as part of the National Primary Drinking Water Regulations enforced by the US Environmental Protection Agency and most states. *See also* drinking water standard.

health effect A modification in the health of an individual as a result of exposure to a chemical, physical, or infectious agent. The meaning of this term is vague, but it is usually thought of in terms of adverse effects on health rather than beneficial effects. Even with this additional specificity, what is meant by the term is not very clear. The opinions of what constitutes an adverse health effect vary widely. Some professionals include small decreases in enzyme activities, whereas others require that clear-cut disease and observable physical pathology be present.

Health Effects Research Laboratory (HERL) The office within the US Environmental Protection Agency's Office of Research and Development responsible for conducting research on the relationships among human exposure to contaminants, internal dose, and health effects. The Health Effects Research Laboratory is located in Research Triangle Park, N.C.

health effects risk analysis The process of estimating the probability that an adverse effect will occur, given a particular set of exposure conditions.

health hazard (1) A cross-connection or potential cross-connection involving any substance that, if introduced into the potable water supply, could cause death or illness, spread disease, or have a high probability of causing such effects. (2) Anything that endangers human health.

heat Energy transfer by virtue of a temperature difference. *See also* energy; work.

heat of adsorption The energy given off during the adsorption of molecules.

heat budget The amount of heat necessary to raise water from the minimum temperature of winter to the maximum temperature of summer. It is usually referred to as the annual heat budget (of a large contained body of water, such as a lake).

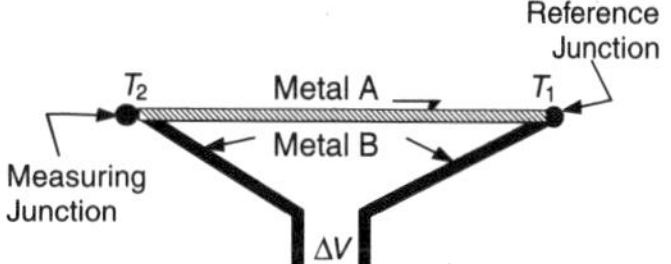

Thermocouple type of heat sensor

heat capacity units In the US customary system, British thermal units per Rankine for a fixed and specified mass. In Système International, joules per kelvin; metres squared kilograms per second squared per kelvin.

heat exchanger A device providing for the transfer of energy as heat between two fluids.

heat flux density units In the US customary system, British thermal units per hour per square foot. In Système International, watts per square metre; kilograms per second cubed.

heat of fusion (water to ice) The energy (not necessarily by heat transfer) released when water changes to ice. Its value is 144 British thermal units per pound mass (335 kilojoules per kilogram), the highest heat of fusion of all substances except liquid ammonia (NH_3). This high value, along with the lower density of ice and the 39.2° Fahrenheit (4° Celsius) temperature of maximum density, means that the subsurface temperature in deep ice-covered lakes does not fall to less than freezing.

heat sensor A device that opens and closes a switch in response to changes in temperature. This device might be a metal contact, a thermocouple that generates a minute electrical current proportional to the difference in heat, or a variable resistor for which the value changes in response to changes in temperature. A heat sensor is also called a temperature sensor.

heat storage well A well used to store heated water for later recovery and use in aquifer-based heating and cooling systems. The relatively good thermal storage properties of rocks make the use of an aquifer as part of a heat pump system in temperate climates feasible.

heat stress An adverse condition that occurs when the body generates more heat than it can release. Symptoms are weakness, dizziness, headache, and nausea.

heat of sublimation The energy that must be supplied (not necessarily by heat transfer) to convert ice directly to vapor. Its value is 1,222 British thermal units per pound mass (2,843 kilojoules per kilogram).

heat transfer An exchange of energy as heat from one fluid to another. It is also called heat exchange.

heat units In the US customary system, British thermal units; foot-pounds force; horsepower hours; kilowatt-hours. In Système International, joules; newton metres; kilograms square metres per second squared.

heat of vaporization The amount of energy (not necessarily by heat transfer) needed to change a given amount of liquid existing at a given temperature into a vapor. The heat of vaporization (evaporation) of water is the highest of all substances—970.3 British thermal units per pound mass (2,257.0 kilojoules per kilogram) for boiling temperatures at the standard atmospheric pressure—and is important in nature because it regulates the evaporation phase of the hydrologic cycle.

heat-exchanger tank A tank used for heating water or sludge, usually equipped with coils to facilitate heat transfer.

heathland Uncultivated open, flat land of poor-quality soils and inferior drainage. Such land is typically covered by shrubs, mostly of the *Ericaceae* family.

heating The process of increasing the temperature of a solid, liquid, or gas.

heating, ventilating, and air conditioning (HVAC) A mechanical system in a building that provides for air circulation for heating, ventilation, and air conditioning.

heat-resistant glass A generic term for commercially manufactured borosilicate glass, e.g., Pyrex® or Chimex®.

heavy metal A metallic element with a specific gravity greater than 5, such as cadmium, copper, lead, and zinc.

heavy rain Rain that is falling at the time of observation with an intensity in excess of 0.30 inches per hour (7.6 millimetres per minute), or more than 0.03 inches (0.8 millimetres) in 6 minutes.

heavy sludge Sludge in fluid form but with a relatively low moisture content.

heavy water Water with an isotope of hydrogen that has an atomic weight of 2.0147 instead of 1.008. It is called dideuto hydrogen oxide or deuterium oxide.

hectare (ha) *See the Units of Measure list.*

hecto (h) A prefix meaning 10^2 in Système International. Its use should be avoided except for measurements of area and volume.

heel The lower zone of an ion exchange bed that is "passed by" in either the softening, deionization, or dealkalization mode or during the application of regenerants. The presence of a heel is usually a result of the configuration of the vessel or the lack of a good underdrain distribution system.

height (*h*) The distance to the top of a structure from some datum plane, frequently the ground surface.

height of dam The difference in elevation between the roadway, walkway, or spillway crest and the lowest part of the excavated foundation along the axis of a fixed dam.

Heinz bodies Inclusion bodies in erythrocytes (red blood cells) resulting from the oxidative damage and precipitation of hemoglobin. They can be produced by chemicals that induce oxidative stress, but they also arise from congenitally abnormal hemoglobins or from deficiencies in enzymes that protect against oxidative stress, reverse the oxidative damage that is induced, or both. *See also* inclusion body.

Hele Shaw model A physical model analog of groundwater flow that is based on the movement of viscous fluid between two closely spaced (transparent) plates.

Helicobacter pylori A pathogenic bacterium that has been shown to be associated with duodenal ulcers in humans, although not all persons who have been found to harbor the organism are affected by ulcers. Transmission probably occurs via the fecal–oral route, which means that waterborne transmission could occur if water directly contaminated by feces from an infected person is ingested. Waterborne transmission was cited in at least one study involving poor-quality water used for drinking. The organism has also been linked to gastritis and gastric carcinoma (cancerous growth).

hemangiosarcoma A malignant tumor that arises from the proliferation of endothelial cells (e.g., blood vessel cells) and fibroblastic tissue (tissue giving rise to fibrous tissue). This type of tumor is produced in humans by vinyl chloride.

hematopoiesis The formation and development of blood cells.

heme ($C_{34}H_{32}N_4O_4Fe$) A deep red pigment that contains reduced (ferrous) iron. Heme is found in red blood cells (hemoglobin). It is also found outside the body in the nonprotein portions of some organic molecules called hemoproteins. In water quality treatment, it may be referred to as heme iron, which is organically bound iron that can cause water to have a pinkish cast. *See* hemoglobin.

heme iron *See* heme.

hemochorial placenta A type of placenta in which the maternal blood comes into direct contact with the chorion (i.e., the outermost nonembryonic membrane surrounding the embryo). This is the type of placenta found in primates, including humans.

hemochromatosis A condition characterized by excess deposition of iron in tissues, particularly the liver, pancreas, and skin. It results in cirrhosis of the liver, diabetes mellitus, and bone and joint changes. A hereditary form exists that is called idiopathic hemochromatosis. Other forms are produced by repeated blood transfusions or consumption of excessive amounts of iron over a long period of time.

hemodialysis The process of purifying a kidney patient's blood by means of a dialysis membrane. *See also* dialysis.

hemoendothelial placenta A placental type in which maternal blood comes in contact with the endothelium of chorionic vessels. Essentially this means that only one layer of cells makes up the placenta, whereas other placental types have as many as six. This type of placenta is found in rats, rabbits, and guinea pigs (animals commonly used in testing for teratogenic effects of chemicals). It is more porous than human placenta.

hemoglobin The respiratory protein of the red blood cells. It transfers oxygen from the lungs to the tissues and carbon dioxide (CO_2) from the tissues to the lungs. It is a conjugated protein consisting of approximately 94 percent globin (protein portion) and 6 percent heme ($C_{34}H_{32}N_4O_4Fe$). It consists of four polypeptides (two α-chains and two β-chains) and four heme groups. It has a molecular weight of 65,000. *See also* heme.

hemolysis The rupturing of red blood cells that sometimes occurs during hemodialysis. Hemolysis may be caused by the presence of chloramines in dialysis water.

hemolytic anemia A decrease in the concentration of erythrocytes (red blood cells) as a result of damage to the structure of the cell membranes. Hemolytic anemia can be induced by a variety of chemicals that induce oxidative stress. The most notable drinking water chemical producing such an effect in vivo is chlorite (ClO_2^-).

henry (H) *See the Units of Measure list.*

Henry's constant (*H*) A coefficient relating the gaseous-phase mole fraction or concentration of a contaminant to the liquid-phase mole fraction or concentration, as expressed in Henry's law. Knowing how Henry's law is being expressed is imperative to determine the proper units for Henry's constant. *See also* air-stripping; Henry's law.

Henry's law A relationship expressing the directly proportional relationship between the liquid-phase and gas-phase concentration of a substance at equilibrium and at a given temperature. Henry's law can be expressed in many forms. This expression is typically used in water treatment to evaluate the potential or extent to which (1) a contaminant will volatilize from the liquid phase to the gaseous phase or (2) a gas will dissolve into the liquid phase.

HEPA filter *See* high-efficiency particulate air filter.

hepatic Pertaining to the liver.

hepatitis An inflammation of the liver that can be produced directly or indirectly by a number of chemical contaminants of drinking water, as well as by infectious agents. A severe case, such as in fulminant viral hepatitis, can be fatal. Damage from milder hepatitis is largely reversible, however, as long as it is not sustained or repeated.

hepatitis A virus (HAV) A ribonucleic acid virus that is also known as infectious hepatitis. It is transmitted by fecal–oral person-to-person contact, by ingestion of contaminated food (e.g., shellfish), by fecally contaminated water, or by contaminated blood products. *See also* waterborne disease.

hepatitis B virus (HBV) A form of hepatitis virus also known as serum hepatitis. It is transmitted through the use of contaminated needles, syringes, and blood products. Hepatitis B virus causes a more severe form of hepatitis than other hepatitis viruses.

hepatitis C virus (HCV) A virus that is transmitted through contaminated blood. Hepatitis C virus is a major cause of hepatitis following transfusions in the United States.

hepatitis D virus (HDV) A circular single-stranded ribonucleic acid virus also known as hepatitis delta agent. The ribonucleic acid of the delta agent requires help from another virus, the deoxyribonucleic acid-containing hepatitis B virus, for transmission. *See also* delta virus.

hepatitis E virus (HEV) A virus that resembles a Norwalk virus or calicivirus in size and structure but causes hepatitis. Transmission is by the fecal–oral route, with fecally contaminated water being the most common vehicle of transmission. *See also* calicivirus; Norwalk virus.

hepatocellular carcinoma A malignant tumor of the liver that is derived from parenchymal cells (i.e., hepatocytes). This is the most frequent type of tumor to be induced in mice by disinfection by-products, including chloroform ($CHCl_3$) and the haloacetic acids. Dichloroacetate ($CHCl_2COO^-$) also produces these tumors in rats.

hepatocyte The parenchymal cell of the liver. These cells contain the bulk of the body's ability to metabolize foreign compounds. They also play a role in the formation of bile and the elimination of chemicals and other normal metabolites into the bile. The liver plays a central role in the coordination of intermediary metabolism in the body as well. Hepatocytes may be damaged by hepatotoxic chemical, physical, or microbial agents in contaminated water.

hepatoma A malignant tumor occurring in the liver. This term can apply to virtually any tumor in the liver, in contrast with the more precise meaning of a hepatocellular carcinoma. *See also* hepatocellular carcinoma.

hepatotoxic Pertaining to a chemical, physical, or microbial agent that damages hepatocytes. The term is sometimes extended to include toxicities to other cell types within the liver.

heptachlor ($C_{10}H_5Cl_7$) A generic name for 1,4,5,6,7,8,8-heptachloro-3a,4,7,7a-tetrahydro-4,7-methanoin dene, a synthetic organic chemical used as an insecticide. It is regulated by the US Environmental Protection Agency. *See also* insecticide; synthetic organic chemical.

```
          Cl
          |
          C
        /   \
    ClC  |   CH ——— CH
     ||  ClCCl |     ||
    ClC  |   CH      CH
        \   /   \   /
          C       C
          |      / \
          Cl    H   Cl
```

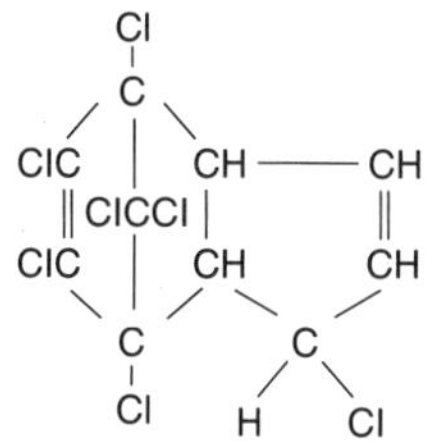

Heptachlor

heptachlor epoxide ($C_{10}H_5Cl_7O$) A degradation product of heptachlor that also acts as an insecticide. This oxidation product of heptachlor occurs in soil and on crops when treatments with heptachlor have been made. It is regulated by the US Environmental Protection Agency. *See also* heptachlor; insecticide.

herbicide A compound, usually a synthetic organic substance, used to stop or retard plant growth.

herd immunity The immunity of a group or community; the resistance of the group to an infectious agent based on the resistance of a high proportion of individuals within the group.

HERL *See* Health Effects Research Laboratory.

hertz (Hz) *See the Units of Measure list.*

heterochromatin Tightly coiled chromatin that contains genetically inactive deoxyribonucleic acid.

heterogeneous Pertaining to a medium for which the characteristics vary in different locations; nonuniform.

heteroskedasticity In regression analysis, the extent to which the standard deviation of the residuals is not constant over the range of the independent variables. A key objective in regression analysis is to explain the variation in the dependent variable by means of independent, causal variables such that the unexplained or residual variation has a random pattern with a mean of zero and a constant variance over the ranges of independent variables. Sometimes a systematic pattern in the residuals can be eliminated by transforming the variables to logs (to remove a percentage change relationship), by including polynomial independent variables (to capture nonlinear relationships), or by including autoregressive terms (to compensate for systematic relationships with the residuals over time). If the heteroskedasticity cannot be removed, inferences about the linear regression model's parameters will not be reliable and the prediction risk of the regression model cannot be accurately estimated. When the variances for the random variables (observations) are identical over all values of the independent variable and are uncorrelated, the errors are specified as homoskedastic.

heterotroph An organism that metabolizes organic compounds as sources of carbon and energy.

heterotrophic Requiring organic compounds as nutrients.

heterotrophic microorganism A bacterium or other microorganism that uses organic matter synthesized by other organisms for energy and growth.

heterotrophic plate count (HPC) A bacterial enumeration procedure used to estimate bacterial density in an environmental sample, generally water. Other names for the procedure include total plate count, standard plate count, plate count, and aerobic plate count.

heuristic knowledge Rules of thumb developed by experience or education and applicable to a particular area of knowledge. The emphasis is on practical solutions to problems that are poorly structured or where insufficient data are available. Generally, heuristic knowledge represents good, or at least workable, solutions to problems, but it may not supply the best solutions in terms of economics, efficiency, or other measures. The "rule of 72" (72/annual interest rate, in percent) for calculating the doubling time of an investment is an example of heuristic knowledge.

HEV *See* hepatitis E virus.

HEX *See* hexachlorocyclopentadiene.

hexachlorobenzene (C_6Cl_6) A synthetic organic chemical with various industrial uses, including use as a fungicide for seeds or as a wood preservative. It is regulated by the US Environmental Protection Agency. *See also* fungicide; synthetic organic chemical.

hexachlorobutadiene ($Cl_2C{:}CClCCl{:}CCl_2$) A solvent for elastomers. *See also* solvent.

hexachlorocyclopentadiene (HEX) (C_5Cl_6) A synthetic organic chemical with various industrial uses, including use as a pesticide or fungicide. It is regulated by the US Environmental Protection Agency. *See also* fungicide; pesticide; synthetic organic chemical.

hexametaphosphate A chemical that is used as a sequestering agent. Normally, a metal (M) hexametaphosphate would have the chemical formula $(MPO_3)_6$, in which the cyclic metal metaphosphate is based on rings of alternating phosphorus and oxygen atoms. The so-called sodium hexametaphosphate, however, is probably a polymer with the formula $(NaPO_3)_n$, in which n is between 10 and 20. *See also* sequestering agent; sodium hexametaphosphate.

hexidecimal A numbering system that uses 16 as its base (rather than 10, which is used by the standard decimal system). The first 10 digits are 0 through 9, and the next six are represented by A, B, C, D, E, and F. Computer memory addresses are conventionally expressed in hexidecimal.

HFSA *See* hollow fiber stripping analysis.

HGL *See* hydraulic grade line.

hierarchical database A database structure in which data are stored based on parent–child or one-to-many relationships with explicit pointers that define the relationship between segments of a record.

hierarchy A group of people or things arranged in order, e.g., class, grade, or rank.

high salting The use of 15 pounds (6.8 kilograms) or more of salt (sodium chloride [NaCl] or potassium chloride [KCl]) to regenerate each 1 cubic foot (0.028 cubic metres) of cation resin. High salting is generally recommended for water with high total hardness and water containing high concentrations of dissolved heavy metals. *See also* hardness.

High Test Hypochlorite® (HTH®) Calcium hypochlorite ($Ca(OCl)_2$), a disinfectant with approximately 70 percent available chlorine. It is sometimes used to disinfect water storage tanks, water lines, and swimming pools.

high-calcium lime A lime containing 95 to 98 percent calcium oxide (CaO).

high-capacity filter A filter that operates at a hydraulic rate higher than usual.

high-density polyethylene (HDPE) A polymer with many industrial applications, including use as a pipe material. Polyethylene ($(H_2C{:}CH_2)_x$) pipe can be used in low-pressure applications for transporting potable water and other liquids. *See also* polymer.

high-efficiency particulate air (HEPA) filter A generic name for air filters used in chemical fume hoods or biological safety cabinets. The filters are designed for removal of a specified percentage of specific sized particles, depending on the requirements of a particular type of application.

high-energy electron beam irradiation A process that uses an electron beam to irradiate a contaminant of concern and modify its structure. This process can be used to convert contaminants that are harmful and difficult to oxidize or difficult to degrade into less harmful by-products. It may potentially also be used to disinfect water on a small scale.

high-frequency ozonation Operation of an ozone (O_3) generator at frequencies equal to or greater than 1,000 cycles per second (1,000 hertz).

high-intensity pulsed ultraviolet light treatment A disinfection method that uses a high-intensity ultraviolet (UV) light source in a pulsed fashion. In a pulsed UV system, alternating current is converted to direct current and is stored in a capacitor. This energy is then released through a high-speed switch to form a pulse of intense radiation.

high-line jumper A pipe or hose connected to a fire hydrant and laid on top of the ground to provide emergency water service for an isolated portion of a distribution system.

high-to-low-dose extrapolation The process of predicting low exposure risks to humans from the measured high-exposure–high-risk data collected in rodents.

high-performance liquid chromatography (HPLC) A technique that is able to separate compounds in the liquid state as they migrate through a chromatographic column. Analytes migrate through the column at a rate based on their relative affinity for the stationary phase versus the mobile phase. The technique has many applications in the analysis of organic compounds that may not be amenable to separation by gas or ion chromatography.

high-pressure fire system A separate high-pressure water system consisting of extra-heavy mains, hydrants, pumps, and appurtenances installed solely for furnishing water for extinguishing fires at pressures suitable for direct hydrant hose streams.

high-pressure liquid chromatography *See* high-performance liquid chromatography.

high-purity oxygen (O_2) A gaseous stream containing oxygen at concentrations greater than approximately 90 percent by weight. High-purity oxygen is produced by selectively liquefying constituents at reduced temperatures to isolate oxygen or by selectively adsorbing all constituents other than oxygen in a gaseous phase.

high-rate clarifier A settling unit operated at a higher hydraulic rate than conventional settling units but still achieving similar levels of particle removal. High-rate clarifiers often employ equipment such as tube or plate settlers to increase the available surface area for settling without increasing the overall area of the unit.

high-voltage electrode During ozonation, the outlet post on a voltage transformer that produces more than 1,000 volts.

high-water line (1) The line of the shore of a river, lake, or sea that is ordinarily reached at high water. (2) Along the seashore, the intersection of the plane of mean high water with the shore.

high-water mark A mark on a structure or natural object, indicating the maximum stage of tide or flood.

high-water-use landscape A landscape with plants and features using water that require 50 to 80 percent of reference evapotranspiration to maintain optimal appearance.

hindered settling The settling of particles in higher concentration than in type I or type II settling. The particles are so close together that interparticle forces hinder the settling of neighboring particles. The particles remain in a fixed position relative to one another, and all settle at a constant velocity. As a result, the mass of particles settles as a zone. *See also* type I settling; type II settling; type III settling.

hindleg paralysis A descriptive symptomatology associated with chemicals or other agents that produce peripheral neuropathy (i.e., damage to the peripheral sensory neurons, motor neurons, or both). Generally, in rodent species used in toxicological testing, the first overt sign of such effects is that the animal begins to drag its hind legs. Hindleg paralysis is produced by such chemicals as n-hexane, methyl butyl ketone, and acrylamide.

histogram A graphic tool for organizing and analyzing data. A modification of a bar chart, it is used to measure distributions that have central tendencies. It provides a quick view of the amount of variation in a set of data.

histograph A map or chart of a river or drainage system on which a series of time lines is placed. Each time line gives the time of transit of water from the given time line to the outlet of the system.

histology The study of the structure of cells and tissues, usually involving microscopic examination of tissue slices.

historic basis An approach for calculating water allotments based on past water consumption.

historical cost The actual cost, without consideration of depreciation or inflation, of constructing a building, facility, or plant at the time of construction.

HIV *See* human immunodeficiency virus.

HK *See* haloketone.

HMI *See* human–machine interface.

HMTA *See* Hazardous Materials Transportation Act.

HOBr *See* hypobromous acid.

HOCl *See* hypochlorous acid.

HOH A variation of the chemical formula for water (H_2O).

holdover storage The portion of the useful storage normally remaining in a reservoir at the end of the drawdown period and held over for use only in a critically dry year.

holiday A void in the coating of a pipe that will allow the passage of electrical current.

hollow cathode lamp A device used in atomic absorption spectrophotometry to produce light from a specific chemical element. Such a lamp contains a cathode constructed from the metal to be determined and an anode in a sealed glass cylinder filled with argon. The cathode emits a line of radiation that is characteristic of the particular metal being determined.

hollow dam A fixed dam, usually of reinforced concrete, consisting essentially of inclined slabs or arched sections supported by transverse buttresses. The load is taken by the slabs or arched sections and transferred to the foundation by the buttresses.

hollow fiber membrane In a crossflow pressure-driven membrane system, a self-supporting membrane with a circular cross-sectional shape having a hollow central bore (typically with an inside diameter of about 1 millimetre or less). Depending on the type of membrane, the flow pattern may be "outside-in," where the feedwater is on the outside of the fiber and the central bore (lumen) carries permeate (or filtrate), or "inside-out," where the flow direction through the membrane is the opposite.

hollow fiber membrane configuration A type of membrane arrangement in a module, permeator, or element where many hollow fiber membranes are installed in a single pressure vessel.

hollow fiber stripping analysis (HFSA) A membrane technique used in the analysis of volatile and semivolatile organic chemicals. Air is recirculated through a water sample, around a bundle of hollow fiber membranes, and through an adsorbent trap. The technique is said to provide faster mass transfer of analytes than closed-loop stripping analysis. A potential application of the hollow fiber stripping analysis technique is in the analysis of taste-and-odor-causing compounds in drinking water. *See also* closed-loop stripping analysis; taste and odor.

hollow fine fiber membrane A type of hollow fiber membrane with a very small diameter (typically about 90 to 95 micrometres for the outside diameter, 40 to 45 micrometres for the inside diameter, with an outer thin skin barrier layer designed for a flow pattern from the outside to the inside) that is used in some reverse osmosis membrane products. *See also* hollow fiber membrane.

hollow-stem auger A large drilling device that looks like a corkscrew but for which the stem (center) of the screw is hollow. Sampling and installation of equipment are routinely conducted in the space of the hollow stem.

Hom model An empirical equation devised by Leonard C. Hom in 1972 describing the killing of microorganisms or the inactivation of viruses by a disinfectant. It is a modification of the Chick–Watson model. The equation for a batch system at constant disinfection concentration or constant ultraviolet irradiation intensity is:

$$r = -k'm[t^{(m-1)}](C^n)N$$

Where:

r = the inactivation rate (organisms killed or inactivated per volume per time)

k' = the rate constant of inactivation (presumed independent of disinfectant concentration)

N = the density of viable organisms at time t, in numbers per volume

t = the time of exposure, in minutes

C = the disinfectant concentration, in milligrams per litre

m = a constant

n = a constant (the coefficient of dilution)

This equation integrates to

$$\ln (N/N_o) = -k'(t^m)(C^n)$$

Where:

N_o = the density of viable organisms at time 0 in numbers per volume

If m = 1, this indicates Chick–Watson kinetics. If m is greater than 1, the plot of ln N/N_o versus time yields a "shoulder." If m is less that 1, "tailing" occurs in the plot of ln N/N_o versus time. *See also* Chick–Watson model; modified Hom model.

home maintenance The inspection of plumbing fixtures for leaks, along with scheduled regular replacement of toilet flappers and ballcocks.

home water treatment unit A treatment device attached to the service connection of an individual dwelling, or at the point of use (e.g., under a sink), to remove various contaminants. Home water treatment units can include such devices as granular activated carbon filters, membrane units, and water softeners. *See also* point-of-use treatment device; Water Quality Association.

homeostasis A state of physiological equilibrium produced by a balance of functions and chemical composition within an organism.

homogeneous Having a uniform structure or composition throughout.

homogeneous solution diffusion (HSD) *See* solution diffusion model.

homogeneous surface diffusion model (HSDM) A mathematical model describing adsorption and desorption in a single-solute and multisolute system with similar-size adsorbents such as granular activated carbon. The solute in this case can be either a specific compound, multiple compounds, or group parameters. For the single-solute system, three equations are needed:

1. $$\frac{\partial C}{\partial t} = -V_s\frac{\partial C}{\partial z} - 3\frac{(1-\varepsilon)}{\varepsilon R}k_f(C - C_s)$$

Where:

C = the concentration of the contaminant in solution, in milligrams per litre or micromoles per litre

t = time, in seconds

V_s = the velocity in pores, in centimetres per second

z = the axial position in the granular activated carbon bed, in centimetres

ε = the porosity of the bed

R = the radius of the granular activated carbon particle, in centimetres

k_f = the liquid film transport coefficient, in centimetres per second

C_s = the concentration of the contaminant at the liquid–particle interface

2. $$\frac{\partial q}{\partial t} = \frac{D_s}{r^2}\left[\frac{\partial}{\partial r}\left(r^2\frac{\partial q}{\partial r}\right)\right]$$

Where:

q = the solute loading on the adsorbate, in milligrams per gram or millimoles per gram

D_s = the surface diffusion coefficient, in centimetres squared per second

r = the radial position within the granular activated carbon particle (radial coordinates), in centimetres

3. $$q_s = KC_s^{1/n}$$

Where:

q_s = the solute loading on the adsorbate at equilibrium, in milligrams per gram or millimoles per gram

K = the Freundlich capacity coefficient, in (milligrams per gram) (litres per milligram)$^{1/n}$ or (millimoles per gram) (litres per milligram)$^{1/n}$

$1/n$ = the Freundlich intensity coefficient

Solving these equations gives the chemical concentration as a function of position and time in a granular activated carbon bed, as well as the granular activated carbon loading as a function of time, position in the granular activated carbon particle, and position in the granular activated carbon bed. *See also* Freundlich isotherm.

homoskedasticity In regression analysis, a condition in which the variances for the random variables (observations) are identical over all values of the independent variable and are uncorrelated. *See also* heteroskedasticity.

hood (1) A ventilation device used to remove gases from a work area, such as in a laboratory or paint booth. (2) A piece of protective equipment worn over the head during sandblasting or painting. If the environment is too harsh, fresh air is supplied to the user through the hood.

hook gauge A pointed, U-shaped hook attached to a graduated staff or vernier scale, used in the accurate measurement of the

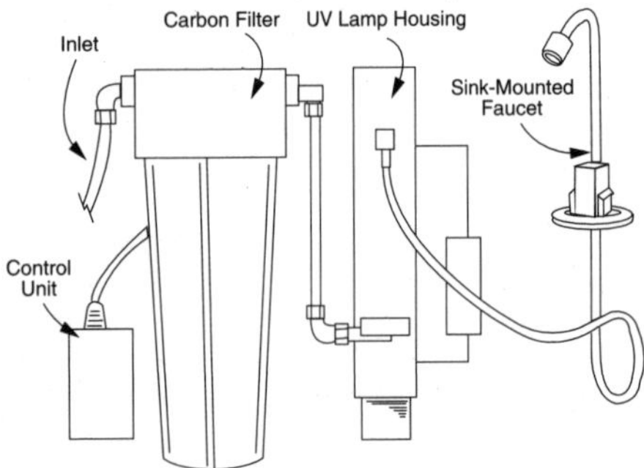

Home water treatment unit incorporating a carbon filter and ultraviolet disinfection system

elevation of a water surface. The hook is submerged and then raised, usually by means of a screw, until the point just makes a pimple on the water surface.

horizon A layer or section of the soil profile, more or less well defined, occupying a position approximately parallel to the soil surface and having characteristics that have been produced through the operation of soil-building processes.

horizontal centrifugal pump A centrifugal pump with the shaft in a horizontal alignment and the impeller rotating in a vertical direction.

horizontal pump (1) A reciprocating pump in which the piston or plunger moves in a horizontal direction. (2) A centrifugal pump in which the pump shaft is in a horizontal position.

horizontal screw pump A pump with a horizontal cylindrical casing in which a runner with radial blades like those of a ship's propeller operates. The pump has a high efficiency at low heads and high discharges and is used extensively in drainage work. Such a pump is also called a wood screw pump.

horizontal well A tubular well pushed approximately horizontally into a water-bearing stratum or under the bed of a lake or stream. Such a well is also called a push well.

horizontal-flow tank A tank or basin, with or without baffles, in which the direction of flow is horizontal.

horizontal-tube distillation Distillation using horizontally oriented evaporation tubes. *See also* multiple-effect distillation.

horsepower (hp) *See the Units of Measure list.*

horsepower·hour (hp·h) *See the Units of Measure list.*

hose bib A location in a water line where a hose is connected. It is often called a faucet.

host A person or other living animal, including birds and arthropods, that provides subsistence or lodging for an infectious agent under natural conditions. Some protozoa and helminths pass successive stages in alternate hosts of different species. In primary or definitive hosts, the parasite attains maturity or passes its sexual stage; in secondary or intermediate hosts, the parasite is in a larval or asexual state. A transport host is a carrier in which the organism remains alive but does not undergo development. In an epidemiologic context, the host may be a certain population group, and the biological, social, behavioral, and other characteristics of this population relevant to health are referred to as host factors.

host resistance Usually, the immunocompetence of an individual to resist invasion by an infectious agent. This term is sometimes applied (incorrectly) to describe resistance to chemical toxicities as well.

hot lime softening A partial softening method that requires adding a lime slurry to water that is at about 212° Fahrenheit (100° Celsius) and then chemically precipitating and removing the calcium and magnesium hardness via sedimentation and filtering. *See also* hot process softening; lime softening.

hot lime–cation softening The combination of a two-stage hot lime softening process followed by hot cation exchange softening for complete hardness removal. *See also* hot process softening.

hot lime–soda softening A method of partially softening water by adding lime and soda ash to chemically precipitate the calcium, magnesium, iron, and silica at a water temperature of about 212° Fahrenheit (100° Celsius). This process also drives off carbon dioxide (CO_2). *See also* hot process softening; lime–soda ash softening.

hot plate An electrical heating unit used to heat solutions.

hot process softening Any of several softening–clarifying processes using lime, lime and soda ash, or lime and cation softening to treat water that is at or near the boiling point. Hot process softening can remove carbon dioxide (CO_2), silica (Si^{4+}), and precipitated magnesium (Mg^{2+}) and is used mainly for boiler feedwater preparation and sulfur mining. *See also* hot lime–cation softening; hot lime–soda softening; hot lime softening.

hot spot An area of soil that, through by survey and analysis or experience, is found to be more corrosive than surrounding soil.

hot water Water that has been heated to a temperature above that of the water in the street main.

hour (h) *See the Units of Measure list.*

house cistern A cistern in which rainwater is stored for household purposes.

house service *See* domestic service.

household cost The cost to a household for a water system to meet a drinking water regulation. Annual household costs of meeting regulations are usually estimated by the US Environmental Protection Agency for proposed and final rules.

household detergent A detergent produced for retail marketing to individuals for use in home laundries. These detergents generally include anionic or nonionic surfactants plus phosphates (PO_4^{-3}) and minor ingredients such as perfumes, brighteners, and bleaches. *See also* detergent; surface active agent.

household equipment *See* residential equipment.

household filter *See* home water treatment unit.

Housing and Urban Development (HUD) The principal federal agency responsible for programs concerned with US housing needs, fair housing opportunities, and improvement and development of US communities. Housing and Urban Development administers the following: Federal Housing Administration mortgage insurance programs that help families become homeowners and facilitate the construction and rehabilitation of rental units; rental assistance programs for lower-income families who otherwise could not afford decent housing; the Government National Mortgage Association mortgage-backed securities program that helps ensure an adequate supply of mortgage credit; programs to combat housing discrimination and affirmatively further fair housing; programs that aid community and neighborhood development and preservation; and programs to help protect the homebuyer in the marketplace.

hp *See* horsepower *in the Units of Measure list.*

hp·h *See* horsepower-hour *in the Units of Measure list.*

HPC *See* heterotrophic plate count.

HPLC *See* high-performance liquid chromatography.

HRT *See* hydraulic residence time.

HSD *See* homogeneous solution diffusion.

HSDM *See* homogeneous surface diffusion model.

HTH® *See* High Test Hypochlorite®.

HU *See* Hazen unit of color.

hub The bell end into which a spigot end is placed with a sealing gasket, gasket and gland, or poured lead joint, depending on the type of pipe. It is also called bell.

HUD *See* Housing and Urban Development.

human consumption As interpreted by the US Environmental Protection Agency, drinking, bathing, showering, cooking, dishwashing, and maintaining oral hygiene. This interpretation has been upheld by the courts; see *United States* v. *Midway Heights County Water District*, 695 F. Supp. 1072, 1074 (E.D. Cal. 1988) ("Midway Heights"). *See also* bathing.

human equivalent dose A dose that, when administered to humans, produces an effect equal to that produced by certain a dose in animals.

human exposure evaluation A component of risk assessment that involves describing the nature and size of a population exposed to a substance and the magnitude and duration of their exposure. The evaluation could concern past exposures, current exposures, or anticipated exposures.

human health risk The likelihood (or probability) that a given exposure or series of exposures may have damaged or will damage the health of individuals experiencing the exposures.

human immunodeficiency virus (HIV) A virus that spreads by blood and sexual contact and leads to the development of acquired immune deficiency syndrome. No evidence exists for casual transmission. Human immunodeficiency virus replicates in T-helper lymphocytes but can also infect a variety of other hematopoietic (blood) cells in the body.

human–machine interface (HMI) A means by which a human operator interacts with a supervisory control and data acquisition (SCADA) system. The human–machine interface of a SCADA system consists of color cathode ray tubes, mapboards, projection video, alarm annunciators, and alarm loggers.

humic acid A brown, polymeric constituent of soils, lignite, and peat. It is not a well-defined compound; rather, it is a mixture of polymers containing aromatic and heterocyclic structures, carboxyl groups, and nitrogen. Aquatic humic acids are major precursors of disinfection by-products. *See also* aquatic humic acid; aromatic heterocycle; aromatic hydrocarbon; disinfection by-product precursor; peat; polymer.

humic material Natural organic matter resulting from partial decomposition of plant or animal matter and forming the organic portion of soil. *See also* humic acid; natural organic matter.

humic substance *See* humic material.

humid (1) Containing or characterized by sensible moisture; moist or damp as applied to the atmosphere. (2) Characterized by precipitation adequate in amount and occurring at such times that agriculture can be carried on without irrigation, and by the possible need for drainage to remove excess moisture; having sufficient precipitation to support a forest vegetation.

humidification The process of increasing the water vapor or moisture content.

humidity The condition of the atmosphere in terms of its content of water vapor.

humin The fraction of the soil organic matter that will not dissolve when the soil is heated with dilute alkali. *See also* humic material.

humoral immune function *See* humoral immunity.

humoral immunity The production of antibodies in lymphocytes or plasma cells following sensitization of an organism to a specific antigen. *See also* cell-mediated immunity; immunity.

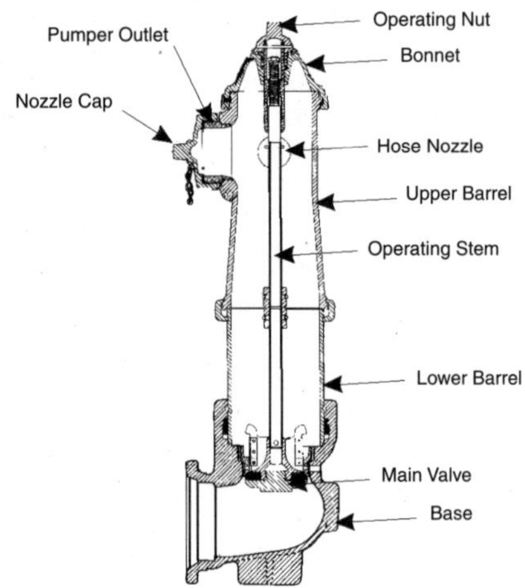

Typical fire hydrant

humus Amorphous, colloidal matter in soil, usually of a dark color, remaining after most of the animal and plant tissues have decomposed.

hundredweight (cwt) *See the Units of Measure list.*

hunting Periodic changes in the controlled variable from one side of the set point to the other as the controller overcorrects first in one direction, then in the other.

HVAC *See* heating, ventilating, and air conditioning.

HX *See* hydrogen halide.

hybrid plant A water treatment plant where two or more separate treatment processes are incorporated synergistically for improved overall performance or cost-effectiveness, e.g., a seawater-desalting plant with distillation and reverse osmosis processes using steam for energy input in such a way as to stabilize the power load requirements at the same time as producing an improved finished water quality by blending the desalted product waters from each process.

hybridization The laboratory technique of matching up and reannealing two single-stranded nucleic acid chains that complement each other. It is used for detecting a chain's complementary sequence or for targeting a specific sequence unique to a particular organism by means of a specific complementary probe.

hydrant A device connected to a water main and provided with the necessary valves and outlets so that a fire hose may be attached for discharging water at a high rate for the purpose of extinguishing fires, washing down streets, or flushing out the water main. A hydrant is also called a fire plug.

hydrant barrel The vertical section of a fire hydrant containing the outlets for discharging water. The valve stem from the bonnet to the valve is located in this section. When the valve is turned off, water drains from this section to keep the hydrant from freezing.

hydrant bonnet The top part of a hydrant connected to the barrel, resembling a bonnet, that contains the operation nut to turn the hydrant on and off. The bonnet can be removed to allow repairs on the internal parts, such as replacement of the valve seat.

hydrant cap A casting with an inside thread designed to fit on the outside thread of a hydrant outlet nozzle and equipped with a special nut that fits a standard hydrant wrench to facilitate removal.

hydrant drain A device for draining the water from the barrel of a hydrant after use to prevent freezing in the barrel. The drain, operated by the main stem, is open when the hydrant is completely closed, allowing the water in the barrel to drain away.

hydrant pitometer A measuring device for determining the velocity of the water discharging freely into the air from one nozzle of a hydrant. It is used in order to determine hydrant flow.

hydrant size The net diameter of the valve seat of the main valve of a hydrant.

hydrant tee A special tee with a 90° elbow, also referred to as a setting tee. Such a tee is used where space is limited. The tee keeps the hydrant close to the water main. It can be installed in the same trench as the main and is anchored to the main.

hydrate A substance formed by combining water with a compound.

hydrated lime Limestone that has been "burned" and treated with water (H_2O) under controlled conditions until the calcium oxide (CaO) portion has been converted to calcium hydroxide ($Ca(OH)_2$). Hydrated lime is quicklime combined with water.

$$CaO + H_2O \rightarrow Ca(OH)_2$$

See also quicklime.

hydration The chemical combination of water and another chemical to form another substance.

hydration water Water that combines with salts when they crystallize. *See also* water of crystallization.

hydraulic Referring to water or other fluids in motion.

hydraulic accumulator A device for storing energy by placing a fluid under pressure in a cylinder. The pressure can be released to accomplish work (e.g., operate brakes, open or close valves) when necessary.

hydraulic bore A standing wave that advances upstream in an open conduit from a point where the flow has suddenly been stopped. The flowing water piles up in the channel against the obstruction that caused the stoppage, and as it reaches a height above the normal water surface approximately equal to its velocity head, the increased depth of water moves upstream in a wavelike shape, that is, a hydraulic bore. *See also* suction wave.

hydraulic classification The rearrangement during fully fluidized backwashing of ion exchange or other media particles according to size. As the result of classification, the smallest particles tend to rise to the top of the bed while the largest tend to sink to the bottom because of increased settling velocity. Hydraulic classification is also called classification or stratification.

hydraulic cleaning Any cleaning that is accomplished via a liquid under pressure and an orifice that provides a high-velocity flow and, consequently, a cleaning action.

hydraulic conductivity (*K*) A coefficient describing the relative ease with which water can move through a permeable medium. It is the constant of proportionality, K, in Darcy's law of linear seepage:

$$Q = -KA \text{ grad}(h)$$

Where:

Q = the total discharge
A = the cross-sectional area of flow
$-\text{grad}(h)$ = the hydraulic gradient, dimensionless
K = hydraulic conductivity

Typical units of hydraulic conductivity are feet per day, gallons per day per square foot, or metres per day (depending on the units chosen for the total discharge and the cross-sectional area). *See also* permeability; permeability coefficient.

hydraulic design basis (HDB) A level of ability of a unit process, or sequence of unit processes that must be accounted for in the process design to meet criteria associated with liquid loading or flow capacity. The hydraulic design basis for clarifier performance may be an overflow rate, whereas the hydraulic design basis for a treatment plant may be the peak flow that must be able to pass through basins and piping.

hydraulic detention time (HDT) The theoretical amount of time required for a given flow to pass through a unit process. It is calculated by dividing the volume of a unit process by the applied flow rate. It is also called the hydraulic residence time.

hydraulic dredge A scow carrying a centrifugal pump that has (1) a suction pipe reaching to the bottom to be excavated and (2) a discharge pipe connecting to a pipeline conveying material to a place of deposit. A hydraulic dredge is also called a sandpump dredge or suction dredge.

hydraulic equation A statement of the law of mass conservation in terms of hydraulic variables. The equation can be expressed as

$$\text{inflow} = \text{outflow} \pm \text{changes in storage}$$

Usually one component of the equation is difficult to measure directly, but knowledge of the remaining components allows an individual to determine the value of the difficult component by solving the equation.

hydraulic element A quantity pertaining to a particular stage of flowing water or to a particular cross section of a conduit or stream channel, such as the depth of water, cross-sectional area of the water, hydraulic radius, wetted perimeter, mean depth of water, velocity, energy head, or friction factor.

hydraulic fill An earth structure or grading operation to which fill material is transported and deposited by means of water pumped through a flexible or rigid pipe.

hydraulic flow net A drawing used for calculating pressures and velocities in a curving stream of water in cases where friction, impact, or eddy losses do not exercise a controlling effect. The hydraulic flow net is established by drawing (1) a system of streamlines so spaced that the rate of flow is the same between each pair of lines, and (2) a system of normal lines so spaced that the distance between the normal lines equals the distance between the streamlines.

hydraulic fracturing A technique of aquifer development in which fluid is injected at pressures that exceed the tensile stress of the aquifer, causing cracks to develop and propagate in the

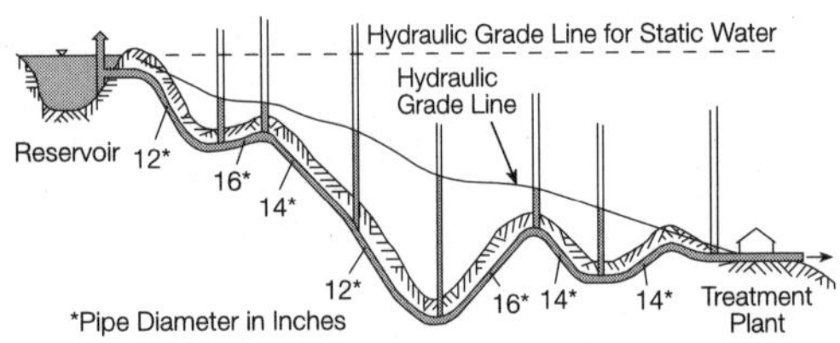

Hydraulic grade line

formation. These cracks serve as conduits for liquid flow to a production well. This technique is used extensively in petroleum recovery and is being investigated as a tool for increasing water production in rock-like aquifers and for contaminant recovery.

hydraulic friction The resistance to flow exerted on the perimeter or contact surface of a body of water moving in a stream channel or conduit caused by the roughness characteristic of the confining surface, which induces turbulence and consequent loss of energy. Energy losses arising from excessive turbulence, impact at obstructions, curves, eddies, and pronounced channel changes are not ordinarily ascribed to hydraulic friction. *See also* roughness coefficient.

hydraulic friction coefficient The ratio of actual discharge in a conduit or open channel to the theoretical discharge if the conduit were frictionless or the open channel free of turbulence. It is also called the roughness coefficient.

hydraulic grade *See* hydraulic grade line.

hydraulic grade line (HGL) A line (hydraulic profile) indicating the piezometric level of water at all points along a conduit, open channel, or stream. In a closed conduit under pressure, artesian aquifer, or groundwater basin, the line would join the elevations to which water would rise in pipes freely vented and under atmospheric pressure. In pipes under pressure, each point on the hydraulic profile is an elevation expressed as the sum of the height associated with the pipe elevation, the pipe pressure, and the velocity of the water in the pipe. In an open channel, the hydraulic grade line is the free water surface. Hydraulic profiles are commonly used to establish elevations through the processes that make up a treatment facility under maximum and minimum flow conditions.

hydraulic gradient The change in static head (pressure) per unit of distance in a pipeline in which water flows under pressure.

hydraulic head The height of the free surface of a body of water above a given point beneath the surface. *See also* head.

hydraulic jetting The process of using water forced through a well screen to suspend fine particles during well development.

hydraulic jump (1) In an open channel, a sudden and usually turbulent passage of water, under conditions of free flow, from a low stage below critical depth to a high stage above critical depth. During this passage the velocity changes from supercritical to subcritical. The hydraulic jump represents the limiting condition of the slope of the surface curve, in which the surface tends to become perpendicular to the stream bed. (2) In a closed conduit, a sudden rise from flow that fills only part of the conduit at a supercritical velocity to flow that fills the conduit under pressure. The depth plus the pressure head downstream from the hydraulic jump equals the high stage obtained for open-channel flow. (3) To dissipate energy in an open channel or at the toe of a spillway section of a dam. (4) A device to promote turbulence. (5) An abrupt rise in water surface that may occur in an open channel when water flowing at a high velocity is retarded.

hydraulic loading The amount of water applied to a given treatment process, usually expressed as volume per unit time or volume per unit time per unit surface area.

hydraulic loss A loss of head attributable to obstructions, friction, changes in velocity, and changes in the form of the conduit.

hydraulic mean depth *See* hydraulic radius.

hydraulic model A flow system operated so that the characteristics of another similar system may be predicted. A model is generally a small-scale reproduction of the prototype, but it may be larger or geometrically distorted. A hydraulic model is used in studies of spillways, stilling basins, flood regulation, or river beds. It may also be called a hydraulic similitude model.

hydraulic permeability The rate of discharge through a unit cross-sectional area of soil normal to the direction of flow when the hydraulic gradient is unity. *See also* permeability coefficient.

hydraulic profile *See* hydraulic grade line.

hydraulic radius The cross-sectional area of a stream of water perpendicular to the flow divided by the length of that part of the water's periphery in contact with the containing conduit; the ratio of area to wetted perimeter. Hydraulic radius is also called hydraulic mean depth.

hydraulic ram A type of pump that uses water as a ram, utilizing kinetic energy or velocity head to force a small portion of water to a higher level by stopping the flow of the volume of water and applying the force to the small portion of the water to be pumped. A hydraulic ram is also called a water ram.

hydraulic reclassification *See* hydraulic classification.

hydraulic residence time (HRT) *See* hydraulic detention time.

hydraulic retention time *See* hydraulic detention time.

hydraulic similitude *See* hydraulic model.

hydraulic slope The slope of the hydraulic grade line. *See also* hydraulic gradient.

hydraulic sluicing The process or operation of moving rock or soil by means of a jet or stream of water.

hydraulic stage *See* hydraulic staging.

hydraulic staging A setup for allowing multiple passes of water between electrodes used in an electrodialysis process, through a sequence of subsequent membranes in a pressure-driven membrane system, or through filters used in a filtration system to achieve further treatment.

hydraulic structure Generally, any major engineering device used to control flow or distribute water. *See also* canal; catch basin; clearwell; detention tank; equalization; headbox; reservoir; sedimentation basin; splitter box; stabilization tank; wet well.

hydraulic turbine A prime mover using the pressure or motion of water for the generation of mechanical energy.

hydraulic turbocharger A pumping device for which two or more turbine impellers rotate on a single shaft and the hydraulic energy of one flow stream is transferred to another flow stream, e.g., in a high-pressure reverse osmosis system where the pressurized waste concentrate stream helps to boost the inlet feedwater pressure.

hydraulic valve A valve operated by means of a hydraulic cylinder or other hydraulic device.

hydraulic-fill dam An earth dam in which all or most of the materials have been transported to the dam and placed in the dam by dredging, sluicing, or pumping during construction. Generally, for greater imperviousness, the fine material in the sluiced earth is segregated and deposited along the center or axis of the dam to form a core.

hydraulics That branch of science or engineering that deals with water or another fluid in motion.

hydrazine (H_2NNH_2) A liquid compound used as a strong reducing agent for transition metals and as an oxidation inhibitor for boiler feedwater and cooling water.

hydric soil A soil that is saturated, flooded, or ponded long enough during the growing season for anaerobic conditions to develop in the upper part.

hydrocarbon An organic compound consisting exclusively of the elements carbon and hydrogen. Hydrocarbons are derived principally from petroleum, coal tar, and plant sources.

hydrocarbon pneumonia An inflammation of the lung with infiltration of fluid caused by aspiration of low-viscosity hydrocarbons into the lung after such hydrocarbons are swallowed. Ingestion of very small amounts (10 millilitres) of such hydrocarbons can be serious and even life-threatening. The lower the viscosity of the hydrocarbon, the more probable that it will produce hydrocarbon pneumonia.

hydrocerussite ($Pb_3(CO_2)(OH)_2$ or $2PbCO_3 \cdot Pb(OH)_2$) The mineral name for a form of basic lead carbonate, described by the formula noted. Hydrocerussite is formed readily on lead surfaces at a pH greater than approximately 8 in the presence of dissolved inorganic carbonate, making it an extremely important solid in controlling lead solubility in drinking water systems. Hydrocerussite tends to be light in color, frequently white, cream, or light bluish gray, and is easily mistaken for calcium carbonate ($CaCO_3$) in the absence of microscopic or chemical examination.

hydrochloric acid (HCl) A water-based solution of hydrogen chloride that is a strong, highly corrosive acid. Hydrochloric acid may be used as a regenerant for cation resin deionization systems operated in the hydrogen (H^+) cycle.

hydrochlorination *See* hypochlorination.

hydrochlorinator *See* hypochlorinator.

hydroclone The basic form of most separators that act on the principle of centrifugal force and are used to remove sand and abrasives from well water. A hydroclone is also called a cyclone precipitator.

hydrodynamic dispersion The phenomenon in which a solute in groundwater flow is mixed with uncontaminated water, resulting in reduced concentrations of the dissolved compound. Hydrodynamic dispersion is roughly analogous to turbulent mixing in open flow systems, although the mechanisms and scale of the mixing are much different in the two cases.

hydrodynamics The study of the motion of, and the forces acting on, fluids.

hydroelectric power Electric power produced by using water power as a source of energy.

hydrofluosilicic acid (H_2SiF_6) A strongly acidic liquid used to fluoridate drinking water.

hydrogen (H) A nonmetallic element having an atomic number of 1. It occurs chiefly in combined form (e.g., in water, hydrocarbons, and other organic compounds). It is also available as a diatomic gas (i.e., H_2). *See also* hydrocarbon.

hydrogen bond The weak attraction between a hydrogen atom carrying a partial positive charge and some other atom with a partial negative charge. Hydrogen bonds occur in polar compounds, such as water, by the attraction of a hydrogen atom of one molecule to two unshared electrons of another molecule. Hydrogen bonds are less than one-tenth as strong as covalent bonds in which electrons are actually shared by a pair of atoms, but they significantly affect properties such as the melting point, boiling point, and crystalline structure of a substance.

hydrogen cycle A cation exchange cycle in which the cation medium is regenerated with acid and all cations in the water are removed by exchange with hydrogen (H^+) ions.

hydrogen halide (HX) A class of inorganic compounds with varying acid strengths, including hydrochloric acid (HCl) and hydrofluoric acid (HF). *See also* hydrochloric acid.

hydrogen ion (1) In aqueous solution, a positively charged combination of a proton and a water molecule (hydrated, H_3O^+). (2) A single proton, usually abbreviated H^+.

hydrogen ion activity *See* hydrogen ion concentration.

hydrogen ion concentration A measure of acidity. It is an important concept in natural waters. The negative logarithm of the hydrogen ion concentration (activity) in moles per litre is operationally defined as the pH. The hydrogen ion concentration can affect the chemical constituents of water, chemical and biological reactivity, water treatment, and water analysis. *See also* pH.

hydrogen peroxide (H_2O_2, H–O–O–H) Strong oxidizing agent used for bleaching and, to some extent, as a disinfectant. Ozone (O_3) in combination with hydrogen peroxide (i.e., PEROXONE), or ultraviolet light in combination with hydrogen peroxide, is used in some advanced oxidation processes to produce hydroxyl radicals (•OH). *See also* advanced oxidation process; disinfectant; hydrogen peroxide/visible–ultraviolet light process; hydroxyl radical; oxidant; PEROXONE process.

hydrogen peroxide/visible–ultraviolet (H_2O_2/Vis–UV) light process One of several advanced oxidation processes that depend largely on the production of hydroxyl radicals (•OH) as the oxidant. The role of the visible light emitted is not known at the time of this writing, but ultraviolet irradiation will split hydrogen peroxide into two hydroxyl radicals, which serve as a powerful oxidant. *See also* ozone–ultraviolet process; PEROXONE process.

hydrogen sulfide (H_2S) A toxic gas produced by the anaerobic decomposition of organic matter and by sulfate-reducing bacteria. Hydrogen sulfide has a very noticeable rotten-egg odor. Respiratory paralysis and death may occur quickly for people exposed to concentrations as low as 0.07 percent by volume in air.

hydrogenesis A process of natural condensation of the moisture in the air spaces in surface soil or rock.

hydrogeochemistry The study of the chemical characteristics of groundwater and surface water in relation to areal and regional geology.

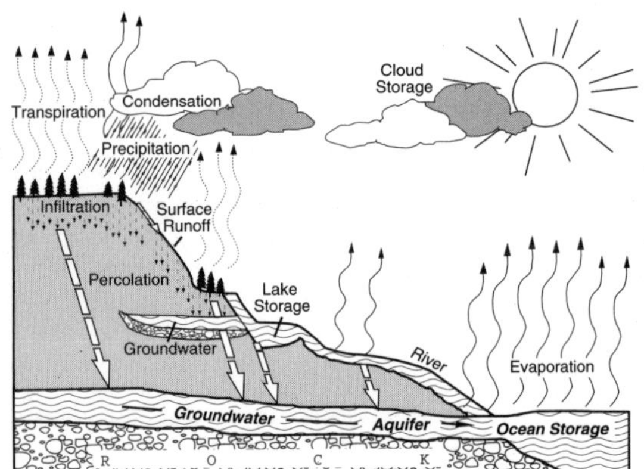

Hydrologic cycle

hydrogeologic condition A condition stemming from the interaction of groundwater and the surrounding soil and rock.

hydrogeologic cycle *See* hydrologic cycle.

hydrogeologist A person who studies and works with groundwater.

hydrogeology *See* geohydrology.

hydrograph A graphic representation of the relationship of the flow, stage, or velocity of a canal or river at a given point as a function of time. Typically it refers to the relationship of discharge and time.

hydrographer (1) A person who makes surveys of bodies of water. (2) In some sections of the country, a person who measures the flow of streams or conduits.

hydrographic Pertaining to the measurement or determination of various characteristics of bodies of water.

hydrographic basin *See* drainage area.

hydrographic survey An instrumental survey made to measure and record physical characteristics of streams and other bodies of water within an area, including such things as location, areal extent and depth, positions and locations of high-water marks, and locations and depths of wells.

hydrography (1) The science of measuring and analyzing the flow of water, precipitation, evaporation, and allied phenomena. (2) The science of measuring and studying oceans, seas, rivers, other waters, and their marginal land areas, as well as the configurations of the underwater surface, including the relief, bottom materials, and coastal structures. In a computer modeling graphic database, these hydrographic features are represented by, for example, water boundaries, stream centerlines, and lake polygons. Such continuous linear features facilitate the structuring of elements into a logical network for modeling purposes.

hydroisobath A contour line on a map that shows the depth to the water table from the ground surface. All points on a particular contour have the same depth. *See also* water-table isobath.

hydroisopleth map A map showing the fluctuation of the water table with respect to time and space.

hydrokinetics The study of the motion of fluids without reference to related forces.

hydrologic area A state or region having similar mean annual flood characteristics.

hydrologic budget An accounting of the inflow to, outflow from, and storage in a hydrologic unit such as a drainage basin, aquifer, soil zone, lake, reservoir, or irrigation project. *See also* available water supply.

hydrologic cycle The natural process recycling water from the atmosphere down to (and through) the Earth and back to the atmosphere.

hydrologic dam station A site at which information applicable to hydrologic studies is collected.

hydrologic inventory The process of (1) evaluating the items of inflow, outflow, and storage and (2) balancing the hydrologic equation. *See also* hydrologic equation.

hydrologic properties Those properties of soils and rocks that control the entrance of water and the capacity to hold, transmit, and deliver water. They include porosity, specific yield, specific retention, permeability, infiltration capacity, and direction of maximum and minimum permeability.

hydrology The science dealing with the properties, distribution, and circulation of water.

hydrolysate The product of hydrolysis.

hydrolysis A chemical reaction in which water molecules react with a substance to form two or more substances.

hydromechanics The science concerned with the equilibrium and motion of fluids and of bodies in or surrounding them.

hydrometric pendulum A device consisting of a metal ball suspended on a cord that is lowered into flowing water. After calibration, measuring the cord's inclination as caused by the current indicates the velocity of the current.

hydrometry (1) The measurement and analysis of the flow of water. (2) The determination of specific gravities of liquids.

hydronium ion *See* hydrogen ion.

hydroperiod The cycle of water depth variation and duration of coverage within a wetland over a period of time.

hydrophilic Having a tendency or affinity to bind with water molecules or absorb water. *See also* hydrophobic.

hydrophilicity The degree to which a substance is hydrophilic. *See also* hydrophilic.

hydrophobic Having a tendency to repel water molecules. *See also* hydrophilic.

hydrophobicity The degree to which a substance is hydrophobic. *See also* hydrophobic.

hydrophone See waterphone.

hydrophyte A plant that grows in saturated soils or water.

hydropneumatic Pertaining to a water system (usually a small one) in which a water pump is automatically controlled (started and stopped) by the air pressure in a compressed-air tank.

hydropneumatic system A system that uses both air and water in its operation. An example is a pressure tank that uses an air chamber to maintain pressure in a well system even when the pump is not operating.

hydropneumatic tank A storage tank of a given volume that is under pressure and thereby establishes a pressure for the water distribution system to which it is connected. Small systems

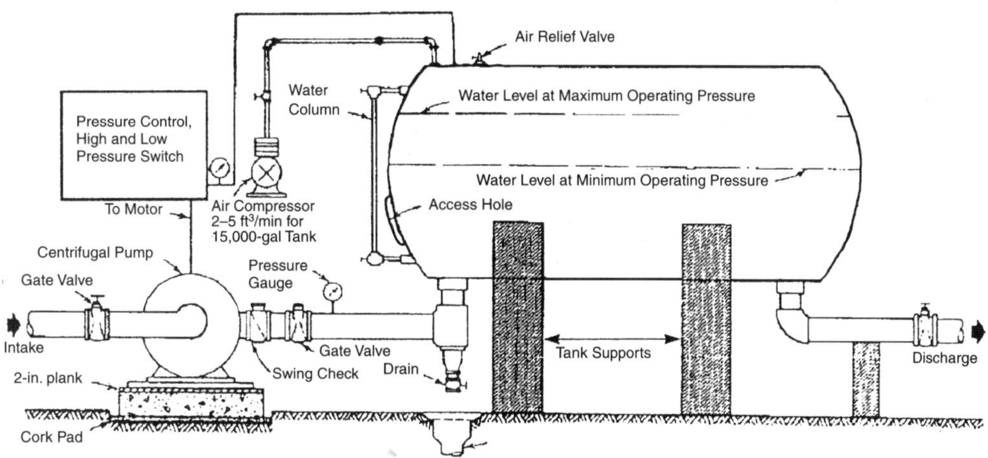

Hydropnuematic water pressure system

often use hydropneumatic tanks instead of elevated storage tanks to establish system pressure.

hydropower *See* hydroelectric power.

hydroscopic Tending to absorb moisture from the atmosphere. Sodium chloride (NaCl) is one hydroscopic substance. This term is identical to the term *hygroscopic*.

hydroscopic water Water that is strongly attracted to the solid particles in a porous medium and held in place by surface tension forces. It typically cannot be removed from a sample except by forcing a phase change, usually by heating the sample in an oven.

hydrosphere The water of the earth, including surface lakes, streams and oceans, underground water, and water in the atmosphere.

hydrostatic head *See* head.

hydrostatic joint A joint of the bell-and-spigot type formed in a large water main by forcing sheet lead into the bell under hydraulic pressure.

hydrostatic level The level or elevation to which the top of a column of water would rise, if afforded opportunity to do so, from an artesian aquifer or basin or from a conduit under pressure.

hydrostatic press A hydrostatic machine for increasing an applied force. It consists of two connected cylinders, one much larger than the other, fitted with watertight pistons. Such a machine is also called a Bramah's press.

hydrostatic pressure (1) The pressure at a specific elevation exerted by a body of water at rest. (2) In the case of groundwater, the pressure at a specific elevation caused by the weight of water at higher levels in the same zone of saturation.

hydrostatic pressure testing The act of determining the pressure that a structure can withstand by exerting a force generated by a fixed water column or pump. For example, a finished water storage reservoir can be tested hydrostatically by filling with water to a predetermined height; the pressure at the base of the tank will then be equivalent to the product of the water depth and the water density. The vessel is then observed, tested for leaks, distortions, mechanical failure, or any combination thereof.

hydrostatic sludge removal The discharge of sludge from hopper-bottomed sedimentation tanks by use of the hydrostatic pressure of the water above the sludge outlet.

hydrostatic test *See* hydrostatic pressure testing.

hydrostatic uplift (1) The upward pressure against the base of an impervious dam transmitted by water in the foundation from the water behind the dam. (2) Pressure on the upper surface of any horizontal joint or crack in a dam. *See also* uplift.

hydrostatics The study of water at rest.

hydrosulfite ($(NaO)_2S_2O_4{\cdot}2H_2O$) A salt used for reducing chlorine residuals. *See also* sodium hydrosulfite.

hydroturbine *See* turbopump.

hydroxide alkalinity Alkalinity caused by hydroxyl ions (OH^-). It is also called hydroxyl alkalinity. *See also* alkalinity.

hydroxide ion (OH^-) The ion formed by an oxygen and a hydrogen atom. This term is used to describe the anionic hydroxyl radical that is partially responsible for the alkalinity of a solution.

hydroxyl The univalent group consisting of one atom of oxygen and one of hydrogen. It occurs in many inorganic compounds that ionize in solution to yield hydroxide ions (OH^-). It is also the characteristic group of alcohols. *See also* alcohol.

hydroxyl radical (·OH) A very strong oxidizing agent that can destroy many organic compounds in water. Various advanced oxidation processes are used to generate hydroxyl radicals. *See also* advanced oxidation process; organic compound.

hydrozone A portion of a landscape area having plants with similar water needs that either are not irrigated or are irrigated by a circuit or circuits having the same schedule.

hydrozoning The design practice of grouping plants by similar water requirements to maximize the potential efficiency of irrigation.

hyetal regions Divisions of the world defined according to rainfall characteristics.

hyetograph (1) A graphical representation of average rainfall, rainfall excess rates, or volumes over specified areas during successive units of time during a storm. (2) A self-registering rain gauge.

hyetography The study of the geographic distribution of rainfall.

hyetology The science that deals with precipitation.

hygrograph A recording hygrometer.

hygrometer (1) An instrument for measuring the relative amount of moisture in the atmosphere. (2) An instrument for determining the dew point. *See also* psychrometer.

hygroscopic *See* hydroscopic.

hygroscopic moisture *See* hydroscopic water.

hygroscopic water *See* hydroscopic water.

hygroscopicity As applied to soil, the ability to absorb and retain moisture.

hyperactivity Any of a large constellation of clinical disorders. Most simply, hyperactivity is any disorder that results in abnormally increased motor activity. *See also* hyperkinetic syndrome.

hypercritical flow *See* shooting flow.

hyperfiltration Filtration using an extremely dense medium capable of separating dissolved ions from a feed stream. This term is sometimes, but not always, used synonymously with *reverse osmosis*. *See also* reverse osmosis.

hyperkinesis Abnormally increased motor activity. It is also referred to as hyperactivity. Hyperkinesis has been associated with lead intake, and drinking water can be a significant contributor to lead burdens.

hyperkinetic syndrome A childhood disorder that usually abates during adolescence. It is characterized by hyperactivity, fidgetiness, impulsiveness, excitability, distractibility, a short attention span, and difficulties in learning and perception. It includes disorders that respond paradoxically to a certain class of stimulant drugs (e.g., Ritalin) by an apparent calming of a child's activity. The syndrome has a variety of etiologies and in some cases is due to overt brain damage. It has been associated with excessive exposure to heavy metals, such as lead, and some food additives. The food association is particularly weak, however.

hypermedia (1) An emerging capability to reference, retrieve, and integrate different forms of geographic data, including vector maps, tabular database information, freeform text, raster images, engineering drawings, standard forms, photographs and other images, or audio records. A geographic information system is often viewed as the index to the variety of data or the "hub" of the multimedia system. (2) Richly formatted documents containing a variety of types of information, such as textual, image, movie, and audio.

hypersensitive animal *See* hypersusceptible animal.

hypersensitivity A condition of having an allergic response to a particular substance. The observed response is very much exaggerated relative to nonallergic individuals. Acute hypersensitivity reactions (type I, or anaphylaxis) can be immediately life-threatening. Type II reactions produce damage through the production of antibodies to tissue antigens, possibly resulting in chronic disease. Chronic toxicities of certain chemicals (e.g., chronic renal injury by mercury) can be mediated by the latter mechanism.

hypersusceptible animal An animal whose sensitivity to a chemical, physical, or microbial agent is greater than average for the species or strain.

hypertension A disease that results in blood pressure that is routinely higher than normal. This term is less commonly applied to describing temporary increases in blood pressure caused by normal physiological states (e.g., a fight-or-flight response). Contaminants in drinking water can influence blood pressure.

hypobromite A shortened name for hypobromite ion (OBr^-). *See also* hypobromite ion.

hypobromite ion (OBr^-) The ionized form of hypobromous acid (HOBr). Ozonation of hypobromite ion produces bromate (BrO_3^-). *See also* bromate ion; hypobromous acid.

hypobromous acid (HOBr) An acid used as a bactericide. During the disinfection of drinking water with chlorine (HOCl) or ozone (O_3), bromide ion (Br^-) is oxidized to hypobromous acid. Hypobromous acid reacts with natural organic matter to form bromine-substituted disinfection by-products. *See also* bactericide; bromide; disinfectant; disinfection by-product; natural organic matter.

hypochlorination Chlorination using solutions of calcium hypochlorite ($Ca(OCl)_2$) or sodium hypochlorite (NaOCl).

hypochlorinator A device used to chlorinate a liquid stream by adding chlorine in the form of liquid sodium hypochlorite (NaOCl, commonly called bleach) or calcium hypochlorite ($Ca(OCl)_2$).

hypochlorite A shortened name for hypochlorite ion (OCl^-). *See also* hypochlorite ion.

hypochlorite ion (OCl^-) The ionized form of hypochlorous acid (HOCl). *See also* hypochlorous acid.

hypochlorite of lime A combination of slaked lime and chlorine in which calcium oxychloride ($Ca(OCl)_2$) is the active ingredient. It is also called chlorinated lime.

hypochlorite solution An aqueous solution of a metallic salt of hypochlorite ion (OCl^-) that is used as a disinfectant and as a bleaching and oxidizing agent. *See also* disinfectant; hypochlorite ion; oxidizing agent.

hypochlorous acid (HOCl) An acid used as a disinfectant and as a bleaching and oxidizing agent. During the disinfection of drinking water with chlorine (Cl_2), hypochlorous acid reacts with natural organic matter and bromide (Br^-) to form disinfection by-products. *See also* bromide; disinfectant; disinfection by-product; natural organic matter; oxidizing agent.

hypolimnion A region of relatively constant temperature below the thermocline and extending to the bottom in a thermally stratified body of water. *See also* epilimnion; thermocline.

hypotension Blood pressure that is lower than normal. *See also* hypertension.

hypothalamo-pituitary-gonadal axis A set of feedback relationships that regulate the activity of hormones involved in sexual development and reproductive functions. These relationships involve a brain region called the hypothalamus, as well as the pituitary gland and the testicles or ovaries. Chemicals that modify interactions between these organs or interfere with the ability of the endocrine system to control function in a variety of other organs are referred to as endocrine disruptors. *See also* endocrine disruptors.

hypothermia A lowering of a person's deep body or core temperature to less than 98.6° Fahrenheit (37° Celsius), usually following exposure to cold water temperatures or ambient air.

hypothesis A supposition, arrived at from observation or reflection, that can be tested (or evaluated) in an analytic investigation. In science, a null hypothesis is put forth in order that it might be refuted; the null hypothesis states that the results observed in a study, experiment, or test are no different from what might have occurred because of chance alone. In environmental epidemiology, the search for causal associations between diseases and environmental exposures involves a series of cycles in which epidemiologists examine existing hypotheses, formulate new hypotheses, and test the acceptability of these new hypotheses. An epidemiologic hypothesis should describe the following with a high degree of specificity: the population and its characteristics, exposure, disease, dose-response relationship, and time-response relationship. *See also* statistically significant.

hypoxia Reduced oxygen tension in tissues. It can have a variety of causes, such as decreased oxygen in the breathing zone, insufficient circulation, or compromised breathing. It can also be produced when chemicals such as carbon monoxide (CO), nitrate (NO_3^-), or chlorite (ClO_2^-) interfere with the oxygen-carrying capacity of the blood.

Hz *See* hertz *in the Units of Measure list.*

I

IAST *See* ideal adsorbed solution theory.
iatrogenic transmission Transmission of a disease or condition by a physician through inadvertent or erroneous treatment.
IC *See* ion chromatography.
ice The solid state of water.
ice gorge The choking of a stream channel caused by the piling up of ice against some obstruction, forming a temporary dam, and the subsequent release of the water impounded behind the dam due to sudden bursting.
ice jam The choking of a stream channel caused by the piling up of ice.
ice pressure Pressure against a dam or other structure located in water caused by the expansion of ice due to a change in temperature.
ice push A sudden increase in the ice pressure.
ICM *See* integrated catchment management.
icosahedral A geometric shape of a particular viral particle, made up by the protein coat and composed of 12 vertices, 20 triangular faces, and 30 edges.
ICP spectroscopy *See* inductively coupled plasma spectroscopy.
ICP–AES *See* inductively coupled plasma spectroscopy.
ICPA *See* inductively coupled plasma spectroscopy.
ICP–MS *See* inductively coupled plasma–mass spectrometry.
I/C/R *See* industrial/commercial/residential.
ICR *See* Information Collection Rule.
icterus A condition characterized by an excess of bile pigments in the blood and tissues that leads to a yellow color of the surface of the skin. It is also known as jaundice.
ID *See* inside diameter.
ID_{50} *See* infectious dose 50.
ideal adsorbed solution theory (IAST) A theory describing a method to evaluate the competitive adsorption of contaminants in similar or different phases, independent of the form of the isotherm (e.g., Freundlich, Langmuir). *See also* competitive adsorption; Freundlich isotherm; Langmuir isotherm.
idiosyncratic reaction A reaction of an individual to a chemical exposure in a way unique to that individual or at least rare within a population. These reactions generally have a genetic basis by which the individual underexpresses a protein that performs some essential function.
IDLH *See* immediately dangerous to life or health.
IE *See* ion exchange.
IEEE *See* Institute of Electrical and Electronics Engineers.
IFI *See* International Fabricare Institute.
igneous rock Rock formed from the cooling and solidification of magma (molten rock).
ihp *See* indicated horsepower.
illuminance units In Système International, lux; candela-steradians per metre squared.
illuviation A process of accumulation by deposition from percolating water of material transported in solution or suspension.
image well An imaginary well used in modeling to simulate the effects of hydraulic boundaries on a pumping or recharge well.
imaginary wells An array of image wells used to simulate the effects of complicated hydraulic boundaries. *See also* image well.
immediate hypersensitivity An acute allergic response also referred to as anaphylaxis. If severe enough, such reactions can be life-threatening.
immediate toxicological effect An adverse effect that develops rapidly after exposure to a chemical. It is usually discussed in relation to effects that have a substantial latent period before they develop.
immediately dangerous to life or health (IDLH) A phrase used to describe an extremely hazardous atmosphere in the workplace that could cause injury or death very quickly.
imminent danger A threatening situation that may cause injury or loss of life if the conditions are not corrected quickly.
immiscible Not capable of being mixed.
immune surveillance The capacity of the immune system to respond by recognizing foreign materials, particularly those that produce infectious disease. This term can also be applied to the immune system's recognition of cells in the body that express specific antigens (e.g., tumor cells). Chronic chemical toxicities are frequently autoimmune effects. The chemical is

metabolized to a reactive form that binds to a protein that is externalized on the exterior of a cell—e.g., on liver cells—and antibodies are formed; the next time an exposure occurs, the liver is damaged by an immune response. This effect occurs with chlorinated hydrocarbons. A syndrome referred to as Goodpasture's syndrome—a glomerulonephritis (i.e., kidney damage)—is produced by exposure to solvents.

immunity (1) A state of increased resistance to infectious disease. Generally, it is the result of the formation of antibodies or special cells (macrophages) to specific antigens expressed by the infectious agent. Immunity can be passive (i.e., the result of medical personnel administering antibodies) or active (i.e., the result of stimulating the individual's own immune system to respond to a particular antigen). (2) With respect to corrosion, the domain of pH and oxidation–reduction potential (variously called E, pE, or EMF) where oxidation of the metal cannot take place. The area of such immunity is sometimes referred to as the area of cathodic protection. *See also* cell-mediated immunity; humoral immunity; immunization.

immunity, acquired *See* acquired immunity

immunity, active humoral *See* active humoral immunity.

immunity, passive humoral *See* passive humoral immunity.

immunization A natural or artificial development of resistance to a specific disease. *See also* infection; pathogenic organism.

immunocompromised Having an inability to respond appropriately to an infection because one or more components of the immune system are not functioning properly. This state can arise in a variety of ways; it may be inherited, induced by chemicals, or caused by infectious agents.

immunodeficiency A lack of one or more immune functions. This term can apply to deficiencies in either cell-mediated or humoral immunity. Immunodeficiency can also arise by a modification of the ability to control and integrate the various arms of the immune system. It can be congenital (i.e., a genetic disorder) or acquired (i.e., arising from viral infection, malnutrition, or a variety of disease states). *See also* acquired immunodeficiency syndrome; cell-mediated immunity; humoral immunity.

immunofluorescence The emission of visible light by a compound that has been irradiated with ultraviolet light. For example, a fluorescent compound (i.e., a fluorescein) can be attached to an antibody. Bacterial, viral, or other antigens that react with the antibody can then be observed by illuminating the sample with ultraviolet light.

immunoglobin The class of proteins to which antibodies (i.e., IgG, IgM, IgA, IgD, IgE) belong.

immunomagnetic assay A method for detecting microorganisms or cells by using antibodies attached to magnetic particles for separation. *See also* antibody.

immunopathology Physically observable damage to organs and cellular elements important to the immune function (i.e., thymus, spleen, lymphocytes, bone marrow, and lymph nodes).

immunosuppression Intentional depression of the function of the immune system, usually for purposes of preventing rejection of transplanted tissues. Obviously, the difference between immunotoxicity and immunosuppression is more a matter of intent. Immunosuppressive drugs are often immunotoxic if their use is not well controlled.

Turbine impeller

immunotoxicity A toxic response of the immune system. It can be manifested as immunosuppression, uncontrolled proliferation (i.e., leukemia and lymphoma), altered host defense mechanisms against pathogens and neoplasia, allergy, or autoimmunity.

impact The striking of one body against another. When particles or streams of water suffer impact, an energy loss or, more accurately, an energy transformation results.

impact fee *See* capacity charge; development impact fee.

impact head A device used in irrigation wherein a type of single-stream rotor uses a lever driven by its own impact on the stream of water to rotate the irrigation nozzle in a full circle or arc. Impact heads produce irrigation areas with large radii and relatively low water application rates, but they do not provide matched water application rates for varying arc patterns (less than a full circle).

impact loss A loss of head in flowing water as a result of the impact of water particles on themselves or on some bounding surface.

impedance The total opposition offered by a circuit to the flow of electric current.

impeller A rotating set of vanes in a pump designed to pump or lift water.

impeller pump Any pump in which the water is moved by the continuous application of power derived from some mechanical agency or medium. *See also* centrifugal pump.

Imperial gallon *See the Units of Measure list.*

impermeable Not allowing, or allowing only with great difficulty, the movement of water; impervious.

impermeable layer A layer not allowing, or allowing only with great difficulty, the movement of water.

impervious Resistant to the passage of water.

impervious bed A bed or stratum through which water will not move.

impervious core A core of an earth dam constructed of puddled or compacted clay or other material, designed to place a water barrier between the upstream and downstream faces.

imperviousness coefficient The ratio, expressed as a decimal or as a percentage, of effectively impervious surface area to the total catchment or tributary area under consideration.

impingement attack *See* erosion corrosion.

impingement corrosion *See* erosion corrosion.

implantation (1) In toxicology, the process by which a conceptus (a blastocyst, or 6–7 day postfertilization of the ovum) attaches to the endometrium. (2) The surgical insertion of a material or device in a tissue, usually for medical purposes. The material can be a drug or hormone. Examples of such devices are heart valves and pacemakers.

impounded water Water that is stored in an artificial basin or dammed ravine by diverting flowing streams or collecting rainfall runoff, as in a reservoir.

impounding dam A barrier constructed across a watercourse to create a reservoir. *See also* dam.

impounding reservoir A reservoir in which surface water is retained for a considerable period of time, ranging from several months upward, and released for use at a time when the ordinary flow is insufficient to satisfy requirements. *See also* reservoir.

impounding reservoir safe yield *See* impoundment safe yield.

impoundment A pond, lake, tank, basin, or other space, either natural or artificial, that is used for storage, regulation, and control of water. *See also* reservoir.

impoundment safe yield The maximum dependable draft that can be made continuously on a source of water supply during a period of years during which the probable driest period or period of greatest deficiency in water supply is likely to occur. For example, impoundment safe yields in the midwestern United States are commonly based on low watershed flows experienced one year in a 50-year period (i.e., a 50-year drought cycle).

impressed-current system A cathodic protection system that uses an outside source of power, converts it to direct current, and injects it into the soil through an anode bed. Such a system is also called a rectifier-ground-bed system.

improvements Buildings, other structures, and other attachments or annexations to land that are intended to remain so attached or annexed, such as sidewalks, trees, drives, tunnels, drains, and sewers.

impulse force The dynamic pressure of a jet or stream in the direction of its motion when its velocity in that direction is entirely destroyed. It is equal and opposite to the reaction force. *See also* reaction force.

impulse pump A type of pump that raises water by periodic application of a force suddenly applied and suddenly discontinued. The hydraulic ram is the most common example of such a pump. *See also* hydraulic ram.

impulse turbine A type of turbine having rotor blades shaped such that the force of fluid jets striking the blades causes a shaft to rotate and transmit energy to a motor or other device. An impulse turbine is sometimes used for energy recovery in a seawater-desalting reverse osmosis system, where pressurized waste concentrate may be used to drive a turbine that recovers energy for the feedwater-pumping device.

impulse–reaction turbine A turbine in which both the impulse and reaction principles are used. *See also* impulse force; reaction force.

IMR *See* infant mortality rate.

IN *See* indicated number.

in. *See* inch *in the Units of Measure list.*

in.^{2} *See* square inch *in the Units of Measure list.*

in.^{3} *See* cubic inch *in the Units of Measure list.*

in situ A Latin term meaning "in its original place." It is often used in the context of making measurements.

in vitro A Latin term meaning "in glass." It refers to tests conducted in a test tube, flask, or other laboratory glassware. *Contrast with* in vivo.

in vitro study A study of chemical effects conducted in tissues, cells, or subcellular extracts from an organism (i.e., not in the living organism).

in vivo A Latin term meaning "in life." It refers to tests conducted in a living organism. *Contrast with* in vitro.

in vivo study A study of chemical effects conducted in intact living organisms. *Contrast with* in vitro study.

inactivation (1) The transformation of a chemical to a less toxic form. This term can also apply more broadly to a variety of means by which the activity of various biomolecules that are used in intracellular, intercellular, and hormonal signaling pathways occurs. For example, acetylcholine is released from nerves to pass a signal to muscles. This acetylcholine is inactivated by an enzyme called acetylcholinesterase. If a chemical interferes with this enzyme, severe poisoning will occur because acetylcholine will not be enzymatically cleaved to the relatively inactive products acetic acid and choline. (2) A process that renders viruses unable to reproduce or infect. Usually, a specific percentage of a population is affected over time. This effect can be accomplished by a variety of methods, e.g., heat, chemicals, or ultraviolet light. In contrast, the term *kill* is applied to other forms of life, such as bacteria, cysts, or algae.

incentive rate A rate that water utilities charge to encourage reductions in water use, particularly where water supplies are limited or uncertain. The incentive rate is typically a higher rate per unit of water for summer use, for summer use in excess of winter use (indoor use), or for consumption in excess of what is considered to be essential or nondiscretionary use irrespective of the time of year. *See also* inclining block rate.

inch (in.) *See the Units of Measure list.*

inches per minute (in./min) *See the Units of Measure list.*

inches per second (in./s) *See the Units of Measure list.*

inch-foot charge for water service *See* fire line service charge.

inch-pound (energy) (in.-lb) *See the Units of Measure list.*

inch-pound (torque) (in.-lb) *See the Units of Measure list.*

incidence The number of instances of illness or infection originating, or persons becoming ill or infected, during a period of time (e.g., the number of *new* cases of disease or infection) in a defined population within a specified period of time. *See also* incidence rate; prevalence.

incidence rate The rate at which people without a disease, illness, or infection develop the condition during a specified period of time. To calculate this rate, divide the number of new cases diagnosed or reported during a defined period of time by the number of people at risk during the time period.

incidence of tumors The percentage of animals with tumors in a toxologic study of a contaminant.

incident command center An area used for the coordination and security of resources and personnel near an emergency or disaster. It is usually organized by a local fire department.

incident rate The number of accidental incidents to which a workforce is exposed in a given time period, typically a year. Incident rates are based on the exposure of 100 full-time workers (40 hours per week, 50 weeks per year), equivalent to 200,000 employee-hours per year. Three incident rates are common:

$$\text{recordable rate} = \frac{\text{number of recordable events}}{\text{total number of productive hours worked during the period covered}} \times 200{,}000$$

$$\text{injury frequency rate} = \frac{\text{number of recordable injuries causing lost days}}{\text{total number of productive hours worked during the period covered}} \times 200{,}000$$

$$\text{severity rate} = \frac{\text{number of lost days}}{\text{total number of productive hours worked during the period covered}} \times 200{,}000$$

incineration Combustion or controlled burning of volatile organic matter in sludge and solid waste. It reduces the volume of the material while producing heat, dry inorganic ash, and gaseous emissions.

incinerator A device used for combustion of solid matter. Incinerators often are used for combusting semiliquid or dewatered materials, such as sludge. In drinking water treatment, incinerators also are used to regenerate or reactivate spent granular activated carbon to restore the porous structure and adsorptive capacity. *See also* regeneration.

incised river (1) A river that has cut its channel through the bed of a valley floor, as opposed to one flowing on a floodplain. (2) A river or stream that has formed its channel by the process of degradation.

inclined plate separator A unit constructed of multiple parallel plates, approximately 2 inches (5 centimetres) apart and oriented at a 45° to 60° angle from the horizontal, to improve settling in a sedimentation basin. The units are placed at the end of a sedimentation basin (across the entire width), and flow travels upward through the plates and exits at the top prior to being discharged from the basin (a configuration called *counterflow*). Cocurrent and cross flow may also be used. The inclined plates provide a much shorter distance for particles to settle prior to being captured and are often used to maintain particle removal at higher flow rates, thereby reducing the need to construct additional basins.

inclining block rate A utility rate structure for which water rates increase for defined blocks of increased water use. (For example, $1.00 may be charged for each 1,000 gallons used per month up to 5,000 gallons, and $1.50 may be charged for each 1,000 gallons used per month between 5,000 and 10,000 gallons.) This rate structure recognizes the higher unit costs of water associated with meeting peak demands. Inclining block rates are sometimes based on cost-of-service analysis, sometimes on the incremental cost of the next unit of supply needed to meet peak demands, and sometimes on the need to conserve

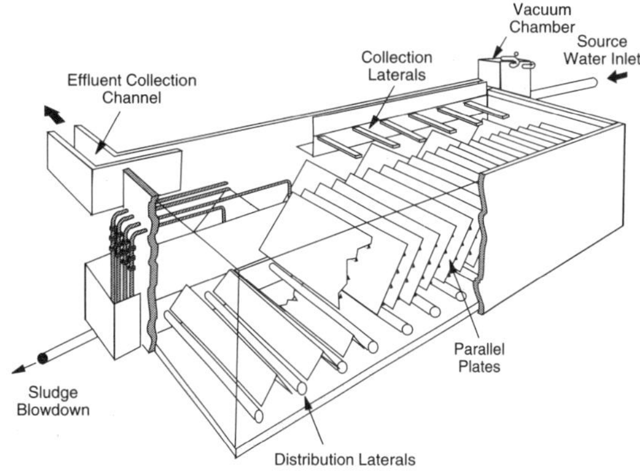

Inclined plate separator

water during peak use periods. Inclining block rates are also known as inverted rates because the rates are the opposite of traditional rates that discount water to high-volume users. *See also* inverted rate.

included gases Gases in isolated interstices in either the zone of aeration or the zone of saturation.

inclusion body An irregular, oval, or round-shaped body occurring in the nuclei or cytoplasm of the cells of the body.

increasing block rate *See* inclining block rate.

incremental cost The cost of water from the most recent (or next) increment of plant capacity and supply, as used in the marginal cost rate-setting concept. Under this rate-setting approach, the charge for all water sold reflects this cost even though the average cost is less. *See also* marginal cost pricing.

incrustant Dense solids formed as a crust on the inside surface of a pipe as a result of hardness and other characteristics of the water carried in the pipe. *See also* hardness; scale; tuberculation.

incubate (1) To maintain microorganisms at a temperature and in an environment favorable to their growth. (2) The holding of disinfection by-product formation potential samples to allow the formation of disinfection by-products. *See also* disinfection by-product formation potential.

incubation (1) The process of placing and holding microbial cultures under conditions favorable for growth. (2) The period of time and conditions necessary for a fertilized ovum to reach full development. (3) The storage of disinfection by-product formation potential samples.

incubation period The time interval between initial contact with an infectious agent and the appearance of the first symptom of the disease.

incubator A heated container that maintains a nearly constant temperature for use in developing biological cultures.

independent variable (1) A variable in an equation that determines the values of other variables. For example, in the equation $y = ax + b$, the variable x is called the independent variable. (2) A variable hypothesized to influence the outcome under study (i.e., to cause a particular outcome or influence its variation). *See also* regression analysis.

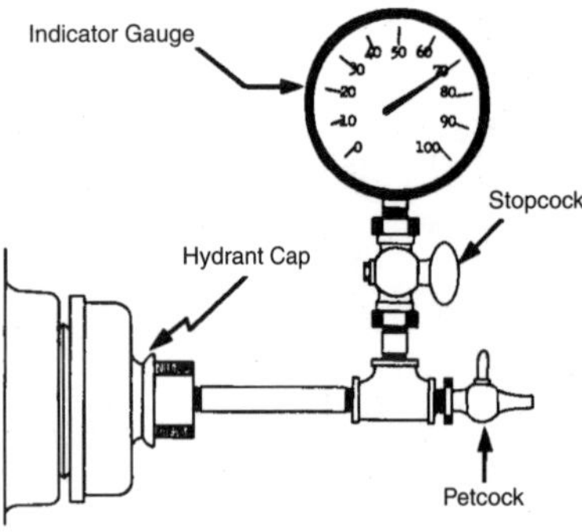

Indicator gauge for hydrant pressure

index (1) An indicator, usually numerically expressed, of the relation of one phenomenon to another. (2) An indicating part of an instrument.

index case The first case of disease or other health condition in a defined population to come to the attention of the epidemiologist.

index organism A microscopic organism for which the presence or absence in bodies of water indicates the presence or absence (and extent) of pollution or some other specific factor under study. An index organism is also called an indicator organism.

Indian tribe According to the Safe Drinking Water Act Section 1401, any Native American tribe having a federally recognized governing body that carries out substantial governmental duties and exercises powers over any area.

indicated horsepower (ihp) A measure of actual work required to compress and deliver a given fluid quantity, including all thermodynamic, leakage, and fluid friction losses.

indicated number (IN) In the testing of bacterial density by the dilution method, the number obtained by taking the reciprocal of the highest possible dilution (smallest quantity of sample) in a decimal series that gives a positive test.

indicating floor stand A device for operating a valve that indicates the extent to which the gate is open.

indicator (1) A substance that gives a visible change, usually of color, at a desired point in a chemical reaction, generally at a specified end point. (2) A device that indicates the result of a measurement. Most indicators in the water utility field use either a fixed scale and movable indicator (pointer), such as a pressure gauge or a movable scale and movable indicator, like those used on a circular flow-recording chart. This type of indicator is also called a receiver.

indicator bacteria In environmental engineering or bacteriology, bacteria used to signal conditions under which enteric pathogenic microbes may be present in treated drinking water.

indicator gauge A gauge that shows—by means of, e.g., an index, pointer, or dial—the instantaneous value of such characteristics as depth, pressure, velocity, stage, discharge, or the movements or positions of water-controlling devices. *See also* indicator; recorder; register.

indicator organism An organism that can be detected with relative ease and specificity and whose presence (or absence) has been shown to be correlated with a specific condition of interest. The condition of interest may be as specific as the potable quality of treated and disinfected drinking water or as broad as the assessment of the general level of pollution in an ecological system. In the latter case, several indicator organisms may be used.

indicator post An aboveground, post-like device attached to an underground valve and provided with (1) a wrench or handwheel for operating the valve and (2) a target designed to show whether the valve is open or shut.

indirect bank protection A kind of works constructed in front of stream banks, rather than directly on them, with the intention of reducing the erosive force of the current either by deflecting the current away or by inducing deposition. An instance of such protection is known as a dike; it may be made of timber or stone.

indirect charge A monetary charge that is a part of a total charge for producing something but is not directly related to production, e.g., overhead.

indirect discharge measurement A determination of the peak discharge (other than by a current meter) on the basis of the slope–area, contracted-opening, culvert, flow-over-dam, or critical-depth method. An indirect measurement also may be based on a survey of high-water marks after a flood.

indirect impedance approach A method of water analysis in which the electrodes of an impedance cell are not in direct contact with a sample. For example, indirect impedance technology can be used to detect certain microorganisms (e.g., coliforms) in water. The signal measured is caused by the carbon dioxide (CO_2) produced during the growth of the microorganisms being absorbed by a potassium hydroxide (KOH) agar (or liquid) around the electrodes. As the carbon dioxide is absorbed by the alkaline potassium hydroxide at pH values greater than 11, it is converted to bicarbonate (HCO_3^-), which in turn leads to a net change in molar conductivity that gives rise to the measured signal.

indirect potable reuse The use of water from streams that have upstream discharges of wastewater.

indirect reuse The use of reclaimed water for nonpotable or potable purposes by first discharging to a water supply source, such as fresh surface water or groundwater, where it mixes, dilutes, and may be transformed before being removed for reuse.

individual water supply A water supply to a single-family domicile. This term has replaced *private water supply* in the context of a supply used only by a single family.

indoor–outdoor rate An incentive rate for water use that an increasing number of water utilities are implementing to induce conservation of irrigation water. The incentive rate is usually set to discourage (1) irrigation beyond the level needed for normal lawn care and (2) other nondiscretionary outdoor uses. The outdoor (summer) rate and the block at which it applies are often based on the cost of meeting peak demands, but sometimes they are based on the need to conserve water during peak-use months. The indoor (winter) rate is sometimes set at less than the actual cost of service during the winter because the difference between the indoor and outdoor rates must be kept large enough to convince customers to reduce their outdoor use. However, the total revenue derived from the respective rates times their volumes must not exceed total revenue requirements.

indoxyl-β-D-glucuronide ($C_{14}H_{15}NO_7$–$C_6H_{13}N$) A chromogenic chemical compound used in bacterial culture media for the specific detection and enumeration of *Escherichia coli.* This compound is hydrolyzed by the enzyme β-glucuronidase to yield a glycone that is converted rapidly to indigo, which is dark blue in color.

induced infiltration *See* induced recharge.

induced recharge The recharge of water to an aquifer from an adjacent water source (e.g., a stream or lake) that occurs when the aquifer water table is lowered by pumping. The difference in head between the water source and aquifer causes water to recharge into the aquifer.

inductance An electrical property by which electrical energy is stored in a magnetic field. It is analogous to inertia in a hydraulic system. Induction has the effect of resisting changes in current flow. It is measured in henrys, or metres squared kilograms per second squared per ampere squared.

induction motor Any motor that relies on magnetic induction to couple energy into the rotor and create the necessary rotor currents.

induction period The period of time required for a specific cause to produce disease; the period of time from the causal action of a factor (e.g., chemical-induced mutation) to the initiation of the disease (e.g., cancer at a specific site). An induction period of many years may be required for cancer. *See also* incubation period; latency period.

inductively coupled argon plasma spectroscopy (ICPA) *See* inductively coupled plasma spectroscopy.

inductively coupled plasma–atomic emission spectroscopy (ICP–AES) *See* inductively coupled plasma spectroscopy.

inductively coupled plasma (ICP) spectroscopy An atomic emission technique that is used in the analysis of metals. Modern inductively coupled plasma instruments have four main components: an inductively coupled plasma source, a spectrometer, sample introduction hardware, and a computer. The inductively coupled plasma source is composed of an argon plasma confined in a radio frequency induction coil. The technique is sensitive for many metals and can allow for the rapid, even simultaneous, determination of metals.

inductively coupled plasma–mass spectrometry (ICP–MS) One of the "hyphenated techniques" in which two powerful instruments are linked. The inductively coupled plasma acts as the ion source for the mass spectrometer. For some elements, the sensitivity is higher than for graphite furnace atomic absorption techniques. Although they can be much faster than graphite furnace units, inductively coupled plasma–mass spectrometry instruments are expensive and, as of this writing, are not yet common in water analysis laboratories. *See also* graphite furnace atomic absorption spectrophotometry.

indurated sand Hardened sand, such as a soft sandstone.

induration A geologic process of hardening of sediments or other rock aggregates through cementation, pressure, heat, or other causes.

industrial consumption The quantity, or quantity per capita, of water supplied in a municipality or district for mechanical, trade, and manufacturing purposes during a given period, generally 1 day. The per capita use is generally based on the total population of the locality, municipality, or district. *Industrial use of water* is a more accurate term.

industrial hygiene The study of health as related to work environments. *See also* environmental health; safety program; toxicology.

industrial recycling The reuse of industrial process water, typically after required treatment.

industrial reuse The use of reclaimed or recycled water for nonpotable industrial purposes.

industrial solvent An organic solvent used for coatings (e.g., paints, varnishes, and lacquers), industrial cleaners, printing inks, extractive processes, and pharmaceuticals. *See also* solvent.

industrial water Water that is withdrawn from a source for sole use in an industrial water system.

industrial water conservation A practice employed to reduce the water consumed during industrial or manufacturing processes. Conservation can include both reducing the water used in a given application and reusing process water to reduce overall consumption.

industrial water use The use of water in industrial activities such as power generation, steel manufacturing, pulp and paper processing, and food processing.

industrial-aid bond A bond issued by a governmental unit, the proceeds of which are used to construct plant facilities for private industrial concerns. Lease payments made by the industrial concern to the governmental unit are used to service the bond.

industrial/commercial/residential (I/C/R) The set of classifications of water utility customers based on service characteristics and demand patterns. These classifications are established to assign water rates that are nondiscriminatory and that meet, as nearly as possible, the cost of providing service to each customer class.

industry water requirement The quantity of water required to produce a unit of product. It is the quantity of water ordinarily used, not necessarily the minimum that could be used.

inelastic demand Demand for a product (water) such that when a price increase (or decrease) is imposed, the resulting percentage change in quantity demanded is less than the percentage change in price. Elasticity is defined as the percentage change in quantity divided by the percentage change in price, which is always a negative term. With inelastic demand, the elasticity ratio is between zero and –1.0. The effect of inelastic demand is that a utility's revenue will increase with rate increases because the percentage increase in price is greater than the percentage decrease in quantity consumed. Water demand is virtually always inelastic.

inert media Synthetic resin beads or water treatment materials that are nonreactive. Inert media are used as natural nonreactive layers to more effectively separate the cation resin from the anion resin in mixed bed deionizers in order to allow each to be regenerated separately.

infant mortality rate (IMR) A measure of the annual rate of deaths for children less than 1 year old. The infant mortality rate is actually a ratio, but the word *rate* is universally used. To calculate this value, divide the number of deaths in a year for children less than 1 year of age by the number of live births in the same year; then multiply this ratio by 1,000 to obtain the infant mortality rate in terms of the number per 1,000 live births.

infection Entry and development or multiplication of an infectious agent (pathogenic microorganism) in humans or animals. Infection may be inapparent (asymptomatic), or it may result in disease.

infectious Capable of causing an infection.

infectious agent An organism capable of causing infection or infectious disease. *See also* pathogenic organism.

infectious disease A clinically apparent disease of humans or animals resulting from infection.

infectious dose The number of organisms that initiates colonization or disease in the host.

infectious dose 50 (ID_{50}) A dose of a given microorganism that, if applied to the total population, will infect 50 percent of the population.

infectious hepatitis An acute viral inflammation of the liver characterized by jaundice, fever, nausea, vomiting, and abdominal discomfort. It may be waterborne.

infectivity titer A method to measure infectious dose through a series of dilutions performed in the laboratory.

infiltrate (1) To filter into. (2) To penetrate, by a liquid or gas, the pores or interstices of a porous material. (3) (archaic) Water containing materials dissolved during passage through a body of soil.

infiltration The flow or movement of water downward through soil.

infiltration capacity The maximum rate at which the soil, when in a given condition, can absorb falling rain or melting snow.

infiltration coefficient The ratio of infiltration to precipitation.

infiltration ditch A ditch that extends into the saturation zone, i.e., a ditch for which the bottom is below the water table in the saturation zone. The water flows from the saturated material by gravity into the ditch, and from there into a sump, well, or canal.

infiltration diversion Diversion of water from a stream by means of perforated pipes or other pervious conduits laid under the stream bed.

infiltration gallery A horizontal subsurface tunnel for intercepting and collecting groundwater by gravity flow.

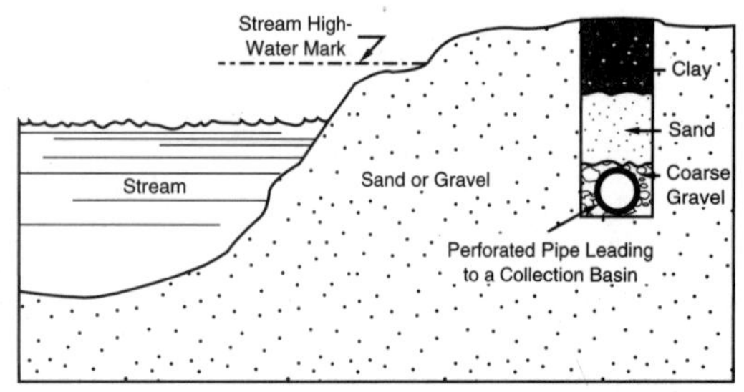

Infiltration gallery

infiltration head The water pressure at the point where subsurface water enters a sewer.

infiltration index A rate of infiltration calculated from records of rainfall and runoff. Several different indexes exist, each calculated by a different method. A common index is the average rainfall intensity for a given storm such that, if this intensity is exceeded, the weight of rainfall will equal the weight of runoff.

infiltration rate (1) The rate at which water enters the soil or other porous material under a given condition. (2) The rate at which infiltration takes place, expressed in depth of water per unit time, usually in inches or centimetres per hour. (3) The rate at which groundwater enters an infiltration ditch or gallery, drain, sewer, or other underground conduit. This value is usually expressed in cubic feet per second, million gallons per day per mile of waterway, cubic metres per second per kilometre, or cubic metres per day per kilometre. Another common method of expression is as an equivalent depth (length) per unit time, which can be multiplied by the area of application to obtain volume per time.

infiltration tunnel A type of infiltration gallery.

infiltration volume The quantity of rainfall, expressed in inches or centimetres depth over a given area, that passes below the root zone of plants and ultimately reaches the zone of saturation. *See also* deep percolation.

infiltrometer Typically a device that consists of two concentric rings driven into the ground. The rings are filled with water, and the time required for water to soak into the soil determines the rate at which water can percolate into the soil. The infiltration rate is usually expressed in inches or millimetres of water per unit time. *See also* lysimeter.

inflation A measurement of a period-to-period (monthly, quarterly, annually) percentage increase in prices for a product or service. To facilitate comparisons and a broad range of applications, inflation for hundreds of products is typically measured by means of price or cost indexes that express each period's price as a ratio with respect to a base period price. Examples of price indices are the consumer price index for a market basket of consumer goods and the *Engineering News Record's* construction cost index that combines the major components of construction.

influence area *See* circle of influence; zone of influence.

influence basin The basin area within which the water table or other piezometric surface is lowered by a withdrawal of water during pumping. *See also* circle of influence; zone of influence.

influent Water or other liquid—untreated or partially treated—flowing into a reservoir, basin, treatment process, or treatment plant. *Contrast with* effluent.

influent seepage A movement of water from the land surface to the water table. This term is commonly used to describe flow from a stream to the adjacent or underlying groundwater body. *See also* influent stream; leakage.

influent stream A stream or stretch of stream that contributes water to the saturation zone. The water surface of such a stream stands at a higher level than the water table or piezometric surface of the groundwater body to which it contributes water. An influent stream is also called a losing stream.

influent water Subsurface water resulting from penetration of surface water or atmospheric precipitation into the lithosphere through large interstices or fractures.

influent weir A weir at the inflow end of a sedimentation basin or channel.

Information Collection Rule (ICR) A rule issued by the US Environmental Protection Agency that requires water utilities serving 10,000 people or more to conduct monitoring in order to gather data for use in developing the Stage 2 Disinfectants–Disinfection By-Products Rule and the Enhanced Surface Water Treatment Rule. The Information Collection Rule was promulgated on May 14, 1996 (*Federal Register*, 61(94):24354–24388).

information collection request A formal request that federal agencies, under the Paperwork Reduction Act, must prepare to obtain clearance from the Office of Management and Budget for any activity that will involve collection of substantially the same information from 10 or more nonfederal respondents. Federal agencies may have to file such a request if information requirements are to be included in a rule (e.g., reporting, monitoring, or record-keeping requirements) or for other information-collecting activities (e.g., conducting studies or surveys, sending out application forms, performing audits).

infrared radiation That portion of the electromagnetic spectrum between the microwave and visible regions. This generally covers wavelengths between 0.76 and 15 micrometres. The interaction of infrared radiation with organic molecules serves as the basis for a technique to identify such molecules. The unit of wavenumber, expressed in units of per centimetre, is often used in infrared spectrometry. The range of greatest interest in the qualitative analysis of organic compounds is 2.5–15 micrometres. *See also* wavenumber.

infrared spectrophotometry An analysis technique in which a sample's absorption of light in the infrared region is measured.

infrared spectroscopy *See* spectroscopy.

infrastructure The basic installations and facilities on which the continuation and growth of a community depends, such as power, roads, wastewater and water plants, and transportation and communication systems.

infusorial Very fine-grained earth that pours.

ingestion The passage of matter such as food, water, or foreign material (ranging from dirt to poisons) into the body by means of the mouth.

inhalation The breathing of vapor, gases, mist, or dust through the mouth or nose.

inhalation exposure Exposure to a chemical or infectious agent via the respiratory tract. The material may be a gas, vapor, fume, or particulate.

inhibition A mechanism for limiting corrosion of a pipe surface. Two general approaches to inhibition exist: precipitation and passivation. With precipitation, insoluble compounds are formed by adjusting water chemistry to cause calcium carbonate ($CaCO_3$) to precipitate on the pipe wall. With passivation, the pipe material itself participates in a reaction to form less soluble metal–hydroxide–carbonate compounds intended to protect the pipe. *See also* passivation.

inhibitor Any chemical substance that is added to a water supply (or solution of any kind) that interferes with ("inhibits") a chemical reaction. Inhibitors are sometimes used to help prevent precipitation or corrosion.

inhibitory toxicity Any demonstrable inhibitory action of a substance on the rate of general metabolism (including rate of reproduction) of living organisms.

in-house Involving the use of water utility staff (full-time, part-time, and temporary employees) to complete a project, as opposed to using a contractor.

initial abstraction retention The total amount of rainfall that may fall without causing a significant amount of direct runoff.

initial compliance period The first full 3-year compliance period that begins at least 18 months after promulgation of a US Environmental Protection Agency drinking water regulation.

initial detention The volume of water on the ground, either in depressions or in transit, at the time active runoff begins.

initial loss In hydrology, rainfall preceding the beginning of surface runoff. It includes interception, surface wetting, and infiltration unless otherwise specified.

initial rain The rain that falls in the beginning of a storm before the depression storage is completely filled. *See also* residual rain.

initial [Br^-]/average [Cl^+] molar ratio *See* bromide-to-FAC ratio.

initiator *See* tumor initiator.

injection technique A technique used to inject a sample (or extract) into a chromatographic instrument (e.g., a gas chromatograph). For example, in gas chromatography, extracts can be injected into a hot injection port (where they are flash vaporized and swept into the chromatographic column) or directly on-column at a "cold" temperature. *See also* gas chromatography.

injection well A well in which fluid is transmitted to a subterranean formation.

injector *See* ejector.

injured bacteria Bacteria that are physiologically damaged because of exposure to adverse physical or chemical conditions.

injury frequency rate *See* incident rate.

in-kind Closely approximating the characteristics of a wetland before it was adversely affected by a regulated activity.

in-kind replacement The provision or management of substitute resources to replace the functional values of the resources lost (e.g., in a restored wetland). Such substitute resources are physically and biologically the same (or nearly the same) as those lost.

in.-lb (energy) *See* inch-pound (energy) *in the Units of Measure list.*

in.-lb (torque) *See* inch-pound (torque) *in the Units of Measure list.*

inlet (1) A surface connection to a drain pipe. (2) A structure at the diversion end of a conduit. (3) The upstream end of any structure through which water may flow. (4) An intake.

inlet control Control of the relationship between headwater elevation and discharge by the inlet or upstream end of any structure through which water may flow.

inlet surface The well screen, perforated sections of casing, or other openings through which water from a water-bearing formation enters a well.

inlet well A chamber that serves as a suction well for pumps in a water-pumping station. *See also* wet well.

inlet zone The initial zone in a sedimentation basin that decreases the velocity of the incoming water and distributes it evenly across the basin.

in-line filtration A treatment process in which coagulant is injected directly into the influent piping to a filter. Coagulation and flocculation occur within the piping, and sedimentation is not provided prior to filtering. *See also* coagulation; direct in-line filtration; flocculation; sedimentation.

in-line mixer *See* static mixer.

in./min *See* inches per minute *in the Units of Measure list.*

inorganic Pertaining to material such as sand, salt, iron, calcium salts, and other mineral materials. Inorganic substances are of mineral origin, whereas organic substances are usually of animal or plant origin and contain carbon. *See also* organic.

inorganic carbon Carbon present in an inorganic compound (e.g., carbon oxides, carbon disulfide [CS_2], and metallic carbonates such as calcium carbonate [$CaCO_3$]). *See also* organic carbon.

inorganic compound A substance that is not derived from hydrocarbons. Examples include sodium chloride (NaCl), ferric hydroxide ($Fe(OH)_3$), and aluminum sulfate ($Al_2(SO_4)_3$).

inorganic contaminant (IOC) An inorganic substance regulated by the US Environmental Protection Agency in terms of compliance monitoring for drinking water. Contained on the agency's list are contaminants as diverse as asbestos, nitrate (NO_3^-), cyanide, and nickel. This abbreviation came into common use in the US Environmental Protection Agency's Phase V drinking water regulations. An inorganic contaminant is sometimes called an inorganic chemical.

inorganic material A chemical substance of mineral origin, or more correctly, not of basically carbon structure.

inorganic matter Any chemical substance that does not arise from the process of living growth, is composed of matter other than plant or animal matter, and does not contain hydrocarbons or compounds of basically carbon structure. Examples are minerals and metals.

inorganic waste Discarded material such as sand, salt, iron, calcium, and other mineral materials that are only slightly affected by the action of organisms. Inorganic wastes are chemical substances of mineral origin, whereas organic wastes are chemical substances of an animal or plant origin.

inotropic action Chemical action to increase the strength of heart contraction. This action is associated with the cardiac glycosides (e.g., digitoxin or digoxin) that are frequently used to treat congestive heart failure. The inotropic effect is usually desired, but recognizing that these are very toxic compounds that occur naturally is also very important.

in-parallel flow A piping arrangement that directs separate streams through two or more units of a treatment system in a balanced manner, with equal flow to each device, so that a higher total flow rate than that from a single unit can be achieved.

in-plant study An evaluation of a unit process or system under full-scale conditions.

input horsepower The total power used in operating a pump and motor. It is equal to the brake horsepower divided by the motor efficiency expressed as a decimal.

in./s *See* inches per second *in the Units of Measure list.*

insanitary Not sanitary; unhealthy; liable to promote disease; contrary to principles known to promote or safeguard health.

insecticide A compound, usually a synthetic organic substance, used to kill insects.

in-series flow A piping arrangement in which the entire effluent flow from one unit of a water treatment system is fed to a second succeeding unit. This arrangement forces the water through multiple treatment units and achieves a greater reduction of contaminants than can be achieved by a single pass of water through a single unit.

insert valve A shutoff valve that can be inserted by a special apparatus into a pipeline while the line is in service under pressure.

inside diameter (ID) (1) The distance between two points on the interior surface of a round pipe or tank as measured along a straight line through the center. (2) The approximate or commercially assigned diameter of a pipe, such as a 6-inch (15-centimetre) pipe.

installation (1) The connecting or setting up and startup operations of any water treatment system. (2) The complete assembly of piping, valves, drain line, pumps, meters, and controls by which such equipment is connected into the water supply system.

instantaneous disinfection by-product concentration [DBP_0] The concentration of disinfection by-products at the time of sampling. In order to determine this amount in a chlorinated sample, the residual chlorine needs to be quenched with a dechlorination agent to stop any further disinfection by-product formation during the time period from sampling to analysis. Routine distribution system monitoring is commonly based on obtaining instantaneous samples. When a disinfection by-product formation potential test is performed on a prechlorinated sample, two parallel samples are collected; one sample is incubated (stored) with a chlorine residual present in order to determine the terminal disinfection by-product concentration [DBP_t], and the other sample is dechlorinated in order to determine the instantaneous disinfection by-product concentration. The disinfection by-product formation potential is equal to [DBP_t] – [DBP_0]. Note that the symbol [DBP_0] is used to represent instantaneous disinfection by-product concentration only in the context of disinfection by-product formation potential testing. *See also* dechlorination agent; disinfection by-product; disinfection by-product formation potential; terminal disinfection by-product concentration.

instantaneous flow rate A flow rate of water measured at one particular instant, such as by a metering device or by the following calculation:

$$Q = AV$$

Where (in any consistent set of units):

Q = the flow rate of water

A = the cross-sectional area of the channel or pipe

V = the velocity of the water at one instant

instantaneous power (*p*) The time rate of transferring or transforming energy. At the terminals of a circuit, instantaneous-power is the rate per unit time at which the energy is transmitted into or out of the circuit. It is often simply called power and is measured in watts, metres squared kilogram per second cubed.

instantaneous total trihalomethane concentration [$TTHM_0$] The concentration of total trihalomethanes at the time of sampling. In order to determine this amount in a chlorinated sample, the residual chlorine needs to be quenched with a dechlorination agent to stop any further trihalomethane formation during the time period from sampling to analysis. Routine distribution system monitoring is normally based on obtaining instantaneous samples. When a trihalomethane formation potential test is performed on a prechlorinated sample, two parallel samples are collected; one sample is incubated (stored) with a chlorine residual present in order to determine the terminal total trihalomethane concentration $[TTHM_t]$, and the other sample is dechlorinated in order to determine the instantaneous total trihalomethane concentration. The total trihalomethane formation potential is equal to $[TTHM_t] - [TTHM_0]$. Note that the symbol $[TTHM_0]$ is used to represent instantaneous total trihalomethane concentration only in the context of trihalomethane formation potential testing. *See also* dechlorination agent; terminal total trihalomethane concentration; total trihalomethanes; trihalomethane formation potential.

Institute of Electrical and Electronics Engineers (IEEE) A large professional organization of electrical and electronics engineers. It is divided into 10 regions, about half of which are in the United States. The organization is also divided into 37 technical societies. It has written many standards on many subjects.

Institution of Water & Environmental Management (IWEM) The organization of professionals in water and environmental management in Great Britain.

instream flow The amount of water in a stream that supports recreational uses, fisheries, aquatic species, and other nonwithdrawal water uses.

instream use A water use that can be carried out without removing the water from its source, as in navigation and recreation.

instrument detection limit The smallest amount of analyte that can be distinguished from an instrument's background signal.

Instrument Society of America (ISA) A professional society that represents the instrumentation and control industries and profession. The group has written standards on many related subjects and conducts other professional activities.

instrumentation (1) The act of developing, equipping with, or using instruments, especially scientific instruments. (2) A set of instruments.

insulated stream A stream (or stretch thereof) that is separated from the saturation zone in the surrounding formation and that neither contributes to nor receives water from that saturation zone.

insulation Material used to separate electric components, such as the resin material surrounding the windings of the stator in an electric motor.

insulator Any of many materials that can be forced to carry electricity only by application of a very strong electrical force.

Insurance Services Office (ISO) A group responsible for rating fire protection service districts within US cities.

intake (1) The works or structures at the head of a conduit into which water is diverted. (2) The process or operation by which water is absorbed into the ground and added to the saturation zone. (3) The flow or rate of flow into a canal, conduit, pump, stack, tank, or treatment process before treatment. *See also* discharge; outfall.

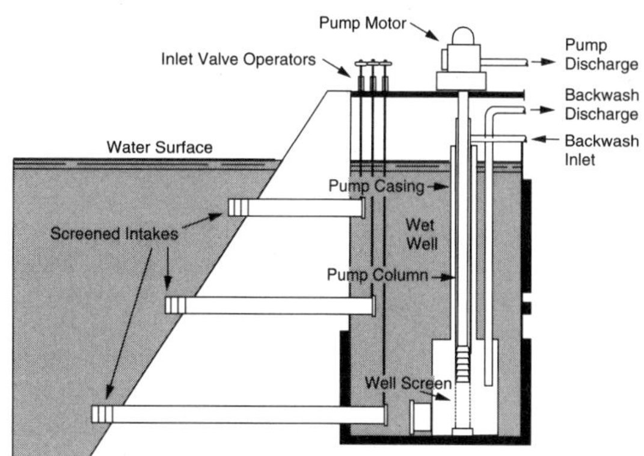

Intake structure with multiple inlet ports

intake area The surface area on which water that eventually reaches an aquifer or groundwater basin is initially absorbed.

intake pipe (1) The pipe at the head of a pipeline or system through which water enters. (2) In water supply, a pipeline conveying water by gravity from a source of supply in a stream or body of water to an intake well.

intake screen A screen placed across the head of, or in, an intake canal or pipeline to prevent fish from entering.

intake section The lower part of the well that lets water in while keeping aquifer materials out.

intake structure A structure or device placed in a surface water source to permit the withdrawal of water from that source.

intake works Structures at the location at which water is taken from a source of supply into a conduit for transportation to other locations.

intangible asset Any element of value pertaining to permanent property of a nonphysical nature, such as a franchise, trademark, patent, copyright, goodwill, or the cost of organizing, developing, and establishing such an asset.

intangible benefit A benefit or advantage of a water resource project that cannot be easily defined or formulated, e.g., a beneficial environmental value.

intangible cost A cost or disadvantage of a water resource project that cannot be easily defined or formulated, e.g., a disadvantageous environmental value.

integral action The control mode that varies the controller's output in proportion to the integral (sum over time) of the process error in the controlled variable. Integral action is also called integral control, reset action, or reset control. *See also* proportional integral derivative control.

integral control *See* integral action.

integrated catchment management (ICM) An approach for coordinated operation and maintenance of either a surface water watershed or an urban stormwater system.

integrated circuit A unified collection of many individual circuits of transistors, resistors, and capacitors that are etched onto a small chip of silicone. Newer chips have millions of components and circuits on a single chip.

integrated exposure assessment A summation over time of the magnitude of human exposure to a toxic chemical via all media.

integrated flow curve A curve representing a summation of all preceding water flows up to a given point in time. *See also* mass diagram.

integrated membrane system *See* dual-membrane system.

integrated resource management (IRM) The use of systematic procedures to allocate and use, in terms of timing and mix, all available assets to meet the objectives of a company or organization.

integrated resources planning (IRP) A comprehensive form of water utility planning that encompasses least-cost analysis of demand management and supply management options, as well as an open and participatory decision-making process, preparation of alternative planning scenarios, and a recognition of the multiple institutions concerned with water resources and the competing policy goals among them. *See also* demand management; supply management.

Integrated Risk Information System (IRIS) A database developed, maintained, and updated by the US Environmental Protection Agency that contains the agency's consensus positions on the potential adverse human health effects of approximately 500 substances. It contains summaries of qualitative and quantitative human health information.

integrated sample *See* depth-integrated sample.

integrated water utility business An organization that incorporates all utility functions—including planning, resource development and conservation goals, a long-term capital improvements program, operation and maintenance budgets, rates and fee schedules, billing, customer services, and so forth—into a cohesive operation that accomplishes the utility mission in a cost-effective manner.

integrating pitot tube A flow element that consists of a rod extending through a pipe, with a number of holes (up to eight) on the upstream side of the rod that sample and average the velocity-induced pressure across the pipe cross section. The pressure is proportional both to the velocity of the fluid in the pipe and to a corresponding flow rate in the pipe. *See also* pitot tube.

integration method (1) A method of measuring and determining the mean velocity along a vertical profile of water flowing in a stream or conduit. A current meter is moved slowly and at a uniform rate of speed from the surface to the bottom and back again one or more times. The total number of revolutions of the meter wheel and the time spent are noted, and the average rotational speed (in revolutions per minute) of the meter wheel is computed. This method allows the velocity of the water throughout the entire vertical section to be integrated or summarized. (2) A method of determining the mean sediment concentration in a vertical profile by lowering a depth-integrating sediment sampler to the streambed and raising it to the surface again. This technique yields a mean sample of the suspended material being transported by the stream in that vertical profile.

integrator A device or meter that continuously measures and calculates (adds) total flows in cubic metres, litres, gallons, million gallons, cubic feet, or some other unit of volume measurement. Such a device is also called a totalizer.

intensity–duration frequency curve A graphic method of representing the statistical probability of the magnitude, length, or both of precipitation events.

Inter-American Association of Sanitary and Environmental Engineering (Asociación Interamericana Ingeniería Sanitaria y Ambiental, AIDIS) An association of environmental engineers and scientists in the United States and Latin America.

interaction The actions of one drug or chemical in modifying the effects of another. The effects can be additive, synergistic (superadditive), or inhibitory. A wide variety of specific mechanisms exist by which interaction can occur. However, these mechanisms are generally classified as chemical (i.e., the chemicals actually react with one another), physiological (i.e., their mechanisms tend to modify functions more or less independent of one another), or pharmacological (i.e., they actually act at the same site of action to modify one another's effects).

interagency agreement An agreement between one or more federal agencies that clarifies, delineates, or both the jurisdictions, responsibilities, or both of the agencies in administering a particular law or regulatory program.

interbasin transfer The movement of water from one watershed to another or one river basin to another, usually involving water rights and intergovernmental relations.

interception (1) The process by which precipitation is caught and held by foliage, twigs, and branches of trees, shrubs, and other vegetation and then lost by evaporation, never reaching the surface of the ground. (2) The amount of precipitation intercepted.

interception loss That portion of precipitation caught by the foliage, twigs, and branches of trees, shrubs, and other vegetation and then lost by evaporation, never reaching the surface of the ground.

interceptometer A rain-collecting device placed under a tree or amid brush and crops. Its catch, or the amount collected, is compared with that of a rain gauge set in the open in order to determine the amount of rainfall loss by interception.

interconnection A physical connection between two water supply systems.

interface (1) A surface that is regarded as the common boundary between two bodies or spaces. (2) In an aquatic system, the area where the water and any solid surface are in contact. (3) In computers and electronics, a point of communication between two or more separate systems. Computers of different manufacture and operating systems use interfaces that enable them to communicate and operate with one another as an integrated system. (4) A physical link between instruments in which analytes are transferred from one component to another, often for the purpose of detecting the analytes, identifying them, or both.

interfacial composite membrane A type of composite membrane made by chemical reactions of substances such as polymers at the support layer surface, creating the membrane's barrier layer.

interference substance Any of several substances with which chlorine reacts. These reactions must be completed before a chlorine residual can be available.

interflow That movement of water of a given density in a reservoir or lake between layers of water of different density. It is usually a result of the inflow of water of either a different temperature or a different sediment or salt content.

Interim Primary Drinking Water Regulations (IPR) The first drinking water regulations promulgated as a result of the Safe Drinking Water Act of 1974. They have now become National Primary Drinking Water Regulations pursuant to the 1986 amendments to the Safe Drinking Water Act. *See also* Primary Drinking Water Regulation.

interlaboratory quality assurance The complete program designed to ensure the reliability of data that a group of analytical laboratories has produced.

interlock An electric switch, usually magnetically operated, used to interrupt all (local) power to a panel or device when the door is opened or the circuit is exposed to service.

intermediate belt That portion of the zone of aeration that lies between the soil–water belt and the capillary fringe.

intermediate groundwater Water in the zone of aeration between fringe and soil water.

intermediate vadose water Subsurface water that lies between the soil–water belt and the capillary fringe.

intermediate water Water lying between the upper surface of the capillary fringe and the soil–water belt.

intermittent chlorination A technique of noncontinuous chlorination used to control biological fouling of surfaces.

intermittent flow An interrupted pattern of water usage that occurs in homes or commercial businesses—as opposed to the steady constant-flow patterns common in industry, such as in factories. This term may also be used to refer to on–off flow patterns of water through treatment units specified in testing standards to simulate customer water use patterns. *See also* continuous flow operation.

intermittent flow operation A method of operation in which supply, treatment, or both is not provided continuously. Intermittent operation may be required if (1) a system having a relatively fixed flow rate cannot meet the variation in demand or (2) a variable-flow system capacity is greater than the variation in demand. For example, well pumps and any associated treatment are often operated intermittently in small systems that have extremely variable demand or in large systems that use wells to meet peak demands in excess of the baseline demand requirement.

intermittent interrupted stream A stream that has intermittent stretches with intervening ephemeral stretches.

intermittent periodic spring A periodic spring that discharges intermittently.

intermittent spring A spring that discharges only during certain periods, ceasing at other periods. Such a spring is also called a temporary spring.

intermittent stream A stream or portion of a stream that flows only in direct response to precipitation. It receives little or no water from springs and no extended supply from melting snow or other sources.

intermittent stretch *See* intermittent stream.

internal drainage *See* underground drainage.

internal friction Friction within a fluid (water) caused by cohesive forces.

internal isolation Isolation, with respect to a potable water system, of a fixture, area, zone, or some combination thereof. Isolation at the fixture means installing an approved backflow preventer at the source of the potential contamination. Area or zone isolation entails confining the potential source within a specific area.

internal water treatment The addition of treatment chemicals to makeup water used in steam generation for boiler operations. Chemicals are often added to prevent scale buildup or corrosive pitting of metal in boiler system components. *See also* external water treatment.

International Classification of Diseases A classification of specific conditions and groups of conditions as determined by an internationally representative group of experts. Every disease entity is assigned a number. The complete list is periodically revised and published by the World Health Organization.

International Drinking Water Supply and Sanitation Decade, 1981 to 1990 A plan that was developed and implemented on November 10, 1980, by the United Nations General Assembly in the hopes of providing clean and adequate sanitation for all people by 1990.

International Fabricare Institute (IFI) An association of dry-cleaning businesses.

International Organization for Standardization (ISO) The specialized international agency for standardization, comprising the national standards bodies of nearly 90 countries. The object of the International Organization for Standardization is to promote the development of standards in the world with a view toward facilitating international exchange of goods and services, as well as developing cooperation in the sphere of intellectual, scientific, technological, and economic activity. International Organization for Standardization standards are in wide use throughout the world, in practically every area of technology, either directly or in the form of identical national standards. The International Organization for Standardization Central Secretariat is located in Geneva, Switzerland.

International Water Supply Association (IWSA) A professional international organization concerned with the public supply of water through pipes for domestic, agricultural, and industrial purposes and the control, provision, and protection of the necessary water resources.

interpolation A technique used to determine values that fall between the marked intervals on a scale.

interpretive report A publication that includes a summary and evaluation of the results of data collection programs according to study objectives or other preestablished goals.

interquartile range The range of data between the twenty-fifth and seventy-fifth percentile values. *See also* box-and-whisker plot.

interrupted stream A stream that contains perennial stretches with intervening intermittent stretches, or intermittent stretches with intervening ephemeral stretches.

interrupted water table A water table that has a pronounced descent along a groundwater dam.

interrupting current The fault current that a fuse or circuit breaker is designed to interrupt before the current becomes destructive.

interspecies extrapolation model A model used to extrapolate from results observed in laboratory animals to humans.

interstate carrier Any vehicle or transport that conveys passengers in interstate commerce.

interstice A very small open space in a rock or granular material. It is also called a void or void space. *See also* pore.

interstitial fibrosis The formation of fibrous material between normal components of a tissue or organ. *See also* fibrosis.

interstitial water (1) Extracellular water within a tissue, i.e., the water found between cells. (2) Subsurface water found in the voids of rocks or in spaces between sediment particles.

interstream groundwater ridge A ridge in the water table between two effluent streams. Percolation toward the surface streams causes a residual groundwater ridge to develop.

intervention epidemiologic study *See* experimental epidemiologic study.

intrahepatic cholestasis A type of liver toxicity. It is most frequently associated with compounds that have steroid-like character. These compounds have to be conjugated and secreted into the bile. Occasionally some of them appear to block the flow of bile, and the person becomes jaundiced. The jaundice is a result of the accumulation of bilirubin.

intrinsic water Ultrapure water that has essentially no ion or mineral content.

intron A noncoding sequence in deoxyribonucleic acid that serves as a spacer region between exons that actually code for a functional protein. In eukaryotic cells a single gene usually consists of several exons that are linked by introns. The intron sequence is cleaved out of the resulting messenger ribonucleic acid prior to the latter being translated into a sequence of amino acids in a protein.

inundation A condition in which water temporarily or permanently covers a land surface.

invasive Tending to spread to healthy tissues or invading cells.

inverse sludge index The reciprocal of the sludge volume index multiplied by 100. It is properly called the sludge density index. *See also* sludge density index.

invert The lowest point of the channel inside a pipe, conduit, or canal.

invertebrate Any animal without a backbone.

inverted capacity The maximum rate at which a well will dispose of water admitted at or near its upper end by discharge through openings at lower levels. Where these lower openings are in a saturation zone, the inverted capacity is approximately equal to the product of the specific capacity and the available pressure head.

inverted rate A utility rate design whereby the unit price of water rises with each successive block, resulting in increasing costs with increasing customer usage. This is the opposite of a declining block rate. *See also* inclining block rate.

inverted well A well in which the movement of water is in the reverse direction of that in an ordinary well. Water is admitted at or near the top of the well and discharged into a permeable bed through one or more openings at lower levels. An inverted well is used during groundwater recharge. *See also* diffusing well.

inverter A device that changes direct electric current to alternating electric current.

investor-owned water utility A public water system owned by an individual, partnership, corporation, or other qualified entity, with the equity provided by shareholders. Regulation may take the form of local or state jurisdiction.

inward-flow turbine A reaction turbine in which the water or steam enters the runners from the outside and flows toward the axis of the runner.

IOC *See* inorganic contaminant.

iodine (I) A nonmetallic element that is the heaviest and least reactive of the naturally occurring halogens. It may be used for disinfection. In both its liquid and vapor forms, iodine is readily adsorbed by activated carbon.

iodine number A surrogate value for the ability of an activated carbon product to adsorb organic substances that have low molecular weights. The iodine number of an activated carbon is equal to the milligrams of iodine at a concentration of 0.02 normal that can be adsorbed on 1 gram of activated carbon.

iodine treatment The addition of salts of iodine to water to prevent goiter, which is caused by an iodine deficiency.

iodometric chlorine test The determination of residual chlorine in water by addition of potassium iodide (KI) and titration of the liberated iodine (I_2) with a standard solution of sodium thiosulfate ($Na_2S_2O_3 \cdot 5H_2O$) with starch solution as a colorimetric indicator.

iodometric method A procedure for determining the concentration of dissolved oxygen in water. It is also known as the Winkler method. The azide modification of this method is commonly used because it is subject to fewer interferences.

ion An atom that is electrically unstable because it has more or fewer electrons than protons. Thus, it is an electrically charged particle. A positive ion is called a cation, and a negative ion is called an anion. In aqueous solution, ions may not actually exist as isolated charged atoms but tend to form a variety of hydrated complexes.

ion balance A type of calculation commonly used as one method to determine the completeness or accuracy of laboratory analyses of a water sample. For ionized solutes in natural waters, the concentrations of cations (typically expressed in milliegivalents per litre or milligrams per litre as calcium carbonate [$CaCO_3$] equivalents) equal the concentrations of anions in the same units.

ion chromatography (IC) A technique for separating substances based on ion exchange. It is commonly used for the analysis of anions and cations in water. Analytes migrate at differing speeds through an ion chromatography column based on their affinity for the stationary phase relative to the strength of the elution solution.

ion exchange (IX or IE) A reversible chemical process in which ions from an insoluble permanent solid medium (the ion exchanger—usually a resin) are exchanged for ions in a solution or fluid mixture surrounding the insoluble medium. The superficial physical structure of the solid is not affected. The direction of the exchange depends on the selective attraction of the ion exchanger resin for the certain ions present and the concentrations of the ions in the solution. Both cation and anion exchange are used in water conditioning. Cation exchange is commonly used for water softening. *See also* anion exchange; cation exchange; deionization; water softening.

ion exchanger A permanent insoluble material (usually a synthetic resin) containing ions that will exchange reversibly with other ions in a surrounding solution. Both cation and anion exchangers are used in water conditioning. The volume of an ion exchanger is measured in cubic feet or cubic metres of exchanger after the exchanger bed has been backwashed and drained and has settled into place. *See also* ion exchange.

ion exclusion chromatography A set of techniques in which chemical separation is based on the size of the compound. Certain materials used in these separations contain cage-like structures of fairly uniform size within their crystals. Retention in a chromatography column depends on molecular size rather than on chemical characteristics of the analytes. Smaller molecules are retained longer in such columns because they can enter the pores more easily than larger molecules.

ion pair Two charged particles in solution joined together by electrostatic attraction. Two ions of opposite charge interact in such a manner that the resulting complex has a distinct chemical nature. Ion pair formation can be of interest to physical chemists who study concentrated salt solutions. Analytic chemists can make use of the formation of ion pairs to extract specific ions from solution. For example, methylene blue active substances can be extracted with an organic solvent after the cationic dye (methylene blue) has formed an ion pair with an anionic substance.

ion product For ionized solutes, the mathematical product of the molar concentration of each ion raised to a power equal to the number of its ions in the solute. For example, for Na_2SO_4, the ion product would be $[Na^+]^2[SO_4^{2-}]$.

ion selectivity *See* selectivity.

ion trap mass spectrometer A type of detector used by analytic chemists and usually interfaced with a gas chromatograph. Ion traps were invented in the 1950s, but were not commercially available until the 1980s. In contrast to other mass spectrometers, in an ion trap, both volatilization and mass analysis occur in the same part of the detector. Variations of radio frequency and direct current voltages allow the detector to dissociate molecules and then isolate and store ions. The detection levels that can be achieved with ion traps are generally much lower than with other types of mass spectrometers.

ion-exchange chromatography *See* high-performance liquid chromatography.

ion-exchange membrane An ion exchange resin cast in flat sheet form and used in electrodialysis and electrodialysis reversal systems. Such a membrane is essentially impermeable to water but readily passes ions. It is sometimes called an electrodialysis membrane. *See also* electrodialysis.

ion-exchange process *See* ion exchange.

ion-exchange resin A bead-like material that removes ions from water. It is used in deionizers. *See also* ion exchange; synthetic ion exchange resin.

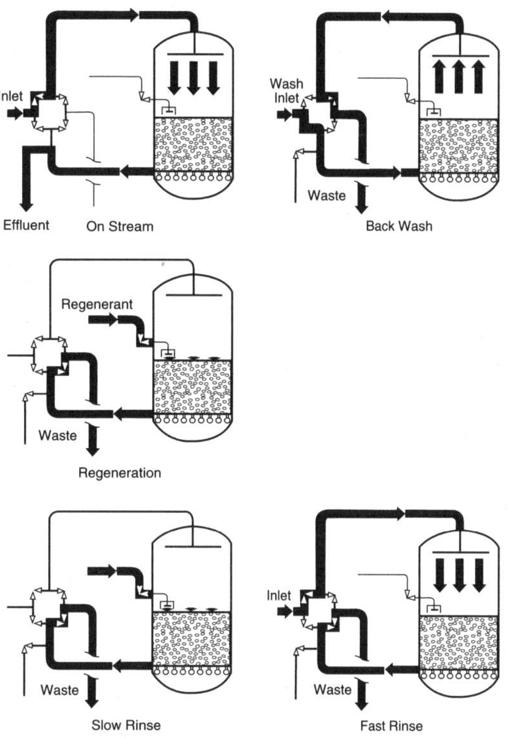

Cycles in ion exchange

ion-exchange softener *See* ion exchange water softener.

ion-exchange treatment The use of ion exchange materials, such as resins or zeolites, to remove undesirable ions from a liquid and to substitute acceptable ions.

ion-exchange water softener A treatment unit used to remove calcium and magnesium from water by means of ion exchange resins.

ion-pair extraction A technique for partitioning out of a solvent (usually water) an analyte based on the behavior of ions in solution. The mechanism for this extraction is the attraction of positively and negatively charged ions to form ion pairs. The ion pairs are more easily extracted into a suitable nonpolar organic solvent. This technique has been used to extract selected metals and organic compounds from aqueous solution. Examples of ion pair reagents are tetraalkylammonium ions, which form complexes with selected inorganic and organic ions.

ion-selective electrode (ISE) A general term used to describe a variety of electrochemical probes used in the determination of specific analytes. Analytic methods are available using a pH meter in conjunction with an ion-selective electrode for the determination of ammonia (NH_3), calcium, dissolved oxygen, fluoride, and other analytes. Electrode potential is measured, often in millivolts, as a function of analyte concentration.

ion-selective membrane In the electrodialysis process, a membrane that allows specific ions to pass through more readily

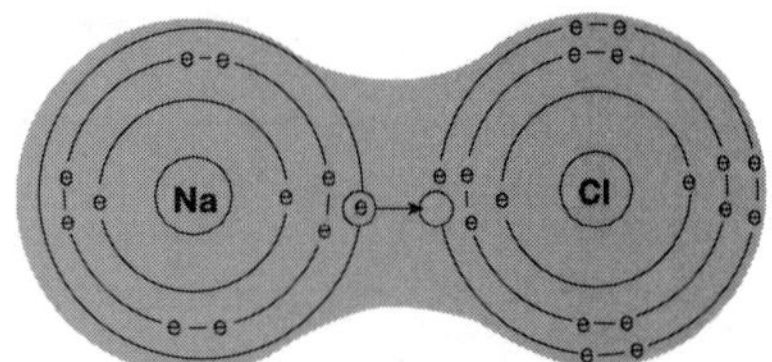

Ionic bonding, sodium chloride (NaCl)

than other ions; especially one that is developed to remove one or more targeted ions more efficiently by having relatively high membrane permselectivity, even between ions of similar charge. An example is an anion-permeable membrane that readily allows bicarbonate (HCO_3^-) and chloride (Cl^-) to pass but has relatively high resistance to sulfate (SO_4^{2-}). *See also* permselective membrane.

ionic bond A type of chemical bond in which electrons are transferred. *See also* covalent bond.

ionic concentration The concentration of any ion in solution, usually expressed in moles per litre.

ionic constant A measure in absolute units of the extent to which a chemical compound or substance in solution will dissociate into ions. *See also* ion.

ionic proportion diagram A graph showing the percentage of major anions (bicarbonate [HCO_3^-], carbonate [CO_3^{2-}], sulfate [SO_4^{2-}], and chloride [Cl^-]) and cations (calcium [Ca^{2+}], magnesium [Mg^{2+}], sodium [Na^+], and potassium [K^+]), expressed in equivalents per litre, in a water body.

ionic strength A measure of the chemical potential of electrolytes in solution. The ionic strength in a solution is defined by

$$\text{ionic strength} = \frac{1}{2}\sum(c_i z_i^2)$$

Where:

c_i = the concentration of the ith ion, in moles per litre

z_i = the valence of the ith ion

The summation takes place over all the different kinds of ions in the solution.

ionic weight The weight of an ion as determined by the sum of the atomic weights of its components.

ionization The splitting or dissociation (separation) of molecules into negatively and positively charged ions.

ionization constant A factor related to the tendency of a particular compound to ionize. It is often expressed as pK_a, the negative logarithm of the ionization constant. The value of pK_a gives information on the charge of a compound relative to the aqueous pH. For example, acetic acid has a pK_a of 4.8, which means that at an aqueous pH less than 4.8, the compound is protonated (i.e., contains a proton [H^+]) and thus is in the acid form. Acetic acid ionizes at an aqueous pH greater than 4.8. An equal proportion of ionized and un-ionized species exist in solution when the pH is equal to the pK_a.

ionize To change or be changed into ions.

IPR *See* Interim Primary Drinking Water Regulations.

IPS *See* iron pipe size.

IRIS *See* Integrated Risk Information System.

IRM *See* integrated resource management.

iron An abundant element (Fe) found naturally in the earth. Dissolved iron is found in most water supplies. An iron concentration higher than 0.3 milligrams per litre causes red stains on plumbing fixtures and on other items in contact with the water. Dissolved iron can also be present in water as a result of corrosion of cast-iron or steel pipes, a frequent cause of red-water problems.

iron bacteria Any of several aquatic bacteria that (1) obtain energy by the oxidation of reduced iron in their habitat from the ferrous (Fe^{2+}) to the ferric (Fe^{3+}) state, or (2) cause the ferrous form to be dissolved or deposited indirectly as a precipitate of hydrated ferric oxide on or in the mucilaginous secretions around their cells or cell sheaths.

iron fouling The accumulation of iron on or within an ion exchange resin bed or filter medium in such amounts that the capacity of the medium is reduced.

iron oxide adsorption A process by which materials are adsorbed through interaction with a synthetic iron oxide (Fe_2O_3) surface applied to granular media. This process may possibly provide a method to remove naturally occurring organic matter from drinking waters.

iron oxide–coated olivine A material composed of a synthetic iron oxide (Fe_2O_3) applied to an inexpensive, mixed iron–manganese silicate (olivine) with physical properties somewhat similar to sand. This material has been successful in adsorbing naturally occurring organic matter found in drinking water.

iron oxide–coated sand A material used in a fixed-bed reactor to remove particles and organic matter. The sand is coated with an iron oxide prepared from a mixture of sodium hydroxide (NaOH) and ferric nitrate ($Fe(NO_3)_3 \cdot 6H_2O$) that is heated to "fix" the coating. The advantage of the iron oxide coating is that the coated sand is positively charged at neutral pH, thereby allowing oxidation–reduction reactions to occur with negatively charged particles and colloidal material and thus improving removal.

iron pipe size (IPS) A pipe designation based on the nominal inside diameter for pipe sizes ⅛ inch and larger in accordance with the American National Standards Institute standard B 36.10 for wrought iron and steel pipe. The outside diameter of each size remains the same, whereas the wall thickness varies as expressed by the schedule number of the pipe.

iron removal The removal of dissolved iron (Fe^{2+}) from water, typically by oxidation to a particulate state and then filtering. Iron is regulated on an aesthetic basis because it can result in a metallic taste and discoloration of fixtures.

iron salt An iron-based coagulant. *See also* coagulant; ferric chloride; ferric sulfate; ferrous salt.

iron vitriol ($FeSO_2$) A common name for copperas; ferrous sulfate.

IRP *See* integrated resource planning.

irradiance units In Système International, watts per metre squared; kilograms per second cubed.

irradiation The process of exposing a material to radiation for microbial inactivation, chemical activation, or creation of hydroxyl radicals (•OH). In industrial applications, process water may be irradiated to provide disinfection without leaving a disinfectant residual that may adversely affect industrial process efficiency. *See also* photon; radiation.

irregular weir A weir with a crest that is not of standard or regular shape.

irreversible toxic effect An adverse, irreversible effect on health induced by chemicals, e.g., the initiation of cancer by production of a mutation. Once established within a cell that has the capability to divide, a mutation is an irreversible effect. Similarly, disruption of physical development produces irreversible disfigurement (i.e., teratogenesis). Another example is the loss of neurons in the central nervous system of an adult mammal, a process considered irreversible because the capability for regenerating neurons does not occur. This finality stands in contrast to the killing of cells within the liver or kidney, where new cells can be generated. Repeated killing of cells in these tissues can, however, have irreversible consequences (e.g., cirrhosis of the liver) if it continues over time.

irrigation The artificial application of water to land to meet the water needs of growing plants not met by rainfall.

irrigation circuit (1) A group of irrigation components, including heads or emitters and pipes, controlled and operated simultaneously by a remote-control valve. (2) The area served by such components.

irrigation controller A mechanical or electronic clock that can be programmed to operate remote-control valves.

irrigation cycle A scheduled sequence of applying water by an irrigation circuit, defined by a start time and duration. Multiple cycles can be scheduled, separated by time intervals, to allow the applied water to infiltrate.

irrigation plan A two-dimensional plan drawn to scale expressing the layout of irrigation components and component specifications. The layout of pipes may be depicted diagrammatically, but the location of irrigation heads and the irrigation schedules should be specified.

irrigation requirement The quantity of water, exclusive of precipitation, that is required for crop production. It includes surface evaporation and other economically unavoidable water waste. *See also* duty of water; water requirement.

irrigation return flow The quantity of water that discharges off an irrigated plot after the water has been used by the crop. This water is frequently contaminated with unacceptable amounts of salt, fertilizer, and pesticides.

irrigation water Water artificially applied to lands to meet the water needs of growing plants. It does not include rainfall.

irrigation zone *See* hydrozone.

ISA *See* Instrument Society of America.

ISE *See* ion-selective electrode.

ISO *See* Insurance Services Office; International Organization for Standardization.

ISO standard *See* International Organization for Standardization.

isobar A line on a map connecting points of the same atmospheric pressure at a given altitude and instant.

isobaric Having equal atmospheric pressure.

isobath An imaginary line on the Earth's surface, as represented on a map, connecting all points having the same vertical distance above a plane of interest, such as the upper or lower surface of an aquifer.

isochion A line of equal snow depth or equal water content of snow, as shown on a map. An isochion is also called an isonival.

isochlor An imaginary line on the Earth's surface, as represented on a map, connecting all points of equal concentration of chlorides. It may be used to represent such points in groundwater, tidal estuaries, and other such bodies of water.

isochron In the graphical presentation of data, a line representing a contour of equal time.

isochrone *See* histograph.

isocratic Pertaining to a stage in the elution of analytes in ion chromatography or high-performance liquid chromatography for which the composition of the mobile phase is held constant during the separation of analytes.

isoelectric point A pH at which the net charge on a compound is neutral. In electrophoretic separation, amphoteric compounds have equal positive and negative charges at the isoelectric point and fail to migrate. This phenomenon is also used in the separation of proteins by precipitation because the solubility of proteins is lowest at the isoelectric point. Good coagulation also occurs at the isoelectric point of the coagulant.

isohyet (1) An imaginary line on the Earth's surface, as represented on a map, connecting all points of equal precipitation. (2) A line or contour representing equal concentration.

isohyetal map A map that shows, through the use of isohyets, the variation and distribution of precipitation occurring over an area during a given period.

isolated interstice A small open space or interstice, usually occurring in lava or other effusive igneous rock (i.e., rock that flowed out on the Earth's surface) of a vesicular (small-hole) texture. *See also* discontinuous interstice.

isolation An extraction technique in which analytes are separated from the matrix for the purposes of increasing the analyte's concentration, minimizing interferences, or both. Examples include liquid–liquid extraction and solid-phase extraction. *See also* liquid–liquid extraction; solid-phase extraction.

isomer A compound that has the same molecular formula but a different chemical structure relative to another compound. For example, 35 isomers have the molecular formula C_9H_{20}. Isomers can have different chemical properties even though the numbers and types of atoms are identical.

isometric view An orientation of a three-dimensional figure such that the length, width, and height characteristics are accurately and dimensionally represented to an observer. For example, the isometric view of a rectangular box is oriented such that one vertical edge occupies the foremost location in a three-dimensional drawing, with the width and length of the box each shown with dimensional accuracy at a 30° angle from horizontal relative to this vertical edge.

isonival *See* isochion.

isopach map A contour map for which the contours are lines of constant thickness. Such a map is used to depict geologic structures such as aquifers.

isopercentile (1) Having equal percentages. (2) A line connecting points of equal percentage of rainfall, usually drawn on a map that shows the annual or monthly rainfall at each rain gauge station as a percentage of the annual long-average figures for that station.

isopiestic line (1) An imaginary line on the Earth's surface, as represented on a map, connecting all points at which water in a water-bearing formation would rise to the same elevation if free to do so; a line connecting all points of equal pressure in a water-bearing formation under pressure; a line connecting all points of equal altitude on the upper surface, or water table, of an unconfined aquifer.

isopiestic map A map showing, by means of isopiestic lines, the shape of the piezometric surface of an aquifer. Such a map is also called a piezometric map. *See also* piezometric surface.

isopleth A line on a chart or map drawn through points of equal value.

isopluvial line (1) A line on a map connecting all points that have the same pluvial index for a given length of storm and a given period of time. (2) An isohyet. *See also* pluvial index.

isopluvial map A map showing isopluvial lines.

***p*-isopropyltoluene ($CH_3C_6H_4CH(CH_3)_2$)** A solvent, also known as para-cymene. *See also* solvent.

isostatic Subject to equal pressure from every side; being in hydrostatic equilibrium.

isotherm (1) A line connecting points having a common temperature, used in a two-dimensional thermal representation. For example, isotherms are shown on weather maps to illustrate regions with common temperatures. In water treatment, isotherms may be used in a cross section of a settling basin to illustrate density currents induced by temperature differences. (2) In adsorption, a relationship that specifies the surface concentration of the material being adsorbed on the adsorbent medium as a function of the concentration of adsorbate in solution at equilibrium. *See also* Freundlich isotherm; Langmuir isotherm.

isothermal (1) Pertaining to the operation of a gas chromatograph with a constant oven temperature, as opposed to temperature programming in which the oven temperature is programmed to increase at a certain rate. The choice of isothermal versus temperature-programming conditions is dependent on how well the analytes can be resolved at a constant temperature or whether a variation in oven temperature is required to facilitate resolution of closely eluting compounds. (2) *See* isothermy. *See also* gas chromatography.

isothermy In limnology, a state in which a lake or pond is at the same temperature throughout and is well mixed. Periods of isothermy occur in spring and autumn in dimictic lakes. A state of isothermy is also called an isothermal.

isotope One of two or more atoms or elements that have the same atomic number (occupy the same position in the periodic table) but that differ in other respects such as atomic weight and number of neutrons in the nucleus. Examples of isotopes are hydrogen (^1H), dueterium (^2H), and tritium (^3H).

isotropic Having the same properties in all directions. This term is used in the context of a medium's elasticity, conduction of heat or electricity, or radiation of heat or light. Cubic (isometric) crystals transmit light equally in all directions, as do liquids, gases, most glasses, and some ion exchange resins. *See also* anisotropic.

iteration Any process of successive approximation used in numerical solution of mathematical equations, including algebraic equations and differential equations.

IWEM *See* Institution of Water & Environmental Management.

IWSA *See* International Water Supply Association.

IX *See* ion exchange.

J

J *See* joule *in the Units of Measure list.*

jacketed pump A pump equipped with jackets around the cylinders, heads, and stuffing boxes through which steam or heat from another source may be forced. These jackets permit the handling of such materials as pitch, resin, and asphalt that are solid when cold but melt on heating.

jacking A method of providing an opening ahead of and slightly larger than a pipe, into which the pipe is forced by means of horizontal jacks.

Jackson candle turbidimeter A device once used to determine turbidity in water. This type of turbidimeter consisted of a standard candle, a calibrated glass tube, and a supporting frame. It was used for many years, but it was not capable of reporting a turbidity less than 25 Jackson turbidity units. The device was included in the 16th edition of *Standard Methods for the Examination of Water and Wastewater* but was eliminated from the 17th edition in 1989. Jackson candle methods were replaced with nephelometric methods of measuring turbidity. *See also* nephelometric turbidity.

Jackson turbidity unit (JTU) An obsolete term for expressing turbidity. *See also* nephelometric turbidity unit *in the Units of Measure list.*

jar test A laboratory procedure for evaluating coagulation and rapid mix, flocculation, and sedimentation processes in a series of parallel comparisons.

jar test apparatus An automatic stirring machine equipped with three to six paddles and a variable-speed motor drive. It is used to conduct jar tests in order to evaluate the coagulation and rapid mix, flocculation, and sedimentation processes. It may also be used to select powdered activated carbon doses for taste and odor control.

jaundice *See* icterus.

jet A stream of water under pressure issuing from an orifice, nozzle, or tube.

jet height The vertical distance to which a jet of water rises above the orifice from which it issues. Such height is always less than the head on the orifice because of air resistance and friction loss in passing through the orifice.

jet injection blending The blending of two or more fluid streams by use of a jet pump. The pump injects fluid under pressure into a flow stream prior to a high-velocity venturi that implements mixing. This process is often used for injecting coagulant chemicals that require high-velocity gradients for proper mixing. *See also* jet pump; velocity gradient; Venturi principle.

jet pump A device that pumps fluids by converting the energy of a high-pressure fluid into that of a high-velocity fluid. It consists of a nozzle that discharges a jet of fluid at high pressure into a venturi tube. The fluid to be pumped enters a chamber just ahead of the Venturi tube, where the high-velocity jet draws the fluid into the tube. *See also* eductor; ejector.

jet washing A method of well development or redevelopment that employs high-velocity, high-pressure water to clean the aquifer or formation outside the well screen.

jetsam Floating discarded material.

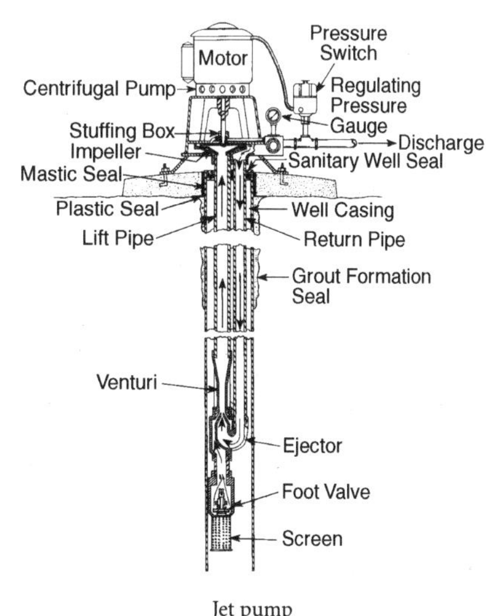

Jet pump

jetted well A shallow well constructed by a high-velocity stream of water directed downward into the ground.

jetting (1) A method of well drilling in which the casing is sunk by driving while the material inside is washed out by a water jet and carried to the top of the casing. It is a very fast method of drilling water supply wells in certain types of soils. (2) A method of sinking piles by means of a water jet. (3) A method of inserting well points by means of a water jet. (4) A method of consolidating unconsolidated backfill by means of a water jet. Jetting is also called jet drilling.

jetty (1) An obstruction built of piles, rock, or other material extending into a stream or into the ocean from the bank or shore, placed so as to induce scouring or bank building or to protect against erosion. (2) A similar obstruction to influence river, lake, or tidal currents or to protect a harbor. (3) A wharf or pier.

jogging The frequent starting and stopping of an electric motor.

Joint Powers Authority (JPA) A type of nonprofit organization frequently established as a separate entity by one or more cities or other public agencies for the specific purpose of borrowing funds for major capital programs with looser restrictions than are common for general obligation or revenue bonds. The Joint Powers Authority typically issues to the public certificates of participation that are equivalent to tax-exempt bonds, and it controls the resulting proceeds to pay for construction. *See also* certificate of participation.

joint (1) A surface of contact between two bodies or masses of material of like or different character or composition. (2) A connection between two lengths of pipe, made either with or without the use of a third part. (3) A length or piece of pipe.

joint clamp A repair mechanism that seals when a rubber gasket is pressed against both the bell and the pipe inserted in the bell. This type of repair is accomplished by placing a ring behind the bell and a follower ring behind the gasket and then compressing the gasket by applying force via bolts between the rings.

joint compound A synthetic compound made of sand and sulfur, plastic, or other materials and used in place of lead or rubber gaskets to produce watertight joints in pipes carrying fluids.

joint spring A spring issuing from joints in a rock formation.

Jonval wheel A parallel- or axial-flow waterwheel.

joule (J) *See the Units of Measure list.*

JTU *See* Jackson turbidity unit.

judicial review A review of a final federal agency action by a federal court. Section 1448 of the Safe Drinking Water Act allows one or more parties to contest one or more provisions of a National Primary Drinking Water Regulation promulgated by the US Environmental Protection Agency by filing a petition for judicial review in the US District Court of Appeals for the District of Columbia.

junction A converging section of a conduit used to facilitate the flow from one or more conduits into a main conduit.

junior rights Water rights that are obtained more recently than older rights and therefore are junior in priority relative to the older, more senior rights. *See also* priority.

jute (1) A coarse, fibrous, rope-like material driven into a run joint to seal the inside so as to contain melted lead that is being poured into the joint. (2) To apply such a material.

juvenile water Water first entering the hydrologic cycle, such as water that enters from geothermal activity deep in the oceans.

K

K *See* kelvin *in the Units of Measure list.*

K *See* distribution constant; hydraulic conductivity.

k *See* kilo.

K_c *See* crop coefficient.

K_F *See* Freundlich adsorption constant.

K. oxytoca *See Klebsiella oxytoca.*

K. pneumoniae *See Klebsiella pneumoniae.*

kaolin A special type of clay, usually high in aluminum content, sometimes used in water treatment.

Kaplan turbine A propeller-type turbine with adjustable blade or runner vanes.

karez A type of infiltration gallery common in the ancient world (ranging from Afghanistan to Morocco) and still in use today. Karezes are hand-dug by skilled workers using methods more than 3,000 years old, but otherwise they are equivalent to modern infiltration galleries. A karez is also known as a quanat, foggara, or falaj. *See also* infiltration gallery.

karotype The chromosomal complement of a cell nucleus. A normal karotype for humans is 23 chromosome pairs that have the distinct morphological characteristics (i.e., appearance) most commonly found in humans.

karst aquifer A type of aquifer formation characterized by carbonate rocks for which a significant portion of the porosity has developed from dissolution of the rocks. The Edwards Aquifer in Texas and the Mammoth Caves in Kentucky are examples of karst aquifers.

karst topography A land surface formed over limestone, dolomite, or gypsum and characterized by sink holes, caves, and underground drainage.

katamorphic zone The zone of rock fracture, where materials break down. It is especially characterized by dissolution, decrease of volume, and softening of materials; these destructive processes result in degeneration. The zone is divided into the belt of weathering and the belt of cementation. *See also* aeration zone; anomorphic zone; weathering zone.

Kb *See* kilobase.

KBEMS *See* knowledge-based energy management system.

KBS *See* knowledge-based system.

kelly In rotary well drilling, a rotating table used in large rotary drilling machines to transmit torque from the machine to the drill string. Torque from a drive table is applied to the square rod to cause the rotary motion. The drive table is chain or gear driven by an engine.

kelvin (K) *See the Units of Measure list.*

Kemmerer sampler A vertical point sampler for a water–suspended-sediment mixture. A "messenger" is slid down a wire to trip the mechanism on the sampler to collect the sample.

Kennison flow nozzle A type of nozzle attached to the end of a pipe in order to measure flow when the pipe is not flowing full.

keratinocyte A cell in the epidermis of the skin that synthesizes keratin. Keratin is the main protein constituent of hair, nails, and other horny tissues. It also forms the organic matrix of teeth on which enamel is formed.

kernicterus A medical condition characterized by severe nervous system symptoms. It arises from high accumulations of bilirubin in the blood. It is also referred to as bilirubin encephalopathy. Kernicterus is produced when a chemical interferes with the elimination of bilirubin in the bile. This elimination process is one of the normal functions of the liver, and its interruption is one manifestation of liver toxicity. Kernicterus can come about because of inhibition of the conjugation reactions that allow bilirubin to be excreted, interference with the secretory mechanism that transfers the conjugated bilirubin into the bile, or interference with the flow of bile. *See also* bilirubin.

ketoacid A class of organic compounds that have both a ketone (R_1–CO–R_2) and carboxylic acid (COOH) functional group in the chemical structure. Some ketoacids are created during the reactions of oxidants used as disinfectants, particularly ozone (O_3), with natural organic matter. *See also* disinfection by-product; ozonation by-product.

$$R_1\text{-}\overset{\displaystyle O}{\overset{\|}{C}}\text{-}R_2\text{-}\overset{\displaystyle O}{\overset{\|}{C}}\text{-OH} \quad \text{or} \quad R\text{-}\overset{\displaystyle O}{\overset{\|}{C}}\text{-}\overset{\displaystyle O}{\overset{\|}{C}}\text{-OH}$$

Ketoacid, where R or R_1 and R_2 are carbon-containing groups

ketone A class of organic compounds in which the carbonyl group (C=O) is attached to two alkyl (hydrocarbon) groups. In industry, ketones are used primarily as solvents. Some ketones are created during the reactions of oxidants used as disinfectants, particularly ozone (O_3), with natural organic matter. *See also* carbonyl; disinfection by-product; ozone by-product; solvent.

$$\begin{array}{c} O \\ \| \\ R_1\text{-}C\text{-}R_2 \end{array}$$

Ketone, where R_1 and R_2 are carbon-containing groups

kg *See* kilogram *in the Units of Measure list.*
KHP *See* potassium hydrogen phthalate.
kHz *See* kilohertz *in the Units of Measure list.*
kill To render microorganisms unable to reproduce, grow, and, in the case of pathogens, infect. In contrast, with respect to viruses, the term *inactivation* is used. *See also* inactivation.
killer cell One of two different types of lymphocytes (white blood cells): the killer T cell, which is derived from the thymus, or the natural killer cells that are produced in bone marrow. The killer cells secrete chemicals that will spontaneously lyse virus-infected cells.
killer T cell A class of T-lymphocytes involved in cell-mediated immunity. *See also* killer cell.
kilo (k) Prefix meaning 10^3, used in Système International.
kilobase (Kb) A set of 1,000 bases (i.e., adenine, thymine, guanine, cytosine, or uracil) of deoxyribonucleic acid or ribonucleic acid.
kilobyte *See the Units of Measure list.*
kilocycle *See* kilohertz *in the Units of Measure list.*
kilograin *See the Units of Measure list.*
kilogram (kg) *See the Units of Measure list.*
kilohertz (kHz) *See the Units of Measure list.*
kilolitre *See the Units of Measure list.*
kilopascal (kPa) *See the Units of Measure list.*
kiloreactive volt-ampere (kvar) *See the Units of Measure list.*
kilovolt (kV) *See the Units of Measure list.*
kilovolt-ampere (kVA) *See the Units of Measure list.*
kilowatt (kW) *See the Units of Measure list.*
kilowatt-hour (kW·h) *See the Units of Measure list.*
kinematic viscosity (ν) A convenient numerical factor of a fluid for use in engineering practice. It is calculated as follows:

$$\nu = \frac{\mu}{\rho}$$

Where:
ν = the kinematic viscosity
μ = the absolute viscosity of the fluid
ρ = the fluid density

In US customary units, kinematic viscosity is expressed in terms of square feet per second. In Système International units, it is expressed in terms of square metres per second. In the application of hydraulic engineering, kinematic viscosity is often used for determining the Reynolds number to classify the flow regimes and to evaluate the friction coefficient for flows in pipes and open channels. At 68° Fahrenheit (20° Celsius), the kinematic viscosity of water is 1.08×10^{-5} square feet per second (1×10^{-6} square metres per second).
kinematic viscosity coefficient The ratio of the coefficient of absolute viscosity of a fluid to the fluid's unit weight.
kinetic energy Energy possessed by a moving body of matter, such as water, as a result of its motion.
kinetic flow factor A standard for measuring the degree of turbulence or tranquillity that prevails in a flowing stream.
kinetic friction coefficient A numerical quantity used as an index of the amount of force necessary to keep a body sliding at a uniform velocity on the surface of another body. It is equal to the ratio of (1) the horizontal force required to slide the body along a horizontal plane surface at a uniform velocity to (2) the weight of the body. It is expressed as a decimal. The coefficient may change with velocity.
kinetic head The theoretical vertical height through which a liquid body may be raised by virtue of its kinetic energy. It is equal to the square of the velocity divided by twice the acceleration due to gravity:

$$\text{kinetic head} = V^2/2g$$

Where (in any consistent set of units):
V = the velocity
g = the acceleration due to gravity *See also* velocity head.
kinetics The study of the relationships between temperature and the motion and velocity of very small particles. Kinetic relationships influence the rate of change in a chemical or physical system and are used in particular to describe the dynamics and rates of chemical reactions.
Kirpich formula An equation used for determining the relationship between rainfall and peak runoff, the time interval for surface runoff to reach the point of flow measurement from the upper reaches of a drainage basin. In equation form,

$$T_c = 0.00013\left(\frac{L^{0.77}}{S^{0.0385}}\right)$$

Where:
T_c = the time of concentration, in hours
L = the maximum length of travel of the water, in feet
S = the slope, equal to H/L
H = the difference in elevation between the most remote point in the basin and the outlet, in feet
KIWA *See* Netherlands Waterworks Testing and Research Institute.
Kjeldahl nitrogen The combination of ammonia nitrogen and organic nitrogen in a water sample. Total Kjeldahl nitrogen is operationally defined by a method that involves digestion of a sample followed by distillation and determination of ammonia (NH_3) in the distillate. The name is derived from J. Kjeldahl, who in 1883 published a method for determining nitrogen in organic matter.
Kjeldahl nitrogen test A standard analytical method used to determine the concentration of ammonia nitrogen and organic nitrogen in a water sample. The Kjeldahl nitrogen test

determines nitrogen in certain organic compounds, but it fails to account for nitrogen in the form of azide ($R(N_3)_x$), azine (–C=N–), azo (–N=N–), hydrazone ($R_2C=N–NH_2$), nitrile (–C≡N), nitro (NO_2–), nitroso (–NO or =NOH), oxime (R_2–C=N–OH), and semicarbazone (H_2N–(CO)–NHN=CR_2). *See also* total Kjeldahl nitrogen.

Klebsiella A genus of lactose-fermenting bacteria belonging to the family Enterobacteriaceae.

Klebsiella oxytoca A genus and species name for a lactose-fermenting bacterium belonging to the family Enterobacteriaceae. It is one of the coliform group of bacteria used as sanitary indicators for drinking water quality.

Klebsiella pneumoniae A genus and species name for a lactose-fermenting bacterium belonging to the family Enterobacteriaceae. It is one of the coliform group of bacteria used as sanitary indicators for drinking water quality.

knockout animal A test animal (e.g., a rodent) in which a gene has been removed to make the animal more or less sensitive to a particular chemical-induced adverse effect. *See also* transgenic animal.

knoll spring A spring occurring on the top of a small knoll or mound. It is produced, wholly or in part, by precipitation of mineral matter from the spring water or by vegetation and sediments blown in by the wind (a method of growth common in arid regions). A knoll spring is also called a mound spring.

knot *See the Units of Measure list.*

knowledge-based energy management system (KBEMS) An expert system designed to manage and minimize energy usage and cost. *See also* expert system.

knowledge-based system (KBS) A set of computer programs that are designed to help people with the solution of tasks involving uncertainty and imprecision and that require judgment and knowledge. Typically the user is not expected to have expertise in the area of knowledge for which the system is designed.

Kolmogorov microscale A length scale indicating the size of eddies below which the kinetic energy of a fluid is dissipated by viscous forces. In equation form:

$$\eta = \left(\frac{\nu^3}{\varepsilon}\right)^{\frac{1}{4}}$$

Where:

η = Kolomogoroff microscale, in feet (metres)

ν = kinematic viscosity, in feet squared per second (metres squared per second)

ε = energy dissipation per unit mass, in feet squared per second cubed (metres squared per second cubed)

Kolmogorov–Smirnov goodness-of-fit test A statistical procedure to determine the goodness of fit of a hypothesized distribution to an observed parametric data set.

kPa *See* kilopascal *in the Units of Measure list.*

Kuch mechanism A mechanism describing the production of red water in distribution systems. Iron (Fe^{2+}) is released from corrosion scales on iron or steel pipe during times of water stagnation. This iron (Fe^{2+}) is oxidized to iron (Fe^{3+}), which then precipitates as ferric hydroxide ($Fe(OH)_3$), causing red water when flow is resumed.

Kutter formula A formula for determining the Chezy resistance factor. In US customary units, it is expressed as follows:

$$C = \frac{41.65 + \frac{0.00281}{s} + \frac{1.811}{n}}{1 + \left(41.65 + \frac{0.00281}{s}\right)\frac{n}{\sqrt{R}}}$$

and in Système International units, it is expressed as

$$C = \frac{23 + \frac{0.00155}{s} + \frac{1}{n}}{1 + \left(23 + \frac{0.00155}{s}\right)\frac{n}{\sqrt{R}}}$$

Where:

C = the Chezy resistance factor, in feet$^{1/2}$ per second (metres$^{1/2}$ per second)

s = slope, dimensionless

n = the Kutter roughness coefficient, dimensionless

R = hydraulic radius, in feet (metres)

See also Kutter roughness coefficient.

Kutter roughness coefficient The roughness coefficient in the formula published in 1869 by E. Ganguillet and W.R. Kutter for determining the Chezy resistance factor in the Chezy formula. *See also* Chezy resistance factor.

kV *See* kilovolt *in the Units of Measure list.*

kVA *See* kilovolt-ampere *in the Units of Measure list.*

kvar *See* kiloreactive volt-ampere *in the Units of Measure list.*

kW *See* kilowatt *in the Units of Measure list.*

kW·h *See* kilowatt-hour *in the Units of Measure list.*

L

L *See* litre *in the Units of Measure list.*

labile Pertaining to a substance that is inactivated by high temperature or radiation; unstable.

laboratory A facility in which water is analyzed for constituents. Many environmental laboratories and larger water utility laboratories are divided into sections based on specialty: microbiology, inorganic chemistry, organic chemistry, and so on.

laboratory information management system (LIMS) A compilation of computer software and hardware designed to aid in the storage and retrieval of laboratory data. It can serve as a tool to boost productivity in use of instruments and personnel. The system allows information on a sample to be tracked from the time of reception in the laboratory to the time the sample is discarded. Data from laboratory analyses are entered either manually or automatically into the system. The data user can retrieve these data in a variety of ways. An LIMS also allows quality assurance and quality control information to be documented. Rapid advances in computer technology have created a possibility of a powerful laboratory information management system functioning in a local area network of microcomputers.

laboratory procedure A mode of conducting laboratory processes and analytical tests consistent with validated standard testing techniques.

laboratory study A laboratory-scale evaluation of whether a particular water is amenable to treatment with specific operations or processes. *See also* pilot-plant study.

laboratory-fortified *See* spiked.

lactation index A measure of the ability of a female experimental animal to provide nutritional support to her offspring during nursing. It is usually defined as the percentage of the offspring alive 4 days after birth (postpartum) that survive to the end of the lactation period (21 days in mice and rats).

lactose A disaccharide (4-β-galactosido-glucose) also known as milk sugar. When hydrolyzed by the enzyme β-galactosidase, it yields the monosaccharides glucose and galactose. Lactose is readily metabolized by coliform bacteria of the genera *Escherichia*, *Klebsiella*, *Enterobacter*, and *Citrobacter*.

lactose negative Lacking the biochemical ability to hydrolyze or ferment lactose. This ability, which is easy to determine, is a key characteristic of the sanitary indicator group bacteria called the coliforms. The lactose fermentation ability is assessed under specified incubation time and temperature conditions and with the use of specific lactose-containing media.

lactose positive Having the biochemical ability to hydrolyze or ferment lactose.

lacustrine Related to, pertaining to, produced by, or formed in a lake or lakes.

lacustrine plain A plain originally formed as the bed of a lake from which the water has disappeared.

lacustrine zone A deep-water zone of a lake or reservoir system.

ladder dredge A continuous chain equipped with buckets and mounted on a scow. It is used to remove sediment from the bottom of a storage reservoir. A ladder dredge is also called a bucket dredge or chain bucket.

lag Any combination of factors that prevents a process from changing instantaneously.

lag growth phase The initial period following bacterial introduction to a food source, during which the population grows slowly as the bacteria acclimate to the new environment.

lagoon (1) Any large holding or detention pond, usually with earthen dikes, used to contain sludges while sedimentation and possibly biological oxidation occur. (2) A shallow, usually small body of water near or communicating with a larger body of water.

lagooning The placement of solid or liquid material in a basin, reservoir, or artificial impoundment for purposes of storage, treatment, or disposal.

laid length The length of pipe measured after that pipe is placed in position. The laid length includes such items as gaskets, space between ends of pipe in a coupling, and inserted bell-and-spigot joints.

lake An inland body of water, fresh or salt, of considerable size (usually more than 50 acres [about 200,000 square metres]) and occupying a basin or hollow on the Earth's surface. It may or may not have a current or single direction of flow. *See also* pond.

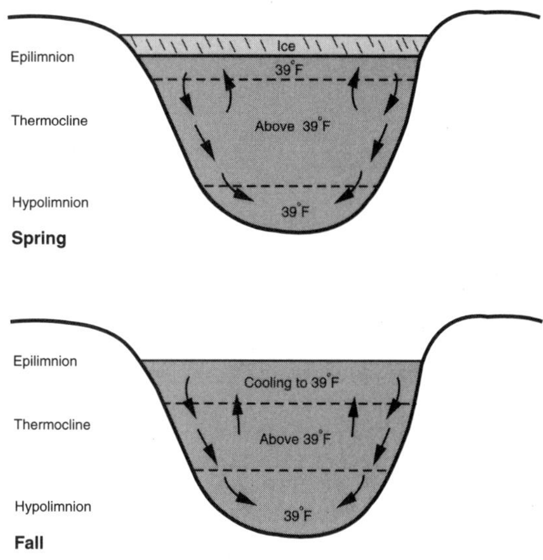

Lake overturn

lake overturn The complete, wind-induced, top-to-bottom circulation of water in deep lakes occurring when the density of the surface water is the same as or only slightly greater than that at the bottom. It is also called lake turnover. *See also* destratification; thermal stratification.

lake wall A low ridge of material formed around lake perimeters on, or just above, the water level that is subject to freezing. The ice on the lake, freezing to the sand, gravel, and boulders on the shallow bottom near the shore, picks up this material, expands shoreward, deposits the picked-up material on the land upon melting, and repeats the process over a period of years.

lambda (λ) *See the Units of Measure list.*

Lamella plate *See* inclined plate separator.

Lamella separator *See* inclined plate separator.

laminar flow The movement of fluid in a particular direction in smooth, continuous, nonturbulent parallel layers that do not mix with each other.

laminar velocity That velocity at which, in a particular channel or conduit and for a particular fluid, laminar flow will always exist.

lamination A formation composed of individual lamina layered upon one another, each one a layer of geologic material that resembles a thin plate. A single geologic unit consisting of alternating layers of shale and sandstone—with each layer thin relative to the thickness of the entire unit and each layer cemented to adjacent layers—is an example of a lamination.

LAN *See* local area network.

land application A disposal technique for sludge generated from water treatment processes. Settled solids, sometimes containing coagulant, can be applied to land as fill material under certain circumstances and within regulatory restrictions.

land batture (1) The land between the channel of a river and a levee that may be inundated at comparatively infrequent intervals. (2) That portion of river bank that is immediately above the shore and is submerged at times.

land disposal Application of raw or treated sludges or solid waste to soils, substrata, or both without production of usable agricultural products. *See also* land treatment.

land drain A drain for drawing off water from land. *See also* groundwater drain.

land drainage The removal of free water both from the land surface and from the soil of the root zone of plants.

land form Multitudinous physical features that, when taken together, make up the surface of the earth. The natural processes involved include weathering, erosion, transportation, deposition, and consolidation.

land information system (LIS) A database describing physical or legal characteristics of land areas. A land information system is sometimes called a land records system. The most common forms of land information systems describe property ownership, land value, tax assessment, and property boundaries.

land pan An evaporation pan located on land.

land rights Rights or obligations in connection with the occupation or ownership of land. *See also* easement; leasehold; right-of-way.

land subsidence A decrease in land surface elevation caused by dewatering and subsequent compaction of the underlying soils as a result of sustained pumping of groundwater.

land treatment Irrigation performed using sludge (not water) on land. Additional treatment is provided by soil, microorganisms, and crops that are grown to use nutrients. Land treatment is rarely used with water treatment sludges. *See also* land application; land disposal.

land use (1) The culture of the land surface that has a determining effect on the broad social and economic conditions of a region and that determines the amount and character of runoff and erosion. Three general classes are recognized: crop, pasture or range, and forest. (2) An existing or zoned economic use of land, such as residential, industrial, farm, or commercial.

landfill A land area upon which water treatment sludge can be applied between layers of earth. Water treatment sludges can be codisposed with other waste materials, such as solid waste, if a maximum water content in the sludge is not exceeded.

landscape area The combination of irrigated area, nonirrigated planted area, water features, hardscape, and natural undeveloped area.

landscape coefficient A factor used to modify reference evapotranspiration and to calculate water requirements for a hydrozone. *See also* reference evapotranspiration.

landscape water requirement A measure of supplemental water needed to maintain the optimal health and appearance of landscape plantings and water features.

landscaping The grading, clearing, and decorative planting of vegetation to improve the natural beauty of a site and to stabilize the soil, thereby preventing runoff.

landslide A movement of a mass of earth, rock, or a mixture of both in a downward direction by sliding. Such sliding is usually along a layer or stratum of soft or slippery material, such as clay, that then becomes wet and loses its cohesiveness.

landslide flow A moderately rapid flowage of large masses of slippery earth material. It occurs along the edges of clay terraces in certain glaciated valleys and in some mountainous regions where shale bedrock occurs. It is also called earth flow.

landslide spring A spring occurring at the base of a landslide.

Langelier index *See* Langelier saturation index.

Langelier saturation index (LSI) The most famous of the calcium carbonate ($CaCO_3$) saturation indexes, the formula for the Langelier saturation index is based on a comparison of the measured pH of a water (pH_a) with the pH the water would have (pH_s) if at saturation with $CaCO_3$ (calcite form) given the same calcium hardness and alkalinity for both pH cases. Many of the other indexes found in the water treatment and corrosion literature are merely approximations. Several versions of the LSI exist in the literature. The version presented here is the one developed by J.R. Rossum and D.T. Merrill for the quadratic equation form of the pH_s equation solution; This material has been updated using newer equilibrium constant and activity coefficient data. (Note that a simplified approximation equation originally formulated by T.E. Larson and A.M. Buswell was presented in the National Interim Primary Drinking Water Regulations amendments.) The basic formula for the Langelier saturation index is:

$$\text{LSI} = \text{pH}_a - \text{pH}_s$$

It is interpreted in the following way: If LSI > 0, water is potentially scale forming (supersaturated with respect to $CaCO_3$). If LSI = 0, water is in equilibrium with $CaCO_3$, not tending to dissolve or precipitate it. If LSI < 0, water will potentially dissolve existing $CaCO_3$ deposits (undersaturated with respect to $CaCO_3$). The quantity pH_s is found from the computed hydrogen ion concentration at hypothetical saturation ($[H^+]_s$) by

$$\text{pH}_s = -\log_{10}([\text{H}^+]_s) - \log_{10} f_m$$

Where:

f_m = the activity coefficient for monovalent ions

The parameter $[H^+]_s$ is computed by the following quadratic equation:

$$[H^+]_s = \frac{-B \pm \sqrt{B^2 - 4AC}}{2A}$$

Where:

$$A = 1 - \frac{[\text{Ca}^{2+}]K_2}{K_s}$$

$$B = K_2\left(2 - \frac{[\text{Ca}^{2+}]\text{Alk}}{K_2}\right)$$

$$C = \frac{K_w K_2}{K_s}[\text{Ca}^{2+}]$$

$[Ca^{2+}]$ = calcium concentration, in milligrams per litre as $CaCO_3$

Alk = alkalinity, in milligrams per litre as $CaCO_3$

K_2 = solubility constant

K_s = equilibrium constant

$K_w = [H^+][OH^-]$

See also Riddick index; Ryznar stability index.

Langmuir equation *See* Langmuir isotherm.

Langmuir isotherm A relationship illustrating the equilibrium between an adsorbent surface and a solute. The graphical presentation illustrates the mass of solute per mass of adsorbent (typically activated carbon) as a function of the concentration of the solute in solution. In equation form,

$$q/Q = bC/(1+bC)$$

Where:

q = the number of moles of adsorbate per mass of adsorbent, in moles per gram

Q = the maximum number of moles adsorbed per mass of adsorbent, in moles per gram

C = the concentration of adsorbate, in milligrams per litre

b = an empirical constant, in litres per milligram

The mass adsorbed, q, is assumed to approach a saturating value, Q, when C becomes very large. The advantages of the Langmuir isotherm relative to other models of adsorption include its simplicity, its foundation in a model with some physical basis, and its ability to fit a broad range of experimental data. The Langmuir adsorption model incorporates an assumption that the energy of adsorption is the same for all surface sites and is not dependent on the degree of coverage. In reality, the energy of adsorption may vary because real surfaces are heterogeneous. In contrast to the Langmuir isotherm, the Freundlich isotherm attempts to account for this variation. *See also* Freundlich isotherm.

lantern ring A metal ring (usually bronze) that is installed over a pump shaft within the packing and is used to help distribute the lubricant for the packing against the rotating shaft.

lapse rate The rate at which atmospheric temperature decreases with altitude. The value is different for wet versus dry air masses.

large meter A water meter that is difficult to move and so is generally tested in place rather than taken to a meter shop for testing. Small meters are generally 5/8 inch to 2 inches, and meters larger than 2 inches are considered large meters. Système International units are not used for these designations.

large water system A public water system serving a population of more than 50,000 people.

Larson ratio (LR) An index developed to describe the corrosion process. Reactive anions, such as chloride (Cl^-) and sulfate (SO_4^{2-}), form strong acids in anodic pits, corroding the exposed metal, whereas bicarbonate (HCO_3^-) and other anions form weak acids and precipitate a protective scale. Rising index values (greater than 0.4) are increasingly corrosive, and falling index values are less corrosive. This index was developed by T.E. Larson. Its formula is

$$LR = \frac{[\text{Cl}^-] + [\text{SO}_4^{2-}]}{[\text{HCO}_3^-]}$$

where the brackets indicate concentration in moles per litre.

LAS *See* linear alkyl sulfonate.

laser particle counter An instrument that uses a laser diode as a light source for the purpose of quantifying particles of various sizes. Different types of particle counters can detect light that is either scattered or blocked by a given set of particles.

latency period A delay between exposure to a disease-causing agent and the appearance of manifestations of the disease, or the period from disease initiation to detection. This term is often used synonymously with *induction period. See also* incubation period; induction period.

latent energy *See* latent heat.

latent heat The amount of heat released or absorbed when a substance changes its physical phase with no change in temperature, e.g., the heat absorbed from surroundings when ice melts into liquid water at the freezing point or the heat released when a gas (steam) condenses into liquid water. The loss or gain of latent heat does not lead to a change in temperature for the melting ice or the condensing water. *See also* heat of fusion; heat of sublimation.

latent period *See* latency period.

lateral (1) Directed toward, coming from, or situated on the side. (2) A ditch, pipe, or other conduit entering or leaving a water main from the side. (3) A secondary conduit diverting water from a main conduit for delivery to distributaries.

lateral erosion The erosion of the sidewalls and side tributaries of the valley of a stream.

lateral-flow spillway A spillway in which the initial and final flows are at approximately right angles to each other. *See also* side-channel spillway.

latitude–longitude coordinate system A global coordinate system in which locations are expressed by geographical coordinates (the geodetic latitude and geodetic longitude) that depict angular measurements relative to the Earth's ellipsoid.

launder trough A trough in a basin designed for evenly distributing or collecting the flow to or from the unit.

laundering weir A sedimentation basin overflow weir. It is a plate with V-notches along the top to ensure a uniform flow rate and avoid short-circuiting.

lavage An irrigation (or washing) of an organ, typically the stomach or intestine.

law *See* public law.

lawn watering The application of water to vegetation and landscaping at a rate sufficient to promote good plant growth.

layer A graphic component of a geographic information system database. Each layer contains a set of homogenous map features registered positionally with respect to other database layers through a common coordinate system. Data are separated into layers based on logical relationships and on the graphic portrayal of sets of features that are organized and stored by the geographical information system software in many, and typically proprietary, ways.

layered bed (1) A multimedia filter bed containing in the same vessel several different filter media (such as anthracite, sand, and garnet) with specific gravities that differ enough to maintain different layers even after backwashing. (2) A single ion exchange bed made up of two or more resins that have bead sizes and densities different enough to maintain different layers after backwashing and that can be regenerated with the same regenerant. For example, a layered bed may have a bed of anion resin on top and cation resin below, with both layers regenerated by salt brine solution.

lb *See* pound *in the Units of Measure list.*

lb/d *See* pounds per day *in the Units of Measure list.*

lb/ft² *See* pounds per square foot *in the Units of Measure list.*

LC *See* lethal concentration; liquid chromatography.

LC_{50} *See* lethal concentration–50 percent.

LCCA *See* Lead Contamination Control Act.

LCD *See* liquid crystal display.

LCR *See* Lead and Copper Rule.

LC–MS *See* liquid chromatography–mass spectrometry.

LC–PB–MS *See* liquid chromatography–particle beam–mass spectrometry.

LD *See* lethal dose.

L/d *See* litres per day *in the Units of Measure list.*

LD_{50} *See* lethal dose–50 percent.

leach To dissolve out by the action of a percolating liquid.

leachate The soluble material dissolved or washed out during leaching.

leaching The dissolution of solids and chemicals into water flowing through a porous sample.

Lead and Copper Rule (LCR) A rule promulgated by the US Environmental Protection Agency on June 7, 1991 (*Federal Register,* 56(110):26460–26564) that set National Primary Drinking Water Regulations for lead and copper.

Lead Contamination Control Act (LCCA) Public Law 100-572, which was passed in 1988 and amended the Safe Drinking Water Act to institute a program to eliminate lead-containing drinking water coolers in schools.

lead (Pb) A metallic element that has had various industrial uses, including in a gasoline additive, in solder and fusible alloys, and in radiation shielding. It is regulated by the US Environmental Protection Agency in water. In addition, it is regulated in air, gasoline, plumbing materials, foods, and house paints. *See also* Lead and Copper Rule; lead-free.

lead A control action used to compensate for lag in a process.

lead calking The driving of lead into a pipe joint after that lead has been poured into the joint and cooled. The lead is tightened into the joint with (1) a blunt, elongated tool known as a calking tool, which is placed against the lead and hit with a hammer, or (2) a similar tool used with either a pneumatic calking gun or hammer. Forcing the lead into and all the way around the joint causes the lead to be pressed against the bell and spigot, creating a watertight seal. The recess in the bell keeps the lead from being forced out of the joint by internal pressure in the pipe. When a lead joint starts seeping, it can usually be sealed by recalking, unless the lead has been forced out of the joint past the recess. *See also* calking tool.

lead-free Meeting the conditions defined in Safe Drinking Water Act, Section 1417 for solder, flux, pipes, and fittings. To be considered lead-free, solder and flux can contain no more than 0.2 percent lead; pipes and fittings no more than 8.0 percent.

lead joint A bell-and-spigot joint in a water pipe sealed by pouring molten lead in the joint after juting (i.e., forcing jute into) the inside of the joint and then calking the lead after cooling to complete the seal and stop leaks. *See also* lead calking; jute.

lead line In hydrographic surveying, a rope, wire, or cable with a weight attached, used for measuring depths of water. A lead line is also called a sounding line.

lead pipe Water piping manufactured and produced from lead metal. Lead was used for service pipe in the past because of the ease of working with it and shaping it. It was sometimes used for small water mains up to 2 inches (5 centimetres) in size. Because of the hazard of lead poisoning, lead pipe can no longer be used. *See also* lead poisoning.

lead poisoning An illness resulting from an accumulation of lead in the body in toxic concentrations, sometimes as a result of the contamination of potable water by lead dissolved from service connections and metal vessels.

lead service *See* lead service line.

lead service line A water line constructed of lead material that was used for the small water line running from the water main to the customer cutoff at the meter box or to the cutoff for the building served. Lead service lines are no longer permitted. *See also* lead poisoning.

LEADSOL (lead solubility) A FORTRAN-language computer program widely used by the US Environmental Protection Agency in research studies on lead solubility in drinking water. The program computes equilibrium lead (Pb^{2+}) solubility at a specified ionic strength and 77° Fahrenheit (25° Celsius) as a function of pH and dissolved inorganic carbon (or alkalinity), orthophosphate, sulfate, and chloride concentrations.

leaf Any flat filter element that has or supports the filter septum.

leak detection The precise locating of underground water leaks in a water system by the use of sounding or sonic devices where leakage is known or suspected. *See also* waterphone.

leak finder A device used for detecting leaks in liquid or gas pipelines. *See also* leakage detector; waterphone.

leak survey A systematic examination of a water system to find leaks.

leakage (1) The uncontrolled loss of water from artificial structures as a result of hydrostatic pressure. (2) The uncontrolled loss of water from one aquifer to another. The leakage may be natural, as through a semi-impervious confining layer, or artificial, as through an uncased well. (3) The presence in treated effluent of the type of ions that the ion exchange process was supposed to remove. Ion removal may be incomplete because of incomplete resin regeneration, excessive service rates, low temperatures, high concentrations of sodium, interfering total dissolved solids in the water being treated, or other factors. This type of leakage is also referred to as slippage.

leakage current The unwanted flow of electric current through liquid passages rather than through membranes and cells.

leakage detector A device or appliance that detects leaks based on the audibility of water flowing through a leak. Most of these devices are marketed under descriptive tradenames. *See also* waterphone.

leaky aquifer An aquifer bounded by a low-permeability layer that can transmit water at sufficient rates to furnish some water to a well in the aquifer.

leasehold Rights to the use of land for which a consideration has been paid, as distinguished from ownership of the land.

least-cost design A facility design that complies with the intent of the specifications at the lowest possible cost.

least-cost planning A utility planning approach that emphasizes a balanced consideration of supply management and demand management options in identifying feasible least-cost alternatives for meeting future water needs. Compared with traditional planning, least-cost planning recognizes that water demand is malleable and that forecast demand does not have to be taken as a given for the planning process. *See also* demand management; supply management.

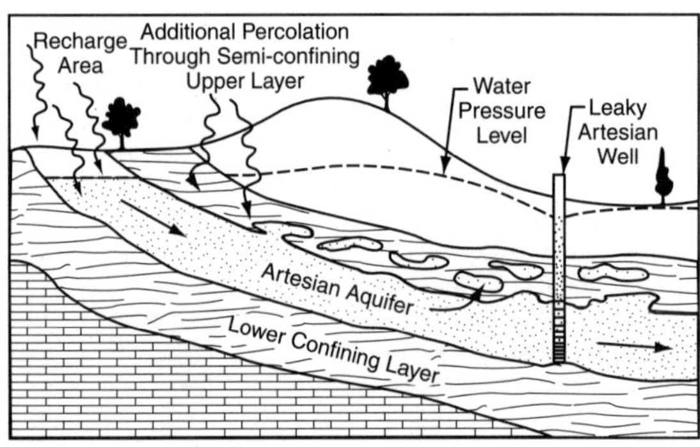

Leaky aquifer

least-squares method A mathematical process used to determine a curve that matches up with observed data points as well as possible, so that the relationship between two or more variables can be approximated by the curve's mathematical formula. This method seeks to minimize the sum of the square of the distances (deviations) of the plotted available data (observations) from the fitted curve.

leather O-rings or gaskets used with piston pumps to provide a seal between the piston and the sidewall. *See also* O-ring.

LED *See* light-emitting diode.

left bank The bank of a stream to the left of an observer who is facing downstream.

legal water level That stage of a body of water where the shoreline defines the riparian boundaries, such as the normal high-water line on a lake.

Legionella A genus of bacteria of the family *Legionellaceae*. As of the time of this writing, it consists of at least 51 serogroups comprising 34 species.

Legionella pneumophila A bacterial species that causes an acute pneumonia (Legionnaires' disease or Legionellosis) that is progressive and sometimes fatal, or a milder form of pneumonic illness (Pontiac fever) that is self-limited (i.e., heals on its own) with respiratory symptoms similar to influenza.

LEL *See* lower explosion limit.

length The greatest dimension along the perimeter in the measurement of a rectangular object.

length of dam The distance between end abutments measured on the top along the center line (axis) of a dam.

length units In the US customary system, feet. In Système International, metres.

LEPC *See* Local Emergency Planning Committee.

lesion A pathological or traumatic discontinuity or loss of function of a part as a result of damage induced by any chemical.

lethal Deadly; fatal.

lethal concentration (LC) A concentration of a chemical in air or water that will kill an organism. Air concentrations tend to be applied to all organisms, whereas those in water usually apply only to aquatic organisms. This term does not apply to ingestion, because in that case the amount (or dose) of a chemical taken in is considered important, not its concentration.

lethal concentration–50 percent (LC_{50}) The concentration of a chemical in air or water that is expected to cause death in 50 percent of test animals living in that air or water.

lethal dose (LD) A dose that would be expected to kill the exposed subject. This is an imprecise term. Estimating a dose that would be expected to kill a certain proportion of exposed individuals (e.g., lethal dose 50) is more precise. This term is most often used in relation to doses likely to be lethal in humans, for whom precise data are rarely available.

lethal dose–50 percent (LD_{50}) The dose of a chemical or other agent that causes death in 50 percent of the treated individuals. To be meaningful, both the dose and route of administration must be stated, as well as how long the test subjects were observed after treatment with the agent.

lethal time The time required for a particular dose to produce a lethal outcome. A more general term would be *latency period.* The lethal time is ordinarily inversely related to the dose administered.

leukemia A progressive and malignant disease of the organs responsible for producing blood cells (i.e., the bone marrow). It is marked by increased numbers of leukocytes and their precursor cells in the blood and bone marrow. It is frequently further characterized by the type of cell (myeloid, lymphoid, or monocytic) involved and the duration of the disease (acute or chronic). In humans, leukemia is primarily associated with benzene (C_6H_6) insults.

leukocytes Cells that are commonly referred to as white blood cells but are actually made up of several different cell types.

leukopenia A reduction in the normal number of leukocytes in the blood. The term is generally applied when the count of leukocytes falls below 5,000 per cubic millimetre because of some chemical insult.

levee A dike or embankment, generally constructed on or parallel to the banks of a stream, lake, or other body of water, intended to protect the land side from inundation by floodwaters or to confine the stream flow to its regular channel.

levee grade (1) The slope of the crown of a levee along the center line. (2) The elevation of the top of a levee.

level control A float device (or pressure switch) that senses changes in a measured variable and opens or closes a switch in response to that change. In its simplest form, this control might be a floating ball connected mechanically to a switch or valve, such as is used to stop water flow into a toilet when the toilet tank (reservoir) is full.

level of detection *See* limit of detection.

Lewis acid Any molecule or ion that can combine with another molecule or ion by forming a covalent bond with two electrons from the second molecule or ion. A Lewis acid is thus an electron acceptor (an electrophile). *See also* Lewis base.

Lewis base A substance that forms a covalent bond by donating a pair of electrons. Such a substance is also called a nucleophile. Neutralization results from a reaction between a Lewis base and a Lewis acid with the formation of a covalent bond. *See also* Lewis acid.

Leydig cell An epithelioid cell that is the endocrine (hormone-secreting) cell of the testis. Such cells are responsible for secreting a number of hormones, testosterone being the most abundant. Haloacetic acids (CX_3COOH, where X = Cl, Br, H, or some combination) have been identified as testicular toxins.

LFL (lower flammable limit) *See* lower explosion limit.

LI *See* Langelier saturation index.

life cycle (1) The series of stages or changes undergone by an organism from fertilization, birth, or hatching to reproduction of the next generation. (2) The sequence of stages of a product from development through production, use, and final disposal.

life cycle cost A method of expressing cost in which both capital costs and operations and maintenance costs are considered for comparing different alternatives. Typically, the amortized annual cost of the capital investment, based on a fixed interest rate and design period, is added to annual operations and maintenance cost to arrive at a total annual cost.

life line rate A utility rate charged to qualifying customers (e.g., senior citizens, the poor, those living on fixed incomes) to ensure they will receive minimum water requirements at a cost lower than that paid by the utility.

life-span The length of an organism's existence, usually expressed as the mean value for the species. Life-span is considered an important variable in extrapolations of risk across species. Treatment periods in toxicological studies are generally considered to apply to the *portion* of the life-span involved in the testing, rather than the absolute time period involved. The life-span used is that of the strain that is being tested rather than the species as a whole, because the life-spans of inbred strains used in laboratory studies are usually significantly shorter than the normal life-spans (excluding predation) seen in wild strains. For example, a 2-year life-span of laboratory rodents is generally considered to be equivalent to a human life-span of 70 years.

lifeline A rope or line attached to a person entering a confined space and by which the person can be raised or lowered.

lifetime exposure The total amount of exposure to a substance that a human would receive in a lifetime (usually assumed to be 70 years).

lift station A structure that contains pumps and appurtenant piping, valves, and other mechanical and electrical equipment for pumping water. A lift station is also called a pump station.

lift-and-turn valve A master control valve for operating and manually regenerating softeners and filters.

ligand A molecule, ion, or atom that is attached to the central atom of a coordination compound, chelate, or other complex. Ligands are also called complexing agents (e.g., in the case of ethylenediaminetetraacetic acid). *See also* chelating agent; ethylenediaminetetraacetic acid.

ligand binding site A site on one chemical that has a specific affinity for a group of other chemicals. In the biomedical sciences, the chemical containing the ligand binding site is usually a protein that functions as a receptor used in passing chemical signals between cells or from the exterior of the cell to its interior. Such sites are common targets for drugs and toxins. Chelating agents also have ligand binding sites with a high affinity for particular metals. Adsorption sites on materials such as activated carbon or resins could also be considered ligand binding sites.

light intensity units In Système International, candelas.

light microscope A compound microscope that uses ordinary visible light to illuminate the specimen. The limit of resolution using visible light is about 0.2 micrometres. *See also* electron microscope.

light obscuration sensor A type of detector used in particle-counting instruments. Particles flowing past a detector block a light source of known intensity. The particle size is measured based on the decrease in light intensity at 180° from the light source. Light obscuration sensors (sometimes called light-blocking sensors) are able to measure a wide range of particle sizes, typically from 1 micrometre to greater than 100 micrometres.

light rain Rain that is falling at the time of observation with an intensity between a trace and 0.10 inches (2.54 millimetres) per hour, or 0.01 inches (0.25 millimetres) in 6 minutes.

light speed *See* speed of light.

light-emitting diode (LED) A semiconductor diode that emits light when current is passed through it. LEDs are commonly used in displays such as calculators and direct readout meters.

lignin An organic substance that, with cellulose, forms the chief part of woody tissue. *See also* fulvic acid; humic acid; tannin.

lignite-based activated carbon Activated carbon derived from combusting lignite, a brownish-black soft coal.

Lim Benyesh–Melnick pool The collective set of 8 pools, consisting of 42 equine antisera, used for typing enteroviruses. The pools have been designed such that a given antiserum appears within one pool, or in two or three pools. An unknown enterovirus may be identified (i.e., typed) based on its neutralization by the pool or pools containing its antiserum.

lime (CaO) A calcined chemical material, calcium oxide. Lime is used in lime softening and in lime–soda ash water treatment, but first it must be slaked to calcium hydroxide ($Ca(OH)_2$). Lime is also called burnt lime, calyx, fluxing lime, quicklime, or unslaked lime. *See also* hydrated lime; lime softening; slake.

lime and soda ash softening *See* lime–soda ash softening.

lime and soda–ash process *See* lime–soda ash softening.

lime clarification The removal of water hardness and turbidity by adding lime to generate carbonate and hydroxide metal precipitates. The solids settle and are removed by manual or mechanical means, generating a clarified effluent with lower hardness and alkalinity than the source water. *See also* alkalinity; hardness.

lime recalcining The reuse of lime sludge by heating to drive off water and carbon dioxide (CO_2), leaving only the calcium oxide (CaO) plus some inert material. Some lime sludge must be wasted periodically to avoid a buildup of inert material. *See also* lime.

lime recovery The process of reclaiming lime by solubilizing magnesium through a carbonation of lime sludge and subsequently separating the solids by mechanical means (e.g., belt filter press or vacuum filter). Source water solids can be separated from the filter cake by flotation followed by lime recalcining of the dewatered calcium carbonate ($CaCO_3$) sludge. *See also* belt filter press; lime recalcining; vacuum filter.

lime scale Hard-water scale formed in pipes and vessels (generally more severe on the hot water side). It contains a high percentage of calcium carbonate ($CaCO_3$) or magnesium carbonate ($MgCO_3$).

lime slaker A mechanical device intended to add sufficient water to quicklime (CaO) to satisfy the latter's affinity for water and to form a slurry. The two basic types of lime slakers are paste (or pug mill) and detention. The paste-type slaker adds water as required to maintain a desired mixing viscosity of about 35 percent calcium hydroxide ($Ca(OH)_2$). The detention-type slaker adds water to maintain a desired ratio with the lime, producing a lime slurry of about 10 percent calcium hydroxide.

lime softened Pertaining to a water from which hardness has been removed by adding lime and precipitating the solids that are composed of metal carbonates and hydroxides.

lime softening The process of removing water hardness by adding lime to precipitate solids composed of metal carbonates and hydroxides. Clarification may or may not also occur.

lime–soda ash method *See* lime–soda ash softening.

lime–soda ash softening A water treatment that makes use of lime softening followed by a reduction of noncarbonate hardness by the addition of soda ash (Na_2CO_3) to form an insoluble precipitate that is removed by filtration. This method of removing hardness by precipitation is sometimes used by municipalities, but it will leave 85 mg/L or more of residual hardness as calcium carbonate ($CaCO_3$). *See also* hot lime–soda softening; lime softening.

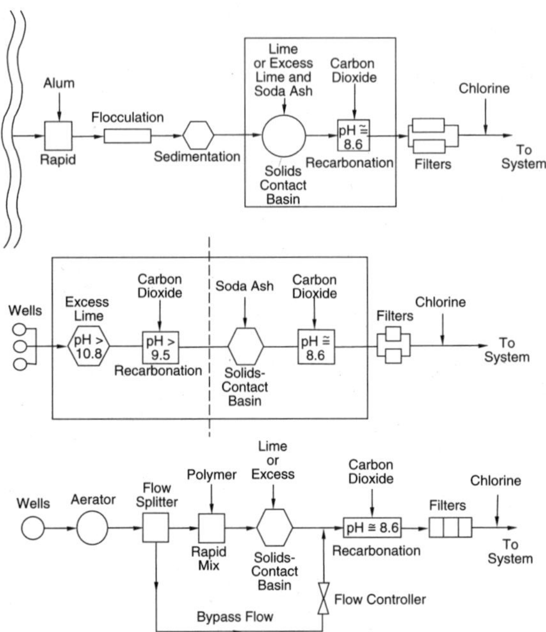

Types of lime–soda ash softening process

limestone A sedimentary rock composed mostly of calcium carbonate ($CaCO_3$) and usually some magnesium carbonate ($MgCO_3$). *See also* calcite.

limestone contactor A treatment device consisting of a bed of limestone through which water is passed to dissolve calcium carbonate ($CaCO_3$). The addition of calcium carbonate to the water decreases corrosivity by increasing the pH, calcium concentration, and alkalinity of the water.

limit of detection (LOD) The minimum concentration that can be detected by a particular analytical method. Limit-of-detection criteria depend on the context, and a different term may be used. For example, the US Environmental Protection Agency defines a method detection limit as the minimum concentration that can be measured and reported with 99 percent confidence that the analyte concentration is greater than zero. The limit of detection is also called the detection limit. *See also* limit of quantitation; method detection limit; practical quantitation limit.

limit of quantitation (LOQ) The concentration level above which the quantification of a substance in a sample is reliable; the region between the limit of detection and the limit of quantification, where detection is reliable but quantification is not. *See also* limit of detection; method detection limit; practical quantitation level.

limit stop Any of several types of diaphragm valves for which adjustments can be made to control flow rates during various processes (backwash, fast rinse, or brine dilution) when brine is pumped in during batch regeneration of the resin in a portable ion exchange tank.

limit switch setting The adjustment of a control device to specify when the device should stop an action, such as a valve operation or door opening at a predetermined location.

limited turf area A prescribed fraction of the landscape area to which turfgrass is restricted.

limiting factor A condition that tends to inhibit a biological, chemical, economic, physical, or social process.

limnetic Relating to or inhabiting the portion of a body of fresh water beyond the outer border of the littoral zone and to a depth that light penetrates.

limnetic community The area of open water in a freshwater lake providing the habitat for fish, phytoplankton, and zooplankton.

limnology The study of fresh waters, especially ponds and lakes, including their biological, geographical, and physical characteristics.

LIMS *See* laboratory information management system.

lin ft *See* linear feet *in the Units of Measure list.*

lindane ($C_6H_6Cl_6$) The legal label name for the γ-isomer of 1,2,3,4,5,6-hexachlorocyclohexane. This pesticide is regulated by the US Environmental Protection Agency. *See also* pesticide.

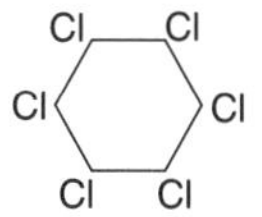

99% gamma isomer

Lindane

line (1) An imaginary trace on the Earth's surface. (2) The distance between two points. (3) A one-dimensional object. A line segment is a direct line between two points. Special forms of lines include: a string, a series of line segments; an arc, a locus of points forming a curve defined by a mathematical function; and a chain, a directed sequence of nonintersecting line segments or arcs with nodes at each end.

line swabbing A method of well development that uses a tool that fits tightly in the well (similar to a brush) and forces water into the formation to remove fines and other blockages near the well.

linear alkyl sulfonate (LAS) (R–C_6H_4–SO_4M, where R is C_{10} or longer and the R group is normal or iso, and M is Na^+ or some other salt) A synthetic detergent specially tailored for its biodegradability. Chemically, linear alkyl sulfonate is a straight-chain type of sulfonate, as opposed to a branch-chain sulfonate type detergent called alkylbenzene sulfonate that is known for its resistance to breakdown by microorganisms. *See also* alkylbenzene sulfonate.

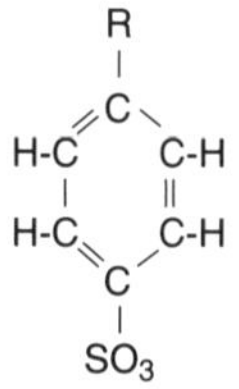

Linear alkyl sulfonate, where R is a straight chain of 10 or more carbon atoms, in which the linear hydrocarbons are normal or iso (branched at the end only)

linear feet *See the Units of Measure list.*

linear polarization A method used to estimate the metal corrosion potential of a fluid stream based on polarization resistance. The linear polarization instrument uses a probe with metal electrodes under a known electrical potential and measures the electrical current flow, considering both anodic and cathodic polarization characteristics, and relates the current flow to a metal corrosion rate (typically millimetres per year or mils per year).

linearity A measure of how closely an instrument measures actual values of a variable through its effective range. This measure is used to determine the accuracy of an instrument.

linearization The modification of a system so that its outputs are approximately linear functions of its inputs.

linearized multistage model (LMS model) A mathematical model in common use to approximate theoretical dose-response relationships for carcinogens. It is based on the assumption that a finite number of irreversible events are involved in the development of cancer. In the model's application to low-dose extrapolation, the assumption is made that the carcinogen being considered acts in an additive way to other causes of cancer in the population at risk. In general, this model implicitly assumes that the chemical or one of its metabolites is responsible for inducing a mutation. Because such an effect is irreversible, the damage is cumulative with dose. Thus, the probability of

Lining

inducing cancer becomes linear at low doses. *See also* log-probit model; logit model; multihit model; multistage model; one-hit model; probit model; Weibull model.

lined canal A canal for which the sides and bottom have been lined or covered with some watertight material to prevent leakage or erosion, improve carrying capacity, or minimize growth of vegetation.

line-to-line voltage For a polyphase circuit, the voltage between a pair of phase conductors, measured in volts.

line-to-neutral voltage For a polyphase circuit, the voltage between a phase conductor and the neutral conductor, measured in volts.

lining (1) A protective covering over all or a portion of the perimeter of a conduit or reservoir, intended to prevent seepage losses, withstand pressure, or resist erosion. In the case of conduits, lining is also sometimes installed to reduce friction losses. (2) A physical method for controlling aquatic plants by placing a permanent lining, such as synthetic rubber, in a water body.

linuron ($C_6H_3Cl_2NHC(O)N(OCH_3)CH_3$) The generic name for the herbicide 3-(3,4-dichlorophenyl)-1-methoxy-1-methylurea. *See also* herbicide.

lipid peroxidation An interaction of oxygen with double bonds present in fatty acids found in lipids. The process is accelerated in the presence of certain metals (e.g., iron, copper, manganese). Lipid peroxides are formed that are in themselves reactive chemicals and that continue to give rise to other reactive chemicals. Lipid peroxidation is thought to contribute to the toxicity of chemicals that induce oxidative stress. Examples include carbon tetrachloride (CCl_4) and paraquat (($CH_3(C_5H_4N)_2CH_3$)$\cdot 2CH_3SO_4$) or ($CH_3(C_5H_4N)_2CH_3 \cdot 2Cl$).

lipids A group of organic compounds that make up fats and other esters that have analogous properties.

lipophilicity The nonpolar character of a chemical substance.

lipoprotein A protein that is strongly associated with lipids through nonpolar interactions. Much of the time these proteins are actually imbedded in cell membranes. In many cases the activity of the protein as an enzyme is dependent on the lipid.

liposome A spherical particle that is found in aqueous media and is formed as a lipid bilayer enclosing an aqueous compartment.

liquefaction The transformation to the liquid state. This term is more commonly used to refer to the changing of gases to liquids rather than the melting of solids to liquids.

liquid A substance that flows freely and takes the shape of its container. A liquid is characterized by free movement of the constituent molecules among themselves, but without the tendency to separate from one another as with gases. The terms *liquid* and *fluid* are often used synonymously, but *fluid* has the broader significance because it includes both liquids and gases. *See also* fluid.

liquid chlorine Elemental chlorine converted to a liquid state by compression and refrigeration of the dry, purified gas. Liquid chlorine is shipped under pressure in steel containers. This term is also sometimes used to refer to hypochlorite (bleach, NaOCl) solutions. *See also* hypochlorite.

liquid chromatography (LC) A technique used in the analysis of organic compounds. It is a separation technique based on the partitioning of an analyte between a liquid mobile phase and a stationary phase, which is typically solid. Liquid chromatography is often practiced as high-performance liquid chromatography. Liquid chromatography techniques allow the analysis of organic compounds that may not be suited to gas chromatographic techniques. Examples include compounds that contain polar functional groups, are thermally unstable, or both. *See also* high-performance liquid chromatography.

liquid chromatography–mass spectrometry (LC–MS) A combined analytical technique that separates compounds in a liquid chromatograph and then detects them in a mass spectrometer. Technological advances have given these instruments a role as powerful tools in the identification of unknown compounds. Although the technique has found applications in research laboratories, at the time of this writing LC–MS is not yet commonly found in many water utility laboratories.

liquid chromatography–particle beam–mass spectrometry (LC–PB–MS) An instrumental technique used in the analysis of organic compounds. Such instruments are more commonly situated in research facilities and, as of the time of this writing, are not yet commonly found in water utility laboratories. The term *particle beam* refers to a type of interface between a liquid chromatograph and a mass spectrometer. One major advantage of a particle beam interface over earlier interfaces is that it produces electron impact spectra. This feature makes it easier to identify unknown compounds because spectral databases typically contain many electron impact spectra. One disadvantage of this type of instrument is an inability to completely ionize nonvolatile compounds at trace concentrations. *See also* electrospray ionization.

liquid crystal display (LCD) A common computer and information display device. It is made up of crystals that are laminated between two pieces of glass. The optical characteristics of the crystals change when voltage is applied across them. New LCD displays consume very little power and are capable of full color and a pixel resolution of 920,000.

liquid oxygen (LOX) A form of oxygen for which the temperature is decreased to an extent that the oxygen is in liquid form. Oxygen in this form can be delivered and stored, thereby relieving the need for on-site generation. *See also* pure oxygen.

liquid scintillation counter An energy-sensitive instrument used in the analysis of radionuclides. A scintillation is a flash of light (photon) produced when an ionizing particle strikes a phosphor. A detector produces a signal that is proportional to the energy deposited in the detector. This behavior gives these instruments an ability to sort, identify, and quantify radiation from different sources. For example, a given instrument can determine the activity of tritium (^3H) and radon (^{222}Rn) in a given sample. In water samples, radon is the radionuclide most commonly determined with a liquid scintillation counter. *See also* scintillation cocktail.

liquid sludge Sludge containing sufficient water (ordinarily more than 85 percent) to permit flow by gravity or pumping.

liquid–liquid extraction (LLE) A class of separation techniques that can be used to isolate or purify a variety of organic and inorganic analytes. It is a commonly used sample preparation technique in water analyses, especially in the analysis of organic compounds. These techniques are based on the partitioning of compounds between two liquids. The trend in laboratory analyses is toward minimizing the use of organic solvents; therefore, many microextraction methods have been developed. Often the partitioning occurs in a separatory funnel or in a sample vial.

liquid–liquid extraction, continuous (CLLE) *See* continuous liquid–liquid extraction.

liquor A solution of one or more chemical substances (gas, solid, or liquid) in water.

LIS *See* land information system.

listing basin A small area in a field formed by plowing or with a special device (called a basin lister) to catch and divert water to ground storage in a semiarid region and to prevent erosion by water that would otherwise constitute surface runoff.

lithium hypochlorite (LiOCl) A dry powder consisting of a combination of lithium and chlorine put together in such a way that, when it is dissolved in water, active chlorine is released. The active chlorine content is usually about 35 percent.

litholitic Derived from rock or from the lithosphere. *See also* lithosphere.

lithology The science dealing with the mineral composition and structure of rocks, especially characteristics of structures that can be studied without high magnification. Lithology also deals with the character of a rock formation or of the rock found in a geological area or stratum, as expressed in terms of structure, composition, color, and texture.

lithosphere That part of the earth that is composed predominantly of rocks (either coherent or incoherent and including the disintegrated rock materials known as soils and subsoils), together with everything in this rocky crust.

litre (L) *See the Units of Measure list.*

litres per day (L/d) *See the Units of Measure list.*

litres per minute (L/min) *See the Units of Measure list.*

litter (1) Vegetative material, such as leaves, twigs, and stems of plants, lying on the surface of the ground in an undecomposed or slightly decomposed state. (2) Solid waste from human activity deposited indiscriminately on land or water.

littoral *See* littoral zone.

littoral zone (1) That portion of a body of fresh water extending from the shoreline lakeward to the limit of occupancy of rooted plants. (2) The strip of land along a shoreline between the high- and low-water levels.

liver A large gland in the upper part of the abdomen that performs many diverse metabolic functions in a higher organism. In mammals, it has two sources of blood flow, one directly from the heart via the hepatic artery and a second from the complex set of vessels that carry newly absorbed materials from the gastrointestinal tract (stomach and intestines) via the portal vein. Consequently, the liver is the first internal organ through which blood flows from the intestine. In terms of toxicology, the liver is the organ most involved in the metabolism of drugs. This metabolism can inactivate an ingested toxic chemical from drinking water or produce reactive intermediates (during phase 1 metabolism) that interact with lipids, proteins, or nucleic acids to induce toxicity, genetic damage, or both, such as mutations or chromosomal abnormalities. Conversely, the liver is frequently involved in detoxification mechanisms (commonly catalyzed by proteins referred to as phase 2 enzymes). The general result of these metabolic reactions is to make a chemical more polar and easier to excrete by the kidney, bile, and intestine. The liver also plays important roles in control of lipid and carbohydrate metabolism and is involved in the metabolism of steroid hormones. It is also the organ that produces and secretes bile.

LLE *See* continuous liquid–liquid extraction.

lm *See* lumen *in the Units of Measure list.*

L/min *See* litres per minute *in the Units of Measure list.*

LMS model *See* linearized multistage model.

ln *See* natural logarithm.

load Almost any quantity, e.g., of electrical power, of people served, or of water carried by a conduit.

load curve A curve that expresses the variation of the load on a water-pumping station or treatment plant over a given period of time, such as a day, week, month, year. The load is usually plotted as the ordinate and time as the abscissa.

load factor The ratio of the average load carried by an operation to the maximum load carried, expressed as a percentage, during a given period of time. *See also* load.

loading The amount of a material carried in a moving medium such as water or air. The loading is also called the mass transport.

loading rate The flow rate per unit area (e.g., gallons per minute per square foot, cubic metres per minute per square metre) of a filter, adsorber, or ion exchange unit through which water passes.

LOAEL *See* lowest-observed-adverse-effect level.

loam Soil that contains 7–27 percent clay, 28–50 percent silt, and less than 52 percent sand.

local area network (LAN) A system for connecting computer hardware and software to provide shared access and data movement among multiple devices within close proximity, normally not exceeding a radius of several miles.

Local Emergency Planning Committee (LEPC) A committee composed of members from entities that provide local emergency services, industries that handle hazardous materials, and government. This group reviews potential hazards and prepares plans to eliminate or respond to emergencies in the community.

local toxic effect An effect that occurs at the site of exposure, e.g., skin reactions to chemicals that are applied topically, lung irritancy induced by inhaled irritant gases, or tumors induced at the site of injection of a carcinogenic chemical.

localized corrosion Corrosion caused by action of local cells, i.e., galvanic cells arising because of nonuniformities between two adjacent areas at a metal surface exposed to an electrolyte.

lockout The application of a lock to the energy sources for a piece of equipment to prevent that equipment from being accidentally energized while it is out of service.

LOD *See* limit of detection.

lodestone *See* magnetite.

LOEL *See* lowest-observed-effect level.

log *See* common logarithm.

log boom A floating structure, usually of timber or logs, used (1) to protect the face of a dam or other structure built in or on the water from damage by wave action or by floating material being dashed against it by the waves, or (2) to deflect floating material away from such a structure. A log boom is also used to control spills in open water.

log growth phase The period during which a bacterial population increases logarithmically with time. The growth rate of the population is a function of the cell division time only if an unlimited supply of food is available.

log Pearson distribution A statistical distribution used in flood frequency analysis to determine the probability that a given flow will occur within a given time interval.

log removal A shorthand term for $\log_{10}$ removal, used in reference to the Surface Water Treatment Rule and the physical–chemical treatment of water to remove, kill, or inactivate pathogenic organisms such as *Giardia lamblia* and viruses. A 1-log removal equals a 90 percent reduction in density of the target organism; a 2-log removal equals a 99 percent reduction; a 3-log removal equals a 99.9 percent reduction; and so on. *See also* common logarithm.

log-mean feed concentration In pressure-driven membrane-desalting processes, the average feed concentration in a concentrate-staged system, considering solute concentration effects throughout the system and varying membrane areas per stage:

$$\text{log-mean feed concentration} = \frac{C_f \ln(C_c/C_f)}{1-(C_f/C_c)}$$

Where:

C_f = feed concentration, in milligrams per litre

C_c = concentrate concentration, in milligrams per litre

For reverse osmosis inorganic solute rejection calculations, the log-mean feed concentration is often preferred over the simple numerical averaging of the feed-concentrate concentration, because it more accurately defines the average system feed concentration. *See also* average feed concentration.

logarithm, common *See* common logarithm.

logarithm, natural *See* natural logarithm.

logarithmic scale A series of intervals (marks or lines), usually placed along one side and/or the bottom of a graph, that represent the range of values of the data and for which the marks or lines are varied logarithmically (as opposed to being equally spaced).

logarithmic transformation A mathematical operation to convert a given number to its logarithm. For example, the logarithmic transform z of the number y can be defined as $z = \log(y)$.

logging, electrical *See* electrical logging.

logit model In toxicology, a tolerance distribution model that assumes a logistic (S-shaped) distribution of the natural logarithms of individual tolerances in the population. This model gives similar results to the log-probit model in the experimental range but gives a flatter tail at a low dose. As a consequence, it tends to predict higher risks at low doses. *See also* linearized multistage model; log-probit model; multihit model; multistage model, one-hit model; probit model; Weibull model.

log-normal distribution A distribution of data in which occurrences of lower values are relatively more abundant compared to occurrences of higher values. Many natural water quality parameters are known to follow a log-normal distribution. A log-normal distribution can be mathematically described by the following equation:

$$f(x) = \frac{1}{x\sigma\sqrt{2\pi}} \exp\left[-\frac{1}{2\sigma^2}(\ln(x)-\mu)^2\right]$$

Where:

σ = population standard deviation

μ = population mean

log-probit model In toxicology, a tolerance distribution model based on the concept that each individual has a threshold dose or exposure at which a toxic effect would be produced. The model assumes a normal distribution and includes location and scaling factors. The data are transformed log 10 on a scale that is adjusted by adding whole integers to ensure that the values representing dose are expressed as positive numbers. These units are known as *probits.* This model is used traditionally in toxicology to estimate the lethal dose—50 percent and other dichotomous (i.e., all-or-nothing) responses. *See also* linearized multistage model; logit model; multihit model; multistage model, one-hit model; probit model; Weibull model.

long pipe Generally, a pipeline for which the length is more than 500 times greater than the diameter. In such pipes, the loss of head caused by entrance into the pipe and by velocity head is negligible and usually disregarded.

long tube A tube inserted in an orifice, having length that is more than three times greater than the diameter.

longitudinal dispersion Movement of a substance forward or backward along the direction of flow as a result of concentration gradients. *See also* Peclet number.

longitudinal epidemiologic study An analytical epidemiologic study in which the time sequence between exposure and disease status can be inferred; i.e., water exposure precedes onset of disease (in contrast to cross-sectional epidemiologic studies, in which disease may influence exposure; for example, disease may increase the need for fluid, thus leading to greater water exposure). Studies can be retrospective or prospective in time.

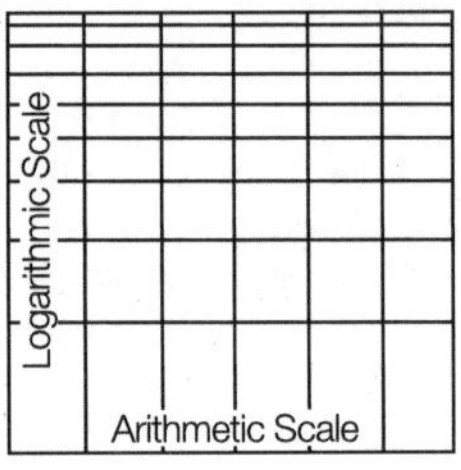

Logarithmic scale

In case-control or retrospective cohort (retrospective) studies, an inference is made about the current disease status of a study population and associated events or experiences in the past. A prospective cohort (prospective) study identifies exposed and unexposed populations and follows these populations to determine their future disease status. *See also* case-control epidemiologic study; cohort; cross-sectional epidemiologic study; prospective study; retrospective analysis.

longitudinal flow A flow pattern in which water travels from the bottom to top (or vice versa) in either a cartridge-type or loose-media tank-type filtration system. Such flow generally produces greater contact time, higher unit capacity, more complete utilization of the medium, and a more uniform water quality. Longitudinal flow is also called axial flow.

longitudinal mixing Dispersion of a substance lengthwise in a stream as it moves downstream.

long-run average cost The total cost of producing a product (e.g., water)—allowing for all factors of production to be changed as desired—divided by the number of units (e.g., 1,000 gallons) produced. In the field of economics, the long run is a period of time long enough so that all costs of production are variable, i.e., all forms of production can be changed, including the plant size, the process, the location, and so forth. By contrast, the short run is a period of time too short to alter all of the factors of production. In the long run, many plant sizes and processes might be devised to produce the required amount of product, each with varying average costs per unit. Each of these alternative plants, once committed, becomes a short-run operation, because all factors of production can no longer be changed. The goal in designing facilities and processes for the long run is to identify which plant configuration, from the array of alternatives evaluated, will minimize the short-run average cost for the expected level of output. *See also* short-run average cost.

long-run marginal cost A value determined by dividing the change in total cost associated with producing the next increment of product (such as the next 1,000 or 1 million gallons of water) by the amount of that increment. Since in the long-run all costs are defined as variable, long-run-marginal cost can be thought of as the change in total cost that results from small variations in the scale of the plant divided by the change in output associated with those changes in plant size. Each scale of plant also has its own short-run average and marginal cost curves. For the optimal plant scale, the short-run marginal cost is equal to the long-run marginal cost at the desired level of output.

long-term animal feeding study *See* carcinogenesis bioassay.

long-term single-element bench-scale test A membrane treatment study specified in the US Environmental Protection Agency's Information Collection Rule as promulgated in May 1996, whereby one spiral wound membrane element (minimum element size 2.5 inches [6.4 centimetres] in diameter and 40 inches [1 metre] in length) with molecular weight cutoff less than 1,000 daltons is evaluated at 75 percent recovery while other operating parameters are held constant for at least 6,600 hours over the course of 1 year. The study is a continuous-flow test using one membrane element and concentrate recycle. *See also* Information Collection Rule.

long-tube evaporator A distillation device in which long tubes for condensate or other steam are configured parallel to the steam flow direction. *See also* multiple-effect distillation; multistage flash distillation.

loop (1) The plumbing network designed to continuously circulate ultrapure-grade water in high-purity water systems between storage and disinfection modes so as to maintain microbiological cleanliness. (2) A plumbing connection used to bypass water around a particular location. It allows a water treatment device to be installed at that location, or it may be used when the treatment system is out of service for any reason. *See also* looping.

looped pooling A process of assigning one water allocation to a group of metered accounts in a looped system. A looped system involves a set of meters in service to the same irrigation main.

looping A distribution system design practice of configuring a piping system in a loop to allow continual flow-through of the treated water without dead ends in which the water can stagnate. Looping promotes good water quality and hydraulic pressure within a distribution system.

loose apron A covering of loose stone or blocks laid on the berm of a river embankment to protect the embankment from erosion. The stone from the apron gradually fails as scour takes place; this process is called launching. *See also* riprap.

loose medium (loose media, pl.) A filter, adsorption, or ion exchange medium (in a tank or bed) that can be expanded during backwashing and rinsing. The term *loose* is used to differentiate from a contained, or "fixed" medium in a tank or a fixed or compressed medium layer in a cartridge filter.

loose-rock dam A dam constructed of rocks without the use of mortar, usually dumped in place without any particular effort at packing, sorting, or arranging. Such a dam is also called a loose rock-fill dam.

LOQ *See* limit of quantitation.

losing stream A stream that loses water by seepage into the ground. A losing stream is also called an influent stream.

loss (1) The cost of a claim against a company. (2) The opposite of profit. Loss occurs when expenses exceed revenues in any accounting period.

loss of head (1) The decrease in energy head between two points resulting from friction, bends, obstructions, expansions, or any other cause. It does not include changes in the elevation of the hydraulic grade line unless the hydraulic and energy grade lines parallel each other. (2) The difference between the total heads at two points in a hydraulic system.

loss-of-head gauge A gauge on a rapid granular filter that indicates the loss of head involved in the filtering operation and

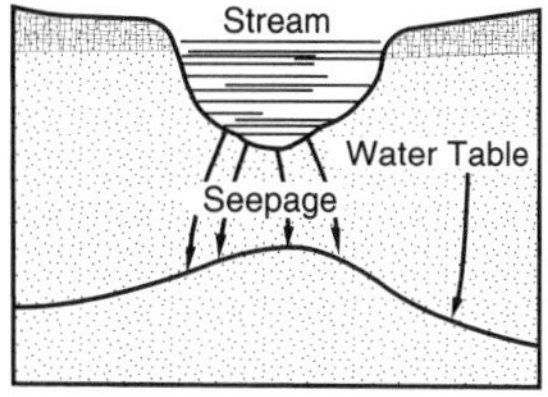

Losing stream

enables the operator to ascertain the need for filter washing. Some gauges are of the indicating–recording type.

loss-of-signal switch Protection for a control loop against overreacting when the loop is receiving no signal. This protection usually locks the control loop at the last value received. Most control signals are represented by a 4–20-milliampere or 1–5-volt signal, with 4 milliamperes and 1 volt representing the minimum value and 20 milliamperes and 5 volts representing the maximum value. If the control signal is lost altogether (0 milliamperes, 0 volts), the control loop must be protected by responding as though it were receiving a 4-milliampere, 1-volt signal.

lost circulation The result of drilling fluid escaping from the borehole into a porous or fractured formation. This fluid loss increases the cost of the drilling operation because the drilling cuttings are not lifted to the surface but instead are forced into the borehole walls. The lost fluid and cuttings can damage the formation near the borehole, so that subsequent well development techniques are ineffective and the well cannot produce economic quantities of water.

lost energy In hydraulics, the heat energy of water flowing in a waterway, produced by friction and lost through absorption in the stream and dissipation in the water or the walls of the waterway.

lost river In geology, a river that, because of acidity, or subterranean flow in karst areas, has lost its trunk, at first only during the driest season and later permanently. The remaining detached tributaries lose themselves in the arid ground.

lost-day rate *See* incident rate.

lotic Pertaining to flowing water bodies, such as rivers.

low duty of water An irrigation water requirement under which a given quantity of water will serve a relatively small area of land.

low-consumption toilet A toilet designed to meet the American National Standards Institute standard A112.19.6, Hydraulic Requirements for Water Closets and Urinals, using 1.6 gallons (6 litres) or less of water per flush cycle.

lower explosion limit (LEL) The lower end of a range in which a substance will be flammable at ambient temperatures. The value is usually represented as a percentage of the gas or vapor in air by volume.

lower flammable limit (LFL) *See* lower explosion limit.

lowest-observed-adverse-effect level (LOAEL) An exposure, dose, or concentration that produces a minimal toxic response. The definition of what constitutes a minimal toxic effect is somewhat arbitrary, but such an effect is ordinarily viewed as a mild, reversible change that does not represent a severe health effect. Although the effect is mild, its relationship to a clearly recognized adverse effect must be recognized. The term has real meaning only when the effects of a chemical have been reasonably well characterized. It should not be applied to results obtained during a single noncomprehensive evaluation of toxic effects. The term has meaning in the derivation of standards only when it serves as the basis for calculating a reference dose. Ordinarily, the calculation of a reference dose from a lowest-observed-adverse-effect level requires additional safety factors relative to a no-observed-adverse-effect level that has been obtained under similar circumstances. *See also* benchmark dose; no-observed-adverse-effect level; reference dose.

lowest-observed-effect level (LOEL) The minimum dose or exposure that leads to some change in an organism. It differs from a lowest-observed-adverse-effect level in that the response has no clear relationship to the production of an adverse health effect. Although it identifies a particular dose level, it is usually not appropriate for use in establishing standards. *See also* benchmark dose; dose; lowest-observed-adverse-effect level; no-observed-adverse-effect level.

low-flow frequency analysis An analysis that uses raw hydraulic data to determine the probability that a particular low-flow design value will be the proper stream-flow-design value for estimating stream flow or impounding reservoir safe yield.

low-head drainage Drainage of water from irrigation lines at the low elevations in an irrigation circuit.

lowland Low-lying land, slightly undulated or flat, principally along coasts and in river valleys.

low-water datum An approximation of the plane of mean low water that has been adopted as a standard reference plane for a limited area and is retained for an indefinite period, even though it may differ slightly from a better determination of mean low water from a subsequent series of observations.

low-water line The line to which the shore of a river, lake, or sea ordinarily recedes at low water.

low-water lunitidal interval The interval between the moon's meridian passage at a given place and the following low water at that place.

low-water regulation Adjustment of low stream flows to a desirable, necessary, or standard condition.

low-water-use plant A plant requiring less than 30 percent of reference evapotranspiration to maintain an optimal appearance.

LOX *See* liquid oxygen.

LR *See* Larson ratio.

LSI *See* Langelier saturation index.

lubricant Usually a natural or synthetic oil or grease (sometimes gas) used to reduce friction and conduct heat away from bearings.

lumen (lm) (1) *See the Units of Measure list.* (2) For a hollow fiber membrane, the central bore of a fiber.

luminance units In Système International, candelas per metre squared.

luminous flux units In Système International, lumens; candelas per steradian.

lunar day The time of the rotation of the Earth with respect to the moon, or the interval between two successive upper transits of the moon over the meridian of a place. The mean lunar day is approximately 24.84 solar hours in length, or 1.035 times as long as the mean solar day.

lux (lx) *See the Units of Measure list.*

L'vov platform An adaptation of a graphite tube used in atomic absorption spectrophotometry. It was named for B.V. L'vov, a pioneer in the field, and is known generically as a stabilized temperature platform furnace. Such a tube contains a shelf onto which a sample is injected. The platform promotes even heating of the sample, which delays atomization of the analyte. This delay minimizes the nonspectral interferences common in graphite furnace work.

lx *See* lux *in the Units of Measure list.*

lye *See* sodium hydroxide.

lymph node A collection point for drainage of tissues by the lymphatic system. The lymph is a clear watery fluid that may contain some red blood vessels but usually contains lymphocytes (white blood cells) and fat. The lymphatic system consists of a system of vessels that come from the blood vessels, converge in lymph nodes, and return to the blood vessels. Frequently, lymph nodes are sites at which metastasizing tumor cells collect. Therefore, lymph nodes are frequently biopsied at the time of cancer surgery to determine whether any signs of metastasis exist. They are examined in all chronic studies of potential drinking water carcinogens. *See also* metastasis.

lymphocyte Essentially a type of white blood cell. The term is applied most specifically to a mononuclear leukocyte that is 7–20 micrometres in diameter and that stains deep blue in the nucleus. A lymphocyte participates in humoral and cell-mediated immune responses. Lymphocytes are involved in inflammatory responses produced by many toxic chemicals. *See also* cell-mediated immunity; humoral immunity.

lymphoma Any neoplastic (i.e., cancerous) disorder of the lymphatic system, including Hodgkin's disease. Because benign lymphomas are relatively rare, the term used alone is usually taken to be synonymous with malignant lymphoma. Lymphoma is a frequent tumor in rodents used in toxicologic studies.

lymphosarcoma Any neoplastic (i.e., cancerous) disorder of the lymphatic system with the exception of Hodgkin's disease; a common tumor type. *See also* lymphoma.

lyophilization An effective method for long-term preservation of bacteria and other microorganisms. It involves removing water from frozen microorganism suspensions by sublimation under reduced pressure.

lysimeter A structure containing a mass of soil and designed to permit the measurement of water draining through the soil. Water is applied by a sprinkling mechanism (rain simulator) that provides a rather uniform sprinkling of water to a prescribed area at prescribed rates and size drops (impact); an auxiliary apparatus consists of a rain gauge (either total or intensity) and a catchment basin or receptacles in which surface runoff rate or total flow is measured.

lysis A process of death of a living microorganism, causing the breakup of the cell wall and resulting in the release of protoplasm from the cell.

M

M *See* molar; moles per litre *in the Units of Measure list;* mega.

M alkalinity *See* alkalinity test.

m *See meta.*

m *See* metre *in the Units of Measure list*; milli.

m T7 medium A type of selective culture medium used for detecting and recovering injured coliform bacteria from treated water, disinfected water, or both. *See also* coliform bacteria.

m³ *See* cubic metre *in the Units of Measure list.*

M⁺ *See* molecular ion.

mA *See* milliampere *in the Units of Measure list.*

MAC *See* maximum allowable concentration.

machine guard Shielding designed to prevent a person from coming into contact with moving parts of equipment.

machine-banded pipe A pipe made of wooden staves held together in a machine and tightly wrapped with wire. It is made in definite lengths in a factory and joined in the field by couplings.

macrobenthos The larger-than-microscopic animals and plants that are fixed or crawl on the bottom of a body of water.

macroinvertebrate Any larger-than-microscopic animal that has no backbone or spinal column.

macroorganism Any plant, animal, or fungal organism visible to the unaided eye.

macrophage Any of a family of phagocytic cells with a small, oval nucleus that occur in the walls of blood vessels and loose connective tissue. These cells are derived from monocytes. Normally immobile, they are activated by inflammation and engulf foreign material, including microbes and their remains.

macrophyte A large aquatic plant (as opposed to small algae such as phytoplankton).

macropore *See* transport pore.

macroporous resin Any of a special grade of ion exchange resins that have a high resistance to oxidation and organic fouling and were developed to provide increased surface area for reactions with high–molecular-weight organic matter. Macroporous resins, which are produced in both anionic and cationic versions, contain high levels (12 percent or more) of divinylbenzene crosslinking, which reduces the swelling of the polymer resin in water.

macroreticular resin Any of a special grade of ion exchange resins (either anionic or cationic) having a rigid polymer porous network that retains its porous structure even after drying. Macroporous, macroreticular, and fixed-pore ion exchangers are, in general, of the same grade and are less subject than other resins to organic fouling.

macroscopic Visible by the eye without the aid of a microscope.

macroscopic organism Any organism big enough to be seen by the eye without the aid of a microscope.

magnesia (MgO) Magnesium oxide that has been specially processed. Magnesia water treatment can be used to modify the pH of water.

magnesium (Mg) One of the elements that make up the Earth's crust as components of many rock-forming minerals, such as dolomite. Magnesium and calcium dissolved in water constitute hardness. The presence of magnesium in water contributes to the formation of scale and the insoluble soap curds that typify hard water.

magnesium hardness The portion of total hardness caused by magnesium compounds, such as magnesium carbonate ($MgCO_3$) and magnesium sulfate ($MgSO_4$).

magnetic affinity cell sorting *See* immunomagnetic assay.

magnetic field strength units In Système International, amperes per metre.

magnetic flowmeter A flow-measuring device in which the movement of water induces an electrical current proportional to the rate of flow.

magnetic flux density units In Système International, teslas; kilograms per second squared per ampere.

magnetic flux units In Système International, webers; metres squared kilograms per second squared per ampere.

magnetic sector hybrid mass spectrometer (MS/MS) A type of mass spectrometer named for the manner in which ions are separated within the mass analyzer. Ions are separated in a strong magnetic field so that ions of a certain mass-to-charge ratio reach the ion collector. Some magnetic sector instruments are capable of high-resolution mass spectrometry and are very expensive.

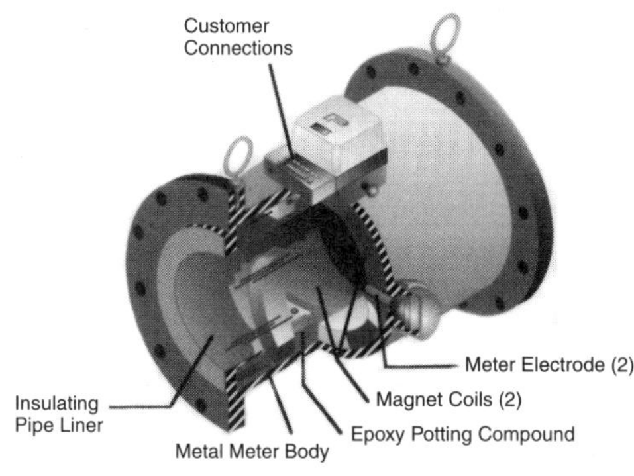

Magnetic flowmeter

magnetic separation A physical treatment process for removing magnetic suspended solids from a liquid by applying a magnetic field.

magnetic stirrer A device used for mixing chemical solutions in the laboratory.

magnetite (Fe_3O_4 or $FeFe_2O_4$) The mineral name for the black, mixed ferrous–ferric iron solid of the formula Fe_3O_4, or $FeFe_2O_4$. The name is indicative of the frequent ability of grains of magnetite solid to be separable from other materials in a scale by the use of a magnet. Magnesium, zinc, manganese, and nickel often substitute for the ferrous iron in the crystal structure. Magnetite is found in layers in many corrosion deposits formed on iron pipe; a relatively high pH, low carbonate concentration, and reducing local environment favor its formation.

main A pipe that transports or distributes water from the supply system to the service lines of a water customer.

main extension Additional distribution piping connected to an existing conduit, line, or main that is already providing service.

main tapping The process of connecting a corporation cock to a water main, including drilling a hole, threading the pipe wall, and tightening the corporation cock. The tap may be made directly to the main or through a tap saddle or sleeve that goes around the pipe and has threads for the corporation stop to be attached. *See also* corporation cock.

main-line meter A water meter installed on a large main of the distribution system.

maintenance Repairs and general upkeep necessary for the efficient operation of physical plants, property, and equipment. Maintenance is not to be confused with replacement or retirement.

maintenance cost The expenses of labor and materials related to upkeep necessary for efficient operation of physical property.

major ions The following anions: bicarbonate (HCO_3^-), carbonate (CO_3^{2-}), chloride (Cl^-), and sulfate (SO_4^{2-}); and cations: calcium (Ca^{2+}), magnesium (Mg^{2+}), potassium (K^+), and sodium (Na^+).

makeup carbon Fresh granular activated carbon that must be added to a column adsorption system after a regeneration cycle or when deemed necessary to bring the total amount of granular activated carbon to specification.

makeup water Treated water added to the water loop of a boiler circuit or cooling tower to make up for the water lost by steam leaks or evaporation.

MAL *See* maximum allowable level.

male end Obsolete term. Use *outside threaded connection* or *spigot end. See also* female end.

male fertility index In toxicity testing, the number of pregnancies that occur per number of confirmed couplings. This index is also referred to as the pregnancy ratio.

male-specific bacteriophage An outdated term used to designate F-specific (male-specific) bacteriophages of *Escherichia coli.* These bacteriophages have their initial attachment point to the bacterial cell somewhere on the F pilus. The F pilus (sex pilus) is a short hair-like projection that is involved in bacterial mating. F-specific deoxyribonucleic acid bacteriophages and F-specific ribonucleic bacteriophages exist.

malignant Very dangerous or virulent, causing or likely to cause death.

managed competition A process of limiting the number of individuals or companies allowed to be eligible for consideration or to bid to provide a service, product, or facility through establishment of prequalifying criteria, such as prescribed financial strength, specific expertise, or proven capability.

managed wetland A wetland that is actively managed by human intervention.

management efficiency In irrigation, a percentage or decimal fraction of the total water applied through irrigation, representing the portion beneficially applied through scheduling, maintenance, and repair of irrigation systems.

manganese (Mn) An abundant element found naturally in the earth. Dissolved manganese is found in many water supplies. At concentrations in excess of 0.05 milligrams per litre, it causes black stains to plumbing fixtures, laundry, and other items in contact with the water.

manganese bacteria Bacteria capable of utilizing dissolved manganese as an energy source and depositing it as hydrated manganic hydroxide.

manganese dioxide (MnO_2) A dark brown or gray-black insoluble compound found in nature as pyrolusite or made synthetically. It is used as an oxidizing agent in water treatment and as a starting material for such permanganate compounds as potassium permanganate ($KMnO_4$). *See also* potassium permanganate.

manganese dioxide-coated pumicite An oxidizing catalyst used to control iron and manganese. The manganese is sacrificial and thus disappears with time.

manganese greensand Greensand that has been processed to incorporate the higher oxides of manganese into its pores and onto its surface. Manganese greensand has a mild oxidizing power and is often used in the oxidation, precipitation, and removal of iron, manganese, hydrogen sulfide (H_2S), or a combination of these three. It is regenerated by solutions of potassium permanganate ($KMnO_4$). *See also* manganese; zeolite.

manganese removal The process of removing manganese from water. Manganese may have adverse health effects at elevated levels and, when oxidized, can form a black solid that can discolor fixtures and clothing.

manganese zeolite A synthetic gel zeolite (sodium alumino silicate) that has been converted to the manganese form by the same process used for manganese greensand and that is used for the same treatment applications as manganese greensand. Manganese zeolite is regenerated with potassium permanganate ($KMnO_4$). *See also* manganese; zeolite.

manganite (MnO(OH)) A form of manganese ore that consists of manganic hydroxide and is used in filters designed to control iron, manganese, or hydrogen sulfide (H_2S). It requires a very high backwash rate because of its very high density (specific gravity 4.3). It is similar to pyrolusite.

manhole *See* personnel access opening.

manifold A pipe with several branches or fittings to allow water or gas to be discharged at several points. In aeration, manifolds are used to spray water through several nozzles.

manmade wetland *See* artificial wetland.

Manning formula A formula for determining the flow velocity in an open channel. In US customary units, the formula is

$$v = \frac{1.486}{n} R^{\frac{2}{3}} S^{\frac{1}{2}}$$

In Système International units, the formula is

$$v = \frac{1}{n} R^{\frac{2}{3}} S^{\frac{1}{2}}$$

Where:

v = the flow velocity, in feet per second (metres per second)
n = the Manning coefficient of channel roughness
R = the hydraulic radius, in feet (metres)
S = the channel slope (for uniform flow) or the energy slope (for nonuniform flow), dimensionless

The energy slope is calculated as $-dH/dx$, where H is the total energy, which is expressed as

$$H = z + y + \frac{v^2}{2g}$$

Where (in any consistent set of units):

z = elevation head
y = water depth
v = velocity
g = gravitational constant
x = distance between any two points

Manning roughness coefficient The roughness coefficient used in the Manning formula for determining the discharge coefficient in the Chezy formula. *See also* Chezy formula; Manning formula.

Manning tables Tabulations of sets of values that satisfy the Manning formula. *See also* Manning formula.

manometer An instrument for measuring pressure. Usually, a manometer is a glass tube filled with a liquid that is used to measure the difference in pressure across a flow-measuring device, such as an orifice or a Venturi meter. The instrument sometimes used to measure blood pressure is a type of manometer.

manual solution feed A method of feeding a chemical solution for small water systems such that the chemical is dissolved in a small plastic tank and transferred to another tank, from which it is fed to the water system using a positive-displacement pump.

map A two-dimensional representation of an area or surface for the purposes of delineating the relative locations of objects or attributes.

map projection A mathematical model used to transform positions on the surface of the Earth, which is curved, onto a flat map surface.

map scale The ratio of units of linear measurement on the map to units of measurement on the Earth. Scale is often stated as a representation fraction, such as 1:2,000, for which one part or unit of measurement on the map is equal to 2,000 parts or identical units of measurement on the Earth. Scale can also be stated in specific measurement units, such as 1 centimetre = 100 metres, for which 1 centimetre on the map is equivalent to 100 metres on the ground.

marble test A laboratory test for calcium carbonate ($CaCO_3$) solubility whereby a water sample is analyzed for alkalinity concentration and then exposed to calcium carbonate powder for at least 3 hours, settled, filtered, and analyzed again. If the alkalinity content increases, decreases, or remains the same, then the sample is undersaturated with calcium carbonate (may be corrosive), supersaturated (may deposit protective scale), or at equilibrium (stable), respectively.

margin of exposure In the US Environmental Protection Agency's proposed cancer guidelines, the means by which degrees of safety will be expressed for those compounds thought not to conform to the assumptions of the linearized multistage model. *See also* linearized multistage model; multistage model.

margin of safety The range between levels of human exposure and those levels shown to produce a measurable toxic response. The concept has generally been applied to chemical exposures for which a toxic threshold exists or may reasonably be thought to exist (for all effects except cancer). *See also* linearized multistage model; multistage model.

marginal cost The change in total (or variable) cost associated with producing the next unit of a product. In the water industry, the next increment of product evaluated is often the next 1,000 or 1 million gallons of water. This term is also often used

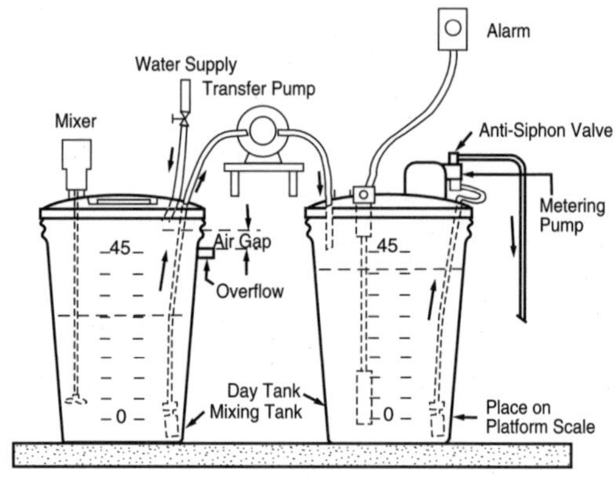

Manual solution feed installation

to refer to the cost of the next increment of supply. The terms *marginal cost* and *incremental cost* are often used interchangeably in water industry practice. *See also* long-run marginal cost; marginal cost pricing.

marginal cost pricing Strictly speaking, the pricing of all water sales at the cost of the next increment of supply, so that water users can decide whether they wish to spend those marginal cost dollars on water or on other products also priced at the margin. Under ideal conditions (rarely achieved in the water industry), marginal cost pricing leads to the most efficient allocation of resources. In the water industry, this term is often applied to the practice of charging a higher price for each increment of water consumed based on the cost to produce that unit. For example, the primary source of water might be surface water, produced and sold for $1.00 per 1,000 gallons. To meet summer demand, remote wells might have to be operated at a cost of $1.20 per 1,000 gallons; in this case water use in excess of the average-month level (or some multiple of the average-month level—say 1.2 times the average, where the production break point is located) is billed at the higher rate. A third increment of supply could result in yet another marginal cost rate block.

marine environment An environment, including the adjacent shoreline and wetlands, where the salinity is greater than 5 parts per thousand (5,000 milligrams per litre) and is in direct contact with waters reaching an ocean.

marine salt An early chemical name for salt.

market basket analysis A regular sampling of foodstuffs for a variety of environmental contaminants. The US Food and Drug Administration conducts national market basket surveys, sampling foodstuffs consumed by the average US family as a basis for estimating the range of exposures to a variety of environmental contaminants. Unfortunately, this approach is not truly representative sampling.

Marsh funnel viscosity The number of seconds required for 1 quart (946 millilitres) of a given fluid to flow through a Marsh funnel. This value is used to determine proper viscosities for drilling fluids.

marsh A tract of soft, wet land, usually vegetated by reeds, grasses, and occasionally small shrubs. *See also* swamp; tidal marsh; wetland.

MAS *See* multiple-address system.

masonry dam A dam constructed of stone set in mortar or in concrete. Masonry may be qualified as rubble masonry, cyclopean masonry, or concrete masonry.

mass The quantity of matter contained in a particle or body regardless of its location in the universe. Mass is constant (unless relativity effects are considered), whereas weight is affected by the distance of a body from the center of the Earth (or of any other planet or satellite). *See also* weight.

mass balance An accounting of all the masses of reactants and products in a reaction or all constituents in water being treated. For a given constituent in water entering a treatment process or blending in multiple flow streams, the mass of a constituent in water entering the system equals its mass exiting the system. The sum of each input stream constituent concentration multiplied by its flow rate equals the sum of the constituent concentrations multiplied by flow rates for all outputs from the process, assuming no mass accumulates in the system (i.e., the system is at steady state).

mass curve *See* mass diagram.

mass density units In the US customary system, pounds mass per cubic foot. In Système International, kilograms per cubic metre.

mass diagram A diagram, curve, or graph plotted with rectangular coordinates and representing a summation (integration) of all preceding quantities up to a point. Each ordinate value is equal to the sum of preceding terms in the series, with the corresponding abscissa value representing elapsed time or some other appropriate variable. Such diagrams are used extensively in storage and regulation studies pertaining to stream flow and water supply systems. A mass diagram is also called a mass curve. *See also* mass runoff; Rippl diagram.

mass loading The total amount of mass of a constituent flowing into a system.

mass movement A unit movement of a portion of the land surface, as in creep, landslide, earth flow, or subsidence.

mass runoff The total volume of runoff over a specified period of time. Successive summations are frequently plotted against time to produce a mass diagram. *See also* mass diagram.

mass spectrometer (MS) An instrument that can aid in the determination of molecular structure. In water analyses, mass spectrometers often serve as detectors after analytes are separated in a gas or liquid chromatograph. Several types of mass spectrometers are available; they are usually classified by the manner in which charged particles are separated within the instrument. For example, the most commonly available mass spectrometers are quadrupole (including ion trap), magnetic sector, and time-of-flight instruments. *See also* ion trap mass spectrometer; magnetic sector hybrid mass spectrometer; quadrupole mass spectrometer; time-of-flight mass spectrometer.

mass spectrometer–mass spectrometer (MS–MS) The combined use of two mass spectrometers, providing a powerful analytical technique. Ion fragments separated in the first mass spectrometer are directed into the second mass spectrometer for the purpose of further fragmentation and separation of masses. This technique is also called tandem mass spectrometry.

mass spectrometer interface The linkage between a mass spectrometer and another instrument. Typically, the other instrument, such a gas or liquid chromatograph, acts as a device to separate analytes. Interfaces are important devices because of the different environments the analytes encounter in each instrument. For example, analytes separated in a gas chromatograph are forced through a chromatographic column via a pressurized carrier gas. Analytes then travel into the vacuum of a mass spectrometer.

mass spectrometer, ion trap *See* ion trap mass spectrometer.

mass spectrometer, magnetic sector hybrid *See* magnetic sector hybrid mass spectrometer.

mass spectrometer, quadrupole *See* quadrupole mass spectrometer.

mass spectrometry A method of chemical analysis in which compounds emerging from a gas chromatograph are fragmented and ionized by bombardment with a beam of electrons. An electromagnetic field separates the ions according to their individual mass-to-charge ratios into a characteristic mass spectrum for each molecule. An analog computer analyzes the spectra and makes it possible to identify molecules even in cases of poor separation on the chromatography column, hence the advantage of mass spectrometry compared to selective chromatograph detectors. *See also* gas chromatography.

mass transfer The amount of molecules of a substance in motion to and across an interface from one phase to another, e.g., an amount (mass) of ozone (O_3) that transfers from air, across the air–water interface, and into water, or the amount of organic material that transfers from water to a solid adsorption surface. The rate and amount of mass transfer can be increased by (1) enlarging the interface boundary by increasing the area of the interface or by rapid renewal or clearance of the interface; (2) increasing the concentration difference (which is the driving force) across the interface boundary; (3) increasing the length of time (contact time) the interface boundary exists; or (4) any combination of these. *See also* Fick's first law of diffusion.

mass transfer coefficient A constant of proportionality that is specific to an individual compound and is used in a mass transfer expression to determine equilibrium conditions between two phases. Mass transfer coefficients are determined experimentally; the units will depend on the nature of the mathematical expression and the phase transfer.

mass transfer zone (MTZ) The place in an adsorption or ion exchange bed where the concentration of adsorbate in solution is changing with depth. The concentration gradient corresponds to the gradual transition of the adsorbent or ion exchange resin from fresh (or virgin) to spent (or exhausted). Aeration or air-stripping columns also have mass transfer zones as the gas or liquid changes from undersaturated to the equilibrium concentration with height. *See also* critical depth.

mass transport *See* loading.

mass units In the US customary system, slugs; pounds mass. In Système International, kilograms. *See also* weight units.

master meter A large instrument (meter) located upstream of other smaller meters and used for water accounting, billing purposes, or both. A utility that does not wholesale water would consider the large meter on the effluent of its treatment plant as a master meter. A wholesale customer that has a large meter at the point of system delivery would consider that to be the master meter.

master plan A facility plan that depicts the major facilities that will be needed for a system in the future based on the anticipated growth over a certain time span. The plan gives a layout of system additions, which can be constructed in increments to meet the demands of growth and demand.

matched precipitation rates Equal water delivery rates of sprinkler irrigation heads of varying arc patterns within an irrigation circuit. Matched precipitation rates are important to achieve a uniform distribution of water.

matching A technique used in epidemiologic studies to make the study group (those exposed to or having the disease being studied) and comparison group (the unexposed or control group) comparable with respect to certain characteristics. Several kinds of matching can be employed, including frequency matching (making sure the frequency distribution of a selected characteristic is similar in the groups being studied) or individual matching (making sure each individual selected for the comparison group is similar, for that chosen characteristic, to each individual enrolled in the study group). In epidemiologic studies, matching is a method used to prevent confounding bias for one or more selected characteristics. For example, if smoking is thought to be a cause of confounding bias in a case–control study, each control selected for inclusion in the study would be matched with a case according to the case's smoking status. For each case who smokes, a control who smokes would be selected; for each case who does not smoke, a control who does not smoke would be selected. Matching cases and controls in this way removes potential study problems due to smoking's association with exposure or disease; thus, smoking will not result in confounding bias in that particular study. Matching for gender, race, or neighborhood is a common practice. However, matching for certain or several characteristics to prevent confounding bias in an epidemiologic study may not always be feasible. Other methods are available to assess and control confounding bias during the analysis. *See also* confounding bias.

material safety data sheet (MSDS) Information on the use, handling, and storage of specific chemicals or products. Material safety data sheets contain mandated types of information concerning physical characteristics, reactivity, required personal protective equipment, and other safeguards.

mathematical model (1) In risk assessment, a model used to perform extrapolations from the effects at one dose to the effects at another dose. (2) Expresssing a phenomenon with a mathematical equation for the purpose of predicting an outcome.

mathematics A branch of science in which shape, quantity, and dependence are studied. Two main areas of mathematics are pure mathematics, dealing with the intrinsic study of mathematical structure, and applied mathematics, dealing with the study of physical phenomena.

mating index In reproductive toxicity testing, the number of confirmed matings per number of females mated per male animal. This index is also referred to as the mating ratio.

matrix diagram A method of organizing large amounts of information into sets of items to be compared. A matrix diagram shows the logical connecting point between any two or more items to illustrate relationships among them.

matrix effect An observation that the sample solution is affecting the performance of an analyte in a given analytical method. For example, solutes—other than the analyte—can have a positive or negative effect on the recovery of the analyte. When this phenomenon is observed experimentally, a matrix effect has occurred. Such observations are important because they can account for bias in an analysis.

matrix modifier A reagent used most commonly in graphite furnace atomic absorption spectrophotometry. The purpose of the matrix modifier is to delay atomization of the analyte metal as long as possible. This delay can result in a more intense signal from the analyte and can minimize interferences in a given determination.

mature river A stream with a slope that is so mild that the tractive forces are barely sufficient to carry debris delivered by the tributaries further downstream.

mature valley A stream valley that has reached a state of development at which cutting of the bottom has practically ceased.

Such a valley has a flat bottom with a low gradient and numerous tributaries of considerable length.

Maui-type well A horizontal water collection tunnel that skims off fresh water. Such wells are common in Hawaii.

maximally tolerated dose (MTD) The highest dose rate at which an experimental animal may be treated without any observable toxic reactions. In the absence of specific signs of toxicity, a decrease in body weight relative to control animals is used as an index. In general, the body weight decrement should be less than 10 percent, although earlier guidelines allowed greater decrements of body weight.

maximum allowable concentration (MAC) A concentration of a chemical in a medium (e.g., air, water, or food) that cannot be legally exceeded. In essence, this term is the standard for that medium.

maximum allowable level (MAL) The maximum concentration of a contaminant in drinking water that a single product is allowed to contribute under American National Standards Institute–NSF International standard 60 and 61 and NSF International standard 61, section 9.

maximum available water The quantity of water that can be readily extracted by a plant and used for normal growth, represented as the difference between field capacity and wilting point.

maximum computed flood The largest momentary flood discharge from a watershed believed possible based on a consideration of meteorologic conditions (such as probable maximum rainfall and snow cover) and geomorphic conditions (such as stream gradients or land slope).

maximum contaminant level (MCL) A value defined under the Safe Drinking Water Act Section 1401(3) as the maximum permissible level (concentration) of a contaminant in water delivered to any user of a public water system. Maximum contaminant levels are the legally enforced standards in the United States.

maximum contaminant level (MCL), primary *See* primary maximum contaminant level.

maximum contaminant level (MCL), secondary *See* secondary maximum contaminant level.

maximum contaminant level goal (MCLG) A federally nonenforceable, health-based goal established by the US Environmental Protection Agency under the Safe Drinking Water Act Section 1412(b)(4) for each contaminant regulated by a national primary drinking water regulation. The maximum contaminant level goal is set at the level at which no known or anticipated adverse effects on human health occur and for which an adequate margin of safety exists.

maximum demand (1) In electricity, the maximum kilowatt load that occurs and persists for a full demand interval during any billing period, usually a month. (2) The maximum flow per hour or maximum flow per day for a water utility.

maximum discharge The maximum rate of flow that a stream, conduit, channel, pipe, pump, or other hydraulic structure is capable of passing. *See also* discharge.

maximum flood flow The maximum rate of floodwater discharge from a given drainage basin resulting from a rainfall of high intensity, melting of snow, breaking of log or ice jams, failure of a structure, or similar conditions causing maximum release.

maximum flow The greatest volume of influent to a treatment plant within a given time period.

maximum mining yield The total volume of groundwater in storage in a particular source that can be extracted and used. *See also* groundwater safe yield.

maximum permissible concentration (MPC) The maximum level of radionuclides that an employee may be exposed to in an average 8-hour work day.

maximum probable precipitation Precipitation of a given amount and duration that can reasonably be expected to occur in a drainage basin.

maximum probable rainfall Precipitation as rain of a given amount and duration that can reasonably be expected to occur in a drainage basin.

maximum pumping rate (MPR) The maximum delivery rate of a pump during a specific demand period. The maximum pumping rate is most significant for systems with limited storage.

maximum residual disinfectant level (MRDL) A level of a disinfectant added for water treatment that may not be exceeded at the consumer's tap without an unacceptable possibility of adverse health effects. For chlorine (HOCl) and chloramines (NH_xCl_y, where $x = 0–2$, $y = 1–3$), a public water system will be in compliance with the federally proposed maximum residual disinfectant level when the running annual average of monthly averages of samples taken from the distribution system, computed quarterly, is less than or equal to the maximum residual disinfectant level. (That is, every 3 months, the average of the previous 12 monthly averages is assessed.) For chlorine dioxide (ClO_2), a public water system will be in compliance when no two consecutive daily samples taken at the entrance to the distribution system exceed the maximum residual disinfectant level. Maximum residual disinfectant levels will be enforceable in the same manner as maximum contaminant levels under Section 1412 of the Safe Drinking Water Act. Convincing evidence exists that addition of a disinfectant is necessary to control waterborne microbial contaminants. Notwithstanding the maximum residual disinfectant levels listed in Section 141.65 of the Disinfectant/Disinfection By-Products Rule (*Federal Register*, July 29, 1994), operators will be able to increase residual disinfectant levels of chlorine or chloramines (but not chlorine dioxide) in the distribution system to a level and for a time necessary to protect public health to address specific microbiological contamination problems caused by such circumstances as distribution system line breaks, storm runoff events, source water contamination, or cross-connections. *See also* Disinfectant/Disinfection By-Products Rule; maximum contaminant level; maximum residual disinfectant level goal.

maximum residual disinfectant level goal (MRDLG) A maximum level of a disinfectant added for water treatment and for which no known or anticipated adverse effect on human health would occur, allowing for an adequate margin of safety. Maximum residual disinfectant level goals are nonenforceable health goals and do not reflect the benefits of adding the chemical for the control of waterborne microbial contaminants. *See also* maximum residual disinfectant level.

maximum storage-replenishment rate analysis An analysis to determine whether a water system can adequately replenish storage facilities during low-demand periods. Such an analysis is also known simply as replenishment analysis.

maximum stream flow A stream's maximum rate of discharge during a specified period.

maximum sustained yield The maximum rate at which groundwater can be withdrawn perennially from a particular source. *See also* groundwater safe yield.

maximum THMFP *See* maximum trihalomethane formation potential.

maximum tolerated dose (MTD) *See* maximally tolerated dose.

maximum trihalomethane formation potential (maximum THMFP) The amount of trihalomethanes formed during a test in which a source or treated water is dosed with a relatively high (and possibly excessive) amount of chlorine (HOCl, typically in excess of what is required to achieve a reasonably low residual) and incubated (stored) under conditions that strongly encourage trihalomethane production (e.g., pH 9.2 or higher, warm water temperature, incubation time of 7 days or longer, or a combination of all three). This test is not a measure of the amount of trihalomethanes that would form under normal drinking water conditions, but rather an indirect measure of the amount of precursors in a sample. In regular testing for trihalomethane formation potential, precursors that can "readily" be converted to trihalomethanes will be converted. Under maximum trihalomethane formation potential conditions, "recalcitrant" precursors will yield trihalomethanes, or "intermediate" disinfection by-products initially formed will be converted to trihalomethanes, or both. Note that the maximum trihalomethane formation potential is not the same thing as the maximum trihalomethane potential. *See also* disinfection by-product precursor; maximum trihalomethane potential; trihalomethane; trihalomethane formation potential.

maximum trihalomethane potential (MTP) A value defined in the 1979 trihalomethane regulation (*Federal Register*, 44(231): 68624–68707) as the maximum concentration of total trihalomethanes that could be produced in a given water containing excess free chlorine (HOCl) after 7 days of incubation at a temperature of 77° Fahrenheit (25° Celsius). In the 1979 regulation, measurement of this parameter could be used in obtaining reduced monitoring requirements. This parameter was designed to measure the maximum amount of total trihalomethanes that could form in a distribution system—as opposed to the maximum trihalomethane formation potential, which represents the level of trihalomethane precursors in a water. The Disinfectant/Disinfection By-Products Rule bases criteria for reduced monitoring considerations on a sample collected in the warmest month at the longest detention time in the distribution system. *See also* maximum trihalomethane formation potential; trihalomethane.

maximum water density The density of water at 39.16° Fahrenheit (3.98° Celsius), i.e., 1.000000 grams per millilitre. Water becomes less dense at both higher and lower temperatures.

maximum water-holding capacity *See* field capacity.

maximum-day demand (MDD) The volume of water consumption used on the highest-consumption day in a year.

maximum-hour demand (MHD) *See* peak-hour demand.

MB *See* megabyte *in the Units of Measure list.*

MBAA *See* monobromoacetic acid.

MBAS *See* methylene blue active substance.

MCAA *See* monochloroacetic acid.

MCL *See* maximum contaminant level.

MCLG *See* maximum contaminant level goal.

MDD *See* maximum-day demand.

MDL *See* method detection limit.

mean *See* arithmetic mean.

mean absolute percentage error An error value determined by one of the tests used to measure the goodness of fit in regression analysis. Calculating this value involves summing the absolute value of the deviation between the predicted value for a series of observations and the actual value and then dividing by the sum of the dependent variable. The mean absolute percentage error in water demand forecasting typically ranges between 3 percent and 10 percent when good forecasting data are available. Note that if the actual plus-or-minus deviations were summed, their sum would equal or be very close to zero and thus would not give any indication about the magnitude of the forecast errors (hence the use of absolute values).

mean annual precipitation The average over a period of years of the annual amounts of precipitation.

mean annual runoff The average over a period of years of the annual amounts of runoff discharged by a stream.

mean depth The average depth of water in a stream channel or conduit, equal to the cross-sectional area divided by the surface width.

mean flow The arithmetic average of the discharge at a given point or station on the line of flow for some specified period of time.

mean range of tide (1) The average difference between the heights of high and low water at any given place. (2) The difference in the elevations of mean high water and mean low water.

mean sea level (MSL) The mean plane about which the tide oscillates; the average height of the sea for all stages of the tide.

mean tide level The plane halfway between mean low water and mean high water in oceans and tidal rivers.

mean velocity The average velocity of a stream flowing in a channel or conduit at a given cross section or in a given reach. It is equal to the discharge divided by the cross-sectional area of the section or by the average cross-sectional area of the reach. Mean velocity is also called average velocity.

mean velocity gradient ($\bar{G}$) A measure of the intensity of mechanical agitation used in rapid mixing and flocculation. It is defined as

$$\bar{G} = \sqrt{\frac{W}{\mu}}$$

Where:

$\bar{G}$ = mean velocity gradient in seconds^{-1}

W = the power input per unit volume of liquid, in foot-pound force per second per foot cubed (watt per metre cubed)

μ = the absolute viscosity of the liquid, in pound force second per foot squared (newton second per metre squared) *See also* flocculation; velocity gradient.

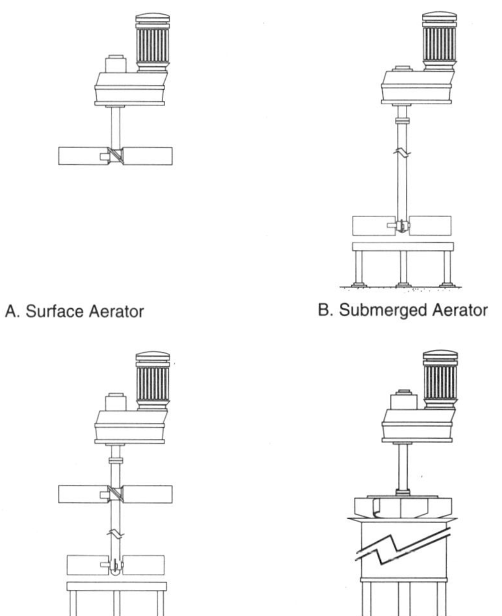

Mechanical aerators

mean velocity position (1) In flow with a free surface, the point on a vertical section at which the actual velocity is equal to the mean velocity. (2) The location of a point along a line or in a cross section where the actual velocity is equal to the mean velocity.

meander One of a series of regular, freely developing sinuous curves, bends, loops, turns, or windings in the course of a stream.

meander belt That part of the valley floor across which the stream shifts its bed naturally as its meanders shift their positions.

meander length The distance along the meander line between the corresponding points at the extreme limits of successive fully developed meanders.

meander line A curve that follows a body of water along the locus of the bank or shoreline at the elevation of mean or ordinary high water.

meander width The amplitude of swing of a fully developed meander, measured from midstream to midstream.

measured variable A factor (flow, temperature) that is sensed and quantified (reduced to a reading of some kind) by a primary element or sensor. *See also* process variable.

mechanical aeration The process of introducing air into a liquid by the mechanical action of paddle, paddle wheel, spray, or turbine mechanism.

mechanical aerator A mechanical device for the introduction of air into a liquid. *See also* mechanical aeration.

mechanical agitation *See* mechanical aeration.

mechanical analysis The operation of determining the distribution of grains of a granular material in accordance with size. The relatively large grains are separated by the use of sieves of various mesh sizes; the material that passes through mesh of one size and is retained on the sieve of the next smaller size is considered to be within one size range or bracket. The distribution is generally expressed in terms of a percentage by weight of materials within each specified size limit. Other methods are used for separating silt and clay.

mechanical control The application of artificial structures, such as terraces, dams, retards, or baffles, for the purpose of reducing erosion. Such control contrasts with vegetative control, in which ground cover is used for such purposes.

mechanical filter A pressure or gravity filter designed to physically separate and remove suspended solids from a liquid by mechanical (physical) means such as straining rather than by chemical means.

mechanical filter respirator A type of respirator that protects against (i.e., filters out) airborne particles, such as dust and mist. It does not give protection from vapors, fumes, or gases.

mechanical joint A flexible connection of two pipes or fittings with a gasket compressed by lugs and bolts.

mechanical pipe joint Any form of flexible joint involving lugs and bolts.

mechanical rake A machine-operated mechanism used for cleaning debris from racks located at the intakes of conduits supplying water to hydroelectric power plants, water supply systems, or other uses.

mechanical seal A mechanical device or assembly that forms a seal between a pump casing and a rotating shaft to prevent leakage along the shaft.

mechanical ventilation Equipment that exchanges the air in a room to remove airborne contaminants.

mechanically cleaned screen A screen equipped with a mechanical cleaning apparatus for removing retained solids.

mechanism of action The entire causal chain of a chemical's impact on an organism's health, from molecular interactions of the chemical with the biological system to describing how the ultimate pathology has to be characterized. *See also* mode of action.

mechanistic equation A mathematical expression derived from an understanding of the underlying mechanics of a phenomenon. For example, if C is the concentration of a reactant that decays according to a first-order chemical reaction, the concentration at any given time t can be predicted by using the first-order chemical reaction model, $C = C_0e^{-kt}$, where C_0 is the initial concentration at time zero and k is the rate constant. In this example, the equation can be called a mechanistic model because it is derived from an understanding of the reaction mechanism.

MED *See* multiple-effect distillation.

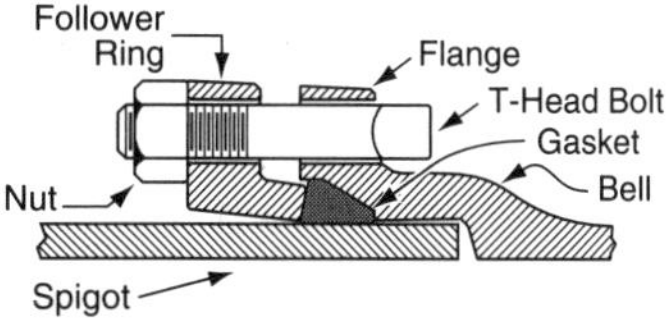

Mechanical joint

media A selected group of materials used in filters and filter devices to form barriers to the passage of certain solids or molecules that are suspended or dissolved in water. The term *media* is commonly used to refer to ion exchange resin products, granular filtration material, or granular adsorption material. *Media* is the plural form of *medium* and refers to more than one type of barrier material.

median A statistical term representing the middle value of a series of numbers when the numbers are arranged in order of their numerical quantity or value. Equal numbers of data points can be found that are greater than or less than the median. For example, in the sequence 1,3,4,7,8,10,12, the median is 7.

median stream flow The rate of discharge for which equal numbers of greater and lesser flow occurrences exist during a specified period.

median tolerance limit In toxicological studies, the concentration of pollutants at which 50 percent of the test animals can survive for a specified period of exposure, usually 96 hours. It is also called the lethal dose 50.

median toxic dose The dose at which 50 percent of test animals exhibit an adverse response. In general, this term is used to quantify the potency of a chemical for producing a nonlethal but clearly adverse effect on an organism.

mediation The process of using a third party to help resolve a dispute via a negotiated resolution.

Mediterranean climate A climate that occurs on the western shores of the populated continents, characterized by moderate temperatures throughout the year, an annual drought, and a rainy season.

medium *See* media.

medium sand Sediment particles having diameters between 0.250 and 0.500 millimetres.

medium-size water system A water system that serves more than 3,300 and less than 50,000 people.

medium-specific concentration (MSC) A risk-weighted concentration limit based on the substance-containing medium (e.g., air, water, soil) and the exposure pathway (e.g., inhalation, ingestion).

medium-water-use plants Plants requiring 30 to 50 percent of reference evapotranspiration to maintain optimal appearance.

meg *See* megger; megohm *in the Units of Measure list.*

mega (M) Prefix meaning 10^6 in Système International.

megabyte (MB) *See the Units of Measure list.*

megacycles *See* megahertz *in the Units of Measure list.*

megahertz (MHz) *See the Units of Measure list.*

megalitre (ML) *See the Units of Measure list.*

megaohm (meg) *See the Units of Measure list.*

megger (meg) An instrument used for checking the insulation resistance on motors, feeders, bus bar systems, grounds, and branch circuit wiring.

melanoma A tumor that is made up of cells containing melanin. In common usage, the term is applied without qualification to malignant melanoma, a very aggressive tumor found in the skin. These tumors commonly arise from dark moles (nevi). Arsenic is associated with skin cancer, but squamous cell carcinomas, not melanomas, are generally formed.

melt water Water that is formed from the melting of snow, rime (white frost), or ice resulting from atmospheric processes.

membrane A natural or synthetic permselective material. Membranes used for water treatment are commonly synthetic organic polymers. In the case of pressure-driven membranes, the polymers are permeable to water (solvent) but reject solutes; for electrodialysis membranes, the polymers are permeable to ions but not to water.

membrane air-stripping A membrane process using a microporous, hydrophobic membrane that is permeable to volatile substances but not to water. The process is driven by a concentration gradient between water on one side of the membranes and air on the other side. Membrane air-stripping can be used for removing volatile organic chemicals.

membrane area The total active area of a permselective membrane.

membrane backflush The process of reversing the flow of water, gas, or both through a membrane from the filtrate side toward the feed side to remove retained substances. This process is typically applicable to microfiltration and some ultrafiltration membranes.

membrane bundle A grouping of hollow fiber or hollow fine membranes placed inside a pressure vessel.

membrane cartridge A spiral wound membrane element. *See also* membrane element.

membrane cleaning The process of removing unwanted substances from membranes by exposing them to one or more cleaning solutions. Typically, cleaning solutions are circulated through the membrane modules without the membranes being physically removed from the treatment unit.

membrane cleaning solution A special chemical solution prepared for membrane cleaning. Typically an acidic solution (e.g., citric acid) is used for carbonate scale and metal hydroxide removal, and alkaline solutions with detergents are used for organic foulant removal. Chelating agents and disinfectants are also commonly used.

membrane concentrate The reject from a membrane filter. Membrane concentrate can be passed through additional membranes to create more product water and higher reject concentrations, but ultimately the concentrate must be disposed of.

membrane configuration The arrangement of membranes in a module, e.g., in a spiral wound, hollow fiber, tubular, or plate-and-frame configuration.

membrane destructive analysis The process of disassembling a membrane element or module for diagnosis, testing, and removal of foulant, scale, or both such that, for practical purposes, the membrane module can no longer be reassembled and put back into service.

membrane distillation A temperature-driven membrane-desalting process whereby a hydrophobic membrane is used for separation of water vapor, which passes through the membrane and condenses to form purified distillate.

membrane element Flat sheet membranes and spacers formed into a spiral wound shape around a central permeate tube and wrapped in fiberglass or tape. One or more such membrane elements are placed inside pressure vessels for operation. *See also* membrane leaf.

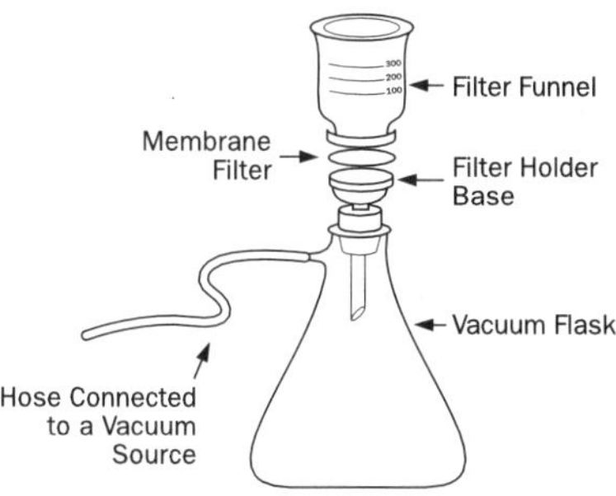

Membrane filter apparatus

membrane element autopsy The disassembly of a spiral wound membrane element or module for diagnostic purposes, commonly to inspect for foulants, scale, or both and to collect samples for analysis. *See also* membrane destructive analysis.

membrane filter (MF) (1) A filtration device used to remove particulate materials (including microorganisms) from a sample of water or other fluid. Membrane filters are used either to ensure that the filtered fluid (filtrate) is free of specific-sized contaminants or to permit analysis of the particulates that were removed from the filtrate. Membrane filters are manufactured to specific nominal pore sizes and are made of a variety of materials depending on the intended application. Membrane filters used in bacteriological analyses are placed on culture media appropriate for detecting the bacteria of interest. (2) A microfiltration or ultrafiltration pressure-driven membrane that rejects particles and produces filtrate.

membrane filter (MF) method *See* membrane filter technique.

membrane filter (MF) technique A microbiological technique used for the analysis of water for the presence of specific bacteria. One of the more common membrane filter techniques is the analysis of water for coliform bacteria, including fecal coliforms and *Escherichia coli*.

membrane filtration Filtration using a microfiltration or ultrafiltration membrane. *See also* membrane filter.

membrane flushing The displacement and removal of concentrated substances from a membrane module following system shutdown.

membrane flux The rate of water passage through a membrane filter, expressed in units of volume per time per membrane area.

membrane flux retention coefficient For some pressure-driven membrane processes, a dimensionless factor that is equal to the fraction of initial productivity remaining after a period of operating time (accounting for membrane compaction). It can be multiplied with the permeate flow rate at a standard condition to calculate permeate flow rate at a given pressure, temperature, or time condition. *See also* compaction.

membrane fouling The loss of membrane flux as a result of material being retained on the surface of the membrane or within the membrane pores. Membrane fouling may be reversible or irreversible. Flux loss by reversible fouling can be recovered by periodic backflushing of the membrane surface. Flux loss by irreversible fouling cannot be recovered; ultimately the production capacity of the membrane becomes limited.

membrane integrity The state of a membrane module such that no imperfections in the membranes or sealing components exist that could compromise the treatment process removal capability beyond rated limits.

membrane leaf For spiral wound membrane elements, the combination of two flat membrane sheets arranged back-to-back with a permeate water carrier between them and sealed on three edges, forming an envelope shape. The open fourth side is attached to a permeate water collection tube. One or more leafs, separated by feed–concentrate spacers, are rolled into spiral wound elements (modules).

membrane life The length of time a membrane performs well enough to remain in service. A membrane life is said to be over when the membrane has an insufficient treatment capability, requires cleaning too frequently, or for some other performance-related reason.

membrane module One or more membrane elements or a membrane bundle housed inside a pressure vessel. For hollow fiber or hollow fine fiber membrane products, a membrane module is sometimes called a permeator.

membrane packing density The number of hollow fiber membranes or the total usable membrane area per unit volume of a membrane module.

membrane permeability coefficient For pressure-driven membrane treatment systems, a measure of the capability of a membrane to allow passage of water (solvent), sometimes expressed in (1) gallons per day of permeate per square foot of membrane area per pound-per-square-inch of net driving force or (2) grams per second per square centimetre per kilopascal. *See also* water transport coefficient.

membrane poisoning Absorption of ions of low mobility into a membrane, resulting in an inability to transport salt ions.

membrane posttreatment Chemical treatment of a membrane after it is manufactured to improve performance.

membrane pretreatment *See* pretreatment.

membrane process A treatment technique using a membrane technology.

membrane product posttreatment (1) The treatment of product water (i.e., permeate or filtrate) from a membrane system to produce finished water meeting quality requirements. (2) The treatment of waste concentrate (i.e., concentrate posttreatment).

membrane reactor A device for chemical reactions using catalysts that are contained on or in a membrane.

membrane reject The material retained or filtered out by the membrane. *See also* membrane concentrate.

membrane selectivity The ability of a membrane to allow passage of only cations, anions, or molecules of a certain size. *See also* dialysis; electrodialysis.

membrane shock treatment A process for disinfection or sterilization of a membrane using intermittent addition of a biocide for a short duration.

membrane softening Removal of water hardness ions (calcium, magnesium) by nanofiltration or, in some cases, by a combination of nanofiltration and reverse osmosis. *See also* nanofiltration; reverse osmosis.

membrane stack In an electrodialysis–electrodialysis reversal system, the vertical arrangement of multiple membrane cell pairs between electrodes.

membrane storage solution A chemical solution used for preventing biological growth and preserving a membrane while not in service. Such a solution is often used when membranes are taken out of service for an extended time period. In some cases, these solutions can be used to protect membranes from freezing.

membrane, ion exchange *See* ion exchange membrane.

m-endo medium A type of selective culture medium used for detecting and recovering coliform bacteria from drinking water and other environmental water samples. *See also* coliform bacteria.

meniscus The curved surface of a column of liquid (water, oil, mercury) in a small tube. When the liquid wets the sides of the container (as with water), the curve forms a valley. When the confining sides are not wetted (as with mercury), the curve forms a hill or upward bulge.

meq *See* milliequivalent *in the Units of Measure list.*

meq/L *See* milliequivalents per litre *in the Units of Measure list.*

mercury (Hg) A metallic element. All inorganic compounds of mercury are highly toxic by ingestion, inhalation, and skin adsorption. Most organic compounds of mercury are highly toxic. Mercury has many industrial uses, e.g., in amalgams, electrical apparatuses, thermometers, or mercury vapor lamps. It is regulated by the US Environmental Protection Agency.

mercury gauge A gauge in which the pressure of a fluid is measured by the height of a column of mercury that the fluid pressure will sustain.

mercury intrusion porosimetry method A method of measuring the void volume of a bed of material by displacing the void space with liquid mercury. This method can be used with materials that have relatively large void spaces between their grains.

mercury vapor ultraviolet (UV) light The ultraviolet light given off as a result of an electron flow through an ionized mercury vapor between electrodes in an ultraviolet lamp. The mercury vapor ultraviolet wavelength that is most destructive to microorganisms in water is 254 nanometres.

merger A process of combining the physical components, managerial components, or both of two or more water systems to form one larger utility.

meromictic Of or pertaining to a lake for which the water is permanently stratified and therefore does not circulate completely throughout the basin at any time during the year.

merry-go-round A sequence of granular activated carbon columns, operated in series, in which the "lead" contactor is sequentially changed as its adsorptive capacity becomes exhausted. In other words, in a sequence of three contactors (labeled 1 through 3 in the initial sequence), column 1 is taken out of service after the granular activated carbon is completely exhausted (i.e., when the effluent concentration equals the influent concentration) and is regenerated. The second column then becomes the lead column, and the sequence becomes 2, 3, 1 after the first column is returned to service. The purpose of this arrangement is to make the most use of the adsorptive capacity of the granular activated carbon and thereby reduce the frequency of regeneration. If operated in parallel, each individual contactor would have to be regenerated at some specific target level, rather than at complete exhaustion. *See also* granular activated carbon; regeneration.

mesh One of the openings or spaces in a screen or woven fabric. The value of the mesh is usually given as the number of openings per inch. This value does not consider the diameter of the wire or fabric; therefore, the mesh number does not always have a definite relationship to the size of the hole.

mesh screen A screen composed of woven fabric of any of various materials.

mesh size The particle size of granular media as determined by the US Sieve Series. Particle size distribution within a mesh series is usually given in the specifications of a particular granular medium, e.g., 10 × 40 granular activated carbon.

mesophile An organism whose growth temperature is in the range of 68° Fahrenheit to 104° Fahrenheit (20° Celsius to 40° Celsius).

mesophreatophyte A phreatophyte that is resistant to neither alkali or drought.

mesophyte A plant that grows under medium or typical conditions of atmospheric moisture supply, as distinguished from one that grows under dry or desert or very wet conditions.

mesotrophic Pertaining to reservoirs and lakes that contain moderate quantities of nutrients and are moderately productive in terms of aquatic animal and plant life.

messenger ribonucleic acid (mRNA) An organic compound that contains the information necessary for the synthesis of proteins that function as structural components or as enzymes in cells. *See also* ribonucleic acid.

***meta* (*m*)** A prefix meaning "beyond." This prefix is used in organic chemistry in naming disubstitution products derived from benzene in which the second substituent atom or functional group is located on the third carbon atom with respect to the first substituent atom. This arrangement is also called the 1,3-position. *See also ortho*; *para.*

A depiction of 1-A-3-B benzene in which B is located in the meta position (on the third carbon) with respect to A

meta-analysis The process of using statistical methods to combine the results of different studies. In epidemiology, a frequent application has been the pooling of results from a number of small, randomized, controlled clinical trials; however, more recently meta-analysis has been used to pool the results of observational epidemiologic studies. Meta-analysis refers to a collection of techniques whereby the results of two or more independent studies are statistically combined to yield a single statistic that is claimed to have important descriptive or inferential properties. Systematic biases of individual epidemiologic studies are not corrected when the studies are pooled to yield a single measure of risk. Two approaches are used to summarize the published research on a given subject. First, a synthetic or aggregative meta-analysis provides a summary estimate of an effect (such as the relative risk), a summary test of statistical significance for the studies, or both. Second, an analytic, comparative, or explanatory meta-analysis examines whether differences among the studies can explain the differences among their results.

metabolism The biotransformation of various chemicals by an organism. These chemicals may be normal substrates for pathways that yield energy or are used for biosynthetic purposes, but the term is also applied to the biotransformation of chemicals, such as drugs and other synthetic chemicals, that are foreign to biological systems.

metabolite The end product of the biotransformation of various chemicals by an organism.

metal An element that forms positive ions when its compounds are in solution and for which the oxides form hydroxides rather than acids with water. Most metals are crystalline solids with metallic luster, are conductors of electricity, and have rather high chemical reactivity. Some metals are quite toxic.

metal hydraulic shoring system A preengineered shoring system composed of aluminum hydraulic cylinders (crossbraces) used in conjunction with vertical rails (uprights) or horizontal rails (wales). Such a system is designed specifically to support the sidewalls of an excavation and prevent cave-ins.

metal oxide semiconductor (MOS) A three-layered gate structure used in the manufacture of a field effect transistor. The configuration enables the device to have a higher power supply voltage fluctuation tolerance and to limit power use.

metal release Products of corrosion that do not remain attached to pipe walls but are released from the pipe to enter the water contacting the pipe. The forms of released metal may be dissolved (ionic), colloidal, particulate, or combinations of all three.

metal solvency The property of a metal that allows it to dissolve in water or another solution. Typical units for a metal dissolved in water are milligrams per litre or micrograms per litre.

metal uptake The absorption or incorporation of a metal by a water as the metal is released from a corroding surface.

metalimnion The middle layer in a stratified lake; another name for the thermocline.

metallothionein A low-molecular-weight protein that has a high content of cysteine ($HSCH_2CH(NH_2)COOH$), a sulfur-containing amino acid. The sulfhydryl groups in metallothionein bind various normal and toxic trace metals. Metallothionein is known particularly for its ability to bind cadmium. The cadmium–metallothionein complex becomes concentrated within the kidney and is largely responsible for the large accumulations of cadmium found within this organ after chronic exposure.

metal-rich lake sediments Bottom sediments in a lake or river containing high metal concentrations.

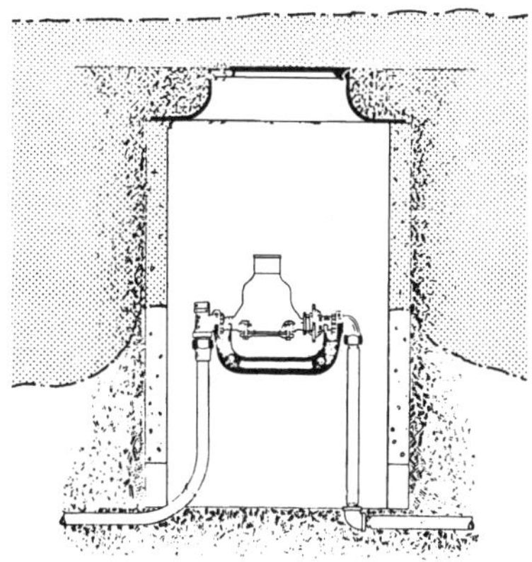

Meter box

metamorphic rock Rock that has been altered from its previous condition through the combined action of pressure and heat.

metamorphic water Water that has been associated with metamorphic rock during the rock's metamorphism.

metastasis A process whereby tumor cells are shed from a tumor and migrate to and colonize other sites within the body. This is a sign of extreme malignancy because these tumor cells are able to avoid all the body's constraints on growth.

metastatic Pertaining to the transfer of disease from one organ or part to another not directly connected with it.

meteoric water Water that is in or derived from the atmosphere. This term is sometimes used to include all subsurface water of external origin, and sometimes to include only water derived by absorption, excluding especially connate ocean water. *See also* interstitial water.

meteorograph An instrument that automatically records on a single sheet the measurements of two or more meteorological elements, such as air pressure, temperature, humidity, or precipitation.

meteorologic elements (1) Six quantities that specify the state of the weather at any given time and place: air temperature, barometric pressure, wind velocity (direction and speed), humidity, cloudiness (types and amounts of clouds), and precipitation. (2) In addition to the preceding quantities, any of the various other meteorological phenomena that distinguish the particular condition of the atmosphere, such as sunshine, visibility, radiation, halos, thunder, mirages, or lightning.

meteorological water Water that is in or derived from the atmosphere. *See also* meteoric water.

meteorology The science concerned with the atmosphere and its phenomena.

meter An instrument for measuring some quantity, such as the amount or rate of flow of liquids, gases, or electric currents. *See also* metre.

meter box The housing or container that encloses a water meter when the latter is installed in a water system. It provides protection for the meter and access to it for reading.

meter coupling A service fitting placed on each side of a meter to connect the service line to the meter, allowing the meter to be easily set and removed. *See also* meter gasket.

meter error The difference between an actual meter reading and a calibrated volume of water (or gas or electricity) put through the meter.

meter gasket A gasket that goes on each end of a meter, between the meter and the meter coupling. It seals when compressed by the meter coupling nut or flange against the meter. *See also* meter coupling.

meter gear train A series of gears within a meter to transmit the pulsations of the disk or piston to the register.

meter interface unit (MIU) An electronic device responsible for communication between a utility office and a meter. The meter interface unit generally reads the meter when given a signal to do so and retransmits the information to the utility. It is used in automatic meter reading and is a hand-held or automatically controlled device to extract the meter reading from the meter.

meter loss The amount of water (or gas or electricity) not accounted for because of a physical malfunctioning of the meter. Meter loss results in a direct loss of revenue for the utility.

meter pit The opening in a vault or enclosure that allows access to a meter and provides protection for it.

meter rate The charge for water based on the quantity used, as measured by water meters.

meter reader A person who visits commercial establishments and residences to read the water meters there, either in person or by remote sensing, for the purpose of determining the quantity of water used since the previous meter reading. This information is used to calculate the water bill. *See also* water bill.

meter reading The periodic noting or reading of the meter register that reflects the usage of water by a customer or the passage of water through the meter.

meter setter A manufactured combination of meter couplings, stops, service pipe, and check valves that enables a meter to be installed merely by connecting the service line and customer plumbing line. A meter setter allows varying depths of services and meter heights with a standard spacing for the meter; it can be used to raise a meter from the initial set location.

meter spud *See* meter coupling.

meter test bench Meter-testing equipment that allows a series of meters to be connected in a line. A measured volume of water at a certain rate of flow is allowed to flow through the meters. The accuracy of the meters can then be determined by comparing the known volume of water that passed through the meter with the amount registered on each meter.

meter vault The concrete structure or housing for a large meter, providing access for maintenance and reading.

meter, water *See* water meter.

metered system A system in which meters are used to measure the flow characteristics at all strategic points on main supply lines, pumping stations, reservoir outlets, connections to other political subdivisions, and at each consumer's service.

metering The process of measuring and recording the quantity of water passing a given point in a system.

metering pump A chemical solution feed pump that adds a measured volume of solution with each stroke or rotation of the pump.

metering system The equipment and processes to determine a customer's consumption of water.

meter-reading system A setup by which a water meter register is accessed and the reading recorded. The system can be manual, with a person reading the register and recording the numbers or entering the numbers in a hand-held device; the reading may be recorded directly to a receiver by direct connection; or the system can be automated by using telephone lines or radio waves to transmit a recording signal back to a central processor or over radio waves to a mobile receiver.

methane (CH_4) A gaseous hydrocarbon that occurs in natural gas and coal gas as a result of decaying vegetation and other organic matter in swamps and marshes. Methane is flammable, as well as asphyxiating. *See also* hydrocarbon.

```
     H
     |
 H — C — H
     |
     H
```

Methane—CH_4

methemoglobin An oxidized form of hemoglobin that is incapable of reversibly binding oxygen. Therefore, this conversion by oxidation decreases the oxygen-carrying capacity of the blood. The conversion of hemoglobin to methemoglobin is responsible for the "blue baby syndrome" produced by various oxidant chemicals, such as nitrite (NO_2^-) and chlorite (ClO_2^-). Excessive conversion of hemoglobin to methemoglobin produces anoxia in tissues and can be life-threatening. *See also* hemoglobin.

methemoglobin diaphorase *See* methemoglobin reductase.

methemoglobin reductase An enzyme responsible for reducing methemoglobin to hemoglobin. The major methemoglobin reductase is actually cytochrome b_5, which requires the reduced form nicotinamide-adenine dinucleotide (most commonly written as NADH) as a cofactor.

methemoglobinemia A condition of increased concentration of methemoglobin within the blood. This condition causes blue-baby syndrome.

method detection limit (MDL) As defined by the US Environmental Protection Agency in 40 CFR Part 136 Appendix B for use in regulatory programs, the minimum concentration of a substance that can be measured and reported with 99 percent confidence that the analyte concentration is greater than zero. In practice, it is a value calculated from the absolute precision of replicate determinations. Many regulatory programs make use of the method detection limit. The concept and application of the method detection limit have been vigorously debated in recent years. The method detection limit is often mistakenly called the minimum detection limit. *See also* limit of detection; limit of quantitation; practical quantitation level; reliable detection level; reliable quantitation level.

method of images A modeling technique to simulate the effects of hydraulic boundaries on pumping wells by the use of image wells. *See also* image well.

method of standard addition A technique sometimes used in atomic absorption to correct for certain nonspectral interferences without removing the interference. Calibration curves are constructed by spiking known amounts of the analyte into the sample. Determinations can then be made using the calibration curve constructed from the spiked sample matrix. One must use this technique with caution because it will not correct for all types of interferences.

methoxychlor ($Cl_3CCH(C_6H_4OCH_3)_2$) A generic name for 2,2-bis(p-methoxyphenyl)-1,1,1-trichloroethane, an insecticide that is regulated by the US Environmental Protection Agency.

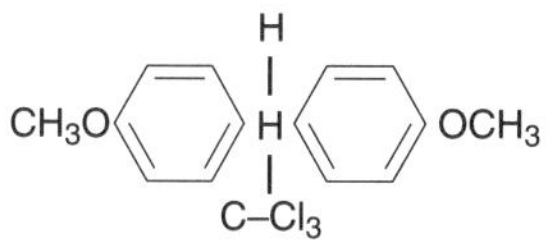

Methoxychlor

methyl ester (R–CO–O–CH_3) Any of a group of fatty esters derived from coconut and other vegetable oils. In addition, diazomethane ($H_2C{=}N^+{=}N^-$) is used to derivatize haloacetic acids, converting them to their corresponding methyl esters, so that they can be analyzed by gas chromatography. *See also* diazomethane; ester.

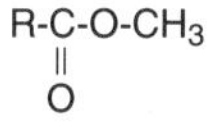

Methyl ester, where R is a carbon-containing group

methyl glyoxal (CH_3COCHO) An aldoketone created during the reaction of ozone (O_3) with natural organic matter. *See also* aldoketone; ozonation by-product.

methyl isobutyl ketone (MIBK) ($(CH_3)_2CHCH_2COCH_3$) A solvent for paints and varnishes. *See also* solvent.

methyl orange ($(CH_3)_2NC_6H_4NNC_6H_4SO_3Na$) A pH indicator dye that can be used in water analysis. In aqueous solutions with pH less than 3.1, the color is red. Solutions for which the pH is greater than 4.4 are yellow in color.

methyl orange alkalinity A measure of the total alkalinity of an aqueous suspension or solution. It is measured by the quantity of standard sulfuric acid (H_2SO_4) required to bring the water to a pH value of 4.5, as indicated by the change in color of methyl orange. It is expressed in units of milligrams per litre of equivalent calcium carbonate ($CaCO_3$).

methyl transferase Any of a group of enzymes that transfer methyl groups to biologically important substrates or to chemicals that are foreign to the body.

methylene blue active substance (MBAS) A substance that reacts with methylene blue as part of an analytical method used to estimate the concentration of anionic surfactants. Methylene blue is a cationic dye that reacts with certain substances to form an ion pair. The ion pair is extracted into an organic solvent such as chloroform ($CHCl_3$). The intensity of the extracted blue color is proportional to the concentration of methylene blue active substances. *See also* ion pair.

methylene blue number The number of milligrams of methylene blue adsorbed by 1 gram of activated carbon in equilibrium with a solution of methylene blue that has a concentration of 1.0 milligram per litre.

methylene chloride *See* dichloromethane.

methyl bromide (CH_3Br) A soil and space fumigant. *See also* fumigant.

2-methylisoborneol (MIB) ($C_{11}H_{20}O$) A common name for 2-exo-hydroxy-2-methylbornane (I), a musty–camphor-smelling chemical produced by certain blue-green algae and *Actinomycetes*. This odorous compound can be perceived at low nanogram-per-litre concentrations. Removal at a treatment plant requires either ozone (O_3) oxidation or activated carbon adsorption. *See also Actinomycetes*; blue-green algae.

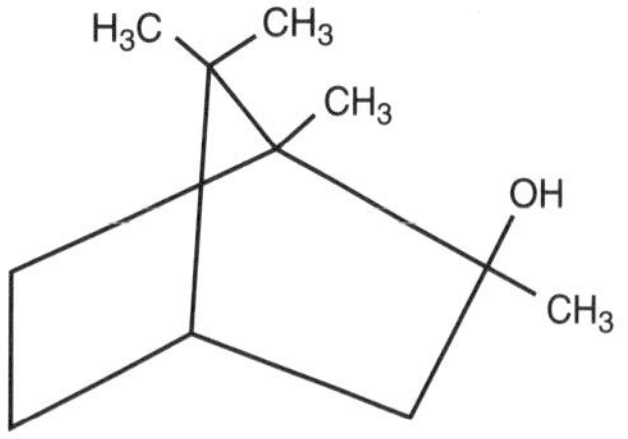

2-methylisoborneol (MIB)

2-methylphenol ($CH_3C_6H_4OH$) The chemical name for ortho-cresol, a synthetic organic chemical derived by fractional distillation of crude cresol from coal tar. This chemical has various industrial uses, e.g., as a disinfectant, fungicide, or insecticide. *See also* coal tar; disinfectant; distillation; fungicide; insecticide; synthetic organic chemical.

methyl-*tert*-butyl-ether (MTBE) ($(CH_3)_3COCH_3$) (1) An organic solvent with moderate polarity, fairly low boiling point, and fairly high vapor pressure. It has become more commonly used as a solvent in liquid–liquid extractions in water analysis laboratories, in part because of its use in specific analytical methods published by the US Environmental Protection Agency. (2) Octane booster for unleaded gasoline. MTBE is the most commonly used fuel oxygenate. MTBE has been detected in groundwater wells as a result of contamination from nearby leaking underground fuel tanks. Surface water supplies are also vulnerable to MTBE contamination, primarily through the use of recreational motorized watercraft. MTBE is highly soluble in water. It is not adsorbed by soil matter, nor does it volatilize readily. In addition, MTBE is not removed by conventional treatment techniques. The US Environmental Protection Agency is looking into the development of a regulation for MTBE in drinking water. In addition to health effects concerns, the aesthetic implications of MTBE in drinking water are of much interest.

4-methylumbelliferyl-β-D-glucuronide (MUG) A fluorogenic compound used in microbiological media for the detection of *Escherichia coli* in drinking water or other environmental samples. It is the substrate for the enzyme β-glucuronidase, an enzyme common to most *Escherichia coli* strains. When 4-methylumbelliferyl-β-D-glucuronide is hydrolyzed by β-glucuronidase, methylumbelliferone is released and fluoresces with a blue color under long-wave (366-nanometre) ultraviolet light. *See also* minimal medium; β-glucuronidase.

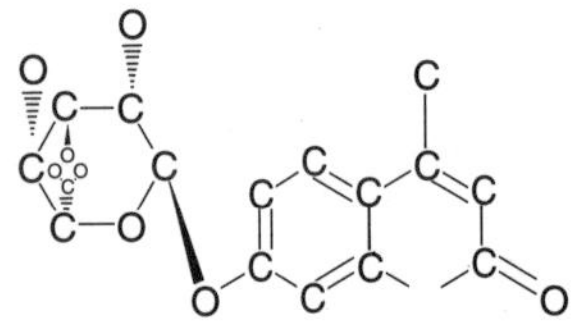

4-methylumbelliferyl-β-D-glucuronide

metolachlor ($(CH_3)C_6H_3(C_2H_5)$-$N(COCH_2Cl)CH(CH_3)CH_2OCH_3$) The common name for the herbicide 2-chloro-*N*-(2-ethyl-6-methylphenyl)-*N*-(2-methoxy-1-methylethyl)acetamide. *See also* herbicide.

metre *See the Units of Measure list.*

metribuzin ($C_8H_{14}N_4OS$) The common name for the herbicide 4-amino-6-(1,1-dimethylethyl)-3-(methylthio)-1,2,4-triazin-5(4*H*)-one. *See also* herbicide.

metric benchmarking A process of systematically tracking multiple operations over a period of time and comparing internal performances against those of similar companies in the same industry.

metric ton (t) *See the Units of Measure list.*

metric units A common name for Système International units.

metrication The conversion of values or expressions to metric (Système International) units.

MeV *See* million electron volts *in the Units of Measure list.*

MF *See* membrane filter; microfiltration.

MF technique *See* membrane filter technique.

MFLOPS *See* millions of floating point operations per second.

mg *See* milligram *in the Units of Measure list.*

mgd *See* million gallons per day *in the Units of Measure list.*

mg/L *See* milligrams per litre *in the Units of Measure list.*

MHD (maximum-hour demand) *See* peak-hour demand.

mho *See the Units of Measure list.*

mHz *See* megahertz *in the Units of Measure list.*

mi *See* mile *in the Units of Measure list.*

mi² *See* square mile *in the Units of Measure list.*

MIB *See* 2-methylisoborneol.

MIBK *See* methyl isobutyl ketone.

MIC *See* microbiologically induced corrosion.

micelle A unit or structure built up from complex molecules in colloids. It may have crystalline properties and is capable of increasing or decreasing in size without changing in chemical nature. The complex molecules have hydrophobic and hydrophilic characteristics.

Michaelis constant An apparent dissociation constant that can be determined by examining the effects of various concentrations of substrate on the rate of the substrate's conversion to product or the extent to which varying the concentrations of a ligand activates a response that is mediated by a receptor. *See also* dissociation constant.

Michaelis–Menten equation A mathematical expression to describe an enzyme-catalyzed biological reaction in which the rate of the reaction is described as a function of the concentration of the reactants. This equation is used in studies of enzyme kinetics and the study of any other saturable reaction; it takes the following general form:

$$v = \frac{V_{max}[S]}{K_m + [S]}$$

Where:

v = rate at any substrate concentration, in micromoles per litre per minute

V_{max} = maximum rate of enzyme activity at saturation, in micromoles per litre per minute

$[S]$ = substrate concentration, in micromoles per litre

K_m = substrate concentration when rate is half the maximum (i.e., $v = 0.5\ V_{max}$), in micromoles per litre

The rates described by this equation are dependent on the amount of enzyme (or catalyst) present in the system and the temperature at which the reaction is taking place.

micro (μ) Prefix meaning 10^{-6} in Système International.

microaerophile A microorganism that grows best under reduced oxygen tension.

microbe *See* microorganism.

microbial activity The activities of microorganisms resulting in chemical or physical changes.

microbial film A gelatinous film of microbial growth attached to or spanning the interstices of a support medium. Such a film is also called biological slime. *See also* extracellular polymeric substance.

microbial growth The activity and growth of microorganisms, such as bacteria, algae, diatoms, plankton, and fungi.

microbials Microbiological contaminants of any sort.

microbiocide A substance that kills or inactivates microorganisms.

microbiological Relating to microorganisms and their life processes.

microbiological analysis The use of various media, techniques, and equipment to determine the density or numbers of microorganisms in a sample.

microbiologically induced corrosion (MIC) Corrosion caused by the activity of microorganisms on a metal surface.

microbiology The study of microscopic organisms and their processes.

microbiota Collectively, the microscopic animals and plants of a region or habitat.

microchemical Pertaining to chemical reactions involving very small quantities of material.

microclimate The local, frequently uniform weather of a specific habitat or place.

microcystin-LR One of more than 50 microcystins, toxins produced by some members of the cyanobacteria (blue–green algae). Microcystin-LR is a hepatotoxin produced by the cyanobacterium *Microcystis aeruginosa* and is one of the most common of the cyanobacterial toxins found in water. It has been responsible for poisoning of both humans and animals. The proposed guideline value for this toxin in water in Canada is

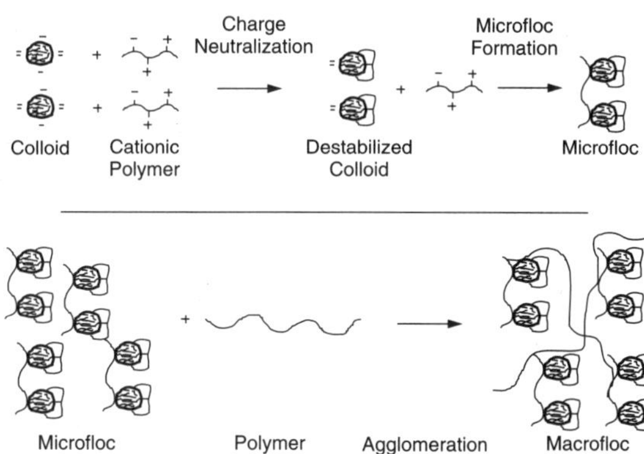

Microfloc and macrofloc formation

0.0015 milligrams per litre; no proposed guideline value exists in the United States at this writing.

microdiameter-depth adsorption system (MIDDAS) A system using "microcolumn" adsorbers containing small adsorbent particles to predict the performance of fixed-bed adsorbers and develop information for scaleup and design. *See also* rapid small-scale column test.

microextraction A liquid–liquid extraction technique in which the water-to-solvent ratio is fairly high. This term is also used to describe extraction systems in which a small volume of organic solvent is used.

microfiltration (MF) A pressure-driven membrane process that separates micrometre-diameter and submicrometre-diameter particles (down to an approximately 0.1-micrometre-diameter size) from a feed stream by using a sieving mechanism. The smallest particle size removed is dependent on the pore size rating of the membrane.

microfloc The initial floc formed immediately after coagulation, composed of small clumps of solids.

microflocculation A process that enhances the effectiveness of particle agglomeration and removal, typically associated with ozone (O_3) treatment. This flocculation phenomenon—in which larger but still microscopic particles are created by the collision of smaller particles—is a result of changes in surface chemistry resulting from ozonation; it can result in improved turbidity removal. *See also* flocculation; ozone.

microflora The microscopic plants present in a volume or area; the flora of a microhabitat.

microgram (μg) *See the Units of Measure list.*

micrograms per litre (μg/L) *See the Units of Measure list.*

microhm *See the Units of Measure list.*

micrometeorology The study of variations in meteorologic conditions over very small areas, such as hillsides, forests, drainage areas of rivers, or individual cities.

micrometre (μm) *See the Units of Measure list.*

micromho *See the Units of Measure list.*

micromole *See the Units of Measure list.*

micron *See the Units of Measure list.*

micron rating A measurement applied to filters or filter media to indicate the particle size for which any larger suspended solids will be removed. As used in the water treatment industry standards, it may be an absolute rating or a nominal rating. *See also* absolute filter rating; nominal filter rating.

micronucleus test A test for the ability of chemicals to damage chromosomes. It essentially involves detection of small remnants of nuclei containing damaged deoxyribonucleic acid that remain for short times after a cell containing chromosomal damage divides. This test is most frequently conducted in mice, where the number of micronuclei observed in polychromatic erythrocytes in the bone marrow are scored.

micronutrient A beneficial element that is needed only in trace amounts. The same element can sometimes be toxic in higher amounts. *See also* trace element.

microorganism A microscopic organism, either plant or animal, invisible to the naked eye. Examples are algae, bacteria, fungi, protozoa, and viruses.

microphyte A microscopic bacterium, plant, or smaller alga.

micropore *See* adsorption pores.

microporous membrane A microfiltration or ultrafiltration membrane with pores that treat water by sieving action.

MICROQL A generalized equilibrium chemistry modeling program derived from the early MINEQL programs, using similar mathematical architecture but programmed in the BASIC computer language. *See also* MINEQL.

microsand enhanced coagulation The use of small sand particles in a coagulation, flocculation, and sedimentation process to improve the settleability of the total particle mix. Particle settling may be hindered when higher coagulant dosages are used for improving organic removal in enhanced coagulation. Using microsand is one method to improve settling. After settling, the sand is separated from the residuals and is recycled to the process. *See also* enhanced coagulation; sand ballasted flocculation–sedimentation.

microscope An instrument used to observe materials and organisms too small to be seen with the unaided eye. Modern microscopes are of the compound variety because they have two sets of lenses, the ocular lenses and the objective lenses. Most laboratory microscopes have three objective lenses: low power, high power, and oil immersion.

microscope, brightfield *See* brightfield microscope.

microscope, electron *See* electron microscope.

microscope, fluorescence *See* fluorescence microscope.

microscope, light *See* light microscope.

microscope, phase contrast *See* phase contrast microscope.

microscope, scanning electron *See* scanning electron microscope.

microscope, transmission electron *See* transmission electron microscope.

microscopic Very small, generally with dimensions between 0.5 and 100 micrometres, and visible only by magnification with a microscope.

microscopic analysis The examination of water to determine the presence and amounts of plant and animal life, such as algae, diatoms, protozoa, and crustacea. Microscopic analysis is sometimes called microscopic examination.

microscopic particulate analysis (MPA) The use of any one of several methods to identify and size particles in water. Several versions of analytical methods are available that can identify *Giardia*, *Cryptosporidium*, algae, nematodes, and other microorganisms. Particles from large volumes of water are isolated onto a cartridge filter with a typical pore size of 1 micrometre. Microscopic particulate analyses are used to assess the performance of water filtration plants, as well as to help identify groundwater that may be under the influence of surface water.

microscopic particulate analysis particle counting *See* microscopic particulate analysis.

microscopy The use of a microscope; observation or investigation using a microscope.

microscopy, atomic force *See* atomic force microscopy.

microscopy, brightfield *See* brightfield microscopy.

microscopy, phase contrast *See* phase contrast microscopy.

microscopy, scanning confocal laser (SCLM) *See* scanning confocal laser microscopy.

microscopy, scanning electron *See* scanning electron microscopy.

microscopy, transmission electron *See* transmission electron microscopy.

microscreen A pretreatment device used to remove fine material such as filamentous algae. Screening media are typically stainless steel or polyester, and media openings are typically 20 to 30 micrometres.

microsiemens (μS) *See the Units of Measure list.*

microsomal enzyme An oxidase that occurs in the diffuse membranous structures within most cells, but primarily in the liver. In a laboratory procedure, when the cell structure is disrupted with homogenization of the tissue, these membranes form small vesicular structures referred to as microsomes. The term *microsomal enzyme* is generally applied to a variety of enzymes that make up the so-called cytochrome P450 superfamily. Other drug-metabolizing enzymes, chemical-metabolizing enzymes, or both are also found in the microsomes. The enzymes are involved in the metabolism of chemicals that are foreign to the body, but they also participate in the metabolism of a variety of natural substrates, most notably steroids. The activities of these enzymes are included in a variety of in vitro assays for genotoxic effects, usually in the form of the so-called S-9 fraction, which is the supernatant liquid obtained following centrifugation of a tissue (usually liver) homogenate at 9,000 times gravity. *See also* genotoxic; microsome.

microsomal inducing agent A chemical or other agent that will increase the synthesis of microsomal enzymes. Most agents do not indiscriminately stimulate the synthesis of all microsomal enzymes, but rather selectively stimulate the synthesis of particular forms that have specificity for certain types of substrates. *See also* microsomal enzyme.

microsome A small vesicular structure that is created upon the laboratory homogenization of cells in a way that empties the cellular contents. Microsomes represent the smooth endoplasmic reticulum, which are membranous structures that are easily observed by electron microscopy. The subcellular fraction that is referred to as microsomes contains the bulk of the enzymes that are involved in the metabolism of foreign chemicals to which the body is exposed.

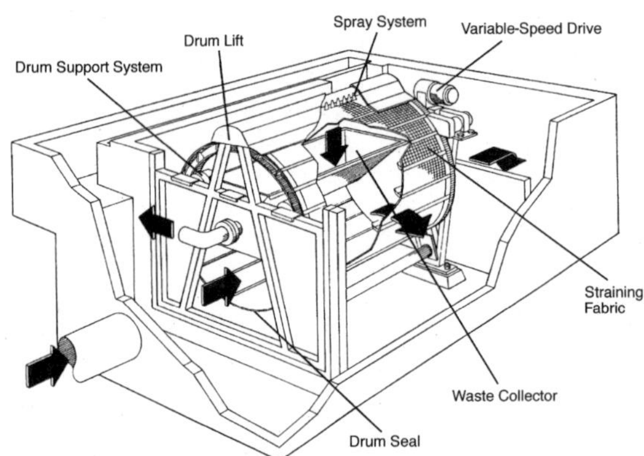

Microstrainer

microsporidia Small, unicellular, obligate intracellular, spore-forming (with spores of about 1–4.5 micrometres) protozoan parasites that are widely distributed in nature and include more than 100 genera and about 1,000 species. They have been recognized as pathogens of insects, fish, birds, and mammals, including humans. Gastrointestinal tract infections are characterized by severe diarrhea, cachexia (general ill health and malnutrition), weight loss, and epigastric pain. Infections have been recognized in immunosuppressed patients (e.g., AIDS patients) and in immunocompetent patients. Species found in humans are *Encephalitozoon cuniculi*; *E. hellem*; *E. intestinalis* (*Septata intestinalis*); *Enterocytozoon bieneusi*; *Nosema conneri*; *Pleistophora* spp.; *Trachipleistophora hominis*; and *Vittaforma corneum*.

microsporidium A "catch-all" genus name for microsporidia that have not yet been classified. *See also* microsporidia.

microstrainer A mechanical device used to remove fine material and particles from water. *See also* microscreen.

microstraining A process that removes very small particles, typically 20 to 30 micrometres across, but sometimes as small as 1 micrometre, often algal-related, from a water source prior to treatment. A screen having small openings, often affixed to a rotating drum, is used for microstraining. The mechanical device is often called a microscreen.

microtubule Any of a variety of microscopic tubular structures found in the cytoplasm of many types of eukaryotic cells. Microtubules are polymers of the protein tubulin and are also called fibrils. They participate in a variety of structural and transport roles within cells. In particular, they are important in the formation of the spindle fibers of mitosis, which are responsible for separation of the chromosomes, an important task during cell division. They extend from the centriole to a special attachment point on each chromosome. They are composed primarily of protein, are contractile, and are involved in the motion of protoplasm (protoplasmic streaming) and transport. Thus, they are important components of cilia and flagella in microorganisms, and they transport organelles from sites of synthesis to functional sites. These proteins are the targets of the effects of various toxic chemicals, such as acrylamide ($CH_2CHCONH_2$) and n-hexane ($CH_3(CH_2)_4CH_3$), that produce a variety of toxicological responses. *See also* organelle.

microwatt (μW) *See the Units of Measure list.*

microwatt-seconds per square centimetre *See the Units of Measure list.*

MIDDAS *See* microdiameter-depth adsorption system.

mid-point The point at or near the middle of a scale or set of experimental data in scientific analyses.

midzonal lesion A lesion indicating that damage to the liver has occurred in the area between the central vein and the portal triad (i.e., portal artery, vein, and bile duct) and not immediately adjacent to either of these structures. *See also* centrilobular lesion; portal lesion.

mil *See the Units of Measure list.*

mil gal *See* million gallons *in the Units of Measure list.*

mild slope (1) A conduit slope less than the critical slope for a particular discharge; a slope for which the depth of flow is greater than Belanger's critical depth and the velocity is less than Belanger's critical velocity. (2) A slope less than the friction slope. A mild slope is also called a flat slope. *See also* Belanger's critical depth; Belanger's critical velocity.

mile *See the Units of Measure list.*

miles per hour (mph) *See the Units of Measure list.*

milk dilution bottle *See* dilution bottle.

milk of lime The lime slurry formed when water is mixed with calcium hydroxide ($Ca(OH)_2$).

milli (m) Prefix meaning 10^{-3} in Système International.

milliampere (mA) *See the Units of Measure list.*

milliequivalent (meq) *See the Units of Measure list.*

milliequivalents per litre (meq/L) *See the Units of Measure list.*

milligram (mg) *See the Units of Measure list.*

milligrams per litre (mg/L) *See the Units of Measure list.*

millilitre (mL) *See the Units of Measure list.*

millimetre (mm) *See the Units of Measure list.*

millimicron (mμ) *See the Units of Measure list.*

millimole (mmol, m*M*) *See the Units of Measure list.*

million electron volts (MeV) *See the Units of Measure list.*

million gallons (mil gal) *See the Units of Measure list.*

million gallons per day (mgd) *See the Units of Measure list.*

millions of floating point operations per second (MFLOPS) A commonly used performance measurement of a computer's operational speed. A higher value corresponds to better performance.

Mills–Reincke phenomenon The coincidence—first pointed out by Hiram F. Mills and J.J. Reincke in 1893–1894, each working independently—between an improvement in the quality of a water supply and a lowering of the death rate in a community from diseases not thought to be directly waterborne.

mils per year (mpy) *See the Units of Measure list.*

min *See* minute *in the Units of Measure list.*

MINEQL (Mineral Equilibrium) A family of equilibrium speciation and modeling computer programs that have evolved for the REDEQL family of programs, with modifications to enhance the efficiency and versatility of the code. The latest version, MINEQL+, was a cooperative effort between the Procter & Gamble Corporation and Environmental Research Software and features an extensive user interface and output management facility for use on IBM-compatible personal computers. *See also* REDEQL.

mineral (1) An inorganic (nonliving) substance that occurs naturally in the earth, has a composition that can be expressed as a chemical formula, and has a set of characteristics (e.g., crystalline structure, hardness) common to all minerals. Examples of minerals are sulfur and salt. Certain organic substances, such as coal, are also sometimes referred to as minerals. The word *mineral* is also used to refer to matter derived from minerals, such as inorganic ions found in water and reported as mineral content. (2) In the water treatment industry, a naturally occurring inorganic cation exchanger formerly used in water softeners—as opposed to the synthetic organic resins used today for water softening.

mineral acid An inorganic acid, especially hydrochloric acid (HCl), nitric acid (HNO_3), and sulfuric acid (H_2SO_4).

mineral acidity Acidity in water caused by the presence of strong inorganic acids (such as hydrochloric acid [HCl], nitric acid [HNO_3], and sulfuric acid [H_2SO_4], as opposed to weak acidity caused by such acids as carbonic acid (H_2CO_3) or acetic acid (CH_3COOH). Mineral acidity is usually expressed in water analysis as free mineral acidity.

Mineral Equilibrium *See* MINEQL.

mineral salt A chemical compound formed by the combination of a mineral acid and a base. Minerals from dissolved rock exist in water in the form of dissolved mineral salts. An excess of mineral salts can give water a disagreeable taste or even be harmful to human health.

mineral spring A spring for which the water contains large quantities of mineral salts, either those commonly occurring in the locality or of a rare or uncommon character.

mineral water (1) Water that is naturally or artificially impregnated with mineral salts or gases (e.g., carbon dioxide, CO_2). (2) Bottled water that contains no less than 250 milligrams per litre total dissolved solids and originates from a protected groundwater source. *See also* bottled mineral water.

mineral-free water Water produced by either distillation or deionization. This term is sometimes found on labels of bottled water as a substitute term for distilled water or deionized water.

mineralization The process of converting organic matter into a mineral or inorganic form. For example, complete oxidation of a chlorinated hydrocarbon results in carbon dioxide (CO_2), water (H_2O), and chloride ions (Cl^-) as the final by-products. Most water treatment oxidation practices, however, result in relatively minor mineralization of organic matter.

miner's salt Another name for mined rock salt.

minimal medium A medium used in a test for total coliform analysis for compliance with the maximum contaminant level for total coliforms and *Escherichia coli* under the Safe Drinking Water Act. The ingredients of the minimal medium are *ortho*-nitrophenol-β-D-galactopyranoside and 4-methylumbelliferyl-β-D-glucuronide. *See also* 4-methylumbelliferyl-β-D-glucuronide; *ortho*-nitrophenyl-β-D-galactopyranoside.

minimize To limit impacts on a wetland by limiting the degree or magnitude of the activity affecting the wetland.

minimum annual flood The smallest of the annual floods during a period of record.

minimum bill rate A rate process whereby the charge to a customer for the initial block of water also recovers billing, collecting, and meter service costs.

minimum day The day during which the minimum-day demand occurs. *See also* minimum-day demand.

minimum detection limit (MDL) *See* method detection limit.

minimum flow (1) The flow occurring in a stream during the driest period of the year. It is also called low flow. (2) The least quantity of influent to a treatment plant within a given time period. *See also* maximum flow.

minimum hour The hour during which minimum-hour demand occurs. *See also* minimum-hour demand.

minimum month The month during which the minimum-month demand occurs. *See also* minimum-month demand.

minimum night flow The amount of water used by an end user or group of end users during the period of lowest demand—typically defined as the hours between midnight and 4:00 A.M.

minimum reporting level (MRL) The lowest concentration of a given analyte that a laboratory feels confident reporting to data users. This value is often used within a laboratory but not in comparison to other laboratories (intralaboratory value). Calculation of the minimum reporting level has not been standardized.

minimum-day demand The least volume per day flowing through the plant for any day of the year.

minimum-hour demand The least volume per hour flowing through a plant for any hour in the year.

minimum-month demand The least volume of water passing through the plant during a calendar month.

mining groundwater The process of extracting groundwater from an aquifer in excess of the natural recharge rate. If mining is conducted long enough, the groundwater resource will be completely depleted. In certain parts of the world, no alternative exists, but proper management can make the resource last for decades or centuries. *See also* groundwater safe yield.

mining yield The amount of groundwater that can be economically withdrawn from an aquifer without regard to replenishment.

minor head loss The energy losses that result from the resistance to flow as water passes through valves, fittings, inlets, and outlets of a piping system.

MINTEQ A family of FORTRAN-language computer programs, most developed by the US Environmental Protection Agency, that combine many of the computational features of the WATEQ programs with the more generalized equation-solving and modeling features of the MINEQL family of programs. The latest version, MINTEQA2, is designed for operation on IBM-compatible personal computers and includes a capability for including adsorption modeling. *See also* MINEQL; WATEQ.

minute (min) *See the Units of Measure list.*

miscible Able to be mixed together or dissolved into each other to produce a homogenous substance.

misclassification bias Bias in an epidemiologic study caused by the erroneous classification of an individual or a characteristic of the individual so that the individual is assigned to a incorrect category of exposure or disease. The probability of misclassification may vary among the groups under study (nonrandom misclassification) or may be the same (random misclassification). Random misclassification usually biases a study toward not observing an association or observing a smaller risk measure than may actually be present, whereas nonrandom bias can result in either higher or lower estimates of risk depending on the distribution of the misclassification error. In environmental epidemiologic studies, misclassification of exposure is often present, and assessing whether misclassification is random or not is important. Random misclassification bias is also called nondifferential misclassification bias by some epidemiologists, and nonrandom misclassification bias is also called differential misclassification bias.

misreplication A mistake in the synthesis of deoxyribonucleic acid (DNA) involving the insertion of an incorrect purine or pyrimidine base in the strand of DNA being synthesized.

missing value In statistics, a missing data point in a set of observations. Various statistical methods are available for estimating missing values.

mist Liquid particles that are suspended in the air by spraying or splashing.

mitigate To moderate or make less severe.

mitigation The act or process of avoiding, minimizing, rectifying, reducing, and compensating for impacts. Compensation covers creation, restoration, or enhancement of wetlands that were or will be lost.

mitigation bank sponsor The entity that develops a mitigation bank. The bank's sponsor might be a different entity from the banker. *See also* mitigation banking.

mitigation banker A person or group of people, a corporation, a government entity, or a consortium that establishes and might also maintain and operate a mitigation bank. The banker might be different from the bank's sponsor. *See also* mitigation banking.

mitigation banking Wetland restoration, creation, or enhancement undertaken expressly to provide compensation credits for wetland losses due to future development activities.

mitochondria Subcellular organelles that are primarily responsible for aerobic metabolism of substrates to yield adenosine triphosphate ($C_{10}H_{16}N_5O_{13}P_3$), the common energy currency within the cell. The mitochondria also have roles in controlling ionic concentrations within the cytosol (especially calcium) in addition to the catabolism of carbohydrates, fats, and amino acids. *See also* organelle.

mitogen A substance that increases the rate of cell division. Such chemicals are of interest primarily because they frequently can act as promoters of cancer.

MIU *See* meter interface unit

mixed bed (1) The intermix of two or more filter or ion exchange products in the same vessel during a service run. The most common use of a mixed bed is in ion exchange systems having 40 percent cations to 60 percent anions in the resin bed, such as for a deionization polisher unit. (2) In filtration, an intermix of two or more media in a single tank, with each stratified into separate layers. *See also* multilayered bed.

mixed media A combination of two or more media products in a single loose-media filtration bed where the products are intermixed rather than stratified in layers. An example is the intermixed use of calcite ($CaCO_3$) and magnesia (MgO) in pH modification or anthracite and sand in a filter for particle removal.

mixed-flow pump A centrifugal pump in which the head is developed partly by centrifugal force and partly by the lift of the vanes on the liquid. This type of pump has a single inlet impeller; the flow enters axially and leaves axially and radially.

mixed-flow turbine A reaction turbine in which the flow enters in a radially inward direction, changes to a direction parallel to the axis of the runner, and in some cases emerges in a more or less radially outward direction.

mixed-function oxidase An enzyme present in the endoplasmic reticulum of many cells that catalyzes reactions important in the metabolism of chemicals normally considered to be foreign to the body. The complex responsible consists of a reductase and a heme protein of the cytochrome P-450 superfamily. *See also* cytochrome P-450.

mixed-media filter A filter that contains media of differing sizes and densities that are intermixed as opposed to being in stratified layers.

mixed-oxidant disinfection Disinfection in which a mixture of oxidants (composed of various chlorine- and oxygen-based compounds, such as ozone (O_3), chlorine dioxide (ClO_2), and short-lived free radicals) is used for disinfection. One technology for generating mixed oxidants uses an electrolytic cell that produces an aqueous mixed-oxidant solution consisting primarily of chlorine (HOCl) with smaller amounts of chlorine dioxide and ozone. *See also* disinfection; electrolytic cell; oxidant; radical.

mixing A process employing energy to create a homogenous mixture from two or more components.

mixing basin (1) A basin or tank in which agitation is applied to water or sludge to increase the dispersion rate of applied chemicals. (2) A tank used for general mixing purposes.

mixing chamber A chamber used to facilitate the mixing of chemicals with liquid or of two or more liquids of different characteristics. It may be equipped with a mechanical device for accomplishing the mixing. *See also* mixing tank.

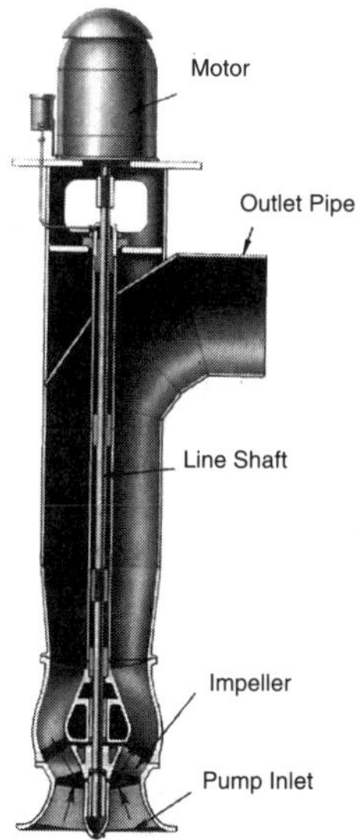

Mixed-flow pump

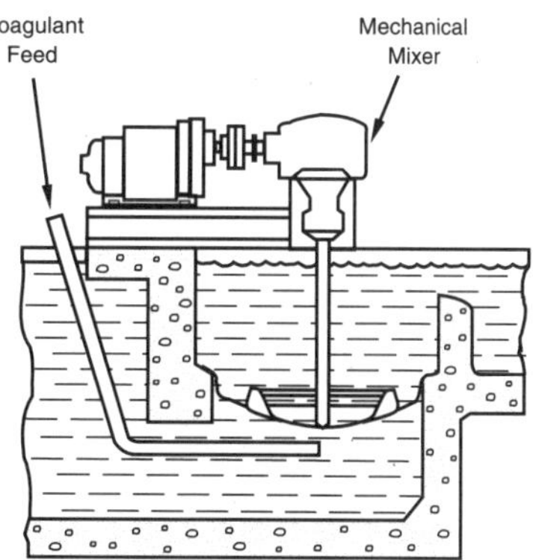

Mechanical mixing chamber

mixing channel A channel provided in a water treatment plant and having hydraulic characteristics or construction features such that chemicals, liquids, or both are thoroughly mixed upon passing through.

mixing tank A tank or channel designed to provide a thorough mixing of chemicals that are introduced into liquids or of two or more liquids that have different characteristics. *See also* mixing chamber.

mixing zone That part of a water body where mixing between the receiving waters and the effluent from a point source (e.g., an industrial discharge) takes place.

mixolimnion The upper zone or layer of a meromictic lake.

mixture Two or more elements, compounds, or both mixed together with no chemical reaction (bonding) occurring.

ML *See* megalitre *in the Units of Measure list.*

mL *See* millilitre *in the Units of Measure list.*

mM *See* millimole *in the Units of Measure list.*

mm *See* millimetre *in the Units of Measure list.*

MMO (minimal medium *ortho*-nitrophenyl-β-D-galactopyransoside) *See* minimal medium.

mmol *See* millimole *in the Units of Measure list.*

MMO-MUG *See* minimal medium.

MMO-MUG technique *See* *ortho*-nitrophenyl-β-D-galactopyransosides; 4-methylumbelliferyl-β-D-glucuronide technique.

mμ *See* millimicron *in the Units of Measure list.*

mobile phase A component of chromatographic separation that transports analytes through the stationary phase. In gas chromatography, the mobile phase is often an inert gas, such as helium. In liquid chromatography, mobile phases can be a single liquid or a combination of several solvents. Differential interaction of the analytes with the mobile and stationary phases causes separation.

mobile water *See* free water.

mobilization A series of initiatives designed to address the fundamental causes of noncompliance with the Safe Drinking Water Act through a coordinated effort by the US Environmental Protection Agency, state drinking water programs, and organizations representing constituencies affected by the Safe Drinking Water Act.

mode In statistics, the most frequently occurring number in a set of numbers. For example, in the set 1, 2, 4, 4, 4, 4, 6, 8, 10, the number 4 is the mode.

mode of action A description of the biological processes a chemical modifies to induce an adverse health effect. *Mode of action* is a less specific term than *mechanism of action.* It is most commonly used to distinguish the effects of carcinogens such as chloroform ($CHCl_3$) that produce cell death and reparative hyperplasia from the effects of carcinogens such as aflatoxin ($C_xH_yO_z$, where x is typically 16 or 17, y is typically 10, 12, or 14, and z is typically 6, 7, or 8) that produce mutations (i.e., are genotoxic). *See also* carcinogen; genotoxic; mechanism of action.

model A mathematical or physical representation of a real-world system, developed to study and understand the behavior of that system. Mathematical models can be used to predict the outcome of a process activity as a result of changes in the variables controlling the process. *See also* mathematical model.

model equation An organized representation of a mathematical model using mathematical symbols, chemical symbols, or both.

model verification The process of comparing the prediction of a mathematical or physical model with the actual outcome of a real-world system.

modeling The act of developing a mathematical or physical model.

moderate rain Rain that is falling at the time of observation with an intensity between 0.11 inches (0.28 centimetres) per hour and 0.30 inches (0.76 centimetres) per hour (i.e., between 0.01 inches [0.025 centimetres] and 0.03 inches [0.076 centimetres] in 6 minutes).

modified Hom model A modified version of the Hom model describing the killing of microorganisms or the inactivation of viruses by a disinfectant. This version is used to describe inactivation during conditions in which the disinfectant is decaying according to a first-order rate, as denoted by the rate constant k^*. The following equation is a very good approximation of the resulting exact solution:

$$\ln(N/N_0) = -kC_0{}^n t^m \left\{ \frac{m[1 - \exp(-nk^* t/m)]}{nk^* t} \right\}^m$$

Where:

N = the number of viable organisms at time t, in numbers per unit volume

N_0 = the original number of organisms, in numbers per unit volume

k = the rate constant of killing or inactivation, in organisms killed or inactivated per unit time

C_0 = the initial concentration of the disinfectant, in milligrams per litre

t = the time of exposure, in minutes

m = a constant to take into account "shoulders" or tailing of the [ln (N/N_0)] versus time curve

n = the coefficient of dilution

k^* = the first-order rate constant for the disappearance of the disinfectant

See also Chick–Watson model; Hom model.

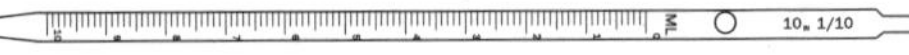

Mohr pipette

modified velocity The velocity as observed after corrections for drift and angularity have been made.

module *See* membrane module.

Mohr pipette A pipette with a graduated stem used to measure and transfer liquids when great accuracy is not required.

moiety (1) An indefinite portion of a sample; one of the portions into which something is divided. (2) A portion of a molecule, e.g., the phenolic moiety.

moisture (1) In hydrology, the amount of water contained in a sample. (2) The percentage by weight of water adsorbed on activated carbon.

moisture adjustment Adjustment of the observed precipitation in a storm via multiplication by the ratio of estimated maximum precipitable water over the drainage area under study to the actual precipitable water, calculated for a particular storm. This procedure is one step in determining the maximum probable precipitation for design purposes; it may lead to a statistically derived design flood.

moisture content The quantity of water present in sludge, usually expressed in terms of a percentage of wet weight.

moisture density The mass of water per unit volume of space occupied by soil, air, and water.

moisture equivalent The quantity of water retained by soil placed in a standardized apparatus when the water content is reduced by means of a constant centrifugal force (1,000 times gravity) acting for a specified period or until a state of capillary equilibrium with the applied force is attained. This value is expressed as a percentage of dry weight. In the reporting of a moisture equivalent, the details of the procedure should be stated.

moisture field deficiency The quantity of water that would be required to restore the soil moisture content to field moisture capacity.

moisture film cohesion In soil stabilization, the resistance of particles to separation as a result of the surface tension of the moisture film surrounding each particle.

moisture gradient The rate of change of the moisture content of soil with depth.

moisture penetration The depth to which moisture penetrates after an irrigation or rain, before the rate of downward movement becomes negligible.

moisture percentage The water content of semiliquid material, such as sludge, expressed as a ratio of the loss in weight after drying at 217.4° Fahrenheit (103° Celsius) to the original weight of the sample. *See also* soil water percentage.

moisture tension (1) A numerical measure of the energy with which water is held in the soil. When expressed as the common logarithm of the head in centimetres of water necessary to produce the suction corresponding to the capillary potential, moisture tension is called pF. (2) The equivalent "negative" or gauge pressure to which water must be subjected in order to be in hydraulic equilibrium, through a porous permeable wall or membrane, with the water in the soil.

moisture-holding capacity *See* field capacity.

mol *See* mole *in the Units of Measure list.*

mol. wt. *See* molecular weight.

molal concentration *See* molality.

molality A measure of concentration defined as the number of moles of solute per litre of *solvent.* Molality is not commonly considered in water treatment. *See also* molarity.

molar (*M*) *See the Units of Measure list.*

molar concentration *See* molarity.

molar energy units In the US customary system, British thermal units per mole. In Système International, joules per mole; metres squared kilograms per second squared per mole.

molar entropy units In the US customary system, British thermal units per mole per kelvin. In Système International, joules per mole per kelvin; metres squared kilograms per second squared per mole per kelvin.

molar heat capacity units In the US customary system, British thermal units per mole per Rankine. In Système International, joules per mole per kelvin; metres squared kilograms per second squared per mole per kelvin.

molarity A measure of concentration defined as the number of moles of solute per litre of solution.

molasses number A value calculated from the ratio of the optical densities of (1) the filtrate of a molasses solution treated with a standard activated carbon and (2) the activated carbon in question. It gives an indication of the volume of large pores in a given activated carbon compared to the standard activated carbon.

mold Multicellular filamentous fungi. *See also* fungus.

mole (mol) *See the Units of Measure list.*

molecular diffusion A process whereby mobile compounds (dissolved or suspended in another compound) move from areas of high concentration to areas of low concentration. Molecular diffusion is described mathematically by Fick's first law of diffusion. *See also* Fick's first law of diffusion.

molecular epidemiology The inclusion of biological markers (cellular, biochemical, or molecular alterations measurable in human tissues, cells, or fluids) in analytical epidemiologic studies. Biological markers may reflect subclinical health effects, define host susceptibility, or serve as indicators of exposure or dose. For example, serum (blood) cholesterol measurements are used in cardiovascular epidemiology, and the identifications of antibodies to various pathogens are used in infectious disease epidemiology. In cancer epidemiology, cellular alterations have been used to identify cancer precursors for screening tests, and deoxyribonucleic acid adducts are being investigated for use in occupational and environmental epidemiologic studies.

molecular ion (M^+) A radical cation formed by the removal of an electron from an analyte in a mass spectrometer. The molecular ion's peak in a mass spectrum is important because it typically represents the molecular weight of the compound. A compound that forms a prominent molecular ion typically has one of the peaks with the highest mass-to-charge ratio. Aromatic compounds tend to have prominent molecular ions because of limited compound fragmentation within a mass spectrometer. Other compounds, such as branched hydrocarbons, tend to fragment extensively and can produce spectra with weak molecular ions.

molecular size An operational definition of the size of a material based on passage through one or several membrane filters of known molecular weight cutoff. For example, the ability of a molecule to undergo diffusion through a membrane, in general, is primarily a function of molecular size and secondarily of molecular shape. *See also* molecular weight cutoff.

molecular substance units In Système International, mole.

molecular weight (mol. wt., MW) The sum of the atomic weights of the atoms in a molecule.

molecular weight cutoff (MWCO) In a pressure-driven membrane process, an approximate characterization of the membrane rejection capability, generally the smallest compounds (in terms of molecular weight) that are rejected.

molecular weight distribution Typically, an expression of the percentage of a complex mixture of organic substances having an apparent molecular weight greater or less than various molecular weight cutoffs (e.g., less than 500, 500–1,000, 1,000–5,000, 5,000–10,000, and greater than 10,000 daltons). *See also* apparent molecular weight; dalton *in the Units of Measure list;* molecular weight cutoff.

molecular weight fractionation The process of determining the distribution of the molecular weight or size of a heterogeneous mixture in liquid form by passing the mixture through a sequence of membrane filters with decreasing molecular weight cutoffs. The amount of a given attribute within a certain molecular weight range can be determined by difference.

molecule The smallest particle of an element or compound that retains all of the characteristics of the element or compound. A molecule is made up of one or more atoms. The helium molecule, for example, has only one atom per molecule. Oxygen molecules (O_2) have two atoms; ozone molecules (O_3) have three atoms. Molecules found in chemical compounds often have many atoms of various kinds.

moles per litre (mol/L, *M*) *See the Units of Measure list.*

molinate (C_2H_5–S–CO–NC_6H_{12}) The common name for the herbicide *S*-ethyl hexahydro-1*H*-azepine-1-carbothioate. *See also* herbicide.

mol/L *See* moles per litre *in the Units of Measure list.*

mollusk Any member of a large phylum (Mollusca) of invertebrate animals, such as clams or snails, with a soft, unsegmented body usually enclosed in a calcareous shell.

moment-of-force units In the US customary system, pounds-force feet. In Système International, newton-metres; metres squared kilograms per second squared.

momentum equation An equation stating, in any consistent set of units, that the impulse (force multiplied by time) applied to a body of water is equal to the momentum (mass multiplied by velocity) acquired by it. The concepts of momentum and impulse, along with energy, are basic to all dynamics.

monitoring (1) The process of tracking information needed to determine exposure limits or medical status. This information includes workplace conditions and employee medical conditions. (2) Routine observation, sampling, and testing of designated locations or parameters to determine the efficiency of treatment or compliance with standards or requirements.

monitoring, compliance *See* compliance monitoring.

monitoring light sensor An indicator light, electrically or electronically activated, attached to a sensor that is positioned in the effluent (product water) stream of a piece of water treatment equipment (deionizer, distiller, reverse osmosis unit, or electrodialysis unit) to detect and signal changes in water quality that might indicate an equipment malfunction. Some lights remain on while water quality is within the desired range and go out if the quality of the water falls into the unacceptable range. Other sensors use red and green light signals. *See also* sensor.

monitoring and reporting (M/R) The monitoring and reporting requirements of the National Primary Drinking Water Regulations. *See also* Primary Drinking Water Regulation.

monitoring well A well drilled primarily for monitoring water levels or water quality in an aquifer. Such wells are distinct and different from production wells that are drilled to produce economic quantities of water for distribution to water users.

monoaldehyde A class of organic compounds that have one aldehyde (HC=O) functional group in the chemical structure (e.g., formaldehyde, CH_2O). Some monoaldehydes are created during the reactions of oxidants used as disinfectants, particularly ozone (O_3), with natural organic matter. *See also* aldehyde; formaldehyde.

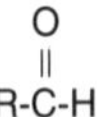

Monoaldehyde, where R is a carbon-containing group

monobromamine (NH_2Br) The brominated analog of monochloramine (NH_2Cl) that can be formed during the chlorination of bromide-containing water. *See also* bromamines.

monobromoacetic acid (MBAA) ($CH_2BrCOOH$) A haloacetic acid containing one bromine atom. It is formed during the chlorination of waters containing moderate amounts of bromide (Br^-), and it may also be formed during the ozonation of bromide-containing waters. *See also* disinfection by-product; haloacetic acid.

monochloramine (NH_2Cl) A chloramine species produced from the mixing of chlorine (HOCl) and ammonia (NH_3). Typically, monochloramine and a small percentage of dichloramine ($NHCl_2$) are formed. Monochloramine is used as a disinfectant, especially for distribution system residual maintenance. *See also* chloramines; dichloramine.

monochloroacetic acid (MCAA) ($CH_2ClCOOH$) A haloacetic acid containing one chlorine atom. Typically, dichlororoacetic acid ($CHCl_2COOH$) and trichloroacetic acid (CCl_3COOH), rather than monochloroacetic acid, are the principal haloacetic acids formed during chlorination of waters containing low amounts of bromide (Br^-). *See also* dichloroacetic acid; disinfection by-product; haloacetic acid; trichloroacetic acid.

monochlorobenzene (C_6H_5Cl) A volatile organic chemical with various industrial uses, including use as a solvent. It is also referred to as chlorobenzene. It is regulated by the US Environmental Protection Agency. *See also* chlorobenzene; solvent; volatile organic compound.

Monochlorobenzene

monochromator A device for isolating monochromatic or narrow bands of radiant energy, such as light, from the source. A monochromator is often used in chemical analysis.

monoclinal spring A contact spring occurring at the surface contact of a monoclinal formation. The water is derived from a pervious water-bearing stratum with a less pervious stratum underneath.

monoclonal antibody An antibody that is specific for a single antigenic determinant.

monocyte A leukocyte (white blood cell) that contains a single nucleus and is phagocytic (i.e., it will engulf foreign material).

Monod equation A mathematical expression first used by J. Monod in describing the relationship between the microbial growth rate and the concentration of a growth-limiting substrate:

$$\mu = \mu_{max}\frac{S}{K_s + S}$$

Where:

μ = the microbial growth rate, in milligrams per litre per hour

S = the concentration of potentially limiting nutrient, in milligrams per litre

μ_{max} = the value of μ when the given nutrient is not limiting, in milligrams per litre per hour

K_s = the concentration of the given nutrient when μ = $0.5\mu_{max}$, in milligrams per litre *See also* Michaelis–Menten equation.

monofilament A single synthetic fiber of continuous length, used in weaving filter cloths.

monofill An ultimate disposal technique for water treatment plant sludge in which the sludge is applied to a landfill dedicated to sludge disposal only (i.e., the landfill does not include municipal solid waste). The monofill is often lined to collect any leachate from the water treatment plant sludge for further processing. When filled, the monofill is "closed" by covering and seeding, as with a typical landfill closure.

monomedia filtration A method of filtration in which particles are removed with a single media type. Monomedia filters are often deeper than conventional filters, up to 6 feet (1.8 metres), and use large-diameter media (from 1.2 to more than 1.8 millimetres in effective size) to reduce head loss. The deeper monomedia can provide additional storage for collected particles. Often used with ozone, monomedia configurations use anthracite or granular activated carbon as a result of their larger particle size compared to sand. *See also* effective size; granular media.

monomer A molecule of low molecular weight capable of reacting with identical or different monomers to form polymers.

monomictic Pertaining to a lake having a single period of free circulation or overturn per year, with consequent disruption of the thermocline. The lake may be either cold monomictic or warm monomictic.

mononuclear phagocyte *See* monocyte.

monovalent *See* univalent.

monovalent ion An ion having a valence charge of 1. The charge can be either positive or negative, e.g., Na^+ or Cl^-.

monsoon A wind blowing with great persistence regularly at a definite season of the year and influenced by seasonal temperature contrasts between land and water masses.

Monte Carlo analysis A method of statistical analysis that involves selecting sample data in proportion to their true or estimated probability of occurrence through a random number generator and using that sample as input to some process (i.e., model) in order to estimate the probability of a particular output from the process. Although Monte Carlo analysis can be applied to a single variable, it is more commonly used to define the attributes of joint probability distributions and multivariate probability distributions. For example, one might wish to define the probability distribution (density function) of total water demand 20 years in the future as it is derived from a functional relationship of population forecasts, current and planned conservation programs, planned changes in water prices, and historical water use per capita. With estimated or known probability distributions for each of these input variables and an equation that links the separate input variables to total water demand, Monte Carlo computer programs are available (for personal computers and mainframes) that will combine the effects of the separate probability distributions into a single probability distribution around the forecast of water demand. This procedure is accomplished by taking hundreds (or thousands) of iterations of sample selections from each of the input variables and then combining them through the functional relationship until a combined distribution takes shape for the projected variable.

monthly flood The maximum flow occurring in a stream during a calendar month.

monthly user charges The charges made to the users of water service through the general water rate structures of the utility to pay for the utility's share of the cost of servicing the water service requirements.

Moody diagram A graphical representation of the relationship among the friction factor, Reynolds number, pipe roughness, and pipe size in pipe flow. The friction factor can be obtained if the other factors are known; this factor is used in the Darcy–Weisbach equation to calculate the head loss in pipe flow. *See also* Darcy–Weisbach formula.

moraine An accumulation of earth, usually with stones, carried and then finally deposited by a glacier.

morbidity rate A measure of both the incidence and prevalence of disease. Use of this term should be avoided. The rate expresses the proportion of a population who are ill with a certain disease in a given period of time. However, one will not necessarily know whether the morbidity includes newly diagnosed cases of disease or all cases diagnosed during the time period. Another disadvantage is that the rate does not take into account such population characteristics as age, race, or gender. Unless standardized, morbidity rates should not be used to compare geographic areas or time periods. *See also* crude rate; incidence; incidence rate; prevalence; prevalence rate; rate difference; rate ratio; standardized mortality rate; standardized rate.

morning-glory spillway *See* glory-hole spillway.

Morrill index The ratio of the time in which 90 percent of a unit volume of liquid passes through a tank to the time in which 10 percent of that unit volume passes through, t_{90}/t_{10}. *See also* dispersion index.

mortality Death.

mortality rate A rate expressing the proportion of a population who die of a specific disease or of all causes during a stated period of time, usually a year. *See also* death rate.

mortgage bond A bond secured by a mortgage against specified properties of an equity, usually its public utilities or other enterprises. If such bonds are primarily payable from enterprise revenues, they are also called revenue bonds.

MOS *See* metal oxide semiconductor.

most advantageous section The cross section of a conduit of such shape and dimensions as to have the smallest wetted perimeter for a given area of cross section.

most probable number (MPN) An estimate of the density of microorganisms in a sample based on certain statistical formulas. Tables of most probable number values for combinations of positive and negative results obtained from the examination of multiple-portion decimal dilution sample aliquots in multiple-tube fermentation procedures (APHA, AWWA, and WEF 1995) are used to interpret the most probable number and the 95 percent confidence limits of a sample. The most probable number is commonly used to estimate the density of coliform bacteria in a water sample. *See also* multiple-tube fermentation test.

mother liquor (1) A residual brine, containing chiefly calcium and magnesium chlorides, obtained after the salt has been crystallized and removed from solution. This term is widely used when salt is produced by use of vacuum pan and gainer operations. In the solar salt evaporation process, the term *bitterns* is often used instead. (2) A solution substantially freed from undissolved matter by a solid–liquid separation process, such as filtration or decanting.

motile Capable of self-propelled movement. This term is sometimes used to distinguish between certain types of organisms found in water.

motor Any electromechanical device for converting electrical energy into mechanical energy. *See also* engine.

motor efficiency The ratio of power delivered by a motor to the power supplied to it during a fixed period or cycle. Electric motor efficiency ratings vary among manufacturers, usually in the range 88.9 to 90.0 percent.

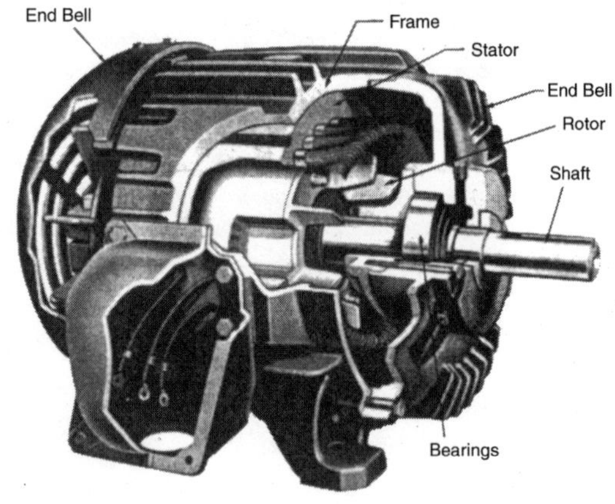

Motor

motor horsepower The horsepower equivalent of the electric power supplied to a motor. *See also* brake horsepower; water horsepower.

motor starter A specially designed switch assembly used to connect and disconnect electrical power from motors. Motor starters may be either manually or electrically operated. If they also contain current-limiting devices (fuses or circuit breakers), they are referred to as combination motor starters.

mottled Spotted or blotched. Teeth can become mottled if excessive amounts of fluoride are consumed during the years of teeth formation.

mottled tooth enamel Tooth enamel that has become discolored because of any of a variety of nutritional disorders and developmental defects. In connection with drinking water, the most common source of mottled tooth enamel is dental fluorosis, whereby excessive fluoride intake results in hypomineralization of the teeth. Mineralization of the tooth structure occurs because removal of the matrix proteins appears to be blocked by excessive fluoride. Thus, the period of a person's highest susceptibility to excessive fluoride is in the early maturation (childhood) stages.

mottling The staining of teeth caused by excessive amounts of fluoride in the water.

mottling of teeth *See* mottled tooth enamel.

mound spring *See* knoll spring.

mouth (1) The exit or point of discharge of a stream into another stream, a lake, or the sea. (2) An opening resembling or likened to a mouth, such as an entrance or exit. (3) The end of a shaft, adit (horizontal mine), drift, entry, or tunnel emerging at the surface. (4) Part of an animal's body where food and liquid intake occurs.

movable bed (1) A streambed made up of materials readily transportable by the stream flow. (2) A bed made up of materials readily movable under the hydraulic conditions established in a tank or chamber.

movable weir (1) A temporary weir that can be removed from the river channel in times of flood. (2) The adjustable weir at the end of a sedimentation tank. *See also* overflow weir.

moving average method A method of obtaining effective dose estimates that minimizes the number of animals used. It involves first range finding by administering the compound in relatively few animals and then moving up or down the dose scale depending on whether a positive or negative response was obtained. For example, in a lethal dose 50 determination, the dose given to a second group of animals would be increased if all animals survived when the first group was treated. On the other hand, if all the animals died, a single animal or a small group would be administered a second lower dose. Once the critical range is identified, more animals could be tested at additional doses spaced appropriately within that range. The number of animals finally employed would depend on how accurate an estimate of the effective dose is necessary. This procedure avoids the possibility that all doses were uniformly lethal or nonlethal in a larger experiment with a fixed number of animals. The statistical method applied to these data as they are developed is called Thompson's moving average interpolation.

moving bed A unique application of granular activated carbon in which a single activated carbon column can approach the efficiency of several columns in series. This is accomplished by the removal of spent activated carbon from one end of the granular activated carbon bed and the addition of fresh granular activated carbon at the other end with little or no interruption in the process.

moving vane A curved surface that moves in a given direction under the dynamic pressure exerted by an impinging jet or stream.

MPA *See* microscopic particulate analysis.

MPC *See* maximum permissible concentration.

mph *See* miles per hour *in the Units of Measure list.*

MPN *See* most probable number.

MPR *See* maximum pumping rate.

mpy *See* mils per year *in the Units of Measure list.*

M/R *See* monitoring and reporting.

MRDL *See* maximum residual disinfectant level.

MRDLG *See* maximum residual disinfectant level goal.

mrem *See the Units of Measure list.*

MRL *See* minimum reporting level.

MS *See* mass spectrometer.

MS–MS *See* mass spectrometer–mass spectrometer.

MSC *See* medium-specific concentration.

MSDS *See* material safety data sheet.

MSF distillation *See* multistage flash distillation.

MSL *See* mean sea level.

MS2 coliphage A ribonucleic acid virus that can replicate only within its bacterial host, *Escherichia coli.*

MTBE *See* methyl-*tert*-butyl-ether.

MTD *See* maximally tolerated dose.

MTF test *See* multiple-tube fermentation test.

MTP *See* maximum trihalomethane potential.

MTZ *See* mass transfer zone.

mu (μ) *See* micro.

muck soil Organic soil derived from the decomposition of peat and containing 20 percent or more mineral material. *See also* peat soil.

mucopolysaccharide A structural component of bacterial cell walls. It is also called cell wall mucopolysaccharide, peptidoglycan, or murein. It is made up of N-acetylglucosamine and N-acetylmuramic acid, with a pentapeptide attached to each molecule of N-acetylmuramic acid. Bridges of amino acids cross-connect the pentapeptides, and each component of the entire cell wall structure is cross-linked by covalent bonds.

MUD *See* municipal utility district.

mud blanket A layer of flocculent material that forms on the surface of a slow granular filter. *See also* schmutzdecke.

mud cake A thin film of drilling fluid that forms on the wall of a borehole during drilling. It is similar to a filter cake. The development of a mud cake is usually a desirable phenomenon during the drilling of water wells. The mud cake is removed when the well is developed for water production.

mud pot A hot spring of limited water flow in which the hot water mixes with clay and undissolved particles, forming a muddy suspension at the spring location.

mud valve A plug valve for draining out sediment, inserted in the bottom of a settling tank.

mudball (1) A clump of granular media that stuck together during backwashing of a rapid granular filter, often caused by uneven distribution of backwash water. Because of their size—from that of a pea up to 1 or 2 inches (2.5 to 5 centimetres) or more in diameter—they sink during backwashing and become very difficult to remove. Surface wash or air wash is usually effective in preventing their formation. (2) A ball of sediment sometimes found in debris-laden flow and channel deposits.

mudball formation The formation of gelatinous, pseudospherical solids of varying diameter in filter media. Mudballs are often formed as a result of poor filter washing, absent or ineffective surface washing, or overdosing of polymeric filter aids. Mudballs must be removed to restore filter efficiency.

muff Fashioned after ear-muffs, muffs protect the wearer from excessive noise levels that could damage a person's hearing. Muffs are assigned different ratings to indicate the levels of noise for which they provide effective protection.

muffle furnace A high-temperature oven used to ignite and burn volatile solids. A muffle furnace is usually operated at temperatures near 1,112° Fahrenheit (600° Celsius).

MUG *See* 4-methylumbelliferyl-β-D-glucuronide; minimal medium.

μg *See* microgram *in the Units of Measure list.*

μg/L *See* micrograms per litre *in the Units of Measure list.*

mulch Any substance spread or allowed to remain on the soil surface to conserve soil moisture and shield soil particles from the erosive forces of raindrops and runoff.

multicollinearity In multiple regression analysis, a condition present when two (or more) independent variables being analyzed to explain the outcomes of a dependent variable are highly correlated with each other. The result is that neither of the coefficients for the independent variables properly measures the independent variable's real causal impact on the dependent variable. The coefficients are said to be biased, leading to incorrect inferences. For example, in a time series forecasting regression model, rainfall and temperature are sometimes highly correlated with each other ($r > 0.8$). When this occurs and both variables are kept in the model, both variables will provide softened measures of their impact on water use. The weaker variable should be omitted or seasonal application of the variables (temperature in summer months, rainfall in winter months) might prove to be without significant multicollinearity.

multicomponent system An engineering system designed to have several components that function together to form the desired end product.

multieffect distillation *See* multiple-effect distillation.

multifilament A number of continuous fiber strands twisted together to form a yarn. Multifilaments are used in weaving filter cloths.

multifunctional medium A single filter or ion exchange medium used to treat water for the removal of more than one constituent. Examples are activated carbon for chlorine removal and sediment filtration, calcite for pH modification and filtering of precipitated iron, or cation resin for reduction of dissolved iron (Fe^{2+})and hardness removal.

multigas monitor A device used to monitor for explosive, toxic, and oxygen-deficient atmospheres. Some monitors can measure specific gases, such as chlorine.

multigeneration study An experiment to determine whether a chemical produces teratogenic or developmental toxicities beyond a single generation. The major concern addressed in these studies is whether a recessive mutation might be produced that would not be detected within the first set of offspring. In general, these studies are limited to two generations, primarily for economic rather than scientific reasons.

multihit model A model used to describe the risk from cancer based on age-specific hazard rates. Multihit models have been developed that allow carcinogenic risks to be calculated at lower doses based on the assumption that the chemical more or less nonspecifically affects a number of events within one or more stages of carcinogenesis. It makes the assumption that the "hits" (interactions of a carcinogen with a cell) fit a Poisson distribution. A multihit model is generally regarded as more restrictive than a multistage model in that it does not account well for dose-response relationships that are linear at low doses but curve upward at higher doses. *See also* linearized multistage model; log-probit model; logit model; multistage model; one-hit model; Poisson distribution; probit model; Weibull model.

multijet meter A current meter for which the velocity of the water passing through the meter causes a turbine wheel to rotate, providing a volume measurement.

multilayered bed A media bed in which more than one filter or ion exchange medium is used in the same vessel, with each medium retaining its stratified position as a layer—even after specified backwashing is performed—because of differences in media densities.

multilevel intake A hydraulic structure by which water may be drawn into or through a dam from various levels at which intake works have been constructed. *See also* intake structure.

multilocus enzyme electrophoresis A laboratory technique for comparing the mobility of enzymes found in isolates of organisms as a means of subtyping the organisms. This technique has been used in epidemiologic studies of many bacterial and fungal diseases. *See also* subtyping.

multimedia filter A filtration device designed to use three or more different types of filter media. The media types usually used are silica sand, anthracite, and garnet sand. This type of filter can sometimes be operated at flow rates higher than 2 gallons per minute per square foot (5 metres per hour).

multimedia filtration *See* depth filtration.

multiple correlation *See* multiple linear regression analysis.

multiple dosing tanks Two or more dosing tanks of equal capacity, each equipped with a dosing device to provide a capability for alternative dosing or rotation of the tanks. *See also* dosing tank.

multiple linear regression analysis A statistical technique that relates one dependent variable with one or more independent variables. The dependent variable depends on or is explained by the independent variables. Multiple linear regression analysis is often referred to as least squares analysis (or something similar) because it mathematically solves for the linear relationship (regression line) between two or more variables that minimizes the sum of the squared deviations of each observation from the regression line. For example, the effect on season runoff (a dependent variable) of such independent variables as antecedent precipitation, groundwater conditions, temperature, and snow depth may be determined through multiple linear regression analysis. The method calculates coefficients (slopes) that define the relationship between the dependent variable and each independent variable. The form of the relationship is typically

$$Y = a + b_1x_1 + b_2x_2 + \cdots + b_nx_n + e$$

Where:

a = an intercept or constant

x_i = the value of the independent variable for the ith period

b_i = the coefficient that defines the relationship between the dependent variable and the ith independent variable

e = the residual or unexplained variation

Various measures are used to define the goodness of fit of the regression equation. *See also* coefficient of determination; F-ratio; regression analysis; standard error of the estimate; t-test.

multiple regression analysis *See* multiple linear regression analysis.

multiple sampler An instrument used for the simultaneous collection of several water-suspended sediment samples of equal or different volumes at each site.

multiple start times The presence of more than one irrigation cycle for a circuit in a given day. Some irrigation controllers can accept programming to allow this feature.

multiple use Use of any resource for more that one purpose. Multiple land use could include grazing of livestock, wildlife production, recreation, watershed, and timber production, whereas a multiple-use reservoir could allow for recreational purposes, fishing, and water supply.

multiple-address system (MAS) A 900-megahertz radio system in which a master radio communicates to a number of remote sites within a specified area of operation.

multiple-arch dam A buttress dam composed of series of inclined arches supported by buttresses or piers. The load is transferred from the arches to the foundation through the buttresses.

multiple-barrier approach The concept of using more than one type of protection or treatment in series in a water treatment process to control contamination and provide overall process reliability, redundancy, and performance. For example, to ensure the safety of drinking water, multiple barriers may include wastewater collection and treatment, protection of water sources, adequate treatment (including disinfection), adequate maintenance, the protection of water quality during storage and distribution, aggressive management, and adequate utility personnel training.

multiple-dome dam A buttress dam in which massive buttresses spaced far apart support a set or series of domes, arched horizontally and longitudinally, on which the water rests. The Coolidge Dam in Arizona is a multiple-dome dam.

multiple-effect distillation (MED) A desalting process in which heated saline water is fed as a thin film on one side of steam-heated tubes in one vessel, causing a portion of the water to vaporize. The hot vapors are introduced on the opposite side of tubes in the following vessel. As the vapors condense to form distillate (product water) in this following vessel, heat is released, aiding in vaporization of saline water on the other side of the tubes in that vessel. The process commonly uses four to eight vessels (effects) operating at successively lower pressures to lower the required boiling temperature point of the saline water.

multiple-effect evaporator A series of single-effect evaporators connected so that the vapor from one effect is the heating medium for the next.

multiple-hearth furnace A furnace consisting of several vertically stacked hearths used for regenerating granular activated carbon (GAC). Dewatered, exhausted GAC is fed into the top hearth and is dried by hot gases. The GAC gradually moves downward through the various hearths until it has been regenerated. This process is carried out in an oxygen-free environment to prevent combustion of the GAC. *See also* regeneration.

multiple-hearth incinerator *See* multiple-hearth furnace.

multiple-tray clarifier A clarifier consisting of a series of trays set one above the other to enhance the collection of suspended solids. It is sometimes equipped with individual sludge scrapers.

multiple-tube fermentation (MTF) *See* multiple-tube fermentation test.

multiple-tube fermentation (MTF) technique *See* multiple-tube fermentation test.

multiple-tube fermentation (MTF) test A microbiological test for estimating the density of bacteria in a sample. Although total coliform bacteria are often the target organisms, the test can be used to detect other bacteria if an appropriate culture medium is used. The test can be used with soil, water, or food samples. It uses a series of tubes (commonly three banks of 5 or 10 tubes each) of liquid culture medium (lactose or other sugar). Each tube receives an inoculum of sample, either as a multiple of 1 millilitre, or as a decimal dilution of the original sample. Following incubation for an appropriate length of time at an appropriate temperature, the tubes are examined for growth, usually indicated by a turbidity increase, a color change in the medium, a

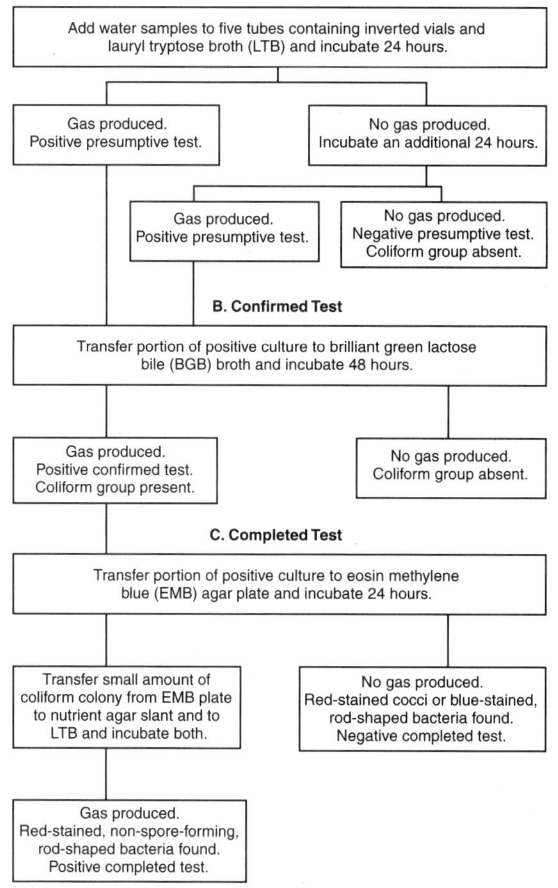
A. Presumptive Test
Add water samples to five tubes containing inverted vials and lauryl tryptose broth (LTB) and incubate 24 hours.
Gas produced. Positive presumptive test.
No gas produced. Incubate an additional 24 hours.
Gas produced. Positive presumptive test.
No gas produced. Negative presumptive test. Coliform group absent.
B. Confirmed Test
Transfer portion of positive culture to brilliant green lactose bile (BGB) broth and incubate 48 hours.
Gas produced. Positive confirmed test. Coliform group present.
No gas produced. Coliform group absent.
C. Completed Test
Transfer portion of positive culture to eosin methylene blue (EMB) agar plate and incubate 24 hours.
Transfer small amount of coliform colony from EMB plate to nutrient agar slant and to LTB and incubate both.
No gas produced. Red-stained cocci or blue-stained, rod-shaped bacteria found. Negative completed test.
Gas produced. Red-stained, non-spore-forming, rod-shaped bacteria found. Positive completed test.

Summary of the multiple-tube fermentation method

production of gas, or another easily observed characteristic. The combination of positive and negative tubes gives a most probable number, as well as a 95 percent confidence interval of the estimate when a table of most probable number values is consulted. *See also* most probable number.

multiple-use reservoir *See* multipurpose reservoir.

multiplexor A device that splits a high-speed communication line into multiple channels supporting multiple devices and recombines multiple channels into a single high-speed channel. Multiplexors maximize the efficiency of a communication line by allowing multiple users to communicate on a single high-speed line.

multiport valve A master control valve used in a filter, deionizer, or water softener to control all the necessary steps in the regeneration process or the backwashing and rinsedown of filters.

multipurpose cadastre An integrated land information system containing legal information (e.g., property ownership or cadastre), physical information (e.g., topography, artificial features), and cultural information (e.g., land use, demographics) in a common and accurate reference framework. The reference framework is typically established with rigorous geodetic and survey control standards, such as the state plane and latitude–longitude coordinate systems. *See also* cadastre.

multipurpose reservoir A reservoir constructed and equipped to provide storage and release of water for two or more purposes, such as flood control, power development, navigation, irrigation, pollution abatement, and domestic water supply.

multipurpose riparian corridor A narrow tract of wetland, immediately adjacent to a river or creek, that provides a variety of functions and uses.

multistage Having more than one stage, as in a multistage pump having more than one pumping impeller.

multistage filtration The use of two or more stages of filtration to improve the efficiency of particle removal. In multistage filtration, a high-rate roughing filter is used in the first stage to remove large particles and reduce particle loading on subsequent filters. The subsequent filters are then used to provide a system with a very high particle removal efficiency.

multistage flash (MSF) distillation A desalting process in which a portion of heated saline water is vaporized (flashed) and condensed in each of several vessels (stages) to form distilled product water. Incoming saline water is heated by a steam heater and thereafter by condensing distillate in each following vessel. The process commonly uses 20 to 40 stages operating at successively lower pressures to lower the required boiling temperature point of the saline water.

multistage model The default (standard) methodology used by the US Environmental Protection Agency for estimating the probability of cancer produced by a chemical at low doses. This model was derived from the Armitage–Doll model, which assumes that cancer develops through a finite number of stages. Progression between stages is assumed to be rare but irreversible. Generally, the stages are produced by mutations in deoxyribonucleic acid sequences that participate in the control of the cell cycle. The transition rate between stages can be variable and can be directly or indirectly affected by the chemical. Some components of the response can also arise spontaneously. However, data are not generally available to estimate potential differential effects on various stages, nor are the mechanisms by which such effects are produced for particular chemicals understood. Therefore, the default methodology usually involves using an estimate of the upper 95 percent bound of some low level of risk based on data obtained from standardized bioassays of the chemical at high doses in experimental animals. *See also* linearized multistage model; log-probit model; logit model; multihit model; one-hit model; probit model; Weibull model.

multistage pump A pump that has more than one impeller. (A single-stage pump has only one impeller.)

multivalent (1) Polyvalent, having a chemical valence usually greater than 2. (2) Represented more that twice in the somatic chromosome number. (3) Having many values, meanings, or appeals. *See also* valence.

multivariate analysis (1) In statistics, any analysis that allows the simultaneous study of two or more dependent variables. (2) In epidemiology, a technique for simultaneous study of the variation in several characteristics, events, or phenomena. Because independent variables are most frequently studied in epidemiology, the term *multiple-variable analysis* has been suggested as being more appropriate than *multivariate analysis. See also* regression analysis.

μM *See* micromole *in the Units of Measure list.*

municipal contract operation Provision of services (e.g., water supply) by contractors, though the municipality retains ownership of all physical facilities.

municipal softening A hardness reduction process performed at municipal central treatment plants. *See also* ion exchange; lime softening; lime–soda ash softening.

municipal utility district (MUD) A special entity authorized by a state government to provide water service, wastewater service, or both to a designated area. The district may authorize bonds to finance the water supply and water facility systems, with repayment from revenue received for the service provided.

municipal water Water that is processed at a central plant to make it potable and then distributed to homes and businesses via water mains. Either public agencies or private companies can be involved in providing municipal water. This general term refers to the common source of water in most urban and suburban areas, as opposed to water obtained from separate proprietary sources such as private wells.

muriatic acid (HCl) Another name for hydrochloric acid.

mushroom valve A valve consisting of a flat disk that raises and lowers without rotation about the valve opening. It is kept in position and on its path of travel by a rod or shaft attached to the disk at right angles to it and extending through the valve opening into a groove or hole that guides its movement. A mushroom valve is also called a poppet valve.

muskeg A bog, usually a sphagnum bog, frequently covered with grassy tussocks, growing in wet, poorly drained boreal (northern) regions, often in areas of permafrost. *See also* bog.

mutagen A material that induces genetic change. *See also* carcinogen; teratogen.

mutagenesis The process of producing a change in the base sequence of deoxyribonucleic acid (DNA) in a cell that is capable of reproducing in the organism. Somatic (i.e., non-germ cell) mutations can also occur, but these are not inheritable. Such mutations can, however, lead to chronic diseases such as cancer. A wide variety of tests are available to determine whether a chemical or one of its metabolites is capable of inducing a DNA change that can be passed on to the progeny of the affected cell. The most commonly applied assays are designed to detect point mutations (e.g., the Ames test, also called the *Salmonella*/microsome assay). Evaluating other genetic end points is also important, however, particularly clastogenic effects (e.g., breaks in chromosomes, changes in chromosome numbers, transfers of parts of chromosomes to one another). The basic mechanisms involved in these effects are potentially quite different, so conducting both types of tests is important in evaluating genotoxic effects. *See also* Ames test; codon; forward mutation; frameshift mutation; genotoxic; germ cell; point mutation; reverse mutation.

mutagenic *See* mutagenesis.

mutagenicity *See* mutagenesis.

mutation An abrupt change in the genotype of an organism, not resulting from recombination. In a mutation, the base sequence of a nucleic acid molecule may undergo qualitative and quantitative alteration or rearrangement. *See also* mutagenesis.

mutual aid plan An important part of emergency preparedness that coordinates assistance from other utilities, contractors, and organizations to come to each other's aid during disasters or emergencies.

mutual water company A cooperative association that is organized to develop and distribute water and that is owned and operated by those receiving water from it.

mutualism A necessary and beneficial interaction between two organisms living in the same environment.

MW *See* molecular weight.

MWCO *See* molecular weight cutoff.

MX *See* 3-chloro-4-(dichloromethyl)-5-hydroxy-2(5H)-furanone.

Mycobacterium A bacterial genus having members characterized as predominantly unicellular, unbranched or branched rods with high lipid content (20–40 percent). These bacteria are often termed acid-fast bacteria because they do not decolorize when rinsed with an acid–alcohol solution in the Gram stain procedure; therefore, their Gram reaction cannot be determined. Based on modern fatty acid analysis technique, however, these organisms have been shown to be gram-negative. The genus *Mycobacterium* includes species that are pathogens for warm-blooded animals and cold-blooded animals.

mycotoxin A chemical substance produced by a fungus and having various adverse health effects. Most commonly, these toxins pose problems to humans when fungi are allowed to contaminate stored foods. The most widely recognized mycotoxins are the aflatoxins produced by *Apergillis flavus* contamination of peanuts. *See also* aflatoxin.

Myer's flood scale A method of expressing the size of a flood (in cubic feet per second) as a numerical percentage of the hypothetical flood given by the product of the square root of the drainage area in square miles and the number 100. The numerical percentage rating is the value on Myer's flood scale.

N

N *See* newton *in the Units of Measure list.*

N *See* normal; normality.

n *See* nano.

n *See* bromine incorporation factor for trihalomethanes; normal.

***n* factor** A value of the roughness factor used in the Manning formula or Kutter formula. *See also* Kutter formula; Manning formula; roughness coefficient.

NA *See* not analyzed; not applicable; not available.

NAA *See* neutron activation analysis.

NADPH–cytochrome P-450 reductase *See* nicotinamide adenine dinucleotide phosphate–cytochrome P-450 reductase.

NAE *See* National Academy of Engineering.

nameplate A durable metal plate found on equipment that lists critical installation information, operating conditions, and the serial number for the equipment.

nano (n) Prefix meaning 10^{-9} in Système International.

nanofiltration (NF) A pressure-driven membrane separation process that generally removes substances in the nanometre size range. Its separation capability is controlled by the diffusion rate of solutes through a membrane barrier and by sieving and is dependent on the membrane type. In potable water treatment, nanofiltration is typically used to remove nonvolatile organics larger than the 200–500-dalton molecular weight cutoff (e.g., natural and synthetic organics, color, disinfection by-product precursors) and multivalent inorganics (for softening). *See also* membrane; pressure-driven membrane.

nanometre (nm) *See the Units of Measure list.*

NAPL *See* nonaqueous phase liquid.

nappe The sheet or curtain of water overflowing a weir or dam. When freely overflowing any given structure, it has a well-defined upper and lower surface.

napthalene ($C_{10}H_8$) A synthetic organic chemical used as a moth repellent and fungicide, as well as in various industrial applications. *See also* fungicide; synthetic organic chemical.

narrow-base terrace A terrace similar to a broad-base terrace in all respects except the width of ridge and channel. The base width of a narrow terrace is usually 4 to 8 feet (1.2 to 2.4 metres). *See also* broad-base terrace.

narrows (1) A contracted part of a stream, lake, or sea. (2) A strait connecting two bodies of water. *See also* water gap.

NARUC *See* National Association of Regulatory Utilities Commissioners.

NAS *See* National Academy of Sciences.

National Academy of Engineering (NAE) A private, honorary organization in Washington, D.C., whose members are elected in recognition of their distinguished and continuing contributions to engineering. Along with the National Academy of Sciences, the National Academy of Engineering advises the federal government through the National Research Council, promotes understanding of the role engineering plays in technological fields, sponsors programs aimed at meeting national needs in the field, and encourages research.

National Academy of Sciences (NAS) A private, nonprofit research organization established by a congressional charter to be the official adviser to the government on scientific and technological matters.

National Association of Regulatory Utilities Commissioners (NARUC) A professional organization of state and federal regulatory commissioners headquartered in Washington D.C., and having jurisdiction over transportation companies and public utilities.

Nappe

National Association of Water Companies (NAWC) The association representing investor-owned and -operated water companies.

National Cancer Institute (NCI) *See* National Toxicology Program.

National Center for Geographic Information and Analysis (NCGIA) A university research consortium formed to evaluate the social, legal, and institutional impacts of the spreading use of geographic information system technology. The center was established in 1988 with a grant from the National Science Foundation and is composed of research groups at the University of California at Santa Barbara, State University of New York at Buffalo, and the University of Maine at Orono.

National Committee for Digital Cartographic Data Standards (NCDCDS) A group sponsored by the federal government and composed of government, university, and industry representatives that propose standards for the structure, content, terminology, and quality of digital cartographic data.

National Conference of State Legislatures (NCSL) A national organization of state legislatures and legislative staff, founded in 1975, that aims to improve the quality and effectiveness of state legislatures; ensure states a strong, cohesive voice in the federal decision-making process; and foster interstate communication and cooperation.

National Contaminant Occurrence Database (NCOD) A repository of information planned by the US Environmental Protection Agency to support the development of possible drinking water regulations. Such a database would, in principle, be linked with both the Unregulated Contaminant Monitoring Rule and the Drinking Water Contaminant Candidate List. *See also* Drinking Water Contaminant Candidate List; Unregulated Contaminant Monitoring Rule.

National Digital Cartographic Database (NDCDB) A collection of digital data maintained and distributed by the US Geological Survey and organized based on the scale and format of the data. The database contains information on transportation, hydrography, boundaries, public land survey system, topography, land use and land cover, and geographic names. Two major types of digital data are produced: elevation data, distributed as digital elevation models, and planimetric and contour data, distributed as digital line graphs.

National Drinking Water Advisory Council (NDWAC) A 15-member council established by the 1974 Safe Drinking Water Act to advise, consult with, and make recommendations to the US Environmental Protection Agency administrator on matters relating to activities, functions, and policies of the agency under the act's terms. Members are appointed by the US Environmental Protection Agency administrator; five are appointed from the general public, five from appropriate state and local agencies, and five from private organizations or groups with an active interest in water hygiene and public water supply. Meeting times and locations are announced in the *Federal Register* and usually include time for public statements.

National Drinking Water Week A week in early May during which an alliance of government, nonprofit organizations, and drinking water purveyors encourages the public to celebrate the water cycle and to become involved in community activities relating to drinking water, wastewater, and other water-related environmental concerns.

National Electrical Code (NEC) A set of legally required rules designed to prevent unsafe installations of electrical systems. The National Electrical Code is developed by the National Fire Protection Association and is updated every 3 years.

National Electrical Manufacturers Association (NEMA) An association of companies that manufacture equipment used for the generation, transmission, distribution, control, and use of electric power. The National Electrical Manufacturers Association's objectives are to maintain and improve the quality and reliability of products; to ensure safety standards in the manufacture and use of products; and to organize and act upon members' interests in terms of productivity, competition from overseas suppliers, energy conservation and efficiency, marketing opportunities, economic matters, and product liability. The association is located in Washington, D.C. It develops product standards covering nomenclature, ratings, performance, testing, and dimensions; participates in developing the National Electrical Code and National Electrical Safety Codes, and advocates acceptance of the codes by state and local authorities; conducts regulatory and legislative analyses on issues of concern to electrical manufacturers; and periodically compiles and issues summaries of statistical data.

National Environmental Laboratory Accreditation Conference (NELAC) A voluntary organization of federal (US Environmental Protection Agency) and state programs established to develop consensus standards for the accreditation of environmental laboratories. Some of these standards received formal approval by NELAC in 1998. State environmental programs are laying the legislative framework for laboratory accreditation, which will eventually replace the current laboratory certification programs required by drinking water regulations. Such nationwide accreditation would prevent laboratories that conduct business in more than one state from having to seek certification from each of the states. NELAC is part of the National Environmental Laboratory Accreditation Program. *See also* National Environmental Laboratory Accreditation Program.

National Environmental Laboratory Accreditation Program (NELAP) The overall program of which the National Environmental Laboratory Accreditation Conference is a part. One of the objectives of the program is to establish nationwide consistency between laboratories performing analyses for compliance with environmental regulations. *See also* National Environmental Laboratory Accreditation Conference.

National Environmental Policy Act (NEPA) A federal law enacted in 1969 (42 USC Section 4321) that sets forth the goals and policies of the federal government directed at protecting and enhancing the quality of the nation's environment. Each federal agency must interpret and exercise its existing authority to achieve specified objectives for environmental preservation and enhancement.

National Environmental Training Association (NETA) A professional organization devoted to serving the environmental trainer and promoting better operation of water treatment and pollution control facilities.

National Fire Protection Association (NFPA) A nonprofit organization that serves as a center for information on fire protection and fire prevention. The association's 270 codes and standards are widely recognized by officials as the basis for laws and regulations.

National Geodetic Reference System (NGRS) A network of monumented geodetic control points in the United States maintained by the National Geodetic Survey.

National Geodetic Vertical Datum of 1929 (NGVD) A geodetic datum derived from a general adjustment of the first-order (highest accuracy) level (elevation) network of the United States and Canada. It was formerly called the Sea Level Datum of 1929.

National Governors Association (NGA) An organization of the governors of the 50 states, Guam, American Samoa, the Virgin Islands, the Northern Mariana Islands, and Puerto Rico, founded in 1908. The National Governors Association serves as a vehicle through which governors influence the development and implementation of national policy and apply creative leadership to state problems; it also keeps the federal establishment informed of the needs and perceptions of states, conducts research programs, and compiles statistics.

National Ground Water Association (NGWA) An association representing groundwater professionals and dedicated to the protection and proper development of groundwater. It was formerly called the National Water Well Association.

National Historic Preservation Act (NHPA) A federal law passed in 1966 (16 USC Section 470) creating the Advisory Council on Historic Preservation to advise Congress and the President of the United States on historic preservation issues. The council is authorized to review and comment in writing on activities licensed or permitted by the federal government that may affect properties listed or eligible for listing on the National Register of Historic Places.

National Inorganics and Radionuclides Survey (NIRS) A survey conducted by the US Environmental Protection Agency to characterize the occurrence of a variety of constituents in community groundwater supplies in the United States, its territories, and its possessions.

National Institute for Occupational Safety and Health (NIOSH) An organization formed under the Occupational Health and Safety Act to study health and safety issues and recommend criteria for health and safety standards. The institute conducts research studies for which employers must submit data.

National Interim Primary Drinking Water Regulation (NIPDWR) A drinking water regulation set by the US Environmental Protection Agency under the 1974 Safe Drinking Water Act, based on the US Public Health Service's 1962 drinking water standards and additional health effects support.

National Marine Fishery Service (NMFS) The federal agency responsible for marine and coastal waters and adjacent wetlands that provide habitat, shelter, and food for marine life. The service provides consultation to the US Army Corps of Engineers during review of the Clean Water Act Section 404 permit applications that impact marine and coastal waters and adjacent wetlands.

National Oceanic and Atmospheric Administration (NOAA) An agency within the US Department of Commerce, Washington, D.C., whose mission is to explore, map, and chart the global ocean and its living resources and to manage, use, and conserve those resources; to describe, monitor, and predict conditions in the atmosphere, ocean, sun, and space environment; to issue warnings against impending destructive natural events; to assess the consequences of inadvertent environmental modification over several scales of time; and to manage and disseminate long-term environmental information.

National Organics Monitoring Survey (NOMS) A national survey conducted by the US Environmental Protection Agency between 1976 and 1977 to determine the frequency of specific organic compounds in drinking water supplies.

National Organics Reconnaissance Survey (NORS) A national survey conducted by the US Environmental Protection Agency to determine the extent of the occurrence of trihalomethanes and other organic chemicals in the United States. It was completed in 1975.

National Pollutant Discharge Elimination System (NPDES) permit The regulatory agency document issued by either a federal or state agency that is designed to control all discharges of pollutants from point sources into US waterways. These permits regulate discharges into navigable waters from all point sources of pollution, including industries, municipal wastewater treatment plants, sanitary landfills, large agricultural feed lots, and return irrigation flows.

National Primary Drinking Water Regulation (NPDWR) *See* Primary Drinking Water Regulation.

National Priorities List (NPL) A list of sites chosen for immediate attention under the Comprehensive Environmental Response, Compensation, and Liability Act of 1980.

National Research Council (NRC) A council organized by the National Academy of Sciences in 1916 to associate the broad community of science and technology with the academy's purposes of furthering knowledge and advising the federal government. Located in Washington, D.C., the National Research Council is the principal operating agency of both the National Academy of Sciences and the National Academy of Engineering in providing services to the government, the public, and the scientific and engineering communities. The council is administered jointly by both of the academies and by the Institute of Medicine.

National Research and Education Network (NREN) The proposed national computer network to be built on the foundation of the National Science Foundation Network. The National Research and Education Network would provide high-speed interconnections between other national and regional networks.

National Risk Management Research Laboratory (NRMRL) A group of offices within the US Environmental Protection Agency's Office of Research and Development responsible for conducting and funding research in support of agency programs that are intended to reduce risk. It includes the Water Supply and Water Resources Division.

National Rural Water Association (NRWA) The association representing rural water districts and systems, particularly those receiving funding from the Farmer's Home Administration.

National Safe Drinking Water Regulation *See* Primary Drinking Water Regulation.

National Safety Council (NSC) An international organization started in 1913 to provide safety professionals of all fields with information expertise and safety assistance.

National Sanitation Foundation (NSF) *See* NSF International.

National Sanitation Foundation maximum allowable level (NSF-MAL) *See* maximum allowable level.

National Science Foundation Network (NSFNET) A computer network funded by the National Science Foundation. It is a high-speed "network of networks" that plays an essential part in academic and research communications.

National Secondary Drinking Water Regulation (NSDWR) *See* Secondary Drinking Water Regulation.

National Society of Professional Engineers (NSPE) A society of professional engineers and engineers-in-training in all fields registered in accordance with the laws of states or territories of the United States or provinces of Canada, as well as qualified graduate engineers, student members, and registered land surveyors. The National Society of Professional Engineers is concerned with social, professional, and ethical matters, as well as programs in public relations, employment practices, ethical considerations, education, and career guidance. It monitors legislative and regulatory actions of interest to the engineering profession.

national standard hose coupling A fire hose coupling that has thread dimensions agreed on by several national associations and approved by the American Standards Association in 1925.

National Technical Information Service (NTIS) A self-supporting federal agency in the US Department of Commerce located in Alexandria, Va., that actively collects and disseminates scientific, technical, engineering, and business-related information generated by the US government and foreign sources. The National Technical Information Service's collection of information exceeds two million works, covering current events, business and management studies, research and development, manufacturing, standards, translations of foreign works, foreign and domestic trade, general statistics, and numerous other areas.

National Toxicology Program (NTP) A program responsible for the conduct of cancer bioassays in experimental animals. The program is associated with the National Institute of Environmental Health Sciences and is located in Research Triangle Park, N.C. It originated from a similar program once housed in the National Cancer Institute.

National Voluntary Laboratory Accreditation Program (NVLAP) A program that certifies independent laboratories' competence to test national standards and publishes lists of certified laboratories for asbestos in air monitoring.

National Water Alliance (NWA) A group of congressional members and other interested people that influences national policies and educates the public regarding water supply.

National Water Resources Association (NWRA) An association that promotes the development, control, conservation, and use of water resources in 17 western states.

National Water Well Association (NWWA) *See* National Ground Water Association.

National Wetlands Inventory (NWI) An inventory of wetlands in the United States conducted by the US Fish and Wildlife Service. In this inventory, a wetland is defined as lands transitional between terrestrial and aquatic systems where the water table is usually at or near the surface or the land is covered by shallow water. For purposes of this classification, wetlands must have one or more of the following three attributes: (1) at least periodically, the land supports predominantly hydrophytes; (2) the substance is predominantly undrained hydric soil; and (3) the substrate is nonsoil and is saturated with water or covered by shallow water at some time during the growing season of each year.

National Wildlife Federation (NWF) A federation of state and territorial conservation organizations and associate members. Located in Washington, D.C., the National Wildlife Federation encourages the intelligent management of the life-sustaining resources of the Earth and promotes greater appreciation of these resources, their community relationship, and their wise use.

nationwide permit (NWP) A permit issued by the US Army Corps of Engineers that authorizes certain defined activities to occur in wetlands or waters of the United States without the need for a detailed individual review. Nationwide permits may be issued for categories of similar activities that individually and cumulatively produce minor environmental impacts.

native and adapted plant A plant indigenous to an area or from a similar climate that requires little or no supplemental irrigation once established.

native landscape A landscape composed of native plant communities.

native salt *See* halite.

natural algal growth Growth of simple plants that contain chlorophyll and require sunlight. Algae occur naturally in all surface waters and grow actively during all seasonal periods when adequate light, water temperature, and inorganic nutrients are available. In ponds, lakes, and reservoirs, massive blooms of planktonic algae are often observed during warm weather periods and may create taste-and-odor problems for water users. *See also* algal bloom; bloom.

natural environment The complex of atmospheric, biological, and geologic characteristics found in an area in the absence of artifacts or influences of a developed technological human culture.

natural filtration The removal of particles or contaminants through in situ soil treatment. When contaminants are discharged into the ground, either by injection or by percolation, natural soil particles may remove these contaminants through filtration or adsorption prior to a location from which the water is recovered (e.g., through a well).

natural flow The flow of a stream as it occurs under natural, as opposed to regulated, conditions. *See also* regulated flow.

natural logarithm (ln) A logarithmic function using the mathematical constant e as its base. The natural logarithm of a real number u is defined as v if the following mathematical condition is satisfied: $e^v = u$. In other words, $\ln u = v$. *See also* common logarithm.

natural logarithmic transformation A mathematical operation to convert a number to its natural logarithm. *See also* natural logarithm.

natural organic matter (NOM) A heterogeneous mixture of organic matter that occurs ubiquitously in both surface water and groundwater, although its magnitude and character differ from source to source. Natural organic matter contributes to the color of a water, and it functions as disinfection by-product precursors in the presence of such disinfectants as chlorine (HOCl). Humic substances (e.g., fulvic acid) represent a significant fraction of natural organic matter in surface water sources. *See also* disinfection by-product precursor; fulvic acid; humic substances.

natural pollution The introduction of substances causing undesirable changes to the environment as a result of non–human-based activities, e.g., the dissolution of arsenic into groundwater from certain subsurface arsenic-containing rocks.

natural replacement The substitution of older, less efficient water fixtures (toilets, faucets, and so forth) with new, more efficient fixtures as the older fixtures break, wear out, or are replaced.

natural resources Those products and features of the Earth that permit it to support life and satisfy people's needs. Land and water are natural resources, as are biological resources on the land and in the water, such as flowers, trees, birds, wild animals, and fish, as well as minerals, such as oil, coal, metals, stone, and sand. Some natural resources are renewable, such as solar, wind, and tidal energy; farmland; forests; fisheries; and surface water. Mineral and fossil fuels are examples of nonrenewable resources. Note that renewable resources may become nonrenewable if they are destroyed by bad management, e.g., farmland losing topsoil or denuding of watersheds.

Natural Resources Conservation Service (NRCS) An agency of the US Department of Agriculture that was formerly the Soil Conservation Service. It has national responsibility for helping farmers, ranchers, and other private landowners develop and carry out voluntary efforts to conserve and protect natural resources. Its key programs include conservation technical assistance, a natural resources inventory, a national cooperative soil survey, snow survey and water supply forecasting, a plant materials program, a small watersheds program, a flood prevention program, the Great Plains Conservation Program, the Resource Conservation and Development Program, a rural abandoned mine program, a water bank program, a Colorado River basin salinity control program, a forestry incentives program, and a farms-for-the-future program.

natural slope The greatest angle to the horizontal assumed by any unsupported granular semisolid or semifluid material. It is also called the angle of repose.

natural softening The replacement of hardness-causing minerals by sodium, potassium, or both as a result of the normal flow of water in the ground. *See also* hardness.

natural sparkling water *See* sparkling water.

natural strainer well A well that taps a water-bearing formation of imperfectly assorted and at least partly incoherent material and that has had its specific capacity increased by the withdrawal of much of the fine-grained part of such material in the immediate vicinity of the screened portion of the well. Such a well is also called a naturally developed well.

natural transformation A mechanism of deoxyribonucleic acid (DNA) exchange between bacteria whereby free or membrane-bound DNA is transferred from one bacterium to another with the use of sex pili. *See also* sex pili.

natural water Water as it occurs in its natural state, usually containing other solid, liquid, or gaseous materials in solution or suspension, in chemical equilibrium with the surroundings.

natural water table A water table in its natural condition and position, not disturbed by artificial additions or extractions of water.

natural watercourse A surface or underground watercourse created by natural processes and conditions.

natural well An abrupt depression in the land surface, not made by humans, that extends into the saturation zone but from which water does not flow to the surface except by artificial processes. It is distinguished from such features as ponds, swamps, lakes, and other bodies of impounded surface water that also extend into the saturation zone in that it has a smaller water surface, is deeper in proportion to its water surface area, and has steeper sides. Although it is not a well in a strict sense, the name has been firmly established as applying to such a feature.

naturally developed well *See* natural strainer well.

naturally soft water Groundwater, surface water, or rainwater sufficiently free of calcium and magnesium salts that no "curd" (precipitated soap) forms and no calcium- or magnesium-based scale forms when the water is heated.

navigable water Any stream, lake, arm of the sea, or other natural body of water that is actually navigable and that, by itself or by its connections with other waters, is of sufficient capacity to float watercraft for the purposes of commerce, trade, transportation, or even pleasure for a period long enough to be of commercial value; any water that has been declared navigable by the Congress of the United States.

NAWC *See* National Association of Water Companies.

NC data *See* noncensored data.

NCDCDS *See* National Committee for Digital Cartographic Data Standards.

NCGIA *See* National Center for Geographic Information and Analysis.

NCH *See* noncarbonate hardness.

NCI (National Cancer Institute) *See* National Toxicology Program.

NCOD *See* National Contaminant Occurrence Database.

NCSL *See* National Conference of State Legislatures.

NCU *See* network communication unit.

NCWS *See* noncommunity water system.

ND *See* not detected.

NDCDB *See* National Digital Cartographic Database.

NDMA *See* N-nitrosodimethylamine.

NDWAC *See* National Drinking Water Advisory Council.

near the first service connection A designation used to indicate any of the 20 percent of all service connections in the entire system that are nearest the water supply treatment facility, as measured by water transport time within the distribution system.

nebulizer A device that reduces a liquid to a fine spray. For example, aqueous samples can be introduced into the flame of an atomic absorption spectrophotometer via a nebulizer. *See also* atomic adsorption spectrophotometer.

neck (1) A narrow stretch of land connecting two larger areas. (2) A narrow body or channel of water between two larger bodies of water.

necropsy The process of taking tissue specimens from animals, usually from a controlled experiment, to be examined for pathological changes.

necrosis A process produced by exogenous influences (chemical or microbial) that causes the death of cells. Necrosis is in contrast to programmed cell death, apoptosis, which occurs naturally. *See also* apoptosis.

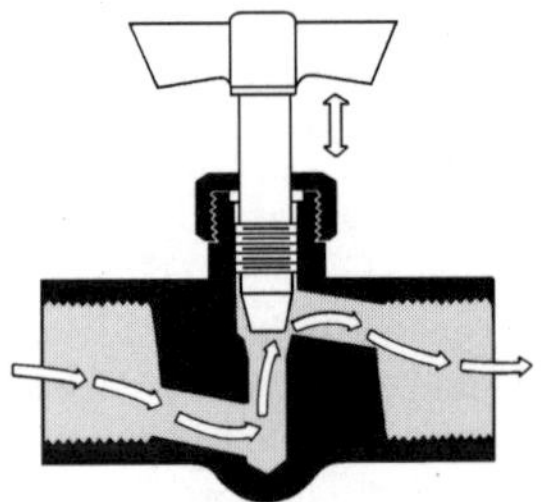

Needle valve

needle A plank, usually made of timber or steel, set on end to close an opening for the control of water. In dams or control structures, a needle may be set in either a vertical or inclined position. *See also* needle dam.

needle dam A movable dam composed of upright pieces or needles, the lower ends of which bear against a sill on the river bottom and the upper ends of which bear against some form of bridge.

needle valve A valve with a circular outlet through which the flow is controlled by means of a tapered needle that extends through the outlet, reducing the area of the outlet as it advances and enlarging the area as it retreats.

needle weir A movable timber weir in which the wooden barrier consists of upright square timbers placed side by side against structural steel frames.

negative Having an electrical charge polarity associated with an excess of electrons or, equivalently, a deficiency of protons.

negative charge An electrical potential inherent to an electron. Anions carry negative charges because they have more electrons than protons.

negative (charge) metal oxide semiconductor (NMOS) A semiconductor incorporating a silicon gate structure that uses negative charges only.

negative confining bed A confining bed that prevents or retards downward movement of groundwater even though the overlying water has sufficient head to produce a resultant downward pressure.

negative head A condition that can develop in a filter bed caused by high head loss. When this occurs, the gauge pressure in the bed can drop to less than atmospheric, often resulting in the release of dissolved gases from the water and the formation of gas bubbles. Negative head is also called negative pressure.

negative pressure A gauge pressure less than the existing atmospheric pressure when the latter is taken as a zero reference.

negative pressure valve A valve that monitors drops in pressure and automatically closes prior to a negative pressure or vacuum being placed on a line or system.

negative sample In a multiple-tube fermentation or membrane filter test, any sample that does not indicate the presence of coliform bacteria.

negative well A shaft or well driven through an impermeable stratum to allow water to drain through to a permeable one. A negative well is also called an absorbing well, dead well, or drain well.

negatively charged filter A filter that has a net negative charge at near-neutral pH values. Some negatively charged filters are composed of fiberglass or nitrocellulose. Virus adsorption to these filters usually requires that water be acidified to approximately pH 3.5, that divalent or trivalent cations be added (e.g., aluminum, Al^{3+}), or both.

negotiated rate A water rate that is sometimes negotiated (downward) with major customers to allow for unique load characteristics, capital facilities provided by the user, technology-specific conditions (e.g., in the oil industry or Silicon Valley), load retention inducement (i.e., to keep a user from leaving the area or sinking a private well), or any other basis that is deemed fair. Rates for reclaimed water are frequently negotiated at less than the potable rates on the basis of facilities provided by the user or costs avoided by the supplier. *See also* economic development rate.

negotiated rulemaking An alternative to the traditional federal agency rulemaking process conducted under the authority of the Negotiated Rulemaking Act of 1990. A proposed rule is negotiated by an advisory committee whose membership includes a representative of each affected interest. Meetings of the committee are open to the public, and information is freely exchanged. New ideas and comments from the public can be offered to the advisory committee for consideration in its negotiations. The goal of committee deliberations is to reach a consensus on the language or issues included in a proposed rule. The negotiated rulemaking process may be initiated at the federal agency's discretion.

Negotiated Rulemaking Act (NRA) Public Law 101-648, enacted in 1990, that establishes a framework for the conduct of negotiated rulemaking and encourages federal agencies to use the process when doing so will enhance the informal rulemaking process.

NELAC *See* National Environmental Laboratory Accreditation Conference.

NELAP *See* National Environmental Laboratory Accreditation Program.

NEMA *See* National Electrical Manufacturers Association.

nematocide An agent that is destructive to soil nematodes. *See also* nematode.

nematode A roundworm belonging to the phylum Nematoda, characterized by a long, cylindrical, unsegmented body and a heavy cuticle. Many nematodes (hookworms and pinworms) are parasites of animals and plants.

neonatal mouse A newborn mouse. Newborn mammals are used in studying a variety of effects of chemicals and other agents during nursing. A specific use of the neonatal mouse is in studying chemical carcinogenesis. Because of their high rates of cell division, young animals can be more sensitive to chemical and physical agents that initiate cancer. The high rates of cell division quickly repair mutations induced by such agents, and the stimulus present for cell division causes the tumors to grow out quickly. On the other hand, the need to metabolize many carcinogens to reactive intermediates may make the neonatal animal less sensitive to a carcinogen, because the ability to metabolize the chemical to an active form is generally less well developed at this age.

neoplasm An abnormal growth or tissue, as a tumor.

NEPA *See* National Environmental Policy Act.

nephelometer *See* nephelometric turbidimeter.

nephelometric turbidimeter An instrument that measures turbidity by measuring the amount of light scattered by particles in a water sample. It is the only instrument approved by the US Environmental Protection Agency to measure turbidity in treated drinking water. It operates by passing light through a sample and then measuring the amount of light deflected (usually at a 90° angle).

nephelometric turbidity A measure of turbidity in a sample as determined by an instrument called a nephelometric turbidimeter. *See also* nephelometric turbidimeter.

nephelometric turbidity unit (ntu) *See the Units of Measure list.*

Nernst equation An equation used to determine the effect of reactant and product concentration on electrochemical cell potential. Conversely, if the cell potential and the concentrations of all but one species are known, the equation can be used to determine the unknown concentration. The equation is applicable to reactions that take place in an electrochemical cell and in solution. In equation form,

$$E_{red/ox} = E^{o}_{red/ox} - \frac{RT}{zF}\ln Q$$

Where:

$E_{red/ox}$ = potential, in volts

$E^{o}_{red/ox}$ = standard potential (in volts), a constant that can be obtained from tables of standard reduction potentials

R = ideal gas constant (about 0.001987 kilocalories per degree-mole)

T = absolute temperature, in degrees kelvin

z = number of electrons transferred in the reaction

F = Faraday constant (23.060 kilocalories per volt)

Q = reaction quotient = {red}/{ox}

{red} = activity of the reductant

{ox} = activity of the oxidant

Nessler tube A glass tube with a flat bottom used in some colorimetric analyses. A common Nessler tube can contain 50 millilitres. These tubes are often sold as matched sets to minimize differences between individual tubes. Lab personnel can visually compare samples and standards by looking directly down the length of the Nessler tube.

Nesslerization A colorimetric technique used in the determination of ammonia (NH_3). The term is derived from Nessler reagent, which is prepared from an aqueous solution of mercuric iodide (HgI), postassium iodide (KI), and sodium hydroxide (NaOH). In this technique, laboratory personnel compare the intensity of a sample to the intensity of standard solutions by visually comparing the solutions or by using an absorption spectrophotometer.

nested polymerase chain reaction The use of nested sets of polymerase chain reaction (PCR) primers for in vitro amplification of a specific target nucleic acid sequence. In a simple nested amplification protocol, the first round of amplification is performed with a single primer pair for 15 to 30 cycles and yields an amplification product; the amplification product from the first round is then transferred to new reaction tube for the second round of amplification, which uses a second primer pair specific for the internal nucleic acid sequence amplified by the first primer pair. Following the second round of amplification for 15 to 30 cycles, the products are detected via an appropriate assay. Nested PCR is also called nested amplification. *See also* polymerase chain reaction.

net available head The difference in pressure elevation between the water in a power conduit before it enters the waterwheel and the first free water surface in the conduit below the waterwheel.

net demand The water demand that is expected to occur in the future after reductions for conservation and natural replacement. It represents the actual demand that should be experienced in the future at the premises of customers.

net driving force *See* net driving pressure.

net driving pressure In a pressure-driven membrane treatment system, the hydraulic pressure differential across the membrane minus the osmotic pressure differential across the membrane. *See also* pressure-driven membrane.

net head The head available in a hydroelectric plant for production of energy after deduction of all losses caused by friction, eddies, entrance, unrecovered velocity head at the mouth of a draft tube, and so on, except those losses chargeable to the turbine. Net head is also called effective head.

net peak flow The total flow at a peak minus the corresponding base flow; that part of the peak discharge that is attributable to direct runoff from precipitation.

net positive suction head (NPSH) A measure of the pressure at the suction side of the pump, including atmospheric pressure and vapor pressure of the liquid being pumped.

net present value (NPV) In the context of cash receipts or expenditures (cash flows) in future periods, a calculated value recognizing that a dollar in hand today is worth more than a dollar to be received in the future because the dollar in hand today has the ability to earn a return either as an interest-earning instrument in a financial institution or from an investment in a capital project. Future cash flows are therefore discounted by the expected return to derive the net present value. The net present value of *n* future values is given by the following formula:

$$\text{NPV} = \sum_{i=1}^{n} \frac{FV_i}{(1 + DR)^i}$$

Where:

n = the total number of time periods (e.g., number of years)

FV_i = the future value received in period i

DR = the discount rate, expressed as a decimal

For example, a single \$100 expenditure that will be made 2 years from today can be shown to have a net present value of \$87.34 if the long-term cost of capital is assumed to be 7.0 percent (\$100/1.07^2 = \$87.34). Net present value calculations are equally applicable to a constant amount or an uneven stream of cash flows over any number of periods.

net rainfall Rainfall minus infiltration and other deductions from surface runoff. Net rainfall is sometimes called excess rainfall.

net revenue requirement In water rate determination, the difference between the total expenditures and the total revenue from nonrate sources. This difference is, then, the amount of net cost that must be recovered or paid for by water rates. Total expenditures include operations and maintenance, revenue-funded capital payments, and (when the utility method is used for defining rate requirements) a return on investment. The nonrate revenues include interest earnings on cash balances, various customer fees, and miscellaneous revenues.

net revenues available for debt service Operating revenues less operations and maintenance expenses but exclusive of depreciation and bond interest. Net revenue available for debt service, as thus defined, is used to compute coverage for revenue-bond issues. Under the laws of some states and the provisions of some revenue-bond indentures, net revenues available for debt service for computation of revenue-bond coverage must be computed on a cash basis rather than in conformity with generally accepted accounting principles. Sometimes indenture provisions permit nonoperating revenue and system-development-charge receipts to be included with operating revenue when one is determining net revenue available for debt service.

NETA *See* National Environmental Training Association.

Netherlands Waterworks Testing and Research Institute (KIWA) The drinking water testing and research organization of the Netherlands.

network (1) A data structure used to model relationships of interconnected linear features, such as roads and water distribution systems. (2) The structure of relationships among the activities, tasks, and events of a project. (3) A total computer system that integrates the functions of hardware and software, as well as the support provided by data communications. (4) A group of sampling locations connected by common objectives and methodologies. (5) A group of pipes in a distribution system.

network analysis The use of analytical procedures to evaluate movement (e.g., of water, sewage, vehicles, or electricity) through an interconnected system based on a number of criteria. Network analysis techniques are usually distinctive for the type of network being evaluated.

network analysis system A computer system used primarily for geographic or statistical analysis of relationships and flows that are dependent on how the segments and nodes in a system, such as a utility network, are connected.

network communication unit (NCU) A network communications operating system or program. It serves to directly control, integrate, or interface computers of the same make, same communications protocol, or both. Computers of different manufacture and having different protocols can be indirectly integrated via a common relational database. The computers are indirectly linked or integrated via the data they exchange. One computer reads the data in, and the other reads it out.

network data model A data storage structure in which records of the same type are grouped in conceptual files, in owner-coupled sets reflecting one-to-many relationships, or in many-to-many relationships.

network protocol A fixed set of rules specifying the format of data exchange between computers. A particular protocol defines how the individual bits are to be arranged in transmitting a message so that the message can be received and interpreted correctly by another node on the network.

network topology The arrangement or layout of the devices and communication lines in a computer network.

networking (1) A method for distributing data process functions through communications facilities. (2) The design of networks.

neural network *See* neural network modeling.

neural network modeling A modeling approach loosely following the functioning of a human brain, in which computer programs using artificial intelligence are developed with dynamically interconnected process modules (or neurons). The interactions between the neurons are defined by using a set of known input parameters and assigned weights to signify the importance of a specific relationship. The process of developing the interactions between the neurons is referred to as training the model; this process can continue during use of the model. This type of model could be useful in developing predictive tools for water quality parameters or as a system tool for automated process control in a treatment plant. Neural network modeling is sometimes referred to as artificial neural network (ANN) modeling.

neuropathy Pathological damage to neurons (nerves). Such pathology can involve different parts of the neuron or be relatively specific to certain types of neurons, depending on the agent or condition that gives rise to the damage. Examples of chemicals that are of some concern in this regard and could occur in drinking water are lead and acrylamide ($CH_2CHCONH_2$).

neurotoxic esterase An enzyme activity of unknown function that is inhibited by chemicals that give rise to a delayed neurotoxicity. Triorthocresyl phosphate and a variety of other organic phosphate compounds produce the effect. Many of these compounds also inhibit acetylcholinesterase, an inhibition activity common to the organophosphate pesticides; this activity can lead to acute neurotoxicity. The acute and delayed neurotoxicity are clearly separable entities, however. For example, not all organophosphate pesticides produce delayed neurotoxicity, and triorthocresyl phosphate is a relatively poor acetylchlolinesterase inhibitor. *See also* delayed neurotoxicity.

neurotoxicant A chemical that induces modifications of nervous system function that give rise to overt symptoms: modified functional activity of the nervous system, actual pathology (damage) to the nervous system, or both.

neutral (1) In electricity, a condition of being uncharged because of a balance between negative and positive charges. The condition of neutrality means that neither an excess nor lack of electrons exists. (2) In water chemistry, the midpoint (neutral) reading of 7.0 on the pH scale, indicating that the solution (water) producing the neutral reading will produce neither an acid nor alkaline reaction. A 7.0 reading on the pH scale means that an equal number of free hydrogen (acidic) ions (H^+) and hydroxide (basic) ions (OH^-) are in the sample.

neutral depth *See* normal depth.

neutralism A condition of minimal interaction between two species living in the same environment.

neutralization (1) A chemical reaction in which water is formed by mutual destruction of the ions that characterize

High Concentration of H^+ Ions	H^+ and OH^- Ions in Balance	High Concentration of OH^- Ions
0—1—2—3—4—5	—6—7—8—	9—10—11—12—13—14
Strong Acid	**Neutral**	Strong Base

Neutral position on the pH scale

acids and bases when both are present in an aqueous solution. Typically, the hydrogen and hydroxide ions react to form water ($H^+ + OH^- \rightarrow H_2O$), and the remaining product is a salt. Neutralization occurs with both inorganic and organic compounds. Neutralization can also occur without water being formed, as in the reaction of calcium oxide and carbon dioxide to form calcium carbonate ($CaO + CO_2 \rightarrow CaCO_3$). Neutralization does not mean that a pH of 7.0 has been attained; rather it means that the equivalence point for an acid–base reaction has been reached. (2) The process by which the cell-attachment protein of a virus is bound by an antibody, thereby inhibiting infection by the virus. *See also* equivalence point.

neutralize *See* neutralization.

neutralizer (1) An alkaline substance, such as calcium carbonate ($CaCO_3$, in the form of calcite) or magnesium oxide (MgO, in the form of magnesia), used to neutralize acidic waters; or an acidic substance, such as acetic acid (CH_3COOH) or dilute hydrochloric acid (HCl), used to neutralize alkaline waters. (2) A calcite or magnesia acid-neutralizing filter used to neutralize acidity, reduce free carbon dioxide (CO_2), or both in water and thereby raise the pH of acidic water.

neutron An uncharged elementary particle that has a mass approximately equal to that of the proton. Neutrons are present in all known atomic nuclei except the lightest nucleus (hydrogen). *See also* electron; proton.

neutron activation analysis (NAA) An instrumental technique that is used for measuring the concentrations of elements in a sample based on a measurement of induced radiation. A sample and standard of known weight of a given element are irradiated with neutrons. Neutron activation analysis can be a very sensitive technique for many elements. Because the neutrons must typically be generated in a nuclear reactor, the technique is unavailable to most analytical chemists.

névé A granular mass, somewhat between snow and ice in form, that occurs in a snow field either as the result of compaction caused by the weight of overlying material or because of alternate thawing and freezing through the day and night of water trickling downward from the surface layers of snow to lower levels. As greater depths are reached, the material becomes more compact, eventually producing solid ice.

New England Water Works Association (NEWWA) A section of the American Water Works Association composed mostly of members in Maine, New Hampshire, Rhode Island, Vermont, and Massachusetts.

new water (1) Groundwater artificially brought to the surface or to the land that, without such diversion, would have run to waste or not have appeared in any known source. (2) Stream flow that has been induced by artificial means. *See also* developed water.

newton (N) *See the Units of Measure list.*

Newtonian flow Flow of a fluid in which the viscosity is independent of shear rate, e.g., water or solutions of coagulants. *See also* non-Newtonian flow.

Newtonian fluid A simple fluid in which the state of stress at any point is proportional to the time rate of strain at that point. The proportionality factor is the viscosity coefficient. In equation form:

$$\tau = \mu \frac{du}{dy}$$

Where:

τ = shear stress, in pound force per square foot (newton per square metre)

μ = coefficient of viscosity, in pound force second per square foot (newton second per square metre)

$\frac{du}{dy}$ = rate of strain or velocity gradient, in seconds^{-1}

NEWWA *See* New England Water Works Association.

NF *See* nanofiltration.

NFPA *See* National Fire Protection Association.

NFR (nonfilterable residue) *See* suspended solids; total suspended solids.

NGA *See* National Governors Association.

N_{Ga} *See* Galileo number.

NGRS *See* National Geodetic Reference System.

NGVD *See* National Geodetic Vertical Datum of 1929.

NGWA *See* National Ground Water Association.

NHPA *See* National Historic Preservation Act.

NiCad battery *See* nickel–cadmium battery.

niche (1) A closely delineated space that is occupied by a population in a biocenosis. (2) The role of a species or population in an ecosystem. *See also* biocenosis.

nickel (Ni) A metallic element used in alloys, in electroplated protective coatings, in alkaline storage batteries, and as a catalyst. It is regulated by the US Environmental Protection Agency.

nickel–cadmium (NiCad) battery A small rechargeable battery used to power laptop or notebook computers and other low-voltage devices. To maintain performance NiCad batteries must be completely discharged prior to being recharged, a process called a deep-cycle discharge and recharge.

nicotinamide adenine dinucleotide phosphate–cytochrome P-450 (NADPH–cytochrome P-450) reductase A flavoprotein enzyme that transfers reducing equivalents from nicotinamide adenine dinucleotide phosphate to cytochrome P450 to reduce the heme iron to the ferrous state. This transfer is an essential step in the function of the cytochrome P450 family of enzymes that participate in the metabolism of many chemicals that are foreign to the body. The enzyme is also referred to as nicotinamide adenine dinucleotide phosphate–cytochrome c reductase because cytochrome c is frequently employed in assays of the enzyme.

95 percent confidence interval *See* box-and-whisker plot; confidence interval.

NIOSH *See* National Institute for Occupational Safety and Health.

NIPDWR *See* National Interim Primary Drinking Water Regulation.

nipple A tubular pipe fitting usually threaded on both ends and less than 12 inches (30 centimetres) long. Pipe longer than this is regarded as cut pipe.

NIRS *See* National Inorganics and Radionuclides Survey.

nitrate (NO_3^-) An oxidized ion of nitrogen. Nitrifying bacteria can convert nitrite (NO_2^-) to nitrate in the nitrogen cycle. Sodium nitrate ($NaNO_3$) and potassium nitrate (KNO_3) are used as fertilizer. The nitrate ion is regulated by the US Environmental Protection Agency. *See also* fertilizer; nitrifying bacteria; nitrite; nitrogen; nitrogen cycle.

nitrate removal The process of removing nitrate (NO_3^-) from water. In drinking water, nitrate is typically removed by an ion exchange process in which nitrate is exchanged for another monovalent ion (e.g., chloride, Cl^-) in a packed bed. Biological denitrification has also been suggested. *See also* biological denitrification; ion exchange.

nitrification The process of formation of nitrate (NO_3^-) from reduced inorganic nitrogen compounds. Nitrification in the environment is carried out primarily by autotrophic bacteria and some chemoorganotrophic bacteria. *See also* autotrophic bacteria; denitrification.

nitrification inhibitor A chemical that slows down the conversion of ammonium ion (NH_4^+) to nitrate nitrogen (NO_3^-).

nitrifying bacteria Bacteria involved in the process of nitrification. The *Nitrosomonas* group of bacteria oxidize ammonia (NH_3) to nitrite (NO_2^-). The *Nitrobacter* group of bacteria oxidize nitrite to nitrate (NO_3^-). Both groups are chemolithotrophic organisms, found together in soils, that obtain energy for carbon dioxide (CO_2) fixation by aerobic oxidation of nitrite and nitrate. Some chemoorganotrophic bacteria—such as isolates of the genera *Vibrio, Bacillus, Mycobacterium,* and *Streptomyces*—can also oxidize ammonia to nitrite or nitrate.

nitrite (NO_2^-) An intermediate oxidized ion of nitrogen. Nitrifying bacteria can convert ammonia (NH_3) to nitrite (NO_2^-) to nitrate (NO_3^-) in the nitrogen cycle. Sodium nitrite ($NaNO_2$) is used in curing meats. The nitrite ion is regulated by the US Environmental Protection agency. *See also* ammonia; nitrate; nitrifying bacteria; nitrogen; nitrogen cycle.

nitrobenzene ($C_6H_5NO_2$) A synthetic organic chemical used as a solvent and in various industrial applications. *See also* solvent; synthetic organic chemical.

Nitrobacter A genus of nitrifying bacteria. Members of this genus typically oxidize nitrite (NO_2^-) to nitrate (NO_3^-).

nitrogen (N) A gaseous element (molecular formula N_2) that constitutes 78 percent of the atmosphere by volume and occurs as a constituent of all living tissues in combined form. Nitrogen is present in surface water and groundwater in the forms of ammonia (NH_3), nitrite (NO_2^-), nitrate (NO_3^-), and organic nitrogen. *See also* ammonia; nitrate; nitrite; organic nitrogen.

nitrogen cycle The conservation of nitrogen in nature through various forms, from living animal matter, to dead organic matter, to various stages of decomposition, to plant life, and again to living animal matter.

nitrogen fertilizer A plant nutrient that can cause algal blooms in sensitive water bodies but is also essential for the development of beneficial life forms. *See also* algal bloom.

nitrogen fixation The use of free nitrogen gas (N_2) in the formation of nitrogen compounds during some forms of biological activity.

nitrogenous Pertaining to chemical compounds (usually organic) that contain nitrogen in combined forms. Proteins and nitrates are nitrogenous compounds.

nitrogen–phosphorus detector A type of ionization detector that is specific to compounds containing nitrogen or phosphorous. Compounds eluting from a gas chromatograph are decomposed on an alkali bead; nitrogen- and phosphorus-containing compounds are preferentially ionized. The current generated is proportional to the concentration of the analyte.

***ortho*-nitrophenyl-β-D-galactopyranoside (ONPG) ($C_{12}H_{15}O_8N$)** A colorless organic compound that is a substrate for the enzyme β-galactopyranosidase produced by lactose-fermenting bacteria. When *ortho*-nitrophenyl-β-D-galactopyranoside is hydrolyzed by active lactose-fermenting bacteria, a yellow-colored compound (*ortho*-nitrophenol) is released, indicating the presence of the bacteria.

ortho-nitrophenyl-β-D-galactopyranoside (ONPG)

***ortho*-nitrophenyl-β-D-galactopyranoside ($C_{12}H_{15}O_8N$), 4-methylumbelliferyl-β-D-glucuronide ($C_{16}H_{16}O_9$) (MMO–MUG) technique** A test for detection of total coliforms and *Escherichia coli* in drinking water that uses the *ortho*-nitrophenyl-β-D-galactopyransoside, 4-methylumbelliferyl-β-D-glucuronide minimal medium. Total coliforms hydrolyze *ortho*-nitrophenyl-β-D-galactopyransoside to release *ortho*-nitrophenol, which gives a yellow color in the medium. Hydrolysis of 4-methylumbelliferyl-β-D-glucuronide by *E. coli* yields a compound (4-methylumbelliferone) that fluoresces when exposed to ultraviolet light (366-nanometre wavelength). *See also* minimal medium.

***N*-nitrosodimethylamine (NDMA) ($(CH_3)_2N_2O$)** A synthetic organic chemical that is a known animal carcinogen and is classified as a probable human carcinogen. It has been used in rocket fuels, as a solvent, and as a rubber accelerator. In addition, it may be formed during drinking water treatment by the reaction of secondary amines (R_1–NH–R_2), nitrite (NO_2^-), and acid. *See also* amine; carcinogen; solvent; synthetic organic chemical.

Nitrosomonas A genus of nitrifying bacteria. Members of this genus typically oxidize ammonia (NH_3) to nitrite (NO_2^-).

nm *See* nanometre *in the Units of Measure list.*

NMFS *See* National Marine Fishery Service.

NMOS *See* negative (charge) metal oxide semiconductor.

NMR *See* nuclear magnetic resonance.

***N*-nitrosodimethylamine** *See under* nitrosodimethylamine.

N,N-diethyl-p-phenylenediamine (DPD) *See under* diethyl-p-phenylenediamine.

NOAA *See* National Oceanic and Atmospheric Administration.

NOAEL *See* no-observed-adverse-effect level.

noble Chemically unreactive, especially toward oxygen, or resistant to chemical action, such as corrosion caused by air, water, or (to a lesser degree) acids. Gold, silver, platinum, palladium, and mercury are said to be noble metals because they do not rust and are resistant to acid damage. Certain gases (e.g., helium, neon, radon) are called noble gases because they are inert (chemically inactive and stable).

noble metal A chemically inactive metal (such as gold); a metal that does not corrode easily and is much scarcer (and more valuable) than the so-called useful metals or base metals. *See also* base metal.

Nocardia Common soil organisms that can survive only in aerobic conditions and can result in foaming in diffused aeration processes.

node (1) A special type of point (zero-dimensional object) that serves as a topological junction or end point and may specify a geometric location. (2) Each computer in a network of computers. Each node has a unique address on the network.

NOEL *See* no-observed-effect level.

noise reduction rating (NRR) Rating given to hearing protection devices to indicate the amount of noise reduction provided to the wearer.

NOM *See* natural organic matter.

nominal diameter An approximate measurement of the diameter of a pipe. Although the nominal diameter is used to describe the size or diameter of a pipe, it is usually not the exact inside diameter of the pipe.

nominal filter rating A filter rating indicating the approximate particle size for which a specified majority of such particles will not pass through the filter. It is generally interpreted as meaning that the filter will retain 85 percent of the particles of the size equal to the nominal filter rating.

nominal pipe size (NPS) The designation of a pipe based on the approximate size of its inside diameter, such as a 6-inch pipe or 24-inch pipe. The inside diameter of a pipe changes with the class of pipe and wall thickness, but all pipes near a certain diameter are referred to as that size of pipe. Note that only US customary units are used for nominal pipe sizes in the United States.

nomograph A graph in which three or more scales are used to solve mathematical problems.

nomographic method A method that uses a graph or other diagram to solve formulas and equations.

NOMS *See* National Organics Monitoring Survey.

nonaqueous phase liquid (NAPL) A liquid in contact or mixed in with water but having distinct boundaries and properties different from water. In a mixture of oil and water, oil is the nonaqueous phase liquid. In contrast, in a mixture of alcohol and water, the alcohol is dissolved in the water, and the distinct boundaries and properties do not exist.

nonartesian well A nonflowing well in which the water does not rise above the zone of saturation.

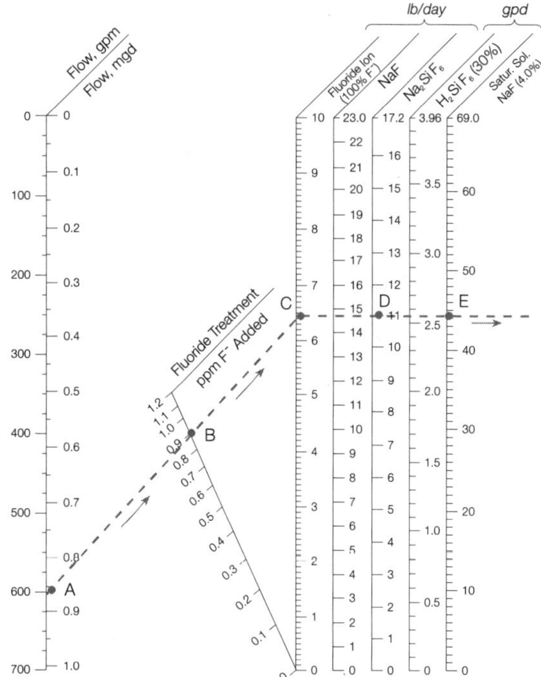

Fluoridation nomograph

nonbinding arbitration A means of mediated conflict resolution such that if the arbitrator's decision is not acceptable to either party, that party may pursue other avenues of resolving the dispute.

nonbiodegradable carbon That portion of the carbon in water that cannot be mineralized by heterotrophic microorganisms. *See also* biodegradable organic carbon.

nonbiological surrogate Any nonbiological indicator analyte—physical or chemical—used as a surrogate for monitoring or measuring for a specific biological agent; for example, measuring removal of turbidity (or particle counts or sizing) during water treatment might be used as a nonbiological surrogate for assessing treatment removal of a biological agent that is not easily monitored, such as *Cryptosporidium parvum* oocysts.

noncarbonate hardness (NCH) The hardness in excess of the carbonate hardness. In equation form,

$$\text{noncarbonate hardness} = \text{total hardness} - \text{alkalinity (carbonate hardness)}$$

when all are expressed as calcium carbonate ($CaCO_3$). *See also* carbonate hardness; hardness; total hardness.

noncensored (NC) data Data that include analytical results below the minimum detection level.

nonchlorine preoxidant An oxidant other than chlorine used to treat source water, e.g., ozone (O_3) or potassium permanganate ($KMnO_4$). *See also* oxidant; ozone; potassium permanganate.

nonclogging impeller An impeller of the open, closed, or semiclosed type designed with large passages for passing large solids.

noncommunity water system (NCWS) A public water system that is not a community water system. *See also* community water system; public water system; nontransient, noncommunity water system; transient, noncommunity water system.

noncompliance analysis The process of comparing data on water quality to desirable quality levels for specific uses. If a given value, usually a concentration, is greater than the prescribed value (or lower in the case of dissolved oxygen) or lies outside the prescribed range, then it is defined as a noncompliant value.

noncondensable Pertaining to gaseous matter that is not liquefied under the existing conditions.

nonconsumptive use Water that is drawn for use but is not consumed, such as water diverted for hydroelectric power generation. It also includes such uses as fishing and boating, where water is still available for other uses at the same location.

noncontributing area (1) In hydrology, a drainage area that, because of physical characteristics or topography, does not contribute surface runoff into a river system. (2) In a determination of drainage basin yields, an area that does not contribute either surface water or groundwater runoff.

nonconventional pollutant Any pollutant that is not listed in a statute or that is poorly understood by the scientific community.

nonculturable state A physiological state in which an organism cannot be grown on laboratory culture media used for its detection and enumeration. This condition can develop for some bacteria (including *Escherichia coli* and other enterics) that are exposed to adverse environmental conditions.

nondegradable Resistant to decomposition or decay by biological means, such as bacterial action, or by chemical means, such as oxidation, heat, sunlight, or solvents.

nondepositing velocity The water velocity that will maintain suspended sediment during movement.

nondetection The inability to detect an analyte because the signal from a detector is indistinguishable from the background signal.

noneroding velocity The water velocity that will maintain silt in movement and at the same time will not scour a stream or canal bed. This velocity depends on the material and age of the bed. *See also* safe velocity.

nonfilterable residue (NFR) *See* suspended solids; total suspended solids.

nonflammable Pertaining to materials that do not ignite or burn easily. This term is used as a rating for certain products, as shown on material safety data sheets, and certain clothes.

nonflowing well A well that does not discharge water at the surface except through the operation of a pump or other lifting device. Such a well is also called a pump well.

nonfunctional landscaping A built landscape, designed for aesthetics rather than to support a practical use or activity.

nongenotoxic Pertaining to a chemical, usually a carcinogen, for which experimental evidence exists to suggest that its toxic effects are not produced by direct interaction with deoxyribonucleic acid (DNA). This particularly poor (but widely used) terminology arises largely because mutation is recognized as an essential part of cancer. In those cases where the cancer is initiated by a chemically induced mutation, the response is considered linear at low doses. If, however, a chemical induces these mutations by some indirect effect (e.g., increasing cell turnover or damaging DNA as a result of inflammatory processes that occur only at high doses), the extra risk of cancer would not be predicted to be linear, and the chemical would be termed nongenotoxic. *See also* epigenetic carcinogen; genotoxic; genotoxic carcinogen.

nongraphic data Digital representations of the characteristics, qualities, or relationships of map features and geographic locations, usually stored in conventional alphanumeric formats. The term *nongraphic* is often used to differentiate those data that do not describe the graphic images of the map features. Nongraphic data are often called textual data or attributes. They are related to geographic locations or graphic elements and are linked to them in the geographic information system through common identifiers or other mechanisms. Often, nongraphic data are managed separately from the graphic data because of their different characteristics or their maintenance and use in other systems. Nongraphic data are sometimes referred to as tabular data.

nonhardening salt A salt containing amounts of such agents as calcium chloride ($Ca(Cl)_2$) or magnesium chloride ($Mg(Cl_2)$) that become soft or liquid by attracting and absorbing moisture from the air and that prevent salt caking and bridging. *See also* bridging.

nonhealth hazard (1) A cross-connection or potential cross-connection involving any substance that generally would not be a health hazard but would constitute a nuisance, or be aesthetically objectionable, if introduced into the potable water supply. (2) Any contamination that would not be a health hazard but may or may not constitute a nuisance.

nonhydric soil A soil that has developed under predominantly aerobic soil conditions.

nonionic Not having an ionic charge.

nonionic flocculant A polyelectrolyte that has no net electrical charge, e.g., unhydrolyzed polyacrylamides.

nonionic polyelectrolyte A polyelectrolyte that forms both positively and negatively charged ions when dissolved in water.

nonionic polymer A polymer that has no net electrical charge.

nonionic resin An ion exchange resin without exchangeable charged ions (e.g., chloride (Cl^-) or hydrogen (H^+)). A nonionic resin either can be a macroreticular, nonionic, acrylic ester type or—without ester cross-linking—could have a less polar structure and take the form of a polystyrene ($(C_6H_5CH{:}CH_2)_n$) backbone cross-linked with divinylbenzene ($C_6H_4(CH{:}CH_2)_2$). A nonionic resin is used to isolate and concentrate natural organic matter. The hydrophobic fraction, defined as the organic compounds adsorbed on the acrylic ester polymer resin at acidic pH, represents the most commonly used operational definition of aquatic humic substances. Organic solutes that do not adsorb on the acrylic ester polymer resin at acidic pH are defined as the hydrophilic fraction. The acrylic ester polymer and polystyrene divinylbenzene have been used in tandem for the isolation of both humic substances and hydrophilic acids. *See also* aquatic humic substance; hydrophilic; hydrophobic; natural organic matter; polymer; resin.

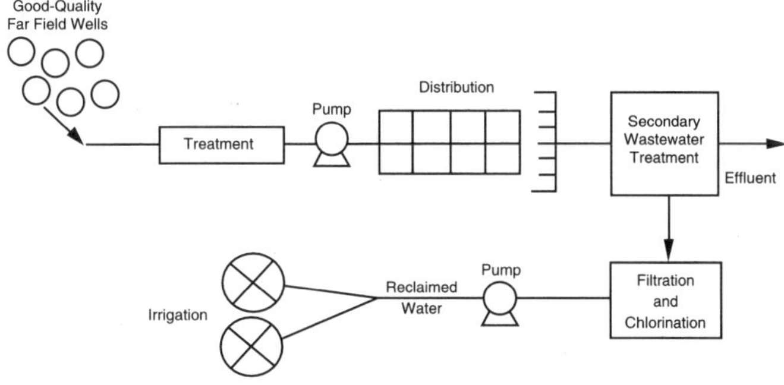

Nonpotable reuse

nonionic surfactant Any of a general family of surfactants for which the entire molecule remains associated in solution. Nonionic molecules orient themselves at surfaces not by an electrical charge, but having separate grease-solubilizing and water-soluble groups within the molecule.

nonisotopic gene probe A specific sequence of nucleic acid (deoxyribonucleic acid or ribonucleic acid) that has been labeled with a nonradioactive molecule or compound for detection after binding to its complementary sequence.

non-Newtonian flow Flow of a fluid having a viscosity that changes with increasing shear rate, e.g., solutions of flocculants. *See also* Newtonian flow.

nonoperating revenue Income of a utility that is not derived from normal operations, such as rents from nonutility facilities, interest on investments or on bank time deposits, and revenue for rendering nonutility services to others. It is also referred to as "other income" in financial statements.

nonparametric data set A set of data in which the individual observations do not follow a normal distribution. Many environmental observations fall in this category by having a significant number of observations that either are not detectable or are at much lower concentrations than the average (mean) value of all the observations. For example, a data set generated by conducting lead or copper corrosion experiment contains high concentrations of the constituents at the beginning of the experiments followed by a prolonged period of very low or nondetectable concentrations. *See also* normal distribution.

nonparametric test A statistical test of significance that is distinguished from parametric tests in that it is applicable with small samples for which the data are not equally spaced, need not be normally distributed, and do not require the population mean, standard deviation, or population proportion parameters that characterize parametric statistics. An example of a data set that requires a nonparametric test is a set of observations in which several of the reported values are indicated as not detectable. Some nonparametric tests (chi-square, for example) have a direct parametric counterpart. When the required assumptions apply, a parametric test will provide greater decision-making sensitivity and should be used; however, when the parametric assumptions are not met, nonparametric tests will avoid incorrect inferences and are clearly superior. In many situations, no parametric approach is available, so the nonparametric tests provide a basis for making statistical inferences that would otherwise not be possible. Frequently used nonparametric tests include the following: the chi-square test (contingency tables), runs test for sequential randomness, Wilcoxon ranked sum test, Mann–Whitney test, Kruskal–Wallis test for independent samples, Friedman test for matched samples, and Spearman's rank correlation test. Each test derives a statistical value from the input data that is then compared to a critical value for the test in order to determine statistical significance. *See also* chi-square distribution.

nonpathogenic Not disease producing.

nonpoint pollution *See* nonpoint waste source.

nonpoint source (NPS) *See* nonpoint waste source.

nonpoint waste source (1) Waste material that enters a water body from overland flow rather than out of a pipe or channel. (2) An unconfined discharge of waste.

nonporous membrane A membrane that acts as a solute barrier without distinct transmembrane openings and uses a diffusion-controlled mechanism for solute separation (as in reverse osmosis or nanofiltration pressure-driven membrane processes).

nonpotable *See* nonpotable water.

nonpotable reuse The use of reclaimed water for nonpotable purposes (i.e., not intended for direct human consumption). Such water is typically used for lawn watering, ornamental and edible food crop irrigation, industrial uses, and other purposes.

nonpotable use Water use for means other than drinking, cooking, or showering and bathing. Typical nonpotable uses include irrigation, maintenance, and some non-food-producing industrial processes (e.g., cooling tower makeup). The level of treatment required for nonpotable water is less than for drinking water treatment. Treated wastewater can be a source for nonpotable water. Because most water consumption is for nonpotable uses, many water-deficient regions are determining that having separate, dual distribution systems (one each for potable and nonpotable water) is cost-effective and necessary to meet water demand. *See also* nonpotable reuse.

nonpotable water Water that may contain objectionable pollution, contamination, minerals, or infective agents and is considered unsafe, unpalatable, or both for drinking.

nonpurgeable organic carbon (NPOC) That fraction of total organic carbon not removed after purging a sample with an inert gas. It is sometimes referred to as nonvolatile total organic carbon, but this is not a recommended term. *See also* dissolved organic carbon; particulate organic carbon; purgable organic carbon; total organic carbon.

nonreferee method An analytical method for testing drinking water featuring speed and practical usefulness rather than high accuracy. It is used for process control or general information rather than settlement of disputed test results.

nonsettleable matter *See* nonsettleable solids.

nonsettleable solids Very small, fine suspended solids, typically colloidal particles of less than 0.1 micrometre in diameter, that will not settle to the bottom of a vessel or water body when the water is calm and nonturbulent in what is considered a reasonable time of about 2 hours.

nonsignificant noncomplier (NSNC) A public water system that violates one or more National Primary Drinking Water Regulations but for which the violation is not serious, is infrequent, or is not persistent enough to meet the US Environmental Protection Agency's definition of a significant noncomplier. *See also* significant noncomplier.

nonsilting velocity The water velocity that will maintain silt in movement.

nontransient, noncommunity water system (NTNCWS) A public water system that is not a community water system but that regularly serves at least 25 of the same people at least 6 months of the year. Examples of such systems include schools, factories, and hospitals that have their own water supplies. *See also* community water system; public water system; transient, noncommunity water system.

nontributary groundwater Underground water in an aquifer that neither contributes to nor draws from a natural surface stream to any measurable degree.

nonuniform Irregular.

nonuniform flow Flow in which the slope, cross-sectional area, and velocity change from section to section in the channel.

nonviable system A water system with technical, financial, or managerial weaknesses that may make it incapable of providing safe drinking water and complying with drinking water regulations on a continuing basis.

nonvolatile matter Material that is only slightly affected by the actions of organisms and is not lost upon ignition of dry solids at 1,022° Fahrenheit (550° Celsius) (e.g., sand, salt, iron, calcium, and other mineral materials). *See also* fixed matter; fixed solids; inorganic waste; volatile matter; volatile solids.

nonvolatile organic carbon (NVOC) *See* nonpurgeable organic carbon.

nonvolatile residue *See* nonvolatile matter.

nonwetland Any area that has sufficiently dry conditions that hydrophytic vegetation, hydric soils, wetland hydrology, or any combination of these is lacking. It includes upland as well as former wetlands that are effectively drained.

no-observed-adverse-effect level (NOAEL) The maximum dose or exposure of a chemical to which an animal or human has been exposed without an effect that is clearly detrimental to health being produced. This exposure level is the most common point of departure for calculating doses or concentrations of a chemical that can be considered safe. *See also* dose.

no-observed-effect level (NOEL) The maximum dose or exposure of a chemical at which no effect can be detected in an animal or human. Such effects need not be adverse; they simply represent physiological adaptation to an exposure. For example, the no-observed-effect level might refer to an effect on an enzyme activity that is not clearly involved in compromising health or producing overt pathology. *See also* dose.

normal (1) A unit of measure for indicating a solution's normality. For example, the equivalent weight of sulfuric acid (H_2SO_4) is 49 (98 [the molecular weight] divided by 2 because two replaceable hydrogen ions are in the acid). Thus, a 1-normal (1*N*) sulfuric acid solution—a solution with a normality of 1—would consist of 49 grams of H_2SO_4 dissolved in each litre of solution. A 2*N* solution would contain 98 grams of H_2SO_4 per litre of solution. (2) A mean or average value established from a series of observations for purposes of comparison of some meteorological or hydrological event. (3) A straight line perpendicular to a surface or to another line. (4) Pertaining to a Gaussian statistical distribution for which the arithmetic and statistical means are equal. (5) Pertaining to hydrocarbon molecules that contain a single unbranched chain of carbon atoms, usually indicated by the prefix *n*- [e.g., *n*-hexane—$CH_3(CH_2)_4CH_3$]. *See also* hydrocarbon; normal distribution; normality.

normal circumstance The soil and hydrology conditions that are normally present, without regard to whether the vegetation has been removed.

normal depletion curve A graphical presentation of the mean of values obtained from a number of groundwater depletion curves.

normal depth (1) The depth of water in an open channel or conduit that occurs when, for a given flow, the velocity is uniform. It is the hypothetical depth under conditions of steady, uniform flow. Under uniform flow conditions, the water depth, flow area, discharge, and the velocity distribution at all sections throughout the channel reach remain unchanged; the water surface line and the channel bottom are parallel to each other. Normal depth is also called uniform depth or neutral depth. (2) The depth of water measured perpendicular to the bed.

normal distribution A statistical distribution that is technically called the standard normal density function but is most frequently referred to as the bell curve. It is mathematically described by its probability density function:

$$f(x) = \frac{1}{\sigma\sqrt{2\pi}}\exp\left[-\frac{1}{2\sigma^2}(x-\mu)^2\right]$$

Where:

σ = population standard deviation
μ = population mean
x = individual datum value

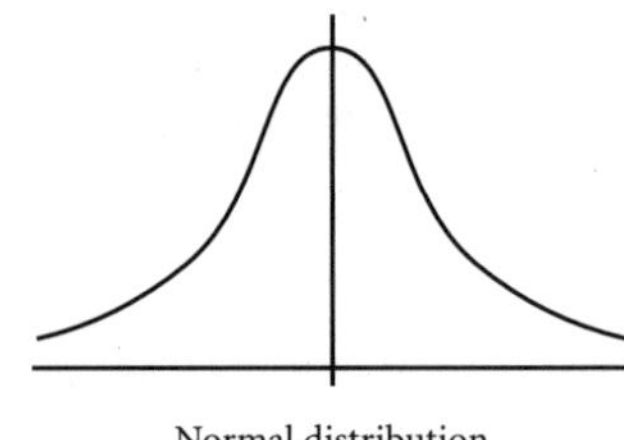

Normal distribution

The normal distribution can take on numerous flat or peaked, narrow or wide appearances, but it is always symmetrical when plotted as a function of the frequency of occurrence. In a normal distribution, the mean and the median are the same number. The shape of the distribution is determined by the mean and standard deviation of a sample taken from a population. All normal distributions are related in a way that makes it possible to use a table of probabilities for one standardized normal distribution to find probabilities for any other distribution in the normal family. The area under the normal distribution can be defined in terms of standard deviations from the mean of the population under consideration. A range of units of 1 standard deviation on both sides of the mean encompasses 68 percent of the area under the normal curve; two standard deviations include 95 percent of the area under the normal curve; and three standard deviations include virtually 100 percent of the area under the curve. The area under the curve is directly related to the probability of occurrence; hence, the normal curve has extensive applicability in large-sample theory for drawing inferences from sample data. A normal distribution is also called a Gaussian distribution.

normal erosion The erosion of land undisturbed by activities of humans or their agents.

normal flow (1) The flow that prevails for the greatest portion of the time; the mean flow or the average flow. (2) Flow at normal depth.

normal flow filtration The flow of the entire feedwater stream in one direction directly through the filter media. The flow is usually normal, or perpendicular, to the media surface area. *See also* crossflow filtration.

normal moisture capacity *See* field capacity.

normal pressure (1) The pressure exerted by any substance against a surface in a direction perpendicular to that surface. (2) The mean atmospheric pressure at any place. (3) The pressure exerted most of the time in a fluid system.

normal solution A solution with a normality of 1. *See also* normal; normality.

normal year A year during which the precipitation or stream flow approximates the average for a long period of record.

normality (*N*) The number of gram-equivalent weights of solute in 1 litre of solution. The equivalent weight of any material is the weight that would react with or be produced by the reaction of 7.999 grams of oxygen or 1.008 grams of hydrogen. Normality is a convenient unit for certain calculations of quantitative analysis. *See also* normal.

normalization For reverse osmosis and nanofiltration membrane processes, a technique whereby performance data are numerically adjusted to a constant set of conditions for comparison of membrane performance at varying times. The goal of normalization is to eliminate the effects of changing operating variables not related to the physical condition of the membrane, such as feedwater quality and temperature, feed pressure, and recovery. *See also* normalized (permeate) flow; normalized salt (solute) passage.

normalized (permeate) flow For reverse osmosis and nanofiltration membrane processes, the (permeate) flow rate through the membrane calculated by numerically adjusting the actual permeate flow rate to constant operating conditions. *See also* normalization.

normalized salt (solute) passage For reverse osmosis and nanofiltration membrane processes, the (solute) salt passage through the membrane calculated by numerically adjusting the actual salt (solute) passage to constant operating conditions. *See also* normalization.

normalized water sales and forecast *See* weather-normalized water sales and forecast.

NORS *See* National Organics Reconnaissance Survey.

Norwalk virus A virus that belongs to the family Caliciviridae, a name derived from the cuplike appearance of the stained particle on electron micrographs. The Norwalk virus causes severe gastroenteritis in humans, mostly children.

NOS *See* not otherwise specified.

nosocomial infection An infection acquired in a medical facility, e.g., a hospital. The infection can appear in a patient either before or after discharge. Infections among staff personnel are included.

not analyzed (NA) A phrase entered into a table of data for a sample that was collected but for which a particular analysis was not performed. The phrase is used because a blank entry might wrongly imply that the parameter was not detected at that site. Because NA is also an abbreviation for "not available" or "not applicable," a footnote to the table should define the abbreviation when first used. *See also* not applicable; not available; not sampled.

not applicable (NA) A phrase used to indicate that a particular analysis is not applicable (or required) for a sample. The phrase is used because a blank entry might wrongly imply that the parameter was not detected at that site. Because NA is also an abbreviation for "not analyzed" or "not available," a footnote to the table should define the abbreviation when first used. *See also* not analyzed; not available.

not available (NA) A phrase used when an interim reporting of data is made and certain data are not yet available. The phrase is used because a blank entry might wrongly imply that the parameter was not detected at that site. Because NA is also an abbreviation for "not analyzed" or "not applicable," a footnote to the table should define the abbreviation when first used. *See also* not analyzed; not applicable.

not detected (ND) A phrase used in reporting test results to indicate that the substance being tested was not detected *at or in excess of the detection limit* by the equipment or method being used for the particular test. This term implies that trace amounts of this substance may be present at a level lower than the detection limit of the test equipment and method. *See also* limit of detection.

not otherwise specified (NOS) A phrase used in shipping regulations for generic classes of substances to which a restriction applies and for which individual members are not listed in the regulations. The class of substance should also be identified by its technical name.

not quantitated (NQ) A phrase sometimes used when a parameter is detected at a trace level lower than the quantitation limit, depending on reporting requirements. Because detection of a signal lower than a quantitation limit may represent a false positive (the signal does not correspond to the analyte of interest), many laboratories simply report "not detected" instead of "not quantitated." *See also* not detected.

not reported (NR) A phrase used when data are being reported and a particular value does not meet quality assurance requirements (and that sample has not been re-analyzed or resampled). The phrase should be entered into the table of data for the corresponding sample because a blank entry might wrongly imply that the parameter was not detected at that site. *See also* quality assurance.

not sampled (NS) A phrase used when data are reported for a routine monitoring program and a particular location was not sampled. The phrase is entered into the table of monitoring data for that location because a blank entry might wrongly imply that the parameter was not detected at that site. *See also* not analyzed.

notch An opening in a dam, spillway, or measuring weir for the passage of water.

notched weir A weir having a substantial width of crest broken at intervals by a notch of known hydraulic characteristics, usually a V-notch. *See also* broad-crested weir; V-notch weir.

notice (1) A word used to call attention to information that is especially significant in understanding and operating equipment or processes safely. (2) A statement from a water utility to consumers about drinking water quality (e.g., a boil water advisory). *See also* boil water advisory; notice of violation; public notification; warning.

notice of proposed rule making (NPRM) An official federal agency notice of a proposed rule that is published in the *Federal Register.*

notice of violation A official notice issued by a primacy agency under the Safe Drinking Water Act to a public water system that the system is in violation of a Primary Drinking Water Regulation.

notifiable disease A disease required by statute to be reported to the appropriate public health authority when the diagnosis is made. In the United States, each state, as well as some localities, requires the reporting or notification of some 40 infectious diseases; the specific notifiable diseases are based on specific legislation in each jurisdiction.

NOX *See Spirillum NOX.*

nozzle (1) A short, cone-shaped tube used as an outlet for a hose or pipe. The velocity of the emerging stream of water is increased by the reduction in cross-sectional area of the nozzle. (2) A short piece of pipe with a flange on one end and a saddle flange on the other end. (3) A side outlet attached to a pipe by such means as riveting, brazing, or welding.

nozzle aerator An aerator consisting of a pressure nozzle through which water is propelled into the air in a fine spray. A nozzle aerator is also called a spray aerator.

nozzle jet The stream of water issuing from the end of a nozzle.

NPDES *See* National Pollutant Discharge Elimination System permit.

NPDWR *See* National Primary Drinking Water Regulations.

NPL *See* National Priorities List.

NPOC *See* nonpurgeable organic carbon.

NPRM *See* notice of proposed rule making.

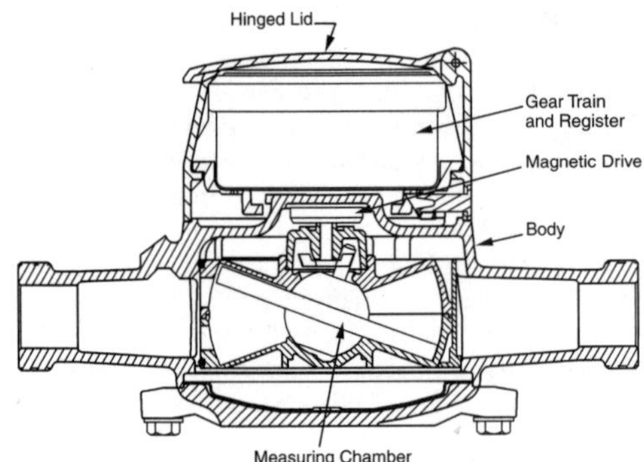

Nutating-disk meter

NPS *See* nominal pipe size; nonpoint waste source.

NPSH *See* net positive suction head.

NPV *See* net present value.

n′(x/y) *See* bromine incorporation factor for haloacetic acids.

NQ *See* not quantitated.

NR *See* not reported.

NRA *See* Negotiated Rulemaking Act.

NRC *See* National Research Council.

NRCS *See* Natural Resources Conservation Service.

NREN *See* National Research and Education Network.

NRMRL *See* National Risk Management Research Laboratory.

NRR *See* noise reduction rating.

NRWA *See* National Rural Water Association.

NS *See* not sampled.

NSC *See* National Safety Council.

NSDWR *See* Secondary Drinking Water Regulation.

NSF *See* NSF International.

NSF International A noncommercial, not-for-profit agency located in Ann Arbor, Mich., and devoted to research, education, and service. Services provided include development of consensus standards, voluntary product testing and certification with policies and practices that protect the integrity of registered trademarks, education and training, and research and demonstration, all relating to public health and the environmental sciences. NSF International was formerly called the National Sanitation Foundation.

NSF-MAL *See* maximum allowable level.

NSFNET *See* National Science Foundation Network.

N_{Sh} *See* Sherwood number.

NSNC *See* nonsignificant noncomplier.

NSPE *See* National Society of Professional Engineers.

NTIS *See* National Technical Information Service.

NTNC *See* nontransient, noncommunity water system.

NTNCWS *See* nontransient, noncommunity water system.

N_{total} *See* total nitrogen.

NTP *See* National Toxicology Program.

ntu *See* nephelometric turbidity unit *in the Units of Measure list.*

nuclear energy Energy obtained from an atom in nuclear reactions.

nuclear magnetic resonance (NMR) A spectrometric technique used in the identification of organic compounds. Absorption of electromagnetic radiation depends on the characteristics of the atomic nuclei present in a compound. Proton (H-1) nuclear magnetic resonance and carbon-13 nuclear magnetic resonance are two versions of the technique, each with advantages in the elucidation of molecular structure. Although nuclear magnetic resonance spectrometry is a powerful analytical tool when used in conjunction with other techniques, the related instruments are typically present in research laboratories rather than water utility laboratories.

nuclear radiation Particulate and electromagnetic radiation emitted from atomic nuclei in various nuclear processes. The important types of nuclear radiation are alpha and beta particles, gamma rays, and neutrons. All types of nuclear radiation are ionizing radiation, capable of producing ions directly or indirectly by their passage through matter, including living tissue.

nuclei *See* nucleus.

nucleic acid Any of numerous large acidic biological polymers that are found concentrated in the nuclei of all living cells. Nucleic acids contain phosphoric acid, sugar, and purine ($C_5H_4N_4$) and pyrimidine ($C_4H_4N_2$) bases. Two types are ribonucleic acid and deoxyribonucleic acid.

nucleic acid probe A segment of ribonucleic acid or deoxyribonucleic acid that has been labeled with an enzyme, antigenic substrate, chemiluminescent moiety, or radioisotope and that can bind with high specificity to a complementary sequence of nucleic acid. Many types of nucleic acid probes have been developed for detection of all types of microorganisms.

nucleus (1) The center of an atom, classically conceived as made up of positively charged particles called protons and uncharged particles called neutrons. (2) Organized and membrane-bound deoxyribonucleic acid. *See also* neutron; proton.

nuisance bacteria A common term for iron and sulfur bacteria, which are associated with water quality problems involving slime formation, colored water, and tastes and odors. The organisms in this group are morphologically and physiologically diverse and may be filamentous or single-celled, autotrophic or heterotrophic, and aerobic or anaerobic.

nuisance species Species of plants that detract from or interfere with a mitigation project, such as most exotic species and those indigenous species whose populations proliferate to abnormal proportions. Nuisance species may require removal through maintenance programs.

null hypothesis *See* hypothesis; statistically significant.

number-average molecular weight (M_n) The weighted average of the corresponding molecular weights of the individual components of natural organic matter (or some other heterogeneous mixture), as illustrated by the following equation:

$$M_n = \frac{\sum W_i(\mathrm{MW})_i}{\sum w_i}$$

Where:

w_i = the mass of the mixture's *i*th component

$(\mathrm{MW})_i$ = the molecular weight of the mixture's *i*th component

Although the measured value of such a basic compositional property is representative of the mixture as a whole, it is not necessarily representative of any component of the mixture. *See also* molecular weight; natural organic matter.

numerator The part of a fraction above the division line. A fraction indicates division of the numerator by the denominator. *See also* denominator.

numerical modeling A technique for mathematically simulating some physical, chemical, or statistical process. *See also* model.

nutating Oscillating or nodding back and forth. Some valve operators utilize a nutating gear motion to generate rotation. The drive gear nutates upon a driven gear with one less tooth, thus producing a rotation. The driven gear advances one tooth for each nutation of the drive gear.

nutating-disk meter An instrument for measuring the flow of a liquid. Liquid passing through a chamber causes a disk to nutate, or roll back and forth. The total number of rolls is then counted.

nutrient An element or compound essential as a raw material for an organism's growth and development. For example, carbon dioxide (CO_2), nitrogen, phosphorous, potassium, and numerous mineral elements are essential plant nutrients, as are water and oxygen.

nutrient pollution Contamination of water resources by excessive inputs of nutrients. In surface waters, excess algal production is the major concern. *See also* algal bloom.

NVLAP *See* National Voluntary Laboratory Accreditation Program.

NVOC *See* nonvolatile organic carbon.

NWA *See* National Water Alliance.

NWF *See* National Wildlife Federation.

NWI *See* National Wetlands Inventory.

NWP *See* nationwide permit.

NWRA *See* National Water Resources Association.

O

O *See* oxygen.
o *See ortho.*
O_3 *See* ozone.
OBL *See* obligate wetlands species.
obligate pathogen A pathogenic organism, such as a virus, that is unable to multiply outside a living host.
obligate plant species (OBL) *See* obligate wetlands species.
obligate upland species (UPL) A plant species that almost always occurs in uplands but that may rarely (less than 1 percent of the time) be found in wetlands (e.g., shortleaf pine).
obligate wetlands species (OBL) A plant species that is nearly always found in wetlands. Its frequency of occurrence in wetlands is 99 percent or more.
OBr^- *See* hypobromite ion.
observation An act of watching the performance of an experiment or computer program. An observation is recorded to identify any change in the status of a variable.
observation bias In epidemiology, systematic error or variation caused by the collection of information on disease in a noncomparable manner from exposed and unexposed groups (for experimental, cohort, or cross-sectional studies) or the collection of information on exposure in a noncomparable manner from cases and controls (for a case–control study). "Blindness" of study participants, investigators, or both and objectivity in obtaining information help minimize this bias. Observation bias is sometimes called information bias. *See also* case-control epidemiologic study; cohort epidemiologic study; cross-sectional epidemiologic study; experimental epidemiologic study.
observation well A well placed near a production well to monitor changes in the aquifer.
observational epidemiologic study A study in which the investigator has no control over the assignment of exposure. The vast majority of epidemiologic studies are observational.
obsolescence The decrease in the value of assets as a result of economic, social, technical, or legal changes. This type of value reduction is an element of depreciation.
occasional storm A rainfall of an intensity that may be expected to occur once in 10 to 25 years. Such a storm is also called an extraordinary storm.
occlusion An absorption process by which one substance is taken in and retained in the interior rather than on the external surface of another substance, sometimes occurring by coprecipitation.
Occupational Safety and Health Act of 1970 Public Law 91-596, enacted December 1970. It is designed to prevent injury and death in industry and is administered at the federal level by the Occupational Safety and Health Administration.
Occupational Safety and Health Administration (OSHA) A federal agency in the Department of Labor that administers safety and health regulations and their enforcement.
Occupational Safety and Health Administration (OSHA) 101 log A form required for every entry on the Occupational Safety and Health Administration 200 log. The 101 supplements the 200 log with greater detail about an event. A state's worker's compensation form or an insurance company report is an acceptable substitute, as long as it includes the same information as the federal 101 form.
Occupational Safety and Health Administration (OSHA) 200 log The federal log of occupational injuries and illnesses, on which the occurrence, extent, and outcome of each case are recorded during the year. It is also required to be posted for employees to review during the month of February.
Occupational Safety and Health Review Commission (OSHRC) A federal commission that reviews and decides the outcome of contested citations and penalties that are issued by the Occupational Safety and Health Administration.
occupational safety A condition of reduced or eliminated dangers, perils, and risks that threaten the safety of a worker performing responsibilities and duties, creating a safe working environment.
Ocean Dumping Act (ODA) A federal act that, as amended (33 U.S.C. Sections 1401, 1445), regulates intentional ocean disposal of materials and authorizes related research. The act is

Occupational saftey notice

administered by the US Environmental Protection Agency, US Army Corps of Engineers, National Oceanic and Atmospheric Administration, and US Coast Guard.

ocean disposal A method of providing ultimate disposal of residuals (e.g., sludge) in which the residuals are discharged into an ocean environment. In general, the disposal is conducted under conditions where it has been determined that the practice will not adversely affect the environmental conditions surrounding the discharge location.

ocean outfall Facilities for the discharge of waters (such as storm drainage, wastewater treatment plant effluent, or membrane treatment waste concentrate) into the ocean.

ocean water Seawater generally having a salinity of approximately 30,000–55,000 milligrams per litre total dissolved solids, depending on location.

oceanic climate A climate in which the meteorological phenomena are primarily influenced by the ocean. Such a climate is characterized by relatively high humidity and fairly uniform temperatures and is also called a marine climate.

OCl^- *See* hypochlorite ion.

OCR *See* optical character recognition.

OD *See* outside diameter.

ODA *See* Ocean Dumping Act.

odds ratio (OR) The odds (chance) of an exposed person having a given disease divided by the odds of an unexposed person having the disease. In case–control studies, the odds ratio is the odds of exposure among cases divided by the odds of exposure among controls. The odds ratio is equivalent to the rate ratio, especially if the disease is rare, e.g., cancer. *See also* case-control epidemiologic study; rate ratio.

odor A quality of something that stimulates the olfactory organ (i.e., its smell). Many compounds have a characteristic odor that serves as an effective means of identification.

odor control The elimination or reduction of odors in a drinking water supply by aeration, algae elimination, superchlorination, activated carbon treatment, and other methods. *See also* odor threshold concentration; threshold odor number.

odor threshold *See* threshold odor.

odor threshold concentration (OTC) The concentration of a chemical compound in water at which the odor detection threshold is surpassed for 50 percent of a group of subjects. The odor detection threshold of a substance is defined as the concentration at which a subject gives a positive response in 50 percent of the cases in which the stimulus is presented. Thus, individual odor detection thresholds are evaluated, and from that data the odor threshold concentration of the compound (for the group of testers) is determined. Individuals differ appreciably in sensitivity to odors. Inter-individual differences in sensitivity show a range of 100 to 10,000 as indicated by threshold determinations. Small parts of the population are even odor-blind (anosmic) to certain compounds. Thus, odor threshold concentrations in the literature may differ depending on the test methods used and the sensitivity of the subjects. *See also* odor; threshold odor number.

odor unit (OU) *See* threshold odor number.

OEM *See* original equipment manufacturer.

off-gas Air or vapor given off or expelled as a by-product or result of an operation or treatment process.

Office of General Counsel (OGC) The office within the US Environmental Protection Agency responsible for the legal affairs of the agency.

Office of Ground Water and Drinking Water (OGWDW) The office within the US Environmental Protection Agency responsible for development and implementation of National Primary Drinking Water Regulations and the Public Water Supply Supervision program.

Office of Information and Regulatory Affairs (OIRA) The office within the Office of Management and Budget responsible for reviewing federal agency regulatory actions.

Office of Information Resources Management (OIRM) The office within the US Environmental Protection Agency responsible for information management and computer services.

Office of Management and Budget (OMB) The agency of the executive branch of the federal government that has the following responsibilities: to assist the president in preparing the budget and in formulating the US government's fiscal program; to supervise and control the administration of the budget; to assist the president by clearing and coordinating department advice on proposed legislation and by making recommendations concerning presidential action on legislative enactments; to assist in developing regulatory reform proposals and programs for paperwork reduction; to assist in considering, clearing, and, where necessary, preparing proposed executive orders and proclamations; to plan and develop information systems that provide the president with program performance data; to plan, conduct, and promote evaluation efforts that assist the president in assessing program objectives, performance, and efficiency; and to keep the president informed of the progress of activities by US government agencies with respect to work proposed, initiated, and completed, together with the relative timing of work among the several agencies of the government, all to the end that the work programs of the executive branch agencies may be coordinated and that the moneys appropriated by Congress may be expended in the most economical manner.

Office of Research and Development (ORD) The office within the US Environmental Protection Agency responsible for conducting and funding research to support agency regulatory programs.

off-peak Pertaining to periods of time when customer demands for water are low.

off-peak power That part of the available load or energy that can be produced at off-peak hours outside the power load curve when the combined primary and secondary load has fallen to less than plant capacity.

Electric potential, measured in volts, is similar to hydraulic pressure.
A. Voltage Measurement

Electric current, measured in amps, is similar to hydraulic flow rate.
B. Current Measurement

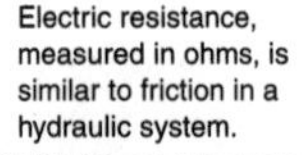

Electric resistance, measured in ohms, is similar to friction in a hydraulic system.
C. Resistance Measurement

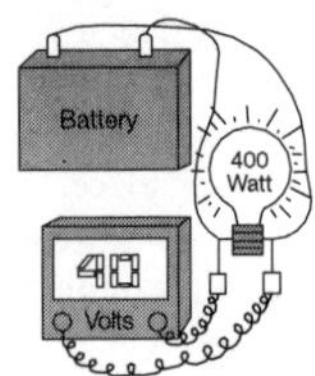

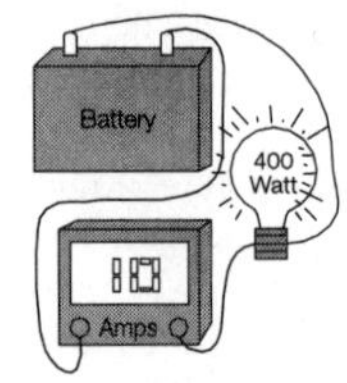

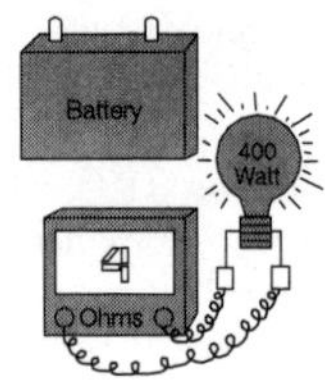

Representation of Ohm's law

off-peak rate A separate, lower-than-average unit charge applied to water delivered during off-peak periods.

offset (1) The difference between the actual value and the desired value (or set point), characteristic of proportional controllers that do not incorporate reset action. This type of offset is also called drift. (2) A combination of elbows or bends that brings one section of a line of pipe out of line with, but into a line parallel with, another section. (3) A pipe fitting in the approximate form of a reverse curve, made to accomplish the same purpose.

offset bend A pipe fitting constructed in a long S-shape (reverse curve) to connect pipes that are not in alignment or to provide a slight alignment change when necessary to go past an obstruction. The use of two offset bends near each other will allow a change in alignment and place the pipe back on the original alignment after going under or around an obstruction.

offset error A small process error that results from proportional control action whenever a load changes.

off-site (1) Not on the same parcel as a wetland that has been adversely affected by a regulated activity. (2) Remote; not nearby.

off-site meter reading (OMR) A process whereby a meter reader, either walking or driving, obtains a reading from the vicinity of the water meter without directly contacting it.

offstream use Water withdrawn from surface or groundwater sources for use at another place.

OGC *See* Office of General Counsel.

ogee A reverse curve shaped like an elongated letter *S*. The downstream faces of overflow dams are often made to have this shape.

OGWDW *See* Office of Ground Water and Drinking Water.

OH⁻ *See* hydroxide ion.

ohm (Ω) *See the Units of Measure list.*

Ohm's law A fundamental law of electricity that relates electromotive force to current and resistance. Ohm's law states that 1 volt of electromotive force will cause 1 ampere of current to flow through 1 ohm of resistance. It is written as

$$E = IR$$

Where:

E = electromotive force, in volts
I = current flow, in amperes
R = resistance, in ohms

oil Any of a wide range of viscous substances that are quite different in chemical nature. Oils derived from animals or from plant seeds or nuts are almost chemically identical with fats, the only difference being one of consistency at room temperature. Petroleum (rock oil) is composed of hydrocarbon mixtures comprising hundreds of chemical compounds. *See also* grease; hydrocarbon.

oil of vitriol Sulfuric acid (H_2SO_4). This term is obsolete.

oil well flooding *See* water flooding.

OIRA *See* Office of Information and Regulatory Affairs.

OIRM *See* Office of Information Resources Management.

old river (1) A river in which all members are graded. (2) A stream system in which all channels, including wet weather rills, are subject to aggradation. It has a wide floodplain, broad meander belt, and gentle slopes.

old valley A stream valley that has reached the last stage of development. Such valleys have very wide, flat bottoms, low and more or less indistinct sidewalls, and stream channels that meander from side to side of the valley bottoms.

oleum A solution of sulfuric acid (H_2SO_4) and sulfur trioxide (SO_3) that is a form of fuming sulfuric acid. The term is used to describe a stage in the production of sulfuric acid.

olfactory fatigue A condition in which a person's nose, after exposure to certain odors, is no longer able to detect the odor.

oligodynamic action The bacteriostatic action exerted by very small amounts of heavy metals, such as copper, silver, and zinc, that deactivates bacteria and creates a hostile environment for the growth of bacterial colonies.

oligonucleotide A synthetically made, single strand of deoxyribonucleic acid, 10 to 25 bases in length, that is complementary to a specific sequence in the target organism's chromosome. An oligonucleotide is usually labeled with a radioactive or nonradioactive molecule and used as a probe for detecting organisms.

oligonucleotide probe A deoxyribonucleic acid or ribonucleic acid probe with fewer than 50 base pairs.

oligotrophic Pertaining to water bodies that are poor in such nutrients as phosphorus, nitrogen, and calcium but have abundant oxygen at all depths. Such waters support little growth of plankton. *Contrast with* eutrophic.

oligotrophic lake A deep, steep-sided lake containing little organic matter and a low level of nutrients. Because of its depth, low level of photosynthesis, limited bacterial action, and narrow littoral zone, the lake's hypolimnion possesses a high level of oxygen. *See also* hypolimnion.

oligotrophic water Water having a low level of nutrients and supporting little organic production.

oligotrophic water quality *See* oligotrophic water.

O&M *See* operations and maintenance.

OMB *See* Office of Management and Budget.

ombrometer A rain gauge.

omega (ω) *See* angular frequency.

omega (Ω) *See* ohm *in the Units of Measure list.*

OMR *See* off-site meter reading

once-through operation A process operation without recycle, reuse, or recirculating flow streams. For example, with once-through cooling systems, the cooling water is used only one time before being discharged to waste.

oncogene A gene that has been altered in expression or by mutation, resulting in disruption of normal control of cell division and thus giving rise to tumors. The corresponding normal genes are referred to as protooncogenes. Activation of oncogenes has been associated with a number of carcinogenic chemicals found in drinking water, including the haloacetic acids.

oncology The study of cancer.

one hundred year (100-year) flood The flood with an average frequency of occurrence of once in 100 years, now usually the design flood for most federal and state water resource agencies.

1,1-dichloroethane *See under* dichloroethane.

1,1-dichloroethylene *See under* dichloroethylene.

1,1-dichloropropene *See under* dichloropropene.

1,1,1-TCP *See* 1,1,1-trichloropropanone.

1,1,1-trichloroacetone *See under* trichloroacetone.

1,1,1-trichloroethane *See under* trichloroethane.

1,1,1-trichloropropanone *See under* trichloropropanone.

1,1,2-trichloroethane *See under* trichloroethane.

1,1,2,2-tetrachloroethane *See under* tetrachloroethane.

1,2-dibromo-3-chloropropane *See under* dibromo-3-chloropropane.

1,2-dichloroethane *See under* dichloroethane.

1,2-dichloroethylene *See under* dichloroethylene.

1,2-diphenylhydrazine *See under* diphenylhydrazine.

1,2,4-trichlorobenzene *See under* trichlorobenzene.

1,2,4-trimethylbenzene *See under* trimethylbenzene.

1,3-dichloropropane *See under* dichloropropane.

1,3-dichloropropene *See under* dichloropropene.

one-hit model An extrapolation model that assumes a linear relationship between the probability of inducing pathology (e.g., cancer) and the dose of a chemical carcinogen. This model implicitly assumes that only a single irreversible change (hit, i.e., mutation) is necessary to produce cancer. Although this assumption may apply to certain tumors that arise in early childhood (i.e., have very short latent periods), it is not consistent with the development of most human cancers. *See also* linearized multistage model; log-probit model; logit model; multihit model; multistage model; probit model; Weibull model.

one-tailed test A statistical procedure to determine the significance of the difference between the respective means of two data distributions. In a one-tailed test, the difference in only one direction (i.e., either greater or smaller) is considered. For example, a one-tailed t-test can be performed to determine whether ferric chloride ($FeCl_3$) is a better coagulant than alum ($Al_2(SO_4)_3 \cdot 14H_2O$) for turbidity removal. In a two-tailed test, differences in either direction are considered. For example, a two-tailed t-test can be performed to evaluate whether membrane units manufactured by one company are any different from those manufactured by another company with respect to *Giardia* removal. *See also* two-tailed test.

one-way analysis of variance A statistical procedure that compares the means of several normally distributed data sets to test the hypothesis that the data sets are statistically similar.

on-line turbidimeter A turbidimeter that continuously samples, monitors, and records turbidity levels in water. *See also* turbidimeter.

on-peak Pertaining to short periods of time when customer demands for water are high.

ONPG *See* minimal medium; *ortho*-nitrophenyl-β-D-galactopyranoside.

ONPG–MUG *See* minimal medium.

on-site (1) On the same parcel at which a wetland has been adversely affected by a regulated activity. (2) At the same location where a given structure exists or activity is taking place.

on-site assessment A method of determining the capability and qualifications of a laboratory to perform specific environmental analyses. The term is used as an integral part of the National Environmental Laboratory Accreditation Program. An assessment team would visit a laboratory and audit various aspects of the operation. Such an assessment would include a review of policies, procedures, records, personnel, analytical instruments, and supplies for the purpose of judging compliance with accreditation standards. *See also* National Environmental Laboratory Accreditation Program.

oocyst A structure that is produced by two coccidian protozoa (i.e., *Cryptosporidium*) as a result of sexual reproduction during the life cycle. The oocyst is usually the infectious and environmental stage, and it contains sporozoites. For the enteric protozoa, the oocyst is excreted in the feces. *See also Cryptosporidium.*

oocyte The germ cell within the ovary.

oogenesis The production of germ cells in a female. In humans this occurs prior to birth when primordial germ cells become segregated from somatic cells. Thus, the female is born with her full complement of germ cells. This is in contrast to the male, for whom sperm are produced throughout adult life. Oogenesis is important in evaluating reproductive toxicology.

opacity The capacity of matter to block the passage of light or other radiant energy, such as heat. One measure of opacity is the percentage of light transmission through a plume. Opacity is the opposite of transparency, and an object with a high degree of opacity is said to be opaque. For example, milk has much higher opacity than drinking water.

open centrifugal pump A centrifugal pump in which the impeller is built with a set of independent vanes.

open channel Any natural or artificial waterway or conduit in which water flows with a free surface.

open conduit In general, an open artificial or natural duct for conveying fluids.

open dumping Disposal of trash in an open area with incidental burning and without covering.

open fire line A fire line or fire system that is connected to a distribution system and contains fire hydrants or fire hose connections through which water can easily be discharged from the system. These systems are usually metered.

open impeller An impeller without attached sidewalls.

open storage Uncovered storage of water in reservoirs that are not enclosed.

open system interconnect (OSI) model A structure for standard interfaces and communication protocols developed and promoted by the International Standards Organization. The model addresses low-level communications (e.g., physical and electrical interfaces, basic network protocols) as well as higher-level communications (e.g., network routing, error correction, communication between application software programs).

open well A well that is large enough to allow a person to enter it and descend in it to the water level. Wells of this class are generally dug wells 3 feet (1 metre) or more in diameter or width.

open-bottomed well *See* open-ended well.

open-channel constriction (1) Any obstruction that reduces the normal cross section of a channel. (2) In a general sense, any type of constriction in an open channel, whether natural or artificial, such as a dam, bridge, or culvert.

open-channel drainage Drainage accomplished by open channels or ditches.

open-channel flow Flow of a liquid with its surface exposed to the atmosphere. The conduit may be an open channel or a closed conduit flowing partly full.

open-ended well A well in which the lower end of the casing at the bottom of the well is left open or unobstructed to permit entrance of water. The casing may have other openings. Such a well is also called an open-bottomed well.

operant conditioning The process of reshaping the behavior of an animal by interactions with its environment. In behavioral toxicology, behavior can be conditioned by providing positive or negative reinforcement to encourage or discourage particular behavior. Interference with the acquisition or extinction of a particular behavioral pattern as a result of positive or negative reinforcement, respectively, is frequently used during tests for the ability of a chemical to interfere with learning.

operating cost The continuing costs involved in running a business, excluding capital costs of acquiring personal and real property required for the business.

operating cycle (1) In filtration, the complete filter cycle consisting of filter service, backwash, and return to service. (2) In ion exchange, the cycle of service run, backwash and regeneration, slow rinse, fast rinse, and return to service. (3) In membrane processes, the cycle of service run, membrane cleaning, and returning to service.

operating expense Expenses necessary for the daily running of an enterprise, rendering of service, and collection of revenue. Such expenses may include the daily costs of maintenance, labor, materials, chemicals, power, and depreciation.

operating floor (1) The floor in a rapid granular filter building on which the operating and indicating devices are generally installed. (2) The floor of a pumping station.

operating nut The end connection of a valve stem, usually a square nut that allows a valve key or operating handle to attach to the valve stem and transmit torque for opening or closing the valve.

operating pressure The manufacturer's specified range of pressure—expressed in pounds per square inch (psi), pascals (Pa), or kilopascals (kPa)—within which a water-processing device or water system is designed to function. A range of 30 psi (210 kPa) to 100 psi (690 kPa) is often indicated. Operating pressure is also called working pressure.

operating pressure differential The operating pressure range for a hydropneumatic system. For example, a system might operate such that, when the pressure drops to less than 40 pounds per square inch (275 kilopascals) the pump will come on and run until the pressure builds up to 60 pounds per square inch (415 kilopascals), at which point the pump will shut off.

operating reserve Funds set aside by water utilities, to allow for the following possibilities: seasonal variations in cash receipts (summer versus winter water sales); cool, wet weather that curbs irrigation water use in summer; unplanned expenditures stemming from natural disasters, major main breaks, and equipment breakdowns; and any other condition that might require a cash payment that is not provided for in the planned revenue for the budget year. Operating cash reserves are separate from reserves that are dedicated for specific purposes, such as connection charges used for capital improvements, capital replacement reserves, or moneys set aside for debt service. For most water utilities, operating cash reserves are targeted at 20 to 30 percent of total annual operating expenditures (excluding major capital programs and debt service).

operating revenue Revenue derived directly from the operations of the utility, either from the sale of water or from closely allied activities.

operating system (OS) The software that interacts directly with the processing unit to control all basic functions of a computer system, such as allocating memory and disk space; loading programs for execution; and controlling input–output operations, user access, and security. The operating system must reside in memory. Many different manufacturers and software companies have written operating system programs.

operating temperature The manufacturer's recommended feedwater or inlet water temperature for a water treatment system.

operation expense Costs incurred in operating source-of-supply facilities, pumping facilities, water purification facilities, and transmission and distribution facilities, as well as costs for customer accounting, customer services, administration, and general operations.

operational salt efficiency *See* salt efficiency.

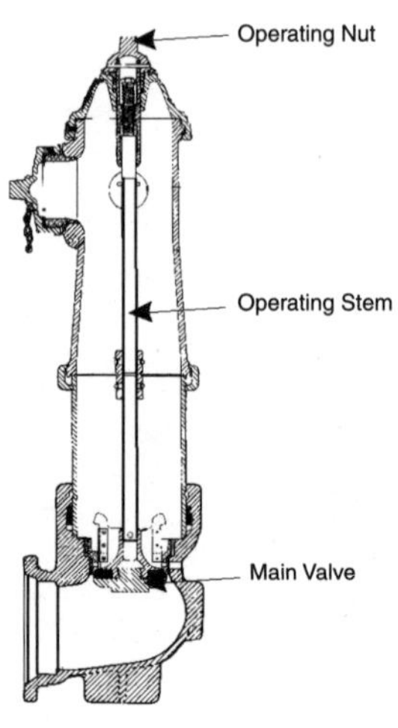

Operating nut on a fire hydrant

operations and maintenance (O&M) The ongoing process of carrying out activities necessary to fulfill the mission of an organization and to keep a system in such condition as to be able to achieve those objectives. Operations represent organized procedures for enabling a system to perform its intended function; maintenance represents organized procedures for keeping the system (equipment, plants, facilities) in such condition that it is able to continue performing its intended function.

operations and maintenance (O&M) cost *See* operations and maintenance expenditures.

operations and maintenance (O&M) expenditures All spending to operate and maintain all water treatment and supply facilities from the source of water through the distribution system. It includes labor, benefits, supplies, rent, utilities, furniture and other minor capital, vehicles, computers, travel, subscriptions, and all other line items of cost that are part of the utility's budget system. Although budgetary distinctions are made, for many utilities this category also includes customer service, computer operations, human resources, public relations, and executive offices expenditures. Some utilities separately classify some of the nonoperating departments as administrative and general expense and allocate part of it to line departments and to capital projects. Operations and maintenance expenditures include neither major capital programs that are funded from debt or from connection fees nor debt service.

operations and maintenance (O&M) manual Bound information, prepared for a particular system, that defines the equipment included in that system and its proper operation.

operator A person who operates the equipment of a water treatment plant.

operator training An educational program for the instruction and development of water treatment plant operators.

opportunistic pathogen An organism that is considered nonpathogenic to immunologically competent individuals but can cause disease in individuals who are immunologically compromised because of existing illness or are immunosuppressed because of medical treatment, old age, or very young age.

optic atrophy Degeneration of the eye, associated with methyl mercury (CH_3Hg) poisoning.

optical character recognition (OCR) The technology that recognizes and converts human-readable symbols or letters into a digital computer input. Scanners are based on this technology. Advanced scanning software provides extremely high accuracy rates.

optical disk technology The equipment and knowledge that provide the capability to store data on optical disk drives. Optical drives now exhibit storage densities up to 20 times those of magnetic media and allow several gigabytes to be stored on a single disk. Most optical drives use a technique in which a laser burns small impressions on a disk surface. The drive reads data by sensing variations in the light reflected from this surface. With this laser technique, however, data can be written to a particular portion of the disk only one time, and no data can be erased. This approach is called "write once read many" (or WORM). Optical drives that can erase and rewrite to the disks are also available but are less common. Optical disk technology is also called optical storage technology.

optical fiber An enclosed bundle of thin, transparent fibers of glass or plastic through which data are transmitted as pulses of light generated by lasers.

optical scanner A peripheral device that digitally encodes information in a raster form from hard-copy maps and documents by optically scanning an image consisting of line work, text, and symbols. The scanner senses variations in reflected light from the surface of the document. *See also* raster image.

optimal point of coagulation The point at which the best floc develops in the shortest time because of selected conditions of pH, mixing, and coagulant–flocculant dose.

optimization of corrosion control For the purposes of the Lead and Copper Rule, the treatment that minimizes the lead and copper levels at users' taps while ensuring that the treatment does not cause the water system to violate any national drinking water regulation. *See also* Lead and Copper Rule.

optimized coagulation Coagulation that occurs under conditions—coagulant type, dose, and pH—that achieve the most effective removal of one or more of the following problems: turbidity, particulates, precursors and disinfection by-products. Normally, optimized conditions are determined by running a series of different coagulant doses and pH levels, varying one parameter at a time. However, achieving optimized conditions for the removal of disinfection by-product precursors may not be practical in high-alkalinity waters. Therefore, the Disinfectant/Disinfection By-Products Rule requires the removal of disinfection by-product precursors by enhanced coagulation rather than optimized coagulation. *See also* coagulation; Disinfectant/Disinfection By-Products Rule; disinfection by-product precursor; enhanced coagulation; jar test.

optimized ribonucleic acid–polymerase chain reaction (optimized RNA–PCR) technology A ribonucleic acid-based polymerase chain reaction protocol that has been optimized in regard to sensitivity and specificity. It is sensitive enough to detect one deoxyribonucleic sequence and specific enough to detect the sequence of interest. It uses a ribonucleic acid strand and reverse-transcriptase (a molecule that allows synthesis of deoxyribonucleic acid from ribonucleic acid) to make a deoxyribonucleic clone that is in turn used for geometric amplification.

optimized RNA–PCR technology *See* optimized ribonucleic acid–polymerase chain reaction technology.

OR *See* odds ratio.

oral Of, through, or by the mouth.

oral–fecal transmission *See* fecal–oral transmission.

ORD *See* Office of Research and Development.

ordinary high-water mark That mark on the shore established by the highest nonflood water level and indicated by such physical characteristics as a clear, natural line impressed on the bank; shelving; changes in the character of soil; destruction of terrestrial vegetation; the presence of litter and debris; or other appropriate means that account for the characteristics of the surrounding areas.

ordinary storm A rainfall of an intensity that may be expected to occur with an average frequency of once in 5 to 10 years.

organelle A membrane-bound structure that forms part of a microorganism and performs a specialized function.

organic Of, relating to, or derived from living organisms (plants or animals); of, relating to, or containing carbon compounds. *See also* inorganic; organic compound.

organic acid (R–COOH) A carbon-containing acid, as opposed to an inorganic (mineral) acid. *See also* acid; amino acid; carboxylic acid; fatty acid.

organic adsorption The removal of organic chemicals by use of a solid adsorbent, such as activated carbon.

organic carbon Carbon derived from living organisms (plants or animals). *See also* dissolved organic carbon; inorganic carbon; total organic carbon.

organic carbon, dissolved *See* dissolved organic carbon.

organic carbon, total *See* total organic carbon.

organic cartridge An air-purifying filter used with a respirator that is designed to remove organic contaminants from inhaled air. *See also* chemical cartridge.

organic chemical A chemical having a carbon–hydrogen structure.

organic chloramines Organic compounds formed during the chlorination or chloramination of water containing organic nitrogen (e.g., N-chlorodimethylamine, $(CH_3)_2NCl$). Unlike inorganic chloramines (e.g., monochloramine (NH_2Cl), dichloramine ($NHCl_2$)), organic chloramines are not considered to be disinfectants. *See also* chloramines; organic compound; organic nitrogen.

organic compound A carbon-containing compound that is derived from living organisms, e.g., a hydrocarbon, alcohol, ether, aldehyde, ketone, carboxylic acid, or carbohydrate. *See also* alcohol; aldehyde; carboxylic acid; ether; hydrocarbon; ketone.

organic contamination Contamination of water, soil, and so on, with an organic compound, such as a volatile organic chemical or synthetic organic chemical. *See also* organic compound; volatile organic compound; synthetic organic chemical.

organic flocculation A method for concentrating viruses from water samples. Beef extract is the organic substance used to form flocs at a low pH (3.5). Viruses adsorbed to the flocs can be recovered by raising the pH to 9.0 in an eluting solution.

organic halogen A halogen-substituted (e.g., chlorine- or bromine-substituted) organic compound. Most disinfection by-products produced during chlorination are organic halogens. *See also* disinfection by-product; halogen; organic compound.

organic material *See* organic matter.

organic matter Material substances derived from living organisms (plants or animals). *See also* natural organic matter.

organic matter degradation The breaking down of organic matter into smaller structural units by hydrolysis, reduction, oxidation, or any combination of these, carried out either chemically or biologically.

organic nitrogen Nitrogen chemically bound in organic molecules, such as proteins, amines ($R_{3-x}NH_x$, where $x = 0$, 1, or 2), and amino acids ($H_2N–R–COOH$).

organic peroxide (R–O–O–R') Any organic compound containing a bivalent O–O group (i.e., the oxygen atoms are univalent). Such compounds are strong oxidizing agents. Some organic peroxides may be created during the reactions of ozone (O_3) with natural organic matter. *See also* hydrogen peroxide.

organic substance A chemical substance of animal or vegetable origin and having carbon in its molecular structure.

organic-free water Water that does not contain organic substances.

organics Organic compounds. *See also* organic compound.

organism Any individual animal or plant having diverse organs and parts that function together as a whole to maintain life and its activities.

organobromine compound Any organic compound containing bromine as a constituent. Some organobromine compounds are formed when bromine reacts with organic substances. Others are synthetic organic chemicals. *See also* bromine-substituted by-products; bromine-substituted organic; synthetic organic chemical.

organochloramine *See* organic chloramine.

organochlorine compound Any organic compound containing chlorine as a constituent. Some organochlorine compounds are formed when chlorine reacts with organic substances. Others are synthetic organic chemicals. *See also* chlorine-substituted by-products; chlorine-substituted organic; synthetic organic chemical.

organochlorine pesticide Any pesticide relating to or belonging to the chlorine-substituted hydrocarbon family.

organogenesis The developmental process whereby cells differentiate into the phenotype associated with their function in particular tissues and form into organs.

organoleptic Affecting or employing one or more of the organs of special sense (e.g., sense of smell or taste).

organophosphorus pesticide A pesticide containing phosphorus as a constituent. Such chemicals damage or destroy cholinesterase, the enzyme required for nerve function in an animal body.

```
RO  O(or S)
  \//
   P
  / \
R'O  O(or S)-R"
```

Organophosphorus pesticide

organotins A family of alkyl (C_nH_{2n+1}) tin compounds widely used as stabilizers for plastics (e.g., dibutyltin compounds—$(C_4H_9)_2SnX_2$, where X is an organic or inorganic group). *See also* inorganic; organic.

orifice An opening (hole) in a plate, wall, or partition. An orifice flange or plate placed in a pipe consists of a slot or a calibrated circular hole smaller than the pipe diameter. The difference in pressure between a point in the pipe upstream and a point at the orifice may be used to determine the flow in the pipe.

orifice box A stilling basin containing a submerged orifice used for measuring the flow of water or other liquids. *See also* orifice; stilling basin.

orifice feed tank A small tank into which a chemical solution is fed from a large chemical storage tank. An orifice feed tank is used for applying the chemical solution to water.

orifice flow Flow through an orifice. For such flow, the inlet is completely submerged, but discharge from the outlet may occur under free-flow conditions, partially or completely submerged, as in a conduit. *See also* sluice flow.

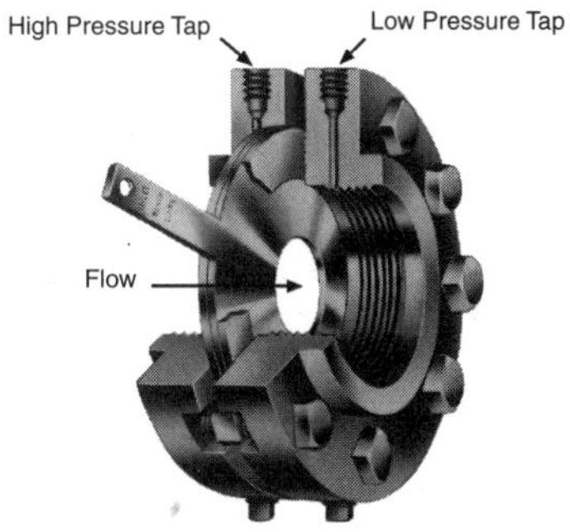

Orifice meter

orifice with full contraction An orifice placed so that filaments of water approach it freely from all directions.

orifice meter A device or plate with an orifice smaller than the diameter of the pipe, installed between flanges to measure the flow of water in the pipe. The velocity is determined by measuring the drop in pressure as water flows through the smaller circular orifice in the plate. The rate of flow is based on the area and velocity for both the pipe and the orifice.

orifice plate A plate containing an orifice. In pipes, the plate is usually inserted between a pair of flanges, and the orifice is smaller in area than the cross section of the pipe.

orifice with suppressed contraction An orifice placed so that some filaments of water must approach it in a direction more nearly parallel to the direction of the jet. If the orifice is flush with a side or a bottom, the contraction on that side of the orifice is wholly suppressed. Rounding the inner edge also suppresses contraction.

original equipment manufacturer (OEM) The company that actually manufactured a component of a preassembled system. All mechanical and electrical systems are made up of components that are manufactured by other equipment manufacturers. Even large companies use components built by someone else and assembled into a complete system. Disk drives and monitors are two examples of such components used in brand name computers.

original interstice An interstice created when the rock in which it occurs came into existence, as a result of the processes by which the rock was formed. Pores in sand or gravel and open pores in sandstone are typical. An original interstice is also called a primary interstice.

O-ring A round, circular, rubber-material gasket shaped like the letter *O* that is used to seal around round objects, such as the ends of two pipes between the bell and spigot; between a valve stem and valve bonnet; or between a fire hydrant steam and a fire hydrant bonnet. O-rings can vary both in diameter of the gasket and in circular length.

orogeny The process or processes of mountain formation, especially the intense deformation of rocks by faulting and folding that, in many mountainous regions, has been accompanied by invasion of molten rock, metamorphism, and organic eruption.

orographic effect The lifting and cooling of air masses as they flow over mountains, leading to condensation and proportional changes in precipitation rate type (e.g., rain to mixed precipitation to snow) with increasing elevation.

orographic precipitation Precipitation caused by the interference of rising land in the path of moisture-laden wind. A horizontal air current striking a mountain slope is deflected upward, and the consequent dynamic cooling associated with the expansion of the air produces precipitation if the air contains sufficient aqueous vapor. *See also* orographic rainfall.

orographic rainfall Rainfall caused by the interference of rising land in the path of moisture-laden wind. *See also* orographic precipitation.

ORP *See* oxidation–reduction (redox) potential.

Orsat method A gas analysis method that uses a portable apparatus to determine mixtures of methane (CH_4), carbon dioxide (CO_2), carbon monoxide (CO), and molecular oxygen (O_2).

***ortho-* (*o*)** (1) A prefix meaning "straight ahead." This prefix is used in organic chemistry in naming disubstitution products derived from benzene (C_6H_6) in which the substituent atoms or functional groups are located on adjoining carbon atoms. It refers to the 1,2-position. (2) A prefix used in inorganic chemistry to mean the most hydrated form of an acid or its salt, in contrast to a less hydrated form that is indicated by the prefix *meta*. For example, H_3PO_4 is orthophosphoric acid; HPO_3 is metaphosphoric acid. *See also meta*; *para*.

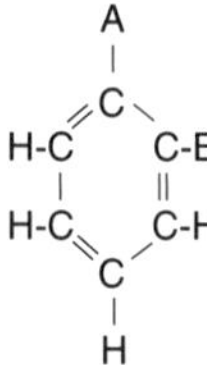

A depiction of 1-A-2-B-benzene where B is in the ortho position (on the second carbon) in respect to A

***ortho*-dichlorobenzene** *See under* dichlorobenzene.

orthokinetic (shear) flocculation A method of agglomerating particles (i.e., making large particles) by the application of external energy to the fluid containing the particles. The agitation or mixing from the external energy results in both spatial and temporal variation in fluid velocity, thus promoting contact between particles that follow the fluid motion. *See also* differential settling; perikinetic flocculation.

***ortho*-nitrophenyl-β-D-galactopyranoside (ONPG)** *See under* nitrophenyl-β-D-galactopyranoside.

orthophosphate (1) A salt that contains phosphorus as PO_4^{3-}. (2) A product of the hydrolysis of condensed (polymeric) phosphates. (3) A nutrient required for plant and animal growth. *See also* nutrient.

***ortho*-tolidine** *See under* tolidine.

***ortho*-(2,3,4,5,6-pentafluorobenzyl)hydroxylamine (PFBHA)** *See under* (pentafluorobenzyl)hydroxylamine.

OS *See* operating system.

oscillation A periodic movement back and forth, or up and down.

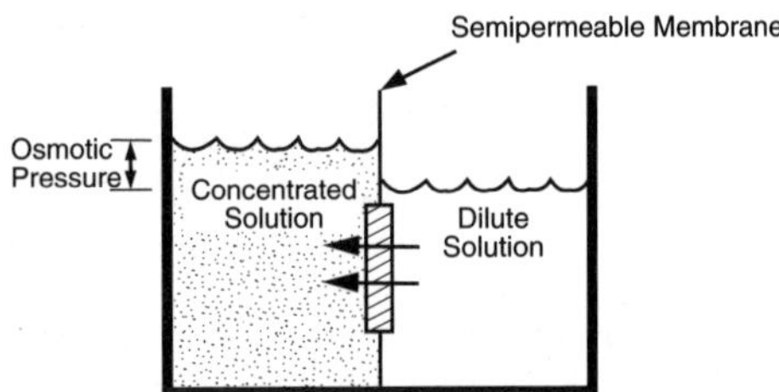

The principle of osmosis

OSHA *See* Occupational Safety and Health Administration.

OSHA 101 log *See* Occupational Safety and Health Administration 101 log.

OSHA 200 log *See* Occupational Safety and Health Administration 200 log.

OSHRC *See* Occupational Safety and Health Review Commission.

OSI model *See* open system interconnect model.

osmometry The measurement of osmotic pressure (the pressure produced by or associated with osmosis). *See also* osmosis; osmotic pressure.

osmosis A natural phenomenon whereby water (or some other solvent) diffuses from the lower-concentration side to the higher-concentration side of a permselective (semipermeable) membrane barrier in a process of equalizing concentrations on both sides.

osmotic pressure The pressure exerted on a solution as a result of osmosis. It is dependent on the molar concentration of the solutes and the temperature of the solution. An approximation of osmotic pressure for a natural water is 1 pound per square inch (6.9 kilopascals) per 100 milligrams per litre total dissolved solids. *See also* osmosis.

osmotic water transport The transfer of water from compartments of lower solute concentration through a semipermeable membrane into compartments of higher solute concentration.

osteosarcoma A tumor derived from cells involved in bone formation. This tumor was associated with fluoride in the National Toxicology Program's study of sodium fluoride (NaF).

OT test *See under* tolidine test.

OTA test *See under* tolidine arsenite test.

OTC *See* odor threshold concentration.

OU *See* threshold odor number.

ounce (oz) *See the Units of Measure list.*

ounce-in. (oz-in.) *See the Units of Measure list.*

outbreak *See* common source epidemic; epidemic.

outcrop (1) That part of a geologic formation or structure that appears at the surface of the earth. (2) Bedrock that is covered only by surficial deposits, such as alluvium.

outfall A conduit through which a treated wastewater flows for ultimate discharge. Outfalls are typically associated with discharges to freshwater or ocean environments.

out-of-kind Pertaining to biological characteristics not closely approximating those of a wetland before it was adversely affected by a regulated or agricultural activity.

out-of-kind replacement In wetland restoration, the act of providing or managing substitute resources to replace the functional values of the resources lost. The substitute resources are physically or biologically different from those lost.

outlet (1) The downstream opening or discharge end of a pipe, culvert, or canal. (2) An opening near the bottom of a dam for draining the reservoir. (3) In plumbing, a discharge opening for water from the water distributing system to a fixture, to atmospheric pressure (except into an open tank that is part of the water supply system), to a boiler or heating system, or to any water-operated device or equipment that requires water to operate but is not a part of the plumbing system.

outlet channel A channel provided to carry water out of or away from a reservoir, lake, or other body of surface water.

outlet control The mechanical means of controlling the relationship between the headwater elevation and the discharge, placed at the outlet or downstream end of any structure through which water may flow.

outlet hose nozzle The small type of hydrant outlet, usually 2.5 inches (62.5 millimetres) in diameter. It is often called a hose connection.

outlet zone The final zone in a sedimentation basin that provides a smooth transition from the settling zone to the effluent piping.

outlier A point in a graph or data set for which the value is suspiciously different from the others in the group. An outlier may result because of experimental errors or because of unexplained factors that are associated with an observation.

outside diameter (OD) The distance from the outside surface of a pipe or cylinder to the opposite outside surface, measured through the center. The outside diameter or dimension of pipe is critical in connecting two pipes or in selecting sleeves, fittings, and repair clamps to be installed on a pipe.

outward-flow turbine A reaction turbine in which the water or steam enters the runners and flows radially outward away from the axis of the runner.

ova Mature female germ cells that develop into new organisms of the same species following fertilization. The singular form is *ovum*.

oven A chamber used to dry, burn, or sterilize materials.

overall efficiency, pump *See* wire-to-water efficiency.

overbank flow That portion of stream flow that exceeds the carrying capacity of the normal channel and overflows one or more adjoining floodplains.

overburden The soil, silt, sand, gravel, and other unconsolidated material that overlies bedrock in a given location.

overchute A flume to carry water, constructed transversely over a canal.

overdevelopment The act of exceeding the economic yield of groundwater from an aquifer. If the transmissibility of an aquifer is limited, excessive withdrawals in a restricted area may cause local drawdown sufficient to make it uneconomical to continue to withdraw water at that rate, even though the physical yield limit for the entire aquifer has not been reached. Such a condition is called local overdevelopment. *See also* mining groundwater.

overdraft (1) The withdrawal of water from a water supply in excess of the amount that a user has a legal right to take.

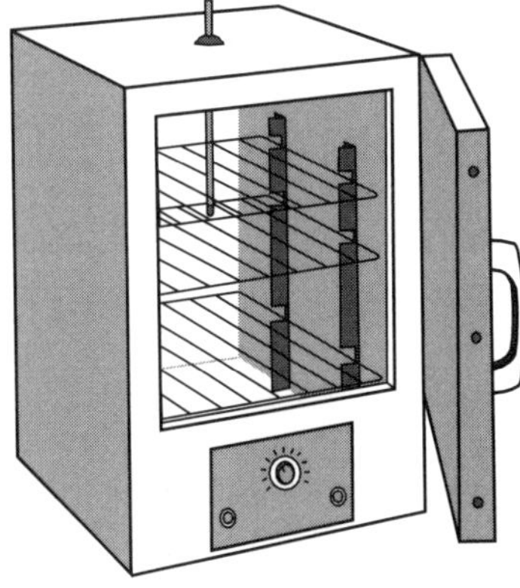

Laboratory oven

(2) The withdrawal of water from an aquifer system in excess of the amount that natural replenishment can sustain without permanent adverse effects on the aquifer water quality or some other damage occurring.

overfall dam A dam constructed with a crest to permit the overflow of water. Such a dam is also called a spillway dam.

overfall weir A device or structure over which water is allowed to flow or waste. *See also* overflow weir.

overfill (1) The part of a dam over which the water spills. (2) The overpouring water.

overflow (1) The water that exceeds the ordinary limits of a containing body, such as the stream banks, the spillway crest, or the ordinary level of a container. (2) To cover or inundate with water or other fluid.

overflow channel An artificial waterway provided to conduct away from the structure the water overflowing from a reservoir, aqueduct, or canal by way of an overflow device provided for that purpose. An overflow channel is also called a spillway channel.

overflow rate A unit process attribute expressing the hydraulic surface loading rate on a treatment basin, such as a settling basin. The units of expression, volume per time per surface area (e.g., gallons per day per square foot or cubic metres per day per square metre) are equivalent to the units of velocity, thereby relating to the settling velocity of the "design" particle to be removed. Particles with settling velocities equal to or greater than the overflow rate will theoretically be removed, as will a fraction of the particles with settling velocities less than the overflow rate.

overflow siphon An outlet installed to divert the liquid above the elevation of the overflow by means of a siphon.

overflow standpipe A pipe that is in a water tank and has the top open and lower than the top of the tank, with the bottom curving sideways and passing through the side of the tank. Water that reaches the top of the pipe overflows through the pipe and out of the tank rather than continuing to rise and place pressure on the top of the tank.

overflow stream In geology, a spillway from a standing water body; any effluent of a lake.

overflow tower A vertical outlet from a water conduit under pressure, inserted on summits to limit the pressure on the pipe by allowing water to be diverted when it reaches a given design elevation.

overflow weir A steel or fiberglass plate designed to evenly distribute flow out of a water treatment vessel. In a sedimentation basin, the weir is attached to the effluent launder. *See also* effluent launder.

overflowed land (1) Tidal marshes that are periodically submerged and uncovered by the rise and fall of the tides. Such marshes are also called tidelands. (2) River bottomlands or floodplains that are temporarily inundated during high stages of streams.

overland flow Precipitation that gathers together and flows in small rivulets at shallow depths (3–30 millimetres). Overland flow eventually reaches defined channels and becomes streamflow.

overland runoff *See* overland flow.

overlay (1) A computerized set, or layer, of data that can be superimposed on another layer. An example would be a digitized map with a superimposed layer showing piping systems, data on age, and type of pipe, and installation and maintenance history. Many geographic information systems have multiple data overlays. (2) A method of detecting viruses. It involves the use of molten agar mixed with virus and host cells and poured onto an agar-containing petri dish.

overload relay A switch device used to sense an overloaded electric motor and disconnect it from the power source before the motor is damaged.

overlying use The use that occurs on top of an aquifer.

overpour head gate A canal head gate through which water passes from the river into the canal by discharging over the crest of a wall or the top edge of a gate or flashboards.

overpumping (1) The extraction by pumping of a quantity of water from a groundwater basin or aquifer in excess of the supply to the basin. Such pumping results in a depletion of the basin. (2) A procedure used in well development to create a high velocity of water entering the well screen. The high velocity is useful to suspend and clear away fine-grained materials from the well that would otherwise reduce the quality and quantity of water it produces. *See also* mining groundwater.

overrun To operate a filter or ion exchange system beyond its predetermined exhaustion point. As a result, the filter or ion exchange system is unlikely to be as effective as it should be, and it will probably need some regeneration to restore capacity (for ion exchange) or a cleansing, backwashing, or media or element replacement to reduce head loss and restore capacity (for a mechanical, adsorption, or neutralization filter).

overshot wheel A waterwheel operated by the weight of water falling into buckets attached to the periphery of the wheel.

overspray The application of water via sprinkler irrigation to areas other than the intended area.

overturn The phenomenon of vertical circulation that occurs in large bodies of water—also known as turnover. It is caused by the increase in density of water at temperatures greater or less than 39.2° Fahrenheit (4° Celsius), the temperature of maximum density. In the spring, as the surface of the water warms toward 39.2° Fahrenheit (4° Celsius), the water increases in density, becomes heavier, and tends to sink, producing vertical currents; in the fall, as the surface water cools toward 39.2° Fahrenheit (4° Celsius) and therefore becomes heavier, it also tends to sink. Wind also creates such vertical currents. *See also* destratification; thermal stratification.

over-and-under baffles A series of physical obstacles within or between unit processes that reduces short-circuiting by requiring the flow to pass over and under the obstacles (called baffles) sequentially. This arrangement often is used in multistage flocculation basins to ensure contact with the mixing devices in each stage. *See also* baffle; end-around baffles; flocculation.

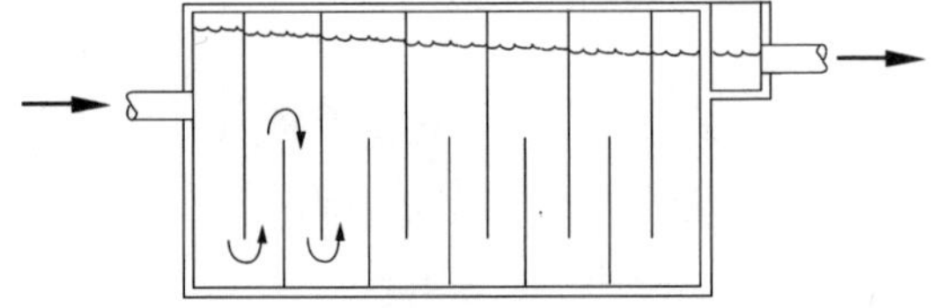

Over-and-under baffles

over-and-under flow pattern A hydraulic regime in which a fluid is passed through a series of chambers or basins to reduce short-circuiting or promote particle contact. The basins are separated by baffles that have openings at the bottom or top, in an alternating fashion, to encourage the flow to traverse the cross-sectional area of the basins. *See also* over-and-under baffles.

over-year storage The accumulation of water in a reservoir during one or several years of greater-than-average supply, the holding of such water for 1 or more years, and the release of it for use during a period of years when the amount of source water supply is insufficient. Over-year storage is also called cyclic storage.

ovicide A chemical compound that kills or inactivates parasite ova. *See also* ova.

ovum *See* ova.

oxalate-carbon equivalent A surrogate for assimilable organic carbon (AOC) that is based on the assumption that the growth of *Spirillum* NOX on oxalate as a defined medium is equivalent to the growth of *Spirillum* NOX on organic carbon in natural water. Oxalate ($OOCCOO^{-2}$) is a good carbon and energy substrate for *Spirillum* NOX; it is used as a reference growth substrate when *Spirillum* NOX is used in the AOC assay. Reporting AOC-NOX in terms of micrograms oxalate-carbon per litre assumes that the growth yield of *Spirillum* NOX on oxalate is equivalent to the growth yield on naturally occurring AOC in a water sample.

oxamyl (($(CH_3)_2N–CO–C(SCH_3):N–O–CO–NH–CH_3$) A common name for methyl *N′*-*N′*-dimethyl-*N*-{(methylcarbamoyl)oxy}-1-thiooxamimidate, a synthetic organic chemical used as an insecticide or nematicide. It is regulated by the US Environmental Protection Agency. *See also* insecticide; nematicide; synthetic organic chemical.

$$(CH_3)_2N–C(=O)–C(SCH_3)=N–O–C(=O)–N(H)–CH_3$$

Oxamyl

oxbow lake An ancient meander that, following a flood and alluviation, formed a neck cutoff that became a lake because the river could not follow. *See also* alluvium.

oxic Pertaining to water that contains dissolved oxygen in solution. An oxic zone in a reservoir, lake, or treatment process is one in which a dissolved oxygen concentration can be measured. *See also* anoxic.

oxidant Any oxidizing agent; a substance that readily oxidizes (removes electrons from) something chemically. Common drinking water oxidants are chlorine (Cl_2), chlorine dioxide (ClO_2), ozone (O_3), and potassium permanganate ($KMnO_4$). Some oxidants also act as disinfectants. *See also* disinfectant; oxidation. *Contrast with* reducing agent.

oxidant demand The quantity of oxidant consumed in a specified time period by reaction with substances present in water that exert an oxidant demand (e.g., natural organic matter, ammonia (NH_3), hydrogen sulfide (H_2S)). The oxidant demand for a given water varies with both contact time and temperature. *See also* chlorine demand.

oxidation (1) A process in which a molecule, atom, or ion loses electrons to an oxidant. The oxidized substance (which lost the electrons) increases in positive valence. Oxidation never occurs alone, but always as part of an oxidation–reduction (redox) reaction. The reduced substance gains electrons and thereby decreases in positive valence. (2) In electrodialysis, a chemical reaction that occurs at an anode and results in the loss of electrons from the anodic material. (3) In ion exchange, a specific attack on the cross-linking of the copolymer of an ion exchange resin by an oxidant (e.g., chlorine, hydrogen peroxide (H_2O_2), or ozone (O_3)) leading to degradation (loss of structure of resin beads) and shortening of the resin life. *See also* oxidation–reduction (redox) reaction.

oxidation by-product A compound that is formed during an oxidation reaction but is not related to the purpose of the oxidation. In water treatment, the most common oxidation by-products are formed at microgram-per-litre concentrations during disinfection by chlorination or ozonation. Many oxidation by-products are formed, but only some have been identified. *See also* disinfection by-product.

oxidation half reaction The electrical potential in volts (with the potential of the hydrogen electrode taken as zero volts) of an oxidation reaction—e.g., $NO_2^- + 2OH^- \rightarrow NO_3^- + H_2O + 2e^-$—equilibrium and under standard conditions with unit activity of the reactants. *See also* reduction half reaction.

oxidation number The number of electrons that must be added or subtracted from an atom in a combined state to convert the atom to the elemental form. For example, in barium chloride ($BaCl_2 = Ba^{2+} + 2\ Cl^-$), the oxidation number of barium is +2 and of chlorine is –1. Many elements can exist in more than one oxidation state. *See also* oxidation state.

oxidation state (1) For simple monatomic substances, simply the net charge on the atom. Thus, the oxidation states of the ferrous (Fe^{2+}) and ferric (Fe^{3+}) ions are +2 and +3, respectively. (2) For polyatomic species, a numeric value arbitrarily selected for each atom (based on some limited rules) so as to make the algebraic sum of the oxidation states equal to the net charge on the molecule or ion. Thus, in the case of the hypochlorite ion (OCl^-), with an oxidation state of –2 assigned to oxygen (which is done for all oxygen-containing compounds except for peroxides), the oxidation state of chlorine is +1.

oxidation treatment The process by which, through the action of living organisms in the presence of oxygen, organic matter is converted into a more stable form.

oxidation–reduction *See* oxidation–reduction chemistry; oxidation half reaction; oxidation–reduction (redox) potential; oxidation–reduction (redox) reaction.

oxidation–reduction (redox) chemistry The study of reactions in which electron transfers occur. *See also* oxidation; reduction.

oxidation–reduction (redox) couple *See* oxidation half reaction; reduction half reaction.

oxidation–reduction (redox) potential (pE) The potential required to transfer electrons from the oxidant to the reductant, used as a qualitative measure of the state of oxidation in treatment systems. It is a dimensionless value. The more positive the value, the more oxidizing the solution; more negative values represent more reducing conditions. Aerobic waters in equilibrium with the atmosphere have a pE value of approximately +13. Anaerobic waters may have a pE of –4. Because electron activity varies with the pH of water, referring to graphs of pE as a function of pH is often useful. In this manner, the range of thermodynamically stable species can be illustrated. *See also* pH.

oxidation–reduction (redox) reaction A reaction in which electron transfers occur. Oxidation and reduction always occur simultaneously, and the substance that gains electrons (and is thus reduced) is termed the oxidant. *See also* oxidation; reduction.

oxidative dealkylation The removal of an alkyl group (C_nH_{2n+1}) as a result of oxidation. For secondary amines (R–NH–R), tertiary amines (R–N(R')–R''), and ethers (R–O–R'), the products are a simpler amine—primary amine (R–NH_2) or secondary amine (R–NHR)—and alcohol (R–OH), respectively, with the alkyl group leaving as an aldehyde (RCHO). Amine compounds (R_{3-x}–NH_x, where $x = 0, 1, 2$) or ether compounds (R–O–R') are metabolized by cytochrome P-450 enzyme systems to produce the products in vivo. *See also* cytochrome P-450.

oxidative deamination The removal of a primary amine (R–NH_2) from an alkyl chain through oxidation of the adjacent carbon. The products are an aldehyde (RCHO) and ammonia (NH_3).

oxidative dehalogenation The removal of a halogen from an organic compound by oxidation of the adjacent carbon–hydrogen bond. It is usually a two-step process, generating a haloalcohol (RX_n–OH, where X = Cl or Br) as an intermediate. The final products are an aldehyde (RCHO) and a haloacid (RX_n–COOH, where X = Cl or Br).

oxide-coated medium A granular medium or resin having a surface upon which is a layer of material that can participate in oxidation–reduction reactions. An oxide-coated medium is generated when the reduced form of a compound such as manganese dioxide (MnO_2), is deposited on the medium. Alternatively, the medium can be specifically designed with the reduced oxide coating in place to initiate the reaction with the water to be treated. *See also* greensand.

oxidizable salt A salt occurring in solution in groundwater—such as ferrous sulfate ($FeSO_4$), ferrous carbonate ($FeCO_3$), or the corresponding salts of manganese—that may be oxidized to other forms and is deposited from solution upon exposure either to air or to dissolved oxygen in surface water.

oxidize (1) To combine with oxygen. (2) To change a compound by increasing the proportion of the electronegative part or to change an element or ion from a lower to higher oxidation state. (3) To remove one or more electrons from an atom, ion, or molecule. *See also* oxidation; oxidation state.

oxidizing agent *See* oxidant.

oxidizing filter A type of filter used to change the valence state of dissolved molecules, making them insoluble and therefore filterable, e.g., a filter that oxidizes ferrous iron (Fe^{2+}), manganous manganese (Mn^{2+}), anionic sulfur (S^{2-}), or a combination of the three by the use of catalytic media (such as manganic oxides) and then filters the oxidized precipitates out of the water.

oxime (R–C(R′)=N–OH) The final product of the reaction of hydroxylamine (H_2NOH) with aldehydes (RCHO) or ketones (R–CO–R′) in which the initial addition to the carbonyl group (C=O) is followed by a subsequent dehydration. The formation of oximes is catalyzed by acids. Use of *ortho*-(2,3,4,5,6-pentafluorobenzyl)hydroxylamine in the analysis of aldehydes and ketones yields a fluorinated oxime that can be detected by a gas chromatography electron capture detector. *See also* aldehyde; carbonyl; electron capture detector; ketone; *ortho*-(2,3,4,5,6-pentafluorobenzyl)hydroxylamine.

R
C=N-OH
R'

Oxime, where R and R′ are carbon containing groups

oxime derivative *See* oxime.

oxoacid A class of organic compounds that have both a carbonyl (C=O) and carboxylic acid (COOH) functional group in the chemical structure. Some oxoacids are aldoacids and others are ketoacids. Some oxoacids (e.g., pyruvic acid, $CH_3COCOOH$; glyoxylic acid, OHCCOOH; ketomalonic acid, HOOCCOCOOH; and oxalacetic acid, $HOOCCOCH_2COOH$) are created during the reactions of oxidants used as disinfectants, particularly ozone (O_3), with natural organic matter. *See also* aldoacid; keto acid.

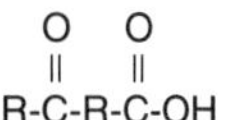

Oxoacid, where R is a carbon containing group

oxychlorine residual The residual concentration of the sum of the oxychlorine species (e.g., chlorine dioxide (ClO_2), chlorite ion (ClO_2^-), and chlorate ion (ClO_3^-)). *See also* chlorine dioxide.

oxychlorine species ($Cl_xO_y^z$) A chemical that contains chlorine and oxygen atoms. Common oxychlorine species in drinking water include chlorine dioxide (ClO_2), the chlorite ion (ClO_2^-), and the chlorate ion (ClO_3^-). *See also* chlorate ion; chlorine dioxide; chlorite ion.

oxygen (O) A nonmetallic gaseous element. Molecular oxygen is O_2 and ozone is O_3. Atmospheric oxygen is the result of photosynthesis. *See also* ozone.

oxygen deficiency A condition in which an atmosphere has less than 19.5 percent oxygen content.

oxygen deficit The difference between the dissolved oxygen level at saturation and the actual dissolved oxygen concentration in water.

oxygen demand The quantity of oxygen used in the oxidation of substances in a specified time, at a specified temperature, and under specified conditions. *See also* biochemical oxygen demand; chemical oxygen demand.

oxygen depletion A state in which oxygen has been used up so that little or none is left.

oxygen, dissolved *See* dissolved oxygen.

oxygen saturation A condition in which a liquid of given chemical characteristics, in equilibrium with the atmosphere, contains the maximum quantity of dissolved oxygen at a given temperature and pressure.

oxygen supply apparatus *See* continuous air supply; self-contained breathing apparatus.

oxygen transfer (1) An exchange of oxygen between a gaseous and a liquid phase. (2) The amount of oxygen absorbed by a liquid compared to the amount fed into the liquid through an aeration or oxygenation device, usually expressed as a percentage. *See also* aeration.

oxygenation capacity In treatment processes, a measure of the ability of an aerator to supply oxygen to a liquid.

oxyhalide An oxygen- and halogen-containing ion. This term is typically used to refer to the following disinfection by-products: chlorite ion (ClO_2^-), chlorate ion (ClO_3^-), and bromate ion (BrO_3^-). *See also* bromate ion; chlorate ion; chlorite ion.

oz *See* ounce *in the Units of Measure list.*

ozonation The process of applying ozone (O_3) to a liquid for disinfection. *See also* ozone.

ozonation by-product A class of compounds formed when ozone (O_3) is used to disinfect water. Aldehydes (RCHO) and aldoacids (HOOCRCHO) are examples of these compounds; they are formed at microgram-per-litre concentrations. Many ozonation by-products are formed, but only some have been identified. *See also* disinfection by-product.

ozonator A device used to apply ozone (O_3) to a liquid. *See also* ozone.

ozone (O_3) An unstable gas that is toxic to humans and has a pungent odor. It is a more active oxidizing agent than oxygen. It is formed locally in air from lightning or in the stratosphere by ultraviolet irradiation; it inhibits penetration of ultraviolet light from the sun to the Earth's surface. It also is produced in automobile engines and contributes to the formation of photochemical smog. For industrial applications, it is usually manufactured at the site of use. It serves as a strong oxidant and disinfectant in the purification of drinking water and as an oxidizing agent in several chemical processes. *See also* disinfectant; oxidant; oxidizing agent; ozonosphere.

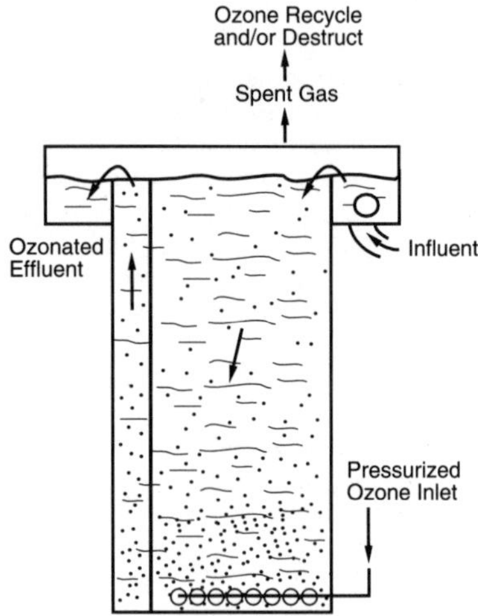

Ozone contactor

ozone (O_3) by-product *See* ozonation by-product.

ozone (O_3) contactor A chamber or basin into which ozone is introduced for the purposes of providing contact with the water to be treated. Because ozone is highly reactive, the contactors are often designed with several chambers into which ozone can be introduced. *See also* ozone.

ozone (O_3) destruction The step by which a component unit of an ozonation system destroys all or some of the ozone present in the off-gas being vented.

ozone (O_3) enrichment A step in the ozonation process in which more ozone is added to a gas that previously contained ozone.

ozone (O_3) generator A mechanical device that produces ozone by applying an electric potential to oxygen. The carrier gas for oxygen can be in the form of either dry air or pure oxygen. *See also* ozone; pure oxygen.

ozone (O_3) half-life The period of time required for 50 percent of a given quantity of ozone to decompose at a specific temperature and pressure.

ozone (O_3) residual The concentration of ozone measured in treated water after its application, typically in units of milligrams per litre. *See also* ozone.

ozone-assisted coagulation A method to improve the coagulation process by using ozone to alter the surface charge on particles. This approach has been reported as a way to reduce coagulant dosages, and it typically works best in low-particle and low–organic content waters with relatively lower coagulant dosages. The dosages used for ozone under these applications may not be sufficient to provide the necessary degree of microbial inactivation, particularly for pathogen cysts such as *Giardia* and oocysts such as *Cryptosporidium*.

Names and Molecular Structure of Ozonation By-Products

Ozonation By-Product	Molecular Structure	Ozonation By-Product	Molecular Structure
Bromate	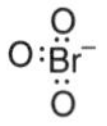 $O{:}\overset{O}{\underset{\ddot{O}}{\ddot{Br}}}{}^-$	***Aldehydes***	
Bromoform	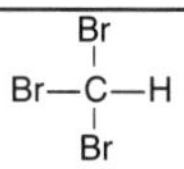 $Br-\overset{Br}{\underset{Br}{C}}-H$	Formaldehyde	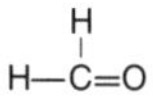 $H-\overset{H}{C}=O$
Dibromoacetonitrile	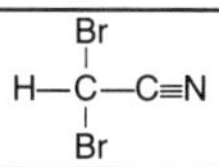 $H-\overset{Br}{\underset{Br}{C}}-C\equiv N$	Acetaldehyde	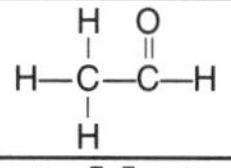 $H-\overset{H}{\underset{H}{C}}-\overset{\overset{O}{\|}}{C}-H$
Cyanogen Bromide	$Br-C\equiv N$	Heptaldehyde	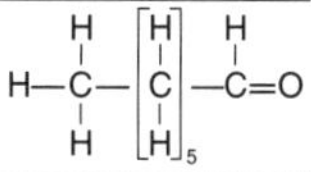 $H-\overset{H}{\underset{H}{C}}-\left[\overset{H}{\underset{H}{C}}\right]_5-\overset{H}{C}=O$
Brominated Haloacetic Acids		Glyoxal	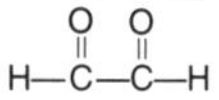 $H-\overset{\overset{O}{\|}}{C}-\overset{\overset{O}{\|}}{C}-H$
Monobromoacetic Acid	$H-\overset{Br}{\underset{H}{C}}-\overset{\overset{O}{\|}}{C}-OH$	Methylglyoxal	$H-\overset{H}{\underset{H}{C}}-\overset{\overset{O}{\|}}{C}-\overset{\overset{O}{\|}}{C}-H$
Dibromoacetic Acid	$Br-\overset{Br}{\underset{H}{C}}-\overset{\overset{O}{\|}}{C}-OH$	***Carboxylic Acids***	
Ketoacids		Formate	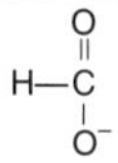 $H-\overset{\overset{O}{\|}}{\underset{O^-}{C}}$
Glyoxylic Acid	$H-\overset{\overset{O}{\|}}{C}-\overset{\overset{O}{\|}}{C}-OH$	Acetate	$H-\overset{H}{\underset{H}{C}}-\overset{\overset{O}{\|}}{\underset{O^-}{C}}$
Pyruvic Acid	$H-\overset{H}{\underset{H}{C}}-\overset{\overset{O}{\|}}{C}-\overset{\overset{O}{\|}}{C}-OH$	Oxalate	$\overset{\overset{O}{\|}}{\underset{O^-}{C}}-\overset{\overset{O}{\|}}{\underset{O^-}{C}}$
Ketomalonic Acid	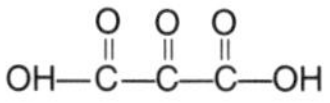 $OH-\overset{\overset{O}{\|}}{C}-\overset{\overset{O}{\|}}{C}-\overset{\overset{O}{\|}}{C}-OH$		

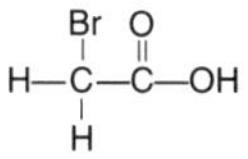

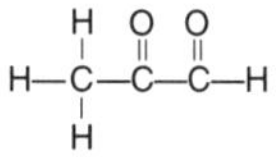

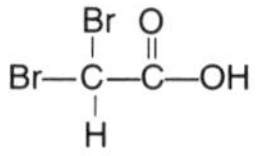

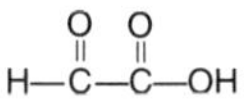

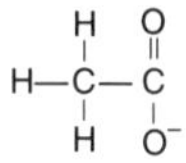

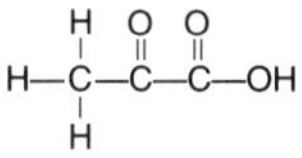

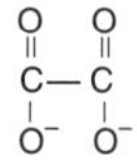

ozone–ultraviolet (O_3–UV) process A process in which ultraviolet light is used together with ozone to oxidize compounds. The ultraviolet light in this advanced oxidation process catalyzes the decomposition of ozone to hydroxyl radicals (•OH), which are very strong oxidizing compounds. *See also* advanced oxidation process; hydroxyl radical.

ozonide A compound that occurs as a by-product of ozonation.

ozonizer A device for producing ozone (O_3) from pure oxygen or air. It consists essentially of two electrodes between which a current of the dry gas is passed. High-voltage electric discharges pass through the air between the electrodes and cause ozone to form. *See also* ozonator.

ozonolysis (1) The oxidation of an organic material by ozone (O_3). (2) The use of ozone as a tool in analytical chemistry to locate double bonds in organic compounds.

ozonosphere A region in the upper atmosphere that contains a relatively high concentration of ozone (O_3) and absorbs certain wavelengths of solar ultraviolet radiation that are not screened out by other substances in the atmosphere.

P

P *See* peta; phosphorus; process and instrumentation diagram.

P *See* active power; delta *P.*

p *See para*; instantaneous power.

P alkalinity *See* alkalinity test.

Pa *See* pascal *in the Units of Measure list.*

P–A test *See* presence–absence test.

PAC *See* powdered activated carbon.

pacing Use of the ratio type of feedforward control for chemical injection.

pack joint A compression fitting or coupling for connecting service lines. The fitting contains a gasket material. Tightening of the end caps compresses the gasket material against the body of the fitting and service line, making a watertight seal. Some fittings contain a mechanical clamp to squeeze against the service line to prevent the fitting from being moved by pressure.

package plant A single, compact unit that contains all of the unit processes necessary for treatment. Package plants are typically installed for small systems serving flows less than 1 million gallons per day (3,785 cubic metres per day).

packed bed A filter or ion exchange bed for which the medium is completely retained so that no bed expansion can occur. No backwash step is used to restratify the media by grain size.

packed bed filter A water treatment unit in which a filtering medium is present with sufficient surface area to promote the removal of contaminants. Packed bed filters include simple rapid granular filters for physical removal of particulates (the granular material is the packing), as well as units with resins for contaminant adsorption or exchange. *See also* adsorption; resin.

packed column A unit designed to transfer contaminants from the liquid to the gaseous phase. The treatment unit, a hollow vertical column, contains packing material that provides surface area over which the liquid to be treated flows, allowing contact between a countercurrent airflow and the liquid and thereby promoting mass transfer. It is also called a packed tower. *See also* air-stripping; degasifier.

packed tower *See* packed column.

packed tower aeration (PTA) A process in which air is introduced into a packed column to promote the transfer of contaminants from the liquid to the gaseous phase. This process is actually misnamed because the contaminants are removed, or stripped, into the gaseous phase and expelled from the column. Thus, a more accurate name would be "packed tower stripping"; however, the common name is "packed tower aeration." *See also* packed column.

packed tower stripping (PTS) *See* packed tower aeration.

packer In well drilling, a device lowered in the borehole or casing that swells automatically or can be expanded by manipulation from the surface at the correct time to produce a watertight joint against the sides of the borehole or the casing, thus entirely excluding water from higher horizons.

packer assembly An inflatable device used to seal a tremie pipe inside a well casing to prevent grout from entering the inside of the conductor casing. *See also* packer; tremie pipe.

packing (1) Specially prepared material placed in a stuffing box around a pump shaft. The packing prevents air from entering the pump and water from leaking from the pump along the shaft. (2) The material placed in a packed tower to provide a very large surface area over which water must pass to attain a high liquid–gas transfer.

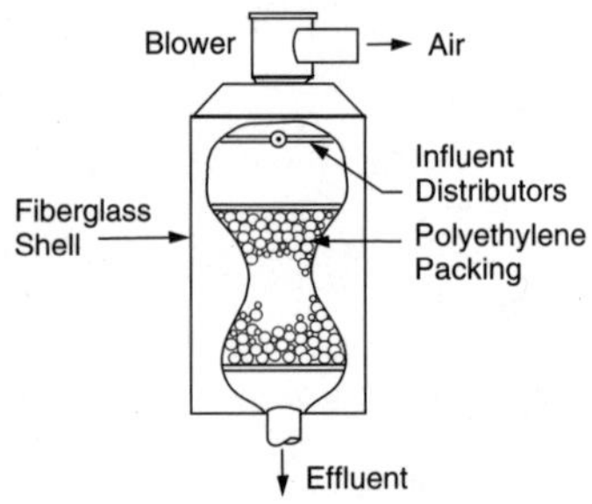

Interior of a packed column

packing gland The designed space between the body of a pump casing and pump shaft where packing material is placed to control the flow of water from the pump along the pump shaft. A follower with bolts in the gland is tightened to compress the packing until the desired amount of water from the pump is released to provide lubrication for the packing against the rotating shaft. *See also* gland.

packing material *See* packing.

PACl *See* polyaluminum chloride.

paddle aerator A device, similar in form to a paddle wheel, that is used in aeration of water.

paddle wheel A waterwheel with paddles or strips of wood or other material attached to its periphery. Such a wheel may be constructed inside a tank, or it may be set in a moving stream or under a falling stream of water that causes the wheel to revolve and generate water power. May also be used in a mechanical flocculator. *See also* flocculation.

paddle-wheel aerator *See* paddle aerator.

PAH *See* polynuclear aromatic hydrocarbon.

PAHO *See* Pan American Health Organization.

PAHS *See* polyaluminum-hydroxysulfate.

paint A uniformly dispersed mixture having a viscosity ranging from a thin liquid to a semisolid paste and consisting of (1) a drying oil, synthetic resin, or other film-forming component, called the binder; (2) a solvent or thinner; and (3) an organic or inorganic pigment. Paints are used (1) to protect a surface from corrosion, oxidation, or other type of deterioration, and (2) to provide decorative effects.

palatability The degree to which water is significantly free of color, turbidity, taste, and odor, is of moderate temperature in summer and winter, and is well-aerated.

palatable Agreeable or pleasant, especially to the sense of taste. *See also* palatability.

Palmer–Bowlus flume A device that is inserted into a pipe or channel to measure flow. The flume is designed with a raised floor elevation that creates higher velocity and thus lower surface elevations in the flume. For a given flume diameter, empirical relationships correlate the depth upstream of the flume to the flow rate.

Palmer–Bowlus meter *See* Palmer–Bowlus flume.

palustrine system The system of (1) all nontidal wetlands dominated by trees, shrubs, persistent emergents, emergent mosses, or lichens, (2) all such wetlands that occur in tidal areas where salinity due to ocean-derived salts is less than 500 milligrams per litre, and (3) wetlands lacking such vegetation but having the following characteristics: an area less than 20 acres (80,000 square metres); active-wave-formed or bedrock shoreline features; a water depth in the deepest part of the basin less than 6.5 feet (2 metres) at low water; and a salinity due to ocean-derived salts that is less than 500 milligrams per litre.

Pan American Health Organization (PAHO) An organization of governments of Western Hemisphere nations united to improve physical and mental health in the Americas. Located in Washington, D.C., the Pan American Health Organization coordinates regional activities combating disease, including exchanges of statistical and epidemiological information, development of local health services, and organization of disease control and eradication programs; encourages development in health systems and technology; provides consulting services; conducts educational courses on public health topics, such as environmental health, food and nutrition, and tropical diseases; and awards fellowships for training health services personnel.

pan coefficient The ratio of evaporation from a large body of water to that measured in an evaporation pan.

pancytopenia A blood disorder in which all the formed elements (i.e., red blood cells, white blood cells, platelets) of the blood are sharply depressed relative to normal. The extreme impact of such depressions is aplastic anemia. This type of response can be observed with high benzene (C_6H_6) exposures. *See also* anemia.

pandemic *See* epidemic.

pannier A long wicker basket containing earth or stones, deposited on wire to serve the same purpose as a fascine. A pannier is also called a gabion. *See also* fascine.

paper chromatography A method for separating or purifying small amounts of mixtures for identification purposes by spotting the mixture at the bottom of a paper strip and allowing a solvent to move and distribute the components along the strip's path.

PAR *See* population attributable risk.

***para* (*p*)** A prefix meaning "opposite." This prefix is used in organic chemistry in naming disubstitution products derived from benzene (C_6H_6) in which the second substituent atom or functional group is attached to the opposite carbon atom with respect to the first substituent atom. It represents the 1,4-position. *See also ortho*; *meta*.

A
C
H-C C-H
H-C C-H
C
B

A depiction of 1-A-4-B-benzene where B is in the para position (attached to the opposite carbon atom) in respect to A

parabolic weir A weir with a notch that is parabolic in shape, with the axis of the parabola vertical.

paracrine Pertaining to one cell affecting another through some signaling process—in contrast with the term autocrine, which applies to self-regulating cells.

***para*-dichlorobenzene** *See under* dichlorobenzene.

parallel misalignment A situation in which the axes of shafts are parallel but not concentric.

parameter A water quality attribute. For example, the presence of certain bacteria, the hardness, and the level of sodium are all parameters.

parasitic bacteria Bacteria that live on other living organisms without serving any useful purpose for those organisms.

paresthesia Localized numbness. Paresthesia can be produced by a variety of chemicals that have local anesthetic effects. It can also be a symptom of permanent sensory nerve damage induced by a chemical.

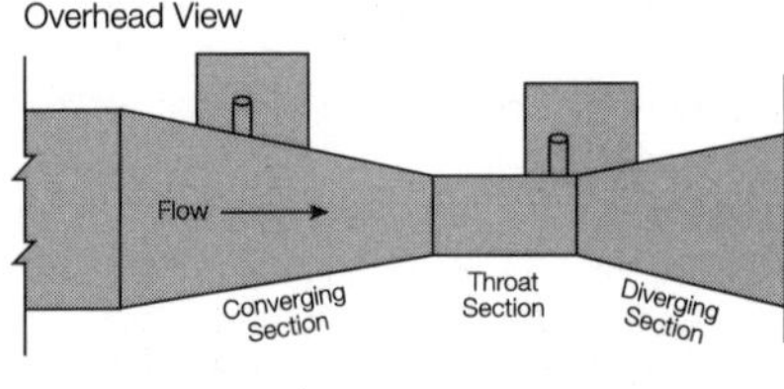

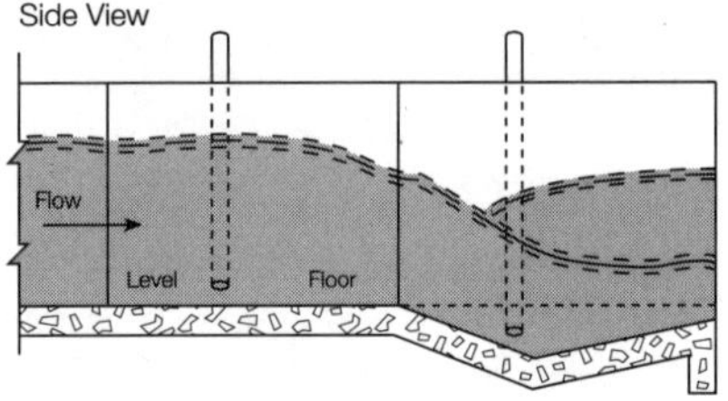

Parshall flume

parcel A map feature depicting land ownership and rights. Parcel boundaries are usually described in narrative form on a deed as metes and bounds or as bearings and distances. In much of the United States, the descriptions may refer to the public land survey system townships, ranges, sections, and aliquot parts (half or quarter sections), which are not usually expressed in Système International units.

parcel identification number (PIN) A numbering scheme for identifying parcels in a computer system. Parcel identification schemes range from using simple sequential numbers to using geocodes that also define locations by incorporating *x* and *y* coordinates for a parcel.

parent material Weathered rock material from which soil is formed.

Pareto chart A graph showing the major causes of the problem under study in descending order.

Parkinson-like syndrome A constellation of diseases that produce symptoms commonly found with Parkinson's disease. In general, chemicals such as manganese destroy dopaminergic activity within the hippocampus region of the brain.

Parshall flume A measuring flume with standard dimensions calibrated for determining the discharge of open channels by measuring the head a specific distance upstream and downstream from the critical area or sill. The surface level is measured in relationship to the bottom or floor level of the flume. The flow is calculated based on this depth or head and on the width of the flume. The downstream water depth does not affect the accuracy of the flow measurement until it exceeds approximately 60 percent of the upstream water depth.

parts per billion (ppb) *See the Units of Measure list.*

parts per million (ppm) *See the Units of Measure list.*

parts per thousand (ppt) *See the Units of Measure list.*

part per trillion (ppt) *See the Units of Measure list.*

partial diversion The taking or removing of water from one location in a natural drainage area and discharging it at another location in the same drainage area.

partial pressure The pressure exerted by each gas independent of the others in a mixture of gases. The partial pressure of each gas is proportional to the amount (percent by volume) of that gas in the mixture.

partial vacuum A space condition in which the pressure is less than atmospheric.

partially suppressed contraction A reduction in the contraction occurring because of the presence of guiding walls that lead the flow to the opening.

participation agreement An agreement in which a developer or distributor pays for the cost of the distribution facilities (such as conduits, pump stations, or treated water reservoirs) required to provide service within that district from the nearest existing available source.

participation charge A contribution of capital required by a utility toward the development of water facilities and offered by those causing the incremental service demands.

particle A very tiny, separate subdivision of matter. Particles suspended in water can vary widely in size, shape, density, and charge. Colloidal and dispersed particles are artificially agglomerated by processes of coagulation and flocculation. *See also* coagulation; flocculation.

particle beam mass spectrometry (PBMS) The use of a particular type of interface in liquid chromatography–mass spectrometry. The interface consists of a nebulizer, a heated desolvation chamber, a set of skimmer lenses, and a probe inlet. The solvent must be removed from the liquid chromatography eluent prior to ionization of the sample compounds in the mass spectrometer source. Particle beam instruments can produce electron impact spectra in which analytes ionize into characteristic fragmentation patterns. This fragmentation improves the ability to identify compounds based on a comparison with large computerized databases of mass spectra. The use of such instruments is considered an improvement over older liquid chromatography–mass spectrometry interfaces; however, it still has a number of shortcomings relative to newer interfaces, such as electrospray. *See also* electrospray ionization; liquid chromatography; mass spectrometer interface; nebulizer; solvent.

particle count The results of a microscopic examination of treated water by a particle counter that classifies suspended particles by number and size. *See also* particle counter.

particle counter An instrument that measures the number of particles within a given size range. These instruments have been used to optimize the performance of filters in water treatment plants.

particle density The mass per unit volume of granular activated carbon, not including the voids between the particles and cracks larger than 0.1 millimetre. It is determined by immersion in mercury and then measuring the displacement of mercury.

particle filtration Filtration of particles having a diameter of 1 micrometre or larger. Particle filtration is typically handled by cartridge filters and media filters.

particle size The average volume or diameter of (1) the particles in a sediment or rock or (2) the grains of a particular mineral that make up a sediment or rock.

particle size distribution A representation of the size distribution of a heterogeneous mixture of particles. The distribution is often given as a percentage of the mixture (based on mass) that is less than or equal to a given size.

particulate A very small solid suspended in water that can vary widely in size, shape, density, and electrical charge. Colloidal and dispersed particulates are artificially gathered together by the processes of coagulation and flocculation. *See also* coagulation; flocculation.

particulate matter Minute separate particles of inorganic matter, organic matter, or both. *See also* particulate; particulate organic carbon.

particulate organic carbon (POC) Organic carbon associated with particulates. In analytical terms, it is that portion of the organic carbon in water that is retained on a filter having a pore diameter of 0.45 micrometre or sometimes larger. For most drinking water sources, the particulate organic carbon fraction represents a very low percentage of the total organic carbon pool. However, some river systems—especially during storm events—may experience an increase in particulate organic carbon as a result of runoff. Conventional water treatment processes are generally efficient in removing particulate organic carbon. *See also* dissolved organic carbon; nonpurgeable organic carbon; particulate; purgeable organic carbon; total organic carbon.

partition coefficient A ratio of the molar concentrations of a solute distributed between two phases. It is often used as an expression of extraction efficiency. It is calculated by dividing the mass fraction of solute in the extraction phase by the mass fraction of the solute in the liquid phase remaining after the liquid has been mixed with the extraction phase.

partly submerged orifice An orifice with its bottom below the water discharge surface and its top above the water discharge surface.

Partnership for Safe Water A volunteer initiative by the US Environmental Protection Agency, the Association of Metropolitan Water Agencies, the Association of State Drinking Water Administrators, the National Association of Water Companies, the American Water Works Association, and the American Water Works Association Research Foundation to optimize water treatment to better protect the public from *Cryptosporidium* oocysts. The program began in September 1995 and commits participating water utilities to a four-phase self-assessment program for optimizing their operations. Phase I of the program consists of the agreement to participate and requires the water supplier to have met the Surface Water Treatment Rule for at least 6 months and to pledge to complete phases II and III of the partnership. Phase II consists of data collection, and Phase III consists of a comprehensive water treatment self-assessment package. Phase IV is a third-party assessment via the US Environmental Protection Agency Composite Correction Program. *See also* Composite Correction Program.

PAS *See* polyaluminum sulfate.

Pa•s *See* pascal-second *in the Units of Measure list.*

pascal (Pa) *See the Units of Measure list.*

pass For a membrane treatment process, a single treatment step or one of multiple membrane treatment steps producing a product stream. For example, a two-pass seawater reverse osmosis desalting system may include seawater reverse osmosis membranes in the first pass, producing permeate (product) that is then repressurized and further desalted by brackish water reverse osmosis membranes in the second pass, yielding a product water of even lower total dissolved solids. Sometimes a membrane system that contains more than one pass is referred to as a permeate-staged system.

passivating film A thin, protective corrosion control layer formed on the surface of a pipe or other material exposed to potable water in the presence of either certain water quality conditions or a corrosion inhibitor. Either may cause the pipe material and the potable supply to interact in such a way that metal compounds are formed on the pipe surface, creating a protective coating of less soluble materials.

passivation A corrosion control technique that causes the pipe materials to create metal/hydroxide/carbonate compounds that form a film on the pipe wall to protect the pipe. *See also* passivating film.

passive circuit A circuit without sources.

passive humoral immunity Immunity attained either naturally by an infant from its mother or artificially by inoculation with specific protective antibodies from immunized animals, convalescent hyperimmune serum, or immune serum globulin from humans.

passivity The phenomenon of an active metal becoming passive, i.e., the state of a metal when its behavior is more noble (resistant to corrosion) than predicted by theory.

Pasteur filter A domestic filter with a filtering medium of unglazed porcelain.

Pasteur pipette A small-diameter glass tube, generally unmarked or ungraduated, with a long tapered tip used for transferring small volumes of liquid culture suspension from one container to another.

pasteurization A heat treatment process originally developed to preserve wine by killing bacteria that could cause the wine to become sour. Pasteurization is now most commonly associated with treating milk to destroy disease-causing organisms that can be transmitted by contaminated milk, including those that cause tuberculosis, Q fever, typhoid fever, paratyphoid fever, bacillary dysentery, diphtheria, scarlet fever, and foot-and-mouth disease. Pasteurization is not a sterilizing process because it does not kill all organisms present in the milk.

pat test *See* wet-shake test.

pathogen *See* infectious agent; pathogenic organism.

pathogenesis The postulated mechanisms by which an etiologic agent produces disease.

pathogenic Biologically harmful.

pathogenic bacteria Bacteria that cause disease in a host organism.

pathogenic organism An organism that can cause disease in a host. *See also* infectious agent.

pathology The study of disease.

pattern diagram A method of visually depicting the relative abundance or absence of dissolved compounds in a water sample.

pattiole *See* particle.

payloader A four-wheeled front-end loader used to load trucks with sand, gravel, or earth material from excavations, such as trenches for water lines.

PBMS *See* particle beam mass spectrometry; performance-based method system.

PBPK modeling *See* physiologically based pharmacokinetic modeling.

PBS *See* phosphate-buffered saline.
PCB *See* polychlorinated biphenyl.
PCE *See* tetrachloroethylene.
pCi *See* picocurie *in the Units of Measure list.*
pCi/L *See* picocuries per litre *in the Units of Measure list.*
PCP *See* pentachlorophenol.
PCR *See* polymerase chain reaction.
PCU *See* color unit *in the Units of Measure list.*
PDD *See* peak-day demand.
PDN *See* predischarge notification.
PDWR *See* Primary Drinking Water Regulation.
PE *See* polyethylene; professional engineer.
Pe *See* Peclet number.
pE *See* oxidation–reduction (redox) potential.
P/E ratio *See* precipitation/evaporation ratio.
PE sample *See* performance evaluation sample.
PE tank *See* portable exchange tank.

peak The highest load carried by an electric generating system or the maximum output of a water treatment plant during any specific period. *See also* peak demand.

peak day The day during which the peak-day demand for water occurs.

peak demand The maximum momentary load (expressed as a rate) placed on a water treatment plant, distribution system, or pumping station. It is usually the maximum average load in 1 hour or less, but it may be specified as instantaneous or for some other short time period.

peak flow Maximum flow.

peak hour The hour during which the peak-hour demand for water occurs.

peak load *See* peak demand.

peak month The month in which the peak-month water demand occurs.

peak operating flow The maximum rate of flow under which a treatment unit is designed to properly function and produce a certain quality of product water.

peak-day demand (PDD) The greatest volume per day flowing through a treatment plant for any day of the year.

peak-hour demand The greatest volume per hour flowing through a treatment plant for any hour in the year.

peaking costs Capital and operating costs incurred specifically to meet peak demands, e.g., the costs of increasing pipe or reservoir storage capacity to serve a particular area, or a capacity addition required to meet growing summer demand for an entire service area. Peaking costs are usually separated from base volume costs through a cost-of-service formula and are allocated to customer classes in proportion to the classes' peaking characteristics. *See also* base–extra-capacity approach to rate structure design; cost-of-service analysis; inclining block rate; marginal cost; peak-load pricing.

peak-load pricing A pricing scheme designed to charge utility customers a higher rate during peak periods of use. In the electric power industry, supply is engaged in increments to meet sales demands that vary during the day and over seasons of the year because electricity cannot economically be stored in significant quantities. Peak loads, therefore, determine the necessary capacity of the system. Increased prices during peak periods can eliminate or shift demand to nonpeak periods and thereby lower the required capacity. In the off-peak periods, costs should be related only to the production of product; in peak periods, capacity costs must also be recovered in the rate charged. A less pronounced situation (with more limited opportunity for shifting demand to off-peak periods) can be found in the water industry, which has the capacity to store its product but only limited time-of-day usage information for peak-load pricing. Consequently, for this industry, pricing during peak periods reflects the cost of treatment, distribution lines, and local storage that are put in place to meet peak demands and are recovered entirely from the peak volumes. Avoided costs of additional capacity required to meet peak demands (rather than actual costs of facilities in place) are also used to price peak demands through the use of inclining block or conservation rate structures. *See also* cost of service; inclining block rate; marginal cost; peaking costs.

peak-month demand The greatest volume of water per calendar month passing through a treatment plant during a given year.

Pearson's correlation matrix A matrix of correlation coefficients between pairs of dependent and independent variables. This type of matrix is calculated to determine the relative dependency of a dependent variable on a number of independent variables.

peat A dark brown or black residue that is produced by the partial decomposition and disintegration of sedges, mosses, trees, and other plants that grow in wet places such as marshes.

peat soil Organic soil that is formed by the accumulation in wet areas of partially decomposed remains of vegetation and that has less than 20 percent mineral material. *See also* muck soil; peat.

Peclet number (Pe) A dimensionless number that expresses the ratio of advective flux to diffusive or dispersive flux in a fluid system. Advective flux is the transport of mass by motion of the host fluid, whereas diffusive or dispersive flux is the transport of mass within the host fluid as a result of concentration gradients. High Peclet numbers are characteristic of diffusion-dominated systems.

pedogenic refractory organic matter (PROM) Refractory natural organic matter originating from soils (as opposed to that originating from aquatic media). *See also* aquagenic organic matter; natural organic matter; refractory organic matter.

PEL *See* permissible exposure limit.

pelagic zone The open waters of a sea or lake, especially where the water is more than 65 feet (20 metres) deep.

pellet reactor A conical-shaped tank that is about half filled with calcium carbonate ($CaCO_3$) granules and in which softening takes place quite rapidly as water passes up through the unit.

pellicular front The even front, developed only in pervious granular material, on which pellicular water depleted by evaporation, transpiration, or chemical action is regenerated by influent seepage.

pellicular water The film of water left around each grain of water-bearing material after gravity water has been drained off. It is also known as the water of adhesion. It is fixed water that can be extracted by root absorption and evaporation but cannot be moved by gravity or by the unbalanced film forces resulting from localized evaporation and transpiration.

pellicular zone The maximum depth from the natural surface down to which evaporation can have its effect. *See also* pellicular water.

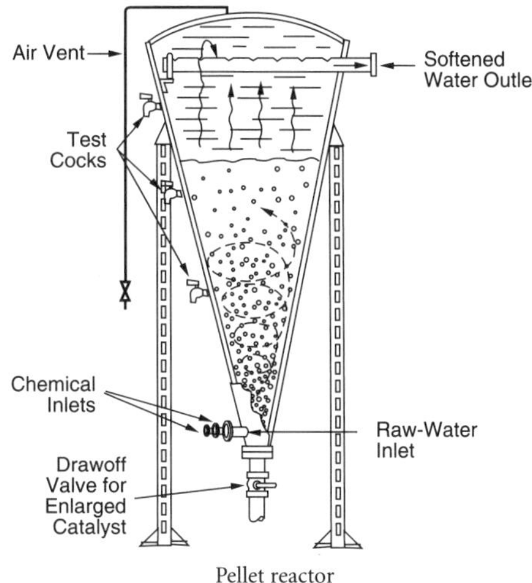

Pellet reactor

Pelton turbine An impulse turbine mounted on a vertical or horizontal shaft and having curved vanes or buckets attached to its periphery that successively receive a jet from one or more nozzles. The jet, on striking the vane or bucket, is turned approximately 180° and leaves the runner with a low absolute velocity. The turbine speed is regulated by closing or deflecting the nozzle. A Pelton turbine is also called a Pelton wheel and is sometimes used as an energy recovery device in a seawater desalting reverse osmosis system.

Pelton wheel *See* Pelton turbine.

Pelton wheel turbine *See* Pelton turbine.

penetrometer An instrument used to estimate a soil's compressive strength. Such an instrument is used to help determine the soil type of an excavation.

penstock The pipeline or conduit that carries water under pressure from the forebay or last free water surface to the turbines in a power-generating facility.

pentachlorophenol (PCP) (C_6Cl_5OH) A synthetic organic chemical used as a fungicide, bactericide, algicide, herbicide, and wood preservative. Unlike some chlorophenols, pentachlorophenol is not a disinfection by-product. Pentachlorophenol is regulated by the US Environmental Protection Agency. *See also* algicide; bactericide; chlorophenols; fungicide; herbicide.

```
        OH
        |
       C
    //   \
Cl-C      C-Cl
   |      ||
Cl-C      C-Cl
    \\   /
       C
       |
       Cl
```

Pentachlorophenol

***ortho*-(2,3,4,5,6-pentafluorobenzyl)hydroxylamine ($C_6F_5CH_2ONH_2$) (PFBHA)** A derivatizing agent in the gas chromatographic determination of carbonyl compounds (i.e., aldehydes and ketones) in aqueous solution. The final product of the reaction of this hydroxylamine with a carbonyl compound is a fluorinated oxime. Fluorine-containing compounds can be detected with good sensitivity by use of an electron capture detector. *See also* carbonyl; derivatization; electron capture detector; oxime.

```
       F
       |
       C
    //   \
F-C       C-CH2-O-NH2
  |       ||
F-C       C-F
    \\   /
       C
       |
       F
```

ortho-(2,3,4,5,6-pentafluorobenzyl)hydroxylamine

***ortho*-(2,3,4,5,6-pentafluorobenzyl)hydroxylamine (PFBHA) ($C_6F_5CH_2ONH_2$) derivatization** A chemical reaction in which *ortho*-(2,3,4,5,6-pentafluorobenzyl)hydroxylamine reacts with a carbonyl compound (i.e., aldehyde or ketone) in aqueous solution to form an oxime. This derivatization step is part of the analytical procedure for the detection of aldehydes and ketones in water. *See also* aldehyde; carbonyl; derivatization, *ortho*-(2,3,4,5,6-pentafluorobenzyl)hydroxylamine; oxime.

peplomer A projection extending from the outer surface of a virus envelope.

per capita Per person. For example, the total demand of a water treatment system divided by the number of people served provides an estimate of the per capita consumption.

per capita use rate An estimate of the water usage in a community, determined by dividing the total water used by the number of persons using it. It is the average amount of water used by a person within a given period of time. It is most commonly expressed in units of gallons (or litres) per capita per day.

percent (%) The fraction of the whole expressed as parts per 100.

percent salt (or specific solute) rejection *See* rejection.

percent recovery *See* recovery.

percent salt (or specific solute) passage *See* salt (or specific solute) passage.

percent saturation The amount or concentration of a substance that is dissolved in a solution divided by the amount or concentration that could be dissolved in the solution and multiplied by 100 to express as a percentage.

percent switch A feature of an irrigation controller that allows percent changes in the duration of programmed irrigation.

percentage reduction The percentage of material removed from water by treatment. It is calculated by dividing the change in concentration through treatment by the concentration prior to treatment and then multiplying by 100 to obtain a percentage.

percentile A point on a frequency distribution indicating what percentage of values are less than or equal to the value being considered. For example, the tenth percentile is the point in a cumulative frequency distribution for which 90 percent of the observations are greater and 10 percent are lesser or equal. Someone who scored in the 78th percentile on a test would have a higher grade than all but 22 percent of those taking the test.

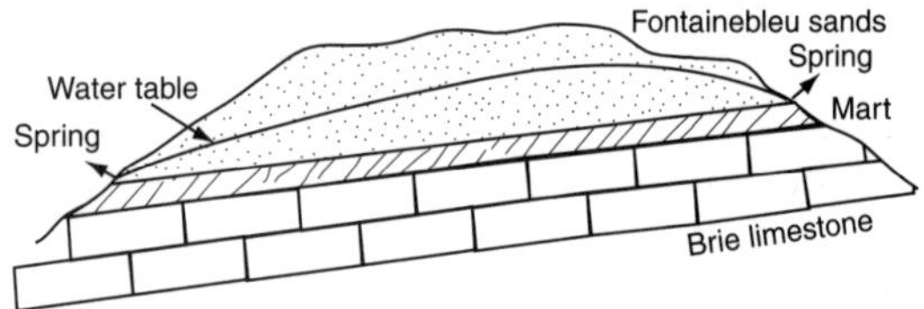

Cross section of a perched aquifer

perched aquifer A small lens of unconfined groundwater separated from an underlying main body of groundwater by an impermeable unsaturated zone.

perched groundwater The water in a perched aquifer. *See also* perched aquifer.

perched spring A spring that has water from a perched aquifer as its source of supply. *See also* perched aquifer.

perched subsurface stream Vadose water that has concentrated in fractures or solution openings and flows toward the water table in a perched aquifer. *See also* vadose water.

perched water Groundwater occurring in a perched aquifer. *See also* perched aquifer.

perched water table The upper water surface of a perched aquifer. *See also* perched aquifer.

perchlorate (ClO_4^-) A salt or ester of perchloric acid ($HClO_4$), e.g., sodium perchlorate, $NaClO_4$. Sodium perchlorate has been used in explosives, in jet fuel, and as an analytical reagent. Perchlorate has been found in a number of drinking water sources as a result of probable contamination from the solid rocket industry. The US Environmental Protection Agency is looking into the development of a regulation for perchlorate in drinking water. *See also* perchloric acid.

perchloric acid ($HClO_4$) A fuming, corrosive strong acid that is the most highly oxidized acid of chlorine and a powerful oxidizing agent when heated.

percolate To move or flow through a porous medium. *See also* infiltrate.

percolating water Water passing through soil or rocks under the influence of gravity but not necessarily flowing downward.

percolation The seeping of water through the soil without a definite channel.

percolation path The course followed by water moving or percolating through any permeable material.

percolation rate The rate of movement of water under hydrostatic pressure through the interstices of the rock or soil, except movement through large openings such as caves.

percutaneous absorption Permeation of the skin by a chemical.

perennial interrupted stream A stream that has perennial stretches with intervening interrupted or ephemeral stretches. *See also* ephemeral; perennial stream.

perennial periodic spring A periodic spring that discharges throughout all seasons of the year and during both dry and wet years.

perennial spring A spring that discharges continuously at all seasons of the year and during both dry and wet years.

perennial stream A stream that flows continuously at all seasons of the year and during both dry and wet years. Such a stream is usually fed by groundwater, and its water surface generally stands at a lower level than that of the water table in the locality.

perennial yield The amount of groundwater that can be withdrawn from an aquifer on a sustained basis without exceeding the natural replenishment rate. *See also* mining groundwater.

perfect fluid A fluid that is incompressible, has a uniform density, and offers no resistance to distorting forces.

perforated baffle wall A wall that segregates unit processes, or chambers within a unit process, and allows flow to pass only through a series or matrix of holes (perforations) in the baffle wall. The perforations promote head loss that, in theory, uniformly distributes the flow into a basin or chamber, thereby minimizing short-circuiting.

perforated casing A well casing that permits the water to enter through holes that have been punched or cut in the casing.

perforated launder A collection trough located at the water surface in a settling basin to allow the flow to be uniformly collected without adversely affecting particle settling. The trough has holes, or perforations, below the water surface to minimize upflow disturbances near the trough.

perforated plate A flat plate with a series of holes, used to control fluid flow distribution and sometimes to create backpressure or dissipate energy.

perforated-casing well A well with a casing that has been perforated to allow the water to enter.

performance-based method system (PBMS) A concept in analytical methods in which characteristics of quality control performance are specified yet still allow the analyst flexibility in terms of how the criteria are achieved. In contrast to methods in which every detail is specified, performance-based methods are developed to allow flexibility in terms of the details.

performance curve A plot of discharge quantity versus total dynamic head for a number of different head and discharge values for a given pump.

performance evaluation An assessment of the quality of data produced by a laboratory for a given analyte. Often this is done as part of a process of laboratory certification. *See also* performance evaluation sample.

performance evaluation (PE) sample A sample used to assess the quality of data a laboratory generates for a given analyte. Typically, these samples are furnished to the laboratory as a blind sample; i.e., the result is unknown to the laboratory being

Perforated baffle wall

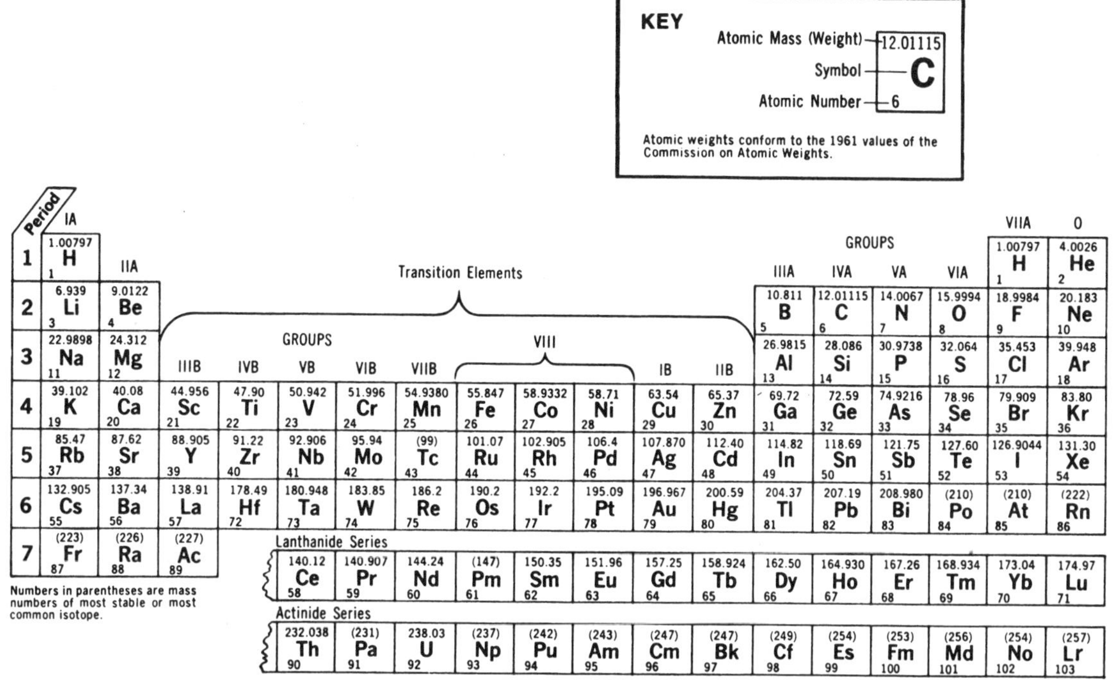

Periodic table of the elements

evaluated. Such samples are often used as part of a laboratory certification process.

performance standard In wetland restoration, a set of specific goals to be reached to ensure that replacement wetlands reach functional maturity before credits are sold.

perigean tide A tide of increased range that occurs when the moon is in perigee (i.e., closest to the Earth in its orbital path).

perikinetic flocculation The contact between particles that occurs as a result of random, or Brownian, motion. *See also* Brownian motion; differential settling; orthokinetic (shear) flocculation.

perimeter The distance around the outer edge of a shape.

period (*T*) (1) For a periodic signal (such as a sinusoidal time function), the minimum time interval after which the same characteristics of the signal recur. This interval is often measured in seconds. (2) The interval required for the completion of a recurring event. (3) Any specified duration of time. (4) A horizontal row of elements in a periodic table.

periodic sample One of a series of grab samples taken at irregular intervals.

periodic spring A spring that has periods of relatively large continuous discharge at more or less regular and frequent intervals. *See also* ebbing-and-flowing spring.

periodic table A chart showing all the chemical elements arranged according to similar chemical properties.

periodical Pertaining to detectable regular or irregular saturated soil conditions or inundation that (1) result from ponding of groundwater, precipitation, overland flow, stream flooding, or tidal influences and (2) occur such that hours, days, weeks, months, or even years pass between events.

peripheral axonopathy Neuron damage that involves the axonal portion of the nerve. Peripheral nerves are those nerves that serve muscles and organs as opposed to nerves within the central nervous system. As a consequence of such damage, the muscle controlled by the nerve loses much of its tone and strength. In general, such effects are diagnosed by virtue of slowed conduction velocities in the peripheral neurons. They can be produced by such chemicals as acrylamide ($CH_2CHCONH_2$) and n-hexane ($CH_3(CH_2)_4CH_3$).

peripheral feed clarifier A settling unit for which the influent is fed from the perimeter and flows inwardly for collection. Peripheral feed clarifiers are typically fed through a surface launder or weir and are often circular. *See also* laundering weir.

peripheral flow Flow of a liquid parallel to the circumference or periphery of a circular tank or other circular structure. Such flow is also called circumferential flow.

peripheral neuropathy Degraded function of a peripheral nerve. This condition is generally associated with a specific pathology. With chemical exposures, the most common neuropathy appears to be the dying-back form, for which the axons of nerves degenerate. *See also* peripheral axonopathy.

peripheral pump A pump having an impeller that develops head by recirculating the liquid through a series of rotating vanes.

peripheral weir An outlet weir extending around the inside portion of the circumference of a circular settling tank, over which the effluent discharges.

periphyton Microscopic plants and animals that are firmly attached to solid surfaces under water, e.g., rocks, logs, and pilings. *See also* plankton.

periportal The region surrounding the branches of the portal vein and hepatic artery that supply blood to the acinus (the organizational unit of the liver). Blood from these vessels flows through the sinusoids that bathe the hepatocytes to the terminal hepatic vein (also known as the central vein). The bile ductules are observed in the same areas as the portal vein and hepatic artery when the liver is seen in an appropriate cross section. Collectively, these three vessels are referred to as the portal triad.

peristaltic pump A positive-displacement pump using a rotary head that squeezes a flexible tube to deliver the pumped liquid. The pump output is determined by the speed of the rotating head. *See also* positive-displacement pump.

perlite (1) A form of volcanic rock that, when processed, yields various grades of filter media. (2) The filter media so obtained.

permafrost Permanently frozen subsoil occurring wherever the temperature remains less than 32° Fahrenheit (0° Celsius) for several years. Permafrost is characteristic of arctic tundra.

permanent dam A dam that is fixed in position and intended for long service.

permanent hardness *See* noncarbonate hardness.

permanent monitoring relief Authority granted under Section 1418 of the Safe Drinking Water Act to state primacy agencies that have a US Environmental Protection Agency–approved source water assessment program to adopt tailored alternative monitoring requirements for public water systems in the state as an alternative to monitoring requirements for chemical contaminants set in the applicable National Primary Drinking Water Regulations.

permanent partial disability A doctor's determination of permanent damage suffered to a portion of the body that will not recover its full abilities.

permanent snow line The line of lowest elevation or the lower limit of a snowfield at any locality on the Earth's surface above which snow accumulates and remains on the ground throughout the entire year.

permanent stream A stream that flows throughout the year. *See also* perennial stream.

permanent total disability A doctor's determination that a worker's injuries will prevent the worker from being employed for life.

permanent water A watering place in the western United States—such as a spring, well, or stretch of stream—where water can be obtained at all seasons of the year and during both dry and wet years.

permanently absorbed water Water that passes downward beyond the root zone and finally reaches the saturation zone, where it becomes an increment to groundwater.

permanently flooded A water regime condition in which standing water covers the land surface throughout the year (but may be absent during extreme droughts).

permeability A measure of the relative ease with which water flows through a porous material. A sponge is very permeable; concrete is much less permeable. Permeability is sometimes called perviousness. *See also* permeability coefficient.

permeability coefficient A coefficient expressing the rate of flow of a fluid through a cross-section of permeable material under a hydraulic gradient or pressure gradient. If the fluid is water, the permeability coefficient is the hydraulic conductivity. A standard coefficient of permeability used in the hydrologic work of the US Geological Survey, known as the Meinzer unit (not a Système International unit), is defined as the rate of flow of water at 60° Fahrenheit (15.6° Celsius), in gallons per day, through a 1-foot cross section under a hydraulic gradient of 100 percent. *See also* hydraulic conductivity.

permeability units (1) For electricity: In Système International, henrys per metre; metre-kilograms per second squared per ampere squared. (2) For groundwater and petroleum engineering: In the US customary system, gallons per day per square foot. In Système International, darcys.

permeable (1) Having communicating (i.e., connected) interstices of capillary or supercapillary size. (2) Having a texture or structure that permits water to move through perceptibly under head differences ordinarily produced in natural or engineered systems.

permeable confining bed A confining bed that produces artesian head in adjacent groundwater not by preventing percolation but by retarding it. Such beds are much more common than impermeable confining beds.

permeable rock A rock having a texture that permits water to move through it perceptibly under the pressure ordinarily found in subsurface water. Such a rock has communicating (i.e., connected) interstices of capillary or supercapillary size.

permeameter A device for measuring the permeability of soils or other material. It usually consists of two reservoirs or tanks connected by a conduit containing the material under investigation. Water is passed from one reservoir, under varying conditions of head, through the connecting conduit.

permeance *See* solvent (water) permeability coefficient.

permeate For a pressure-driven membrane treatment process, the portion of the feed solution that passes through the membrane. For potable water membrane treatment systems, the permeate is often referred to as the product flow stream.

permeate flux The permeate flow rate per unit active membrane area. It is commonly expressed in units of gallons per day per square foot or cubic metres per hour per square metre. *See also* flux.

permeate staging For a pressure-driven membrane process, a membrane module arrangement for which the permeate from one stage is further processed in a following stage, usually with an additional pumping step prior to the following stage. *See also* array; hydraulic staging; pass.

permeation The passage of water and associated substances through a membrane or other porous medium. *See also* dialysis; leaching; percolation; reverse osmosis.

permeator A membrane treatment device for which hollow fiber membranes are housed in a pressure vessel.

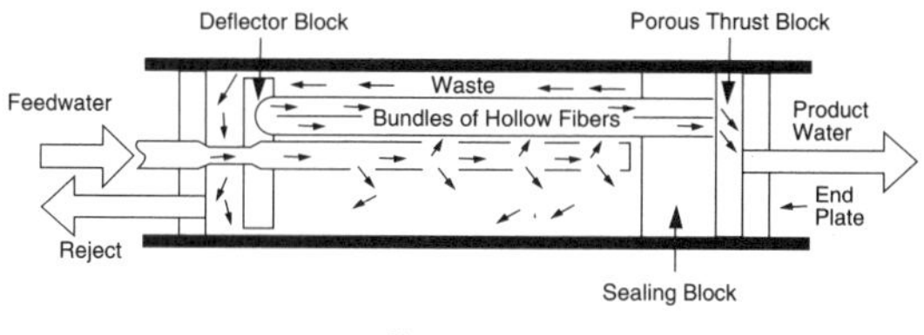

Permeator

permissible dose The dose of a chemical that may be received by an individual without the expectation of a significantly harmful result. *See also* dose; no-observed-adverse-effect level.

permissible exposure limit (PEL) The maximum concentration of a substance that a worker may be exposed to in each 8-hour work day for 30 years and not experience related health problems.

permissible velocity The highest velocity at which water may be carried through a structure, canal, or conduit without excessive damage.

permit-required confined space A confined space that requires a checklist to be completed before employees may enter. *See also* confined space.

permitting authority The district engineer of the US Army Corps of Engineers or other such individual as may be designated by the secretary of the US Army to issue or deny permits under Section 404 of the Clean Water Act; or the state director (or delegated representative) of a permit program approved by the US Environmental Protection Agency under Section 404(g) and Section 404(h) of the Clean Water Act. A permit is required under Section 404 of the Clean Water Act to discharge dredged or fill material into the navigable waters of the United States.

permittivity units In Système International, farads per metre; seconds to the fourth amperes squared per cubic metre per kilogram.

permselective material *See* permselective membrane.

permselective membrane A membrane used for separation of constituents in a fluid based on differences in one or more constituent properties, such as diffusion rate, solubility, electrical charge, or size and shape. A permselective membrane is also known as a semipermeable membrane.

peroxidase Any of a family of enzymes capable of catalyzing the breakdown of peroxides.

peroxide Any compound containing two oxygen atoms united together into a bivalent –O–O– group. A peroxide readily releases abnormally active atomic oxygen and is therefore a strong oxidizing agent. Hydrogen peroxide (H_2O_2) is an example.

peroxisome A subcellular organelle (or microbody) that contains peroxidase, catalase, D-amino acid oxidase, and the capability for β-oxidation of fatty acids. β-oxidation of fatty acids in peroxisomes differs from that in mitochondria in that hydrogen peroxide (H_2O_2) is produced as an obligatory product.

peroxisome proliferator Any of a group of chemicals that are capable of increasing the numbers of peroxisomes in particular tissues. A rather diverse group of chemicals possess this property, and most members of the class are capable of inducing tumors, particularly in the livers of rodents. Of particular interest in drinking water are the disinfection by-product trichloroacetic acid (CCl_3COOH) and the plasticizer di(2-ethylhexyl) phthalate ($C_6H_4(COOCH_2CH(C_2H_5)C_4H_9)_2$). Marked differences occur in the responses in different animal species to these chemicals, and the question of whether such chemicals would produce cancer in humans is very controversial.

peroxisome proliferator activated receptor (PPAR) Any of a family of proteins involved in the signal transduction pathway that leads to increased synthesis of peroxisomes and associated proteins. Although peroxisome proliferators are clearly necessary for the response, many of them appear not to interact directly with this type of protein; instead they seem to act by increasing the level of some endogenous factor. The peroxisome proliferator activated receptor interacts with other nuclear receptors that are important in development, including the retinoic acid receptor and thyroid hormone receptor. Despite the very weak activity of peroxisome proliferators to induce synthesis of peroxisomes in human and other mammalian cells, humans do possess the peroxisome proliferator activated receptors. This raises the possibility that the induction of peroxisomes may not be a prerequisite for the tumorigenic activity of these chemicals.

PEROXONE A combination of ozone (O_3) and hydrogen peroxide (H_2O_2).

PEROXONE process An advanced oxidation process in which hydrogen peroxide (H_2O_2) is added to ozone (O_3) to catalyze the formation of hydroxyl radicals ($\cdot OH$), which are strong oxidizing agents. This process is often used to oxidize taste-and-odor-causing compounds. *See also* advanced oxidation process; hydroxyl radical; ozone.

persistence The tendency of a refractory, nonbiodegradable material to remain essentially unchanged after being introduced into the environment.

person An individual, corporation, company, association, or partnership; a municipality; or a state, federal, or tribal agency.

person-to-person spread The conveyance of an infectious disease from one person to another by direct transmission because of close proximity either within a family or with the population at large. *See also* transmission of infection.

personal protection equipment (PPE) Such equipment as hard hats, gloves, goggles, and steel-toed shoes used by workers to prevent potential injuries from workplace hazards.

personnel access opening The opening in a vault or caisson to allow maintenance and inspection personnel to enter and perform their duties. Such openings were formerly called manholes.

PERT *See* program evaluation review technique.

pervaporation A membrane process used primarily in industrial applications for separation of mixtures of dissolved solvents. In this process, transport through the membrane is induced by the difference in partial pressure between the liquid feed solution and the permeate vapor. Commonly, the permeate is under a vacuum produced by a vacuum pump or by a cooling and condensing of the permeate vapor.

pervious *See* permeable.

pervious bed A bed or stratum that contains voids through which water will move under ordinary hydrostatic pressure.

perviousness *See* permeability.

pesticide Any substance or chemical applied to kill or control pests. *See also* algicide; herbicide; insecticide; rodenticide.

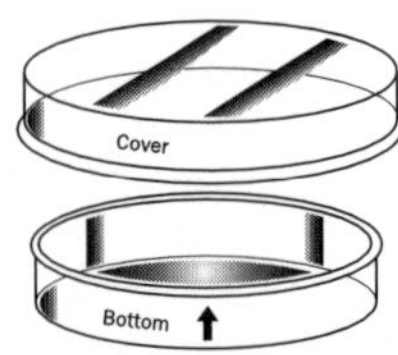

Petri dish

pet cock A small valve or faucet used to drain a cylinder or fitting.

peta (P) A prefix meaning 10^{15} in Système International.

petition for review A legal action taken under Section 1448 of the Safe Drinking Water Act to contest one or more provisions of a National Primary Drinking Water Regulation issued by the US Environmental Protection Agency.

petri dish A shallow glass or plastic dish with vertical sides, a flat bottom, and loose-fitting cover, used for growing microbiological cultures.

petrochemical A product or component that arises primarily from the chemical processing of petroleum and natural gas hydrocarbons.

petroleum derivative A chemical substance derived from petroleum (e.g., a chemical formed when gasoline breaks down in contact with groundwater).

petrology A science the deals with the origin, history, occurrence, structure, chemical composition, and classification of rocks.

PF *See* power factor; protection factor.

pF A numerical measure of the energy with which water is held in the soil. It is expressed as the common logarithm of the head, in centimetres of water, necessary to produce the suction corresponding to the capillary potential. *See also* moisture tension.

PFBHA *See ortho*-(2,3,4,5,6-pentafluorobenzyl)hydroxylamine.

PFBHA derivatization *See ortho*-(2,3,4,5,6-pentafluorobenzyl)hydroxylamine derivatization.

PFBOA An abbreviation sometimes used for *ortho*-(2,3,4,5,6-pentafluorobenzyl)hydroxylamine. PFBHA is the more appropriate abbreviation, however. *See also ortho*-(2,3,4,5,6-pentafluorobenzyl)hydroxylamine.

PFR *See* plug-flow reactor.

pfu *See* plaque-forming unit *in the Units of Measure list.*

pg *See* picogram *in the Units of Measure list.*

PGP *See* programmatic state general permit.

pH A measure of the acidity or alkalinity of a solution, such that a value of 7 is neutral; lower numbers represent acidic solutions and higher numbers, alkaline solutions. Strictly speaking, pH is the negative logarithm of the hydrogen ion concentration (in moles per litre). For example, if the concentration of hydrogen ions is 10^{-7} moles per litre, the pH will be 7.0. As a measure of the intensity of a solution's acidic or basic nature, pH is operationally defined relative to standard conditions that were developed so that most can agree on the meaning of a particular measurement. The pH of an aqueous solution is an important characteristic that affects many features of water treatment and analysis.

pH meter A sensitive voltmeter used to measure the pH of liquid samples.

pH of saturation (pH_s) The pH at which water is saturated with calcium carbonate ($CaCO_3$).

pH Redox Equilibrium Equation (PHREEQE) A general-purpose equilibrium chemistry modeling computer program written in FORTRAN and developed by the US Geological Survey. This flexible program can be used to perform calculations for aqueous speciation and solid saturation, as well as to model precipitation, dissolution, oxidation–reduction, titration, and mixing reactions. Versions that work on IBM-compatible and Macintosh personal computers are available from various sources.

pH shock A short-term depression of pH levels caused by a sudden influx of acid runoff into a water body. The acid runoff could result from an intense rainfall of low pH or from meltwater. A pH shock is also called an acid shock.

phage *See* bacteriophage.

phagocyte Any cell that engulfs bacteria, other cells, or foreign material. Usually the material is digested within the phagocyte. Macrophages and polymorphonuclear lymphocytes (white blood cells) are examples of phagocytes.

pharmaceutical grade water The collective term for six types of water as defined by the *US Pharmacopeia*: (1) purified water, (2) water for injection, (3) bacteriostatic water for injection, (4) sterile water for inhalation, (5) sterile water for injection, and (6) sterile water for irrigation.

pharmacodynamics The quantitative (i.e., kinetic) relationships between the amount of a chemical (or drug) at its active site within the body and the molecular, biochemical, and cellular and tissue responses the chemical produces within the body.

pharmacokinetics The formal mathematical description of processes involved in the absorption, distribution, metabolism, and elimination of chemicals in an organism. In medicine, pharmacokinetic models are used to determine the required dosing regimen of a drug to maintain blood levels in the therapeutic range. In toxicology and risk assessment, pharmacokinetic models provide a means of adjusting external doses to an organism to the effective dose that is delivered to the site within the body at which the adverse effect is produced. With the inclusion of species-specific rate constants, pharmacokinetic models can be used for interspecies comparisons of effective doses of a chemical for a given external dose.

pharmacological antagonism A situation in which two chemicals act through the same receptor but one of the chemicals acts as an activator of the receptor (agonist) and the other acts as an antagonist. *See also* antagonism.

phase (1) A distinct state of matter. The three physical states are solid, liquid, and gas. In addition, colloids are said to be the dispersed phase and liquids are said to be the continuous phase. (2) An attribute of the type of electrical service being used. Water treatment utilities utilize three-phase power for 2 horsepower (1.5 kilowatts) and greater; single-phase power is used for smaller loads. (3) An indication of the time relationships of two or more alternating electrical currents' oscillations. Two or more signals with the same time relationships (i.e., matching peaks and valleys on a visual plot of the signal over time) are said to be in phase; if the timing is different, the signals are said to be out of phase. The phase in a circuit is determined by interaction among the resistance, inductance, and capacitance components of the circuit.

phase angle (θ) A measure of the progress of a periodic wave in time and space from a chosen instant or position. It is obtained by multiplying the phase by 360° or by 2π radians.

phase contrast microscope A modified light microscope that enhances the visibility of unstained objects that differ only very slightly from the background in terms of the refractive index.

phase contrast microscopy The examination of microorganisms using a phase contrast microscope.

phase I reaction An initial reaction of a chemical being metabolized by the body. Phase I reactions frequently result in the production of reactive intermediates that are capable of interacting with and altering the structure and function of such macromolecules as proteins, nucleic acids, or lipids that make up biological membranes.

Phase I Rule A rule promulgated by the US Environmental Protection Agency on July 8, 1987 (*Federal Register,* 52(130):25690–25717), that set maximum contaminant level goals and National Primary Drinking Water Regulations for eight volatile organic chemicals (trichloroethylene, carbon tetrachloride, 1,1,1-trichloroethane, vinyl chloride, 1,2-dichloroethane, benzene, 1,1-dichloroethylene, and *para*-dichlorobenzene).

phase II reaction A secondary reaction that tends to detoxify reactive metabolites produced by phase I reactions. Phase II reactions are frequently conjugation reactions with sulfate, glucuronides, glutathione ($C_{10}H_{17}O_6N_3S$), cysteine ($HSCH_2CH(NH_2)COOH$), or other biochemicals. Occasionally, however, further processing of metabolites produced by phase II reactions can lead to reactive metabolites, e.g., the cysteine conjugates that arise from processing of the glutathione conjugate of such chemicals as tetrachloroethylene ($Cl_2C{:}CCl_2$) or trichloroethylene ($CHCl{:}CCl_2$).

Phase II Rule A rule promulgated by the US Environmental Protection Agency on January 30, 1991 (*Federal Register,* 56(20):3526–3597), that set maximum contaminant level goals and National Primary Drinking Water Regulations for 26 synthetic organic chemicals (*ortho*-dichlorobenzene, *cis*-1,2-dichloroethylene, *trans*-1,2-dichloroethylene, 1,2-dichloropropane, ethylbenzene, monochlorobenzene, styrene, tetrachloroethylene, toluene, xylenes, alachlor, atrizine, carbofuran, chlordane, 1,2-dibromo-3-chloropropane, 2,4-dichlorophenoxyacetic acid, ethylene dibromide, heptachlor, heptachlor epoxide, lindane, methoxychlor, polychlorinated biphenyls, toxaphene, 2-(2,4,5-trichlorophenoxyl propionic acid, acrylamide, epichlorohydrin) and 7 inorganic chemicals (asbestos, cadmium, chromium, mercury, nitrate, nitrite, selenium).

Phase V Rule A rule promulgated by the US Environmental Protection Agency on July 17, 1992 (*Federal Register,* 57(138):31776–31849) that set maximum contaminant level goals and National Primary Drinking Water Regulations for 18 synthetic organic chemicals (dichloromethane, 1,2,4-trichlorobenzene, 1,1,2-trichloroethane, dalapon, dinoseb, diquat, endothall, endrin, glyphosate, oxamyl, picloram, simazine, benzo(a)pyrene, di(2,ethylhexyl)adipate, di(2-ethylhexyl)phthalate, hexachlorobenzene, hexachlorocyclopentadiene, 2,3,7,8-tetrachlorodibenzo-*p*-dioxin) and 5 inorganic chemicals (antimony, beryllium, cyanide, nickel, sulfate, thallium).

phenanthroline method A colorimetric procedure used to determine the concentration of iron in water by using 1,10-phenanthroline ($C_{12}H_8N_2{\cdot}H_2O$) after the iron is reduced to the Fe^{2+} state with boiling hydrochloric acid (HCl) and hydroxylamine ($NH_2OH{\cdot}HCl$).

phenolic compound A hydroxy derivative of benzene (C_6H_6). The simplest phenolic compound is hydroxy benzene, C_6H_5OH (phenol).

phenolphthalein ($C_{20}H_{14}O_4$) An organic compound used as an acid-base indicator. Phenolphthalein solutions gradually change from colorless for a pH less than 8.3 to pink in solutions of pH greater than 10 and then to dark red.

phenolphthalein alkalinity The alkalinity in a water sample as measured by the amount of standard acid required to lower the pH to a level of 8.3. If phenolphthalein is used, the solution changes from pink to clear at this pH. Phenolphthalein alkalinity is expressed in units of milligrams per litre as calcium carbonate ($CaCO_3$).

phenotype A cell characteristic that arises for a particular cell from expression of a particular protein or group of proteins. The phenotype may reflect properties that are determined by changes in expression seen during normal differentiation and development of a particular tissue, or it can reflect changes in the characteristics of the wild-type cell as the result of a mutation (i.e., a change in the genotype). For example, liver and kidney cells are of a different phenotype. Tumor cells have different phenotypes than the normal cells from which they are derived.

phosphate A salt or ester of phosphoric acid (H_3PO_4). *See also* orthophosphate; phosphorus.

phosphate buffer The chemical(s) used to buffer water that is used in microbiological analyses of drinking water and other environmental water samples.

phosphate concentration The concentration of phosphates in drinking water or another source, normally presented in terms of phosphate ion (PO_4^{3-}) or phosphorus (P). The phosphate ion is a component of some corrosion inhibitors that are added to drinking water. Hence, the phosphate ion can be measured at the point of entry into the transmission system and at key locations within the distribution system to determine the adequacy of corrosion control.

phosphate-buffered saline (PBS) Phosphate-buffered water used for washing bacteria in the microsuspension mutagenicity test method. *See also* phosphate buffer.

phosphorus (P) An essential chemical element and nutrient for all life forms. It occurs in orthophosphate, pyrophosphate, tripolyphosphate, and organic phosphate forms. Each of these forms, as well as their sum (total phosphorus), is expressed in terms of milligrams per litre elemental phosphorus. *See also* nutrient.

phosphorus fertilizer A plant nutrient, composed of phosphorus, that can cause algal blooms in sensitive water bodies. *See also* algal bloom.

phosphorus removal The precipitation of soluble phosphorus from water by coagulation and subsequent flocculation and sedimentation.

photochemical Pertaining to chemical reactions that depend on light.

photochemical oxidation Changes in chemicals caused by light.

photochemically driven reaction A chemical reaction (e.g., photosynthesis, photolytic decomposition, photochemical oxidation) in which absorption of radiant energy (photons) induces or modifies chemical changes. *See also* photochemical oxidation; photolytic decomposition; photosynthesis.

photodegradation The breakdown of a compound dispersed in the atmosphere or present on an exposed solid surface into simpler components by the action of solar radiation.

photogrammetric digitizing A technique used to compile new maps from aerial photographs. It is similar to manual digitizing in its approach, but the table digitizer is replaced by a photogrammetric instrument, such as an analytical stereoplotter. Photogrammetric digitizing is most often used to record very precise and accurate digital planimetric features and elevation data from stereophotography.

photogrammetric mapping The process of compiling maps by measurement of controlled aerial photos.

photogrammetry The science of making reliable measurements by the use of photographs (usually aerial) in surveying and mapmaking.

photoionization detector (PID) A detector used in gas chromatography, typically for the analysis of aromatic compounds. A photoionization detector elicits an electrical signal from compounds with double bonds and can be fairly sensitive to conjugated double bonds, such as in aromatic compounds.

photolysis *See* photolytic decomposition.

photolytic decomposition Chemical decomposition that is driven by photons of sunlight or ultraviolet light.

photolyzed Pertaining to something that has undergone photolytic decomposition.

photometer An instrument used to measure the intensity of light transmitted through a sample or the degree of light absorbed by a sample.

photomultiplier tube Tube used to magnify the primary photocurrent by the ejection of several electrons for each electron striking a dynode. The magnification goes through several successive stages operating at increasingly higher voltages. These tubes are used in instruments like spectrophotometers and liquid scintillation counters.

photon The basic unit (quantum) of electromagnetic radiation. Light waves, gamma rays, x rays, and so on consist of photons. Photons are discrete concentrations of energy that have no rest mass and move at the speed of light. They are emitted when electrons move from one energy state to another, as in an excited atom. *See also* irradiation; radiation.

photon emitter A radionuclide that emits photons. Photon emitters and beta particles make up the radiological measurement of gross beta radiation that is regulated by the US Environmental Protection Agency. *See also* beta particle; gross beta particle activity; photon; radiological contaminant.

photooxidant An oxidant that participates in a photochemical reaction involving oxidation–reduction chemistry. *See also* oxidant; oxidation–reduction (redox) reaction; photochemically driven reaction.

photosensitization The process of increasing the sensitivity of a cell or organism to the influence of light. Some drugs or chemicals are referred to as photosensitizers because they increase the adverse effects of ultraviolet light, visible light, or both. Two basic types of reactions can be produced: photoallergic or phototoxic. In humans, such effects most often involve the skin or the eyes.

photosynthesis The synthesis of complex organic materials, especially carbohydrates, from carbon dioxide (CO_2), water, and inorganic salts under the following conditions: with sunlight as the source of energy, with the aid of a catalyst such as chlorophyll, and with the simultaneous liberation of oxygen.

photosynthetic bacteria Bacteria that obtain their energy for growth from light by photosynthesis.

phototoxicity Biological damage caused by photon-driven chemical reactions. Phototoxic reactions are divided into two broad classes: those that are dependent on reactions with oxygen (i.e., photodynamic reactions) and those that do not require oxygen. The former generally result in the formation of reactive free radicals that are responsible for the damage. In the latter case, the chemical facilitates the reaction of a photon with cellular biochemicals. The methoxypsoralens, certain polycyclic aromatic hydrocarbons, and selected drugs are chemicals known to induce various types of phototoxicity. *See also* free radical.

photovoltaic Pertaining to the process by which some materials convert incident light to an electromotive force.

photovoltaic cell A device used to convert light energy to electrical energy. In such a cell, light falling on a semiconductor surface releases electrons to a collector electrode, so that the output current produced by the cell is very nearly proportional to the intensity of illumination.

photozone of seawater The region in the ocean into which light will penetrate.

phreatic Pertaining to that layer of soil or rock in which a water table exists.

phreatic cycle A period of rise and succeeding period of decline for a water table. The most common kinds of phreatic cycles are daily, annual, and secular.

phreatic decline A downward movement of the water table.

phreatic divide *See* groundwater divide.

phreatic fluctuation *See* phreatic cycle.

phreatic line The upper boundary of (1) the water table in soils or (2) seepage water in earth dams, levees, and dikes. It is the line of atmospheric pressure, and it lies between the capillary zone and the saturation zone.

phreatic rise An upward movement of the water table.

phreatic surface A surface that represents the position to which water would rise in wells anywhere in an unconfined aquifer. A water-table contour map is a method of depicting the phreatic surface of an unconfined aquifer.

phreatic water Water beneath the water table in an unconfined aquifer.

phreatic water discharge Discharge of water from the saturation zone directly onto the land surface, into a body of surface water, or into the atmosphere by means of springs, wells, infiltration galleries, or infiltration tunnels and other subterranean channels. Such discharge is also called groundwater discharge.

phreatic wave A wavelike movement of the water table in the direction of the latter's slope. It assumes the form of a groundwater mound or ridge. It is produced by a considerable addition of water to the water table in a relatively short time and over a relatively small area, and it flattens out as it progresses to lower levels. A phreatic wave is also called a groundwater wave.

phreatophyte A type of plant with very long roots and extensive root systems that draws its water from the water table or other permanent groundwater supplies. Examples of phreatophytes are willows and salt cedars. Excessive phreatophyte growth is undesirable in areas where water is scarce because these plants can consume large quantities of water.

PHREEQE *See* pH redox equilibrium equation.

pH_s *See* pH of saturation.

phthalate ($C_6H_4(COOR)_2$) Any of a class of synthetic organic chemicals with various industrial uses, including use as a plasticizer. The compound di(2-ethylhexyl) phthalate ($C_6H_4(COOCH_2CH(C_2H_5)C_4H_9)_2$ is regulated by the US Environmental Protection Agency. *See also* di(2-ethylhexyl) phthalate; plasticizer; synthetic organic chemical.

```
        H
        |    O
       C     ||
H-C       C-C-O-R
 |         || O
 |         || ||
H-C       C-C-O-R
       C
        |
        H
```

Phthalate

phylogeny The study of comparative evolutionary relationships among organisms.

physical adsorption The accumulation or concentration of substances (adsorbate) at a surface or interface with another substance (adsorbent) by physical attraction and interaction.

physical analysis An examination of a sample for parameters that represent the collective properties of the sample. For example, in the examination of water, examinations of appearance, conductivity, turbidity, temperature, and color represent physical analyses.

physical stability A measure of the ability of an ion exchanger or filter medium to resist breakdown caused by the physical forces to which the exchanger or medium is subjected during use. Such forces include crushing, attrition, or forces caused by high temperatures.

physical treatment Any water treatment process involving only physical means of solid–liquid separation, e.g., centrifugation, clarification, distillation, filtration, flocculation solely by agitation, and heat treatment.

physical–chemical treatment Treatment of water by unit processes other than those based on microbiological activity. Unit processes commonly included under this heading are precipitation with coagulants, flocculation with or without chemical flocculants, filtration, adsorption, chemical oxidation, air-stripping, ion exchange, membrane treatment, and several others.

physiognomy The general outward appearance of a community as determined by the life form of the dominant species, e.g., grassland, forest.

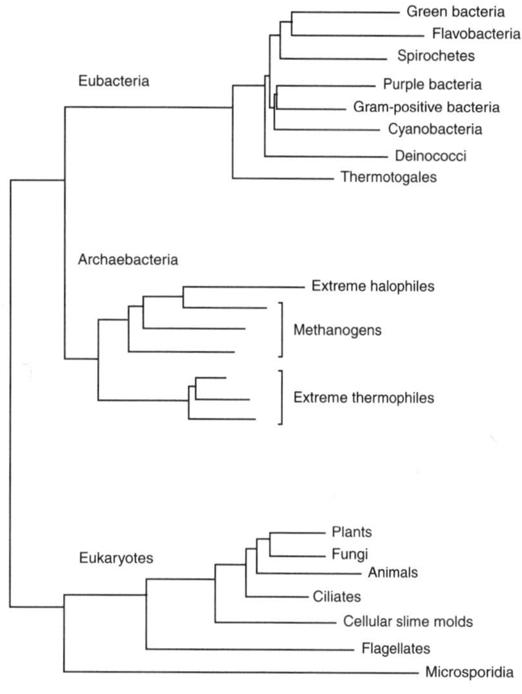

Phylogeny

physiographic balance A condition of equilibrium between surface topography and vegetal cover on the one hand, and precipitation and the accompanying processes of erosion on the other. It is said to exist in a given locality when topography and cover have become so adjusted to precipitation and erosion that changes in the topography and cover are imperceptible over a considerable period of time, even though they might be taking place.

physiographic province Terrain having a pattern of natural relief features or landforms—including drainage, geomorphological characteristics, and topography—that differs significantly from that of adjacent zones. Also called *physiographic regions.*

physiological antagonism A circumstance in which chemicals or drugs activate different physiological systems to produce opposing effects. For example, many of the effects of chemicals that activate the parasympathetic nervous system (a portion of the autonomic nervous system that uses acetylcholine as the major neurotransmitter) are counteracted by chemicals that activate the sympathetic system (a portion of the autonomic nervous system that uses catecholamines as neurotransmitters). A simpler example may be the antagonistic effects of stimulants and depressants. *See also* antagonism.

physiologically based pharmacokinetic (PBPK) modeling A very particularized way of modeling how a chemical, its metabolites, or both are handled within the body. These models contain specific information on blood flows, as well as transfer rates (frequently described by the partition coefficients) of each chemical species between the external environment (e.g., air concentration) to the blood plasma and tissues. They can also include kinetic parameters related to the metabolism and excretion of each chemical form from the body and any specialized binding of the chemical within each compartment. Many believe these models provide a more accurate basis than other models for extrapolating exposure and dose information across species. These models are very data-intensive, however, relative to the compartmental models that have frequent application in the use of drugs. *See also* biologically based dose response modeling; compartmental model.

phytometer A device for measuring transpiration, consisting of a vessel containing soil in which one or more plants are rooted and sealed so that water can escape only by transpiration from the plant. *See also* lysimeter.

phytoplankton Collectively, all the microscopic plants that live suspended in the water of aquatic habitats and are unattached.

pi (π) The ratio of the circumference of a circle to the diameter of that circle, approximately equal to 3.14159, or about 22⁄7.

PICl *See* polyiron chloride.

picloram ($C_6H_3O_2N_2Cl_3$) A common name for 4-amino-3,5,6-trichloropicolinic acid, a systemic herbicide that has been used as defoliant. It is regulated by the US Environmental Protection Agency. *See also* defoliant; herbicide.

Picloram

pico Prefix meaning 10^{-12} in Système International.

picocurie (pCi) *See the Units of Measure list.*

picocuries per litre (pCi/L) *See the Units of Measure list.*

picogram (pg) *See the Units of Measure list.*

Picornaviruses The smallest of the ribonucleic acid viruses. The *Picornaviruses* are subdivided into four genera: *Enterovirus, Cardiovirus, Rhinovirus,* and *Aphthovirus.*

picosecond (ps) *See the Units of Measure list.*

picture element (pixel) The smallest indivisible element that makes up an image. In raster processing, data are represented spatially on a matrix of grid cells by picture elements, which are assigned values for image characteristics or attributes. The higher the resolution, the higher the pictures element count for a given area. *See also* raster image.

PID *See* proportional integral derivative control; photoionization detector.

piedmont alluvial deposit A series of alluvial fans built up by streams originating from a mountain range.

piedmont alluvial plain A series of alluvial fans that merge into one another. A piedmont alluvial plain is also called a compound alluvial fan.

Pielkenroad separator A settling unit containing parallel corrugated plates in a portion of the unit, at a fixed angle, to increase the available surface area and thereby enhance particle separation. The flow is conducted upward through the plates, with the clarified effluent being discharged at the top and the collected solids being discharged at the bottom. The corrugated plates reportedly enhance sludge thickening and are less costly to construct than tube settlers. *See also* inclined plate separator; tube settler.

piestic interval The difference in altitude between two isopiestic lines (lines of equal pressure).

piezometer An instrument for measuring pressure head in a conduit, tank, or soil by determining the location of the free water surface.

piezometric head The elevation of water (with respect to a datum) plus the pressure head expressed in units of length; above a datum, the total head at any cross section minus the velocity head at that cross section. It is equivalent to the elevation of the water surface in open-channel flow, and it gives the elevation of the hydraulic grade line at any point. *See also* pressure head.

piezometric map A map showing, by means of isopiestic lines, the shape of the piezometric surface of an aquifer.

piezometric surface A surface that represents the position to which water would rise in wells anywhere in an aquifer. A groundwater-level contour map is a method of depicting the piezometric surface of an aquifer. *See also* isopiestic line.

piezometric surface map A contour map of the imaginary surface to which the water in an artesian aquifer will rise. Such a map is also called a pressure surface map.

piezometric tube A tube open at both ends, with one end that is placed into a liquid flow system (e.g., an aquifer or pipeline) to measure the static water level in the system. The height of water in the tube measures the energy in the liquid due to pressure (pressure head) and elevation (elevation head). *See also* piezometer.

pig *See* polypig.

pigging The process of forcing an in-line scraper or polypig through a water line by the force of moving water or flush water to remove scale, sand, and other foreign matter from the interior surface of the pipe.

pilot filter A small tube containing the same media as plant filters and through which flocculated plant water is continuously passed, with a recording turbidimeter continuously monitoring the effluent. The amount of water passing through the pilot filter before turbidity breakthrough can be correlated to the operation of the plant filters under the same coagulant dose.

pilot hole A test well.

pilot plant A small-scale treatment process designed to simulate a full-scale treatment unit or sequence of units. It operates at a lower flow and is used to evaluate the impact of variations in operating criteria on treatment performance. A pilot plant is operated in a continuous-flow manner but uses sufficiently small quantities of water that the effluent can be wasted, thus

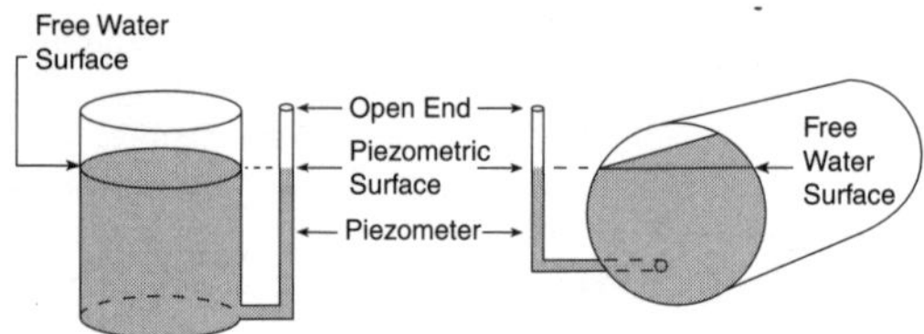

Piezometer and piezometric surface

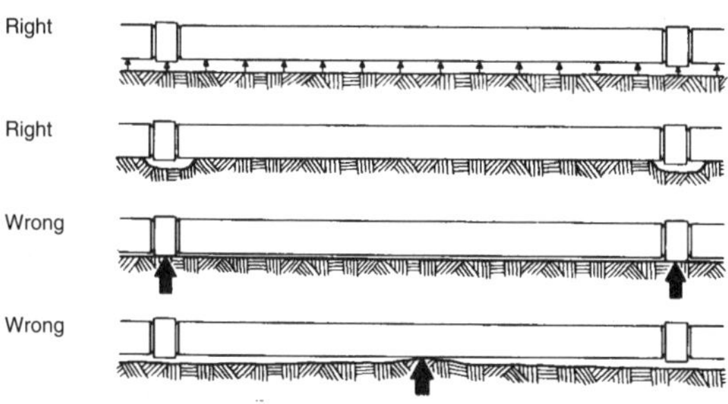

Good and bad pipe bedding

providing the distinct benefit of simulating the process without being required to generate a potable water for consumption.

pilot project An undertaking of a special unit of work or research (as in a school, laboratory, or plant) organized and carried out as a trial unit for experimentation, testing, or seeking a solution to a specific problem.

pilot-plant study Evaluation, on a scale larger than laboratory scale but smaller than full scale, of the amenability of water to treatment by particular operations or processes.

pilot-plant test data Information regarding water quality and operating parameters obtained from a small-scale, continuous-flow unit constructed to simulate the behavior of a full-scale treatment unit.

PIN *See* parcel identification number.

pinch point An area on equipment or a point in a procedure where an employee may have a portion of clothes or body caught between two pieces of equipment or between equipment and the working surface.

pipe A conduit that diverts or conducts water from one location to another.

pipe bedding The surface, material, or both on which pipe rests in a ditch.

pipe class The working pressure rating, including allowances for surges, of a specific pipe for use in water distribution systems. This term is used for cast-iron, ductile iron, asbestos–cement, and glass pipe, as well as some plastic pipe.

pipe corrosion The destruction of a pipe as a result of a chemical reaction with its surroundings. Pipe corrosion is generally a physicochemical interaction between a pipe's material and its environment that results in an alteration of the pipe material's properties. Internal pipe corrosion refers to destruction associated with the inside of the pipe, which is in contact with the substance being conveyed, e.g., potable water. Internal pipe corrosion may be uniform over the surface or it may form pits and tubercles. External pipe corrosion refers to destruction from the outside of the pipe toward the interior. External pipe corrosion is often associated with stray current or certain soil conditions. *See also* tubercle.

pipe cutter A mechanical tool with rolling blades that indent a pipe as the tool rotates around and cuts the pipe. Such a tool is used for cast-iron pipe, with a smaller cutter used for copper tubing.

pipe diameter The nominal or commercially designated inside diameter of a pipe, unless otherwise stated. *See also* inside diameter. *See also* inside diameter.

pipe finder An instrument for locating underground metal water pipes by magnetic attraction or electronic detection.

pipe fitting A set of connections, appliances, and adjuncts designed to be used in connecting pipes. Examples are elbows and bends to alter the direction of a pipe; tees and crosses to connect a branch with a main; plugs and caps to close an end; and bushings, diminishers, or reducing sockets to couple two pipes of different dimensions.

pipe gallery (1) Any conduit in which pipes are to be located, usually of a size to allow a person to walk through. (2) A gallery provided in a treatment plant for the installation of the conduits and valves and for a passageway to provide access to them.

pipe gauge A number that defines the thickness of the sheet used to make steel pipe. The larger the number, the thinner the pipe wall.

pipe grade The slope or fall of the pipe in the direction of flow.

pipe loop An experimental apparatus consisting of several feet of pipe complete with joints, elbows, and connections for flow-through testing of corrosion and corrosion control.

pipe offset A casting in the form of a reverse curve, designed for the continuation of a line of pipe parallel to its beginning. *See also* offset.

pipe saw (1) A pneumatically or hydraulically powered tool that is self-propelled on a chain or rollers and has a saw blade that saws the pipe as the machine goes around the pipe. Such a tool is used to cut large-diameter cast-iron and ductile iron pipe. (2) A hand-held gasoline-powered saw that rotates a grinding blade at high speed and will saw or grind pipe. The pipe can be rotated or the saw can be moved all the way around the pipe. This type of saw can be used for ductile iron, large cast-iron, steel, and concrete pipe.

pipe schedule A classification in a sizing system of arbitrary numbers that specifies the inside diameter and outside diameter for pipe. This term is used for steel, wrought iron, and some types of plastic pipe. It is also used to describe the strength of some types of plastic pipe.

pipe-lateral system A filter underdrain system using a main pipe (header) with several smaller perforated pipes (laterals) branching from it on both sides. *See also* filter bottom.

pipe-to-soil potential An electrical potential (driving force expressed in volts) that any two connected metals develop when in

Pipe fittings for ductile iron pipe

contact with a common medium, such as soil. Traditionally, pipe-to-soil potential is the potential measured between an underground metal, such as pipe, and a copper–copper sulfate electrode.

pipeline A collection of pipes jointed to provide a conduit through which fluids flow.

Piper trilinear diagram A graphical method for showing the content of major ions in water by plotting the percent equivalents of anions and cations along the sides of two triangles and using a rhombus (equilateral parallelogram) to represent the total ionic constituents.

pipette A slender glass or plastic tube (sometimes graduated) used to measure and transfer small volumes of liquids (usually less than 25 millilitres). *See also* volumetric pipette.

piping (1) The action of water passing through or under a dam and carrying some of the finer soil material to the surface at the downstream toe. Such action may result in excessive leakage or even failure because the increased porosity of the material caused by removal of the fines allows an increase in the velocity of the water; in turn more and larger-sized material is removed. (2) The result of such action. (3) The system of pipes, incorporated fittings, and appurtenances that is provided to carry any fluid in a structure or plant. (4) The installation or laying of a system of pipes.

piping system A system of pipes, fittings, and appurtenances within which a fluid flows.

***p*-isopropyltoluene** *See under* isopropyltoluene.

piston A sliding metal cylinder or disk that reciprocates in a tubular housing to move water or apply force to water.

piston pump A reciprocating pump in which the cylinder is tightly fitted with a reciprocating piston.

piston-type pump *See* piston pump.

pitless adapter A tube with a waterproof cap that is attached to a well casing above the frostline to protect the well discharge pipe and pump from freezing. The tube permits power wiring to enter but is otherwise waterproof.

pitometer A device operating on the principle of the Pitot tube, principally used for determining the velocity of a flowing fluid at various points in a water distribution system and to ascertain waste, leakage, or clogging of pipes.

pitometer survey A survey conducted with a pitometer to determine the velocity of flow of water flowing at various points in a distribution system and to ascertain waste, leakage, or clogging of pipes.

Pitot tube A device for measuring the velocity of flowing fluid by using the velocity head of the stream as an index of velocity. One type of Pitot tube consists essentially of an orifice that is held so as to point upstream and that is connected with a tube in which the impact pressure due to the incoming water's velocity head may be observed and measured. A Pitot tube may also be constructed to have both an upstream and a downstream orifice; or it may have an orifice pointing upstream to measure the velocity head or pressure and piezometer holes in a coaxial tube to measure the static head or pressure. In either of these latter cases, the difference in pressure is the index of velocity.

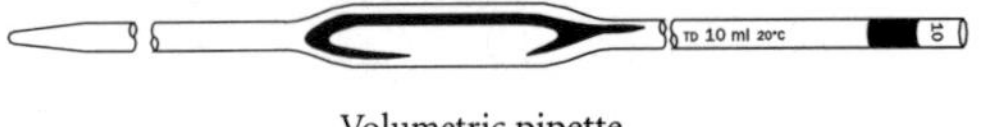

Volumetric pipette

Pitot-integrating tube A flow element that consists of a rod extending through the pipe, with a number of holes (up to eight) on the upstream side of the rod that sample and average the velocity-induced pressure across the pipe cross section. The pressure is proportional both to the velocity of the fluid in the pipe and to a corresponding flow rate in the pipe. *See also* Pitot tube.

pitting corrosion Highly localized corrosion resulting in deep penetration at only a few spots.

pixel *See* picture element.

pK_a The negative common logarithm of the equilibrium constant (K) for the dissociation of an acid. The weaker an acid, the larger its pK_a. Thus, at 77° Farenheit (25° Celsius), for sulfuric acid (H_2SO_4, a strong acid), the pK_a is approximately –3.0; for acetic acid (CH_3COOH, a weak acid), the pK_a equals 4.76; and for boric acid (H_3BO_3, a very weak acid), the pK_a equals 9.24. When pH equals pK_a, the acid is 50 percent deprotonated. *See also* acid; dissociation.

pK_x A common logarithmic equilibrium constant for which x is replaced by an appropriately descriptive letter (e.g., pK_a for acid).

PL *See* public law.

placeholder A requirement included in a proposed regulation that may be replaced by a different requirement defined at a later date.

placental barrier The cellular and basement membranes that prevent a free diffusion of chemicals from the mother's blood to the fetal circulation. It is not an absolute barrier. Its structure differs significantly in different species and during different times in pregnancy. Up to six layers of cells may be present, and placenta are classed according to the number of layers present.

placental transfer The rate at which a chemical or other agent moves from the maternal blood into the fetus.

plain angle units In the US customary system, degrees. In Système International, radians.

plain sedimentation The sedimentation of suspended matter without the use of chemicals or other special means.

plain settling tank A tank or basin in which water containing settleable solids is retained for a sufficient time, and in which the velocity of flow is sufficiently low, to remove by gravity a part of the suspended matter. *See also* sedimentation basin.

plan view A diagram or photo showing an object or facility as it would appear when one is looking down on it from above. *See also* elevation view; isometric view.

planimetric map A map that shows the horizontal position on the earth's surface of natural and cultural physical entities, such as roads, buildings, and water bodies that are visible, identifiable, and able to be compiled into map features through photogrammetric or surveying procedures.

plankton (1) Small, usually microscopic, plants (phytoplankton) and animals (zooplankton) in aquatic systems. The chief constituents of phytoplankton are unicellular algae. Zooplankton consist of protozoa, small crustaceans, and various invertebrates' larvae. They are used as food by higher aquatic life forms such as fish. (2) The aggregate of passively floating, drifting, weakly motile organisms in a body of water. *See also* periphyton; phytoplankton; zooplankton.

planktonic Pertaining to plankton.

planned reuse Water reuse implemented by a prepared system that was designed in advance (e.g., golf course irrigation with treated wastewater) rather than unintentional reuse (such as using a water body contaminated with wastewater as a drinking water source).

plant community The plant populations existing in a shared habitat or environment.

plant factor For a water treatment plant, the quantity of the production output over a given period of time divided by the plant's rated capacity, usually expressed as a percentage. A plant factor is sometimes also referred to as a plant operating factor.

plant hydraulic capacity The level of flow into a plant in excess of which the system will be hydraulically overloaded.

plant operating factor *See* plant factor.

plant tap A water-sampling site upstream of the entrance to the distribution system but downstream of treatment.

plant water requirement The amount of irrigation needed to replace moisture depleted from the soil around plant roots as a result of evapotranspiration.

planting plan A two-dimensional plan drawn to scale that shows the layout of landscape plantings and plant specifications.

plant-scale study A full-scale evaluation of whether or not a particular water is amenable to treatment with particular operations or processes.

plaque A clear circular zone seen in a bacterial lawn resulting from in vitro virus infection.

plaque-forming unit (pfu) *See the Units of Measure list.*

plasma (1) Any of several types of high-energy discharges of hot (3,000–10,000 kelvin) ionized gas. Plasma is applied in certain spectroscopic instruments for the analysis of metals. For example, an inductively coupled plasma is energized by a radio frequency coil and resulting magnetic field. The plasma is often cooled with large volumes of an inert gas such as argon. The plasma makes a very good atomizer for metals analysis and can be a powerful instrument when interfaced with atomic emission or mass spectrometric detectors. (2) The fluid portion of the blood in which the particulate compounds are suspended.

plasma membrane The lipid structure that separates the contents of a cell from the cell's external environment. Such membranes serve the purpose of excluding materials from the cell and provide a means of maintaining cellular contents, including various substrates, at a sufficient concentration to maintain normal physiological function.

plasma protein A protein found in the blood. The two major classes found in mammalian blood are albumin and globulins. As observed in therapeutics and toxicology, plasma proteins bind certain chemicals. Such binding serves as a buffer, preventing rapid diffusion of chemicals to cells where they produce their effects but also serving to limit the rate at which chemicals are removed from the body (i.e., the protein acts as a temporary sink for the chemicals). Thus, chemicals that are heavily bound to plasma proteins will have their initial or peak effects blunted, but the duration of the chemical effects might well be prolonged as a result.

plasma water The aqueous phase of the blood. *See also* plasma.

plasmid A cytoplasmic, autonomously replicating chromosomal element found in bacteria. *See also* antibiotic resistance.

plasmid deoxyribonucleic acid (plasmid DNA) A circular strand of deoxyribonucleic acid that can exist in the cytoplasm of a bacterial cell and replicate independent of bacterial genomic deoxyribonucleic acid. Genes encoded in the plasmid strand are accessory genes, such as antibiotic resistance genes, and are not needed by the organism for survival under normal conditions.

plastic Any high-molecular-weight polymer material (usually synthetic) that during its manufacture or processing can be extruded, molded, cast, drawn, or laminated into objects of all sizes and shapes by application of heat or pressure, by chemical condensation, or by casting during polymerization of monomers, and that can retain the new shape under conditions of use. *See also* polymer.

plastic fluid A fluid, in laminar flow, that requires a minimum shear stress (called the yield stress) before velocity gradients can be established.

plastic media Media that are made of plastic and are used within a packed bed or packed tower process. The media promote mass transfer or serve as attachment points for microorganisms. They are often in the shape of hollow spheres with varying geometric patterns to increase their surface area. *See also* packed bed filter; packed tower aeration.

plastic pipe An identification used for nonmetallic pipe manufactured from different polymers, such as polyvinyl chloride, $(CH_2{:}CHCl)_n$; polyethylene, $(H_2C{:}CH_2)_n$; or polybutylene, $(C_4H_8)_n$.

plastic soil A soil that is easily molded and shaped without cracking or losing volume.

plasticizer An organic compound added to a high-molecular-weight polymer to facilitate processing, to increase the flexibility, and to increase toughness of the final product by internal modification of the polymer molecule. Among the more important plasticizers are nonvolatile organic liquids and low-melting solids (e.g., phthalate, adipate). *See also* adipate; phthalate; polymer.

plate Any flat-surface filter element. Plates are usually found in horizontal plant filters.

plate count An enumeration of bacteria in a sample. An appropriate volume of sample, or diluted sample, is placed in a sterile petri dish with a culture medium. The sample is then incubated at an appropriate temperature for some specified time period (usually 24 hours to 48 hours, but some types of samples and media are incubated for several days to several weeks). Each bacterial colony that develops is considered to have developed from 1 colony-forming unit, which may be a single viable cell or a clump of viable cells. All colonies on the plate are counted and adjusted for dilution, and the colony count recorded in colony-forming units per millilitre. *See also* heterotrophic plate count.

plate press *See* plate-and-frame filter press.

plate screen A screen composed of one or more perforated plates through which a liquid flows. *See also* perforated baffle wall.

plate settler *See* inclined plate separator; Pielkenroad separator.

plate-and-frame filter press A mechanical device used to dewater sludge. The square plates are fitted with a filter fabric into which sludge is introduced, and the plates are pressed together, in accordion fashion, to dewater the sludge. After the pressing

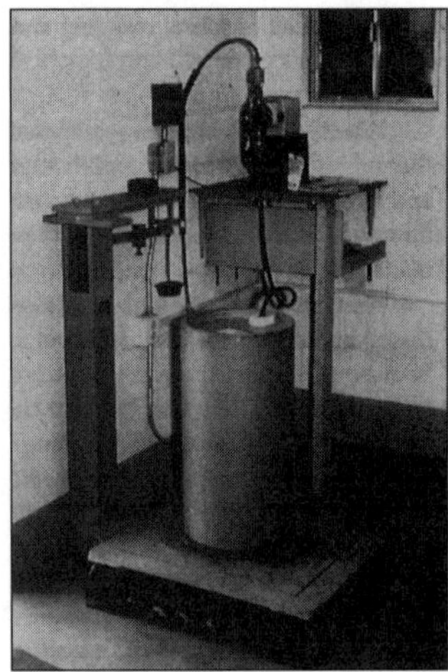

Platform scale

cycle, the plates are opened and the dewatered residual falls into a hopper. In some instances the dewatered sludge may have to be manually dislodged from the filter fabric. Although somewhat labor-intensive, plate-and-frame filter presses can achieve a very high solids content in the dewatered sludge, potentially up to 50 percent.

plate-and-frame membrane configuration A type of membrane arrangement in which flat sheets of pressure-driven membranes are placed between porous separation plates. The arrangement is designed for a pressure-differential driving force across the membrane.

platelet A disk-shaped cell lacking a nucleus. Such cells are found in the blood of mammals and play an important role in the coagulation (clotting) of blood.

platelet aggregation The process by which platelets accumulate rapidly at the site of injury within blood vessels. The platelets attach to exposed collagen fibers and, through an active process, recruit the adhesion of other platelets to initiate a clot. During this process a factor is exposed on the membrane of the platelet that augments thrombin formation, further accelerating clot formation.

platform furnace *See* L'vov platform.

platform scale A weight-measuring device typically used for relatively heavy objects. The scale is located at grade elevation so that the object to be weighed can be rolled into place and does not have to be lifted. A truck scale is an example of a platform scale.

platinum–cobalt (Pt–Co) color unit *See the Units of Measure list.*

platinum–cobalt method A procedure used to determine the amount of color in water by comparing the colors of standards made up with specified platinum–cobalt solutions to the color in the sample.

playa (1) A lake bed found in arid or desert regions in the lowest part of an enclosed valley with centripetal or inward drainage; a bolson. The lake is usually dry, except after heavy rainstorms, when it may be covered by a thin sheet of water that quickly evaporates. (2) A shore, strand, beach, or bank of a river.

playa deposit A deposit of sedimentary material laid down in a playa.

PLC *See* programmable logic controller.

plerotic water Subsurface water occupying the saturation zone, from which wells and springs are fed. In a strict sense, this term refers only to water below the water table. Plerotic water is also called groundwater or phreatic water.

ploidy The number of copies of chromosome complements that are contained in a cell.

PLSS *See* public land survey system.

plug (1) A fitting for the bell end of cast-iron pipes to close the opening. (2) A fitting that has an exterior pipe thread and a projecting head by which the fitting is screwed into the opening of another fitting. (3) The movable part of a tap, cock, faucet, plug, or valve. (4) A clogging of a pipe. (5) A specific quantity, or "chunk," of water moving through a containing structure. *See also* plug flow.

plug cock A shutoff valve in which the liquid passageway is a hole in a rotatable plug fitted into the valve body. Rotation of the plug through a right angle stops or starts the flow.

plug flow A flow regime in which no longitudinal dispersion (mixing in the direction of flow) occurs. The input, or plug, travels through a pipe or unit process in such a fashion that the entire mass or volume is discharged at exactly the theoretical hydraulic retention time of the unit (calculated as volume of the vessel divided by the flow rate). *See also* hydraulic detention time; longitudinal dispersion; plug.

plug valve A valve in which the movable control element is a cylindrical or conical plug, in contrast to a flat disk.

plug-flow reactor (PFR) A reactor in which the hydraulic behavior is such that the residence time of a given input, or plug, is exactly equal to the theoretical hydraulic retention time. *See also* hydraulic detention time.

plug-flow sampling A technique of collecting samples in which grab samples are taken at downstream intervals in a flowing stream so that the same plug of water is always being sampled.

plugging In a membrane treatment process, the depositing of solids in the membrane module, causing a physical flow restriction that negatively affects performance.

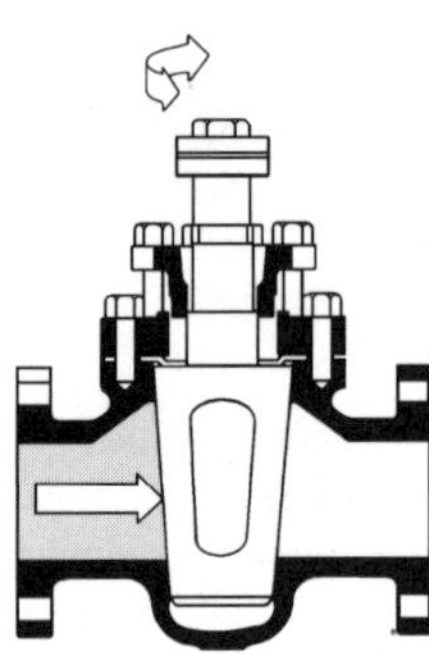

Plug valve

plugging factor An empirical measure of the plugging characteristics of membrane feedwater based on passing the water through a membrane filter test apparatus containing a 0.45-micrometre pore diameter filter (commonly a cellulosic type) in a dead-end filtration mode at a constant feed pressure (typically 30 pounds per square inch [207 kilopascals]) for a specified duration (typically 15 minutes) and observing the flow rate decline. The plugging factor is calculated based on the measured time required for 500 millilitres of filtrate to pass through the filter at the beginning of the test filtration period and the time for collection of 500 millilitres of filtrate at a later time. It is computed as [1 – (initial time in seconds/final time in seconds)] × 100, expressed as a percentage.

plumbing (1) The pipes, fixtures, and other apparatuses inside a building for bringing in the water supply and removing liquid- and waterborne wastes. (2) The installation of the foregoing pipes, fixtures, and other apparatuses.

plumbing fixture A receptacle that receives water and discharges wastewater into a drainage system (e.g., a sink or toilet).

plumbing system A building's distributing pipes for water supply; fixtures and fixture traps; soil, waste, and vent pipes; building drain and building sewer; and stormwater drainage pipes; along with the associated devices, appurtenances, and connections within and adjacent to the building.

plumbosolvation The actual dissolution of lead into drinking water, as opposed to the electrochemical deterioration of the lead-bearing material.

plume The pattern formed as polluted water extends downstream from the pollution source (analogous to smoke drifting downwind from a smokestack).

plunger pump A reciprocating pump having a plunger that does not come in contact with the cylinder walls, but enters and withdraws from the cylinder through packing glands. Such packing may be inside or outside the cylinder, according to the design of the pump. The packing is in the casing rather than around the plunger or piston.

pluripotent stem cell A primitive stem cell that has not yet committed to a final developmental pathway and can still give rise to several different cell types. Such cells can be recognized in some organs and tissues, but their identity in some tissues (e.g., the liver) are controversial as of the time of this writing. Stem cells are believed to be important in carcinogenesis because cancer is assumed to evolve only from stem cells in which a mutation has been inserted or in which control over the rate of division has been lost. *See also* stem cell.

plutonic water Water derived from deep in the Earth's magma by volcanic activity.

pluvial index An indicator of the total rainfall that may be expected to be exceeded during a given period, such as 1 day or 2 days, in a given locality during a given number of years. *See also* rainfall index.

pluviometric coefficient The ratio of the actual mean rainfall in a particular month to a calculated mean rainfall, expressed as a percentage and based on the assumption that the annual rainfall is evenly distributed throughout the year.

PMOS *See* positive (charge) metal oxide semiconductor.

PMP *See* probable maximum precipitation.

PMR *See* proportional mortality ratio.

PNAH *See* polynuclear aromatic hydrocarbon.

pneumatic Powered or moved by air pressure or compressed air.

pneumatic controller A device for controlling the mechanical movement of another object (e.g., a valve stem) by variations in pneumatic pressure connected to the controller.

pneumatic ejector A device by which a material is ejected or educted into a pipe through reduced air pressure created by a venturi. *See also* venturi.

pneumatic hammer *See* water hammer.

pneumatic pumping Pumping by means of an air-lift pump.

pneumatic system A system in which air pressure is used as the primary motive force.

pneumatic tank A pressurized holding tank that is part of a closed water system (such as for a household well system) and is used to create a steady flow of water and avoid water surges created by the pump turning on and off.

POC *See* particulate organic carbon; purgeable organic carbon.

pocket spring A spring located at the lower edge of a deposit of pervious material filling an irregular depression of the bedrock.

pocket storage The volume of water, expressed as a depth for the entire water surface of the area, that is required to fill natural depressions, large or small, to their overflow levels. Pocket storage is also called depression storage.

POE *See* point of entry.

pOH A measure of a solution's hydroxyl ion (OH^-) concentration.

poietin Any of a variety of hormones that control the numbers of various cell types found in the peripheral blood (e.g., the numbers of red blood cells or white blood cells).

point of application The location, in relation to other processes, where a specific treatment chemical is introduced.

point dilution method A technique for measuring water discharge by adding a known concentration of a tracer compound at one location and measuring the concentration of that tracer at a downstream location. A simple mixing calculation allows an estimate to be made of the total discharge between the two points based on how much the tracer was diluted.

point of diminishing returns The point at which additional increments of coagulant result in an "insignificant" removal of disinfection by-product precursors. A requirement in the Disinfectant/Disinfection By-Products Rule states that surface water systems using conventional treatment must enhance their coagulation processes to remove disinfection by-product precursors. Because some waters contain natural organic matter that is less amenable to removal by coagulation than for typical surface waters, a utility can perform a series of jar tests to determine a point of diminishing returns that can be employed in requesting alternative performance criteria. The rule establishes a protocol for determining the point of diminishing returns for a particular source water. *See also* Disinfectant/Disinfection By-Products Rule; disinfection by-product precursor; enhanced coagulation; natural organic matter.

point of disinfectant application The point where disinfectant is applied and where water downstream of that point is not subject to recontamination by surface water runoff.

point of entry (POE) The location at which a product to be delivered, such as potable water, enters the delivery system, such as a distribution system.

point gauge A sharp-pointed rod attached to a graduated staff or vernier scale, used for measuring the elevation of the surface of water. The point is lowered until the tip barely touches the water and forms either a streak (in flowing water) or a meniscus jump (in still water). A point gauge can also be used in a still well, and it can operate on an electric current with a buzzer or light activated by contact with the water.

point mutation A change in the base sequence at a single site on deoxyribonucleic acid. Point mutations usually represent deletions or base-pair substitutions. Deletions are particularly destructive if they cause a number of bases not divisible by three to be lost. In this case, the mutation is referred to as a frame-shift mutation. *See also* frame-shift mutation; mutagenesis.

point rainfall The rainfall rate at a single station, in contrast to the average rate for a region.

point source (PS) A discharge that comes out of the end of a pipe—as opposed to runoff or a discharge from a field or similar source, which is called a nonpoint source. *See also* nonpoint waste source.

point of use (POU) The location at which a product, such as potable water, reaches the end user.

point-of-entry (POE) treatment Water treatment at the inlet to an entire building or facility. (Outside faucets at the location may be excepted from such treatment.)

point-of-entry (POE) treatment device A treatment device applied to the drinking water entering a house or building for the purpose of reducing contaminants in the drinking water distributed throughout that house or building. *See also* Water Quality Association.

point-source emitter A drip-irrigation component that delivers water from a single orifice at a predictable rate, usually measured in gallons or litres per hour.

point-of-use (POU) treatment Water treatment at a single outlet or limited number of water outlets in a building, but for less than the whole building or facility. Point-of-use treatment is often used to treat water for drinking and cooking only.

point-of-use (POU) treatment device A treatment device applied to a small subset of a building's taps (usually one) for the purpose of reducing contaminants in drinking water at those taps. *See also* Water Quality Association.

poise (Pa·s) *See the Units of Measure list.*

poised river A river or stream that, as a whole, maintains its slopes, depths, and channel dimensions without any raising or lowering of its bed. Such a river condition may be only temporary from a geologic point of view, but it may be accepted as having considerable stability for practical engineering purposes.

poison *See* toxicant.

Poisson distribution In statistics, a frequency distribution for a population that is randomly dispersed and for which the mean and the variance are equal. *See also* arithmetic mean; variance.

polar Having distinct positively and negatively charged ends.

polar organic by-product A disinfection by-product that is a polar organic compound. Polar compounds have a positive and negative charge in separate positions on the molecule. An example of a polar organic by-product is a compound with a carboxyl (COOH) group, such as haloacetic acids (CX_3COOH, where X = Cl, Br, or H, or a combination) or carboxylic acids (R-COOH). *See also* disinfection by-product; organic compound.

polar organic compound A compound with a molecular structure composed of one or more functional groups that make the structure unbalanced in terms of electron placement. Polar organic compounds are characterized by high water solubility, and they are generally difficult to extract from water.

polar solvent A solvent for which the molecules are polarized. In other words, the electrostatic charge is relatively positive on one side of the molecule and, consequently, relatively negative on the other because of an unequal sharing of electrons in the covalent bonding between the atoms. Polar solvents separate the structural units of the solute by surrounding each ion with a cluster of solvent molecules pointing opposite charges toward the positive or negative solute ions. Water and methanol (CH_3OH) are examples of highly polar solvents.

polarity reversal In an electrodialysis reversal membrane process, the periodic change in electrode polarity of the electrodes, whereby the demineralizing compartments become concentrate compartments and the concentrate compartments become demineralizing compartments. *See also* electrodialysis reversal.

polarization (1) The concentration or depletion of ions in the thin boundary layer adjacent to an ion-rejecting membrane. (2) In the electrodialysis process, the point at which the current density is great enough to dissociate water molecules, forming hydrogen ions (H^+) and hydroxide ions (OH^-). *See also* concentration polarization.

polarize (1) To make something polar. (2) To disrupt the corrosion process by developing a barrier on an anodic or cathodic surface.

polarography An electrochemistry technique, generally performed in aqueous solution, whereby current is measured as a function of potential applied to an electrolytic cell.

pole One end of a magnet: the north or south pole.

pole shader A copper bar circling the laminated iron core inside the coil of a magnetic starter.

poliovirus A member of the genus *Enterovirus*. It is responsible for an acute infectious disease involving the central nervous system.

polished water *See* intrinsic water.

polisher A treatment stage placed at the end of other treatment to bring the water to a more highly conditioned and more perfect state, e.g., a mixed bed of ion exchange media installed as the final treatment step in the deionization process to remove last traces of undesirable ions.

polishing filter A filter installed for use after the primary water treatment stage to remove any traces of undesirable matter or to "polish" the water. *See also* polisher.

pollutant Dredged spoil, solid waste, incinerator residue, wastewater, garbage, wastewater sludge, munitions, chemical wastes, biological materials, or radioactive materials not covered by the Atomic Energy Act; heat; wrecked or discarded equipment; or rock, sand, cellar dirt, or industrial, municipal, or agricultural waste discharged into water.

pollution (1) A condition characterized by the presence of harmful or objectionable material in water. (2) The presence in a natural body of water of any material that diminishes the optimal economic use of the water body by the population it serves and that has an adverse effect on the surrounding environment.

pollution control The implementation of a program to reduce or eliminate substances that contaminate another substance (e.g., air, water, soil) after the contamination is verified by comparison with a standard.

pollution plume A mass of water that contains undesirable concentrations of some pollutant. Water at the periphery of the mass may also contain the pollutant but at acceptable concentrations and so would not be considered part of the plume. Plumes are often elliptical or teardrop shaped, with the long axis aligned with the direction of flow.

pollutional index A criterion by which the degree of pollution of a stream or other body of water may be measured, such as bacterial density, plankton level, benthos level, biochemical oxygen demand, or dissolved oxygen concentration.

pollutional load The quantity of material that is carried in a body of water and exerts a detrimental effect on some subsequent use of that water.

poly wrap A thin, loose plastic sheet or bag that is placed around ductile iron pipe or iron fittings to separate the material from the surrounding soil, providing corrosion protection for the pipe or fitting.

polyacrylic acid (($H_2C{:}CHCOOH)_x$) A major chemical component of some types of scale inhibitors (antiscalants). Polyacrylic acid-based scale inhibitor is sometimes added during pretreatment for membrane desalting systems.

polyaluminum chloride (PACl) Al $(OH_x(Cl)_y(SO_4)_z$ with *y* typically from 1.5 to 2.5 and *z* from 0 to 1.5 A hydrolyzed form of aluminum chloride ($AlCl_3$) that is used for coagulation, typically in low-turbidity waters. As a result of its polymeric form, lower dosages can be used compared to metal coagulants. *See also* coagulant; coagulation.

polyaluminum sulfate (PAS) ($Al(OH)_x(Cl)_y(SO_4)_z$, with *y* typically from 1 to 2.5 and *z* from 0 to 1.5) A hydrolyzed form of aluminum sulfate ($Al_2(SO_4)_3$) that is used for coagulation, typically in low-turbidity waters. As a result of its polymeric form, lower dosages can be used compared to metal coagulants. *See also* coagulant; coagulation.

polyaluminum-hydroxysulfate (PAHS) A hydrolyzed form of aluminum sulfate (aluminum-to-hydroxide ratio = 1.5) that includes water of hydration. It is used for coagulation, typically in low-turbidity waters. As a result of its polymeric form, lower dosages can be used compared to metal coagulants. *See also* coagulant; coagulation.

polyamide membrane A membrane with polyamide-based barrier-layer composition. Such membranes are commonly used for hollow fiber membrane permeators or for thin film composite flat-sheet membranes formed into spiral wound elements. *See also* flat-sheet membrane; hollow fiber membrane.

polyaromatic hydrocarbon *See* polynuclear aromatic hydrocarbon.

polybutylene ($C_4H_8)_n$ A thermoplastic polymer, such as butyl rubber, of butylene liquefied petroleum gas—isobutene, butene-1, or butene-2. Polybutylene is manufactured in various degrees of elasticity, strength, and stability. It is used for films, coatings, pipes, tubing, fittings, and many other services. *See also* plastic; plastic pipe.

polycarbonate ($(-C_6H_4C(CH_3)_2C_6H_4OCOO-)_n$) A thermoplastic polymer resin that is a linear polyester of carbonic acid (H_2CO_3). Polycarbonate is a transparent, nontoxic, noncorrosive, heat resistant, high-impact-strength plastic. It is generally stable but may be subject to attack by strong alkalies and some organic hydrocarbons. It can be molded, extruded, or thermoformed, and it is commonly used for numerous services, such as nonbreakable windows, household appliances, tubing, piping, and cartridge filter sumps. *See also* plastic; plastic pipe.

polychlorinated biphenyl (PCB) ($C_{12}X_{10}$, where two or more of the substituent atoms are chlorine and the remainder are hydrogen) Any of a class of aromatic compounds containing two benzene nuclei with two or more substituted chlorine atoms. These compounds are highly toxic. Because of their persistence and ecological damage via water pollution, their manufacture was discontinued in the United States in 1976. They are regulated by the US Environmental Protection Agency. *See also* aromatic hydrocarbon.

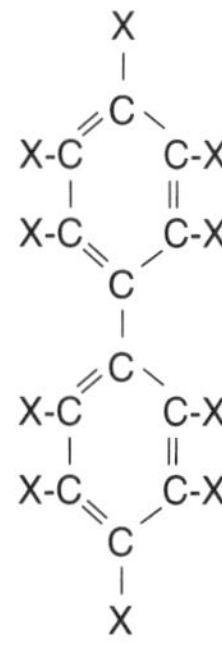

Polychlorinated biphenyl, where two or more substituent atoms (X) are chlorine and the remainder are hydrogen

polydispersity A measure of the range of molecular weights of the components of natural organic matter in a water sample. For example, a sample with low polydispersity contains substances that occupy a narrow size fraction and do not possess molecular weights that vary by orders of magnitude. *See also* molecular weight distribution; natural organic matter.

polyelectrolyte A complex polymeric compound, usually composed of synthetic macromolecules that form charged species (ions) in solution. Water-soluble polyelectrolytes are used as flocculants; insoluble polyelectrolytes are used as ion exchange resins. *See also* coagulant; polymer.

polyelectrolyte flocculant A polymeric organic compound used in water treatment to induce or enhance the flocculation of suspended and colloidal solids and thereby facilitate sedimentation or the dewatering of sludges.

polyethylene (PE) A tough thermoplastic polymer $(-CH_2CH_2-)_x$ of ethylene ($CH_2{:}CH_2$) that resists chemicals and absorbs very little moisture. Polyethylene can vary from soft and flexible to hard and rigid depending on the pressures and catalysts used during manufacturing. Low-density polyethylene has a melting

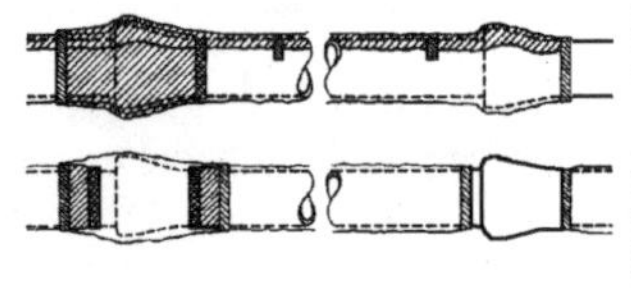
Method A: One length of polyethylene tube fro each length of pipe, overlapped at joint.
Method B: Separate pieces of polyethylene tube for barrel of pipe and for joints. Tube over joints overlaps tube encasing barrel.

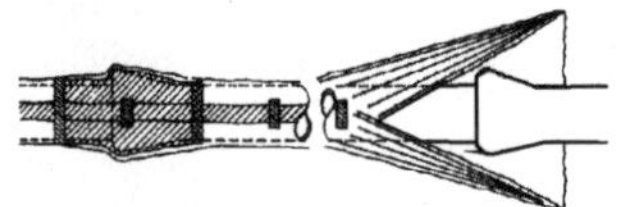
Method A: One length of polyethylene tube fro each length of pipe, overlapped at joint.

Polyethylene encasement of ductile iron pipe

point of about 240° Fahrenheit (115° Celsius) and a tensile strength of 1,500 pounds per square inch (10,345 kilopascals); high-density polyethylene melts at 270° Fahrenheit (135° Celsius) and has a tensile strength of 4,000 pounds per square inch (27,580 kilopascals). Polyethylene is commonly used for tubing and piping, food packaging, garment bags, and molded plastic products. *See also* plastic; plastic pipe.

polyethylene encasement An 8-mil (0.2-millimetre) polyethylene film placed around gray or ductile cast-iron pipe to prevent corrosion.

polygon (1) A closed, two-dimensional figure with three or more sides and intersections. (2) An enclosed geographic area, such as a land parcel or political jurisdiction.

polygon overlay A software function that generates derivative polygons based on the union, intersection, or subtraction of polygons in multiple geographic information system data layers. *See also* polygon.

polyiron chloride (PICl) A prepolymerized, hydrolized form of iron used as a coagulant. *See also* polyaluminum chloride.

polymer A synthetic organic compound with high molecular weight and composed of repeating chemical units (monomers). Polymers may be polyelectrolytes (such as water-soluble flocculants), water-insoluble ion exchange resins, or insoluble uncharged materials (such as those used for plastic or plastic-lined pipe). *See also* coagulant; plastic pipe.

polymer aging A process in which a polymer is stored in a diluted form and slowly mixed to allow the polymer to develop certain characteristics that are required for its intended purpose. Polyelectrolyte polymers typically are aged in day tanks, in which the concentrated polymer is mixed with dilution water and is gradually fed over the course of 1 to several days. *See also* day tank; polymer.

polymer bridging A flocculation mechanism in which previously formed particles are connected by a polymer, thus creating a larger particle that can more readily be removed by physical processes (e.g., settling). *See also* bridging; polymer.

polymer coagulant Any of a group of coagulants that are polymeric in nature and of varying molecular weights and charges. A polymer coagulant is used to destabilize particles in suspension for subsequent removal by physical processes. Most polymer coagulants that are used as the primary, or sole, coagulants are positively charged.

polymerase chain reaction (PCR) An assay or protocol that allows exponential amplification (cloning) of a specific deoxyribonucleic acid sequence from the original sequence (template).

polyneuropathy A disease that involves several nerves (as opposed to a single nerve).

polynomial equation A mathematical equation in which various exponential forms of the independent variable appear simultaneously. For example, $y = ax + bx^2 + cx^3$ is a polynomial equation.

polynuclear aromatic hydrocarbon (PAH) (C_xH_y) A complex aromatic ring system built up by fusing two (or more) rings—usually of the benzenoid type (C_6H_6)—to one or more common sides. For example, naphthalene ($C_{10}H_8$) is a fusion of two rings; anthracene ($C_{14}H_{10}$) and phenanthrene ($C_{14}H_{10}$) are fusions of three rings; pyrene ($C_{16}H_{10}$) and naphthacene ($C_{18}H_{12}$) are fusions of four rings; and benzopyrene ($C_{20}H_{12}$) is a fusion of five rings.

Napthalene, an example of a polynuclear aromatic hydrocarbon (other examples have more rings fused together)

polyoxyethylene (20) sorbitan monooleate *See* polysorbate 80.

polyphosphate Any of a broad family of inorganic phosphorus compounds that are commonly referred to as molecularly dehydrated phosphates or condensed phosphates, including hexametaphosphate, tripolyphosphate ($P_3O_{10}{}^{5-}$), and pyrophosphate ($P_2O_7{}^{5-}$), among others. Unlike orthophosphate ($PO_4{}^{3-}$), for which four oxygen atoms always surround a single central phosphate, polyphosphates are arranged as polymeric chains or occasionally rings that vary in their phosphorus-to-oxygen ratio and chain length with different commercial formulations. Inorganic polyphosphates may be protonated like weak acids when they are not coordinated with metal ions, and they tend to break down (hydrolyze) into orthophosphate groups plus molecules of shortened chain length over time and under extreme temperature and pH conditions.

polypig A flexible polyurethane cleaning swab that is flushed through a distribution line with water to scrape, remove foreign matter, and assist in flushing and cleaning water mains. *See also* scraper.

polyploidy A condition in which a cell contains more than two full sets of homologous chromosomes (e.g., triploid, three sets; tetrapolid, four sets). This condition is not necessarily abnormal in certain tissues (e.g., the liver), but it may result in observable anomalies if such cells do not normally appear in a particular tissue or organism.

polypropylene (C_3H_6)$_n$ A thermoplastic polymer of propylene resembling polyethylene and used for making molded and

Polystyrene resin beads

extruded plastic products, such as water pipe, tubing, and fittings. *See also* plastic; plastic pipe; polyethylene.

polysorbate 80 The generic name for polyoxyethylene(20) sorbitan monooleate. In general, polysorbates are nonionic surfactants obtained by esterification of sorbitol ($C_6H_8(OH)_6$) with one or three molecules of a fatty acid (e.g., oleic acid ($CH_3(CH_2)_7CH{:}CH(CH_2)_7COOH$)) under conditions that cause splitting out of water from the sorbitol, leaving sorbitan. Polysorbates are used as surfactants, emulsifying agents, dispersing agents, foaming and defoaming agents, and so on. Polysorbate 80 is used in bacterial media to disperse microorganisms that would grow in clumps. *See also* esterification; fatty acid; nonionic; surfactant; surface-active agent.

polystyrene ($C_6H_5CH{:}CH_2)_n$ Polymerized styrene. Polystyrene forms the skeletal structure of many ion exchange resin beads.

polystyrene resin The most common type of resin used in ion exchange processes.

polysulfone ($(–C_6H_4SO_2C_6H_4OC_6H_4C(CH_3)_2C_6H_4O–)_n$) A synthetic thermoplastic polymer. Derivation is from the condensation of bis-phenol A ($(CH_3)_2C(C_6H_4OH)_2$) and dichlorophenyl ($Cl_2C_6H_3$) sulfone. (A sulfone is an organic compound containing an $–SO_2$ group.) *See also* polysulfone membrane.

polysulfone membrane A composite membrane with a polysulfone-based barrier layer composition.

polytetrafluoroethylene (PTFE) ($(C_2F_4)_n$) A polymer of tetrafluoroethylene. It is used as a nonstick coating. *See also* polymer.

polyunsaturated Pertaining to a fat or oil that is based at least partly on fatty acids having two or more double bonds per molecule. *See also* fatty acid.

polyurea Urea ($CO(NH_2)_2$) that is used in the preparation of plastics, resins, and so on. *See also* polyurea membrane.

polyurea membrane A composite membrane with a polyurea-based barrier layer composition.

polyvinyl alcohol (PVA) ($(–CH_2CHOH–)_x$) A water-soluble synthetic polymer made by alcoholysis (a chemical reaction between an alcohol and another organic compound, analogous to hydrolysis) of polyvinyl acetate ($(–CH_2CH(OOCCH_3)–)_x$). *See also* polymer.

polyvinyl chloride (PVC) ($(–H_2CCHCl–)_x$) An artificial polymer made from vinyl chloride monomer ($CH_2{:}CHCl$) and frequently used in pipes, sheets, and vessels for transport, containment, and treatment in water facilities. *See also* polymer.

polyvinyl chloride (PVC) pipe A type of pressure pipe for water distribution manufactured from polyvinyl chloride compound. This type of pipe is found in plumbing and irrigation systems.

pond A body of water of limited size, either naturally or artificially confined and usually smaller than a lake.

ponded stream A section or stretch of stream flowing through an area that has been warped upward or uplifted at a geological period subsequent to the time when the stream channel was originally established.

pool (1) A small and rather deep body of relatively quiet water, as a pool in a stream. (2) A small body of standing or stagnant water; a puddle.

pool cushion A pool of water maintained at the toe of an overflow dam or other structure to receive and spread the impact of water that falls with high velocity over the structure.

pool spring A spring that occurs as a pool of water in the bottom of a surface depression.

pooling A process whereby water allocations from several different irrigation accounts are reassigned to better fit actual watering needs.

poorly buffered Having minimal ability to resist changes in pH.

poorly drained Pertaining to a condition in which water is removed from the soil so slowly that the soil is saturated periodically during the growing season or remains wet for long periods (greater than 7 days).

poppet valve *See* mushroom valve.

population In ecology, a group of individuals of the same type, particularly of the same species, within a community. Populations have certain characteristics not shown by either individual organisms or the community as a whole. Characteristics of a population include age distribution, birth and death rates (natality and mortality, respectively), density (size in relation to unit space), dispersion (distribution of the individuals within the area), growth rate, and sex ratio.

population attributable risk (PAR) The fraction of the disease burden (e.g., cancer) that could be eliminated if an exposure were absent.

population at risk A population subgroup that is more likely to be exposed to a chemical, or is more sensitive to a chemical, than is the general population.

population-based fee An annual operating fee determined by the size of the population being served by a community water system.

pore A very small open space in a rock or granular material. *See also* interstice.

pore diffusion The mechanism by which molecules pass through a pore in a medium for subsequent removal by adsorption. This term is typically associated with diffusion through granular activated carbon pores, in which the limiting step in adsorption may be associated with the ability of a molecule of a given size to pass through (or be excluded by) a pore of similar, or smaller, size. *See also* adsorption; granular activated carbon.

pore space An open space in rock or granular material. *See also* interstice.

pore velocity The fluid velocity in each pore of a porous medium. *See also* actual groundwater velocity.

pore volume The volume of the internal void spaces in a granule that are smaller than 0.1 millimetre across and large enough to allow access to helium. It is measured as the difference in the volumetric displacement by a measured quantity of granules in mercury as compared to the same measurement in helium at standard conditions.

pore water Water in interstices of granular permeable rocks, as in sedimentary and alluvial deposits.

porosity The ratio of void volume to bulk volume in a sample of rock, sediment, or filter pack material.

porous Full of pores through which water, light, and so on may pass.

portable exchange (PE) tank A tank containing up to 2 cubic feet (0.06 cubic metres) of ion exchange products or filter media, rented to homeowners or business clients with the bed fully regenerated and ready for use. Portable exchange tanks do not have the valving controls required for regeneration. Upon exhaustion (as indicated by predetermined calendar days, a meter, or a monitoring device), the tanks are returned to a central regeneration plant, where the resin or other media in each tank are reprocessed and restored for reuse. Portable exchange tanks may be available with water-softening or deionization resins, mixed ion exchange media, manganese zeolite, activated alumina, or activated carbon. Portable exchange tanks are used for both household and commercial applications.

portal lesion Necrotic damage to the liver in the area surrounding the portal vein.

positive Having an electrical charge polarity associated with an excess of protons (or, equivalently, a deficiency of electrons).

positive bacteriological sample A water sample for which a positive total coliform result is obtained by the use of any of the approved total coliform analytic techniques specified under the US Environmental Protection Agency's Coliform Rule, promulgated June 29, 1989, or any such techniques approved subsequent to promulgation of the rule. *See also* Total Coliform Rule; total coliform test.

positive charge The electrical potential associated with a proton. A cation is an atom that has lost one or more electrons, leaving it with more protons than electrons; hence, cations carry positive charges.

positive (charge) metal oxide semiconductor (PMOS) A metal oxide semiconductor incorporating a silicone gate structure that uses only positive charges.

positive confining bed A confining bed that prevents or retards upward movement of groundwater so that the underlying water has a higher static level than the overlying water, creating an upward pressure. Artesian head is maintained by confining beds of this class. *See also* permeable confining bed.

positive head The energy possessed per unit mass of a fluid due to the fluid's elevation above some datum. Positive head is also called elevation head.

positive rotary pump A type of displacement pump consisting essentially of elements rotating in a pump case in which they closely fit. *See also* rotary pump.

positive sample In the context of a multiple-tube fermentation or membrane filter test, any sample that contains coliform bacteria.

positive-displacement meter A quantity meter that captures definite volumes of water in a chamber with a piston or disk and discharges the water, counting operations as the water passes through the meter.

positive-displacement pump A pump in which a measured quantity of water is collected or entrapped and forced to the discharge side by a reciprocating piston, gear, or rotary vane.

positively charged filter A filter that has a net positive or only slight negative charge at pH values near neutrality. Such filters may be composed in part of asbestos or aluminum hydroxide gels. They adsorb viruses in water at pH values near neutrality without the need for special water treatment.

positron A particle having a mass equal to that of an electron but with an opposite charge of equal magnitude.

possible precipitation For a given area, the depth and duration of precipitation that can be reached but not exceeded under known meteorological circumstances.

post hydrant A fire hydrant that rises above the sidewalk surface.

postammoniation The addition of ammonia (NH_3) to a treated water containing free chlorine (HOCl), following the final treatment process in a plant, to produce a chloramine (NH_xCl_y, where $x = 0–2$, $y = 1–3$) residual. *See also* chloramines.

postchloramination The application of chloramines (NH_xCl_y, where $x = 0–2$, $y = 1–3$) to water after filtration. Postchloramination can be used for secondary disinfection after primary disinfection has been accomplished with a stronger disinfectant (e.g., chlorine (HOCl), ozone (O_3), or chlorine dioxide (ClO_2)). *See also* chloramination.

postchlorination (1) The application of chlorine (HOCl) to water after filtration. (2) The application of chlorine subsequent to any treatment. The treatment should be specified, e.g., postsedimentation chlorination.

postdisinfection The addition of a disinfectant following the final treatment process in a plant, often following filtration.

postozonation The addition of ozone (O_3) following the final treatment process in a plant, often following filtration, to provide disinfection. Recently, however, a more accepted treatment practice is to follow ozone with a granular medium, such as a filter, that is biologically active in order to remove biodegradable compounds produced during ozonation. *See also* biologically active filter.

posttreatment An additional treatment step following a primary treatment process, such as permeate posttreatment for corrosion control or disinfection following membrane desalting.

potable Safely drinkable.

potable reuse The reuse of reclaimed water for potable purposes. *See also* direct reuse; indirect reuse.

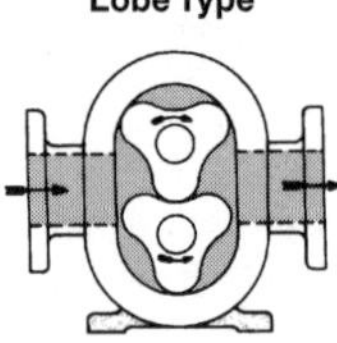

One type of positive-displacement pump

potable water Water that is safe and satisfactory for drinking and cooking.

potable water approval An approval status conferred on a water supply by a regulatory agency, signifying that the treated drinking water meets applicable standards for potable water.

potamology The branch of hydrology that pertains to streams; the science of rivers.

potassium (K) An alkali metal. It is an essential element in plant growth and in animal and human nutrition. Potassium occurs in all soils. *See also* alkali metals.

potassium acid phthalate *See* potassium hydrogen phthalate.

potassium chloride (KCl) A colorless potassium salt that can be used as a regenerant in cation exchange water softeners.

potassium cycle The use of potassium chloride (KCl) salt instead of sodium chloride (NaCl) salt in the regeneration of cation ion exchange water softeners. The potassium ion (K^+) becomes the exchangeable ion rather than the sodium ion (Na^+).

potassium hydrogen phthalate (KHP) ($HOOCC_6H_4COOK$) An analytical standard used for total organic carbon analysis (2.1254 grams of anhydrous potassium hydrogen phthalate per litre of solution equals 1.00 milligram of carbon per millilitre). Potassium hydrogen phthalate is also used as a standard for chemical oxygen demand analyses. (It has a theoretical chemical oxygen demand of 1.176 milligrams of oxygen per milligram.) *See also* chemical oxygen demand; total organic carbon.

potassium permanganate ($KMnO_4$) A substance in the form of dark purple crystals used as an oxidant in drinking water treatment. Potassium permanganate is used for taste-and-odor control and for iron and manganese removal. Unlike other oxidants, potassium permanganate has not been associated with the production of disinfection by-products. *See also* disinfection by-product; oxidant.

potential (1) Electrostatic potential, the gradient (derivative) of which gives the electric field strength. (2) Velocity potential, the gradient of which gives the velocity in a fluid flow. (3) Gravitational potential, the derivative of which yields the gravitational force. (4) The work required to move a unit mass against a unit force from one specified point to another, yielding force in a mechanical field. (5) In soil water, a quantity synonymous with *head.* It is what makes water move through soil, as a voltage difference causes an electric current to flow through a resistance. Head equals the elevation of any point in question with respect to an arbitrary datum plus the pressure of water in the soil pores per unit weight of water (i.e., assuming that the velocity of flow is small and that the water is not changing temperature). Below the water table (piezometric surface), the pressure head (the second term in the preceding sentence) is positive. Going down X feet or metres into the soil below the water table, the elevation head decreases by X feet or metres, but the pressure head increases by X feet or metres (pressure per unit weight $= X$); thus, no change in head occurs in a geostatic sense, and the groundwater does not flow. Above the water table (capillary zone), the pressure head is negative and is equal (in absolute value) to the distance from the piezometric surface to the point in the capillary zone in question times the unit weight of water. Above the capillary zone, in the vadose zone, where the soil becomes unsaturated, the pressure begins to be controlled by suction, or a negative pressure in the pore water caused by the formation of air bubbles with high surface tension. (6) The possibility of the creation of disinfection by-products when a disinfectant reacts with precursor organic matter. *See also* formation potential; precursor.

potential energy Energy possessed by a body of matter as a result of its position or condition.

potential evaporation The evaporation that would occur under given climatic conditions if unlimited moisture were available to evaporate.

potential evapotranspiration The evapotranspiration that would occur if unlimited soil moisture were present.

potential head The height of any particle of water above a specified datum. Potential head is also called elevation head.

potential-determining ion A charged species that establishes the electrostatic attraction or repulsion among suspended particles. *See also* coagulant.

potentiation The effect of one chemical to increase the effect of another chemical.

potentiometer An instrument in which electrical potential (voltage) is measured. It is a variable resistor of electricity that measures voltage.

potentiometric surface A surface that represents the level to which water will rise in a tightly cased well. If the head varies greatly with depth, then multiple potentiometric surfaces may exist. The water table is a particular potentiometric surface for an unconfined aquifer.

potentiometrical procedure Any laboratory procedure that measures a difference in electric potential (voltage) to indicate the concentration of a constituent in water.

pothole A depressional wetland commonly found in the upper Midwest (North and South Dakota and western Minnesota), as well as a similar wetland found elsewhere.

POTW *See* publicly owned treatment works.

POU *See* point of use.

pound (lb) *See the Units of Measure list.*

pounds per day (lb/d) *See the Units of Measure list.*

pounds per square foot (lb/ft^2) *See the Units of Measure list.*

pounds per square inch (psi) *See the Units of Measure list.*

pounds per square inch absolute (psia) *See the Units of Measure list.*

pounds per square inch gauge (psig) *See the Units of Measure list.*

Pourbaix diagram A diagram showing thermodynamically stable species of a metal as a function of potential (relative to the standard hydrogen electrode and pH).

powdered activated carbon (PAC) Activated carbon composed of fine particles and providing a large surface area for adsorption. Powdered activated carbon is typically added as a slurry on an intermittent or continuous basis to remove taste-and-odor-causing compounds or trace organic contaminants and is not reused. *See also* granular activated carbon.

power *See* active power; apparent power; complex power; exponent; instantaneous power; reactive power.

power capacity of a stream The total hydroelectric power that, according to estimates, can be developed on a stream during a stipulated portion of the time, such as 50 percent of the time.

Powdered activated carbon (right) versus granular activated carbon

The term is generally used in studies of prospective hydroelectric power, and it is flexible in the application. In estimates of power capacity, the feasibility of storage, proportion of total runoff and of total fall that can be used, and other factors must be taken into consideration.

power factor (PF) For sinusoidal signals, the cosine of the voltage phase angle minus the current phase angle.

$$PF = \cos(\theta_v - \theta_i)$$

Where:

PF = the power factor

θ_v = the voltage phase angle, in radians

θ_i = the current phase angle, in radians

power function A mathematical equation in which several independent variables appear as a multiplicative term and the independent variables contain different exponents. For example, $y = ab^m c^n d^o$ is a power function.

power head An actuating mechanism at the power end of a deep-well pump that transmits the power for lifting the water.

power plant A facility designed to produce electricity from an energy source, such as fossil fuel (e.g., burning coal), wind, solar energy, or hydroelectric energy.

power pump A pump that requires energy other than manual energy for its operation.

power requirement The rate of energy input needed to operate a piece of equipment, a treatment plant, or other facility or system. The form of energy may be electricity, fossil fuel, hydroelectric power, or some other form or combination of forms.

power site Any site where a power plant has been built, is being built, or is contemplated.

power units In the US customary system, horsepower; British thermal units per hour; foot-pounds force per minute. In Système International, watts; volt-amperes; metres squared kilograms per second cubed.

POX *See* purgeable organic halogen.

POXFP *See* purgeable organic halogen formation potential.

PPAR *See* peroxisome proliferator activated receptor.

ppb *See* parts per billion *in the Units of Measure list.*

PPE *See* personal protection equipment.

ppm *See* parts per million *in the Units of Measure list.*

ppt *See* precipitate. *See also* parts per thousand; parts per trillion *in the Units of Measure list.*

PQL *See* practical quantitation level.

practical quantitation level (PQL) The lowest concentration that can be quantified within specified limits of precision and accuracy during routine laboratory operations. The practical quantitation level is intended to represent the lowest concentration achievable by good laboratories under practical and routine laboratory conditions. Practical quantitation levels are set by the US Environmental Protection Agency based on interlaboratory studies and applications of statistics. They are usually set at a level between 5 and 10 times the minimum detection limit. *See also* limit of detection; limit of quantitation; method detection limit; reliable detection level; reliable quantitation level.

practical quantitation limit *See* practical quantitation level.

practical turf area A landscape design and management concept promoting turf only in those areas of the landscape that are functional, as well as the efficient management of supplemental irrigation required for those areas.

PRD-1 phage A bacterial virus that infects *Salmonella typhirmurium*, also known as a bacteriophage. *See also* bacteriophage.

preaeration The use of an aeration device, such as a packed tower aerator or tray aerator, prior to a sequence of treatment processes to (1) increase the dissolved oxygen concentration in the water, (2) reduce the concentration of a contaminant by transferring it from the liquid to the vapor phase, or (3) oxidize a dissolved contaminant so that it can be removed by subsequent physical removal process. Preaeration has historically been used for taste-and-odor removal, as well as oxidation of iron and manganese prior to filtration.

preammoniation Chloramination in which ammonia (NH_3) is added before the chlorine (Cl_2). This order of chemical addition minimizes free chlorine contact time in order to control the formation of disinfection by-products. *See also* chloramination; disinfection by-product.

prechloramination Chloramination of source water prior to other unit processes (e.g., before coagulation). The location where the chloramine (NH_xCl_y, where $x = 0–2$, $y = 1–3$) is added should be specified to avoid confusion, e.g., source water chloramination, prefiltration chloramination. *See also* chloramination.

prechlorination Chlorination of source water prior to other unit processes (e.g., before coagulation). The location where the chlorine (Cl_2) is added should be specified to avoid confusion, e.g., source water chlorination, prefiltration chlorination. *See also* chlorination.

precipitable water The total water vapor contained in an atmospheric column of unit cross-sectional area, expressed in terms of depth of water of the same cross-sectional area.

precipitate (ppt) (1) A substance separated from a solution or suspension by a chemical reaction. (2) To form such a substance.

precipitated iron *See* ferric iron.

precipitation (1) The total measurable supply of water received directly from clouds as rain, hail, or sleet, usually expressed as depth in a day, month, or year (i.e., daily, monthly, or annual precipitation, respectively). (2) The process by which atmospheric moisture is discharged onto a land or water surface.

(3) The process by which small particles settle out of a liquid or gaseous suspension by gravity. (4) The process of particle formation during a chemical reaction. (5) In the context of corrosion, the shifting of chemical equilibrium to cause the formation of a solid protective coating, usually calcium carbonate ($CaCO_3$) on interior pipe surfaces.

precipitation gauge A device for catching and measuring the depth of precipitation. *See also* rain gauge.

precipitation mass curve A graph showing accumulated precipitation as a function of time. *See also* mass diagram.

precipitation oscillation The long-term changes in average precipitation (such averages being taken as the mean of a period of 50 to 100 years) that occur over periods ranging from less than 1 year to more than several thousand years. Although no actual measurements exist of the precipitation over such periods, evidence based on studies of tree rings, geological deposits, and changes in vegetation indicates that such changes do occur.

precipitation rate The amount of precipitation occurring in a unit of time, generally expressed in inches per hour or centimetres per hour.

precipitation/evaporation (P/E) ratio The ratio of monthly precipitation to monthly evaporation, used in evapotranspiration studies.

precipitative softening A unit process by which the dissolved minerals in water, particularly calcium and magnesium, are removed during lime (CaOH) softening (which causes a pH increase) through the formation of a precipitate. For the Disinfectant/Disinfection By-Products Rule, precipitative softening can be used for the removal of disinfection by-product precursors (a process referred to as enhanced softening). *See also* Disinfectant/Disinfection By-Products Rule; disinfection by-product precursor; enhanced softening; ion exchange; lime softening; precipitate.

precision A measure of the agreement among replicate measurements. *See also* accuracy; bias.

precoat A very fine granular filter medium, such as diatomaceous earth, applied (usually by slurry) to a retaining membrane or fabric surface prior to a service run. At the end of each service run, the precoat medium is rinsed off and disposed of prior to application of a new precoat to the filter septum.

precoat filtration A process designed to remove particulates by applying the water to be treated to a membrane or fabric coated with a very fine granular medium, such as diatomaceous earth. *See also* diatomaceous earth; precoat.

precoating The process of applying a precoat to a support surface called a septum. *See also* precoat.

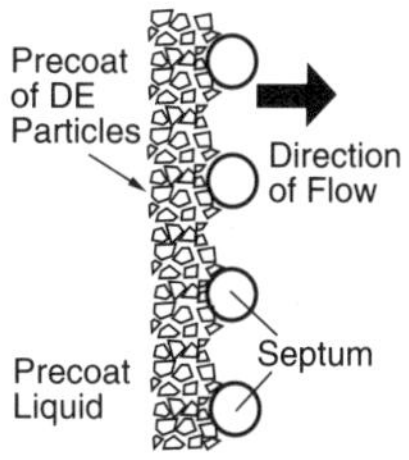

Diatomaceous earth precoat

precursor A compound or mixture that can be converted to a specific substance. For example, upon disinfection, disinfection by-product precursors are converted to disinfection by-products. *See also* disinfection by-product precursor.

precursor compound *See* precursor.

precursor limited Pertaining to a situation in which the concentration of precursor material in a solution (e.g., water) is limiting. For example, during chlorination of effluent from granular activated carbon treatment, the concentration of disinfection by-product precursors is the limiting factor to disinfection by-product formation. *See also* disinfection by-product precursor.

predator–prey interaction A natural process controlling living populations. In every society or community of organisms in its natural state, predators survive by eating prey.

prediction An act of attempting to foretell the value of a dependent variable from known values of independent (or predictor) variables.

predictive equation A mathematical equation used for developing predictions of dependent variables for known values of independent variables.

predischarge notification (PDN) Notification to the US Army Corps of Engineers prior to discharge of fill material, as required under a Clean Water Act Section 404 Nationwide Permit.

predisinfection Application of a disinfectant prior to a treatment step. This term typically refers to source water disinfection. Better usage is to specify where the disinfectant is added, e.g., source water disinfection, prefiltration disinfection.

preformed chloramine A concentrated solution of chloramine (NH_xCl_y, where $x = 0–2$, $y = 1–3$) formed outside of the treatment train. Preformed chloramines are added to the water directly, rather than separately adding the chlorine (Cl_2) and ammonia (NH_3) to the water. *See also* chloramines.

preliminary filter A filter used in a water treatment plant for the partial removal of turbidity prior to final filtration. Such filters are usually of the rapid type, and their use allows final filtration at a more rapid rate or reduces or eliminates the necessity of other preliminary treatment of the water. A preliminary filter is also called a contact filter, contact roughing filter, or roughing filter.

preliminary treatment Any physical, chemical, or mechanical process used before the main water treatment processes. Preliminary treatment can include screening, presedimentation, and chemical addition. It is also called pretreatment, although this term is often confusing.

preloading A method of applying a compound to a unit process prior to putting the process into service. For example, granular activated carbon can be preloaded with natural organic matter to determine the latter's impact on the removal of specific synthetic organic compounds. Preloading can be deliberate, as in the preceding example, or accidental, as in an attempt to adsorb a specific organic compound on granular activated carbon in the presence of naturally occurring background organic matter.

premarking facilities The process of painting personnel access openings, valves, and so on with distinct symbols or patterns so that the facilities can easily be identified in aerial photography and captured during photogrammetric compilation.

premise piping Piping that is located on the property of the water customer and is designed to serve the needs of that particular customer or group of customers. Premise piping includes the service line and home or building plumbing. It does not include a water transmission or distribution line that crosses a customer's property on an easement.

premises isolation The prevention of backflow into a public water system from a user's premises by installation of a suitable backflow preventer at the user's connection.

preneoplastic lesion A group of cells, usually clonal in origin, that are recognized as being precursors to tumors. However, such groups of cells do not inevitably lead to tumors.

preoxidation Application of an oxidant prior to a water treatment step. Better usage is to specify where the oxidant is added, e.g., source water oxidation or prefiltration oxidation.

preozonation Application of ozone (O_3) prior to a water treatment step. This term is typically used to refer to source water ozonation (whereas *intermediate ozonation* and *postozonation* are used to refer to ozonation of settled and filtered water, respectively). *Source water ozonation* is a better term. *See also* ozonation.

preplumbed installation An installation that allows domestic water treatment equipment to be installed easily because the necessary bypass and valves are already in place. An example would be a new home that already has all of the plumbing needed for installing a water treatment device.

prescriptive Pertaining to water rights that are acquired by diverting water and putting it to use in accordance with specified procedures. These procedures include filing a request (with a state agency) to use unused water from a stream, river, or lake.

presedimentation A preliminary treatment process used to remove gravel, sand, and other gritty material from the source water before the water enters the main treatment plant. This treatment is usually performed without the use of coagulating chemicals.

presedimentation impoundment A large earthen or concrete basin used for sedimentation of source water prior to further treatment. Such a basin is also useful for storage and for reducing the impact of source water quality changes on water treatment processes.

presence–absence method *See* presence–absence test.

presence–absence reporting For drinking water compliance monitoring, the reporting of coliform bacteria test results as "coliforms present" or "coliforms absent," regardless of whether the results were obtained from a quantitative procedure (such as the membrane filter test or the most probable number test) or the original presence–absence test. *See also* presence–absence test.

presence–absence (P–A) test A qualitative microbiological test indicating the presence or absence of coliform bacteria in water. A standard 100-millilitre sample volume is inoculated into a bottle of presence–absence medium, the inoculated medium is incubated for 24 to 48 hours, and the results are recorded. Any amount of gas, acid, or both constitutes a positive presumptive test and requires confirmation. Acid production from lactose fermentation is indicated by a distinct yellow color in the medium; if gas is also being produced, gentle shaking of the bottle results in foaming of the medium. Bottles of inoculated medium that do not change color, do not produce gas, or both are considered as indicating an absence of coliform bacteria. *See also* presence–absence reporting.

preservation *See* preserved wetland.

preservation, sample *See* sample preservation.

preservative In water treatment, a substance added to a water sample to maintain certain constituents in the form in which they are found in the water body, e.g., dissolved metals in solution.

preserved wetland A specified wetland area for which prohibition in perpetuity of specific uses and activities is established in a binding legal agreement.

presettling A process of sedimentation applied to a liquid before some other treatment.

press filter *See* filter press.

pressure The force pushing on a unit area. Water pressure is normally measured in pounds per square inch, kilopascals, or feet or metres of head.

pressure aquifer *See* confined aquifer.

pressure control A switch that operates based on changes in pressure. Such a switch is usually a diaphragm pressing against a spring. When the force on the diaphragm overcomes the spring pressure, the switch is actuated (activated).

pressure correction factor For some pressure-driven membrane processes, a dimensionless coefficient that is dependent on net driving pressures across the membrane. This factor can be multiplied by permeate flow at a standard condition to calculate the permeate flow rate at a given net driving pressure condition.

pressure differential The difference in pressure between two points in a water system. The difference may be caused by a difference in elevation, a pressure drop resulting from water flow, or both.

pressure differential meter Any flow-measuring device that creates and measures a difference in pressure proportional to the rate of flow. Examples include the Venturi meter, orifice meter, and flow nozzle.

pressure drop (1) A decrease in water pressure that occurs as the water flows. Pressure drop may occur for several reasons: internal friction between the molecules of water, external friction between the water and the walls of the piping system, or rough areas in the channel through which the water flows. (2) The difference between the inlet and outlet water pressure during water flow through a water treatment device, such as water conditioner. This type of pressure drop is abbreviated ΔP and is measured in pounds per square inch or pascals gauge pressure. *See also* delta *P*; head loss.

pressure energy Energy associated with a body because of the body's location in a pressure system, usually with respect to atmospheric pressure.

pressure filter (1) An enclosed vessel having a vertical or horizontal cylinder of iron, steel, wood, or other material containing granular media through which liquid is forced under pressure. (2) A mechanical filter for partially dewatering sludge. *See also* filter press; plate-and-frame filter press.

pressure filtration A filtration process in which the media are completely enclosed and the operation takes place under pressure via pumping of the water to be treated. Pressure filtration is often used in applications for which insufficient elevation is

Pressure relief valve

available to support gravity filtration, such as in industrial settings or swimming pool filters.

pressure gauge A device having a metallic sensing element for registering the pressure of solids, liquids, or gases. The device may be graduated to register pressure in any units desired. *See also* piezometer.

pressure granular filter A granular filter placed in a cylindrical steel pressure vessel. The water is forced through the medium by pumping.

pressure grouting The placing of grout (a watery mixture of either portland cement and water, or of various organic chemicals, that hardens in place) under pressure in void spaces in a structure, in soil or rock adjacent to a structure, or in and around underground pipes to strengthen the structure or make it more watertight.

pressure head A measurement of the amount of energy in water due to water pressure.

pressure intensity *See* pressure.

pressure loss *See* head loss; pressure drop.

pressure potential The amount of work a unit mass of fluid could perform against pressure forces to change from its current position to an arbitrary reference position.

pressure reducer A control valve that opens to allow flow if the downstream pressure is less than a certain value and that closes when the set pressure is reached. A pressure reducer ensures that the downstream pressure does not become too high. It is used on house services where the distribution pressure is high and in other situations that require reductions from higher-pressure planes to lower-pressure planes.

pressure regulator A device for controlling pressure in a pipeline or pressurized tank, such as a pressure-regulating valve or a pump drive-speed controller. *See also* regulator.

pressure relief cone The depression in the piezometric (potentiometric) surface of a confined aquifer that forms around a well through which water is being extracted at a given rate. This cone is analogous to the pumping depression cone in the water table of an unconfined aquifer that forms around a well through which water is being extracted. *See also* cone of depression.

pressure relief valve A valve that opens automatically if the water pressure exceeds a certain level.

pressure strainer A device that is used in a pressure pipe and has two or more isolated compartments holding removable screens or strainers. Such a device is removable so that screens can be made accessible for cleaning without interfering with the flow.

pressure surface An imaginary surface that everywhere coincides with the static level of the water in an aquifer. *See also* isopiestic line; piezometric surface.

pressure tank A tank that is used in connection with a water distribution system for a single household, for several houses, or for a portion of a larger water system; is airtight and holds both air and water; and in which the air is compressed and the pressure is transmitted to the water.

pressure test A measurement of pressure with a pressure gauge at a point in the distribution system. When no flow is occurring in the system, the pressure is the static pressure; pressure during flow is called residual pressure. Measuring the static pressure and residual pressure at a fire hydrant while measuring the flow of another fire hydrant nearby allows one to determine the amount of fire flow available at the flowing hydrant at various residual pressures. Usually the flow is determined for a residual pressure of 20 psi (137.8 kilopascals), which is the minimum desired pressure during fire flow.

pressure units In the US customary system, pounds per square inch. In Système International, pascals; kilograms per metre per second squared.

pressure vessel A device designed to contain pressure, such as a housing used for pressure filters or membranes.

pressure zone An area within a distribution system of a domestic or municipal water supply in which the pressure in the mains is maintained within certain specified limits.

pressure surface map A contour map of the imaginary surface to which the water in an artesian aquifer will rise. Such a map is also called a piezometric surface map.

pressure vacuum breaker (PVB) assembly A backflow protection device to prevent water from being drawn back into a water supply when the line is closed. The assembly opens to the atmosphere, thus preventing a vacuum in the line, such as an irrigation line.

pressure-compensating emitter A drip-irrigation emitter designed to deliver water at a consistent flow rate under a range of operating pressures.

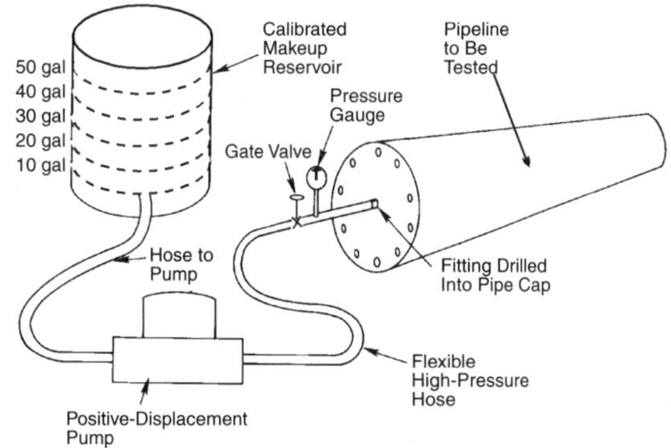

Pressure test of pipeline

pressure-driven membrane A membrane that uses a pressure differential across the membrane, from the feed side to the permeate or filtrate side, as the driving force.

pressure–momentum curve A curve derived from calculations of flow in an open channel and showing the variation of the sum of the forces due to pressure and momentum flux with depth of flow.

pressure-reducing valve A flow valve used to allow water to flow from a higher-pressure plane to a lower-pressure plane so as not to exceed a set maximum pressure in the lower-pressure plane.

pressure-regulating valve (PRV) A valve designed to maintain a set pressure on the downstream side of the valve regardless of pressure changes on the source side.

pressure-swing adsorption A process that uses an adsorbent (e.g., zeolite) to separate individual components of a gas mixture. The process employs a sequencing batch control to alternately pressurize and depressurize the vessels that contain the adsorbent. Pressure-swing adsorption can be used to separate oxygen from air to form pure oxygen. *See also* pure oxygen; zeolite.

pressure-type vacuum breaker *See* pressure vacuum breaker assembly.

prestressed Pertaining to pipe that has been reinforced with wire strands (which are under tension) to give the pipe an active resistance to loads or pressures on it.

prestressed concrete pipe A type of reinforced concrete pipe placed in compression by a highly stressed, closely spaced helical wire winding. The reinforcement increases a concrete pipe's ability to withstand tension forces.

presumptive test The first major step in the multiple-tube fermentation test. This step presumes (indicates) the presence of coliform bacteria based on gas production in a nutrient broth after incubation. *See also* completed test; confirmed test.

presystemic elimination The process by which a substance is metabolized or immediately secreted back into the bile by the liver before it reaches the systemic circulation. The extent to which this occurs is frequently referred to as the first-pass effect. Presystemic elimination is particularly marked for certain chemicals that are introduced to the body via the gastrointestinal tract.

pretreated Pertaining to water that has received pretreatment.

pretreatment One or more actions or processes occurring immediately before water enters a downstream treatment unit process. Better usage is to specify the type of subsequent treatment, e.g., presedimentation or prefiltration.

prevailing wind Wind that regularly comes from the same compass point or direction, most frequently observed during a given period (e.g., day, month, season, or year).

prevalence The number of people in a population who have a disease, illness, or infection at a given point or period in time. Prevalence depends on how many people have become ill in the past and the duration of their illness. Whereas incidence measures the rate at which new illness occurs, prevalence measures the residual of such illness at a given time. *See also* incidence.

prevalence index (1) A weighted average measure of the sum of the frequency of occurrences of all species along a single transect. (2) The average of the prevalence index of all sample transects through the study area.

prevalence rate The number of people with a disease, illness, or infection (i.e., the existing number of cases) at a particular time divided by the total population at risk at that time.

prevalence study An epidemiological study that examines the relationships between diseases and exposures as they exist in a defined population at a particular point in time.

preventive maintenance program A specific, ongoing plan to inspect, monitor, and service—at scheduled and specified intervals—equipment, motors, pumps, valves, vehicles, and so on to ensure efficient operation of such equipment and prolong its productive life.

previously disturbed soil Soil that has been excavated, exhumed, or worked on prior to the present excavation.

Price current meter A wheel-like current meter with a series of conical cups fastened to a flat framework through which a pin extends. The pin sits in the framework of the meter, and the cups are rotated around it in a horizontal plane by the flowing water. The number of revolutions is registered by an acoustical or electrical device. The velocity of the water may be computed based on the number of revolutions.

price elasticity (of demand) The degree of responsiveness of the demand for a good (such as water or oil) relative to a change in its price.

primacy The responsibility for ensuring that a law is implemented, as well as the authority to enforce that law and related regulations.

primacy agency The agency that has the primary responsibility for administering and enforcing regulations.

Primary Drinking Water Regulation (PDWR) A regulation defined in Section 1401 of the Safe Drinking Water Act that (1) applies to public water systems, (2) specifies contaminants that may have any adverse effect on the health of persons, (3) specifies for each contaminant a maximum contaminant level or treatment technique, and (4) contains criteria and procedures to ensure compliance.

Primary Drinking Water Standard *See* Primary Drinking Water Regulation.

primary maximum contaminant level Maximum contaminant levels for various water quality indicators that have been established to protect public health.

primary In electrical systems, the high-voltage side of a step-down transformer. *See also* secondary.

primary cell A cell taken directly from living tissue and used to start a cell culture. *See also* cell culture.

primary element (1) The part of a pressure differential meter that creates a signal proportional to the water velocity through the meter. (2) A hydraulic structure used to measure flows. In open channels, weirs and flumes are primary elements or devices. In pipes or pressure conduits, Venturi meters and orifice plates are the primary elements.

primary interstice *See* original interstice.

primary opening *See* original interstice.

primary power Power that is developed at a hydroelectric power plant and is always available and dependable for carrying a given load. Its amount is governed by the minimum amount of water available at any time, including that derived from storage. Primary power is also called firm power.

primary production Production of biomass by green plants, whether on land or in water.

primary sedimentation The first of a sequence of settling steps (typically two steps) designed to remove large particles. Primary sedimentation can be conducted with or without the use of coagulants.

primary treatment A level of treatment associated with the removal of particles, typically by settling, with or without the use of coagulants. Dissolved contaminants are not significantly removed in primary treatment.

prime To fill a pump casing or siphon with water to remove the air. Most pumps must be primed before start-up or they will not pump any water.

primer A short synthetic strand of deoxyribonucleic acid (DNA) used in polymerase chain reaction technology. The synthetic stands of DNA bind to their complementary sequences on the original double-stranded DNA sequence. Two primers are used in polymerase chain reaction technology, one upstream and one downstream of the target sequence. This bracketing of the target sequence determines the length of clones (copies of the target sequence).

priming (1) The first filling with water of a canal, reservoir, or other structure built to contain water. (2) The action of filling a pump casing or siphon with water to remove the air and allow the pump or siphon to start.

primitive water Water that has been trapped in the interior of the lithosphere (in either molecular or dissociated condition) since the formation of the Earth. *See also* interstitial water.

principal isotope *See* isotope.

prior appropriation A doctrine of water law that allocates the right to use water on a first-come–first-serve basis.

prioritization matrix A diagramming tool that a decision maker uses to prioritize issues, product service characteristics, tasks, and so forth based on known weighted criteria.

priority The status of a water user when water rights are based on the doctrine of prior appropriation. Under this doctrine, those who have been using the water longer have higher priority (i.e., greater rights to use the water) than parties who started using the water more recently.

priority contracting The process of signing agreements with other utilities or outside contractors prior to disasters or emergencies for services, products, or both, allowing the signer to be first to receive those products or services in times of disaster.

prism (1) The liquid, mobile volume of water flowing in a specified segment of a stream. (2) A piece of glass for dividing light into individual wavelengths.

pristine Unchanged by civilization; fresh and clean.

pristine watershed A water basin or valley that is clean and unchanged by civilization.

private use of water The use of water solely by those entities directly associated with the agency engaged in the development and distribution of that water.

private utility An enterprise owned by private individuals or by a corporation and operated for the purpose of rendering utility service.

private water Streams or other bodies of surface water that are not publicly owned. *See also* public water.

private water supply A water supply not available to the general public because it is located on or has outlets on private property to which the public does not have access or legal right of entry. *See also* individual water supply; privately owned water utility.

privately owned water utility A public water system owned by one or more private investors. *See also* investor-owned water utility.

privatization The sale to private investors of government-owned equity in nationalized industry or other commercial enterprises. *Contrast with* contract operation and management.

probability *See* probability value.

probability distribution A graphical distribution that is based on a frequency distribution (also called a histogram) and that reflects the frequency of actual sample data falling into selected intervals of occurrence (e.g., 0 to 2.0, 2.1 to 4.0, 4.1 to 6.0). In frequency distributions, the highest frequency usually falls near the center of the range of actual values, near the mean. A frequency distribution can, however, have many shapes, e.g., symmetrical (normal) or with the distribution skewed (tailing) to the right or left; concentrated near the mean or spread out; or high peaked or relatively flat. The probability distribution concept derives from the inference that large or repeated samples of data will generate relatively consistent recurring patterns or frequency distributions that can be interpreted to show the expected frequency (probability) of occurrence for any of the data values, individually or cumulatively. A number of different types of probability distributions are frequently used in business, social science, medicine, and engineering applications, e.g., the normal distribution, binomial distribution, Poisson distribution, and the gamma distributions that include the exponential and chi-square distributions. Probability distributions are also referred to as density functions. *See also* chi-square distribution; normal distribution; Poisson distribution.

probability plot A form of graphical representation of data in which one of the axes follows a particular probability distribution.

probability (*p*) value The probability that a difference at least as large as the measured difference between two groups under study could have arisen at random, assuming the groups were drawn from the same population and the measured characteristic obeys the normal probability distribution. For example, the statement $p < 0.05$ means that at most, one chance in twenty exists that a difference as big as that observed would have occurred if the null hypothesis (that the groups represent no real difference) were true. Investigators may set their own values, but $p < 0.05$ is frequently used to assess statistical significance. *See also* confidence interval; hypothesis; normal distribution; statistically significant.

probable carcinogen A chemical for which the weight of available scientific evidence would suggest that the chemical poses a carcinogenic hazard for humans. This is a qualitative judgment, not a quantitative one. The meaning of the term somewhat depends on the agency making the determination (e.g., the International Agency for Research on Cancer or the US Environmental Protection Agency). Once the US Environmental Protection Agency makes such a determination, it regulates the chemical as a carcinogen. Frequently, such a determination has resulted in the assignment of a letter or number ranking to summarize the weight of evidence. The evidence considered includes epidemiological data to suggest that the chemical produces cancer in exposed human populations. Negative epidemiological data, however, are not generally considered helpful unless the study had sufficient power to detect relatively small increases in cancer risk by virtue of a very high exposure or unusually large and well-defined population to study. The ability of the chemical to induce cancer in more than one species of experimental animal is most frequently the critical determination. The evidence is further strengthened if the chemical produced tumors at multiple sites in experimental animals and particularly if it produced tumors at sites in the animal at which the background rate of cancer was particularly low. Certain in vitro data will heighten the concern that the chemical is capable of inducing cancer in humans, but rarely can that information alone be the basis of indicting the chemical as a human carcinogen. In particular, strong evidence that the chemical can induce point mutations in deoxyribonucleic acid or damage chromosomes at less than cytotoxic doses adds significantly to other evidence that the chemical is capable of inducing cancer in vivo. *See also* point mutation.

probable maximum precipitation (PMP) The best judgment of meteorologists of the realistic upper limit of precipitation that can occur at a location. This value is used in the design of large dam spillways.

probe method *See* electrode method.

probing In subsurface investigations, the act of pushing or driving a rod down through the soil as far as it will penetrate with the equipment at hand. *See also* sounding.

probit model In toxicology, a model for which the response values on a dose-response curve are expressed in terms of standard deviations. If the data are normally distributed, this type of model has the benefit of rendering a linear dose-response curve, which facilitates comparisons of curves produced by different chemicals. Conventionally, 5 is added to all the standard deviations to avoid negative numbers (i.e., otherwise the mean would be zero). *See also* linearized multistage model; log-probit model; logit model; multihit model; multistage model; one-hit model; Weibull model.

problem area wetland A wetland that is difficult to identify because it may lack indicators of wetland hydrology, hydric soils, or both, or because its dominant plant species are more typical of nonwetlands.

procarcinogen A chemical that can be modified by chemical reaction or by enzymatic action to become a chemical that is capable of initiating cancer. Most generally, metabolism of the procarcinogen results in the formation of a reactive chemical that interacts with the pyrimidine or purine bases found in nucleic acids. Binding of the chemical to these bases results in the misreading or misrepair of the deoxyribonucleic acid template and, following cell division, a cell containing a mutation.

process control The use of industrial control computers to monitor and control such plant facilities as pump stations, regulating valves, reservoirs, and filter plants as well as to coordinate and maintain telemetry, radio, and communications equipment.

Process control center

process decision program chart A procedure to map out conceivable contingencies and events that may occur in any implementation plan and to identify feasible countermeasures.

process error The difference between the controlled variable and the controller set point. Also known as *deviation*.

process flowchart A visual representation of all the steps in a process under study. It is a tool that helps develop a clear and common vision of all the elements in a process.

process and instrumentation diagram (P&ID) A plan schematically illustrating the manner in which the components of a unit process, or group of unit processes, interrelate and how the process will be controlled. Process and instrumentation diagrams include such information as monitoring devices (e.g., flowmeters), alarms, and control devices (e.g., switches). Standards for process and instrumentation diagrams are issued by the Instrument Society of America. *See also* Instrument Society of America.

process lag A time delay in a process response to an adjustment. Process lag is caused by the inertia of the process. A contributing factor, along with dead time, is the time required for a controlled variable to respond to a change in value of the final control element.

process safety management Implementation of an Occupational Safety and Health Administration program designed to protect workers and communities from exposure to extremely hazardous chemicals in the workplace. A review of the entire chemical storage, handling, mixing, and feeding process is conducted and evaluated.

process train An independent treatment process or group of processes in series. Commonly, multiple trains are placed in parallel in a treatment facility to subdivide the overall treatment capacity or for some other purpose.

process variable A physical or chemical quantity that is usually measured and controlled in the operation of a water treatment plant or industrial plant. *See also* measured variable.

process water Water used in a manufacturing or treatment process or in the actual product manufactured. Examples include water used for washing, rinsing, direct contact, cooling, solution makeup, chemical reactions, or gas scrubbing in industrial and food-processing operations. In many cases, water is specifically treated to produce the quality of water needed for the process.

product In chemistry, a substance resulting from a chemical reaction. The products of a reaction are shown on the right-hand side of a chemical equation. *See also* reactant.

product staging In reverse osmosis, the practice of using some of the product water from the first stage of treatment as feed-water for the second stage.

product stewardship The act of responsibly assessing the environmental impact of a product to ensure that the product is used properly and does not damage the environment. *See also* ecology.

product water Water that has passed through a water treatment plant, i.e., water for which all the treatment processes are completed or finished. This water is the product from the water treatment plant and is ready to be delivered to the consumers. It is also called finished water.

product water dispensing rate That amount of product water available from the full-open dispensing outlet of a water treatment unit for a specified period of time.

production rate The amount of product water the system produces per minute or (especially for reverse osmosis) per 24-hour period.

productivity (1) In biology, the rate of production of biomass resulting from a biological activity in a given area or volume. (2) In business, the ability to bring into existence—by intellect, creative ability, or a work program—goods and services that have exchange value.

professional engineer (PE) A person registered to practice engineering in one or more disciplines in accordance with the laws of a state or territory of the United States.

proficiency testing (PT) A program of providing environmental samples to a laboratory for analysis. This term is equivalent to the term *performance evaluation* in current laboratory certification programs. It is used in National Environmental Laboratory Accreditation Conference standards. The respective compositions of proficiency testing samples are unknown to the analyst, and the results must be reported to the sample provider. *See also* National Environmental Laboratory Accreditation Conference; performance evaluation.

profile (1) A vertical section of the surface of the ground, underlying strata, or both along any fixed line. (2) A graphical representation of elevation plotted as a function of distance. (3) In open channel hydraulics, a plot of water surface elevation as a function of channel distance; the hydraulic grade line. (4) A longitudinal section along a pipeline, conduit, or stream.

program evaluation review technique (PERT) A time–event network analysis system in which the various events in a project or program are identified, with a planned time established for each. These events are placed in a network diagram showing the relationships of each event to the other events.

programmable logic controller (PLC) A control device used for sequential control of processes or functions. Older programmable logic controllers did not possess the control capability of current units, which are indistinguishable from process control computers. Programmable logic controllers are easily configured or programmed and have a full range of control and data display functions, including erasable programmable read-only memory. *See also* remote terminal unit.

programmable read-only memory (PROM) chip A computer chip manufactured, configured, or customized for a special application or need. Once such a chip's memory is programmed (or "burned"), it cannot be changed.

programmatic state general permit (PGP) A general permit that is issued by the US Army Corps of Engineers to a state delegating certain general wetland permitting authority. Programmatic state general permits operate in conjunction with state and local regulatory programs and are intended to reduce unnecessary duplication resulting from federal and state or local programs. The US Army Corps of Engineers issues programmatic state general permits for dredge and fill activities only where existing state, regional, or local programs provide the same or higher levels of environmental protection as the federal regulatory program provides. The US Army Corps of Engineers maintains control over the aspects of Clean Water Act Section 404 permitting that are not covered by a general permit.

Programming in Logic *See* PROLOG.

progressive cavity pump A positive-displacement pump using a screw or auger to lift liquids, usually sludge or wastewater. It is a low-lift, high-volume, wide-range-of-flow, nonclogging pump.

project flood The design flood chosen for a particular water resources project.

projected savings An estimate of the amount of water that will not be used because both customers and suppliers are implementing efficiency practices.

prokaryote A cellular organism for which the nucleoplasm is not surrounded by a membranous envelope.

proliferation The growth or production by multiplication of parts, as in budding or cell division.

PROLOG A high-level computer language used mainly for artificial intelligence applications. It manipulates knowledge rather than numbers.

PROM *See* pedogenic refractory organic matter.

PROM chip *See* programmable read-only memory chip.

prometon ($(H_7C_3HN)_2C_3N_3OCH_3$) The common name for the herbicide 2,4-bis(isopropylamino)-6-methoxy-*s*-triazine. *See also* herbicide.

promoter *See* tumor promoter.

promoter region A sequence within deoxyribonucleic acid to which a transcription factor can bind to increase the synthesis of a particular messenger ribonucleic acid. This binding ordinarily increases the rate of synthesis of the protein coded for by the messenger ribonucleic acid.

promotion (1) Advancement in rank, position, or the next higher stage or grade, usually accompanied by an increase in honor, pay, prestige, and other rewards. (2) In oncology the process by which a chemical provides a selective advantage to cells that have been initiated for cancer. This selective advantage (i.e., an increase in the rate of cell division or a decrease in the rate at which these populations of cells die relative to normal cells) accelerates the development of tumors. *See also* tumor initiator.

promulgate To announce final federal agency action on a rule making by publication of the final rule in the *Federal Register*; to publish officially.

propagate (1) To transmit or spread from place to place. (2) To cause to increase or multiply.

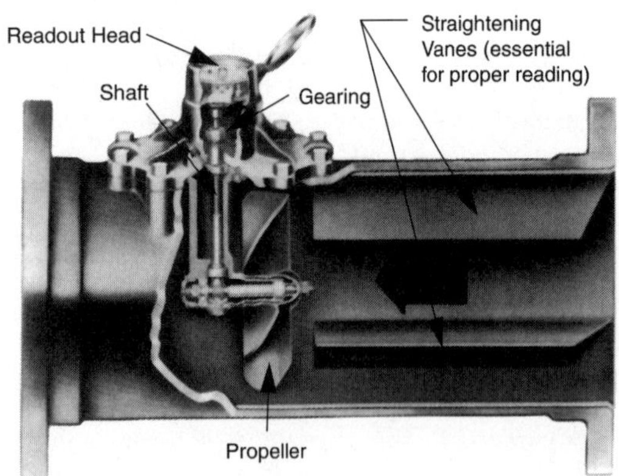

Propeller meter

propeller meter An instrument for measuring the quantity of fluid flowing past a given point. The flowing stream turns a propeller-like device, and the number of revolutions are related directly to the volume of fluid passed.

propeller pump A centrifugal pump that develops most of its head by the propelling or lifting action of the vanes on the liquids. Such a pump is also called an axial-flow pump.

propeller-type impeller An impeller of the straight axial-flow type.

propeller-type turbine A turbine of the axial-flow type that is fitted with a runner having large blades similar to those of the propeller of ship, and that makes use of both the kinetic and pressure energy of the water.

proportional Having or occurring in the same ratio. For example, the fractions ½ and 2⁄4 are proportional to each other.

proportional action A type of control mode that varies the controller's output in proportion to the error of the controlled variable. *See also* proportional integral derivative control.

proportional band The gain of a proportional action controller. It is defined as a percentage of the controlled variable's change that will cause the output to change from minimum to maximum (100 percent).

proportional control *See* proportional action.

proportional counter A radiation detection instrument suitable for determining alpha activity at the alpha operating plateau, as well as beta plus alpha activity and a little gamma activity at a higher operating voltage. A proportional counter is extremely sensitive for making low-level radiation measurements because the sample may be introduced into the gas-flow counting chamber for radioassay.

proportional integral derivative (PID) control Control based on the deviation between the actual output and desired output according to the proportion of deviation, the integral of deviation, and the derivative of deviation. *See also* derivative action; integral action; proportional action.

proportional flowmeter Any flowmeter that diverts a small portion of the main flow and measures the flow rate of that portion as an indication of the rate of the main flow. The rate of the diverted flow is proportional to the rate of the main flow.

proportional mortality rate *See* proportional mortality ratio.

proportional mortality ratio (PMR) The ratio of the number of deaths observed in the study population to the number of deaths expected in the study population, generally multiplied by 100 for convenience in making comparisons. In the proportional mortality ratio, the expected number of deaths in the study population (usually age, time, and cause specific) is computed on the basis of the proportion of that cause of death in the standard or general population. This ratio is used frequently in occupational epidemiology when a standard mortality ratio cannot be computed. Empirically, in occupational studies, the interpretation of a proportional mortality ratio is similar to the interpretation of a standard mortality ratio. *See also* standardized mortality ratio.

proportional weir A special type of weir in which the discharge through the weir is directly proportional to the head of the water being discharged.

proposed rulemaking A formal proposal issued by a federal agency for a rule announced via publication in the *Federal Register.*

propping agent A granular material, such as sand, that is used to keep conduits formed by hydraulic fracturing from collapsing. Propping agents are required to prevent the fractures from collapsing when the pressure is released. *See also* hydraulic fracturing.

prospective study An epidemiological study that examines the development of disease in a group of people determined to be free of the disease at present.

protean (1) Very changeable; taking on different shapes and forms. (2) Any of a group of insoluble derived proteins that are the first products of the action of water, dilute acids, or enzymes on proteins.

protection factor (PF) The ratio of airborne contaminant concentration in the environment to the concentration in air that has passed through a respirator.

protective coating A covering on one material of another material that has greater resistance to corrosion.

protective system Engineered procedures, shielding, or both established or set up to protect workers in an excavation from

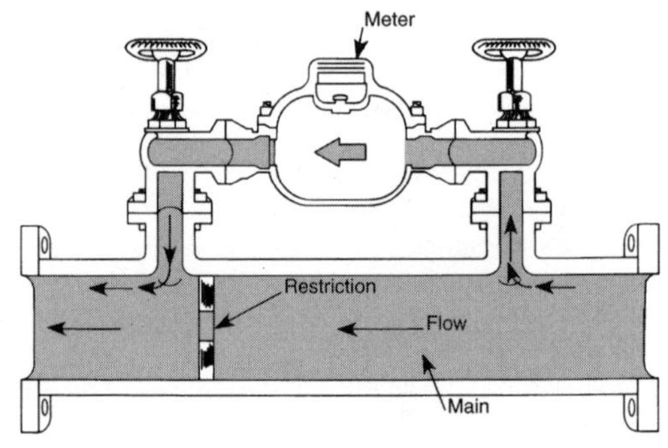

Proportional flow meter

cave-ins or from material breaking off the sides of a trench or adjacent structures.

protein Any of numerous naturally occurring, extremely complex substances that consist of amino acid residues joined by peptide bonds, containing the elements carbon, hydrogen, nitrogen, oxygen, usually sulfur, and occasionally other elements (such as phosphorus or iron). Many essential biological compounds (e.g., enzymes, hormones, or immunoglobulins) contain proteins.

protein code The amino acid sequence of a functioning protein. The building blocks of proteins are amino acids, and it is the amino acid sequence and its length (i.e., number of amino acids) that determine the protein's function.

proteolysis Hydrolysis of proteins or peptides into simpler compounds.

protocol (1) The plan, including the methodology, of a scientific study or experiment. (2) A set of formats and procedures governing the exchange of information between systems. (3) A set of rules that define a function. In synchronous communications, blocks of information are structured to adhere to a particular protocol understood by devices on the network. A network protocol is a fixed set of rules specifying the format of data exchange. A protocol defines how the individual bits are to be arranged in transmitting a message so that the message can be received and interpreted correctly by another device or node on the network.

proton One of the three classical elementary particles of an atom (along with neutrons and electrons). The proton is a positively charged particle located in the nucleus of an atom. The number of protons in the nucleus of an atom determines the atomic number of that element. *See also* electron; neutron.

proton nuclear magnetic resonance (proton NMR) An absorption spectrometric technique used in the characterization of organic compounds. It is based on the absorption of radio frequency radiation by protons in the presence of a magnetic field. This technique allows a nuclear magnetic resonance spectrum—a plot of the absorption frequency versus intensity—to be generated. The peak area of the plot is proportional to the number of protons in the compound; the pattern of peaks in the spectrum depends on the chemical environment of protons in a compound.

protoplasm Cell material inside the cell membrane. Protoplasm is composed largely of proteins along with significant quantities of fats and carbohydrates; it is responsible for all the metabolic activities of the cell.

prototype (1) A natural or full-scale version of a device or system that previously had only been simulated by a model. (2) A trial version of a product for testing before a final version is brought to market.

protozoa Small, one-celled organisms, including amoebae, ciliates, and flagellates.

protozoan cyst An environmentally stable life-cycle stage of protozoan parasites. Such cysts are excreted through feces into the environment. *See also* cyst; *Giardia*; oocyst.

proximate carcinogen The form of a carcinogen (usually an electrophilic or radical cation intermediate) that actually reacts with the purine and pyrimidine bases within nucleic acids to initiate the carcinogenic response.

Prussian blue A blue paste or liquid (often supported on a paper, such as carbon is supported on carbon paper) used to show an area of contact between two pieces of equipment. The Prussian blue rubs off at the points of contact. For example, Prussian blue is used to determine whether gate valve seats fit properly.

PRV *See* pressure-regulating valve.

PS *See* point source.

pseudocholinesterase Any of a broad class of nonspecific enzymes that catalyze the hydrolysis of the esteratic bond in acetylcholine. These enzymes are found in high concentrations within the blood. The term *pseudocholinesterase* was given to these enzymes to differentiate them from the true cholinesterases found in nerves and red blood cells that are actually responsible for terminating the action of acetylcholine. Pseudocholinesterases can play an important role in terminating the activity of a number of chemicals that have similar properties to acetylcholine.

Pseudomonas A genus of aerobic bacteria belonging to the family Pseudomonadaceae, characterized as gram-negative motile rods with either monotrichous (having a single flagellum at one end of the cell), amphitrichous (having a tuft of flagella at one end of the cell), or lophotrichous (having a single flagellum or a tuft of flagella at each end of the cell) polar flagellation.

Pseudomonas aeruginosa A gram-negative, aerobic, motile bacterial species, 1.5 to 3 micrometres long by 0.5 micrometres wide, that often produces a bluish green, soluble diffusible pigment called pyocyanin and a yellowish green fluorescent pigment. This organism is found in a variety of environments, generally aquatic, and is pathogenic for humans. It has been associated with infections of wounds and burns, middle-ear infections (swimmer's ear), endocarditis, pneumonia, meningitis, gastrointestinal infections, and generalized systemic infection.

Pseudomonas fluorescens A gram-negative, aerobic, motile bacterial species that is commonly found in aquatic environments and produces a diffusible yellowish green fluorescent pigment. The numerous strains of *Pseudomonas fluorescens* do not appear to be pathogenic to humans.

***Pseudomonas fluorescens* strain P17** A bacterial strain used in bioassay procedure for determining the concentration of assimilable organic carbon in drinking water. *See also* assimilable organic carbon.

P17 *See Pseudomonas fluorescens* strain P17.

psi *See* pounds per square inch *in the Units of Measure list.*

psia *See* pounds per square inch absolute *in the Units of Measure list.*

psig *See* pounds per square inch gauge *in the Units of Measure list.*

psychrometer An instrument or hygrometer used to determine the relative humidity and vapor tension of the atmosphere. It usually consists of two thermometers, with one bulb left dry and the other encased in cloth or a cloth wick that can be saturated with water. Evaporation from the wet bulb causes the temperature to fall to less than that of the air. The relative humidity and the vapor pressure can be determined based on this temperature difference and specially prepared psychrometric tables.

psychrophilic bacteria Bacteria whose optimal temperature for growth is 54° to 64° Fahrenheit (12° to 18° Celsius).

psychrophilic range The optimal temperature range for growth of psychrophilic bacteria, generally accepted as 54° to 64° Fahrenheit (12° to 18° Celsius).

PT *See* proficiency testing.

PTA *See* packed tower aeration.

Pt–Co *See* platinum–cobalt color unit *in the Units of Measure list.*

PTFE *See* polytetrafluoroethylene.

PTS *See* packed tower stripping.

public land survey system (PLSS) A reference scheme for recording property ownership by township, range, section, and aliquot parts (half or quarter sections) in the United States. The public land survey system was laid out during the settlement of the country, dividing land areas into townships of thirty-six 1-square-mile sections.

public landscape Public street medians, parks, recreational facilities, and landscapes next to a public facility.

public law (PL) A bill or joint resolution (other than for amendments to the US Constitution) passed by both houses of the US Congress and approved by the president of the United States. Bills and joint resolutions that are vetoed by the president but overridden by Congress also become public law.

public notification Disclosure to the public of when and how public water systems are in violation of the Safe Drinking Water Act, as per the requirements set October 27, 1987 (*Federal Register*, 52(208):41534,41550) by the US Environmental Protection Agency. For an acute violation, public notification is required by TV and radio within 72 hours, a general circulation newspaper within 14 days, and by mail or hand delivery within 45 days. If no newspaper of general circulation is available, posting or hand delivery is required. Notification by mail or hand delivery is repeated quarterly as long as the system is in violation. *See also* consumer confidence report.

public use of water The use of water that is provided by an agency engaged in the development and distribution of water and is offered—up to the full capacity of the agency's water system—to all consumers who can be served under that system and who may apply for this service.

public utility A business enterprise, in the form of a public service corporation that performs an essential public service and is regulated by the federal, state, or local government.

public water Water that is open to public recreational use.

public water supply *See* public water system.

Public Water Supply Supervision (PWSS) program The program created and administered by the US Environmental Protection Agency's Office of Ground Water and Drinking Water to implement and enforce Safe Drinking Water Act regulations in the United States.

public water system (PWS) As defined in Section 1401(4) of the Safe Drinking Water Act, as a system for providing to the public water for human consumption through pipes or other constructed conveyances, if such system has at least 15 service connections or regularly serves at least 25 individuals daily at least 60 days out of the year. Such a system includes (1) any collection, treatment, storage, and distribution facilities under control of the operator of such system and used primarily in connection with such system, and (2) any collection of pretreatment storage facilities not under such control but that are used primarily in connection with such system. A public water system is either a community water system or a noncommunity water system. On Aug. 5, 1998, the US Environmental Protection Agency published guidance regarding interpretation of the definition of *public water system* (*Federal Register*, 63:41939–41946). *See also* community water system; noncommunity water system; transient noncommunity water system.

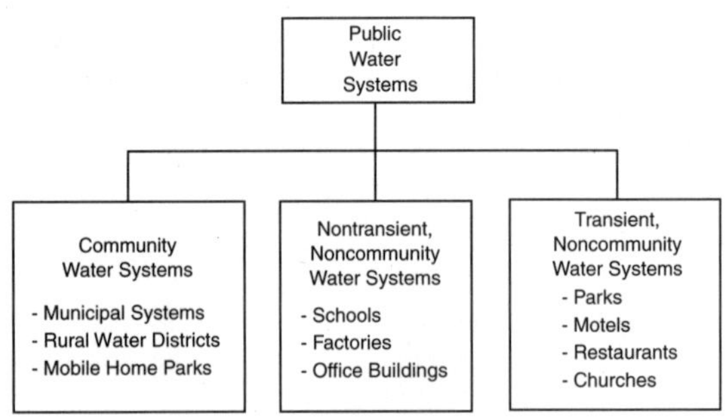

Classification of public water systems

public water system inventory A listing maintained by the US Environmental Protection Agency that categorizes all public water systems in the United States that are subject to Safe Drinking Water Act regulations.

public water use The use of water to support public facilities, such as public buildings and offices, parks, golf courses, picnic grounds, campgrounds, and ornamental fountains.

public works Structures for public use, such as dams, airports, and buildings, paid for by government funds.

publicly owned treatment works (POTW) A wastewater treatment facility owned by a municipality or local government authority.

publicly owned water utility A water system owned by a municipal or quasi-municipal (e.g., special district) government.

puddle A small pool of water, usually a few inches in depth and from several inches to several feet in its greatest dimension.

puddle clay Clay with certain adhesive properties that make it suitable for use in forming a watertight lining or backing for a reservoir embankment or other structure used to hold back water.

puddled core An earth dam core constructed of puddle clay, compacted clay, or other fine material. *See also* impervious core.

pulmonary test A medical test required by the Occupational Safety and Health Administration's respiratory standard and given to employees before they are allowed to wear respirators on the job. This test measures the volume of air that a person's lungs can bring into the body.

pulse bed *See* moving bed.

pulse sensor clearance The gap or distance over which the sensor in an instrument or meter will pick up the pulse signal. For a magnetic sensor, a pulse or signal that has to travel more than this distance will not be sensed or detected.

pumicite A stable, natural, glassy aluminum silicate mineral derived from volcanic ash that is used as a water treatment filtration medium. *See also* manganese dioxide-coated pumicite.

pump A mechanical device for driving fluid flow, for raising or lifting a fluid, or for applying pressure to a fluid.

pump amperage The electrical current flow, at a given voltage that must be applied to a pump to produce a certain flow at a specific head. This current flow is expressed in amperes.

pump bowl The submerged pumping unit in a well, including the shaft, impellers, and housing.

pump capacity The amount or rate of flow a pump can deliver. Pump capacity varies with the head or pressure and is usually expressed at a point near the highest efficiency. *See also* rated capacity.

pump center line An imaginary line through the center of a pump.

pump characteristics curve A graphic representation of the performance of a pump in relation to the rate of flow against the total head, with the efficiency of the pump shown at points along the curve. These curves are generally established from actual run tests of the pumps. The shut-off head at which the pump will no longer deliver water and the rated capacity at the most efficient point are included with the curve. From these curves a range of heads and delivery amounts can be determined, allowing one to select a pump for a specific pump application and for the most efficient operating range.

pump curve *See* pump characteristic curve.

pump efficiency The ratio of the pump output (i.e., water horsepower) to the pump horsepower, expressed as a percentage. If the motor is included with the pump as a unit, then the ratio of the pump output to the horsepower required by the motor gives a wire-to-water efficiency. One must know the efficiency of the motor to calculate the pump efficiency to determine the wire-to-water efficiency. *See also* wire-to-water efficiency.

pump horsepower The power (in units of horsepower) that must be delivered to a pump shaft in order to achieve a certain rate of flow of water at a certain pressure. *See also* pump kilowatts.

pump house A shelter for housing a small water pump.

pump kilowatts The Système International equivalent of pump horsepower.

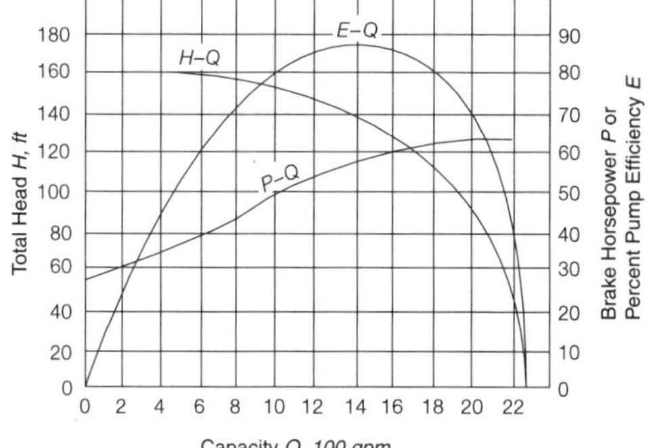

Pump characteristics curve

pump output The flow rate times the total head—with the product expressed in units of horsepower or kilowatts—that a pump will deliver to water at a specific pressure or head. Pump output is also known as water horsepower or water kilowatts. *See also* water horsepower; water kilowatts.

pump overall efficiency *See* wire-to-water efficiency.

pump packing The material placed in the designed space between the pump housing and the pump shaft to prevent excessive loss of water along the pump shaft. The packing is inserted in rings and is compressed by a follower gland with bolts to reduce leakage. A small amount of water provides lubrication of the packing against the rotating shaft, preventing overheating and wear of the shaft.

pump pit A dry well or chamber, below ground level, in which a pump is located.

pump primer A vacuum pump attached to the suction end of a pump, used for priming the pump automatically.

pump priming The act of filling a pump casing with water to replace the air and provide a liquid to which the pump may apply a force. Pumps for which the suction side is set below the water level are said to have a positive suction head and can be primed by venting the top of the pump casing, allowing water to replace the air.

pump rod The line of rods of a reciprocating well pump, connecting the piston in the working barrel with the power head.

pump setting A measure of the energy added to a system by a pump. It is equal to the height of the pump head above the pump location or, for a well pump, the total length of the discharge column.

pump slip The percentage of water that is taken into the suction end of a pump but is not discharged because of clearances in the moving unit.

pump stage The number of impellers in a centrifugal pump; e.g., a single-stage pump has one impeller, and a two-stage pump has two impellers.

pump station A structure containing relatively large pumps and appurtenant piping, valves, and other mechanical and electrical equipment for pumping water. A pump station is also called a lift station. *See also* pump house.

pump strainer A device placed on the inlet of a pump to strain out suspended matter that might clog the pump.

pump stroke The lineal distance traveled by the piston or plunger of a reciprocating pump through one-half of its cycle of movement.

pump submergence The vertical distance of a pump inlet or suction below the water level in a pump pit or afterbay.

pump valve An opening through which water enters or leaves the cylinders of a displacement pump.

pump well A well that does not discharge water at the surface except through the operation of a pump or other lifting device. A pump well is also called a nonflowing well.

pumpage The total quantity of liquid pumped in a given interval, usually 1 day, 1 month, or 1 year.

pumped storage Water pumped by a power plant into a reservoir during off-peak periods of plant operation. Such water is later used to develop hydroelectric power during periods of peak demand. The reservoir is usually at a considerable elevation above the power plant; thus, the stored water provides a high head, so only a relatively small amount of storage space is needed. This method of supplying peak power often results in a considerable improvement in the load factor of the main plant and a consequent savings in equipment costs. *See also* pumped-storage plant.

pumped-storage plant A hydroelectric power plant in which peak-load power capacity is produced by water pumped into storage reservoirs during off-peak periods by using the plant's turbine-driven generators as motor-driven pumps. The power needed for pumping is obtained from the system when excess power-producing capacity is available; it would be wasted were it not used for this purpose. *See also* pumped storage.

pumper outlet nozzle The large type of hydrant outlet, usually 4.5 inches (11.4 centimetres) in diameter. Such an outlet is often called a pumper connection. It is also sometimes called a steamer connection because of its use in connection with steam fire engines years ago.

pumping The application of force to a liquid to move or pressurize the liquid or to lift it from a lower elevation to a higher elevation.

pumping depression area The surface area overlying the area of influence or pressure relief cone created by a pumping operation. *See also* area of influence; pressure relief cone.

pumping depression cone The depression, roughly conical in shape, produced in a water table or other piezometric surface by the extraction of water from a well at a given rate. *See also* cone of depression; cone of influence; drawdown; pressure relief cone.

pumping groundwater level The depth to water in a production well as observed when the well is being pumped. The difference between this depth and the static (nonpumping) depth to water is called the drawdown. Typically the measurement is made after 1 hour of pumping during which the specific capacity of the well is measured. *See also* cone of depression; drawdown.

pumping head The sum of the static head and friction head on a pump discharging a given quantity of water.

pumping level The elevation at which the water level stands in a well when the well is being pumped at a given rate.

pumping line The discharge pipe from a pump.

pumping sampler An apparatus for sampling a water–sediment mixture by withdrawing it through a pipe or hose. The intake of the pipe or hose is placed at the desired sampling point.

pumping station *See* pump station.

pumping trough A hydrodynamic barrier to seawater intrusion in coastal areas created by operating a line of wells near the coast to extract brackish water from an aquifer. These wells are designed to stabilize the location of a saltwater wedge in the aquifer.

pumping water level (pwl) The vertical distance from the center line of the pump discharge to the level of the free pool of the upper level storage or water supply tank while water is being drawn from the tank.

pump-valve cage A casting containing a number of ports for water valves on the suction end of a reciprocating pump.

pure culture *See* axenic culture.

pure oxygen An oxygen source that is at least 90 percent oxygen. Pure oxygen for process applications is either produced on-site or shipped in liquid form.

pure water A term that has no real meaning unless the word *pure* is defined by some standard, such as for, say, reagent grade water, distilled water, deionized water, reverse osmosis treated water, or activated carbon adsorption treated water. *See also* reagent grade water.

purge water The water used in cleaning and disinfecting a water line or water storage facility. It is purged, wasted, and replaced with clean potable water.

purge-and-trap method A combination of isolation techniques used in the analysis of volatile organic compounds. Typically, an inert gas is bubbled through a water sample to purge volatile analytes into the headspace. Analytes are then adsorbed from the gaseous phase onto a solid trap. Compounds are then desorbed onto the head of a chromatographic column for separation and subsequent analysis.

purgeable organic carbon (POC) The fraction of total organic carbon removed from an aqueous solution by gas stripping under specified conditions. Purgeable organic carbon is also referred to as volatile organic carbon, which includes the carbon content of chemicals that are volatile under the specified stripping conditions of the test. *See also* dissolved organic carbon; nonpurgeable organic carbon; particulate organic carbon; total organic carbon; volatile organic compound.

purgeable organic halogen (POX) A surrogate measurement of the total quantity of purgeable (fairly volatile) halogen-substituted organic material in a water sample. It is the volatile fraction (under conditions of the test) of the total organic halogen. The presence of purgeable halogen-substituted organic molecules in source water is typically due to the presence of volatile organic chemicals (e.g., industrial solvents), whereas in finished water it is typically due to the presence of volatile disinfection by-products (e.g., trihalomethanes). *See also* disinfection by-product; halogen-substituted organic material; surrogate measurement; total organic halogen; volatile organic compound.

purgeable organic halogen formation potential (POXFP) The amount of purgeable organic halogen (POX) formed during a test in which a source or treated water is dosed with a relatively high amount of disinfectant (normally chlorine) and is incubated (stored) under conditions that maximize purgeable organic halogen production. This value is not a measure of the amount of purgeable organic halogen that would form under normal drinking water treatment conditions, but rather an indirect measure of the amount of purgeable organic halogen precursors in a sample. If a water has a measurable level of purgeable organic halogen prior to the formation potential test (e.g., from volatile organic chemicals in the water being tested), then the formation potential equals the POX concentration measured at the end of the test minus the initial value ($(POX_t) - (POX_0)$). *See also* disinfection by-product formation potential; purgeable organic halogen.

purification The act of removing objectionable matter from water by natural or artificial methods.

purified water A *United States Pharmacopeia* grade water that is produced from water meeting US Environmental Protection Agency standards for potable water, has microbiological content under control, and is free of foreign substances. *United States Pharmacopeia* grade water has various stipulated ions

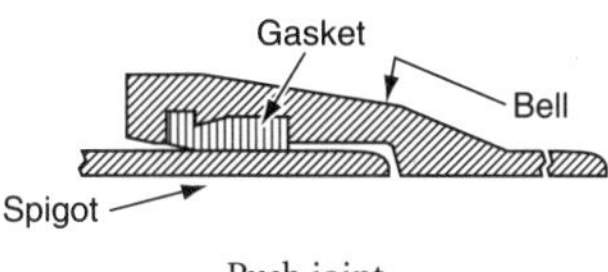

Push joint

and gases removed, and it has a 1.0-megaohm-centimetre resistivity measurement.

purveyor, water *See* water purveyor.

push joint A type of gasket joint in bell-and-spigot pipe. The gasket is placed in a groove in the bell end and is sealed when compressed against the pipe and bell. This compression occurs when the spigot end of the pipe is inserted or pushed into the bell.

push well A tubular well pushed approximately horizontally into a water-bearing stratum or under the bed of a lake or stream. A push well is also called a horizontal well.

putative carcinogen A chemical that is thought to be responsible for the induction of cancer.

putrefaction Biological decomposition of organic matter associated with anaerobic conditions, resulting in the production of ill-smelling and bad-tasting products.

putrescibility The relative tendency of organic matter to undergo decomposition in the absence of oxygen.

PVA *See* polyvinyl alcohol.

***p*-value** *See* probability value.

PVB assembly *See* pressure vacuum breaker assembly.

PVC *See* polyvinyl chloride.

PVC pipe *See* polyvinyl chloride pipe.

pwl *See* pumping water level.

PWS *See* public water system.

PWSS program *See* Public Water Supply Supervision program.

pycnometer A device for measuring densities of liquids, consisting of a container of known capacity that is filled with the liquid and weighed. The difference in mass compared to the mass of the empty container indicates the density.

pyrochromatogram A chromatogram generated from the pyrolysis of a sample. Such a chromatogram illustrates the fragments produced upon rapid heating of complex molecules.

pyrogen A substance (often of unknown origin) that produces fever when introduced into the human body. Because they are chemically and physically stable, pyrogens are not necessarily destroyed by conditions that kill bacteria.

pyrolusite (MnO_2) *See* manganese dioxide.

pyrolysis The destructive distillation of organic compounds in an oxygen-free environment, converting the organic matter into gases, liquids, and char.

pyrolysis gas chromatography–mass spectrometry (pyrolysis GC–MS) A technique of introducing components of a sample to a gas chromatograph–mass spectrometer. It involves rapid heating of a sample so that the sample decomposes, and the products are analyzed by a combined gas chromatograph–mass spectrometer. High-molecular-weight substances have been analyzed by this technique.

pyrolysis GC–MS *See* pyrolysis gas chromatography–mass spectrometry.

Q

Q *See* reactive power.
QA *See* quality assessment.
QALY *See* quality-adjusted life year.
QA–QC *See* quality assessment; quality control.
QC *See* quality control.

quadrupole mass spectrometer A device used in the detection and identification of compounds. A quadrupole type of mass spectrometer is a fairly compact instrument and is commonly interfaced with a gas or liquid chromatograph. The term *quadrupole* refers to the four parallel metal rods that act as a filter for ions formed in the ion source portion of the mass spectrometer. Alternating radio frequency and direct current fields on the rods allow separation of ions by permitting only substances of a certain mass-to-charge ratio to pass entirely through the quadrupole. This type of device might better be called a quadrupole filter.

quagga mussel A bivalve (*Dreissena bugensis*) that multiplies rapidly in fresh water and can clog intake pipes. It has been identified in the Great Lakes and is similar in appearance to the zebra mussel. *See also* zebra mussel.

qualification test A set of tests and verifications performed to validate the conformance of water treatment equipment to a specific standard.

qualified right A water right that is not absolute but must be shared with others.

qualitative Pertaining to kind, type, or direction, as opposed to size, magnitude, or degree.

qualitative assay A test measurement resulting in a non-numerical result (e.g., positive versus negative, presence versus absence).

quality The presence of value for the customer; the degree of excellence that a thing possesses.

quality assessment (QA) A process of measuring and evaluating quality. Often this assessment involves evaluating such information as control charts, replicate measurements, spiked samples, and standard reference materials. *See also* certified reference material; control chart; spiked sample.

quality assurance (1) An overall system of management functions designed to provide assurance that a specified level of quality is being obtained. It can be thought of as being composed of quality control (QC) and quality assessment (QA). (2) The management of products, services, and production or delivery processes to ensure the attainment of operational performance, product, or both in keeping with quality requirements. *See also* quality assessment; quality control.

quality circle A small group of people, usually from similar units and organizational levels, who meet regularly for the purpose of improving productivity, morale, product, profit, and quality of performance.

quality control (QC) A system of functions carried out at a technical level for the purpose of maintaining and documenting quality. It includes such features as personnel training, standard operating procedures, and instrument calibrations.

quality design team In total quality management, a small group of people—working at any and every appropriate level—that formulates, communicates, promotes, and guides the process; plans organizational activities; and approves principal staff improvement teams. *See also* total quality management.

quality system The overall program of quality assurance policies and quality control procedures used by a laboratory. Quality systems are in place to ensure that analyses are of high and documented quality. This term is used in National Environmental Laboratory Accreditation Conference standards. *See also* National Environmental Laboratory Accreditation Conference.

quality-adjusted life year (QALY) A year of potential life adjusted based on a consideration of the quality of life at various ages. In epidemiology, an adverse effect of exposure is sometimes measured in terms of the number of years of potential life lost, a public health measure of the premature deaths that may result from exposure. The method of calculation assigns equal value to each year of life lost, but it has often been proposed that an indicator of the quality of life should be considered because certain years of life may be more valuable than others, both to the individual and to society. This requires that the years of potential life lost be adjusted. However, epidemiologists have neither reached a general agreement on which years of life are more valuable nor devised an accepted standard for weighting those years in the calculation of years of potential life lost. Because the measure of a quality-adjusted life year can vary from investigator to investigator, knowing how the quality has been determined in calculating quality-adjusted life years lost is important.

QualServ A program offered jointly by the American Water Works Association and the Water Environment Federation to help water, wastewater, and joint water–wastewater utilities improve performance and increase customer satisfaction on a continuing basis. The program also serves as a set of tools to help utilities hone their competitiveness.

quanat A type of infiltration gallery common in the ancient world (ranging from Afghanistan to Morocco) and still in use today. Quanats are hand-dug by skilled workers using methods over 3,000 years old, but otherwise they are equivalent to modern infiltration galleries. A quanat is also called a karez, foggara, or falaj. *See also* infiltration gallery.

quantification The measurement of the quantity of a substance. This term is preferred over *quantitation.*

quantitation The measurement of the quantity of a substance. The term *quantification* is preferred.

quantitative Pertaining to size, magnitude, or degree.

quantitative assay A test measurement resulting in a numerical result.

quantity-of-electricity units In Système International, coulombs; ampere seconds.

quantity-of-heat units In the US customary systems, British thermal units; horsepower hours; foot-pounds force. In Système International, joules; metres squared kilograms per second squared.

quarterly user charges The charges made to the users of water service through the general water rate structures of the utility for the utility's share of the cost of servicing the water service requirements.

quartile confidence interval In statistics, an interval or range that is estimated to enclose one-quarter of the values that a particular random variable might assume. A quartile represents one-quarter of a population or sample. The four quartiles are reported as lower, second, third, and upper for a population or sample.

quartz jacket A clear, pure, fused quartz tube used to protect the high-intensity ultraviolet lamps in ultraviolet systems. It usually retards less than 10 percent of the ultraviolet radiation dose. A quartz jacket is also called a quartz sleeve.

quartz sleeve *See* quartz jacket.

quaternary amine (R_4N^+) A class of organic compounds of nitrogen that may be considered as derived from the ammonium ion (NH_4^+) by replacing all four of the hydrogen atoms with alkyl groups. *See also* strong base cation exchanger.

quaternary ammonium salt ($R_4N^+X^-$) A type of organic nitrogen compound in which the molecular structure includes a central nitrogen atom joined to four organic groups (R) as well as to an acid radical (X). An ammonium salt in which the nitrogen atom shares its four valence electrons with carbon atoms is usually formed by the reaction between a tertiary amine (R–N(R′)–R″) and an alkylating agent (R‴X). They are all cationic surface-active coordination compounds and tend to be adsorbed on surfaces. Quaternary ammonium salts have many industrial uses, including use as a disinfectant, cleanser, and sterilizer. Cationic quaternary ammonium compounds adsorb on the cell walls of microbes and react chemically with the negative charges carried by the cell walls. Quaternary ammonium salts provide the exchange sites on certain anion exchange resins. *See also* strong base anion exchanger.

quench To cool a material suddenly or halt a process or reaction abruptly.

quenching agent A substance used to inhibit a chemical reaction. For example, ascorbic acid ($C_6H_8O_6$) can be used as a reducing agent to quench the chlorine (HOCl) present in a treated water sample.

query To selectively retrieve data from a database. Having a computer list all items at a particular location that have not been replaced in the last year is an example. Most queries are accomplished with structured query language.

questionnaire (1) A form used for information gathering, inviting open-ended feedback from either a targeted or random group. It has a question-and-answer format and may allow for comments. The compiled results provide information on a range of opinions but have no statistical validity. (2) A formally developed set of questions used to solicit information about the individuals in a study for use in epidemiologic analyses. Information is collected to help assess various exposures, risk factors, and confounding characteristics. The questionnaire is usually developed by a survey statistician and is designed to collect information in an unbiased manner. Objectivity is sought in obtaining information; i.e., objective (closed-ended) answers rather than subjective (open-ended) answers are solicited. A questionnaire can be either self-administered or administered by a trained interviewer, but it must be administered in a consistent fashion according to established guidelines. Whenever possible, interviewers are unaware of the exposure or disease status of a study participant, and study participants are unaware of the associations being studied. Most often the actual study participants are questioned, but sometimes information about individuals in the study must be obtained from a spouse or another individual familiar with the study participant.

queue A temporary holding place for data. The data are usually processed on a "first-in–first-out" basis. A computer print queue is a good example. The print buffer stores, or queues, the data, which are then, in a job-by-job order, output to a printer.

quicklime Another name for calcium oxide (CaO), which is used in water softening and stabilization. *See also* lime.

quick-operating valve A valve that has a revolving plug and a lever attached to it to facilitate sudden opening or closing.

R

R *See* gas constant.

r *See* roentgen *in the Units of Measure list.*

R factor *See* R plasmid.

R plasmid A plasmid that encodes for antibiotic resistance. An R plasmid is also called an R factor. *See also* plasmid.

***r* squared (r^2) value** *See* coefficient of determination.

RAA *See* running annual average.

rabble arm A metal arm attached to the center column in a multiple-hearth incinerator that rotates, moving material to be incinerated around each hearth (i.e., chamber). It conducts the material to an opening through which the material falls downward to the hearth below. *See also* multiple-hearth furnace.

rack rake A hand-operated or mechanically operated rake designed to remove trash and debris lodged on a rack or screen.

rad *See* rad; radian *in the Units of Measure list.*

radial drainage pattern A drainage pattern formed by streams running radially outward from a central high area or radially inward to a central low area.

radial flow Flow that moves across a basin or turbine from the center to the outside edge, or vice versa. *See also* radial inward flow; radial outward flow.

radial gate A pivoted crest gate, the face of which is usually a circular arc with the center of curvature at the pivot about which the gate swings when opening. *See also* tainter gate.

radial to impeller Perpendicular to the impeller shaft. The flow of water is forced by centrifugal force to the outer edge of the impeller with discharge away from the shaft.

radial inward flow Flow of water or another liquid from the periphery to the center in a circular tank or water turbine. The inlet or inlets are located at the periphery and the outlet or outlets at the center.

radial outward flow Flow of water or another liquid from the center to the periphery in a circular tank or water turbine. The inlet or inlets are located at the center and the outlet or outlets at the periphery.

radial well A very wide, relatively shallow well with horizontal drilled screen points at the bottom. Radial wells produce large quantities of water.

radial-arm collector well A well composed of a relatively large-diameter vertical shaft with smaller-diameter horizontal shafts extending radially outward from the central shaft. The horizontal shafts are screened to allow groundwater to enter the well. These wells are sometimes called Ranney wells.

radial-flow tank A circular tank in which the direction of flow is from the center to the periphery, or vice versa.

radian (rad) *See the Units of Measure list.*

radiance units In the US customary system, British thermal units per hour per square foot per steradian. In Système International, watts per square metre per steradian.

radiant flux units In the US customary system, British thermal units per hour. In Système International, watts; metres squared kilograms per second cubed.

radiant intensity units In the US customary system, British thermal units per hour per steradian. In Système International, watts per steradian.

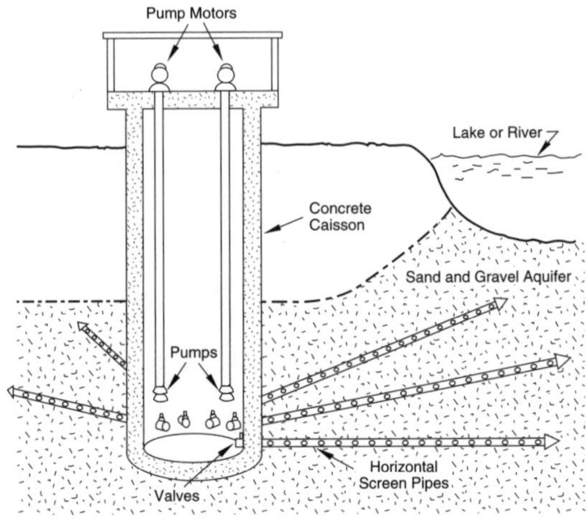

Radial well

radiation Energy in the form of electromagnetic waves. Radiation is emitted from matter in the form of photons, each having an associated electromagnetic wave that has a frequency and wavelength. The various forms of radiation are characterized by their respective wavelengths, and together they form the electromagnetic spectrum, the components of which include cosmic rays, gamma rays, x-rays, ultraviolet rays, visible light rays, infrared (heat) radiation, microwave radiation, and radio and electric rays. Radiation might also be called irradiation, radiant energy, or light. *See also* photon.

radiation adsorbed dose *See* rad *in the Units of Measure list.*

radiation concentration guide The amount of any specific radioisotope that is acceptable in water for continuous consumption. This value is equivalent to what was formerly called the maximum permissible concentration.

radiation protection guide (RPG) A set of tables that indicate maximum exposure levels for workers and provide guidance for the efforts that should be taken to minimize worker exposure to radiation.

radical A group of atoms, acting as a single atom, that go through chemical reactions without being changed. Some examples are bicarbonate (HCO_3^-), hydroxide (OH^-), and sulfate (SO_4^{2-}). *See also* free radical; hydroxyl radical.

radical scavenger Any substance that is added or naturally present in a system or mixture and that consumes or inactivates free radicals. For example, during advanced oxidation processes, hydroxyl radicals can be efficiently trapped by the bicarbonate ion (HCO_3^-), which is a common component of natural waters. In some laboratory experiments, *tert*-butanol ($(CH_3)_3COH$) has been used as a hydroxyl radical scavenger. *See also* advanced oxidation process; hydroxyl radical; radical.

radioactive Pertaining to a material with an unstable atomic nucleus that spontaneously decays or disintegrates, producing various forms of radiation.

radioactive contaminant A radioactive material that is present in activities or places where it may spoil experiments, harm life, or make products or equipment unsuitable or unsafe for use.

radioactive decay The decrease in activity of any radioactive material with the passage of time caused by the spontaneous emission from the atomic nuclei of either alpha or beta particles, sometimes accompanied by gamma radiation; the spontaneous radioactive transformation of one nuclide into a different nuclide or into a different energy state of the same nuclide. Every decay process has a specific half-life. *See also* half-life.

radioactive fallout Airborne particles of radioactive nuclei that result from nuclear explosions and settle out of the atmosphere, often thousands of miles away from the place of detonation.

radioactive solution gauging A method of measuring the flow of water by introducing a constant flow of a radioactive solution of known concentration for a sufficient lapse of time at one section of the water conduit and then determining the resulting degree of dilution of this solution at another section downstream. *See also* dilution gauging.

radioactive spring A spring in which the water has an abnormally high and easily measurable radioactivity.

radioactive waste Water or any other materials—including spent nuclear reactor fuel, work clothes, or tools—that contain radioisotopes.

radioactivity The emission of radiation (alpha, beta, or gamma rays or a combination) from some elements (e.g., radium, radon, uranium, thorium) as a result of the spontaneous disintegration of the nuclei of the atoms of these elements.

radioactivity shielding Any material or obstruction that absorbs radiation and thus tends to protect personnel or material from the effects of exposure to radioactivity. A considerable thickness of high-density material may be needed for nuclear radiation, especially in the case of gamma radiation.

radioactivity units In the US customary system, picocuries. In Système International, becquerels.

radiochemical analysis That specialty in chemical analysis dealing with radioactive substances.

radiofocus assay *See* radioimmunofocus assay.

radioimmuno assay (RIA) A technique that uses radioactive labeled antibodies for detection of viral antigens in blood.

radioimmunofocus assay (RIFA) A type of assay used for viruses that are difficult to grow in tissue culture and do not produce cytopathic effects in tissue culture, e.g., hepatitis A. The presence of virus in cell culture is detected by the addition of an isotopically labeled antibody into the cell culture flask. If the virus is present in the cell culture, the antigen–antibody complex that has been formed can be visualized by using X-ray film. *See also* cytopathic effect.

radioisotope A radioactive isotope of an atom, usually artificially produced.

radiological contaminant A radioactive substance or type of radiation (e.g., radon, radium, uranium, beta particles, photon emitters, alpha particles) regulated or proposed to be regulated by the US Environmental Protection Agency. *See also* alpha particle; beta particle; photon emitter; radioactive; radium; radon; uranium.

radionuclide A material with an unstable atomic nucleus that spontaneously decays or disintegrates, producing radiation.

Radionuclide Rule A rule developed by the US Environmental Protection Agency to establish National Primary Drinking Water Regulations for radionuclides. The proposed rule covering radon, uranium, radium-226, radium-228, alpha emitters, and beta and photon emitters was published July 18, 1991 (56 *Federal Register* 33050–33127). As of the time of this writing, a final rule is under development.

radium (Ra) A naturally occurring radioactive element (in the form of radium-226 or radium-228) created in the decay of the uranium and thorium series. Radium can be removed from water by cation exchange softening.

radium-226 + radium-228 (Ra-226 + Ra-228) The sum of the naturally occurring radioactive isotopes of radium. The interim regulation for radium by the US Environmental Protection Agency is for the sum of the isotopes. The Radionuclide Rule proposes separate standards for each isotope. *See also* isotope; Radionuclide Rule; radium.

radius The distance from the center of a circle to its edge; one half of the diameter.

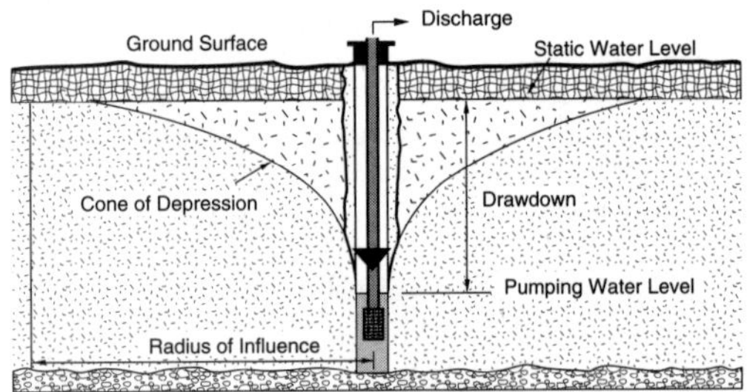

Radius of influence

radius of influence (of a well) The distance from a pumping well beyond which the drawdown caused by the well is negligible or unobservable. The value depends on the aquifer's hydraulic properties and the pumping rate of the well. *See also* cone of depression; drawdown.

radon (Rn) A gaseous radioactive element that derives from the radioactive decay of radium. This radionuclide is to be regulated by the US Environmental Protection Agency. Radon can be removed from water by aeration. *See also* radioactive decay; radium.

rafter-and-strut framed dam A fixed dam, usually low in height, that is built of squared timbers with a plank facing and is not filled with rock. The bents or buttresses are framed like the letter *A* lying on its side, with its apex upstream, one side forming a sill and the other the support for the deck. The deck usually makes an angle of less than 30° with the horizontal. Such a dam is also called an A-frame timber dam.

rain Particles of liquid water that have become too large to be held by the atmosphere. Such particles are generally of a diameter greater than 0.02 inches (0.5 millimetres), and they usually fall to the earth at velocities greater than 10 feet (3 metres) per second in still air.

rain gauge A device for catching and measuring the depth of rainfall. Various kinds and sizes of rain gauges exist, most of them employing the principle of catching the rainfall in a collector that has a larger area than the measuring compartment, so that a given depth of water in the latter represents a considerably smaller depth of rainfall catch. A rain gauge is also called a pluviometer.

rain shutoff device A device connected to an irrigation controller that overrides scheduled irrigation when significant precipitation has been detected.

rainfall (1) A fall of rain; precipitation in the form of water. (2) The amount of rain, usually expressed in inches or millimetres depth of water on an area, that reaches the surface of the earth. The term sometimes also includes other forms of atmospheric precipitation, such as snow and dew, but technically only the term *precipitation* should be used in this broader scope.

rainfall area The geographic surface extent of a rainfall.

rainfall distribution coefficient A coefficient obtained for a storm by dividing the maximum rainfall at any point within the storm area by the mean rainfall in the drainage area or basin.

rainfall effectiveness The ability of rainfall to promote vegetation, governed by the amount of rain absorbed by the soil and thus made available to the plant roots. Controlling factors are the total amount and character of the rainfall, the rate of fall and seasonal distribution, the losses caused by runoff and evaporation, and the character and condition of the soil and of the vegetation.

rainfall frequency analysis A frequency analysis of point rainfall values made and published by the National Weather Service for rainfall durations of 5 minutes to 24 hours and for return periods ranging from 2 years to 100 years. *See also* return period.

rainfall index The average rainfall intensity in excess of which the volume of rainfall will equal the volume of observed runoff. The index represents the combined effects of interception and depression storage, as well as of infiltration. *See also* pluvial index.

rainfall infiltration (1) The process by which rainfall passes downward through the surface of the soil or rock. (2) That portion of the rainfall that passes downward from the surface into the soil or rock. Such infiltration is equal to the total precipitation less the losses caused by interception by vegetation, retention in the depressions on land surfaces, evaporation from all moist surfaces, and surface runoff.

rainfall intensity *See* rainfall rate.

rainfall intensity curve A curve that expresses the relationship between rates of rainfall and their duration. Each curve depicts the probability that a particular rainfall rate–duration combination will occur in any given year. The probabilities are expressed as return periods in years. A return period of 100 years means that the probability that a rainfall rate–duration combination will fall on the curve or lie above the curve in any given year is 1 percent. A return period of 2 years means that the probability that a rainfall rate–duration combination will fall on the curve or lie above the curve in any given year is 50 percent.

rainfall penetration The depth below the ground surface reached by a given quantity of infiltrating rainfall. The maximum extent of penetration is limited by the depth to groundwater.

rainfall province An area characterized by its similarity in terms of the distribution and seasonal concentration of precipitation.

rainfall rate The amount of rainfall occurring in a unit of time, generally expressed in inches or millimetres per hour.

rainfall simulator A device that provides simulated rainfall, primarily for use in determining infiltration on small areas. Some instruments permit the application of droplets of controlled size, simulating natural rainfall; others use sprays. A rainfall simulator is sometimes called a rain maker.

rainfall year A continuous 12-month period during which a complete annual cycle occurs. The start date for the cycle is arbitrarily selected based on the hydrologic data being studied. For example, rainfall is often quite low in mid-summer, and a hydrologist might select July 1 as the start of each rainfall year for a particular analysis. The US Geological Survey uses October 1 as the start of its annual cycle for the publication of stream flow records. *See also* climatic year; water year.

rain-fed stream A stream in which the water comes primarily from surface runoff resulting from rainfall.

ramping Controlling stream flow so that flow changes are gradual and thus do not adversely impact property and safety downstream.

random error In epidemiologic studies, an error that is governed by chance, as opposed to systematic error or bias. *See also* statistically significant.

random number Any number on a list of nonrepetitive numbers that does not follow any sequence or algorithm.

range The spread from minimum to maximum values over which an instrument is designed to measure. *See also* effective range; span.

range of values The difference between the highest and the lowest values in a database.

rangeability The ratio of the maximum flow rate to the minimum flow rate under which a meter can operate.

Ranney collector A water collector constructed as a dug well, 12 to 16 feet (3.5 to 5 metres) in diameter, that has been sunk as a caisson near the bank of a river or lake. Screens are driven radially and approximately horizontally from this type of well into the sand and gravel deposits underlying the river. *See also* radial-arm collector well.

rapid bench-scale membrane test (RBSMT) A membrane treatment study specified in the US Environmental Protection Agency's Information Collection Rule as promulgated in May 1996, whereby two membranes with molecular weight cutoffs less than 1,000 daltons are evaluated at four different recoveries (30, 50, 70, and 90 percent) while other operating parameters are held constant during each quarter of 1 year. The study uses a batch test employing flat-sheet membrane coupons in tangential flow cells and concentrate recycle. *See also* Information Collection Rule.

rapid granular filter A filter in which particulates are removed by granular media through which water flows, typically by gravity. A rapid granular filter typically operates at design filtration rate of approximately 2 gallons per minute per square foot (5 metres per hour) or higher. Some form of pretreatment typically is used to improve performance either by removing particles prior to filtration or by making them more amenable to removal by filtration. The term *rapid* was assigned when the filters were first being applied to contrast them with slow sand filters, which operate as the sole treatment step at filtration rates at least 10 times lower than those for rapid granular filters. *See also* slow sand filter.

rapid granular filtration The removal of particles by a rapid granular filter.

rapid mixing The process of quickly mixing a chemical solution uniformly through the water.

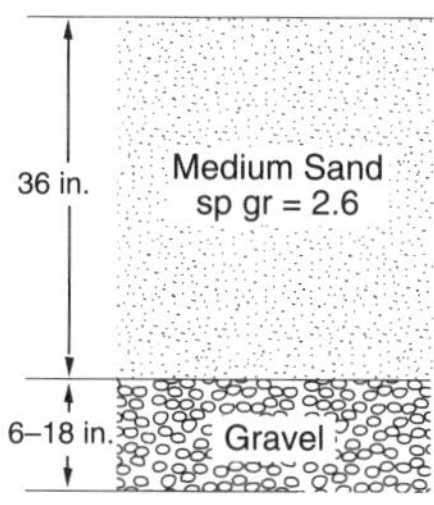

Rapid granular filter media

rapid sand filter A type of granular filter for which the granular material is sand. *See also* rapid granular filter.

rapid small-scale column test (RSSCT) A procedure used to predict the removal of contaminants by granular activated carbon adsorption at operating times much shorter than for pilot- or full-scale facilities. Breakthrough behavior of a contaminant can be predicted in a matter of days or weeks with such a test, as opposed to months for large-scale facilities. In this procedure, water is applied continuously to a small-scale column containing crushed granular activated carbon. The column typically has an inside diameter between 0.15 and 0.6 inches (4 and 15 millimetres). The length of the column depends on the empty bed contact time and the flow-through velocity but is typically between 4 and 10 inches (10 and 25 centimetres). *See also* empty-bed contact time; granular activated carbon.

rapids A part of a river where the current is turbulent and swift.

raster image A format for storing, processing, and displaying graphic data in which graphic images are stored as values for uniform grid cells or picture elements (pixels). *See also* picture element.

raster-to-vector conversion The process of identifying dots that represent line work, text, or symbols; separating these dots from the background; and converting them to a connected vector representation. This vector representation allows the information to be manipulated, edited, attributed, and stored for use in a geographic information system. *See also* geographic information system.

ratchet rate A type of water rate for which the charge for the highest rate block the customer reaches during a rate period applies to all water volume, as opposed to charging for each block at the rate designated for that block. This method, which is rarely used in the United States, is intended to deter high water use through the "retroactive" application of the highest block rate.

rate *See* water rate.

rate base The total of all allowable assets to be multiplied by an allowable rate of return to derive the earnings to a utility's owners. The allowable assets usually include the net book value of the utility plant in service, construction work in progress, and working capital (accounts receivable and inventories less accounts payable). Contributions in aid of construction are deducted from the total asset base because a return on contributed assets is not allowed to the owners. *See also* rate of return.

rate covenant A legal commitment by a bond issuer to maintain utility rates at certain levels necessary to generate a specified level of revenue, a specified ratio of revenue to debt service, or specified interest. Rate covenants can be designated on the basis of a ratio of either gross revenues or net revenue to debt service.

rate, crude *See* crude rate.

rate difference In epidemiology, the absolute difference between two rates, e.g., the incidence rate of morbidity or mortality for an exposed population minus the incidence rate for an unexposed population. This value has the same units of measure as the rates being subtracted. *See also* incident rate.

rate impact analysis An analysis of any program for its impact on water rates. This term is usually restricted to major programs amenable to a number of alternative approaches. For example, as many as a dozen new sources of water might be employed to meet a particular set of long-term water demand requirements. These alternatives could have widely differing costs per unit of supply. Hence, a rate impact analysis is conducted to determine which alternatives have the least impact on rates. This term also applies in the case of a high-cost single project; the intent of the analysis is simply to show the impact of the project on rates, with alternative combinations of sources of funds (e.g., rates versus connection charges, debt funding versus a pay-as-you-go approach).

rate making The process of analyzing rate requirements and setting rates for a future period. The term is somewhat archaic; current terminology includes setting rates, updating rates, and evaluating rate requirements. The process includes an analysis of total water revenue requirements, a cost-of-service analysis to determine who pays, identification of a fair return on assets invested (for investor-owned utilities), and identification of specific rates that will provide the required revenues in a manner fair to all ratepayers.

rate ratio In epidemiology, the ratio of two rates (e.g., the incidence rate of morbidity or mortality for an exposed population divided by the incidence rate for an unexposed population). *See also* relative risk.

rate of return The level of profit provided to water utility owners or stockholders as a return on their investment. In the utility method of rate development, which is used for most investor-owned or governmental utilities regulated by public service agencies, a return on investment or rate of return on investment is allowed on the total allowable rate base of the utility. The rate base generally includes the net assets of the firm; in some situations, projected assets for a test year are also allowed. This return on investment to the owners or stockholders can vary based on time, area, and the practices of the regulatory agency in a given area. The adequacy of the rate of return is usually based on cost-of-capital considerations; the rate is usually in the area of 10 percent. In the utility method, the return on capital is included in the rate requirements along with operations and maintenance costs and depreciation on assets. *See also* cash requirements method; utility method.

rate schedule The set of different rates and charges assigned to the various customer classes and customers.

rate stabilization fund Money set aside to provide a reasonable level of utility rate stability when major capital expenditures are planned. A very large rate impact can result from such expenditures even when the capital program is funded from debt. An increasingly popular approach for providing rate stability is to ramp up water rates 3 to 5 years prior to the rate impact, to maintain the excess receipts during the ramping period in a rate stabilization fund, and to use the stabilization fund during the early years of higher rate requirements until the current rates are in line with current rate requirements. This concept can be used not only to ramp up to a higher rate requirement, but also to bridge the gap between two or more major revenue-funded (not debt-funded) capital programs. Stabilization funds are also used to maintain debt-service coverage requirements, because the withdrawal of stabilization funds can generally be treated as revenues for debt-service coverage.

rate, standardized *See* standardized rate.

rate structure analysis Any of an entire array of analyses conducted to determine how rates are applied in terms of fixed monthly charges and water volume rates, as well as how the rates are applied to each customer group and area of the utility's total service area. Monthly service charges are typically designed to recover fixed customer costs, by meter size, that do not vary with volume. Volume charges can recover total costs related to the volume of water used in a single unit rate, or they can be broken down into base costs related to average load conditions and extra-capacity costs related to water use in excess of average use in order to allocate costs based on the peaking characteristics of different customer segments. The rate structure could be tiered based on volume to encourage conservation or to recover costs according to the incremental expenses associated with peak usage. No single method of designing a rate structure exists. In the design of a rate structure, the utility's policies and attitudes regarding growth, conservation, who pays, and fairness issues are every bit as important as conventional cost-of-service analysis. *See also* base–extra-capacity approach to rate structure design; cost-of-service analysis; inclining block rate.

rate tariff *See* rate schedule.

rate-controlling step The slowest elementary process in a sequence. The products of a sequence of processes can be formed no faster than the rate of the slowest step in the sequence. Therefore, if one of the steps in a sequence is much slower than all the others, the rate of the overall reaction will be limited by, and be exactly equal to, the rate of this slowest step.

rated capacity (1) In filtration or adsorption, the manufacturer's statement regarding the expected number of days the equipment will be in service or the expected volume of product water delivered before backwashing and rinsedown are expected to occur. (2) In softening or ion exchange, a value as stated by the manufacturer, given in relation to the period between regenerations: the expected number of days the equipment will be in service, the expected volume of product water delivered, or the weight of total hardness removed. The capacity of an ion exchange system varies, within limits, with the amount of regenerant used. *See also* rated softening capacity.

rated in-service life As specified by the manufacturer of a treatment unit, the total volume of treated water delivered or the length of time, based on water flow rates, the unit will be in operation before servicing (regenerating, cleaning, or replacement) of the unit is expected to be necessary.

rated pressure drop In softening or filtration, the expected pressure drop across a treatment unit as stated by the equipment manufacturer or obtained under test conditions. Rated pressure drop is based on the rated service flow rate for clean water at 60° Fahrenheit (15.5° Celsius) using a freshly regenerated softener or a backwashed filter. The general limit is 15 pounds per square inch (105 kilopascals) at the specified service flow rate for the unit.

rated service flow The manufacturer-specified maximum and minimum flow rates at which a particular piece of water treatment equipment will continuously produce the desired quality of water.

rated softening capacity As provided by a water softener manufacturer, the expected level of total hardness (in milligrams per litre or grains per gallon as calcium carbonate ($CaCO_3$)) that will be removed between regenerations at the specified flow rate and for the specified amount of regenerant (usually sodium chloride, NaCl). The capacity of an ion exchange resin to remove hardness increases, within limits, with higher regeneration salt dosages; therefore, the rated softener capacity must be related to the weight of salt required for each regeneration. *See also* rated capacity.

rate-determining step *See* rate-controlling step.

rate-of-flow controller *See* filter rate-of-flow controller.

rating In electricity, the limit of power, current, or voltage at which a material or piece of equipment is designed to operate under specific conditions.

rating agency An organization that ranks a utility's debt instruments based on an evaluation of the relative credit quality of that utility. The agency makes its initial rating after conducting a detailed evaluation of the financial status of the utility. The rating process often includes a live presentation by the utility and a "question-and-answer" session with the utility. The two most prominent rating agencies for municipal finance are Standard & Poor's Corporation and Moody's Investors Service. *See also* financial risk assessment.

ratio The relative size of two quantities expressed as the quotient of one divided by the other. A ratio may be expressed using colons (e.g., 1:2, 3:7), or it may be expressed as a fraction (e.g., 1/2, 3/7) or written out (e.g., 1 to 2, 3 to 7).

ratio control A feedforward control mode that causes the controller's output to always have a fixed ratio (e.g., 5 times, 1.5 times, 2.3 times) to the controller's input signal. *See also* pacing.

rational method In hydrology, the application of a simple formula to estimate the peak flow from a small drainage area. This method is used for the design of drainage facilities. The formula is

$$Q_p = C\,I\,A$$

Where:

Q_p = the peak runoff rate, in cubic feet per second

C = the runoff coefficient, in cubic feet per second per acre-inch hour

I = the average rainfall intensity lasting for a critical period of time (t_c), in inches per hour

t_c = the critical time period, called the time of concentration, in hours

A = the size of the drainage area, in acres

The runoff coefficients are determined empirically or read from a table that relates land use descriptions to values of the runoff coefficient. The formula should not be used for areas greater than 200 acres (810,000 square metres). Système International units are not used.

rational model A rate law to describe the decline in organism numbers during disinfection. It takes the following form:

$$r = -kC^nN^x$$

Where:

r = the rate of change of microorganism density, in number per unit time

C = the disinfectant residual concentration, in milligrams per litre

N = the initial density of microorganisms, in number per unit volume

k, n, x = constants

The constant x allows a description of the shoulders and tailing off of a curve of microorganism density versus time. *See also* Chick–Watson model; Hom model; modified Hom model.

Ra-226 + Ra-228 *See* radium-226 + radium-228.

raw water Water from the supply source prior to treatment. *See also* source water.

RBSMT *See* rapid bench-scale membrane test.

RCI *See* Riddick index.

RCRA *See* Resource Conservation and Recovery Act.

RDL *See* reliable detection level.

RDMS See relational database management system.

RDX *See* cyclonite.

Re *See* Reynolds number.

reach A portion of a canal or river in which the hydraulic conditions remain the same.

reactance The effect of inductance or capacitance that resists changes in current or voltage, respectively, in an alternating current system. It is measured in ohms.

reactant Any of the chemicals brought together in a chemical reaction. The chemical reactants are shown on the left-hand side of a chemical equation, with the reaction products on the right-hand side. *See also* product.

reaction force As determined by Newton's third law of motion, to every action (force) an equal and contrary reaction (force) always occurs, or the mutual actions of two bodies are always equal and oppositely directed along the same straight line.

reaction order An expression of the relationship between the rate of a reaction and the concentrations of the reactants. For example, a zero-order reaction does not depend on the concentration of the reactant, whereas in a single-order (first-order) reaction the reaction rate is directly proportional to the concentration of one of the reactants. In a double-order (second-order) reaction the rate of reaction is directly proportional to the product of two reactants or the square of the concentration of a single reactant.

reactivate To remove the adsorbed materials from spent activated carbon and restore the activated carbon's porous structure so that the activated carbon can be used again. The reactivation process is similar to that used to initially activate carbon. *See also* adsorption; granular activated carbon.

reactivation *See* reactivate.

reactive intermediate In toxicology, an unstable metabolite of a chemical that is reactive with biological molecules. Of particular concern are forms that bind to deoxyribonucleic acid in such a way that bases are mispaired and result in a cell containing a mutation following cell division. *See also* reactive metabolite.

reactive metabolite In toxicology, an intermediate that is produced in the metabolism of a compound and that readily reacts with such biological molecules as lipids, proteins, nucleic acids, or a combination. Such metabolites might be radicals if homolytic scission of a chlorine–carbon bond results. For example, oxygen is frequently inserted across double bonds to form an epoxide ($-(C_2O)-$). Acid chloride, episulfonium ions ($R-S^+-$

Reagent bottle

(C_2R_4)), carbanions ($R_3C{:}^-$), and carbenes ($R_2C{:}$) are other examples of reactive metabolites. *See also* reactive intermediate.

reactive power (*Q*) For a passive circuit, the exchange of stored energy between an inductor and a capacitor, reversing for each cycle. For sinusoidal quantities in a two-wire circuit, reactive power is the product of the voltage, the current, and the sine of the phase angle between them. For nonsinusoidal quantities, it is the sum of all harmonic components. In a polyphase circuit, reactive power is the sum of the reactive powers of the individual phases. Reactive power is customarily measured in volt-amperes-reactive.

reactor A vessel in which a chemical reaction or process (e.g., coagulation, flocculation, adsorption) takes place.

reactor–clarifier A device in which both flocculation and particle separation occur. The "reaction" occurs when water to which coagulants have been applied is flocculated in a conically shaped compartment, with mechanical mixing, in the center of the clarifier. The flocculated water flows outward into the settling zone, in which solids settle and clarified water is collected at the surface. The advantage of using reactor–clarifiers—as opposed to a separate flocculator and settling tank—is that area requirements for treatment can be reduced.

readily accessible equipment Equipment that can be cleaned and inspected without the need for tools. *See also* accessible equipment.

readily removable Capable of being taken away from a water treatment unit without the use of tools. *See also* removable.

reaeration Generally, a process that dissolves oxygen into water. For example, reaeration might involve the introduction of air through forced-air diffusers into the lower layers of a reservoir. As the air bubbles form and rise through the water, oxygen from the air dissolves into the water and replenishes the dissolved oxygen. The rising bubbles also cause the lower waters to rise to the surface, where oxygen from the atmosphere is transferred to the water. This latter phenomenon is sometimes called surface reaeration. *See also* aeration.

reagent A substance used in chemical analysis or synthesis. Reagents can be either chemicals or solutions that participate in a chemical reaction.

reagent blank A mixture of reagents used in an analytical method for checking the reagents' purity.

reagent bottle A bottle made of borosilicate glass fitted with a ground glass stopper, used to store standard chemical solutions.

reagent grade water Very high-purity water produced to meet the standards outlined by the American Society for Testing and Materials in their standard D1193-77, Standard Specification for Reagent Water. Four grade levels, types I through IV, are specified, and three levels of maximum total bacterial count, types A through C, are listed depending on intended use. Reagent grade water is used for chemical analysis and physical laboratory testing.

reagent water *See* reagent grade water.

real density The density of the skeleton of an activated carbon granule, as determined by measuring the amount of helium displaced by a known volume and mass of granular activated carbon.

reasonable use In the context of water rights, a water use that is acceptable in general terms.

reasonable use doctrine A water use concept stating that the user may use a reasonable amount of water. The water must not be taken in such quantity as to deny another user of the water, nor may it be taken to intentionally deny another user the water.

reauthorization An action taken by the US Congress to authorize federal funding for implementation of an existing federal law.

recalcination The heating of lime sludge to drive off water and carbon dioxide (CO_2), leaving only the calcium oxide (CaO) or quicklime plus some inert material and enabling the lime to be reused. Some lime sludge must be wasted periodically to avoid a buildup of inert material. *See also* lime; quicklime.

recarbonation The introduction of carbon dioxide (CO_2) into the water, after precipitative softening using excess lime for magnesium removal, to lower the pH of the water. *See also* excess-lime treatment.

receiver A device that indicates the value of a measurement. Most receivers in the water utility field use either a fixed scale and movable indicator (pointer), as on a pressure gauge, or a moving chart with a movable pen, as on a circular flow-recording chart. A receiver is also called an indicator.

receiving water All distinct bodies of water that receive runoff or wastewater discharges, such as streams, rivers, ponds, lakes, and estuaries.

receptor A specific area on the surface of a host cell that serves as a point for the attachment of viruses.

recessed impeller pump A centrifugal pump that has an impeller located in the end of the casing away from the suction inlet and in which the liquid does not go around the impeller. The vanes, which are located on only one side of the impeller, cause the liquid to rotate and apply force to the water as it leaves the outlet. The space between the inlet and impeller accommodates the passage of solids through the pump.

recession velocity *See* retreat velocity.

recharacterization In membrane treatment, the process of testing the performance of a membrane under specified conditions for comparison with performance at a previous time.

recharge (1) To add water to the groundwater supply via precipitation; by infiltration from surface streams, lakes, reservoirs, and snowmelt; or by utility pumping. (2) The process of such addition.

recharge area An area with downward components of hydraulic gradient in an aquifer. Water can infiltrate into deeper parts of an aquifer in a recharge area.

recharge basin A basin excavated to provide a way for water to soak into the ground at rates exceeding the natural infiltration rate.

recharge mound A rise of the water table that occurs beneath recharge basins during infiltration of water into the aquifer.

recharge rate The quantity of water per unit time that replenishes or refills an aquifer.

recharge water Water that replenishes the zone of saturation of an aquifer.

recharge well A well designed to allow water to flow *into* an aquifer to recharge the groundwater reservoir.

recharging *See* regeneration.

reciprocal A quantity inversely or oppositely related to another. For example, 1,000 ohms of resistance is the reciprocal of $1/1{,}000$ siemens of conductance; the fraction $4/3$ is the reciprocal of $3/4$.

reciprocating flocculator A device that imparts slow mixing to water by raising and lowering rods in an alternating fashion. Cone-shaped devices are mounted onto the rods and impart energy to the water, thereby creating a velocity gradient for flocculation. A reciprocating flocculator has the advantage of having the bearings out of the water and is also called a walking beam flocculator. *See also* flocculation.

reciprocating positive-displacement pump A pump in which motion and pressure are applied to a fluid by a reciprocating piston in a cylinder discharging a known quantity or volume of liquid with each stroke of the piston.

reciprocating pump A pump that applies pressure and motion to a fluid by the back-and-forth movement of a piston or plunger in a cylinder. An example is a windmill or pump jack that uses a sucker rod to apply the movement in the cylinder.

recirculation (1) In water treatment, the continuous operation of a transfer pump to keep water flowing through one or more unit processes at a rate in excess of the water flow through the entire plant in order to improve the operation of the unit processes. (2) In crossflow membrane filtration systems, the recycling of a portion of the reject stream to maintain a desirable flow across the membrane while the system is in operation.

recirculation of filter backwash A practice of returning filter backwash water to the head of the plant.

recirculation flow rate The flow rate at which retentate or concentrate is recirculated in a recirculation loop in some membrane processes. *See also* recirculation loop.

recirculation loop A flow in some membrane processes by which retentate or concentrate is returned to the suction side of the recirculation or recycle pump, blended with source or pretreated water, and returned as feed to the membrane process for treatment. The recirculation loop produces increased flow rates and velocities on the feed–retentate (concentrate) side of the membranes to minimize concentrations, material deposits, or both at the membrane surface.

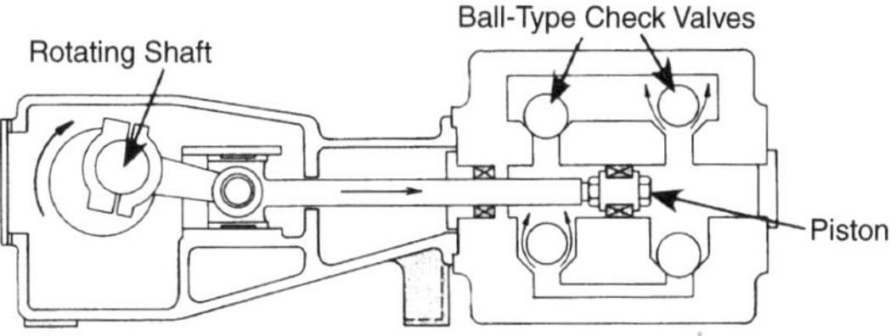

Double-action reciprocating pump

reclaim To recover something useful.

reclaimed brine That portion of a previously used brine solution used to regenerate a batch of cation resin for portable exchange softener tanks. Brine that still measures at least 30 percent saturation and is low in total hardness can be reused in the first stages of the next cation batch regeneration. *See also* sweet brine.

reclaimed wastewater Wastewater that has been treated and recovered for useful purposes.

reclaimed water *See* reclaimed wastewater.

reclaimed wetland A restored wetland. This term is often used in other parts of the world to refer to a wetland that was damaged by filling or draining.

reclassification *See* hydraulic classification.

recombinant deoxyribonucleic acid A strand of deoxyribonucleic acid that has had genes exchanged, relocated, or deleted from their original sequence.

recommended maximum contaminant level (RMCL) As established by the 1974 Safe Drinking Water Act, the contaminant concentration at which no known or anticipated human health effects occur, allowing for an adequate margin of safety. The recommended maximum contaminant level was renamed the maximum contaminant level goal (MCLG) by the 1986 Safe Drinking Water Act amendments. *See also* maximum contaminant level goal.

record of determination (ROD) A written record, maintained by a drinking water primacy agency, of a determination made by that agency concerning a public water system's compliance with applicable provisions of a National Primary Drinking Water Regulation.

record retention A comprehensive schedule that determines the necessary period of time for retaining records and that controls the ultimate disposition of records at the appropriate time.

recordable rate *See* incident rate.

recorder A device that creates a permanent record, on a paper chart or on magnetic tape, of the changes of some measured variable.

recovery In a membrane water treatment system, the fraction of the feedwater that is converted to permeate, filtrate, or product. In equation form,

$$\text{recovery}(\%) = \frac{Q_p}{Q_f} \times 100$$

Where:

Q_p = permeate, filtrate, or product flow rate or volume

Q_f = feedwater flow rate or volume

For microfiltration and ultrafiltration systems, recovery is often calculated as follows:

$$\text{recovery}(\%) = 1 - \frac{\text{volume of water wasted}}{\text{volume of source water used}} \times 100$$

Recovery is sometimes called permeate recovery, product water recovery, feedwater recovery, or conversion.

recovery method A procedure for aquifer evaluation that measures how quickly water levels return to normal after pumping.

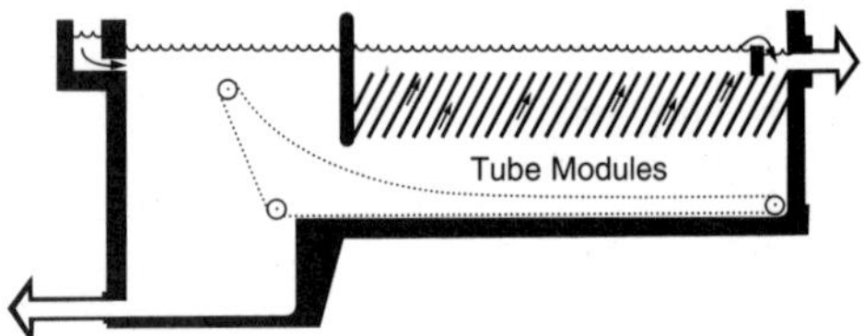

Rectangular clarifier with tube settlers

recovery rate *See* recovery.

recreation A beneficial use of a multipurpose impoundment or reservoir project for an outdoor activity, e.g., boating or fishing.

rectangular clarifier A treatment unit that is used for particle separation by settling and is rectangular in shape. Flow is introduced on one of the sides of shorter length and is removed on the opposite side.

rectifier A device that changes alternating electric current to direct current.

rectifier–ground–bed system *See* impressed-current system.

rectilinear flow Uniform flow in a horizontal direction.

recycle To use something over again.

red border review The formal mechanism by which US Environmental Protection Agency senior management (usually assistant regional administrators and the general counsel) review and approve regulations before presentation to the US Environmental Protection Agency administrator or other approving official. The red border review period generally lasts 3 weeks.

red tide Coastal surface waters of a reddish color caused by the presence of large numbers of red-pigmented, toxin-producing microorganisms. *Gymnodinium breve* and *Cochlodinium* spp., among other organisms, have been implicated in red tide events.

red water Rust-colored water. Such color is usually due to the presence of precipitated ferric iron (Fe^{3+}) salts or to dead microorganisms that depended on iron (Fe^{2+}) and manganese (Mn^{2+}). *See also* red tide.

REDEQL redox equilibrium An early generalized computer program for computing aqueous speciation and precipitation or dissolution reactions, written originally in the FORTRAN language for mainframe computers. The program was originally developed at the California Institute of Technology, and later development was supported by the US Environmental Protection Agency.

redox *See* oxidation–reduction chemistry; oxidation–reduction (redox) potential; oxidation–reduction (redox) reaction.

redox chemistry *See* oxidation–reduction chemistry.

redox couple *See* oxidation–reduction (redox) couple.

Redox Equilibrium *See* REDEQL.

redox potential (pE) *See* oxidation–reduction (redox) potential.

redox reaction *See* oxidation–reduction (redox) reaction.

reduced monitoring A provision in the US Environmental Protection Agency's Lead and Copper Rule allowing small and medium systems that meet the lead copper action level during each of two consecutive 6-month monitoring periods to reduce the number of samples, in accordance with the rule, and the frequency to once per year. *See also* Lead and Copper Rule.

reduced pressure zone (RPZ) device A backflow device that is typically installed on high-hazard installations, such as plating plants.

reduced-instruction-set computer (RISC) A computer with an operating system that executes complex commands very rapidly in one processing cycle, in contrast to conventional processing unit architectures that require many machine cycles to complete complex instructions. Although these computers offer increased performance for many applications, they require large main memories and special compilers to translate program source code into machine language. *See also* complex-instruction-set computer.

reduced-pressure principle backflow-prevention assembly (RPBA) In its approved form, an assembly consisting of two independently acting approved check valves together with a hydraulically operating, mechanically independent pressure differential relief valve located between the check valves and below the first check valve. These three items are located between two tightly closing, resilient-seated shutoff valves as an assembly and are equipped with properly located resilient-seated test cocks.

reducing agent Any chemical that decreases the positive valence of an ion. *Contrast with* oxidant.

reduction The counterpart of the process of oxidation. Reduction may involve (1) acceptance of one or more electrons by an atom or ion, (2) removal of oxygen from a compound, or (3) addition of hydrogen to a compound. *See also* oxidation.

reduction half reaction The electrical potential in volts (with the potential of the hydrogen electrode taken as zero volts) of a reduction reaction—e.g., $ClO^- + H_2O + 2e^- \rightarrow Cl^- + 2OH^-$—at equilibrium and under standard conditions with unit activity of the reactants. *See also* oxidation half reaction.

reductive Pertaining to reduction.

reductive dehalogenation Scission of a carbon–chlorine bond under reductive conditions, resulting in the formation of both a chlorine radical and a carbon-centered free radical.

reference A physical or chemical quantity for which the value is known exactly. A reference is used to calibrate or standardize instruments.

reference dose (RfD) A dose (or exposure) that has been determined to have a low probability of producing a harmful effect. It is arrived at by applying a series of uncertainty factors to a dose that produced no adverse effect or minimal adverse effects in studies conducted in animals or humans. The uncertainties adjust for uncertain differences in susceptibility between animals and humans or within species differences in animals. Additional safety factors are frequently applied to adjust for the adequacy of the data (e.g., to adjust for study duration) or the severity of the effect that the chemical produces. *See also* benchmark dose; lowest-observed-adverse-effect level; no-observed-adverse-effect level.

reference evapotranspiration (ET_o) The evapotranspiration of a broad expanse of well-watered, cool-season grass 4 to 6 inches (10 to 15 centimetres) tall.

refractory organic matter A stable by-product resulting from bacterial decomposition of a wide number of biochemical compounds essentially derived from cells of dead organisms.

refrigerator A cabinet that contains a heat exchanger and is used to store chemical solutions and preserve water samples, typically at temperatures near the freezing point of water.

reg neg *See* regulatory negotiation.

regenerant A chemical solution used to restore an exhausted bed of ion exchange resin to the fully ionic (regenerated) form necessary for the desired ion exchange to again take place effectively. Regenerants are also used in a similar fashion to restore catalyst media (e.g., for iron and manganese removal) for reuse. *See also* regeneration.

regenerate To restore the properties of a granular medium used for removing contaminants by surface interactions. For example, ion exchange resins are regenerated by passing through the medium a solution that exchanges ions of similar charge for those that have been removed. After regeneration, the medium is put back into service to remove the target contaminants.

regeneration (1) The periodic restoration of an ion exchange resin back to a usable form by employing a regenerant to displace ions removed during the treatment process. Ion exchange resins are regenerated by reversing the exchange reaction between the exchanging ions and the ions that are removed during treatment and retained on the resin. The regenerant typically contains a high concentration of the exchanging ions to drive the exchange reaction in a reverse direction from the normal service cycle. This process is also called recharging or rejuvenation. Catalyst media are recharged similarly. (2) The removal of the adsorbed materials from spent granular activated carbon, thus restoring the activated carbon's porous structure so it can be used again. The regeneration process is similar to that used to activate carbon. The process is usually called reactivation. *See also* reactivate.

regeneration cycle The sequence of steps—including backwash, application of regenerant, dilution, and fresh or deionized water rinse—necessary to regenerate an ion exchange bed or oxidizing filter.

regeneration efficiency In the regeneration of ion exchange resin, the degree to which regenerant ions are utilized on an equivalent basis to those ions replaced. *See also* regenerate; regeneration.

regeneration level The quantity of regenerant, usually expressed in pounds per cubic foot (kilograms per cubic metre) of ion exchanger bed or in pounds (kilograms) per regeneration, used in the regeneration cycle of an ion exchange system. The regeneration level may also be called the salt dosage.

regeneration rate The flow rate per unit area at which a regeneration solution is passed through an ion exchange bed of resin.

regeneration tank A large vessel, either gravity operated (open) or pressurized (closed), in which batches of ion exchange resin used in portable exchange tanks are regenerated. Some filter media, such as manganese zeolite, are also reprocessed in this manner.

regeneration water All of the water consumed in the regeneration steps: backwash, regenerant application (brining), dilution, and rinse. Source water or partially treated water may be used for rinsedown.

regional office An office of the US Environmental Protection Agency responsible for overseeing the implementation of regulations for several states or territories. US states and territories are divided among 10 regional offices (see table). *See also* US Environmental Protection Agency.

regional watershed The entire watershed within a region, including the land mass for which the runoff drains into the aquatic ecosystem.

register (1) A memory circuit that is contained in a computer's central processing unit. Registers are made up of system addresses for storing programming instructions, data, or addresses for other data used during the execution of a program. (2) The dial(s) on the face of a water meter for indicating the total amount of water that has passed through the meter.

regolith The unconsolidated mantle of soil material and weathered rock overlying solid rock.

regression analysis A statistical procedure in which an empirical relationship between an independent variable and a dependent variable can be determined such that the average tendency of the observed values and the average tendency of the predictive values given by the empirical equation will be identical. This procedure attempts to find the best mathematical model, given the available data, to describe a dependent variable as a function of an independent variable or, in the case of multiple regression analysis, more than one independent variable. The most common form of

USEPA regional offices

Region	Office Address	States Included
1	1 Congress Street Boston MA 02114 (617) 918-1990	Connecticut, Massachusetts, Maine, New Hampshire, Rhode Island, Vermont
2	290 Broadway New York NY 10007 (212) 637-3000	New Jersey, New York, Puerto Rico, Virgin Islands
3	1650 Arch Street Philadelphia PA 19103 (215) 814-2900	Delaware, Maryland, Pennsylvania, Virginia, West Virginia, District of Columbia
4	Atlanta Federal Center 61 Forsyth Street, SW Atlanta GA 30303 (404) 562-9900	Alabama, Florida, Georgia, Kentucky, Mississippi, North Carolina, South Carolina, Tennessee
5	77 W. Jackson Street Chicago IL 60604 (312) 353-2000	Illinois, Indiana, Ohio, Michigan, Minnesota, Wisconsin
6	1445 Ross Ave., Suite 1200 Dallas TX 75202 (214) 655-2200	Arkansas, Louisiana, New Mexico, Oklahoma, Texas
7	901 N. 5th Street Kansas City, KS 66101 (913) 551-7003	Iowa, Kansas, Missouri, Nebraska
8	One Denver Place 999 18th St., Suite 500 Denver CO 80202-2466 (303) 312-6312	Colorado, Montana, North Dakota, South Dakota, Utah, Wyoming
9	75 Hawthorn Street San Francisco CA 94105 (415) 744-1500	Arizona, California, Hawaii, Nevada, American Samoa, Guam, Trust Territories of the Pacific
10	1200 Sixth Ave. Seattle WA 98101 (206) 553-1200	Alaska, Idaho, Oregon, Washington

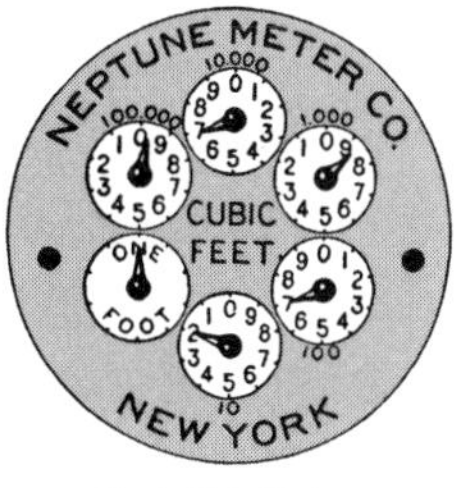

Circular Meter Register Reading in Cubic Feet

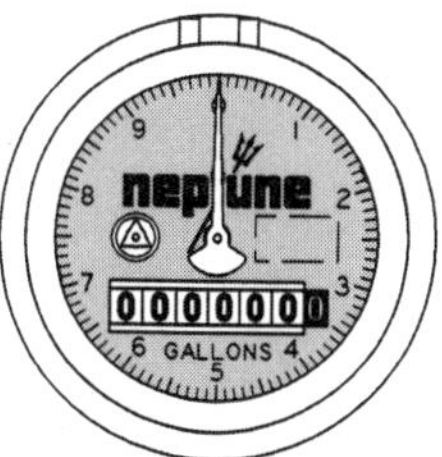

Straight Meter Register Reading in Gallons

Water-meter registers

model is linear regression analysis. In epidemiology, the dependent variable is usually the probability or odds that a disease for which the variation is sought will be explained or accounted for by the influence of one or more independent variables (i.e., one or more characteristics or events that may cause the disease or contribute to its variation). The logit model, a linear model for the natural logarithm of the odds of disease, is often used. *See also* coefficient of determination; logit model; multiple linear regression analysis; natural logarithm.

regression coefficient In regression analysis, a derived coefficient (usually found through the method of least squares) that expresses the change in the dependent variable associated with a change in one or more independent variables. The regression coefficient is also referred to as the slope of the relationship between the variables, as the derivative (in bivariate analysis), or as the partial derivative (in multiple regression). *See also* multiple linear regression analysis; regression analysis.

regrowth A posttreatment increase in the concentration of culturable heterotrophic bacteria in distribution water as measured by a plate count or heterotrophic plate count method. The terms *aftergrowth* and *regrowth* are often used interchangeably. *See also* aftergrowth.

regulated flow (1) The controlled movement of water in a conduit, through an orifice, or through an open channel across a weir for such purposes as flow measurement or control. (2) Surface water flow in natural watercourses that is modified by outlet controls in a dam.

regulation A rule or order issued by a federal, state, or local government agency having the force of law.

regulator (1) An employee of a federal, state, or local regulatory agency. (2) A device for controlling flow, movement, pressure, and so forth. *See also* flow regulator; pressure regulator.

regulatory agency A government organization or body that creates and promulgates enforceable regulations and standards.

regulatory agenda An agenda prepared by the US Environmental Protection Agency and published in the *Federal Register* that lists all of the final, proposed, and anticipated rulemakings since the previous update.

regulatory flexibility analysis (RFA) An analysis required by the Regulatory Flexibility Act of 1980. A federal agency (e.g., US Environmental Protection Agency) must prepare a regulatory flexibility analysis for all rules that are subject to the notice and comment procedures of the Administrative Procedure Act and that have a significant impact on a substantial number of small entities. The Regulatory Flexibility Act defines a small entity as any city, county, town, township, village, school district, or special district that has a population of fewer than 50,000 people. The regulatory flexibility analysis identifies regulatory alternatives for small entities and describes the objectives and legal basis of a rule; the benefits derived; the characteristics of the affected industry with analysis of demographics, cost, competitive effects, and exemptions or allowances; the record-keeping, reporting, or other compliance requirements; and other regulations that may duplicate, overlap, or conflict with the proposed rule.

regulatory framework The technical and legal structure within which new regulations are developed.

regulatory guidance letter (RGL) A letter issued by a regulatory agency to provide guidance or clarification on one or more issues related to the implementation of a rule.

regulatory impact analysis (RIA) A study of the economic impacts of a proposed or final rule, as required under Executive Order 12291. This executive order classifies federal agency proposed and final rules into two categories: major and nonmajor. Major rules are those having significant impact on society, either in terms of costs (more than $100 million annually) or in terms of effects on employment, international trade, or an industrial sector. A regulatory impact analysis is prepared for major rules, and an economic impact analysis is prepared for nonmajor rules. *See also* economic impact analysis.

regulatory negotiation (reg neg) A federal agency rulemaking procedure authorized by the Negotiated Rulemaking Act of 1990 whereby an advisory committee is formed under the Federal Advisory Committee Act to negotiate the provisions of proposed rule. *See also* negotiated rulemaking.

regulatory stage One of a series of rules and regulations—frequently sequenced in time—that are part of an overall regulatory package. For example, the Disinfectant/Disinfection By-Products Rule has Stage 1 and Stage 2 regulations. *See also* Disinfectant/Disinfection By-Products Rule.

reject *See* reject stream.

reject solution *See* reject stream.

reject staging A method used to raise the production rates of a reverse osmosis system by using the reject water from the first stage as the feedwater for the second or succeeding stage of the process.

reject stream For a pressure-driven membrane process, the concentrated solution containing substances that do not pass through the membrane. The reject stream is also called the concentrate or the concentrate stream. For some reverse osmosis systems, it is also called the brine or the brine stream.

reject water *See* reject stream.

rejection In a pressure-driven membrane process, a measure of the membrane's ability to retard or prevent passage of solutes and other contaminants through the membrane barrier. In equation form,

$$\text{rejection \%} = \frac{C_f - C_p}{C_f} \times 100$$

Where:

C_f = feedwater concentration, in milligrams per litre

C_p = permeate (product) concentration, in milligrams per litre

rejection rate *See* rejection.

rejuvenation *See* regeneration.

relational database A database structure commonly used in a geographic information system in which data is stored based on two-dimensional tables where multiple relationships between data elements can be defined and established in an ad hoc manner. This structure allows great flexibility in the range of queries that can be handled and is particularly useful in situations where nonstandard inquiries are common. *See also* geographic information system.

relational database management system (RDMS) A system for managing information via a database that is organized and accessed according to relationships between data items. In its simplest conception, a relational database is actually a collection of data files that "relate" to one another through one or more common links.

relative density The ratio of the density of a specific substance to the density of another substance that is used as a standard. The standard for comparison with liquids is pure water at 39.2° Fahrenheit (4° Celsius). The standard for gases is air at normal temperature and pressure. Relative density is also called specific gravity. *See also* specific gravity.

relative humidity The actual amount of water vapor in a given volume of air at a definite temperature divided by the maximum amount of water vapor that would be present if the air were saturated with water vapor at that temperature expressed as a percentage.

relative risk The ratio of the risk of disease or death among an exposed population to the risk among an unexposed population. A value of 1 indicates no association or no increased risk caused by the exposure or risk factor being studied; a value greater than 1 indicates an increased risk, provided the association is not subject to systematic or random error. For example, a relative risk of 2.0 in a study of cancer risk in a population exposed to a drinking water contaminant indicates that the risk of cancer is doubled among those exposed to the contaminant compared to those not exposed. *See also* odds ratio; rate ratio; risk indicator.

relative source contribution The contribution of drinking water to human exposure of a contaminant, relative to all other sources of exposure. For example, a contaminant for which drinking water contributes 20 percent of the total exposure would have a relative source contribution of 20 percent, or 0.2.

relative standard deviation (RSD) The measure of the variability of a population (i.e., the standard deviation) divided by the mean of the population, often expressed as a percentage.

relay A device activated by an electrical or physical condition to cause some other device in an electrical circuit to operate.

reliable detection level (RDL) As proposed in the early 1990s by the American Chemical Society and the US Environmental Protection Agency, the lowest level at which qualitative decisions should be made about a single sample result. The reliable detection level has been proposed to be set equal to twice the value of an interlaboratory method detection limit. At the reliable detection level, the probability of false positives and false negatives is equal and low. *See also* limit of detection; limit of quantitation; method detection limit; practical quantitation level; reliable quantitation level.

reliable quantitation level (RQL) As proposed in the early 1990s by the American Chemical Society and the US Environmental Protection Agency, the lowest level at which regulatory decisions should be made on a single sample result. The reliable quantitation level has been proposed to be set equal to four times the interlaboratory method detection limit. If used by regulatory programs, the RQL could replace the practical quantitation limit concept. *See also* limit of detection; limit of quantitation; method detection limit, practical quantitation limit; reliable detection level.

relief A land surface's elevations or inequalities, considered collectively.

Components of a remote electronic meter reading system

reliquefaction The return of a gas to the liquid state, e.g., a condensation of chlorine gas to return it to its liquid form by cooling.

reluctance A measure of the opposition presented to magnetic flux in a magnetic circuit, analogous to resistance in an electric circuit.

rem *See the Units of Measure list.*

remote communication Communications among devices that are physically remote from the central computer site and rely on commercial, private, or government communication facilities, such as a telephone system, fiber optic channels, microwave transmission, or satellites for the transmission of data.

remote electronic meter reading (REMR) The process by which the output from an encoded register meter is captured through a probe. The probe is attached to a hand-held data entry terminal and applied to a receptacle on the outside of the customer's premises. Sometimes the signals are transmitted directly to the utility via telephone lines, cable systems, or radio frequencies. *See also* automatic meter reading.

remote metering *See* remote electronic meter reading.

remote sensing The analysis and interpretation of images gathered through techniques that do not require direct contact with the subject. A discipline that evolved from photogrammetry, remote sensing of the earth's resources uses aerial or space photographs, electronic scanners, and other devices to collect data about the earth's surface and subsurface.

remote terminal unit (RTU) A microcomputer-based field device that interconnects the supervisory control and data acquisition system's master station computer to water system instrumentation and control relays. The remote terminal unit also performs logical control functions. *See also* programmable logic controller.

remote-control valve An electric solenoid valve that is wired to an irrigation controller and controls the flow of water to an irrigation circuit.

removable Capable of being taken away from a water treatment equipment unit by the use of simple tools, such as a screwdriver, pliers, or open-end wrench. *See also* readily removable.

REMR *See* remote electronic meter reading.

renal Pertaining to the kidney.

renal clearance The rate at which a substance is removed from the blood. In pharmacokinetics, all clearance terms are expressed in units of volume per unit time, corresponding to the volume of blood from which a chemical has been completely removed in the specified time interval. This particular measure is arbitrary because chemicals are not generally completely removed from any given volume of blood in a single pass.

renal excretory mechanism Either of two general mechanisms by which chemicals are removed from the body by the kidney. The first is filtration in renal glomeruli, and the second is tubular secretion. Tubular secretion may involve facilitated or active transport of the chemical from the interstitial space surrounding the renal tubule into the lumen of the tubule. The main determining factor for a chemical's renal excretion rate by the kidney is the chemical's polarity. If the chemical carries a charged group at physiological pHs, it will be effectively filtered or secreted and very efficiently excreted by the kidney. If the chemical is non-polar, it will be reabsorbed from the kidney tubule by passive diffusion as water is removed from the tubule because of decreases in osmotic pressure caused by the active transport of sodium out of the tubule. The removal of water out of the tubule creates a concentration gradient for the chemical to be readsorbed.

renal function The ability of a person's kidneys to efficiently form urine and secrete waste products from the body and to reabsorb or salvage such nutrients as glucose or amino acids. *See also* renal excretory mechanism.

renal tubule The microscopic structure that leads from the glomerulus, or site of filtration of the blood in the kidney, to the urine-collecting ducts. The ultrafiltrate of blood becomes concentrated as sodium is actively transported out of the tubule just downstream from the glomerulus (i.e., within the proximal tubule); this transport process in turn leads to a massive reabsorption of water because of the reduced osmotic pressure with the tubule. In this same region various useful substrates (e.g., glucose, amino acids, and low-molecular-weight proteins) are reabsorbed. The renal tubule is an important target of many chemicals that are toxic to the kidney.

renal tumor A tumor of the kidney.

renewable resource *See* natural resources.

renewable water supply *See* available fresh water.

Reovirus A respiratory enteric orphan (hence "reo") virus. A Reovirus is a double-stranded ribonucleic acid virus and may be isolated in the absence of known disease. Reovirus is known as an orphan virus because, at this writing, it is not associated with any disease. It is found in the respiratory and enteric tracts of humans and many animals.

repeat compliance period Any subsequent compliance period after the initial compliance period.

repeat sample A sample taken to confirm the presence of a contaminant that was detected in a previous sample.

repeatability control The accuracy and repeatability of a number of consecutive output measurements for the same input value under the same operating conditions over the full-scale operating range of the instrument or field device. For example, good repeatability control would be present when a 50 percent signal produces a 50 percent valve open position under all operating conditions.

replaceable component *See* disposable component.

replicate sample (1) Any of multiple samples taken simultaneously. (2) Any of multiple samples taken at the same place sequentially at specified intervals over a specific period of time.

representative sample (1) A portion of water or material that is, as nearly as possible, identical in content and consistency to that in the larger body of water or material being sampled. (2) In water treatment equipment manufacturing and testing, a typical production line sample that exhibits the essential features corresponding to and equivalent to (within plus or minus 10 percent) the other production units of that model.

representative site A location chosen to be typical of a given environment.

reproductive test A toxicological procedure designed to determine whether a chemical affects reproductive function. Such tests include measures of both male and female reproductive function. At the most elemental level, these tests attempt to determine whether the chemical affects the ability of sperm to fertilize and the ability of the female to conceive and maintain a pregnancy. *See also* reproductive toxicity.

reproductive toxicity Any effect of a chemical that would prevent normal reproduction of a species. In humans and other mammals this includes effects on the development and competency of sperm to fertilize an oocyte (or egg), on the development of an oocyte that is fertile, and on the capacity of the female to maintain pregnancy. Reproductive toxicities can also relate to the ability to perform sexually. Such failure could be related to neurotoxicity.

required fire flow The minimum fire flow or amount of water to be delivered from a fire hydrant at 20 pounds per square inch (138 kilopascals) near a building or structure, as established by city fire prevention and protection codes. This flow is generally based on the size, use, and construction method of the facility to be protected.

research (1) Investigation or experimentation aimed at the discovery and interpretation of facts, revision of accepted theories or laws in light of new facts, or practical application of such new or revised theories or laws. The first two types of investigations are often referred to as basic research, whereas the third is often referred to as applied research. (2) The use of information-gathering techniques to confirm assumptions or hunches, to clarify questions for which limited or contradictory information is available, or to reorient beliefs.

reservoir An impounded body of water or controlled lake in which water can be collected and stored. *See also* impoundment.

reservoir of infection Any person, animal, arthropod, plant, soil, or substance in which an infectious agent normally lives and multiplies, on which that agent depends primarily for survival, and where it reproduces so that it can be transmitted to a susceptible host.

reservoir routing For design and planning purposes, methods for evaluating how a flood wave is altered as it passes through an impounding reservoir.

reservoir safe yield *See* impoundment safe yield.

reservoir sedimentation The accumulation of sediment in impoundment reservoirs, eventually resulting in a reduction of both storage space and safe yield.
reservoir surcharge Water that is above the spillway height in a reservoir.
reservoir tank *See* storage tank.
reservoir trap efficiency The ability of an impounding reservoir to retain sediment, expressed as the percentage of incoming sediment retained in the impoundment.
reset control *See* integral action.
residential equipment Small water-processing equipment that has been designed primarily for home use and for intermittent household water flow rates up to 12 gallons (45 litres) per minute. *See also* Water Quality Association.
residential retrofit A modification or change of a residential service to replace the existing service or components, such as replacement of a lead service with a copper or plastic service, changing a tapered curb stop with a ball valve, or raising a meter in the meter box.
residential water conservation The wise use of domestic water in the home and on the premises by families and individuals to reduce overall water use. The process may be voluntary or mandatory.
residential water use The portion of a community water system used in the home and on the premises by families and individuals for drinking, cooking, cleaning, and other domestic purposes.
residual chlorine The concentration of free available chlorine remaining after a given contact time under specified conditions.

Types of residual chlorine and their effectiveness

Type	Chemical Abbreviation	Estimated Effectiveness Compared With HOCl
Hypochlorus acid	HOCl	1
Hypochlorite ion	OCl^-	1/100
Trichloramine*	NCl_3	†
Dichloramine	$NHCl_2$	1/80
Monochloromine	NH_2Cl	1/150

* Commonly called nitrogen trichloride.
† No estimate; possibly more effective that dichloramine.

residual disinfectant concentration (*C*) The concentration of a disinfectant after a given contact time. In the Surface Water Treatment Rule, disinfection credit is based on achieving specified $C \times T$ values, where C is the concentration of the disinfectant in milligrams per litre determined before the water reaches the first customer, and T is the corresponding contact time in minutes. *See also* $C \times T$; contact time; disinfectant; disinfection; Surface Water Treatment Rule.
residual flow control A method of controlling a chlorine feed rate based on the residual chlorine concentration after the feeder.
residual pressure The pressure on a water main while water is flowing. In general, it is the static pressure minus the pressure loss due to line loss or friction.
residual rain Rain falling near the end of a storm at a rate less than infiltration capacity. *See also* initial rain.
residual of regression The magnitude of the difference between an actual observed value and the value predicted by the regression equation. The magnitude of the summation of the squares of these residual values determines the goodness of fit of the regression equation. *See also* goodness of fit.

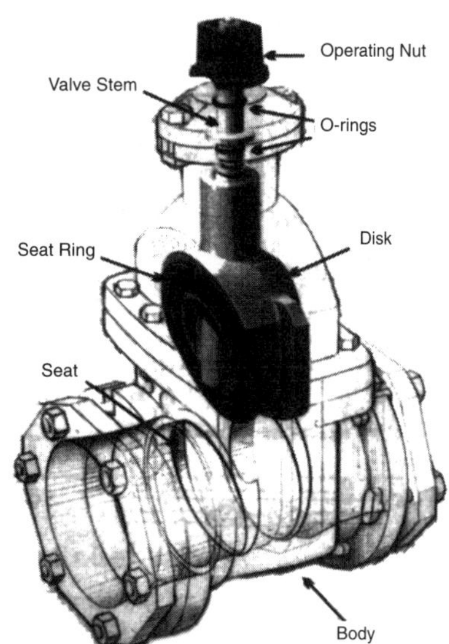

Resilient-seat gate valve

residuals Any gaseous, liquid, or solid by-product of a treatment process that ultimately must be disposed of. For example, in a fixed-bed filter for removing particles from water, both the filter backwash water and the solids in the backwash water are residuals. A residual is often called a sludge.
residuals blowdown *See* sludge blowdown.
residuals cake *See* sludge cake.
residuals conditioning *See* sludge conditioning.
residuals dewatering *See* sludge dewatering.
residuals disposal *See* sludge disposal.
residuals lagoon *See* sludge lagoon.
residuals management The practice of coordinating the handling and disposal of solid material, liquid streams, gaseous emissions, or any combination thereof generated in a treatment process. Residuals management is important because specific regulatory requirements for the methods of disposal for specific types of materials often exist. The costs associated with these regulatory requirements can often significantly affect the selection of the process or processes to be used to remove contaminants.
residuals processing *See* sludge processing.
residuals-drying bed *See* sludge-drying bed.
residue A material that remains after gases, liquids, and some solids have been removed, often by heating the sample for a specified period of time at a specified temperature.
resilience The ability of a system (e.g., an ecosystem) to recover from or resist stress.
resilient-seat gate valve A gate valve that uses a flexible solid disk for closure, such that the disk presses against the body of the valve in the flowway.

Respirator

resin In the water-processing industry, an ion exchange resin product (usually in the form of specifically manufactured organic polymer beads) used in softening and other ion exchange processes to remove dissolved salts from water. *See also* ion exchanger.

resin degradation *See* degradation.

resin regeneration *See* regenerate; regeneration.

resistance A measure of a material's ability to impede the flow of electricity. Resistance is measured in ohms. *See also* Ohm's law.

resistivity A measure of the ability of a solution or any other bulk material to retard electrical current. Resistivity is inversely proportional to conductance and is commonly used in highly purified water treatment systems to quantify the degree of process water demineralization. Resistivity is measured in ohm-metres.

resistivity meter A battery-powered instrument used to measure the average resistance to electrical current flow in a medium.

resonance A mathematical concept used to describe or express the true chemical structure of certain compounds that cannot be accurately represented by any one valence–bond structure. The resonance concept was originally applied to aromatic compounds, such as benzene (C_6H_6), for which many approximate structures exist, none of which is completely satisfactory. This concept indicates that the actual molecular structure lies somewhat between these various approximations. *See also* aromatic hydrocarbon; aromaticity; benzene.

Resource Conservation and Recovery Act (RCRA) An amendment to the Solid Waste Disposal Act that created a "cradle-to-grave" regulatory system for hazardous waste. Enacted in 1976 (Public Law 94-580), the act requires generators, transporters, and disposers to maintain written records of waste transfers, and it requires the US Environmental Protection Agency to establish standards, procedures, and permit requirements for disposal.

resource recovery The reuse of natural resources consumed in a process. For example, waste heat from a boiler can be used to heat water, or waste steam can be used to power turbine generators.

respirable size particulate An airborne contaminant that, when inhaled, is small enough to go deep into the lungs.

respiration The process by which a living organism takes in oxygen from the air or water; uses it in oxidation, thereby producing energy; and gives off the products of oxidation, especially carbon dioxide (CO_2). Breathing is an example of respiration.

respirator A mask or filter that prevents inhalation of airborne contaminants.

response The behavior of a system or any of its parts in reaction to a disturbance.

restoration of wetlands Actions performed to establish nontidal wetlands on former wetland sites.

restored wetland A wetland returned from a disturbed or altered condition to a previously existing natural or altered condition by human action (e.g., fill removal).

restructuring The process by which a small system that is unable to achieve and maintain compliance with the drinking water regulations may choose to delegate some or all of its responsibilities to another entity, such as a public utility, satellite management organization, regional water authority, or unit of local government. This process may be achieved by a change of ownership, by receivership or regulatory takeover where provided by law, by contracting for those services that the small system cannot carry out on its own, or by forming voluntary partnerships with nearby water systems.

retained water The water remaining in a soil after gravity drainage.

retainer gland A device that holds a mechanical component in place, such as packing around a pump shaft. *See also* gland; packing gland.

retentate For a pressure-driven membrane process, the portion of feed solution that is rejected by the membrane. Retentate is also called concentrate.

retention (1) The ability of a membrane to retard the passage of a substance through the membrane barrier. (2) That part of the precipitation falling on a drainage area that does not escape as surface stream flow during a given period. It is the difference between total precipitation and total runoff during the period, and it represents evaporation, transpiration, subsurface leakage, infiltration, and, when short periods are considered, temporary surface or underground storage on the area.

retention capacity The maximum precipitation retention capability under given conditions. *See also* retention.

retention coefficient A dimensionless measure of the degree of membrane retention of a substance, calculated by dividing the substance concentration in the permeate by its concentration in the retentate and subtracting the ratio from 1.

retention period *See* detention time.

retention pond A basin, usually enclosed by an artificial dike, that is used for temporary storage of water treatment plant residuals.

retention time *See* detention time.

reticulocyte A newly synthesized red blood cell (erythrocyte). The name comes from the reticulum (network) that displays basophilic staining with hemotoxylin ($C_{16}H_{14}O_6$).

retinopathy Any disease of the retina (excluding inflammatory conditions, such as those that would be produced by infectious diseases that affect the eye). For example, methyl mercury (CH_3Hg) causes a retinopathy that affects vision.

retreat velocity The mean velocity of water flowing in a stream channel or conduit immediately downstream from a structure over or through which such water flows. Retreat velocity is also called recession velocity.

retrofit (1) A process of changing, altering, or adjusting plumbing fixtures to save water. (2) To perform such a process.

retrofit kit Prepackaged hardware and information for distribution in a plumbing retrofit program. Kits usually contain replacement showerheads or flow restrictors, toilet tank dams or displacement bags, instructions, and literature on water conservation.

retrogression (1) A lowering of the bed level of a river. (2) A drop in the water level at the same discharge. *See also* aggradation; stream.

retrospective analysis In epidemiology, an analysis in which an association about exposure and disease is determined from historical or past information about the study participant's exposure characteristics, risk factors, events, or experiences, the disease having already occurred in the study population. A case-control study is sometimes referred to as a retrospective study or analysis. A cohort study can be either retrospective and prospective in nature. An ecologic study is almost always retrospective. *See also* case-control epidemiologic study; cohort study; ecologic epidemiologic study.

return current path The metallic connection between the anode and cathode of an electrochemical cell.

return flow *See* return water.

return offset In plumbing, a double offset installed to return the pipe to its original alignment. *See also* offset.

return on investment (ROI) A measure of an owner or shareholder's profit from an investment. A number of return-on-investment concepts have relevance in public utilities, e.g., the return on assets, which measures net profit as a percentage of net assets; the discounted cash flow–return on investment approach, which derives the return rate that equates discounted cash flows for expenditures and receipts over a project life (identical to the internal rate of return function in spreadsheet programs); and the return on investment on the rate base that is allowed for investor-owned and regulated utilities. All of the return-on-investment concepts measure profit or after-tax cash flows as a return on investment, either for the utility as a total entity or for a selected project within the utility. *See also* rate of return.

return period A way of expressing the probability that a particular value of a hydrologic variable will be equaled or exceeded in a given time period. For example, if a flow of a particular magnitude has a return period of 100 years, then the probability in any single year that this magnitude will be equaled or exceeded is 1 percent (a relatively rare occurrence). If the flow of a particular magnitude has a return period of 2 years, then the probability in any single year that that magnitude will be equaled or exceeded is 50 percent (a relatively common occurrence). The numerical values of the probabilities are obtained as the ratio of the time base to the return period. Thus, an event with a return period of 100 years occurs with probability of 0.01 (1 year/100 years = 0.01).

return seepage Water that percolates from canals and irrigated areas to underlying strata, raising the groundwater level, and that eventually returns to natural channels.

return water (1) Water that has been diverted from a stream or other body of water for irrigation purposes and that, not having been consumed by evaporation or transpiration, passes directly back to a stream or other body of water or downward to the water table and ultimately reaches a surface stream or other body of water. (2) Water that is diverted from a stream for industrial plant use and returned to the stream.

reuse The application of appropriately treated wastewater to a constructive purpose. *See also* reclaim; recycle; water reuse.

reuse water Wastewater treated to be made useful.

revenue requirement An amount of income needed by a utility for a specific purpose. The total revenue requirements of a utility are the sum of all the costs for operations and maintenance, for revenue-funded capital expenditures, for debt service on debt-funded capital expenditures, and for a return on investment (for investor-owned utilities) that are required to provide service to its customers. The revenue requirements are met primarily from water rates after nonrate revenues are subtracted from the total revenue requirements. Nonrate revenues consist of such items as interest earnings on fund balances, taxes, connection and installation charges, and use of cash reserves.

revenue stream All sources of revenue to a utility, including rates and charges, connection charges, taxes, miscellaneous income from rentals and minor property sales, and any other cash receipts. Proceeds from loans or bond issues are not included in the revenue stream because they are considered a direct source of funds for major capital programs rather than a revenue.

revenue-funded capital expenditures Capital expenditures in a utility that are paid for from rate revenue (often called user charges). These expenditures are contrasted with debt-funded expenditures, which are paid for from the proceeds of a debt instrument, such as general obligation bonds, revenue bonds, assessment district bonds, or certificates of participation (lease bonds). Capital programs that are revenue funded are usually smaller improvement programs, replacement programs, furniture and fixtures, vehicles, and computers. However, some utilities prefer to not use debt funding except for very large projects. These utilities accumulate cash reserves from rates or user charges over a period of years and then pay cash for major facilities. *See also* debt-funded expenditure.

revenue-neutral rate A new rate or (frequently) a series of rates by rate block that will produce the same revenue as the rates that are being replaced, typically at a reduced volume of sales. This concept is most frequently employed when inclining block rates are implemented to replace a single unit rate or to replace existing block rates with different break points or higher prices. It is, however, applicable to any price change. The primary consideration is that the new rates will have a price response (elasticity) such that reduced volume and lower total revenue could result, depending on the elasticity of demand. Evaluations of the potential responses to higher rates are usually made when rate structures are changed or existing rates are increased. The goal in rate design is to have a revenue-neutral or higher (if additional revenue is required) response.

reverse curve A curve shaped like an elongated *S*.

reverse deionization The use of the anion exchange resin ahead of the cation exchange resin (the reverse of the usual order) in a deionization system.

reverse flow Flow in a direction opposite to the normal flow.

reverse mutation A mutation that returns an organism to the wild (i.e., natural) phenotype. Reverse mutation assays are frequently used in mutagenesis testing because of the ease with which reverse mutations can be detected. Organisms that have spontaneously mutated or have been genetically engineered to have a specific mutation at a specific locus within a gene to make them dependent on the presence of a particular nutrient are tested after they have been treated with the test chemical to

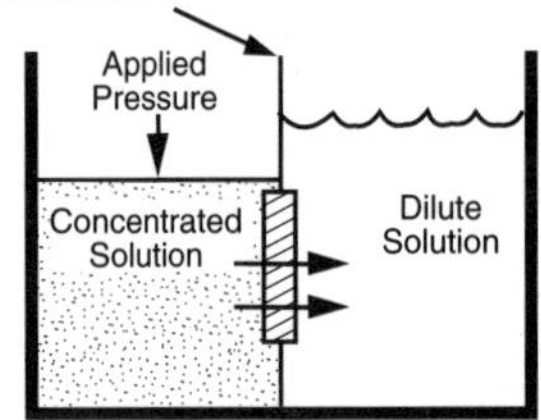

The principle of reverse osmosis

determine whether they are able to grow in the absence of the particular nutrient. *See also* Ames test; mutagenesis.

reverse osmosis (RO) A pressure-driven membrane separation process that removes ions, salts, and other dissolved solids and nonvolatile organics. The separation capability of the process is controlled by the diffusion rate of solutes through a membrane barrier and by sieving; it is dependent on the membrane type. In potable water treatment, reverse osmosis is typically used for desalting, specific ion removal, and natural and synthetic organics removal. It is no longer commonly called hyperfiltration. *See also* desalting; membrane; membrane softening; nanofiltration; pressure-driven membrane; ultrafiltration.

reverse osmosis membrane A synthetic membrane used for reverse osmosis separation. *See also* membrane; reverse osmosis.

reverse phase technique The liquid chromatographic analysis of nonpolar compounds. Analytes are separated based on differences in how they adsorb onto a nonpolar stationary phase. A polar solvent elutes the compounds from the chromatographic column. Reverse phase techniques are the most common type of liquid chromatography. *See also* liquid chromatography.

reverse transcriptase An enzyme that is coded for by retroviruses and is able to make a complementary deoxyribonucleic acid chain on the ribonucleic acid template and then convert this deoxyribonucleic acid chain to the double-stranded form. *See also* polymerase chain reaction.

reverse transcriptase–polymerase chain reaction (RT–PCR) A method of detection for ribonucleic acid (RNA) viruses. RNA is reverse transcribed to a complementary strand of deoxyribonucleic acid that is then amplified for detection by standard polymerase chain reaction methodology. *See also* c-deoxyribonucleic acid; probe hybridization; polymerase chain reaction.

reverse transcription The process by which a complementary deoxyribonucleic acid chain is created on a ribonucleic acid template and then converted to the double-stranded form.

reversed phase chromatography *See* high-performance liquid chromatography.

reversible toxic effect An effect that will disappear when exposure to a chemical is removed.

reversing current A common feature of the flow of a tidal waterway, such that the flow is alternately upstream (flood current) and downstream (ebb current).

reversion The hydrolytic phenomenon common to all the higher polymeric or condensed phosphates, in which they revert to the uncondensed form, orthophosphate (PO_4^{3-}), in aqueous media.

revetment A facing of stone, concrete, or other material placed to protect an embankment or shore structure against erosion by wave action or currents, normally extending from top of the bank to the thalweg or beyond the deepest part of the channel.

revetment wall A wall constructed along the toe of an embankment to protect the slope against erosion.

revolutions per minute (rpm) *See the Units of Measure list.*

revolutions per second (rps) *See the Units of Measure list.*

revolving distributor A movable distributor having a rotary motion. *See also* distributor.

revolving screen A screen or rack in the form of a cylinder or a continuous belt that is revolved mechanically. The screenings are removed either manually or by water jets or automatic scrapers. *See also* traveling screen.

revolving vanes Vanes attached to an axis around which they move as a result of dynamic pressure tangential to the circumference of revolution.

Reynolds critical velocity That velocity in a conduit or channel at which flow changes from laminar to turbulent and at which friction ceases to be proportional to the first power of the velocity and becomes proportional to a higher power, nearly the square.

Reynolds number (Re) A dimensionless numerical quantity used to characterize the type of flow in a hydraulic structure, where resistance to motion depends on the viscosity of the liquid in conjunction with the resisting force of inertia. The Reynolds number is equal to the ratio of inertia forces to viscous forces, i.e., the product of a characteristic velocity of the system (e.g., the mean, surface, or maximum velocity) and a characteristic linear dimension (such as diameter or depth) divided by the kinematic viscosity of the liquid. All units must be consistent so that the combinations will be dimensionless. The number is chiefly applicable to closed systems of flow, such as pipes or conduits where no free water surface exists, or to bodies fully immersed in the fluid so that the free surface need not be considered. For pipes,

$$\mathrm{Re} = \frac{Vd}{\nu}$$

Where:

V = the velocity of the water
d = the pipe diameter
ν = the kinematic viscosity of the water

See also kinematic velocity.

RFA *See* regulatory flexibility analysis.

RfD *See* reference dose.

RGL *See* regulatory guidance letter.

RHA *See* Rivers and Harbors Act.

rhizoid In the lower plants, one of the rootlike filaments (unicellular or multicellular) that serve for attachment and absorption.

rhizoplane The external surface of roots and the debris and soil particles adhering to the roots.

RIA *See* radioimmuno assay; regulatory impact analysis.

ribbon test A soils test that helps indicate the type of soil in an excavation. It involves rolling the soil between the hands to an approximately ⅛-inch (3-millimetre) thickness to determine whether the soil holds together.

ribonucleic acid (RNA) A polymeric chemical based on sequences of nucleotides of four bases—guanine ($C_5H_5N_5O$), uracil ($C_4H_4O_2N_2$), cytosine ($C_4H_5N_3O$), and adenine ($C_5H_5N_5$)—bound through ribose and phosphate linkages. Ribonucleic acid is classically divided into three major types; ribosomal ribonucleic acid (rRNA), which provides the structure for protein synthesis; messenger ribonucleic acid (mRNA), which carries the template from which protein is synthesized; and transfer ribonucleic acid (tRNA), which recognizes the codon for a specific amino acid and brings the amino acid into the appropriate site on the template. *See also* codon.

ribonucleic acid–polymerase chain reaction (RNA–PCR) *See* polymerase chain reaction.

Riddick index (RCI) An empirical index, proposed by T. Riddick in 1944, that includes a number of solution water quality parameters and attempts to relate them to solution corrosiveness. The index is most appropriate for soft, surface water supplies of the northeastern and northwestern United States. Its formula is:

$$\text{RCI} = \frac{75}{\text{Alk}}\left\{[CO_2] + 0.5(\text{hardness} - \text{alk}) + [Cl^-] + \frac{2[NO_3^-]\left(\frac{10}{[SiO_2]}\right)(\text{DO} + 2)}{\text{DO}_{sat}}\right\}$$

Where:

$[CO_2]$ = carbon dioxide concentration, in milligrams per litre
hardness = hardness in milligrams per litre as $CaCO_3$
alk = alkalinity, in milligrams per litre as $CaCO_3$
$[Cl^-]$ = chloride ion concentration, in milligrams per litre
$[NO_3^-]$ = nitrate ion concentration, in milligrams per litre as N
$[SiO_2]$ = silica concentration, in milligrams per litre
DO = dissolved oxygen, in milligrams per litre
DO_{sat} = dissolved oxygen saturation, in milligrams per litre

The meaning of the index value is generally taken as follows: RCI less than 5, extremely noncorrosive; RCI between 6 and 25, noncorrosive; RCI between 26 and 50, moderately corrosive; RCI between 51 and 75, corrosive; RCI between 76 and 100, very corrosive; and RCI greater than 100, extremely corrosive. *See also* Langelier saturation index; Ryznar stability index.

ridge (1) A raised line or strip; something raised in a continuous line above the adjacent surface. (2) To form a ridge.

RIFA *See* radioimmunofocus assay.

right of access A well-founded legal claim to approach, enter on, or be on the bank of a stream or other body of water. One who seeks the right to use or divert water in or from the stream or lake must have the right of access before the other right may be lawfully exercised.

right bank The right-hand bank of a stream when the observer is facing downstream.

right of entry The authority granted under the Safe Drinking Water Act, as amended, for the US Environmental Protection Agency administrator or a designated representative, upon presenting appropriate credentials and a written notice, to enter any establishment, facility, or other property of a supplier of water or other person subject to a National Primary Drinking Water Regulation.

right-angle drive An assembly placed on a vertical line-shaft turbine pump to allow the pump to be driven by an internal combustion engine having a shaft at right angles to the pump shaft.

right-to-know *See* hazard communication.

right-of-way A right to pass over another person's land. *See also* easement.

rigid conduit A conduit with a cross-sectional shape that cannot be distorted sufficiently to change its vertical dimension more than 0.1 percent without causing injurious cracks.

rill A very small stream. A rill is also called a rivulet or streamlet.

ring nozzle A nozzle in which the internal diameter at the smaller end is suddenly reduced by a ring.

Ringer's solution A physiological saline solution used for in vitro maintenance of cells, tissues, or organs. It is sometimes supplemented with glucose and used as a temporary blood replacement for persons in shock to avoid vascular collapse.

rinse The step in the regeneration process for a softener or ion exchanger in which fresh water is passed through the bed of resin to remove any excess or spent regenerant prior to the unit being placed back into service. *See also* fast rinse; regeneration; slow rinse.

riparian Pertaining to or situated on the banks of a lake or river.

riparian doctrine *See* riparian water right.

riparian land Land that abuts on the banks of a stream or other natural body of water.

riparian right *See* riparian water right.

riparian vegetation Vegetation growing on the banks of a stream or other body of water.

riparian water right A legal right ensuring that the owner of land abutting on a stream or other natural body of water will have the use of such water. This type of water right originated in the common law that allowed each riparian owner to require

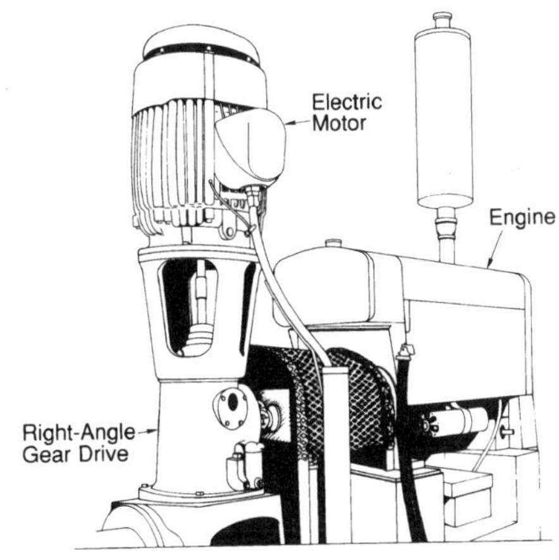

Right-angle drive

the waters of a stream to reach his or her land "undiminished in quantity and unaffected in quality" except for minor domestic uses. It has been abrogated in a number of the western states and greatly modified in others. In general, it now allows each riparian owner to make a reasonable use of the water on his or her riparian land, with the extent of such use being governed by the reasonable needs and requirements of other riparian owners and the quantity of water available. This legal concept is also called the riparian doctrine.

riparian zone The border or banks of a stream. Although this term is sometimes used interchangeably with *floodplain*, the riparian zone is generally regarded as relatively narrow compared to a floodplain. In a riparian zone, the duration of flooding is generally much shorter and the timing less predictable than in a river floodplain.

riparians Those whose property holdings are along the shores of a water body.

Rippl diagram A mass diagram for the study of storage from the net yield of a source of water supply during a continuous time period, generally of a long extent.

riprap Broken stone or boulders placed compactly or irregularly on dams, levees, dikes, or similar embankments for protection of earth surfaces against the action of waves or currents.

riser pipe (1) A steel pipe used in jet well drilling to carry water under pressure to the well point where unconsolidated material is loosened. (2) A particular pipe in an ion exchange system. In co-current systems, the riser is the central (internal) pipe that carries the processed water from the bottom of the resin bed into the service lines or directs backwash water to the bottom of the ion exchanger bed. In countercurrent systems, the riser pipe distributes the regenerant to the bottom of the bed for the upflow regeneration process.

rising rate In a solids-contact unit, the rate of operation in the plane at the separation line. This rate is typically expressed in units of gallons per minute per square foot or metres per hour.

rising time The time necessary for removal, by flotation, of suspended or aggregated colloidal substances. *See also* dissolved air floatation.

risk (1) An individual's assessment of possible events that could lead to harm and danger or exposure to uncomfortable situations. (2) A utility's assessment in the workplace of dangers or harm to employees, equipment, or both. (3) The probability that a particular adverse event will occur. *See also* health effect.

risk analysis The determination or prediction of possible harm, danger, or occurrence of an adverse event. *See also* health effects risk analysis; water demand risk analysis.

risk characterization The integration of the following to determine an estimate of risk: information for a chemical, microbe, or process that can produce harm (hazard identification); the expected dose-response relationships; and the extent and distribution of exposure(s) in the population of concern.

risk communication The act of providing information on risks in terms that are understood by an intended audience, e.g., a client, a judge or jury, an impacted population, or the public as a whole.

risk estimate A description of the probability that organisms exposed to a specified dose of chemical will develop an adverse response (e.g., cancer).

risk extrapolation In risk assessment, the process of extending information beyond the scope of available data. The two most common extrapolations made in risk assessment are (1) that effects produced in experimental animals will also be produced in humans, and (2) that effects observed at high doses in experimental animals or humans extend to lower doses based on some assumptions about the mechanism of action. Such assumptions are frequently necessary to make decisions in the near term. However, once such decisions are made, reversing them with new data is frequently difficult. *See also* linearized multistage model; multihit model; multistage model; one-hit model; probit model; Weibull model.

risk factor A physical condition, chemical, microbial agent, or behavior that is found to be associated with the development of a particular disease state. The term *risk factor* is often used to refer to both risk indicators and the cause associated with the development of a specific disease. *See also* causality; risk indicator.

risk indicator In epidemiology, the term is preferred in place of *risk factor*. If a disease is more common among those with a certain characteristic than among those without it, and if the characteristic occurs or is present prior to the disease, the characteristic is a risk indicator for the disease. Knowledge of risk indicators can be used to help identify high-risk populations for preventive interventions. Age, gender, occupation, and smoking are risk indicators for many diseases. *See also* risk factor.

risk management The process of weighing policy alternatives, selecting the most appropriate regulatory action, and integrating the results of risk assessment with engineering data and social, economic, and political concerns to reach a decision. *See also* risk analysis.

risk management plan (RMP) A plan intended to prevent and minimize the impact of accidental releases of hazardous substances. A risk management plan must be prepared according to US Environmental Protection Agency (USEPA) regulations if any regulated chemical is present at a facility in quantities that exceed the threshold listed in USEPA regulations; see 40 Code of Federal Regulations 68.

Risk Reduction Engineering Laboratory (RREL) An organization that—under the Office of Research, US Environmental Protection Agency, Phase I reorganization—became the National Risk Management Research Laboratory. The Risk Reduction Engineering Laboratory no longer exists. *See also* National Risk Management Research Laboratory.

risk-specific dose The dose associated with a specified risk level.

river A large stream of water that serves as the natural drainage channel for a drainage basin of considerable area. This term is a comparative one and relates to basin size rather than water volume. *See also* stream.

river basin The area drained by a river and its tributaries. *See also* drainage area.

river bend A sudden shift in a river's direction.

river forecasting The process of predicting the river stage and discharge based on hydrology and meteorology, including research into forecasting methods. In some countries, the term *hydrometeorology* is used with this limited meaning.

river gauge A device for measuring a river stage, i.e., for indicating the height of the water surface above a specific point. Types in common use include the staff gauge, water-level recorder, and wire-weight gauge.

river lake *See* drainage area.

river piracy A condition in the geological development of a stream system whereby one stream, by reason of more rapid development through headward erosion, pushes its headwater divide into the drainage basin of an adjoining stream and ultimately diverts such stream or one or more of its tributaries into its own channel. River piracy is sometimes called river capture.

river safe yield The maximum dependable draft that can be made continuously on a source of water supply during a period of years during which the probable driest period or period of greatest deficiency in water supply is likely to occur. It is frequently expressed in minimum stream flows during the minimum 7-day period occurring with a 10-year frequency.

river sleeve A long sleeve placed over other pipe joints to prevent injury to joints laid on a river bottom or underwater.

river system A principal stream and all its tributaries.

river terrace A terrace or bench that borders a stream and was formed by previous meanderings of the stream in a geologic period when the stream's present channel was at a higher elevation or when the land level was at a lower elevation. *See also* alluvial terrace.

river training The construction of engineering works including artificial plantations, with or without the construction of embankments built along a river or a section of river, in order to direct or lead the flow into a prescribed channel. The training of a river for navigation involves developing a waterway of required depth, either by construction of a series of locks and dams or, if the slope is sufficiently flat, straightening and confining the flow.

river valley A valley formed primarily by the erosive activities of a river.

riverbank filtration A process of collecting water in an infiltration gallery located within a bank along a river to allow the river water to pass through the soil in the riverbank. Riverbank filtration provides particle removal, as well as partial removal of organic compounds. *See also* infiltration gallery.

riverbed The bottom of a river below the usual water surface. *See also* streambed.

riverine *See* riparian.

Rivers and Harbors Act (RHA) Legislation, enacted in 1899 (33 U.S.C. 403), that regulates certain activities in navigable waters of the United States. For example, construction of any dam or dike across any navigable water of the United States is prohibited in the absence of congressional consent and approval by the US Army Corps of Engineers; construction or alteration of any navigable water of the United States is prohibited unless authorized by the US Army Corps of Engineers; excavating from or depositing material into such waters is prohibited unless approved by the US Army Corps of Engineers.

Riveted steel pipe being installed in Philadelphia in 1907

riverwash Barren, usually coarse-textured, alluvial soil in and along waterways that is exposed during low water levels and is subject to shifting during flood periods.

riveted steel pipe Steel pipe made from sheets of steel curved into a cylindrical shape with the edges riveted together to form a seam. The seams may form a spiral running around the circumference of the pipe.

rivulet *See* rill.

RMCL *See* recommended maximum contaminant level.

RMP *See* risk management plan.

rms value *See* root-mean-square value.

Rn *See* radon.

RNA *See* ribonucleic acid.

RO *See* reverse osmosis.

rock In groundwater hydrology, all formations in the lithosphere, whether solid or granular, consolidated or unconsolidated.

rock flowage zone That part of the lithosphere in which all rocks are under stresses exceeding their elastic limits. In this zone, the rocks undergo permanent deformation and are said to flow; hence, in this zone interstices are absent or insignificant.

rock flowage-and-fracture zone That part of the lithosphere lying between the zone of rock fracture and the zone of rock flowage. In this zone, the strongest rocks behave as in the rock fracture zone, and the weakest rocks as in the rock flowage zone. *See also* rock flowage zone; rock fracture zone.

rock formation A geological formation normally composed of solid rock material. In geologic usage, this term may be applied to all earth materials, whether consolidated or unconsolidated.

rock fracture zone The upper part of the lithosphere, in which rocks are under stresses less than those required to close their interstices by deformation of the walls of the interstices.

rock salt *See* halite.

rock-fill dam A dam composed of loose rock supporting a watertight face or containing a watertight core of impervious earth, concrete, steel sheet pile, or some combination of these materials, or backed by an earth embankment.

rock-filled crib dam A timber dam consisting of a series of cribs or rectangular cells made of square or round timbers drift-bolted together and filled with broken rock or boulders, with an upstream facing and deck covered with heavy planks to provide watertightness.

ROD *See* record of determination.

rod float A rod or staff designed to float in a practically vertical position in flowing water and used for observing the time taken to traverse a measured distance or to indicate the direction of flow.

rodenticide Any substance or chemical used to kill or control rodents.

roentgen (R) *See the Units of Measure list.*

roentgen equivalent man (person) *See* rem *in the Units of Measure list.*

ROI *See* return on investment.

roiliness A milky appearance of water, possibly caused by fine suspended particles of silt or clay.

rolled-earth dam A dam in which selected material of proper moisture content is placed in thin layers and compacted by rolling. The resulting fill attains a very high density and is relatively impervious, but it may contain a line of saturation that remains within the base of the structure.

roller (1) An instance of rolling water in a generally circular path outside a separation zone. It is a form of eddy, and is frequently observed, for example, in stilling basins. (2) A heavy cylinder used for compacting embankments or smoothing surfaces.

roller gate A hollow, horizontal, cylindrical crest gate with spur gears at each end meshing with an inclined rack anchored to a recess in the end pier or wall.

roller-bearing gate A crest gate provided with roller bearings that transmit the water pressure against the gate to a track placed in the structure, thereby reducing the friction incurred in opening and closing.

roller-way face A face on low overflow dams designed to reduce erosion. A roller-way face usually has a slope of 12 to 18 units horizontal to 1 unit vertical and is toughened to offer frictional resistance.

rolling dam A movable dam consisting of a large steel cylinder gate placed across the waterway between piers or abutments and arranged to be rolled up clear of the stream flow on an inclined rack track.

rolling-up curtain weir A movable timber weir in which the wooden barrier consists of a curtain that is composed of horizontal lathes increasing in thickness downward and connected by watertight hinges. This type of weir is rolled up from the bottom.

rollway (1) The overflow portion of a dam. (2) An overflow spillway.

roof weir A gate type of bear-trap crest gate, improved by the addition of a vertical lip to the upstream leaf and equipped with rollers bearing on the downstream leaf.

root That portion of a dam or diversion weir that extends into the natural ground surface at each end.

root of an equation A value of a variable that satisfies a mathematical equation. An equation may have multiple roots.

root zone That part of the soil that is invaded by roots of plants.

root-mean-square (rms) value The square root of the time average value of the square of a quantity. For a periodic function, the rms value is the square root of the average of the square of the function taken throughout one period. For sinusoidal functions, the root-mean-square (rms) value of the signal is the amplitude divided by the square root of 2:

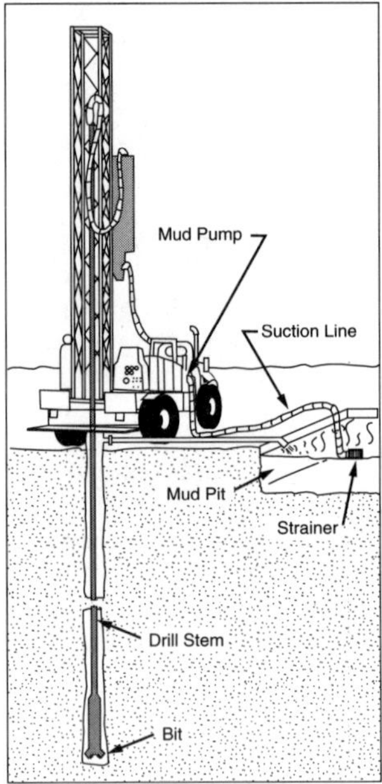

Rotary drilling equipment

$$A_{\text{rms}} = \frac{A}{\sqrt{2}}$$

Where:

A_{rms} = the root-mean-square amplitude

A = the amplitude

The root-mean-square value is also called the effective value.

rotameter A device used to measure the flow rate of gases and liquids. The gas or liquid being measured flows vertically up a tapered calibrated tube. Inside the tube is a small ball- or bullet-shaped float (it may rotate) that rises or falls depending on the flow rate. One can read the flow rate on a scale behind or on the tube by looking at the middle of the ball or at the widest part or top of the float.

rotameter control valve A device usually consisting of a pressure-regulating valve, flowmeter, and flow rate valve to control and measure the flow rate. *See also* rotameter.

rotary drilling A common hydraulic well-drilling method that uses a rotating drill pipe with a hard-tooled drill bit attached at the bottom. A fluid (drilling mud) is forced down through the drill pipe and then forced up again between the drill pipe and the well hole, carrying rock chippings (cuttings) up with the mud.

rotary evaporation An analytical technique used to concentrate solutes in a sample. In this approach, a water sample held in a rotating flask is boiled under reduced pressure, and the solutes are concentrated while the water is distilled off. Rotary evaporation is a fairly gentle concentration technique.

rotary pump A type of displacement pump consisting essentially of elements rotating in a pump case in which they closely fit. The rotation of these elements alternately draws in and discharges the water being pumped. Such pumps act with neither suction nor discharge valves, operate at almost any speed, and do not depend on centrifugal forces to lift the water.

rotary surface washer A device used to clean the surface of filter media, typically used in the first stages of a backwash cycle. A rotary surface washer has arms connected radially to a mounted rod and rotates in a plane parallel to the surface of the media. The rotating arms, typically located less than 6 inches (15 centimetres) above the surface of the media to be cleaned, rotate as a result of water pressure exerted when water flows into the arms and is forced out, downward into the media surface, through small openings along one side of each arm. *See also* backwash; stationary surface washer.

rotary valve A valve that (1) consists of a casing more or less spherical in shape and a gate that turns on trunnions through 90° when opening or closing, and (2) has a cylindrical opening of the same diameter as that of the pipe it serves.

rotation The motion of turning around on an axis or center; rotary motion.

rotavirus A member of the Reoviridae family associated with acute enteritis in infants and young children.

Rotifera The class of minute, multicelled aquatic animals.

rotor The rotating part of an alternating current electrical motor, or of a turbine or propeller meter.

roughing filter *See* preliminary filter.

roughness A measure of the resistance to fluid flow of a channel, pipe, or other conduit, as a result of the conduit's fabrication, scale formation, biological growth, or other causes. *See also* roughness coefficient.

roughness coefficient A factor—as in the Chezy, Darcy–Weisbach, Hazen–Williams, Kutter, Manning, and other formulas—for computing the average velocity of water flow in a conduit or channel. It represents the effect of the roughness of the confining material on the energy losses in the flowing water. *See also* Chezy open-channel formula; Darcy–Weisbach formula; Hazen–Williams formula; Kutter formula; Manning formula; *n* factor.

roughness factor *See* roughness coefficient.

round robin test A procedure whereby samples containing known amounts of certain constituents are analyzed by various laboratories to establish the competence of those laboratories.

rounded-crest weir (1) A weir having a crest that is convex upward in the direction of flow over the weir. (2) A weir having a crest with a flat center and rounded corners.

roundworm *See* nematode.

route of administration The method of introducing a drug into the body. The most frequent route is the administration of a capsule or tablet by the oral route. However, drugs are provided in forms suitable for injection, inhalation, topical introduction (via the skin), rectal introduction, and introduction by many other routes. *See also* gavage.

route of exposure The pathway by which a chemical is introduced to the body, e.g., inhalation of a fume, vapor, or gas.

routine sample A sample required by the National Primary Drinking Water Regulations to be taken at regular intervals to determine compliance with maximum contaminant levels. *See also* maximum contaminant level.

routing (1) The derivation of an outflow hydrograph for a given reach of a stream from known values of upstream inflow, the speed of a water wave moving through the stream, and the storage properties of the stream. (2) Estimating the flood at a downstream point from the inflow at an upstream point. *See also* flood routing; flow routing.

RPBA *See* reduced-pressure principle backflow-prevention assembly.

RPG *See* radiation protection guide.

rpm *See* revolutions per minute *in the Units of Measure list.*

rps *See* revolutions per second *in the Units of Measure list.*

RPZ device *See* reduced pressure zone device.

RQL *See* reliable quantitation level.

RREL *See* Risk Reduction Engineering Laboratory.

RS 232-C A recommended standard of the Electronics Industries Association that defines a method for distributing data by wire. This standard is used in asynchronous transmission in which such peripheral devices as terminals, printers, or plotters are connected with individual lines to the host computer.

RSD *See* relative standard deviation.

RSI *See* Ryznar stability index.

RSSCT *See* rapid small-scale column test.

RT–PCR *See* reverse transcriptase–polymerase chain reaction.

RTU *See* remote terminal unit.

rubber sheeting A process that geometrically adjusts map features to "force" a digital map to fit a designated base. This term is used for this process because the map is mathematically stretched to fit the base, given a set of known coordinate values for which the locations are defined on both the base map and the map to be rubber sheeted. The process uses mathematical operations to minimize distortion.

rubble dam A dam constructed of rocks laid in place without mortar.

rule A principle, regulation, or policy governing conduct, action, procedure, arrangements, and so on.

rule of continuity A physical rule stating that the flow *Q* that enters a system must also be the flow that leaves the system. Mathematically, this rule is generally stated as $Q_1 = Q_2$ or, because *Q* equals the product of cross-sectional area (*A*) and flow velocity (*V*), as $A_1 V_1 = A_2 V_2$.

rule of correlative right A water rights doctrine specifying that rights are not absolute but depend on the rights of others as well.

rulemaking Activity of a federal, state, or local regulatory agency to prepare a new regulation.

Rules, Safe Drinking Water Act *See* Safe Drinking Water Act Rules.

run (1) A watercourse; a small stream. (2) One of a series of tests. (3) To become fluid; to melt. (4) To flow as a fluid, as a running stream. (5) To discharge or flow out, as a running faucet. (6) A length of pipe that is made up of more than one piece of pipe. (7) In a tee or cross fitting, the end of the fitting opposite the filling entrance. (8) An execution of a computer program.

runner The revolving part of a turbine.

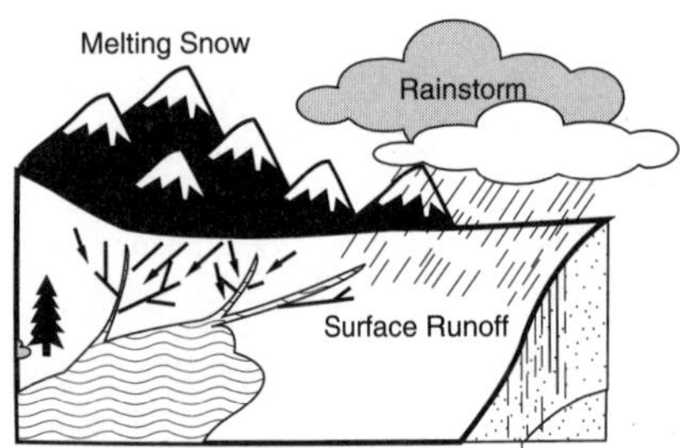

Runoff

running annual average (RAA) A method of calculating a running 12-month average concentration of a contaminant, as used to determine compliance with a maximum contaminant level for some contaminants. For example, if compliance monitoring for a contaminant is required quarterly, the running annual average would be the average of the current quarter's monitoring results along with the results of the three previous quarters.

runoff (1) That portion of the earth's available water supply that is transmitted through natural surface channels. (2) The total quantity of such runoff during a specified time. (3) In the general sense, that portion of the precipitation that is not absorbed by the deep strata but finds its way into the streams after meeting the persistent demands of evapotranspiration, including interception and other losses. (4) The discharge of water in surface streams, usually expressed in inches or (centimetres) depth on the drainage area or as a volume in such terms as cubic feet, acre-feet, or cubic metres. (5) That part of the precipitation that runs off the surface of a drainage area and reaches a stream or other body of water or a drain or sewer.

runoff coefficient (1) The ratio of the maximum rate of runoff to the uniform rate of rainfall with a duration equaling or exceeding the time of concentration that produced this rate of runoff. (2) The ratio of the depth of runoff from the drainage basin to the depth of rainfall. *See also* Kirpich formula.

runoff cycle That portion of the hydrologic cycle between incident precipitation over land areas and subsequent discharge through stream channels or direct return to the atmosphere through evapotranspiration.

runoff distribution curve A graph showing the typical distribution of runoff from a drainage basin in terms of the percentage of the total runoff, expressed as the average discharge in a given time interval.

runoff measurement A method of estimating maximum probable flood flows that will occur from the drainage basin above any point. *See also* rational method.

runoff percentage The amount of runoff expressed as a percentage of total rainfall on a given area. *See also* runoff coefficient.

runoff rate The volume of water running off per unit time from a surface, expressed in inches (centimetres) depth per hour, in cubic feet (cubic metres) per second, in cubic feet (cubic metres) per second per square mile (kilometre), or in other units.

runoff volume The total quantity or volume of runoff during a specified time. It may be expressed in acre-feet (cubic metres), in inches (millimetres) depth over the drainage area, or in other units.

run-of-stream plant A hydroelectric plant that operates on water obtained from the natural flow of the stream without storage or without an appreciable amount of pondage. Such a plant is also called a streamflow plant or run-of-river plant.

rural water survey (RWS) A survey of the quantity, quality, and availability of rural drinking water supplies conducted by the US Environmental Protection Agency as required by the Safe Drinking Water Act of 1974 (Public Law 93-523).

rust A reddish corrosion product occasionally found in water. Rust is formed as a result of electrochemical interaction between iron and atmospheric oxygen in the presence of moisture. It is also called ferric oxide.

rusty water *See* red water.

RWS *See* rural water survey.

Ryznar index *See* Rzynar stability index.

Ryznar stability index (RSI) An index indicating whether a water has a tendency to corrode (dissolve precipitated $CaCO_3$) or to precipitate in the form of calcium carbonate ($CaCO_3$).

$$RSI = 2pH_s - pH_a$$

Where:

pH_s = the pH of the water if it were saturated and at equilibrium with $CaCO_3$

pH_a = the actual pH

Ryznar stability index values between 6.0 and 7.0 indicate slight scale formation or corrosion. Values less than 6.0 or greater than 7.0 indicate varying degrees of scale-forming conditions or corrosive ($CaCO_3$ dissolving) conditions, respectively. *See also* Langelier saturation index; Riddick index.

S

S *See* siemens *in the Units of Measure list*; sulfur.
S *See* complex power.
s *See* second *in the Units of Measure list.*
|S| *See* apparent power.
s^{-1} *See* velocity gradient.
S. sonnei *See Shigella sonnei.*
SAB *See* Science Advisory Board.
SAC *See* spectral absorption coefficient.
sacrificial anode An anode made of suitable metal placed in a water heater tank or water storage tank to protect the tank from corrosion.
sacrificial anode system A cathodic protection system in which the driving voltage for the protective current is generated by a galvanic corrosion cell, with the protected structure serving as the cathode.
saddle (1) A steel or concrete structure used for supporting a pipe or penstock laid above the surface of the ground. (2) A depression in a ridge. (3) An assembly of circumferential metal straps on a pipe where a connection is to be installed.
safe Pertaining to a condition of exposure under which a "practical certainty" exists that no harm will result for exposed individuals.
safe bearing capacity The load, usually expressed in tons per square foot (kilograms per square metre), that a given soil or foundation can safely support without appreciable settlement or movement.
safe drinking water Water that does not contain harmful bacteria or toxic materials or chemicals. Water may have taste-and-odor problems, color, and certain mineral problems and still be considered safe for drinking. *See also* potable water.
Safe Drinking Water Act (SDWA) Public Law 93-523, enacted Dec. 16, 1974, establishing Title XIV of the US Public Health Service Act, codified generally as 42 U.S.C. 300f-300j-11. It required the US Environmental Protection Agency to set national primary (health-related) drinking water regulations that were the first to apply to all public water systems, as defined by the act, in the United States. *See also* Safe Drinking Water Act Amendments of 1986; Safe Drinking Water Act Amendments of 1996.
Safe Drinking Water Act Amendments of 1986 Public Law 99-339, which was enacted June 16, 1986, amending Title XIV of the US Public Health Service Act, commonly known as the Safe Drinking Water Act. It required the US Environmental Protection Agency to regulate a list of 83 specific contaminants, to regulate an additional 25 contaminants every 3 years, to specify criteria for deciding when filtration of surface supplies is required, to require disinfection of all public water supplies with some exceptions for groundwater that meets as yet unspecified criteria, and additional provisions, including new programs for wellhead protection and sole source aquifer. *See also* Safe Drinking Water Act; Safe Drinking Water Act Amendments of 1996; wellhead protection.
Safe Drinking Water Act Amendments of 1996 Public Law 104-182, which made substantial revisions to the Safe Drinking Water Act, including elimination of the 1986 requirement that 25 contaminants be regulated every 3 years; addition of a new process for listing and regulating new contaminants; revision of the standard-setting process to include consideration of costs and benefits and competing health risks; addition of new programs for source water assessment, local source water petitions, and source water protection grants; mandatory regulation of filter backwash water recycle, radon, and arsenic; revision of requirements for unregulated contaminant monitoring and a national occurrence database; creation of a new state revolving loan fund for drinking water; new provisions regarding small system variances, treatment, technology, and assistance centers; and development of capacity development and operator certification guidelines by the US Environmental Protection Agency. *See also* Safe Drinking Water Act; Safe Drinking Water Act Amendments of 1986.
Safe Drinking Water Act Rules Regulations developed and promulgated by the US Environmental Protection Agency under the authority of the Safe Drinking Water Act.
Safe Drinking Water Hotline A US Environmental Protection Agency telephone hotline that provides assistance and regulatory

knowledge to the regulated community (public water systems) and to the public on the regulations and programs developed in response to the Safe Drinking Water Act Amendments of 1986. Hours are from 9:00 A.M. to 5:30 P.M. Eastern time, Monday through Friday. The phone numbers are (800)426-4791 or (703)527-5190.

Safe Drinking Water Information System (SDWIS) A computerized information system established and maintained by the US Environmental Protection Agency that contains information on public water systems and compliance with National Primary Drinking Water Regulations. The Safe Drinking Water Information System is accessible via the Internet at http://www.epa.gov/enviro/html/sdwis/sdwis_ov.html.

safe velocity The fluid velocity that will maintain solids in movement and, at the same time, will not scour the conduit in which the fluid is being carried. *See also* noneroding velocity.

safe yield, groundwater *See* groundwater safe yield.

safe yield, impounding reservoir *See* impoundment safe yield.

safe yield, river *See* river safe yield.

safety audit A review of hazards in the environment or workplace, with a view toward reducing them to an acceptable level.

safety belt A belt used to secure a person who is performing an activity and who requires fall protection. This type of belt was phased out by regulations beginning in 1997. *See also* safety harness.

safety can A container used for safe storage of such flammables as gasoline. Such a can has a spring-loaded cap to allow release of internal pressure upon exposure to high temperatures. Laid on its side, this can should not leak more than 3 drops per minute.

safety factor (1) A numerical factor by which a dose or exposure that is known to produce minimal effects or no effect in an experimental situation (usually involving experimental animals or humans) is multiplied to help prevent adverse effects in a more susceptible species or individual. Safety factors are frequently applied to the determination of "safe" doses or exposures to chemicals. In general, however, the term *uncertainty factor* is favored, because it more precisely indicates why these factors are used. (2) An amount of water added to demand projections to protect against unforeseen changes in water demand and supply. *See also* lowest-observed-adverse-effect level; no-observed-adverse-effect level; uncertainty factor.

safety harness A harness worn by employees requiring fall protection to distribute the shock of a fall evenly on the body. In contrast, a safety belt directs all of the fall energy to the waist of the wearer. *See also* safety belt.

safety program A set of written procedures and guidelines that instruct workers on how to carry out their job duties safely and avoid hazards.

safety shoes Shoes that have steel toes and are shanked to prevent damage to the foot. Such shoes must meet American National Standards Institute Standard Z41.

safety shower A shower that delivers a drenching flow to the user. A safety shower is usually located near hazardous chemicals.

safety switch Generally a switch that disables or disconnects a circuit, piece of equipment, or particular control function. Some safety switches, however, cause the functions to freeze in the last setting. Although most safety switches are of the circuit breaker type, some are contained in the software of a control program that requires inputs from other circuits. If the input is not received, the equipment is shut down or remains at the last setting that was received. A safety switch is also called a disconnect switch.

safety valve A valve that automatically opens when prescribed conditions, usually of pressure, are exceeded in a pipeline or other closed receptacle containing liquids or gases. A safety valve prevents such conditions from being exceeded and causing damage.

sal marina The Spanish name for sea salt.

salimeter A hydrometer that measures the percentage of salt as sodium chloride (NaCl) in brine or other salt solutions. A 100 percent reading on a salimeter is about 26.4 percent salt by weight at 60° Fahrenheit (15.5° Celsius). Some use the term *salinometer* to refer to a salimeter but the two devices are different; a salinometer measures electrical conductivity, not density. *See also* salinometer.

saline Pertaining to or containing salt.

saline contamination Contamination of water by intrusion of salt water. *See also* saline water intrusion.

saline soil A nonalkaline soil with the conductivity of the saturation extract greater than 4 millimhos per centimetre at 77° Fahrenheit (25° Celsius), the exchangeable sodium percentage less than 15, and the pH usually less than 8.5.

saline solution Usually, any solution with the total dissolved solids ranging from 15,000 to 30,000 milligrams per litre, e.g., a solution of sodium chloride (NaCl) at this concentration, and usually also containing other salts. Saline solution may also be called saline water.

saline spring A spring in which the water contains a considerable quantity of sodium chloride (NaCl) or other common minerals.

saline water Water containing relatively high concentration of salts or dissolved solids, generally at least 1,000 milligrams per litre total dissolved solids. *See also* brackish water; brine.

saline water intrusion The movement of saline groundwater into a formerly freshwater aquifer as a result of pumping in that aquifer. Saline water intrusion typically occurs in coastal areas where the source of saline water is the nearby ocean. Passive saline water intrusion refers to the slow intrusion caused by a general lowering of the freshwater potentiometric surface. Active saline water intrusion refers to a much more rapid intrusion caused by a lowering of the freshwater potentiometric surface below the local mean sea level.

saline–alkaline soil A soil with the exchangeable sodium percentage greater than 15, the pH usually 8.5 or lower, and the conductivity of the saturation extract greater than 4 millimhos per centimetre at 77° Fahrenheit (25° Celsius).

salinity (1) The relative concentration of dissolved salts in water. (2) The total dissolved solids in salt water after all carbonates and organic matter have been oxidized and all bromide and iodide have been replaced with chloride. This type of salinity is usually reported in grams per kilogram or parts per thousand. Salinity in parts per thousand empirically is equal to 0.03 plus the value of 1.805 multiplied by chlorinity in parts per thousand. *See also* chlorinity; chlorosity.

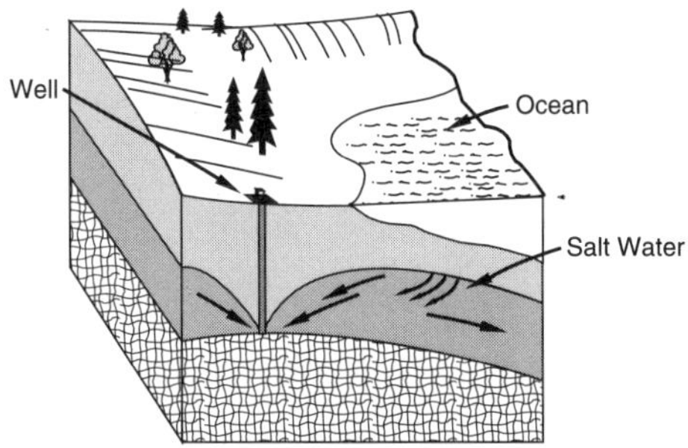

Saline water intrusion

salinity wedge *See* saltwater wedge.

salinization The process of salts accumulating in soil.

salinometer An instrument for determining the salt concentration (salinity) of brine water by measuring the electrical conductivity of the solution. A salinometer is sometimes called a salt gauge. *See also* salimeter.

Salmonella A genus of enteric bacteria belonging to the family Enterobacteriaceae. *Salmonella* are gram-negative, rod-shaped, non-spore-forming, facultative anaerobes that are morphologically indistinguishable from the coliform bacteria. Most species are actively motile by means of peritrichous flagella. All members of the *Salmonella* group are pathogenic to some degree in humans, primarily causing gastrointestinal illnesses ranging from mild to severe (typhoid fever is an example of the latter). They may also cause disseminated systemic infections. The natural hosts for these organisms are lower animals (both warm- and cold-blooded), especially rodents and birds, that provide a reservoir for human disease. *See also* typhoid fever; waterborne disease.

***Salmonella* microsome test** A mutagenesis test designed to detect reverse mutations induced by chemicals in one or more tester strains of *Salmonella* bacteria. *Salmonella* microsomal or S-9 fractions are frequently added to the incubation to enable detection of chemicals that have to be metabolically activated to induce mutations. The enzymes that mediate such metabolism in mammals are most often found in the endoplasmic reticulum. Upon homogenization of tissues, the endoplasmic reticulum is reduced to vesicular structures referred to as microsomes. *See also* Ames test; microsome; reverse mutation; S-9 fraction.

salmonellosis *See* waterborne disease.

salometer *See* salimeter.

salt A compound resulting from acid–base mixtures.

salt block Evaporated salt or fine rock salt that is mechanically compressed into dense blocks, weighing about 50 pounds (23 kilograms) each, that are sometimes used in residential water softeners.

salt cake Sodium sulfate (Na_2SO_4) that is only 90–99 percent pure (i.e., contains 1–10 percent substances other than sodium sulfate). Salt cake is made by heating rock salt with sulfuric acid (H_2SO_4), producing muriatic acid (HCl) and salt cake.

salt dosage *See* regeneration level.

salt efficiency The hardness removal capacity of a water softener unit, calculated as the milligrams (grains) of hardness removed divided by the in kilograms (pounds) of salt that is used to achieve that amount of hardness reduction. Operational salt efficiency refers to the performance, in terms of salt efficiency, of a water softener under conditions of actual or simulated long-term use (6 months or more) in a household where the amount of water usage typically varies from day to day.

salt flux *See* solute (salt) flux.

salt mass transfer coefficient *See* solute (salt) permeability coefficient.

salt method A method of measuring the discharge of water flowing through closed passages under pressure. A salt solution of known strength is injected at a known rate into the water at one point, and samples are taken of the water at a downstream point and analyzed. Knowing the quantity of salt in the solution, the solution's rate of injection, and the quantity of salt in the resulting mixture, one can compute the quantity of discharge by a mass balance. This method is frequently used in measuring water flow through hydraulic turbines. *See also* dilution method.

salt (or specific solute) passage For a pressure-driven membrane treatment process, the fraction of feedwater salt or specific solute that passes through the membrane and becomes part of the permeate (product). In equation form,

$$\text{salt (or specific solute) passage (\%)} = \frac{C_p}{C_f} \times 100$$

Where:

C_p = permeate concentration, in milligrams per litre

C_f = feedwater concentration, in milligrams per litre

It is also equal to 100 percent minus the percentage salt or specific solute rejection.

salt passage correction factor For some pressure-driven membrane processes, a dimensionless coefficient that is dependent on the net driving pressure and concentration gradient across the membrane. This factor can be used to calculate salt passage for a given net driving pressure and concentration gradient by multiplication with salt passage at a standard condition.

salt permeability coefficient (constant) *See* solute (salt) permeability coefficient (constant).

salt rejection *See* rejection.

salt splitting An ion exchange process in which neutral salts in water are converted to their corresponding acids or bases. A strong base anion exchanger resin (R) can convert a salt solution to caustic (i.e., base), e.g.,

$$NaCl + \overline{ROH} = \overline{RCl} + NaOH$$

A strong acid cation exchanger can convert a salt to acid, e.g.,

$$NaCl + \overline{HR} = \overline{NaR} + HCl$$

salt spring A saline spring.

salt transport coefficient *See* solute (salt) permeability coefficient.

salt water Water containing relatively high concentration of salts or dissolved solids, usually over 1,000 milligrams per litre. *See also* ocean water; saline water.

salt-affected soil Soil that has been adversely modified for the growth of most crop plants by the presence of certain types of exchangeable ions or soluble salts. Soils having an excess of salts, exchangeable sodium, or both, fit in this category.

saltation The transportation of clastic (broken or fractured) sediments in air or water by intermittent leaps or bounds.

salting out The act of decreasing the solubility of a solute by addition of a salt. This phenomenon has been applied in analytical chemistry as a way to increase the extraction efficiency of certain analytes. For example, liquid–liquid extraction methods for the analysis of haloacetic acids (CX_3COOH, where X = Cl, Br, or H in various combinations) in water recommend the addition of large amounts of salt to improve the extraction efficiency.

salt-splitting capacity test A test that measures the performance of a used ion exchange resin to determine the capacity of the used resin versus the standard rated capacity of the resin when fresh. This allows a calculation of the degree of exhaustion of the resin.

salt-velocity method A means of measuring water velocity by injecting salt at one point and observing the time required for the salt solution to reach another point. *See also* salt method.

saltwater encroachment The invasion of a body of fresh water by a body of salt water. *See also* saline water intrusion.

saltwater intrusion *See* saline water intrusion.

saltwater system A system of water mains, separate from the regular distribution system, that conveys salt water for fighting fires.

saltwater underrun The more rapid movement of the ocean water along the bottoms of tidal estuaries, making the bottom water saltier than the top water. *See also* saltwater wedge.

saltwater wedge A salinity intrusion that occurs in certain tidal waterways and has the distinguishing characteristic of being a stratum of salt water underflowing a stratum of comparatively fresh water. The thickness of the stratum of salt water is greatest at the entrance and least at some point upstream, giving the stratum a wedge shape. *See also* saline water intrusion; saltwater underrun.

salvage water right A legal right to appropriate and use waters that would otherwise go to waste.

salvaged water That part of a particular stream or other water supply that, as a result of artificial work is not lost from the supply and therefore is retained within the supply and made available for use. Salvaged water is already in the area and is saved and restored to the usable supply within the area by artificial means.

SAMP *See* special area management plan.

sample (1) A fraction of a population or material tested or analyzed in order to determine the composition, nature, or properties, or a combination of all three. (2) To collect a sample.

sample bottle A wide-mouth glass or plastic bottle used for taking microbiological and chemical water samples.

sample point A location or apparatus in a water distribution system that will allow a water sample to be collected for testing.

sample preservation Treatment given to a sample to maintain its original composition.

sample split Part of a larger sample.

sampler A person or device that collects a portion of a substance for the purpose of analysis.

sampling (1) A process of collecting a portion of a substance for analysis. A goal of sampling is to obtain a portion of the substance that is representative of the whole. (2) In statistics, a selection of a subset of the entire population of subjects or products for analysis such that the measured values obtained collectively represent the entire population within certain confidence limits. In general, a method of selecting individual samples in an unbiased way must be found. The most common approach is to randomly select the samples for analysis.

sampling iron A metal frame used during the collection of water samples from rivers and lakes. A sampling iron is designed to hold sampling bottles of different sizes.

sampling vertical A vertical line from the bottom of a water body to the surface, along which one or more samples are collected.

San Dimas flume A special form of flume for measuring the flow of debris-laden water in open conduits.

SANCHO model A computer model to simulate the behavior of chlorine (HOCl), biodegradable organic carbon, and biomass in distribution system pipes. This model can predict the spatial fluctuations of the modeled parameter under a steady-state condition.

sanctuary Territory set apart and maintained in a natural state for the preservation of plants and animals.

sand (1) A soil particle between 0.05 and 2.0 millimetres in diameter. Chemical composition does not influence the definition, only size. (2) Any one of five soil separates: very fine sand, fine, medium, coarse, or very coarse.

sand ballasted flocculation–sedimentation A flocculation and sedimentation process that is enhanced by the use of sand. During the flocculation process, the sand attaches to other particles, often formed through the addition of coagulants. These particles then settle rapidly in the sedimentation process. The sand is then recovered from the residuals and is recycled to the beginning of the process. *See also* microsand enhanced coagulation.

sand bar A ridge of sand built up to the surface or near the surface of a river or along a beach.

sand barrier A layer of gravel around the curb of a dug well.

sand boil The violent washing action in a granular water filter caused by an uneven distribution of backwash water.

sand drying bed A device used to separate water and solids in residuals. The residuals are applied to the surface of a layer of sand, ranging in depth between 6 and 9 inches (15 and 23 centimetres), supported by 12 inches (30 centimetres) of a gravel underdrain system. The dewatering process occurs by two

Sample bottle

mechanisms: (1) gravity drainage through the sludge cake and sand filter, and (2) air drying from the surface of the sludge cake by evaporation. Sand drying beds are simple to operate but require large land areas to provide effective dewatering. *See also* sludge bed.

sand ejector A portable device for transporting sand by water under pressure.

sand filter The oldest and most basic filtration process, which generally uses two grades of sand (coarse and fine) for turbidity and particle removal. A sand filter can serve as a first-stage roughing filter or prefilter in more complex processing systems. *See also* granular media; slow sand filter.

sand gate A sluice gate used for sluicing out sand and sediment collected at a sand trap in an irrigation canal.

sand grain bridging The formation of bridges by sand grains during well construction and development. Such bridging can result in inefficient wells and damage to pumps when the bridges break and sand particles pass through the pump.

sand interceptor A detention chamber designed to remove sand from a conduit.

sand pump A long cylindrical tube fitted with a valve at the bottom and open at the top, used for raising the mud that accumulates in the bottom of a boring during the sinking process. A sand pump is also called a shell pump or sludger.

sand trap An enlargement of a conduit carrying untreated water that allows the water velocity to slow down so that sand and other grit can settle.

sand washer *See* sand-washing machine.

sand-pump dredge *See* suction dredge.

sandstone Sedimentary rock, usually consisting of quartz sand united by some cement, such as silica (SiO_2) or calcium carbonate ($CaCO_3$).

sand-washing machine A device used for washing the sand in a slow sand filter, as well as the sand from a rapid granular filter. *See also* slow sand filter.

sandy Containing a large fraction of sand. This term may be applied to any one of the soil classes that contains a large percentage of sand.

sanitary drinking fountain A drinking fountain that delivers water in a stream in such a manner that (1) the individual drinks from the stream without coming in contact with the equipment, and (2) the unused water flows only into the waste basin.

sanitary engineer *See* environmental engineering.

sanitary landfill A dedicated location for disposal of solid waste or dried, dewatered residuals on land. After it reaches a specified elevation, the landfill is decommissioned and a cover material, such as soil or grass, is placed on the surface.

sanitary seal Any device or system that creates a protective union between two mechanical or process system elements.

sanitary survey An on-site review of a water utility's water source, facilities, equipment, and operations and maintenance records for the purpose of evaluating the system's adequacy in producing and distributing safe drinking water.

sanitation The improvement of environmental conditions favorable to health; measures designed to prevent disease through an environmental health program that includes provisions for safe food and drinking water, satisfactory quality of indoor and outdoor air, proper collection and treatment of domestic and industrial wastewater, and proper management of industrial and agricultural runoff and solid and hazardous wastes.

saprophyte An organism living on dead or decaying organic matter. Saprophytes help the natural decomposition of organic matter in water.

saprophytic Living on dead or decaying matter.

saprophytic bacteria Bacteria that live on dead organic matter.

SAR *See* sodium adsorption ratio.

SARA *See* Superfund Amendments and Reauthorization Act.

satellite management Operations and maintenance services provided through contracts with one or more other water systems.

saturable kinetics A feature of a chemical process when that process is independent of the substrate concentration (i.e., of zero order). This term is commonly used in toxicology when a process's capacity for clearing a compound by a metabolic or excretory pathway is exceeded. Doses of an administered toxic compound that exceed these capacities will produce exaggerated responses relative to lower doses. However, if a metabolite is responsible for the effect rather than a toxic compound, the metabolite will flatten (i.e., make less exaggerated) the relationship between external dose and the measured response. *See also* reaction order.

saturate (1) To fill all the empty spaces between soil particles with a liquid. (2) To form a solution of the highest possible concentration under a given set of physical conditions in the presence of an excess of the solute. (3) To fill something to capacity.

saturated (1) Pertaining to a material that can absorb no more of a second material. Saturated soil has its void spaces completely filled with water, so any water added will run off and not soak in. (2) Pertaining to a chemical in which all available valence bonds of an atom (especially carbon) are attached to

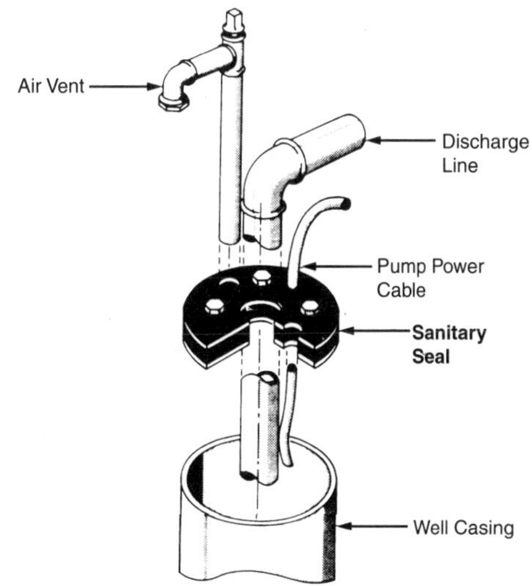

Components of a sanitary seal

other atoms. The straight-chain alkanes ($R–CH_2–R'$) are typical saturated compounds. *See also* alkane; unsaturated; valence.

saturated air Air containing as much water vapor as possible at a given temperature and pressure. *See also* relative humidity.

saturated liquid A liquid that contains as much of a solute as it can retain in the presence of an excess of that solute at a given temperature.

saturated rock A rock that has all its interstices or void spaces filled with water.

saturated soil A soil sample for which all the pore space is occupied by some liquid. Typically, the liquid is water, but it could be some other liquid or a mixture of liquids. In the latter case, a modifier is usually added to the term, e.g., a gasoline-saturated soil.

saturated solution A solution that contains the maximum amount of the dissolved substance (solute) that it can normally hold at this temperature. *See also* supersaturated.

saturated zone The portion of the subsurface where the pore space is completely occupied by water at or in excess of atmospheric pressure.

saturation The condition of a liquid when the liquid has taken into solution the maximum possible quantity of a given substance at a given temperature and pressure.

saturation capacity The maximum capacity of an ion exchange column, usually expressed in terms of kilograms of exchanged material per cubic foot of exchange resin (kilograms per cubic metre).

saturation extract The extract from a sample of soil that has been saturated with water.

saturation index (SI) A numerical value obtained by calculation from a water analysis. Such an index is intended to make predicting the scale-forming or scale-dissolving tendencies of water possible. *See also* Langelier saturation index; Riddick index; Ryznar stability index.

saturation line A horizontal line on a cross section of an earth dam, dike, or levee that marks the uppermost limit of water flow through the dam and subsoil. Material below the line of saturation will be saturated with water. *See also* phreatic line.

saturation pH The pH at which a water is saturated with a compound, normally calcium carbonate ($CaCO_3$).

saturation point The point at which a solution can no longer dissolve any more of a particular chemical. Precipitation of the chemical will occur beyond this point.

saturation zone That portion of the lithosphere in which the functional interstices of permeable rocks are saturated with water under hydrostatic pressure.

saturator A piece of equipment that feeds a sodium fluoride (NaF) solution into water for fluoridation. A layer of sodium fluoride is placed in a plastic tank and water is allowed to trickle through the layer, forming a constant-concentration solution that is fed to the water system. A saturator can also be used for feeding lime ($Ca(OH)_2$).

sausage dam A low dam constructed of cylinders of loose rock formed by wire wrapping. The cylinders are laid in either a horizontal or a vertical position. Such dams are not impervious and are used primarily to reduce erosion or retard debris flows.

SAV *See* submerged aquatic vegetation.

SCADA *See* supervisory control and data acquisition.

scale A coating or precipitate deposited on surfaces. Examples include oxide flakes forming on metal surfaces, carbonate (CO_3^{2-}) precipitates depositing in piping systems or water heaters, sulfate (SO_4^{2-}) scale precipitating in desalting systems, or silica (SiO_2) scale forming in boilers. Waters that contain carbonates or bicarbonates (HCO_3^-) of calcium or magnesium are especially likely to cause scale when heated. Scale is also called hard water scale.

scale inhibitor *See* scale prevention compound.

scale prevention compound A chemical additive that prevents the formation and buildup of a scale that can be caused by supersaturation of a compound in the bulk water. For example, the addition of sodium polyphosphate ($Na_{n+2}P_nO_{3n+1}$) can reduce precipitation and scaling associated with supersaturation of calcium carbonate ($CaCO_3$) in potable water. Scale prevention compounds are often used to prevent scaling in boilers. They are also called scale inhibitors or antiscalants.

scaling The deposition of scale on a surface. *See also* scale.

scanning A technique for capturing and converting analog data to digital form. A scanner uses an optical laser or other electronic device to scan an existing map and convert its images to digital format. Most scanners produce digital data in raster image format, recording a value of dark (e.g., representing a line or symbol) or light (no line or symbol) for each grid cell or picture element (pixel) of the scan. *See also* picture element; raster image.

scanning confocal laser microscope (SCLM) A special laser light source microscope that keeps illumination and detection confined to the same spot in a specimen at any one time. Only what is in focus is detected, registered electronically, and contributes to generation of a complete optical image of the specimen. To build up an image, the point probe is scanned over the field of view and the final image is generated electronically from a serial signal derived from the output of one or more photomultipliers.

scanning confocal laser microscopy The practice of using a scanning confocal laser microscope.

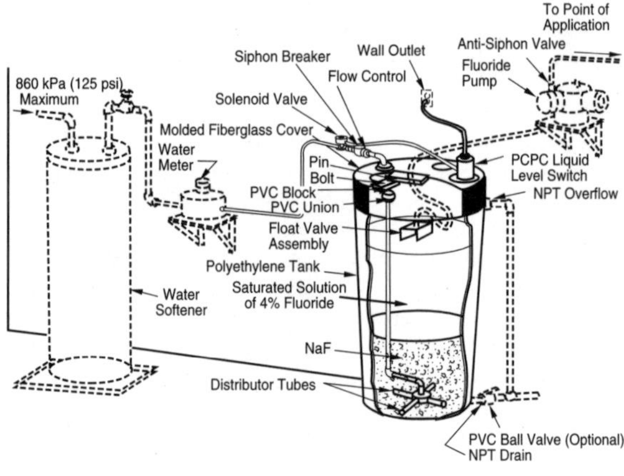

One type of saturator

scanning electron microscope (SEM) A type of electron microscope that operates at a much lower voltage than a transmission electron microscope and for which the electron beam operates as a probe by being deflected across the surface of the specimen by deflection coils within the bore of the final lens. The whole scanning electron microscope can be regarded as a slow-scan TV system with the microscope column acting as a very high-resolution camera tube and the specimen as the camera tube target. *See also* transmission electron microscope.

scanning electron microscopy The practice of using a scanning electron microscope.

scatter diagram A graphical depiction of the relationship between one variable and another, used to screen for possible cause-and-effect relationships.

scavenger A polymer matrix or ion exchanger that is used specifically to remove organic species from the feedwater before the water is to pass through a deionization process.

scavenger well A well that pumps brackish water from beneath a freshwater lens to control saline water intrusion.

scavenging A process in which free radicals are consumed or inactivated by a radical scavenger. *See also* free radical; radical; radical scavenger.

SCBA *See* self-contained breathing apparatus.

SCD *See* Soil Conservation District; streaming current detector.

SCFM *See* standard cubic feet per minute *in the Units of Measure list.*

schedule, pipe *See* pipe schedule.

schema The set of characteristics of computer files that contain nongraphic attributes. The schema holds information such as the names of data elements, the size of element fields in bytes or columns, the data element format (e.g., alpha, integer, binary), and other data required by the software to process attribute data.

schmutzdecke The layer of solids and biological growth that forms on top of a slow sand filter, allowing the filter to remove turbidity effectively without chemical coagulation. *See also* biofilm; slow sand filtration.

Schumberger array *See* dipole array.

Science Advisory Board (SAB) An independent advisory board of scientists and engineers initially established in 1973 as part of the US Environmental Protection Agency's Office of Research and Development and moved to the US Environmental Protection Agency administrator's office in 1976. Congress provided statutory authorization for the Science Advisory Board in 1978 under the Environmental Research, Demonstration, and Development Authorization Act (42 U.S.C. Section 4365). The Science Advisory Board reviews and provides advice and comments on the scientific adequacy of proposed criteria, documents, standards, limitations, or regulations issued by the US Environmental Protection Agency. The Safe Drinking Water Act requires the US Environmental Protection Agency to request comments from the Science Advisory Board before proposing any maximum contaminant level goal or National Primary Drinking Water Regulation.

scientific notation A means of expressing any number as a number between 1 and 10 (excluding 10 itself) multiplied by a power of 10. For example, 1,300,000 would be 1.3×10^6; 0.065 would be 6.5×10^{-2}.

scintillation cocktail A solution used in the analysis of radionuclides. These solutions can be prepared in the laboratory or purchased commercially. They are often composed of one or more organic solvents containing a scintillator. A method for the determination of radon in water makes use of a scintillation cocktail as both an extraction solvent and a medium containing a scintillator. Particles from the radioactive decay of radon interact with a scintillator that then emits photons. Those photons are then detected in a liquid scintillation counter. *See also* scintillator.

scintillation counter, liquid *See* liquid scintillation counter.

scintillator A substance that produces flashes of photons upon interaction with radioactive particles. A liquid scintillator is typically an organic compound dissolved in an organic solution (scintillation cocktail). Radioisotopes, such as radon-222, carbon-14, and tritium emit beta particles that interact with the scintillator, which in turn emits photons. These photons are detected and counted by an instrument, such as a liquid scintillation counter. *See also* scintillation cocktail.

SCLM *See* scanning confocal laser microscope.

scoop wheel A type of pump consisting of a series of flat vanes revolving in a curved channel. A scoop wheel is practically a reverse undershot waterwheel and is used for drainage pumping.

scour (1) The action of a flowing liquid as it lifts and carries away the material on the sides or bottom of a waterway, conduit, or pipeline. (2) The enlargement of a flow section of a waterway through the action of the fluid in motion carrying away the material composing the boundary. *See also* scouring velocity.

scouring sluice An opening in a gate-controlled dam through which accumulated silt, sand, and gravel may be ejected.

scouring velocity The minimum velocity necessary to dislodge stranded material from the boundary of a waterway, conduit, or pipeline by a fluid in motion.

scraper (1) A device for insertion in pipelines that is pushed or hauled through by some method or device (such as water pressure, rope, or cable) to remove accumulated organic or mineral deposits. Scrapers are used principally in pipes too small for access by persons and are of various designs and sizes. (2) A device used in the bottom of a sedimentation tank to move settled sludge to a discharge port. (3) A blade used to separate accumulated sediment from filter or screen surfaces. *See also* polypig; squeegee.

screen (1) A device with openings, generally of uniform size, used to retain or remove suspended or floating solids in flowing water to prevent them from entering an intake or passing a given point in a conduit. The screening element may consist of parallel bars, rods, wires, grating, wire mesh, or perforated plate, and the openings may be of any shape, although they are usually circular or rectangular. (2) A sieving device used to segregate granular material, such as sand, crushed rock, and soil, into various sizes. *See also* surface screen; well screen.

screen chamber A chamber in which screens are installed.

screen copy device A device used to generate a small-format hard copy of a geographic information system display screen.

screen size *See* mesh size. *See also* geographic information system.

screened well A well into which water enters through one or more screens.

screening A pretreatment method using coarse screens to remove large debris from the water to prevent clogging of pipes or channels leading to the treatment plant and to protect pumps.

screenings (1) Material removed from liquids by screens. (2) Broken rock, including the dust, of a size that will pass through a given screen, depending on the character of the stone.

screenings dewatering The removal of a large part of the water content of waste screenings by draining or by mechanical means.

screenings shredder A device that shreds screenings.

screw conveyer A device that axially transports liquid–solid mixtures by means of a rotating screw-type shaft that progressively moves the mixture between the screw threads along the interface between the screw and a conveyor belt.

screw impeller The helical impeller of a screw pump.

screw pump A special type of rotary positive-displacement pump in which the fluid is carried between screw threads on one or more rotors and is displaced axially as the screws rotate. Because screw pumps employ a low internal velocity, they offer a number of advantages in applications where liquid agitation is objectionable or where the specific gravity, viscosity, or both of the fluid to be pumped are relatively high.

screwed pipe A pipe provided with threaded ends and connected by threaded couplings.

screw-feed pump A pump with either horizontal or vertical cylindrical casing, in which a runner with radial blades like those of a ship's propeller operates. *See also* horizontal screw pump; vertical screw pump.

SCS *See* Natural Resources Conservation Service.

S-curve hydrograph A graph showing the summation of the ordinates of a series of unit hydrographs spaced at unit rainfall duration intervals. Such a graph represents the hydrograph of a unit rate of rainfall excess indefinitely continued. If the ordinate values of an S-curve are a percentage of the total unit hydrograph volume at any given time, with the abscissa in units of time, the S-curve is known as a summation graph.

SDE *See* simultaneous distillation extraction.

SDI *See* silt density index; sludge density index.

SDS *See* simulated distribution system test.

SDS-DBP *See* simulated distribution system disinfection by-product concentration.

S&DSI *See* Stiff-and-Davis stability index.

SDS-THM *See* simulated distribution system trihalomethane concentration.

SDWA *See* Safe Drinking Water Act.

SDWIS *See* Safe Drinking Water Information System.

SDWR *See* Secondary Drinking Water Regulation.

S_e *See* standard error of the estimate.

SEA *See* size exclusion analysis.

sea level In general, the surface of the sea used as a reference for elevation; a curtailed form of the term *mean sea level.*

seal Packing gland material or a mechanical device that fits around a pump shaft and prevents either air intake to the pump or water leakage.

sea-level datum A determination of mean sea level that has been adopted as a standard datum for elevation. The sea level is subject to some variations from year to year; however, because the permanency of any datum is of prime importance in engineering work, a sea-level datum should, in general, be maintained indefinitely after adoption even though it may differ slightly from later determinations of mean sea level based on a longer series of observations. The sea level used for the US Coast and Geodetic Survey level net is officially known as the sea level datum of 1929 (the year refers to the last general adjustment of the net). The datum itself can be considered to be an approximation based on observations of the tide at various tide stations along the coasts of the United States over a number of years. *See also* mean sea level; National Geodetic Vertical Datum of 1929.

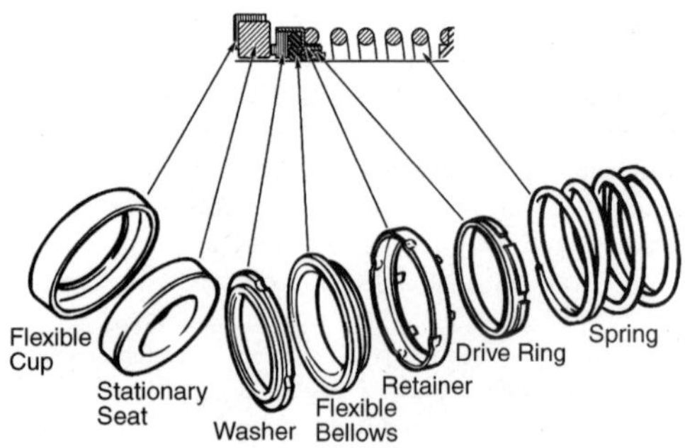

Parts of a mechanical seal

season A division of the year generally determined by some annually recurrent natural phenomenon, such as the state of vegetation or the meteorological conditions.

seasonal Related to, characteristic of, or depending on the season or seasons.

seasonal analysis Analysis of seasonal water use by a water utility. Monthly water consumption, particularly for single-family residential accounts, tends to follow a highly consistent pattern from year to year. In areas that have snowy winters, winter consumption represents indoor water use, and summer consumption in excess of winter use is primarily used for irrigation. Spring and fall consumption reflect the lower irrigation requirements during periods of moderate temperatures. Rainfall will affect water consumption primarily during summer months but also during hot weather in the spring and fall. A very useful method of quantifying the seasonal pattern of water use is to calculate a seasonal index. Most statistical programs use a moving average method to determine the seasonal influence of each month. The most frequently used method is to compute a centered 13-month moving average of the time series data that is centered on the middle, i.e., the seventh month. Because the 13 months exceed the period of a year, the seasonal or monthly pattern is removed in the average. Then each month is expressed as a ratio with respect to the moving average, and the ratios for all the January data, all February data, and so on are averaged to derive a typical index (or ratio with respect to the average) for each month. The sum of all the monthly ratios equals 12. The seasonal index can be used for seasonally adjusting water sales—removing the influence of season (by dividing actual sales by the index)—or for forecasting future monthly sales (by multiplying the moving average by the index).

seasonal depletion Withdrawal of groundwater or surface water from a source at a rate in excess of the rate of supply during a given season, but not in excess of the average supply over a secular cycle. (Withdrawal at a rate in excess of the average rate of supply over a secular cycle is termed *cyclic depletion.*)

seasonal index *See* seasonal analysis.

seasonal rate A seasonally based water rate, with the summer rate higher than the winter rate, sometimes established either to recover the additional cost associated with peak summer water production or as an inducement for customers to conserve during the peak-use season. The summer rate can be applied to all summer water use or only for summer use in excess of winter use. The latter approach is preferred by most rate economists because it allows for a relatively large summer surcharge (e.g., to induce conservation) without driving down the winter rate simply to maintain annual revenues at the same level as before the change. Seasonal rates can also include an inclining block design. *See also* inclining block rate; revenue-neutral rate.

seasonal recovery The general rise of the elevation of the water table caused by the addition of water during or after the wet season. *See also* seasonal depletion.

seasonal storage Storage of water in a reservoir during that portion of the year when an excess or surplus occurs in the source of supply.

seasonal water table A water table that rises and falls substantially with the seasons as a result of either natural causes (seasonal rainfall or fluctuations in the levels of nearby rivers) or human activity (seasonal irrigation). *See also* water table.

seawater *See* ocean water; salt water.

seawater intrusion The invasion of a body of fresh water by seawater. Both coastal surface water and groundwater can be impacted by seawater intrusion. Seawater intrusion in surface water supplies is common during drought, high-tide, or low-river-flow conditions. *See also* saline water intrusion.

SEBST *See* single-element bench-scale test.

Secchi disk A circular metal plate, 8–12 inches (20–30 centimetres) in diameter, that is painted in black and white quadrants. A Secchi disk is used to measure the clarity of water and determine visible light extinction coefficients.

second(s) *See the Units of Measure list.*

second feet *See the Units of Measure list.*

secondary Pertaining to the low-voltage side of a step-down transformer. *See also* primary.

secondary carcinogen A chemical that requires metabolism to produce cancer. Generally, such chemicals are metabolized to a reactive form that interacts covalently with nucleophilic centers in deoxyribonucleic acid. (This is an older term.)

Secondary Drinking Water Regulation (SDWR) A US Environmental Protection Agency nonenforceable regulation for a contaminant that may adversely affect the taste, odor, or appearance of water. These effects may cause an adverse impact on the public welfare, e.g., by a substantial number of people discontinuing use of the water provided by a public water system. Secondary Drinking Water Regulations establish secondary maximum contaminant levels.

Secondary Drinking Water Standards *See* Secondary Drinking Water Regulations.

secondary interstice An opening in rock developed by processes that affected the rock after it was formed.

secondary maximum contaminant level (SMCL) A nonenforceable numerical limit set by the US Environmental Protection Agency for a contaminant on the basis of aesthetic effects to prevent an undesirable taste, odor, or appearance that would have an adverse impact on the public welfare. *See also* Secondary Drinking Water Regulation.

secondary opening *See* secondary interstice.

secondary soil Soil developed through the action of weathering and other processes on materials, originating from previously existing soils and rock debris that have been eroded from their former locations and redeposited by the transporting agents; transported soil.

second-foot day *See the Units of Measure list.*

section (1) A view of something as it would appear if cut through by an intersecting plane. (2) *See the Units of Measure list.*

sectionalizing valve A large valve installed in a pipeline to shut off flow in a section of the pipeline for the purpose of inspection or repair. Such valves are usually installed in the main pipelines.

sector gate A roller type of crest gate in which the roller is constructed as a sector of a cylinder instead of as a complete cylinder (i.e., the roller has a cross section that resembles a piece of pie).

secular cycle A period of time that includes a group of years during which the precipitation is, in general, considerably greater than average, as well as a group of years during which it is, in general, considerably less than average.

sedentary soil Soil formed in place without removal from the site of the original rock from which it was formed. Sedentary soil is also called residual soil.

Sedgwick–Rafter cell A special microscope cell used in the Sedgwick–Rafter method. It is about 50 millimetres long, 20 millimetres wide, and 1 millimetre deep. It holds 1 millilitre of sample to allow counting of the organisms.

Sedgwick–Rafter filter A cylindrical funnel used as a filter for concentrating organisms in the Sedgwick–Rafter method for the quantitative determination of organisms that are microscopic but larger than bacteria in water. *See also* Sedgwick–Rafter method.

Sedgwick–Rafter method A method for the quantitative determination of organisms in water that are microscopic but larger than bacteria. This method involves a low-magnification (less than 200 times) microscopic examination of water. A Sedgwick–Rafter cell is often used in the quantification of phytoplankton in water. *See also* Sedgwick–Rafter cell; Sedgwick–Rafter filter.

sediment (1) Solid material settled from suspension in a liquid. (2) Mineral or organic solid material that is being transported or has been moved from its site of origin by air, water, or ice and has come to rest on the ground surface either above or below sea level. (3) Inorganic or organic particles originating from weathering, chemical precipitation, or biological activity. (4) Any solid phase settling out of a liquid phase, e.g., deposits in rivers and lakes or sludge in clarifiers.

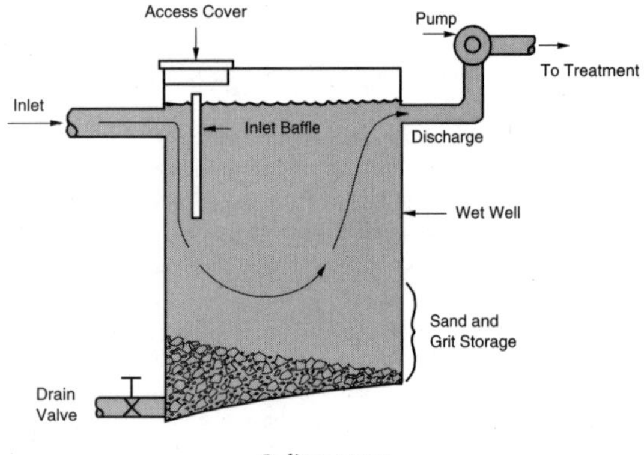

Sediment trap

sediment concentration The ratio of the weight of the sediment in a water–sediment mixture to the total weight of the mixture. The sediment concentration is sometimes expressed as the ratio of the volume of sediment to the volume of mixture. It is dimensionless and is usually expressed as a percentage for high values of concentration and in parts per million for low values.

sediment discharge Usually the mass, but sometimes the volume, of sediment passing a stream transect in a unit of time.

sediment discharge curve A graphical presentation of the relationship between the stage of a river and the sediment discharge.

sediment erosion Detachment of sediment particles by water, wind, ice, or gravity.

sediment load The quantity of sediment that passes a cross section of a stream or river in a specified period of time.

sediment source area That area of a drainage basin that yields a large amount of eroded material.

sediment station A river section where samples of suspended load are taken either each day or periodically.

sediment transport curve A graph showing the relationship between the water discharge, in cubic feet (metres) per second, and the amount of sediment in transport in a stream channel, in tons (kilograms) per day.

sediment trap A device, often a simple enlargement in cross-sectional area, placed in a conduit to arrest, by deposition, the sand or silt carried by the water. A sediment trap usually includes a means for ejecting the settled material. A sediment trap is also called a sand trap. *See also* sand trap.

sediment yield The total amount of suspended sediment and bed load outflow from a drainage basin, usually expressed in terms of mass or volume per unit time.

sedimentary rock Consolidated rock formed from material that was either deposited in layers from suspension or precipitated from solution.

sedimentation A treatment process using gravity to remove suspended particles.

sedimentation basin A basin or tank in which water is retained to allow settleable matter, such as floc, to settle by gravity. A sedimentation basin is also called a settling basin, settling tank, settler, or sedimentation tank.

sedimentation compartment That portion of a water treatment tank used as a settling tank; the settling section of a chemical flocculation and sedimentation unit.

sedimentation tank *See* sedimentation basin.

sedimentation test A test in which a soil sample and a jar of water are used to determine the soil's composition and characteristics for excavations.

seep (1) A more or less poorly defined area where water oozes from the earth in small quantities. (2) To appear or disappear, as water or other liquid, from a poorly defined area of the ground surface. (3) *See* seepage.

seep water Water that has passed through or under a levee or dam.

seepage The slow movement of water or other fluid downward through unsaturated soil, rocks, artificial structures, or other porous media. It is to be distinguished from percolation, which is the predominant type of movement of water in saturated material.

seepage area An area where discharged groundwater appears at the ground surface, often a sizable area. Water in seepage areas may pond and begin to move as a current of flowing water.

seepage boil A concentrated outcropping of seepage landward of a levee or downstream of a dam.

seepage face The rock or soil face at which a seepage area occurs.

seepage loss The loss of water through a dam by capillary action and slow movement.

seepage spring A spring in which the water discharges from numerous small openings in permeable material. A seepage spring is also called a filtration spring.

seepage velocity The velocity of water flowing in a porous medium, obtained by dividing the total discharge by the product of the cross-sectional area and porosity. *See also* actual groundwater velocity; pore velocity.

segmental degeneration A neuropathologic condition characterized by loss of myelin from neurons (nerve cells) but with the axon (the portion of the nerve that carries electrical signals away from the cell body of a nerve) largely spared.

seiche An oscillation of the water surface of a lake or other large landlocked body of water as a result of unequal atmospheric pressure, wind, or other cause that sets the surface in vibration. The amplitude of the oscillation may be 1 foot (30 centimetres) or more, and the period may reach several hours. *See also* storm surge.

seize-up A mechanical condition that occurs when an engine overheats and a part expands to the point where the engine will not run. Seize-up is also called freezing.

select backfill Backfill material of a specified quality, often imported from a different location to meet a particular specification.

selection bias In epidemiology, systematic error caused by differences in characteristics among those selected for study. Selection bias occurs only as a result of poor study design, and it must be prevented because it cannot be controlled during the analysis. Preventing selection bias requires exposed and unexposed groups to be selected by the investigator without knowledge of disease, or cases and controls must be selected without knowledge of exposure.

selective adsorption A method used in gas chromatography to separate organic compounds so that their concentrations can be determined.

selective ion electrode A device used in the electrochemical analysis of specific ions in solution. For example, pH (i.e., hydrogen ion), fluoride, and calcium can be determined in water by the use of electrodes designed for the particular ion.

selective ion exchanger An ion exchange medium that shows selectivity, e.g., a chelating ion exchange resin that will remove only gold ions from solution. *See also* selectivity.

selective ion probe *See* selective ion electrode.

selective membrane A sheet of material that is preferentially selective to the passage of either cations or anions in solution.

selective toxicity Adverse effects that apply to specific organisms. This concept has been used in the development of various pesticides and drugs to disrupt metabolic pathways so as to render the target organism uniquely sensitive to the substance's toxic effects. For example, penicillin displays selective toxicity to bacteria because it inhibits the synthesis of the cell wall; animals do not synthesize cell walls, so penicillin is not toxic to animals.

selectivity (1) The relative capacity of an ion exchange resin to remove different ions of similar charge (i.e., co-ions). (2) For pressure-driven membrane processes, the relative permeability of a solute through the membrane. (3) The ability of an analytical test method for a microorganism to encourage growth of the target organism while retarding the development of nontarget organisms.

selectivity band The respective region or zone within an ion exchanger or adsorption medium bed where individual ions or substances accumulate and are removed from the water in the order of their individual respective preferences for the medium. Because different substances each have different affinities or selectivity preferences for the treatment medium, they are removed in different zones of the medium bed. The removal zone for the ion or substance with the lowest selectivity moves downward through the bed most quickly, and the zone for the highest selectivity substance moves downward through the bed least quickly. *See also* chromatographic behavior; mass transfer zone; selectivity.

selenium (Se) A nonmetallic element that occurs in certain soils. This inorganic chemical has various industrial uses, including in electronics and as a trace element in animal feed. Selenium is regulated in drinking water by the US Environmental Protection Agency.

self-backwashing filter A filter whose hydraulic design allows backwashing with the filtered water from operating filters without pumping. Filtered water effluent weirs are located at an elevation higher than the surface of the backwash gullet, allowing the filtered water to flow upward through the filter being backwashed by gravity. *See also* gullet.

self-cleansing velocity The minimum velocity in a water supply intake necessary to keep solids in suspension, thereby preventing them from depositing and from subsequently creating a nuisance in the form of stoppages.

self-contained breathing apparatus (SCBA) A full-face respirator connected to a tank that contains a 30- or 60-minute air supply of grade D air. *See also* grade D air.

self-maintaining system An ecosystem that can perform all of its natural ecological functions without human intervention or dependence on engineered structures.

seltzer water *See* soda water; bottled sparkling water.

SEM *See* scanning electron microscope; standard error of the mean.

SEMI *See* Semiconductor Equipment and Materials International.

semiarid Neither entirely arid nor strictly humid, but intermediate, with a tendency toward an arid character. This term may be applied to dry farming country in which many crops grow without irrigation, but in which far better yields result from irrigation. *See also* semihumid.

semiarid climate A climate characterized by 10 to 20 inches (25 to 50 centimetres) of annual rainfall.

semibatch reactor A reactor in which one of the reactants is added continuously but the other component was added in the initial batch of water. For example, in a semibatch ozone reactor, ozone (O_3) is continuously added to a static volume of liquid.

semibolson A topographic basin that has an inward-flowing drainage system, is more or less longitudinal in shape, and is drained by an intermittent stream.

Semiconductor Equipment and Materials International (SEMI) The group that has set the accepted standards for electronics-grade purified water.

semiconductor A material that can be easily forced to carry electricity only under certain circumstances.

semiconfined aquifer *See* leaky aquifer.

semidiurnal Having a duration of approximately one-half of a tidal day. The predominating type of tide throughout the world is semidiurnal, with two high waters and two low waters each tidal day.

semihumid Neither entirely arid nor strictly humid, but intermediate, with a tendency toward a humid character. Sufficient moisture to raise all crops may be present, but irregularity of precipitation during the year makes providing irrigation facilities essential to raising better crops. *See also* semiarid.

semihydraulic-fill dam An earth dam for which, during construction, some of the material was transported to the dam by some means other than water and some of it was moved into place by the action of water.

seminal vesicle A structure that secretes seminal fluid, the fluid that provides nutrients to sperm and serves as a carrier for sperm upon ejaculation.

semiperched Pertaining to an overlying body of groundwater that has a greater pressure head than an underlying body of water but is not separated from the underlying body by unsaturated rock.

semiperched water Groundwater that has a greater pressure head than an underlying body of groundwater from which the first type of groundwater is not completely separated hydraulically. Semiperched water can occur in layered aquifer systems where an upper aquifer is in contact with a lower aquifer and the lower aquifer has less water pressure than the upper aquifer. In such cases, the upper aquifer is said to be semiperched. The upper aquifer can have greater water pressure because its net recharge exceeds that of the lower aquifer. *See also* semiperched water table.

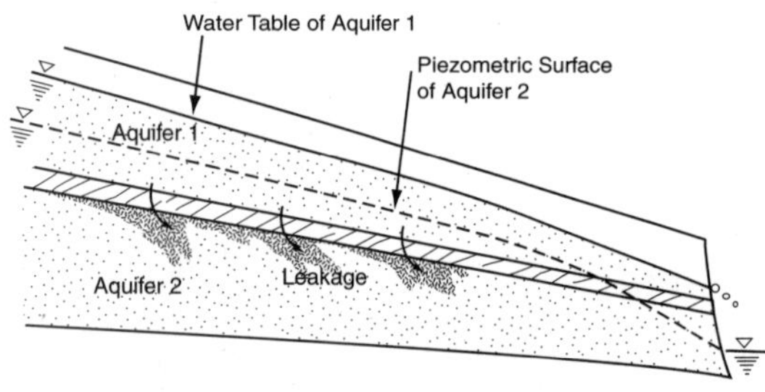

Semiperched water

semiperched water table The upper surface of a body of semiperched water. This is the water table of the hydraulically connected underlying water, too. *See also* semiperched water.

semipermanent snow line The line of lowest elevation or the lower limit of a snow field at any locality on the ground surface above which snow accumulates and remains on the ground continuously for a considerable portion of the year.

semipermeable membrane *See* permselective membrane.

semivariable costs Those water utility costs that are not completely fixed with respect to treated water volume but that do not vary directly with volume, either. Examples are indirect labor, overtime premiums, portions of chemical and energy costs, and costs that are reassigned to meet peak demands. In the short run, most of the costs of operating and maintaining a water utility are fixed; i.e., they do not vary with the volume of water produced. Examples are facility rent or debt service, leases, insurance, supplies, and so forth. In most utilities, labor is also a fixed (or programmed) cost for a given scope of work. A few costs vary directly with volume, such as some chemicals and some pumping power costs. *See also* long-run average cost.

senescent Having aged or grown old, in a physiological sense. This term is used to describe the aging process of cells or whole organisms.

sensitivity (1) The smallest amount of a chemical or microorganism that can be discriminated from a blank value with a specified degree of reliability. (2) The responsiveness of an organ or organism to external stimuli. (3) In analytical testing, the lowest practical detection level (not necessarily the minimum detection level). (4) In terms of a test method for a microorganism, the likelihood that the test result will be positive when the target organism is present. (5) The capacity of a detector to respond to a class of compounds or a specific analyte. In some methods, sensitivity is defined as the concentration of an analyte that produces a specific detector response. *See also* limit of detection; limit of quantitation; method detection limit; practical quantitation level; reliable detection level; reliable quantitation level.

sensitivity analysis An analysis of the sensitivity of the dependent variable in a mathematical expression as a function of variations in the value of any independent variables or coefficients associated with the independent variables. For example, in the following expression:

$$Y = ax_1 + bx_2$$

a sensitivity analysis would determine how much the value of Y is affected by changes in the values of a and b or by changes in the values of x_1 and x_2. In hydraulics, a sensitivity analysis might be conducted to determine the effect of small changes in the assumed values for known heads, consumptions, and pipe or pump characteristics on a hydraulic model's solutions. Sensitivity analysis can be done automatically by a computer or by hand.

sensitivity control A measure of the smallest change in input that a device can detect and react to.

sensitivity level The lowest concentration that can be detected and quantified by a test method. *See also* limit of detection; limit of quantitation; method detection limit; practical quantitation level; reliable detection level; reliable quantitation level; sensitivity.

sensitization The initial exposure of an individual to an antigen. This exposure stimulates the synthesis of antibodies that will precipitate an immune response with subsequent exposures. This term is most frequently applied to circumstances in which the secondary exposure leads to hypersensitivity reactions. Sensitization also appears to be involved in autoimmune disorders that appear after chronic exposures to certain toxic chemicals.

sensor An instrument that measures (senses) a physical condition or variable of interest. Floats and thermocouples are examples of sensors. A sensor is also called a primary element.

separation line In a solids-contact unit, the interface of the clarified liquid and the sludge blanket.

separation process An operation—possibly including more than one unit process—in which the various components of a mixture are isolated. Some important separation processes are evaporation, distillation, drying, gas adsorption, sedimentation, solvent extraction, pressure extraction, adsorption, and filtration.

separation technique Any of a variety of analytical and preparatory techniques for the purpose of distinguishing between solutes or solvent. Extraction, chromatography, and distillation are three examples of separation techniques.

separator In desalting, a device designed to remove entrapped droplets from a vapor stream. A separator is also called a demister entrainment unit. *See also* spacer.

sepralator A spiral wound membrane element or cartridge used in crossflow membrane systems.

septic Pertaining to a condition produced by bacteria in water when all oxygen supplies are depleted. If this condition is severe in a body of water, the bottom deposits produce hydrogen sulfide (H_2S), the deposits and water turn black and give off foul odors, and the water will have a greatly increased chlorine demand.

septic system An on-site system designed to treat and dispose of domestic wastewater. A typical septic system consists of a tank that receives wastes from a residence or business and a system of tile lines or a pit for disposal of the liquid effluent that remains after the solids settle out in the tank.

septic tank A tank (usually underground) into which the solid matter of household wastewater flows and is held to settle and for some degree of decomposition by bacteria. Septic tanks are common in rural areas where no municipal wastewater system is available. A proper septic tank facility must also include an adequate tile field for disposal of the liquid. *See also* tile field.

septicemia A systemic disease associated with the presence and persistence of pathogens or their toxins in the blood.

septicity A condition characterized by growth of anaerobic organisms.

septum (1) Any permeable material that supports filter media. (2) The lining of a bottle cap that can be pierced to draw a sample. (3) The covering on an inlet port to a chromatograph or other piece of analytical equipment that can be pierced to inject a sample.

sequent depths The depths before and after a hydraulic jump. Sequent depths are also called conjugate depths.

sequential composite sample A sample obtained either by continuously pumping water from the liquid stream being sampled or by mixing equal volumes of water collected from the liquid stream being sampled at regular time intervals.

sequential control Process control implemented via a series of discrete steps.

sequester To keep a substance (e.g., iron or manganese) in solution through the addition of a chemical agent (e.g., sodium hexametaphosphate) that forms chemical complexes with the substance. In the sequestered form, the substance cannot be oxidized into a particulate form that will deposit on or stain fixtures. Sequestering chemicals are aggressive compounds with respect to metals, and they may dissolve precipitated metals or corrode metallic pipe materials. *See also* complexation; hexametaphosphate; sodium hexametaphosphate.

sequestering A chemical reaction in which certain chemicals (sequestering or chelating agents) "tie up" other chemicals, particularly metal ions, so that the chemicals no longer react. Sequestering agents are used to prevent the formation of precipitates or other compounds.

sequestering agent (1) A chemical that forms complexes with metallic ions in solution so that the metallic ions may no longer be precipitated. For example, calcium soap precipitates are not produced from hard water treated with sodium hexametaphosphate. (2) Any agent that prevents an ion from exhibiting its usual properties because of close combination with an added material. *See also* hexametaphosphate; sequester; sodium hexametaphosphate.

sequestration *See* sequestering.

serial correlation A statistical condition in which errors for different periods of time series data are correlated. Serial correlation is very common in time series analyses, such as those dealing with monthly water consumption. A tendency exists for errors (residuals) in one period to be correlated with errors in preceding periods. If these patterns are predictable (i.e., if they can be defined with a regression coefficient just like any other explanatory variable), then a coefficient should be determined for the serial correlation effect because including that coefficient will make the coefficients of the other independent variables more true to their actual causal influences. The standard test used to determine whether serial correlation exists in a time series is the Durbin–Watson test. The computed value of the Durbin–Watson test can range from 0, indicating extreme positive serial correlation, to 4.0, indicating extreme negative serial correlation. A Durbin–Watson value of 2.0 indicates no serial correlation. Generally, it is desirable for the Durbin–Watson value to be within a range of 1.7 to 2.3. The effect of serial correlation is usually demonstrated either by including the dependent variable as an independent variable lagged by one or more periods or by including an autoregressive variable as an independent variable. *See also* autocorrelation; autoregressive term; Durbin–Watson test.

serial filtration The arrangement of two or more filtering steps, one following the other, in order to remove increasingly finer particulates at each stage and to provide for filtration of all sizes of suspended solids. Cartridge-style units often employ this method, using depth prefilters (compressed fibers) followed by surface filtration with a micromembrane cartridge element.

serious violation An Occupational Safety and Health Administration's classification assigned to a workplace hazard that is deemed dangerous to a worker's well-being.

seroepidemiologic study A study based on serological testing of a change in the serum level of specific antibodies. Through such a study, asymptomatic infections and carriers can be detected in addition to clinical cases.

serological typing *See* serotype analysis.

serotype analysis An analysis in which microorganisms can be identified based on serological procedures that detect the presence of specific characteristic antigens. Such an analysis can be performed for any microorganism for which specific antibodies can be produced. Serotype analysis is used to identify particular groups within a large group of closely related organisms, such as bacteria in the genus *Salmonella*.

Sertoli cell A cell that mechanically supports developing spermatids and is thought to be involved in providing metabolic signals that trigger critical steps in the development of sperm.

serum The aqueous phase remaining in a blood sample following clotting. *See also* plasma; plasma water.

service The small water line—including such components as the meter, corporation stop, curb stop, and meter couplings—that provides a facility to move water from a water main to or into a building.

service age The period of time between the date when a new physical unit of property was put into service and the date when the physical unit is studied for decommissioning or rehabilitation. *See also* service life.

service box A valve box used with a corporation cock or curb cock.

service charge A fixed monthly charge that most utilities incorporate as part of their rate structures. This charge is intended to recover the costs of meter reading and servicing, accounting, billing and collection, customer service, and any other costs that the utility believes are directly related to the customer, as contrasted with the costs of producing and distributing water. Service charges are almost always allocated based on meter size or number of equivalent meters. Sometimes monthly service charges are added to the cost to provide service by water consumption block and are then identified as a fixed monthly minimum charge or a standby charge. *See also* base–extra-capacity approach to rate structure design.

service clamp (1) A clamp used to repair a pinhole or leak on a service line. (2) A full-circle repair clamp with an outlet tap for a corporation stop, used for repairing a leak on a water line near a service connection.

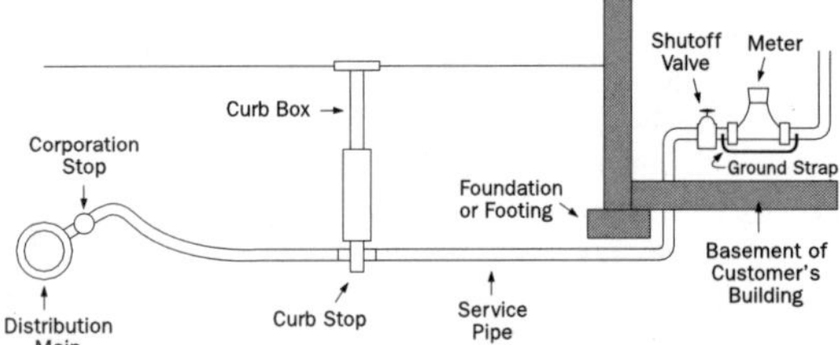

Small service connection with the meter located in a basement

service connection That portion of the service line from the utility's water main to and including the curb stop at or adjacent to the street line or the customer's property line, including valves and fittings but not the curb box.

service flow The rate at which a given water-processing system can deliver product water, expressed in gallons (litres) per minute. The rating may be for intermittent peak flow or constant flow.

service life The period of time between the date when a new physical unit of property was first put into service as a part of the property and the date when the unit was retired, or is anticipated to be retired, from service. *See also* service age.

service line The pipe and all appurtenances that run between the utility's water main and the customer's place of use, including fire lines.

service line sample A 1-litre sample of water that has been collected in accordance with Section 141.86(b)(3) of the Code of Federal Regulations and was standing for at least 6 hours in a service line prior to sampling.

service meter A water meter installed on a consumer's service line.

service outlet A device for releasing water at a dam for downstream uses.

service pipe The pipeline extending from the water main to the building served or to the consumer's system.

service reservoir A reservoir supplying a given district and capable of storing water during hours of low demand for use when demands exceed the capacity of the treatment plant(s) providing water to the district.

service run That portion of the operating cycle of a water-processing system during which the water is actually being treated. The other portion of the cycle is regeneration or cleaning. For example, that portion of the cycle in which the hard water supply is passed through a regenerated and rinsed bed of ion exchange material, thereby producing softened water, is a service run.

service unit Typically, a portable exchange tank. *See also* portable exchange tank.

sessile Pertaining to animals (e.g., Asiatic clams) that live attached to a substratum, such as the shell of another animal or a rock.

set The direction of a current of water.

set point The position at which a process controller is set. The set point is the same as the desired value of the process variable.

***S*-ethyl di-*N,N*-propylthiocarbamate** *See under* ethyl di-*N,N*-propylthiocarbamate.

setting A point or instruction for a controller, operator, or transmitter at which an action is to occur or specific rate is to be maintained. Examples include a fully open or closed valve position that triggers a limit switch to stop the travel of a valve operator; a level in a tank that causes a pump to come on or turn off; or a chemical feed rate to be maintained at a certain flow of water. The setting may be fixed or variable based on changing conditions, such as the positions of a valve based on a certain flow rate through the valve.

setting tee *See* hydrant tee.

settleability The tendency of suspended solids to settle.

settleability test A determination of the settleability of solids in a suspension by measuring the volume of solids settled out of a measured volume of sample in a specified interval of time, usually reported in millilitres per litre.

settleable solids The material settling out of a suspension within a defined period of time. *See also* settleability test.

settling *See* sedimentation.

settling basin *See* sedimentation basin.

settling chamber A chamber in which water containing settleable solids is retained to allow a part of the suspended matter to be removed by gravity, e.g., the settling portion of a combination flocculating–settling tank. *See also* sedimentation basin.

settling solids Solids that settle in a tank or chamber constructed for the purpose of removing this fraction of suspended solids. *See also* settleable solids.

settling tank *See* sedimentation basin.

settling time The time necessary for the removal of suspended or colloidal substances by gravitational settling, aggregation, or precipitation.

settling tube (1) A tall, vertical, transparent tube used to simulate the settling behavior of aqueous suspensions of particulates in the full-scale settling tanks used in water treatment. (2) An inclined tube placed in a settling tank (as part of a bank of such tubes) to improve solid–liquid separation. *See also* tube settler.

settling velocity The velocity at which subsidence and deposition of the settleable suspended solids in water will occur. *See also* standard fall velocity; Stokes's law.

settling zone The zone in a sedimentation basin that provides a calm area so that the suspended matter can settle.

severity rate *See* incident rate.

sewage *See* wastewater.

sex pili Thin filaments consisting of two protein fibers protruding from the surface of gram-negative bacteria.

SFC *See* supercritical fluid chromatography.

SFE *See* supercritical fluid extraction.

shading A physical method for controlling aquatic plants by limiting the amount of sunlight reaching the bottom of the water body.

shaft A cylindrical piece of metal connected to rotating machine parts to transmit power or motion for such equipment as an electric motor or centrifugal pump.

shaft spillway *See* glory-hole spillway.

shale A fissile rock formed by the consolidation of clay, mud, or silt. Shale has a finely laminated or stratified structure and is composed of minerals essentially unaltered since deposition.

shallow well A well sunk in easily penetrated ground to a point that is below the water table but usually less than about 30 feet (10 metres) in depth.

shallow-depth sedimentation A modification of the traditional sedimentation process using inclined tubes or plates to reduce the distance the settling particles have to travel to be removed. *See also* inclined plate separator; Pielkenroad separator; tube settler.

shallow-water deposit A deposit on the bed of the ocean made in water of a depth of about 100 fathoms (600 ft; 300 metres) or less.

sharp-crested weir A weir having a crest—usually consisting of a thin plate (generally metal)—so sharp that the edge over which the water passes is only a line.

sharp-edged orifice An orifice with sharp edges, such that the water passing through touches only the line of the edge. Such an orifice is also called a standard orifice.

sharps Objects that can puncture or cut the skin, such as needles, glassware, and scalpels, and that have come in contact with bodily fluids in laboratory, dental, or medical settings.

shear gate A pivoted gate, without guides, that is closed by water pressure and seating lugs that force the gate against a seat. This type of gate closes in one direction only and is generally used on drain outlet pipes.

shear vane *See* vane shear.

shearing action The action of water flowing at high rates that splits and separates particle agglomerations and prevents the formation of floc deposits during the coagulant feed and filtration process. This action may also tear away from the filter grains any previous deposits or suspended matter.

sheet erosion The gradual uniform removal of a section of ground surface without the formation of rills or gullies. This type of erosion may occur when water flows in a sheet down a sloping surface and removes material from the surface.

sheet flow Flow in a relatively thin sheet of generally uniform thickness.

sheet pile A structural element of timber, concrete, or steel of rectangular sheet shape rather than circular or square that is driven, jetted, or otherwise embedded on end into the ground to support a load or compact the soil.

sheet piling The portion of a shoring system that keeps the earth in place.

sheeting *See* sheet piling.

shell pump *See* sand pump.

shellfish Any aquatic animal with a shell, as a shelled mollusk or crustacean, especially an edible one, such as a clam, lobster, or shrimp.

shelter (1) A small housing providing protection against the elements. (2) A louvered box having openings on all sides and bottom to permit free flow of air and fitted with instruments to measure some condition, such as temperature, humidity, or air quality.

Sherwood number (N_{Sh}) The ratio of the actual mass transfer across the water boundary (as measured by the mass transfer coefficient) to the theoretical mass transfer across the water boundary (as calculated by the diffusion coefficient), divided by the thickness of the liquid film. In equation form,

$$N_{Sh} = \frac{k_L}{D \times d}$$

Where:

N_{Sh} = the Sherwood number, dimensionless
k_L = the mass transfer coefficient across the liquid boundary, in length per time
D = the diffusion coefficient, in length squared per time
d = the thickness of the "stagnant" water boundary, in length

Ideally, the Sherwood number should be 1, but 1.5 is more common.

Shewhart control chart A process control chart used to illustrate statistical quality control. The concept was developed in the 1930s by Walter A. Shewhart. Although these graphs can be of several varieties, a typical Shewhart chart is a plot of a series of replicated measurements compared with a mean value and control limits.

shield A structure built on-site in accordance with Occupational Safety and Health Administration's No. 1926.650 excavation standard or by manufacturers that specialize in excavation safety systems. Shields are able to withstand the forces that are imposed by the surrounding earth to prevent cave-ins and thereby protect workers.

shield system *See* shield.

shielding (1) The separation and insulation of metal parts of a pipe joint by means of a special fitting that will not conduct electric current. The fitting prevents corrosion caused by galvanic action between two different metals. (2) A protective cover or barrier that prevents transmittance of heat or radiation to or from a component of water treatment equipment.

Shigella A genus of gram-negative, rod-shaped, facultatively anaerobic bacteria belonging to the family Enterobacteriaceae. These organisms are intestinal pathogens of humans and other primates, causing bacillary dysentery. *See also* waterborne disease.

Shigella sonnei A bacterial species that causes dysentery. This species may be waterborne and is characterized by slow lactose fermentation. *See also Shigella*; waterborne disease.

shigellosis An acute enteric disease characterized by diarrhea, fever, nausea, and sometimes vomiting, cramps, and tenesmus (urge to defecate). Although some cases have watery diarrhea, typically stools contain blood and mucus. The infectious dose of the bacteria responsible is low for humans; 10–100 bacteria are known to have caused disease. In outbreaks, secondary attack rates can be as high as 40 percent. The only significant reservoir of the pathogen is people. Mild and asymptomatic infections occur, and illness is usually self-limiting, lasting 4 to 7 days. The illness can also be severe; the severity of illness and the case fatality rate depend on age and nutrition of host and *Shigella* serotype. *Shigella sonnei* and *Shigella fexneri* are frequently identified as etiologic agents for waterborne outbreaks in the United States; occasionally *Shigella boydii* is identified. *Shigella dysenteriae* accounts for most isolates from developing countries and is often associated with serious illness; case fatality rates are as high as 20 percent. Worldwide, 600,000 deaths are estimated per year to be caused by shigellosis, primarily in children under 10 years of age. *See also* waterborne disease.

shim A thin steel plate, sometimes wedge shaped, used to level pumps and motors by raising the base or feet.

Shipek sampler A device used to collect samples of the top layer of bottom sediments that are undisturbed.

shock load The arrival at a water treatment plant of source water containing unusual amounts of algae, colloidal matter, color, suspended solids, turbidity, or other pollutants.

shooting flow Flow with a mean velocity greater than the critical velocity and depth less than critical depth. For this type of flow, the Froude number is greater than 1 and the velocity is greater than the speed of a gravity wave. Shooting flow is also called supercritical flow. *See also* Froude number; gravity wave.

shore The land bordering any body of water. This term is sometimes used to represent the bank of a stream.

shoreline The line along which the land surface meets the water surface of a lake, sea, or ocean. Strictly speaking, it is not a line but rather a narrow strip or area embracing that part of the land surface that comes in contact with wave action both above and below the surface of the water. The term does not apply on tidal flats or marshes that are inundated by the tides; it applies essentially to strips where the land surface has an appreciable slope toward the water. *See also* coastline.

shoring A mechanical, metal hydraulic or timber system that is able to withstand the forces that are imposed by the surrounding earth to prevent cave-ins and thereby protect workers. A shoring system is composed of wales, sheeting, cleats, uprights, and crossbraces. *See also* metal hydraulic shoring system; timber shoring system.

shoring jack A hydraulic jack that is placed between sheeting and extended to brace against the faces of an excavation to prevent cave-ins.

shoring system *See* shoring.

short circuit *See* short-circuiting.

short nipple A nipple with a length only a little greater than that of its two threaded lengths, or somewhat longer than a close nipple. A short nipple always has some unthreaded shoulder between the two threads. *See also* close nipple.

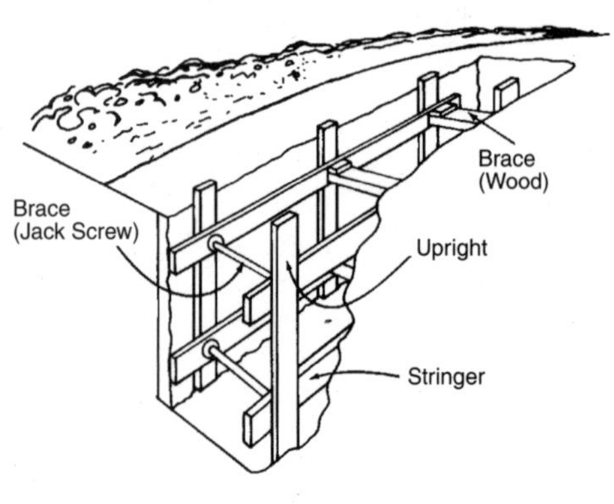

Shoring

short pipe Usually, a pipeline for which the length is less than 500 times the diameter. In such pipes, the effects of entrance and velocity head may be significant and should be considered.

short-circuiting A hydraulic condition in a basin in which the actual flow time of water through the basin is less than the design flow time (i.e., less than the tank volume divided by the flow).

short-run average cost The total cost of producing a product (water) with existing facilities divided by the number of units produced. In economics, the short run is a period of time too short to alter all of the factors of production. By contrast, in the long run, all factors of production can be changed. *See also* long-run average cost.

short-run marginal cost The change in total (or variable) cost associated with using existing facilities to produce the next increment of product, such as the next 1,000 or 1 million gallons of water. The term *incremental cost* is often used to mean the same thing as the short-run marginal cost. Short-run costs contrast with long-run costs in that, in the long run, all costs are variable, including those for the plant scale and the technology employed. *See also* long-run marginal cost.

short-term exposure limit (STEL) The maximum exposure concentration that a worker can be exposed to in a continuous 15-minute period, with a maximum of four exposure periods per day and at least 60 minutes between each exposure period. This limit is not to exceed the standard time-weighted average for that particular hazard. These limits are set by the Occupational Safety and Health Administration. *See also* time-weighted average.

short-term test A biological test that can be used to screen for effects of chemicals that may be beneficial or harmful. In toxicology, short-term tests are developed to be predictive of particular adverse health effects. The most frequently applied short-term tests in toxicology are mutagenesis tests. Such tests are used to determine whether a chemical produces mutation in bacteria. Positive results suggest, but do not prove, that such chemicals may produce cancer. Unfortunately, such tests predict only carcinogenic effects of mutagenic chemicals. Other mechanisms are now known to be important with a variety of chemical carcinogens that are associated with drinking water. *See also* Ames test.

shortwave ultraviolet *See* germicidal ultraviolet.

shot drill A core drill for which the cutting edge consists of a short length of steel tube with one or more notches. Under the weight of the drill rod, the shot—chilled steel balls about 1⁄16 inch (1.6 millimetres) in diameter—is fed down with the water inside the drill rods, rolls under the bit, and wears out an annular ring similar to that cut by a diamond drill. Shot drills are limited to vertical holes.

shoulder ditch A ditch constructed above a cut slope to convey surface water runoff and reduce erosion or cutting of the slope. *See also* berm ditch.

shrub Usually, woody vegetation greater than 3 feet (1 metre) but less than 20 feet (6 metres) tall, including multistemmed, bushy shrubs and small trees and saplings. (Woody seedlings less than 3 feet [1 metre] tall are considered part of the herbaceous layer.)

shrub wetland A wetland area dominated by woody vegetation less than 20 feet (6 metres) tall. The species include true shrubs, young trees, and trees or shrubs that are small or stunted because of environmental conditions.

shrunk joint A joint secured in place by shrinking of a larger pipe on a smaller one.

shutoff A valve that can close off the source of water or a section of a distribution system in a water system.

shutout The closing of valves in a water distribution system to isolate and stop flow to a certain area or section of a water main. The shutout may be made using one valve for a dead-end system, two valves on a single main, or multiple valves when a connection is involved.

shutter (1) A type of flashboard that swings open by water pressure. (2) A swinging crest gate that may be opened quickly under water pressure. *See also* flashboard.

shutter weir A movable weir consisting of a row of large panels hinged at the bottom and inclined slightly downstream toward the top when the weir is closed.

SI *See* saturation index; Système International d'Unites *in the Units of Measure list.*

side contraction A contraction of the nappe, or reduction in width of water overflowing a weir, brought about by the detachment of the sides of the nappe or jet of water while the water passes over the sides of the weir.

side slope The slope of the sides of a canal, dam, or embankment. Custom has sanctioned the naming of the horizontal distance first, with "1.5 to 1" (or, frequently, "1½:1") meaning a distance of 1.5 units horizontal to 1 unit vertical. A better and more strictly correct form of expressing such a slope—one that is not subject to misinterpretation—would be "1 on 1½."

side water depth (SWD) The depth of water measured along a vertical interior wall.

side-channel spillway A spillway in which the initial and final flow are approximately at right angles to each other. The water, after passing over the spillway weir or ogee crest, is carried away by a channel running essentially parallel to the crest. A side-channel spillway is also called a lateral-flow spillway.

side-flow weir A diverting weir constructed on the side of a channel or conduit, usually at right angles to the center line of the main channel. *See also* diverting weir; overfall weir; overflow weir.

siderite ($FeCO_3$) The mineral name for the ferrous carbonate solid. Siderite is frequently found in layers of corrosion deposits on cast-iron pipes, as well as in encrustations in groundwater well screens and pumps. It is a member of the calcite structural group of minerals that includes calcium carbonate ($CaCO_3$), magnesium carbonate ($MgCO_3$), and manganese carbonate ($MnCO_3$). Formation of ferrous carbonate usually requires high inorganic carbonate concentrations and a near-neutral to alkaline pH, as well as a relatively low oxidation–reduction potential in the local environment.

sidewall (1) The vertical, curved, or inclined wall that supports the arch or top of a masonry conduit. (2) The wall at the side of a structure.

siemen (S) *See the Units of Measure list.*

sievert (Sv) *See the Units of Measure list.*

sigma (σ) *See* standard deviation; surface tension.

signal conditioning The act of processing the form or mode of a signal so as to make it intelligible to or compatible with a given device, such as a data transmission line.

signal transduction The conversion of a signal to different modalities. For example, the electrical signal of a neuron is converted to a chemical signal at a synapse between two nerves. Interactions at the postsynaptic site can generate a second electrical signal, or they can initiate processes within a cell to activate enzymes that generate secondary signals (which are frequently referred to as second messengers). This cascade of events eventually triggers a response on the part of the targeted cell.

significant digits *See* significant figures.

significant figures Digits that may be interpreted as accurate in a number. Significant figures are used because all instruments have limitations in terms of how precisely they can provide an accurate measured value. For example, if experimental or mathematical conditions dictate that the number 28,645 must be rounded off to two significant figures, the number will be expressed as 29,000 (where the 2 and the 9 are the significant figures). If the number 0.0046783 must be rounded off to three significant figures, it will be expressed as 0.00468.

significant noncomplier (SNC) A community water system; nontransient, noncommunity water system; or transient, noncommunity water system that has serious, frequent, or persistent National Primary Drinking Water Regulation violations. The criteria that classify a system as a significant noncomplier are set by the US Environmental Protection Agency for each regulated contaminant. *See also* community water system; nonsignificant noncomplier; nontransient, noncommunity water system; transient, noncommunity water system.

silica (SiO_2) A common name for silicon dioxide. Silica occurs widely in nature as sand, quartz, flint, and diatomite. It has various industrial uses, including in the manufacture of glass and ceramics and in water filtration. *See also* activated silica.

silicates A group of compounds containing oxygen and silica (SiO_2). Silicates are considered anodic corrosion inhibitors, combining with the free metal released at the anode site of corrosion activity and forming an insoluble metal–silicate compound. Silicates can also be used to sequester soluble iron and manganese present in source waters to help prevent oxidation and the formation of red and black water.

siliceous gel zeolite *See* gel zeolite.

sill (1) The base on which a supporting structure, such as a bent of a trestle, rests. Usually a sill is made of wood or concrete and rests directly on the ground surface. (2) The timber, steel, or concrete base, located at the bottom of a lift-gate opening, on which the gate rests when closed. (3) The timber, steel, or concrete base, affixed to the crest of a dam or spillway, to which gates or flashboards are attached or against which they are supported. (4) A low concrete or masonry dam in a small mountain stream, designed to retard the downward cutting of the stream in relatively soft rock or unconsolidated material. (5) A low wall used on an apron or in a stilling basin to assist in energy dissipation and in controlling the water currents within and leaving the stilling basin. (6) An intrusive sheet of igneous rock of an approximately uniform thickness and a much greater lateral extent, forced between level or gently inclined beds. (7) A structure built underwater across deep pools of a river, with the aim of correcting the depth of the river.

silt Detritic sedimentary rock that has noncemented grains of particle size between fine sand and clay (i.e., between 0.002 and 0.05 millimetres in equivalent diameter).

silt density index (SDI) An empirical measure of the plugging characteristics of membrane feedwater based on passing the water through a membrane filter test apparatus containing a 0.45-micrometre pore diameter filter (commonly a cellulosic type) in a dead-end filtration mode at a constant feed pressure (typically 30 pounds per square inch [207 kilopascals]) for a specified duration (15 minutes most common) and observing the flow rate decline. The silt density index is calculated based on the measured time required for collection of a 500-millilitre filtrate sample at the beginning of the test filtration period (t_i) and the time for collection of a 500-millilitre filtrate sample at a later time (t_f). For a test period T, the SDI value (which is dimensionless) is calculated as follows:

$$SDI_T = \frac{1 - \frac{t_i}{t_f}}{T} \times 100$$

Where:

t_i = initial time to collect 500-millilitre sample, in seconds

t_f = final time to collect 500-millilitre sample, in seconds

T = filtration test period, in minutes

siltation The accumulation of silt (small soil particles having a diameter between 0.002 and 0.05 millimetres) in an impoundment.

silting *See* siltation.

siltstone A very fine-grained, consolidated rock composed predominantly of particles of silt grade.

silver (Ag) A metallic element that has various industrial uses, e.g., in the manufacture of photographic chemicals and in electronic equipment, table cutlery, jewelry, and dental amalgams. It is regulated in drinking water by the US Environmental Protection Agency.

silvex ($Cl_3C_6H_2OCH(CH_3)COOH$) A generic name for 2-(2,4,5-trichlorophenoxy)propionic acid. Silvex is an herbicide and plant growth regulator that is regulated by the US Environmental Protection Agency. *See also* herbicide; 2-(2,4,5-trichlorophenoxy)propionic acid.

Cl, CH$_3$, Cl, –O–CH, O, Cl, C, OH

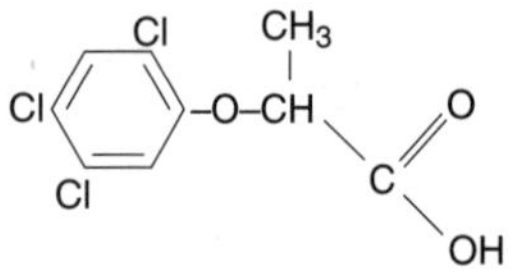

Silvex

simazine ($ClC_3N_3(NHC_2H_5)_2$) A generic name for 2-chloro-4,6-bis(ethylamino)-*s*-triazine, an herbicide that is regulated by the US Environmental Protection Agency. *See also* herbicide.

Cl, C, N, N, C_2H_5NHC, $CNHC_2H_5$, N

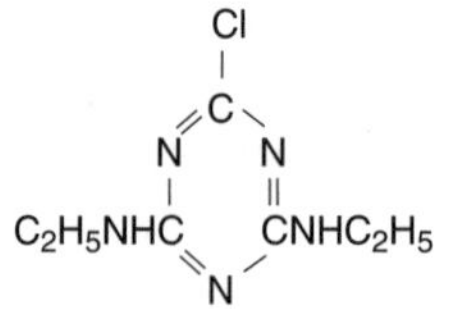

Simazine

similitude A condition that exists when the attributes of a model and prototype unit are equal. For example, geometric similitude exists between model and prototype if the ratios of all corresponding dimensions in the model and prototype are equal. Hydraulic similitude is difficult to achieve.

simple electrolyte Any compound, such as a coagulant, that forms charged species (ions) in solution. *See also* polyelectrolyte.

simple surge tank A surge tank that does not have a restricted orifice to create a head differential between the conduit and the tank.

simple water budget A water budget used in landscaping on the premise that no change in water storage in the soil occurs. It is used to determine the amount of water required to irrigate landscape plants to prevent wilting. The irrigation amount is determined as the product of a reference evapotranspiration rate, the irrigated area, and a conversion factor. *See also* reference evapotranspiration.

simulate To reproduce the action of some process, usually on a smaller scale.

simulated distribution system (SDS) *See* simulated distribution system test.

simulated distribution system (SDS) test A type of test in which a treated water is dosed with a typical treatment plant level of the disinfectant normally applied and is incubated (stored) under conditions that simulate disinfection by-product production in a distribution system. The pH, temperature, and contact time should match the values found in the actual distribution system. This test does not measure the amount of disinfection by-product *precursors* in a sample; that parameter is measured in a disinfection by-product formation potential test. However, an SDS test should predict the disinfection by-product concentration at a consumer's tap if the water's time of travel to that tap is the same as the contact time in the test. *See also* disinfection by-product; disinfection by-product formation potential; uniform formation conditions.

simulated distribution system disinfection by-product (SDS-DBP) concentration The amount of disinfection by-products formed during a test in which a treated water is dosed with a typical treatment plant level of disinfectant (normally but not always chlorine) and is incubated (stored) under conditions that simulate disinfection by-product production in a distribution system. The pH, temperature, and contact time should match the values found in the actual distribution system. The SDS-DBP level is not a measure of the amount of disinfection by-product *precursors* in a sample; that parameter is measured in a disinfection by-product formation potential test. However, the SDS-DBP concentration should be similar to the disinfection by-product concentration at a consumer's tap if the water's time of travel to that tap is the same as the contact time in the test. *See also* disinfection by-product; disinfection by-product formation potential; uniform formation condition.

simulated distribution system trihalomethane (SDS-THM) concentration The amount of trihalomethanes formed during a test in which a treated water is dosed with a typical treatment plant level of disinfectant (normally but not always chlorine) and is incubated (stored) under conditions that simulate trihalomethane production in a distribution system. The pH, temperature, and contact time should match the values found in the actual distribution system. The SDS-THM concentration is not a measure of the amount of trihalomethane *precursors* in a sample; that parameter is measured in a trihalomethane formation potential test. However, the SDS-THM level should predict the trihalomethane concentration at a consumer's tap if the water's time of travel to that tap is the same as the contact time in the test. *See also* trihalomethane; trihalomethane formation potential; uniform formation conditions.

simulated water quality Water quality values generated by a computer simulation of a treatment process or a physical simulation of a distribution system. Computer simulations of a treatment process use predictive equations to determine the changes in water quality. Physical simulation of a distribution system involves holding a batch of finished water for an equivalent time (compared to the distribution system) under conditions similar to that expected within the distribution system.

simulation An act of reproducing the effects of a system by conducting mathematical manipulation of predictive equations or by using a physical model intended to represent the system on a smaller scale.

simulation model *See* simulation program.

simulation program A computer program that has a collection of various equations describing different processes of a system. A simulation program can be used to study the approximate effects of changing certain operating conditions on the outcome of the system.

simultaneous distillation extraction (SDE) An analytical technique for extracting and concentrating trace organic compounds. A water sample and a solvent are heated separately and are then condensed together. The steam and steam-distillable organic compounds mix with the solvent vapor. The condensate separates into two phases; from each phase the solvent and extracted organic compounds are recovered for subsequent concentration and analysis. *See also* organic compound; solvent.

single probe A probe, usually 4–4.5 feet (1.2–1.4 metres) in length, used to measure the resistivity of soil around the tip or point of the probe.

single-action pump A reciprocating pump in which the suction inlet admits water to only one side of the plunger or piston and the discharge is intermittent.

single-arch dam A curved masonry dam that depends principally on arch action for stability. A single-arch dam is also called an arch dam.

single-centrifugal pump A centrifugal pump in which the suction inlet admits water to one side of the impeller.

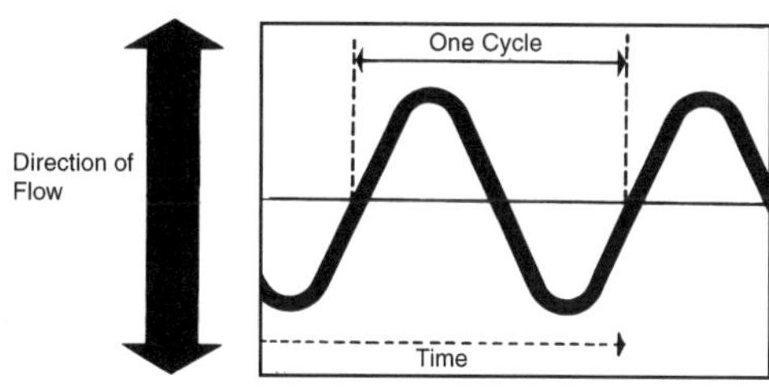

Single-phase power

single-effect evaporator An evaporator in which the liquid is subjected to only one evaporating step. *See also* evaporator; multiple-effect evaporator.

single-element bench-scale test (SEBST) A membrane treatment study specified in the US Environmental Protection Agency's Information Collection Rule as promulgated in May 1996, whereby two spiral wound membrane elements (minimum element size 2.5 inches [6.4 centimetres] in diameter and 40 inches [1 metre] in length) with molecular weight cutoffs less than 1,000 daltons are evaluated at 75 percent recovery while other operating parameters are held constant for 4 weeks during each quarter of 1 year. The study for each membrane is a continuous-flow test using a single element and concentrate recycle. *See also* Information Collection Rule.

single-family structure A building that was constructed as a residence for one family and is currently used as either a residence or a place of business.

single-main system A water system for which one main supplies both potable and fire-fighting water.

single-pan balance A balance used to make quick, accurate mass measurements. The material to be weighed is placed on the pan, and counterweights, located on arms (beams) beneath the pan, are adjusted to balance the material, thus indicating the weight. A single-pan balance is also known as a beam balance.

single-phase circuit An electric circuit energized by a single voltage source. An alternating current single-phase circuit usually consists of two conductors.

single-phase power Alternating current power in which the current flow peaks in each direction only once per cycle.

single-phasing (of a motor) The loss of a power phase to an electrical motor in a three-phase power supply. This condition will cause the motor to be underpowered and to overheat.

single-stage pump A pump that has only one impeller. (A multistage pump has more than one impeller.)

single-stage recirculation A process used in a multiple (2–6) membrane reverse osmosis system in which a portion of the concentrate stream is split off and routed back to the inlet and mixed with the feedwater. This approach increases the flow across the membrane without increasing the amount of feedwater and increases the overall recovery rate. The pump capacity will affect the amount of water to be recirculated and the recovery rate. *See also* recirculation.

single-stage system A reverse osmosis system in which the water is passed through the membrane(s) only once by a single high-pressure pump.

single-stroke deep-well pump A reciprocating, power-driven, deep-well pump having a single pump rod connecting the reciprocating mechanism to the power head with the cylinder plunger.

single-suction impeller An impeller with one suction inlet.

sink (1) A depression in the land surface, especially one having a central playa or saline lake with no outlet. (2) One of the hollows in limestone regions, often connected to a cavern or subterranean passage, so that waters running into it are lost. (3) In hydromechanics, a theoretical singular point in the fluid medium at which fluid matter is annihilated at a constant rate; an infinitesimal orifice or exit; a negative source. (4) A place into which anything flows; e.g., ice is a heat sink. *See also* sinkhole.

sinker bar A heavy bar of round iron that makes up the mass in a string of well-boring tools.

sinkhole A depression in local topography that connects to an underlying aquifer or underground watercourse. Sinkholes can become very large; entire homes have been lost when a sinkhole suddenly increases in size following large recharge events when the unconsolidated material in the sinkhole is washed into the aquifer.

sinuosity The ratio between the thalweg length and the linear distance through the air between two points on a river. *See also* thalweg.

sinusoidal signal A physical quantity (e.g., voltage or current) that is varying in time as

$$A\sin(\omega t + \theta) \text{ or } A\cos(\omega t + \theta)$$

Where:

A = the amplitude or maximum amplitude of the quantity, in volts or amperes depending on the signal

ω = the angular frequency, in radians per second

t = the time, in seconds

θ = the phase angle of the sinusoidal time function, in radians in both US customary and Système International units

siphon (1) A hydraulic process in a closed system in which a low enough pressure created at the siphon discharge permits a fluid to flow upward and be transferred across a higher elevation as long as the ultimate discharge point is at a lower elevation than the hydraulic grade of the system at the siphon starting point. For example, a siphon will transport water up and over the edge of a container as long as the discharge point is below the water surface in the container and the system is closed (i.e., no air gaps occur in the siphon). (2) To create such a condition.

siphon spillway An enclosed spillway that uses the operating principle of a siphon. The top of the spillway inlet or air vent is constructed at the elevation at which spilling from an impoundment is to cease, and the top of the inside of the spillway tube is placed at the maximum elevation the water in an impoundment is to attain. When the water level in an impoundment somewhat exceeds the elevation of the spillway crest, the air is exhausted in the tube, siphoning begins, and the discharge through the spillway is greatly accelerated. When the water level in the impoundment falls below the top of the inlet or air vent, the siphon is broken and discharge ceases. Siphon spillways have several advantages. First, because siphon spillways have such a large capacity for a small rise in water level, their use allows a considerable reduction in the length of the spillway. Second, a siphon spillway can be constructed around a dam or through a dam without deep excavations. Third, a siphon spillway allows for a rise in the water level in an impoundment prior to discharging water.

siphonage The movement of liquid from a higher to a lower level by action of a siphon. Shutting off and draining a distribution main can create conditions for siphonage to occur from either a building or leak in the same shutout. *See also* siphon.

site area The total area of a development site, including building footprints, roadways, and parking areas.

16SrRNA probe A nucleic acid probe against a ribosomal 16SrRNA gene sequence. 16SrRNA gene sequences are found in all bacteria and are used in phylogenetic, evolutionary, and diagnostic studies because they accumulate mutations at a slow, constant rate over time and therefore serve as molecular clocks.

size exclusion analysis (SEA) An analytical technique to determine the molecular weight and molecular weight distribution of a sample. The method involves the separation of molecules according to molecular size. *See also* molecular size; molecular weight distribution.

skeletal fluorosis A condition of brittle bones that arises from excessive deposition of fluoride in bones. The skeletal changes that occur consist of combined osteosclerosis and osteomalacia.

skeletonization The practice of deleting certain minor pipes and facilities from a system model. Ideally, this creates a manageable model that decreases execution time without a loss of overall accuracy in model performance.

skewness *See* skewness of data.

skewness of data A statistical condition in which the data in a distribution are not symmetrical about the mean value. If more data points have larger values than smaller values, the database is called positively skewed or skewed to the right. If more data points have smaller values, the data set is said to be skewed to the left or negatively skewed.

skimmer well A type of infiltration gallery, common on volcanic islands, for which the pumps are designed to produce very little drawdown so as not to mix the underlying salt water with the freshwater lens.

skimming (1) The process of diverting water from the surface of a stream or conduit by means of a shallow overflow in order to avoid diversion of sand, silt, or other debris or material carried as bottom load. (2) The process of diverting water from any chosen elevation in a reservoir by means of outlets at different elevations or by any other skimming device in order to obtain the most palatable drinking water.

skimming weir A weir on a tank or reservoir with an adjustable crest elevation to afford a means of restricting the depth of overflow so that a layer of water of limited depth is removed or skimmed off. *See also* overflow weir.

skin absorption The passage of a chemical through the skin. In general, polar chemicals pass through the skin very poorly, but uncharged, lipophilic chemicals can readily pass through the skin as long as they are applied to the skin in a solubilized form.

skin of water The thin layer of water at the free surface that, because of the different molecular conditions that exist at this interface, is denser and develops a somewhat membranous character, allowing it to resist slight upward or downward pressures. *See also* surface tension.

slack water (1) In tidal waters, the state of a tidal current when the current's velocity is at a minimum, especially the moment when a reversing current changes direction and its velocity is zero. The relation of the time of slack water to the tidal phases varies in different localities. In some cases slack water occurs near the times of high and low water, whereas in other localities the slack water may occur midway between high and low water. (2) In streams, a place where very little current exists.

slake To add water to quicklime (calcium oxide, CaO) to form calcium hydroxide ($Ca(OH)_2$) that can then be used in a softening or stabilization process. *See also* quicklime.

slaked lime *See* hydrated lime.

slaker The part of a quicklime feeder that mixes the quicklime (CaO) with water to form hydrated lime [calcium hydroxide, $Ca(OH)_2$].

sleeve (1) A pipe fitting for uniting two pipes of the same nominal diameter in a straight line. (2) A tube into which a pipe is inserted.

Slichter method A method of measuring the velocity of groundwater flowing through granular material, developed by Charles S. Slichter. It consists essentially of driving sets of perforated pipes into the material at known distance intervals, charging the groundwater passing the upstream pipe with a strong electrolyte (ammonium chloride, NH_4Cl, is very suitable), and noting the time involved in the passage of the charged water to another perforated pipe downstream by determining the electrical resistance of the water in the latter pipe at intervals of time. When the resistance is plotted against time, the point of inflection of the resulting curve is taken as the time of passage between the pipes, and the velocity is computed by dividing the distance between the two pipes by the time of passage. The method is repeated for multiple pipes across the cross section of granular material.

slickens (1) The thin layer of extremely fine silt deposited after a general flood over the valley of a large, silt-bearing river, such as the Nile, Sacramento, or Mississippi rivers.

slide gate A device on a dam to control the release of water.

sliding gate A gate valve, or any type of gate, that slides directly on its bearings or seats while opening or closing without the use of rollers or other devices to reduce friction.

sliding-panel weir A movable timber weir in which the barrier consists of wooden panels that slide in grooves and are placed between pairs of immovable frames.

slime (1) Any of numerous substances of viscous organic nature that are usually formed from microbiological growth and that attach themselves to other objects, forming a coating. (2) The coating of biomass (humus, schmutzdecke, sluff) that accumulates on the top of slow sand filters. *See also* biofilm; schmutzdecke; slow sand filter.

slip (1) In a pump, the percentage of water taken into the suction end that is not discharged because of clearances in the moving unit. (2) In geology, a movement that dislocates a rock or soil mass.

slip joint An inserted joint in which the end of one pipe is slipped into the flared or swaged end of an adjacent pipe.

slipoff slope A meandering stream's bank that is not eroded by stream action and is being built up gradually.

slippage *See* leakage.

slope The ratio of vertical distance to horizontal distance, or "rise over run." *See also* grade; side slope.

slope area discharge measurement A determination, by indirect measurement during a field survey, of the peak discharge of a reach of channel, as well as the high-water marks, usually after a flood has passed. Discharge is usually computed by the Manning formula, modified to account for nonuniform flow. *See also* Manning formula.

slope gauge A gauge used to determine the slope of a water surface. *See also* sloping gauge.

slope wash The process by which erosional debris are removed from sloping surfaces by natural runoff of water that is not concentrated in well-defined channels.

slope-discharge curve A graph of discharge in a stream as a function of the slope of the water surface. The slope of the water surface is determined as the ratio of the difference in water elevation at two gauges to the distance between the two gauges. A slope-discharge curve is also called a slope-discharge diagram.

sloping A means to protect workers from cave-in during an excavation by angling the sides of the excavation. Sloping prevents the faces of the excavation from sloughing off and burying the workers. The slope of the sides depends on the type of soil in which the excavation is being performed. *See also* natural slope.

sloping gauge A staff gauge that is used to indicate the elevation of the water surface in a stream channel, conduit, reservoir, or tank and is graduated to such a scale that the graduations represent vertical distances above a specific datum plane. Such a gauge is usually installed on a flat sloping bank where increased accuracy in reading the gauge is desirable. A sloping gauge is also called an inclined gauge or slope gauge.

sloping wave A translatory wave or decrease in depth of the water surface in an open channel caused by a sudden change in conditions of flow. This wave is usually discernible only as a lowering of the water surface, and it acts in accordance with the same principles that apply to an abrupt wave. *See also* hydraulic jump.

sloping-benching A series of steps intended to provide protection to workers from a cave-in during an excavation. This arrangement prevents the faces of the excavation from sloughing off and burying the worker. The number of benches depends on the type of soil in which the excavation is being performed.

slot A narrow opening.

slot opening The size of the holes in a well screen that lets water pass from an aquifer into a well.

slotted lateral pipe An underdrain lateral pipe that has many tiny openings that appear as though the pipe had been partially cut around the perimeter with a hacksaw. *See also* filter underdrain.

slough (1) A small muddy marshland or tidal waterway that usually connects other tidal areas. (2) A tideland or bottom land creek; a side channel or inlet, as from a river or a bayou.

This type of slough may be connected at both ends to a parent body of water. (3) To shed, cast off, drop off, separate from, or become detached. In reference to biofilms developed on materials in contact with drinking water, biofilm material that sloughs from granular filtration media, or from the surfaces of distribution pipes and storage tanks, may result in increased densities of bacteria (including coliforms) in the water.

sloughing The process of shedding, casting off, dropping off, separating, or detaching. Biofilms or other types of films applied to or developed on a surface may become detached by a process of sloughing due to physical or chemical action. Sloughing of aquatic biofilms usually occurs because of the shearing action of turbulent water flow.

slow rinse In ion exchange, that portion of the rinsing stage that usually follows the introduction of the regenerant and during which the rinse water (deionized, decationized, softened, or untreated water) passes through the resin at the same flow rate as the regenerant. *See also* fast rinse.

slow sand filter A filter for the purification of water in which water, without previous treatment, is passed downward through a filtering medium consisting of a layer of sand 24 to 40 inches (0.6 to 1 metre) thick. The filtrate is removed by an underdrainage system, and the filter is cleaned by scraping off the clogged sand and eventually replacing the sand. A slow sand filter is characterized by a slow rate of filtration, commonly 3–6 million gallons per day per acre of filter area (0.1–0.2 metres per hour). Its effectiveness depends on the biological mat, or schmutzdecke, that forms in the top few millimetres. *See also* schmutzdecke.

slow sand filtration (SSF) The use of a slow sand filter.

sludge A term that is being replaced by the term *residuals*. It has the following meanings: (1) The accumulated solids separated from a liquid, such as water, during processing. (2) Organic deposits on the bottoms of streams or other bodies of water. (3) The removed material resulting from chemical treatment, coagulation, flocculation, sedimentation, or flotation (in which case the sludge is called *float*) of water. (4) Any solid material containing large amounts of entrained water and collected during water treatment. *See also* residuals; settleable solids.

sludge bed An area comprising natural or artificial layers of porous material on which water treatment sludge is dried by drainage and evaporation. A sludge bed may be open to the atmosphere or covered, usually with a greenhouse-type superstructure. Such a bed is also called a sludge drying bed.

sludge blanket A zone in which sludge accumulates and concentrates. Sludge blankets are often associated with the region in a settling basin in which sludge deposits and collects.

sludge blowdown The controlled withdrawal of sludge from a solids-contact basin to maintain the proper level of settled solids in the basin.

sludge cake Semi-dry solids composed of sludge that sticks together, typically after being pressed by a dewatering device. Sludge cakes typically contain solids concentrations of 20 to 25 percent, and sometimes higher. *See also* sludge dewatering.

sludge collector A mechanical device for scraping the sludge on the bottom of a settling tank to a sump from which it can be drawn.

sludge compartment A separate compartment incorporated in a settling process.

sludge concentration Any process of reducing the water content of sludge.

sludge conditioning A chemical or thermal process in which the dewaterability of sludge is enhanced. Sludges are often conditioned with polymers that promote the release of bound water. The resulting solids form a better sludge cake. *See also* polymer; sludge cake; sludge dewatering.

sludge density index (SDI) A measure of the degree of compaction for 1,000 millilitres of sludge after settling in a graduated container takes place for 30 minutes. This index is expressed in units of millilitres per gram. It is also called the sludge volume index.

sludge dewatering The process of removing a part of the water in sludge by any method (mechanical or physical), such as draining, evaporation, pressing, vacuum filtration, centrifuging, exhausting, passing between rollers, acid flotation, or dissolved air flotation, with or without heat. Sludge dewatering involves reducing the sludge from a liquid to a spadable condition rather than merely changing the density of the liquid (concentrating) or drying (as in a kiln). Mechanical means are typically more costly and operator-intensive, but they require less land than physical means. *See also* belt filter press; centrifuge; sand drying bed; sludge bed; sludge lagoon.

sludge disposal The ultimate discharge of collected solids to the environment. Typical disposal options include land disposal (i.e., landfilling), incineration, and ocean disposal.

sludge dryer A device for removal of a large percentage of moisture from sludge or screenings by heat.

sludge drying bed *See* sand drying bed; sludge bed.

sludge filter A device in which wet sludge, usually conditioned by a coagulant, is partly dewatered by means of vacuum or pressure.

sludge index *See* sludge density index.

sludge lagoon A basin or pond in which liquid sludge is discharged. The solids settle to the bottom and the separated liquid is discharged or treated. Sludge lagoons essentially allow a sedimentation process for long periods (e.g., 30 to 60 days or longer). Collected sludge can be either physically removed or allowed to accumulate until the lagoon is permanently taken out of service.

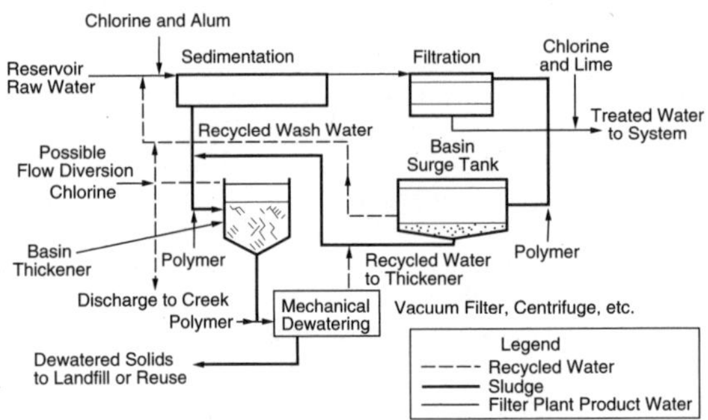

Schematic for a mechanical system for sludge dewatering

sludge moisture content The quantity of water present in sludge, usually expressed in terms of the percentage of wet weight.
sludge pressing The process of dewatering sludge by subjecting it to pressure, usually within a cloth fabric through which the water passes and in which the solids are retained. *See also* belt filter press; plate-and-frame filter press.
sludge processing Collection, treatment, and disposal of sludge. *See also* sludge conditioning.
sludge solids Dissolved and suspended solids in sludge.
sludge thickener A tank or other piece of equipment designed to concentrate water treatment sludges. The principal mechanism is gravity settling of sludge solids.
sludge thickening A process of removing water from accumulated solids prior to dewatering. Thickened sludge has a solids concentrations ranging from less than 2 percent to 5 or 6 percent. Sludge is often thickened by less mechanically intensive processes than in dewatering, such as the use of settling tanks, to reduce the volume to be treated in more mechanical and cost-intensive (but less land-intensive) dewatering processes.
sludge volume index *See* sludge density index.
sludge wasting The process of removing sludge from a unit process.
sludge zone The bottom zone of a sedimentation basin. This zone receives and stores the settled particles.
sludge-blanket clarifier *See* solids-contact basin.
sludge-drying bed A process used to separate water and solids in sludge. The sludge is applied to a filtering medium, such as sand, with an underdrain system to collect the filtered water. The sludge-dewatering process occurs by two mechanisms: (1) gravity drainage through the sludge cake and sand filter, and (2) air drying from the surface of the sludge cake by evaporation. Sludge-drying beds are simple to operate but require large land areas to provide effective dewatering. *See also* sand-drying bed; sludge bed.
sludger *See* sand pump.
slug (1) A temporary, abnormally high concentration of an undesirable substance that appears in the product or distributed water. For example, a slug of iron rust might appear because of the shearing action of a high-demand flow that loosens a previously deposited iron precipitate. (2) *See the Units of Measure list.*
slug dose A single applied addition of a material to a process. In tracer testing, a slug dose of the tracer chemical may be added; the hydraulic characteristics of the basin being evaluated are determined based on the mass of tracer recovered in the effluent over time. *See also* bolus dose.
slug test An aquifer test, conducted in wells, in which a known volume of tracer (the "slug") is added to the groundwater to determine hydraulic properties of the aquifer or overlying soils. *See also* auger hole test.
sluggish stream A stream in which the peaks of a flood form more slowly than usual because of the decrease in the slope as the stream system ages or as the flow is reduced or retarded by withdrawal or storage in upstream reaches.
sluice (1) A conduit for carrying water at high velocity. (2) An opening in a structure for passing debris. (3) To cause water to flow at high velocities for wastage, for such purposes as excavation or ejecting debris.

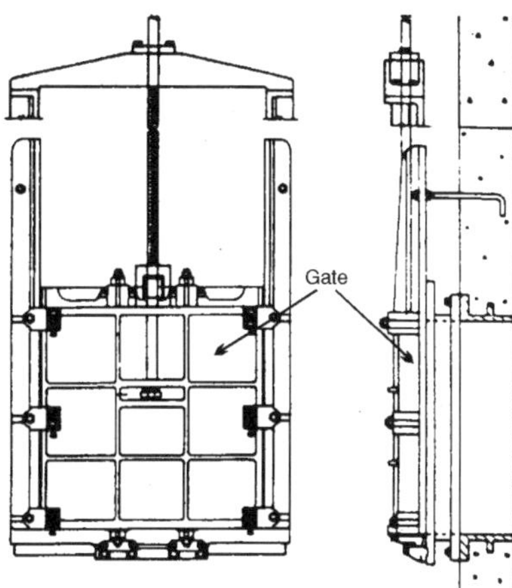

Sluice gate

sluice flow That part of the total discharge that flows through sluices. Sluice flow is also called orifice flow.
sluice gate A gate used for sluicing. Such a gate is constructed to slide vertically and is fastened into or against the masonry of dams, tanks, or other structures.
sluiceway (1) An opening in a diversion dam, usually adjacent to the head gate. A sluiceway may consist of an open panel in the dam or a gate near the base of the dam. (2) A culvert or controlled drainage opening in a levee or dam.
sluicing The operation of moving earth, sand, gravel, or other material by means of flowing water.
slump A type of landslide involving the downward slipping of a mass of rock or unconsolidated material of any size, moving as a unit or as several subsidiary units, usually with backward rotation.
slurry A mixture of liquid and solids that can be pumped and transported by liquid-handling equipment. Drilling mud and wet concrete are examples of slurries.
slurry recirculation unit A center-feed clarifier with flocculation zones built into a central compartment. A motor-driven impeller can recirculate within the reactor up to five times the source water inflow rate, allowing a specified slurry concentration to develop. The recirculation allows a higher slurry concentration to be obtained, thus dampening source water quality or hydraulic loadings that can cause floc carryover or operational difficulty in maintaining proper slurry concentrations.
slush ice An unfrozen mixture of water and ice.
SMA *See* Snow Mountain agent.
small-bore piping Copper or stainless-steel pipe with a diameter of 0.5 inch (13 millimetres). Such pipe is commonly found in pump-assisted hot water central heating systems.
Small Systems Technology Initiative A program developed by the US Environmental Protection Agency's Office of Drinking

Water to promote the use of drinking water treatment technologies for small systems (those serving fewer than 3,300 people).

small water system A public water system serving fewer than 3,300 people.

SMCL *See* secondary maximum contaminant level.

SMCRA *See* Surface Mining Control and Reclamation Act.

Smith–Heiftje background correction An instrumental technique of compensating for interferences in atomic absorption spectrophotometry. The technique measures the difference between signals obtained from a hollow cathode lamp before and during a high-current pulse. This technique can be used for analytes in which the spectral line is in either the ultraviolet or visible region.

smooth nozzle A nozzle in which the internal diameter is gradually reduced from its large end to its small end.

SMSA *See* standard metropolitan statistical area.

snake (1) A fiber yarn rope that is placed around a pipe joint and covered with mud to hold molten lead being poured into a joint until the lead solidifies. A vee is formed in the rope on the top side of the pipe to allow the lead to be poured into the joint. (2) A flexible wire with a cutter, sometimes used to clear clogged pipes.

SNARL *See* suggested no-adverse-response level.

SNC *See* significant noncomplier.

S-9 fraction The supernatant liquid produced after a cell homogenate is spun in a centrifuge at a force 9,000 times that of gravity. Typically this preparation is made from the liver of a rodent that has been pretreated to induce enzymes that metabolize chemicals to reactive intermediates. It is incorporated into a variety of in vitro tests (e.g., the Ames test). *See also* Ames test.

snow (1) Precipitation of water from the atmosphere in the form of branched hexagonal crystals or stars, often mixed with simple ice crystals. Snow falls more or less continuously from a solid cloud sheet. These crystals may fall either separately or in coherent clusters forming snowflakes. (2) The accumulation of fallen snow.

snow bin A large box used to measure the depth of snowfall. The box is usually 5 to 10 feet (1.5 to 3 metres) square, protected against eddying air currents along the top of its sides by screens, and lined with some waterproofing material. The snow is allowed to melt, with such melting sometimes being aided by means of rock salt sprinkled in the bottom of the bin or by a heating apparatus, such as an electric lamp, heating element, or a coal oil lamp installed in the box. The melted snow drains into a measuring tank.

snow board A flat board, usually about 18 by 24 inches (0.45 by 0.6 metres) and covered with white flannel, used to measure the depth of snowfall. The board is laid horizontally on the ground or snow surface; the white surface reflects the radiant heat of the sun and is very similar in texture to snow. The depth of snowfall is measured directly as the snow is caught on the board.

snow course A line on a drainage area, laid out and permanently marked, along which the snow is sampled at definite distances or stations at appropriate times during a snow survey to determine the snow's depth, water equivalent, and density.

snow cover The accumulated snow and ice on the surface of the ground at any time. *See also* snowpack.

snow gauge A device for measuring snow depth. *See also* snow scale.

Snow Mountain agent (SMA) A Norwalk-like viral agent isolated from an outbreak of acute gastrointestinal illness that occurred at a ski resort near Granby, Colo., in 1976. The virus is 27–32 nanometres in size and is morphologically similar to, but antigenically distinct from, Norwalk agent. Like Norwalk virus, at present it cannot be cultivated in vitro. It has been positively identified with both waterborne and foodborne disease outbreaks, and airborne spread by aerosolized droplets from infected persons is suspected, as well. It is related to the calicivirus group and the Norwalk virus.

snow residuum Accumulated snow on the ground surface at the end of any period, or that snow's water equivalent.

snow ripening The process, occurring during early stages of melting, by which snow crystals tend to become granular, spaces between the crystals become filled with water, and the water content of the snow tends to become uniform at all depths. When the snow becomes still more dense and begins to lose water, it is sometimes said to be overripe.

snow sample A core taken from the snow mantle on a snow course by a snow sampler during the course of a snow survey. From this core, the depth, water equivalent, and density of the snow mantle may be determined.

snow sampler A device used in securing snow samples, consisting essentially of a set of light, jointed, metal tubes for taking samples and a spring scale graduated to read directly the corresponding depth of water contained in the snow sample.

snow sampling The process or operation, in connection with a snow survey, of observing the depth and determining the density or water content of snow existing at given times at given locations or along given snow courses on a drainage basin.

snow scale A device, usually a rod, bearing markings so that snow depth can be read directly from a distance without the snow cover being disturbed.

snow stake A wooden stick, either driven into the ground or held upright by guy wires, used in regions of deep snowfall to indicate the depth of snow. The depth is read directly from graduations marked on the stake. *See also* snow scale.

snow storage Storage of water on a drainage basin in the form of snow and ice.

snow survey The process or operation of determining the depth, water content, and density of snow at various selected points on a drainage basin in order to ascertain the amount of water stored in the form of snow so that subsequent runoff can be forecasted.

snow trap An opening made in a dense forest to allow snow to fall to the ground, thus trapping the snow.

snowfall (1) The rate at which snow falls, usually expressed in terms of inches (centimetres) of snow depth per 6-hour period. (2) The actual depth of snow on the ground after a snowstorm. (3) A snowstorm.

snowfield (1) An area, usually at high elevation or in polar latitudes, where snow accumulates and remains on the ground during the entire year. (2) An unbroken expanse of snow.

snowpack A field of naturally packed snow that provides a steady supply of water and is also useful in furnishing water power. This term is used in the mountain regions of western United States. Snow pack is also called snow storage or, sometimes, snow cover.

soap The water-soluble reaction product of a fatty acid ester (RCOOR′) and an alkali (usually sodium hydroxide, NaOH) that produces suds when used with water for washing or cleaning purposes. Some soaps are mild disinfectants. Common soaps, such as sodium and potassium soaps, are soluble but can be converted to insoluble calcium and magnesium soap curds (bathtub ring) by the presence of hardness ions in the water. *See also* hardness.

SOC *See* synthetic organic chemical.

Society for Epidemiologic Research An association of epidemiologists and biostatisticians that sponsors the *American Journal of Epidemiology.*

socket pipe In pipe fitting, a cast-iron pipe that is provided with a socket at one end and a spigot at the other. The sockets of wrought pipes are couplings and are screwed over the ends on the outside diameter.

soda ash (Na_2CO_3) A common name for commercial sodium carbonate.

soda water Water that has been impregnated with carbon dioxide (CO_2) so that it will be effervescent when not under pressure. Soda water is the same thing as seltzer water and sparkling water. *See also* bottled sparkling water.

sodium (Na) A metallic element found abundantly in compounds in nature but never existing alone. Sodium compounds are highly soluble and do not form curds when used with soaps or detergents. Many sodium compounds are used in the water treatment industry. Notable are the uses of sodium chloride (NaCl) as a regenerant in the cation exchange water-softening process and sodium hypochlorite (NaOCl) as a disinfectant.

sodium adsorption ratio (SAR) An expression of the relative activity of sodium ions in exchange reactions with soil, indicating the sodium or alkali hazard to soil. It is calculated from the following expression:

$$SAR = \frac{[Na^+]}{\sqrt{0.5([Ca^{2+}] + [Mg^{2+}])}}$$

Where:

$[Na^+]$ = sodium ion concentration
$[Ca^{2+}]$ = calcium ion concentration
$[Mg^{2+}]$ = magnesium ion concentration

All concentrations are expressed in milliequivalents per litre. This term is used in guidelines for the interpretation of the quality of water for irrigation. In addition, interactions between soil ions and dissolved organic carbon may be characterized by the value of the sodium adsorption ratio. Calcium is able to act as a bridge between organic matter and clay and produce insoluble organic matter complexes, thereby increasing the retention capacity of the soil. When calcium-based ameliorants are applied to soil, the added calcium will change the sodium adsorption ratio of the solution in contact with the soil, thereby reducing clay dispersion, which will reduce the desorption of ions and molecules from these surfaces.

sodium aluminate ($Na_2Al_2O_4$) A substance sometimes used as an auxiliary coagulant for the removal of fine turbidity and color in soft, low-pH water. *See also* coagulant.

sodium aluminosilicate *See* gel zeolite.

sodium arsenite ($NaAsO_2$) A salt used in testing for residual chlorine. It is also known as sodium metaarsenite.

sodium bicarbonate ($NaHCO_3$) A substance that can be added to drinking water to increase alkalinity and provide additional buffering intensity. It is generally supplied as a powder and is used as part of a corrosion control strategy or in the coagulation treatment process. It is also called baking soda.

sodium bisulfite ($NaHSO_3$) A chemical additive used as a disinfectant, preservative, and dechlorination or reducing agent.

sodium bromide (NaBr) A salt of sodium and bromine. When chlorine is applied to a water solution of NaBr, gaseous bromine (Br_2) is released.

sodium carbonate (Na_2CO_3) A salt used in water treatment to increase the alkalinity or pH value of water or to neutralize acidity. It is also called soda ash.

sodium chloride (NaCl) The chemical name for common table salt. Sodium chloride is also widely used for regeneration of ion exchange water softeners and in some dealkalizer systems. *See also* common salt.

sodium cycle The cation exchange water-softening process in which sodium ions in the resin are exchanged for hardness ions in the water. Sodium chloride (NaCl) is commonly used for resin regeneration.

sodium fluoride (NaF) A dry chemical used in the fluoridation of drinking water. It is commonly used in saturators.

sodium hexametaphosphate A substance that has a molecular ratio of 1.1 parts of sodium monoxide (Na_2O) to 1 part of phosphorus pentoxide (P_2O_5), with a guaranteed minimum of 76 percent P_2O_5. Several specialized compositions are available. In general, it is used as a sequestering, dispersing, and deflocculating agent. It is soluble in water and insoluble in organic solvents. It is also called sodium polyphosphate or glassy sodium phosphate. *See also* hexametaphosphate; sequestering agent.

sodium hydrosulfite ($Na_2S_2O_4$) A crystalline salt that is a strong reducing agent and the main ingredient in several resin cleansers used to clean iron-fouled ion exchange resin beds.

sodium hydroxide (NaOH) A strong caustic chemical used in treatment processes to neutralize acidity, increase alkalinity, or to raise the pH value. It is also known as caustic soda, sodium hydrate, lye, or white caustic.

sodium hypochlorite (NaOCl) A substance, commonly known as liquid bleach, that is used for disinfection as an alternative to chlorine gas, especially where safety concerns over storage of the gas exist. When the hypochlorite is added to water, it hydrolyzes to form hypochlorous acid (HOCl), the same active ingredient that occurs when chlorine gas (Cl_2) is used. *See also* chlorine; hypochlorous acid.

sodium hyposulfite ($Na_2S_2O_4$) A strong reducing agent. *See also* sodium metabisulfite.

sodium metabisulfite ($Na_2S_2O_5$) A dry commercial form of sodium bisulfite. *See also* sodium hydrosulfite.

sodium silicate (Na_2SiO_3) A caustic solution of silica used in water treatment to prevent corrosion or, after a treatment known as activation, to assist coagulation. *See also* activated silica.

sodium silicofluoride (Na_2SiF_6) A dry chemical used in the fluoridation of drinking water. It is derived from hydrofluosilicic acid.

sodium thiosulfate ($Na_2S_2O_3$) A reducing agent used to remove residual chlorine from solution. *See also* reducing agent.

sodium tripolyphosphate (STP or STPP) ($Na_5P_3O_{10}$) A water-soluble chelating agent for certain metals in solution. It is also called sodium triphosphate or pentasodium tripolyphosphate.

sodium 2-(parasulfophenylazo)-1,8-dihydroxy-3,6-naphthalene disulfonate (SPADNS) method A colorimetric method for the determination of fluoride in water. It is based on the reaction of fluoride with zirconium and the sodium 2-(parasulfophenylazo)-1,8-dihydroxy-3,6-naphthalene disulfonate (SPADNS) reagent. The intensity of the dye complex is inversely proportional to the concentration of fluoride.

soft In water, having low hardness (low concentrations of calcium, magnesium, or both). *See also* hard water; hardness; soft water.

soft water Water having a relatively low concentration of calcium and magnesium ions. Generally, soft water exhibits a hardness within the range of 0 to 75 milligrams per litre (0 to 4.4 grains per gallon) as calcium carbonate ($CaCO_3$). *See also* hard water.

softened water Any water that has been processed in some manner to reduce the total hardness to 17 milligrams per litre (1 grain per gallon) or less, expressed as calcium carbonate ($CaCO_3$) equivalent. *See also* hard water; zero soft water.

softening *See* water softening.

soil The unconsolidated, naturally occurring mineral or organic material on the immediate ground surface, capable of supporting plant growth. It extends from the surface to 6 inches (15 centimetres) below the depth at which properties produced by soil-forming processes can be detected. These properties differ from those found in the underlying unconsolidated material, and they consist of the interactions among living organisms, climate, and other natural processes acting on soil and soil parent material.

soil air The gases in the interstices of the aeration zone that open directly or indirectly to the surface and hence are connected with the atmosphere. Soil air is also called ground air or subsurface air.

soil alkalinity The pH value of soil extract. The soil is alkaline when the pH is greater than 7.0 and acidic when less than 7.0.

soil amendment Organic and inorganic materials added to soils to improve texture, nutrients, moisture-holding capacity, and infiltration rates.

soil box A device used to determine the resistivity of a confined volume of soil. Such boxes are manufactured by several companies; however, one can be constructed conveniently. A soil box measures 4 cm × 4 cm × 4 cm and two of the opposite sides are of stainless steel.

soil classification The process of identifying soil types based on different characteristics in the excavation, as required by the Occupational Safety and Health Administration's 1926.650 excavation standard.

soil colloid Soil particles that constitute the physical fraction of soils generally considered as being less than 2 micrometres in diameter. These particles are usually clay minerals, although other minerals may also be present with diameters less than 2 micrometres. These particles are mainly composed of silica (SiO_2), alumina (Al_2O_3), iron oxide (most commonly Fe_2O_2), and combined water.

soil conditioning Alteration of the physical nature of a soil by addition of solid agents to improve water movement.

soil conservation district (SCD) A local government entity within a defined water or soil protection area that provides assistance to farmers and other local residents in conserving natural resources, especially soil and water. This type of entity is also called the soil and water conservation district in some areas.

Soil Conservation Service (SCS) *See* Natural Resources Conservation Service.

soil corrosion Corrosion of underground materials resulting from soil conditions.

soil creep The slow movement, usually over relatively short distances, of a mass of soil acting under the force of gravity. This term is usually applied to movement that is much smaller in magnitude and extent than a landslide.

soil discharge As applied to groundwater, the direct discharge of water through evaporation. The water discharged is, for the most part, lifted by capillarity from the zone of saturation nearly to the surface, where evaporation takes place.

soil erodibility A measure of the soil's susceptibility to raindrop impact, runoff, and other erosional processes.

soil horizon A layer or section of the soil profile, more or less well defined, occupying a position approximately parallel to the soil surface and having characteristics that have been produced through the operation of soil-building processes.

soil improvement The addition of soil amendments (additives) to planted areas.

soil infiltration rate The maximum rate at which a soil, in a given condition at a given time, can absorb water.

soil moisture Pellicular water in the soil zone. It is divided into available and unavailable moisture—the former being water easily abstracted by roots of plants, the latter being water held so firmly by adhesion and other forces that it cannot usually be absorbed by plants rapidly enough to produce growth. *See also* pellicular water.

soil moisture content The volumetric water content of a soil sample; the ratio of water volume in a sample to total volume of the sample.

soil organic matter The organic fraction of the soil. It includes plants and animal residues in various stages of decomposition, cells and tissues of soil organisms, and substances synthesized by the soil population. Soil organic matter is commonly considered to be those organic materials that accompany the soil after passage through a 2-millimetre pore diameter sieve.

soil pipe A standard type of bell-and-spigot cast-iron pipe of limited strength.

soil porosity The percentage of the soil (or rock) volume that is not occupied by solid particles, including all pore space filled with air and water. The total porosity may be calculated from the following formula:

$$n = \left(1 - \frac{\gamma_d}{\gamma_w G}\right) \times 100$$

Where:

n = the percent pore space

γ_d = the unit weight (weight per unit of volume) of the soil after all free water has been removed from the pores of the soil (γ is the universal symbol in soil mechanics for unit weight or weight density; the subscript d symbolizes dry)

γ_w = the unit weight of water (62.4 pounds per cubic foot in US customary units, 9.81 kilonewtons per cubic metre in Système International units)

G = specific gravity of the solid particles in the soil

soil priming The amount of saturation locally required to produce runoff.

soil profile The layerlike arrangement of soil horizons, often with differing colors, consistencies, and textures formed by natural weathering processes in soil material.

soil resistivity An indication of difficulty with which a soil conducts electrical current; the average electrical resistance of a volume of soil.

soil and rock deposit classification The classification of a soil and rock deposit by a competent person as either type A, type B, or type C on the basis of such properties as unconfined compressive strength, whether the soil is fissured, whether the soil has previously been disturbed, and so on. *See also* type A soil; type B soil; type C soil.

soil salinity *See* saline-alkaline soil.

soil saturation The filling of all the pore space in a soil with water. Soil below the depth of the water table is said to be saturated. *See also* duration of inundation.

soil sealant A chemical or physical agent that plugs porous soils and prevents leaching or percolation.

soil solution The aqueous liquid phase of the soil and its solutes, the latter consisting of (1) ions dissociated from the surfaces of the soil particles and (2) other soluble materials.

soil structure The arrangement of soil particles into aggregates.

soil tank A tank filled with soil and used for measuring evaporation and evapotranspiration. A soil tank is also called a lysimeter or vegetation tank. *See also* lysimeter.

soil texture The proportions of soil particles (sand, silt, and clay) in a soil profile.

soil water Water in the aeration zone, immediately below the surface of the ground. Such water may be discharged into the atmosphere in appreciable quantities by plant transpiration or soil evaporation.

soil water percentage The percentage of water in soil based on the weight of material oven dried at 230° Fahrenheit (110° Celsius). It is the ratio, expressed as a percentage, of (1) the weight of a given volume of soil under specified conditions minus the weight of such soil after it has been dried in an oven at a temperature of 230° Fahrenheit (110° Celsius) until its weight becomes constant, to (2) the weight of such volume of oven-dried soil.

soil water table The upper surface of a body of soil water.

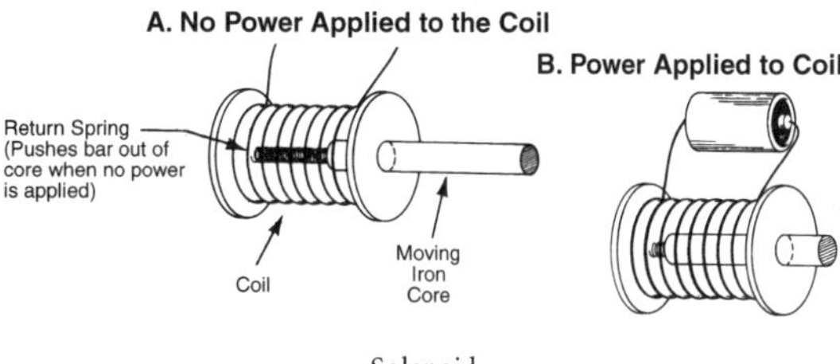

Solenoid

soils report A report by a soils engineer indicating soil type(s), soil depth, uniformity, infiltration rates, and pH for a given site.

soil-saving dam A dam, usually small, constructed to impound or retard surface runoff temporarily and bring about deposition of a substantial portion of the soil being carried away by stormwater runoff. Such a dam is also called a detention dam.

soil-water zone The portion of the unsaturated zone where plants can use soil water in their metabolic processes.

solar distillation A desalting process in which distillation takes place in a device that uses solar energy as the heat source (i.e., a solar still). This process is also called solar humidification. *See also* solar still.

solar energy Energy imparted from the sun.

solar humidification *See* solar distillation.

solar irradiation The use of the sun's energy to initiate or catalyze a reaction.

solar salt Common salt that is produced by solar evaporation in shallow ponds or lagoons and is used in water softener regeneration.

solar still A device used for solar humidification or solar distillation, consisting of a saline water basin under a transparent cover that allows solar energy to pass through, resulting in a temperature rise and evaporation followed by the subsequent condensation of vapors and collection of the distillate.

solder A metal or metallic alloy that is melted and used to join metallic surfaces. Many solders contain lead and tin. Historically, 50:50 tin:lead solder was used to join copper piping. The US Environmental Protection Agency now requires that low-lead solders be used when the solder is exposed to potable water.

sole source aquifer (1) The single water supply available in an area. (2) A US Environmental Protection Agency designation for a water source requiring protection.

solenoid A magnetically operated mechanical device (electric coil). Solenoids can operate small valves or electric switches.

solenoid shutoff valve An electrical device operated by a magnetic coil to make a valve either open or close to allow or shut off water flow. This type of valve is used extensively for flow control and direction on many water-processing systems.

solid Matter that is in its most highly concentrated form (i.e., the atoms or molecules are more closely packed than in gases or liquids) so it is more resistant to deformation. The normal condition of the solid state is a crystalline structure (the orderly arrangement of the constituent atoms of a substance in a framework called a lattice). All solids can be melted by heat and thereby converted to a liquid.

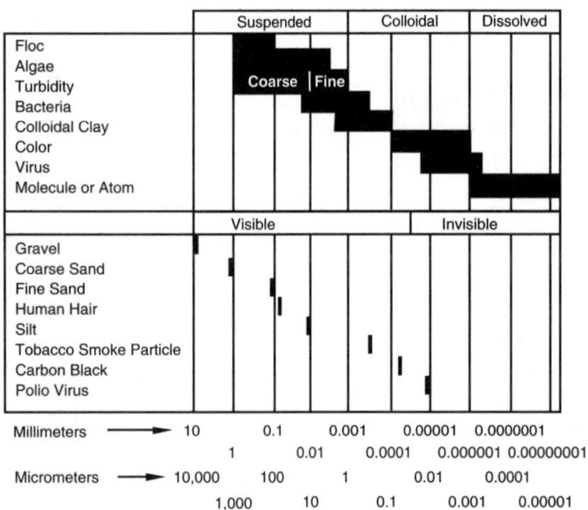

Size range of solids

solid angle units In Système International, steradians.

solid sleeve (1) A pipe fitting for uniting two pipes of the same nominal diameter in a straight line. (2) A tube into which a pipe is inserted.

Solid Waste Disposal Act (SWDA) Public Law 89-272, Title II, enacted in 1965, which focused on research, demonstrations, and training. It provided for the federal government to share with the states the costs of making surveys of waste disposal practices and problems, as well as the costs of developing waste management plans. The Solid Waste Disposal Act was so extensively amended by the Resource Conservation and Recovery Act of 1976 (Public Law 94-580) that it is commonly referred to by the latter act's title rather than by its own official title. The Solid Waste Disposal Act–Resource Conservation and Recovery Act and amendments are codified in 42 U.S.C. 6901–6991k.

solid–liquid separation Any process for removing suspended solids from water.

solid-phase extraction A technique in which analytes are isolated from solution by adsorbing onto a solid matrix. The matrix is often a chromatographic material coated onto resin beads or a membrane filter. Analytes are typically desorbed from the solid matrix with an elution solvent. *See also* solid-phase microextraction.

solid-phase extraction disk A disk, similar in size and shape to standard filter paper but made of various hydrophobic compounds, that is used to adsorb (extract) organic compounds from a water sample. The disk is then extracted (desorbed) with a solvent, such as methylene chloride (CH_2Cl_2), for subsequent analysis. Use of such a disk serves as a substitute for liquid–liquid extraction to lessen the volume of waste solvents generated. *See also* liquid–liquid extraction.

solid-phase microextraction (SPME) A separation technique used in the analysis of organic compounds. Analytes are adsorbed onto the stationary phase of a small section of a fused silica fiber. Analytes can be desorbed into an organic solvent or directly into an instrument for further analysis.

solids Matter suspended or dissolved in water. As of the time of this writing, the term *solids* is used instead of the term *residue*. In addition, the terms *suspended solids* and *dissolved solids* are used to refer to nonfilterable residue and filterable residue, respectively. Different categories of solids are defined based on the analytic method of determination. *See also* dissolved solids; fixed solids; settleable solids; suspended solids; total solids; volatile solids.

solids disposal Any process for ultimate disposal of sludges by incineration, landfilling, soil conditioning, or other means.

solids, dissolved *See* dissolved solids.

solids, fixed *See* fixed solids.

solids flux The application of solids per unit time and area, typically expressed in units of pounds per square foot per second (kilograms per square metre per second). Solids flux is a design criterion frequently used in sedimentation to define the limiting amount of solids that can be applied to a given area in a specified time to permit the necessary thickening to take place or to govern the underflow of solids in the bottom of a clarifier.

solids, suspended *See* suspended solids.

solids, total *See* total solids.

solids, volatile *See* volatile solids.

solids-contact basin A unit process in which both flocculation and particle separation occur. Coagulated water is passed upward through a solids blanket, allowing flocculation and particle separation to take place in a single step. The solids blanket is typically 6 to 10 feet (2 to 3 metres) below the water surface, and clarified water is collected in launder troughs along the top of the unit. Solids are continually withdrawn from the solids blanket to prevent undesired accumulation. *See also* launder trough.

solids-contact clarifier *See* solids-contact basin.

solids-contact process A process combining coagulation, flocculation, and sedimentation in one treatment unit, in which the flow of water is vertical. *See also* solids-contact basin.

solubility The ability or tendency of one substance to blend uniformly with another. Solids vary from 0 to 100 percent in their degree of solubility in liquids, depending on the chemical nature of the substances. To the extent that they are soluble, solids lose their crystalline form and become molecularly or ionically dispersed in the solvent to form a true solution. Liquids and gases are often said to be miscible in other liquids and gases rather than soluble. *See also* miscible; solvent.

solubility coefficient *See* gas solubility coefficient.

solubility product constant (K_{sp}) In a saturated solution at a specified temperature, the equilibrium constant of the dissolution reaction of a solid in water. For example, if solid B_mA_n dissolved in water, it would become m B^{x+} and n A^{y-} ions (where x and y are the valence of the ions). In this case, K_{sp} would equal $[B^{x+}]^m[A^{y-}]^n$. If the solid were $Mg(OH)_2$, then $x = 2, y = 1, m = 1$, and $n = 2$; hence, K_{sp} would be $[Mg^{2+}]^1[OH^{1-}]^2$, or more commonly $[Mg^{2+}][OH^-]^2$. *See also* equilibrium constant; saturated; solubility.

solubilization To treat a material so that it becomes more readily dissolved.

solute A substance dissolved in a solvent.

solute (salt) flux For a membrane separation process, the solute (salt) mass passing through the membrane in the permeate per

Solution, composed of a solute and a solvent

unit area and time. Solute (salt) flux is commonly expressed in pounds per square foot per second (grams per square metre per second).

solute (salt) mass transfer coefficient *See* solute (salt) permeability coefficient.

solute (salt) permeability coefficient A factor describing the flow of solute (salt) through a membrane for a diffusion-controlled, pressure-driven membrane process, such as reverse osmosis. The coefficient is dependent on the specific membrane type and characteristics. It is equal to the solute (salt) flow rate through a unit area of membrane (the solute or salt flux) divided by the solute (salt) concentration gradient across the membrane. It is typically expressed in units of feet or metres per second, depending on the system of units used. Sometimes this term is referred to as the solute (salt) transport coefficient or solute (salt) mass transfer coefficient. In equation form,

$$B = F_s/(C_1 - C_2)$$

Where:

B = solute permeability coefficient, in feet per second (metres per second)

F_s = solute flux, in pounds per second per square foot (grams per second per square metre)

$C_1 - C_2$ = solute concentration gradient across the membrane, in pounds per cubic foot (grams per cubic metre)

solute transport The movement of a given dissolved constituent in a fluid, or from one phase to another. Solute transport can describe the movement of a contaminant in groundwater, or it can describe the movement of a contaminant from a liquid to a solid phase (e.g., granular activated carbon adsorption).

solute transport coefficient *See* solute (salt) permeability coefficient.

solution A mixture in which one or more substances (solutes) are dissolved into another substance (solvent), usually a liquid, in such a way that the solute is equally distributed (homogeneous) throughout the solvent in the form of either molecules (as in a sugar solution) or ions (as in a salt solution).

solution channel An opening in solid rock created by the dissolution, usually by water, of material that formerly occupied the opening and the carrying away of the material in solution. A solution channel is also called a solution opening.

solution diffusion model For some pressure-driven membranes (e.g., a reverse osmosis membrane), a mathematical model that describes the transport of solvent (water) and solute (salt) through a permselective (semipermeable) membrane. For a membrane type with given solvent permeability, solute permeability characteristics, and membrane thickness, the solvent flux is directly proportional to the net driving force across the membrane and the solute flux is directly proportional to the concentration gradient across the membrane. *See also* net driving pressure.

solution feeder A feeder for dispensing a concentrated chemical solution to water at a rate that is controlled manually or automatically.

solution opening *See* solution channel.

solvent A liquid substance that dissolves another substance (the solute) to form a solution.

solvent flux *See* flux; flux rate (solvent, water).

solvent (water) mass transfer coefficient *See* solvent (water) permeability coefficient.

solvent (water) permeability coefficient A factor describing the flow of solvent (water) through a membrane for a pressure-driven membrane process such as reverse osmosis. The coefficient is dependent on the specific membrane type and characteristics. It is equal to the solvent (water) flow rate through a unit area of membrane (the solvent flux) divided by the net driving pressure across the membrane. This term is sometimes referred to as permeance, solvent (water) transport coefficient, or solvent (water) mass transfer coefficient. In equation form,

$$A = \frac{F_w}{\text{NDP}}$$

Where:

A = solvent permeability coefficient, in gallons per day per square foot per pound per square inch (cubic metres per hour per square metre per kilopascal)

F_w = solvent flux, in gallons per day per square foot (cubic metres per day per square metre)

NDP = net driving pressure (hydraulic pressure differential across the membrane minus the osmotic pressure differential across the membrane), in pounds per square inch (kilopascals)

solvent (water) transport coefficient *See* solvent (water) permeability coefficient.

somatic (1) Of or pertaining to the body. (2) Pertaining to any of the cells of an organism that become differentiated into the tissues, organs, and so on; pertaining to nonreproductive cells or tissues.

somatic bacteriophage A bacteriophage that attaches to receptor sites on the bacterial cell surface, as opposed to an F-specific or male-specific bacteriophage that attaches only to bacteria that have F-pili. *See also* male-specific bacteriophage.

somatic coliphage A coliphage that enters bacterial cells by attachment to the bacterial cell wall. *See also* coliphage.

sonic flowmeter A device for measuring flow rates across fluid streams by either Doppler-effect measurements or time-of-transit determination. In both types of flow measurement, the displacement of the portion of the flowing stream carrying the sound waves is determined, and the flow rate is calculated from the effect on sound wave characteristics.

sonic logging A geophysical logging technique that determines the properties of a formation by measuring its response to high-energy sound waves.

sonoscope *See* waterphone.

sorbent A solid material, such as activated carbon or ion exchange resin, that is used to concentrate dissolved compounds. *See also* activated carbon; adsorbent; adsorption; ion exchange resin.

sorption A surface phenomenon involving absorption, adsorption, or a combination of the two. This term is often used when the specific mechanism is not known. *See also* absorption; adsorption.

sounding (1) In hydrography, the process of measuring the depth of water with a lead line or by other means. (2) In subsurface investigations, the act of pushing or driving a rod down through the soil as far as it will penetrate with the equipment at hand. Such soundings are most useful in soft materials to determine the depth of a solid stratum. This type of sounding is also called probing.

sounding line In hydraulic surveying, a rope, wire, or cable with a weight attached, used for measuring depths of water. A sounding line is also called a lead line.

sounding rod A portable staff gauge for measuring the depth of a liquid in a vessel or stream. For deep, quickly flowing streams, a sounding wire is used.

sounding stick A portable staff gauge for measuring the depth of a liquid in a vessel or stream. *See also* sounding rod.

sounding tube A pipe or tube used for measuring the depths of water.

sounding wire A flexible wire with weight attached at the bottom, used for determining the depth of water. A sounding wire is also called a lead line or log line.

sour brine Brine that contains a high concentration of calcium, magnesium, or other substances that would interfere with the brine's use or reuse for effective regeneration of exhausted ion exchange resin.

source contribution *See* relative source contribution.

source of infection The person, animal, object, or substance from which an infectious agent passes to a host. *See also* reservoir of infection.

source water The supply of water for a water utility. Source water is usually treated before distribution to consumers, but some groundwaters are of such a quality that they can be distributed untreated. This term is preferred over *raw water.*

Source Water Assessment Program (SWAP) A US Environmental Protection Agency program established by the 1996 Safe Drinking Water Act amendments, requiring states to develop a comprehensive program to identify the areas that supply public tap water, inventory contaminants and assess water system susceptibility to contamination, and inform the public of the results.

source-of-funds analysis An analysis often undertaken when major expenditures are planned in a utility to determine "who pays," especially if new capacity is planned. The purpose of the analysis is to match revenue requirements with the available sources of funds. For new capacity, the concept of "growth pays for growth" is often used, in which case connection charges are established or increased to pay for the portion of new facilities constructed to serve new demand. Major capital programs are usually funded from debt (rather than directly from rates), and the debt service is paid for from rates for facilities serving existing customers or from connection charges for facilities serving new customers. The source-of-funds analysis is usually confined to rates, debt (with many alternatives), and connection charges, but it could also be applied to the rate structure to determine how, for example, to pay for new peak capacity.

sp gr *See* specific gravity.

sp ht *See* specific heat.

spacer For a membrane treatment unit, a material that is designed to allow controlled fluid flow and is used to separate membrane layers, such as a feed–concentrate or permeate spacer for a spiral wound membrane or a concentrating compartment or demineralizing compartment spacer for an electrodialysis membrane stack.

spadable sludge Residuals that can readily be forked or shoveled from a drying bed. Such sludge is ordinarily less than 75 percent moisture.

SPADNS method *See* sodium 2-(parasulfophenylazo)-1,8-dihydroxy-3,6-naphthalene disulfonate method.

span The scale or range of values across which an instrument is designed to measure. *See also* range.

sparger A perforated pipe or porous fitting on a pipe in an aerator or ozone (O_3) contact compartment through which the air or ozone-containing air is introduced into the water. A sparger allows for the diffusion of the air or ozone into the water.

sparingly soluble salt A salt that is relatively insoluble in water and may precipitate when moderately concentrated even at ambient temperatures. For membrane desalting processes, the concentrations of sparingly soluble salts—such as calcium carbonate ($CaCO_3$), calcium sulfate ($CaSO_4$), strontium sulfate ($SrSO_4$), and barium sulfate ($BaSO_4$)—are commonly evaluated and provisions made to avoid their precipitation in the system.

spark source mass spectroscopy (SSMS) A type of mass spectrometry in which ions representative of the sample material are produced by a pulsed radio frequency spark source. A spark between the conducting sample electrodes is produced by application of a pulsed radio frequency. Sample electrodes may be made directly from the sample material if that material is a conductor or semiconductor, whereas nonconducting materials must be mixed with pure conductors, such as silver or graphite, and compressed. Ions produced in the source are accelerated into the analyzer portion of the mass spectrometer. This technique offers high sensitivity and comprehensive elemental coverage, as well as linear response for many elements. For example, this technique has been used to analyze the boron content of different steel samples. *See also* mass spectrometry.

sparkling bottled water *See* bottled sparkling water.

sparkling water Carbonated water for which the carbon dioxide (CO_2) content comes from the same source as the water itself. *See also* bottled sparkling water.

spatial (1) Pertaining to an object, dataset, or activity that has a geographic component or is related to a location in a coordinate system. Spatial analysis and spatial modeling are based on geographical relationships of data. (2) In biology or ecology, pertaining to the occurrence of or relationships among microorganisms, plants, or animals living in the same habitat, as in the spatial distribution of a particular species.

spatial analysis The use of analytical techniques associated with the study of the locations of geographical entities together with their spatial dimensions.

spatial variation Changes in water quality characteristics among various locations in a water body or among water bodies.

SPC An abbreviation for standard plate count, an obsolete term. *See also* heterotrophic plate count.

SPC system *See* state plane coordinate system.

SPE method *See* solid-phase microextraction.

special aquatic site A geographic area, large or small, possessing special ecological characteristics in terms of productivity, habitat, wildlife protection, or other important and easily disrupted ecological values.

special area management plan (SAMP) A type of comprehensive plan authorized by the Coastal Zone Management Act amendments of 1980, providing for natural resource protection balanced against reasonable economic growth. Special area management plans are developed by the US Army Corps of Engineers, often in coordination with state and local planning efforts, and they identify areas either suitable for development or in need of protection.

special A specific section of pipe constructed for a special purpose or location that meets specifications but is not of a standard length or standard shape. A special may have nonstandard connection outlets or degree of bend. *See also* pipe fitting.

special casting Any of numerous fittings of various shapes and materials made to serve specific purposes.

speciation The chemical makeup of a group of compounds. For example, for disinfection by-products, the speciation may be limited to chlorine-substituted by-products or may alternatively include a mixture of chlorine- and bromine-substituted species.

species A group of related individuals with common chromosomic number and structure, hereditary morphology, physiological characteristics, and way of life, separated from neighboring groups by a barrier that is generally sexual in nature and occupying a definable geographic area. Members of the same species are capable of reproduction producing offspring of the same species.

species composition The kinds and numbers of species occupying an area.

species difference The presence in different animal species of varying degrees of sensitivity to the effects of a chemical, physical, or biological agent. Such differences can be related to differences in the metabolism of a chemical or by different species, or they may simply reflect differences in intrinsic sensitivity. *See also* selective toxicity.

species diversity The number and relative abundance of species in an area.

species population Similar organisms residing in a defined space at a certain time, taken as a group.

specific absorption The capacity of water-bearing material to absorb water after the free water has been removed. Specific absorption is equal to the specific yield of the material, except in cases when such material has suffered compaction from the burden of overlying material after groundwater has been removed, in which case allowance must be made for the effect of such compaction. *See also* free water; specific yield.

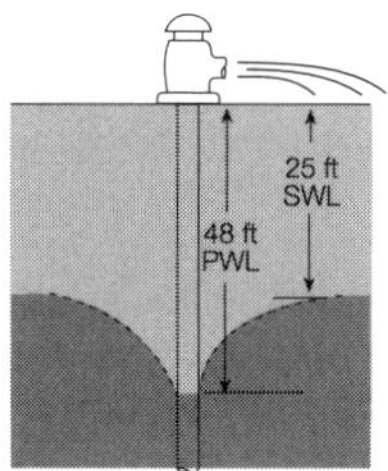

Well Output = 115 gpm
Drawdown = 48 ft. – 25 ft. = 23 ft.
Specific Capacity = Output/Drawdown
115 gpm/23 ft. = 5 gpm/ft.

Specific capacity

specific capacity A measurement of the well yield per unit length of drawdown. *See also* drawdown.

specific capacity test A testing method for determining the adequacy of an aquifer or well by measuring the specific capacity. *See also* specific capacity.

specific conductance The conductivity of a solution as measured using a standard cell with a 1-centimetre width. Specific conductance is expressed in units of micromhos or microsiemens at 77° Fahrenheit (25° Celsius). *See also* conductivity.

specific discharge The discharge per unit area. This value is often used to define flood magnitudes. In porous media and filtration, specific discharge is the discharge per unit area of the filter or porous media.

specific energy (1) The energy with regard to the channel bed as a datum, or the energy measured above the channel bed. In equation form,

$$E = Y + \frac{V^2}{2g}$$

Where (in any consistent set of units):

E = the specific energy
Y = water depth
V = mean velocity in a channel
g = gravitational constant

(2) The total energy per unit weight of a substance. *See also* critical velocity.

specific energy units In the US customary system, British thermal units per pound mass. In Système International, joules per kilogram; square metres per second squared.

specific entropy units In the US customary system, British thermal units per pound mass per rankine. In Système International, joules per kilogram per kelvin; square metres per second squared per kelvin.

specific flux For pressure-driven membrane processes, the permeate (water) flux divided by the net driving pressure. In the US customary system, the units are gallons per square foot per day per pounds per square inch; in Système International, they are grams per square metre per second per kilopascal. Specific

Specific gravities of various solids and liquids

Substance	Specific Gravity
Solids	
Aluminum (20°C)	2.7
Steel (20°C)	7.8
Copper (20°C)	8.9
Activated carbon*†	0.13–0.45 (avg. 0.19)
Lime*†	0.32–0.80
Dry alum*†	0.96–1.2
Soda ash*†	0.48–1.04
Coagulant aids (polyelectrolytes)*†	0.43–0.56
Table salt*†	0.77–1.12
Liquids	
Liquid alum (36°Bé, 15.6°C)	1.33
Water (4°C)	1.00
Fluorosilicic acid (30%, –8.1°C)	1.25–1.27
Sulfuric acid (18°C)	1.83
Ferric chloride (30%, 30°C)	1.34

* Bulk density used to determine specific gravity.
† Temperature and/or pressure not given.

flux is commonly temperature corrected to a standard or constant temperature to negate water temperature effects for performance analysis, in which case it is sometimes called the temperature-corrected specific flux. Specific flux is equal to specific productivity divided by the active membrane area. *See also* flux; net driving pressure; specific productivity; temperature correction factor.

specific gravity (sp gr) The ratio of the density of a substance to a standard density. For solids and liquids, the density is compared to the density of water at 39.2° Fahrenheit (4° Celsius) (i.e., 1 kilogram per litre). For gases the density is compared to the density of air at standard temperature and pressure (i.e., 1.2 grams per litre).

specific heat (sp ht) (1) The ratio of the heat capacity of a substance to the heat capacity of water; or the quantity of heat required for a 1° temperature change in a unit mass of material. Specific heat is commonly expressed in units of calories per gram per degree Celsius or British thermal units per pound per degree Fahrenheit. (2) The quantity of heat required to increase the temperature of a unit mass of granular activated carbon through a given interval of temperature, divided by the corresponding quantity for water.

specific humidity The ratio of the mass of water vapor to the total mass of the mixture of air and water vapor.

specific ion meter A sensitive voltmeter used to measure the concentration of specific ions, such as fluoride, in water. Electrodes designed specifically for each ion must be used.

specific level The level of the water surface in a river at any particular site for a given discharge. The specific level may change with time, depending on changes in the river section at the site.

specific productivity For pressure-driven membrane processes like reverse osmosis, the permeate production rate divided by the net driving pressure. Specific productivity is commonly expressed in units of gallons per day per day per pounds per square inch or cubic metres per second per kilopascal. It is commonly temperature corrected, in which case it is sometimes called the temperature-corrected specific productivity. *See also* permeate; net driving pressure; specific flux; temperature correction factor.

specific resistance (SR) (1) A measure of the resistance of a sludge to dewatering under specified conditions. (2) The capacity for resisting the flow of electrical current. In the case of liquids, such as water, specific resistance is the resistance of a 1.0-cubic-centimetre cube, which is the resistance offered by the liquid between two electrode plates 1 centimetre square and placed 1 centimetre apart. Specific resistance is the reciprocal of specific conductance and is expressed in units of ohm-centimetres. One hundred milligrams per litre of sodium chloride (NaCl) dissolved in water causes a specific resistance of 4,716 ohm-centimetres, whereas a 1,000-milligrams-per-litre solution of NaCl in water results in a specific resistance of 500 ohm-centimetres. *See also* specific conductance.

specific resistance to filtration (SRF) A measure of the filterability of a sludge, based on the amount of time required for the sludge to pass through a filter paper with a known pore opening size under a specific applied pressure.

specific retention The quantity of water retained against the pull of gravity by rock or earth after the material has been saturated and allowed to drain completely to a remote body of mobile water by way of continuous capillary interstices. Specific retention is expressed as the ratio of the volume of water that the rock or earth material will retain against gravity to the volume of the material itself. The sum of specific yield and specific retention is the porosity of the sample. *See also* porosity; specific yield.

specific speed A speed or velocity of revolution, expressed in revolutions per minute, at which the runner of a given type of turbine would operate if it were so reduced in size and proportion that it would develop 1 horsepower under a 1-foot head (not a Système International unit). This quantity is used in determining the proper type and character of turbine to install at a hydroelectric power plant under given conditions.

specific ultraviolet absorbance (SUVA) The ultraviolet absorbance at 254 nanometres (measured in units of per metre) divided by the dissolved organic carbon concentration (in milligrams per litre). Typically, a specific ultraviolet absorbance less than 3 litres per metre-milligram corresponds to largely nonhumic material, whereas a specific ultraviolet absorbance in the range of 4–5 litres per metre-milligram corresponds to mainly humic material. Because humic materials are more easily removed through coagulation than nonhumic substances, higher specific ultraviolet absorbance values should indicate a water that is more amenable to enhanced coagulation. *See also* dissolved organic carbon; enhanced coagulation; humic material; ultraviolet absorbance at 254 nanometres.

specific volume units In the US customary system, gallons per pound. In Système International, cubic metres per kilogram.

specific weight Weight of a substance per unit volume. It is equal to the density times the gravitational constant.

specific weight units In the US customary system, pounds force per cubic foot. In Système International, newtons per cubic metre.

specific well yield The maximum rate at which a well will yield water under a stipulated set of conditions, such as a given drawdown, pump and motor, or engine size. Specific well yield may be expressed in terms of cubic metres per second, gallons per minute, cubic feet per second, or similar units.

specific year flood A stream flow that is equaled or exceeded, on the average, once in a designated period, e.g., 10 years, 50 years, or 100 years. *See also* flood frequency.

specific yield The quantity of water that a unit volume of permeable rock or soil, after being saturated, will yield when drained by gravity. Specific yield may be expressed as a ratio or as a percentage by volume. The sum of specific yield and specific retention is the porosity of the sample. *See also* porosity; specific retention.

specifications Precise standards of performance for construction work, materials, and manufactured products. Specifications make it possible to express expected values when work or items are purchased or contracted for, and they provide means of determining conformance with expectations after purchase or construction.

specificity In the context of an analytical test method for a microorganism, the ability of the method to select and distinguish the microorganism under consideration from all others in the same environment (sample).

spectral absorption coefficient (SAC) The ultraviolet absorbance of a sample divided by the spectrophotometer cell path length. Ultraviolet absorbance measurements are typically reported in units of per centimetre or per metre. *See also* spectroscopy; ultraviolet absorption.

spectrograph An instrument for photographing or producing a representation of a spectrum.

spectrometer A chemical analytical instrument used in spectroscopy. *See also* spectroscopy.

spectrometry *See* spectroscopy.

spectrophotometer A photometer that uses a diffraction grating or a prism to control the light wavelengths used for specific analysis.

spectrophotometry *See* spectroscopy.

spectroscopy A technique used in chemical analyses that is based on the principle that many substances, when crossed by a beam of light, allow a unique and well-defined fraction of that light to pass or emit a well-defined fraction of radiation when returning from an atomic vapor state to their fundamental state. The characteristic wavelength pattern of the absorbed or emitted light can be used to identify the particular substance with great certainty. The quantity of the light absorbed or emitted is proportional to the concentration of the substance. Spectroscopy is one of the most frequently used analytical methods for water analysis. *See also* atomic absorption spectroscopy; emission spectroscopy.

spectrum A representation of electromagnetic radiation over a range of wavelengths.

speed coefficient The ratio of the linear velocity of the runner periphery of a waterwheel to the spouting velocity of the water under a prescribed head. *See also* spouting velocity.

speed of light In a vacuum, 1.86×10^5 miles per second (2.99×10^8 metres per second).

speed units In the US customary system, feet per second. In Système International, metres per second.

sperm *See* spermatozoon.

spermatid An intermediate cell in the development of sperm. It is derived from a secondary spermatocyte by fission (meosis) and develops into a spermatozoon.

spermatocyte An intermediate cell in the development of sperm. It is derived from spermatogonia and develops into a spermatid.

spermatogenesis The orderly process of producing sperm, beginning with the development of spermatogonia and proceeding through primary and secondary spermatocyte and spermatid stages before production of mature spermatozoa.

spermatogonia The stem cell for sperm, found attached to the basal lamina of the seminiferous tubule.

spermatotoxic Causing the death of mature spermatozoa.

spermatozoon A mature male germ cell; the specific output of the testes.

spermhead abnormality A change in the structure of the head of sperm. Because a normal spermhead has well-defined genetic determinants and a complex structure, production of spermhead abnormalities can be used as the basis of a short-term in vivo test for genotoxic chemicals. The effects can be differentiated from simple spermatotoxic effects by the timing of the appearance of the abnormalities following exposure. This timing corresponds to the stage of spermatogenesis in which the effect is produced. If the effects are delayed relative to exposure, the effect is likely to reflect a genotoxic effect. *See also* genotoxic.

sphagnum Soft moss found mainly in bogs on the surface.

sphericity A measure of the bead roundness or "whole bead" count of beads in an ion exchange resin product or other bead-form absorbent or filter medium.

spigot The end of a pipe, fitting, or valve that is inserted into the bell end.

spiked Pertaining to a sample fortified with a known amount of a substance to evaluate the accuracy (recovery) of an analytical method as part of a laboratory quality assessment and quality control program. The term *fortified* (or *laboratory-fortified*) is being used more commonly to refer to a spiked sample.

spiked sample A sample to which a known amount of a substance has been added.

spill An intentional or unintentional release of compounds into the environment or workplace.

spill water Water released from an impoundment because the impoundment lacks sufficient storage capacity.

spillway A device used to release water from a dam.

spillway channel An artificial waterway provided for conducting away from the structure the water overflowing from a reservoir, aqueduct, or canal by way of an overflow device. A spillway channel is also called an overflow channel.

spillway chute An open conduit conducting water, usually at supercritical velocity, from a reservoir to the waterway downstream from a dam. A spillway chute is also called a spillway trough.

spillway dam A dam constructed with a crest to permit the overflow of water. A spillway dam is also called an overfall dam.

spillway lip That part of the spillway crest where overflow starts.

spillway trough *See* spillway chute.

spillway tunnel A tunnel used as an outlet channel for a spillway.

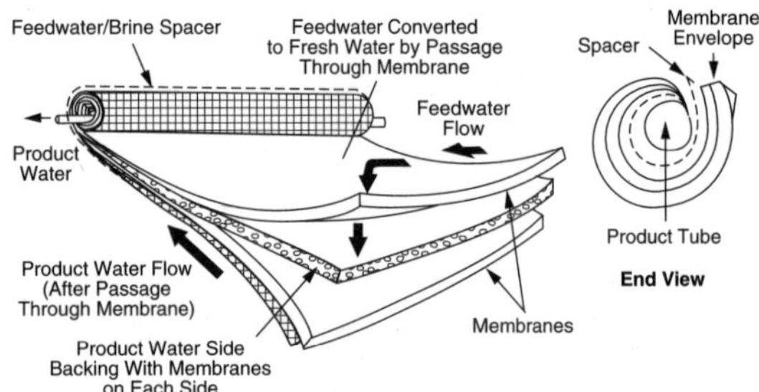

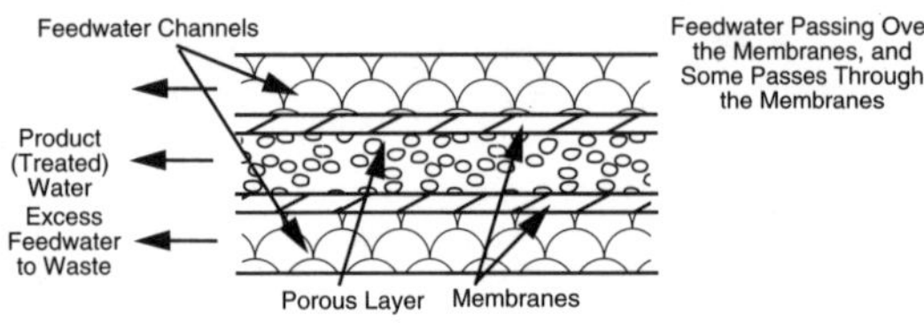

Details of a spiral wound membrane

spiral riveted pipe Steel pipe made from sheets of steel curved into a cylindrical shape with the edges riveted together to form a seam around the pipe.

spiral welded pipe Steel pipe made from sheets of steel curved into a cylindrical shape with the edges welded together. The seams may form a spiral running around the circumference of the pipe with welding along the seam.

spiral wheel An axial-flow turbine modified with runner blades having helical surfaces and usually mounted on a horizontal or slightly inclined shaft.

spiral wound Having a construction configuration very common for one style of reverse osmosis membrane and cartridge filter element. In reverse osmosis membranes, the membrane sheets are assembled in layers around a perforated mandrel product water tube, with coarse mesh spacer screens between the layers, to form a complete module element. In cartridge filter elements, the filtration material, such as fiber cord, is continuously wound around a perforated mandrel core tube.

spiral wound membrane *See* spiral wound.

Spirillum A genus of bacteria characterized as rigid, helical cells (1.4–1.7 micrometres in diameter by 14–60 micrometres in length) that are aerobic or microaerophilic, gram-negative, motile by means of large bipolar tufts of flagella, and found in stagnant, freshwater environments.

Spirillum NOX **(NOX)** A bacterial strain used in a bioassay procedure to estimate oxalate ($OOCCOO^{-2}$)–carbon equivalents of the assimilable organic carbon concentration in a water sample. *See also* assimilable organic carbon; oxalate–carbon equivalent.

Spirillum **sp.** Any species of bacteria belonging to the genus *Spirillum.*

splash goggles Safety goggles with shatterproof lenses designed to provide a tight covering around the eyes, protecting them from chemicals and flying particles.

spline interpolation The construction of a smooth curve by fitting simple curves represented by mathematical formulas to connect a series of points. (If the simple curves are line segments, the process is called linear interpolation.)

split addition The multiple addition of a coagulant, flocculant, or both at several locations along a flow path to achieve cumulative benefits.

split sample A single sample that separated into at least two parts such that each part is representative of the original sample.

split treatment A water treatment scheme whereby a portion of the potential feedwater to a treatment process bypasses the process and combines with the product water from the treatment process, producing a blended product water. Split treatment generally is considered when treating all of the water with a specific process would have negative affects (such as increased corrosivity) or when product water quality goals can be met without treating all of the water with the process and the bypass would reduce overall production costs. Split treatment is typically used in ion exchange softening, such that one train will be designed for virtually complete hardness removal and the output will then be blended with bypass water to meet finished water hardness goals.

split-case pump A centrifugal pump with the suction and discharge ports parallel but on opposite sides of the pump. The pump is split along the shaft so that the top half of the casing can be removed.

splitless injector *See* split–splitless injector.

split-stream treatment *See* split treatment.

split–splitless injector A device that is commonly used on a gas chromatograph and allows two types of injection techniques. The split injection technique allows a portion of a sample to be directed to the gas chromatographic column and the remainder of the sample sent to waste. The splitless injection technique directs the entire sample onto the chromatographic column.

splitter box (1) A division box that splits the incoming flow into two or more streams. (2) A device for splitting and directing discharge from a head box to two separate points of application.

SPME *See* solid-phase microextraction.

spoil Excavated material, such as soil from the trench of a water main.

spore A propagative unit that is typically unicellular, is often uninucleate, and may be formed with (sexually) or without (asexually) a change in ploidy. Most types of nonmotile spores are dormant and are more resistant to environmental change than are vegetative cells. The principal groups of spore-forming bacteria are those in the genera *Bacillus* and *Clostridium.* Bacterial spores are asexual; are formed within the vegetative cell (one spore per cell); and are much more resistant to heat, desiccation, and many bactericidal chemicals than are vegetative cells. Some spores can behave as gametes. *See also* cyst; gameteogenesis; ploidy; sporocyst; sporozoite.

spore removal surrogate technique In drinking water treatment, the use of indigenous *Bacillus* spores, or of *Bacillus subtilis* spores added as a spike, to evaluate the effectiveness of the coagulation, flocculation, and filtration process for removal of particulates from drinking water. Spore removal is used as a surrogate for the removal of specific parasite cysts or oocysts

(*Giardia* and *Cryptosporidium*) because detecting the spores is simpler than the sample processing and analysis for the parasite cysts or oocysts.

sporocyst A cyst structure enclosing one or more sporozoites of organisms that belong to the order Coccidia and class Sporozoa and are intracellular parasites of invertebrate and vertebrate animals, and in the case of malaria, in white and red blood cells of vertebrates.

Sporozoa One of four classes of the subphylum Plasmodroma. The members of the Sporozoa are all parasites and are characterized by a lack of organelles of locomotion and by the production of spores at the end of their life cycle. Furthermore, many species exhibit both asexual and sexual reproduction. *See also* organelle.

sporozoite A motile individual contained within a spore of Sporozoa, a major class of the phylum Protozoa.

spouting velocity The theoretical velocity (V) of water issuing from an orifice or other opening under a given head (H) when the effect of friction is eliminated.

$$V = \sqrt{2gH}$$

Where (in any consistent set of units):
g = the gravitational constant
H = head

Sprague–Dawley rat A strain of albino rat commonly used for toxicological studies.

spray aerator An aerator consisting of a pressure nozzle through which water is propelled into the air in a fine spray. A spray aerator is also called a nozzle aerator.

spray dryer An evaporator that is used to concentrate and dry solids. Commonly, feedwater (such as salt water) is atomized and sprayed with a hot gas (air) stream into a drying chamber, where evaporation and drying of the solids takes place; a filter then separates the solids from the gas stream. A spray dryer is sometimes used in zero-discharge wastewater treatment facilities and for concentrate disposal in some small-scale desalting systems. *See also* zero discharge.

spray head A sprinkler irrigation nozzle installed on a riser that delivers water in a fixed pattern. Flow rates of spray heads are high relative to the area covered by the spray pattern.

spray tower A tower built around a spray aerator to keep the wind from blowing the spray and to prevent the water from freezing during cold temperatures.

spreader A wood, timber, concrete, or masonry plate or wall set, with its upper edge level, in a channel or ditch. A spreader is intended to spread the flow evenly over the channel floor. A spreader is also called a sill.

spreading (1) The process by which a small amount of a liquid, as a drop, when placed on a solid that it wets, will spread out as a film. (2) A process of diverting water from streams when the supply exceeds the demand, and carrying it—either by means of ditches or by allowing it to flow broadly—to areas of absorbent material, where it sinks to the zone of saturation.

spring A concentrated discharge of groundwater appearing at the ground surface as a current of flowing water.

spring breakup Fragmentation of the ice cover on water bodies, pieces of which are carried by currents.

spring line (1) The theoretical center of a pipeline. (2) The guideline for laying a course of bricks.

spring tide The tide as it occurs when its range is a maximum, on the days following the new and full moon.

spring water Water obtained from an underground formation from which water flows naturally to the surface, or would flow naturally to the surface if it were not collected underground. *See also* spring.

spring-fed intermittent stream A stream, or stretch of a stream, that flows only at certain times when it receives water from springs. As with most intermittent streams, the intermittent character of spring-fed intermittent streams is generally due to fluctuations of the water table such that stream channels stand a part of the time below the water table and a part of the time above it.

sprinkle In meteorology, a light rain of scattered drops.

sprinkler system In fire protection, a network of overhead piping provided with systematically spaced sprinkler heads and connected to a suitable water supply, arranged so that actuation of fusible elements in the heads or other heat-sensitive devices causes the system to discharge water over a fire starting at any point.

spur (1) An obstruction of stone, timber, brushwood, or earth constructed from the bank of a channel and projecting into the channel to train the flow. (2) A similar structure for protecting the seashore from erosion.

spur terrace A short terrace used to collect or divert runoff.

squamous cell carcinoma A malignant tumor derived from squamous cell epithelium (a particular cell in the skin).

SQL *See* Structured Query Language.

square foot (ft^2) *See the Units of Measure list.*

square inch (in.^{2}) *See the Units of Measure list.*

square metre (m^2) *See the Units of Measure list.*

square mile (mi^2) *See the Units of Measure list.*

squeegee The metal blades attached to the scraper mechanism in a tank bottom.

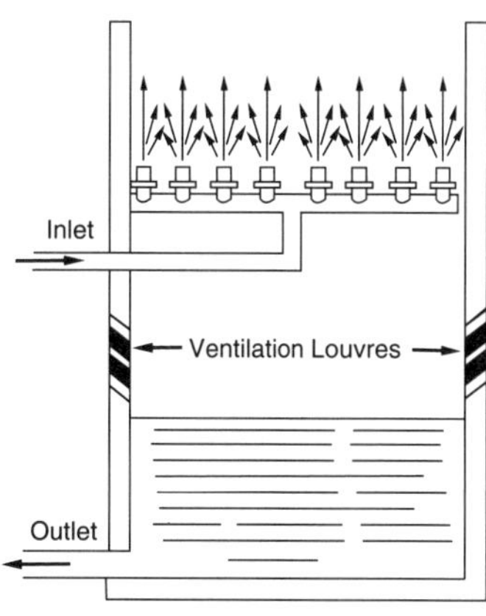

Spray aerator

SR *See* specific resistance.
sr *See* steradian *in the Units of Measure list.*
SRB *See* sulfate-reducing bacteria.
SRF *See* specific resistance to filtration; state revolving loan fund.
ss *See* stainless steel.
SSF *See* slow sand filtration.
SSMS *See* spark source mass spectrometry.
stability (1) The ability of any substance to resist change. (2) The ability of an engineering structure, such as a dam or retaining wall, to resist movement when loads are applied to it. (3) The ability of an ion exchange product or filter medium to withstand physical and chemical degradation in cycle-after-cycle operations. (4) The resistance of a density-stratified or thermally stratified body of water to mixing or overturning. *See also* thermal stratification.
stability index A numerical value that indicates the degree to which waters may cause scale or corrosion. *See also* Langelier saturation index; Riddick index; Ryznar stability index.
stabilization A water treatment process intended to reduce the corrosive or scale-forming tendencies of water.
stabilization tank A basin used in water treatment to adjust the pH of the water being treated. Following lime softening, for example, the pH typically is adjusted downward (i.e., stabilized) to an acceptable range for distribution and consumption. *See also* recarbonation.
stabilize (1) To produce a water that is exactly saturated with calcium carbonate ($CaCO_3$). (2) To prevent soil movement or the overturning of a structure. *See also* Langelier saturation index; Riddick index; Ryznar stability index.
stabilized channel An earth channel or canal in which, over a period of time, no appreciable erosion or deposition of silt or sediment occurs.
stabilized temperature platform furnace An instrument that furnishes the conditions needed to provide a consistent analytical process in atomic absorption spectrophotometry. Examples of these features include an L'vov platform, matrix modification, and background correction. These conditions minimize nonspectral interferences and allow many types of samples to be treated in a similar manner. *See also* atomic absorption spectrophotometric method; L'vov platform.
stable Resistant to change.
stack In demineralization, a basic electrodialysis operating unit consisting of membranes, separator, electrodes, and other appurtenances necessary to make a complete operating unit.
staff gauge A graduated scale (vertical unless otherwise specified), on, e.g., a plank, metal plate, pier, or wall, used to indicate the height of a liquid surface above a specified point or datum plane.
stage (1) The elevation of a water surface above its minimum or above or below an established low-water plane or datum of reference. (2) One of many steps in the operation of an evaporator, filter, compressor, or pump, each of which is operated at different conditions of pressure. Such a stage is also called an effect. *See also* electrical stage; gauge height; hydraulic staging; regulatory stage.
stage, electrical *See* electrical stage.
stage, hydraulic *See* hydraulic staging.
stage, regulatory *See* regulatory stage.
stage treatment Any treatment in which similar processes are used in series or stages.
stage–area curve A graphical representation of the relationship between reservoir depth and reservoir surface area.
stage–discharge curve A graphical representation of the relationship between surface water storage levels and surface water releases from a body of water or wetland.
stage–discharge relation The relation between gauge height and discharge of a stream or conduit at a gauging station. This relation is shown by the rating curve or rating table for such station.
stagnation A condition that occurs when water is motionless, or nearly motionless, and does not flow in a stream or move in a lake. Similar no-flow or low-flow conditions can occur in a finished water storage reservoir or in a distribution system piping network.
stainless steel (ss) A chromium alloy steel (usually 10 to 25 percent chromium) that is resistant to rusting and corrosion. Austenitic stainless steel contains 16 percent or more chromium and 7 percent or more nickel, as well as 2 percent silicon for stress-corrosion-resistant stainless steel. Ferritic stainless steel contains chromium and cannot be hardened by heat treatment. Martensitic stainless-steel contains chromium and can be hardened by heat treatment.
stainless-steel band A full-circle water main repair band made from stainless steel and placed around the main to repair holes or leaks. A gasket material, attached to the inside of the band, is pressed against the water main to form the seal as the band is tightened by bolts connected to each end of the band. The stainless steel provides a measure of protection against corrosive soils.
stake dam A dam composed of brush held in an advantageous place by staggered rows of stakes, used to retard the velocity of surface water flow on a slope.
stakeholder A group, organization, person, or agency who has an interest in, decision-making responsibility for or authority over a process and who is affected by or will benefit from the outcome of a process.
stale water Water that has not flowed recently and may have picked up tastes and odors from distribution or storage facilities.
standard (1) A recommended practice in the manufacturing of products or materials or in the conduct of a business, art, or profession. Such standards may or may not be used as (or called) specifications. (2) A document that specifies the minimum acceptable characteristics of a product or material, issued by an organization that develops such documents (e.g., an American Water Works Association standard). (3) A numerical contaminant limit set by a regulatory agency (e.g., a US Environmental Protection Agency maximum contaminant level). *See also* standard method(s).
standard addition, method of *See* method of standard addition.
standard, analytic *See* analytic standard.
standard atmospheric pressure *See* standard pressure.
standard bioassay for carcinogens *See* carcinogenesis bioassay.
standard biochemical oxygen demand Biochemical oxygen demand as determined under a standard laboratory procedure for 5

days at 68° Fahrenheit (20° Celsius), usually expressed in milligrams of oxygen per litre.

standard cell potential (ΔE^o) The voltage measured in an electrochemical cell under standard conditions (i.e., 1-molal concentration—which is essentially the same as 1 molar—for all dissolved materials; 1 atmosphere (101.3-kilopascal) pressure for gases; and the most stable form of solids at 77° Fahrenheit (25° Celsius). When a cell operates under standard conditions, its ΔE^o depends only on the chemical nature of the reactants and products. The standard cell potential can be taken as a quantitative measure of the tendency of reactants in their standard states to form products in their standard states. In short, ΔE^o represents the driving force of the chemical reaction. For example, in a zinc–copper cell, the relevant equations are as follows:

$$Zn^{2+} + 2e^- = Zn, E^o = -0.76 \text{ volt;}$$
$$\text{or } Zn = Zn^{2+} + 2e^-, E^o = +0.76 \text{ volt}$$
$$Cu^{2+} + 2e^- = Cu, E^o = +0.34 \text{ volt}$$
$$Zn + Cu^{2+} = Cu + Zn^{2+}, \Delta E^o = 1.10 \text{ volts}$$

standard cubic feet per minute (SCFM) *See the Units of Measure list.*

standard deviation A measure of the dispersion of a data series about its mean. The standard deviation is calculated as the square root of the squared deviations of all values in the series from the mean of the series. The more spread out the distribution of the values included, the higher the standard deviation. If the data series (e.g., persons, firms, farms, trees, or some other series) includes the totality of items in the data series (population data), the standard deviation is denoted as σ and is given by the following formula:

$$\sigma = \sqrt{\frac{\sum(x_i - \bar{x})^2}{N}}$$

Where:

x_i = the value of the *i*th item in the data set
$\bar{x}$ = the mean of the data set
N = the population size of the data set

If the data are drawn from a sample of the population, the standard deviation is usually denoted as *S* and is given by the formula

$$S = \sqrt{\frac{\sum(x_i - \bar{x})^2}{n-1}}$$

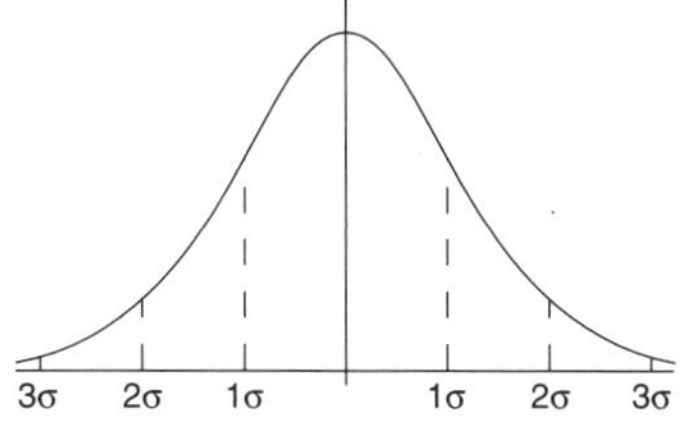

Standard deviation

Where:

x_i = the value of the *i*th item in the data set
$\bar{x}$ = the mean of the data set
n = the sample size of the data set

The $(n - 1)$ term in the denominator reflects the loss of one degree of freedom resulting from the use of sample data rather than whole-population data. In the context of probabilities, the standard deviation enables taking account of the variation in a data series, thereby helping in drawing sound conclusions from the available information. For example, in a normal probability distribution (a bell-shaped frequency curve), one standard deviation on each side of the mean value encompasses about 68 percent of the area under the curve—i.e., there is a 68 percent probability in repeated samples that the true mean value will lie within ±1 standard deviation of the sample mean. Similarly, a 95 percent probability exists that the true mean is within ±2 standard deviations of the sample mean; and the probability that the true mean is within ±3 standard deviations is virtually 100 percent. The standard deviation is often called the standard error because its derivation stems from deviations (or errors) from a predicted or estimated mean. *See also* frequency distribution; normal distribution; probability distribution; standard error of the coefficients; standard error of the estimate; standard error of the mean.

standard electrode potential (E^o) A measure of the tendency of a given half-reaction (e.g., a metal losing electrons to become an ion or an ion accepting electrons to become a metal) to proceed for a reactant or product in its standard state. The standard cell potential is the sum of the two half-cell potentials, one associated with each of the half-reactions in the cell. It is also referred to as the standard half-cell potential. *See also* standard cell potential.

standard error *See* standard deviation.

standard error of the coefficient A statistical measure of the dispersion around regression analysis coefficients. In regression analysis, a coefficient is calculated for each independent variable included in the analysis. Each coefficient is derived from the best fit given by the method of least squares. Because each coefficient is presumed to be from a random sample and unbiased, it has a probability distribution around it. In a normal probability distribution (i.e., a bell curve) the area under the curve is directly related to the number of standard errors above and below the mean value. The mean ±1 standard error encompasses about 68 percent of the area under the curve; 2 standard errors equates to about 95 percent, and 3 standard errors equates to 100 percent of the area under the curve. These percentages of area under the curve reflect the probability that, in repeated samples, the true population mean will be included within the probability range defined by the number of standard errors about the population mean. *See also* frequency distribution; probability distribution; standard deviation; t-test.

standard error of the estimate ($S_{Y.123...}$) A statistical measure of the dispersion of regression analysis point estimates. In regression analysis, point estimates are made of the dependent variable for any given values of the independent variables. The point estimates are made from sample data and consequently have a probability distribution around them. If the errors are randomly distributed (with a mean of zero and a constant variance over the range of independent variable values), the actual dependent variable values will fall within a probability range defined by the point estimate plus or minus a number of standard errors that relate to selected probabilities. One standard error relates to about 68 percent probability, 2 standard errors to about 95 percent probability, and 3 standard errors to virtually 100 percent probability. The general formula for computing the standard error of the estimate in multiple regression is as follows:

$$S_{Y.123...} = \sqrt{\frac{\text{unexplained variation}}{\text{number of degrees of freedom}}}$$

Where:

$S_{Y.123...}$ = the error of the forecast of the dependent variable Y when independent variables 1, 2, 3, ... are used to make the forecast

Unexplained variation is the sum of the differences between the point estimates derived from the regression and the actual values of Y given the independent variable values. Unexplained variation is equal to $\Sigma(Y_i - Y')^2$, which is the sum of the squared differences between the actual values of the dependent variable (Y_i) and the forecasted value (Y'). The number of degrees of freedom is calculated as the total sample size less the number of coefficients estimated in the analysis. Therefore, the larger the sample size, the better the estimate of the population standard error. The general formula for the standard error of the estimate with two independent variables would be

$$S_{Y.12} = \sqrt{\frac{\sum(Y_i - Y')^2}{n-3}}$$

where n is the sample size (of number of observations in a regression analysis). Note that, with two independent variables, three degrees of freedom are lost because one degree of freedom is lost for each independent variable and one for the estimate of the intercept term, the constant in $Y = a + bx_1 + cx_2$. An additional degree of freedom would be lost for each additional independent variable that might be added.

standard error of the mean ($\sigma_{\bar{x}}$) A statistical value computed by dividing the standard deviation of a sample drawn from a population by the square root of the sample size. This value provides an estimated standard deviation for the distribution of repeated sample means that could be drawn from a population—hence the term *standard error of the mean.* The formula is as follows:

$$\sigma_{\bar{x}} = \sqrt{\frac{\sigma_x}{n}}$$

Where:

$\sigma_{\bar{x}}$ = the estimated standard deviation of the mean of sample means

σ_x = the standard deviation of the sample taken

n = the number of observations

standard fall diameter The diameter of a sphere that has a specific gravity of 2.65 and has the same standard fall velocity as the particle in question.

standard fall velocity The average rate of fall that a particle would attain if falling alone in quiescent distilled water of infinite extent and at a temperature of 75.2° Fahrenheit (24° Celsius). *See also* fall velocity; settling velocity; Stokes's law.

standard fire stream A stream of water delivered at a rate of 250 gallons per minute (0.016 cubic metres per second) from a 1⅛-inch (28.6-millimetre) smooth nozzle.

standard free energy (G^o) The Gibbs free energy of a reactant or a product in its standard state. *See also* Gibbs free energy; standard free-energy change.

standard free-energy change (ΔG^o) The free-energy change that accompanies the conversion of reactants in their standard states to products in their standard states. If ΔG^o for a chemical reaction is negative, the reactants in their standard states will be converted spontaneously to products in their standard states. If ΔG^o is positive, this conversion will not be spontaneous; however, the corresponding reverse reaction will be. When ΔG^o is positive, some products can be formed, but not in concentrations as great as for the standard state. *See also* Gibbs free energy; standard cell potential.

standard method(s) (1) An assembly of analytical techniques and descriptions commonly accepted in water and wastewater treatment. (2) A shorthand way of referring to *Standard Methods for the Examination of Water and Wastewater,* published jointly by the American Public Health Association, the American Water Works Association, and the Water Environment Federation. (3) Validated methods published by professional organizations and agencies covering specific fields or procedures. The publishers include, among others, the American Public Health Association, American Public Works Association, American Society of Civil Engineers, American Society of Mechanical Engineers, American Society for Testing and Materials, American Water Works Association, United States Bureau of Standards, United States of America Standards Institute (formerly American Standards Association), United States Public Health Service, Water Environment Federation, and US Environmental Protection Agency.

standard metropolitan statistical area (SMSA) A land area classification that was introduced in the United States to provide an appropriate unit for statistical descriptions. Both a city and its surrounding areas are included in an SMSA because of extensive interactions between these areas. To qualify as a standard metropolitan statistical area, an area must meet certain criteria with respect to size, population density, and social and economic interactions and conditions. The US Bureau of the Census sets these criteria.

standard monitoring Initial monitoring required by the Safe Drinking Water Act.

standard National Weather Service gauge A rain gauge developed and used at regular and cooperative stations of the National Weather Service. It consists of a circular collector, 8 inches (203 millimetres) in diameter, with a funnel-shaped bottom that drains into a cylindrical brass measuring tube. The tube has a cross-sectional area equal to one-tenth that of the collector plus the area of the measuring stick, so that 1 inch (25.4 millimetres) of rainfall on a collector will fill the measuring tube to a depth of 10 inches (254 millimetres).

standard National Weather Service pan A land evaporation pan developed and used by the National Weather Service. It is circular in shape, 4 feet (1.2 metres) in diameter, 10 inches (254 millimetres) in depth, and set on a framework of timbers laid crosswise with the top 12 inches (305 millimetres) from the ground. The water surface in the pan is maintained during the course of observation between 1.5 and 2.5 inches (38.1 and 63.5 millimetres) from the top of the pan.

standard orifice *See* sharp-edged orifice.

standard plate count (SPC) *See* heterotrophic plate count.

standard pressure Atmospheric pressure at sea level under standard conditions of temperature, 273.1 kelvin. *See also* atmospheric pressure.

standard pumper-connection fire hydrant A standard fire hydrant with two hose connections and a 4-inch (10-centimetre) outlet for connection to a pumper truck that can take suction from the fire hydrant and pump water to increase pressure for fire fighting.

standard reference material *See* certified reference material.

standard sample An aliquot of finished drinking water that is examined for the presence of coliform bacteria. *See also* coliform bacteria.

standard short tube A tube with a diameter about one-third its length. Such a tube is also called a standard tube.

standard solution A solution with an accurately known concentration, used in the laboratory to determine the properties of unknown solutions.

standard temperature and pressure (STP) A standard set of conditions used in developing certain relationships in chemistry: a temperature of 273.1 kelvin (0° Celsius) and a pressure of 101.3 kilopascals (1 atmosphere).

standard test conditions A set of specified variables that are fixed for performance testing or comparison, e.g., for factory testing of reverse osmosis membranes at a constant feedwater quality, temperature, pressure, and recovery to measure permeate flow rate and salt rejection or salt passage at a specified operating time.

standard tube *See* standard short tube.

standardization (1) The procedure necessary to bring a preparation to a specified or known concentration, e.g., preparation and adjustment of a standard solution in volumetric analysis. (2) The process of comparing an instrument or device with a standard to determine the relationship between results obtained with the instrument or device and those obtained with the standard. *See also* rating.

standardize (1) In wet chemistry, to find out the exact strength of a solution by comparing it with a standard of known strength. This information enables one to adjust the strength of the solution by adding more water or more of the substance dissolved. (2) To set up an instrument or device to read a standard. This procedure allows one to adjust the instrument so that it reads accurately or to apply a correction factor to the readings.

standardized monitoring framework A compliance monitoring framework developed by the US Environmental Protection Agency and intended to standardize monitoring cycles. It is based on a 9-year compliance cycle. Each 9-year cycle consists of 3 three-year compliance periods. All compliance cycles and periods are based on a calendar year. The framework was first applied in the US Environmental Protection Agency Phase II Rule. *See also* Phase II Rule.

standardized mortality rate (SMR) A summary measure of the death rate that a population would have if it had a standard age structure. Morbidity can also be expressed in this manner. The standardized mortality rate is also called the age-adjusted death rate.

standardized mortality ratio (SMR) The ratio, multiplied by 100, of the number of deaths observed in the study population to the number of deaths (usually age, time, and cause specific) expected if the study population had the same specific rates as the standard or general population. This concept is used frequently in occupational epidemiology. A standardized mortality ratio of 175 indicates that mortality is 75 percent greater than would be expected. Morbidity can also be expressed in this manner. *See also* proportional mortality ratio.

standardized rate A weighted averaging of a characteristic-specific rate, e.g., the crude mortality rate, according to a standard distribution of age or other selected characteristics of the population. Standardization is necessary when one is comparing crude rates for two or more populations that differ with respect to a characteristic (confounding variable), such as age, that independently influences the risk of death. Adjustment methods are classified as either direct or indirect. *See also* crude rate.

standardized rate ratio A rate ratio in which both rates in the ratio (i.e., both the numerator and denominator rates) have been standardized to the same population distribution.

standby plant An electric power plant that is maintained in a condition to produce power in case of emergencies, such as breakdowns occurring in the regular generating stations or transmission lines.

standing crop The biota present at a selected point in time in an environment.

standing sample A 1-litre sample collected from the kitchen or bathroom cold-water faucets of targeted sample sites, representing water standing in the interior piping for at least 6 hours. Standing samples will often contain higher levels of a contaminant that has been picked up because of prolonged exposure to a pipe wall or faucet. A standing sample is also known as a first-draw tap sample by the US Environmental Protection Agency. *See also* first-draw residential lead sample.

standing water level The water level in a nonpumping well, whether the well is within or outside the influence area of pumping wells. If the well is outside the influence area, this term is equivalent to *static water level*; if the well is within the influence area, the standing water level is one elevation on the pumping depression cone. *See also* cone of depression.

standpipe A high tank, usually small in diameter compared to height, for holding water. This water is used to maintain pressure in a water supply system and as storage for fire protection.

start-action notice A notice that initiates work on a rule or related action and establishes a work group within the US Environmental Protection Agency. The office within the US Environmental Protection Agency developing the rule, referred to as the lead office, prepares the start-action notice at the outset of its effort to develop a regulation. The notice provides brief, descriptive information about the rule to be developed, alerts other agency offices to the lead office's intention to develop a rule, and serves as the mechanism for reaching internal agreement on the necessary review steps for the new rule.

starter A device used to start up large motors gradually to avoid severe mechanical shock to a driven machine and to prevent disturbance to the electric lines (which would cause dimming and flickering of lights).

startup The activity associated with initiating the operation of a unit process or treatment facility.

state (1) The agency of the state government that has jurisdiction over public water systems. During any period when a state has not been delegated primary enforcement responsibility, the designation *state* will refer to the regional administrator of the US Environmental Protection Agency. (2) The characteristics or description of a system. As associated with thermodynamics, on a macroscopic scale, the state of a system is determined when it can be accurately characterized or described (to the limits of experimental measurement ability) in terms of all of the properties of thermodynamic interest. Measurable properties of matter that describe the macroscopic state of a system are called state variables (e.g., pressure, temperature, volume, density, refractive index, and magnetic susceptibility). Time-dependent processes taking place within a system characterize a transient state. Time-independent processes can be either dynamic or static. If dynamic, they characterize a steady state. If static, they characterize an equilibrium state. Equations of state (e.g., the ideal gas law) describe a quantitative relationship that exists between intensive (mass-independent) system parameters (e.g., pressure, temperature) and the extensive (mass-dependent) system parameters (e.g., volume, number of moles of a substance). *See also* state function.

state of the art The most advanced level of development of a device, procedure, process, technique, or science at any particular time.

state function A system property that has some definite value for each state and is independent of the manner in which the state is reached.

State Plane Coordinate (SPC) System A system of *x–y* coordinates for each state maintained by the National Geodetic Survey and commonly used in geographic information system technology in the United States. One of two map projections—Transverse Mercator (for states with a north–south orientation) or Lambert Conformal (for states with an east–west orientation)—is usually used to define one or more specific zones for each state. The geographic extent of the zones is limited to a distance of 158 miles (254 kilometres) (east–west in the Transverse Mercator or north–south in the Lambert Conformal) to limit the amount of distortion that accumulates because of the curvature of the Earth. *See also* geographic information system.

state revolving loan fund (SRF) A mechanism whereby federal funds are provided to states to capitalize a state loan for the purpose of providing low-interest loans. For example, a state revolving loan fund was established under the federal Clean Water Act and the federal Safe Drinking Water Act whereby the federal government provided capitalization grant money to states to provide low-interest loans to communities for constructing wastewater treatment facilities and water treatment facilities, respectively. In general, states must meet specific US Environmental Protection Agency requirements to receive capitalization grants for these revolving loan funds and provide a specified percentage of matching funds.

state water quality certification A certification issued by a state water quality agency under Section 401 of the Clean Water Act.

static Fixed in position; resting; without motion.

static discharge head The difference in height between the pump center line and the level of the discharge free water surface.

static groundwater level The level of groundwater in a well when the well pump is not operating and the water level is not changing (i.e., the level at equilibrium). Depending on aquifer properties, it might take anywhere from several minutes after shutoff to many days after shutoff to reach this level.

static head When water is not moving, the vertical distance from a specific point to the water surface. *See also* dynamic pressure head; static pressure.

static mixer A device designed to produce turbulence and mixing of chemicals with water by means of fixed sloping vanes within the unit, without the need for any application of power. The energy required is a function of the head loss through the unit.

static pressure When water is not moving, the vertical distance from the specific point of interest to the water surface. The static pressure is the static head multiplied by the specific weight of water. *See also* dynamic pressure head; static head.

static suction head The difference in elevation between the pump center line and the free water surface of the reservoir feeding the pump. In the measurement of static suction head, the piezometric surface of the water at the suction side of the pump is higher than the pump; otherwise, static suction lift is what is being measured. *See also* static suction lift.

static suction lift The difference in elevation between the pump center line and the free water surface of the liquid being pumped. In a static suction lift measurement, the piezometric surface of the water at the suction side of the pump is lower than the pump; otherwise static suction head is what is being measured. *See also* static suction head.

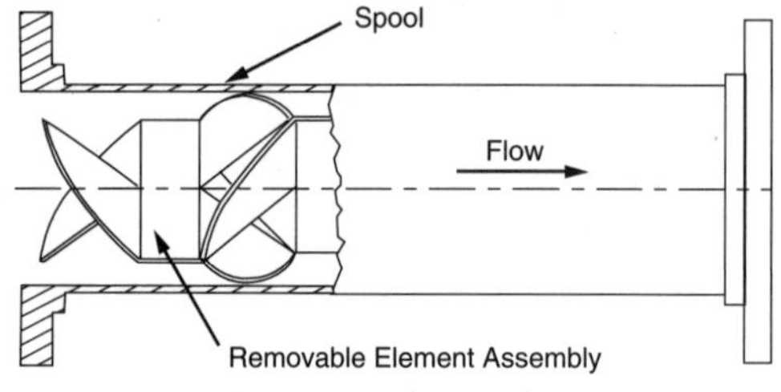

Static mixer

static system A system or process in which the reactants are not flowing or moving.

static water depth The vertical distance from the center line of the pump discharge down to the surface level of the free pool while no water is being drawn from the pool or water table.

static water level (SWL) The water level in a well measured when no water is being taken from the aquifer by pumping.

static water system A water system in which, at the moment, water is not moving.

station rating curve A curve showing the relationship between gauge height and discharge of a stream or conduit at a given gauging station. A station rating curve is a graphical representation of a rating or discharge table.

stationary phase A substance that remains in place during chromatographic separations. Typically, stationary phases are bonded to particles or to the interior wall of capillary tubes. Stationary phases interact with analytes and a mobile phase to allow analytes to be separated. A number of mechanisms, such as adsorption or ion exchange, can account for the interaction between analytes and the stationary phase. *See also* gas chromatography; high-performance liquid chromatography.

stationary surface washer A stationary system used to clean the media surface in a filter. Unlike a rotary surface washer, a stationary surface washer is a fixed pipe grid with small holes that produce high-velocity water or air jets to clean the media surface at the beginning of a backwash cycle. *See also* rotary surface washer.

statistical analysis A set of techniques used to analyze a set of observations from a population in order to draw inferences about the population.

statistical power The probability, at a specified confidence level, that the null hypothesis will be rejected when it is false or that a study will demonstrate a statistically significant result when a difference or an epidemiologic association actually exists in the population. *See also* hypothesis; statistically significant.

statistical significance test A statistical procedure to test the null hypothesis. An estimate is made of the probability of the observed difference or epidemiologic association under conditions of the null hypothesis in an experiment or random sample of a given size. From this estimate, the statistical significance can be stated, usually as a *p* (for probability) value. Statistical significance does not mean the association is of biological significance or that bias or systematic error was not present. Neither does it mean that random error or chance can be completely ruled out as the explanation for the observed result. Most epidemiologists feel that strict reliance on statistical significance testing is not appropriate. *See also* bias; confidence interval; hypothesis; probability value.

statistically significant Pertaining to a situation in which, at a specified confidence interval and for a valid null hypothesis, an observed difference or association is not likely due to chance alone. *See also* confidence interval; hypothesis; probability value.

statistics A branch of mathematics dealing with the collection and analysis of data with a view toward drawing inferences on the population from which the data are collected.

stator (1) The nonrotating or stationary part of an alternating current electrical motor. (2) A fixed baffle in a mixing chamber or vessel to help promote the mixing of a solution.

Stauwerke gate An automatic crest gate operated by counterweights placed either above or below the water surface.

steady flow A type of flow in which the properties of flow do not vary with time. Equilibrium flow to a well is a type of steady flow. *See also* steady uniform flow.

steady nonuniform flow A flow in which the quantity of water flowing per unit of time remains constant at every point along the conduit but the velocity varies along the conduit because of changes in the hydraulic characteristics.

steady operating flow rate The flow rate in gallons (litres) per minute at which a water-processing filter or ion exchanger will deliver its rated capacity. For water softeners, this flow is based on delivering softened water from an incoming source water having 342 mg/L (20 grains per gallon) total hardness as calcium carbonate ($CaCO_3$).

steady state A condition in which the input energy equals the output energy. This term may apply to any continuous physical, chemical, or biological process.

steady uniform flow A flow in which the velocity and the quantity of water flowing per unit of time remain constant. *See also* steady flow.

steady-state analysis A pipe network analysis that assumes that consumptions, supplies, storage-tank levels, and certain other variables are unchanging for one specific point in time.

steam distillation extraction *See* simultaneous distillation extraction.

steam power Any type of energy or power generated or developed through the use of a steam engine. This term is commonly applied to electrical energy or power generated by the use of steam engines or turbines to drive electric generators.

steam pump A pump operated by a steam engine, in which the steam and water cylinders placed in the same machine are considered a single unit.

steam turbine A prime mover in which the pressure or motion of steam on vanes is used for the generation of mechanical power.

steam vacuum pump A displacement pump in which steam admitted to the cylinder is condensed by a stream of water, creating a vacuum that serves to draw the water into the cylinder. When the cylinder is full, or nearly so, steam under full boiler pressure is admitted, forcing the water out through the outlet valve.

steel An iron-base alloy malleable under certain conditions, containing up to 2 percent carbon, and having varying qualities of hardness, elasticity, and strength according to composition and the type of heat treatment given to the alloy.

steel bolted tank A cylindrical tank that is constructed with curved steel plates bolted together, with gasket material between plates at the bolted areas.

steel pipe Pipe manufactured out of rolled steel sheets or plates with welded longitudinal or spiral seams. For water applications the interior can be lined and a coating applied to the exterior for corrosion protection.

steep slope (1) A conduit slope greater than the critical slope for a particular discharge; the slope for which the depth of flow

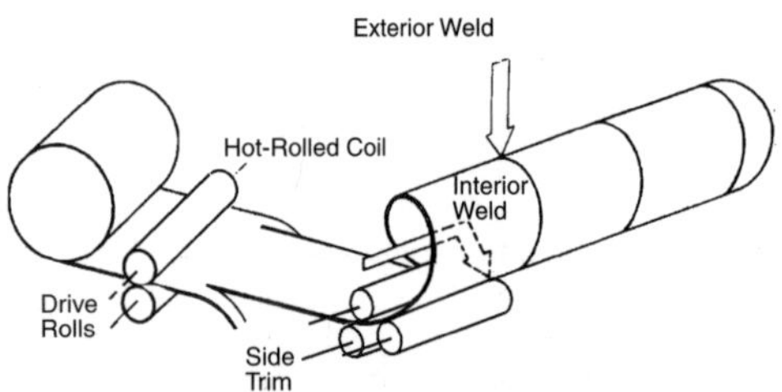

Process for making spiral-seam steel pipe

is less than Belanger's critical depth and the velocity is greater than Belanger's critical velocity. (2) A slope greater than the friction slope. *See also* Belanger's critical velocity.

steering committee A standing group within the US Environmental Protection Agency (USEPA), with representation from each assistant administrator and the Office of General Counsel, that coordinates and integrates the USEPA's regulatory development activities. The committee's key functions are to approve start-action notices; to charter and monitor the progress of staff-level work groups, especially regarding cross-media or interoffice problem solving; and to ensure, when appropriate, that significant issues are resolved or elevated to top agency management. USEPA's regional offices participate in steering committee activities through regional regulatory contacts who coordinate reviews in their regions and facilitate rule-related activities for the regional administrators.

STEL *See* short-term exposure limit.

stem cell A cell that still possesses the capability of differentiating. Stem cells are generally nondifferentiated cells that do not yet possess all of the characteristics of the functional cell of an organ. *See also* stromal cell.

stemflow Rainfall or snowmelt fed to the ground down the trunks or stems of plants.

step dose Any of the doses applied under conditions such that the chemical is fed continuously at progressively larger or smaller dosages. One can use a step dose in tracer testing to determine the hydraulic characteristics of the basin being evaluated by comparing the amount of time required for the concentration of the tracer in the effluent to reach a specified percentage of the influent concentration.

step testing The measurement of water flow in successive sections of a water supply network. A large increase in water flow between successive readings suggests that a leak may exist in the last section closed down.

stepwise multiple linear regression analysis A statistical procedure to develop a regression equation consisting of several independent variables to predict dependent variables. The independent variables are entered into the equation one at a time in order of their relative importance in terms of explaining the dependent variable. *See also* multiple linear regression analysis.

stepwise regeneration The process of regenerating ion exchange resin beds several times with the same regenerant but at a higher concentration each time. This method is usually used to avoid calcium sulfate ($CaSO_4$) precipitation when sulfuric acid (H_2SO_4) is employed as a regenerant for cation bed units of deionizer systems being used to decationize unsoftened feedwater.

steradian (sr) *See the Units of Measure list.*

stereoplotter A device used in photogrammetric mapping to digitize the horizontal position and elevation of selected points and features visible in aerial photographs.

sterilization The process of destroying all forms of microbial life on and in an object by physical or chemical means. *See also* disinfection; pasteurization.

sterilize To apply physical treatment (heat, ultraviolet irradiation, gamma irradiation, or filtration) or chemical treatment to kill or remove all life forms in a liquid, in or on a material, or on a surface. *See also* disinfection; pasteurization.

sterilized water *See* bottled sterile water.

Stiff-and-Davis stability index (S&DSI) An index, generally applicable to waters with total dissolved solids greater than 10,000 milligrams per litre, that indicates whether a water is in equilibrium with calcium carbonate ($CaCO_3$).

$$S\&DSI = pH_a - pH_s$$

Where:

pH_a = the actual pH of the water

pH_s = the pH of saturation, the pH of the water if it were saturated and at equilibrium with calcium carbonate

The pH_s value is calculated based on the concentrations of calcium and alkalinity, as well as a specified constant that is dependent on ionic strength and temperature. Stiff-and-Davis stability index values greater than, less than, or equal to zero indicate the tendency of a water to deposit calcium carbonate, dissolve it, or be at equilibrium with it, respectively. They are used for scale-control calculations, such as for reverse osmosis membrane concentrate streams. *See also* Langelier saturation index; Riddick index; Ryznar stability index.

still water A portion of a stream or basin in which no apparent current exists.

stilling basin A structure or excavation that reduces the velocity or turbulence of flowing or falling water.

stilling well A pipe, chamber, or compartment with one or more comparatively small inlets connected with a main body of water. The purpose of a stilling well is to dampen waves or surges while permitting the internal water level to rise and fall with the major fluctuations of the main body of water. A stilling well is used with water-measuring devices to improve the accuracy of measurement.

stirred sludge volume index A measure of the settling characteristics of a solid suspension determined by evaluating the settled volume of the suspension after a prescribed time for a given initial concentration. The suspension is slowly stirred by a mechanical apparatus; the container is often a cylindrical column, such as a graduated cylinder. The index is expressed in millilitres of settled sludge per gram of solids in suspension.

stochastic Pertaining to actions for which the results occur from probabilistic events.

stochastic process A process involving a random variable dependent on a parameter, usually time.

stoichiometric Pertaining to the proportions in which chemicals combine to form compounds, as well as the mass relations in chemical reactions.

stoichiometric equation A symbolic representation of a chemical equation in which the proportions of various constituents are also represented.

stoichiometry The mathematical and theoretical study of how elements combine in predetermined quantities to form compounds.

Stokes's law A formula for calculating the rate of fall of particles through a liquid medium. In the laminar flow region, the rate at which a spherical particle will fall when suspended in a liquid medium varies directly with the square of the particle's diameter, the density of the particle, and the viscosity of the fluid. In equation form,

$$V_s = \frac{g(\rho_s - \rho)d^2}{18\mu}$$

Where (in any consistent set of units):

V_s = the settling velocity
g = the gravity constant
ρ_s = the density of the particle
ρ = the density of the fluid (water)
d = the diameter of the particle
μ = the absolute viscosity

stomacher A mechanical, electric motor–driven device used for thoroughly mixing a food or environmental sample with sterile water or other diluting fluid prior to analysis of the sample.

stoney gate A crest gate used in connection with very large openings in hydraulic structures, such as spillways on dams. This type of gate is carried by a set of rollers that travel vertically in grooves in masonry piers and are independent of the gate and piers. The gate moves vertically by rolling along the rollers; this design eliminates friction due to water pressure to a great extent, and it reduces the power required to open and close the gate.

stop box A receptacle that provides access to the service shutoff valve located on a service line, usually between the main and the meter. Quite often the stop box is located on the property line or between the curb and the sidewalk. A stop box is also referred to as a curb stop or valve box. *See also* curb stop and box.

stop gate (1) A hand-placed gate, usually made of metal, used to divert or block flow in an open channel. (2) A form of lock gate that is raised vertically.

stop log A log, plank, cut timber, or a steel or concrete beam fitting into end guides between walls or piers to close an opening in a dam or conduit to the passage of water. Stop logs usually are handled or placed one at a time.

stop plank A removable plank (usually wooden, but sometimes steel) that is placed in a groove or rack to block off the flow of a liquid from one compartment or channel to another.

stop valve A large valve installed in a pipeline to shut off flow in a section to permit inspection or repair. Such valves are usually installed in the main lines. A stop valve is also called a sectionalizing valve.

stopcock A small valve for stopping or regulating the flow of a fluid through a pipe or burette.

storage The impounding of water, in either surface reservoirs or underground reservoirs, for future use. Storage differs from pondage and regulation of stream flow in that the latter refer to more or less temporary retention of the water, whereas storage involves retention for much longer periods.

storage capacity The maximum volume of water available for use from a water storage tank, e.g., the amount available from a reverse osmosis or distiller water storage tank.

storage capacity curve A curve expressing the relationship between the volume of a space and the upper level of elevation of the material occupying the space. In the case of a reservoir, the curve depicts the relationship between the water surface elevation in the reservoir and the volume of water below that elevation. A storage capacity curve is also called a capacity curve.

storage coefficient (1) For surface waters, a coefficient devised by Allen Hazen to express the relationship of storage capacity in a reservoir to the mean annual flow of a stream above the dam forming the reservoir. The coefficient is the ratio of reservoir storage capacity to the annual water volume entering the reservoir. A reservoir with a large storage coefficient (greater than 1) can store multiple years worth of flow, whereas one with a small coefficient must pass some fraction of the inflow over the dam every year. (2) For groundwater, the volume of water released from storage per unit decline in peizometric head per unit area of aquifer. In the US customary system, the storage coefficient would be the volume of water (in cubic feet) released from storage in a vertical column of aquifer having a base of 1 square foot when the piezometric surface falls 1 foot. In Système International, it would be the volume of water (in cubic metres) released from storage in a vertical column of aquifer having a base of 1 square metre when the piezometric surface falls 1 metre. The value of this storage coefficient is approximately equal to the specific yield in an unconfined aquifer. *See also* specific yield.

storage equation An axiom stating that the volume of inflow equals the volume of outflow plus or minus the change in storage. The storage equation is also called the continuity equation.

storage gallery A water-collecting gallery that is located in a water-bearing formation and can be used for either conveying or storing groundwater.

storage ratio The ratio of the net available storage of an impounding reservoir to the annual mean flow of the stream feeding it. Usage of this term must be limited to large reservoirs.

storage reservoir A reservoir in which surface water is retained for a considerable period of time. *See also* impounding reservoir; reservoir.

Storage and Retrieval of US Waterways Parametric Data (STORET) A repository of waterway parametric data, including information on ambient, intensive survey, effluent, and biological water quality of the waterways within and contiguous to the United States. This program was first developed by the US Public Health Service in 1964 to collect and disseminate basic information on chemical, physical, and biological quality of the nation's waters. Today, the US Environmental Protection

Agency manages the program, which is available on the agency's mainframe computer. Public access to this program can be obtained by subscription to a US Environmental Protection Agency user account through the National Technical Information Service.

storage tank A compartment used to accumulate the product water from a water treatment unit so that sufficient quantity, pressure, or both are available for intermittent periods of higher-flow-rate water use. A storage tank is also called a clearwell.

STORET *See* Storage and Retrieval of US Waterways Parametric Data.

storm Usually, an occurrence of such phenomena as rain, snow, hail, and wind, although wind storms may be described in accordance with the material that is carried in suspension in the air, such as dust or sand. This term is often used in connection with a meteorologic phenomenon that is either unusual or of great magnitude, rate, or intensity.

storm center The center of the area covered by a storm, especially the place of lowest pressure in a cyclonic storm or the place where wind velocities approach zero.

storm distribution pattern The manner in which the depth of rainfall varies from station to station throughout an area.

storm drainage Surface movement of water resulting from storms.

storm flow That portion of the precipitation that leaves the drainage area in a comparatively short time on or near the surface after the occurrence of precipitation. Storm flow is also called excess rainfall or surface runoff.

storm overflow Some portion of flow caused by stormwater.

storm runoff That portion of the total runoff that reaches the point of measurement within a relatively short time after the occurrence of precipitation. Storm runoff is also called direct runoff.

storm seepage The rainfall that infiltrates into the surface soil and moves away from the area on which it falls through the upper soil horizons at a rate much in excess of normal groundwater seepage. *See also* infiltration; subsurface runoff.

storm surge A rise or piling up of water against the shore, produced by wind stress and atmospheric pressure differences in a storm. *See also* storm tide; wind tide.

storm tide A wind tide caused by storm winds, as in the case of abnormally high or low water in tidal bays and rivers as a result of winds and freshwater runoff. *See also* wind tide.

stormwater Water that is collected as runoff from a rainfall event. Separate collection facilities and piping are often designed to prevent stormwater from overloading sanitary wastewater collection systems.

stormwater runoff That portion of the rainfall over a given area that finds its way to natural or artificial drainage channels.

stovepipe well casing A well casing made of two thin sheets of steel, one fitting closely over the other for half its length. When casing segments are fitted together in several lengths, a double-thickness casing is formed, which is jacked into the borehole. The material at the bottom of the hole is removed by a sand bucket or bailer. This method of well driving is called the California stovepipe method and is used for wells of reasonable depth in relatively loose soil with few boulders.

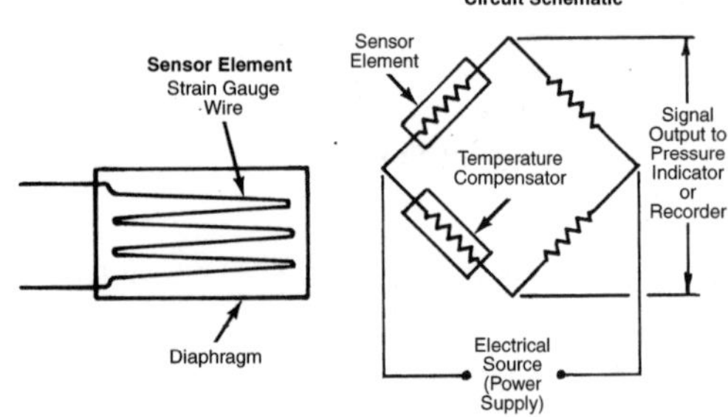

Strain gauge

STP *See* sodium tripolyphosphate; standard temperature and pressure.

STPP *See* sodium tripolyphosphate.

straightening vanes Horizontal vanes mounted on the inside of fluid conduits to reduce swirl ahead of a flowmeter.

straight-flow pump A pump in which the suction and discharge pipes and the pump are all in line, with the water not changing flow direction as it passes through.

straightway valve A valve through which the fluid passes without deviation. Such valves offer the least resistance to flow.

strain difference A difference in the response of a different strain of the same species to a stimuli or insult. Such differences are genetically determined but may have multiple determinants (i.e., multiple genes could contribute to such differences).

strain gauge A device that uses a change in the electrical resistance of a wire under strain to measure pressure.

strainer A filtration device in which particles are separated from a fluid stream by sieving. The removal rate is dependent on the size and shape of the strainer openings and the physical characteristics of the particles.

strainer head A perforated device inserted in one type of underdrain of a rapid granular filter, through which the filtered water is collected and through which the wash water is distributed when the filter is washed. A strainer head is also called a filter head. *See also* filter bottom.

strainer system A network system equipped with strainers to collect filtered water in the underdrain of a rapid granular filter.

strainer well A tube well in which a strainer is interposed at the level where water-bearing strata must be tapped to prevent sand and debris from entering the well.

strategic load management The use of certain techniques by water utilities to alter the level or pattern of demand to improve overall system efficiency. Strategic load management includes methods that affect the level and timing of peak and off-peak water use.

strategic quality planning A process of using multilevel, multidisciplinary, multifaceted teams to change, challenge, and improve an organization as a whole.

stratification (1) The formation of separate mixing layers in a body of water, usually characterized by different temperature regimes and usually called thermal stratification. (2) Collectively, the existence of beds or laminae. Such stratification is also called lamination. (3) The layering of granular media on the basis of grain size during backwashing (fluidization). (4) An arrangement into classes, groups, or strata. (5) An arrangement or deposition of sedimentary material in layers; of sedimentary rock in strata marked by change in color, texture, dimension of particles, and composition. (6) A method of analyzing epidemiologic data to assess and control confounding bias. The population sample is separated for analysis into smaller groups based on specified criteria, such as age, gender, or other characteristics. For example, in a study of the possible association of bladder cancer and drinking water exposure, an appropriate analysis would be to stratify on smoking status to determine whether risks differ for smokers and nonsmokers. *See also* bedding; hydraulic classification; hypolimnion; mixolimnion; thermocline.

stratified bed A bed in which two ion exchangers of different classes and different densities have been placed in the same column (bed), e.g., a weak base anion resin on top of a strong base anion exchanger or, in cation exchange systems, a weak acid on top of a strong acid resin. *See also* hydraulic classification.

stratified data Data that are grouped (or stratified) according to specified criteria for the purpose of analyzing how the criteria may affect the data. In epidemiology, heath effects of interest are studied among various groups or strata (e.g., age, gender, or smoking status) to determine the risks of disease in each stratum (e.g., are risks greater among the elderly or young, males or females, smokers or nonsmokers?). Stratified data are commonly used to assess and control confounding bias. *See also* confounding bias.

stratum A layer that is characterized by certain unifying attributes, characteristics, or properties distinguishing it from adjacent layers.

straw man A document issued by the US Environmental Protection Agency to the public that presents initial thoughts and concepts for a new rule for discussion. A straw man rule is not a formal proposal; it merely serves to initiate public discussion on issues associated with a regulation under development by the US Environmental Protection Agency.

stray current Direct electrical current traveling through the earth around an existing underground structure. When the current uses a water main as the path of travel and exits, metal from the main will be removed. Thus, the current will cause pitting, corrosion, or both. *See also* stray current corrosion.

stray current corrosion Corrosion that is caused by stray currents from some external source.

streak line A locus of the temporary location of all particles that have passed through a given point.

stream A course of running water usually flowing in a particular direction in a definite channel and discharging into some other stream or body of water. In the law of water rights, an important distinction exists between a stream and water that appears on the surface in a diffused state with no permanent source of supply or regular course for any considerable time.

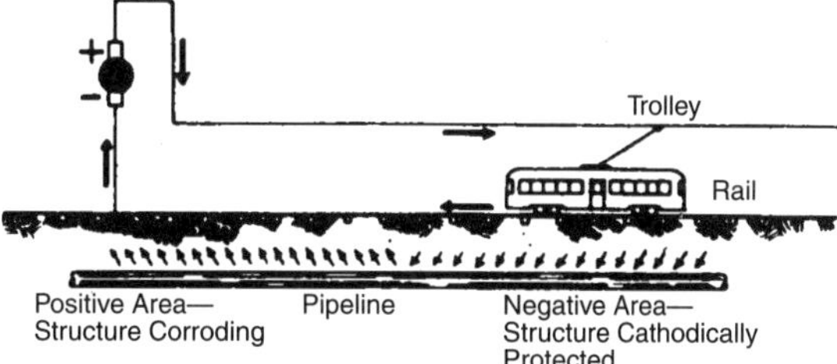

Stray current corrosion

Surface water may at times collect and flow through a land depression or gully, but a stream usually flows even though it may be dry temporarily. *See also* channel; river.

stream adjustment The natural process involving the changes in the courses of streams, extending over long periods of time, by which such courses develop definite and sustained relations to the rock structure in which they are located. This process involves a continued tendency on the part of the streams to seek out and erode channels in rocks of weaker structure.

stream aggradation *See* aggrading river.

stream banks The side slopes of a natural channel between which the normal flow of a stream is confined. *See also* shore.

stream biota The collective animal and plant life of a stream.

stream degradation *See also* degrading river.

stream discharge The rate of flow, or volume of water per unit time, flowing in a stream at a given place.

stream gauging The measurement of the velocity of a stream of water in a channel or an open conduit, as well as the cross-sectional area of the water, for the purpose of determining the discharge. Measurements are usually obtained in conjunction with a record of stage to give a stage–discharge relation for the channel. Stream gauging is also called gauging. *See also* discharge measurement.

stream gradient The general slope, or rate of change in vertical elevation per unit of horizontal distance, of the water surface of a flowing stream.

stream rotor A sprinkler irrigation head that delivers rotating streams of water in arcs or full circles. Some types use a gear mechanism and water pressure to generate a single stream or multiple streams. Stream rotors have relatively low precipitation rates, and multiple stream rotors provide matched precipitation for varying arc patterns.

stream system A principal stream and all its tributaries.

stream upflow The groundwater moving parallel to and discharging into a surface stream within and through the porous streambed.

streambed The bottom of a stream below the usual water surface. The streambed is the area that is kept practically bare of vegetation by the washing effect of stream water from year to year.

stream bed permeability The capacity of the bottom of a watercourse to transmit water to and from underground layers.

streamflow The volume of water per unit time flowing in a stream and usually measured at a stream gauge.

streamflow depletion The amount of water that flows into a valley or onto a particular land area, minus the water that flows out of the valley or off the particular land area.

streamflow record A tabulation of the flow of a stream. Such a record may include daily, monthly, annual, and instantaneous extremes of discharge.

streamflow regulation Control of the quantity or quality of water in a stream by regulating the release of impounded water.

streamflow wave A traveling wave caused by a sudden increase of flow. *See also* flood wave.

streaming current A current gradient generated when a solution or suspension containing electrolytes, polyelectrolytes, or charged particles passes through a capillary space, as influenced by adsorption and electrical double layers. This phenomenon is used in monitoring and controlling coagulation and flocculation processes. *See also* electrokinetic potential; electrophoretic mobility; streaming current detector; zeta potential.

streaming current detector (SCD) An on-line instrument used to adjust coagulant dosage at a water treatment plant. Within the instrument a streaming current is generated by charged particles in the water; this current indicates the degree of destabilization of the particles. Cost savings in coagulant usage have been reported when streaming current detectors have been used to control the dosage, especially when the quality of the source water is highly variable.

streaming flow Turbulent flow with a mean velocity less than Belanger's critical velocity. Streaming flow is also called subcritical flow. *See also* Belanger's critical velocity.

streamlet A rivulet or rill.

streamline A line that is everywhere parallel to the instantaneous direction of motion in a fluid. Thus, if u, v, and w are the velocity components in Cartesian coordinates, the differential equations of the streamlines are

$$\frac{dx}{u} = \frac{dy}{v} = \frac{dz}{w}$$

The streamline field is customarily constructed in such a way that the spacing of the streamlines at a given point is inversely proportional to the speed of the motion at this point.

streamline flow A type of fluid flow in which a continuous steady motion of the particles occurs, with the motion at a fixed point always remaining constant. *See also* laminar flow.

stress (1) In ecology, a stimulus or series of stimuli of such magnitude as to tend to disrupt the balanced state of an organism or an ecosystem. (2) Good or bad physical or mental tension experienced by an individual. (3) The action on a body by any system of balanced forces such that deformation results. Such stress is usually measured in units of pounds per square inch or kilopascals.

stress corrosion Corrosion that acts on metal at points of tensile stress, working, or vibration wear. Corrosion can be accelerated by stress, either residual internal stress in the metal or externally applied stress. Residual stresses can occur during fabrication and assembly and can be caused by exposure of the pipe to certain environmental conditions. Welding can cause high stress and be a source of trouble. Both process piping and water distribution piping can be subject to cracking as a result of stress corrosion.

stress corrosion cracking Cracking of water pipes due to stress corrosion.

stressor A specific mental or physical event that causes stress.

string wound element A cartridge-style filter element constructed by continuous spiral winding of natural or synthetic yarn around a preformed product water tube core and then a building up of layers to form a depth-type filter element.

strip To remove gases from water by passing large volumes of air through the water. *See also* air-stripping; packed tower aeration; stripping.

strip cropping A crop production system that involves planting alternating strips of row crops and close-growing forage crops. The forage strips intercept and slow runoff from the less protected row crop strips.

stripping A process of transferring a target compound from the liquid phase to the gas phase. *See also* air-stripping; packed tower aeration.

stromal cell A cell that is involved in establishing the structural matrix of an organ as opposed to one that serves the characteristic function of the organ (i.e., a differentiated cell). *See also* stem cell.

strong acid An acid that has a great tendency to transfer a proton to another molecule. The quantitative measure of acid strength is the acid dissociation constant, i.e., the equilibrium constant for a chemical reaction of the following form (in which A represents an anion):

$$HA + H_2O = H_3O^+ + A^-$$

For example, hydrochloric acid (HCl) has a great tendency to lose a proton and is considered a strong acid. *See also* acid.

strong acid cation exchanger A cation exchange resin with an exchange site (i.e., active group, usually sulfonic acid—R–SO_2OH) capable of splitting neutral salts [e.g., sodium chloride, NaCl; magnesium sulfate, $MgSO_4$; calcium nitrate, $Ca(NO_3)_2$] to form their corresponding free acids (hydrochloric acid, HCl; sulfuric acid, H_2SO_4; nitric acid, HNO_3). *See also* strong base anion exchanger; sulfonic acid.

strong acid ion exchange Ion exchange using a cationic resin that contains exchangeable functional groups derived from a strong acid, such as sulfuric acid (H_2SO_4).

strong base anion exchanger An anion exchange resin with an exchange site [i.e., active group, usually quaternary amine, $-N(CH_3)_3{}^+$] capable of splitting neutral salts (e.g., sodium chloride, NaCl; calcium sulfate, $CaSO_4$; potassium nitrate, KNO_3) to form their corresponding free bases [sodium hydroxide, NaOH; calcium hydroxide, $Ca(OH)_2$; potassium hydroxide, KOH]. *See also* strong acid cation exchanger.

strong base ion exchange Ion exchange using an anionic resin that contains exchangeable functional groups derived from a quaternary ammonium compound.

structure Any of the improvements to land that are used or expected to be used in the operation of a utility, including buildings, utility lines, tanks, driveways, walks, and fences.

structure contour A line passing through all points that are on the upper or lower surface of any geologic formation or aquifer and that have the same elevation above a given datum.

Structured Query Language (SQL) A data query language developed by IBM for use with several of its mainframe database management packages. An American National Standards Institute SQL standard has been developed and is adhered to by many relational database management systems. Many developers of geographic information system software have also adopted Structured Query Language for performing database queries. *See also* geographic information system.

Student's t-test *See* t-test.

Student's t distribution *See* t-test.

styrene monomer ($C_6H_5CH{:}CH_2$) A chemical compound also called vinylbenzene. Styrene monomers are used in the manufacture of polystyrene, synthetic rubbers, thermoplastics, and resins. Styrene is regulated in drinking water by the US Environmental Protection Agency. *See also* polystyrene; resin.

Sub-Part H utilities Water utilities that are subject to the US Environmental Protection Agency's Surface Water Treatment Rule, codified in the Code of Federal Regulations as Title 40, Part 141, Sub-Part H. Such utilities use surface water or groundwater under the direct influence of surface water.

subacute Literally, something less than acute. This term is particularly ambiguous and is used to describe toxicity tests involving multiple doses given over an extended time (e.g., from 5 to 30 days). The term implies a single dose, but that implication is incorrect.

subaqueous Existing, taking place, or formed in or under water.

subaqueous pipe A pipe that is submerged in water.

subarid Neither entirely arid nor strictly humid, but intermediate, with a slight tendency to be arid. *See also* semiarid.

subartesian well A nonflowing well in which the water rises by hydrostatic pressure above the saturation zone but not to the land surface.

subcapillary interstice An interstice, smaller than a capillary interstice, in which molecular attraction spans the entire space and water is held immovable against gravity by the forces of adhesion alone. Such water is pellicular and hygroscopic. *See also* pellicular water; supercapillary interstice.

subcapillary opening An opening smaller in size than a capillary opening.

subchronic Of intermediate duration. This term is usually used to describe studies or levels of exposure between 5 and 90 days.

subchronic toxicity test A toxicity test that involves a substantial treatment period. Most commonly this term is used to represent a 90-day study in rodents.

subcritical flow *See* streaming flow.

subcritical fluidization backwash A procedure to clean a filter in which backwash water is added at a rate that does not completely fluidize the bed; i.e., the particles are not freely supported by the liquid. Under a subcritical fluidization regime, the media are in greater contact, potentially "scouring" difficult-to-dislodge particles. When combined with air, this flow regime can implement a collapse pulsing backwash. Subcritical fluidization may be followed by full fluidization to allow the scoured particles to be released from the media bed for complete cleaning. *See also* collapse pulsing backwash; full fluidization backwash.

subdrain A drain constructed beneath a lined conduit, such as a canal or tunnel, or beneath a dam or other structure. The purpose of a subdrain may be to prevent groundwater from entering the conduit, especially during construction, by providing the water with a method of escape, to prevent passage of water through to the downstream face of an earth dam, or to prevent or reduce uplift pressure in the case of a masonry dam.

sublimation The direct passage of a substance from solid to vapor without appearing in the intermediate (liquid) stage.

submarine spring A freshwater spring that occurs in the ocean.

submerged aquatic vegetation (SAV) Aquatic types of vegetation, such as sea grasses, that cannot withstand excessive drying and therefore live with their leaves at or below the water surface. Submerged aquatic vegetation provides an important habitat for young fish and other aquatic organisms.

submerged crib A water treatment plant intake built of masonry, timber, or metal and resting on the bed of a waterway, with its top below normal water level, to protect the exposed ends of intake pipes and offer minimum resistance to floating ice and debris.

submerged jet flocculation A novel flocculation process in which coagulated water is introduced upward into a reactor chamber though a center nozzle. A perforated plate separates the chamber from the settling unit and creates turbulence that results in hydraulic mixing. The reported advantages of such a unit are low head loss, low electrical energy use, easy operation, and production of a low-turbidity effluent.

submerged orifice An orifice discharging entirely underwater.

submerged outlet An outlet entirely covered by water.

submerged pipe A pipeline laid either on the bed or bottom or in an open or covered trench under the bed of a stream or other body of surface water.

submerged soil Soil that is underwater or saturated with water.

submerged spillway A spillway in which the water level downstream stands at an elevation higher than the crest of the spillway.

submerged tube A square tube with a length not exceeding 3.5 times the distance of the center line of the tube below the hydraulic gradient, and with both ends submerged.

submerged weed An aquatic plant, such as pondweed, that grows entirely beneath the surface of water.

submerged weir A weir that, when in use, has the water level on the downstream side at an elevation equal to or higher than the weir crest. The rate of discharge is affected by the tailwater. A submerged weir is also called a drowned weir.

submergence (1) The condition of a weir when the elevation of the water surface on the downstream side is equal to or higher than that of the weir crest. (2) The ratio, expressed as a percentage, of the height of the water surface downstream from a weir above the weir crest to the height of the water surface upstream above the weir crest. The distances upstream or downstream from the crest at which such elevations are measured are important but have not been standardized. (3) The depth of flooding over a pump suction inlet.

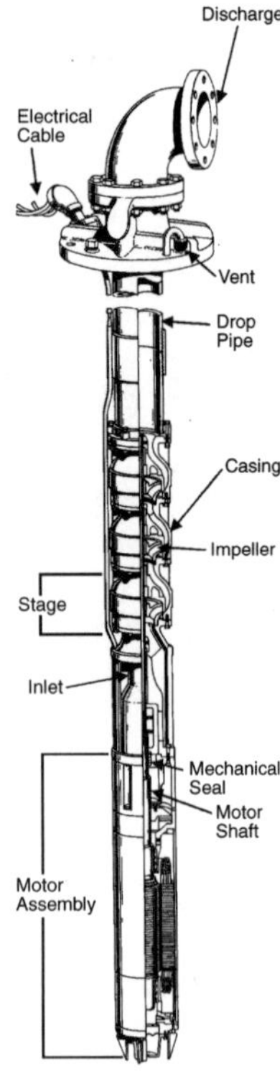

Submersible pump

submergent vegetation Aquatic plants that live completely beneath the surface of water.

submersible pump A pump designed to fit inside the well casing and to operate below the water level in a drilled well.

submetering Separate metering of a portion of water use associated with a metered water service connection.

submicrometre filter A cartridge-type membrane filter used in fine particle separation applications to remove particulates less than 1 micrometre in size.

subnatant liquid The liquid that remains beneath the surface of floating solids after flotation.

subpermafrost water Water beneath the permafrost.

subset A group composed only of some, but not all, elements in a set.

subsidence tank A tank or basin in which water or another liquid containing settleable solids is retained for a sufficient time, and in which the velocity of flow is sufficiently low, for a part of the suspended matter to be removed by gravity. *See also* sedimentation basin.

subsidence velocity The vertical rate at which a solid particle settles in a liquid under the influence of gravity. *See also* settling velocity; Stokes's law.

subsiding basin *See* subsidence tank.

subsoil The layer of soil below the arable layer.

subsoil drain A land drain constructed at a sufficient depth to intercept water from the subsoil.

subsoil drainage The removal of surplus or excess groundwater from the soil mass. Subsoil drainage is accomplished by means of natural drainage or by such artificial means as drains placed under the soil surface.

substation An electric power-switching station, generally accompanied by a transformer.

substrate (1) A substance used by organisms for growth. (2) A substance acted upon by an enzyme. (3) The basic—and usually most abundant—component of a medium, or a sample of that component.

substratum (1) A part, substance, or element that underlies and supports another; a foundation. (2) In botany, the base or material to which a plant is attached and from which it gets nutrients; the substrate. This term is loosely used to refer to the subsoil.

subsurface Pertaining to, formed, or occurring underneath the ground surface.

subsurface air *See* soil air.

subsurface drainage The removal of excess water from the soil or from prepared beds by a system of underdrains. *See also* subsoil drainage.

subsurface drainage check A gate or other structure placed in a deep drainage ditch or covered drain and by use of which the drainage discharge can be controlled to maintain a constant groundwater level for subirrigation.

subsurface drip irrigation The application of water via buried pipe and emitters, with flow rates measured in gallons (litres) per hour.

subsurface filter A gallery, with openings in its sides and bottom, extending generally horizontally into a water-bearing formation to collect water.

subsurface float A submerged body attached by a line to a surface float. Movement of the surface float indicates movement of the subsurface float. A subsurface float is used for observing the time taken to traverse a measured distance or to indicate the direction of flow. Such a float is also called a double float. *See also* surface float.

subsurface irrigation Irrigation by means of underground porous tile or its equivalent.

subsurface runoff (1) Groundwater runoff from a temporary zone of saturation in the soil. The runoff occurs so rapidly that it cannot be distinguished on the stream hydrograph from overland runoff and is therefore included in direct runoff. (2) Water that enters the soil but returns to the surface or appears in channels at a lower level without entering the water table in the zone of saturation.

subsurface velocity The velocity of water flowing in an open channel or conduit at any point below the water surface. Such velocity may be greater or less than the surface velocity.

subsurface water Water in the lithosphere. Subsurface water may be in a liquid, solid, or gaseous state. It comprises suspended water and groundwater. *See also* groundwater.

subsurface water basin A pervious formation, with sides and bottom composed of relatively impervious material, in which groundwater is held or retained. Such a formation is also called a groundwater basin.

subterranean stream (1) A stream flowing through large interstices or openings in rock, such as caves or caverns. The stream must have a fairly high discharge and velocity and must not be a mere trickle passing through a small crack or fissure. (2) A well-defined body of groundwater having a measurable, though small, velocity and flowing in a definite direction and confined in a permeable formation by formations less permeable.

subterranean water Water that occurs in open spaces within the rock materials of the Earth's crust. *See also* groundwater; subsurface water.

subtyping A process of using immunologic (or other) characteristics to separate and identify subtypes of bacteria in the same genus and species from each other.

subvariable spring A spring for which the difference between the maximum and minimum levels of discharge is at least 25 percent, but no more than 100 percent, of the average discharge.

success criteria Specific parameters that can be quantitatively compared to site-specific data and from which the degree of progress is determined.

suction dredge A scow carrying a centrifugal pump that has (1) a suction pipe reaching to the bottom to be excavated, and (2) a discharge pipe connecting to a pipeline conveying material to a place of deposit. A suction dredge is also called a hydraulic dredge or sand-pump dredge.

suction head (1) The head between the center line of a pump and the level of water on the suction side. When the head is positive above the center line, it is called suction head; otherwise it is sometimes called suction lift. (2) The head less than atmospheric pressure in a piping system.

suction lift The amount of head a pump located above the suction level must add to water to lift the water into the pump inlet or to the center line of the pump. The amount of lift is limited by atmospheric pressure and is generally in the 20–25-feet (6.1–7.6-metre) range, depending on the elevation above sea level.

suction lift pump A pump set above the surface of the body of water that supplies the pump. This arrangement requires the water to be lifted from such surface to the pump cylinder or casing.

suction pipe The inlet pipeline of a pump; the suction side.

suction piping A pipe or system of pipes that carries water into the inlet of a pump.

suction pit A walled pit in which the suction pipe or inlet openings of a pump are placed. A suction pit is sometimes called a sump or wet well.

suction valve A valve located on the suction line of a pump and used to close the suction line or pump from the source.

suction wave A wave that advances upstream in an open conduit from a point where the flow has suddenly been increased, as by the sudden opening of a gate. The water flowing through the opening consists of the normal flow in the conduit plus an additional quantity of flow that has been accelerated by the increase in slope in the water surface. This phenomenon is the opposite of the hydraulic bore; the front of the wave faces downstream instead of upstream and has a flatter slope as it advances upstream. *See also* hydraulic bore.

sudden contraction A reduction in the cross-sectional area of a stream channel, conduit, or other hydraulic structure that occurs abruptly or over a distance that is small compared with the dimensions of the channel or conduit, such as depth or diameter.

sugar of iron ($FeSO_4$) A common name for ferrous sulfate.

suggested no-adverse-response level (SNARL) An estimate of the contaminant concentration beneath which no adverse response is anticipated if the contaminant is ingested over a specified time period, assuming a dose level, body weight, daily water intake, and safety factor. The no-observed-adverse-effect level and the lowest-observed-adverse-effect level are used in place of the suggested no-adverse-response level to formulate a health advisory. *See also* dose; Health Advisory; lowest-observed-adverse-effect level; no-observed-adverse-effect level.

sulfate (SO_4^{2-}) An inorganic ion that is widely distributed in nature. It may be present in natural waters in concentrations ranging from a few to several thousand milligrams per litre.

sulfate reduction Chemical reduction of the sulfate ion (SO_4^{2-}) by anaerobic bacteria to form sulfide (S^{2-}) and hydrogen sulfide (H_2S). This reaction can occur in oxygen-deficient waters, which can be present in distribution system dead ends, storage reservoirs, and hot water tanks.

sulfate-reducing bacteria (SRB) Bacteria characterized by their ability to reduce sulfate under anaerobic conditions, producing hydrogen sulfide (H_2S). Sulfate is used as an electron acceptor for the oxidation of organic materials. Sulfate-reducing bacteria are generally found in anaerobic soils and sediments, as well as in areas of drinking water systems where appropriate conditions exist. These organisms are involved in pipe corrosion. The sulfate-reducing bacteria of most importance belong to the genera *Thiobacillus* and *Desulfovibrio*. *See also Desulfovibrio*; electron acceptor; *Thiobacillus*.

sulfide (S^{2-}) An anion often present in groundwater, especially in hot springs. Its common presence in wastewaters comes partly from the decomposition of organic matter, sometimes from industrial wastes, but mostly from the bacterial reduction of sulfate (SO_4^{2-}). Acidification of solutions of soluble sulfide compounds leads to the evolution of hydrogen sulfide (H_2S), a foul-smelling gas that can be poisonous. *See also* sulfate reduction; hydrogen sulfide.

sulfonic acid ($-SO_2OH$) A specific acidic group that forms the exchange site (i.e., active group) in certain cation exchange resins and gives these resins their ion exchange capability. *See also* strong acid cation exchanger.

sulfotransferase Any of a class of conjugating enzymes that transfer sulfate (SO_4^{2-}) to form esters (R–CO–O–R') with hydroxyl (R–OH) or free amino groups (R–NH_2) on an organic compound. This esterification serves to increase the polarity of the compound, facilitating the compound's elimination in the urine.

sulfoxidation A process by which thioethers (R–S–R') are oxidized (usually by microsomal monooxygenases) to sulfoxides (R–(SO)–R').

sulfur (S) A nonmetallic element that occurs either free or combined (especially in sulfide (S^{2-}) and sulfate (SO_4^{2-}) compounds). Sulfur exists in several forms, including as yellow crystals. *See also* sulfate; sulfide.

sulfur bacteria *See Thiobacillus.*
sulfur compound joint A bell-and-spigot joint that was sealed with a melted sulfur compound rather than lead to reduce cost and eliminate the need to calk the joint. Because of sulfur-reducing bacteria in many soils, the life of a sulfur compound joint has been found to be short, and these joints often must be reworked or joint clamps installed to stop leaks. *See also* joint clamp.
sulfur compound jointing material Any of the sulfur compounds used instead of lead to seal a bell-and-spigot joint because they offer cost savings and the ability to seal the joint without calking. The compounds are subject to attack by sulfur bacteria and have been the cause of numerous joint leaks in systems where they were used. At the height of their popularity, as many as 50 such products were available; all contained 50 to 65 percent sulfur. Sulfur compound jointing materials are sometimes referred to as mineral lead.
sulfur dioxide (SO_2) An extremely effective oxidizing and reducing agent. It has many industrial uses, including use as a food additive (for inhibition of browning, of enzyme-catalyzed reactions, or of bacterial growth). Sulfur dioxide is also used as a dechlorination agent in drinking water facilities. *See also* dechlorination agent; oxidant; reducing agent.
sulfur spring A spring in which the water contains certain compounds of sulfur, usually in gaseous form. Such springs are usually identified by the odor of the hydrogen sulfide (H_2S) that emanates from them.
sulfuric acid (H_2SO_4) A very strong, corrosive, and hazardous acid used as a regenerant for the cation stage of an ion exchange deionization system. Sulfuric acid is also used occasionally to lower the pH of highly alkaline water. When higher concentrations of sulfuric acid are combined with high concentrations of calcium, calcium sulfate ($CaSO_4$) crystals precipitate, creating tenacious fouling of media particles. Sulfuric acid was once commonly called oil of vitriol.
sulfur-reducing agent A sulfur-containing compound used as a reducing agent (e.g., sulfur dioxide, SO_2; sodium bisulfite, $NaHSO_3$; sodium sulfite, Na_2SO_3; sodium thiosulfate, $Na_2S_2O_3$). *See also* sodium bisulfite; sodium thiosulfate; sulfur dioxide.
sum of five haloacetic acids (HAA5) The sum of the concentrations, in milligrams per litre, of five haloacetic acid compounds: monochloro-, dichloro-, trichloro-, monobromo-, and dibromoacetic acid. At least initially, the Disinfectant/Disinfection By-Products Rule regulates the sum of these five species; data were not available on the occurrence and control of the other four haloacetic acid species during the rulemaking process. *See also* Disinfectant/Disinfection By-Products Rule; haloacetic acid.
sum of six haloacetic acids (HAA6) The sum of the concentrations, in milligrams per litre, of six haloacetic acid compounds, including the five species in HAA5 as well as bromochloroacetic acid. The Information Collection Rule requires monitoring for these six species in order to develop data on the occurrence and control of these compounds for future disinfection by-product rulemaking efforts. *See also* Information Collection Rule; sum of five haloacetic acids.
summation graph A graph showing the summation of the ordinates of a series of unit hydrographs spaced at unit rainfall duration intervals, such that the ordinates of the S-curves are expressed as the percentage of the total unit hydrograph volume and the abscissas expressed in units of time. *See also* S-curve hydrograph.

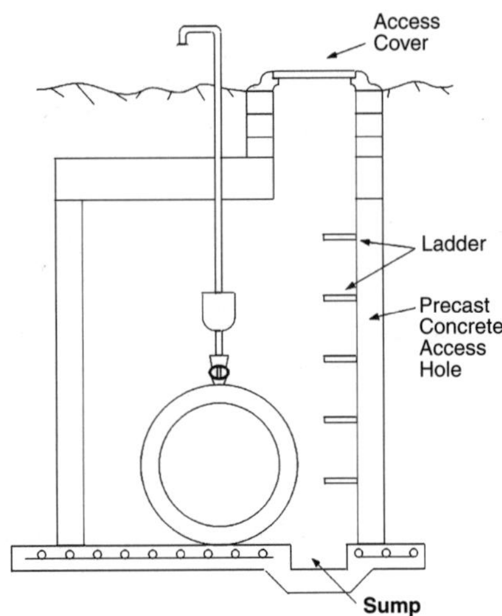

Sump in an access hole

sump A tank or pit that receives drainage and stores it temporarily and from which the discharge is pumped or ejected.
sump pump A mechanism used for removing water from a sump or wet well. It may be energized by air, water, steam, or an electric motor. Ejectors and submerged centrifugal pumps, controlled either by a float or manually, are often used for this purpose.
sunk costs Those costs that have already either been incurred or committed. Sunk costs are not incremental costs and, as a result, no longer play a part in any decision-making process. For example, if $1.0 million has been spent on a new process or facility before it is determined that another process offers a far better way to achieve a goal, the sunk costs for the abandoned process should play no role in the decision-making process regarding the benefit–cost ratio or return on investment for the new process.
supercapillary interstice An interstice that is larger than a capillary interstice, in which water is not held at an appreciable height above a water table or hydrostatic pressure level, and through which the movement of water may be turbulent. *See also* subcapillary interstice.
supercapillary percolation Percolation through that portion of a porous medium where the size of the pore space is too large to generate appreciable capillary forces. *See also* capillary force; percolation.
superchlorination The addition of excess amounts of chlorine to a water supply to speed chemical reactions or ensure disinfection within a short contact time. The chlorine residual following superchlorination may be high enough to make the water unpalatable, so dechlorination is commonly employed before the water is used. *See also* dechlorination.

supercritical Pertaining to a fluid held in excess of its critical temperature and pressure. Such a substance acts physically like a gas yet chemically like a liquid.

supercritical flow *See* shooting flow.

supercritical fluid chromatography (SFC) A separation technique that has been used in the analysis of relatively polar, nonvolatile, or thermally labile compounds. It has proven to be complementary to gas and liquid chromatography. The mobile phase is a supercritical fluid, e.g., carbon dioxide (CO_2) that allows the chromatographic separation of analytes at low temperatures.

supercritical fluid extraction (SFE) A technique used in the analysis of solid matrices. Water and air samples can be analyzed indirectly by isolation of compounds onto an adsorbent phase and subsequent desorption of analytes by supercritical fluid extraction. The technique involves a fluid, such as carbon dioxide (CO_2), held in excess of its critical pressure and temperature. Such a supercritical substance acts physically like a gas yet chemically like a liquid. Supercritical fluid extraction takes advantage of the fluid's solvating power, the extent of which depends on the density of the fluid.

Superfund Amendments and Reauthorization Act (SARA) Public Law 99-499, enacted in 1986, that enlarged and reauthorized the Comprehensive Environmental Response, Compensation, and Liability Act of 1980, commonly known as Superfund. The original legislation and amendments are codified generally as 42 U.S.C. 9601–9675.

supernatant Located above or on top of something else.

supernatant liquid The liquid situated over the surface of settled sediment. *See also* subnatant liquid.

supernatant liquor *See* supernatant liquid.

superoxide [$M(O_2)_x$, where $x = 1$ or 2 depending on the valence of metal M] A compound characterized by the presence of the superoxide anion (O_2^-) in its structure. *See also* superoxide anion.

superoxide anion (O_2^-) A highly reactive form of oxygen that is produced when molecular oxygen is reduced by a single electron (i.e., when it gains one electron). Superoxide anion is a common product of many different biological oxidations.

supersaturated Pertaining to a solution in which the solvent (e.g., water) contains more dissolved matter (solute) than is present in a saturated solution of the same components at an equivalent temperature. *See also* saturated solution.

supervisory control and data acquisition (SCADA) A computer-monitored alarm, response, control, and data acquisition system used by drinking water facilities to monitor their operations.

supplemental water supply source A water source furnishing additional supplies when water needs to be obtained from more than just a primary source.

supplied-air respirator A respirator, full faced or half faced, with a continuous supply of grade D breathing air and with a 5-minute escape bottle serving as a backup supply. *See also* grade D air.

supplier of water Any entity that owns or operates a public water system.

supply curve A representation of the amount of a product that a firm will offer in the market at different prices. In the economics of pure competition, the prices at which a firm will offer its product are based on the firm's marginal cost curve for the product over a feasible range of output with a specified plant and production process. In a nonprofit monopoly situation, which is typical in the drinking water business, the supply curve is represented not by the marginal cost curve but by the average cost curve. The regulated monopoly will price not for profit maximization but rather to ensure that average revenue is equal to average cost.

supply and demand The combination of (1) the total amount of water available to a utility or water provider from all sources (supply) and (2) the total water use of all customers on the system (demand). Both supply and demand are frequently measured for peak-hour, peak-day, peak-month, and peak-season amounts, as well as for annual amounts.

supply line A conduit between a source of water supply and a distribution system.

supply management The use of specific measures by water utilities to enhance their capability to supply water, including capacity additions as well as efficiency improvements in the water supply infrastructure. Supply management includes the use of water audits, leak detection and repair, metering, source protection, conservation, and source development that meets least-cost and other planning criteria.

support media bed In filtration or ion exchange, a bed of material of a specific graded particle size (such as gravel) used as a subfill to support the primary medium bed. This bed improves the collection of processed water and promotes a more uniform distribution of the backwash water.

suppressed contraction A condition existing in an orifice or weir when one or both sides, bottom, or all three are flush with the sides of the orifice or sides or crest of the weir. Such a condition eliminates the contraction in cross-sectional area for water passing through the orifice or over the weir.

suppressed weir A weir with one or both sides flush with the channel of approach. This arrangement prevents contraction of the nappe adjacent to the flush side. The suppression may occur at one end or both ends of the weir.

suppression A technique used in ion chromatography as a way of decreasing the conductivity of the eluant and increasing the conductivity of the analyte. Because conductivity detectors are commonly used in ion chromatography, the use of suppression has

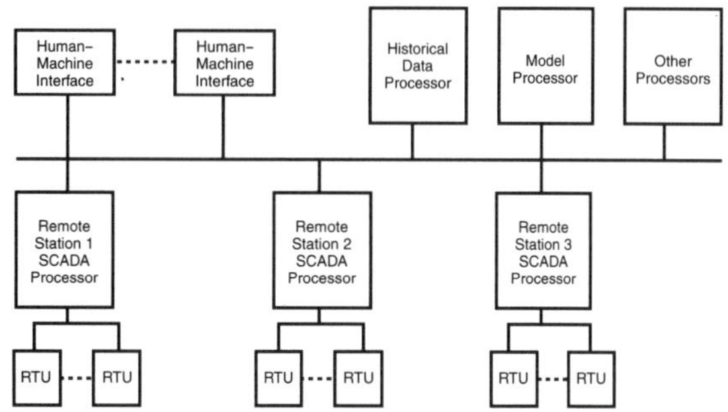

Distributed SCADA system controlling three remote pumping stations

dramatically improved the sensitivity of ion chromatographic analyses. *See also* conductivity detector; ion chromatography.

suprapermafrost water Groundwater above permafrost, forming what is called the suprapermafrost layer.

surcharge pricing A method of charging for water use whereby customers are charged higher rates during times of peak usage. This method is used to pay for the extra costs associated with providing large amounts of water for short periods of time, such as during the summer months. Surcharge pricing can be used in combination with other pricing mechanisms, such as a flat rate, decreasing block rate, or increasing block rate.

surcharge storage Storage available above an established reservoir level. Such storage is sometimes used to control floods.

surface aeration The absorption of air through the surface of a liquid.

surface area For granular activated carbon, an area value empirically determined by the Brunauer-Emmett-Teller method. Such surface area is generally expressed in square metres per gram of activated carbon. *See also* Brunauer-Emmett-Teller method.

surface curve (1) The longitudinal profile assumed by the surface of a stream of water flowing in an open channel or conduit. Surface curves are usually catalogued into 12 classifications depending on the slope of the conduit bed and whether the depth of flow is greater than, less than, or between the normal depth and the critical depth. The 12 classifications of the surface curve are used to describe the possible gradually varied flows in an open channel: three for the mild slope (M1, M2, M3), three for the steep slope (S1, S2, S3), two for the critical slope (C1, C3), two for the horizontal slope (H2, H3), and two for the adverse slope (A2, A3). (2) The hydraulic grade line for an open conduit or for a closed conduit flowing partly full. *See also* hydraulic grade line.

surface detention That part of the rain that remains on the ground surface during rainfall and either runs off or infiltrates after the rain ends. Surface detention does not include depression storage. The detention depth increases until discharge reaches an equilibrium in which the rate of supply equals the surface runoff.

surface diffusion An intraparticle transport mechanism for adsorbed particles. Once adsorbed, particles can diffuse along a surface according to a concentration gradient of previously adsorbed particles.

surface drag Resistance to fluid flow due to viscous shear in the boundary layer. Surface drag is also called skin friction.

surface drainage (1) The removal of surplus or excess surface water collecting on land. Surface drainage is accomplished by natural means or by such artificial means as levees, open ditches, and terracing. (2) Runoff.

surface evaporation Evaporation from the surface of a body of water, moist soil, snow, or ice. *See also* evapotranspiration.

surface filtration Filtration that occurs at a surface layer (as opposed to within the depth of the filter bed as occurs with granular media filtration) of a filter. Surface filtration is accomplished by passing the material to be filtered over a grating, screen, diatomaceous earth layer, biofilm layer, sieve, or membrane fabric with micrometre-sized holes. The size of the filter holes determines which particles will pass through and which will be filtered out (the determinants of the straining mechanism). *See also* depth filtration; diatomaceous earth filtration; slow sand filtration.

surface float A float on a water surface. Observation of the float indicates the flow direction, and observation of the time the float takes to traverse a measured distance indicates the velocity. *See also* subsurface float.

surface loading rate A unit process criterion expressing the rate at which the hydraulic surface of a treatment basin will receive a flow of water. The units of expression, volume per time per surface area, are equivalent to the units of velocity, thereby relating to the settling velocity of the "design" particle to be removed. Particles with settling velocities greater than the surface loading rate theoretically will be removed, whereas only a portion of the particles with settling velocities less than the surface loading rate will be removed. *See also* overflow rate.

surface microlayer The thin zone, usually several molecules thick, at the surface of a body of water within which physical processes are modified because of contact with the atmosphere, surface tension, and other surface phenomena. *See also* surface tension.

Surface Mining Control and Reclamation Act (SMCRA) Public Law 95-87 (30 U.S.C. 1201–1328), which seeks to protect the environment from the adverse effects of surface coal-mining operations. The Surface Mining Control and Reclamation Act requires permits for surface coal mining and reclamation.

surface overflow rate *See* surface loading rate.

surface pressure An inward pull acting on the free surface of a liquid. Surface pressure differs with differing curvature of the liquid surface. It increases with increasing convexity and decreases with increasing concavity of the surface of the liquid.

surface profile The longitudinal profile assumed by the surface of a stream of water flowing in an open channel.

surface pump A mechanical device for removing water from a sump or wet well.

surface runoff (1) That portion of the runoff of a drainage basin that has not percolated beneath the surface after precipitation. (2) The water that reaches a stream by traveling over the soil surface or by falling directly into the stream channels, including not only the large permanent streams but also the tiny rills and rivulets.

surface screen A device for excluding debris from a surface water intake.

surface slope The inclination of a water surface, expressed as the change in elevation per unit of surface length; the sine of the angle that the water surface makes with the horizontal. The tangent of that angle is ordinarily used, with no appreciable error resulting except for the steeper slopes.

surface soil The soil's uppermost part that is ordinarily moved in tillage, or its equivalent in uncultivated soils. Surface soil ranges in depth from 3 to 10 inches (7.5 to 25 centimetres) and is frequently designated as the plow layer.

surface storm flow That portion of the storm flow that is induced by gravity to move over the surface of the ground.

surface supply Water from streams and snow runoff, as distinguished from subterranean sources or groundwater.

surface tension (σ) The strength with which a liquid forms a relatively tough "skin" or film on its surface. Surface tension is caused by the attraction between the molecules of the liquid, and it causes water molecules to stick together and form drops.

surface tension units In the US customary system, pounds force per foot. In Système International, newtons per metre; kilograms per second squared.

surface wash (1) A supplementary method of washing the filtering medium of a rapid granular filter by applying water under pressure at or near the surface of the sand by means of a system of stationary or rotating jets. (2) The surface runoff draining into a ditch or drain. *See also* rotary surface washer; sheet erosion; stationary surface washer.

surface washing *See* filter agitation.

surface water All water on the surface, as distinguished from subsurface water or groundwater.

surface water drain A drain constructed to carry surface water.

surface water inlet An inlet providing entrance for surface water into a drain that is located below the ground surface.

surface water safe yield *See* impoundment safe yield.

Surface Water Treatment Rule (SWTR) The common name for the US Environmental Protection Agency regulation promulgated June 29, 1989, that set maximum contaminant level goals for *Giardia lamblia*, viruses, and *Legionella*, as well as National Primary Drinking Water Regulations for public water systems using surface water sources or groundwater under the direct influence of surface water. These regulations include (1) criteria under which filtration (including coagulation and sedimentation, as appropriate) are required and procedures by which the states are to determine which systems must install filtration, and (2) disinfection requirements. *See also* groundwater under the direct influence of surface water.

surface weed *See* floating weed.

surface-active agent A substance—such as a detergent, wetting agent, or emulsifier—that, when added to water, lowers surface tension and increases the "wetting" capabilities of the water. Reduced surface tension allows water to spread and to penetrate fabrics or other substances, enabling them to be washed or cleaned. Surface-active agents are sometimes called surfactants or wetting agents. *See also* surface tension.

surface-fed intermittent stream A stream or stretch of a stream that flows during a protracted period while receiving water from a surface source, usually the gradual and long-continued melting of snow in a mountainous area or other cold tributary area. This term may arbitrarily be restricted to streams or stretches of stream that flow continuously during periods of at least 1 month.

surfactant *See* surface-active agent.

surficial Of or relating to a surface. For example, the surficial velocity is the fluid velocity at the surface of the liquid.

surge A sudden increase in the movement or pressure of water in a pipe or container, usually caused by opening or closing of a valve, starting or stopping of a pump, or trapping of air in a water main. *See also* water hammer.

surge block A plug that is lowered and raised in a well to create pressure shocks and to dislodge fine materials from the well screen. Surge blocks are used to develop wells for water production.

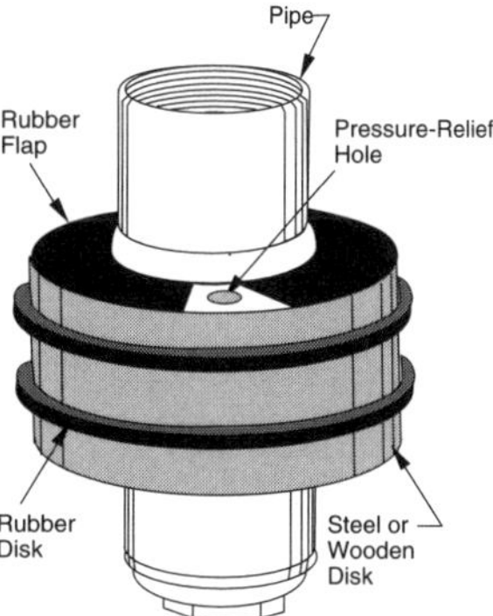

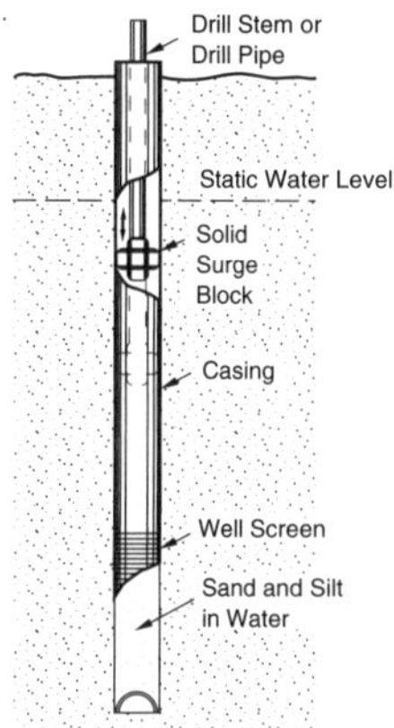

Surge block

surge chamber *See* surge tank.

surge pressure The pressure caused by a sudden movement of water or liquid from such causes as a directional change in the flow, the starting or stopping of a pump, and the opening or closing of a valve or fire hydrant. *See also* water hammer.

surge suppressor A device, such as a ball valve or cone valve, that opens and closes slowly with the turning on and off of a pump to minimize the surges with changes in water motion.

surge tank A tank used to attenuate the pressure increase or decrease resulting from a sudden change in fluid momentum. A surge tank accumulates water in response to a pressure surge and supplies water in instances of moderate negative gauge pressure, thereby preventing damage to pumping and piping equipment. Surges result when sudden changes in a flow regime occur, such as when pumping equipment is stopped or valves are closed too quickly. *See also* water hammer.

surger A device to develop a well resembling a disk that fits tightly in the well with a one-way valve. The surger is alternately pushed down into the well and then pulled up like a piston. It causes rapid hydraulic pressure changes and, somewhat like a syringe, pulls water and drilling cuttings from the formation into the well for removal. Once the well is fully developed and no residual drilling cuttings flow into the well, the surger is removed and the production pump is placed into the well.

surging and bailing A well development method that uses a surger to force drilling cuttings into the well and a bailer to remove the suspension of cuttings and water that flow into the well during use of the surger. *See also* bailer; surger.

surplus The amount by which the assets of a private utility exceed the total of the utility's liabilities, reserves, and capital stock; in a publicly owned utility, the amount by which the assets exceed the total liabilities, reserves, and municipal equity.

surrogate compound A substance that is not usually found in the environment but is chemically similar to an environmental contaminant. Such similarities allow the use of surrogate compounds as internal standards in organic analytical procedures. Surrogate compounds are used, for example, to determine recoveries and adjust the retention times of gas chromatographic methods.

surrogate measurement An analytical measurement for a nonspecific or indicator parameter that can be used in lieu of performing tests for specific contaminants. For example, dissolved organic halogen measurements can be made instead of (or in addition to) tests for specific halogen-substituted disinfection by-products. Similarly, total coliform testing can be used as a surrogate measurement for pathogenic microorganisms in water (i.e., coliform densities can provide an indication of the degree of pollution of a water). *See also* total coliform; total organic halogen.

surveillance sample A water sample analyzed by a regulatory agency to evaluate the ability of a public water system to meet drinking water regulations. Surveillance samples taken at the discretion of the regulatory agency and without objection from the water supplier may be counted as monitoring samples.

survey meter A portable instrument, such as a Geiger counter or ionization chamber, used to detect nuclear radiation and to measure the dose rate.

surveying A branch of applied mathematics that provides techniques for determining the area of any portion of the Earth's surface, the lengths and directions of the bounding lines, the contour of the surface, and an accurate delineation of the whole on paper.

suspected carcinogen A formal term applied by the International Agency for Research on Cancer or by the US Environmental Protection Agency to indicate that limited evidence exists that a chemical could produce cancer in humans. Giving a chemical this label entails using a "weight of evidence" approach that can be based on some combination of tests in experimental animals and results from human epidemiological studies. *See also* weight-of-evidence approach.

suspended growth Biological growth that is uniformly dispersed and mixed in a liquid medium for the purposes of removing target compounds and contaminants. *Contrast with* biofilm.

suspended load The part of a liquid's sediment load that is in suspension.

suspended matter *See* suspended solids.

suspended sediment *See* suspended solids.

suspended solids Solid organic and inorganic particles that are held in suspension by the action of flowing water and are not dissolved. Suspended solids are retained on a standard glass fiber filter or a 0.45-micrometre pore diameter membrane filter after filtration of a well-mixed sample. The residue is dried at 217° to 221° Fahrenheit (103° to 105° Celsius). Suspended solids are also known as suspended matter or suspended sediment. *See also* solids; total suspended solids.

suspended water Subsurface water that partially occupies interstices in the zone of aeration. Suspended water is composed of hygroscopic, pellicular, mobile, and fringe water.

suspended water zone *See* unsaturated zone.

suspended-frame weir A movable timber weir in which the structural steel frames, in times of flood, can be raised from the river and supported by an overhead bridge.

sustainability A decision-making concept describing development that meets the needs of the present without compromising the ability of future generations to meet their own needs.

sustained yield A continual periodic or annual yield of plants or plant material from an area.

Sutro weir A weir with at least one curved side and a horizontal crest, formed so that the head above the crest is directly proportional to the discharge.

SUVA *See* specific ultraviolet absorbance.

Sv *See* sievert *in the Units of Measure list.*

SVI *See* sludge density index.

swabbing The process of lowering and raising a surge block or line swab to develop a well for water production.

swamp A flat, wet area that is usually or periodically covered by standing water and that supports a growth of grasses, shrubs, and trees. In contrast to a bog, the organic soil in a swamp is thin and readily permeated by roots and nutrients.

swamp gas *See* methane.

SWAP *See* Source Water Assessment Program.

SWD *See* side water depth.

SWDA *See* Solid Waste Disposal Act.

sweat joint A joint for copper pipe or tubing with fittings, created by heating of the joint until melted solder runs into the joint between the tubing and fitting, making a seal and connection as the solder cools.

sweep flocculation *See* sweep-floc coagulation.

sweep-floc coagulation A process of overdosing coagulant to ensure that the hydrolyzed chemical will form sufficient precipitate to settle rapidly and "sweep" particles out of the suspension. Most conventional water treatment plants operate in the sweep-floc coagulation, or sweep flocculation, mode. *See also* adsorption-destabilization; bridging; double-layer compression.

sweet brine Brine that contains sufficient sodium or potassium content and is relatively low in calcium, magnesium, or other interfering substances such that it is effective for use or reuse in regenerating exhausted ion exchange resin. *See also* reclaimed brine; sour brine.

swelling The expansion of certain ion exchange resins when they are converted into a specific ionic state (either exhaustion or regeneration, depending on the resin). This expansion is reversible. Some exchangers will expand as they exhaust. Cation exchange water-softening resins will generally swell when they are exhausted (loaded with hardness ions) and will shrink when regenerated with heavier salt dosages of 10 to 15 pounds of sodium chloride (NaCl) or potassium chloride (KCl) per cubic foot (160 to 240 kilograms per cubic metre) of resin. Standard cation softening resin [8 percent polystyrene $((C_6H_5CHCH_2)_n)$/divinyl benzene $(C_6H_4(CH{:}CH_2)_2)$], in the calcium form, will shrink about 5 percent in volume when treated with a 25 percent salt-brine solution.

swirl Rotary motion of a flow superimposed on the forward motion.

switch A device to manually disconnect electrical equipment from the power source.

switchgear A heavy, constructed panel for a large motor or other equipment using high voltage, generally 1,200 to 7,200 volts. The device has circuit breakers and other protective features along with low-voltage transformers for control power. Frequent starting and stopping of the equipment should be limited because of high-voltage starting requirements and excessive wear to the equipment. Relatively new switchgear may have start limit controls—protective devices that prevent the starting of motors more than a limited number of times over a set time period to prevent overheating and damage.

SWL *See* static water level.

SWTR *See* Surface Water Treatment Rule.

symbiosis The mutually beneficial cohabitation of two dissimilar organisms. For example, lichens, which grow on the surfaces of rocks, are examples of mutual beneficial relationships between algae and fungi.

symbol A graphical element used to represent a feature as a point on a map. A symbol is generally stored once in a library or file and is referenced multiple times within a graphic drawing. A geographic information system is equipped to display a wide variety of symbols representing points, lines, areas, and facilities. *See also* geographic information system.

symmetrical membrane A type of membrane having a structure that is relatively uniform throughout.

syncarcinogenesis The administration of two carcinogens resulting in either additive or synergistic (i.e., more than additive) effects. Two types of syncarcinogenesis are recognized: that which occurs when the chemicals are administered together, and that which occurs when they are sequentially administered.

synchronous communication Communication occurring at timed intervals regulated by pulses from a computer clock. It is commonly used in data communications to synchronize a transmitter and receiver. Synchronous transmission of data between computers eliminates the need for start and stop bits. *See also* asynchronous communication.

synchronous motor An electric motor that is designed to run at synchronous speed. *See also* synchronous speed.

synchronous speed The speed of rotation of the magnetic field in an alternating current electric motor. This speed is a function of the frequency of the applied power and the number of poles in the motor.

synclinal spring A contact spring occurring at the outcrop of a pervious water-bearing stratum in a syncline.

syncline A folded geologic formation where the limbs of the fold dip (slope downward) toward the axis.

syndet *See* synthetic detergent.

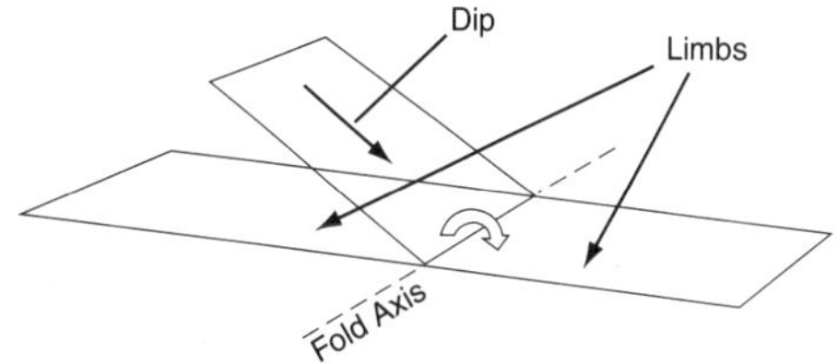

Simple syncline

synecology The study of the interrelationships between organism groups in a natural community and their environment.

Synedra A species of diatom found in eutrophic fresh surface waters.

synergism A phenomenon in which the effects of two or more organisms or substances acting together are greater than the sum of their individual effects—often expressed as "2 + 2 = 5."

synoptic Varying with distance; longitudinal.

synthetic detergent A manufactured cleaning agent. Detergents can be classified as anionic, cationic, or nonionic.

synthetic ion exchange resin A manufactured ion exchange resin, commonly made with cross-linked polymers having exchangeable functional groups. *See also* ion exchange.

synthetic membrane A manufactured membrane, commonly made with manufactured polymers.

synthetic organic chemical (SOC) An organic compound that is commercially made. Some synthetic organic chemicals are contaminants in drinking water and are regulated by the US Environmental Protection Agency. The regulated synthetic organic contaminants include volatile organic chemicals, pesticides, herbicides, polychlorinated biphenyls, selected treatment chemicals (e.g., acrylamide), and polynuclear aromatic hydrocarbons. *See also* acrylamide; herbicide; pesticide; Phase I Rule; Phase II Rule; Phase V Rule; polychlorinated biphenyl; polynuclear aromatic hydrocarbon.

synthetic organic compound (SOC) *See* synthetic organic chemical.

synthetic organic contaminant (SOC) *See* synthetic organic chemical.

synthetic organic matter *See* synthetic organic chemical.

synthetic resin *See* synthetic ion exchange resin.

synthetic unit hydrograph A unit hydrograph developed for an ungauged drainage area, based on known physical characteristics of the basin.

system (1) A combination of several pieces of equipment integrated to perform a single function. For example, a water treatment system employs various processes—such as coagulation, sedimentation, filtration, and disinfection—with the common objective of producing potable water. (2) An assemblage or combination of things, or a blend of functionally compatible components, to achieve objectives that have been properly determined.

system analysis *See* network analysis.

system atlas An up-to-date graphical representation of a water distribution system depicting such facilities as mains, service connections, fire hydrants, valves, tanks, and pump stations, along with utility property, easements, and rights-of-way. These facilities, as shown in a systematic manner, become geographic locations to form a map book or atlas used in the operation and maintenance of the distribution system.

system development charge A contribution of capital toward existing or planned future backup plant facilities that are needed to meet the service needs of the customers to whom such fees apply. Various terms are used to describe these charges in the water treatment industry, but the basic intent of the charges is to provide funds for use in financing all or a portion of the capital improvements necessary to serve new customers. *See also* capacity charge.

system development charge facilities Those facilities, or a portion of those facilities, that have been identified as being required for new customer growth. The cost of the facilities will be recovered in total or in part through a system development charge.

system efficiency *See* hardware efficiency.

system head curve A plot of the head loss in a water distribution system as a function of discharge in the system. Hydraulic engineers use system head curves to match pumps to hydraulic systems and to operate systems efficiently. Typically pumps and hydraulic systems operate efficiently over a narrow range of discharges and added heads, and the two parts must be carefully matched to avoid wasting energy.

system leakage The quantity of water that goes through a distribution system but cannot be accounted for. The number is derived by subtracting the amount of water that is measured by meters and billed to customers from the water that is leaving treatment plants or well fields. The percentage varies greatly depending on how well the system is maintained. System leakage is also called unaccounted-for water.

system loss *See* system leakage.

system with a single service connection A system that supplies drinking water to consumers via a single service line.

systematic error *See* bias.

Système International (SI) *See the Units of Measure list.*

systemic Acting within the body as a result of being absorbed and distributed by the blood.

systemic effect An effect observed at sites distant from the entry point of a chemical as a result of the chemical's absorption and distribution into the body. Chemicals that act locally on epithelial structures, such as the skin or lining of the stomach or intestine, would not be considered as producing systemic effects.

systemic toxicity Toxicity that occurs as a result of absorption of a chemical into the body. *See also* systemic effect.

systems analysis A systematic approach to assessing a problem, arriving at alternatives, and projecting the probable consequences of employing those alternatives in order to facilitate making correct decisions.

T

T *See* disinfectant contact time; tera; tesla.

T *See* absolute temperature; period.

t *See* metric ton *in the Units of Measure list.*

T cell A lymphocyte (white blood cell) that is influenced by or passes through the thymus gland (i.e., a thymus-dependent lymphocyte). T-cells are of different types; some act to suppress whereas others assist the stimulation of antibodies within B-lymphocytes. They have additional immune system functions, as well, and are involved in cell-mediated immunity. Some are capable of killing tumor cells or cells from transplanted tissues.

table (1) A condensed collection and arrangement of related numbers or other items assembled in some definite order, such as a table of logarithms. (2) A surface, such as a water table.

tagged and locked out Unavailable for use as a result of safety measures. This phrase describes a safety precaution to prevent the power to such electrical equipment as a pump motor from being turned on while maintenance is being performed on the equipment or to prevent operation of the equipment for other reasons. A tag is placed on the switchgear or starter to provide notice, and a lock is placed on the switchgear or starter to prevent its operation.

tagout To communicate to workers that a piece of equipment is out of service and not to be energized until the warning tag is removed.

tail bay The part of a canal lock immediately below the tail gates.

tail gate The gate at the low-level end of a lock.

tail pipe The suction pipe of a pump.

tailed phage The most numerous and most varied of all viruses. Tailed phages are identified by morphology, physicochemical properties, nucleic acid content, resistance to physical and chemical agents, serology, and nucleic acid hybridization reactions.

tailgate safety meeting A safety meeting regularly held by members of the construction industry around the tailgate of a truck.

tailrace A channel that conducts water from a waterwheel; an afterbay.

tailwater Water discharged at the downstream end of a structure.

tainter gate A crest gate for which the face is a section of a cylinder that rotates about a horizontal axis downstream from the gate. The water pressure against the gate is concentrated in the axis; this arrangement reduces friction in raising and lowering the gate. A seal is placed along the sides and bottom of the gate face for watertightness. The gate is raised and lowered by winches or hoists attached to cables or chains fastened to the bottom edge of the gate and lying against its water face; this allows a vertical lifting force to be applied. A tainter gate can be closed under its own weight. Such gates were formerly called canal locks.

tall fescue A hybridized, cool-season turfgrass characterized by deeper root systems and more drought tolerance than bluegrass.

tangential flow filtration *See* crossflow filtration.

tangential turbine A turbine that depends on the impulse of moving water or steam delivered at high velocities against a set of vanes or buckets placed on the periphery of a wheel that is rotated by the impulse of the steam or water. Such a turbine is also called an impulse turbine.

tangible asset Permanent property of a physical nature, such as lands, buildings, mineral deposits, wells, reservoirs, plant equipment of all kinds, utensils, furnishings, rolling stock, or merchandise, intended for immediate use and permanent improvements. Tangible assets are also called physical assets.

tank A structure or container used to hold water for such purposes as aeration, disinfection, equalization, holding, sedimentation, treatment, mixing, dilution, feeding, or other handling of chemical additives.

tannin A colored compound formed in water when plant matter degrades.

tap The connection of a service line to a water line by means of a corporation stop. A tap is installed by drilling and threading the water main for the corporation stop or by using a tap saddle that attaches around the water main and provides threads for the corporation stop. *See also* corporation cock.

tap fee *See* system development charge.

tap saddle A device that encircles a water main and provides an outlet to connect a corporation cock.

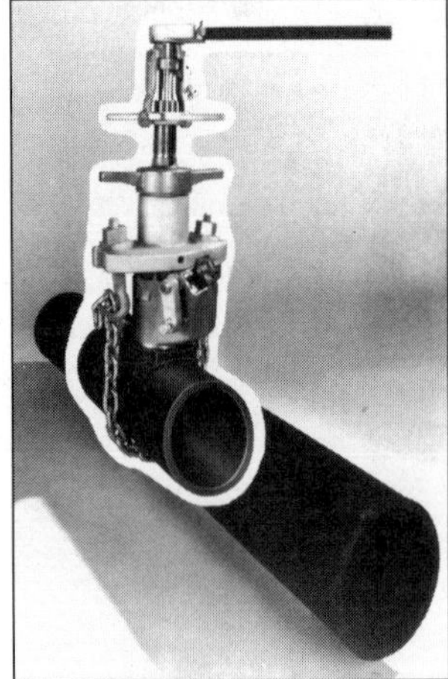

Tapping machine

tape gauge A gauge to measure the distance to a water surface. It consists of a tagged or indexed chain, tape, or other line attached to a weight that is lowered to touch the water surface. When the weight touches the surface, the gauge height is then read on a graduated staff or opposite an index. Opposite ends of the tape may be attached to a float and counterbalance and the tape arranged to hang on a pulley. The gauge height may then be read on a graduated staff or opposite an index.
tapping The drilling and threading of a hole in a water line with the insertion of a corporation cock to allow the connection of a water service line. *See also* corporation cock.
tapping machine A tool that can be attached to a water main to drill and thread a hole in the water main and insert a corporation cock under pressure. *See also* corporation cock.
tapping sleeve A split sleeve used in making a wet connection where a single branch line is to be tapped into a water main under pressure.
tapping valve A valve with flanges that can be connected to a tapping tee and a drilling machine to drill a hole in a main through the valve. The drill is withdrawn and the valve closed, allowing removal of the drilling machine and then connection of the new water main.
tare weight The weight of a container when empty.
target compound A chemical for which an analytical protocol has been developed for detection. *See also* broad-screen analysis.
target concentration The chosen concentration in the effluent of a treatment process to designate when the treatment system must be stopped.
target organ A bodily organ in which a chemical produces its effects. In therapeutics, the effect may be a desirable effect, e.g., to alleviate symptoms of disease. In toxicology, the target organ is the organ in which adverse effects are produced.
taste (1) To ascertain the flavor of a substance by taking a portion into the mouth; to perceive or recognize by the sense of taste. (2) The act of tasting. Taste is one of the special senses that perceives and distinguishes the sweet, sour, bitter, or salty quality of a dissolved substance. It is mediated by taste buds on the tongue. Strictly speaking, *taste* refers to the four basic tastes just discussed, whereas *flavor* refers to the blend of taste and smell sensations evoked by a substance in the mouth. *See also* flavor; odor.
taste and odor The combination of sensations perceived by the mouth (i.e., by taste buds on the tongue) and nose. *See also* taste.
taste threshold The minimum concentration of a chemical or biological substance that can just be tasted. *See also* threshold odor; threshold odor number.
TAW *See* Technical Advisory Workgroup.
TBAA *See* tribromoacetic acid.
TBM *See* tribromomethane.
TBS *See* tetrapropylene-benzene sulfonate.
TC *See* total carbon; total coliform.
T_c *See* Kirpich formula.
TCA *See* trichloroacetate; 1,1,1-trichloroethane; 1,1,2-trichloroethane.
TCAA *See* trichloroacetic acid.
TCD *See* thermal conductivity detector.
TCDD *See* 2,3,7,8-tetrachlorodibenzo-*p*-dioxin.
TCE *See* trichloroethylene.
$TCID_{50}$ *See* tissue culture infectious dose 50.
TCLP *See* toxicity characteristic leaching procedure.
TCM *See* total catchment management; trichloromethane.
TCP *See* trichlorophenol.
1,1,1-TCP *See* 1,1,1-trichloropropanone.
TCP–IP *See* transmission control protocol–Internet protocol.
TCR *See* Total Coliform Rule.
tcu *See* true color unit *in the Units of Measure list.*
TDH *See* total dynamic head.
TDS *See* total dissolved solids.
Technical Advisory Workgroup (TAW) A committee of the American Water Works Association's Water Utility Council charged with evaluating and developing comments on one or more drinking water regulatory issues as well as draft and proposed drinking water regulations issued by the US Environmental Protection Agency.
technical assistance A one-to-one approach whereby any trained person or entity provides information and support to water systems in an effort to provide safe, adequate, and palatable drinking water.
technical summary document (TSD) A document prepared by the US Environmental Protection Agency and the US Army Corps of Engineers that summarizes the findings of a field study and literature review of the natural characteristics and functions of a wetland area. Preparation of a technical summary document is the first step in the advance identification process. The technical summary document is provided to the federal, state, and local review agencies for their consideration and recommendations. *See also* advance identification.
technological change A major or sweeping change in knowledge in such fields as applied science and engineering, e.g., the change from vacuum tubes to transistors to integrated circuits or the change from mixed-media filtration to membrane technology.
technological obsolescence A condition in which some thing or process continues to work but in common practice has been

replaced by a newer version. A chemical laboratory that does not use new kinds of analytical instrumentation would suffer from technological obsolescence.

technology Specifically, the branch of knowledge that deals with industrial arts, applied science, engineering, and so on. Technology generally encompasses the sum of the ways in which members of a social group or society provide themselves with the material objects of their civilization.

technology transfer The sharing or transfer of new techniques, processes, or equipment by one agency or country with or to another. Technology can also be transferred from one advanced industrial use to another, e.g., aerospace instrumentation to water quality monitoring.

tee A fitting in the shape of a letter *T* that has an opening perpendicular to the run, allowing a three-way connection.

TEF *See* toxicity equivalency factor.

telemetering The use of devices and attachments for complete measuring, transmitting, receiving, recording, and integrating data from a distance by electrical translating means.

telemetry The use of an electrical link between a transmitter and a receiver. A telephone line commonly serves as the electrical link.

telemetry system A system designed to transmit signals to a more centralized location for control of, and data acquisition from treatment processes and distribution facilities. Typical means of signal transfer include telephone lines, radio waves, and microwaves.

telephone dial-inbound A type of automatic meter-reading system in which meter interface units at customers' premises call into the utility office to provide meter-reading data.

telephone dial-outbound A type of automatic meter-reading system in which a central computer polls meter interface units over the telephone lines, typically on the telephone company's secondary circuit.

television inspection The inspection of inaccessible or hazardous locations by the use of closed-circuit television. This term usually applies to truck-mounted, semiportable equipment used to inspect inside buried water lines.

telltale (1) A water-level indicator installed in a reservoir. (2) A small overflow pipe that indicates, by dripping, when a tank or cistern is full.

TEM *See* transmission electron microscopy.

temperature (1) The thermal state of a substance with respect to its ability to communicate heat to its environment. (2) A measure of the thermal state on some arbitrarily chosen numerical scale, e.g., Celsius, kelvin, or Fahrenheit. *See also* heat.

temperature, absolute *See* absolute temperature.

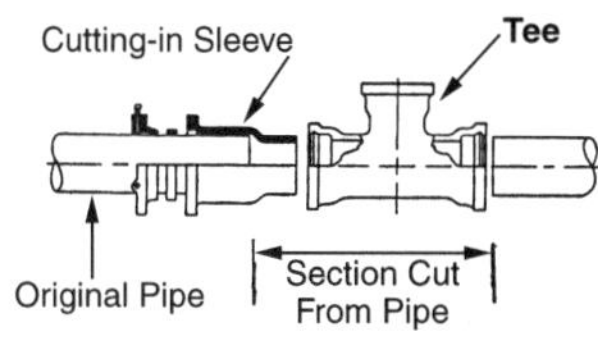

A tee being installed

temperature approach The lowest temperature difference between the two fluids within a discrete-surface heat exchanger or between air and water in a cooling tower.

temperature correction factor For some pressure-driven membrane processes, a dimensionless coefficient accounting for effects of process fluid temperature variations. Such a factor can be used to calculate permeate flow rate at a given temperature by multiplication with permeate flow rate at a standard temperature.

temperature degree A unit on a scale to measure the hotness or coldness of anything, usually measured on a thermometer.

temperature sensor A device that opens and closes a switch in response to changes in temperature. This device might be a metal contact, a thermocouple that generates a small electrical current proportional to the difference in heat, or a variable resistor for which the value changes in response to changes in temperature. A temperature sensor is also called a heat sensor.

temperature units In the US customary system, degrees Celsius for chemistry and degrees Fahrenheit for engineering. In Système International, kelvin.

temperature-driven membrane A hydrophobic membrane that uses a thermal gradient across the membrane as the driving force. A portion of the liquid on the feed side vaporizes and passes through the membrane to the permeate side, where it condenses. Such a membrane can be used for water desalting, in which case the process is called membrane distillation.

temporal Transitory; lasting only for a time; transient.

temporal variation Change in some condition (e.g., water quality) that occurs over time.

temporarily absorbed water Water that reaches only the belt of soil water or the intermediate saturation zone, from which it escapes either by evaporation from film surfaces or by root absorption and transpiration from vegetation.

temporary dam A dam constructed of readily available materials and equipment in such a manner as to be nonpermanent and to perform as a cofferdam.

temporary hardness *See* carbonate hardness.

temporary partial disability An injury that does not result in death or permanent injury yet allows the employee to temporarily perform alternate duties that accommodate the disability until the employee is able to return to normal duties.

temporary spring A spring in which a discharge occurs only during certain periods and ceases at other periods. Such a spring is also called an intermittent spring.

temporary total disability An injury that does not result in death or permanent injury yet prevents the employee from performing duties. A full day of work after the day of the injury must be missed to fall under this Occupational Safety and Health Administration classification.

Ten State Standards A reference manual providing guidance on the design of water treatment plants, prepared under the supervision of representatives from 10 states, including Illinois, Indiana, Iowa, Michigan, Minnesota, Missouri, New York, Ohio, Pennsylvania, and Wisconsin, as well as the province of Ontario, Canada.

tender-option bond A bond that can be redeemed by the bondholder on a date or dates prior to the stated maturity date. Such a bond is also known as a demand bond.

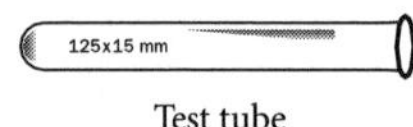

Test tube

tensiometer A term used for two distinctly different devices: (1) A device used to measure the soil moisture pressure (tension suction) in the unsaturated zone. (2) A device to measure the interfacial (surface) tension of a liquid. *See also* surface tension; water tension.

tension (1) The French word for *voltage*. This French term is often found in print materials dealing with ozonation technology. (2) In English, a term usually used to mean high voltage, as in a high-tension transformer or high-tension lines. (3) The degree of stiffness or tautness.

TEQ (toxicological equivalency) *See* hazard index.

tera (T) A prefix meaning 10^{12} in Système International.

teratogen A chemical that is capable of producing structural abnormalities of the organs and skeleton that are observed at birth. *See also* carcinogen; developmental toxicities; mutagen; teratogenic effect.

teratogenesis The process of producing structural abnormalities of the skeleton or soft tissues during development.

teratogenic effect The inducement of malformations in utero. This effect is included in developmental toxicities. *See also* developmental toxicities; teratogen.

teratogenicity The capacity of a physical or chemical agent to cause teratogenesis in offspring. *See also* teratogenesis.

teratology The study of physical and chemical agents that induce developmental toxicities.

terbacil ($C_9H_{13}O_2N_2Cl$) The common name for the herbicide 3-*tert*-butyl-5-chloro-6-methyluracil. *See also* herbicide.

terbufos ($(CH_3CH_2O)_2P(S)\text{-}S\text{-}CH_2\text{-}S\text{-}C(CH_3)_3$) The common name for the soil insecticide and nematicide *S*-[[(1,1-dimethylethyl)thio]methyl]-*O*,*O*-diethyl phosphorodithioate. *See also* insecticide; nematicide.

term bond A bond with the entire principal maturing on one date.

terminal disinfection by-product concentration (DBP_t) The concentration of disinfection by-products present at the end of a disinfection by-product formation potential test at time *t*. This value equals the sum of the initial amount of disinfection by-products (i.e., the concentration present at the start of the test) and the disinfection by-product formation potential of the sample. *See also* disinfection by-product formation potential.

terminal half-life The time obtained by pharmacokinetic analyses for the slowest elimination half-life for a chemical or drug in the body. It is usually measured in the blood, but it can also be deduced from excretion rates into the urine or other body fluids (e.g., saliva). The terminal half-life is based on removal of the drug from the compartment for which the drug has the highest affinity. For example, highly fat-soluble compounds accumulated in fat deposits within the body are removed very slowly from that compartment; this removal would eventually dictate blood levels of the compound if the external exposure were removed. *See also* half-life; pharmacokinetics.

terminal head loss The head loss in a filter at which water can no longer be filtered at the desired rate because of the suspended matter filling the voids in the filter and greatly increasing the resistance to flow.

terminal settling velocity The maximum rate of sedimentation of a suspended particle that is not hindered in any way. *See also* Stokes's law.

terminal temperature The maximum temperature attained by the feedwater in a distillation device.

terminal total trihalomethane concentration ($TTHM_t$) The arithmetic sum of trihalomethane concentrations present at the end of a trihalomethane formation potential test at time *t*. This value equals the sum of the initial amount of total trihalomethanes (i.e., the concentration present at the start of the test) and the total trihalomethane formation potential of the sample. *See also* trihalomethane formation potential.

terminal trihalomethane concentration (THM_t) The concentration of a specific trihalomethane present at the end of a trihalomethane formation potential test at time *t*. This value equals the sum of the initial amount of the trihalomethane (i.e., the concentration present at the start of the test) and the trihalomethane formation potential of the sample. *See also* trihalomethane formation potential.

terrace (1) A flat, level (or nearly level) area of land bounded on at least one side by a definite steep slope rising upward from it and on the other sides by downward slopes. (2) A low embankment or ridge of earth constructed across a slope to control surface runoff and minimize soil erosion.

terrace spring A spring that has built up at its mouth a series of terraces or basins through deposition of material carried out of the earth in solution.

terrain conductivity curve A graph produced following surface geophysical measurement of the conductivity of earth materials by the use of electromagnetic waves. These measurements are especially useful for locating gravel, saline intrusions, cavities in the subsurface, contaminated plumes of groundwater, and bedrock topography.

terrestrial Growing, living, or peculiar to the land, as opposed to the aquatic environment.

tesla (T) *See the Units of Measure list.*

test In chemistry, a laboratory procedure used to determine, e.g., the concentration of constituents in water.

test pit An open pit dug to permit the examination of the underlying material for suitability as a foundation or to test the current condition of buried water system facilities, such as valves and pipe.

test tube A slender glass or plastic tube with an open top and rounded bottom. Test tubes are used for a variety of tests.

test year The annualized period for which costs are to be analyzed and rates established.

tested well capacity The maximum rate at which a well is known to have yielded water without appreciable increase in drawdown. If the well has been tested with the water level drawn down to the intake, the tested capacity is the total capacity.

testicle The organ in which sperm are produced. This organ also has a number of secretory functions that are, at least in part, responsible for secondary sex characteristics, including behavior that would be examined in a full evaluation of testicular function. A variety of testing procedures are applied to assess the effects of chemicals on male reproductive function.

testing A process of examining something for the purpose of gaining further knowledge about it.

2,3,7,8-tetrachlorodibenzo-*p*-dioxin (TCDD) ($C_{12}H_4O_2Cl_4$) A chlorinated hydrocarbon that occurs as an impurity in the herbicide 2-(2,4,5-trichlorophenoxy)propionic acid. In addition, other environmental sources for this contaminant exist. 2,3,7,8-tetrachlorodibenzo-*p*-dioxin is highly toxic and persistent. It is regulated by the US Environmental Protection Agency.

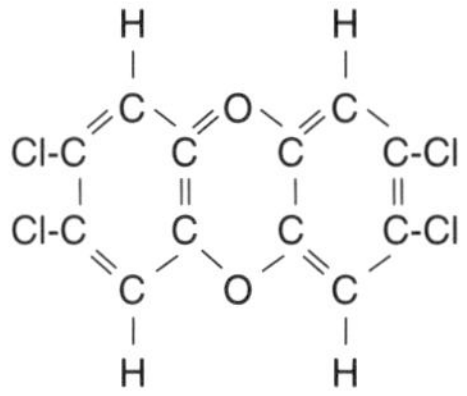

```
        H           H
        |           |
     //C\   //O\   //C\
Cl-C      C      C      C-Cl
   |      ||     |      ||
Cl-C      C      C      C-Cl
     \\C/   \O/   \\C/
        |           |
        H           H
```

2,3,7,8-tetrachlorodibenzo-*p*-dioxin

1,1,2,2-tetrachloroethane ($CHCl_2CHCl_2$) A volatile organic compound with various industrial uses, e.g., as a solvent. *See also* solvent; volatile organic compound.

tetrachloroethylene ($Cl_2C{=}CCl_2$) The chemical name for perchloroethylene, a volatile organic compound used as a dry-cleaning or vapor-degreasing solvent. Tetrachloroethylene is regulated by the US Environmental Protection Agency. *See also* solvent; volatile organic compound.

tetrapropylene-benzene sulfonate (TBS) ($RC_6H_4SO_3^-$) A petrochemical product belonging to the class of synthetic detergents known as alkyl aryl sulfonates. It is one of the most useful surface-active products developed.

tetrasomy A condition in which a cell contains two additional chromosomes of a particular type in an otherwise diploid cell. Whereas the normal diploid cell has two of each type of chromosome, a tetrasomic cell has an additional pair of one of the chromosomes.

texture The disposition or manner of arrangement of the particles or small constituents of a body or substance.

TFC (thin film composite) membrane *See* composite membrane.

TH *See* total hardness.

thallium (Tl) A metallic element with miscellaneous industrial uses, including in mercury alloys, rodenticides, and photoelectric applications. Thallium is regulated by the US Environmental Protection Agency. *See also* rodenticide.

thalweg (1) The line following the lowest part of a valley, whether underwater or not. The thalweg is usually the line following the deepest part or middle of the bed or channel of a river. (2) The middle or chief navigable channel of a waterway; the thread of a stream.

thatch The buildup of organic material at the base of turfgrass leaf blades. Thatch repels water and reduces infiltration capacity.

thematic mapping Portrayal of some geographic variable or "theme," such as land use, geology, or population distribution, by shading or symbolizing areas or map features to represent the value or class of that theme.

theoretical discharge The discharge over a weir or through an orifice, tube, or pipeline as computed from such variables as cross-sectional area, head, and slope, with a corrective factor or coefficient applied to account for losses due to frictional resistance, contractions, and similar occurrences.

theoretical model A theoretical representation of the working principles of a system. *See also* mathematical model; system.

theoretical pump displacement A pump manufacturer's published rating of a pump, expressed in units of gallons per minute (cubic metres per second) per 100 revolutions per minute.

theoretical velocity The velocity that water or another liquid under a given head would attain in passing through an orifice, conduit, or other structure if its flow were not reduced by friction or other losses.

therapeutic index The ratio of the dose required to produce a toxic or lethal effect to the dose required to produce a nonadverse or therapeutic response.

thermal conductivity A measure of the ability of a substance to conduct heat. Mathematically, thermal conductivity is the ratio of the rate of heat flow to the rate of temperature change in the particular substance.

thermal conductivity detector (TCD) A device used as a nonselective detector (i.e., one that responds to almost all compounds) in gas chromatography. Such a device operates based on a comparison of resistance values resulting from changes in the thermal conductivity of a heated source. As analytes in the eluant flow past the heated source, changes in the thermal conductivity are detected. Although not as sensitive as many other gas chromatographic detectors, this device is relatively rugged and simple to operate.

thermal conductivity units In the US customary system, British thermal units per hour per degree Fahrenheit per foot. In Système International, watts per metre per kelvin; metre-kilograms per second cubed per kelvin.

thermal log A type of geophysical log that measures the temperature of groundwater. Flow paths can be inferred from this type of log, as can groundwater velocities. *See also* geophysical log.

thermal ozone destructor A unit in an ozonation system that employs high temperature to destroy excess ozone (O_3).

thermal power Any type of energy or power generated or developed through the use of heat energy. In common practice, thermal power is electrical energy or power generated by the use of steam engines or turbines to drive electric generators.

thermal regeneration A process using combustion in the absence of oxygen for regenerating, or reactivating, the surface area of granular activated carbon to restore the activated carbon's adsorptive properties. The combustion typically takes place in two stages: the first, at a lower temperature, to drive off adsorbed organics, and the second, at a higher temperature, to regenerate the surface. *See also* granular activated carbon; reactivation; regeneration.

thermal spring A spring that discharges water having a temperature in excess of the normal local groundwater. Hot springs and warm springs are types of thermal springs. The water in a thermal spring is usually highly mineralized.

thermal stratification The formation of layers of different temperatures in bodies of water as a result of water density differences. *See also* destratification; epilimnion; hypolimnion; metalimnion; thermocline.

thermal transfer plotter A peripheral device that makes graphic plots of computer data using a heating element to melt dots of waxy ink from a film substrate to a specially treated paper. Most thermal transfers require three or four passes to plot a color image. One primary color or black is plotted on each pass.

thermal-convection storm A storm, caused by local inequalities in the temperature, in which the rainfall is intense, short-lived, and limited to only a small area.

thermistor A thermally sensitive electrical resistor.

thermocline In a thermally stratified lake, the layer below the epilimnion and above the hypolimnion. The thermocline is the stratum in which a rapid rate of decrease in temperature occurs with depth. *See also* destratification; epilimnion; hypolimnion; metalimnion; thermal stratification.

thermocouple A heat-sensing device made of two conductors of different metals joined at their ends. An electric current is produced when a difference in temperature occurs between the ends.

thermoelectric power Electricity produced by heat when water is used for steam production.

thermograph An instrument designed to make an automatic record of temperature. The thermometric element is, most commonly, either a bimetallic strip or a metal tube filled with a liquid. In all types, the pen writes the record of temperature on a ruled sheet that is wrapped around a cylinder revolved by enclosed clockwork; from the trace thus made, the instantaneous behavior of the temperature may be determined.

thermophone An instrument for determining temperatures at different depths in a reservoir, in masonry dams, or in other structures.

thermoplastic Any of numerous materials, such as certain synthetic resins and plastics, that soften or fuse when heated and harden and become rigid when cooled, and that can usually be remelted and cooled time after time with no appreciable chemical change.

thermoplastic pipe casing Plastic pipe used as riser pipe in well construction. Thermoplastic pipe casing is routinely used in all waters, but it is especially useful in highly corrosive waters. It is composed of polyvinyl chloride, acrylonitrile butadiene styrene, and rubber-modified polystyrene.

thermoset Any of numerous plastics and synthetic resins that, once solidified, will not resoften or fuse when heated. Thermoset materials may decompose at high temperatures but will not soften or melt.

theta (θ) *See* phase angle.

thickener Any process equipment or process that, after gravity sedimentation, increases the concentration of solids in sludges with or without the use of chemical flocculants. *See also* sludge thickener.

Thiessen polygon A geometric construction for determining the zone within which data taken at a rain-gauge station are applicable in a network of gauging stations. The points of location of rain gauges on a map are joined by straight lines and the lines' perpendicular bisectors are drawn, forming a polygon around each rain-gauge station.

thin film composite (TFC) membrane *See* composite membrane.

thin-layer chromatography A chemical separation technique in which the stationary phase is supported on a glass plate. Analytes are separated in two dimensions by partitioning between a liquid mobile phase and a solid stationary phase. *See also* paper chromatography.

Thiobacillus A genus of small, single-celled bacteria that can create gaseous hydrogen sulfide (H_2S) and the resulting "rotten egg" odor in water supplies.

thiram ($((CH_3)_2NCS)_2S_2$) The common name for tetramethylthiuram disulfide. It is used as a fungicide, seed protectant, and animal repellent. Thiram has been shown to react with nitrite (NO_2^-) under acidic conditions to produce the carcinogen *N*-nitrosodimethylamine ($(CH_3)_2N_2O$). *See also* fungicide; *N*-nitrosodimethylamine.

THM *See* trihalomethane.

THM precursor *See* trihalomethane precursor.

THMFP *See* trihalomethane formation potential.

THM_t *See* terminal trihalomethane concentration.

thread of a stream The line equidistant from both of a stream's edges at the ordinary level of the water in the stream. In some cases, this term has been construed to mean the median line of the main channel of the stream. *See also* thalweg.

thread test A test for measuring the cohesiveness of soil. Soil that can be rolled to a thread that is 2 inches (5 centimetres) long and ⅛ inch (3 millimetres) in diameter without tearing is considered cohesive. *See also* ribbon test.

3-chloro-4(dichloromethyl)-5-hydroxy-2(5H)-furanone *See under* chloro-4(dichloromethyl)-5-hydroxy-2(5H)-furanone.

3-log removal 99.9 percent removal or inactivation for *Giardia* cysts, as required by the Surface Water Treatment Rule under the US Environmental Protection Agency's Safe Drinking Water Act. *See also* log removal.

303d listing A list called for by Section 303(d) of the Clean Water Act, which requires states to identify and develop a priority listing of waters in the state that are not "fishable, swimmable" and to develop total maximum daily loads for them, with oversight from the US Environmental Protection Agency. *See also* total maximum daily load.

three-phase circuit A system of alternating current supply comprising three or four conductors. For a sinusoidal symmetrical and balanced three-phase circuit, the same magnitude of root-mean-square voltage exists between any two conductors.

three-phase power Alternating current power in which the current flow reaches three peaks in each direction per cycle.

threshold The dose rate or other stimulus level at which a biological response first occurs. For example, a nerve membrane must undergo a certain amount of depolarization before it will initiate an action potential (i.e., a nerve impulse). In toxicology and risk assessment, certain types of adverse effects are thought to have thresholds, i.e., doses beneath which the adverse effect will not be produced.

threshold limit value (TLV) An exposure limit established for chemicals by the American Conference of Governmental

Industrial Hygienists. These values are the most widely accepted limits for chemicals encountered in the workplace.

threshold odor The minimum odor of a water sample that can just be detected after successive dilutions with odorless water. The threshold odor is also called the odor threshold.

threshold odor concentration *See* odor threshold concentration.

threshold odor number (TON) A numerical value based on the greatest dilution of a sample with odor-free water that yields a definitely perceptible odor. For example, if odor were perceptible in 1:1, 2:1, and 3:1 dilutions of a sample with odor-free water, but not in a 4:1 dilution, the threshold odor number would be 3. *See also* odor threshold concentration.

threshold substance *See* trace substance.

threshold treatment Prevention of the deposition of calcium carbonate ($CaCO_3$) in water, after softening treatment, by the addition of a small quantity of sodium hexametaphosphate. *See also* sodium hexametaphosphate.

thrombosis The formation of a clot of blood factors, mainly platelets and fibrin, that causes vascular obstruction.

throttle valve A valve installed to reduce the flow through a pipeline. A throttle valve may be any one of a number of types of valves designed for this purpose.

throughfall In a vegetated area, the precipitation that falls directly to the ground or the rainwater or snowmelt that drops from twigs or leaves. *See also* stemflow.

throughput volume The volume of water passed through an ion exchange resin bed or water treatment system before the exchanger or system reaches exhaustion.

thrust A force resulting from water under pressure and in motion. Thrust pushes against fittings, valves, and hydrants and can cause couplings to leak or entirely separate.

thrust anchor A block of concrete, often a roughly shaped cube, cast in place between a fitting and the undisturbed soil at the side or bottom of the pipe trench and tied to the fitting with anchor rods. A thrust anchor resists vertical thrust.

thrust block *See* thrust anchor.

thumb penetration A test that can be used to estimate the unconfined compressive strength of cohesive soils. Type A soils can be indented by the thumb with great effort. Type C soils can easily be penetrated several inches by the thumb and can be molded by light finger pressure. This test should be performed on undisturbed soil as soon as possible to prevent drying of the sample.

thunderstorm A local and short-lived atmospheric disturbance accompanied by electrical phenomena and heavy showery precipitation, often with gusts of wind and sometimes with hail. *See also* thermal-convection storm.

thyroid nodule A focal area of hyperplasia within the thyroid. Some of these lesions have the potential for neoplastic transformation (i.e., tumor development).

TIC *See* total inorganic carbon.

tidal Pertaining to periodic water level fluctuations due to the action of lunar (moon) and solar (sun) forces upon the rotating Earth.

tidal bore A wave of water having a nearly vertical front advancing upstream as a result of high tides in certain estuaries. *See also* bore.

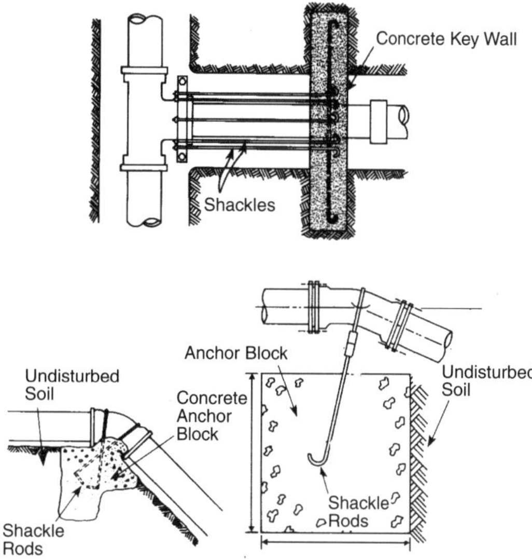

Various types of thrust anchors

tidal current A current brought about or caused by tidal forces.

tidal day A time interval between tides; 24.84 hours. This term should not be confused with *tidal interval. See also* tidal interval.

tidal efficiency A barometric efficiency of an aquifer, expressed as the ratio of the groundwater elevation change in an aquifer to the change in oceanic tide elevation. This value can be used to estimate water reserves and hydraulic performance of wells in aquifers that have oceanic suboutcrops.

tidal flat *See* tidal marsh.

tidal flood interval The time between the transit of the moon over the meridian of a place and the time of the following tidal flood.

tidal flood strength The tidal flood current at the time of maximum velocity.

tidal inlet A transverse channel through a barrier bar or beach that permits seawater to sweep in and out of a lagoon.

tidal interval The average usual interval of time between tides. For diurnal tides, this value amounts to 12.42 hours from high water to high water. *See also* tidal day.

tidal marsh A low, flat marshland traversed by interlacing channels and tidal sloughs and usually inundated by tides. This term is usually applied to marshlands with brackish or salt water.

tidal range The height difference between consecutive low and high tide levels.

tidal river A river in which flow and water surface elevation are affected by the tides. Such effects usually occur in the lower stretch near the mouth, where the gradient is very flat. In some streams, the effects may extend 100 miles (160 kilometres) or more upstream from the mouth.

tidal table A table published by the US Coast and Geodetic Survey that gives daily predictions of the times and heights of the tide. These predictions are usually supplemented by tidal differences and constants by means of which additional predictions can be made for numerous other places.

tidal water The water that rises and falls in a predictable and measurable rhythm or cycle as a result of the gravitational pulls of the moon and sun. Tidal waters end where the rise and fall of the water surface can no longer be practically measured in a predictable rhythm because of masking hydrologic, wind, or other effects.

tidal water level The altitude reached by a tidal surface, usually fitting in one of the following categories:

- High water (HW): The highest level reached by a rising tide.
- Higher high water (HHW): The higher of the two high waters of any lunar day. The single high water occurring daily during periods when the tide is diurnal is considered a higher high water.
- Higher low water (HLW): The higher of the two low waters of any lunar day.
- Low water (LW): The lowest level reached by a falling tide.
- Lower high water (LHW): The lower of the two high waters of any lunar day.
- Lower low water (LLW): The lower of the two low waters of any lunar day. The single low water occurring daily during periods when the tide is diurnal is considered a lower low water.
- Mean high water (MHW): The mean level of high waters over a long period. When a tide is semidiurnal or mixed, only the higher high water values are included. In this case, mean high water is the same as mean higher high water.
- Mean higher high water (MHHW): The mean level of the higher high waters over a long period.
- Mean low water (MLW): The mean level of the low waters over a long period. When the tide is semidiurnal or mixed, only the lower low water values are included. In this case, mean low water is the same as mean lower low water.
- Mean lower low water (MLLW): The mean level of lower low water over a long period.

tidal wave (1) The general rise and fall of the surface of the ocean that is generated by gravity effects of the sun and moon and that travels around the Earth each day. (2) An extremely large wave caused by a seismic disturbance or a great storm, often causing overflow of low-lying lands not usually inundated by ordinary wave or tidal action. *See also* bore; tsunami.

tidal-current curve A graphical representation of the flow of tidal current. Time is shown on the abscissa and velocity of the current is shown on the ordinate, with the flood velocity considered positive and the ebb velocity negative. In general, a tidal-current curve approximates a cosine curve.

tide (1) The periodic rising and falling of water that results from the gravitational attraction of the moon and sun acting on the rotating Earth. This term should not be used to designate the horizontal movement of the water; the correct term in that case is *tidal current.* (2) A flood in a stream. This meaning appears in colloquial usage in the southeastern United States. *See also* tidal current.

tide epoch The phase or time of occurrence of high water, calculated from a fixed time origin.

tide gauge (1) A staff gauge that indicates the height of the tide. (2) An instrument that automatically registers the rise and fall of the tide. In some instruments, the registration is accomplished by printing the heights at regular time intervals; in others by a continuous graph in which the height of the tide is shown on the ordinate of the curve and the corresponding time on the abscissa.

tide generator A device for generating tides in hydraulic models.

tideland (1) Land subject to overflow by water during the higher stages of the tide. (2) A coastal area situated above the mean low tide and below the mean high tide, alternately uncovered and covered by the ebb and flow of ordinary daily tides. *See also* tidal water level.

Tier 1 violation A violation of a National Primary Drinking Water Regulation that is considered more serious so has more extensive public notification requirements than a Tier 2 violation. Exceeding a maximum contaminant level, violating a treatment technique requirement, and violating a variance or exemption schedule are Tier 1 violations, which are further classified as acute or nonacute. Acute violations involve an acute risk to human health and have the most extensive public notification requirements, including television and radio announcements. *See also* acute violation.

Tier 2 form The annual reporting form for the Superfund Amendments and Reauthorization Act Title III for hazardous chemicals. The report form includes the following information: owner's name and address, location of facility, chemical descriptions, Chemical Abstracts Service number, physical and health hazards, quantities, types of storage and locations, certification and verification by authorized representative or owner. This form, along with any fees, must be returned by March 1 of each year.

Tier 2 violation A violation of a National Primary Drinking Water Regulation that is considered less serious than a Tier 1 violation. Violating a monitoring requirement, not using an approved test procedure, and issuing a variance or exemption are Tier 2 violations, and all must be reported to the public.

tiered rate *See* inclining block rate.

TIGER system *See* Topological Integrated Geographic Encoding and Referencing system.

tight junction A joining of cells so tightly together that chemicals must penetrate cell membranes to pass through the structure. For example, the endothelial cells that make up the capillary walls in the brain have tight junctions and so form part of the blood–brain barrier. Tight junctions in the epithelium of the skin also limit the absorption of polar material or high-molecular-weight materials through this organ. Similar junctions are found in the intestinal epithelium.

tile ceramic pipe.

tile drainage The removal of surplus groundwater by means of buried pipes. Water enters through the unsealed joints or through perforations in the pipe. Water sometimes enters the drain tile through surface inlets.

tile field A system of open-jointed tile, usually laid on a rock fill, used for dispersing wastewater effluents into the ground. *See also* subsurface irrigation.

tile filter bottom A type of underdrainage system placed under a filter bed to provide a collecting system for the filtered water. *See also* filter bottom.

tile underdrainage A system of tile drains laid in covered trenches underground, in most cases with open joints, to collect and carry off excess groundwater.

tillage Plowing, seedbed preparation, and cultivation practices.

tilting dam An overflow dam constructed in such a way that a top section will tilt and allow the passage of excess water. Sections are set to tilt at various elevations of the impounded water. That is, the sections that tilt are set at different elevations; one tilts at a time, depending on the desired level at which water should start spilling.

tilting gate A crest gate, hinged at the top or bottom, that (1) automatically tilts or drops as water rises and flows over it as a result of an increase in discharge, and (2) resumes its original position when the flow is reduced.

timber dam A dam constructed of framed timbers. The dam may be of cellular or crib construction, filled with rock to give it stability; or it may attain such stability through its design.

timber shoring system Wood members of the shoring system. Minimum sizes of members are specified for use in different types of soil. Members that are to be selected on the basis of this information are the cross braces, the uprights, and the wales (horizontal rails), where wales are required. Six tables of information are available, two for each soil type.

time composite A composite sample consisting of several equal-volume samples taken at specified times. *See also* composite sample.

time of concentration (T_c) *See* Kirpich formula.

time to filter The time required in a capillary suction test to filter a fixed volume of sludge suspension through a porous membrane. The time is an indication of the filterability of the sludge sample. *See also* capillary suction test.

time of flow The time required for water flowing in a stream to travel from a given point to some other downstream point. *See also* travel time.

time lag (1) The time elapsing between the occurrence of corresponding changes in discharge or water level at two points in a river. (2) The time between the beginning, or center of mass, of rainfall to the peak, or center of mass, of runoff. (3) The time between the beginning of snow melt and the start of the resulting runoff. (4) The time required for processes and control systems to respond to a signal or to reach a desired level. (5) The time from the start of a study of biological growth until a biological culture begins to grow. *See also* time of flow.

time management A systematic method of allocating the time required to perform specific operations or tasks or to achieve a specific goal, with a view toward finding more efficient methods.

time series A data set collected over a long period of time at (usually) equal time intervals.

time series analysis In statistics, the use of a body of analytical methods for data collected over a period of time. Annual, quarterly, monthly, or daily data are forms of time series data. Such data contrast with data related to a single point in time, such as the diameter of a tree related to its height or the amount of water use per day for a number of families based on household size. Time series data are generally composed of seasonal patterns, cycles, trends, and irregular movements (random errors). Seasonal patterns are identified with seasonal indexes, from which cycles are isolated by means of moving averages and regression analysis; trends are most frequently identified by means of regression analysis. Regression analysis, exponential smoothing, and autoregressive integrated moving-average methods (such as the Box–Jenkins method) may be used to forecast time series data. *See also* autoregressive integrated moving-average forecasting methods.

time of travel The determination, usually by modeling, of the time in years for groundwater recharge to travel from a certain field point to the wellhead.

time units In the US customary system or Système International, seconds.

time-of-flight mass spectrometer A type of mass spectrometer in which the ions are separated based on their velocity through a drift tube portion of an instrument. Ions entering the drift tube have the same kinetic energy; therefore, the velocities of different ions depend on their masses. Because velocity through the instrument is inversely proportional to the ion's mass, lighter ions reach the end of the detector faster than heavier ions. Mass is calculated by measuring the time it takes for ions of a given mass to travel the distance of the drift tube. These instruments are often larger than other popular types of mass spectrometers. Because of the very fast acquisition of data within time-of-flight instruments, these devices can be useful for the analysis of short-lived chemical species. *See also* mass spectrometer.

timer A device for automatically starting or stopping a machine or other device at a given time.

time-weighted average (TWA) The average value of a parameter (e.g., the concentration of a chemical in air) that varies over time.

time-weighted average (TWA) concentration A concentration of a toxic airborne material, weighted typically for an 8-hour exposure period.

tissue A group of similar cells in the body.

tissue culture Growth of tissue cells in vitro on artificial media for experimental research.

tissue culture infectious dose 50 ($TCID_{50}$) The concentration of virus at which 50 percent of the cells in a tissue culture become infected.

titer (1) The concentration of a substance in solution as determined by titration. (2) The concentration of a substance (chemical or microbial) based on a series of dilutions to a defined end point.

titrate To add a chemical solution of known strength to a sample drop-by-drop until a certain color change, precipitate, or pH change in the sample is observed (i.e., until the end point is reached).

titration A process of adding known volumes of a reagent solution to a sample. Specific chemical reactions are monitored until an end point is reached. Titrations are commonly used in water analyses for determining hardness, alkalinity, chlorine residual, and many other parameters.

titrimetric method Any laboratory procedure that uses titration to determine the concentration of a constituent in water.

TKN *See* total Kjeldahl nitrogen.

TLV *See* threshold limit value.

TNCWS *See* transient, noncommunity water system.

TOBr *See* total organic bromine.
TOC *See* total organic carbon.
TOCl *See* total organic chlorine.
TOD *See* total oxygen demand.
toe (1) The downstream edge at the base of a dam. (2) The line where a natural or fill slope intersects the natural ground. (3) The lowest edge of a backslope of a cut where it intersects the roadbed or bench.
toilet dam A flexible insert placed across the bottom of a toilet's reservoir tank to hold back water when the toilet is flushed, thereby not using the entire volume of water in the reservoir for each toilet flush. Use of a toilet dam is a water conservation technique.
tolerance An organism's capacity for enduring or being less responsive (i.e., adapting) to the influence of a drug or poison, particularly when such capacity is acquired by continued use of a substance or exposure to unfavorable environmental factors.
***ortho*-tolidine ($(C_6H_3(CH_3)NH_2)_2$)** *See ortho*-tolidine arsenite test; *ortho*-tolidine reagent; *ortho*-tolidine test.
***ortho*-tolidine arsenite (OTA) test** An approximate technique, largely obsolete, for differentiating between free available chlorine, combined available chlorine, and color caused by interfering substances. It involves using *ortho*-tolidine reagent, sodium arsenite ($NaAsO_2$) reagent, and colorimetric standards.
***ortho*-tolidine reagent** A solution prepared from *ortho*-tolidine crystals and used to determine colorimetrically the concentration of available chlorine.
***ortho*-tolidine (OT) test** An approximate technique for determining residual chlorine in water by using *ortho*-tolidine reagent and colorimetric standards. This test is used for routine measurement; however, its accuracy is affected by interfering substances and color, making the test largely obsolete.
toluene ($C_6H_5CH_3$) A common name for methylbenzene, a volatile organic compound with miscellaneous industrial applications, including use as an admixture in aviation gasoline; as a solvent, diluent, or thinner; and in explosives (i.e., trinitrotoluene, or TNT ($CH_3C_6H_2(NO_2)_3$)). Toluene is regulated by the US Environmental Protection Agency. *See also* solvent; volatile organic compound.
TON *See* threshold odor number; total organic nitrogen.
ton *See the Units of Measure list.*
tonne *See the Units of Measure list.*
too numerous to count An expression used to indicate that the total number of bacterial colonies exceeds 200 on a 47-micrometre pore diameter membrane filter used for coliform detection.
top contraction (1) An abrupt or gradual drop in the roof of a conduit. (2) The extent of the drop in the water surface caused by a constriction or contraction.
topographic divide The line that follows the ridges or summits forming the exterior boundary of a drainage basin and that separates one drainage basin from another. A topographic divide is also called a drainage divide.
topographic feature A map feature that defines elevation information. Such features are often called third-dimension values or *z* values. Topographic data are usually compiled from aerial photographs at relatively large scales and at high levels of positional accuracy. The features recorded typically include contour lines, point elevations, and elevation values.

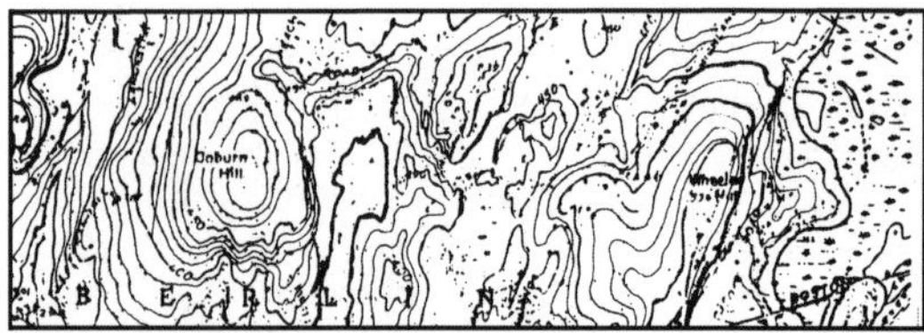

Topographic map

topographic geology The branch of geology that relates the shape of the land surface to the characteristics of the underlying soil and rocks.
topographic map A map that shows the elevation of the land in a specified area.
topographic mapping The process of delineating on maps the position and elevation of natural and constructed features of a place or region.
topography The configuration of a surface, including its relief and the position of its natural and artificial features.
topologic Pertaining to a geographic data structure in which the inherent spatial connectivity and adjacency relationships of the geographic features are implicitly stored and maintained.
Topological Integrated Geographic Encoding and Referencing (TIGER) line file A record for each line or feature segment (road, railroad, hydrographic feature, and so forth). The file is a flat American Standard Code for Information Interchange file, meaning that it is a simple text file containing one record for each line segment. The format actually contains several different files, each with a different record type, linked to the line segment by a record identification code. *See also* Topological Integrated Geographic Encoding and Referencing system.
Topological Integrated Geographic Encoding and Referencing (TIGER) system A major digital database that was developed by the Bureau of the Census and the US Geological Service in preparation for the 1990 Decennial Census of the United States. The system is the first comprehensive digital street map of the United States, containing digital data at a scale of 1:100,000 for every street and road in the nation. The system also contains the range of address numbers located along each section of every street in the 345 largest urban areas, all the railroads and names of their operating companies, all significant hydrographic features and their associated names, and other information. The system's line file is a flat American Standard Code for Information Interchange file that contains a record for each street segment or other feature. The segments make up the census geographical areas, such as census tracts. The system's line file can be used to construct two forms of data in a geographic information system: a logical street center-line network with address ranges, and census area polygons for demographic analysis. *See also* geographic information system; Topological Integrated Geographic Encoding and Referencing line file.
topological structure The collective properties that define the relative positioning of data elements in space with respect to

one another but without reference to actual distances. These properties are used (e.g., in a geographic information system) to define such relationships as polygon boundary segments, connectivity between points, networks, and area adjacencies. *See also* geographic information system.

topology (1) A branch of mathematics that deals with non-metric spatial relationships, such as connectedness and adjacency. (2) In a geographic information system, the data that deal with topological relationships among point, line, and area objects (e.g., identifying which streets meet at an intersection, which streets bound a block, or which blocks front on a street). *See also* geographic information system.

topping Work performed to raise the level of the top of a levee. The results of such work are usually intended to be only temporary. However, topping is meant to create permanent results to increase flood protection when levees are below grade and a threat of overtopping exists.

topsoil The soil layer moved during cultivation.

torpedo sinker A type of weight attached to the bottom of a log line for observing depths in rapid streams.

torr *See the Units of Measure list.*

tortuosity The actual length of a sinuous flow path in a porous medium divided by the distance between the two ends of the path.

tortuous flow A type of fluid flow in which an unsteady motion of the particles occurs in a tortuous or meandering channel. *See also* turbulent flow.

tortuous path (1) A water flow path through channels that are constricted and marked by repeated twists, bends, and winding turns. (2) In an electrodialysis system, a water flow path in which spacers, turbulence promoters, or cross-traps are used to produce turbulence in the flow stream. *See also* turbulence.

total acidity The total of all forms of acidity in a solution, including mineral acidity, carbon dioxide (CO_2), and acid salts. *See also* acidity.

total alkalinity *See* alkalinity.

total available chlorine The total chlorine, present as chloramines, other chlorine-containing species, or both that is present in a water and is still available for disinfection, oxidation of organic matter, or both. Note that the *total chlorine* may include organic chloramines that do not have disinfection capabilities; such species are not considered part of the total available chlorine. In most drinking water applications, the total chlorine and total available chlorine are quite similar; in chlorinated wastewater, however, organic chloramines may represent a significant fraction of the total chlorine. Thus, for drinking water, the more commonly used terms are *free available chlorine* for chlorine and *total* (or *combined*) *chlorine* for chloramines, when chloramines are the dominant species, with or without some free chlorine present. *See also* chloramines; free available chlorine; organic chloramines; total chlorine.

total body clearance The rate at which a chemical is removed from the body, expressed in terms of volume cleared per unit time. *See also* clearance.

total bromine-substituted trihalomethanes *See* total trihalomethane bromine.

total capacity The maximum rate at which a well or spring will yield water by pumping after the water stored in the well or spring has been removed. Total capacity is the rate of yield when the water level is drawn down to the screen under conditions of constant discharge pressure. *See also* capacity.

total carbon (TC) A quantitative measure of the concentration of both total inorganic carbon and total organic carbon in water, as determined instrumentally, frequently by chemical oxidation to carbon dioxide (CO_2) and subsequent infrared detection in a carbon analyzer.

total catchment management (TCM) An integrated plan for the collection, treatment, and disposal of runoff from a drainage basin.

total chlorine The total concentration of the chlorine in a water, including the combined chlorine (i.e., inorganic and organic chloramines) and the free available chlorine. *See also* chloramines; combined chlorine; free available chlorine; organic chloramines.

total chlorine residual The total amount of chlorine residual present after a given contact time in a water sample, regardless of the type of chlorine. *See also* residual chlorine; total chlorine.

total chlorine-substituted trihalomethanes *See* total trihalomethane chlorine.

Total Coliform Rule (TCR) A rulemaking of the US Environmental Protection Agency that sets National Primary Drinking Water Regulations for total coliforms, fecal coliforms, and *Escherichia coli.* The rule was promulgated June 29, 1989 (54 *Federal Register* 27544–27568) and amended Jan. 15, 1991 (56 *Federal Register* 1556–1557).

total coliform techniques *See* total coliform test.

total coliform test Any of several tests for the presence of indicator bacteria in water. The test used may be the multiple-tube fermentation test, the membrane filter test, the presence–absence test, or an enzyme test based on *ortho*-nitrophenol-β-D-galactopyranoside, 4-methylumbelliferyl-β-D-glucuronide, or another substrate for which a specific test has been defined for detection of coliform bacteria. *See also* total coliforms.

total coliforms (TC) The group of bacteria used as warm-blooded animal fecal pollution indicator organisms of drinking water quality. Total coliforms are regulated by the US Environmental Protection Agency. *See also* coliform group bacteria; Total Coliform Rule.

total concentration The concentration of a given constituent in water.

total dissolved phosphorus The total phosphorus content of material that will pass through a filter of a specific pore size.

total dissolved solids (TDS) The weight per unit volume of solids remaining after a sample has been filtered to remove suspended and colloidal solids. The solids passing the filter are evaporated to dryness. The filter pore diameter and evaporation temperature are frequently specified.

total dissolved solids (TDS) test A test used to measure the total dissolved solids in water by first filtering out any undissolved solids and then evaporating the filtered water to dryness. The residue that remains represents the total dissolved solids. *See also* total dissolved solids.

total dynamic discharge head The total dynamic head plus the dynamic suction head or minus the dynamic suction lift.

total dynamic head (TDH) The difference in height between the hydraulic grade line on the discharge side of the pump and the hydraulic grade line on the suction side of the pump. This head is a measure of the total energy that a pump must impart to the water to move it from one point to another.

total hardness (TH) The total of the amounts of divalent metallic cations, principally calcium hardness and magnesium hardness, expressed in terms of calcium carbonate ($CaCO_3$) equivalent. *See also* carbonate hardness; noncarbonate hardness.

total head The total energy per unit weight of water in a hydraulic system. Total head is composed of pressure head, velocity head, and elevation head. The various heads are roughly analogous to the internal, kinetic, and potential energy of the water at the point of measurement in the system.

total heterotrophic bacteria Those bacteria detected or enumerated by the heterotrophic plate count method. *See also* heterotrophic plate count.

total inorganic carbon (TIC) The sum of the inorganic carbon species in water (i.e., carbonate, CO_3^{2-}; bicarbonate, HCO_3^-; and dissolved carbon dioxide, CO_2).

total Kjeldahl nitrogen (TKN) The total concentration of organic nitrogen and ammonia nitrogen, i.e., nitrogen with a valence of –3 (N^{3-}). The amount of organic nitrogen can be obtained by separately measuring the ammonia nitrogen and subtracting that value from the total Kjeldahl nitrogen. *See also* Kjeldahl nitrogen; Kjeldahl nitrogen test; organic nitrogen.

total matter The sum of all suspended and dissolved matter in a water sample.

total maximum daily load (TMDL) A quantitative assessment of water quality problems, contributing sources, and load reductions or control actions needed to restore and protect bodies of water. Some surface waters in the United States do not meet the national goal of "fishable, swimmable," as specified in the Clean Water Act (CWA), in spite of nationally required levels of pollution control technology having been implemented for many pollution sources. Section 303(d) of the CWA addresses these waters that are not "fishable, swimmable" by requiring states to identify the waters and to develop total maximum daily loads for them, with oversight from the US Environmental Protection Agency. *See also* 303d listing.

total nitrate plus nitrite The sum of nitrate nitrogen (NO_3^--N) and nitrite nitrogen (NO_2^--N)concentrations, which represents the total oxidized inorganic nitrogen. Total nitrate plus nitrite, in addition to nitrate and nitrite as separate chemicals, is regulated by the US Environmental Protection Agency. *See also* nitrate; nitrite.

total nitrogen (N_{total}) A measure of the complete nitrogen content in water. In water, the forms of nitrogen of greatest interest are nitrate (NO_3^-), nitrite (NO_2^-), ammonia (NH_3), and organic nitrogen. All these forms of nitrogen, as well as nitrogen gas, are biochemically interconvertible and are components of the nitrogen cycle. The total nitrogen content of water can be determined by measuring nitrate, nitrite, ammonia, and Kjeldahl nitrogen. Alternatively, total nitrogen can be measured by evaporating a sample to dryness and performing an elemental analysis. *See also* ammonia; elemental analysis; nitrate; nitrite; nitrogen; nitrogen cycle; organic nitrogen; total Kjeldahl nitrogen.

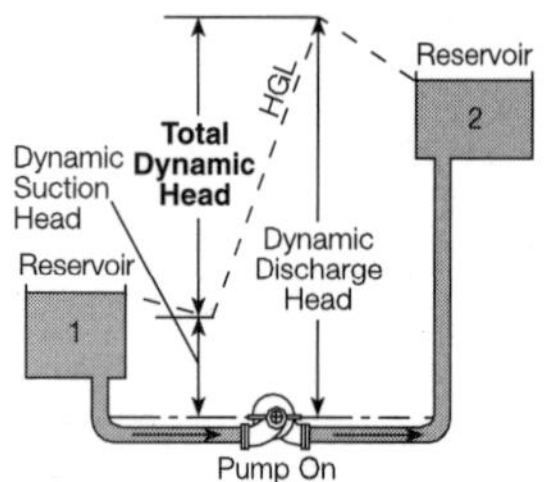

Total dynamic head

total organic bromine (TOBr) A surrogate measurement of the total quantity of bromine-substituted organic compounds in a water sample. Total organic bromine is a component of total organic halogen. In source water, bromine-substituted organic molecules are typically present because of the existence of synthetic organic chemicals, whereas in finished water they typically occur as disinfection by-products and high-molecular-weight, partially halogen-substituted aquatic humic substances. *See also* aquatic humic substance; disinfection by-product; surrogate measurement; synthetic organic chemical; total organic halogen.

total organic carbon (TOC) A measure of the concentration of organic carbon in water, determined by oxidation of the organic matter into carbon dioxide (CO_2). Total organic carbon includes all the carbon atoms covalently bonded in organic molecules. Most of the organic carbon in drinking water supplies is dissolved organic carbon, with the remainder referred to as particulate organic carbon. In natural waters, total organic carbon is composed primarily of nonspecific humic materials. Total organic carbon is used as a surrogate measurement for disinfection by-product precursors, although only a small fraction of the organic carbon will react to form these by-products. Quantitatively, total organic carbon is determined by removing interfering inorganic carbon, such as bicarbonate (HCO_3^-), and oxidizing the organic carbon to carbon dioxide. Typically, the carbon dioxide is then measured with a nondispersive infrared detector. *See also* disinfection by-product precursor; dissolved organic carbon; particulate organic carbon; total inorganic carbon.

total organic chlorine (TOCl) A surrogate measurement of the total quantity of chlorine-substituted organic compounds in a water sample. Total organic chlorine is a component of total organic halogen. In source water, chlorine-substituted organic molecules are typically present because of the existence of synthetic organic chemicals, whereas in finished water they typically occur as disinfection by-products and high-molecular-weight, partially halogenated aquatic humic substances. *See also* aquatic humic substance; disinfection by-product; surrogate measurement; synthetic organic chemical; total organic halogen.

total organic halogen (TOX) A surrogate measurement of the total quantity of halogen-substituted organic material in a water sample. The presence of halogen-substituted organic molecules in source water is typically caused by synthetic organic chemicals, whereas in finished water it is typically caused by disinfection by-products and high-molecular-weight, partially halogenated aquatic humic substances. Total organic halogen is measured by the dissolved organic halogen test and is also called adsorbable organic halogen, carbon adsorbable organic halogen, or dissolved organic halogen. *See also* aquatic humic substance; disinfection by-product; halogen-substituted organic material; surrogate measurement; synthetic organic chemical.

total organic halogen formation potential (TOXFP) The amount of total organic halogen formed during a test in which a source or treated water is dosed with a relatively high amount of disinfectant (normally chlorine) and is incubated (stored) under conditions that maximize total organic halogen production. This value is not a measure of the amount of organic halogen that would form under normal drinking water treatment conditions; rather, it is an indirect measure of the amount of organic halogen precursors in a sample. If a water has a measurable level of organic halogen prior to the formation potential test (e.g., from synthetic organic chemicals in the source water), then the formation potential equals the terminal value measured at the end of the test minus the initial value. *See also* disinfection by-product formation potential; total organic halogen.

total organic nitrogen (TON) A measure of the organic nitrogen in water. Organic nitrogen is defined functionally as organically bound nitrogen in the trinegative state (N^{3-}). Analytically, organic nitrogen and ammonia can be determined together; the combination may be referred to as Kjeldahl nitrogen. *See also* total Kjeldahl nitrogen.

total oxygen demand (TOD) A quantitative measure of all oxidizable material in a sample of water, as determined instrumentally by measuring the depletion of oxygen after high-temperature combustion. *See also* total organic carbon.

total ozone demand The total amount of ozone (O_3) gas that must be mixed with a liquid, solid, or gas in order to satisfy all the ozone oxidation requirements. *See also* ozonation.

total particulate phosphorus The total phosphorus content of material retained on a filter of a specific pore size.

total phosphorus The sum of all phosphorus forms in a sample.

total pumping head A measure of the energy increase imparted to each unit weight of liquid as it is pumped; the algebraic difference between the total discharge head and the total suction head.

total quality management (TQM) A comprehensive management approach based on the fundamental concept of continuous improvement developed primarily by W. Edwards Deming. Principle components are understanding and striving to meet customer expectations and needs; focusing on work processes that produce a service or product; and encouraging, respecting, and valuing employee contributions.

total residual chlorine *See* total chlorine residual.

total residuals content The total amount of solids contained in a water sample. It includes both suspended and dissolved material. *See also* total solids.

total residue *See* total solids.

total runoff The accumulated volumetric runoff from a drainage area for a definite period of time, such as a day, month, or year, or for the duration of a particular storm.

total solids The concentration of all suspended and dissolved solids in a liquid. This value was formerly called residue or total residue. It is measured by evaporation of a sample and

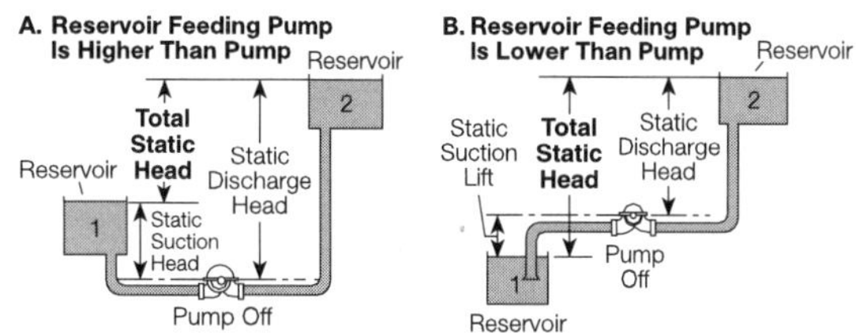

Total static head

subsequent drying in an oven at a defined temperature. Total solids includes total suspended solids and total dissolved solids. *See also* total dissolved solids; total suspended solids.

total solids content *See* total residuals content.

total static head The total height that the pump must lift the water when moving it from one point to another; the vertical distance from the suction free water surface to the discharge free water surface.

total suspended solids (TSS) A measure of all suspended solids in a liquid. A well-mixed sample is filtered through a standard glass fiber filter, and the residue retained on the filter is dried to a constant weight at 217° to 221° Fahrenheit (103° to 105° Celsius). The increase in the weight of the filter represents the total suspended solids. *See also* suspended solids.

total trihalomethane bromine (TTHM-Br) The molar sum of bromine in individual trihalomethanes. This value represents the numerator in both the bromine incorporation factor and in bromide utilization calculations. For example, in 1 mole of bromoform ($CHBr_3$), 3 moles of bromine are present (i.e., three bromine atoms per molecule of bromoform); thus, TTHM-Br = 3 for this example. *See also* bromide utilization; bromine incorporation factor; total trihalomethane chlorine.

total trihalomethane chlorine (TTHM-Cl) The molar sum of chlorine in individual trihalomethanes. For example, in 1 mole of chloroform ($CHCl_3$), 3 moles of chlorine are present (i.e., three chlorine atoms per molecule of chloroform); thus, TTHM-Cl = 3 for this example. *See also* total trihalomethane bromine.

total trihalomethanes (TTHM) The sum of the four chlorine- and bromine-containing trihalomethanes (i.e., chloroform, bromodichloromethane, dibromochloromethane, and bromoform). The US Environmental Protection Agency regulates the sum of these four species on a weight concentration basis. *See also* trihalomethane.

total volatile solids (TVS) A measure of the volatile solids in a liquid. When the residue from a total solids test (dried at 217–221° Fahrenheit, or 103–105° Celsius) is ignited to constant weight at 932±122° Fahrenheit (500±50° Celsius), the remaining solids represent the fixed total solids, and the weight lost on ignition is the total volatile solids. Likewise, the residue from a total dissolved solids test (dried at 356° Fahrenheit, or 180° Celsius) or a total suspended solids test (dried at 217–221° Fahrenheit, or 103–105° Celsius) can be ignited to constant weight at 932±122° Fahrenheit (500±50° Celsius) in order to determine the volatile dissolved solids or the volatile suspended solids. *See* *also* fixed solids; total dissolved solids; total solids; total suspended solids; volatile solids; volatile suspended solids.

total water cycle management (TWCM) *See* total water management.

total water management The management of water resources with a comprehensive approach balancing resources, demands, and environmental issues. Total water management considers water supply, water quality and treatment, storage, conveyance, potential use of alternative water supplies (such as water reuse or desalting of saline waters), conservation and demand-side management, effects of water users, and environmental needs and concerns.

total well capacity The maximum rate at which a well will yield water by pumping after the water stored in the well has been removed. Total well capacity is the rate of yield when the water level in the well is drawn down to the top of the screen.

totalizer (1) A precipitation gauge used for catching and holding the total precipitation that occurs over a considerable period of time, such as several months or a year. Such a device is usually used in remote and inaccessible locations that can be visited only infrequently. (2) A device or meter that continuously measures and calculates (adds) total flows in cubic metres, gallons, million gallons, cubic feet, or some other unit of volume measurement. Such a device is also called an integrator.

tower crib A water treatment plant intake built on the bottom of a river or lake and extending above high water for the purpose of allowing water to be drawn at different levels.

TOX *See* total organic halogen.

toxaphene (approximate formula, $C_{10}H_{10}Cl_8$) A generic name for technical chlorinated camphene (content of combined chlorine equal to 67–69 percent). Toxaphene is an insecticide that is regulated by the US Environmental Protection Agency. *See also* insecticide.

TOXFP *See* total organic halogen formation potential.

toxic Poisonous.

toxic pollutant A material that is contaminating the environment and can cause death, disease, or birth defects in organisms that ingest or absorb them. The quantities and length of exposure necessary to cause these effects can vary widely.

toxic substance A chemical element or compound, such as lead, radon, benzene (C_6H_6), dioxin, and numerous others, that has toxic properties following either ingestion, inhalation, or absorption into the body. There is considerable variation in the degree of toxicity among the various toxic substances and in the exposure levels that induce toxicity. *See also* toxicity.

Toxic Substances Control Act (TSCA) Public Law 94-469, enacted Oct. 11, 1976 (15 U.S.C. 2601 *et seq.*), authorizing the US Environmental Protection Agency to screen existing and new chemicals used in manufacturing and commerce to identify potentially dangerous products or uses that should be subject to federal control.

toxic waste Waste that, when it comes in contact with a biological entity, causes an adverse response.

toxicant A substance that causes harmful effects to organisms on contact, inhalation, or ingestion. A toxicant is also called a poison.

toxicity The total adverse effect of a poisonous substance to which the body is exposed, typically by way of the mouth, skin, or lungs.

toxicity assessment A characterization of the toxicological properties and effects of a chemical, including all aspects of its absorption, metabolism, excretion, and mechanism of action, with special emphasis on establishment of dose-response characteristics.

toxicity characteristic leaching procedure (TCLP) An extraction procedure used to determine whether a solid waste is hazardous under the Resource Recovery and Conservation Act. The toxicity characteristic leaching procedure was promulgated by the US Environmental Protection Agency on March 29, 1990 (55 *Federal Register* 11799–11877). The method for this procedure appears in 40 CFR part 261, Appendix II.

toxicity equivalency factor (TEF) A means of estimating toxicity and normalizing the potency of a mixture of chemicals that induce toxicity by interacting with a common receptor molecule, allowing one to arrive at an estimate of the hazards these chemicals pose. The concept has been applied almost exclusively to mixtures of dioxin and dibenzofurans and has little application to other chemicals or chemical classes because the receptors involved in inducing toxicity are poorly understood. In the case of the dioxins and dibenzofurans ($C_{12}H_8O$), the toxicity equivalency factor is determined by the relative affinity of each component to the Ah receptor with 2,3,7,8-tetrachlorodibenzo-*p*-dioxin as the standard. A problem with this concept is that it fails to deal with the differing pharmacokinetic parameters of the different dioxins. *See also* Ah receptor; hazard index; 2,3,7,8-tetrachlorodibenzo-*p*-dioxin.

toxicodynamics The qualitative and quantitative description of the response produced following delivery of a chemical dose to its active site within the body. *See also* pharmacodynamics.

toxicokinetics The quantitative description of the absorption, distribution, and clearance of a chemical for which the expected effects are adverse rather than beneficial. *See also* pharmacokinetics.

toxicology The study of the quantitative effects of chemicals on biologic tissue, particularly in terms of definite harmful actions and degrees of safety. *See also* acute; chronic; environmental health; potable water approval; product stewardship.

toxin A poisonous substance that comes from another organism.

Toxoplasma gondii *See* toxoplasmosis.

toxoplasmosis A systemic disease caused by *Toxoplasma gondii*, an intracellular coccidian protozoan of felines. Mammals and birds are infected worldwide. The definitive hosts are cats and other felines that acquire infection mainly from eating infected mammals or birds. Only felines harbor the parasite in the intestinal tract, where the sexual stage of the parasite's life cycle occurs. Oocysts shed in feces by cats remain infective in water or moist soil for long periods of time. Intermediate hosts, including sheep, cattle, goats, rodents, swine, and chickens, may carry infective cysts that may remain viable for long periods in tissue, especially muscle and brain tissue. Human infections may be acquired from raw or undercooked meat, raw milk containing cysts, or other foods or water contaminated with feline feces. Children become infected by ingesting infective oocysts from soil in sandboxes and play areas where cats have defecated. The infection is not directly transmitted person-to-person except in utero and is rarely acquired by blood transfusion. Infections are common in humans and frequently asymptomatic. Acute disease may cause symptoms resembling infectious mononucleosis. A primary infection during early pregnancy may lead to fetal infection, with death of the fetus or other serious effects, including brain damage. Later in pregnancy, material infection results in mild or subclinical fetal disease with delayed manifestations.

2,4,5-TP *See* 2-(2,4,5-trichlorophenoxy)propionic acid.

T-phage A tailed coliphage. *See also* coliphage.

TQM *See* total quality management.

trace (1) A very small quantity of a constituent. (2) The amount of rainfall or other form of precipitation that occurs when the quantity is so small that it cannot be measured in the rain gauge. (3) The chart record made by a self-registering instrument.

trace element An element essential to plant nutrition, animal nutrition, or both in a trace concentration of 1 percent or less of the total nutrition budget, e.g., iron, manganese, zinc, copper, potassium, or sodium. Such elements are also called micronutrients.

trace metal A metal that is present in relatively small concentrations in water. *See also* trace element.

trace of precipitation An amount of precipitation too small to be measured in a gauge, usually less than 0.005 inches (0.125 millimetres).

trace of rain An amount of rain too small to be measured.

trace substance A substance that is found during water analysis in a concentration high enough to be detected but too low to be quantified accurately by standard testing methods. A trace substance is sometimes referred to as a threshold substance.

tracer (1) A foreign substance mixed with or attached to a given substance for subsequent determination of the location or distribution of the foreign substance. (2) An element or compound that has been made radioactive so that it can be easily followed (traced) in biological and industrial processes. Radiation emitted by the radioisotope pinpoints its location.

tracer study A study using a substance that can readily be identified in water (such as a dye) to allow one to determine the distribution and rate of flow in a basin, pipe, groundwater, or stream channel.

tracer test A type of flow test in which a tracer is used to determine flow paths, arrival times, and dispersion effects in water systems.

train *See* process train.

trans Pertaining to an isomer for which a given atom or organic functional group is positioned on the opposite side of the carbon axis or backbone. *See also cis*; isomer.

transbasin diversion The deliverance of water from its natural basin into another basin.

***trans*-1,2-dichloroethylene** *See under* dichloroethylene.

transceiver A device that is capable of both transmitting and receiving information or data. When the device is simultaneously transmitting and receiving, it utilizes collision-detection logic.

transcription The process of transferring the sequence information from deoxyribonucleic acid into messenger ribonucleic acid. The messenger ribonucleic acid transcript is processed and used as a template for the synthesis of proteins.

transducer An electromechanical device that senses some varying condition (such as temperature or pressure) measured by a primary sensor and converts it to an electrical or other signal for transmission to some other device (a receiver) for processing or decision making.

transect Any imaginary line along which samples are taken at specified distance intervals. Transect sampling is usually done on large bodies of water, such as rivers and lakes.

transfer pipette *See* volumetric pipette.

transfer pump A pump specifically designed to convey water or chemical solutions from one tank to another.

transformation Acquisition by a cell of the property of uncontrolled growth.

transformer A device used with alternating current to increase current while decreasing voltage, or to decrease current while increasing voltage.

transgenic animal A test animal (e.g., a rodent) in which a gene has been added to alter the response of the animal to a chemical to be tested. *See also* knockout animal.

transient analysis A pipe network analysis that models pressure waves created by sudden changes in system flow rates.

transient, community water system *See* transient water system.

transient, noncommunity water system (TNCWS) A noncommunity water system that does not serve at least 25 of the same people over 6 months of the year in a nonresidential area such as a campground, motel, or gas station. *See also* noncommunity water system; nontransient, noncommunity water system.

transient system *See* transient water system.

transient water system (TWS) A public water system that serves a transient population. *See also* transient, noncommunity water system.

transit time (1) In a sonic transit time flowmeter, the time duration for the sonic pulse to travel from the transmitter to the

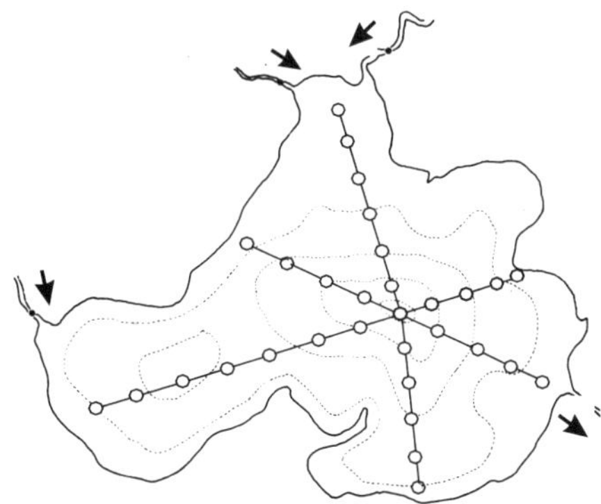

Transect sampling plan

receiver. (2) The time for a segment of water to flow from one location to another.

transition A short section of a conduit used as a conversion section to unite two conduits having different hydraulic elements.

transition zone The relatively thin layer between the upper and lower layers of the water in a reservoir where the temperature changes readily with the depth. *See also* thermocline.

translation The synthesis of protein using messenger ribonucleic acid as a template of information that dictates the sequence in which amino acids are assembled.

translatory wave A moving or advancing wave or series of waves that tend to overtake each other and form a single larger wave. A translatory wave is caused by any sudden change in the conditions of flow.

transmembrane flux Flux through a membrane. *See also* flux.

transmembrane pressure (1) For a pressure-driven membrane process, the net driving force across the membrane. (2) For microfiltration and ultrafiltration with negligible osmotic pressure differential across the membrane, the hydraulic pressure differential from the from feed side to permeate side.

transmissibility coefficient An outdated form of the term *aquifer transmissivity. See also* aquifer transmissivity; permeability coefficient.

transmission The process or act of passing along or conveying something, such as moving water from one point to another via pipelines and conduits.

transmission constant *See* hydraulic conductivity; permeability coefficient.

transmission control protocol–Internet protocol (TCP–IP) A complex set of rules and standards that control communications on the Internet, allowing data exchange with other computers. The rules were developed by the US Department of Defense.

transmission electron microscope A microscope in which the image is formed by an electron beam transmitted through the specimen, as opposed to the image being formed by an electron beam reflected (incident) from the specimen. *See also* scanning electron microscope.

transmission electron microscopy (TEM) Use of a transmission electron microscope to visually examine structures too fine to be resolved with ordinary, or light, microscopes, or to study surfaces that emit electrons. *See also* transmission electron microscope.

transmission of germicidal ultraviolet (UV) irradiation The percentage of light at a wavelength of 253.7 nanometres transmitted through water.

transmission of infection Any mechanism by which an infectious agent is spread directly or indirectly from a source or reservoir of infection to a person. These mechanisms are as follows: (1) Direct transmission of an infectious agent to a receptive portal of entry, usually by touching, kissing, biting, sexual activities, or direct projection of droplet spray onto the conjunctiva or mucous membranes by sneezing, coughing, and so on. (2) Indirect transmission in which a vehicle serves as an intermediate means by which an infectious agent is transported. The agent may or may not have multiplied or developed in or on the vehicle. Vehicles may include contaminated inanimate materials or objects (fomites, such as toys, diapers, or soiled clothes), water, food, milk, or biological products (such as blood or tissue). (3) Indirect transmission by a vector from either (a) the carriage of an infectious agent by a crawling or flying insect through the soiling of its feet or proboscis or by passage of organisms through the agent's gastrointestinal tract, or (b) injection of salivary gland fluid during biting, or regurgitation or deposition of feces or other material that may penetrate the skin through a bite wound or after scratching. (4) Indirect transmission by airborne dissemination of microbial aerosols suspended in the indoor and outdoor air environment to the respiratory tract. (Note: droplets and other large particles promptly settle and are not considered airborne.)

transmission lag The dead time specifically associated with transmitting the measured value of a variable between its measuring instrument and the receiving instrument, such as a controller or recorder.

transmission line A pipeline that transports source water to a water treatment plant or treated water to a distribution grid system.

transmission main A water main that transports water from the main supply or source to a distant area where the water is distributed through distribution lines.

transmissivity A measure of the ability to transmit water. *See also* aquifer transmissivity.

transmittancy The ability of water to transmit or convey ultraviolet energy.

transmitter The part of a pressure differential meter that measures the signal from the primary element and sends another signal to the receiver.

transpiration The process by which water vapor is lost to the atmosphere from living plants.

transpiration ratio The ratio of the weight of water lost to the atmosphere from a plant to the weight of dry plant substance produced.

transport capacity A measure of the ability of a stream to transport a suspended load. Transport capacity is expressed in terms of the total weight of the particles.

transport competency A measure of the ability of a stream to transport sediment at a given velocity. Transport competency is expressed in terms of the dimensions of the particles that may be transported.

transport pore In granular activated carbon, a pore larger than the largest adsorption pores. Transport pores function as a diffusion path to transport organic and inorganic adsorbates. Adsorption does not occur in these locations, even at near-saturated conditions.

transport process An exchange of heat, mass, or momentum between materials or objects.

transportation In geology, the movement of detached soil material across the land surface or through the air. Transportation may be brought about by running water, wind, or gravity. *See also* erosion.

transported soil Soils formed by the consequent or subsequent weathering of materials transported and deposited by some agency, such as water, air, or ice.

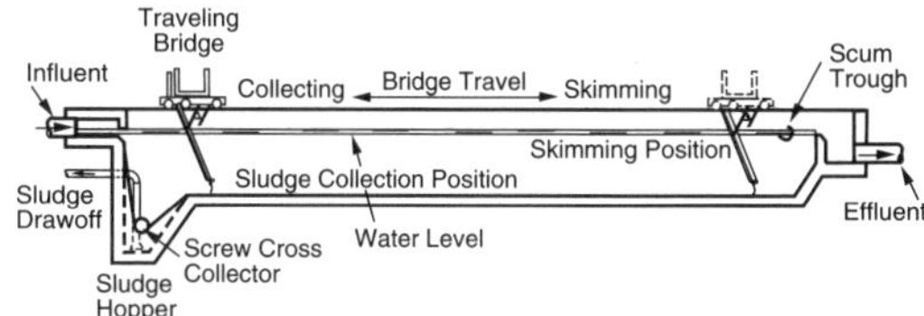

Traveling bridge sludge collector

transporting erosive velocity A flow velocity that is high enough not only to maintain sediment in movement but also to scour the bed of a stream or canal.

transverse flow An operating condition, in some types of pressure-driven membrane water treatment processes, where the feedwater and resulting concentrate or retentate pass tangentially across the surface of the membrane and the permeate passes in a perpendicular direction through the membrane barrier. With hollow fiber membranes, transverse flow commonly means that pressurized feed passes tangentially on the outside of the membrane fibers and that the permeate flows outside in, or shell side to lumen, and is collected from the inner lumen of the fibers (opposite the permeate flow direction, as in a crossflow hollow fiber membrane). *See also* crossflow; hollow fiber membrane; lumen.

transverse Mercator projection A map projection that is used to establish zones in the State Plane Coordinate System. *See also* State Plane Coordinate System.

trap A device used to prevent a material flowing or carried through a conduit from reversing its direction of flow or movement or from passing a given point.

trapezoidal weir A weir with a trapezoid-shaped notch.

trash Debris that may be removed from reservoirs by coarse racks.

trash gate A gate over which floating trash may be skimmed off or through which trash removed from racks may be discharged.

trash screen A screen installed or constructed in a waterway to collect and prevent the passage of trash.

trashrack *See* bar screen.

trauma An injury that was caused by a force outside the body. For example, hitting the hand with a hammer would be considered a trauma injury.

travel time The time for a water particle to travel from one location to another along a flow line.

traveling bridge sludge collector A mechanism that collects sludge from the bottom of a rectangular sedimentation basin by moving a bridge horizontally across tracks on the top walls of the basin. The bridge supports a suction-type sludge collection mechanism that extends to the bottom of the basin. The sludge is suctioned out into a channel that runs parallel to the basin at the surface.

traveling screen A trash screen that revolves such that water always passes through a clean area of the screen.

traveling wave A translatory wave in an open channel traveling downstream because of a sudden increase of flow, or traveling upstream when the rate of flow is suddenly diminished or stopped. *See also* bore; tidal bore.

tray aerator A device used to transfer target compounds from the liquid to the gaseous phase by cascading water over sequential trays that create turbulence in the liquid, thereby promoting mass transfer. *See also* stripping.

treated water Water that has been subjected to treatment processes.

treatment plant The central portion of water facilities that contains various treatment processes exclusive of collection or distribution of water.

treatment plant residue An accumulation of solid material removed from water during treatment.

treatment system A combination of various unit processes with a common objective of producing potable water. *See also* potable water.

treatment technique (TT) requirement A requirement of the National Primary Drinking Water Regulations that specifies, for a particular contaminant, one or more specific treatment techniques that lead to a reduction in the level of the contaminant sufficient to achieve compliance with the requirements of 40 Code of Federal Regulations Part 141. *See also* best available technology.

tree diagram A graphical method of mapping out in increasing detail, e.g., the full range of paths and tasks that need to be accomplished in order to achieve a goal and related subgoals or how a drinking water regulation will affect various categories of water utilities.

tree-forest cover Trees and shrubs that cover the soil, protecting it from variations in humidity and temperature, as well as from erosion.

trellis drainage pattern A stream pattern that is characterized by many more or less right-angle bends and is commonly developed in regions of folded geological structure. The rectilinear arrangement resembles a trellis.

tremie A box or frame of wood or metal used for depositing concrete underwater. Its upper section forms a hopper above water to receive the concrete, and it may be moved laterally or vertically by any suitable device, such as a traveling crane.

tremie pipe A small pipe inserted into the annular space between a borehole and well casing to place filter pack materials, cements, and grouts.

trench An excavation made for installing pipes and masonry walls, as well as for other purposes. A trench is distinguished from a ditch in that the opening is temporary and is eventually backfilled.

trench box A box that can be stacked, laid end to end with other such boxes, and interlocked to provide protection on deep excavations. Trench boxes are fabricated by manufacturers that specialize in excavation safety systems. They must be able to withstand the forces that are imposed from the surrounding earth to prevent cave-ins and thereby protect workers.

trench fill revetment In the United States, a covering of loose stone or blocks laid on the berm of a river embankment to protect the embankment from erosion. *See also* loose apron; riprap.

trench shield *See* shield.

triangular weir A weir having a notch that is triangular in shape. *See also* V-notch weir.

triazine herbicides ($R_3C_3N_3$) A family of herbicides based on a symmetrical triazine structure, where there are a variety of attached groups (R). The triazine structure is composed of a six-membered ring with alternating carbon and nitrogen atoms, where every other carbon–nitrogen bond is a single bond and the other bonds are double bonds. The R groups are attached to the carbon atoms. *See also* atrazine; cyanazine; herbicide; prometon; simazine.

tribromamine (NBr_3) The bromine-substituted analog of trichloramine (NCl_3) that can be formed during the chloramination of bromide-containing water. *See also* bromamines.

tribromoacetic acid (TBAA) (CBr_3COOH) The bromine-substituted analog of trichloroacetic acid that may be formed during the chlorination or ozonation of bromide-containing water. As of the time of this writing, research suggests that this haloacetic acid is relatively unstable in water. *See also* haloacetic acid.

tribromomethane (TBM) ($CHBr_3$) The chemical name for bromoform. *See also* bromoform.

tributary A stream or other body of water—surface or underground—that contributes its water to another and larger stream or body of water.

tributary groundwater Water below the earth's surface that is hydrologically or physically connected to natural stream water so as to affect the first water's flow, whether in movement from or to that stream.

trichloramine (NCl_3) A compound that is formed by the chlorination of water containing ammonia (NH_3). This compound is also called nitrogen trichloride. Trichloramine is the last in a sequence of compounds formed by the continual addition of ammonia—first monochloramine (NH_2Cl), then dichloramine ($NHCl_2$), and finally trichloramine. Thus, this species is not normally detected in chloraminated waters. Trichloramine has little disinfection capacity and is a source of taste and odors. Upon additional chlorination, trichloramine will be converted to elemental nitrogen gas, and any additional chlorine (HOCl) will be measured as free chlorine. *See also* breakpoint chlorination; chloramines.

trichloroacetate (TCA) (CCl_3COO^-) A salt of trichloroacetic acid. At drinking water pH levels, trichloroacetic acid is present as the trichloroacetate ion. *See also* trichloroacetic acid.

trichloroacetic acid (TCAA) (CCl_3COOH) A haloacetic acid containing three chlorine atoms. Typically, this acid and dichloroacetic acid ($CHCl_2COOH$) are the principal haloacetic acids formed during chlorination, except in waters containing moderate amounts of bromide ion (Br^-). Trichloroacetic acid is also commercially prepared for miscellaneous industrial uses, including use as an herbicide. *See also* dichloroacetic acid; haloacetic acid; herbicide.

1,1,1-trichloroacetone *See under* trichloropropanone.

trichloroacetaldehyde *See* chloral hydrate.

1,2,4-trichlorobenzene ($C_6H_3Cl_3$) A volatile organic compound used as a solvent and an insecticide. Trichlorobenzene is regulated by the US Environmental Protection Agency. *See also* insecticide; solvent; volatile organic chemical.

trichloroethane (TCA) *See* 1,1,1-trichloroethane; 1,1,2 trichloroethane.

1,1,1-trichloroethane (TCA) (CH_3CCl_3) A volatile organic compound used as a solvent and a pesticide. 1,1,1-trichloroethane is regulated by the US Environmental Protection Agency. *See also* pesticide; solvent; volatile organic compound.

1,1,2-trichloroethane (TCA) ($CHCl_2CH_2Cl$) A volatile organic compound used as a solvent. 1,1,2-trichloroethane is regulated by the US Environmental Protection Agency. *See also* solvent; volatile organic compound.

trichloroethene *See* trichloroethylene.

trichloroethylene (TCE) ($CHCl{=}CCl_2$) A volatile organic compound that has been used as a metal degreaser, as a solvent, in dry cleaning, and as a fumigant. Trichloroethylene (also called trichloroethene) is regulated by the US Environmental Protection Agency. *See also* fumigant; solvent; volatile organic compound.

trichloromethane (TCM) ($CHCl_3$) The chemical name for chloroform. *See also* chloroform.

trichlorophenol (TCP) A chlorophenol with three chlorine atoms. One of the trichlorophenols is 2,4,6-trichlorophenol ($C_6H_2Cl_3OH$), which is a fungicide, herbicide, and defoliant. In addition, 2,4,6-trichlorophenol can be formed during the chlorination of water. However, this disinfection by-product is typically detected at low (or sub) microgram-per-litre levels. *See also* chlorophenols; defoliant; disinfection by-product; fungicide; herbicide.

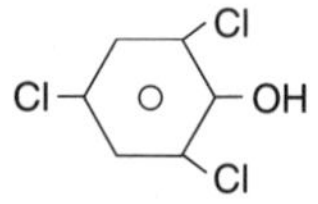

2,4,6-trichlorophenol

2-(2,4,5-trichlorophenoxy)propionic acid (2,4,5-TP) ($Cl_3C_6H_2OCH(CH_3)COOH$) An herbicide and plant growth regulator that is regulated by the US Environmental Protection Agency. It is also called silvex. *See also* herbicide.

1,1,1-trichloropropanone (1,1,1-TCP) (CCl_3COCH_3) A halogen-substituted ketone containing three chlorine atoms and commonly called 1,1,1-trichloroacetone. Typically, this compound is the principal halogen-substituted ketone formed during chlorination, except in waters containing moderate amounts of bromide (Br^-). It is converted to chloroform ($CHCl_3$) during base-catalyzed hydrolysis. *See also* base-catalyzed hydrolysis; haloketone.

trigger point An instantaneous condition for a demand-initiated regeneration water softener or in a valve control cycle when the regeneration step is about to be started.

trihalomethane (THM) Any of numerous organic compounds named as derivatives of methane (CH_4) in which three halogen atoms (chlorine, bromine, iodine, singly or in combination) are substituted for three of the hydrogen atoms. Trihalomethanes are formed during the disinfection of water with free chorine. Because of their carcinogenic potential and other possible health effects, these compounds are regulated by the US Environmental Protection Agency. *See also* disinfection by-product; total trihalomethanes.

trihalomethane (THM) precursor A substance that can be converted into a trihalomethane during disinfection. Typically, most of these precursors are constituents of natural organic matter. In addition, the bromide ion (Br^-) is a precursor material. Trihalomethane formation potential tests are used to indirectly measure the level of trihalomethane precursors in a sample. *See also* bromide; natural organic matter; trihalomethane; trihalomethane formation potential.

trihalomethane formation potential (THMFP) The amount of trihalomethanes formed during a test in which a source or treated water is (1) dosed with a relatively high amount of disinfectant (normally chlorine) to produce a residual at the end of the test of about 3 milligrams per litre and (2) incubated or stored under conditions that maximize THM production (e.g., neutral to alkaline pH, warm water temperature, contact time of 4 to 7 days). This value is not a measure of the amount of trihalomethanes that would form under normal drinking water treatment conditions, but rather an indirect measure of the amount of trihalomethane precursors in a sample. If a water has a measurable level of trihalomethanes prior to the formation potential test (e.g., in a chlorinated sample), then the formation potential equals the terminal concentration measured at the end of the test ($TTHM_t$) minus the initial concentration ($TTHM_0$). *See also* disinfection by-product precursor; trihalomethane; trihalomethane precursor; uniform formation condition.

1,2,4-trimethylbenzene ($C_6H_3(CH_3)_3$) A volatile organic compound with various industrial uses, e.g., in the manufacture of dyes and pharmaceuticals. *See also* volatile organic chemical.

triple-expansion steam pump A direct-connected steam pump in which the steam is exhausted from the high-pressure cylinder to an intermediate cylinder, where it expands, and is then further exhausted to a third cylinder.

triple-media filter A filter containing three media—anthracite coal, supported by sand, supported by garnet—that stratify after backwashing according to their respective specific gravities. Triple-media filtration *may* have improved filtration properties over dual-media filters in some applications. A triple-media filter is sometimes called a mixed-media filter. *See also* dual-media filtration.

triplex pump A reciprocating pump with three single-acting cylinders placed next to each other in line, all connected with the same suction and discharge line, with valves arranged so that the intake and discharge through the pump are continuous.

tripolyphosphate A water-soluble chelating agent for certain metals in solution. As sodium tripolyphosphate, the formula is $Na_5P_3O_{10}$. *See also* sodium tripolyphosphate.

trisodium phosphate ($Na_3PO_4 \cdot 12H_2O$) (TSP) An alkaline chemical additive used in some membrane-cleaning solutions.

trivalent Having a valence of three.

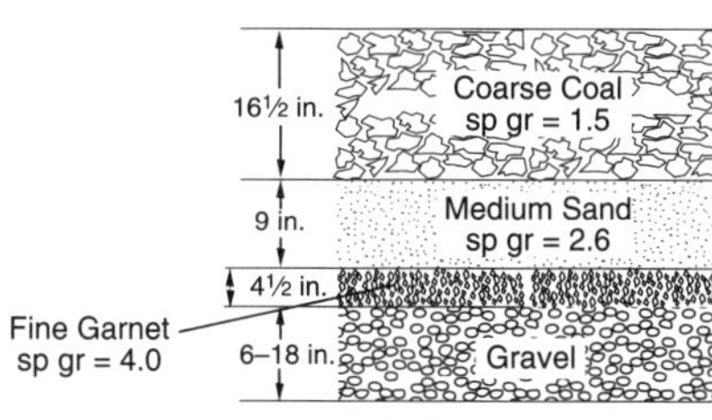

Triple-media filter

trivalent ion An ion having a valence charge of three. The charge can be positive or negative.

troph *See* trophozoite.

trophic level Any of the stages of the food chain at which an organism feeds.

trophic status An indicator of the ability of a lake to support plant growth as measured by phosphorus content, algal abundance, and depth of light penetration. *See also* eutrophic; mesotrophic; oligotrophic.

trophozoite The vegetative state or phase of a protozoan life cycle in which asexual reproduction occurs. A trophozoite is sometimes called a troph.

tropic tide A tide occurring semimonthly when the effect of the moon's maximum declination is greatest.

troubleshooting (1) A process of finding ways and methods to resolve disputes or impasses. (2) A process of discovering and eliminating the cause of trouble in, e.g., mechanical equipment, distribution systems, treatment plants, or power lines.

trough A structure, usually with a length several times its transverse dimensions, used to hold or transport water or other liquids.

true cohesion In soil stabilization, the resistance of particles to being pulled apart, resulting from the attraction between molecules at the point of contact.

true color The color of water from which the turbidity has been removed.

true color unit (tcu) *See the Units of Measure list.*

true groundwater velocity The actual or field velocity of groundwater percolating through water-bearing material. This velocity is equal to the volume of groundwater passing through a unit cross-sectional area per unit time divided by the effective porosity. True groundwater velocity is also called actual groundwater velocity, effective groundwater velocity, or field groundwater velocity.

true monthly mean precipitation The weighted mean precipitation for each month for a large area, derived from the study of an isohyetal map of the area. The weighted precipitation is the product of an isohyetal value and the area represented by this value. The sum of the weighted precipitations for the entire area is divided by the total basin area to determine the value of the mean precipitation for the whole area. *See also* isohyet.

true power Power available to do useful work. *See also* active power; complex power; instantaneous power; reactive power.

truncation of data The act of reducing the number of significant figures after a decimal point because of inaccuracies in the measurement technique. *See also* significant figure.

trunk main A large pipe serving as a supply main or feeder main in a water distribution system.

trunnion A roller device, placed under ton containers of chlorine, for example, to hold them in place.

TS *See* total solids.

TSCA *See* Toxic Substances Control Act.

TSD *See* technical summary document.

TSP *See* trisodium phosphate.

TSS *See* total suspended solids.

tsunami A gravity wave caused by an underwater seismic disturbance, such as sudden faulting, landsliding, or volcanic activity. *See also* tidal wave.

TT requirement *See* treatment technique requirement.

t-test The test most often used in regression analysis to provide a measure of the goodness of fit for individual independent variables that enter the regression equation. The t-test, which is often called *Student's t-test,* is related to the concept of normal probability. It uses specific numbers of standard errors to define an area under a probability curve that approaches a normal curve when the sample size is about 50 observations. For example, a span of 2 standard errors on either side of the sample mean encompasses about 95 percent of the area under a normal curve, leaving about 5 percent of the area in the tails. In this context, including in a regression analysis an independent variable that has a t value of 2.0 means that about a 5 percent probability exists that the variable could be significant in the regression equation by chance (i.e., not be a true causal influence). Consequently a t value of 2.0 is usually used to determine whether variables should be included in or excluded from regression equations. A t-test can be either a one-tailed test or a two-tailed test. *See also* one-tailed test; regression analysis; standard error of the coefficient; two-tailed test.

TTHM *See* total trihalomethanes.

TTHM-Br *See* total trihalomethane bromine.

TTHM-Cl *See* total trihalomethane chlorine.

($TTHM_0$) *See* instantaneous total trihalomethane concentration.

($TTHM_t$) *See* terminal total trihalomethane concentration.

tube (1) A closed conduit used to convey fluids. A tube may take many different forms, e.g., converging, diverging, uniform diameter, or special shape. (2) An instrument for determining the velocity and quantity of flow. *See also* pitot tube.

tube settler A unit constructed of parallel tubes that are typically arranged in a honeycomb fashion and are approximately 2 inches (5 centimetres) in width oriented at a 45° to 60° angle from horizontal. Tube settlers are used to improve settling in a sedimentation basin. The units are placed at the end of the sedimentation basin (across the entire width) and flow travels upward through the tubes and exits at the top, prior to being discharged from the basin. The inclined tubes provide a much shorter distance for particles to settle prior to being captured, resulting in a low overflow rate, and they are often used to maintain particle removal at higher flow rates, thereby reducing the need to construct additional basins. *See also* inclined plate separator; Pielkenrod separator; shallow-depth sedimentation.

tube settling A shallow-depth sedimentation process that uses a series of inclined tubes. *See also* tube settler.

tube well A well constructed by pushing or hammering a tube into the ground. The bottom of the tube is closed and tapered like an arrow. Holes in the side of the tube allow water to enter the tube and come to the surface.

tubercle (1) A small knob or button of rust formed on the inside of an iron pipe. (2) An attached deposit of metallic salts in pipes, frequently resembling a bunch of grapes.

tubercular corrosion A type of corrosion that occurs when pitting corrosion products build up at the anode over the top of the underlying pit. In iron and steel pipes, tubercles are made up of rust or iron oxides plus other compounds. These tubercles may appear as individual nodules or may be so plentiful that they grow together to form a continuous tuberculated surface. Copper pipe may also form tubercles that are generally smaller in size than those for iron.

tuberculation (1) The formation of tubercles in pipe. (2) Localized corrosion at scattered locations resulting in knoblike mounds.

tubing (1) Flexible pipe of small diameter, usually less than 2 inches (5 centimetres). (2) A special grade of high-test pipe fitted with couplings and fittings of special design.

tubular membrane A membrane installed within a porous tube that provides membrane support. Feed flow passes through the tube, and permeate passes through the membrane barrier in a perpendicular direction.

tubular spring A spring that issues through more or less rounded openings, such as solution passages in limestone or gypsum or natural channels in basaltic lava.

tubular well Normally, a well of small diameter installed with a strainer or sand point without gravel packing. The slots in the strainer are selected to suit the formation penetrated so that the well will not pump excessive amounts of sand after development.

tumbling basin A basin holding a cushion of water, designed to absorb energy from vertically dropping water.

tumor (1) A growth of tissue in which the multiplication of cells is not fully under control and has the potential for progression to cancer. (2) Broadly speaking, a swelling that signals inflammation.

tumor incidence The fraction of animals having a tumor of a certain type.

tumor initiator A physical or chemical agent that is capable of producing an irreversible event that carries with it a probability of a tumor being produced. Ordinarily, such agents act by interacting with deoxyribonucleic acid to produce a mistake in its replication so that a mutation arises in a critical site. Such mutations alter the cell's response to growth control mechanisms. Because a mutation is an irreversible event, the effects of a tumor initiator are demonstrably irreversible in appropriately designed experiments. Certain chemicals will stimulate the growth of such altered cells. These latter types of chemicals are referred to as tumor promoters. *See also* carcinogenic; tumor promoter.

tumor promoter (1) In toxicology, a chemical that increases the growth rate of clones of cells that develop into tumors. (2) In molecular biology, a region within a deoxyribonucleic acid sequence to which a protein binds to increase the synthesis of messenger ribonucleic acid from particular genes. This region is also called the promoter region. *See also* carcinogenic; messenger ribonucleic acid; tumor initiator.

Tubercular corrosion in a pipe

tumorigenic response An effect elicited by a physical, chemical, or microbial agent that results in the formation of tumors. *See also* carcinogenic; tumor.

tundra Level or undulating plainland without trees, as characteristic of Arctic regions.

tuning The act of adjusting the proportional band, integral factor, derivative factor, or any combination of the three in an automatic controller.

turbid Having a cloudy or muddy appearance.

turbidimeter An instrument that measures the amount of light scattered by suspended particles in a water sample, with a standard suspension used as a reference.

turbidimetry (1) The measurement of turbidity based on the ratio of the intensity of light scattered at a fixed angle by the particles in suspension to the intensity of the incident light. (2) The depth at which a target disappears beneath the layer of turbid medium. *See also* nephelometric turbidity unit *in the Units of Measure list*; nephelometric turbidimeter.

turbidity (1) A condition in water caused by the presence of suspended matter, resulting in the scattering and absorption of light. (2) Any suspended solids that impart a visible haze or cloudiness to water and can be removed by treatment. (3) An analytical quantity, usually reported in nephelometric turbidity units, determined by measurements of light scattering. The turbidity in finished water is regulated by the US Environmental Protection Agency. *See also* color; nephelometric turbidity unit *in the Units of Measure list.*

turbidity breakthrough Passage of turbidity all the way through a filter such that a measurable change in turbidity can be detected. In the filtration process, a filter ripening period occurs in which turbidity declines, followed by an operating time when the turbidity is typically less than the goal value and is relatively stable, followed by a steady increase. Under ideal conditions, terminal head loss should precede breakthrough, thereby protecting the filtered water quality from a sharp increase in particulates. *See also* terminal head loss; turbidity.

turbidity unit *See* nephelometric turbidity unit *in the Units of Measure list.*

turbine Any of various machines having a rotor, usually with vanes or blades, driven by the pressure, momentum, or reactive thrust of a moving fluid, such as steam, water, hot gases, or air. The moving fluid occurs either in the form of free jets or as a fluid passing through and entirely filling a housing around the rotor.

turbine efficiency (1) The ratio of energy converted into useful work to the energy supplied to the turbine. (2) The energy difference between the water in a wheel pit and the tailwater.

turbine meter A meter for measuring flow rates by measuring the speed at which a turbine spins in water, indicating the velocity at which the water is moving through a conduit of known cross-sectional area.

turbine pump A series of small-diameter centrifugal pumps (called bowls) that are connected together and driven by a single motor through one shaft. Using a series of bowls is necessary to provide the head because of the limited diameter of the bowls, especially in a deep well application.

turbopump A pumping device in which one or more turbine impellers rotate on a shaft with one or more pump impellers. The hydraulic energy of a pressurized flow stream passing through the turbine impeller(s) is transferred through the common shaft to the flow stream being pumped. For example, in a high-pressure reverse osmosis system, a feedwater pump having an energy recovery turbine installed on the same shaft as the pumping impellers and motor is a turbopump. The pressurized waste concentrate flow stream passes through the energy recovery turbine and lowers the required energy input from the drive motor.

turbulence A pattern of water flow characterized by cross currents and eddies that mix the water (as opposed to a streamlined or laminar flow pattern). Turbulence may be caused by excessive flow rates, by curves or rough surfaces in the flow channel, or by turbulence promoters (such as baffles) purposely created to mix the water. Turbulence significantly increases pressure drop in a system. *See also* Kolmogoroff microscale; laminar flow.

turbulence promoter A device inserted into a channel to improve the mixing characteristics and increase the turbulence and velocity gradient. Typical turbulence promoters include baffles, spiral wires, balls, spacers, and static mixers. *See also* baffle; static mixer; velocity gradient.

turbulent flow A flow regime characterized by random motion of the fluid particles in the transverse direction, as well as motion in the axial direction. Turbulent flow occurs at high Reynolds numbers and is the type of flow most common in industrial fluid systems. It is a type of flow in which forces caused by inertia are more significant than forces caused by viscosity and in which adjacent fluid particles are more or less random in motion. *See also* Reynolds number.

turbulent velocity That velocity of water flowing in a conduit in excess of which the flow will always be turbulent and beneath which the flow may be either turbulent or laminar, depending on circumstances.

turfgrass Hybridized grasses that, when regularly mowed, form a dense growth of leaf blades and roots.

turn-down ratio The ratio of the design range of an instrument to the range of acceptable accuracy or precision. *See also* effective range.

turned joint A joint used in wooden and other pipes in which the inside of the bell (or enlarged end of one length of pipe) is turned or bored on a lathe to an accurate fit, the outside of the spigot (or end of the joining length of pipe) is also turned to a fit, and the two lengths of pipe are driven together by a wooden ram or other tool. The contacting surfaces may be painted or greased, and a small amount of cementing material added to the outside. A turned joint is also called a turned-and-bored joint.

turnkey construction A construction process in which the owner contracts with an outside consultant who is given the responsibility to design, construct, and start up a facility. Upon completion the project is turned over to the owner for operations and maintenance.

turnkey system (1) A system for which components from various vendors are assembled by an integrator who installs the system and provides necessary data and applications to perform a set of functions. The integrator generally assumes responsibility for the performance of all individual components of the system. (2) A system that is immediately operational upon installation.

turn-on (or turn-off) charges Fees assessed by a utility for activating (or deactivating) water service to a customer. Emergency turnoffs normally are exempted from such charges.

turnover *See* overturn.

TVS *See* total volatile solids.

TWA concentration *See* time-weighted average concentration.

TWCM *See* total water cycle management.

twin dosing tanks Two dosing tanks of equal capacity, each equipped with a dosing device so that they may be used alternately. *See also* dosing tank.

2-methylisoborneol *See under* methylisoborneol.

2-methylphenol *See under* methylphenol.

2,2-dichloropropane *See under* dichloropropane.

2-(2,4,5-trichlorophenoxy)propionic acid *See under* (trichlorophenoxy)propionic acid.

2,3,7,8-tetrachlorodibenzo-*p*-dioxin *See under* tetrachlorodibenzo-*p*-dioxin.

2,4-D *See* 2,4-dichlorophenoxyacetic acid.

2,4-dichlorophenol *See under* dichlorophenol.

2,4-dichlorophenoxyacetic acid *See under* dichlorophenoxyacetic acid.

2,4-dinitrophenol *See under* dinitrophenol.

2,4-dinitrotoluene *See under* dinitrotoluene.

2,4,5-TP *See* 2-(2,4,5-trichlorophenoxy)propionic acid.

2,6-dinitrotoluene *See under* dinitrotoluene.

two-bed A pairing of cation and anion exchange tanks, typically operating in series. A two-bed is best used for the deionization of relatively high volumes of water and is capable of producing product water with resistivity of up to 1 megohm-centimetre.

two-tailed test A statistical procedure to determine whether the differences between two sets of data are significant. In a two-tailed test, differences in either direction (i.e., greater or smaller) are considered. *See also* t-test; one-tailed test.

TWS *See* transient water system.

type A soil Cohesive soil with an unconfined, compressive strength of 1.5 tons per square foot (144 kilopascals) or greater. Examples of cohesive soils are clay, silty clay, sandy clay, clay loam, and, in some cases, silty clay loam and sandy clay loam. Cemented soils such as caliche and hardpan are also considered type A. However, no soil is type A if any of the following apply: (1) the soil is fissured; (2) the soil is subject to vibration from heavy traffic, pile driving, or similar effects; (3) the soil has been previously disturbed; (4) the soil is part of a sloped, layered system where the layers dip into the excavation on a slope of four horizontal to one vertical or greater; or (5) the material is subject to other factors that would require it to be classified as a less stable material. *See also* soil and rock deposit classification.

type B soil (1) Cohesive soil with an unconfined compressive strength greater than 0.5 ton per square foot (48 kilopascals) but less than 1.5 tons per square foot (144 kilopascals). (2) Granular cohesionless soil, including angular gravel (similar to crushed rock), silt, silt loam, sandy loam, and, in some cases, silty clay loam and sandy clay loam. (3) Previously disturbed soil, except those soils that would otherwise be classed as type C soil. (4) Soil that meets the unconfined compressive strength or cementation requirements for type A but is fissured or subject to vibration. (5) Dry rock that is not stable. (6) Material that is part of a sloped, layered system where the layers dip into the excavation on a slope less steep than four horizontal to one vertical, but only if the material would otherwise be classified as type B. *See also* soil and rock deposit classification.

type C soil (1) Cohesive soil with an unconfined compressive strength of 0.5 ton per square foot (48 kilopascals) or less. (2) Granular soil, including gravel, sand, and loamy sand. (3) Submerged soil or soil from which water is freely seeping. (4) Submerged rock that is not stable. (5) Material in a sloped, layered system where the layers dip into the excavation or a slope of four horizontal to one vertical or steeper. *See also* soil and rock deposit classification.

type I settling The settling of discrete, nonflocculent particles in a dilute suspension. The particles settle as separate units, and no apparent interaction occurs between particles. Such settling is also called free settling.

type II settling The settling of flocculent particles in a dilute suspension. The particles flocculate during settling; thus, they increase in size and settle at a faster velocity as time passes. Such settling is sometimes called flocculent settling.

type III settling The settling of a suspension intermediate in concentration between type II and type IV. The particles are so close together that interparticle forces hinder the settling of neighboring particles. The particles remain in fixed positions relative to each other, and all settle at a constant velocity. As a result, the mass of particles settle as a zone. Such settling is also called zone settling or hindered settling.

type IV settling The settling of particles that are at such a high concentration that the particles touch each other and settling can occur only by the compression of the compacting mass. Such settling is also called compression settling.

typhoid fever A serious infectious disease caused by the bacterium *Salmonella typhi*. It was one of the most widespread and important of all bacterial diseases until the advent of adequate water treatment processes, including disinfection. The disease, although generally thought of as a gastrointestinal infection, is actually a generalized infection of the body, and symptoms may include frontal headache, lack of appetite, nosebleed, muscular weakness, and diarrhea. Death can occur if the illness is not diagnosed and treated aggressively. Typhoid fever is generally acquired through contaminated drinking water, and epidemics still persist, primarily in undeveloped and developing countries where drinking water treatment and wastewater treatment are inadequate and the level of sanitation in the general population remains at a low level. Typhoid fever is sometimes just called typhoid. *See also* waterborne disease.

U

U pump A reciprocating piston pump in which the valves are placed in the piston and the flow through the cylinder is in one direction with no reversal.
uc *See* uniformity coefficient.
UCA *See* utilities communication architecture.
UCMR *See* Unregulated Contaminant Monitoring Rule.
UEL *See* upper explosion limit.
UF *See* ultrafiltration; uncertainty factor.
UFBV *See* unit filter backwash volume.
UFC *See* uniform formation conditions.
UFRV *See* unit filter run volume.
UFW *See* unaccounted-for water.
UIC program *See* Underground Injection Control Program.
ultimate disposal The final release of biologically and chemically stable residuals into the environment.
ultimate standard water demand A schedule of the requirements of water—in terms of both quantity and time of need—for a particular purpose and, in some cases, for a particular locality. The purpose may be irrigation of various specified crops, municipal or domestic water supply, plant transpiration, or power.
ultracentrifugation Centrifugation at speeds creating forces that will sediment macromolecules, such as proteins and viruses, thus allowing them to be separated from the fluid that contains them.
ultrafiltration (UF) A pressure-driven membrane process that separates submicron particles (down to a 0.01-micrometre size or less) and dissolved solutes (down to a molecular weight cutoff of approximately 1,000 daltons) from a feed stream by using a sieving mechanism that is dependent on the pore size rating of the membrane. *See also* molecular weight cutoff.
ultra-low-flush toilet A water-conserving or low-water-use fixture for elimination of human waste matter. It uses about 1.5 gallons (6 litres) per flush. Some models improve performance by using compressed air to assist in flushing the waste.
ultrapure water High-purity water that has extremely low organic and inorganic content and a resistivity of approximately 18 megohms. Ultrapure water is commonly used for industrial applications, e.g., as semiconductor rinse water. *See also* 18-megohm water.
ultrasonic flowmeter A device that measures the flow rate of a fluid in a channel or pipe by transmitting and receiving sound waves that cross diagonally through the flow path. The length of time required for the wave to cross the flow is proportional to the flow velocity. This velocity, together with the cross-sectional area, can be used to calculate a flow rate.
ultrasonic nebulizer A device used in such atomic spectrometry techniques as inductively coupled plasma spectroscopy. Ultrasonic nebulizers are sample introduction devices that produce aerosol particles more efficiently than conventional nebulizers, resulting in improved sensitivity in, e.g., inductively coupled plasma analyses.
ultraviolet (UV) *See* ultraviolet irradiation.
ultraviolet (UV) absorbance at 254 nanometres (UV-254 or UV_{254}) A measure of water's ultraviolet absorption at a wavelength of 254 nanometres. This value is an indirect measure of compounds containing double bonds (including, but not limited to, aromatic compounds). Therefore, this measurement has

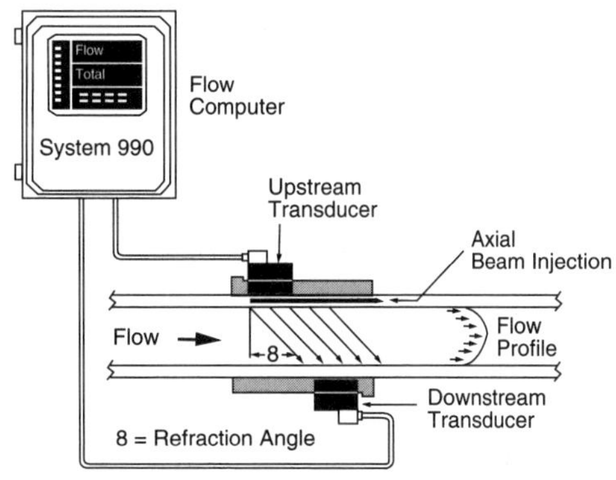

Ultrasonic flowmeter

been considered representative of the humic content of natural organic matter, as well as acting as a surrogate for disinfection by-product precursors. *See also* aromatic hydrocarbon; disinfection by-product precursor; humic material; natural organic matter; specific ultraviolet absorption; ultraviolet irradiation.

ultraviolet (UV) absorber A substance that absorbs ultraviolet radiation. Ultraviolet absorbers are added to plastic products (e.g., in plastic tanks and fittings) and rubber products to make them less likely to decay as a result of absorbing ultraviolet rays.

ultraviolet (UV) absorption Retention by a substance of incident ultraviolet radiation, followed by a compensatory change in the energy state of the substance's molecules. For example, the ultraviolet component of sunlight is absorbed as the light passes through glass and some organic compounds, with the radiant energy being transformed into thermal energy.

ultraviolet (UV) chamber An area where the water is irradiated with ultraviolet rays to disinfect the water or produce hydroxyl radicals (•OH). *See also* advanced oxidation processes; germicidal ultraviolet; hydroxyl radical; ultraviolet light disinfection.

ultraviolet (UV) demand The amount of ultraviolet irradiation required to inactivate certain microorganisms.

ultraviolet (UV) detector A device used in the analysis of both organic and inorganic compounds. An ultraviolet detector generates radiation (in the UV range) that interacts with certain types of compounds. The difference in radiation is detected and transformed into an electrical signal within an instrument. This signal, when calibrated, is related to particular compounds at a particular concentration. When used in water analyses, the device is often interfaced with an instrument such as a high-performance liquid chromatograph or ion chromatograph. *See also* high-performance liquid chromatography; ion chromatography.

ultraviolet (UV) dosage The amount of disinfectant (germicidal) ultraviolet irradiation delivered to the organisms in water being disinfected. Dosage is calculated as the product of ultraviolet intensity and contact time. It is measured in watt-seconds per square centimetre. US customary units are not used.

ultraviolet (UV) extinction A process of eliminating or reducing ultraviolet radiation. The extinction (A) of ultraviolet irradiation can be determined by

$$A = \varepsilon bc$$

Where:

ε = the molar extinction coefficient

b = the path length of the absorption cell, in centimetres

c = the molar concentration of the ultraviolet-absorbing substance, in moles per litre

The value of the molar extinction coefficient is characteristic of the absorbing functional group [e.g., carbon–carbon double bond, carbonyl group (C=O)] in a molecule. *See also* ultraviolet absorption.

ultraviolet (UV) irradiation Radiation in the region of the electromagnetic spectrum that includes wavelengths from 10 to 390 nanometres. *See also* radiation.

ultraviolet (UV) light Radiation having a wavelength shorter than 390 nanometres (the shortest wavelength of visible light) and longer than 10 nanometres (the longest wavelength of x-rays). This wavelength puts ultraviolet light at the invisible violet end of the light spectrum. Ultraviolet light may be used as a disinfectant or to create hydroxyl radicals (OH). *See also* advanced oxidation processes; germicidal ultraviolet; hydroxyl radical; ultraviolet light disinfection.

ultraviolet (UV) light disinfection A process for inactivating microorganisms by irradiating them with ultraviolet light. The ultraviolet waves disrupt the metabolic activities of the organisms, rendering them inactive and incapable of reproduction. The ultraviolet light does not leave a disinfectant residual, however, so a form of chlorine disinfection must be applied if a residual is desired. To allow the irradiation to reach the organisms effectively, the water to be disinfected must be relatively free of particles, as in a filtered water. *See also* germicidal ultraviolet.

ultraviolet (UV) light spectroscopy An analytical technique of determining the electronic structure of a molecule through that molecule's absorption of energy in the ultraviolet region of the spectrum.

ultraviolet (UV) ray *See* ultraviolet irradiation.

ultraviolet oxidation (UVOX) A process used to oxidize target compounds in water by irradiating them with ultraviolet light, either with or without the addition of hydrogen peroxide (H_2O_2). The ultraviolet light causes double bonds in organic compounds to vibrate, eventually liberating electrons and changing the molecular structure. When H_2O_2 is used, ultraviolet light irradiation results in the formation of hydroxyl radicals (•OH), which are very strong oxidizing agents. This is called an advanced oxidation process. *See also* advanced oxidation process; hydroxyl radical.

ultraviolet–visible (UV–Vis) irradiation Radiant energy in both the ultraviolet and visible wavelengths.

unaccounted-for water (UFW) Water use that does not go through meters (such as that lost from leaks) and thus is not accounted for by the utility. *See also* system leakage.

unamortized credit A credit entirely accounted for in the current account period.

unamortized debit A debit entirely accounted for in the current account period.

unbilled sale The estimated amount due a utility for services rendered but not yet billed.

unbuffered Pertaining to a solution lacking the solutes that would resist a change in pH.

uncertainty analysis *See* complete uncertainty analysis.

uncertainty factor (UF) A factor (or combination of factors) applied by the US Environmental Protection Agency in deriving a reference dose to account for differences in response to toxicity within the human population and between humans and animals. For example, quantitative data describing the dose–response relationships for toxic chemicals in humans are rare. Uncertainty factors are intended to compensate for such factors as intra- and interspecies variability, the small number of animals tested compared with the size of the exposed population, sensitive subpopulations, and possible synergistic effects. For instance, to avoid the possibility that humans are more sensitive than the species in which the toxicological data were obtained, the no-observed-adverse-effect level in animals is multiplied by 0.1 or some other value to provide an estimate

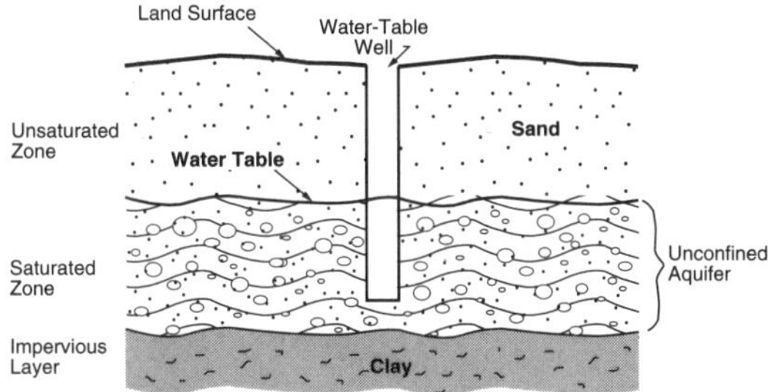

Unconfined aquifer

that would be more protective of humans. *See also* no-observed-adverse-effect level; reference dose; toxicity.

unconfined aquifer An aquifer for which the upper boundary of flow is a water table. *See also* aquifer.

unconfined compressive strength The load per unit area at which a soil will fail in compression. It can be determined by laboratory testing or estimated in the field by using a pocket penetrometer, thumb penetration tests, or other methods. *See also* thumb penetration.

unconfined groundwater The water in an aquifer where a water table exists.

unconformity spring A contact spring at the surface outcrop of the boundary between two different geologic formations, the upper one being water bearing and the lower one being impervious or less pervious.

unconsolidated formation A formation of sediment that is loosely arranged or unstratified (not in layers) or for which the particles are not cemented together (soft rock). Such a formation can occur either at the ground surface or at a depth below the surface. *See also* consolidated formation.

unconsolidated material Loosely arranged, uncemented material, such as soil or sand.

unconstrained demand The demand that would be experienced if not for natural replacement and conservation.

uncontrolled storage Water storage that is not controlled by gates or other control devices. An example is water above a spillway, which is not gated; hence, the water level elevation cannot be controlled.

underbed A layer of gravel or grout used to fill the bottom of a filter or softener tank, usually in a system with a manifold and lateral underdrain design. *Underbed* is not the same per se as the media support bed, which is a layer of granular material (e.g., gravel) used to support filtering media. *See also* filter underdrain.

undercut-slope bank A stream bank eroded by stream action.

underdrain *See* filter underdrain.

underflow (1) The filtered water collected after filtration. (2) The wash water distributed through the filter for backwash. (3) The solids collected at the bottom of a tank.

underflow conduit A permeable formation that underlies a surface stream channel, has a bottom and sides that are more or less definitely limited by formations of relatively low permeability, and contains groundwater that percolates downstream.

underflow flow That movement of water of a given density along or near the bottom of an impoundment or tank beneath water of a lesser density.

underground drainage Natural or artificial evacuation of waters beneath the surface of the Earth. Underground drainage is also called internal drainage.

Underground Injection Control (UIC) Program A regulatory program of the US Environmental Protection Agency established by the Safe Drinking Water Act. It is intended to protect underground sources of drinking water from contamination by injection of wastes.

underground injection control Any means of regulating waste disposal by underground injection.

underground runoff Water flowing toward stream channels after infiltration into the ground.

underground storage tank (UST) A tank situated underground and used for holding, e.g., industrial chemicals, gasoline, or other petroleum products.

underground utility (1) A subsurface system of pipes and conduits or a subsurface development for conveying water, wastewater, drainage, gas, power, or communications or for protecting electric wires or cables. (2) Any buried system.

underground water Water that occurs in the lithosphere. *See also* groundwater.

underground watercourse A known and defined subterranean channel that was created by natural conditions and contains flowing water. An underground watercourse is clearly distinct from an aquifer in that the flow moves in a channel, not through a porous medium. This type of channel is sometimes called an underground river.

undershot head gate A canal head gate at which the water passes from the river into the canal through gate openings that are formed, when the gates are raised, between the sill of the gate opening and the lower edge of the gate.

undershot wheel A waterwheel operated by the impact of flowing water against paddles or vanes on the periphery while the paddles or vanes are partly or wholly submerged in the moving stream of water.

undeveloped landscape *See* unlandscaped natural area.

Unfunded Mandate Reform Act Public Law 104-4, which was enacted March 22, 1995, and is intended to end the imposition, in the absence of full consideration by Congress, of federal mandates on state, local, and tribal governments without adequate federal funding in a manner that would displace other essential state, local, and tribal government priorities, unless Congress deliberately chooses to impose the mandate.

uniform corrosion Corrosion that results in an equal amount of material loss over an entire pipe surface.

uniform depth *See* normal depth.

uniform flow A type of flow for which the properties of flow do not change with respect to position. Laminar flow in a rectangular channel is uniform flow, whereas radial flow to a well is nonuniform.

uniform formation conditions (UFCs) Conditions found in a test protocol that is used to evaluate disinfection by-product formation in a distribution system and can be uniformly applied for any water. This protocol stands in contrast to the simulated distribution system test, which is based on site-specific conditions. The uniform formation conditions include an incubation time of 24 ± 1 hours, a temperature of 68° ± 1.8° Fahrenheit (20.0 ± 1.0° Celsius), a pH of 8.0 ± 0.2, and a 24-hour free chlorine residual of 1.0 ± 0.3 milligrams per litre. These values were based on typical (average) conditions used in US distribution systems. The conditions for determining disinfection by-product precursors (formation potential) are different. *See also* disinfection by-product; disinfection by-product precursor; formation potential; simulated distribution system.

uniform particle size The particle size distribution screen sizing for ion exchanger and filtration media as established by US Mesh Standards.

uniform rate A utility pricing structure in which the price per unit is constant as consumption increases. This rate is also known as a flat fee or a uniform block rate. *See also* flat fee.

uniformity coefficient (uc) A measure of how well a sediment is graded. The coefficient may be calculated as follows:

$$uc = \frac{d_{60}}{d_{10}}$$

Where:

d_{60} = the grain size, in millimetres, for which 60 percent (by weight) of the sediment's grains are finer, as shown on the sediment's grain size distribution curve

d_{10} = the grain size, in millimetres, for which 10 percent (by weight) of the sediment's grains are finer, as shown on the sediment's grain size distribution curve

uninterruptible power supply (UPS) A computer power supply (usually battery operated) that is used in the event of a commercial power failure. Most businesses isolate the computer by supplying power through the batteries at all times. Other businesses use motor-driven generators to supply power. The motors normally use commercial power, and if they fail they automatically switch over to a battery supply. These systems are extremely reliable and provide good clean power, free from the voltage fluctuations common to commercial power grids.

union (1) A device used to connect pipes. A union commonly consists of three pieces: the thread end, fitted with exterior and interior threads; the bottom end, fitted with interior threads and a small exterior shoulder; and the ring, which has an inside flange at one end while the other end has an inside thread like that on the exterior of the thread end. When a union is in use, a gasket is placed between the thread and bottom ends, which are drawn together by the ring. Gaskets are often supplanted by ground joints. Unions are used extensively because they permit connection with little disturbance of the pipe positions. (2) A number of people, societies, states, or the like associated or joined together for some common purpose, e.g., workers joined together in a labor or trade union.

unit cancer risk An estimate of the lifetime risk caused by each unit of exposure in the low-exposure region.

unit cost The cost of a unit of product or service, for example, the cost of pumping 1 million gallons of water.

unit filter backwash volume (UFBV) The design volume of water used to backwash a filter. The unit backwash volume is determined by multiplying the backwash rate by the time of filter backwash. *See also* backwash; filter.

unit filter run volume (UFRV) The volume of water that is processed through a filter in a single filter run between backwashes. The purpose of determining the unit filter run volume is to compare the throughput of a filter under different filtration rates. Although filter run times will decrease as filtration rate increases, the unit filter run volume may stay the same or increase because of the higher filtration rate.

unit hydrograph The discharge-versus-time relationship at the outlet for a particular watershed that results from 1 inch (25.4 millimetres) of water (rainfall excess) uniformly applied over the entire watershed at a uniform rate during a specified period of time. A unit hydrograph is also called a unitgraph.

unit hydrograph method A procedure, originated by L.K. Sherman, for determining the rates of surface runoff in a drainage basin based on observations of rainfalls and the corresponding observed hydrographs of surface runoff from the same basin. This method is based on the concept that in a given drainage basin, surface runoff from rainfall occurring in a unit of time will produce hydrographs of approximately equal abscissa values and with ordinate values varying with the quantity of rainfall minus infiltration and other subtractions.

unit power The number of watts produced from a theoretical 1-metre-diameter waterwheel or turbine operating under a 1-metre head. US customary units are not used.

unit risk The risk associated with a particular dose or dose rate of an agent (i.e., chemical, microbe, or physical agent) or circumstance (e.g., risk per mile driven in an automobile or per mile flown in airplanes). In the former case, the units are usually exposed in terms of dose rate, such as milligrams per kilogram per unit body weight for a lifetime.

unit speed The speed of a theoretical 1-metre-diameter waterwheel or turbine operating under a 1-metre head. US customary units are not used.

unit water use The quantity of water used at an industrial plant per unit of production. It may be expressed, e.g., in gallons (cubic metres) per pound (kilogram) or per unit of product.

***United States Pharmacopeia* (USP)** The official publication for drug product standards, including six water quality standards for pharmaceutical uses. The *United States Pharmacopeia* was established by the US Congress in 1884 to control makeup of drugs. *See also* pharmaceutical grade water.

unitgraph *See* unit hydrograph.

univalent Having a valence of one. This term is equivalent to *monovalent.*

universal gravitational constant (G) In the US customary system, 3.434×10^{-8} pounds force per square foot per slug squared. In Système International, 6.67×10^{-11} newton metres squared per kilogram squared.

universal soil loss equation A formula developed by the US Department of Agriculture for predicting rates of soil loss by sheet erosion:

$$A = (R)(K)(LS)(C)(P)$$

Where:

A = the average annual soil loss, in tons per acre

R = the rainfall factor

K = a soil erodibility factor

LS = a slope length and steepness factor
C = a cropping and management factor
P = the supporting conservation factor

For information on the details of these factors, contact the Natural Resources Conservation Service (formerly the Soil Conservation Service) of the US Department of Agriculture in your county or state.

Universal Transverse Mercator (UTM) A map projection and plane coordinate system based on 60 north and south trending zones, each 16° of longitude wide, that circle the globe. The Universal Transverse Mercator grid is commonly used in geographic information system technology in the United States. *See also* geographic information system.

UNIX A portable operating system for computers, developed at the University of California at Berkeley and at AT&T Bell Laboratories. Originally developed for scientific applications on large processors, UNIX is becoming increasingly popular on multipurpose computers as a multiuser, multitasking operating system.

unlandscaped natural area The portion of a development site where existing plant communities have not been removed or replaced.

unmetered rate A charge for water assessed monthly, bimonthly, quarterly, or annually based on amount of land and size of structure rather than on the total amount of water delivered to the premises. This type of rate is also called a flat fee. *See also* flat fee.

unplanned reuse Unintentional water reuse; reuse without a system designed and constructed for reuse purposes—for example, a community's use of a river as a public drinking water supply source where a significant portion of the river flow stream is derived from treated wastewater discharges from upstream communities.

unreasonable risk to health (URTH) A contaminant concentration set by the US Environmental Protection Agency or the state primacy agency under the Safe Drinking Water Act at or beneath which the risk to human health is considered reasonable. A utility's drinking water may not exceed unreasonable-risk-to-health concentrations if the utility wishes to receive a variance or exemption under the Safe Drinking Water Act.

unregulated contaminant A contaminant that the US Environmental Protection Agency requires water systems to monitor but for which the agency does not have a maximum contaminant level goal or maximum contaminant level. *See also* maximum contaminant level; maximum contaminant level goal.

Unregulated Contaminant Monitoring Rule (UCMR) A rulemaking being developed by the Office of Ground Water and Drinking Water of the US Environmental Protection Agency (USEPA). The UCMR is intended to satisfy Section 1445 of the Safe Drinking Water Act, which required that USEPA issue by August 1999, and every 5 years thereafter, a list of not more than 30 unregulated contaminants to be monitored by public water systems for inclusion in the National Drinking Water Occurrence Database. An unregulated contaminant is a contaminant for which a National Primary Drinking Water Regulation has not been established. *See also* National Drinking Water Occurrence Database.

unsanitary Not sanitary; unhealthy; liable to promote disease; contrary to principles known to promote or safeguard health. *See also* insanitary.

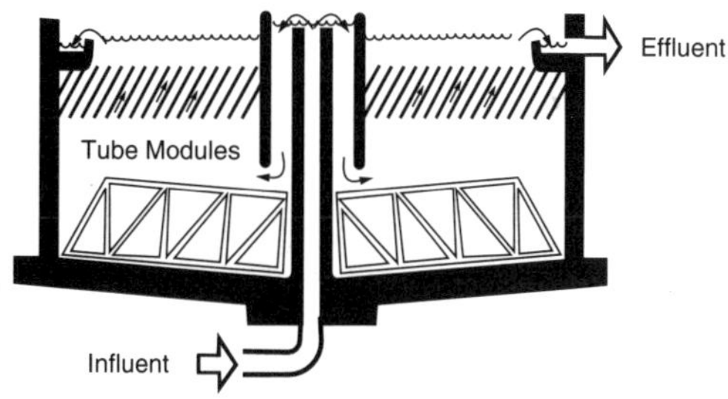

Upflow clarifier with tube settlers

unsaturated (1) In the context of a chemical compound, not having all available valence bonds along the alkyl chain satisfied. In such compounds, the extra bonds usually form double or triple bonds (chiefly with carbon). An unsaturated compound (e.g., ethylene, C_2H_4) has fewer hydrogen atoms than the corresponding saturated compound (e.g., ethane, C_2H_6). (2) *See* unsaturated zone. *See also* alkane; saturated; valence.

unsaturated zone The zone that is situated between the land surface and the water table and that contains, in the pore space, water at less than atmospheric pressure, as well as air.

unslaked lime *See* lime.

unstable (1) Readily decomposing or changing otherwise in chemical composition or biological activity. (2) Corrosive or scale forming.

unsteady flow Flow that occurs when, at any point in the flow field, the magnitude or direction of the specific discharge changes with time. Nonequilibrium flow to a well or varying flow from a constant displacement pump is unsteady flow. Unsteady flow is also called transient flow or nonsteady flow.

unsteady nonuniform flow Flow in which the velocity and the quantity of water flowing per unit time at every point along the conduit vary with respect to time and position.

unwater To remove water from an excavation or from within a cofferdam so that work within the unwatered area can be conducted in dryness.

upflow A pattern of water flow in which a solution (usually water or regenerant) enters at the bottom of the vessel or column and flows out at the top of the vessel or column during any phase of a treatment unit's operating cycle. This term is used to describe ion exchange system, filtration, or granular activated carbon adsorber flow patterns or water flow through filter media. An ion exchange system can have upflow during the treatment cycle and downflow during regeneration. Upflow is also called countercurrent operation. *See also* countercurrent operation.

upflow clarifier A unit that combines flocculation and settling in a single tank. The flocculation portion of the tank is designed to provide the necessary mixing for good floc formation, while the sedimentation portion acts as a true upflow-type clarifier with the surface overflow rate controlling particle removal. A sludge or solids blanket is not involved. *See also* sludge blanket clarifier; solids-contact clarifier; upflow contact clarifier.

upflow coagulation Coagulation achieved by passing liquid, to which coagulating chemicals may have been added, upward through a blanket of settling sludge.

upflow contact clarifier A unit process in which both flocculation and particle separation occur. Coagulated water is passed upward through a solids blanket, allowing flocculation and particle separation to take place in a single step. The sludge blanket is typically 6 to 10 feet (2 to 3.5 metres) below the water surface, and clarified water is collected in launder troughs along the top of the unit. Solids are continually withdrawn from the sludge blanket to prevent undesired accumulation. *See also* launder trough; sludge blanket clarifier; solids-contact clarifier; upflow clarifier.

upflow filter A gravity or pressure filtration system in which the water flows upward, generally first through a coarse medium and then through a fine medium, before discharging.

upflow softening A pattern of water flow used in softeners whereby the water flows upward through the ion exchange bed. The media are restricted in movement, usually because they are in the form of a packed bed. The regeneration brine usually flows downward in such systems.

upflow tank (1) *See* upflow contact clarifier. (2) A vertical-flow tank.

UPL *See* obligate upland.

upland (1) High land, especially at some distance from the ocean. (2) Ground elevated above the lowlands between hills or along rivers.

uplift The upward pressure against the base of an impervious dam, transmitted by water in the foundation from the water behind the dam; or pressure on the upper surface of any horizontal joint or crack in a dam. Such pressure tends to offset the weight of the dam and in some cases to overturn it; it reduces the dam's stability or resistance to overturning.

upper distributor The piping arrangement inside and at the top of softeners and filters to distribute the incoming water more uniformly over the resin or filter media bed. In small domestic units this distributor also distributes the brine for regeneration.

upper explosion limit (UEL) The maximum concentration of a substance in the air that will burn or detonate when ignited. The upper explosion limit is expressed as a percentage of vapor or gas in the atmosphere by volume.

upper-bound estimate An estimate of a value not likely to be lower than the true value. Such an estimate is often used in risk assessment.

upright The vertical part of a timber shoring system used for excavations. The upright braces the sheet piling that prevents cave-ins. *See also* timber shoring system.

UPS *See* uninterruptible power supply.

upstream In the opposite direction of a flow.

uptake The rate of reversible or irreversible buildup of a compound or element in an organism through inhalation, ingestion, absorption, or a combination of the three, with subsequent assimilation, utilization, clearance, or a combination of the three. *See also* clearance.

upwelling An upward flow of water from a subsurface current.

uranium (U) A metallic element with three natural radioactive isotopes (i.e., U-234, U-235, and U-238). Uranium is required to be regulated by the US Environmental Protection Agency by November 2000, as stipulated in a Nov. 18, 1996, court ruling. *See also* isotope; radioactive.

Urban and Regional Information Systems Association (URISA) An international professional organization for geographic information system users. *See also* geographic information system.

urban runoff Surface drainage from streets, parking areas, and other developed areas of cities and towns.

URISA *See* Urban and Regional Information Systems Association.

URTH *See* unreasonable risk to health.

US Agency for International Development (USAID) An agency within the US government's executive branch that administers US foreign economic and humanitarian assistance programs in the developing world, central and eastern Europe, and the independent states of the former Soviet Union. The agency functions under an administrator to conduct a worldwide network of programs in more than 100 countries.

US Army Corps of Engineers (COE, Corps) The US Army command responsible for managing the Army's real property, performing the full cycle of real property activities (requirements, programming, acquisition, operations, maintenance, and disposal); managing and executing engineering, construction, and real estate programs for the Army and the US Air Force; and performing research and development in support of these programs. The US Army Corps of Engineers manages and executes civil works programs, including research and development, planning, design, construction, operations, and maintenance, as well as real estate activities related to rivers, harbors, and waterways; administers laws for the protection and preservation of navigable waters and related sources, such as wetlands; and assists in recovery from natural disasters.

US Conference of Mayors (USCM) A nonprofit organization founded in 1932 and headquartered in Washington, D.C. Membership consists of cities with populations over 30,000, represented by their mayors. The US Conference of Mayors promotes improved municipal government by cooperation between cities and the federal government; provides educational information, technical assistance, and legislative services to cities; conducts research programs; and compiles statistics.

US customary units *See the Units of Measure list.*

US Environmental Protection Agency (USEPA) An independent agency of the executive branch of the US government, created Dec. 2, 1970, to permit coordinated and effective government action on behalf of the environment. It administers the implementation of environmental laws to abate and control pollution systematically through monitoring, standard setting, enforcement, and research activities. *See also* regional office.

US Environmental Protection Agency water treatment plant (WTP) model A computer model used by the US Environmental Protection Agency in developing the Disinfectant/Disinfection By-Products Rule. The empirically based model simulates disinfection by-product formation, removal of natural organic matter, inorganic water quality changes, and disinfectant decay in water treatment processes. This model is part of the Disinfection By-Product regulatory assessment model. *See also* disinfectant; Disinfectant/Disinfection By-Products Rule; disinfection by-product; disinfection by-product regulatory assessment model; natural organic matter.

US Fish and Wildlife Service (FWS) A bureau within the US Department of the Interior headquartered in Washington,

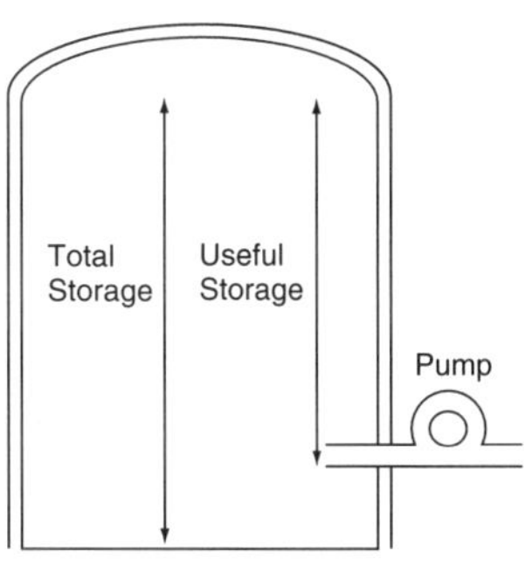

Useful storage

D.C., responsible for migratory birds, endangered species, certain marine mammals, inland sport fisheries, and specific fishery and wildlife research activities. Its mission is to conserve, protect, and enhance fish and wildlife and their habitats for the continuing benefit of the people of the United States.

US General Accounting Office (GAO) *See* General Accounting Office.

US Public Health Service (USPHS) An agency within the US Department of Health and Human Services responsible for promoting the protection and advancement of the nation's physical and mental health by coordinating with the states to set and implement national health policy and pursue effective intergovernmental relations; generating and upholding cooperative international health-related agreements, policies, and programs; conducting medical and biomedical research; sponsoring and administering programs for the development of health resources and the prevention and control of diseases and alcohol and drug abuse; providing resources and expertise to the states and other public and private institutions in the planning, direction, and delivery of physical and mental health care services; and enforcing laws to ensure the safety and efficacy of drugs and to protect against impure and unsafe foods, cosmetics, medical devices, and radiation-producing projects.

USAID *See* US Agency for International Development.

USCM *See* US Conference of Mayors.

useful life The amount of time during which conduits, equipment motors, and so on can perform their assigned functions.

useful storage Water storage that is readily available for discharge into a distribution system, such as water in an elevated storage tank or in a ground storage tank that can be pumped into the system. Water in a ground storage tank below the suction level of the pump would be storage but not useful storage.

USEPA *See* US Environmental Protection Agency.

USEPA WTP model *See* US Environmental Protection Agency water treatment plant model.

user The product water consumer. Users are also called customers.

user charge A fee paid by a utility customer for a specific service, such as a laboratory fee for a test run, a tap fee, or a meter charge. *See also* rate making; rate structure analysis.

user fee The periodic (monthly, bimonthly, quarterly, or other) charges made to the user of water service through the water utility's rate structure. *See also* water bill.

USGAO *See* General Accounting Office.

USP *See United States Pharmacopeia.*

USP grade water *See* pharmaceutical grade water.

USPHS *See* US Public Health Service.

UST *See* underground storage tank.

utilities communication architecture (UCA) Standards for data communications for the utility industry.

utility An organization—either created by a state or other government agency through legislative action or owned by an individual, partnership, or corporation—with the primary purpose of providing a designated area with, e.g., potable water, gas, electricity, or wastewater service at reasonable costs so that the people of the area can improve their health, safety, and welfare. *See also* water utility.

utility belt A leather or nylon belt with pouches, pockets, and loops to hold tools and parts.

utility line Any underground or overhead transmission line, pipe, cable, or wire for the conveyance of public or private water, wastewater, or natural gas or the transmission of electrical, radio, or telecommunications service.

utility method A method of accounting used by most investor-owned and regulated utilities for rate-making analysis (as opposed to the cash requirements method). The major features of this method are the use of accrual accounting, the inclusion of depreciation in operating expenses, and the inclusion of a return on investment as part of the net revenue requirement that must to be recovered from water rates. *See also* cash requirements method; net revenue requirement; rate of ruturn; return on investment.

utility oven A laboratory oven used primarily to dry labware and chemicals prior to weighing or to sterilize labware.

utilization factor (1) The ratio, usually expressed decimally, of the water flowing in a stream that is used for power development to the quantity that is available for use. The latter quantity is limited by the flow as indicated by the flow-duration curve and also, at higher stages, by the capacity of the waterwheels. (2) The factor representing the ratio of net power to gross power in a power-generating scheme.

UTM *See* Universal Transverse Mercator.

U-tube contactor An ozone contactor designed in a vertical, U-tube configuration. Detention time is provided by the volume available in the U-tube. This type of contactor reduces surface area requirements and allows higher concentrations of ozone to remain in solution as a result of the pressure created by the vertical column of water.

U-tube manometer A device for measuring gauge pressure or differential pressure by means of a U-shaped transparent tube partly filled with a liquid of known specific gravity, often water. A pressure slightly greater or less than atmospheric may be measured by connecting one leg of the "U" to the pressurized space and observing the height of liquid while the other leg is open to the atmosphere. Similarly, a small differential pressure may be measured by connecting both legs to separate pressurized locations (e.g., high- and low-pressure regions across an orifice or venturi). The liquid rises in one leg and drops in the other; the difference between the levels in the two legs is proportional to the difference in pressures.

UV An abbreviation for *ultraviolet.*

UV absorber *See* ultraviolet absorber.

UV absorption *See* ultraviolet absorption.

UV chamber *See* ultraviolet chamber.

UV demand *See* ultraviolet demand.

UV dosage *See* ultraviolet dosage.

UV extinction *See* ultraviolet extinction.

UV/H_2O_2 *See* hydrogen peroxide/visible–ultraviolet light process.

UV irradiation *See* ultraviolet irradiation.

UV light *See* ultraviolet light.

UV light disinfection *See* ultraviolet light disinfection.

UV light spectroscopy *See* ultraviolet light spectroscopy.

UV-254, UV_{254} *See* ultraviolet absorbance at 254 nanometres.

UV–Vis irradiation *See* ultraviolet–visible irradiation.

UVOX *See* ultraviolet oxidation.

V

V *See* volt *in the Units of Measure list.*

V *See* apparent volume of distribution.

VA *See* volt-ampere *in the Units of Measure list*; Department of Veterans Affairs.

vaccine A preparation of virus used to immunize a host. The virus may be live-attenuated, mutant, or inactivated.

vaccine-strain A type of virus from a specific source, maintained in successive cultures.

vacuum A space in which the pressure is far less than normal atmospheric pressure so that the remaining gases do not affect processes being carried on in the space. In a water system, a vacuum is created when water is discharged from a low point in a closed system without a point for air to enter the system. Such a vacuum can result in backflow from connections to the system.

vacuum breaker A mechanical device that prevents backflow due to a siphoning action created by a partial vacuum. A vacuum breaker allows air into the piping system, breaking the vacuum.

vacuum deaeration The use of special equipment operating under a vacuum to remove dissolved gases from liquid.

vacuum filter A device used to remove water from sludge by applying suction through a filter fabric. A partially submerged, horizontal cylindrical drum covered by a fine filter fabric rotates continuously through an open container filled with sludge. A partial vacuum is applied to the filter fabric, thus drawing solids to the surface of the filter fabric and allowing water (i.e., filtrate) to pass through the fabric. The filtrate is discharged or recycled for further treatment, and the solids are scraped from the fabric in the nonsubmerged zone for ultimate disposal. *See also* filter press.

vacuum filtration Filtration by means of a vacuum filter. This is one of the oldest mechanical dewatering techniques in continuous use. In municipal softening, this process is used to separate water from lime sludge for sludge disposal. *See also* vacuum filter.

vacuum freezing A form of desalination using a vacuum to help cool and fast-freeze source water having a high total dissolved solids content. This process separates the solids by concentrating them in the portion of the water that does not freeze or that freezes last, in a manner similar to what occurs in the cloudy centers of ice cubes.

vacuum pan An airtight container used to produce granulated water softener salt by a process involving the evaporation, in a partial vacuum, of brine turned to steam.

vacuum pump A pump used to provide a partial vacuum needed for such filtering operations as the membrane filter test.

vacuum relief valve A valve that admits air to relieve vacuum conditions and that permits accumulated air to escape. Such a valve is also called an air-and-vacuum valve or a vacuum valve.

vacuum valve *See* vacuum relief valve.

vadose Pertaining to liquids in the Earth's crust down to the groundwater level.

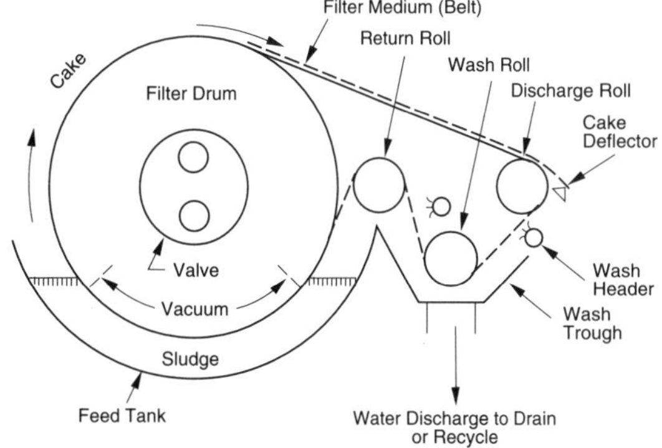

Sludge-dewatering vacuum filter (belt type)

vadose water All suspended water in the aeration zone, including water in the capillary fringe at the bottom, soil water at the top, and intermediate waters.

vadose zone The unsaturated portion of the soil column between the land surface and the water table. A better term is *unsaturated zone. See also* unsaturated zone.

vadose-water discharge Discharge of soil water not derived from the zone of saturation.

valence An integer representing (1) the number of hydrogen atoms with which one atom of an element (or one radical) can combine (in which case the valence is negative) or (2) the number of hydrogen atoms the atom or radical can displace (in which case the valence is positive). For example, calcium has a valence of +2 and chloride has a valence of −1.

validation (1) In water treatment, a determination upon testing that a representative sample of water treated by a particular equipment model has met the requirements of a specified standard. (2) In modeling, the process of testing a mathematical equation used in a model with an independent data set (i.e., with data that were not used in developing the equation). *See also* certification.

valley fill The sedimentary deposits laid down in a valley after the time of its formation.

valley spring A spring occurring on the side of a valley at the outcrop of the water table.

valley storage (1) The volume of water below the water surface profile. (2) The natural storage capacity or volume occupied by a stream in flood after that stream has overflowed its banks. This storage includes the quantity of water within the main channel (channel storage) and that which has overflowed its banks (lateral storage).

valley train Outwashed material laid down by a stream flowing from a glacier.

value The magnitude of a measurement; e.g., the pH of a water sample has a certain value, such as 6.5.

value engineering A process by which an engineering design is subjected to intense peer review by a designated group of experts in the field with the purpose of challenging design concepts and assumptions. The results are then reviewed by the design team. Costs are developed for alternative concepts and are compared with the initial design, thereby potentially reducing the overall cost of the project.

valve A mechanical device installed in a water distribution system or treatment plant to close off or regulate the flow of water in the system. *See also* ball valve; butterfly valve; check valve; cone valve; corporation stop; curb stop; gate valve; pet cock.

valve box A housing that encloses the operating nut of a gate valve and extends to the ground surface, allowing an access opening for an operating or valve key to be inserted and connected to the operating nut so that the valve may be opened and closed.

valve-exercising program The periodic and scheduled inspection and operation of valves in a water distribution system to maintain the valves in working condition for use in an emergency.

valve key A metal wrench with a socket to fit a valve nut and with a long handle for operating a gate valve from a distance of several feet.

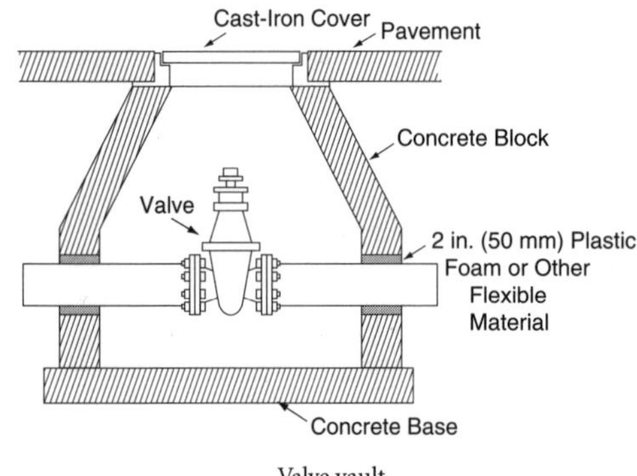

Valve vault

valve stem The rod by means of which a valve is opened or closed. The rod lifts and pushes down the gate.

valve tower A hollow cast-iron or masonry tower built within a reservoir and equipped with drawoff pipes at different levels for withdrawing water for supply purposes.

valve vault A concrete structure with an access opening that completely houses a valve, allowing access to the valve for maintenance, inspection, and operation. The structure is usually constructed just beneath ground or street level, with a round or rectangular frame and lid for entry into the vault.

valving The designated arrangement of irrigation circuits and the corresponding remote-control valves.

van der Waals's forces Weak attractive forces acting between molecules. These attractive forces between colloidal particles allow the coagulation process to take place.

van Dorn sampler A messenger-operated water-suspended sediment point sampler used to collect samples at a depth of 6.5 feet (2 metres) or greater. The long axis of the sampling cylinder can be either horizontal or vertical as the cylinder is lowered.

vanadium (V) A metallic element with various industrial uses, including in the manufacture of alloy steels.

vane shear A device used in determining a soil's unconfined compressive strength. Two cross blades attached to a vertical rod are forced into the soil and turned like a valve wrench. A torque reading is taken from the indicator once the soil fails. A second test indicates the soil's remolded strength.

vapor (1) The gaseous form of any substance for which the usual form is a liquid or a solid. (2) Visible particles of moisture suspended in air, such as mist or fog. *See also* evaporation.

vapor blanket The layer of air that overlies a body of water and that, because of its proximity to the water, has a higher content of water vapor than the surrounding atmosphere.

vapor compression (VC) *See* vapor compression distillation.

vapor compression (VC) distillation A desalting process in which heated saline water is introduced as a fine spray on the outside walls of heated tubes in a chamber for which the pressure is lower than required for vaporization, causing some of the water to vaporize. The tubes are heated by the compression of the vapor collected from the chamber and subsequent condensation of distillate (product water) inside the tubes from compressor discharge. Typically the vapor is compressed by using a mechanically driven compressor or a steam jet thermocompressor. Units of one to four stages are common.

vapor density The relative weight of a gas or vapor as compared with some specific standard, such as air.

vapor pressure The pressure at which a substance is in a state of balance between its liquid and solid states.

vaporimeter An instrument for measuring the volume or pressure of a vapor.

vaporization The process by which a substance (e.g., water) changes from the liquid or solid state to the gaseous state.

vapor-pressure deficit The difference between actual vapor pressure and the vapor pressure of a saturated atmosphere, for given conditions of temperature and pressure.

VAR *See* volt-ampere-reactive *in the Units of Measure list.*

variability Variation that occurs within the elements or members of a data set or population and is usually irreducible through further measurement. The variability of a factor (e.g., variation of drinking water intake among different people) cannot be accurately characterized by a single value, but must be described by one or more measures derived from relevant data (e.g., average, median, percentiles, geometric mean) as well as by the degree of variation (e.g., variance, geometric standard deviation, absolute deviation, dispersion, skewness).

variable Any characteristic, attribute, event, or mathematical term, that can have different values and for which measurements (quantitative or otherwise) are made in an epidemiologic study or other type of study or experiment.

variable costs Those costs of production that vary directly with production. In the water industry, expenses for energy and chemicals are variable costs. Variable costs contrast with fixed costs, which do not change with the volume of product produced. *See also* breakeven analysis.

variable, dependent *See* dependent variable.

variable, independent *See* independent variable.

variable, measured *See* measured variable.

variable, process *See* process variable.

variable spring A spring for which the difference between the maximum and minimum discharge is greater than the value of the average discharge.

variable-displacement pump A pump for which the discharge is inversely proportional to the total head delivered by the pump. For example, a centrifugal pump is a variable-displacement pump.

variable-frequency drive (VFD) A motor speed controller that uses adjustment of the applied power frequency to affect motor speed control. This type of controller is used with induction motors.

variable-speed drive (VSD) *See* variable-frequency drive.

variable-speed pump A pump that can be operated at a variable speed to change the flow rate delivered in the process. Variable-speed pumps are driven by motors that can draw variable current, thereby modulating the speed and changing the flow rate. Variable-speed pumps are often used to reduce pump cycling (i.e., on–off operation) and to allow continuous operation of downstream processes.

variance (1) A temporary relief from compliance with a maximum contaminant level that may be granted to a public water system by a state primacy agency under the Safe Drinking Water Act. A variance may be granted only if (a) the system cannot meet the maximum contaminant level in spite of the application of best available technology, treatment techniques, or other means (taking costs into consideration) because of the characteristics of the source water, and (b) the variance will not result in an unreasonable risk to health. A system may also be granted a variance from a specified treatment technique if it can show that, because of the nature of the source water, such treatment is not necessary to protect public health. (2) In statistics, a measure of the variation in a group of observations or a data set. The variance is (a) the sum of the squares of deviations from the mean divided by the number of degrees of freedom or (b) the standard deviation squared. *See also* exemption; standard deviation; unreasonable risk to health; variability.

varied flow A type of flow occurring in a stream or conduit having a variable cross section or slope. When the discharge is constant, the velocity changes with each change of cross section and slope.

varve A distinct band that represents the annual deposit of sedimentary materials, regardless of origin. This band usually consists of two layers: (1) a thin, dark-colored layer of clay laid down in the fall and winter and (2) a thick, light-colored layer of silt and fine sand laid down in the spring and summer.

vascular system A well-developed system of conducting tissue in plants to transport water, mineral salts, and foods. *See also* cardiovascular system.

vault (1) An arched structure, usually made of bricks, concrete, or stones, framing a ceiling or roof over a hall, room, or other wholly or partially enclosed construction. (2) An underground chamber used by a utility in its distribution system to house pumps, meters, and so on.

VC distillation *See* vapor compression distillation.

vector (1) A quantity having both magnitude and direction. (2) An organism that serves to transmit disease from one source to another. *See also* vector data.

vector data Data composed of *x–y* coordinates. In a geographic information system, these coordinates represent locations on the Earth. These data take the form of single points when graphically displayed. Vector data can define polygons, objects, and other complex entities that can be manipulated or displayed on the basis of their attributes. *See also* geographic information system.

vector sum The algebraic sum of two or more vector quantities (i.e., quantities having both magnitude and direction).

vector-borne transmission Transmission of an infectious agent to a human by an insect or animal that carries the agent. The agent may or may not multiply in the vector. *See also* fecal–oral transmission; transmission of infection.

vegetal discharge As applied to groundwater, the discharge of water through the physiologic functioning of plants. The water may be taken into roots of plants directly from the saturation zone or from the capillary fringe, which in turn is supplied from the saturation zone. The water is discharged from the plants by transpiration.

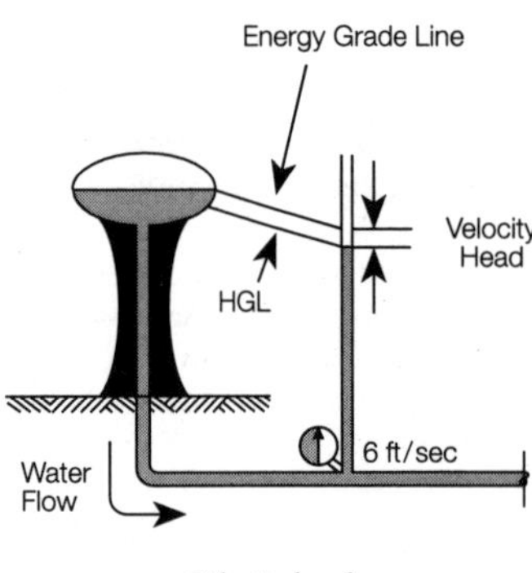

Velocity head

vegetation The sum total of macrophytes that occupy a given area.

vegetation tank A soil-filled tank used for measuring evaporation and evapotranspiration. Such a tank is also called a soil tank. *See also* lysimeter.

vegetative control Non-point-source pollution control practices that involve plants (vegetative cover) to reduce erosion and minimize the loss of pollutants.

vehicle-borne transmission Transmission of infectious or other etiologic agents to a human host through contaminated materials, such as water, food, diapers, or bedding, in which the agent may or may not have multiplied. *See also* fecal–oral transmission; transmission of infection; vector-borne transmission.

velocity The time rate of change of position of a body. Velocity represents the rate of movement of water flowing in a pipe, conduit, drainage way, or basin. It is expressed in units of distance per time, such as feet (metres) per second.

velocity of approach The mean velocity in a conduit immediately upstream from a weir, dam, Venturi tube, or other structure.

velocity coefficient A numerical factor, always less than unity, that expresses the ratio between the actual velocity issuing from an orifice or other hydraulic structure or device and the theoretical velocity that would occur if no friction losses were caused by the orifice, structure, or device. The square of the velocity coefficient is a measure of the efficiency of a structure as a waterway.

velocity gradient (*G*) A measure of the mixing intensity in a water process. The velocity gradient, which is expressed in units per second, is dependent on the power input, the viscosity, and the reactor volume. Very high velocity gradients (greater than 300 per second) are used for complete mixing and dissolution of chemicals in a coagulation process, whereas lower values (less than 75 per second) are used in flocculation to bring particles together and promote agglomeration. *See also* coagulation; flocculation; mean velocity gradient; orthokinetic flocculation.

velocity head The theoretical vertical height through which a liquid body may be raised by its kinetic energy. Velocity head is equal to the square of the velocity divided by twice the acceleration of gravity. *See also* head.

velocity meter A water meter using a rotor with vanes (such as a propeller) and operating on the principle that the vanes move at about the same velocity as the flowing water.

velocity profile The relationship between the velocity of fluid flowing adjacent to the conduit wall or membrane surface and that flowing at a distance from the wall or surface.

velocity units In the US customary system, feet per second. In Système International, metres per second.

velocity–area method A method used to determine the discharge of a stream or any open channel by measuring the velocity of the flowing water at several points within the cross section of the stream and summing up the products of these velocities and their respective fractions of the total area.

velocity-rod correction The correction to be applied to the velocity value given by a velocity rod in order to convert that value into mean velocity.

vena contracta The most contracted sectional area of a stream, jet, or nappe issuing through or over an orifice or weir notch. It occurs downstream from the plane of such notch or orifice.

Venn diagram A drawing that employs circles to represent logical relationships between, and operations on, sets by the inclusion, exclusion, or intersection of the circles.

ventilate To provide an exchange of fresh air in a confined space, room, or building, usually by means of a fan or blower.

ventilator A piece of equipment, such as a fan or blower, that provides air exchange to a building, room, or pit.

Venturi flume A flow-measuring device in which the flow is calculated based on the difference in elevation in different sections of the flume. The flume has a restricted cross-sectional area at its center, causing a change in velocity that results in changes in flow depth that can be related to flow on the basis of hydraulic principles. This nonmechanical method of flow measurement is reliable and effective as long as the geometry of the flume is well understood and is calibrated. *See also* Parshall flume.

Venturi meter An instrument for measuring fluid flow rate in a piping system by use of a nozzle section to increase velocity and produce a head difference that can be measured and mathematically converted to a flow rate value. A Venturi meter is also used in a filter rate-of-flow controller. *See also* filter rate-of-flow controller.

Venturi principle The phenomenon in a Venturi device whereby the constricted throat causes an increase in flow velocity and a decrease in pressure. The phenomenon can be used to measure flow or to draw other fluids into the flow through a small inlet.

Venturi tube A type of primary element used in a pressure differential meter that measures flow velocity based on the amount of pressure drop through the tube.

versenate Disodium dihydrogen salt of versene $(HOOCCH_2)_2NCH_2CH_2N(CH_2COOH)_2$. Versenate can be used in water analysis to detect water hardness or to colorimetrically measure hardness quality. *See also* ethylenediaminetetraacetic acid.

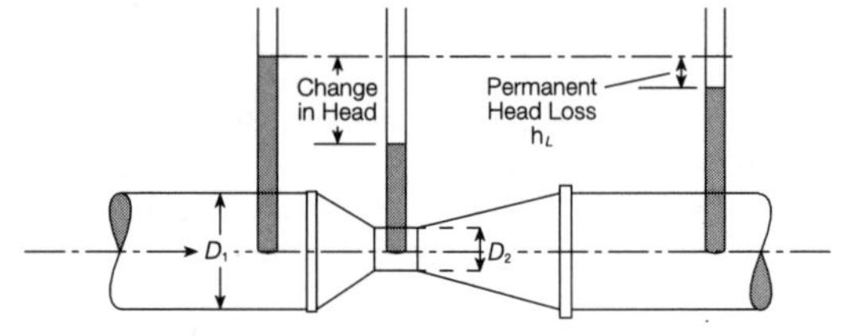

Venturi meter

vertex A point defining the beginning or end of a line feature or defining a change in alignment of the feature. A vertex is defined by a pair of *x–y* coordinates.

vertical mowing Mechanical cultivation of turfgrass by the use of vertically oriented knives to penetrate cored material, thatch, or underlying soil.

vertical pump (1) A reciprocating pump in which the piston or plunger moves in a vertical direction. (2) A centrifugal pump in which the pump shaft is in a vertical position.

vertical screw pump A pump that is similar in shape, characteristics, and use to a horizontal screw pump but that has the axis of its runner in a vertical position. *See also* horizontal screw pump.

vertical survey control monument A monument or benchmark that provides the control structure from which elevations can be measured, as used in compilation of topographic data. When aerial photography is the source for compilation, these monuments or other vertical control points are marked for clear visibility on the photographs. Specific densities of vertical control points must be available to achieve particular levels of accuracy.

vertical turbine pump A centrifugal pump, commonly of the multistage diffuser type, in which the pump shaft is mounted vertically.

vertical velocity curve A curve showing the relation between depth and velocity, at a given point and along a vertical line, of water flowing in an open channel or conduit. Such a curve is also called a mean velocity curve.

vertical-flow tank A sedimentation tank in which water enters near the bottom, rises vertically, and flows out at the top.

vertical-tube distillation Distillation using vertically oriented evaporation tubes. *See also* multiple-effect distillation.

very fine sand Sediment particles having diameters between 0.062 and 0.125 millimetres.

very poorly drained Pertaining to a soil condition in which water is removed from the soil so slowly that free water remains at or on the surface during most of the growing season.

very small system A public water system serving fewer than 500 people.

very, very small system A public water system serving 25 to 100 people.

Veterans Administration *See* Department of Veterans Affairs.

VFD *See* variable-frequency drive.

viability (1) An indication of whether a water system has the technical, financial, and managerial capabilities to provide drinking water that complies with federal and state drinking water regulations. (2) The ability to live, grow, and develop.

viability index One of several parameters used in the study of the toxicology of a chemical on the reproductive system. This index is defined as the number of offspring that survive 4 days past birth (in rodents).

viable (1) Capable of living. (2) Acceptable or beneficial.

viable-nonculturable (VNC) A classification used to describe a physiological adaptation exhibited by many culturable bacteria as a response to environmental conditions. The organisms remain alive as measured by one or more viability parameters (nutrient uptake, enzyme activity, photosynthesis respiration, energy charge) but fail to grow in vitro on culture media.

Vibrio A genus of gram-negative, straight or curved rod shaped, facultatively anaerobic bacteria belonging to the family Vibrionaceae. These organisms are primarily aquatic bacteria found in seawater and fresh water and in association with aquatic animals. This genus contains several species that are pathogenic for humans, as well as for marine vertebrates and invertebrates.

Vibrio cholerae A bacterial pathogen that causes cholera. Epidemic strains of *Vibrio cholerae* serovar O:1 (O1) may be divided into the classic and eltor biovarieties (biovars) and further divided into the Ogawa and Inaba subtypes. *Vibrio cholerae* is subdivided on the basis of the O antigens, and the O1 serovars produce cholera toxin that is regulated by chromosomal genes. In general, O1 serovars produce cholera toxin, but nontoxigenic O1 serovars also occur. Transmission of toxigenic *Vibrio cholerae* O1 to humans is by food or water that has been contaminated by fecal material from infected persons, and in epidemic outbreaks the disease rapidly becomes endemic. The natural reservoir of *Vibrio cholerae* O1, long thought to be humans, is unknown, but recent evidence has suggested that it may be the aquatic environment.

***Vibrio cholerae* El Tor (eltor)** *See Vibrio cholerae.*

video inspection An inspection of inaccessible or hazardous locations using closed-circuit television. This term is usually applied to truck-mounted, semiportable equipment used to inspect the insides of large-diameter water mains, pipes, and intakes. Video inspection is also called television inspection.

viewshed The area that is visible from a specific location, e.g., from a ridge top or building location. Obstructions to sight caused by terrain and other physical features reduce the viewshed.

vintage rate A program in which customers are classified and customer rates and charges are based on the date or period when a customer connects to and first obtains service from the utility system. Such rates and charges can include user rates; customer contributions of capital for system development, main extension, connection fees; or charges for ancillary services rendered. The concept has been used during periods of rising average costs to reflect the higher costs associated with serving new customers.

vinyl chloride (CH_2:CHCl) A vinyl monomer. Vinyl compounds are the basis of a number of important plastics. This volatile organic compound is regulated by the US Environmental Protection Agency. *See also* monomer; plastic; volatile organic chemical.

violation A failure by a water utility to meet all of the National Primary Drinking Water Regulations and all state and local regulations. Utilities that have failed to meet these regulations are said to be "in violation."

viradel method A method used for concentrating viruses from large volumes of water by using electropositive cartridge filters. The virus is eluted from the filter cartridge by an organic flocculation concentration procedure.

viral Pertaining to or caused by a virus.

viral gastroenteritis *See* waterborne disease.

VIRALT A semianalytical groundwater flow and contaminant transport (viral-transport) model (i.e., a computer program) commissioned by the US Environmental Protection Agency to predict virus concentrations in groundwater from various sources. It simulates the fate of viruses in groundwater flow systems and is used to help define wellhead protection areas. This program can be used to calculate capture zones and contaminant breakthrough curves for pumping wells.

virgin granular activated carbon (virgin GAC) Granular activated carbon that has not been used in a prior treatment application. Virgin granular activated carbon typically has the highest adsorption capacity. Though granular activated carbon can subsequently be regenerated or reactivated to restore its adsorptive properties, a small amount of the adsorption capacity is lost in each reactivation cycle.

virtual safe dose A dose for which the estimate of risk is low enough that virtually no one in a population is expected to be affected by the exposure. An officially stated virtual safe dose will usually reflect a definite point from a specified mathematical model for estimating low-level risks (e.g., one in a million lifetime extra risk from a given exposure to a chemical), but agencies may differ on the point. For example, some might accept one in a million as the low-level risk, whereas others may accept as high as one in 10,000.

virulence The ability of an infectious agent to produce pathologic effects.

virus (1) A minute organism not visible by light microscopy. A virus is an obligate parasite dependent on nutrients inside cells for its metabolic and reproductive needs. It consists of a strand of either deoxyribonucleic acid or ribonucleic acid, but not both, separated by a protein covering called a capsid. (2) A computer program that replicates itself through incorporation into other programs.

viscosity A measure of the capacity of a substance to internally resist flow. Viscosity is the ratio of the shear stress rate to the rate of shear strain. Viscosity is expressed in units of stress-time.

$$\mu = \tau\left(\frac{du}{dy}\right)$$

Where:

μ = viscosity, in pound force seconds per foot squared (newton seconds per metre squared or pascal seconds)

τ = shear stress, in pound force per foot squared (newtons per metre squared)

$\frac{du}{dy}$ = rate of shear strain, rate of angular deformation, or velocity gradient, in 1/second

Water at 68° Fahrenheit (20° Celsius) has a viscosity of 0.0208 pound force seconds per foot squared (1.002×10^{-3} pascal-seconds, or 1.002 centipoise). *See also* Newtonian flow; non-Newtonian flow.

viscous Having a sticky quality.

viscous flow A type of fluid flow in which a continuous steady motion of the particles exists, with the motion at a fixed point always remaining constant. Viscous flow is also called streamline flow. *See also* laminar flow.

vital dye A dye or stain that is capable of penetrating living cells or tissues and not inducing immediate evident degenerative changes. A vital dye is also called a vital stain.

vital stain *See* vital dye.

vital statistic Tabulated information about births, deaths, marriages, and so on, used for demographic purposes.

vitrified clay pipe A pipe made from various clays or combinations of clays that are shaped, dried, and fired to a point where the glass-forming components fuse to form a bond between the crystalline grains. Such pipe is used for drains at water treatment plants and for wastewater lines.

vitriol *See* sulfuric acid.

VNC *See* viable-nonculturable.

V-notch weir A triangular weir. In US customary units, the flow through a V-notch weir is given by

$$Q = 2.54\ H^{2.5}$$

Where:

Q = the flow, in cubic feet per second

H = the head, in feet

In Système International units,

$$Q = 1.38\ H^{2.5}$$

Where:

Q = the flow, in cubic metres per second

H = the head, in metres

VOC *See* volatile organic compound.

void The percentage by volume of the interstices to total bed volume. *See also* void space; void volume.

void ratio The ratio of the volume of voids to the volume of particles in a mass of particles, sand, ion exchange resin, or other porous medium.

void space A pore or open space in rock or granular material, not occupied by solid matter. It may be occupied by air, water, or other gaseous or liquid material. A void space is also called an interstice or a void.

void volume The volume occupied by the interstitial spaces between the particles of ion exchangers, filter media, or other granular materials in a bed or column. Void volume is often expressed as a percentage of the total volume occupied by the medium bed. *See also* void ratio.

volatile Capable of turning to vapor (evaporating). This term should be associated with a temperature and pressure because they determine whether or not a chemical is volatile. A chemical may be volatile under one set of conditions but not another.

volatile constituent A sample component that is readily lost by evaporation. Volatile constituents include dissolved gases as well as substances with low boiling points.

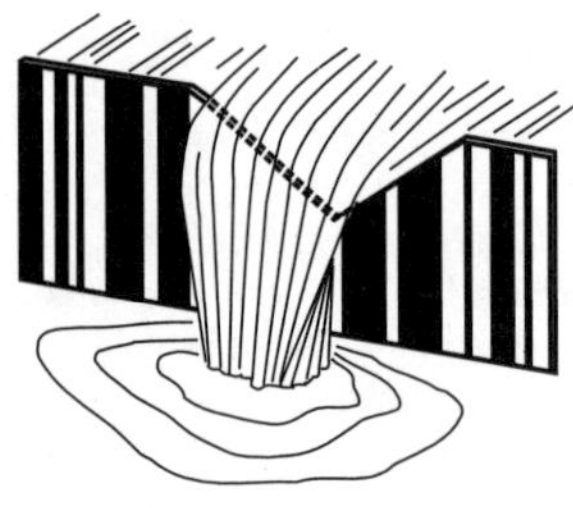

90° V-notch weir

volatile liquid A liquid that easily vaporizes or evaporates at room temperature.

volatile matter The apparent loss of matter from a residue ignited at 1,022±77° Fahrenheit (550±25° Celsius) for a period of time sufficient to reach a constant weight of residue, usually 10–15 minutes. *See also* volatile solids.

volatile organic carbon *See* purgeable organic carbon.

volatile organic compound (VOC) A class of organic compounds that includes gases and volatile liquids. Many volatile organic chemicals are used as solvents. A number of these compounds are regulated by the US Environmental Protection Agency. *See also* organic compound; solvent; volatile.

volatile organic contaminant *See* volatile organic compound.

volatile residue *See* volatile solids.

volatile solids (VS) In the laboratory analysis of the solids content of a substance (such as water), the portion of the total suspended solids, dissolved solids, or both that become expulsed or driven off after heating or burning of a given sample of the substance at a specified temperature and for a specified time. *See also* fixed solids; volatile matter.

volatile suspended solids (VSS) That fraction of suspended solids, including organic matter and volatile inorganic salts, that will ignite and burn when heated to 1,022° Fahrenheit (550° Celsius) for 60 minutes in an oxygen-containing atmosphere.

volatilization Loss of a substance through evaporation.

volcanic spring A spring with water derived at considerable depths and brought to the surface by volcanic forces.

volcanic water Water furnished by lava flows.

volt (V) *See the Units of Measure list.*

voltage The electrical pressure available to cause a flow of current (amperage) when an electrical circuit is closed. Voltage is measured in volts. *See also* electromotive force.

voltage-driven membrane An electrodialysis membrane.

volt-ampere (VA) *See the Units of Measure list.*

volt-ampere-reactive (VAR) *See the Units of Measure list.*

voltmeter An instrument for the measurement of a potential difference, in volts or in fractions or multiples thereof, between two points.

voltmeter-ohmmeter-ammeter A test instrument having a number of different ranges for measuring voltage, current, and resistance.

volume (1) The amount of space occupied in three dimensions; cubic contents or cubic magnitude. (2) The quantity, strength, or loudness of sound. *See also* decibel *in the Units of Measure list.*

volume charge That portion of a water rate that is related to the volume of water used. The volume charge can be a single unit charge for all volume used, or it can be increased or decreased at selected block levels. *See also* inclining block rates; rate structure analysis.

volume units In the US customary system, cubic feet. In Système International, cubic metres.

volume per volume (v/v) The initial volume of the solute divided by the final volume of the solution. When multiplied by 100, the value becomes the percentage by volume. *See also* solute; solution.

volumetric Based on the volume of some factor. For example, volumetric titration is a means of measuring unknown concentrations of water quality indicators in a sample by determining the volume of titrant or liquid reagent needed to complete particular reactions. *See also* gravimetric.

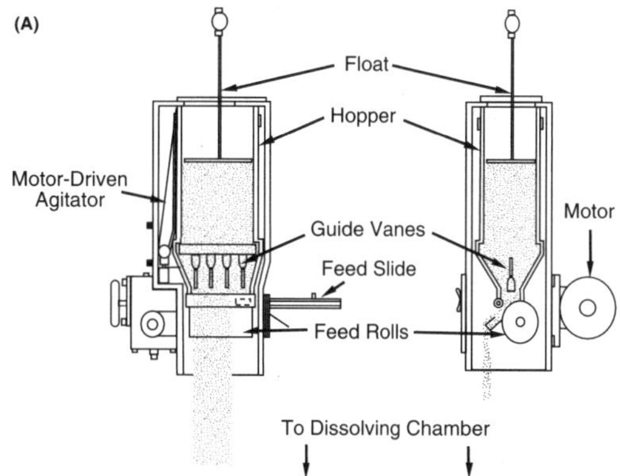

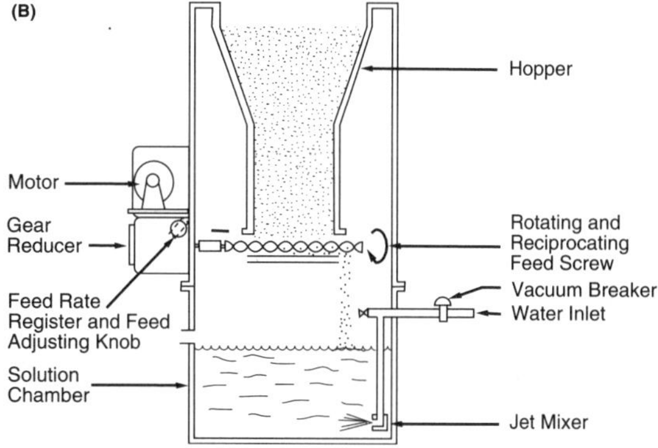

Two types of volumetric feeders: (A) Roll-type and (B) Screw type

volumetric analysis Any of a number of quantitative analyses based on the addition of known volumes of a reagent solution until a chemical equivalence point is reached. Water samples are commonly titrated with standardized solutions until a given end point is observed. The volume of standardized solution required is proportional to the concentration of the analyte.

volumetric dry feeder *See* volumetric feeder.

volumetric feeder A chemical feeder that adds specific volumes of dry chemical. *See also* gravimetric feeder.

volumetric flask A squat bottle with a long, narrow neck, used to prepare fixed volumes of solution. Each flask is calibrated for a single volume only.

volumetric pipette A pipette calibrated to deliver a single volume only, also known as a transfer pipette.

volute A spiral casing for a centrifugal pump designed so the pump rotation or speed will be converted to pressure in a smooth transition as the water leaves the impeller and is discharged from the pump.

volute pump A centrifugal pump with a casing made in the form of a spiral or volute as an aid to the partial conversion of

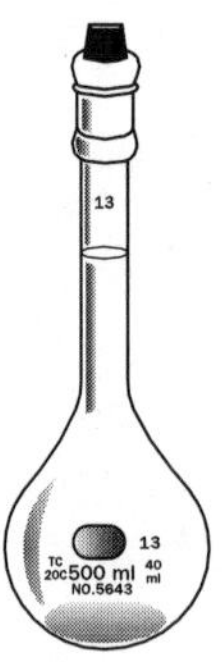

Volumetric flask

the velocity energy into pressure head as the water leaves the impellers.

vortex A revolving mass of water that forms a whirlpool. This whirlpool is caused by water flowing out of a small opening in the bottom of a basin or reservoir. A funnel-shaped opening is created downward from the water surface.

vortex pump A pump with a recessed impeller that creates a negative gauge pressure, thereby drawing liquid into the pump cavity and passing the liquid to the discharge. Solids are drawn into the vortex of the swirling liquid and are discharged by centrifugal force from the open pump chamber, seldom touching the impeller. Although vortex pumps (which are also called recessed impeller pumps) are inefficient, they are used to pump sludges or slurries (e.g., fluids with high solids concentrations) in instances when having the impeller contact the material and directly transport the fluid is undesirable.

vortex-shedding meter A flowmeter in which fluid velocity is determined from the frequency at which vortices are generated by an obstruction in the flow or a bend in a pipe. *See also* vortex.

VS *See* volatile solids.

VSD *See* variable-speed drive.

VSS *See* volatile suspended solids.

vulnerability The susceptibility or likelihood that a public water system or source water would be contaminated or otherwise adversely affected by an activity, practice, or unexpected emergency condition.

vulnerability assessment An evaluation of the likelihood that a water system could be contaminated by a particular chemical that is covered by the National Primary Drinking Water Regulations. Vulnerability assessments are conducted by the state primacy agency and serve as the basis for the state to allow a water system to lessen monitoring frequencies for certain contaminants.

v/v *See* volume per volume.

W

W *See* watt *in the Units of Measure list.*

w *See* width.

wadi A valley, ravine, or watercourse that is dry except during the rainy season. This term is usually used in reference to northern African and Arabian locales. *See also* intermittent stream.

waiver A document that is issued by the primacy agency and that permits a water system either not to monitor or to perform reduced monitoring for a specific contaminant.

waiver for filtration A document issued by the primacy agency stating that a water source can be distributed to users without prior filtration.

wale A horizontal member of a timber shoring system for excavation. The wale is located behind the uprights and sheeting piling that serve as the vertical retainers of the system. Cross-braces are between the wales across the excavation. *See also* timber shoring system.

warm spring A thermal spring with water having a temperature lower than 98° Fahrenheit (37° Celsius).

warning An indication of a hazard level between the levels indicated by the terms *caution* and *danger. See also* caution; danger; notice.

wash (1) The flow of a body of water. (2) The floodplain of an intermittent stream. This meaning is encountered principally in the arid regions of the United States and is usually applied to sand, gravel, and boulder deposits. (3) Loose or eroded surface material of the earth transported and deposited by running water. (4) A gully. (5) A marsh, bay, fen, or inlet. (6) The rough water above a breakwater or jetty that is caused by a wave breaking on it. (7) To clean a unit process.

wash boring (1) The operation of drilling or boring a hole through soft material with a wash drill. (2) A hole bored by means of a wash drill.

wash drill A pipe from which water is discharged under pressure to liquefy soils and lift them to the surface.

wash load As related to transport processes, that part of the suspended load of a stream that is composed of particle sizes not generally found in shifting portions of the streambed. Wash load is also called fine sediment load.

wash water Water that is used to clean a unit process. Wash water is typically identified as backwash water and is associated with the wastewater resulting from the cleaning of filter media to remove attached particles. *See also* filter backwash.

washout valve A valve installed in a low point or depression on a pipeline to allow drainage of the line. Such a valve is also called a blowoff valve.

wash-water rate The rate at which wash water is applied to a rapid granular filter during the washing process. This rate is usually expressed as the rate of rise of water in the filter in inches per minute, centimetres per minute, gallons per minute per square foot, metres per hour, or cubic metres per square metre per minute.

wash-water tank An elevated tank at a rapid granular filtration plant, into which water is pumped at a rate such that the tank will be filled between washings and set at a height such that the wash water will have a pressure of about 15 pounds per square inch (102 kilopascals) at the underdrain.

wash-water trough A trough placed above filter media to collect backwash water and carry it to the drainage system.

waste (1) Something that is superfluous or rejected; something that can no longer be used for its original purpose. (2) Any

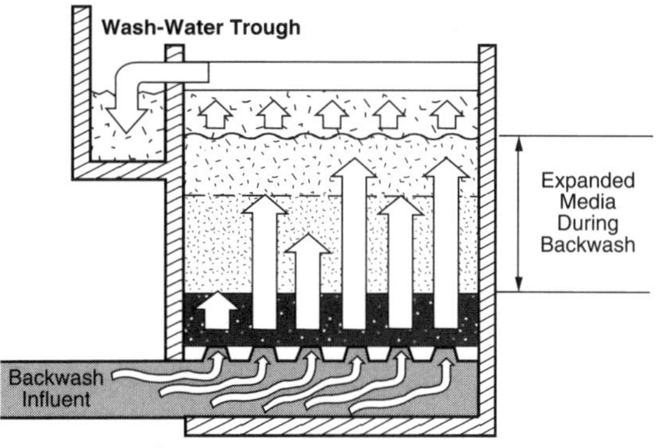

Wash-water trough

solid, liquid, or gaseous emission of no known use that results from human activity.

waste gate A gate inserted in a canal for wasting surplus water.

waste metering A process of measuring water flowing into an isolatable portion of a water supply pipe network over time. Increases in the flow rate over time can indicate the presence of leaks.

waste-heat evaporator An evaporator that uses the heat given off by another device, such as a diesel engine, a gas turbine, or an exhaust gas flue, as its source of heat for the feedwater.

wastewater The used water and water-carried solids from a community (including used water from industrial processes) that flow to a treatment plant. Stormwater, surface water, and groundwater infiltration may also be included in the wastewater that enters a wastewater treatment plant. The term *sewage* usually refers to household wastes, but it has been replaced by the term *wastewater.*

wastewater renovation Treatment of wastewater for reuse.

wastewater reuse *See* water reuse.

wasteway A channel used to convey water discharged into it from a spillway, escape, or sluice.

wasteweir A structure installed in a canal or open conduit to allow spilling or wasting of excess water from other sources that might reach the canal and cause damage by overflowing the banks. The water being spilled or wasted goes to a drainage system or nearby watercourse. A wasteweir usually consists of a short section of canal with a level crest, protected against erosion or cutting, with an elevation that is below the top of the canal bank but above the normal water surface in the canal. *See also* spillway.

WATEQ (Water Equilibrium) A family of computerized chemical equilibrium speciation programs developed by the US Geological Survey. The original WATEQ was written in the PL/1 language, as was its successor, WATEQ2. Later variants include WATEQF and WATEQ4F, which were written in the FORTRAN language, the latter for operation on personal computers. These programs take analytical water chemistry data and compute equilibrium aqueous speciation and saturation states of solids of interest, as well as various other parameters, such as ion balance errors and ion ratios. *See also* WATEQX.

WATEQX A FORTRAN version of the WATEQ family of programs, modified to make the thermodynamic database and computational structure more generalized to enable better user control and modification without recompilation of data. WATEQX was specifically developed with strict adherence to the American National Standards Institute language standard and can be run on both IBM-compatible personal computers and mainframes. *See also* WATEQ.

water (1) A transparent, odorless, tasteless compound of hydrogen and oxygen, H_2O. At a pressure of 1 atmosphere (101.3 kilopascals), water freezes at 32° Fahrenheit (0° Celsius) and boils at 212° Fahrenheit (100° Celsius). Water, in a more or less impure state, constitutes rain, oceans, lakes, rivers, and other such surface water bodies as well as groundwater. It contains 11.188 percent hydrogen and 88.812 percent oxygen by weight. It may exist as a solid, liquid, or gas. As normally found in the lithosphere, hydrosphere, and atmosphere, it may have other solid, gaseous, or liquid materials in solution or suspension. (2) To wet, supply, or irrigate with water.

water allowance *See* water budget.

water analysis An examination of constituents in water for chemical, physical, or biological characteristics. Such analyses can be quantitative or qualitative.

water association A nonprofit, nontaxing organization formed to provide drinking water related services for a specific area.

water audit A thorough examination of the accuracy of water agency records or accounts (volumes of water) and system control equipment. Water managers can use audits to determine their water distribution system efficiency. The overall goal is to identify and verify water and revenue losses in a water system.

water balance A method used to estimate evaporation from a water surface, based on a measurement of the continuity of water. This method uses the following equation:

$$E = I - O - S$$

Where (in any consistent set of units):

E = evaporation
I = inflow
O = outflow
S = the change in the lake, reservoir, or impoundment volume

water banking A process whereby unused water allocations are kept for use in future water allocations.

water bill A statement issued to a customer on a regular basis detailing the amount of money owed for water supplied during a described period of time based on a predetermined rate.

water bloom A prolific growth of plankton, including blue-green algae, that may occur and be so dense that it imparts a greenish, yellowish, or brownish color to water near the surface of a lake, pond, impoundment, or reservoir. A water bloom often causes tastes and odors in water. *See also* algal bloom.

water body Water impounded in a natural or artificial basin or moving in a definable watercourse, such as a river or stream.

water budget The quantity of water needed to maintain plants and other features in an ornamental landscape.

water budget approach A method of establishing water efficiency standards by describing limits on water consumption for irrigated landscapes.

water charger A device for filling the water passage of a pump for the purpose of priming the pump. *See also* pump primer.

water closet A flushable toilet.

water column (1) The water above the valve in a set of pumps. (2) A measure of head or pressure in a closed pipe or conduit. *See also* head.

water of compaction Water furnished by a reduction of pore space that has resulted from compaction of sediments.

water compensation The quantity of water that must be allowed to pass diversion works to satisfy the needs of holders of prior rights on the stream.

water of condensation Water formed by condensation of water vapor from the atmosphere, from soil, and from air in rock interstices, as well as water vapor arising from the magmosphere or from the volcanic forces of the lithosphere.

water conditioning *See* water treatment.

water conservation Promotion of the efficient use of water through the economically or socially beneficial lessening of water withdrawals, water use, or water waste. Conservation can forestall future supply capacity needs and can be implemented on the supply side as well as on the demand side. It can consist of both temporary measures used during emergencies and permanent measures used to improve long-term efficiency.

water consumption The quantity, or quantity per capita, of water supplied in a municipality or district for a variety of uses or purposes during a given period. Water consumption is usually taken to include all municipal use of water plus the quantity wasted, lost, or otherwise unaccounted for. *See also* unaccounted-for water.

water content In plant ecology, the water of the soil or habitat. Physiologically, water content is the available water supply; physically, it is the total amount of soil water. It is also called moisture content.

water correction Treatment—other than filtration, softening, and chlorination—used to stabilize water to reduce its tendency to corrode pipes, form incrustations, and so on. *See also* corrosion inhibitor.

WATER CO$T model A computer software program developed for the US Environmental Protection Agency to determine the capital and operational costs for a treatment system of a capacity greater than 1 million gallons per day (3,785 cubic metres per day).

water of crystallization The water that combines with salts when they crystallize. Water of crystallization combines in a definite quantity and is a molecular constituent of the crystalline compound.

water cycle *See* hydrologic cycle.

water of dehydration Water that was once in chemical combination with certain minerals but has been released by later chemical changes.

water demand A schedule of the water requirements for a particular purpose, as for irrigation, power, municipal supply, plant transpiration, or storage.

water demand risk analysis The risk associated with forecasting water demands. Inaccurate forecasts could lead to the design of facilities that are much larger than needed or are grossly inadequate to provide the desired level of service. A number of elements in a water demand forecast have an inherent amount of error built in: population and account growth, consumption per account by account group, the response to the utility's price policy, weather, and more. The utility forecaster will want to design facilities to meet demand, say, 90 percent of the time, which requires an assessment of the potential range of demand. Usually, a confidence interval can be constructed around each of the contributing variables; this interval can be used, with little difficulty, to produce mean or average forecasts. However, combining the probabilities of all the variables to predict a joint probability distribution for all variables simultaneously is a more complex process. Monte Carlo and simulation programs are now readily available that permit this kind of risk analysis to be applied to water demand and water supply forecasts over extended periods. *See also* financial risk; Monte Carlo analysis.

water of dilation Water in excess of water of saturation held by sedimentary material in an inflated state.

water district A quasi-governmental taxing entity formed to provide and finance services related to drinking water for a specific area.

water engine (1) An engine operated to raise water. (2) An engine propelled by water.

Water Engineering Research Laboratory (WERL) An office within the US Environmental Protection Agency's Office of Research and Development that was responsible for research support for the agency's water programs. The office was eliminated when the agency was reorganized in the late 1980s.

water equivalent of snow The depth of water that would result if an accumulation of snow, at a point or over an area, were reduced to water by melting.

Water Environment Federation (WEF) A technical society, formerly known as the Water Pollution Control Federation, representing chemists, biologists, ecologists, geologists, operators, educational and research personnel, industrial wastewater engineers, consultant engineers, municipal officials, equipment manufacturers, and university professors and students dedicated to the enhancement and preservation of water quality and resources. It seeks to advance fundamental and practical knowledge concerning the nature, collection, treatment, and disposal of domestic and industrial wastewaters, as well as the design, construction, operation, and management of facilities for these purposes. It disseminates technical information and promotes good public relations and regulations that improve water quality and the status of individuals working in this field.

water feature A fountain, pool, water sculpture, canal, channel, waterfall, constructed pond or lake, or other element using water as part of its design composition.

water filtration plant (WFP) A facility designed to treat water by means of filtration. In the most generic sense of this term, a facility that provides potable water and uses a filtration process, with or without sedimentation, is often called a water filtration plant. *See also* filtration.

water flooding A process in underground mining (e.g., in oil recovery) in which oil or a mineral from underground formations is replaced by an infusion of water to bring the underground substance to the surface for recovery. Water flooding is also known as oil well flooding.

water flux *See* flux; flux rate.

water gap (1) A pass in a mountain ridge through which a stream flows. (2) A constricted section of a stream valley where the stream bank material has been more resistant to erosion than the other material in the locality and has therefore confined the erosive action of the stream to narrower lateral limits. A water gap is also called a narrows, although the latter term is sometimes applied to a small gap.

water glass ($Na_2O{\cdot}xSiO_2$) The common name of a sodium silicate substance used for corrosion control in potable waters or as an ingredient in the manufacture of synthetic gel zeolite. *See also* corrosion control.

water hammer The phenomenon of pressure oscillation that occurs in pipes when a valve is opened or closed very rapidly, creating a sound similar to someone hammering on a pipe. When a valve position is changed quickly, the water pressure in a pipe increases and decreases in a very quick sequence, potentially causing serious damage to the system. The speed of the pressure wave created by an instantaneous shutoff of a system is given as

$$C = \sqrt{\frac{E_v}{2\rho}}$$

Where (in any consistent set of units):
C = the speed of the pressure wave
E_v = the bulk modulus of elasticity of water
ρ = the density of water
At 68° Fahrenheit (20° Celsius) and 1 atmosphere (101.3 kilopascals), C is 4,866.5 feet per second (1,483.3 metres per second).

water harvesting A process of capturing and using water runoff from storms on-site.

water hole (1) A natural hole or hollow containing water. (2) A hole in the dry bed of an intermittent stream. (3) A spring in a desert. (4) A pool, pond, or small lake.

water horsepower (WHP) The amount of power applied by a pump to actually add energy to water. In US customary units, water horsepower is computed as

$$\text{WHP} = \frac{Qh}{3{,}956}$$

Where:
WHP = the water horsepower, in horsepower
Q = the discharge, in gallons per minute
h = the added head, in feet
In Système International units, water horsepower is computed as

$$\text{WHP} = Q\gamma h$$

Where:
WHP = the water horsepower, in newton-metres per second
Q = the discharge, in cubic metres per second
γ = the specific weight of the liquid, in newtons per cubic metre
h = the added head, in metres
The water horsepower supplied by a pump is always less than the electric horsepower required by the pump because of mechanical inefficiencies and friction.

water of hydration Water that has been chemically combined with a substance to form a hydrate and that can then be removed (as by heating) without essentially changing the chemical composition of the substance.

water of imbibition The amount of water that a rock can contain above the saturation zone.

Water Industry Coordinating Committee (WICC) A committee consisting of representatives of the American Water Works Association, Association of Metropolitan Water Agencies, National Association of Water Companies, National Rural Water Association, and National Water Resources Association, formed to coordinate the regulatory and legislative activities of the respective organizations.

Water Industry Data Base (WIDB) A database of water utility information collected and maintained by the American Water Works Association. For information on data availability, call the Water Industry Data Base manager.

Water Industry Technical Action Fund (WITAF) A fund created by the Water Industry Coordinating Committee and administrated by the American Water Works Association for the purpose of funding technical studies related to developing or proposed US Environmental Protection Agency regulations and the activities of the Technical Advisory Workgroups. *See also* Technical Advisory Workgroup; Water Industry Coordinating Committee.

water of infiltration That part of surface water or precipitation that infiltrates into the upper parts of the lithosphere through small capillary pores of the soil and interstices in the rock.

water for injection (WFI) *See* pharmaceutical grade water.

water intake (1) The total gross quantity of water withdrawn from supply in a given interval, usually a day, month, or year. Water intake is also called pumpage. (2) The physical structure capable of removing water from a source, such as a lake. *See also* intake.

water jacket An outer casing that holds water or through which water flows and circulates to absorb heat and cool the interior of the mechanism or machinery that the casing surrounds.

water law The body of laws dealing with water—either surface water or groundwater—in its natural state, its maintenance in that state, and its use for various purposes. *See also* call; conditional water right; junior rights; riparian water right.

water level (1) The free water surface of a body of water. (2) The elevation of the free water surface of a body of water above or below any datum. The water level is also called the gauge height or stage.

water loss (1) In any water system, that portion of water that leaves the system without being used as intended. (2) In water supply and water power, that portion of the precipitation that reaches the surface of the ground or vegetation and is returned to the atmosphere by evaporation and plant transpiration. (3) In flood hydrology, that part of the storm rainfall that does not appear as runoff for the duration of the flood. *See also* unaccounted-for water.

water main The water pipe, located beneath the ground, from which domestic water supply is delivered to the service pipe leading to specific premises.

water management (1) The combination of regulations, laws, and policies and their implementation and enforcement by the

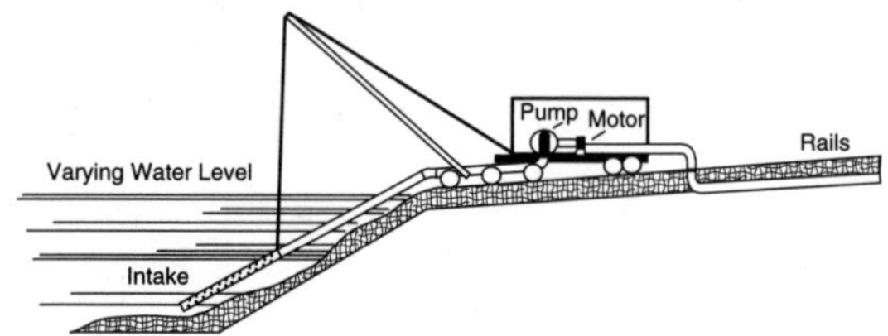

Moveable pump-driven water intake

private sector and various levels of government by which a water body is used, preserved, and enhanced for the benefit of present and succeeding generations. (2) The monitoring, planning, and administration of water resources for various purposes. Where these purposes include water supply, navigation, flood control, or irrigation, then the design, financing, operation, maintenance, and evaluation of performance of projects for these purposes are also included.

water mass transfer coefficient *See* solvent (water) permeability coefficient.

water meter An instrument, mechanical or electronic, used for recording (in cubic feet, gallons, or cubic metres) the quantity of water passing through a particular pipeline or outlet. In water-processing systems, meters may initiate certain functions, such as automatically starting the regeneration cycle in an ion exchange system.

water meter corrosion Corrosion of a water meter as a result of one of the following causes: excessive wearing of mechanical parts, erosion resulting from excessive velocity through the meter, or direct connection of dissimilar metals (i.e., erosion corrosion, galvanic corrosion).

water metering The measurement of water supplied to consumers by the installation of a meter at each consumer's service. *See also* metered system.

water mining A process of withdrawing groundwater accumulated in the ground at a rate faster than it is being replenished. Water mining is also called depletion. *See also* groundwater recharge.

WATER model A computer software program developed for the US Environmental Protection Agency to determine the capital and operational cost for a treatment system of a capacity not exceeding 1 million gallons per day (3,785 cubic metres per day).

Water for People (WFP) A nonprofit organization founded in 1991 that works to receive, administer, and expend funds to help developing countries organize and implement drinking water and sanitation projects that improve the health and welfare of the population. It also provides related technical referral and technical literature distribution services, as well as training programs, technical assistance services, and exchange of information.

water permeability coefficient *See* solvent (water) permeability coefficient.

water plant *See* drinking water treatment plant.

water pocket (1) A water hole in the bed of an intermittent stream. (2 The hole formed in the bed by a stream falling over a nearly vertical face.

water pollution The addition into water of harmful or objectionable materials and substances in large enough quantities to adversely affect the water's usefulness.

Water Pollution Control Federation (WPCF) The former name of the Water Environment Federation.

water power Any type of energy or power that can be or has been developed through use of the energy in falling or moving water. In common practice, this term is applied to electric energy or power generated by waterwheels used to drive electric generators.

water pressure plane A zone in a distribution system that covers a specific area and maintains pressure within a certain range either by pumping or by keeping the water within certain levels in an elevated tank.

water privilege (1) The right to use water; the right to use running water to turn machinery. (2) A stream or body of water capable of being used to drive machinery.

water processing *See* water treatment.

water provider A city, town, private water company, or public water wholesaler responsible for the direct or indirect distribution of water to its customers.

water purification *See* water treatment.

water purveyor An agency or person that supplies water (usually potable water) to someone else.

water quality The chemical, physical, and biological characteristics of water with respect to its suitability for a particular purpose. The same water may be acceptable for one purpose or use but unacceptable for another, depending on its characteristics and the requirements for the particular use. *See also* drinking water regulation; water quality criteria.

Water Quality Association (WQA) An organization of individuals and firms engaged in the manufacture, assembly, distribution, or retail selling of water treatment equipment, supplies, and services. The association promotes the acceptance and use of industry equipment, products, and services; provides activities, programs, and services designed to improve economy and efficiency within the industry; and conducts expositions and certification and equipment validation programs.

water quality certification (WQC) Certification by the state in which a discharge originates or would originate, under Section 401 of the Clean Water Act (33 U.S.C. Section 1341), that the discharge will comply with the applicable effluent limitations and water quality standards established by the state. Such certification is required for any applicant of a federal permit or license to conduct any activity that may result in a discharge of a pollutant into waters of the United States.

water quality criteria Scientific standards on which a decision or judgment may be based concerning the suitability of water of a specific quality to support a designated use.

water quality determinant *See* water quality parameter.

water quality guideline A numerical concentration limit or narrative statement recommending a designated water use.

water quality monitoring The process of measuring water quality characteristics, usually at regular intervals over a long period of time, at the same location.

water quality monitoring network A series of locations that have been chosen according to the objectives of the program and at which biota, sediment, and water samples are collected, usually at regular time intervals over a long period of time.

water quality objective A numerical concentration or statement of criteria established for water at a specified site based on the water's designated uses.

water quality parameter A measurable physical, chemical, radiological, or biological characteristic of an aquatic environment. Such a parameter is also called a water quality determinant or water quality variable. *See also* drinking water standard; primary maximum contaminant level; secondary maximum contaminant level.

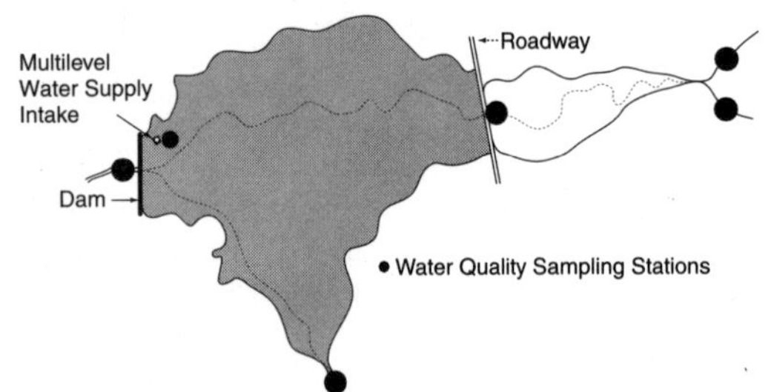

Water quality sampling stations for a reservoir

water quality regulation (1) A regulation issued by a government agency that applies to a surface water based on intended or designated use, typically setting one or more water quality standards. (2) A regulation issued by a government agency that applies to drinking water quality, typically setting one or more drinking water quality standards. *See also* drinking water regulation; Primary Drinking Water Regulation; Secondary Drinking Water Regulations.

water quality sampling station A designated location on a water body where biota, sediment, or water are sampled for the purpose of physical, chemical, radiological, or biological analysis.

water quality standard (1) A numerical limit set by a government agency for a constituent or pollutant that applies to a surface water, typically based on the water's intended use. (2) A numerical limit set by a government agency for a constituent or pollutant that applies to drinking water. *See also* drinking water standard; Primary Drinking Water Regulation; Secondary Drinking Water Regulations.

water quality variable *See* water quality parameter.

water ram A device for lifting water, powered by the water hammer or impulse of water produced by periodically stopping the flow of water in the supply pipeline. A water ram is also called a hydraulic ram. *See also* water hammer.

water rate Generally, all of the components of the rates for water on a water bill. A water rate includes the monthly service charge if applicable, including a minimum charge; the water rate per unit of water consumed, which could be a single unit rate or an increasing or decreasing block rate; and any surcharges, such as those to recover costs of providing water to elevated areas or for summer use. A water rate does not include fees for turning services on or off, connection charges, or assessments. *See also* assessment district; capacity charge; rate making; water bill.

water reclamation The treatment of wastewater to produce reclaimed water that is discharged to the environment or directly to a water reuse system. *See also* reclaimed wastewater.

water recycle The reuse of reclaimed water for an application in which the water was already originally used. This term is often associated with industrial water reuse.

water regain *See* water retention.

water requirement (1) Water needed for process makeup. (2) The water needed by crops for normal growth under field conditions.

water resources Water in various forms—such as groundwater, surface water, rain, snow, ice, clouds, and reclaimed or reused water—that is potentially useful for some purpose.

Water Resources Development Act (WRDA) A series of public laws that authorize federal spending for US Army Corps of Engineers water projects. The Water Resources Development Act of 1986 (P.L. 99-662) authorized $16 billion in total spending for construction of 263 water projects. The Water Resources Development Act of 1988 (P.L. 100-676) authorized 16 new water projects with an estimated federal cost of $1.7 billion. The Water Resources Development Act of 1990 (P.L. 101-640) authorized 26 new US Army Corps of Engineers water projects, and the Water Resources Development Act of 1992 (P.L. 102-580) authorized 23 new US Army Corps of Engineers water projects with an estimated federal cost of $1.56 billion.

water retention The amount of water, expressed as a percentage of the wet weight of an ion exchanger, retained within the resin bead and on the surface of fully swollen and drained ion exchange media. Water retention is also called water regain.

water reuse The reuse of reclaimed water for one or more purposes, generally only as part of a planned reuse scheme. *See also* direct reuse; indirect reuse; nonpotable reuse; planned reuse; potable reuse; unplanned reuse. *See also* water recycle.

water right A legal right to use the water of a natural stream or water furnished through a ditch or canal for general or specific beneficial use. *See also* water law.

water right acquired by prescription A legal right to use water gained by one's own efforts or actions, acquired through the continued use of the water over a long period of time. *See also* call; conditional water right; junior rights; riparian water right.

water rights A body of law that determines water ownership. *See also* appropriation doctrine; English rule; prior appropriation; water law.

water sale Delivery of water to consumers at a predetermined rate or price.

water of saturation Water that completely fills the interstices of rock or of earth when the particles are in contact. *See also* saturated zone.

water saver A toilet that meets the standard of 3.5 gallons (13 litres) per flush cycle as defined by American Society of Mechanical Engineers–American National Standards Institute Standard A112.19.6. A water saver should not be confused with a low-flow toilet using no more than 1.6 gallons (6 litres) per flush cycle. *See also* ultra-low-flush toilet.

water service pipe In plumbing, the pipe from the water main or other source of water supply to the premises being served.

water service schedule A statement required from water supply companies by state regulatory commissions, setting forth the rates to be charged to consumers.

water softener (1) A chemical compound that, when introduced into water used for cleaning or washing, will counteract the effects of the hard water minerals (calcium and magnesium) and produce the effect of softened water. For example, detergent additives and polyphosphates are water softeners. *See also* sodium hexametaphosphate. (2) A pressurized water treatment device in which hard water is passed through a bed of cation exchange media (either inorganic or synthetic organic) for the purpose of exchanging calcium and magnesium ions for sodium or potassium ions, thus producing a softened water that is more desirable for laundering, bathing, and dishwashing. This cation exchange process was originally called zeolite water softening. Most modern water softeners use a sulfonated bead form of styrene ($C_6H_5CH{:}CH_2$)/divinyl benzene ($C_6H_4(CH{:}CH_2)_2$) cation resin. *See also* hard water.

water softener salt Salt suitable for regenerating residential and commercial cation exchange water softeners. The most commonly used salt for this purpose is sodium chloride (NaCl) in crystal or pelletized form. Rock grade salt should be 96–99 percent NaCl; evaporated salt should be greater than 99 percent NaCl. Potassium chloride (KCl) may also be used for the regeneration cycle in the cation exchange process, thus minimizing amount of sodium added to both the softened water and the spent regenerant water going to the drain.

water softening The removal of calcium and magnesium ions, which are the principal causes of hardness in water. The cation exchange resin method is most commonly used for residential and commercial water treatment. In municipal and industrial water treatment, the process can be lime softening or lime–soda ash softening, called precipitative softening. *See also* ion exchange; lime–soda ash softening; lime softening; water softener.

water solubility The maximum concentration of a chemical compound that can result when the compound is dissolved in water. If a substance is water soluble, it can very readily disperse through the environment.

water source The basic origin of a water, either a surface source (such as a lake, river or reservoir) or a subsurface source (such as a well). After treatment and pumping via pipelines, the treated and pumped water becomes potable water. For water-desalting treatment, seawater or saline water is the source.

water spreading The artificial application of water to lands for the purpose of storing it in the ground for subsequent withdrawal by pumps for crops. *See also* spreading.

water standard A criterion of water quality established as a basis of control for various water use classifications. *See also* water quality standard.

water still A device used to produce distilled water by evaporation and condensation of water.

water stop (1) A valve installed in a water service pipe for control of the flow of water to a building. A water stop is also called a curb stop and box, meter stop, or cutoff. (2) A flexible material installed in the construction joint of a concrete structure to prevent leakage at the joint.

water storage Water that is contained in tanks, reservoirs, clearwells, and the distribution system, and is readily available for use. This can include water for a specific purpose, such as fire protection.

water of supersaturation Water that is in excess of water of saturation in a granular material in which the particles have lost contact—as in the case of quicksand, plastic clay, or flowing mud—and are more or less separated by water. *See also* water of dilation.

water supplier A person who owns or operates a public water system.

water supply (1) In general, the sources of water for public or private uses. (2) The furnishing of a good quality potable water under satisfactory pressure for domestic, commercial, industrial, and public service, as well as an adequate quantity of water under reasonable pressure for fire fighting.

water supply development The planning, design, and construction of sources of water for public or private uses.

water supply facility The works, structures, equipment, and processes required to supply and treat water for domestic, industrial, and fire use.

water supply source A river, brook, stream, lake, reservoir, impoundment, spring, or aquifer from which a supply of water is or can be obtained. The use of the term *raw water* for source water is discouraged.

water supply system (1) Collectively, all property involved in a water utility, including land, water source, collection systems, dams and hydraulic structures, water lines and appurtenances, pumping system, treatment works, and general properties. (2) In plumbing, the water distribution system in a building or complex of buildings, including appurtenances.

Water Supply and Water Resources Division (WSWRD) The division within the US Environmental Protection Agency's Office of Research and Development responsible for funding and conducting research on drinking water issues other than those pertaining to health effects. Health effects research is the responsibility of the Health Effects Research Laboratory. *See also* Health Effects Research Laboratory.

water system Collectively, all of the property involved in the *operation* of a water utility, including land, water lines and appurtenances, pumping stations, treatment plants, and general property.

water system appurtenance A structure, device, or appliance—other than a pipe or conduit—that is used in connection with a water distribution system, such as a valve, hydrant, corporation cock, or service.

water system master plan A plan developed or updated by many water utilities every 5 years or so to identify the facilities and capacities necessary to meet current and future water demands. Such a plan includes assessments of account growth by customer type and area, as well as related water demands; supply sources and alternatives; transmission, treatment, distribution, and storage facility requirements; compliance with water quality regulations through the treatment plants and the distribution system; fire flow requirements; hydraulic evaluation; and any other issues that deal with the parts or the whole of the water system. The water system is often modeled (simulated) and evaluated against hydraulic criteria to identify any deficiencies that might emerge at key nodes and milestone points, usually at 5- or 10-year increments up through buildout. Water rates are often evaluated as a part of the master plan, particularly if conservation goals are part of the long-run supply plan. *See also* capital improvement plan.

water table The surface in an unconfined aquifer where the water pressure is atmospheric. The water table is determined by measuring the water level in shallow wells installed a few feet into the zone of saturation. *See also* piezometric surface map.

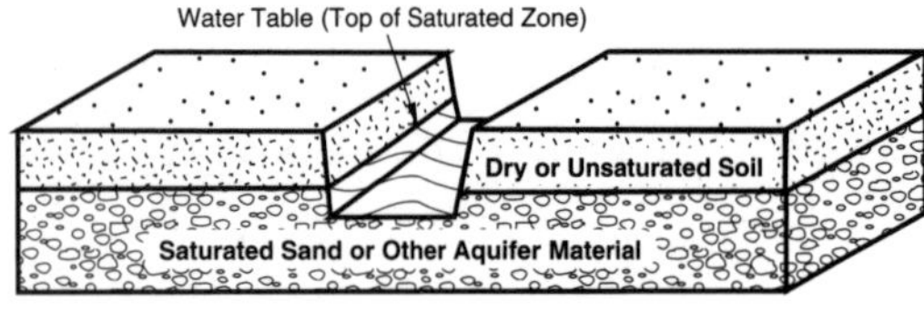

Water table

water table aquifer An aquifer confined only by a lower impermeable layer. *See also* unconfined aquifer.

water tension The negative water pressure in unsaturated porous materials. The negative water pressure represents the energy per unit weight of water that must be exerted on the soil column to remove water. This energy is required to overcome both the capillary forces that "suck" water into the porous material and the adsorptive forces between the water and porous materials. In contrast, a saturated material has water at positive pressure that represents the energy per unit weight of water that can be recovered by draining. Tension is also called suction. To remove water from unsaturated materials, suction—i.e., negative pressure—must be applied; to force water into saturated materials, pressure must be applied.

water tightness test A hydrostatic test for newly installed piping in a distribution system. It involves pressurizing the line up to a specific pressure and determining the rate at which water is lost while the pressure is maintained over a length of time.

water tower A tower containing a tank in which water is stored, normally for providing local storage in a distribution system where ground-level storage would provide inadequate pressure. *See also* ground storage; standpipe.

water transport coefficient *See* solvent (water) permeability coefficient.

water treatment (1) The act of removing contaminants from source water by addition of chemicals, filtration, and other processes, thereby making the water safe for human consumption. (2) The act of adjusting water quality to satisfy the requirements of any end use. *See also* potable water.

water treatment facility *See* drinking water treatment facility.

water treatment plant (WTP) *See* drinking water treatment plant.

water treatment plant waste Residuals resulting from the treatment of water. Waste may take the form of a liquid (e.g., filter backwash water), solid (e.g., dried sludge cake), or gas (e.g., off-gas from air-stripping or activated carbon regeneration furnaces). *See also* activated carbon regeneration; air-stripping; filter backwash.

water treatment works (WTW) *See* drinking water treatment plant.

water tunnel (1) In a hydraulic laboratory, a device similar to a wind tunnel that uses water for the working fluid. (2) In water supply systems, a tunnel used to transport water.

water usage fee A fee based on the quantity of water used by each customer of a community water system.

water use A classification of the utilization of waters in natural watercourses for such purposes as potable water supply, recreation and bathing, fish culture, industrial processes, waste assimilation, transportation, and power production.

water utility The collective set of features of a system that provides potable water to the public, including the source water facilities, treatment plant, and water distribution network. This term is typically used to designate all aspects of a given system. A less preferred term is *water works*.

water vapor The gaseous form of water; molecules of water present as a gas in an atmosphere of other gases. Water vapor moves from higher to lower vapor pressure regions to maintain vapor pressure equilibrium. It is also called aqueous vapor.

water waste survey A survey undertaken to locate, measure, and control locations where water leaks away or is in any way wasted.

Water and Wastewater Equipment Manufacturer's Association (WWEMA) A trade organization of manufacturers of equipment for water, wastewater, or industrial waste treatment or disposal plants. The group is based in Washington, D.C.

water well An artificial excavation that derives water from the interstices of the rocks or soil that it penetrates.

water wing A wall on the bank of a river adjoining a bridge abutment to protect the abutment's foundation from the action of the current. A water wing is sometimes referred to as a bund or a wing dike. *See also* wing wall.

water works *See* water utility.

water year A continuous 12-month period during which a complete annual cycle occurs, arbitrarily selected from the presentation of data relative to hydrologic or meteorologic phenomena. The US Geological Survey uses the period October 1 to September 30 in the publication of its records of stream flow. A water year is also called a climatic year.

water-based disease A disease for which the pathogen spends an essential part of its life in water or is dependent on aquatic organisms for completion of its life cycle. Examples include primary meningoencephalitis (caused by *Naegleria*), schistosomiasis (caused by *Schistosoma*), and dracontiasis (caused by *Dracunculus*).

water-based pathogens A grouping of pathogenic microorganisms that are naturally present in water, such as *Legionella* and *Mycobacterium* (which cause respiratory disease), blue-green algae (which produce toxins), and certain parasites, such as *Naegleria*.

water-bearing bed A porous, water-bearing formation. Such a formation is also called a water-bearing deposit. *See also* aquifer.

water-bearing deposit *See* water-bearing bed. *See also* aquifer.

water-bearing formation A geologic formation that contains water. Usually, a geologic formation that contains and transmits water is called an aquifer. Strictly speaking, however, a geologic formation that is a poor transmitter of water but still contains water could be considered a water-bearing formation. An example might be a clay-rich bed of material that forms the confining layer in a confined aquifer system. The clay-rich layer contains water but transmits it slowly. *See also* aquifer.

water-bearing medium A geologic formation that will supply water to wells or drains at a rate sufficient to justify extraction of the water. Such a formation is also called a water-bearing stratum. *See also* aquifer.

water-bearing stratum *See* water-bearing medium.

waterborne pathogen Any of a group of pathogenic microorganisms that are excreted or shed by infected humans, infected animals, or both in their feces and are then acquired by ingestion. Waterborne organisms may be transmitted directly by ingestion of water or indirectly by, e.g., contamination of food during washing or processing or transmission through the skin or the anal area during recreational use of water (as in hot tubs). *See also* fecal–oral transmission.

waterborne disease A disease transmitted through the ingestion of contaminated water. Water acts as a passive carrier of the infectious agent or chemical. Water is only one mode of transmission, and epidemiological investigation is necessary to identify water as the source of infection. Cholera (caused by *Vibrio chloerae*) and typhoid fever, the classic waterborne diseases, continue to occur but are important primarily in the developing countries of the world. Outbreaks of these diseases are occasionally reported in the United States and other more industrialized countries. During the period 1971–1992, 684 waterborne outbreaks were reported in the United States, resulting in 164,158 cases of illness, 1,170 hospitalizations, and 12 deaths. Approximately 50 percent of these outbreaks were classified as acute gastrointestinal illness of undetermined etiology, characterized by such symptoms as abdominal cramps, nausea, vomiting, and diarrhea occurring 24 to 48 hours after consumption of contaminated water. Protozoan (*Giardia*, *Cryptosporidium*, and occasionally *Entamoeba histolytica*), bacterial (*Campylobacter*, *Shigella*, *Salmonella*, *Yersinia*), toxigenic (*Escherichia coli*), or viral (Norwalk agent, rotavirus) pathogens were identified as the cause of 19 percent, 14 percent, and 8 percent of the outbreaks, respectively. Acute chemical poisonings (copper, fluoride, and nitrate were the most frequently identified) accounted for 9 percent of the outbreaks. *See also* fecal–oral transmission.

waterborne disease outbreak A significant occurrence of acute infectious illness epidemiologically associated with the ingestion of water from a public water system that is deficient in treatment, as determined by the appropriate local or state agency.

waterborne illness *See* waterborne disease.

water-break An arrangement of obstructions placed in a ditch or ravine to reduce wash.

watercourse (1) A running stream of water. (2) A natural or artificial channel for the passage of water.

watercourse bed The part of a watercourse that carries water at ordinary stages.

water-demand curve A curve of the summation of water demands over an assumed period of time.

water-dependent activity An activity that requires access or reasonable proximity to water to fulfill its basic purpose.

water-distributing pipe In plumbing, a pipe that conveys water from the service pipe to the plumbing fixtures and other outlets.

water-efficient landscape An ornamental landscape that minimizes water requirements and consumption through proper design, installation, and management.

waterfall A very steep or perpendicular descent of the water of a stream.

water-flow formula One of a number of formulas for determining velocity or discharge of water in a conduit or channel, e.g., the Chezy, Darcy–Weisbach, Hazen–Williams, Kutter, and Manning formulas. *See also* Chezy open-channel formula; coefficient; Darcy–Weisbach formula; Hazen–Williams formula; Kutter formula; Manning formula.

water-hammer arrester A device used on a service pipe to protect plumbing fixtures from the effects of water hammer. *See also* water hammer.

waterhead *See* head.

water-level gauge A gauge, recording or otherwise, that indicates the water level in a well, surface body of water, reservoir, or other receptacle.

water-level recorder A device for producing, graphically or otherwise, a record of the rise and fall of a water surface with respect to time. Such a device is also called a water-stage register.

waterlogged (1) Saturated with water. (2) In a condition such that the groundwater stands at a level that is detrimental to plant growth. Waterlogged land may result from excessive irrigation or seepage, coupled with inadequate drainage.

waterlogged soil A soil so continuously wet as to drive out all gases and in which normal upland plants cease to grow.

waterlogged tank A tank (as in a domestic water well pumping system) in which too much water has accumulated and has replaced some of the air in the tank's air cushion, causing a disruption in the normal pressure pattern needed for pumping and uniform water flow.

waterlogging The accumulation of excessive moisture in the soil within the zone or depth desirable for favorable root development of plants.

water-meter load factor The ratio, expressed as a percentage, of the annual registration of a meter in cubic feet (cubic metres) to the discharge capacity of the meter, operating continuously, in cubic feet per year (cubic metres per year).

waterphone A device used to locate points of leakage in an underground water pipe system under pressure.

water-pressure engine An engine similar to a steam engine but using water under pressure instead of steam.

water-producing zone A porous, water-bearing geologic formation; an aquifer. *See also* aquifer; saturated zone.

water-related disease A disease transmitted by contaminated water through ingestion, inhalation of aerosols, dermal contact, poor sanitation and hygiene, or insects that breed in water. *See also* water-based disease; waterborne disease; water-vectored disease; water-washed disease.

water-retaining capacity The quantity of water retained against the pull of gravity by rock or earth that has become saturated and has been allowed to drain completely to a remote body of mobile water by way of continuous capillary interstices. *See also* specific retention.

water-right value The monetary value of the right to use water from a given source. This value may be expressed in dollars per unit rate of flow or per unit of volume diverted during a year.

waters of the United States The sum total of the following (as specified in 33 Code of Federal Regulations 328.a): (1) All waters that are currently used, or were used in the past, or may be susceptible to use by the United States in interstate or foreign commerce, including all waters subject to the ebb and flow of the tide. (2) All interstate waters, including interstate wetlands. (3) All other waters, such as intrastate lakes, rivers, streams (including intermittent streams), mudflats, sandflats, wetlands, sloughs, prairie potholes, wet meadows, playa lakes, or natural ponds, for which use, degradation, or destruction could affect interstate or foreign commerce, including any such waters (a) that are or could be used by interstate or foreign travelers for recreation or other purposes, (b) from which fish and shellfish are or could be taken and sold in interstate or foreign commerce, or (c) that are used or could be used for industrial purpose by industries in interstate commerce. (4) All impoundments of waters otherwise defined as waters of the United States under this definition. (5) Tributaries of waters identified under headings (1) to (4) above. (6) The territorial seas. (7) Wetlands adjacent to waters (other than waters that are themselves wetlands).

water-saving device *See* water-saving fixture.

water-saving fixture A plumbing fixture or other appliance specifically designed to lessen water use when installed.

water-saving kit A set of devices sometimes distributed by utilities to reduce customer water usage.

watershed (1) The drainage basin area contained within the bounds specified by a divide and above a specified point on a stream. In water supply engineering, a watershed is also called a catchment area; in river control engineering, it is called a drainage area, a drainage basin, or a catchment area. (2) The divide between drainage basins.

watershed boundary The high point on a watershed that directs the flow of water in a certain direction. This boundary is also called the watershed divide.

watershed divide *See* watershed boundary.

watershed management A combination of planning, design, construction, operation, and government programs for the highest and best uses of a drainage area.

watershed planning (1) The process of planning the use and treatment of land and water that will most nearly meet all the objectives of soil and water conservation from the standpoint of both the individual farms and all the residents of the watershed. (2) The process of considering the interrelated uses of water from several watersheds to meet regional water supply requirements as well as local watershed objectives. This process may be part of state or national water resource planning.

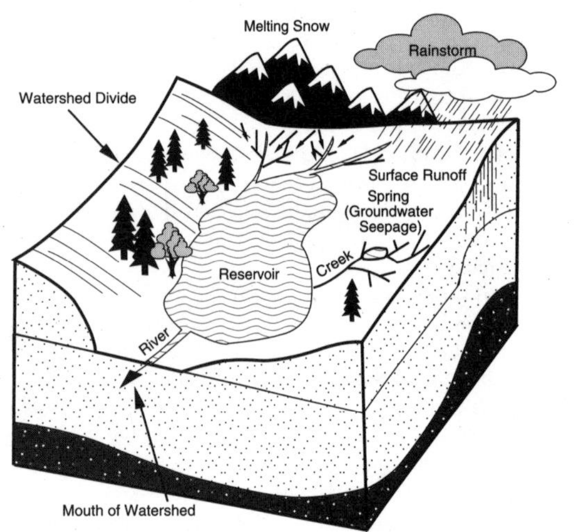

Schematic of a typical watershed

watershed sanitation (1) The removal of actual sources of pollution on a watershed, as well as the prevention of potential sources. (2) Measures undertaken to minimize unavoidable pollution or to improve the quality of water for drinking.

watershed survey A technical review of a drainage area, usually for purposes of planning, design, and watershed control programs.

watersoaked Completely filled with water. This term is applied to a porous body that has its pores and other openings completely filled with water. *See also* waterlogged; waterlogged soil.

waterspout A conduit or orifice from which water is spouted.

water-stage register *See* water-level recorder.

water-stage transmitter A device designed to transmit electrically a reading of the water level as it changes, for recording on a mechanically actuated chart.

water-table contour A line connecting all points on a water table that have the same elevation above a given datum. *See also* isopiestic line.

water-table decline A downward movement of the water table, also called phreatic decline. *See also* mining groundwater.

water-table depression cone *See* circle of influence; cone of depression; drawdown; well cone of influence; zone of influence.

water-table fluctuation The alternate upward and downward movement of a water table. Such movement is also called phreatic fluctuation.

water-table gradient The rate of change of altitude per unit of horizontal distance in the water table at a given place and in a given direction. If the direction is not mentioned, it is generally understood to be the direction along which the maximum rate of change occurs. Where the rate of change is uniform between two points, the gradient is equal to the ratio of the altitude difference between the two points to the horizontal distance between them.

water-table isobath An imaginary line on the ground surface or a line on a map connecting all points at which the vertical distance above a certain water table is the same.

water-table map A contour map of the upper surface of an unconfined groundwater body.

water-table outcrop A place where the water table intersects the surface of the ground.

water-table profile (1) A line formed by the intersection of a water table with a vertical plane. (2) A graphical representation of such a line.

water-table rise Upward movement of a water table.

water-table slope *See* water-table gradient.

water-table spring A spring that occurs at the intersection of a water table and the surface.

water-table stream A concentrated groundwater flow at the water table in a formation or structure of high permeability.

water-table wave A wavelike movement of the water table in the direction of its slope. *See also* groundwater wave.

water-table well A well in which the source of supply is free groundwater in the saturation zone below a water table.

water-tank indicator A wood or metal strip bearing measurement marks and numbers, or a dial and needle (used with a float), to indicate the depth of water in a tank.

watertight Impermeable to water unless sufficient pressure occurs to cause rupture. This term applies to water treatment equipment and materials of a certain level of precision of construction and fit.

water-vectored disease A disease transmitted by insects that breed in water, like malaria-carrying mosquitoes, or insects that bite near water. Examples of such diseases include yellow fever, dengue, filariasis, malaria, and onchocerciasis.

water-washed disease A disease closely related to poor hygienic habits and sanitation. The availability of a sufficient quantity of water is felt to be more important than water quality in controlling these diseases.

waterway Any natural or artificial channel or depression in or under the ground surface that provides a course for water flowing either continuously or intermittently.

waterwheel A wheel so arranged with floats, buckets, or other such equipment that it may be turned by flowing water. The energy of moving water causes the wheel to revolve either by the weight of water dropping from a higher to a lower level and hitting a portion of the wheel or by dynamic pressure due to a change in direction or velocity of a stream of water striking a part of the wheel. A waterwheel is also called simply a wheel.

waterworks *See* water utility.

waterworks aluminum Aluminum sulfate ($Al_2(SO_4)_3 \cdot 14\ H_2O$). *See also* alum.

watt (W) *See the Units of Measure list.*

wave front In a granular activated carbon adsorber, the activated carbon loading gradient that exists in the critical bed depth. This gradient corresponds to the gradual transition of the activated carbon from fresh (or virgin) to spent (or exhausted). *See also* mass transfer zone.

wave velocity (1) The velocity at which the front of an increment of flow moves downstream. (2) The speed at which an individual wave form advances. *See also* celerity.

wavenumber A measure of energy used in infrared spectrometry that is inversely proportional to wavelength by the following equation:

$$\text{wavenumber (in units of per centimetre)} = \frac{1}{\text{wavelength (in micrometres)}} \times 10^4$$

The wavenumber is often the preferred quantity in infrared spectrometry because it is proportional to the vibrational energy of a molecule.

waves of pollution Successive masses of polluted water occurring at intervals in a moving body of water.

Wb *See* weber *in the Units of Measure list.*

weak acid An acid that results in a relatively low concentration of hydrogen ions (H^+) from ionization. For example, acetic acid (CH_3COOH) is a weak acid. *See also* acid; strong acid.

weak acid cation exchanger Those cation exchange products with functional groups that are not capable of splitting neutral salts to form corresponding free acids. Weak acid cation exchange resins have a much higher (three to four times higher) regeneration efficiency than their strong acid counterparts, but they can exchange only cations that are associated with alkalinity. The cations associated with strong acids, such as sulfate (SO_4^{2-}), chloride (Cl^-), and nitrate (NO_3^-), cannot be removed with a weak acid cation exchanger. *See also* free acid form; strong acid cation exchanger.

weak base anion exchanger Those anion exchange products with functional groups that are not capable of splitting neutral salts to form corresponding free bases. Weak base anion exchange resins have a much higher (three to four times higher) regeneration efficiency than their strong base counterparts, but they can exchange only mineral acid anions, such as sulfate (SO_4^{2-}), chloride (Cl^-), and nitrate (NO_3^-). The anions associated with weak acids, such as carbonates (CO_3^{2-}), bicarbonates (HCO_3^-), silicates, and organic acids, cannot be removed with a weak base anion exchanger. *See also* free base form; strong base cation exchanger.

wearing ring A ring installed between the casing and impeller of a centrifugal pump to reduce the space and slip of water, improving the efficiency of the pump.

weather (1) The present climate; one phase in the succession of phenomena for which the complete cycle, recurring with some degree of uniformity every year, constitutes the climate of any locality; the meteorological condition of the atmosphere defined by the measurement of the six meteorological elements: air temperature, barometric pressure, wind velocity, humidity, clouds, and precipitation. (2) To disintegrate or decompose rock at or near the surface by physical and chemical action of atmospheric agencies.

weather-normalized Statistically accounting for departures of weather variables from normal. Multiple regression analysis is frequently used in water demand forecasting to identify and quantify the causal influences of numerous variables on water demand. Weather is almost always a significant variable in such analyses, especially if monthly data are analyzed. As it does with other variables, regression analysis derives a coefficient (partial derivative) for each weather variable, stating how much water demand changes per unit of change in the weather variable. These coefficients can then be applied to historical periods to restate water demand on a normalized basis, i.e., as if the actual weather had been normal. This is accomplished by substituting normal weather (long-run average) into the demand (regression) equation in place of actual weather data. Similarly, forecasts for future periods can be expressed in terms of normal weather by applying the regression coefficients to normal (long-run average) weather for these periods. The regression coefficients may further be used to predict the sensitivity of water demand to abnormal weather in a given period for peaking or revenue-planning analyses. *See also* multiple linear regression analysis.

weathering A geological transformation process of alteration, decomposition, and disintegration caused by the chemical and physical action of atmospheric agents on minerals and rocks, at the level of or near the surface of the lithosphere.

weathering zone The top layer of the lithosphere in the katamorphic zone, extending from the surface of the ground to the groundwater table. In this belt, the physical and chemical processes are destructive, resulting in a breaking down of the material. The zone is characterized by dissolution, softening, and a decrease in volume of its material. Below the belt of weathering lies the belt of cementation. *See also* aeration zone; anomorphic zone; katamorphic zone.

web browser A computer program that requests, retrieves, and displays electronic files from the World Wide Web.

weber (Wb) *See the Units of Measure list.*

Weber number A numerical quantity used as an index to characterize the type of flow in a hydraulic structure where surface tension influences motion phenomena as a result of tension in the elastic skin, particularly when considerable curvature of this skin exists, in conjunction with the resisting force of inertia. The Weber number is useful in certain studies of surface waves, formation of drops and air bubbles, entrainment of air in flowing water, and other related phenomena. In equation form

$$W_e = \frac{\rho V^2 L}{\sigma}$$

Where:

W_e = the Weber number

ρ = the fluid density

V= the characteristic velocity of the system (e.g., the mean, surface, or maximum velocity, depending on the situation)

L = the characteristic length (e.g., diameter or depth, depending on the situation)

σ = the surface tension

These terms must be expressed in consistent units so that the Weber number will be dimensionless.

weep hole An opening formed during the construction of retaining walls, aprons, canal linings, and foundations to permit drainage of water collecting behind and beneath such structures. Such drainage is needed to reduce hydrostatic head.

WEF *See* Water Environment Federation.

Weibull model A low-dose extrapolation model based on a distribution of sensitivities in a population (i.e., based on the idea that some level of exposure exists beneath which individuals would not respond with an adverse effect—a threshold). The Weibull model expresses the probability of an adverse effect at low doses by establishing a dose-response relationship that considers a distribution of thresholds in a population. It generally provides estimates of risk that are intermediate between those provided by the linearized multistage model and multihit models. *See also* linearized multistage model; log-probit model; logit model; multihit model; multistage model; one-hit model; probit model; threshold.

weight (wt) A quantity of force associated with an object, depending on the context. Considerable confusion exists in the use of this term. In commercial and everyday use, the term *weight* nearly always means mass, e.g., in the context of a person's weight. This nontechnical use of the term *weight* in everyday life will probably persist. In science and technology, a body's weight refers to the force that, if applied to the body, would give it an acceleration equal to the local acceleration of free fall (with *local* usually referring to a location with respect to the surface of the Earth). In this context, the local acceleration of free fall is represented by the symbol g (commonly referred to as the gravitational constant or the acceleration of gravity). The term *force of gravity* is preferred instead of *weight*. Because of its dual usage, the term *weight* should be avoided in technical practice, except under circumstances in which its meaning is completely clear. When the term is used, knowing whether mass or force is intended is important. Note, however, that on the surface of the Earth, weight is often expressed in mass units. For example, a value in terms of milligrams per litre is conventionally considered weight per unit volume. *See also* gravitational constant; mass.

weight concentration ratio In ultrafiltration, the ratio of the initial weight of the feedwater to the weight of the reject water remaining at any time during the process.

weight per volume (w/v) The weight of the solute divided by the final volume of the solution. If a dilute solution is prepared in water, the density of the solution will be approximately 1 gram per millilitre. Thus, the weight-per-volume value can be readily converted to a weight-per-weight (or percent by weight) basis. Units of milligrams per litre represent an example of a weight-per-volume unit for a trace level of a solute. *See also* solute; solution; weight per weight.

weight per weight (w/w) The weight of the solute divided by the final weight of the solution. When multiplied by 100, the value becomes the percent by weight. Units of parts per million represent an example of a weight-per-weight unit for a trace level of a solute. *See also* parts per million *in the Units of Measure list*; solute; solution.

weight units In the US customary system, pounds force. In Système International, newtons. *See also* mass units.

weighted average A calculated average of a set of observations in which some observations are given more or less importance than others by assignment of weighting factors. A weighting factor greater than 1 is assigned to increase the importance of an observation, whereas a weighting factor less than 1 is assigned to signify less importance.

weighted monthly mean precipitation *See* true monthly mean precipitation.

weight-of-evidence approach An examination of all the available data to make a scientific judgment of whether a chemical is carcinogenic in humans. It includes data obtained from epidemiological analysis, where cause can be narrowed to a few agents or a defined mixture of agents; animal data in which concordance across species is examined; and mechanistic data, which may provide insight into processes that induced the tumor at the high doses normally tested. This information is then examined in light of human exposures to determine if the chemical could possibly or even probably be a human carcinogen.

weighting agent A material, such as bentonite, added to low-turbidity waters to provide additional particles for good floc formation.

weir A dam-like device with a crest and some side containment of known geometric shape placed perpendicular to flow to measure or control the flow rate of water in a channel.

weir box A box installed in a narrow open channel upstream from a weir to provide an enlargement of the waterway, for the purpose of reducing the velocity of approach to the weir.

weir diameter The length of a line from one edge of a circular clarifier's circular weir to the opposite edge, passing through the center of the circle formed by the weir. Many circular clarifiers have circular weirs within the outside edges of the clarifiers. All the water leaving such a clarifier flows over this weir.

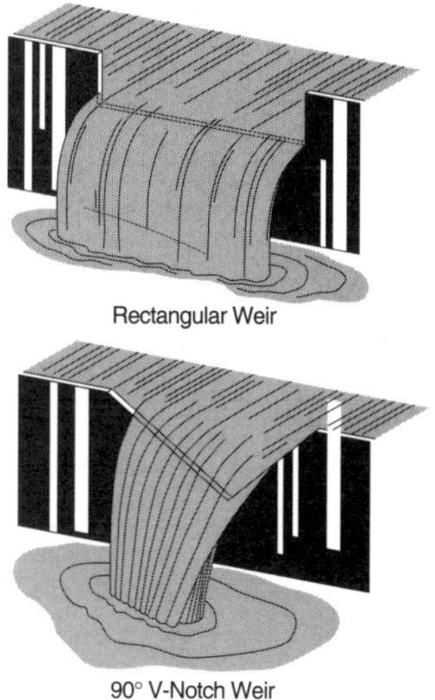

Two types of weirs

weir head (1) The vertical distance from the crest of a weir to the water surface upstream of the weir. (2) The energy head of the water measured above the height of the crest of the weir.

weir loading *See* weir loading rate.

weir loading rate (1) In a solids contact unit, the rate—in gallons per minute per foot (cubic metres per minute per metre) of weir length—at which clarified or treated liquid is leaving the unit. (2) A guideline used to determine the length of weir needed on settling tanks and clarifiers in treatment plants. Such a value is used by operators to determine whether weirs are hydraulically (flow) overloaded. Maximum weir loading rates are often specified to avoid undesirable performance resulting from poor outlet hydraulics. For example, high weir loading rates in a sedimentation basin can cause high, localized vertical velocities, resulting in poor particle settling. *See also* overflow weir.

weir overflow rate *See* weir loading weir.

weir teeth The portion of a notched weir plate that is designed to maintain a constant level behind the weir. Water flows through the notches, or teeth, and a constant level is maintained behind the weir. *See also* alligator teeth; V-notch weir.

welded steel pipe Steel pipe made from sheets of steel curved into a cylindrical shape with the edges welded together. Field joints may be flanged, screwed, or welded. The seams may form a spiral running around the circumference of the pipe with welding along the seam. *See also* riveted steel pipe.

well (1) An artificial excavation that derives water from the interstices of the rocks or soil that the excavation penetrates. (2) A shaft or hole into which water may be conducted by ditches to drain water away from a construction area.

well capacity The maximum rate at which a well will yield water under a stipulated set of conditions, such as a given drawdown, pump and motor, or engine size. Well capacity may be expressed in terms of cubic feet per second, gallons per minute, cubic metres per minute, or other similar units. *See also* circle of influence; cone of depression; zone of influence.

well casing The nonperforated riser pipe that connects a well to the surface.

well cone of influence The depression, roughly conical in shape, produced in a water table or other piezometric surface by the extraction of water from a well at a given rate. The volume of the cone will vary with the rate and duration of withdrawal of water. Such a cone is also called a cone of depression. *See also* circle of influence; cone of depression; drawdown; zone of influence.

well curb A concrete or masonry parapet built around a well at ground level to prevent drainage of surface water into the well.

well development The process of cleaning the fines and drilling residue from a new well to improve its yield and quality for subsequent use in water production.

well field A group of wells treated as a single entity for administrative, production, and treatment purposes. Hydraulic models frequently treat several wells in the same geographic area as a single well with a discharge equivalent to that of all the wells in the well field.

well field placement The process of arranging wells to achieve a desired output from an aquifer.

well flooding (1) A process of diverting water from streams when the supply exceeds the demand and allowing it to pass into the saturation zone through a well constructed for that purpose. As a result, the water may be stored for future use. (2) The overtopping of a well by floodwaters. *See also* flood spreading; spreading.

well hydrograph A graphical representation of the fluctuations of the water surface in a well (plotted along the ordinate) against time (plotted along the abscissa).

well infiltration area The area of a water-bearing formation that contributes to a well.

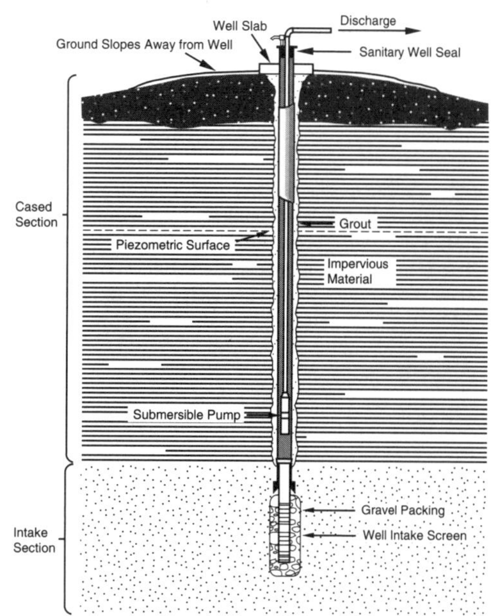

Parts of a typical well

well intake The well screen, perforated sections of casing, or other openings through which water from a water-bearing formation enters a well.

well interference The additional drawdown observed when two or more wells are operating in the same aquifer. Neglecting the effect of other wells can lead to inadequate suction head at a well's pump because the expected drawdown was computed based on only the one well operating.

well isolation zone A surface or zone of restricted land use surrounding a water well or well field that supplies a public water system. Such restricted land uses prevent contaminants due to forbidden activities from moving toward and reaching the water well or well field. *See also* wellhead protection area.

well log A record of the thickness and characteristics of the soil, rock, and water-bearing formations encountered during the drilling (sinking) of a well.

well monitoring The process of measuring, by on-site instruments or laboratory methods, the quality of water in a well.

well mouth The orifice or opening at the upper end of a well, generally at or near the surface, that provides access to the interior of the well.

well plug A watertight and gastight seal installed in a borehole or well to prevent movement of fluids. Wells are plugged when they are abandoned to prevent the introduction of pollutants into the subsurface.

well point A pointed hollow tube driven into the ground with a small screened portion near the end. Well points are used primarily for construction dewatering and water quality sampling, but they can also supply production quantities of water in appropriate regions.

well record A concise statement (tabular or otherwise) containing all the available data regarding the hydraulic performance of a well. In some sections of the United States, this term is applied only to the record of fluctuations of the water level in a well, along with the date when each change in the water level elevation was observed and measured.

well recovery The decrease in drawdown of the water table (return to predevelopment conditions) when a pumping well is shut off.

well screen A sleeve with slots, holes, gauze, or wire wrap placed at the end of a well casing to allow water to enter the well. The screen prevents sand from entering the water supply.

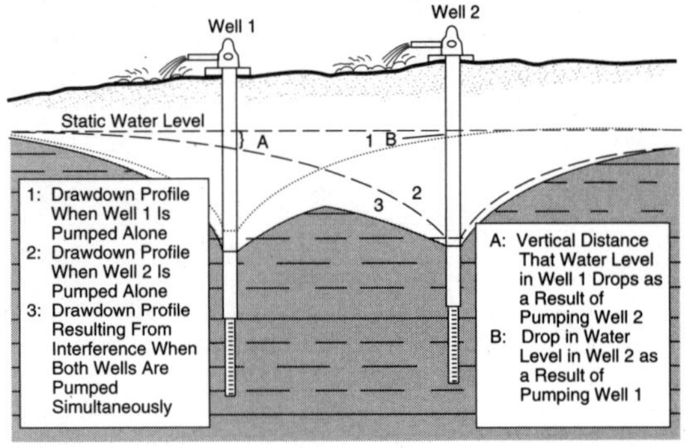

Well interference as a result of overlapping cones of depression

well seepage area *See* well infiltration area.

well shooting The act of firing a charge of nitroglycerin or other high explosive in the bottom of a well to increase the flow of water.

well strainer A type of well screen manufactured from woven screen of varied gauge wrapped on a casing perforated with holes. This type of screen is used in well points that are driven into the ground with percussion for shallow domestic wells or for dewatering purposes. *See also* well point.

well water Water obtained from a well built into or drilled into the zone of saturation. *See also* groundwater.

wellhead The surface appurtenance (top) of a particular well. The wellhead is the location of the pump motor (unless the motor and pump are both submersible), the concrete slab that surrounds the well casing, and any plumbing dedicated to that particular well. A group of wells that are geographically close may share a storage tank, but each well has its own wellhead. A group of wells is often called a well field. *See also* well field.

wellhead protection A process of guarding against potential groundwater contamination for a specific groundwater source.

wellhead protection area (WHPA) The surface and subsurface area that surrounds a water well or well field supplying a public water system and through which contaminants may move toward and reach the water well or well field. Frequently a 1-year, 5-year, or 10-year time of contaminant travel is used to define the size of the protection area, and this area has restricted land-use activity to protect the quality of the water serving the water supply well. *See also* well isolation zone.

wellhead protection program (WHPP) A state program authorized by Section 1428 of the Safe Drinking Water Act to protect wellhead areas from contaminants that may have any adverse effects on the health of persons. The state wellhead protection program, at a minimum, must (1) specify the duties of state agencies, local governmental entities, and public water supply systems with respect to the development and implementation of the wellhead protection program; (2) for each wellhead, determine the wellhead protection area based on all reasonably available hydrogeologic information on groundwater flow, recharge, and discharge, as well as other information the state deems necessary to adequately determine the wellhead protection area; (3) identify within each wellhead protection area all potential anthropogenic sources of contaminants that may have any adverse effect on the health of persons; (4) describe a program that contains, as appropriate, technical assistance, financial assistance, implementation of control measures, education, training, and demonstration projects to protect the water supply within wellhead protection areas from such contaminants; (5) include contingency plans for the location and provision of alternative drinking water supplies for each public water system in the event of well or wellfield contamination by such contaminants; and (6) include a requirement that consideration be given to all potential sources of such contaminants within the expected wellhead area of a new water well that serves a public water supply system. *See also* wellhead protection area.

well-point method A method of removing water from the ground by the use of well points.

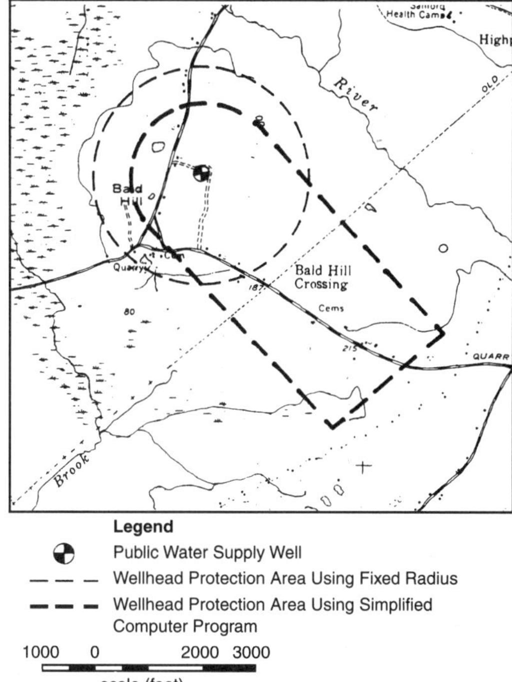

Wellhead protection area laid out on a topographic map

well-point system A system of well points, header pipe, pumps, valves, discharge pipe, and appurtenances installed for removing water from the ground.

Wenner array *See* dipole array.

WERL *See* Water Engineering Research Laboratory.

Western blot A procedure used to detect proteins in which the proteins are first electrophoretically separated and then transferred to a paper matrix such as nitrocellulose. Subsequent probing of the proteins immobilized on such a solid support membrane with antibodies or specific ligand reagents allows individual components of a complex mixture to be analyzed.

wet chemistry analysis The use of laboratory procedures to analyze a sample of water such that liquid chemical solutions are used instead of, or in addition to, laboratory instruments.

wet connection A connection made to a water main under pressure by a pipe-tapping machine without an interruption in service.

wet scrubber A device installed to remove dust from a dry chemical feeder by means of a continuous water spray.

wet soil Soil that contains significantly more moisture than moist soil but in such a range of values that cohesive material will slump or begin to flow when vibrated. Granular material that would exhibit cohesive properties when moist will lose those cohesive properties when wet.

wet volume capacity The maximum liquid volume that a unit process is designed to hold. Additional volume, or freeboard, is often provided so that the process does not overflow.

wet well A basin that is used to collect a specified volume of water and therefore permit the continuous operation of unit processes. For example, a pumping wet well is used to ensure that a minimum volume is available to be pumped to subsequent unit processes. The level in the wet well may vary, and the pumping rate may be changed, to respond to needed changes in the flow rate and to permit continuous plant operation.

wet-bulb depression The difference in degrees between the current or concurrent temperatures of the dry-bulb and wet-bulb thermometers of a psychrometer. *See also* psychrometer.

wet-bulb temperature The temperature of air as measured by the wet-bulb thermometer in a psychrometer. This temperature will be lower than that measured by the dry-bulb thermometer in inverse proportion to the humidity. Wet-bulb and dry-bulb temperatures are the same when the air is saturated. *See also* dry-bulb temperature; relative humidity.

wet-bulb thermometer One of two similar thermometers of a psychrometer, the bulb of which is moistened by means of a special wick. (The other thermometer is a dry-bulb thermometer.) *See also* wet-bulb temperature.

wetland An area that is inundated or saturated by surface water or groundwater at a frequency and duration sufficient to support, and that under normal circumstances does support, a prevalence of vegetation typically adapted for life in saturated soil conditions. Wetlands generally include swamps, marshes, bogs, and similar areas.

wetland boundary The line along the ground at which a shift from a wetland to a nonwetland area occurs, as determined on the basis of a qualified person's judgment.

wetland creation The conversion of a persistent upland or shallow water area into a wetland through human activity.

wetland determination The process by which an area is identified as a wetland or nonwetland.

wetland enhancement *See* enhanced wetland.

wetland function The physical, chemical, and biological processes or attributes of a wetland without regard to their importance to society.

wetland hydrology In general terms, permanent or periodic inundation or prolonged soil saturation sufficient to create anaerobic conditions in the soil.

wetland indicator status The exclusiveness with which a plant species occurs in wetlands. *See also* facultative plant; facultative upland plant; facultative wetland plant; obligate wetland species.

wetland value A wetland process or attribute that is valuable or beneficial to society.

wet-line correction In wire sounding to measure water depths, the correction applied to that part of the line below the water surface, where large vertical angles are induced because of high water velocities, great depth, insufficient sounding weight, or any combination. *See also* air-line correction.

wetness index A numerical quantity, usually expressed as a percentage, calculated as the ratio of the annual runoff of a stream or drainage basin for a given year to the average runoff over a considerable period of years.

wet-salt saturator tank A type of brine tank in which the saturated brine is always above the undissolved salt level in the bottom of the tank. This type of tank is used on large commercial

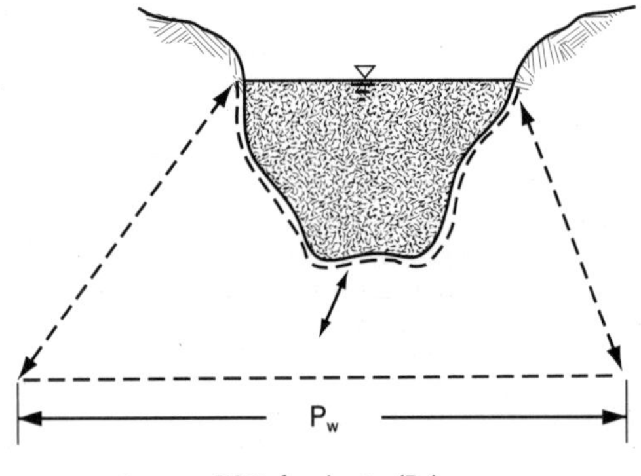

Wetted perimeter (P_w)

water softeners and older manual residential softeners. Most automatic, home-sized water softeners now use dry-salt saturator tanks. *See also* dry-salt saturator tank.

wet-shake test A test used to help determine the soil type in an excavation. A person performs this test by taking a soil patty approximately ⅛ to ¼ inch (3 to 6 millimetres) thick in the palm of his or her hand. The person wipes the surface clear of any water, then slaps the back of his or her hand. If water comes to the surface, the soil is considered to be mostly cohesive silt or sand. An absence of water indicates that the soil is mostly cohesive clay. The wet-shake test is also called the pat test.

wetted perimeter The length of wetted contact between a stream of flowing water and its containing conduit or channel, measured in a plane at right angles to the direction of flow. *See also* hydraulic radius.

wetting A process by which a liquid is absorbed by a solid surface and forms a liquid film that wets the surface.

wetting agent *See* surface-active agent.

WFI (water for injection) *See* pharmaceutical grade water.

WFP *See* water filtration plant; Water for People.

wheel *See* waterwheel.

wheel pit An enclosure in which waterwheels are located.

wheeling charge Fees assessed by a utility or company for the use of its facilities to another utility, district, company, or organization for transmission of electricity, gas, oil, water, and so forth that are owned by the transmitter. Charges for use of transmission facilities, volumes to be transmitted, and times when such transfers can occur are set forth in use agreements or contracts.

Whipple grid A micrometer reticule grid used along with a stage micrometer with a standardized, accurately ruled scale for calibration of a microscope. The grid is placed in the eyepiece of the microscope. A Whipple grid is also called a Whipple square.

whirl vortex The whirling or circular motion that is generated in water passing out of a small opening or orifice in the bottom of a basin or other receptacle and that creates a funnel-shaped opening downward from the water surface, extending ultimately below the bottom of the basin. *See also* vortex.

white blood cell count The number of white blood cells present in a specified volume of blood. This count can be broken down for various types of white blood cells. *See also* lymphocyte.

white paper A document prepared as an argument concerning a particular issue on behalf of an organization. It may establish a more argumentative tone than a position paper.

WHO *See* World Health Organization.

wholesale contract An agreement whereby water is sold by one utility to another utility and is delivered at one or more major points for resale to individual customers by the purchaser.

wholesome water Water that is safe and palatable for human consumption. *See also* potable water.

WHP *See* water horsepower.

WHPA *See* wellhead protection area.

WHPP *See* wellhead protection program.

WICC *See* Water Industry Coordinating Committee.

wicket dam A movable dam consisting of a sill and a number of wickets or rectangular panels of wood or iron, hinged on the sill and held up nearly vertical by a hinged prop that has a footing on the sill and can be tripped, dropping the wickets flat on the sill.

WIDB *See* Water Industry Data Base.

wide-crested weir A weir having a substantial width of crest in the direction parallel to the direction of water flow over it. The nappe is supported on the crest for an appreciable length, and the weir produces no bottom contraction of the nappe. Such a weir is also called a broad-crested weir.

width (*w*) The smaller of the two dimensions in the measurement of a rectangular object. *See also* length.

Wilcoxon rank-sum test A nonparametric statistical test used for comparing two groups of data that fall within certain limits (e.g., the number of animals that died within certain limits of time). The data in the two groups and the sum of the ranks (i.e., the sum of the rank orders of each data point in the group) found in each group are compared.

Wild and Scenic Rivers Act (WSRA) A public law enacted in 1968 (16 U.S.C. Section 1271 *et seq.*) to protect rivers selected for preservation because of their "outstandingly remarkable scenic, recreational, geologic, fish and wildlife, historic, cultural, or other similar values."

wild variable Any input to a feedforward controller that is not impacted, even indirectly, by the output from the controller. An example would be the flow signal used for pacing of chemicals in a water plant. *See also* feedforward control.

wilderness A wild and uncultivated region (e.g., a desert or a forest) that is inhabited only by wild animals and is generally unchanged by humans.

wildlife habitat Part of the terrestrial ecosystem on which wildlife can be supported.

willful violation A knowing breach of a standard or failure to correct known hazards that results in an Occupational Safety and Health Administration citation. A repeat of a willful violation can result in a fine of up to $70,000.

wind deposit Soil deposited by the action of the wind. A wind deposit is also called an aeolian deposit.

wind direction The point of the compass from which the wind blows (not that toward which it is moving).

wind gap A narrow opening through a range of hills or mountains, originally formed as a water gap but from which the original stream that formed it has been diverted.

wind gauge An instrument for measuring the force or velocity of wind; an anemometer. *See also* anemometer.

wind pump A pump operated by the force of the wind rotating a multiple-blade propeller. A wind pump is sometimes called a windmill.

wind setup In lakes and reservoirs, a deviation from a still water surface elevation caused by the transport of surface water by winds. *See also* seiche.

wind tide In any body of water, a deviation from a still water surface elevation caused by the transport of surface water by winds.

wind wave (1) A wave formed and growing in height under the influence of wind. (2) Loosely, any wave generated by wind.

windage Water carried from a stripping tower by the wind. This water is not available for recirculation.

windbreak A line of trees, shrubs, or other vegetation usually planted perpendicular or nearly so to the principal wind direction to protect such things as soil, crops, homesteads, and roads against the effects of wind, such as wind erosion and the drifting of soil and snow.

windings The wiring in an electric motor laid around the stator and embedded in an insulating resin material.

windrow sludge composting A method of composting sludge by applying the sludge in rows, with space between the rows to allow for placing, mixing, and removing the composted sludge. A special machine is used to periodically intermix and turn over the sludge to promote stabilization. *See also* composting.

wing dam A wall, crib, row of piles, stone jetty, or other barrier projecting outward from the shore or bank into a stream or other body of water for such purposes as protecting the shore or bank from erosion, arresting sand movement along the shore, or concentrating the low flow of a stream into a smaller channel. A wing dam is also called a groin, jetty, or spur.

wing screen A screen for which the screening elements are set in radial planes of curved vanes rotating on a horizontal axis.

wing wall (1) The wall of a dam structure extending from the headwall downstream a distance equal to the length of the apron, constructed to prevent sloughing of banks or channels and to direct and confine overfall. (2) The wall extending from a culvert headwall, intended to prevent sloughing of embankment into the waterway.

Winkler titration An iodometric titration method for volumetrically determining dissolved oxygen in water. It is used both to determine dissolved oxygen and to calibrate other methods of determining dissolved oxygen. Several modifications of the procedure are available to account for certain interferences.

wire dam A dam constructed of rock held together by wire mesh, used principally in flood protection works. Such a dam is sometimes called a wire-wrapped dam.

wire sounding A process of measuring water depth by using a calibrated wire with a sounding weight on the bottom.

wire-mesh screen A screen made of a wire fabric attached to a metal frame. Such a screen is usually part of a motorized system so that it can move continuously through the water and be automatically cleaned with a water spray. A wire-mesh screen removes finer debris from the water than a bar screen.

wire-to-water efficiency The efficiency of a pump and motor together. This value is also called the overall efficiency.

wire-weight gauge A river gauge in which a weight suspended on wire is lowered to the water surface from a bridge or other overhead structure to allow the distance from a point of known elevation on the bridge to the water surface to be measured.

WITAF *See* Water Industry Technical Action Fund.

withdrawal In pharmacology and therapeutics, a syndrome that occurs following discontinuation of the use of a drug that imparts some physical dependence with repeated use (i.e., addiction). The symptoms of withdrawal are specific to the nature of the drug. For example, barbiturates produce a different complex of symptoms than heroin.

wood screw pump A pump that has a horizontal cylindrical casing and in which a runner with radial blades like those of a ship's propellor operates. *See also* horizontal screw pump.

wood-based granular activated carbon Granular activated carbon that is made from wood as a raw material. This substance has a larger external pore structure than comparable products that reportedly can assist in maintaining a biologically active population, particularly under colder temperatures. *See also* biologically enhanced activated carbon.

work In a mechanical sense, an energy transfer across a system boundary in an organized form such that its sole effect is equivalent to the raising of a mass. *See also* energy.

work hour An industrial time unit of 60 minutes (1 hour) of work by one person for a previously agreed upon rate of compensation. This term has replaced the term *man hour*.

work order A written order authorizing and directing the performance of a certain task and issued to the person who is to direct the work. Items of information shown on the order are the nature and location of the job, specifications of the work to be performed, and the work order number to be referred to in reports of the amount of labor, materials, and equipment used.

work in progress Construction work undertaken but not yet completed.

work units In the US customary system, foot-pounds force. In Système International, joules; metres squared kilograms per second squared.

workable sludge Sludge that can readily be forked or shoveled from a sludge-drying bed. Workable sludge ordinarily has less than 75 percent moisture.

working barrel The metal tube or pump cylinder that is fastened to the lower end of the dropline of a deep well and that contains the valves and pistons of a reciprocating pump.

working pressure *See* operating pressure.

working pressure head The actual head of water flowing at any point in a conduit; the vertical height from the center line of a conduit to the hydraulic grade line.

works A group or assemblage of physical devices and structures for any of a variety of useful purposes, e.g., a water treatment plant.

World Health Organization (WHO) A specialized agency of the United Nations with primary responsibility for international health matters and public health. Created in 1948, the World Health Organization promotes the development of comprehensive health services, the prevention and control of diseases, the improvement of environmental conditions, the development of health-related human resources, the coordination and development of biomedical and health services research, and the planning and implementation of health programs. The organization is headquartered in Geneva, Switzerland.

World Wide Web (www) A worldwide series of linked computer files residing in machines scattered throughout the Internet. The World Wide Web is accessed via browser software that allows users to view graphics and text. The World Wide Web is hypertext based. It is sometimes called simply the *web* or W^3.

worst-case scenario A set of assumptions that represent the worst possible outcome of a circumstance, theory, or estimate.

WPCF *See* Water Pollution Control Federation.

WQA *See* Water Quality Association.

WQC *See* water quality certification.

WRDA *See* Water Resources Development Act.

wrought pipe Pipe made by welding long steel plates with butt or lap welds or by piercing and drawing a seamless tube from a billet.

wrought-iron pipe Pipe made of true wrought iron.

WSRA *See* Wild and Scenic Rivers Act.

wt *See* weight.

WTP *See* drinking water treatment plant.

WTW *See* drinking water treatment plant.

w/v *See* weight per volume.

w/w *See* weight per weight.

WWEMA *See* Water and Wastewater Equipment Manufacturer's Association.

www *See* World Wide Web.

wye (1) A pipe branching off a straight main run at an angle of 45°. (2) A polyphase electric circuit for which the phase difference is 120° and that, when drawn, resembles the letter *Y*.

wye strainer A screen shaped like the letter *Y*. The water flows in at the top of the *Y*, and the debris in the water is removed in the bottom two branches of the *Y*.

X

X *See* halogen.

x bit map (XBM) A simple computer image format. X bit maps only appear in black and white. They are found in HyperText Mark-up Protocol Language documents.

X ray (1) Electromagnetic radiation with a very short wavelength (0.01 to 12 nanometres), shorter than for ultraviolet radiation. (2) An image created by short-term exposure of an object to X rays, used in spectrometry analysis and medical therapy.

XBM *See* x bit map.

xenobiotic A synthetic product not formed by natural biosynthetic processes; a foreign substance or poison. The production of many novel xenobiotics has introduced into the environment many compounds that microorganisms normally do not encounter and are therefore not prepared to biodegrade. Many of these compounds are toxic to living systems, and their presence in aquatic and terrestrial habitats often has serious ecological consequences, including major kills of indigenous biota. *See also* anthropogenic; biodegradation; biota; synthetic organic chemical.

Xeriscape A landscape concept to describe attractive landscaping, using plants with low water demand, to conserve water. The concept was developed by the Denver Water Department. The word comes from the Greek *xeros* meaning dry.

Xeriscaping A landscaping practice that involves selecting trees and shrubs that require small amounts of water.

xerophreatophyte A phreatophyte that is able to resist drought.

xerophyte A plant structurally adapted to growth in an area of limited water supply, such as a desert.

X-linked *See* cross linkage.

X-ray diffraction A technique used in the identification of crystalline compounds. This technique is also used to study the structure of crystals. Within a laboratory instrument, X rays bombard a sample and interact with the layers of atoms composing a crystal. The ordered structure of a crystal scatters the X rays and produces a diffraction pattern characteristic of the substance. The traditional way of recording diffraction patterns is on a photographic surface. This information can be used to empirically identify the crystalline substance.

X-ray fluorescence A technique used for the identification and semiquantitative analysis of solids and liquids. Following absorption of X rays, excited atoms in a sample emit fluorescent X rays that are characteristic of the atoms.

***x–y* coordinates** The elements of a Cartesian coordinate system that specify the location of a point on a plane. Most maps are based on a geographic reference grid that indicates the horizontal (*x*) and vertical (*y*) coordinates of all features included on the map. The grid is defined by an origin point, and coordinates are defined as increments from this origin along the *x* and *y* axes. The reference grid may be an arbitrary local coordinate system, or it may be based on a global reference grid (e.g., latitude–longitude) that accurately references m ap features to absolute positions on the Earth's surface.

xylene ($C_6H_4(CH_3)_2$) The common name for dimethylbenzene, a commercial mixture of three isomers, *ortho-*, *meta-*, and *para*-xylene. Xylenes are used in aviation gasoline and protective coatings, and they are used as solvents and insecticides. These volatile organic compounds are regulated by the US Environmental Protection Agency. *See also* insecticide; isomer; solvent; volatile organic compound.

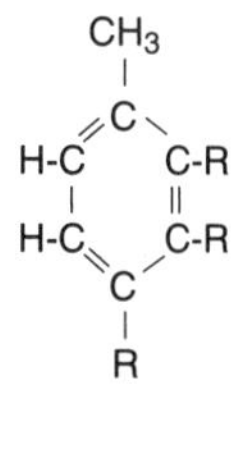

Xylene, where one of the "R" groups is another methyl (CH_3) group and the other two Rs are hydrogen atoms. The other methyl group can be in either the *ortho, meta,* or *para* position.

Y

Y connection (1) A pipe fitting with three branches positioned in one plane in the pattern of the letter *Y*. (2) More generally, any junction of three pipes or open channels forming roughly, in plan view, the letter *Y*.

yard (yd) *See the Units of Measure list.*

yard meter A water service separate from the domestic service and used only for irrigation purposes. An irrigation water rate, without charges for wastewater, applies for a yard meter.

yd *See* yard *in the Units of Measure list.*

yd³ *See* cubic yard *in the Units of Measure list.*

yeast A broad group of fungal microorganisms capable of causing fermentation.

yellow water Water that is colored as a result of the presence of iron. The iron can originate either from the source water or from corrosion reactions with iron piping. Yellow water is sometimes referred to as red water.

Yersinia A genus of bacteria in the family Enterobacteriaceae and characterized as gram-negative, fermentative, facultatively anaerobic, rod-shaped bacteria. The most important species is *Yersinia enterocolitica*, toxigenic strains of which are human pathogens. All species of *Yersinia* are widely distributed in the aquatic environment, but few appear to be pathogenic for humans.

Yersinia enterocolitica A species of gram-negative, rod-shaped bacteria that has been implicated in a few outbreaks of waterborne yersiniosis, as well as in foodborne outbreaks. Swine, and possibly wild rodents, appear to be natural reservoirs of these organisms and may cause frequent contamination of water. This organism is able to grow well at low temperatures (including refrigerator temperatures) and may adapt well to the aquatic environment. Thus, isolation of this species from the environment—from refrigerated food samples or from clinical samples—involves using cold enrichment, followed by plating on one or more selective enrichment media. (Cold enrichment involves the use of low-temperature—39° Fahrenheit (4° Celsius)—incubation. These bacteria are better able than others to grow at low temperatures; hence, they grow to increased numbers while other bacteria do not.)

yersiniosis An acute enteric disease caused by *Yersinia pseudotuberculosis* (with symptoms including abdominal pain mimicking appendicitis) or *Yersinia enterocolitica* (with symptoms including diarrhea, fever). Bloody diarrhea is seen in 10 to 30 percent of children infected with *Yersinia enterocolitica*; adults often report joint pain. *Yersinia pseudotuberculosis* primarily causes zoonosis, with humans as incidental hosts. *Yersinia enterocolitica* has been recovered from a wide variety of animals, with the pig being the principal reservoir. About two-thirds of *Yersinia enterocolitica* cases occur among infants and children. *See also* waterborne disease; zoonosis.

yield (1) The quantity of water, expressed as a rate of flow, that can be collected for a given use or uses from surface water or groundwater sources on a watershed. The yield may vary based on the use proposed, the plan of development, and economic considerations. (2) Total runoff. (3) The stream flow in a given interval of time derived from a unit area of watershed, usually expressed in cubic feet per second per square mile (cubic metres per second per square kilometre). (4) An expression of the amount of material ultimately produced by a given process, e.g., volume or weight of sludge produced. (5) The net amount of a desired product from a chemical reaction.

yield curve A graphical depiction of the series of annual returns to investors.

yield per foot (metre) The amount of water produced per foot (metre) of drawdown. As applied to groundwater, the yield per foot (metre) is the same as the specific capacity of a well. As applied to a surface reservoir or impoundment, the yield per foot (metre) is the volume of water produced for each foot (metre) of decline in storage depth. *See also* circle of influence; cone of depression; drawdown; specific capacity; zone of influence.

yoke (1) A designed meter setter providing proper spacing, alignment, and bracing of the inlet and outlet pipes, allowing simplified setting and removal of meters. (2) A 45° branch in the shape of a letter *Y*.

young river In geological terms, a river that is actively eroding its channel.

young valley A valley that is in its earliest stages of development. Such a valley is usually V-shaped in cross section, with a narrow bottom, relatively steep slopes, and poorly developed small tributaries. *See also* youthful valley.

youthful valley A stream valley that is still in the early stages of its development. Such a valley is usually V-shaped in cross section, has a narrow bottom with a steep gradient, has few tributaries, and is of relatively short length.

Z

z coordinate The third element of a point's location in space in a Cartesian coordinate system. The horizontal positions are identified by *x* and *y* coordinates. The *z* coordinate identifies the vertical position, or elevation, of a point or feature. *See also* *x–y* coordinates.

zebra mussel A bivalve that has become a serious pest, because it blocks intake pipes. *See also* quagga mussel.

Zeeman background correction An instrumental technique of compensating for interferences in graphite furnace atomic absorption spectrophotometry. In this technique, a strong magnetic field divides spectral lines into polarized components. Background correctors based on the Zeeman effect are independent of the analyte and wavelength used in the determination.

zeolite Any of numerous hydrated sodium alumina silicates, either naturally occurring mined products or synthetic products, with ion exchange properties. Zeolites were formerly used extensively for residential and commercial water softening but have been largely replaced by synthetic organic cation resin ion exchangers. Modified zeolites, such as manganese greensand and synthetic manganese zeolites, are still used as catalyst/oxidizing filter media for the removal of iron, hydrogen sulfide, and manganese. *See also* cation exchange resin; manganese greensand; manganese zeolites.

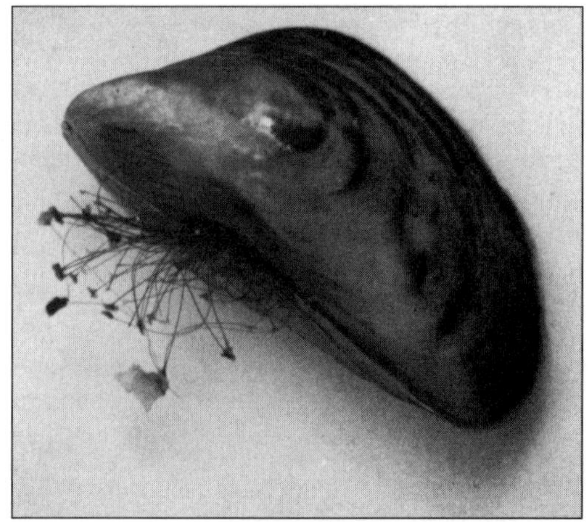

Zebra mussel

zeolite filter In water softening, a treatment unit designed to remove certain chemical constituents from water by base exchange. Zeolite is the exchange medium. *See also* zeolite process.

zeolite process A process of softening water by passing it through a zeolite substance that exchanges other constituents for hardness constituents in the water.

zeolite softening The removal of calcium and magnesium hardness from water by a base exchange process using natural or synthetic zeolites. Since the introduction of synthetic organic cation exchange resins, zeolite softening has generally been replaced by cation exchange softening. Zeolite softening was also called base exchange. *See also* cation exchange.

zero discharge (1) A complete recycling of water in a process. (2) Discharge of essentially pure water from a process. (3) Discharge of a treated effluent containing no substance at a concentration higher than that found normally in the local environment. (4) A condition whereby a facility discharges no process liquid effluent, typically through water reuse and recycling and, in some cases, concentrating and drying of liquid waste residuals into solid or semisolid waste products. This type of zero discharge is commonly associated with an industrial facility practicing nearly complete liquid recycling and reuse within the facility.

zero discharge water A condition in which a discharge limit is applied to manufacturing and commercial establishments such that only normal human sanitary wastewaters may be discharged to the municipal sewage system. No other types of wastewater, such as water used in manufacturing processes, may enter the municipal sewage system. Other types must nonetheless be recycled, and the resulting waste product from such water must be taken to an alternative approved disposal facility.

zero point of charge The pH corresponding to a surface charge of zero on a particle. Many particles found in natural waters have surface charges dependent on the solution pH and can exhibit both positive and negative charges (i.e., can be amphoteric). *See also* electrophoretic mobility; streaming current; electrokinetic potential.

zero soft water Water produced by the cation exchange process and having less than 1.0 grain per gallon (17.1 milligrams per litre) as calcium carbonate ($CaCO_3$).

zero-base budget A system of budgeting in which resource allocations, from zero to the the top, are determined within the total budget in terms of achieving organization objectives, with opportunities and risks taken into account.

zero-coupon bond A bond purchased at a large discount relative to its value at maturity. The bond then earns interest that is reinvested, helping the bond grow to its stated value at maturity. For example, a 5-year $1,000 zero-coupon bond earning interest of 8.0 percent per year would be initially sold for $680.58 and redeemed 5 years later for $1,000. The retention of interest makes a zero-coupon bond a powerful earning instrument. Although the interest is not paid out to the bondholders as it is earned, it is taxable in the year it is earned. A zero-coupon bond is also called a capital-appreciation bond. *See also* deep-discount bond.

zero-moisture index The amount of precipitation *uniformly* distributed throughout the year that would supply all the water needed for maximum evaporation and transpiration. Annual precipitation in excess of the zero-moisture index will produce surface runoff and recharge to the groundwater system. Annual precipitation less than the zero-moisture index will require moisture stored in the soil to be used to satisfy the evaporation–transpiration requirement. The zero-moisture index is so called because annual precipitation at this value would not change the soil moisture content during the year.

zero-order kinetics *See* reaction order.

zeta meter *See* zeta potential meter.

zeta plus filter *See* electropositive filter.

zeta potential (ζp) *See* electrokinetic potential.

zeta potential meter A device used to measure the zeta potential of a water sample. This type of meter is used to control the dosage of coagulants in water treatment and is also called an electrophoretic mobility apparatus.

zinc (Zn) A metallic element that is an essential growth element in terms of nutrition. Zinc is used industrially in alloys, electroplating, batteries, and fungicides. *See also* fungicide; zinc orthophosphate.

zinc orthophosphate ($Zn_3(PO_4)_2$) A chemical used in corrosion control that reacts with the surface of a pipe to form a protective coating. *See also* corrosion; corrosion inhibitor.

zinkgeriesel The granular zinc precipitate that results from corrosion of galvanized pipe.

zone of aeration The soil or rock located between the ground surface and the top of the water table.

zone of influence The area around a well or well field that experiences a measurable drawdown during pumping of the well or group of wells. *See also* circle of influence; cone of depression; drawdown; well cone of influence.

zone of saturation That part of an aquifer that has all available pore spaces filled with water.

zone settling *See* type III settling.

zoning A civil process that specifies which types of land use are allowable in specific locations.

zoonosis An infection or disease transmissible under natural conditions from vertebrate animals to humans.

zooplankton Microscopic animals living unattached in aquatic ecosystems. They include single-celled animals and small crustacea.

ζp *See* electrokinetic potential.

zwitterion A molecule in which the positive and negative groups are equally ionized. At the isoelectric point—the pH at which the net charge of a molecule in solution is zero—amino acids exist almost entirely in the zwitterion state. In the zwitterion structure of an amino acid, the acidic proton from the carboxyl group has been appropriated by the basic nitrogen of the amino group. *See also* amino acid; isoelectric point; zero point of charge.

Zwitterion structure of an amino acid

Units of Measure

acre A unit of area. This is not a Système International unit.

acre-foot (acre-ft) A unit of volume. One acre-foot is the equivalent amount or volume of water covering an area of 1 acre that is 1 foot deep. This is not a Système International unit.

ampere (A) In Système International, that constant current that, if maintained in two straight parallel conductors of infinite length or negligible cross section and placed 1 metre apart in a vacuum, would produce a force equal to 2×10^{-7} newton per metre of length (adopted at the 9th General Conference on Weights and Measures in 1948).

ampere-hour (A·h) A unit of electric charge equal to 1 ampere flowing for 1 hour. This is not a Système International unit.

angstrom (Å) A unit of length equal to 10^{-10} metre. This is not a Système International unit.

atmosphere (atm) A unit of pressure equal to 14.7 pounds per square inch (101.3 kilopascals) at average sea level under standard conditions. This unit is not a Système International unit.

bar A unit of pressure defined as 100 kilopascals. This unit is not a Système International unit.

barrel (bbl) A unit of volume, frequently 42 gallons for petroleum or 55 gallons for water. This is not a Système International unit.

baud A measure of analog data transmission speed that describes the modulation rate of a wave, or the average frequency of the signal. One baud equals 1 signal unit per second. If an analog signal is viewed as an electromagnetic wave, one complete wavelength or cycle is equivalent to a signal unit. The term *baud* has often been used synonymously with *bits per second*. The baud rate may equal bits per second for some transmission techniques, but special modulation techniques frequently deliver a bits-per-second rate higher than the baud rate.

becquerel (Bq) In Système International, the becquerel is the activity of a radionuclide decaying at the rate of one spontaneous nuclear transition per second.

billion electron volts (BeV) A unit of energy equivalent to 10^9 electron volts.

billion gallons per day (bgd) A unit for expressing the volumetric flow rate of water being pumped, distributed, or used. This is not a Système International unit.

binary digits (bits) per second (bps) A measure of the data transmission rate. A binary digit is the smallest unit of information or data, represented by a binary "1" or "0."

British thermal unit (Btu) A unit of energy. One British thermal unit is the quantity of heat required to raise the temperature of 1 pound of pure water 1° Fahrenheit. This is not a Système International unit.

bushel (bu) A unit of volume. This is not a Système International unit.

caliber (1) The diameter of a round body, especially the internal diameter of a hollow cylinder. (2) The diameter of a bullet or other projectile, or the diameter of a gun's bore. In US Customary units, usually expressed in hundreths or thousandths of an inch and typically written as a decimal fraction (e.g., 0.32). In Système International units, expressed in units of millimeters.

calorie (gram calorie) A unit of energy. One calorie is the amount of heat necessary to raise the temperature of 1 gram of pure water at 15° Celsius by 1° Celsius. This unit is not a Système International unit.

candela (cd) In Système International, a unit of luminous intensity. One candela is the luminous intensity, in a given direction, of a source that emits monochromatic radiation of frequency 540×10^{12} hertz and that has a radiant intensity in that direction of $1/683$ watt per steradian (adopted by the 16th General Conference on Weights and Measures in 1979).

candle A unit of light intensity. One candle is equal to 1 candela. This unit is not a Système International unit. Candelas are the preferred units.

candlepower A unit of light intensity. One candlepower is equal to 1 candela. This unit is not a Système International unit. Candelas are the preferred units.

centimetre (cm) A unit of length defined as one-hundredth of a metre.

centipoise A unit of absolute viscosity equivalent to 10^{-2} poise. This is not a Système International unit. *See also* poise.

chloroplatinate (Co–Pt) unit (cpu) *See* color unit.

cobalt–platinum unit *See* color unit.

color unit The unit used to report the color of water. Standard solutions of color are prepared from potassium chloroplatinate (K_2PtCl_6) and cobaltous chloride ($CoCl_2 \cdot 6H_2O$). Adding the following amounts in 1,000 millilitres of distilled water produces a solution with a color of 500 color units: 1.246 grams potassium chloroplatinate, 1.00 grams geobaltous chloride, and 100 millilitres concentrated hydrochloric acid (HCl).

colony-forming unit (cfu) A unit of expression used in enumerating bacteria by plate-counting methods. A colony of bacteria develops from a single cell or a group of cells, either of which is a colony-forming unit.

coulomb (C) In Système International, a quantity of electricity or electric charge. One coulomb is the quantity of electricity transported in 1 second by a current of 1 ampere, or about 6.25×10^{18} electrons. Coulombs are equivalent to ampere-seconds.

coulombs per kilogram (C/kg) A unit of exposure dose of ionizing radiation. *See also* roentgen.

cubic feet (ft^3) A unit of volume equivalent to a cube with a dimension of 1 foot on each side. This is not a Système International unit.

cubic feet per hour (ft^3/h) A unit for indicating the rate of liquid flow past a given point. This is not a Système International unit.

cubic feet per minute (ft^3/min, CFM) A unit for indicating the rate of liquid flow past a given point. This is not a Système International unit.

cubic feet per second (ft^3/s, CFS, cfs) A unit for indicating the rate of liquid flow past a given point. This is not a Système International unit.

cubic inch ($in.^3$) A unit of volume equivalent to a cube with a dimension of 1 inch on each side. This is not a Système International unit.

cubic metre (m^3) A unit of volume equivalent to a cube with a dimension of 1 metre on each side.

cubic yard (yd^3) A unit of volume equivalent to a cube with a dimension of 1 yard on each side. This is not a Système International unit.

curie (Ci) A unit of radioactivity. One curie equals 37 billion disintegrations per second, or approximately the radioactivity of 1 gram of radium. This is not a Système International unit.

cycles per second (cps) A unit for expressing the number of times something fluctuates, vibrates, or oscillates each second. These units have been replaced by hertz. One hertz equals 1 cycle per second.

dalton (D) A unit of weight. One dalton designates ¹⁄₁₆ the weight of oxygen-16. One dalton is equivalent to 0.9997 atomic weight unit, or nominally 1 atomic weight unit. *See also* atomic weight unit *in the main text.*

darcy (da) The unit used to describe the permeability of a porous medium (e.g., the movement of fluids through underground formations studied by petroleum engineers, geologists or geophysicists, and groundwater specialists). A porous medium is said to have a permeability of 1 darcy if a fluid of 1-centipoise viscosity that completely fills the pore space of the medium will flow through it at a rate of 1 cubic centimetre per second per square centimetre of cross-sectional area under a pressure gradient of 1 atmosphere per centimetre of length. In Système International units, 1 darcy = 9.87×10^{-13} square metres.

day (d) A unit of time equal to 24 hours.

decibel (dB) A dimensionless ratio of two values expressed in the same units of measure. It is most often applied to a power ratio and defined as decibels = 10 $\log_{10}$ (actual power level/reference power level), or dB = 10 $\log_{10} (W_2/W_1)$, where W is the power level in watts per square centimetre for sound. Power is proportional to the square of potential. In the case of sound, the potential is measured as a pressure, but the sound level is an energy level. Thus, dB = 10 $\log_{10} (p_2/p_1)^2$ or dB = 20 $\log_{10} (p_2/p_1)$, where p is the potential. The reference levels are not well standardized. For example, sound power is usually measured above 10^{-12} watts per square centimetre, but both 10^{-11} and 10^{-16} watts per square centimetre are used. Sound pressure is usually measured above 20 micropascals in air. The reference level is not important in most cases because one is usually concerned with the difference in levels, i.e., with a power ratio. A power ratio of 1.26 produces a difference of 1 decibel.

decilitre (dL) A unit of volume defined as one-tenth of a litre. This unit is often used to express concentration in clinical chemistry. For example, a concentration of lead in blood would typically be reported in units of micrograms per decilitre.

degree (°) A measure of the phase angle in a periodic electrical wave. One degree is ¹⁄₃₆₀ of the complete cycle of the periodic wave. Three hundred sixty degrees equals 2π radians.

degree Celsius (°C) A unit of temperature. The degree Celsius is exactly equal to the kelvin and is used in place of the kelvin for expressing Celsius temperature (symbol t) defined by the equation $t = T - T_0$, where T is the thermodynamic temperature in kelvin and $T_0 = 273.15$ kelvin by definition.

degree Fahrenheit (°F) A unit of temperature on a scale in which 32° marks the freezing point and 212° the boiling point of water at a barometric pressure of 14.7 pounds per square inch. This is not a Système International unit.

degree Kelvin (K) *See* kelvin.

dram (dr) Small weight. Two different drams exist: the Apothecary's dram (equivalent to 1/3.54 gram) and the Avoirdupois dram (equivalent to 1/1.17 gram).

dyne A unit of force in the centimetre–gram–second metric system adopted in 1881, defined as that force that will impart to a free mass of 1 gram an acceleration of 1 centimetre per second squared. This is not a Système International unit.

einstein A unit of measure expressing the total energy of 1 mole of light photons, which depends on the wavelength. In equation form,

$$E = \frac{hc}{\lambda}$$

Where:

E = total energy per photon, in joules
$h = 6.63 \times 10^{-34}$ joule-seconds per photon
$c = 3.0 \times 10^8$ metres per second
λ = wavelength, in metres

As an example, for $\lambda = 254 \times 10^{-9}$ metres, 1 einstein is 7,870 watt-minutes or 472.2 kJ.

electron volt (eV) In the US customary system, a unit of energy commonly used in the fields of nuclear and high-energy physics. One electron volt is the energy transferred to a charged particle with single charge when that particle falls through a potential of 1 volt. An electron volt is equal to 1.6×10^{-19} joule.

equivalents per litre (eq/L) In Système International, an expression of concentration equivalent to normality. The normality of a solution (equivalent weights per litre) is a convenient way of expressing concentration in volumetric analyses. *See also* normality *in the main text.*

farad (F) In Système International, the unit of electrical capacitance. One farad is the capacitance of a capacitor between the plates of which a difference of potential of 1 volt appears to be present when the capacitor is charged by a quantity of electricity equal to 1 coulomb. Farads are equivalent to seconds to the fourth amperes squared per metre squared per kilogram.

fathom A unit of length equivalent to 6 feet, used primarily in marine measurements. This is not a Système International unit.

feet (ft) The plural form of a unit of length (the singular form is *foot*). This is not a Système International unit.

feet board measure (fbm) A unit of volume. One board foot is represented by a board measuring 1 foot by 1 foot by 1 inch thick (144 cubic inches). A board measuring 0.5 feet by 2 feet by 2 inches thick would equal 2 board feet. This unit is not a Système International unit.

feet per hour (ft/h) A unit for expressing the rate of movement. This is not a Système International unit.

feet per minute (ft/min) A unit for expressing the rate of movement. This is not a Système International unit.

feet per second (ft/s, fps) A unit for expressing the rate of movement. This is not a Système International unit.

feet per second squared (ft/s^2) A unit of acceleration (rate of change of linear motion). For example, the acceleration caused by gravity is 32.2 ft/s^2 at sea level. This is not a Système International unit.

feet squared per second (ft^2/s) A unit used in flux calculations. This is not a Système International unit.

fluid ounce (fl oz) A unit for expressing volume, equivalent to $1/128$ of a gallon. This is not a Système International unit.

foot A unit of length, equivalent to 12 inches. This is not a Système International unit. *See also* US customary system of units.

foot of water (39.2° Fahrenheit) A unit for expressing pressure or elevation head. This is not a Système International unit.

foot per second per foot (ft/s/ft; s^{-1}) A unit for expressing velocity gradient. This is not a Système International unit.

foot-pound, torque A unit for expressing the energy used in imparting rotation, often associated with the power of engine-driven mechanisms. This is not a Système International unit.

foot-pound, work A unit of measure of the transference of energy when a force produces movement of an object. This is not a Système International unit.

formazin turbidity unit (ftu) A turbidity unit appropriate when a chemical solution of formazin is used as a standard to calibrate a turbidimeter. If a nephelometric turbidimeter is used, nephelometric turbidity units and formazin turbidity units are equivalent. *See also* nephelometric turbidity unit.

gallon (gal) A unit of volume, equivalent to 231 cubic inches. This is not a Système International unit. *See also* Imperial gallon.

gallons per capita per day (gpcd) A unit typically used to express the average number of gallons of water used by the average person each day in a water system. The calculation is made by dividing the total gallons of water used each day by the total number of people using the water system. This is not a Système International unit.

gallons per day (gpd) A unit for expressing the discharge or flow past a fixed point. This is not a Système International unit.

gallons per day per square foot (gpd/ft^2, gfd, gsfd) A unit of flux equal to the quantity of liquid in gallons per day through 1 square foot of area. It may also be expressed as a velocity in units of length per unit time. In pressure-driven membrane treatment processes, this unit is commonly used to describe the volumetric flow rate of permeate through a unit area of active membrane surface. In settling tanks, this rate is called the overflow rate. This is not a Système International unit.

gallons per flush (gal/flush) The number of gallons used with each flush of a toilet. This is not a Système International unit.

gallons per hour (gph) A unit for expressing the discharge or flow of a liquid past a fixed point. This is not a Système International unit.

gallons per minute (gpm) A unit for expressing the discharge or flow of a liquid past a fixed point. This is not a Système International unit.

gallons per minute per square foot (gpm/ft^2) A unit for expressing flux, the discharge or flow of a liquid through a unit of area. In a filtration process, this unit is commonly used to describe the volumetric flow rate of filtrate through a unit of filter media surface area. It may also be expressed as a velocity in units of length per unit time. This unit is not a Système International unit.

gallons per second (gps) A unit for expressing the discharge or flow past a fixed point. This is not a Système International unit.

gallons per square foot (gal/ft^2) A unit for expressing flux, the discharge or flow of a liquid through each unit of surface area of a granular filter during a filter run (between cleaning or backwashing). This is not a Système International unit.

gallons per square foot per day *See* gallons per day per square foot.

gallons per year (gpy) A unit for expressing the discharge or flow of a liquid past a fixed point. This is not a Système International unit.

gamma (γ) A symbol used to represent 1 microgram. Use of this symbol should be avoided. The preferred symbol is μg.

gigabyte A unit of computer memory. One gigabyte equals 1 megabyte times 1 kilobyte, or 1,073,741,824 bytes (roughly 1 billion bytes).

gigalitre (GL) A unit of volume defined as 1 billion litres.

grad A unit of angular measure equal to $1/400$ of a circle. This is not a Système International unit.

grain (gr) A unit of weight. This is not a Système International unit.

grains per gallon (gpg) A unit sometimes used for reporting water analysis concentration results in the United States and Canada. This is not a Système International unit.

gram (g) A fractional unit of mass. One gram was originally defined as the weight of 1 cubic centimetre or 1 millilitre of water at 4° Celsius. Now it is $1/1{,}000$ of the mass of a certain block of platinum–iridium alloy known as the international prototype kilogram, preserved at Sèvres, France.

gram molecular weight The molecular weight of a compound in grams. For example, the gram molecular weight of carbon dioxide (CO_2) is 44.01 grams. *See also* mole.

gray (Gy) In Système International, the unit of *absorbed* ionizing radiation dose. One gray, equal to 100 rad, is the absorbed dose when the energy per unit mass imparted to matter by ionizing radiation is one joule per kilogram. *See also* rad; rem; sievert.

hectare (ha) A unit of area equivalent to 10,000 square metres. This is not a Système International unit.

henry (H) In Système International, the unit of electric inductance, equivalent to metres squared kilograms per second squared per ampere squared. One henry is the inductance of a closed circuit in which an electromotive force of 1 volt is produced when the electric current in the circuit varies uniformly at a rate of 1 ampere per second.

hertz (Hz) In Système International, the unit of measure of the frequency of a periodic phenomenon in which the period is one second, equivalent to $second^{-1}$. Hertz units were formerly expressed as cycles per second.

horsepower (hp) A standard unit of power. This is not a Système International unit. *See also* US customary system of units.

horsepower-hour (hp·h) A unit of energy or work. This is not a Système International unit.

hour (h) An interval of time equal to one twenty-fourth of a day. This is not a Système International unit.

hundredweight (cwt) A unit of weight. This is not a Système International unit.

Imperial gallon A unit of volume used in the United Kingdom, equivalent to the volume of 10 pounds of fresh water. This unit is not a Système International unit.

inch (in.) A unit of length. This is not a Système International unit.

inch of mercury (32° Fahrenheit) A unit of pressure or elevation head. This is not a Système International unit.

inch-pound (in.-lb) A unit of energy or torque. This is not a Système International unit.

inches per minute (in./min) A unit of velocity. This is not a Système International unit.

inches per second (in./s) A unit of velocity. This is not a Système International unit.

joule (J) In Système International, the unit for energy, work, or quantity of heat, equivalent to metres squared kilograms per second squared. One joule is the work done when the point of application of a force of 1 newton is displaced a distance of 1 metre in the direction of the force (1 newton-metre).

kelvin (K) In Système International, the unit of thermodynamic temperature. One kelvin is 1/273.16 of the thermodynamic temperature of the triple point of water (adopted at the 13th General Conference on Weights and Measures in 1967). No degree sign (°) is used. Zero kelvin is absolute zero, the complete absence of heat.

kilobyte A unit of measurement for digital storage of data in various computer media, such as hard disks, random access memory, and compact discs. One kilobyte is 1,024 bytes.

kilocycle *See* kilohertz.

kilograin A unit of weight equivalent to 1,000 grains.

kilogram (kg) In Système International, the unit of mass. One kilogram is equal to the mass of a certain block of platinum–iridium alloy known as the international prototype kilogram (nicknamed Le Grand K), preserved at Sèvres, France, and adopted at the 1st and 3rd General Conferences on Weights and Measures in 1889 and 1901. A new standard is expected early in the 21st century.

kilohertz (kHz) A unit of frequency equal to 1,000 hertz, or 1,000 cycles per second.

kilolitre A unit of volume equal to 1,000 litres or 1 cubic metre.

kilopascal (kPa) A unit of pressure equal to 1,000 pascals.

kiloreactive volt-amperes (kvar) A unit of reactive power equal to 1,000 volt-amperes reactive.

kilovolt (kV) A unit of electrical potential equal to 1,000 volts.

kilovolt-ampere (kVA) A unit of electrical power equal to 1,000 volt-amperes. This is not a Système International unit.

kilowatt (kW) A unit of electrical power equal to 1,000 watts.

kilowatt-hour (kW·h) A unit of energy or work. This is not a Système International unit.

knot The unit of speed used in navigation. One knot is equal to 1 nautical mile (6,080.20 feet) per hour. This is not a Système International unit.

lambda (λ) A symbol used to represent 1 microlitre. Use of this symbol should be avoided. The preferred symbol is µL.

linear feet (lin ft) A unit of distance in feet along an object. This is not a Système International unit.

litre (L) A unit of volume. One litre of pure water weighs 1,000 grams at 4° Celsius at 1 atmosphere of pressure.

litres per day (L/d) A unit for expressing a volumetric flow rate past a given point.

litres per minute (L/min) A unit for expressing a volumetric flow rate past a given point.

lumen (lm) In Système International, the unit of luminous flux equivalent to candela-steradian. One lumen is the luminous flux emitted in a solid angle of 1 steradian by a point source having a uniform intensity of 1 candela.

lux (lx) In Système International, the unit of illuminance. One lux is the illuminance intensity given by a luminous flux of 1 lumen uniformly distributed over a surface of 1 square metre. One lux is equivalent to 1 candela-steradian per metre squared.

megabyte (MB) A unit of computer memory storage equivalent to 1,048,576 bytes.

megahertz (mHz) A unit of frequency equal to 1 million hertz, or 1 million cycles per second.

megalitre (ML) A unit of volume equal to 1 million litres. This is not a Système International unit.

megaohm (megohm) A unit of electrical resistance equal to 1 million ohms. This is the unit of measurement for testing the electrical resistance of water to determine its purity. The closer water comes to absolute purity, the greater its resistance to conducting an electric current. Absolutely pure water has a specific resistance of more than 18 million ohms across 1 centimetre at a temperature of 25° Celsius. *See also* ohm.

metre (m) In Système International, a unit of length. One metre is the length of the path traveled by light in a vacuum during a time interval of 1/299,792,458 second (adopted by the 17th General Conference on Weights and Measures in 1983).

metres per second per metre (m/s/m; s^{-1}) A unit for expressing velocity gradient.

metric system A system of units started in about 1900 based on three basic units: the metre for length, the kilogram for mass, and the second for time—the so-called MKS system. Decimal fractions and multiples of the basic units are used for larger and smaller quantities. The principle departure of Système International from the more familiar form of metric engineering units is the use of the newton as the unit of force instead of the kilogram-force. Likewise, the newton instead of kilogram-force is used in combination units including force, for example, pressure or stress (newton per square metre), energy (newton-metre = joule), and power (newton-metre per second = watt). *See also* Système International.

metric ton (t) A unit of weight equal to 1,000 kilograms.

mho A unit of electrical conductivity in US customary units equal to 1 siemens, which is a Système International unit. *See also* siemens.

microgram (μg) A unit of mass equal to one-millionth of a gram.

micrograms per litre (μg/L) A unit of concentration for dissolved substances based in their weights.

microhm A unit of electrical resistance equal to one-millionth of an ohm.

micrometre (μm) A unit of length equal to one-millionth of a metre.

micromho A unit of electrical conductivity equal to one-millionth of an mho. This is not a Système International unit. *See also* microsiemens.

micromhos per centimetre (μmho/cm) A measure of the conductivity of a water sample, equivalent to microsiemens per centimetre. Absolutely pure water, from a mineral content standpoint, has a conductivity of 0.055 micromhos per centimetre at 25° Celsius. This is not a Système International unit.

micromolar (μM) A concentration in which the molecular weight of a substance (in grams) divided by 10^6 (i.e., 1 μmol) is dissolved in enough solvent to make 1 litre of solution. *See also* micromole; molar.

micromole (μmol) A unit of weight for a chemical substance, equal to one-millionth of a mole. *See also* mole.

micron (μ) A unit of length equal to 1 micrometre. Micrometres are the preferred units.

microsiemens (μS) A unit of conductivity equal to one-millionth of a siemens. The microsiemens is the practical unit of measurement for conductivity and is used to approximate the total dissolved solids content of water. Water with 100 milligrams per litre of sodium chloride (NaCl) will have a specific resistance of 4,716 ohm-centimetres and a conductance of 212 microsiemens per centimetre. Absolutely pure water, from a mineral content standpoint, has a conductivity of 0.055 microsiemens per centimetre at 25° Celsius.

microwatt (μW) A unit of power equal to one-millionth of a watt.

microwatt-seconds per square centimetre (μ-s/cm^2) A unit of measurement of irradiation intensity and retention or contact time in the operation of ultraviolet systems.

mil A unit of length equal to one-thousandth of an inch. This is not a Système International unit.

mile (mi) A unit of length, equivalent to 5,280 feet. This is not a Système International unit.

miles per hour (mph) A unit of speed. This is not a Système International unit.

milliampere (mA) A unit of electrical current equal to one-thousandth of an ampere.

milliequivalent (meq) A unit of weight equal to one-thousandth the equivalent weight of a chemical.

milliequivalents per litre (meq/L) A unit of concentration for dissolved substances based on their equivalent weights.

milligram (mg) A unit of mass equal to one-thousandth of a gram.

milligrams per litre (mg/L) The unit used in reporting the concentration of matter in water as determined by water analyses.

millilitre (mL) A unit of volume equal to one-thousandth of a litre.

millimetre (mm) A unit of length equal to one-thousandth of a metre.

millimicron (mμ) A unit of length equal to one-thousandth of a micron. This unit is correctly called a nanometre.

millimolar (mM) A concentration in which the molecular weight of a substance (in grams) divided by 10^3 (i.e., 1 mmol) is dissolved in enough solvent to make 1 litre of solution. *See also* millimole; molar.

millimole (mmol) A unit of weight for a chemical substance, equal to one-thousandth of a mole. *See also* mole.

million electron volts (MeV) A unit of energy equal to 10^6 electron volts. This unit is commonly used in the fields of nuclear and high-energy physics. *See also* electron volt.

million gallons (mil gal, MG) A unit of volume equal to 10^6. This is not a Système International unit.

million gallons per day (mgd) A unit for expressing the flow rate past a given point. This is not a Système International unit.

mils per year (mpy) A unit for expressing the loss of metal due to corrosion. Assuming the corrosion process is uniformly distributed over the test surface, the corrosion rate of a metal coupon may be converted to a penetration rate (length per time) by dividing the unit area of metal loss by the metal density (mass per volume). The penetration rate, expressed as mils per year, describes the rate at which the metal surface is receding because of the corrosion-induced metal loss. *See also* mil.

minute (min) A unit of time equal to 60 seconds. This is not a Système International unit. This unit is not to be confused with the navigational term *minute* as in *minute-of-arc.*

molar (*M*) A unit for expressing the molarity of a solution. A 1-molar solution consists of 1 gram molecular weight of a compound dissolved in enough water to make 1 litre of solution. A gram molecular weight is the molecular weight of a compound in grams. For example, the molecular weight of sulfuric acid (H_2SO_4) is 98. A 1-molar, or 1-mole-per-litre, solution of sulfuric acid would consist of 98 grams of H_2SO_4 dissolved in enough distilled water to make 1 litre of solution. *See also* molarity *in main text.*

mole (mol) In Système International, the amount of substance that contains as many elementary entities as atoms in 0.012 kilogram.
moles per litre (mol/L, *M*) A unit of concentration for a dissolved substance.
mrem An expression or measure of the extent of biological injury that would result from the absorption of a particular radionuclide at a given dosage over 1 year.
nanometre (nm) A unit of length defined as 10^{-12} metre.
nephelometric turbidity unit (ntu, NTU) A unit for expressing the cloudiness (turbidity) of a sample as measured by a nephelometric turbidimeter. A turbidity of 1 ntu is equivalent to the turbidity created by a 1:4,000 dilution of a stock solution of 5.0 millilitres of a 1.000-gram hydrazine sulfate ($(NH_2)_2 \cdot H_2SO_4$) in 100 millilitres of distilled water solution plus 5.0 millilitres of a 10.00-gram hexamethylenetetramine ($(CH_2)_6N_4$) in 100 millilitres of distilled water solution that has stood for 24 hours at 25 ± 3° Celsius.
newton (N) In Système International, the unit of force. One newton is equivalent to 1 kilogram-metre per second squared. It is that force that, when applied to a body having a mass of 1 kilogram, gives it an acceleration of 1 metre per second squared. The newton replaces the unit kilogram force, which is the unit of force in the metric system.
ohm (Ω) In Système International, the unit of electrical resistance, equivalent to metres squared kilograms per second cubed per ampere squared. One ohm is the electrical resistance between two points of a conductor when a constant difference of potential of 1 volt, applied between these two points, produces in this conductor a current of 1 ampere, with this conductor not being the source of any electromotive force.
100 cubic feet (Ccf) A unit of volume. This is not a Système International unit.
ounce (oz) A unit of force, mass, and volume. This is not a Système International unit.
ounce-inch (ounce-in., ozf-in.) A unit of torque. This is not a Système International unit.
parts per billion (ppb) A unit of proportion, equal to 10^{-9}. This expression represents a measure of the concentration of a substance dissolved in water on a weight per weight basis or the concentration of a substance in air on a weight per volume basis. One litre of water at 4° Celsius has a mass equal to 1.000 kilogram (specific gravity equal to 1.000, or 1 billion micrograms. Thus, when 1 microgram of a substance is dissolved in 1 litre of water with a specific gravity of 1.000 (1 microgram per litre), this would be one part of substance per billion parts of water on a weight per weight basis. This terminology is now obsolete, and the term *micrograms per litre* (µg/L) should be used for concentrations in water.
parts per million (ppm) A unit of proportion, equal to 10^{-6}. This terminology is now obsolete, and the term *milligrams per litre* (mg/L) should be used for concentrations in water. *See also* parts per billion.
parts per thousand (ppt) A unit of proportion, equal to 10^{-3}. This terminology is now obsolete, and the term *grams per litre* (g/L) should be used for concentrations in water. *See also* parts per billion.
parts per trillion (ppt) A unit of proportion, equal to 10^{-12}. This terminology is now obsolete, and the term *nanograms per litre* (ng/L) should be used for concentrations in water. *See also* parts per billion.
pascal (Pa) In Système International, the unit of pressure or stress equivalent to newtons per metre per second squared. One pascal is the pressure or stress of 1 newton per square metre.
pascal-second (Pa·s) A unit of absolute viscosity equivalent to kilogram per second per metre cubed. The viscosity of pure water at 20° Celsius is 0.0010087 pascal-second.
picocurie (pCi) A unit of radioactivity. One picocurie represents a quantity of radioactive material with an activity equal to one-millionth of one-millionth of a curie, i.e., 10^{-12} curie. This is not a Système International unit.
picocuries per litre (pCi/L) A radioactivity concentration unit. This is not a Système International unit.
picogram (pg) A unit of mass equal to 10^{-12} gram or 10^{-15} kilogram.
picosecond (ps) A unit of time equal to one-trillionth (10^{-12}) of a second.
plaque-forming unit (pfu) A unit expressing the number of infectious virus particles. One plaque-forming unit is equivalent to one virus particle.
platinum–cobalt (Pt–Co) color unit (PCU) *See* color unit.
poise A unit of absolute viscosity, equivalent to 1 gram mass per centimetre per second. This is not a Système International unit.
pound (lb) In US customary units, a unit used to represent either a mass or a force. This can be a confusing unit because two terms actually exist, *pound mass* (lbm) and *pound force* (lbf). One pound force is the force with which a 1 pound mass is attracted to the Earth. In equation form,

$$\text{pounds force} = (\text{pounds mass})\left(\frac{\text{local acceleration due to gravity}}{\text{standard acceleration due to gravity}}\right)$$

One pound mass, on the other hand, is the mass that will accelerate at 32.2 feet per second squared when a 1 pound force is applied to it. As an example of the effect of the local acceleration due to gravity, at 10,000 feet (3,300 metres) above sea level, where the acceleration due to gravity is 32.17 feet per second squared (979.6 centimetres per second squared) instead of the sea level value of 32.2 feet per second squared (980.6 centimetres per second squared), the force of gravity on a 1 pound mass would be 0.999 pounds force. On the surface of the Earth at sea level, pound mass and pound force are numerically the same because of the acceleration due to gravity being applied to an object, although they are quite different physical quantities. This may lead to confusion. Neither pound unit is a Système International unit.
pounds per day (lb/d) A unit for expressing the rate at which a chemical is added to a water treatment process. This is not a Système International unit.
pounds per square foot (lb/ft²) A unit of pressure. This is not a Système International unit.
pounds per square inch (psi) A unit of pressure. This is not a Système International unit.

pounds per square inch absolute (psia) A unit of pressure reflecting the sum of gauge pressure and atmospheric pressure. This is not a Système International unit. *See also* absolute pressure *in main text.*

pounds per square inch gauge (psig) A unit of pressure reflecting the pressure measured with respect to that of the atmosphere. The gauge is adjusted to read zero at the surrounding atmospheric pressure. This is not a Système International unit. *See also* gauge pressure *in the main text.*

radian (rad) In Système International, the unit of measure of a plane angle between two radii of a circle that cut off on the circumference an arc equal in length to the radius. This unit is also used to measure the phase angle in a periodic electrical wave. Note that 2π radians is equivalent to 360°.

radians per second (rad/s) A unit of angular frequency.

rad (radiation absorbed dose) A unit of adsorbed dose of ionizing radiation. Exposure of soft tissue or similar material to 1 roentgen results in the absorption of about 100 ergs (10^{-5} joules) of energy per gram, and this was defined by the International Commission on Radiation Units and Measurements in 1953 as 1 rad. This is not a Système International unit. *See also* gray; rem; sievert.

rem (roentgen equivalent man [person]) A unit of *equivalent* dose of ionizing radiation, developed by the International Commission on Radiation Units and Measurements in 1962 to reflect the finding that the biological effects of ionizing radiation were dependent on the nature of the radiation as well as other factors. For X and gamma radiation, the weighting factor is 1; thus, 1 rad equals 1 rem. For alpha radiation, however, 1 rad equals 20 rem. This is not a Système International unit. *See also* gray; rad; sievert.

revolutions per minute (rpm) A unit for expressing the frequency of rotation, or the number of times a fixed point revolves around its axis in 1 minute.

revolutions per second (rps) A unit for expressing the frequency of rotation, or the number of times a fixed point revolves around its axis in 1 second.

roentgen (r) The quantity of electrical charge produced by X or gamma radiation. One roentgen of exposure will produce about 2 billion ion pairs per cubic centimetre of air. It was first introduced at the Radiological Congress held in Stockholm as the special unit for expressing exposure to ionizing radiation. It is now obsolete. *See also* gray; rad; rem; sievert.

second (s) In Système International, the duration of 9,192,631,770 periods of radiation corresponding to the transition between the two hyperfine levels of the ground state of the cesium-133 atom (adopted at the 13th General Conference on Weights and Measures in 1967).

second feet A unit of flow equivalent to cubic feet per second. This is not a Système International unit.

second-foot day A unit of volume. One second-foot day is the discharge during a 24-hour period when the rate of flow is 1 second foot (i.e., 1 cubic foot per second). This is not a Système International unit. In ordinary hydraulic computations, 1 cubic foot per second flowing for 1 day is commonly taken as 2 acre-feet. The US Geological Survey now uses the term *cfs day* (cubic feet per second day) in its published reports.

section A unit of area in public land surveying. One section is a land area of 1 square mile. This is not a Système International unit.

siemens (S) In Système International, the derived unit for electrical conductance, equivalent to seconds cubed amperes squared per metre squared per kilogram. One siemens is the electrical conductance of a conductor in which a current of 1 ampere is produced by an electric potential difference of 1 volt.

sievert (Sv) In Système International, the unit of *equivalent* ionizing radiation dose, adopted by the General Conference on Weights and Measures in 1977. One sievert is the dose equivalent when the adsorbed dose of ionizing radiation multiplied by the dimensionless factors Q (quality factors) and N (product of any other multiplying factors) stipulated by the International Commission on Radiological Protection is 1 joule per kilogram. One sievert is equal to 100 rem. *See also* gray; rad; rem.

slug In the US customary system, the base unit of mass. A slug is a mass that will accelerate at 1 foot per second squared when 1 pound force is applied. This is not a Système International unit.

square foot (ft^2) A unit of area equivalent to that of a square 1 foot on a side. This is not a Système International unit.

square inch ($in.^2$) A unit of area equivalent to that of a square 1 inch on a side. This is not a Système International unit.

square meter (m^2) A unit of area equivalent to that of a square 1 metre on a side.

square mile (mi^2) A unit of area equivalent to that of a square 1 mile on a side. This is not a Système International unit.

standard cubic feet per minute (SCFM) A unit for expressing the flow rate of air. This unit represents cubic feet of air per minute at standard conditions of temperature, pressure, and humidity (32° Fahrenheit, 14.7 pounds per square inch absolute, and 50 percent relative humidity). This is not a Système International unit.

steradian (sr) In Système International, the unit of measure of a solid angle which, having its vertex in the center of a sphere, cuts off an area on the surface of the sphere equal to that of a square with sides of length equal to the radius of the sphere.

Système International d'Unités (SI) The international system of units of measure as defined by the periodic meeting of the General Conference on Weights and Measures (CGPM). This system is sometimes called the International Metric System or Le Système International d'Unités. Système International, as it is called, is a rationalized selection of units from the metric system with seven base units for which names, symbols, and precise definitions have been established. Many derived units are defined in terms of the base units, with symbols assigned to each and, in some cases, given names, e.g., the newton (N). The great advantage of Système International is its establishment of one and only one unit for each physical quantity—the metre for length, the kilogram (not the gram) for mass, the second for time, and so on. From these elemental units, units for all other mechanical quantities are derived. Another advantage is the ease with which unit conversions can be made, as few conversion factors need to be invoked.

The tables that follow list some especially important units.

Base Système International Units

Quantity	Unit	Abbreviation
length	metre	m
mass	kilogram	kg
time	second	s
electric current	ampere	A
thermodynamic temperature	kelvin	K
amount of substance	mole	mol
luminous intensity	candela	cd

Supplementary Système International Units

Quantity	Unit	Abbreviation
plane angle	radian	rad
solid angle	steradian	sr

Derived Système International Units With Special Names

Quantity	Unit	Abbreviation	Equivalent-Units Abbreviation
frequency (of a periodic phenomenon)	hertz	Hz	s^{-1}
force	newton	N	$kg{\cdot}m/s^2$
pressure, stress	pascal	Pa	N/m^2
energy, work, quantity of heat	joule	J	N•m
power, radiant flux	watt	W	J/s
quantity of electricity, electric charge	coulomb	C	A•s
electric potential, potential difference, electromotive force	volt	V	W/A
electrical capacitance	farad	F	C/V
electrical resistance	ohm	Ω	V/A
electrical conductance	siemens	S	A/V
magnetic flux	weber	Wb	V•s
magnetic flux density	tesla	T	Wb/m^2
inductance	henry	H	Wb/A
luminous flux	lumen	lm	cd•Sr
luminance	lux	lx	lm/m^2
activity (of a radionuclide)	becquerel	Bq	disintegrations/s
absorbed ionizing radiation dose	gray	Gy	J/kg
ionizing radiation dose equivalent	sievert	Sv	J/kg

Some Common Derived Units of Système International

Quantity	Unit	Abbreviation
absorbed dose rate	grays per second	Gy/s
acceleration	metres per second squared	m/s^2
angular acceleration	radians per second squared	rad/s^2
angular velocity	radians per second	rad/s
area	square metre	m^2
concentration (amount of substance)	moles per cubic metre	mol/m^3
current density	amperes per square metre	A/m^2
density, mass	kilograms per cubic metre	kg/m^3
electric charge density	coulombs per cubic metre	C/m^3
electric field strength	volts per metre	V/m
electric flux density	coulombs per square metre	C/m^2
energy density	joules per cubic metre	J/m^3
entropy	joules per kelvin	J/K
exposure (X and gamma rays)	coulombs per kilogram	C/kg
heat capacity	joules per kelvin	J/K
heat flux density irradiance	watts per square metre	W/m^2
luminance	candelas per square metre	cd/m^2
magnetic field strength	amperes per metre	A/m
molar energy	joules per mole	J/mol
molar entropy	joules per mole per kelvin	J/(mol•K)
molar heat capacity	joules per mole per kelvin	J/(mol•K)
moment of force	newton-metre	N•m
permeability (magnetic)	henrys per metre	H/m
permittivity	farads per metre	F/m
power density	watts per square metre	W/m^2
radiance	watts per square metre per steradian	$W/(m^2{\cdot}sr)$
radiant intensity	watts per steradian	W/sr
specific heat capacity	joules per kilogram per kelvin	J/(kg•K)
specific energy	joules per kilogram	J/kg
specific entropy	joules per kilogram per kelvin	J/(kg•K)
specific volume	cubic metres per kilogram	m^3/kg
surface tension	newtons per metre	N/m
thermal conductivity	watts per metre per kelvin	W/(m•K)
velocity	metres per second	m/s
viscosity, absolute	pascal-second	Pa•s
viscosity, kinematic	square metres per second	m^2/s
volume	cubic metre	m^3
wave number	per metre	m^{-1}

tesla (T) In Système International, the unit of magnetic flux density, equivalent to kilograms per second squared per ampere. One tesla is the magnetic flux density given by a magnetic flux of 1 weber per square metre.

ton A unit of force and mass defined as 2,000 pounds. This is not a Système International unit.

tonne (t) A unit of mass defined as 1,000 kilograms. This is not a Système International unit. A tonne is sometimes called a metric ton.

torr A unit of pressure. One torr is equal to 1 centimetre of mercury at 0° Celsius. This is not a Système International unit.

true color unit (tcu) A unit of color measurement based on the platinum–cobalt color unit. This unit is applied to water samples in which the turbidity has been removed. One true color unit equals 1 color unit. *See also* color unit.

turbidity unit *See* nephelometric turbidity unit.

US customary system of units A system of units based on the yard and the pound, commonly used in the United States and defined in "Unit of Weights and Measures (United States Customary and Metric): Definitions and Tables of Equivalents," *National Bureau of Standards Miscellaneous Publication MP 233*, Dec. 20, 1960. Most of the units have a historical origin from the United Kingdom, e.g., the length of a king's foot for the length of 1 foot, the area a team of horses could plow in a day—without getting tired—for an acre, the load a typical horse could lift in a minute for horsepower, and so forth. No organized method of multiples and fractions is involved. *See also* Système International.

volt (V) In Système International, the unit of electrical potential, potential difference, and electromotive force, equivalent to metres squared kilograms per second cubed per ampere. One volt is the difference of electric potential between two points of a conductor, carrying a constant current of 1 ampere, when the power dissipated between these points is equal to 1 watt.

volt-ampere (VA) A unit used for expressing apparent power and complex power.

volt-ampere-reactive (VAR) A unit used for expressing reactive power.

watt (W) In Système International, the unit of power and radiant flux, equivalent to metres squared kilograms per second cubed. One watt is the power that gives rise to the production of energy at the rate of 1 joule per second. Watts represent a measure of active power and instantaneous power.

weber (Wb) In Système International, the unit of magnetic flux, equivalent to metres squared kilograms per second squared per ampere. One weber is the magnetic flux that, linking a circuit of one turn, produces in the circuit an electromotive force of 1 volt as the magnetic flux is reduced to zero at a uniform rate in 1 second.

yard (yd) A unit of length equal to 3 feet. This is not a Système International unit.

Table of Conversion Factors

Conversions*		Procedure		
From	**To**	**Multiply Number of**	**By**	**To Obtain Number of**
acres	square feet (ft^2)	acres	43,560	ft^2
acres	square metres (m^2)	acres	4,047	m^2
acre-feet (acre-ft)	gallons (gal)	acre-ft	325,851	gal
ampere-hours (A-h)	coulombs (C)	A-h	3,600	C
angstroms (Å)	metres (m)	Å	1×10^{-10}	m
atmospheres (atm), standard	pascals (Pa)	atm	101,325	Pa
barns (b)	square metres (m^2)	b	1×10^{28}	m^2
barrels (bbl)	cubic metres (m^3)	bbl	0.21	m^3
barrels (bbl), oil	gallons (gal)	bbl	42	gal
barrels (bbl), water	gallons (gal)	bbl	55	gal
bars	pascals (Pa)	bars	100,000	Pa
billion gallons per day (bgd)	gigalitres per day (GL/d)	bgd	3.785	GL/d
billion gallons per day (bgd)	cubic metres per day (m^3/d)	bgd	3,785,000	m^3/d
British thermal unit (Btu), 39° Fahrenheit	joule (J)	Btu	1,059.67	J
bushels (bu)	cubic metres (m^3)	bu	0.03524	m^3
caliber	metres (m)	caliber	0.00025	m
calories (cal), 15° Celsius	joules (J)	cal	4.1858	J
centipoise (dynamic viscosity)	pascal-seconds (Pa•s)	centipoise	0.001	Pa•s
coulombs (C)	ampere-hours (A-h)	C	0.000237	A-h
cubic feet (ft^3)	cubic metres (m^3)	ft^3	0.02832	m^3
cubic feet per hour (ft^3/h)	cubic metres per hour (m^3/h)	ft^3/h	0.02832	m^3/h
cubic feet per minute (ft^3/min, cfm)	cubic metres per minute (m^3/min)	ft^3/min	0.02832	m^3/min
cubic feet per minute (ft^3/min, cfm)	cubic metres per second (m^3/s)	ft^3/min	0.0004719	m^3/s
cubic feet per second (ft^3/s, cfs)	cubic metres per second (m^3/s)	ft^3/s	0.02832	m^3/s
cubic inches ($in.^3$)	cubic metres (m^3)	$in.^3$	0.00001639	m^3
cubic inches ($in.^3$)	cubic millimetres (mm^3)	$in.^3$	16,390	mm^3
cubic inches ($in.^3$)	litres (L)	$in.^3$	0.01639	L
cubic metres (m^3)	cubic feet (ft^3)	m^3	35.31	ft^3
cubic yards (yd^3)	cubic metres (m^3)	yd^3	0.7646	m^3
curies (Ci)	becquerels (Bq)	Ci	3.7×10^{10}	Bq
darcys (da)	square metres (m^2)	da	9.8692×10^{-13}	m^2
degrees (°), angle	radians (rad)	°	0.0174	rad
drams (dr), avoirdupois	grams (g)	dr	0.855	g
drams (dr), apothecary	grams (g)	dr	0.282	g
dynes	newtons (N)	dynes	1×10^{-5}	N

* Any conversion can be reversed by using the reciprocal of the conversion factor. For example, cubic feet (ft^3) can be converted to gallons (gal) by multiplying by 1/0.1337.

Conversions*		Procedure		
From	To	Multiply Number of	By	To Obtain Number of
electron volts (eV)	joules (J)	eV	1.6022×10^{-19}	J
fathoms	metres (m)	fathoms	1.8288	m
feet (ft)	metres (m)	ft	0.3048	m
feet board measure (fbm)	cubic inches (in.^{3})	fbm	144	in.^{3}
feet board measure (fbm)	cubic metres (m^3)	fbm	0.00236	m^3
feet per hour (ft/h)	metres per hour (m/h)	ft/h	0.31	m/h
feet per minute (ft/min)	metres per second (m/s)	ft/min	0.005080	m/s
feet per second (ft/s)	metres per second (m/s)	ft/s	0.3048	m/s
feet per second squared (ft/s^2)	metres per second squared (m/s^2)	ft/s^2	0.3048	m/s^2
fluid ounces (US)	cubic metres (m^3)	fluid ounces	2.9574 ¥ 10^{-5}	m^3
foot-pounds (ft-lb) torque or work	joules (J)	ft-lb	1.36	J
gallons (gal)	cubic metres (m^3)	gal	0.003785	m^3
gallons (gal)	litres (L)	gal	3.785	L
gallons (gal)	cubic feet (ft^3)	gal	0.1337	ft^3
gallons, Canadian liquid	cubic metres (m^3)	Canadian gallons	0.004546	m^3
gallons, UK liquid	cubic metres (m^3)	UK gallons	0.004546	m^3
gallons per day (gpd)	cubic metres per day (m^3/d)	gpd	0.003785	m^3/d
gallons per day (gpd)	litres per day (L/d)	gpd	3.785	L/d
gallons per day per square foot (gpd/ft^2)	metres per day (m/d)	gpd/ft^2	0.04075	m/d
gallons per day per square foot (gpd/ft^2)	millimetres per second (mm/s)	gpd/ft^2	0.0004716	mm/s
gallons per hour (gph)	litres per second (L/s)	gph	0.001052	L/s
gallons per minute (gpm)	cubic metres per second (m^3/s)	gpm	0.0000631	m^3/s
gallons per minute per square foot (gpm/ft^2)	metres per hour (m/h)	gpm/ft^2	2.44	m/h
gallons per square foot (gal/ft^2)	litres per square metre (L/m^2)	gal/ft^2	40.7	L/m^2
grade	radians	grade	0.01571	radians
grains (gr)	grams (g)	gr	0.06480	g
grains (gr)	kilograms (kg)	gr	6.480×10^{-5}	kg
grains per gallon (gpg)	milligrams per litre (mg/L)	gpg	17.1	mg/L
grays (Gy)	radiation absorbed dose (rad)	Gy	100	rad
hectares (ha)	square metres (m^2)	ha	10,000	m^2
Henry's coefficient, atmospheres (H_{atm})	Henry's coefficient, dimensionless (H_u)	H_{atm} at 20°C	7.49×10^{-4}	H_u
Henry's coefficient, atmosphere cubic metres per mole (H_m)	Henry's coefficient, dimensionless (H_u)	H_m	41.6	H_u
Henry's coefficient, atmosphere litres per milligram (H_D)	Henry's coefficient, dimensionless (H_u)	H_D	55,600 divided by the molecular weight of the substance	H_u
horsepower (hp)	foot-pounds per minute (ft-lb/min)	hp	33,000	ft-lb/min
horsepower (hp)	kilowatts (kW)	hp	0.7457	kW
horsepower (hp)	watts (W)	hp	745.7	W
horsepower-hours (hp-h)	joules (J)	hp-h	2,680,000	J
hours (h)	seconds (s)	h	3,600	s
hundredweight, short	kilograms	hundredweight	45.36	kg
Imperial gallons (gal (Imp))	gallons (gal)	gal (Imp)	1.2	gal
Imperial gallons (gal (Imp))	litres (L)	gal (Imp)	4.53	L
inches (in.)	metres (m)	in.	0.0254	m
inch of mercury (in. Hg), 32° Fahrenheit	pascals (Pa)	in. Hg	3,386	Pa
inches per minute (in./min)	millimetres per second (mm/s)	in./min	0.423	mm/s
inch-pounds (in.-lb)	joules (J)	in.-lb	0.113	J

* Any conversion can be reversed by using the reciprocal of the conversion factor. For example, cubic feet (ft^3) can be converted to gallons (gal) by multiplying by 1/0.1337.

Conversions*		Procedure		
From	**To**	**Multiply Number of**	**By**	**To Obtain Number of**
joules (J)	calories (cal)	J	4.2	cal
joules (J)	foot-pounds (ft-lb)	ft-lb	0.735	ft-lb
joules (J)	newton-metres (N•m)	J	1	N·m
kilocalories (kcal), mean	joules (J)	kcal	4,190	J
kilogram force (kgf)	newtons (N)	kgf	9.807	N
kilogram mass (kgm)	kilograms (kg)	kgm	1.00	kg
kilojoules (kJ)	British thermal units (Btu)	kJ	0.944	Btu
kilopascals (kPa)	feet (ft) of hydraulic head	kPa	0.3346	ft of head
kilopascals (kPa)	pounds per square inch (psi)	kPa	0.145	psi
kilovolt-amperes (kVA)	watts (W)	kVA	0.001	W
kilowatts (kW)	foot-pounds per minute (ft-lb/min)	kW	44,250	ft-lb/min
kilowatts (kW)	horsepower (hp)	kW	1.341	hp
kilowatts (kW)	watts (W)	kW	1,000	W
kilowatt-hours (kW•h)	joules (J)	kW•h	3,600,000	J
knots, international	metres per second (m/s)	knots	0.5144	m/s
litres (L)	cubic metres (m^3)	L	0.001	m^3
lux (lx)	lumens per square metre (lm/m^2)	lx	1.00	lm/m^2
metric tons (t)	kilograms (kg)	t	1,000	kg
mhos	siemens (S)	mhos	1	S
microns (μ)	metres (m)	μ	1×10^{-6}	m
miles (mi), US statute	metres (m)	mi	1,609.3	m
miles per hour (mph)	kilometres per hour (km/h)	mph	1.609	km/h
miles per hour (mph)	metres per second (m/s)	mph	0.4470	m/s
milligrams per litre (mg/L)	grains per gallon (gpg)	mg/L	0.0585	gpg
mils	micrometres (μm)	mils	25.4	μm
ounce, US fluid (oz)	cubic metres (m^3)	oz	0.00002957	m^3
ounce-inches (oz-in.) torque	joules (J)	oz-in.	0.00706	J
ounces-force, avoirdupois (ozf)	newtons (N)	ozf	0.278	N
ounces-mass, avoirdupois (ozm)	kilograms (kg)	ozm	0.0285	kg
pascals (Pa)	feet (ft) of hydraulic head	Pa	0.0003346	ft of head
pascals (Pa)	newtons per metre squared (N/m^2)	Pa	1.00	N/m^2
pascals (Pa)	pounds per square inch (psi)	Pa	0.0001450	psi
pascal-seconds (Pa•s)	poise	Pa•s	10	poise
picocuries per litre (pCi/L)	becquerels per cubic metre (Bq/m^3)	pCi/L	0.027	Bq/m^3
poise (absolute viscosity)	pascal-seconds (Pa•s)	poise	0.1	Pa•s
pounds mass (lbm), avoirdupois	kilograms (kg)	lbm	0.4536	kg
pounds force (lbf)	newtons (N)	lbf	4.448	N
pounds per square foot (lb/ft^2)	pascals (Pa)	lb/ft^2	47.88	Pa
pounds per square inch (psi)	kilopascals (kPa)	psi	6.895	kPa
pounds per square inch (psi)	pascals (Pa)	psi	6,895	Pa
radiation absorbed dose (rad)	grays (Gy)	rad	0.01	Gy
roentgen (R)	coulombs per kilogram (C/kg)	R	0.000258	C/kg
roentgen equivalent man (person) (rem)	sieverts (Sv)	rem	0.01	Sv
sections	square metres (m^2)	sections	2,590,000	m^2
sieverts (Sv)	roentgen equivalent man (person) (rem)	Sv	100	rem

* Any conversion can be reversed by using the reciprocal of the conversion factor. For example, cubic feet (ft^3) can be converted to gallons (gal) by multiplying by 1/0.1337.

Conversions*		Procedure		
From	To	Multiply Number of	By	To Obtain Number of
slugs	kilograms (kg)	slugs	14.594	kg
square feet (ft^2)	square metres (m^2)	ft^2	0.0929	m^3
square feet per second (ft^2/s)	square metres per second (m^2/s)	ft^2/s	0.093	m^2/s
square yards (yd^2)	square metres (m^2)	yd^2	0.8361	m^2
tons	kilograms (kg)	tons	907	kg
tons	pounds (lb)	tons	2,000	lb
tonnes (t)	kilograms (kg)	t	1,000	kg
torrs, cm Hg, 0° Celsius	pascals (Pa)	torrs	133.32	Pa
yards (yd)	metres (m)	yd	0.9144	m

*Any conversion can be reversed by using the reciprocal of the conversion factor. For example, cubic feet (ft^3) can be converted to gallons (gal) by multiplying by 1/0.1337.

Sources of Information

American Society for Testing and Materials. 1992. *Standard Practice for Use of the International System of Units (SI)*. Publication E380-92. Philadelphia: ASTM.

American National Standards Institute/American Water Works Association. C111/A21.11. 1995. *Rubber-Gasket Joints for Ductile Pipe and Fittings.* Denver, Colo.: AWWA.

American Public Health Association, American Water Works Association, and Water Environment Federation. 1998. *Standard Methods for the Examination of Water and Wastewater,* 20th ed. Washington, D.C.: APHA.

American Water Works Association. 1989. *Manual M11, Steel Pipe—A Guide for Design and Installation.* Denver, Colo.: AWWA.

American Water Works Association. 1992. *Centrifugal Pumps and Motors: Operation and Maintenance.* Denver, Colo.: AWWA.

American Water Works Association. 1993. *Drinking Water Handbook for Public Officials* Denver, Colo.: AWWA.

American Water Works Association. 1995a. *Basic Science Concepts and Applications.* 2nd ed. Denver, Colo.: AWWA.

American Water Works Association. 1995b. *Water Quality,* 2nd ed. Denver, Colo.: AWWA.

American Water Works Association. 1995c. *Water Sources,* 2nd ed. Denver, Colo.: AWWA.

American Water Works Association. 1995d. *Water Transmission and Distribution,* 2nd ed. Denver, Colo.: AWWA.

American Water Works Association. 1995e. *Water Treatment,* 2nd ed. Denver, Colo.: AWWA.

American Water Works Association. 1997. *Manual M12: Simplified Procedures for Water Examination.* Denver, Colo.: AWWA.

American Water Works Association. 1999. *Water Quality and Treatment,* 5th ed. New York: McGraw-Hill.

Auerbach, J. 1994. Costs and Benefits of Current SDWA Regulations. *Jour. AWWA,* 86(2): 69–78.

Barrett et al. 1991. *Manual of Design for Slow Sand Filtration.* Denver, Colo.: AWWARF and AWWA.

Brock, T.D., and M.T. Madigan. 1991. *Biology of Microorganisms.* Englewood Cliffs, N.J.: Prentice-Hall.

De Marsily, G. 1986. *Quantitative Hydrogeology.* San Diego, Calif.: Academic Press.

Farm Chemicals Handbook: Section C. Pesticide Dictionary. 1984. Willoughby, Ohio: Meister Publishing.

Franzini, J.B. 1997. *Fluid Mechanics With Engineering Applications.* New York: McGraw-Hill.

Gerstberger, eds. 1994. *Instrumentation and Computer Integration of Water Utility Operations.* Denver, Colo.: AWWARF and Japan Water Works Association.

Gotoh, K., J.K. Jacobs, S. Hosoda, R.L. Gerstberger, eds. 1994. *Instrumentation and Computer Integration of Water Utility Operations.* Denver, Colo.: AWWARF and Japan Water Works Association.

Harrison, J.F., and W. McGowan. 1993. *WQA Glossary of Terms.* J. Grove, ed. Lisle, Ill.: Water Quality Association.

Hawley, G.G. 1981. *The Condensed Chemical Dictionary,* 10th ed. New York: Van Nostrand Reinhold.

Heath, R.C. 1989. *Basic Ground-Water Hydrology.* US Geological Survey Water-Supply Paper 2220. Reston, Va.: US Geological Survey.

Krasner, S.W., C.J. Hwang, and M.J. McGuire. 1981. Development of a Closed-Looped Stripping Technique for the Analysis of Taste- and Odor-Causing Substances in Drinking Water. In *Advances in the Identification and Analysis of Organic Pollutants in Water.* Vol. 2. Ann Arbor, Mich.: Ann Arbor Science.

Krasner, S.W., M.J. McGuire, J.G. Jacangelo, N.L. Patania, K.M. Reagan, and E.M. Aieta. 1989. The Occurrence of Disinfection By-Products in U.S. Drinking Water. *Jour. AWWA,* 81(8): 41–53.

Mackenthun, K.M., and W.M. Ingram. 1967. *Biological Associated Problems in Freshwater Environments.* Cincinnati, Ohio: US Department of the Interior, Federal Water Pollution Control Administration.

Mazur, A., and B. Harrow. 1969. Biochemistry: *A Brief Course.* Philadelphia: W.B. Saunders.

Moeller, D.W. 1992. *Environmental Health.* Cambridge, Mass.: Harvard University Press.

Najm, I.N., and S.W. Krasner. 1995. Effects of Bromide and NOM on By-Product Formation. *Jour. AWWA,* 87(1): 106–115.

National Weather Service (NWS). 1961. Frequency Atlas of the United States. Technical Paper No. 40. Washington, D.C.: US Department of Commerce.

Pirbazari, M., B.N. Badriyha, J.F. Green, and R.J. Miltner. 1989. Analytical Techniques for Assessment of Treatment Processes for Aldicarb and Its Derivatives. In *Proc. 1989 AWWA WQTC.* Denver, Colo.: AWWA.

Salvato, J.A. 1992. *Environmental Engineering and Sanitation,* 4th ed. New York: John Wiley & Sons.

Schenck, K.M., J.R. Meier, H.P. Ringhand, and F.C. Kopfler. 1990. Recovery of 3-Chloro-4-(dichloromethyl)-5-hydroxy-2(5H)-furanone from Water Samples on XAD Resins and the Effect of Chlorine on Its Mutagenicity. *Environ. Sci. Technol.,* 24: 863–867.

Texas Water Commission. 1989. *The Underground Subject.* Pamphlet C89-05.

Texas Water Utilities Association. *Manual of Water Utility Operations* 1986. Austin, Tex.: TWUA.

Thakker, D.R., M. Nordqvist, H. Yagi, W. Levn, D. Ryan, P. Thomas, A.H. Conney, and D.M. Jerina. 1979. Comparative Metabolism of a Series of Polycyclic Aromatic Hydrocarbons by Rat Liver Microsomes and Purified Cytochrome P-450. In *Polynuclear Aromatic Hydrocarbons.* Edited by P.W. Jones and P. Leber. Ann Arbor, Mich.: Ann Arbor Science Publishers.

US Dept. of the Interior, Bureau of Reclamation 1981. *Water Measurement Manual.* Denver, Colo.: US Government Printing Office.

US Environmental Protection Agency. 1991. *Protecting Local Ground-Water Supplies Through Wellhead Protection* Washington. D.C.: USEPA.

US Environmental Protection Agency and Association of Metropolitan Water Agencies. 1989. *Disinfection By-Products in United States Drinking Water.* Los Angeles, Calif.: James M. Montgomery Consulting Engineers, Inc. and Metropolitan Water District of Southern California.

US Public Health Service. *Algae in Water Supplies.* 1962. Washington, D.C.: USPHS.

Verschueren, K. 1983. *Handbook of Environmental Data on Organic Chemicals.* 2nd edition. New York: Van Nostrand Reinhold Company, Inc.

Von Huben, H., ed. 1999. *Water Distribution Operator Training Handbook,* 2nd ed. Denver, Colo.: American Water Works Association.

Wylie, E.B., and V.L. Streeter. 1983. *Fluid Transients.* Ann Arbor, Mich.: FEB Press.

Illustration Credits

Altitude valve: GA Industries, Inc., Cranberry Township, Pa.
Axial flow pump: Ingersoll-Dresser Pump Co., Liberty Corner, N.J.
Ball valve: A.Y. McDonald Mfg. Co., Dubuque, Iowa
Bimetal couple: Dr. J.R. Myers, JRM Associates, Franklin, Ohio 45005
Cadastral map: Courtesy of Denver Water
External cathodic protection using sacrificial anodes: Farwest Corrosion Control Co., Gardena, Calif.
Check valve: Valve Manufacturers Assoc. of America, Washington, D.C.
Chlorine cylinder: Chlorine Institute, Inc., Washington, D.C.
Cipolletti weir: US Dept. of the Interior, Bureau of Reclamation
Closed impeller: ITT-Goulds Pumps, Seneca Falls, N.Y.
Coke-tray aerator: General Filter Company, Ames, Iowa
Compound meter: Courtesy of Schlumberger Industries Water Division
Container valve: Chlorine Institute, Inc., Washington, D.C.
Ball-style corporation cock: A.Y. McDonald Mfg. Co., Dubuque, Iowa
Curb stop and box: Ford Meter Box Co., Wabash, Ind.
Cyclone separator: LAKOS Filtration Systems, Claude Laval Corp., Fresno, Calif.
Deep-well pump: Ingersoll-Dresser Pump Co., Liberty Corner, N.J.
Diatomaceous earth filtration system: Celite Corp., Lompoc, Calif.
Point-of-use distillation system: Courtesy of the Water Quality Association
Domestic filter unit: Courtesy of the Water Quality Association
Compression fittings: Ford Meter Box Company, Wabash, Ind.
Flush hydrant: Mueller Co., Decatur, Ill.
Full-face shield: Thomas Scientific, Inc., Swedesboro, N.J.
Gate valve: Valve Manufacturers Assoc. of America, Washington, D.C.
Gear pump: Hydraulic Institute, Parsippany, N.J.
Globe valve: Valve Manufacturers Assoc. of America, Washington, D.C.
Home water treatment unit: Courtesy of the Water Quality Association
Hydrant: U.S. Pipe and Foundry Co., Birmingham, Ala.
Hydropneumatic water pressure system: From *Environmental Engineering and Sanitation,* 4th ed., J.A. Salvato, Copyright © 1992 by John Wiley & Sons, Inc. Reprinted by permission of John Wiley & Sons, Inc.
Turbine impeller: Ingersoll-Dresser Pump Co., Liberty Corner, N.J.
Inclined plate separator: Courtesy of Infilco Degremont Inc., Richmond, Va.
Cycles in ion exchange: Infilco Degremont Inc., Richmond, Va.
Leaky aquifer: Texas Water Commission 1989
Lining: Stevens Geomembranes, Holyoke, Mass.
Magnetic flowmeter: ABB Fischer & Porter, Warminster, Pa.
Mechanical aerators: Philadelphia Mixers, Palmyra, Pa.
Microstrainer: U.S. Filter, Warren, N.J.
Mixed-flow pump: Ingersoll-Dresser Pump Co., Liberty Corner, N.J.
Needle valve: Valve Manufacturers Assoc. of America, Washington, D.C.
Nutating-disk meter: Schlumberger Industries Water Division, Tallassee, Ala.
Occupational safety notice: Courtesy of U.S. Pipe and Foundry Company
Operating nut on a fire hydrant: U.S. Pipe and Foundry Co., Birmingham, Ala.
Orifice meter: Bristol-Babcock Division, Acco Industries Inc., Watertown, Conn.
Laboratory oven: Barnstead/Thermolyne Corp.
Perched aquifer: From De Marsily, G. 1986. *Quantitative Hydrogeology.* San Diego, Calif.: Academic Press.
Permeator: U.S. Filter, Warren, N.J.
Good and bad pipe bedding: J-M Manufacturing Co., Inc., Livingston, N.J.
Pipe fittings for ductile iron pipe: U.S. Pipe and Foundry Co., Birmingham, Ala.
Plug valve: Valve Manufacturers Assoc. of America, Washington, D.C.
Polystyrene resin beads: U.S. Filter, Warren, N.J.
One type of positive-displacement pump: Hydraulic Institute, Parsippany, N.J.
Pressure relief valve: Watts Regulator Co., North Andover, Mass.
Double action reciprocating pump: Hydraulic Institute, Parsippany, N.J.
Rectangular clarifier: U.S. Filter, Microfloc Products, Sturbridge, Mass.
Register: Schlumberger Industries Water Division, Tallassee, Ala.

Components of a remote electronic meter reading system: ABB Water Meters, Inc., Ocala, Fla.

Resilient-seat gate valve: Mueller Co., Decatur, Ill.

Riveted steel pipe being installed in Philadelphia in 1907: American Iron and Steel Institute, Washington, D.C.

Parts of a mechanical seal: Aurora Pump, North Aurora, Ill.

Details of a spiral wound membrane: Dow Chemical Co., Midland, Mich.

Spray aerator: Infilco Degremont Inc., Richmond, Va.

Tapping machine: Mueller Co., Decatur, Ill.

Various types of thrust anchors: J-M Manufacturing Co., Inc., Livingston, N.J.

Traveling bridge sludge collector: Courtesy of F.B. Leopold Company, Inc.

Tubercular corrosion in a pipe: Girard Industries, Houston, Tex.

Ultrasonic flowmeter: Controlotron Corp., Hauppauge, N.Y.

Upflow clarifier with tube settlers: U.S. Filter, Microfloc Products, Sturbridge, Mass.

Venturi meter: Public Works®, © 1999 Public Works Journal Corporation

Wellhead protection area laid out on a topographic map: US Envrionmental Protection Agency

Zebra mussel: Fred Snyder, Ohio Sea Grant, Port Clinton, Ohio

Suggestion Form

Please photocopy or detach and mail to Publishing Group, Books, American Water Works Association, 6666 West Quincy Avenue, Denver, CO 80235-3098, or fax to (303) 794-7310. Please print clearly. Thank you.

I suggest a change in *The Drinking Water Dictionary* on page _____ in line _____ of the

entry entitled ____

I recommend that the line read ____________________________________

____

____

____

Please provide contact information for follow-up, if needed:

Name ____

Telephone Number ____

Fax Number ____

Email Address ____

ABOUT THE AUTHORS

The leading organization of water management professionals in the United States, the **American Water Works Association** (AWWA) empaneled a committee of 23 experts to compile authoritative definitions and develop this dictionary. This special Technical Advisory Committee (TAC) has worked since 1994 creating this unequaled reference. **James M. Symons,** Cullen Distinguished Professor Emeritus of Civil Engineering at the University of Houston, served as Project Supervisor and Editor. AWWA has its headquarters in Denver, Colorado.

REFERENCE
DO NOT CIRCULATE

REF TD 208 .D75 2001

The drinking water

WITHDRAWN

ASHLAND COMMUNITY COLLEGE
3 3631 1095299 8